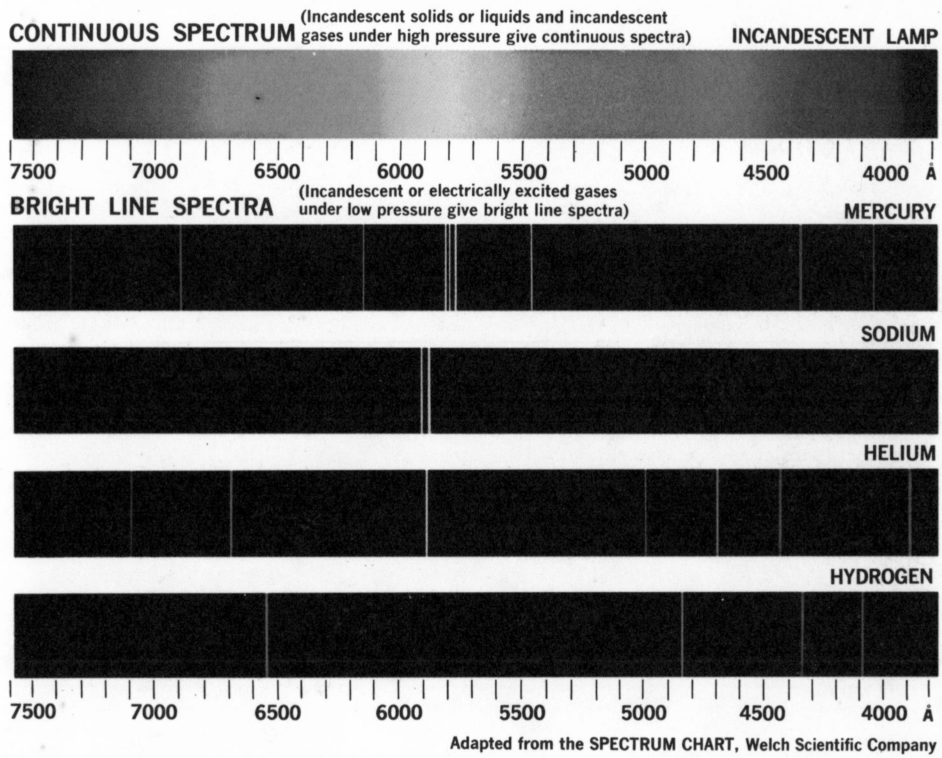

Physical Data Often Used[a]

Acceleration due to gravity	9.80 m/s^2
Average earth-moon distance	3.84×10^8 m
Average earth-sun distance	1.49×10^{11} m
Average radius of the earth	6.37×10^6 m
Density of air	1.29 kg/m^3
Density of water (20°C and 1 atm)	1.00×10^3 kg/m^3
Mass of the earth	5.99×10^{24} kg
Mass of the moon	7.36×10^{22} kg
Mass of the sun	1.99×10^{30} kg
Standard atmospheric pressure	1 atm = 1.013×10^5 Pa

[a] These are the values of the constants as used in the text.

PHYSICS

FOR SCIENTISTS & ENGINEERS

Second Edition

VOLUME I

SAUNDERS COLLEGE PUBLISHING COMPLETE PACKAGE FOR TEACHING WITH PHYSICS FOR SCIENTISTS AND ENGINEERS SECOND EDITION
by Raymond A. Serway

INSTRUCTOR'S SOLUTIONS MANUAL
H. Leap, J.R. Gordon, L. Hmurcik, and R.A. Serway

STUDY GUIDE
with Computer Exercises
R.A. Serway and J.R. Gordon

COURSEWARE DISK
to Accompany Study Guide with Computer Exercises
D. Oliver

PRINTED TEST BANK
Georgia Institute of Technology

COMPUTERIZED TEST BANK
for Apple II and IBM

OVERHEAD TRANSPARENCIES

PHYSICS
FOR SCIENTISTS & ENGINEERS
Second Edition

VOLUME I

Raymond A. Serway
James Madison University

SAUNDERS GOLDEN SUNBURST SERIES

Saunders College Publishing
Philadelphia New York Chicago San Francisco
Montreal Toronto London Sydney
Tokyo Mexico City Rio de Janeiro Madrid

Address orders to:
383 Madison Avenue
New York, NY 10017

Address editorial correspondence to:
West Washington Square
Philadelphia, PA 19105

Text Typeface: Caledonia
Compositor: Progressive Typographers
Acquisitions Editor: John Vondeling
Developmental Editor: Lloyd Black
Project Editor: Sally Kusch
Copyeditor: Charlotte Nelson
Art Director: Carol Bleistine
Text Design: Edward A. Butler
Layout Artist: Dorothy Chattin
Cover Design: Lawrence R. Didona
New Text Artwork: J & R Technical Services, Inc.
Production Manager: Tim Frelick
Assistant Production Manager: JoAnn Melody

Cover Credits: Mel Di Giacomo/Bob Masini/THE IMAGE BANK,
Department of Energy

To my wife, Elizabeth Ann, and children, Mark, Michele, David and the most recent light in my life, Jennifer Lynn, for their love and understanding.

Physics for Scientists and Engineers 2/e, VOLUME I ISBN 0-03-012244-9

6789 032 987654321

CBS COLLEGE PUBLISHING
Saunders College Publishing
Holt, Rinehart and Winston
The Dryden Press

PREFACE

This two-volume textbook is intended for a course in introductory physics for students majoring in science or engineering. The book is an extended version of *Physics for Scientists and Engineers* in that Volume II includes five additional chapters covering selected topics in modern physics. This material on modern physics has been added to meet the needs of those universities which choose to cover the basic concepts of quantum physics and its application to atomic, molecular, solid state, and nuclear physics as part of their curriculum.

The entire contents of the text could be covered in a three-semester course, but it is possible to use the material in shorter sequences with the omission of selected chapters and sections. The mathematical background of the student taking this course should ideally include one semester of calculus. If that is not possible, the student should be enrolled in a concurrent course in introduction to calculus.

A number of changes and improvements have been made in preparing the second edition of this text. Many of these changes are in response to comments and suggestions offered by users of the first edition and reviewers of the manuscript. The following represent the major changes in the second edition:

1. The order of topics has been changed slightly so that the chapters on wave motion and sound now follow the material on Newtonian mechanics, and hence precede the chapters dealing with electricity and magnetism.

2. An extensive amount of rewriting was done on the chapters concerned with rotational dynamics. Chapter 6 was rewritten so as to emphasize application of Newton's second law to circular motion and motion through a viscous medium. Most of the material concerning the Universal Law of Gravity now appears in Chapter 14.

3. The book contains many new thought questions requiring verbal answers; all questions are now located near the end of each chapter following the summaries. New problems have also been added to the book, and many of the original problems have been revised and edited. Furthermore, all end-of-chapter problems are now designated as problems. (The previous separation into exercises and problems caused some confusion.) About two thirds of the problems are keyed to specific sections, and an attempt has been made to place the more difficult problems near the end of a section group. The remaining problems, labeled "General Problems," are not keyed to any section. The more challenging problems are preceded by a bullet (•). A number of problems requiring the use of either a programmable calculator or a computer have been added to those chapters which are appropriate for such computations.

4. The artwork in the book is substantially new, as is the physical layout. A second color has been added to improve the clarity and effectiveness of the figures. Airbrushing has been used in those figures where three-dimensional effects were required. Color has also been used to highlight important statements, marginal notes, and equations. Many photographs have been added to the text.

5. A number of worked examples have been added to the text. Many of the worked examples are now followed by exercises with answers, which are extensions of the worked examples.

6. As an additional motivational component, the book now contains many essays written by guest authors. These essays cover a wide variety of topics describing many current applications in physics, and some of the more exciting recent discoveries. The essays are optional reading for the student, but an attempt has been made to locate them in related chapters.

Objectives

The main objectives of this introductory physics textbook are twofold: to provide the student with a clear and logical presentation of the basic concepts and principles of physics, and to strengthen an understanding of the concepts and principles through a broad range of interesting applications to the real world. In order to meet these objectives, emphasis is placed on sound physical arguments. At the same time, I have attempted to motivate the student through practical examples which demonstrate the role of physics in other disciplines.

Coverage

The material covered in this book is concerned with fundamental topics in classical physics. The book is divided into three parts: Part I (Chapters 1–15) deals with the fundamentals of Newtonian mechanics and the physics of fluids; Part II (Chapters 16–18) covers wave motion and sound; Part III (Chapters 19–22) is concerned with heat and thermodynamics; This order of presentation differs from the first edition in that the chapters on wave motion and sound now precede the treatment of electricity and magnetism, which is included in Volume II.

Features

Most instructors would agree that the textbook selected for a course should be the student's major "guide" for understanding and learning the subject matter. Furthermore, a textbook should be easily accessible and should be styled and written for ease in instruction. With these points in mind, I have included many pedagogic features in the textbook which are intended to enhance its usefulness to both the student and instructor. These are as follows:

Organization The book is divided into three parts: mechanics, wave motion and sound, and heat and thermodynamics. Each part includes an overview of the subject matter to be covered in that part and some historical perspectives.

Style As an aid for rapid comprehension, I have attempted to write the book in a style that is clear, logical, and succinct. The writing style is somewhat informal and relaxed, which I hope students will find appealing and enjoyable to read. New terms are carefully defined, and I have tried to avoid jargon.

Worked Examples A large number of worked examples of varying difficulty are presented as an aid in understanding concepts. In many cases, these examples will serve as models for solving the end-of-the-chapter problems. The examples are set off with colored bars for ease of location, and most examples are given titles to describe their content.

Worked Example Exercises As an added feature of this second edition, many of the worked examples are followed immediately by exercises with answers. These exercises are intended to make the textbook more interactive with the student, and to test immediately the student's understanding of problem-solving techniques. The exercises represent extensions of the worked examples and are numbered in case the instructor wishes to assign them for homework.

Problems An extensive set of problems is included at the end of each chapter. Answers to odd-numbered problems are given at the end of the book in a section which is shaded at the edges for ease of location. For the convenience of both the student and instructor, about two thirds of the problems are keyed to specific sections of the chapter. The remaining problems, labeled "General Problems," are not keyed to specific sections. In general, the problems within a given section are presented such that the straightforward problems are first, followed by problems of increasing difficulty. I have also included a small number of challenging problems which are marked with a bullet (•). In my opinion, assignments should consist mainly of the keyed problems so as to help build self-confidence in students.

Calculator/Computer Problems Numerical problems that can be best solved with the use of programmable calculators or a computer are given in a selected number of chapters. These will be useful in those courses where the instructor wishes to put programming skills to practice.

Units The international system of units (SI) is used throughout the text. The British engineering system of units (conventional system) is used only to a limited extent in the chapters on mechanics, heat, and thermodynamics.

Previews Most chapters begin with a chapter preview, which includes a brief discussion of chapter objectives and content.

Thought Questions A list of questions requiring verbal answers is given at the end of each chapter. Some questions provide the student with a means of self-testing the concepts presented in the chapter. Others could serve as a basis for initiating classroom discussions. Answers to most questions are included in the Student Study Guide With Computer Exercises that accompanies the text.

Summaries Each chapter contains a summary which reviews the important concepts and equations discussed in that chapter.

Guest Essays As an added motivational feature in this second edition, I have included a number of essays written by guest authors. Most of these essays cover topics of current interest to scientists and engineers and are intended as supplemental readings for the student.

Special Topics Many chapters include special topic sections which are intended to expose the student to various practical and interesting applications of physical principles. Most of these are considered optional, and as such are labeled with an asterisk (*).

Important Statements and Equations Most important statements and definitions are set in color for added emphasis and ease of review. Important equations are shaded in color for review or reference.

Marginal Notes Comments and marginal notes set in color are used to locate important statements, equations, and concepts in the text.

Illustrations The readability and effectiveness of the text material and worked examples are enhanced by the large number of figures, diagrams, photographs, and tables. A second color is used to add clarity to the artwork. For example, vectors are color-coded, and curves in *xy*-plots are drawn in color. Three-dimensional effects are produced with the use of airbrushed areas, where appropriate.

Mathematical Level Calculus is introduced gradually, keeping in mind that a course in calculus is often taken concurrently. Most steps are shown when basic equations are developed, and reference is often made to mathematical appendices at the end of the text. Vector products are introduced later in the text where they are needed in physical applications. The dot product is introduced in Chapter 7, "Work and Energy." The cross product is introduced in Chapter 11, which deals with rotational dynamics.

Appendices and Endpapers Several appendices are provided at the end of the text. Most of the appendix material represents a review of mathematical techniques used in the text, including scientific notation, algebra, geometry, trigonometry, differential calculus, and integral calculus. Reference to these appendices is made throughout the text. Most mathematical review sections include worked examples and exercises with answers. In addition to the mathematical reviews, the appendices contain tables of physical data, conversion factors, atomic masses, and the SI units of physical quantities, as well as a periodic chart. Other useful information, including fundamental constants and physical data, mathematical symbols, the Greek alphabet, and standard abbreviation of units appears on the endpapers.

Ancillaries The ancillaries available with this text include an Instructor's Solutions Manual, a Printed Test Bank containing over 1200 multiple choice questions, a Computerized Test Bank, a Student Study Guide with Computer Exercises, a Courseware Disk Software package to accompany the Study Guide, and a set of overhead transparencies.

The Student Study Guide With Computer Exercises is a unique student aid in that it combines the value of a problem-solving oriented study guide with a select group of integrated and interactive computer exercises. Most chapters in the study guide contain a list of objectives, skills necessary for that unit, a review and summary of important concepts, a list of equations and their meanings, answers to most end-of-chapter questions, and several programmed exercises that test the student's understanding of concepts and methods of problem solving. The Study Guide also includes the option of using a select group of computer programs (presented in special computer modules) that are interactive in nature. That is, the student's input will have direct and immediate effect on the output. This feature will enable students to work through many challenging numerical problems, and experience the power of the computer in scientific computations. The computer exercises contained in the study guide direct the student's use of the programs contained on the Courseware Disk, which is available upon adoption of the Study Guide.

Teaching Options

This book is structured in the following sequence of topics: Volume I includes classical mechanics, matter waves, and heat and thermodynamics; Volume II includes electricity and magnetism, light waves, optics, relativity, and modern physics. This presentation is a more traditional sequence and differs from that in the first edition in that the subject of matter waves is presented before electricity and magnetism. Some instructors may prefer to cover this material after completing electricity and magnetism (after Chapter 34). The chapter on relativity was placed near the end of the text because this topic is often treated as an introduction to the era of "modern physics." If time permits, instructors may choose to cover Chapter 39 in Volume II after completing Chapter 14, which concludes the material on Newtonian mechanics.

For those instructors teaching a two-semester sequence, some sections and chapters could be deleted without any loss in continuity. I have labeled these with asterisks (*) in the Table of Contents and in the appropriate sections of the text. For student enrichment, some of these sections or chapters could be given as extra reading assignments. The guest essays could also serve the same purpose.

Acknowledgments

Both editions of this textbook were prepared with the guidance and assistance of many professors who reviewed part or all of the manuscript. I wish to acknowledge the following scholars and express my appreciation for their suggestions, criticisms, and encouragement:

Elmer E. Anderson, University of Alabama
Wallace Arthur, Farleigh Dickinson University
Duane Aston, California State University at Sacramento
Richard Barnes, Iowa State University
Marvin Blecher, Virginia Polytechnic Institute and State University
William A. Butler, Eastern Illinois University
Don Chodrow, James Madison University
Clifton Bob Clark, University of North Carolina at Greensboro
Lance E. De Long, University of Kentucky
Jerry S. Faughn, Eastern Kentucky University
James B. Gerhart, University of Washington
John R. Gordon, James Madison University
Herb Helbig, Clarkson University
Howard Herzog, Broome Community College
Larry Hmurcik, University of Bridgeport
William Ingham, James Madison University
Mario Iona, University of Denver
Karen L. Johnston, North Carolina State University
Brij M. Khorana, Rose-Hulman Institute of Technology
Carl Kocher, Oregon State University
Robert E. Kribel, Auburn University
Fred Lipschultz, University of Connecticut
Francis A. Liuima, Boston College

Charles E. McFarland, University of Missouri, Rolla
Bruce Morgan, U.S. Naval Academy
Clem Moses, Utica College
Curt Moyer, Clarkson University
A. Wilson Nolle, The University of Texas at Austin
Thomas L. O'Kuma, San Jacinto College North
George Parker, North Carolina State University
William F. Parks, University of Missouri, Rolla
Philip B. Peters, Virginia Military Institute
Joseph W. Rudmin, James Madison University
James H. Smith, University of Illinois at Urbana-Champaign
Edward W. Thomas, Georgia Institute of Technology
Gary Williams, University of California, Los Angeles
George A. Williams, University of Utah
Earl Zwicker, Illinois Institute of Technology

I am especially grateful to Henry Leap for writing and solving many of the new problems that appear in this second edition, for his assistance in reorganizing the problem sets, and in the preparation of the Instructor's Manual with Solutions that accompanies the text. I am indebted to John R. Gordon for his many contributions during the development of this text, his continued encouragement and support, and for his expertise in writing the Student Study Guide With Computer Exercises. I am grateful to David Oliver for developing the computer software which accompanies the Student Study Guide With Computer Exercises. I wish to thank William Ingham, William McFarland, Dorn Peterson, and Joseph W. Rudmin for suggesting new problems for the text. I am grateful to Mario Iona for making many excellent suggestions for improving the figures in the text. I thank Lloyd Black, Jerry Faughn, Henry Leap, and Jim Lehman for locating and/or providing many excellent photographs. I thank Roger W. Clapp, Jr., for checking solutions to all the new problems. I appreciate the assistance of my wife, Elizabeth, and my sons Mark and David in proofreading the galleys, and their cutting and pasting work in preparing the final product. I thank Agatha Brabon, Linda Delosh, Mary Thomas, Georgina Valverde, and Linda Miller for an excellent job in typing various stages of the original manuscript. During the development of this textbook, I have benefited from valuable discussions with many people including Subash Antani, Gabe Anton, Randall Caton, Don Chodrow, Jerry Faughn, John R. Gordon, William Ingham, David Kaup, Len Ketelsen, Henry Leap, H. Kent Moore, Charles McFarland, Frank Moore, Clem Moses, William Parks, Dorn Peterson, Joe Rudmin, Joe Scaturro, Alex Serway, John Serway, Giorgio Vianson, and Harold Zimmerman. Special recognition is due to my mentor and friend, Sam Marshall, a gifted teacher and scientist who helped me sharpen my writing skills while I was a graduate student.

During the preparation of the last five chapters of this text, I benefited from valuable discussions with Don Chodrow, John R. Gordon, and Joe Rudmin. These individuals, together with Elmer Anderson, Larry Hmurcik, and Clem Moses, kindly provided most of the exercises and problems in these chapters. I am most grateful for their contributions and interest in this work.

I am most grateful to the professional staff at Saunders College Publishing for their outstanding work and skills in transforming the manuscript into a most attractive textbook. I am especially indebted to Lloyd Black, Developmental Editor, and Sally Kusch, Project Editor, who kept track of all the details and worked so diligently on the project. I thank John Vondeling, Associate

Publisher, for his great enthusiasm for the project, his friendship, and his confidence in me as an author. I am most appreciative of the excellent artwork by J & R Technical Services, Inc., Tom Mallon, Linda Maugeri, and Larry Ward, and the excellent design work by Edward A. Butler and Carol Bleistine.

I am most grateful to the hundreds of students at Clarkson University who used this text in manuscript form during its development and for the supportive environment provided by both Clarkson University and James Madison University. I also wish to thank the many users of the first edition who submitted suggestions and pointed out errors. With the help of such cooperative efforts, I hope to have achieved my main objective; that is, to provide an effective textbook for the student.

And last, I thank my wonderful family for their continued patience and understanding. The completion of this enormous task would not have been possible without their endless love and faith in me.

Raymond A. Serway
James Madison University
Harrisonburg, Virginia
April 18, 1986

TO THE STUDENT

I feel it is appropriate to offer some words of advice which should be of benefit to you, the student. Before doing so, I will assume that you have read the preface, which describes the various features of the text that will help you through the course.

How To Study

Very often instructors are asked "How should I study physics and prepare for examinations?" There is no simple answer to this question, but I would like to offer some suggestions based on my own experiences in learning and teaching over the years.

First and foremost, maintain a positive attitude towards the subject matter, keeping in mind that physics is the most fundamental of all natural sciences. Other science courses that follow will use the same physical principles, so it is important that you understand and be able to apply the various concepts and theories discussed in the text.

Concepts and Principles

It is essential that you understand the basic concepts and principles *before* attempting to solve assigned problems. This is best accomplished through a careful reading of the textbook before attending your lecture on that material. In the process, it is useful to jot down certain points which are not clear to you. Take careful notes in class, and then ask questions pertaining to those ideas that require clarification. Keep in mind that few people are able to absorb the full meaning of scientific material after one reading. Several readings of the text and notes may be necessary. Your lectures and laboratory work should supplement the text and clarify some of the more difficult material. You should reduce memorization of material to a minimum. Memorizing passages from a text, equations, and derivations does not necessarily mean you understand the material. Your understanding of the material will be enhanced through a combination of efficient study habits, discussions with other students and instructors, and your ability to solve the problems in the text. Ask questions whenever you feel it is necessary.

Study Schedule

It is important to set up a regular study schedule, preferably on a daily basis. Make sure to read the syllabus for the course and adhere to the schedule set by your instructor. The lectures will be much more meaningful if you read the corresponding textual material before attending the lecture. As a general rule, you should devote about two hours of study time for every hour in class. If you are having trouble with the course, seek the advice of the instructor or students who have taken the course. You may find it necessary to seek further instruction from experienced students. Very often, instructors will offer review sessions in addition to regular class periods. It is important that you avoid the practice of delaying study until a day or two before an exam. More often than not, this will lead to disastrous results. Rather than an all night study session, it is better to briefly review the basic concepts and equations, followed by a good night's rest. If you feel in need of additional help in understanding the concepts, preparing for exams, or in problem-solving, we suggest that you acquire a copy of the student study guide which accompanies the text, which should be available at your college bookstore.

Use the Features

You should make full use of the various features of the text discussed in the preface. For example, marginal notes are useful for locating and describing important equations and concepts, while important statements and definitions are highlighted in color. Many useful tables are contained in appendices, but most are incorporated in the text where they are used most often. Appendix B is a convenient review of mathematical techniques. Answers to odd-numbered problems are given at the end of the text, and answers to end-of-chapter questions are provided in the study guide. Exercises (with answers), which follow some worked examples, represent extensions of those examples, and in most cases you are expected to perform a simple calculation. Their purpose is to test your problem-solving skills as you read through the text. An overview of the entire text is given in the table of contents, while the index will enable you to locate specific material quickly. Footnotes are sometimes used to supplement the discussion or to cite other references on the subject. Many chapters include problems that require the use of programmable calculators or computers. These are intended for those courses that place some emphasis on numerical methods. You may want to develop appropriate programs for some of these problems even if they are not assigned by your instructor.

After reading a chapter, you should be able to define any new quantities introduced in that chapter, and discuss the principles and assumptions that were used to arrive at certain key relations. The chapter summaries and the review sections of the study guide should help you in this regard. In some cases, it will be necessary to refer to the index of the text to locate certain topics. You should be able to correctly associate with each physical quantity a symbol used to represent that quantity and the unit in which the quantity is specified. Furthermore, you should be able to express each important relation in a concise and accurate prose statement.

Problem Solving

R.P. Feynman, Nobel laureate in physics, once said, "You do not know anything until you have practiced." In keeping with this statement, I strongly advise that you develop the skills necessary to solve a wide range of problems. Your ability to solve problems will be one of the main tests of your knowledge of physics, and therefore you should try to solve as many problems as possible. It is essential that you understand basic concepts and principles before attempting to solve problems. It is good practice to try to find alternate solutions to the same problem. For example, problems in mechanics can be solved using Newton's laws, but very often an alternative method using energy considerations is more direct. You should not deceive yourself into thinking you understand the problem after seeing its solution in class. You must be able to solve the problem and similar problems on your own.

The method of solving problems should be carefully planned. A systematic plan is especially important when a problem involves several concepts. First, read the problem several times until you are confident you understand what is being asked. Look for any key words that will help you interpret the problem, and perhaps allow you to make certain assumptions. Your ability to interpret the question properly is an integral part of problem solving. You should acquire the habit of writing down the information given in a problem, and decide what quantities need to be found. You might want to construct a table listing quantities given, and quantities to be found. This procedure is sometimes used in the worked examples of the text. After you have decided on the method you feel is appropriate for the situation, proceed with your solution.

I often find that students fail to recognize the limitations of certain formulas or physical laws in a particular situation. It is very important that you understand and remember the assumptions which underlie a particular theory or formalism. For example, certain equations in kinematics apply only to a particle moving with constant acceleration. These equations are not valid for situations in which the acceleration is not constant, such as the motion of an object connected to a spring, or the motion of an object through a fluid.

Experiments

Physics is a science based upon experimental observations. In view of this fact, I recommend that you try to supplement the text through various types of "hands-on" experiments, either at home or in the laboratory. These can be used to test ideas and models discussed in class or in the text. For example, the common "Slinky" toy is excellent for studying traveling waves; a ball swinging on the end of a long string can be used to investigate pendulum motion; various masses attached to the end of a vertical spring or rubber band can be used to determine their elastic nature; an old pair of Polaroid sunglasses and some discarded lenses and magnifying glass are the components of various experiments in optics; you can get an approximate measure of the acceleration of gravity by dropping a ball from a known height by simply measuring the time of its fall with a stopwatch. The list is endless. When physical models are not available, be imaginative and try to develop models of your own.

Scientific Method

All that has been said can be summarized in an approach called the scientific method. The scientific method, which is used in all branches of science, consists of five steps:

1. Recognize the problem.
2. Hypothesize an answer.
3. Predict a result based upon the hypothesis.
4. Devise and perform an experiment to check the hypothesis.
5. Develop a theory which links the confirmed hypothesis to previously existing knowledge.

Someone once said that there are only two professions in which people truly enjoy what they are doing: professional sports and physics. I suspect that this is an exaggeration, but it is true that both fields are exciting and stretch your skills to the limit. It is my sincere hope that you too will find physics an exciting and enjoyable experience, and that you will profit from this experience, regardless of your chosen profession.

Welcome to the exciting world of physics.

Whatever trouble life holds for you, that part of your lives which you spend finding out about things, things that you can tell others about, and that you can learn from them, that part will be essentially a gay, a sunny, a happy life, not untouched by rivalry, maybe not even untouched by an occasional regret that somebody else thought of something that you should have thought of first, but on the whole, one of those nobler parts of the human experience. This makes it true that the life of the scientist is, along with the life of the poet, soldier, prophet, and artist, deeply relevant to man's understanding of his situation and his view of his destiny.

J. ROBERT OPPENHEIMER
Uncommon Sense,
Boston, Birkhauser, 1984.

CONTENTS OVERVIEW

CONTENTS

° These sections are optional.

LIST OF TABLES

PART I
Mechanics

Facts which at first seem improbable will, even in scant explanation, drop the cloak which has hidden them and stand forth in naked and simple beauty.
GALILEO GALILEI

Nature and nature's laws lay hid in the night. God said: "Let Newton be"; and all was light.
ALEXANDER POPE

Physics, the most fundamental physical science, is concerned with the basic principles of the universe. It is the foundation upon which the other physical sciences—astronomy, chemistry, and geology—are based. The beauty of physics lies in the simplicity of the fundamental physical theories and in the manner in which just a small number of fundamental concepts, equations, and assumptions can alter and expand our view of the world around us.

The myriad physical phenomena in our world are a part of one or more of the following five fundamental divisions of physics:

1. Mechanics, which is concerned with the motion of material objects
2. Thermodynamics, which deals with heat, temperature, and the behavior of a large number of particles
3. Electromagnetism, which involves the theory of electricity, magnetism, and electromagnetic waves
4. Relativity, which is a theory describing particles moving at very high speeds
5. Quantum mechanics, a theory dealing with the behavior of particles at the submicroscopic level

The first part of this textbook deals with mechanics, sometimes referred to as classical mechanics or newtonian mechanics. This is an appropriate place to begin an introductory text since many of the basic principles used to understand mechanical systems can later be used to describe such natural phenomena as waves and heat transfer. Furthermore, the laws of conservation of energy and momentum introduced in mechanics retain their importance in the fundamental theories that follow, including the theories of modern physics.

The first serious attempts to develop a theory of motion were provided by the Greek astronomers and philosophers. Although they devised a complex model to describe the motions of heavenly bodies, their model lacked correlation between such motions and the motions of objects on earth. This lack of universality was recognized much later, following a number of careful astronomical investigations by Copernicus, Brahe, and Kepler in the 16th century. In the same period, Galileo attempted to relate the motion of falling bodies and projectiles to the motion of planetary bodies, and Sevin and Hooke were studying forces and their relation to motion. The theory of mechanics reached its peak in 1687 when Newton published his *Mathematical Principles of Natural Philosophy.* This elegant theory, which remained unchallenged for more than 200 years, was based on contributions made by Galileo and others, together with Newton's hypothesis of universal gravitation.

Today, mechanics is of vital importance to students from all disciplines. It is highly successful in describing the motions of material bodies, such as the planets, rockets, and baseballs. In the first part of the text, we shall describe the laws of mechanics and examine a wide range of phenomena that can be understood with these fundamental ideas.

1
Introduction: Physics and Measurement

Physics is a fundamental science concerned with understanding the natural phenomena that occur in our universe. It is a science based on experimental observations and quantitative measurements. The main objective of the scientific approach is to develop physical theories based on fundamental laws that will predict the results of some experiments. Fortunately, it is possible to explain the behavior of many physical systems with a limited number of fundamental laws. These fundamental laws are expressed in the language of mathematics, the tool that provides a bridge between theory and experiment.

Whenever a discrepancy arises between theory and experiment, new theories and concepts must be formulated to remove the discrepancy. Many times a theory is satisfactory under limited conditions; a more general theory might be satisfactory without such limitations. A classic example is Newton's laws of motion, which accurately describe the motion of bodies at normal speeds but do not apply to objects moving at speeds comparable to the speed of light. The special theory of relativity developed by Albert Einstein (1879–1955) successfully predicts the motion of objects at speeds approaching the speed of light and hence is a more general theory of motion.

Classical physics, developed prior to 1900, includes the theories, concepts, laws, and experiments in three major disciplines: (1) classical mechanics, (2) thermodynamics (heat transfer, temperature, and the behavior of a large number of identical particles), and (3) electromagnetism (the study of electric and magnetic phenomena, optics and radiation).

Galileo Galilei (1564–1642) made the first significant contributions to classical mechanics through his work on the laws of motion in the presence of constant acceleration. In the same era, Johannes Kepler (1571–1630) used astronomical observations to develop empirical laws for the motion of planetary bodies.

The most important contributions to classical mechanics were provided by Isaac Newton (1642–1727), who developed classical mechanics as a systematic theory and was one of the originators of the calculus as a mathematical tool. Major developments in classical physics continued in the 18th century. However, thermodynamics and electricity and magnetism were not well understood until the latter part of the 19th century, principally because the apparatus for controlled experiments was either too crude or unavailable until then. Although many electric and magnetic phenomena had been studied earlier, it was the work of James Clerk Maxwell (1831–1879) that provided a unified theory of electromagnetism. In this text we shall treat the various disciplines of classical physics in separate sections; however, we will see that the disciplines of mechanics and electromagnetism are basic to all the branches of classical and modern physics.

A new era in physics, usually referred to as *modern physics,* began near the end of the 19th century. Modern physics developed mainly because of the discovery that many physical phenomena could not be explained by classical physics. The two most important developments in this modern era were the theories of relativity and quantum mechanics. Einstein's theory of relativity completely revolutionized the traditional concepts of space, time, and energy. Among other things, this theory corrected Newton's laws of motion for describing the motion of objects moving at speeds comparable to the speed of light. The theory of relativity also assumes that the speed of light is the upper limit of the speed of an object or signal and shows the equivalence of mass and energy. The formulation of quantum mechanics by a number of distinguished scientists provided a description of physical phenomena at the atomic level.

Scientists are constantly working at improving our understanding of fundamental laws, and new discoveries are being made every day. In many research areas, there is a great deal of overlap between physics, chemistry, and biology. The many technological advances in recent times are the result of the efforts of many scientists, engineers, and technicians. Some of the most notable recent developments are (1) unmanned space missions and manned moon landings, (2) microcircuitry and high-speed computers, and (3) nuclear energy. The impact of such developments and discoveries on our society has indeed been great, and it is very likely that future discoveries and developments will be exciting, challenging, and of great benefit to humanity.

1.1 STANDARDS OF LENGTH, MASS, AND TIME

The laws of physics are expressed in terms of basic quantities that require a clear definition. For example, such physical quantities as force, velocity, volume, and acceleration can be described in terms of more fundamental quantities that in themselves are defined in terms of measurements or comparison with established standards. In mechanics, the three fundamental quantities are length (L), time (T), and mass (M). All other physical quantities in mechanics can be expressed in terms of these.

Obviously, if we are to report the results of a measurement to someone who wishes to reproduce this measurement, a standard must be defined. It would be meaningless if a visitor from another planet were to talk to us about a length of 8 "gliches" if we do not know the meaning of the unit glich. On the other hand, if someone familiar with our system of measurement reports that a wall is 2.0 meters high and our unit of length is defined as 1.0 meter, we then know that the height of the wall is twice our fundamental unit of length. Likewise, if we are told that a person has a mass of 75 kilograms and our unit of mass is defined as 1.0 kilogram, then that person is 75 times as massive as our fundamental unit of mass.[1]

In 1960, an international committee established rules to decide on the latest set of standards for these fundamental quantities. The system that was established is an adaptation of the metric system, and it is called the **International System** (SI) of units. In this system, the units of mass, length, and time are the kilogram, meter, and second, respectively. Other standard SI units established by the committee are those for temperature (the *kelvin*), electric

[1] The need for assigning numerical values to various physical quantities through experimentation was expressed by Lord Kelvin (William Thomson) as follows: "I often say that when you can measure what you are speaking about, and express it in numbers, you should know something about it, but when you cannot express it in numbers, your knowledge is of a meager and unsatisfactory kind."

current (the *ampere*), and luminous intensity (the *candela*). These six fundamental units are the basic SI units. In the study of mechanics, however, we will be concerned only with the units of mass, length, and time.

Mass

The SI unit of mass, the **kilogram,** is defined as the mass of a specific platinum-iridium alloy cylinder kept at the International Bureau of Weights and Measures at Sèvres, France.

This mass standard was established in 1901, and there has been no change since that time because platinum-iridium is an unusually stable alloy. The Sèvres cylinder is 3.9 centimeters in diameter and 3.9 centimeters in height. A duplicate is kept at the National Bureau of Standards in Gaithersburg, Md.

Length

Before 1960, the standard for length, the *meter,* was defined as the distance between two lines on a specific platinum-iridium bar stored under controlled conditions. This standard was abandoned for several reasons, a principal one being that the limited accuracy with which the separation between the lines on the bar can be determined does not meet the present requirements of science and technology. Until recently, the meter was defined as 1 650 763.73 wavelengths of orange-red light emitted from a krypton-86 lamp. However, in October 1983, the meter was redefined as follows:

One meter—the distance traveled by light in vacuum during a time of 1/299 792 458 second.

(*Left*) The National Standard Kilogram No. 20, an accurate copy of the International Standard Kilogram kept at Sèvres, France, is housed under a double bell jar in a vault at the National Bureau of Standards. (Courtesy National Bureau of Standards, U.S. Dept. of Commerce) (*Right*) The primary frequency standard (atomic clock) at the National Bureau of Standards. When operated as a clock, this device keeps time with an accuracy of about 3 millionths of a second per year. (Courtesy National Bureau of Standards, U.S. Dept. of Commerce)

In effect, this latest definition establishes that the speed of light in vacuum is 299 792 458 meters per second.

Time

TABLE 1.1 Mass of Various Bodies (Approximate Values)

	Mass (kg)
Milky Way Galaxy	7×10^{41}
Sun	2×10^{30}
Earth	6×10^{24}
Moon	7×10^{22}
Shark	1×10^{4}
Human	7×10^{1}
Frog	1×10^{-1}
Mosquito	1×10^{-5}
Bacterium	1×10^{-15}
Hydrogen atom	1.67×10^{-27}
Electron	9.11×10^{-31}

Before 1960, the standard of time was defined in terms of the *mean solar day.*[2] Thus, the *mean solar second,* representing the basic unit of time, was originally defined as $(\frac{1}{60})$ $(\frac{1}{60})$ $(\frac{1}{24})$ of a mean solar day. Time that is referenced to the rotation of the earth about its axis is called *universal time.*

In 1967, the second was redefined to take advantage of the high precision that could be obtained using a device known as an *atomic clock.* In this device, the frequencies associated with certain atomic transitions (which are extremely stable and insensitive to the clock's environment) can be measured to an accuracy of one part in 10^{12}. This is equivalent to an uncertainty of less than one second every 30 000 years. Such frequencies are highly insensitive to changes in the clock's environment. Thus, in 1967 the SI unit of time, the *second,* was redefined using the characteristic frequency of a particular kind of cesium atom as the "reference clock":

One second—the time required for a cesium-133 atom to undergo 9 192 631 770 vibrations.

This new standard has the distinct advantage of being "indestructible" and widely reproducible.

The orders of magnitude (approximate values) of various masses, lengths, and time intervals are presented in Tables 1.1 to 1.3. Note the wide range of these quantities.[3] You should study these tables and get a feel for what is meant by a kilogram of mass, for example, or by a time interval of 10^{10} seconds. Systems of units commonly used are the SI or *mks* system, in which the units of

TABLE 1.2 Approximate Values of Some Measured Lengths

	Length (m)
Distance from earth to most remote known quasar	1.4×10^{26}
Distance from earth to most remote known normal galaxies	4×10^{25}
Distance from earth to nearest large galaxy (M 31 in Andromeda)	2×10^{22}
Distance from earth to nearest star (Proxima Centauri)	4×10^{16}
One lightyear	9.46×10^{15}
Mean orbit radius of the earth	1.5×10^{11}
Mean distance from earth to moon	3.8×10^{8}
Mean radius of the earth	6.4×10^{6}
Typical altitude of orbiting earth satellite	2×10^{5}
Length of a football field	9.1×10^{1}
Length of a housefly	5×10^{-3}
Size of smallest dust particles	1×10^{-4}
Size of cells of most living organisms	1×10^{-5}
Diameter of a hydrogen atom	1×10^{-10}
Diameter of an atomic nucleus	1×10^{-14}

[2] A solar day is the time interval between successive appearances of the sun at the highest point it reaches in the sky each day.

[3] If you are unfamiliar with the use of powers of ten (scientific notation), you should review Section B.2 of the mathematical appendix at the back of this book.

TABLE 1.3 Approximate Values of Some Time Intervals

	Interval (s)
Age of the universe	5×10^{17}
Age of the earth	1.3×10^{17}
Average age of a college student	6.3×10^{8}
One year	3.2×10^{7}
One day (time for one revolution of earth about its axis)	8.6×10^{4}
Time between normal heartbeats	8×10^{-1}
Peroid[a] of audible sound waves	1×10^{-3}
Period of typical radio waves	1×10^{-6}
Period of vibration of an atom in a solid	1×10^{-13}
Period of visible light waves	2×10^{-15}
Duration of a nuclear collision	1×10^{-22}
Time for light to cross a proton	3.3×10^{-24}

[a] Period is defined as the time interval of one complete vibration.

mass, length, and time are the kilogram (kg), meter (m), and second (s), respectively; the *cgs* or Gaussian system, in which the units of length, mass, and time are the centimeter (cm), gram (g), and second, respectively; and the British engineering system (sometimes called the conventional system), in which the units of length, mass, and time are the foot (ft), slug, and second, respectively. Throughout most of this text we shall use SI units since they are almost universally accepted in science and industry. We will make some limited use of conventional units in the study of classical mechanics.

Some of the most frequently used prefixes for the various powers of ten and their abbreviations are listed in Table 1.4. For example, 10^{-3} m is equivalent to 1 millimeter (mm), and 10^3 m is 1 kilometer (km). Likewise, 1 kg is 10^3 g, and 1 megavolt (MV) is 10^6 volts.

TABLE 1.4 Some Prefixes for Powers of Ten

Power	Prefix	Abbreviation
10^{-18}	atto	a
10^{-15}	femto	f
10^{-12}	pico	p
10^{-9}	nano	n
10^{-6}	micro	μ
10^{-3}	milli	m
10^{-2}	centi	c
10^{3}	kilo	k
10^{6}	mega	M
10^{9}	giga	G
10^{12}	tera	T
10^{15}	peta	P
10^{18}	exa	E

1.2 DENSITY AND ATOMIC MASS

Any piece of matter tends to resist any change in its motion. This property of matter is called *inertia*. The word *mass* is used to describe the amount of inertia associated with a particular body.

A fundamental property of any substance is its **density** ρ (Greek letter rho), defined as *mass per unit volume* (a table of the letters in the Greek alphabet is provided at the back of the book):

$$\rho \equiv \frac{m}{V} \tag{1.1}$$

For example, aluminum has a density of 2.70 g/cm^3, and lead has a density of 11.3 g/cm^3. Therefore, a piece of aluminum of volume 10 cm^3 has a mass of 27.0 g, while an equivalent volume of lead would have a mass of 113 g. A list of densities for various substances is given in Table 1.5.

The difference in density between aluminum and lead is due, in part, to their different *atomic weights;* the atomic weight of lead is 207 and that of aluminum is 27. However, the ratio of atomic weights, $207/27 = 7.67$, does not correspond to the ratio of densities, $11.3/2.70 = 4.19$. The discrepancy is due to the difference in atomic spacings and atomic arrangements in their crystal structures.

TABLE 1.5 Densities of Various Substances

Substance	Density ρ (kg/m^3)
Gold	19.3×10^3
Uranium	18.7×10^3
Lead	11.3×10^3
Copper	8.93×10^3
Iron	7.86×10^3
Aluminum	2.70×10^3
Magnesium	1.75×10^3
Water	1.00×10^3
Air	0.0013×10^3

All ordinary matter consists of atoms, and each atom is made up of electrons and a nucleus. Practically all of the mass of an atom is contained in the nucleus, which consists of protons and neutrons. Thus we can understand why the atomic weights of the various elements differ. The mass of a nucleus is measured relative to the mass of an atom of the carbon-12 isotope (this isotope of carbon has six protons and six neutrons).

The mass of ^{12}C is defined to be 12 atomic mass units (u), where $1\ u = 1.66 \times 10^{-27}$ kg. In these units, the proton and neutron have masses of about 1 u. More precisely,

$$\text{mass of proton} = 1.0073\ u$$
$$\text{mass of neutron} = 1.0087\ u$$

The mass of the nucleus of ^{27}Al is *approximately* 27 u. In fact, a more precise calculation shows that the nuclear mass is always slightly *less* than the combined mass of the protons and neutrons making up the nucleus. The processes of nuclear fission and nuclear fusion are based on this mass difference.

One mole of any element (or compound) consists of Avogadro's number, N_A, of molecules of the substance. Avogadro's number is defined so that one mole of carbon-12 atoms has a mass of exactly 12 g. Its value has been found to be $N_A = 6.02 \times 10^{23}$ molecules/mole. For example, one mole of aluminum has a mass of 27 g, and one mole of lead has a mass of 207 g. Although the two have different masses, one mole of aluminum contains the same number of atoms as one mole of lead. Since there are 6.02×10^{23} atoms in one mole of any element, the mass per atom is given by

Atomic mass

$$m = \frac{\text{atomic weight}}{N_A} \tag{1.2}$$

For example, the mass of an aluminum atom is

$$m = \frac{27\ \text{g/mole}}{6.02 \times 10^{23}\ \text{atoms/mole}} = 4.5 \times 10^{-23}\ \text{g/atom}$$

Note that 1 u is equal to N_A^{-1} g.

EXAMPLE 1.1. How Many Atoms in the Cube? A solid cube of aluminum (density 2.7 g/cm³) has a volume of 0.2 cm³. How many aluminum atoms are contained in the cube?

Solution: Since the density equals mass per unit volume, the mass of the cube is

$$\rho V = (2.7\ \text{g/cm}^3)(0.2\ \text{cm}^3) = 0.54\ \text{g}.$$

To find the number of atoms, N, we can set up a proportion using the fact that one mole of aluminum (27 g) contains 6.02×10^{23} atoms:

$$\frac{6.02 \times 10^{23}\ \text{atoms}}{27\ \text{g}} = \frac{N}{0.54\ \text{g}}$$

$$N = \frac{(0.54\ \text{g})(6.02 \times 10^{23}\ \text{atoms})}{27\ \text{g}} = 1.2 \times 10^{22}\ \text{atoms}$$

1.3 DIMENSIONAL ANALYSIS

The word *dimension* has a special meaning in physics. It usually denotes the physical nature of a quantity. Whether a distance is measured in units of feet or meters or furlongs, it is a distance. We say its dimension is *length*.

The symbols that will be used to specify length, mass, and time are L, M, and T, respectively. We will often use brackets [ ] to denote the dimensions

TABLE 1.6 Dimensions of Area, Volume, Velocity, and Acceleration

System	Area (L^2)	Volume (L^3)	Velocity (L/T)	Acceleration (L/T^2)
SI	m^2	m^3	m/s	m/s^2
cgs	cm^2	cm^3	cm/s	cm/s^2
British engineering (conventional)	ft^2	ft^3	ft/s	ft/s^2

of a physical quantity. For example, in this notation the dimensions of velocity, v, are written $[v] = L/T$, and the dimensions of area, A, are $[A] = L^2$. The dimensions of area, volume, velocity, and acceleration are listed in Table 1.6, along with their units in the three common systems. The dimensions of other quantities, such as force and energy, will be described as they are introduced in the text.

In many situations, you may be faced with having to derive or check a specific formula. Although you may have forgotten the details of the derivation, there is a useful and powerful procedure called *dimensional analysis* that can be used to assist in the derivation or to check your final expression. This procedure should be used whenever an equation is not understood and should help minimize the rote memorization of equations. Dimensional analysis makes use of the fact that *dimensions can be treated as algebraic quantities.* That is, quantities can be added or subtracted only if they have the same dimensions. Furthermore, the terms on each side of an equation must have the same dimensions. By following these simple rules, you can use dimensional analysis to help determine whether or not an expression has the correct form, because the relationship can be correct only if the dimensions on each side of the equation are the same.

To illustrate this procedure, suppose you wish to derive a formula for the distance x traveled by a car in a time t if the car starts from rest and moves with constant acceleration a. In Chapter 3, we shall find that the correct expression for this special case is $x = \frac{1}{2}at^2$. Let us check the validity of this expression from a dimensional analysis approach.

The quantity x on the left side has the dimension of length. In order for the equation to be dimensionally correct, the quantity on the right side must also have the dimension of length. We can perform a dimensional check by substituting the basic units for acceleration, L/T^2 and time, T, into the equation. That is, the dimensional form of the equation $x = \frac{1}{2}at^2$ can be written as

$$L = \frac{L}{\cancel{T^2}} \cdot \cancel{T^2} = L$$

The units of time cancel as shown, leaving the unit of length.

A more general procedure of dimensional analysis is to set up an expression of the form

$$x \propto a^n t^m$$

when n and m are exponents that must be determined and the symbol $\propto$ indicates a proportionality. This relationship is only correct if the dimensions of both sides are the same. Since the dimension of the left side is length, the dimension of the right side must also be length. That is,

$$[a^n t^m] = L$$

Since the dimensions of acceleration are L/T^2 and the dimension of time is T,

$$(L/T^2)^n T^m = L$$

or

$$L^n T^{m-2n} = L$$

Since the exponents of L and T must be the same on both sides, we see that $n = 1$ and $m = 2$. Therefore, we conclude that

$$x \propto at^2$$

This result differs by a factor of 2 from the correct expression, which is given by $x = \frac{1}{2}at^2$.

EXAMPLE 1.2. Analysis of an Equation
Show that the expression $v = v_0 + at$ is dimensionally correct, where v and v_0 represent velocities, a is acceleration, and t is a time interval.

Solution: Since

$$[v] = [v_0] = L/T$$

and the dimensions of acceleration are L/T^2, the dimensions of at are

$$[at] = (L/T^2)(T) = L/T$$

and the expression is dimensionally correct. On the other hand, if the expression were given as $v = v_0 + at^2$, it would be dimensionally *incorrect.* Try it and see!

EXAMPLE 1.3. Analysis of a Power Law
Suppose we are told that the acceleration of a particle moving in a circle of radius r with uniform velocity v is proportional to some power of r, say r^n, and some power of v, say v^m. How can we determine the powers of r and v?

Solution: Let us take a to be

$$a = kr^n v^m$$

where k is a dimensionless constant. With the known dimensions of a, r, and v, we see that the dimensional equation must be

$$L/T^2 = L^n(L/T)^m = L^{n+m}/T^m$$

This dimensional equation is balanced under the conditions

$$n + m = 1 \quad \text{and} \quad m = 2$$

Therefore, $n = -1$ and we can write the acceleration

$$a = kr^{-1}v^2 = k\frac{v^2}{r}$$

When we discuss uniform circular motion later, we shall see that $k = 1$.

1.4 CONVERSION OF UNITS

Sometimes it is necessary to convert units from one system to another. Conversion factors between the SI and conventional units of length are as follows:

1 mile = 1609 m = 1.609 km 1 ft = 0.3048 m = 30.48 cm
1 m = 39.37 in. = 3.281 ft 1 in. = 0.0254 m = 2.54 cm

A more complete list of conversion factors can be found in Appendix A. Units can be treated as algebraic quantities that can cancel each other. For example, suppose we wish to convert 15.0 in. to centimeters. Since 1 in. = 2.54 cm (exactly), we find that

$$15.0 \text{ in.} = (15.0 \cancel{\text{in.}})\left(2.54 \frac{\text{cm}}{\cancel{\text{in.}}}\right) = 38.1 \text{ cm}$$

Can you perform the conversion? (Photo Ohio Department of Transportation)

EXAMPLE 1.4. The Density of a Cube
The mass of a solid cube is 856 g and each edge has a length of 5.35 cm. Determine the density ρ of the cube in SI units.

Solution: Since 1 g = 10^{-3} kg and 1 cm = 10^{-2} m, the mass, m, and volume, V, in SI units are given by

$$m = 856\,\cancel{g} \times 10^{-3}\ \text{kg}/\cancel{g} = 0.856\ \text{kg}$$

$$V = L^3 = (5.35\,\cancel{cm} \times 10^{-2}\ \text{m}/\cancel{cm})^3$$

$$= (5.35)^3 \times 10^{-6}\ \text{m}^3 = 1.53 \times 10^{-4}\ \text{m}^3$$

Therefore

$$\rho = \frac{m}{V} = \frac{0.856\ \text{kg}}{1.53 \times 10^{-4}\ \text{m}^3} = 5.60 \times 10^3\ \text{kg/m}^3$$

1.5 ORDER-OF-MAGNITUDE CALCULATIONS

It is often useful to compute an approximate answer to a given physical problem even where little information is available. Such results can then be used to determine whether or not a more precise calculation is necessary. These approximations are usually based on certain assumptions, which must be modified if more precision is needed. Thus, we will sometimes refer to an *order of magnitude* of a certain quantity as the power of ten of the number that describes that quantity. Usually, when an order-of-magnitude calculation is made, the results are reliable to within a factor of 10. If a quantity increases in value by three orders of magnitude, this means that its value is increased by a factor of $10^3 = 1000$.

The spirit of attempting order-of-magnitude calculations, sometimes referred to as "guesstimates" or "ball-park figures," is given in the following quotation: "Make an estimate before every calculation, try a simple physical argument . . . before every derivation, guess the answer to every puzzle. Courage: no one else needs to know what the guess is."[4]

[4] E. Taylor and J.A. Wheeler, *Spacetime Physics*, San Francisco, W.H. Freeman, 1966, p. 60.

EXAMPLE 1.5. The Number of Atoms in a Solid
Estimate the number of atoms in 1 cm^3 of a solid.

Solution: From Table 1.2 we note that the diameter of an atom is about 10^{-10} m. Thus, if in our model we assume that the atoms in the solid are solid spheres of this diameter, then the volume of each sphere is about 10^{-30} m^3 (more precisely, volume $= 4\pi r^3/3 = \pi d^3/6$, where $r = d/2$). Therefore, since 1 $cm^3 = 10^{-6}$ m^3, the number of atoms in the solid is of the order of $10^{-6}/10^{-30} = 10^{24}$ atoms.

A more precise calculation would require knowledge of the density of the solid and the mass of each atom. However, our estimate agrees with the more precise calculation to within a factor of 10. (This same approach should be used for Problem 21.)

EXAMPLE 1.6. How Much Gas Do We Use?
Estimate the number of gallons of gasoline used by all U.S. cars each year.

Solution: Since there are about 200 million people in the United States, an estimate of the number of cars in the country is 40 million (assuming one car and five people per family). We must also estimate that the average distance traveled per year is 10 000 miles. If we assume a gasoline consumption of 20 mi/gal, each car uses about 500 gal/year. Multiplying this by the total number of cars in the United States gives an estimated total consumption of 2×10^{10} gal. This corresponds to a yearly consumer expenditure of over 20 billion dollars! This is probably a low estimate since we haven't accounted for commercial consumption and for such factors as two-car families.

1.6 SIGNIFICANT FIGURES

All real measurements of quantities have some degree of inaccuracy. Whenever a physical quantity is measured, both the value and the precision of the measured quantity are important. For example, if observer A measures the speed of an object to be 5.38 m/s to a precision of 1%, the result could be expressed (5.38 ± 0.05) m/s. Therefore the true value lies between 5.33 m/s and 5.43 m/s. On the other hand, if an independent measurement is made on the same object by observer B, with a precision of only 3%, a value of (5.25 ± 0.16) m/s should be reported. In either case, all three digits in the measured value are significant; however, the last digit is uncertain to some degree. This uncertainty will depend on many factors, such as the quality of the instruments used, experimental technique, and human error. The following rule should be followed when reporting the accuracy of a measurement: *The last figure in the measurement should be the first uncertain figure.* For example, it would be wrong to claim that the speed of an object is 5.384 m/s if the digit 8 is uncertain.

The following is a good rule of thumb to use as a rough guide in determining the number of significant figures that can be claimed. *When multiplying several quantities, the number of significant figures in the final answer is the same as the number of significant figures in the least accurate of the quantities being multiplied. The same rule applies to division.* For example, if we were to perform the multiplication 3.60×5.387, the result would be 19.4 and not 19.3932. We can only claim three significant figures since the less accurate number, 3.60, contains three significant figures.

The presence of zeros in an answer may also be misinterpreted. For example, suppose the mass of an object is measured to be 1500 g. This value is ambiguous because it is not known whether the last two zeros are being used to locate the decimal point or whether they represent significant figures in the measurement. In order to remove this ambiguity, it is common to use scientific notation to indicate the number of significant figures. In this case, we would express the mass as 1.5×10^3 g if there are two significant figures in the measured value and 1.50×10^3 g if there are three significant figures.

Finally, *when numbers are added (or subtracted), the number of decimal places in the result should equal the smallest number of decimal places of any*

Could this be the result of poor data analysis? (Photo Mill Valley, CA, University Science Books, 1982)

term in the sum. For example, if we wish to compute 123 + 5.35, the answer would be 128 and not 128.35. When performing a series of calculations, it is best to round off your answer at the very end of the problem.

Throughout this book, *we shall generally assume that the given data are precise enough to yield an answer having three significant figures.* Thus, if we state that a jogger runs a distance of 5 m, it is to be understood that the distance covered is 5.00 m. Likewise, if the speed of a car is given as 23 m/s, its value is understood to be 23.0 m/s.

EXAMPLE 1.7. The Area of a Rectangle

A rectangular plate has a length of (21.3 ± 0.2) cm and a width of (9.80 ± 0.10) cm. Find the area of the plate and the uncertainty in the calculated area.

Solution:

$$\text{Area} = \ell w = (21.3 \pm 0.2)\ \text{cm} \times (9.80 \pm 0.10)\ \text{cm}$$

$$\approx (21.3 \times 9.80 \pm 21.3 \times 0.10 \pm 9.80 \times 0.2)\ \text{cm}^2$$

$$\approx (209 \pm 4)\ \text{cm}^2$$

Note that the input data were given only to three significant figures, so we cannot claim any more in our result. Furthermore, you should realize that the uncertainty in the product (2%) is approximately equal to the sum of the uncertainties in the length and width (each uncertainty is about 1%).

1.7 MATHEMATICAL NOTATION

Many mathematical symbols will be used throughout this book, some of which you are surely aware of, such as the symbol = to denote the equality of two quantities.

The symbol $\propto$ is used to denote a proportionality. For example, $y \propto x^2$ means that y is proportional to the square of x.

The symbol $<$ means *less than,* and $>$ means *greater than.* For example, $x > y$ means x is greater than y.

The symbol $\ll$ means *much less than,* and $\gg$ means *much greater than.*

The symbol $\approx$ is used to indicate that two quantities are *approximately equal* to each other.

The symbol $\equiv$ means *is defined as.* This is a stronger statement than a simple $=$.

It is convenient to use a symbol to indicate the change in a quantity. For example, Δx (read delta x) means the *change in the quantity* x. (It does not mean the product of Δ and x). For example, if x_i is the initial position of a particle and x_f is its final position, then the *change in position* is written

$$\Delta x = x_f - x_i$$

We will often have occasion to sum several quantities. A useful abbreviation for representing such a sum is the Greek letter Σ (capital sigma). Suppose we wish to sum a set of five numbers represented by x_1, x_2, x_3, x_4, and x_5. In the abbreviated notation, we would write the sum

$$x_1 + x_2 + x_3 + x_4 + x_5 \equiv \sum_{i=1}^{5} x_i$$

where the subscript i on a particular x represents any one of the numbers in the set. For example, if there are five masses in a system, m_1, m_2, m_3, m_4, and m_5, the *total* mass of the system $M = m_1 + m_2 + m_3 + m_4 + m_5$ could be expressed

$$M = \sum_{i=1}^{5} m_i$$

Finally, the *magnitude* of a quantity x, written $|x|$, is simply the absolute value of that quantity. The magnitude of x is *always positive,* regardless of the sign of x. For example, if $x = -5$, $|x| = 5$; if $x = 8$, $|x| = 8$.

A list of these symbols and their meanings is given on the back endsheet.

1.8 SUMMARY

Mechanical quantities can be expressed in terms of three fundamental quantities, *mass, length,* and *time,* which have the units *kilograms* (kg), *meters* (m), and *seconds* (s), respectively, in the SI system. It is often useful to use the *method of dimensional analysis* to check equations and to assist in deriving expressions.

The **density** of a substance is defined as its *mass per unit volume.* Different substances have different densities mainly because of differences in their atomic masses and atomic arrangements.

The number of atoms in one mole of any element or compound is called **Avogadro's number,** N_A, which has the value 6.02×10^{23} atoms/mole.

QUESTIONS

1. What types of natural phenomena could serve as alternative time standards?
2. The height of a horse is sometimes given in units of "hands." Why do you suppose this is a poor standard of length?
3. Express the following quantities using the prefixes given in Table 1.4: (a) 3×10^{-4} m, (b) 5×10^{-5} s, (c) 72×10^2 g.
4. Does a dimensional analysis give any information on constants of proportionality that may appear in an algebraic expression? Explain.

5. Suppose that two quantities A and B have different dimensions. Determine which of the following arithmetic operations *could* be physically meaningful: (a) $A + B$, (b) A/B, (c) $B - A$, (d) AB.

6. What accuracy is implied in an order-of-magnitude calculation?

7. Apply an order-of-magnitude calculation to an everyday situation you might encounter. For example, how far do you walk or drive each day?

PROBLEMS

Section 1.2 Density and Atomic Mass

1. Calculate the density of a solid cube that measures 5 cm on each side and has a mass of 350 g.
2. A solid sphere is to be made out of copper, which has a density of 8.93 g/cm^3. If the mass of the sphere is to be 475 g, what radius must it have?
3. A hollow cylindrical container has a length of 800 cm and an inner radius of 30 cm. If the cylinder is completely filled with water, what is the mass of the water? Assume 1.0 g/cm^3 as the density of water.
4. Calculate the mass of an atom of (a) helium, (b) iron, and (c) lead. Give your answers in atomic mass units and in grams. The atomic weights are 4, 56, and 207, respectively, for the atoms given.
5. A small particle of iron in the shape of a cube is observed under a microscope. The edge of the cube is 5×10^{-6} cm long. Find (a) the mass of the cube and (b) the number of iron atoms in the particle. The atomic weight of iron is 56, and its density is 7.86 g/cm^3.

Section 1.3 Dimensional Analysis

6. Show that the expression $x = vt + \frac{1}{2}at^2$ is dimensionally correct, where x is a coordinate and has units of length, v is velocity, a is acceleration, and t is time.
7. The displacement of a particle when moving under uniform acceleration is some function of the time and the acceleration. Suppose we write this displacement $s = ka^mt^n$, where k is a dimensionless constant. Show by dimensional analysis that this expression is satisfied if $m = 1$ and $n = 2$. Can this analysis give the value of k?
8. The square of the speed of an object undergoing a uniform acceleration a is some function of a and the displacement s, according to the expression given by $v^2 = ka^ms^n$, where k is a dimensionless constant. Show by dimensional analysis that this expression is satisfied if $m = n = 1$.
9. Suppose that the displacement of a particle is related to time according to the expression $s = ct^3$. What are the dimensions of the constant c?
10. (a) One of the fundamental laws of motion states that the acceleration of an object is directly proportional to the resultant force on it and inversely proportional to its mass. From this statement, determine the dimensions of force. (b) The newton (N) is the SI unit of force. According to your results for (a), how can you express a newton using the SI fundamental units of mass, length, and time?

Section 1.4 Conversion of Units

11. Convert the volume 8.50 in.^{3} to m^3, recalling that 1 in. = 2.54 cm and 1 cm = 10^{-2} m.
12. A rectangular building lot is 100.0 ft by 150.0 ft. Determine the area of this lot in m^2.
13. An object in the shape of a rectangular parallelepiped measures 2.0 in. $\times$ 3.5 in. $\times$ 6.5 in. Determine the volume of the object in m^3.
14. A creature moves at a speed of 5 furlongs per fortnight (not a very common unit of speed). Given that 1 furlong = 220 yards and 1 fortnight = 14 days, determine the speed of the creature in m/s. (The creature is probably a snail.)
15. A solid piece of lead has a mass of 23.94 g and a volume of 2.10 cm^3. From these data, calculate the density of lead in SI units (kg/m^3).
16. Estimate the age of the earth in years using the data in Table 1.3 and the appropriate conversion factors.
17. The proton, which is the nucleus of the hydrogen atom, can be pictured as a sphere of whose diameter is 3×10^{-13} cm having a mass of 1.67×10^{-24} g. Determine the density of the proton in SI units and compare this number with the density of lead, 1.14×10^4 kg/m^3.
18. Using the fact that the speed of light in free space is about 3.00×10^8 m/s, determine how many miles a pulse from a laser beam will travel in one hour.
19. Radio waves are electromagnetic and travel at a speed of about 3.0×10^8 m/s in free space. Use this fact and the data in Table 1.2 to determine the time it would take an electromagnetic pulse to make a round trip from the earth to Proxima Centauri, the star nearest the earth.
20. The mean radius of the earth is 6.37×10^6 m, and that of the moon is 1.74×10^8 cm. From these data calculate (a) the ratio of the earth's surface area to that of the moon and (b) the ratio of the earth's volume to that of the moon. Recall that the surface area of a sphere is $4\pi r^2$ and the volume of a sphere is $\frac{4}{3}\pi r^3$.
21. The mass of a copper atom is 1.06×10^{-22} g, and the density of copper is 8.9 g/cm^3. Determine the number of atoms in 1 cm^3 of copper and compare the result with the estimate in Example 1.5.
22. Aluminum is a very lightweight metal, with a density of 2.7 g/cm^3. What is the weight in pounds of a solid sphere of aluminum of radius 50 cm? The result might surprise you. (*Note:* A 1-kg mass corresponds to a weight of 2.2 pounds.)

Section 1.5 Order-of-Magnitude Calculations

23. Estimate the total number of times the heart of a human beats in an average lifetime of 70 years. (See Table 1.3 for data.)
24. Estimate the number of Ping–Pong balls that can be packed into an average-size room (without crushing them).
25. Soft drinks are commonly sold in aluminum containers. Estimate the number of such containers thrown away each year by U.S. consumers. Approximately how many tons of aluminum does this represent?
26. Approximately how many raindrops fall on a 1-acre lot during a 1-in. rainfall?
27. Determine the approximate number of bricks needed to face all four sides of an average-size home.
28. Estimate the number of piano tuners living in New York City. This question was raised by E. Fermi, a world-famous nuclear physicist of the 1940s.

Section 1.6 Significant Figures

29. A particular hamburger chain advertises that it has sold more than 50 billion hamburgers. Estimate how many pounds of hamburger meat have been used by the restaurant chain and how many head of cattle were required to furnish the meat.
30. Determine the number of significant figures in the following numbers: (a) 23 cm (b) 3.589 s (c) 4.67×10^3 m/s (d) 0.0032 m.
31. Calculate (a) the circumference of a circle of radius 3.5 cm and (b) the area of a circle of radius 4.65 cm.
32. Carry out the following arithmetic operations: (a) the sum of the numbers 756, 37.2, 0.83, and 2.5; (b) the product 3.2×3.563; (c) the product $5.6 \times \pi$.
33. If the length and width of a rectangular plate are measured to be (15.30 ± 0.05) cm and (12.80 ± 0.05) cm, find the area of the plate and the uncertainty in the calculated area.
34. The *radius* of a solid sphere is measured to be (6.50 ± 0.20) cm, and its mass is measured to be (1.85 ± 0.02) kg. Determine the density of the sphere in kg/m^3 and the uncertainty in the density.

2
Vectors

Mathematics is the basic tool used by scientists and engineers to describe the behavior of physical systems. Physical quantities that have both numerical and directional properties are represented by vectors. Some examples of vector quantities are force, velocity, and acceleration. This chapter is primarily concerned with vector algebra and with some general properties of vectors. The addition and subtraction of vectors will be discussed, together with some common applications to physical situations. Discussion of the products of vectors will be delayed until these operations are needed.[1]

Vectors will be used throughout this text, and it is imperative that you master both their graphical and algebraic properties.

2.1 COORDINATE SYSTEMS AND FRAMES OF REFERENCE

Many aspects of physics deal in some form or other with locations in space. For example, the mathematical description of the motion of an object requires a method for describing the position of the object. Thus, it is perhaps fitting that we first discuss how one describes the position of a point in space. This is accomplished by means of coordinates. A point on a line can be described with one coordinate. A point in a plane is located with two coordinates, and three coordinates are required to locate a point in space.

A coordinate system used to specify locations in space consists of

1. A fixed reference point O, called the origin
2. A set of specified axes or directions
3. Instructions that tell us how to label a point in space relative to the origin and axes.

One convenient coordinate system that we will use frequently is the *cartesian coordinate system*, sometimes called the *rectangular coordinate system*. Such a system in two dimensions is illustrated in Figure 2.1. An arbitrary point in this system is labeled with the coordinates (x, y). Positive x is taken to the right of the origin, and positive y is upward from the origin. Negative x is to the left of the origin, and negative y is downward from the origin. For example, the point P, which has coordinates (5, 3), may be reached by first going 5 meters to the right of the origin and 3 meters above the origin. Similarly, the point Q has coordinates $(-3, 4)$, corresponding to going 3 meters to the left of the origin and 4 meters above the origin.

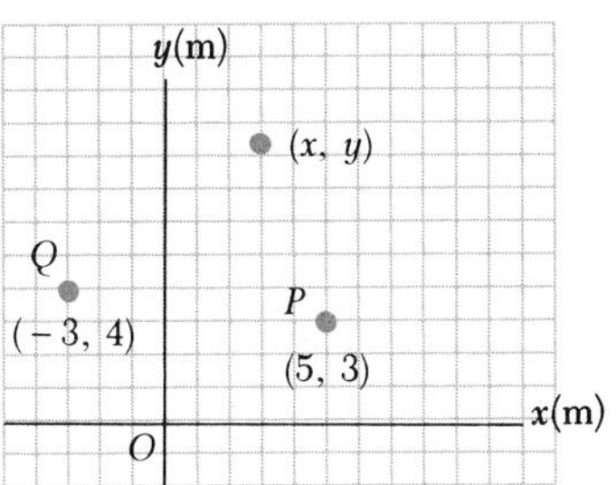

Figure 2.1 Designation of points in a cartesian coordinate system. Any point is labeled with coordinates (x, y).

[1] The dot, or scalar, product is discussed in Section 7.3, and the cross, or vector, product is introduced in Section 11.1.

Sometimes it is more convenient to represent a point in a plane by its *plane polar coordinates*, (r, θ), as in Figure 2.2a. In this coordinate system, r is the distance from the origin to the point having cartesian coordinates (x, y) and θ is the angle between r and a fixed axis, usually measured counterclockwise from the positive x axis. From the right triangle in Figure 2.2b, we find $\sin\theta = y/r$ and $\cos\theta = x/r$. (A review of trigonometric functions is given in Appendix B.4.) Therefore, starting with plane polar coordinates, one can obtain the cartesian coordinates through the equations

$$x = r\cos\theta \tag{2.1}$$

$$y = r\sin\theta \tag{2.2}$$

Furthermore, it follows that

$$\tan\theta = y/x \tag{2.3}$$

and

$$r = \sqrt{x^2 + y^2} \tag{2.4}$$

You should note that these expressions relating the coordinates (x, y) to the coordinates (r, θ) apply only when θ is defined as in Figure 2.2a, where positive θ is an angle measured *counterclockwise* from the positive x axis. If the reference axis for the polar angle θ is chosen to be other than the positive x axis, or the sense of increasing θ is chosen differently, then the corresponding expressions relating the two sets of coordinates will change.

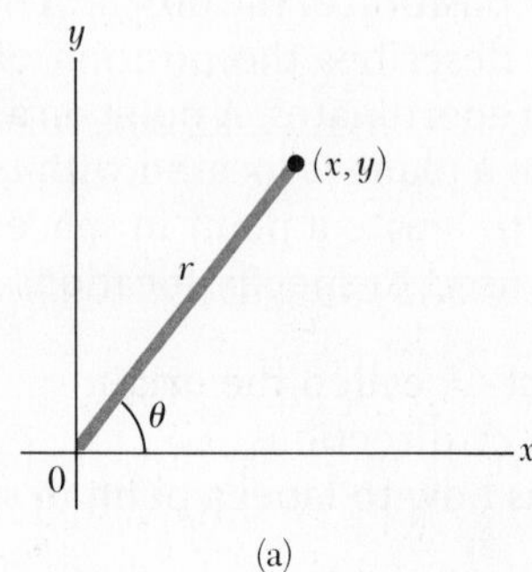

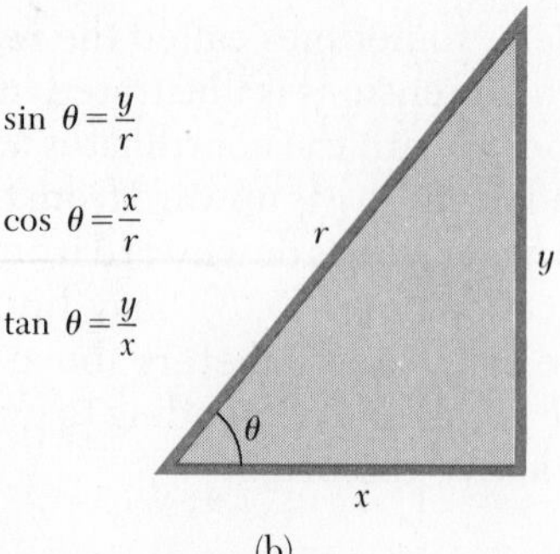

Figure 2.2 (a) The plane polar coordinates of a point are represented by the distance r and the angle θ. (b) The right triangle used to relate (x, y) to (r, θ).

EXAMPLE 2.1. Polar Coordinates

The cartesian coordinates of a point are given by $(x, y) = (-3.5, -2.5)$ meters as in Figure 2.3. Find the polar coordinates of this point.

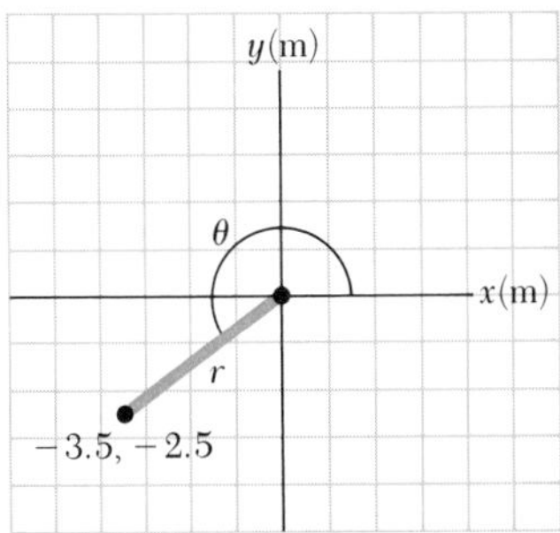

Figure 2.3 (Example 2.1).

Solution:

$$r = \sqrt{x^2 + y^2} = \sqrt{(-3.5)^2 + (-2.5)^2} = 4.3 \text{ meters}$$

$$\tan \theta = \frac{y}{x} = \frac{-2.5}{-3.5} = 0.714$$

$$\theta = 216°$$

Note that you must use the signs of x and y to find that θ is in the third quadrant of the coordinate system. That is, $\theta = 216°$ and not $36°$.

2.2 VECTORS AND SCALARS

The physical quantities that we shall encounter in this text can be placed in one or the other of two categories: they are either scalars or vectors. A scalar is a quantity that is completely specified by a number with appropriate units. That is,

> A **scalar** has only magnitude and no direction. On the other hand, a **vector** is a physical quantity that must be specified by both magnitude and direction.

The number of apples in a basket is an example of a scalar quantity. If you are told there are 38 apples in the basket, this completely specifies the required information; no direction is required. Other examples of scalars are temperature, volume, mass, and time intervals. The rules of ordinary arithmetic are used to manipulate scalar quantities.

Force is an example of a vector quantity. To describe completely the force on an object, we must specify both the direction of the applied force and a number to indicate the force's magnitude. When the motion of an object is described, we must specify both how fast it is moving and the direction of its motion.

Another simple example of a vector quantity is the **displacement** of a particle, defined as the *change in the position* of the particle. Suppose the particle moves from some point O to the point P along a straight path, as in Figure 2.4. We represent this displacement by drawing an arrow from O to P, where the tip of the arrow represents the direction of the displacement and the length of the arrow represents the magnitude of the displacement. If the particle travels along some other path from O to P, such as the broken line in Figure 2.4, its displacement is still $\mathbf{OP}$. The vector displacement along any indirect path from O and P is defined as being equivalent to the displacement for the direct path from O to P. Thus, *the displacement of a particle is completely known if its initial and final coordinates are known.* The path need not be specified. In other words, *the displacement is independent of the path*, if the end points of the path are fixed.

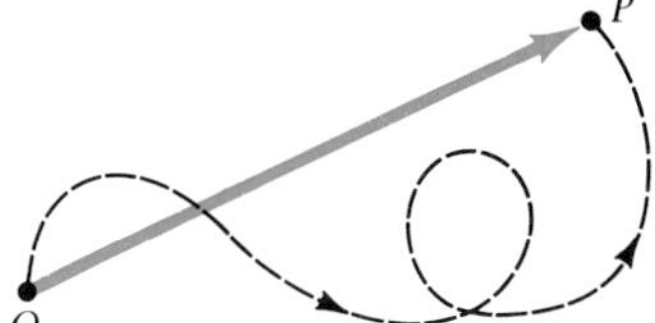

Figure 2.4 As a particle moves from O to P along the broken line, its displacement vector is the arrow drawn from O to P.

If the particle moves along the x axis from position x_i to position x_f, as in Figure 2.5, its displacement is given by $x_f - x_i$. As mentioned in Chapter 1, we use the Greek letter delta (Δ) to denote the *change* in a quantity. Therefore, we write the change in the position of the particle (the displacement)

Definition of displacement along a line

$$\Delta x = x_f - x_i \tag{2.5}$$

From this definition, we see that Δx is positive if x_f is greater than x_i and negative if x_f is less than x_i. For example, if a particle changes its position from $x_i = -3$ units to $x_f = 5$ units, its displacement is 8 units.

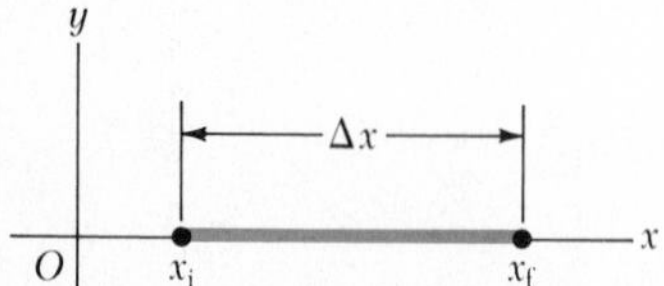

Figure 2.5 A particle moving along the x axis from x_i to x_f undergoes a displacement $\Delta x = x_f - x_i$.

There are many physical quantities in addition to displacement that are vectors. These include velocity, acceleration, force, and momentum, all of which will be defined in later chapters. In this text, we will use boldface letters, such as **A**, to represent an arbitrary vector. Another common method for vector notation that you should be aware of is to use an arrow over the letter: $\vec{A}$. The magnitude of the vector **A** is written A or, alternatively, |**A**|. The magnitude of a vector has physical units, such as cm for displacement or m/s for velocity, as discussed in Chapter 1. Vectors combine according to special rules, which will be discussed in later sections.

2.3 SOME PROPERTIES OF VECTORS

Equality of Two Vectors Two vectors **A** and **B** are defined to be equal if they have the same magnitude and point in the same direction. That is, **A** = **B** only if $A = B$ *and* they act along parallel direction. For example, all the vectors in Figure 2.6 are equal even though they have different starting points. This property allows us to translate a vector parallel to itself in a diagram without affecting the vector. In fact, any true vector can be moved parallel to itself without affecting the vector.

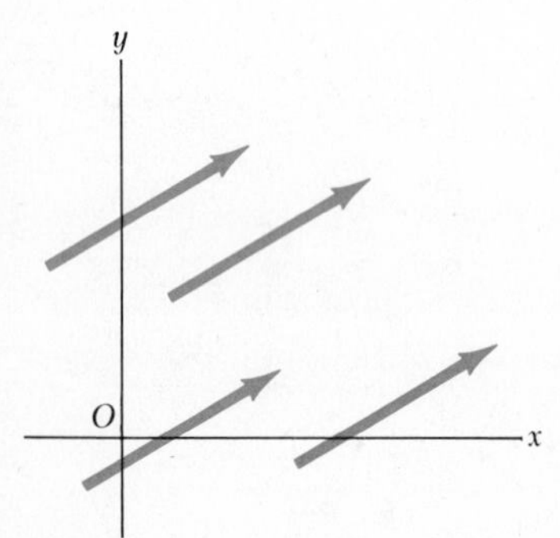

Figure 2.6 Four representations of the same vector.

Addition When two or more vectors are added together, *all* vectors involved *must* have the same units. For example, it would be meaningless to add a velocity vector to a displacement vector since they are different physical quantities. Scalars also obey the same rule. For example, it would be meaningless to add time intervals and temperatures.

The rules for vector sums are conveniently described by geometric methods. To add vector **B** to vector **A**, first draw vector **A**, with its magnitude represented by a convenient scale, on graph paper and then draw vector **B** to

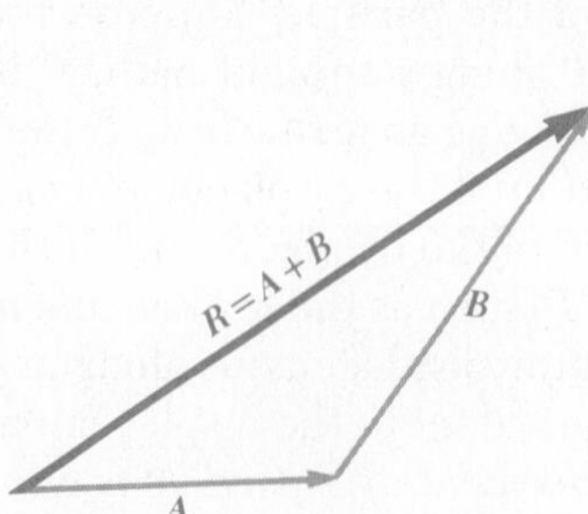

Figure 2.7 When vector **A** is added to vector **B**, the resultant **R** is the vector that runs from the tail of **A** to the tip of **A** to the tip of **B**.

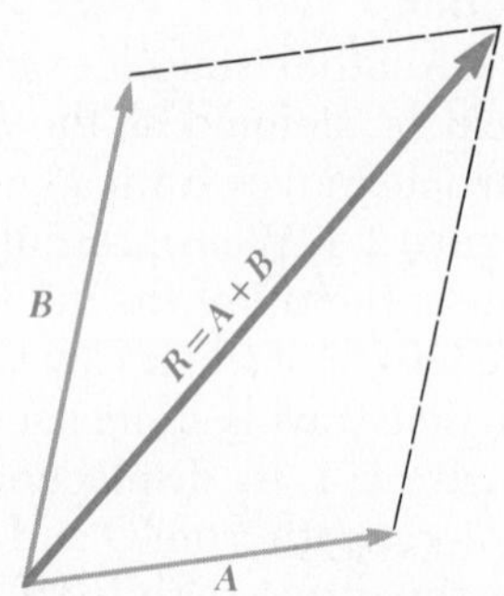

Figure 2.8 This construction shows that **A** + **B** = **B** + **A**. Note that the resultant **R** is the diagonal of a parallelogram with sides **A** and **B**.

the same scale with its tail starting from the tip of **A**, as in Figure 2.7. The *resultant vector* **R** = **A** + **B** is the vector drawn from the tail of **A** to the tip of **B**. This is known as the *triangle method of addition*. An alternative graphical procedure for adding two vectors, known as the *parallelogram rule of addition*, is shown in Figure 2.8. In this construction, the tails of the two vectors **A** and **B** are together and the resultant vector **R** is the diagonal of a parallelogram formed with **A** and **B** as its sides.

When two vectors are added, the sum is independent of the order of the addition. This can be seen from the geometric construction in Figure 2.8 and is known as the **commutative law of addition:**

Commutative law

$$\mathbf{A} + \mathbf{B} = \mathbf{B} + \mathbf{A} \tag{2.6}$$

If three or more vectors are added, their sum is independent of the way in which the individual vectors are grouped together. A geometric proof of this for three vectors is given in Figure 2.9. This is called the **associative law of addition:**

Associative law

$$\mathbf{A} + (\mathbf{B} + \mathbf{C}) = (\mathbf{A} + \mathbf{B}) + \mathbf{C} \tag{2.7}$$

Thus we conclude that *a vector is a quantity that has both magnitude and direction and also obeys the laws of vector addition* as described in Figures 2.7 to 2.10.

Geometric constructions can also be used to add more than three vectors. This is shown in Figure 2.10 for the case of four vectors. The resultant vector sum **R** = **A** + **B** + **C** + **D** is *the vector that completes the polygon*. In other words, **R** is *the vector drawn from the tail of the first vector to the tip of the last vector*. Again, the order of the summation is unimportant.

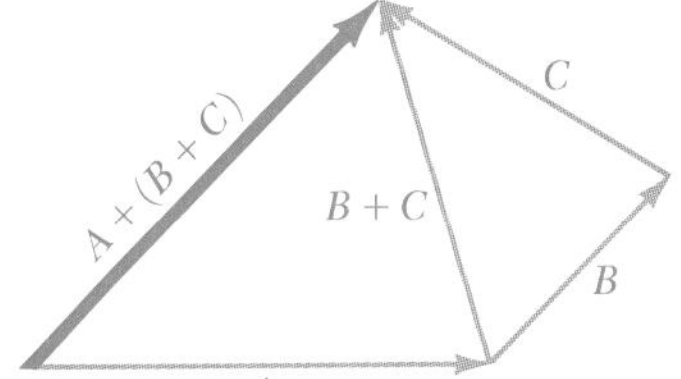

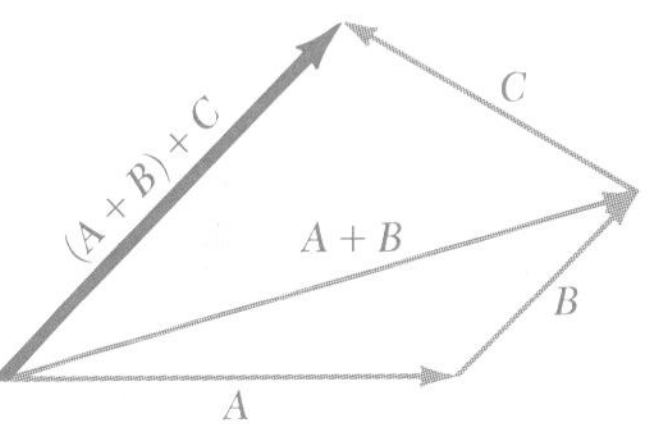

Figure 2.9 Geometric constructions for verifying the associative law of addition.

Negative of a Vector The negative of the vector **A** is defined as the vector that when added to **A** gives zero. That is, **A** + (−**A**) = 0. The vectors **A** and −**A** have the same magnitude but point in opposite directions.

Subtraction of Vectors The operation of vector subtraction makes use of the definition of the negative of a vector. We define the operation **A** − **B** as vector −**B** added to vector **A**:

$$\mathbf{A} - \mathbf{B} = \mathbf{A} + (-\mathbf{B}) \tag{2.8}$$

The geometric construction for subtracting two vectors is shown in Figure 2.11.

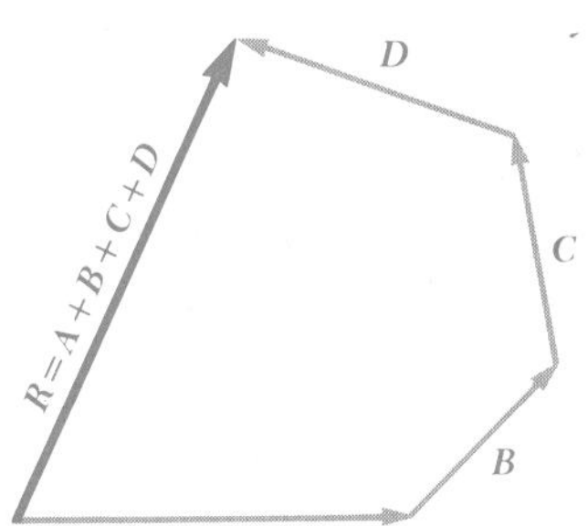

Figure 2.10 Geometric construction for summing four vectors. The resultant vector **R** completes the polygon.

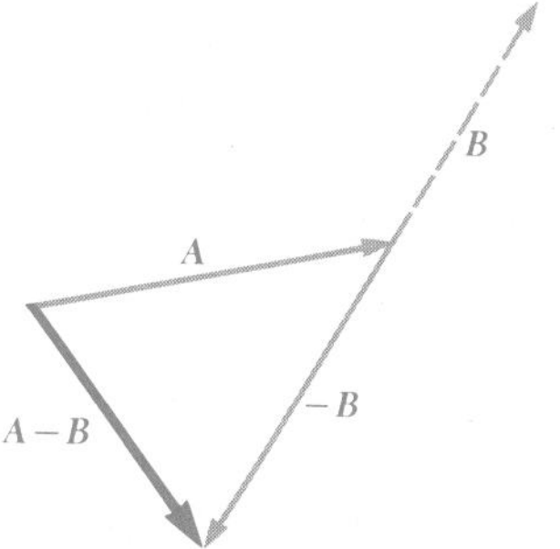

Figure 2.11 This construction shows how to subtract vector **B** from vector **A**. The vector −**B** is equal to and opposite the vector **B**.

Multiplication of a Vector by a Scalar If a vector **A** is multiplied by a positive scalar quantity m, the product $m\mathbf{A}$ is a vector that has the same direction as **A** and magnitude mA. If m is a negative scalar quantity, the vector $m\mathbf{A}$ is directed opposite **A**. For example, the vector $5\mathbf{A}$ is five times as long as **A** and points in the same direction as **A**. On the other hand, the vector $-\frac{1}{3}\mathbf{A}$ is one third the length of **A** and points in the direction opposite **A** (because of the negative sign).

EXAMPLE 2.2. A Vacation Trip

A car travels 20.0 km due north and then 35.0 km in a direction 60° west of north, as in Figure 2.12. Find the magnitude and direction of the car's resultant displacement.

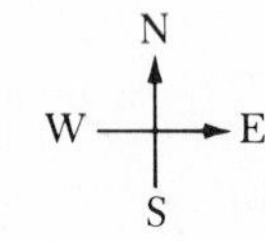

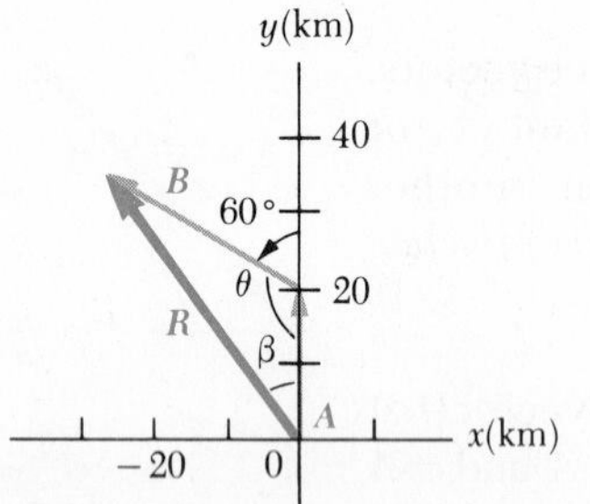

Figure 2.12 (Example 2.2) Graphical method for finding the resultant displacement $\mathbf{R} = \mathbf{A} + \mathbf{B}$.

Solution: The problem can be solved geometrically using graph paper and a protractor, as shown in Figure 2.12. The resultant displacement $\mathbf{R}$ is the sum of the two individual displacements **A** and **B**.

An algebraic solution for the magnitude of R can be obtained using the law of cosines from trigonometry as applied to the obtuse triangle (Appendix B.4). Since $\theta = 180° - 60° = 120°$ and $R^2 = A^2 + B^2 - 2AB\cos\theta$, we find that

$$R = \sqrt{A^2 + B^2 - 2AB\cos\theta}$$

$$= \sqrt{(20)^2 + (35)^2 - 2(20)(35)\cos 120°} = 48.2 \text{ km}$$

The direction of $\mathbf{R}$ measured from the northerly direction can be obtained from the law of sines from trigonometry:

$$\frac{\sin\beta}{B} = \frac{\sin\theta}{R}$$

$$\sin\beta = \frac{B}{R}\sin\theta = \frac{35}{48.2}\sin 120° = 0.629$$

or

$$\beta = 39°$$

Therefore, the resultant displacement of the car is 48.2 km in a direction 39° west of north.

2.4 COMPONENTS OF A VECTOR AND UNIT VECTORS

The geometric method of adding vectors is not the recommended procedure in situations where high precision is required or in three-dimensional problems. In this section, we describe a method of adding vectors that makes use of the *projections* of a vector along the axes of a rectangular coordinate system. These projections are called the **components** of the vector. Any vector can be completely described by its components.

Consider a vector **A** lying in the xy plane and making an arbitrary angle θ with the positive x axis, as in Figure 2.13. The vector **A** can be expressed as the sum of two other vectors $\mathbf{A}_x$ and $\mathbf{A}_y$, called the **vector components** of **A**. The vector component $\mathbf{A}_x$ represents the projection of **A** along the x axis, while $\mathbf{A}_y$ represents the projection of **A** along the y axis. From Figure 2.13, we see that $\mathbf{A} = \mathbf{A}_x + \mathbf{A}_y$. We will often refer to the magnitudes of $\mathbf{A}_x$ and $\mathbf{A}_y$, namely A_x and A_y, as the **components** of **A**. The components of a vector are numbers with

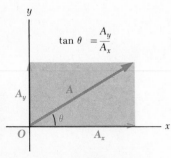

Figure 2.13 Any vector **A** lying in the xy plane can be represented by its rectangular vector components $\mathbf{A}_x$ and $\mathbf{A}_y$, where $\mathbf{A} = \mathbf{A}_x + \mathbf{A}_y$.

units that can be positive or negative. The component A_x is positive if $\mathbf{A}_x$ points along the positive x axis and is negative if $\mathbf{A}_x$ points along the negative x axis. The same is true for the component A_y.

From Figure 2.13 and the definition of the sine and cosine of an angle, we see that $\cos\theta = A_x/A$ and $\sin\theta = A_y/A$. Hence, the rectangular components of $\mathbf{A}$ are given by

$$A_x = A\cos\theta$$

and

$$A_y = A\sin\theta \tag{2.9}$$

Components of the vector $\mathbf{A}$

These components form two sides of a right triangle the hypotenuse of which has a magnitude A. Thus, it follows that the magnitude of $\mathbf{A}$ and its direction are related to its rectangular components through the expressions

$$A = \sqrt{A_x^2 + A_y^2} \tag{2.10}$$

Magnitude of $\mathbf{A}$

and

$$\tan\theta = \frac{A_y}{A_x} \tag{2.11}$$

Direction of $\mathbf{A}$

To solve for θ, we can write $\theta = \tan^{-1}(A_y/A_x)$, which is read "$\theta$ equals the angle the tangent of which is the ratio A_y/A_x." *Note that the signs of the rectangular components A_x and A_y depend on the angle θ.* For example, if $\theta = 120°$, A_x is negative and A_y is positive. On the other hand, if $\theta = 225°$, both A_x and A_y are negative. Figure 2.14 summarizes the signs of the components when $\mathbf{A}$ lies in the various quadrants.

II	I
A_x negative A_y positive	A_x positive A_y positive
A_x negative A_y negative	A_x positive A_y negative
III	IV

Figure 2.14 The signs of the rectangular components of a vector $\mathbf{A}$ depend on the quadrant in which the vector is located.

If you choose reference axes or an angle other than what is shown in Figure 2.13, the components of a vector must be modified accordingly. In many applications it is more convenient to express the components of a vector in a coordinate system having axes that are not horizontal and vertical, but still perpendicular to each other. Suppose a vector $\mathbf{B}$ makes an angle θ with the x' axis defined in Figure 2.15. The rectangular components of $\mathbf{B}$ along these axes are given by $B_{x'} = B\cos\theta$ and $B_{y'} = B\sin\theta$, as in Equation 2.9. The magnitude and direction of $\mathbf{B}$ are obtained from expressions equivalent to Equations 2.10 and 2.11. Thus, we can express the components of a vector in *any* coordinate system that is convenient for a particular situation.

The components of a vector, such as a displacement, are different when viewed from different coordinate systems. Furthermore, the components of a vector can change with respect to a fixed coordinate system if the vector changes in magnitude, orientation, or both.

Vector quantities are often expressed in terms of unit vectors. **A unit vector** is *a dimensionless vector one unit in length* used to specify a given *direction*. Unit vectors have no other physical significance. They are used simply as a convenience in describing a direction in space. We will use the symbols $\boldsymbol{i}$, $\boldsymbol{j}$, and $\boldsymbol{k}$ to represent unit vectors pointing in the x, y, and z directions, respectively. Thus, the unit vectors $\boldsymbol{i}$, $\boldsymbol{j}$, and $\boldsymbol{k}$ form a set of mutually

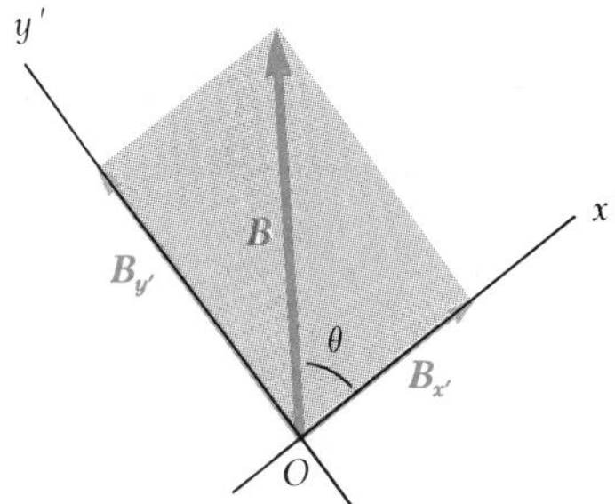

Figure 2.15 The vector components of a vector $\mathbf{B}$ in a coordinate system that is tilted.

perpendicular vectors as shown in Figure 2.16a, where the magnitude of the unit vectors equals unity or where $|i| = |j| = |k| = 1$.

Consider a vector $\boldsymbol{A}$ lying in the xy plane, as in Figure 2.16b. The product of the component A_x and the unit vector $\boldsymbol{i}$ is the vector $A_x\boldsymbol{i}$ parallel to the x axis with magnitude A_x. Likewise, $A_y\boldsymbol{j}$ is a vector of magnitude A_y parallel to the y axis. Thus, the unit-vector notation for the vector $\boldsymbol{A}$ is written

$$\boldsymbol{A} = A_x\boldsymbol{i} + A_y\boldsymbol{j} \tag{2.12}$$

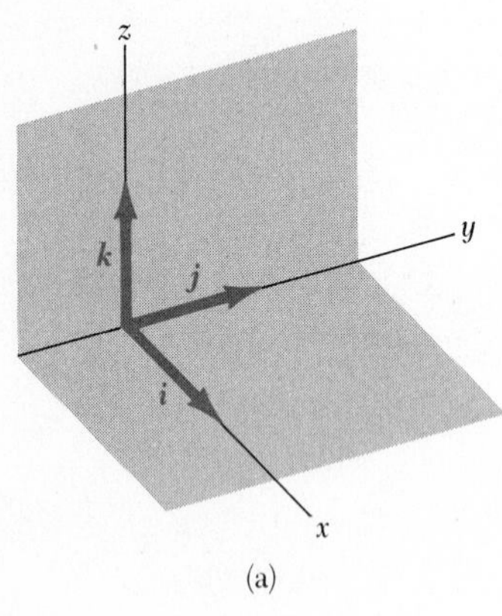

(a)

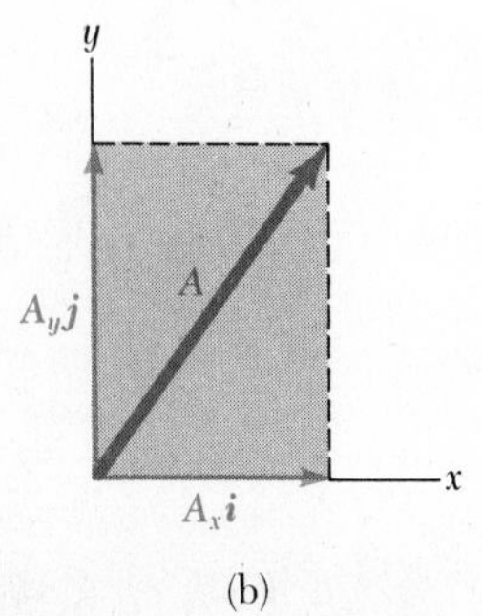

(b)

Figure 2.16 (a) The unit vectors $\boldsymbol{i}$, $\boldsymbol{j}$, and $\boldsymbol{k}$ are directed along the x, y, and z axes, respectively. (b) A vector $\boldsymbol{A}$ lying in the xy plane has component vectors $A_x\boldsymbol{i}$ and $A_y\boldsymbol{j}$ where A_x and A_y are the rectangular components of $\boldsymbol{A}$.

The vectors $A_x\boldsymbol{i}$ and $A_y\boldsymbol{j}$ are the component vectors of $\boldsymbol{A}$. These should not be confused with A_x and A_y, which we shall always refer to as the components of $\boldsymbol{A}$.

Now suppose we wish to add vector $\boldsymbol{B}$ to vector $\boldsymbol{A}$, where $\boldsymbol{B}$ has components B_x and B_y. The procedure for performing this sum is to simply add the x and y components separately. The resultant vector $\boldsymbol{R} = \boldsymbol{A} + \boldsymbol{B}$ is therefore given by

$$\boldsymbol{R} = (A_x + B_x)\boldsymbol{i} + (A_y + B_y)\boldsymbol{j} \tag{2.13}$$

Thus, the rectangular components of the resultant vector are given by

$$R_x = A_x + B_x$$

$$R_y = A_y + B_y \tag{2.14}$$

The magnitude of $\boldsymbol{R}$ and the angle it makes with the x axis can then be obtained from its components using the relationships

$$R = \sqrt{R_x^2 + R_y^2} = \sqrt{(A_x + B_x)^2 + (A_y + B_y)^2} \tag{2.15}$$

and

$$\tan\theta = \frac{R_y}{R_x} = \frac{A_y + B_y}{A_x + B_x} \tag{2.16}$$

The procedure just described for adding two vectors $\boldsymbol{A}$ and $\boldsymbol{B}$ using the component method can be checked using a geometric construction, as in Figure 2.17. Again, you must take note of the *signs* of the components when using either the algebraic or the geometric method.

The extension of these methods to three-dimensional vectors is straightforward. If $\boldsymbol{A}$ and $\boldsymbol{B}$ both have x, y, and z components, we express them in the form

$$\boldsymbol{A} = A_x\boldsymbol{i} + A_y\boldsymbol{j} + A_z\boldsymbol{k} \tag{2.17}$$

$$\boldsymbol{B} = B_x\boldsymbol{i} + B_y\boldsymbol{j} + B_z\boldsymbol{k} \tag{2.18}$$

The sum of $\boldsymbol{A}$ and $\boldsymbol{B}$ is given by

$$\boldsymbol{R} = \boldsymbol{A} + \boldsymbol{B} = (A_x + B_x)\boldsymbol{i} + (A_y + B_y)\boldsymbol{j} + (A_z + B_z)\boldsymbol{k} \tag{2.19}$$

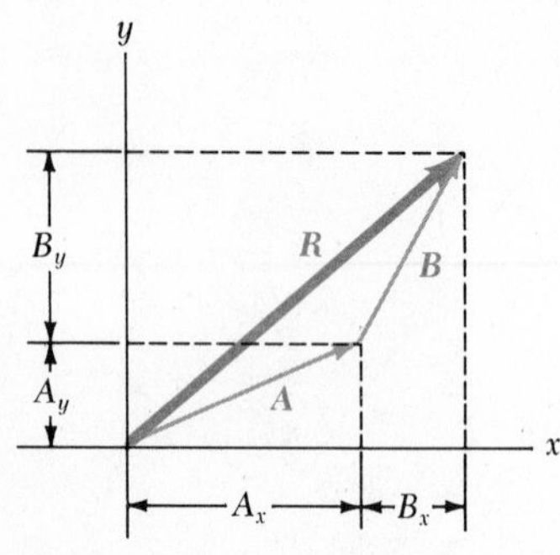

Figure 2.17 Geometric construction showing the relation between the components of the resultant $\boldsymbol{R}$ of two vectors and the individual vector components.

Thus, the resultant vector also has a z component given by $R_z = A_z + B_z$. The same procedure can be used to sum up three or more vectors.

When two or more vectors are to be added, the following step-by-step procedure is recommended:

1. Select a coordinate system.
2. Draw a sketch of the vectors to be added (or subtracted), with a label on each vector.
3. Find the x and y components of all vectors.
4. Find the resultant components (the algebraic sum of the components) in both the x and y directions.

5. Use the Pythagorean theorem to find the magnitude of the resultant vector.
6. Use a suitable trigonometric function to find the angle the resultant vector makes with the x axis.

EXAMPLE 2.3. The Sum of Two Vectors
Find the sum of two vectors $\boldsymbol{A}$ and $\boldsymbol{B}$ lying in the xy plane and given by

$$\boldsymbol{A} = 2\boldsymbol{i} + 2\boldsymbol{j} \quad \text{and} \quad \boldsymbol{B} = 2\boldsymbol{i} - 4\boldsymbol{j}$$

Solution: Note that $A_x = 2$, $A_y = 2$, $B_x = 2$, and $B_y = -4$. Therefore, the resultant vector $\boldsymbol{R}$ is given by

$$\boldsymbol{R} = \boldsymbol{A} + \boldsymbol{B} = (2 + 2)\boldsymbol{i} + (2 - 4)\boldsymbol{j} = 4\boldsymbol{i} - 2\boldsymbol{j}$$

or

$$R_x = 4, \; R_y = -2$$

The magnitude of $\boldsymbol{R}$ is given by

$$R = \sqrt{R_x^2 + R_y^2} = \sqrt{(4)^2 + (-2)^2} = \sqrt{20} = 4.47$$

Many examples in this text will be followed by an exercise. The purpose of these exercises is to test your understanding of the example by asking you to do a calculation or answer some other question related to the example. Answers to these exercises will be provided at the end of the exercise, when appropriate. Here is your first exercise, related to Example 2.3.

Exercise 1 Find the angle θ that the resultant vector $\boldsymbol{R}$ makes with the positive x axis.
Answer: 333°.

EXAMPLE 2.4. The Resultant Displacement
A particle undergoes three consecutive displacements given by $\boldsymbol{d}_1 = (\boldsymbol{i} + 3\boldsymbol{j} - \boldsymbol{k})$ cm, $\boldsymbol{d}_2 = (2\boldsymbol{i} - \boldsymbol{j} - 3\boldsymbol{k})$ cm, and $\boldsymbol{d}_3 = (-\boldsymbol{i} + \boldsymbol{j})$ cm. Find the resultant displacement of the particle.

Solution:

$$\begin{aligned} \boldsymbol{R} &= \boldsymbol{d}_1 + \boldsymbol{d}_2 + \boldsymbol{d}_3 \\ &= (1 + 2 - 1)\boldsymbol{i} + (3 - 1 + 1)\boldsymbol{j} + (-1 - 3 + 0)\boldsymbol{k} \\ &= (2\boldsymbol{i} + 3\boldsymbol{j} - 4\boldsymbol{k}) \text{ cm} \end{aligned}$$

That is, the resultant displacement has components $R_x = 2$ cm, $R_y = 3$ cm, and $R_z = -4$ cm. Its magnitude is

$$\begin{aligned} R &= \sqrt{R_x^2 + R_y^2 + R_z^2} = \sqrt{(2)^2 + (3)^2 + (-4)^2} \\ &= 5.39 \text{ cm} \end{aligned}$$

EXAMPLE 2.5. Taking a Hike
A hiker begins a trip by first walking 25 km due southeast from her base camp. On the second day, she walks 40 km in a direction 60° north of east, at which point she discovers a forest ranger's tower. (a) Determine the rectangular components of the hiker's displacements for the first and second days.

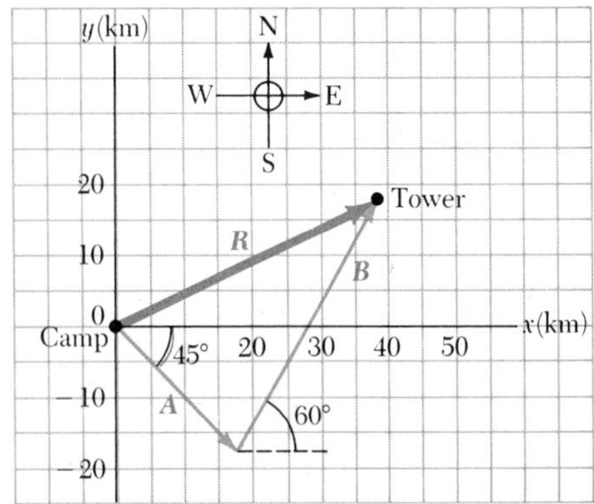

Figure 2.18 (Example 2.5) The total displacement of the hiker is the vector $\boldsymbol{R} = \boldsymbol{A} + \boldsymbol{B}$.

If we denote the displacement vectors on the first and second days by $\boldsymbol{A}$ and $\boldsymbol{B}$, respectively, and use the camp as the origin of coordinates, we get the vectors shown in Figure 2.18. Displacement $\boldsymbol{A}$ has a magnitude of 25.0 km and is 45° southeast. Its rectangular components are

$$A_x = A\cos(-45°) = (25 \text{ km})(0.707) = 17.7 \text{ km}$$

$$A_y = A\sin(-45°) = -(25 \text{ km})(0.707) = -17.7 \text{ km}$$

The negative value of A_y indicates that the y coordinate had decreased for this displacement. The signs of A_x and A_y are also evident from Figure 2.18. The second displacement, $\boldsymbol{B}$, has a magnitude of 40.0 km and is 60° north of east. Its rectangular components are

$$B_x = B\cos 60° = (40 \text{ km})(0.50) = 20.0 \text{ km}$$

$$B_y = B\sin 60° = (40 \text{ km})(0.866) = 34.6 \text{ km}$$

(b) Determine the rectangular components of the hiker's total displacement for the trip.

The resultant displacement for the trip, $\boldsymbol{R} = \boldsymbol{A} + \boldsymbol{B}$, has components given by

$$R_x = A_x + B_x = 17.7 \text{ km} + 20.0 \text{ km} = 37.7 \text{ km}$$

$$R_y = A_y + B_y = -17.7 \text{ km} + 34.6 \text{ km} = 16.9 \text{ km}$$

In unit-vector form, we can write the total displacement $\boldsymbol{R} = (37.7\boldsymbol{i} + 16.9\boldsymbol{j})$ km.

Exercise 2 Determine the magnitude and direction of the total displacement.
Answer: 41.3 km, 24.1° north of east from the base camp.

*2.5 FORCE

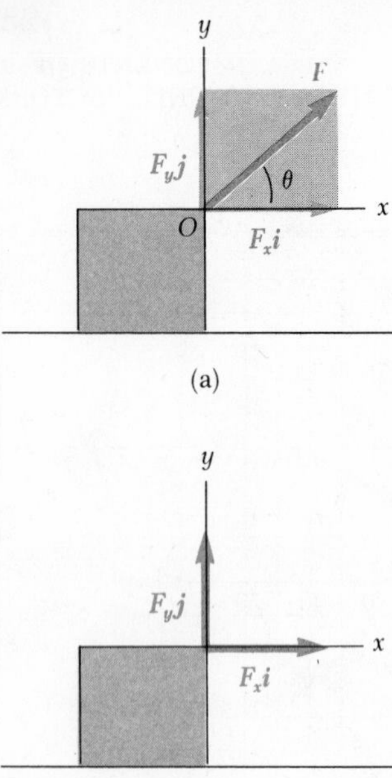

Figure 2.19 (a) The force $\boldsymbol{F}$ acting on the object has components F_x and F_y. (b) The vector sum of the forces $F_x\boldsymbol{i}$ and $F_y\boldsymbol{j}$ is equivalent to the force $\boldsymbol{F}$ shown in (a).

Force is an important concept in all branches of physics. If you push or pull an object in a certain direction, you exert a force on that object. The force of gravity exerted on every body on the earth (the weight of the body) is a common force that we experience in our everyday activities. For example, in order to lift an object from the ground, one must exert an upward force that is greater than the weight of the object. Any force on an object is specified completely by its *magnitude, direction*, and point of application. Force is more fully discussed in Chapter 5; this section merely describes how forces can be treated algebraically. The method of replacing a force by its components is emphasized, since this often simplifies the description of the behavior of a system under the influence of several forces.

The SI unit of force is the newton N,[2] whereas the conventional unit of force is the more familiar pound (lb). The conversion between the two units is 1 N = 0.2248 lb or 1 lb = 4.448 N. For example, if you weigh 125 lb, your weight in newtons is approximately 556 N.

Suppose a force $\boldsymbol{F}$ acts on an object at the point O at an angle θ relative to the horizontal, as in Figure 2.19a. The rectangular components of $\boldsymbol{F}$ are F_x and F_y, where $F_x = F\cos\theta$ and $F_y = F\sin\theta$. The vector sum of the component vectors in Figure 2.19b is equivalent to the original force $\boldsymbol{F}$. That is, *any force* $\boldsymbol{F}$ *can be represented by its rectangular components, provided that the force acting along each component originates at the same point as* $\boldsymbol{F}$.

As a numerical example, consider a single force of magnitude 7 N acting on an object at an angle of 30° to the horizontal, as in Figure 2.20. Because $F = 7$ N, its rectangular components are

$$F_x = F\cos\theta = (7\text{ N})(\cos 30°) = 6.06\text{ N}$$

$$F_y = F\sin\theta = (7\text{ N})(\sin 30°) = 3.50\text{ N}$$

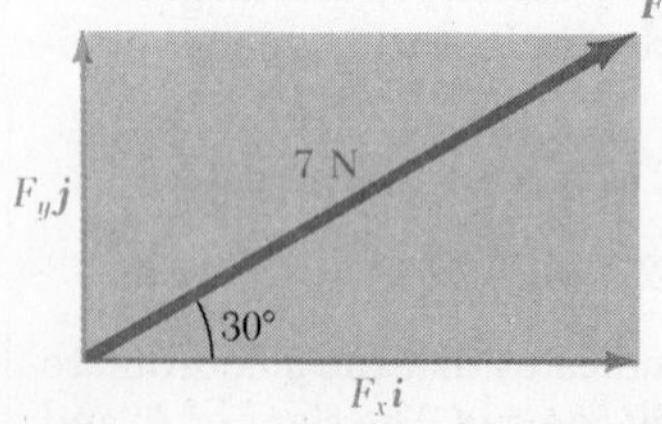

Figure 2.20 The rectangular components of the 7-N force are F_x and F_y.

Hence, we can express $\boldsymbol{F}$ in unit-vector form:

$$\boldsymbol{F} = F_x\boldsymbol{i} + F_y\boldsymbol{j} = (6.06\boldsymbol{i} + 3.50\boldsymbol{j})\text{ N}$$

Now consider two forces acting on an object as in Figure 2.21a. Suppose you want to find the resultant force on the object, that is, you wish to know

[2] The newton is defined in Chapter 5.

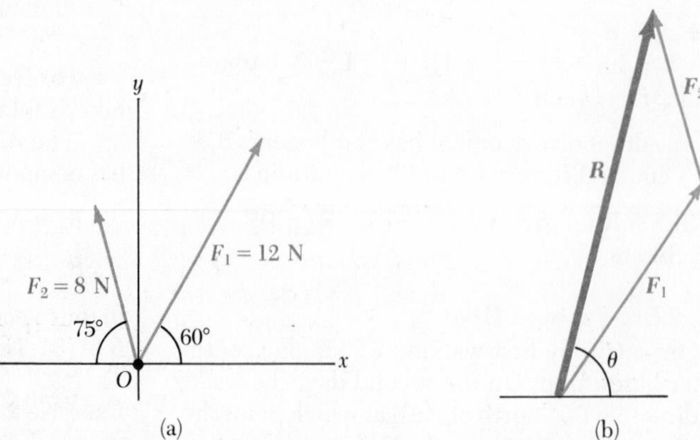

Figure 2.21 (a) Two forces acting on an object at the origin. (b) Graphical method for obtaining the resultant force $\boldsymbol{F}$.

what *single force would be equivalent to the two forces shown.* The x and y components of the 12-N force are given by

$$F_{x1} = F_1 \cos 60° = (12 \text{ N})(0.50) = 6.00 \text{ N}$$

$$F_{y1} = F_1 \sin 60° = (12 \text{ N})(0.866) = 10.4 \text{ N}$$

Likewise, the components of the 8-N force are

$$F_{x2} = F_2 \cos(105°) = (8 \text{ N})(-0.259) = -2.07 \text{ N}$$

$$F_{y2} = F_2 \sin 105° = (8 \text{ N})(0.966) = 7.73 \text{ N}$$

Note that the component F_{x2} is negative because the component vector is directed along the negative x axis. We are using the usual sign conventions of analytical geometry, in which x components to the right are positive and those to the left are negative. Likewise, y components upward are positive and those downward are negative. Adding the x and y components gives the components of the resultant force $\boldsymbol{R} = \boldsymbol{F}_1 + \boldsymbol{F}_2$:

$$R_x = F_{x1} + F_{x2} = 6.00 \text{ N} - 2.07 \text{ N} = 3.93 \text{ N}$$

$$R_y = F_{y1} + F_{y2} = 10.4 \text{ N} + 7.73 \text{ N} = 18.1 \text{ N}$$

In unit-vector form, $\boldsymbol{R}$ can be expressed

$$\boldsymbol{R} = (3.93\boldsymbol{i} + 18.1\boldsymbol{j}) \text{ N}$$

The magnitude and direction of R are given by

$$R = \sqrt{R_x^2 + R_y^2} = \sqrt{(3.93 \text{ N})^2 + (18.1 \text{ N})^2} = 18.5 \text{ N}$$

$$\theta = \tan^{-1} \frac{R_y}{R_x} = \tan^{-1}\left(\frac{18.1 \text{ N}}{3.93 \text{ N}}\right) = 77.7°$$

You should check these results against the graphical solution shown in Figure 2.21b.

If you have difficulty keeping track of the various forces and their components, it is suggested that you set up a table similar to the one shown here, which summarizes the above calculations. This procedure is especially useful when dealing with three or more forces.

Force	F_x (x Component)	F_y (y Component)
12 N	6.00 N	10.4 N
8 N	−2.07 N	7.73 N
Resultant R	$R_x = 3.93$ N	$R_y = 18.1$ N

Finally, suppose you wish to determine the magnitude and direction of another force $\boldsymbol{F}$, which when applied to the body will make the resultant force zero. This can easily be calculated by first finding the resultant $\boldsymbol{R}$ of the original forces and then applying the condition that $\boldsymbol{R} + \boldsymbol{F} = 0$, or $\boldsymbol{F} = -\boldsymbol{R}$. That is, $\boldsymbol{F}$ must be equal in magnitude to the resultant of the original forces and in the opposite direction. For instance, the third force $\boldsymbol{F}$ that must be applied to the body given in Figure 2.21 to make the resultant force zero is given by

$$\boldsymbol{F} = -\boldsymbol{R} = (-3.93\boldsymbol{i} - 18.1\boldsymbol{j}) \text{ N}$$

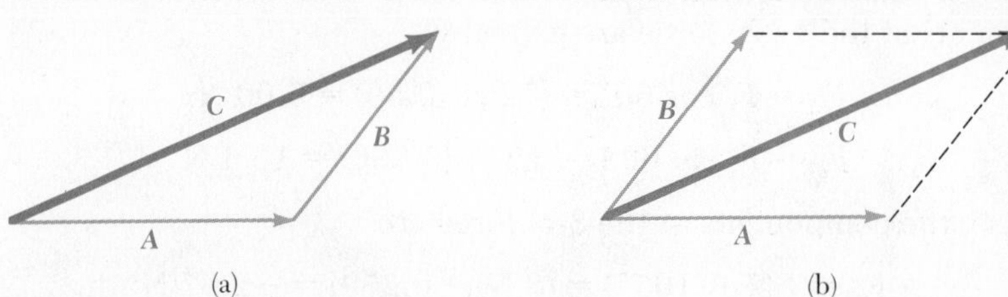

Figure 2.22 (a) Vector addition using the triangle method. (b) Vector addition using the parallelogram rule.

2.6 SUMMARY

Vectors are quantities that have both magnitude and direction and obey the vector law of addition. **Scalars** are quantities that have only magnitude.

Two vectors $\boldsymbol{A}$ and $\boldsymbol{B}$ can be added using either the triangle method or the parallelogram rule. In the triangle method (Figure 2.22a), the vector $\boldsymbol{C} = \boldsymbol{A} + \boldsymbol{B}$ runs from the tail of $\boldsymbol{A}$ to the tip of $\boldsymbol{B}$. In the parallelogram method (Figure 2.22b), $\boldsymbol{C}$ is the diagonal of a parallelogram having $\boldsymbol{A}$ and $\boldsymbol{B}$ as its sides.

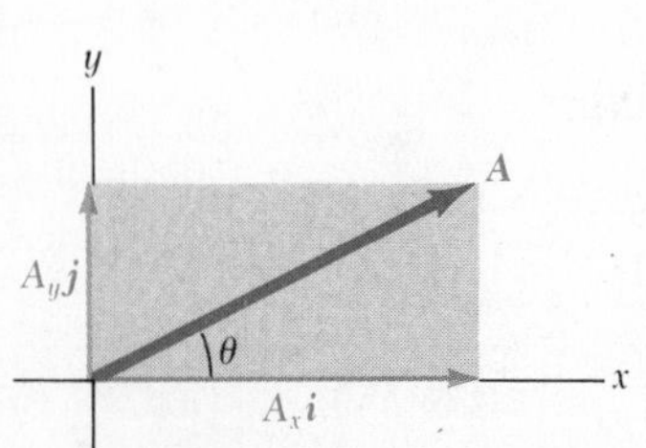

Figure 2.23 The x and y components of a vector $\boldsymbol{A}$ are A_x and A_y.

The x component, A_x, of the vector $\boldsymbol{A}$ is equal to its projection along the x axis of a coordinate system as in Figure 2.23, where $A_x = A \cos \theta$. Likewise, the y component, A_y, of $\boldsymbol{A}$ is its projection along the y axis, where $A_y = A \sin \theta$. The resultant of two or more vectors can be found by resolving all vectors into their x and y components, finding the resultant x and y components, and then using the Pythagorean theorem to find the magnitude of the resultant vector. The angle that the resultant vector makes with respect to the x axis can be found by use of a suitable trigonometric function.

If a vector $\boldsymbol{A}$ has an x component equal to A_x and a y component equal to A_y, the vector can be expressed in unit-vector form as $\boldsymbol{A} = A_x\boldsymbol{i} + A_y\boldsymbol{j}$. In this notation, $\boldsymbol{i}$ is a unit vector pointing in the positive x direction and $\boldsymbol{j}$ is a unit vector in the positive y direction. Since $\boldsymbol{i}$ and $\boldsymbol{j}$ are unit vectors, $|\boldsymbol{i}| = |\boldsymbol{j}| = 1$.

QUESTIONS

1. A book is moved once around the perimeter of a table of dimensions 1 m × 2 m. If the book ends up at its initial position, what is its displacement? What is the distance traveled?
2. If $\boldsymbol{B}$ is added to $\boldsymbol{A}$, under what condition does the resultant vector have a magnitude equal to $A + B$? Under what conditions is the resultant vector equal to zero?
3. Can the magnitude of a particle's displacement be greater than the distance traveled? Explain.
4. The magnitudes of two vectors $\boldsymbol{A}$ and $\boldsymbol{B}$ are $A = 5$ units and $B = 2$ units. Find the largest and smallest values possible for the resultant vector $\boldsymbol{R} = \boldsymbol{A} + \boldsymbol{B}$.
5. A vector $\boldsymbol{A}$ lies in the xy plane. For what orientations of $\boldsymbol{A}$ will both of its rectangular components be negative? For what orientations will its components have opposite signs?
6. Can a vector have a component equal to zero and still have a nonzero magnitude? Explain.
7. If one of the components of a vector is not zero, can its magnitude be zero? Explain.
8. If the component of vector $\boldsymbol{A}$ along the direction of vector $\boldsymbol{B}$ is zero, what can you conclude about the two vectors?
9. If $\boldsymbol{A} = \boldsymbol{B}$, what can you conclude about the components of $\boldsymbol{A}$ and $\boldsymbol{B}$?
10. Can the magnitude of a vector have a negative value? Explain.
11. If $\boldsymbol{A} + \boldsymbol{B} = 0$, what can you say about the components of the two vectors?
12. Which of the following are vectors and which are not: force, temperature, the volume of water in a can, the ratings of a TV show, the height of a building, the velocity of a sports car, the age of the universe?
13. Under what circumstances would a vector have components that are equal in magnitude?
14. Is it possible to add a vector quantity to a scalar quantity? Explain.
15. Two vectors have unequal magnitudes. Can their sum be zero? Explain.

PROBLEMS

Section 2.1 Coordinate Systems and Frames of Reference

1. Two points in the xy plane have cartesian coordinates (2.0, −4.0) and (−3.0, 3.0), where the units are in m. Determine (a) the distance between these points and (b) their polar coordinates.
2. A point in the xy plane has cartesian coordinates (−3.0, 5.0) m. What are the polar coordinates of this point?
3. The polar coordinates of a point are $r = 5.50$ m and $\theta = 240°$. What are the cartesian coordinates of this point?
4. Two points in a plane have polar coordinates (2.50 m, 30°) and (3.80 m, 120°). Determine (a) the cartesian coordinates of these points and (b) the distance between them.

Section 2.2 Vectors and Scalars and Section 2.3 Some Properties of Vectors

5. A pedestrian moves 6 km east and 13 km north. Find the magnitude and direction of the resultant displacement vector using the graphical method.
6. Vector $\boldsymbol{A}$ is 3 units in length and points along the positive x axis. Vector $\boldsymbol{B}$ is 4 units in length and points along the negative y axis. Use graphical methods to find the magnitude and direction of the vectors (a) $\boldsymbol{A} + \boldsymbol{B}$, (b) $\boldsymbol{A} - \boldsymbol{B}$.
7. Vector $\boldsymbol{A}$ is 6 units in length and at an angle of 45° to the x axis. Vector $\boldsymbol{B}$ is 3 units in length and is directed along the positive x axis ($\theta = 0$). Find the resultant vector $\boldsymbol{A} + \boldsymbol{B}$ using (a) graphical methods and (b) the law of cosines.
8. A person walks along a circular path of radius 5 m, around one half of the circle. (a) Find the magnitude of the displacement vector. (b) How far did the person walk? (c) What is the magnitude of the displacement if the circle is completed?
9. A particle undergoes three consecutive displacements such that its *total* displacement is zero. The first displacement is 8 m westward. The second is 13 m northward. Find the magnitude and direction of the third displacement using the graphical method.
10. Each of the displacement vectors $\boldsymbol{A}$ and $\boldsymbol{B}$ shown in Figure 2.24 has a magnitude of 3 m. Find graphically (a) $\boldsymbol{A} + \boldsymbol{B}$, (b) $\boldsymbol{A} - \boldsymbol{B}$, (c) $\boldsymbol{B} - \boldsymbol{A}$, (d) $\boldsymbol{A} - 2\boldsymbol{B}$.

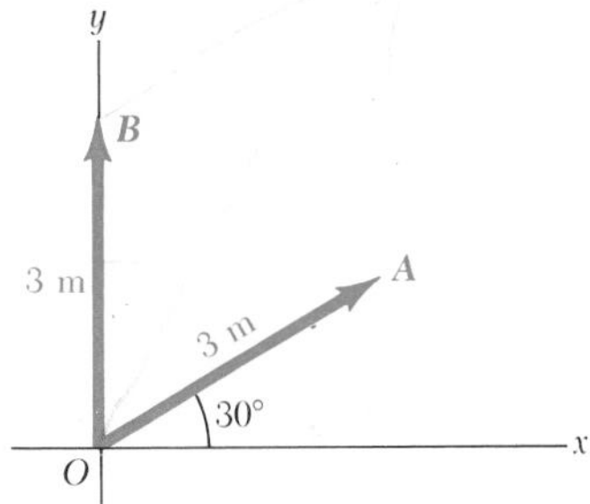

Figure 2.24 (Problems 10 and 25).

Section 2.4 Components of a Vector and Unit Vectors

11. A vector has an x component of −25 units and a y component of 40 units. Find the magnitude and direction of this vector.
12. A displacement vector $\boldsymbol{A}$ makes an angle θ with the positive x axis, as in Figure 2.13. Find the rectangular components of $\boldsymbol{A}$ for the following values of A and θ: (a) $A = 8$ m, $\theta = 60°$; (b) $A = 6$ ft, $\theta = 120°$; (c) $A = 12$ cm, $\theta = 225°$.
13. Vector $\boldsymbol{A}$ lies in the xy plane. Construct a table of the signs of the x and y components of $\boldsymbol{A}$ when the vector lies in the first, second, third, and fourth quadrants.
14. A displacement vector lying in the xy plane has a magnitude of 50 m and is directed at an angle of 120° to the positive x axis. What are the rectangular components of this vector?
15. Find the magnitude and direction of the resultant of three displacements having components (3, 2) m, (−5, 3) m, and (6, 1) m.
16. Vector $\boldsymbol{A}$ has x and y components of −8.7 cm and 15 cm, respectively; vector $\boldsymbol{B}$ has x and y components of 13.2 cm and −6.6 cm, respectively. If $\boldsymbol{A} - \boldsymbol{B} + 3\boldsymbol{C} = 0$, what are the components of $\boldsymbol{C}$?
17. Two vectors are given by $\boldsymbol{A} = 3\boldsymbol{i} - 2\boldsymbol{j}$ and $\boldsymbol{B} = -\boldsymbol{i} - 4\boldsymbol{j}$. Calculate (a) $\boldsymbol{A} + \boldsymbol{B}$, (b) $\boldsymbol{A} - \boldsymbol{B}$, (c) $|\boldsymbol{A} + \boldsymbol{B}|$, (d) $|\boldsymbol{A} - \boldsymbol{B}|$, (e) the direction of $\boldsymbol{A} + \boldsymbol{B}$ and $\boldsymbol{A} - \boldsymbol{B}$.
18. Three vectors are given by $\boldsymbol{A} = \boldsymbol{i} + 3\boldsymbol{j}$, $\boldsymbol{B} = 2\boldsymbol{i} - \boldsymbol{j}$, and $\boldsymbol{C} = 3\boldsymbol{i} + 5\boldsymbol{j}$. Find (a) the sum of the three vectors and (b) the magnitude and direction of the resultant vector.
19. Obtain expressions for the position vectors with polar coordinates (a) 12.8 m, 150°; (b) 3.3 cm, 60°; (c) 22 in., 215°.
20. Vectors $\boldsymbol{A}$ and $\boldsymbol{B}$ have components $A_x = -5.0$ cm, $A_y = 1.1$ cm, $A_z = -3.5$ cm and $B_x = 8.8$ cm, $B_y = -6.3$ cm, $B_z = 9.2$ cm. Determine the components of the vectors (a) $\boldsymbol{A} + \boldsymbol{B}$, (b) $\boldsymbol{B} - \boldsymbol{A}$, (c) $3\boldsymbol{B} + 2\boldsymbol{A}$. (d) Express the vector $\boldsymbol{B} - \boldsymbol{A}$ in unit-vector notation.
21. A particle undergoes the following consecutive displacements: 3.5 m south, 8.2 m northeast, and 15.0 m west. What is the resultant displacement?
22. A quarterback takes the ball from the line of scrimmage, runs backward for 10 yards, then sideways parallel to the line of scrimmage for 15 yards. At this point, he throws a 50-yard forward pass straight downfield perpendicular to the line of scrimmage. What is the magnitude of the football's resultant displacement?
23. An airplane flies from city A to city B in a direction due east for 800 miles. In the next part of the trip the airplane flies from city B to city C in a direction 40° north of east for 600 miles. What is the resultant displacement of the airplane between city A and city C?
24. A particle undergoes three consecutive displacements. The first is to the east and has a magnitude of 25 m. The second is to the north and has a magnitude of 42 m. If the resultant displacement has a magnitude

of 38 m and is directed at an angle of 30° north of east, what are the magnitude and direction of the third displacement?

25. Find the x and y components of the vectors $\boldsymbol{A}$ and $\boldsymbol{B}$ shown in Figure 2.24. Derive an expression for the resultant vector $\boldsymbol{A} + \boldsymbol{B}$ in unit-vector notation.

26. A particle undergoes two displacements. The first has a magnitude of 150 cm and makes an angle of 120° with the positive x axis. The *resultant* displacement has a magnitude of 140 cm and is directed at an angle of 35° to the positive x axis. Find the magnitude and direction of the second displacement.

27. The vector $\boldsymbol{A}$ has x, y, and z components of 8, 12, and -4 units, respectively. (a) Write a vector expression for $\boldsymbol{A}$ in unit-vector notation. (b) Obtain a unit-vector expression for a vector $\boldsymbol{B}$ one fourth the length of $\boldsymbol{A}$ pointing in the same direction as $\boldsymbol{A}$. (c) Obtain a unit-vector expression for a vector $\boldsymbol{C}$ three times the length of $\boldsymbol{A}$ pointing in the direction opposite $\boldsymbol{A}$.

28. Two vectors are given by $\boldsymbol{A} = -2\boldsymbol{i} + \boldsymbol{j} - 3\boldsymbol{k}$ and $\boldsymbol{B} = 5\boldsymbol{i} + 3\boldsymbol{j} - 2\boldsymbol{k}$. (a) Find a third vector $\boldsymbol{C}$ such that $3\boldsymbol{A} + 2\boldsymbol{B} - \boldsymbol{C} = 0$. (b) What are the magnitudes of $\boldsymbol{A}$, $\boldsymbol{B}$, and $\boldsymbol{C}$?

29. A vector $\boldsymbol{A}$ has a magnitude of 35 units and makes an angle of 37° with the positive x axis. Describe (a) a vector $\boldsymbol{B}$ that is in the direction opposite $\boldsymbol{A}$ and is one fifth the size of $\boldsymbol{A}$, and (b) a vector $\boldsymbol{C}$ that when added to $\boldsymbol{A}$ will produce a vector twice as long as $\boldsymbol{A}$ pointing in the negative y direction.

30. A vector $\boldsymbol{A}$ has a positive x component of 4 units and a negative y component of 2 units. What second vector $\boldsymbol{B}$ when added to $\boldsymbol{A}$ will produce a *resultant* vector three times the magnitude of $\boldsymbol{A}$ directed in the positive y direction?

31. A vector $\boldsymbol{A}$ has a negative x component 3 units in length and a positive y component 2 units in length. (a) Determine an expression for $\boldsymbol{A}$ in unit-vector notation. (b) Determine the magnitude and direction of $\boldsymbol{A}$. (c) What vector $\boldsymbol{B}$ when added to $\boldsymbol{A}$ gives a resultant vector with no x component and a negative y component 4 units in length?

32. A particle moves in the xy plane from the point (3, 0) m to the point (2, 2) m. (a) Determine a vector expression for the resultant displacement. (b) What are the magnitude and direction of this displacement vector?

Section 2.5 Force

33. Find the magnitude and direction of a force having x and y components of -5 N and 3 N, respectively.

34. A 40-N force is applied at an angle of 30° to the horizontal. What are the x and y components of this force?

35. Two 25-N forces are applied to an object as shown in Figure 2.25. Find the magnitude and direction of the resultant force.

36. Three forces are given by $\boldsymbol{F}_1 = 6\boldsymbol{i}$ N, $\boldsymbol{F}_2 = 9\boldsymbol{j}$ N, and $\boldsymbol{F}_3 = (-3\boldsymbol{i} + 4\boldsymbol{j})$ N. (a) Find the magnitude and direction of the resultant force. (b) What force must be added to these three to make the resultant force zero?

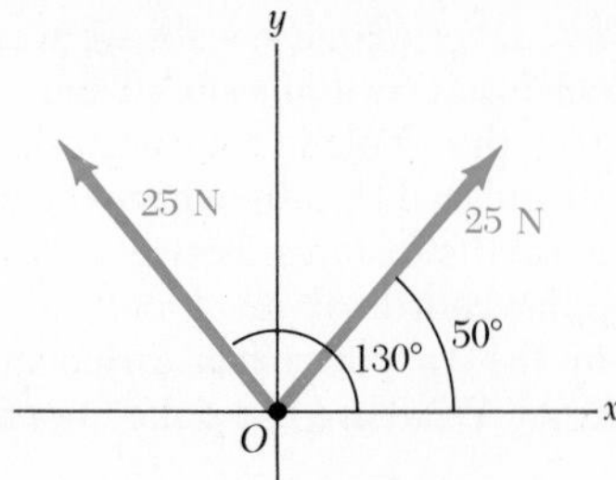

Figure 2.25 (Problem 35).

37. Three forces act at the point O as shown in Figure 2.26. Find (a) the x and y components of the resultant force and (b) the magnitude and direction of the resultant force.

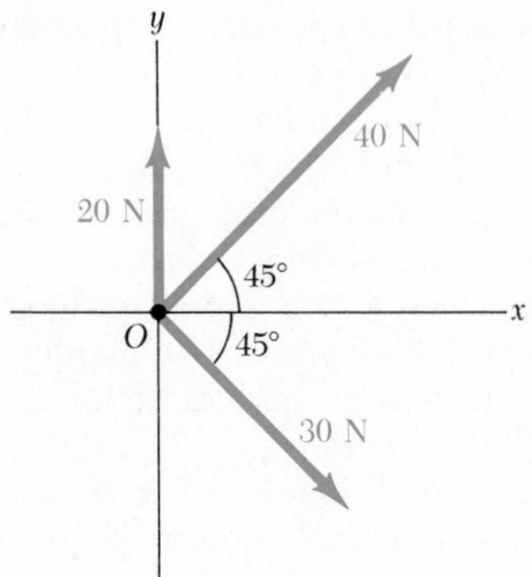

Figure 2.26 (Problem 37).

GENERAL PROBLEMS

38. A vector is given by $\boldsymbol{R} = 2\boldsymbol{i} + \boldsymbol{j} + 3\boldsymbol{k}$. Find (a) the magnitudes of the x, y, and z components, (b) the magnitude of $\boldsymbol{R}$, and (c) the angles between $\boldsymbol{R}$ and the x, y, and z axes.

39. A person going for a walk follows the path shown in Figure 2.27. The total trip consists of four straight-

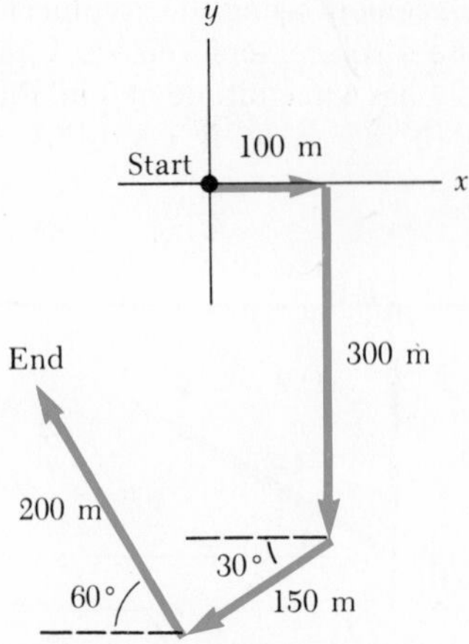

Figure 2.27 (Problem 39).

line paths. At the end of the walk, what is the person's resultant displacement measured from the starting point?

40. Two people pull on a stubborn mule as seen from the helicopter view shown in Figure 2.28. Find (a) the single force which is equivalent to the two forces shown, and (b) the force that a third person would have to exert on the mule to make the net force equal to zero.

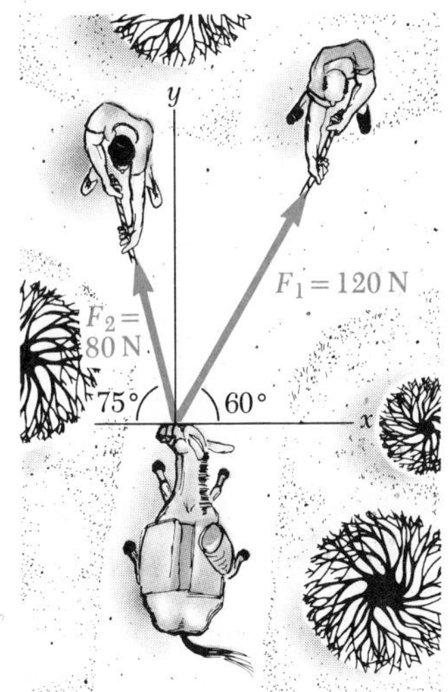

Figure 2.28 (Problem 40).

41. A particle moves from a point in the xy plane having cartesian coordinates $(-3, -5)$ m to a point with coordinates $(-1, 8)$ m. (a) Write vector expressions for the position vectors in unit-vector form for these two points. (b) What is the displacement vector? (See Problem 44 for definition.)

42. A rectangular parallelepiped has dimensions a, b, and c, as in Figure 2.29. (a) Obtain a vector expression for the face diagonal vector $\boldsymbol{R}_1$. What is the magnitude of this vector? (b) Obtain a vector expression for the body diagonal vector $\boldsymbol{R}_2$. What is the magnitude of this vector?

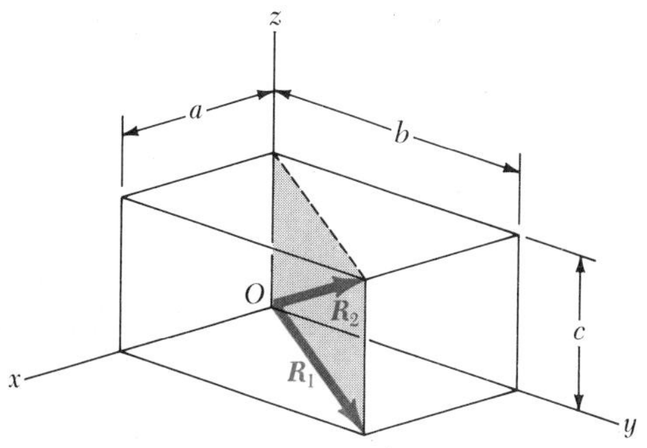

Figure 2.29 (Problem 42).

43. A point P is described by the coordinates (x, y) with respect to the normal cartesian coordinate system shown in Figure 2.30. Show that (x', y'), the coordinates of this point in the rotated $x'y'$ coordinate system, are related to (x, y) and the rotation angle α by the expressions

$$x' = x \cos \alpha + y \sin \alpha \text{ and } y' = -x \sin \alpha + y \cos \alpha.$$

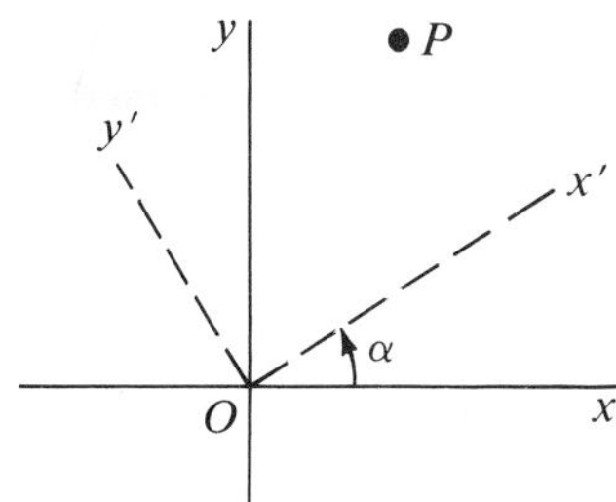

Figure 2.30 (Problem 43).

44. (a) Show that a point lying in the xy plane and having coordinates (x, y) can be described by the position vector $\boldsymbol{r} = x\boldsymbol{i} + y\boldsymbol{j}$. (b) Show that the magnitude of this vector is $r = \sqrt{x^2 + y^2}$. (c) Show that the displacement vector for a particle moving from (x_1, y_1) to (x_2, y_2) is given by $\boldsymbol{d} = (x_2 - x_1)\boldsymbol{i} + (y_2 - y_1)\boldsymbol{j}$. (d) Plot the position vectors $\boldsymbol{r}_1$ and $\boldsymbol{r}_2$ and the displacement vector $\boldsymbol{d}$, and verify by the graphical method that $\boldsymbol{d} = \boldsymbol{r}_2 - \boldsymbol{r}_1$.

3
Motion in One Dimension

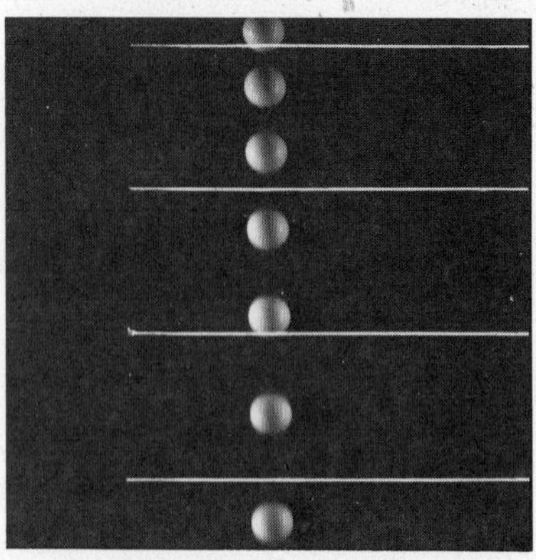

Dynamics is concerned with the study of the motion of an object and the relation of this motion to such physical concepts as force and mass. It is convenient to describe motion using the concepts of space and time, without regard to the causes of the motion. This portion of mechanics is called *kinematics.* In this chapter we shall consider motion along a straight line, that is, one-dimensional motion. In the next chapter we shall extend our discussion to two-dimensional motion. Starting with the concept of displacement discussed in the previous chapter, we shall define velocity and acceleration. Using these concepts, we shall proceed to study the motion of objects undergoing constant acceleration. The subject of *dynamics,* which is concerned with the causes of motion and relationships between motion, forces, and the properties of moving objects, will be discussed in Chapters 5 and 6.

From everyday experience we recognize that motion represents the continuous change in the position of an object. The movement of an object through space may be accompanied by the rotation or vibration of the object. Such motions can be quite complex. However, it is sometimes possible to simplify matters by temporarily neglecting the internal motions of the moving object. In many situations, an object can be treated as a *particle* if the only motion being considered is one of translation through space. An idealized particle is a mathematical point with no size. For example, if we wish to describe the motion of the earth around the sun, we can treat the earth as a particle and obtain reasonable accuracy in a prediction of the earth's orbit. This approximation is justified because the radius of the earth's orbit is large compared with the dimensions of the earth and sun. On the other hand, we could not use the particle description to explain the internal structure of the earth and such phenomena as tides, earthquakes, and volcanic activity. On a much smaller scale, it is possible to explain the pressure exerted by a gas on the walls of a container by treating the gas molecules as particles. However, the particle description of the gas molecules is generally inadequate for understanding those properties of the gas that depend on the internal motions of the gas molecules, namely, rotations and vibrations.

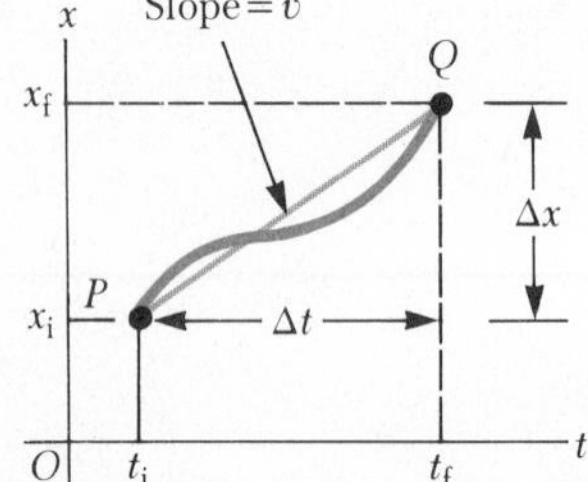

Figure 3.1 Position-time graph for a particle moving along the x axis. The average velocity $\bar{v}$ in the interval $\Delta t = t_f - t_i$ is the slope of the straight line connecting the points P and Q.

3.1 AVERAGE VELOCITY

The motion of a particle is completely known if its position in space is known at all times. Consider a particle moving along the x axis from point P to point Q. Let its position at point P be x_i at some time t_i, and let its position at point Q be x_f at time t_f. (The indices i and f refer to the initial and final values.) At times other than t_i and t_f, the position of the particle between these two points may vary as in Figure 3.1. Such a plot is often called a *position-time graph.* In the time interval $\Delta t = t_f - t_i$, the displacement of the particle is $\Delta x = x_f - x_i$. (Recall that the displacement is defined as the change in the position of the particle, which equals its final minus its initial position value.)

The x-component of the **average velocity** of the particle, $\bar{v}$, is defined as the ratio of its displacement, Δx, and the time interval, Δt:

$$\bar{v} \equiv \frac{\Delta x}{\Delta t} = \frac{x_f - x_i}{t_f - t_i} \quad (3.1)$$

Average velocity

From this definition, we see that the average velocity has the dimensions of length divided by time, or m/s in SI units and ft/s in conventional units. The average velocity is independent of the path taken between the points P and Q. This is true because the average velocity is proportional to the displacement, Δx, which in turn depends only on the initial and final coordinates of the particle. It therefore follows that if a particle starts at some point and returns to the same point via any path, its average velocity for this trip is zero, since its displacement along such a path is zero. The displacement should not be confused with the distance traveled, since the distance traveled for any motion is clearly nonzero. Thus, average velocity gives us no details of the motion between points P and Q. (How we evaluate the velocity at some instant in time is discussed in the next section.) Finally, note that the average velocity in one dimension can be positive or negative, depending on the sign of the displacement. (The time interval, Δt, is always positive.) If the coordinate of the particle increases in time (that is, if $x_f > x_i$), then Δx is positive and $\bar{v}$ is positive. This corresponds to a velocity in the positive x direction. On the other hand, if the coordinate decreases in time ($x_f < x_i$), Δx is negative and hence $\bar{v}$ is negative. This corresponds to a velocity in the negative x direction.

The average velocity can also be interpreted geometrically by drawing a straight line between the points P and Q in Figure 3.1. This line forms the hypotenuse of a triangle of height Δx and base Δt. The slope of this line is the ratio $\Delta x/\Delta t$. Therefore, we see that the *average* velocity of the particle during the time interval t_i to t_f is equal to the "slope" of the straight line joining the initial and final points on the space-time graph. (The word *slope* will often be used when referring to the graphs of physical data. Regardless of what data are plotted, the word *slope* will represent the ratio of the change in the quantity represented on the vertical axis to the change in the quantity represented on the horizontal axis.)

EXAMPLE 3.1. Calculate the Average Velocity

A particle moving along the x axis is located at $x_i = 12$ m at $t_i = 1$ s and at $x_f = 4$ m at $t_f = 3$ s. Find its displacement and average velocity during this time interval.

Solution: The displacement is given by

$$\Delta x = x_f - x_i = 4\text{ m} - 12\text{ m} = -8\text{ m}$$

The average velocity is

$$\bar{v} = \frac{\Delta x}{\Delta t} = \frac{x_f - x_i}{t_f - t_i} = \frac{4\text{ m} - 12\text{ m}}{3\text{ s} - 1\text{ s}} = -\frac{8\text{ m}}{2\text{ s}} = -4\text{ m/s}$$

Since the displacement and average velocity are negative for this time interval, we conclude that the particle has moved to the left, toward decreasing values of x.

3.2 INSTANTANEOUS VELOCITY

The velocity of a particle at any instant of time, or at some point on a space-time graph, is called the **instantaneous velocity.** This concept is especially important when the average velocity in different time intervals is *not constant.*

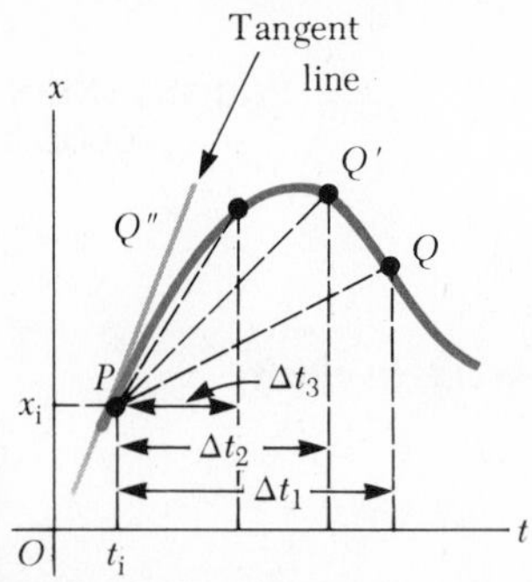

Figure 3.2 Position-time graph for a particle moving along the x axis. As the time intervals starting at t_i get smaller and smaller, the average velocity for that interval approaches the slope of the line tangent at P. The instantaneous velocity at P is defined as the slope of the tangent line at the time t_i.

Consider the motion of a particle between the two points P and Q on the space-time graph shown in Figure 3.2. As the point Q is brought closer and closer to the point P, the time intervals ($\Delta t_1, \Delta t_2, \Delta t_3, \ldots$) get progressively smaller. The average velocity for each time interval is the slope of the appropriate dotted line in Figure 3.2. As the point Q approaches P, the time interval approaches zero, but at the same time the slope of the dotted line approaches that of the line tangent to the curve at the point P. The slope of the line tangent to the curve at P is defined to be the *instantaneous velocity* at the time t_i. In other words,

the instantaneous velocity, v, equals the limiting value of the ratio $\Delta x/\Delta t$ as Δt approaches zero[1]:

$$v \equiv \lim_{\Delta t \to 0} \frac{\Delta x}{\Delta t} \tag{3.2}$$

In the calculus notation, this limit is called the *derivative* of x with respect to t, written dx/dt:

Definition of the derivative

$$v \equiv \lim_{\Delta t \to 0} \frac{\Delta x}{\Delta t} = \frac{dx}{dt} \tag{3.3}$$

The instantaneous velocity can be positive, negative, or zero.

When the slope of the space-time graph is positive, such as at the point P in Figure 3.3, v is positive. At point R, v is negative since the slope is negative. Finally, the instantaneous velocity is zero at the peak Q (the turning point), where the slope is zero. *From here on, we shall usually use the word* velocity *to designate instantaneous velocity.*

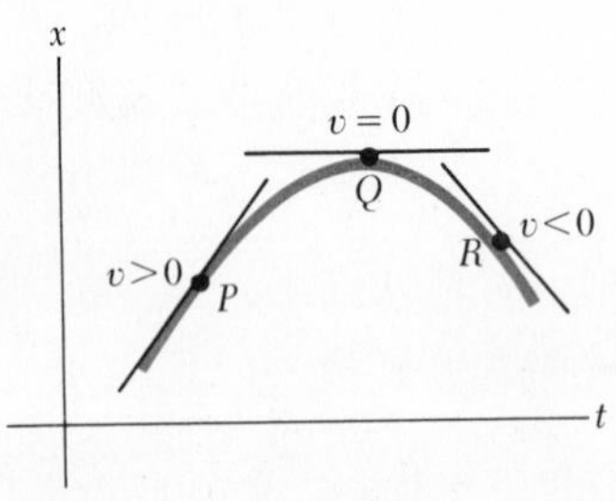

Figure 3.3 In the position-time graph shown here, the velocity is positive at P, where the slope of the tangent line is positive; the velocity is zero at Q, where the slope of the tangent line is zero; and the velocity is negative at R, where the slope of the tangent line is negative.

The *instantaneous speed* of a particle is defined as the magnitude of the instantaneous velocity vector. Hence, by definition, *speed* can never be negative.

It is also possible to find the displacement of a particle if its velocity is known as a function of time using a mathematical technique called integration. Because this procedure may not be familiar to many students, the topic is treated in Section 3.6 for general interest and for those courses that cover this material.

[1] Note that the displacement, Δx, also approaches zero as Δt approaches zero. However, as Δx and Δt become smaller and smaller, the ratio $\Delta x/\Delta t$ approaches a value equal to the *true* slope of the line tangent to the x versus t curve.

EXAMPLE 3.2. Average and Instantaneous Velocity
A particle moves along the x axis. Its x coordinate varies with time according to the expression $x = -4t + 2t^2$, where x is in m and t is in s. The position-time graph for this motion is shown in Figure 3.4. Note that the particle first moves in the negative x direction for the first second of motion, stops instantaneously at $t = 1$ s, and then heads back in the positive x direction for $t > 1$ s. (a) Determine the displacement of the particle in the time intervals $t = 0$ to $t = 1$ s and $t = 1$ s to $t = 3$ s.

In the first time interval, we can set $t_i = 0$ and $t_f = 1$ s. Since $x = -4t + 2t^2$, we get for the first displacement

$$\begin{aligned}\Delta x_{01} &= x_f - x_i \\ &= [-4(1) + 2(1)^2] - [-4(0) + 2(0)^2] \\ &= -2 \text{ m}\end{aligned}$$

Likewise, in the second time interval we can set $t_i = 1$ s and $t_f = 3$ s. Therefore, the second displacement in this interval is

$$\begin{aligned}\Delta x_{13} &= x_f - x_i \\ &= [-4(3) + 2(3)^2] - [-4(1) + 2(1)^2] \\ &= 8 \text{ m}\end{aligned}$$

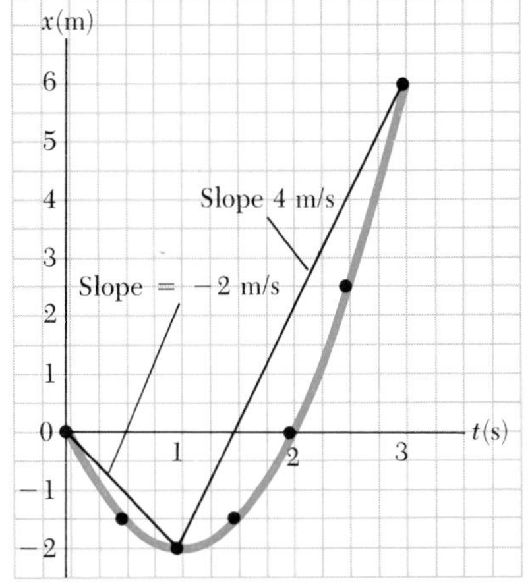

t(s)	x(m)
0	0
0.5	−1.5
1	−2
1.5	−1.5
2	0
2.5	2.5
3	6

Figure 3.4 (Example 3.2) Position-time graph for a particle having an x coordinate that varies in time according to $x = -4t + 2t^2$. Note that $\bar{v}$ is *not* the same as $v = -4 + 4t$.

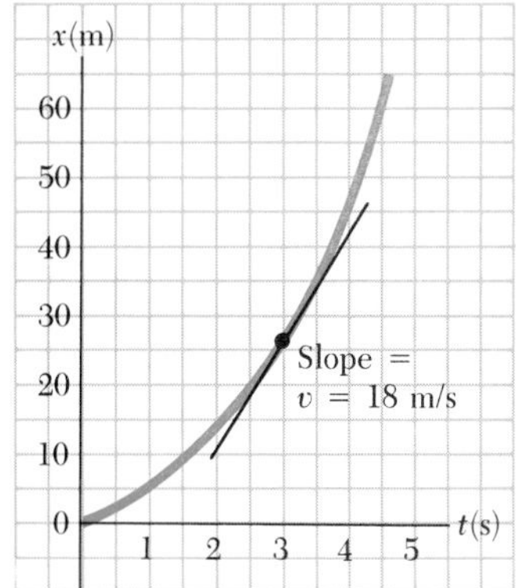

Figure 3.5 (Example 3.3) Position-time graph for a particle having an x coordinate that varies in time according to $x = 3t^2$. Note that the instantaneous velocity at $t = 3$ s equals the slope of the line tangent to the curve at this point.

Note that these displacements can also be read directly from the position-time graph (Fig. 3.4).

(b) Calculate the average velocity in the time intervals $t = 0$ to $t = 1$ s and $t = 1$ s to $t = 3$ s.

In the first time interval, $\Delta t = t_f - t_i = 1$ s. Therefore, using Equation 3.1 and the results from (a) gives

$$\bar{v}_{01} = \frac{\Delta x_{01}}{\Delta t} = \frac{-2\text{ m}}{1\text{ s}} = -2\text{ m/s}$$

Likewise, in the second time interval, $\Delta t = 2$ s; therefore

$$\bar{v}_{13} = \frac{\Delta x_{13}}{\Delta t} = \frac{8\text{ m}}{2\text{ s}} = 4\text{ m/s}$$

These values agree with the slopes of the lines joining these points in Figure 3.4.

(c) Find the instantaneous velocity of the particle at $t = 2.5$ s.

By measuring the slope of the position-time graph at $t = 2.5$ s, we find that $v = 6$ m/s.[2] (You should show that the velocity is -4 m/s at $t = 0$ and zero at $t = 1$ s.) Do you see any symmetry in the motion? For example, does the speed ever repeat itself?

EXAMPLE 3.3. The Limiting Process

The position of a particle moving along the x axis varies in time according to the expression $x = 3t^2$, where x is in m and t is in s. Find the velocity at any time.

Solution: The position-time graph for this motion is shown in Figure 3.5. We can compute the velocity at any time t by using the definition of the instantaneous velocity. If the initial coordinate of the particle at time t is $x_i = 3t^2$, then the coordinate at a later time $t + \Delta t$ is

$$x_f = 3(t + \Delta t)^2 = 3[t^2 + 2t\,\Delta t + (\Delta t)^2]$$
$$= 3t^2 + 6t\,\Delta t + 3(\Delta t)^2$$

Therefore, the displacement in the time interval Δt is

$$\Delta x = x_f - x_i = 3t^2 + 6t\,\Delta t + 3(\Delta t)^2 - 3t^2$$
$$= 6t\,\Delta t + 3(\Delta t)^2$$

The average velocity in this time interval is

$$\bar{v} = \frac{\Delta x}{\Delta t} = 6t + 3\,\Delta t$$

To find the instantaneous velocity, we take the limit of this expression as Δt approaches zero. In doing so, we see that the term $3\,\Delta t$ goes to zero, therefore

$$v = \lim_{\Delta t \to 0} \frac{\Delta x}{\Delta t} = 6t\text{ m/s}$$

Notice that this expression gives us the velocity at *any* general time t. It tells us that v is increasing linearly in time. It is then a straightforward matter to find the velocity at some specific time from the expression $v = 6t$. For example, at $t = 3$ s, the velocity is $v = 6(3) = 18$ m/s. Again, this can be checked from the slope of the graph at $t = 3$ s.

The limiting process can also be examined numerically. For example, we can compute the displacement

[2] We could also use the rules of differential calculus to find the velocity from the displacement. That is, $v = \frac{dx}{dt} = \frac{d}{dt}(-4t + 2t^2) = 4(-1 + t)$ m/s. Therefore, at $t = 2.5$ s, $v = 4(-1 + 2.5) = 6$ m/s. A review of basic operations in the calculus is provided in Appendix B.6.

and average velocity for various time intervals beginning at $t = 3$ s, using the expressions for Δx and $\bar{v}$. The results of such calculations are given in Table 3.1. Notice that as the time intervals get smaller and smaller, the average velocity more nearly approaches the value of the instantaneous velocity at $t = 3$ s, namely, 18 m/s.

TABLE 3.1 Displacement and Average Velocity for Various Time Intervals for the Function $x = 3t^2$ (the intervals begin at $t = 3$ s)

Δt (s)	Δx (m)	$\Delta x/\Delta t$ (m/s)
1.00	21	21
0.50	9.75	19.5
0.25	4.69	18.8
0.10	1.83	18.3
0.05	0.9075	18.15
0.01	0.1803	18.03
0.001	0.018003	18.003

3.3 ACCELERATION

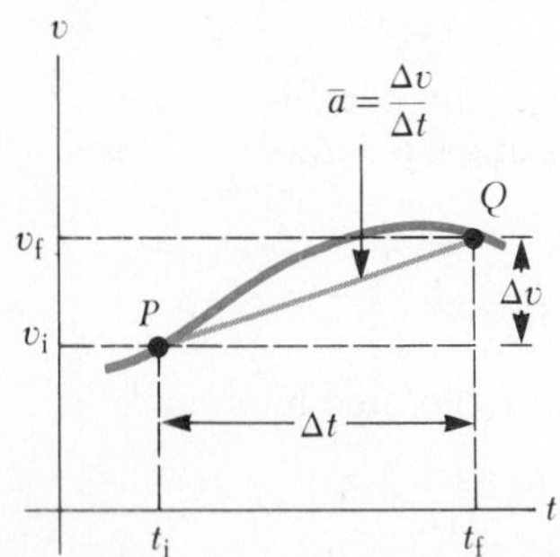

Figure 3.6 Velocity-time graph for a particle moving in a straight line. The slope of the line connecting the points P and Q is defined as the average acceleration in the time interval $\Delta t = t_f - t_i$.

When the velocity of a particle changes with time, the particle is said to be *accelerating*. For example, the velocity of a car will increase when you "step on the gas." The car will slow down when you apply the brakes. However, we need a more precise definition of acceleration than this.

Suppose a particle moving along the x axis has a velocity v_i at time t_i and a velocity v_f at time t_f, as in Figure 3.6.

The **average acceleration** of the particle in the time interval $\Delta t = t_f - t_i$ is defined as the ratio $\Delta v/\Delta t$, where $\Delta v = v_f - v_i$ is the *change* in velocity in this time interval:

$$\bar{a} \equiv \frac{v_f - v_i}{t_f - t_i} = \frac{\Delta v}{\Delta t} \tag{3.4}$$

Acceleration has dimensions of length divided by (time)^2, or L/T^2. Some of the common units of acceleration are meters per second per second (m/s^2) and feet per second per second (ft/s^2).

In some situations, the value of the average acceleration may be different over different time intervals. It is therefore useful to define the **instantaneous acceleration** as the limit of the average acceleration as Δt approaches zero. This concept is analogous to the definition of instantaneous velocity discussed in the previous section. If we imagine that the point Q is brought closer and closer to the point P in Figure 3.6 and take the limit of the ratio $\Delta v/\Delta t$ as Δt approaches zero, we get the instantaneous acceleration:

Definition of instantaneous acceleration

$$a \equiv \lim_{\Delta t \to 0} \frac{\Delta v}{\Delta t} = \frac{dv}{dt} \tag{3.5}$$

That is, the instantaneous acceleration equals the derivative of the velocity with respect to time, which by definition is the slope of the velocity-time graph. Again you should note that if a is positive, the acceleration is in the positive x direction, whereas negative a implies acceleration in the negative x direction. *From now on we shall use the term* acceleration *to mean instantaneous acceleration.* Average acceleration is seldom used in physics.

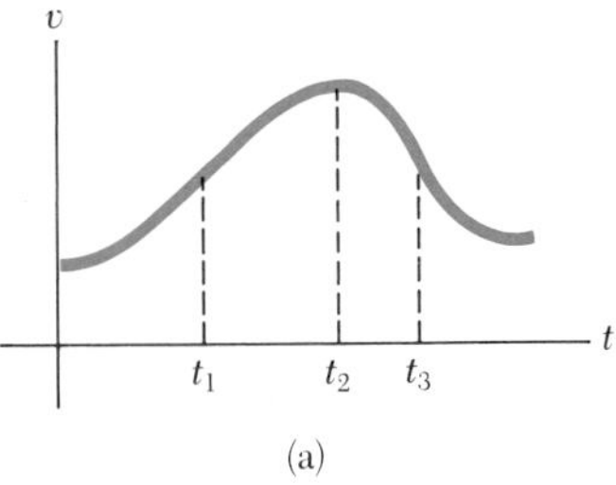

(a)

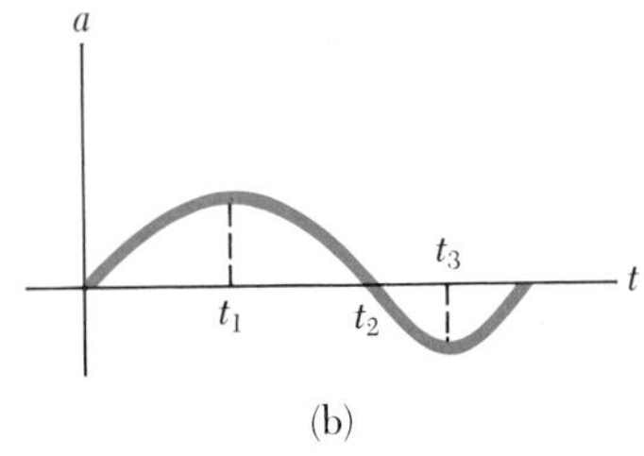

(b)

Figure 3.7 The instantaneous acceleration can be obtained from the velocity-time graph (a). At each instant, the acceleration in the a versus t graph (b) equals the slope of the line tangent to the v versus t curve.

Since $v = dx/dt$, the acceleration can also be written

$$a = \frac{dv}{dt} = \frac{d}{dt}\left(\frac{dx}{dt}\right) = \frac{d^2x}{dt^2} \tag{3.6}$$

That is, the acceleration equals the *second derivative* of the coordinate with respect to time.

Figure 3.7 shows how the acceleration-time curve can be derived from the velocity-time curve. In these sketches, the acceleration at any time is simply the slope of the velocity-time graph at that time. Positive values of the acceleration correspond to those points where the velocity is increasing in the positive x direction. The acceleration reaches a maximum at time t_1, when the slope of the velocity-time graph is a maximum. The acceleration then goes to zero at time t_2, when the velocity is a maximum (that is, when the velocity is momentarily not changing and the slope of the v versus t graph is zero). Finally, the acceleration is negative when the velocity in the positive x direction is decreasing in time.

EXAMPLE 3.4. Average and Instantaneous Acceleration

The velocity of a particle moving along the x axis varies in time according to the expression $v = (40 - 5t^2)$ m/s, where t is in s. (a) Find the average acceleration in the time interval $t = 0$ to $t = 2$ s.

The velocity-time graph for this function is given in Figure 3.8. The velocities at $t_i = 0$ and $t_f = 2$ s are found by substituting these values of t into the expression given for the velocity:

$$v_i = 40 - 5t_i^2 = 40 - 5(0)^2 = 40 \text{ m/s}$$

$$v_f = 40 - 5t_f^2 = 40 - 5(2)^2 = 20 \text{ m/s}$$

Therefore, the average acceleration in the time interval $\Delta t = t_f - t_i = 2$ is given by

$$\bar{a} = \frac{v_f - v_i}{t_f - t_i} = \frac{(20 - 40) \text{ m/s}}{(2 - 0) \text{ s}} = -10 \text{ m/s}^2$$

The negative sign is consistent with the fact that the slope of the line joining the initial and final points on the velocity-time graph is negative.

(b) Determine the acceleration at $t = 2$ s.

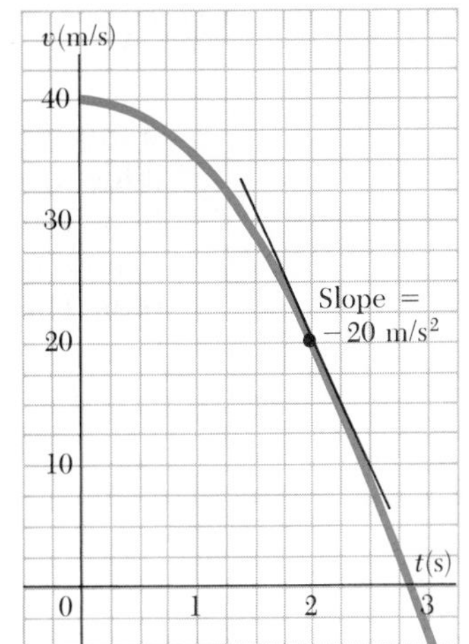

Figure 3.8 (Example 3.4) The velocity-time graph for a particle moving along the x axis according to the relation $v = (40 - 5t^2)$ m/s. Note that the acceleration at $t = 2$ s is equal to the slope of the tangent line at that time.

The velocity at time t is given by $v_i = (40 - 5t^2)$ m/s, and the velocity at time $t + \Delta t$ is given by

$$v_f = 40 - 5(t + \Delta t)^2 = 40 - 5t^2 - 10t\,\Delta t - 5(\Delta t)^2$$

Therefore, the change in velocity over the time interval Δt is

$$\Delta v = v_f - v_i = [-10t\,\Delta t - 5(\Delta t)^2] \text{ m/s}$$

Dividing this expression by Δt and taking the limit of the result as Δt approaches zero, we get the acceleration at *any* time t:

$$a = \lim_{\Delta t \to 0} \frac{\Delta v}{\Delta t} = \lim_{\Delta t \to 0} (-10t - 5\,\Delta t) = -10t \text{ m/s}$$

Therefore, at $t = 2$ s, we find that

$$a = -10(2) = -20 \text{ m/s}$$

This result can also be obtained by measuring the slope of the velocity-time graph at $t = 2$ s. Note that the acceleration is not constant in this example. Situations involving constant acceleration will be treated in the next section.

So far we have evaluated the derivatives of a function by starting with the definition of the function and then taking the limit of a specific ratio. Those of you familiar with the calculus should recognize that there are specific rules for taking the derivatives of various functions. These rules, which are listed in Appendix B.6, enable us to evaluate derivatives quickly.

Suppose x is proportional to some power of t, such as

$$x = At^n$$

where A and n are constants. (This is a very common functional form.) The derivative of x with respect to t is given by

$$\frac{dx}{dt} = nAt^{n-1}$$

Applying this rule to Example 3.3, where $x = 3t^2$, we see that $v = dx/dt = 6t$, in agreement with our result of taking the limit explicitly. Likewise, in Example 3.4, where $v = 40 - 5t^2$, we find that $a = dv/dt = -10t$. (Note that the derivative of any constant is zero.)

3.4 ONE-DIMENSIONAL MOTION WITH CONSTANT ACCELERATION

If the acceleration of a particle varies in time, the motion can be complex and difficult to analyze. A very common and simple type of one-dimensional motion occurs when the acceleration is constant, or uniform. Because the acceleration is constant, the average acceleration equals the instantaneous acceleration. Consequently, the velocity increases or decreases at the same rate throughout the motion.

If we replace $\bar{a}$ by a in Equation 3.4, we find that

$$a = \frac{v_f - v_i}{t_f - t_i}$$

For convenience, let $t_i = 0$ and t_f be any arbitrary time t. Also, let $v_i = v_0$ (the initial velocity at $t = 0$) and $v_f = v$ (the velocity at any arbitrary time t). With this notation, we can express the acceleration

$$a = \frac{v - v_0}{t}$$

Velocity as a function of time

$$v = v_0 + at \qquad \text{(for constant } a\text{)} \qquad (3.7)$$

This expression enables us to predict the velocity at *any* time t if the initial velocity, acceleration, and elapsed time are known. A graph of velocity versus time for this motion is shown in Figure 3.9a. The graph is a straight line the slope of which is the acceleration, a, consistent with the fact that $a = dv/dt$ is a constant. From this graph and from Equation 3.7, we see that the velocity at any time t is the sum of the initial velocity, v_0, and the change in velocity, at. The graph of acceleration versus time (Fig. 3.9b) is a straight line with a slope of zero, since the acceleration is constant. Note that if the acceleration were negative (a decelerating particle), the slope of Figure 3.9a would be negative.

Because the velocity varies linearly in time according to Equation 3.7, we can express the average velocity in any time interval as the arithmetic mean of the initial velocity, v_0, and the final velocity, v:

$$\bar{v} = \frac{v_0 + v}{2} \qquad \text{(for constant } a\text{)} \qquad (3.8)$$

Note that this expression is only valid when the acceleration is constant, that is, when the velocity varies linearly with time.

We can now use this result and Equation 3.1 to obtain the displacement as a function of time. Again, we choose $t_i = 0$, at which time the initial position is $x_i = x_0$. This gives

$$\Delta x = \bar{v}\,\Delta t = \left(\frac{v_0 + v}{2}\right) t$$

or

$$x - x_0 = \tfrac{1}{2}(v + v_0)t \qquad \text{(for constant } a\text{)} \qquad (3.9)$$

We can obtain another useful expression for the displacement by substituting Equation 3.7 into Equation 3.9:

$$x - x_0 = \tfrac{1}{2}(v_0 + v_0 + at)t$$

$$x - x_0 = v_0 t + \tfrac{1}{2}at^2 \qquad \text{(for constant } a\text{)} \qquad (3.10)$$

Finally, we can obtain an expression that does not contain the time by substituting the value of t from Equation 3.7 into Equation 3.9. This gives

$$x - x_0 = \tfrac{1}{2}(v_0 + v)\left(\frac{v - v_0}{a}\right) = \frac{v^2 - v_0{}^2}{2a}$$

Velocity as a function of displacement

$$v^2 = v_0{}^2 + 2a(x - x_0) \qquad \text{(for constant } a\text{)} \qquad (3.11)$$

A position-time graph for motion under constant acceleration assuming positive a is shown in Figure 3.9c. Note that the curve representing Equation 3.10 is a parabola. The slope of the tangent to this curve at $t = 0$ equals the initial velocity, v_0, and the slope of the tangent line at any time t equals the velocity at that time.

If motion occurs in which the acceleration is *zero*, then we see that

$$\left.\begin{aligned} v &= v_0 \\ x - x_0 &= vt \end{aligned}\right\} \text{when } a = 0$$

That is, when the acceleration is zero, the velocity is a constant and the displacement changes linearly with time.

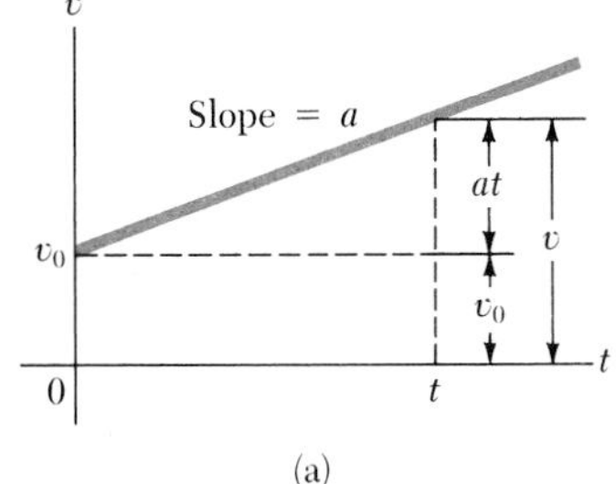

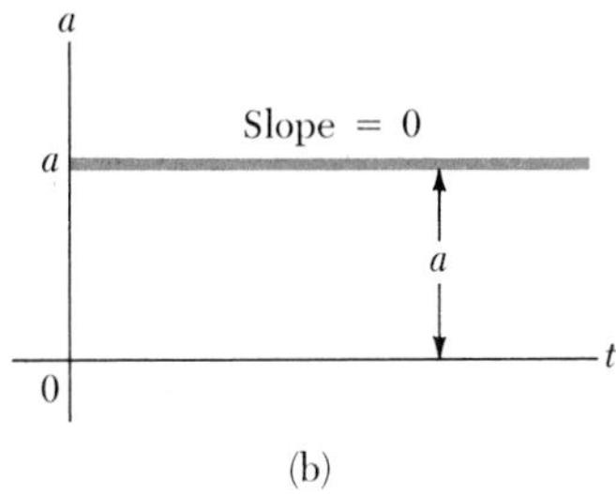

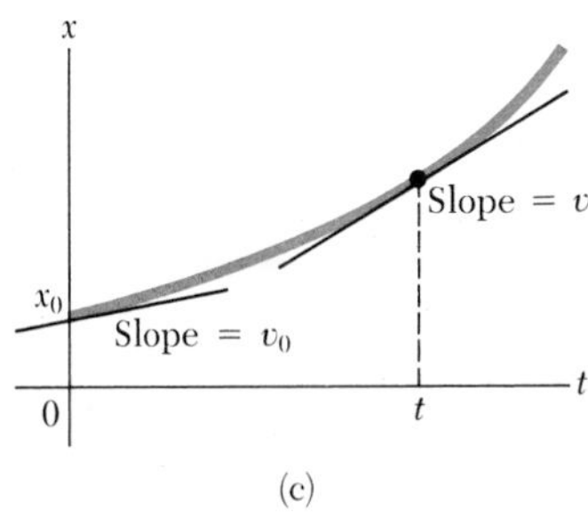

Figure 3.9 A particle moving along the x axis with uniform acceleration a; (a) the velocity-time graph, (b) the acceleration-time graph, and (c) the space-time graph.

TABLE 3.2 Kinematic Equations for Motion in a Straight Line Under Constant Acceleration

Equation	Information Given by Equation
$v = v_0 + at$	Velocity as a function of time
$x - x_0 = \frac{1}{2}(v + v_0)t$	Displacement as a function of velocity and time
$x - x_0 = v_0 t + \frac{1}{2}at^2$	Displacement as a function of time
$v^2 = v_0^2 + 2a(x - x_0)$	Velocity as a function of displacement

Note: Motion is along the x axis. At $t = 0$, the position of the particle is x_0 and its velocity is v_0.

Equations 3.7 through 3.11 are five *kinematic expressions that may be used to solve any problem in one-dimensional motion with constant acceleration.* Keep in mind that these relationships were derived from the definition of velocity and acceleration, together with some simple algebraic manipulations and the requirement that the acceleration be constant. It is often convenient to choose the initial position of the particle as the origin of the motion, so that $x_0 = 0$ at $t = 0$. In such a case, the displacement is simply x.

The four kinematic equations that are used most often are listed in Table 3.2 for convenience.

The choice of which kinematic equation or equations you should use in a given situation depends on what is known beforehand. Sometimes it is necessary to use two of these equations to solve for two unknowns, such as the displacement and velocity at some instant. For example, suppose the initial velocity, v_0, and acceleration, a, are given. You can then find (1) the velocity after a time t has elapsed, using $v = v_0 + at$, and (2) the displacement after a time t has elapsed, using $x - x_0 = v_0 t + \frac{1}{2}at^2$. You should recognize that the quantities that vary during the motion are velocity, displacement, and time.

You will get a great deal of practice in the use of these equations by solving a number of exercises and problems. Many times you will discover that there is more than one method for obtaining a solution.

EXAMPLE 3.5. The Supercharged Sportscar

A certain automobile manufacturer claims that its superdeluxe sportscar will accelerate uniformly from rest to a speed of 87 mi/h in 8 s. (a) Determine the acceleration of the car.

First note that $v_0 = 0$ and the velocity after 8 s is 87 mi/h = 128 ft/s. (It is useful to note that 60 mi/h = 88 ft/s exactly.) Because we are given v_0, v, and t, we can use $v = v_0 + at$ to find the acceleration:

$$a = \frac{v - v_0}{t} = \frac{128 \text{ ft/s}}{8 \text{ s}} = 16 \text{ ft/s}^2$$

(b) Find the distance the car travels in the first 8 s.

Let the origin be at the original position of the car, so that $x_0 = 0$. Using Equation 3.9 we find that

$$x = \tfrac{1}{2}(v_0 + v)t = \tfrac{1}{2}(128 \text{ ft/s})(8 \text{ s}) = 512 \text{ ft}$$

(c) What is the velocity of the car 10 s after it begins its motion, assuming it continues to accelerate at the rate of 16 ft/s²?

Again, we can use $v = v_0 + at$, with $v_0 = 0$, $t = 10$ s, and $a = 16$ ft/s². This gives

$$v = v_0 + at = 0 + (16 \text{ ft/s}^2)(10 \text{ s}) = 160 \text{ ft/s}$$

which corresponds to 109 mi/h.

EXAMPLE 3.6. Accelerating an Electron

An electron in a cathode ray tube of a TV set enters a region where it accelerates uniformly from a speed of 3×10^4 m/s to a speed of 5×10^6 m/s in a distance of 2 cm. (a) How long is the electron in this region where it accelerates?

Taking the direction of motion to be along the x axis, we can use Equation 3.9 to find t, since the displacement and velocities are known:

$$x - x_0 = \tfrac{1}{2}(v_0 + v)t$$

$$t = \frac{2(x - x_0)}{v_0 + v} = \frac{2(2 \times 10^{-2} \text{ m})}{(3 \times 10^4 + 5 \times 10^6) \text{ m/s}}$$

$$= 8 \times 10^{-9} \text{ s}$$

(b) What is the acceleration of the electron in this region?

To find the acceleration, we can use $v = v_0 + at$ and the results from (a):

$$a = \frac{v - v_0}{t} = \frac{(5 \times 10^6 - 3 \times 10^4)\ \text{m/s}}{8 \times 10^{-9}\ \text{s}}$$

$$= 6.2 \times 10^{14}\ \text{m/s}^2$$

We also could have used Equation 3.11 to obtain the acceleration, since the velocities and displacement are known. Try it! Although a is very large in this example, the acceleration occurs over a very short time interval and is a typical value for such charged particles in acceleration.

3.5 FREELY FALLING BODIES

It is well known that all objects, when dropped, will fall toward the earth with nearly constant acceleration. There is a legendary story that Galileo Galilei first discovered this fact by observing that two different weights dropped simultaneously from the Leaning Tower of Pisa hit the ground at approximately the same time. Although there is some doubt that this particular experiment was carried out, it is well established that Galileo did perform many systematic experiments on objects moving on inclined planes. Through careful measurements of distances and time intervals, he was able to show that the displacement of an object starting from rest is proportional to the square of the time the object is in motion. This observation is consistent with one of the kinematic equations we derived for motion under constant acceleration (Eq. 3.10). Galileo's achievements in the science of mechanics paved the way for Newton in his development of the laws of motion.

Galileo Galilei (1564–1642), an Italian physicist and astronomer, investigated the motion of objects in free fall (including projectiles) and the motion of an object on an inclined plane, established the concept of relative motion, and noted that a swinging pendulum could be used to measure time intervals. Following his invention of the telescope, Galileo made several major discoveries in astronomy. He discovered four moons of Jupiter and many new stars, investigated the nature of the moon's surface, discovered sun spots and the phases of Venus, and proved that the Milky Way consists of an enormous number of stars. (Courtesy of AIP Niels Bohr Library)

You might want to try the following experiment. Drop a coin and a crumpled-up piece of paper simultaneously from the same height. In the absence of air resistance, both will experience the same motion and hit the floor at the same time. (In a real experiment, air resistance cannot be neglected.) In the idealized case, where air resistance *is* neglected, such motion is referred to as *free fall.* If this same experiment could be conducted in a good vacuum, where air friction is truly negligible, the paper and coin would fall with the same acceleration, regardless of the shape of the paper. On August 2, 1971, such an experiment was conducted on the moon by astronaut David Scott. He simultaneously released a geologist's hammer and a falcon's feather, and in unison they fell to the lunar surface. This demonstration would have surely pleased Galileo!

Acceleration due to gravity $g = 9.80\ \text{m/s}^2$

We shall denote the *acceleration due to gravity* by the symbol **g**. The magnitude of g decreases with increasing altitude. Furthermore, there are slight variations in g with latitude. The vector **g** is directed downward toward the center of the earth. At the earth's surface, the magnitude of g is approximately $9.80\ \text{m/s}^2$, or $980\ \text{cm/s}^2$, or $32\ \text{ft/s}^2$. Unless stated otherwise, we shall use this value for g when doing calculations.

Definition of free fall

When we use the expression *freely falling object,* we do not necessarily refer to an object dropped from rest. A freely falling object is any object moving freely under the influence of gravity, *regardless* of its initial motion. Objects thrown upward or downward and those released from rest are all falling freely once they are released. Furthermore, it is important to recognize that any freely falling object experiences an acceleration directed *downward.* This is true regardless of the initial motion of the object.

> An object thrown upward (or downward) will experience the same acceleration as an object released from rest. Once they are in free fall, all objects will have an acceleration downward, equal to the acceleration due to gravity.

If we neglect air resistance and assume that the gravitational acceleration does not vary with altitude, then the motion of a freely falling body is equivalent to motion in one dimension under constant acceleration. Therefore our kinematic equations for constant acceleration can be applied. We shall take the vertical direction to be the y axis and call y positive upward. With this choice of coordinates, we can replace x by y in Equations 3.7, 3.9, 3.10, and 3.11. Furthermore, since positive y is upward, the acceleration is negative (downward) and given by $a = -g$. The negative sign simply indicates that the acceleration is downward. With these substitutions, we get the following expressions:[3]

Kinematic equations for a freely falling body

$$v = v_0 - gt \tag{3.12}$$

$$y - y_0 = \tfrac{1}{2}(v + v_0)t \tag{3.13}$$

$$y - y_0 = v_0 t - \tfrac{1}{2}gt^2 \tag{3.14}$$

$$v^2 = v_0^2 - 2g(y - y_0) \tag{3.15}$$

(for constant a, $a = -g$)

You should note that *the negative sign for the acceleration is already included in these expressions.* Therefore, when using these equations in any free-fall problem, you should simply substitute $g = 9.80\ \text{m/s}^2$.

Consider the case of a particle thrown vertically upward from the origin with a velocity v_0. In this case, v_0 is *positive* and $y_0 = 0$. Graphs of the displace-

[3] One can also take y positive downward, in which case $a = +g$. The results will be the same, regardless of the convention chosen.

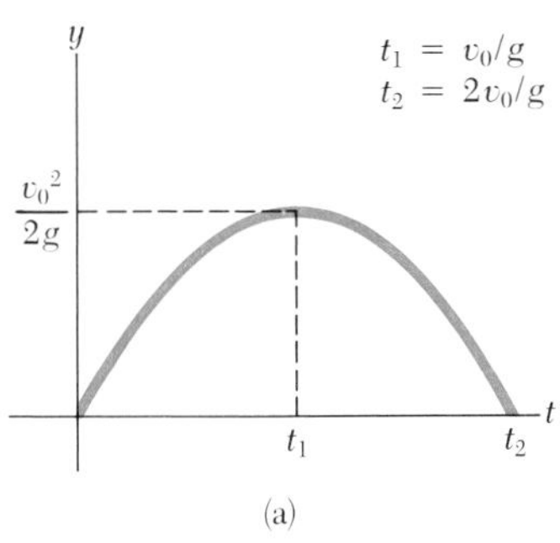

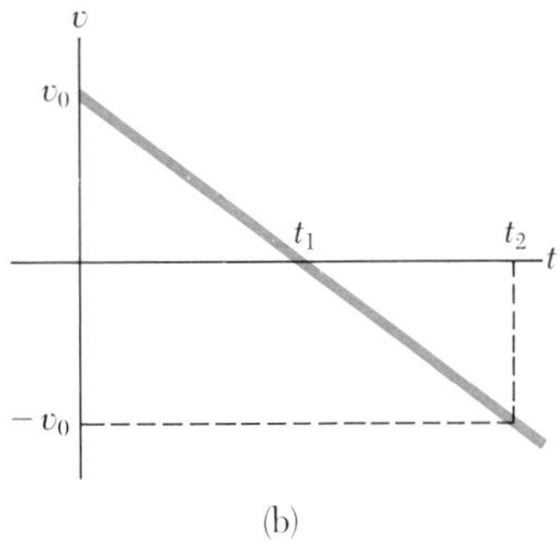

Figure 3.10 Graphs of (a) the displacement versus time and (b) the velocity versus time for a freely falling particle, where y and v are taken to be positive upward. Note the symmetry in the curves about $t = t_1$.

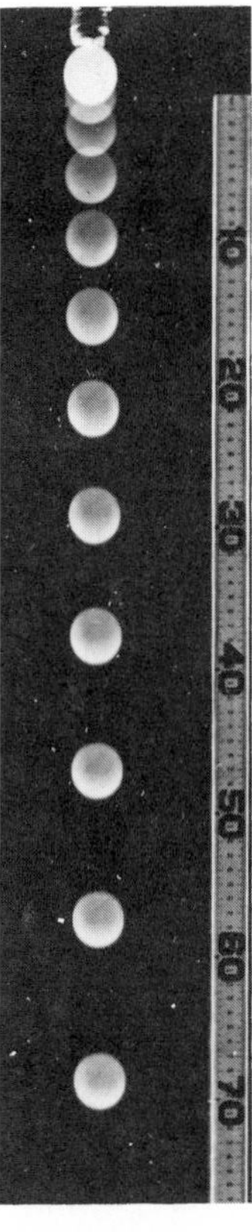

A multiflash photograph of a freely falling ball. The time interval between flashes is (1/30) s and the scale is in cm. Can you determine g from these data?

ment and velocity as functions of time are shown in Figure 3.10. Note that the velocity is initially positive, but decreases in time and goes to zero at the peak of the path. From Equation 3.12, we see that this occurs at the time $t_1 = v_0/g$. At this time, the displacement has its largest positive value, which can be calculated from Equation 3.14 with $t = t_1 = v_0/g$. This gives $y_{max} = v_0^2/2g$.

At the time $t_2 = 2t_1 = 2v_0/g$, we see from Equation 3.14 that the displacement is again zero, that is, the particle has returned to its starting point. Furthermore, at time t_2 the velocity is given by $v = -v_0$. (This follows directly from Equation 3.12.) Hence, there is symmetry in the motion. In other words, both the displacement and the magnitude of the velocity repeat themselves in the time interval $t = 0$ to $t = 2v_0/g$.

In the examples that follow, we shall, for convenience, assume that $y_0 = 0$ at $t = 0$. Notice that this does not affect the solution to the problem. If y_0 is nonzero, then the graph of y versus t (Fig. 3.10a) is simply shifted upward or downward by an amount y_0, while the graph of v versus t (Fig. 3.10b) remains unchanged.

EXAMPLE 3.7. Look Out Below!

A golf ball is released from rest from the top of a very tall building. Neglecting air resistance, calculate the position and velocity of the ball after 1, 2, and 3 s.

Solution: We choose our coordinates such that the starting point of the ball is at the origin ($y_0 = 0$ at $t = 0$) and remember that we have defined y to be positive upward. Since $v_0 = 0$, Equations 3.12 and 3.14 become

$$v = -gt = -(9.80 \text{ m/s}^2)t$$

$$y = -\tfrac{1}{2}gt^2 = -\tfrac{1}{2}(9.80 \text{ m/s}^2)t^2$$

where t is in s, v is in m/s, and y is in m. These expressions give the velocity and displacement at any time t after the ball is released. Therefore, at $t = 1$ s,

$$v = -(9.80 \text{ m/s}^2)(1 \text{ s}) = -9.80 \text{ m/s}$$

$$y = -\tfrac{1}{2}(9.80 \text{ m/s}^2)(1 \text{ s})^2 = -4.90 \text{ m}$$

Likewise, at $t = 2$ s, we find that $v = -19.6$ m/s and $y = -19.6$ m. Finally, at $t = 3$ s, $v = -29.4$ m/s and $y =$ 44.1 m. The minus signs for v indicate that the velocity vector is directed downward, and the minus signs for y indicate displacement in the negative y direction.

Exercise Calculate the position and velocity of the ball after 4 s.
Answer: -78.4 m, -39.2 m/s.

EXAMPLE 3.8. Not a Bad Throw for a Rookie

A stone is thrown from the top of a building with an initial velocity of 20 m/s straight upward. The building is 50 m high, and the stone just misses the edge of the roof on its way down, as in Figure 3.11. Determine (a) the time needed for the stone to reach its maximum height, (b) the maximum height, (c) the time needed for the stone to return to the level of the thrower, (d) the velocity of the stone at this instant, and (e) the velocity and position of the stone at $t = 5$ s.

Solution: (a) To find the time necessary to reach the maximum height, use Equation 3.12, $v = v_0 - gt$, noting that $v = 0$ at maximum height:

$$20 \text{ m/s} - (9.80 \text{ m/s}^2)t_1 = 0$$

$$t_1 = \frac{20 \text{ m/s}}{9.80 \text{ m/s}^2} = 2.04 \text{ s}$$

(b) This value of time can be substituted into Equation 3.14, $y = v_0 t - \frac{1}{2}gt^2$, to give the maximum height as measured from the position of the thrower:

$$y_{max} = (20 \text{ m/s})(2.04 \text{ s}) - \tfrac{1}{2}(9.80 \text{ m/s}^2)(2.04 \text{ s})^2 = 20.4 \text{ m}$$

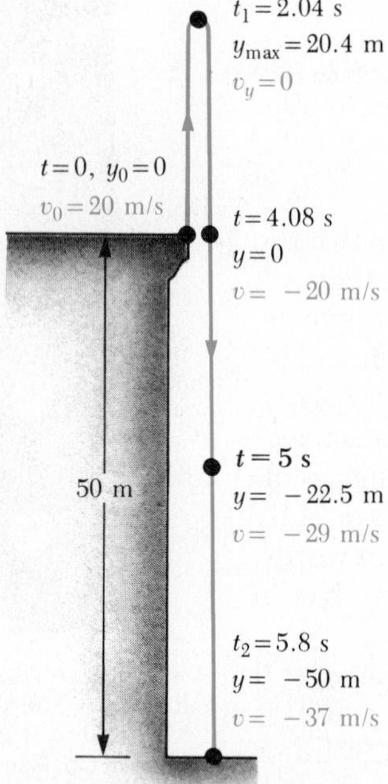

Figure 3.11 (Example 3.8) Position and velocity versus time for a freely falling particle thrown initially upward with a velocity $v_0 = 20$ m/s.

(c) When the stone is back at the height of the thrower, the y coordinate is zero. From the expression $y = v_0 t - \frac{1}{2}gt^2$ (Eq. 3.14), with $y = 0$, we obtain the expression

$$20t - 4.9t^2 = 0$$

This is a quadratic equation and has two solutions for t. (For some assistance in solving quadratic equations, see Appendix B.2.) The equation can be factored to give

$$t(20 - 4.9t) = 0$$

One solution is $t = 0$, corresponding to the time the stone starts its motion. The other solution is $t = 4.08$ s, which is the solution we are after.

(d) The value for t found in (c) can be inserted into $v = v_0 - gt$ (Eq. 3.12) to give

$$v = 20 \text{ m/s} - (9.80 \text{ m/s}^2)(4.08 \text{ s}) = -20.0 \text{ m/s}$$

Note that the velocity of the stone when it arrives back at its original height is equal in magnitude to its initial velocity but opposite in direction. This indicates that the motion is symmetric.

(e) From $v = v_0 - gt$ (Eq. 3.12), the velocity after 5 s is

$$v = 20 \text{ m/s} - (9.80 \text{ m/s}^2)(5 \text{ s}) = -29.0 \text{ m/s}$$

We can use $y = v_0 t - \frac{1}{2}gt^2$ (Eq. 3.14) to find the position of the particle at $t = 5$ s:

$$y = (20 \text{ m/s})(5 \text{ s}) - \tfrac{1}{2}(9.80 \text{ m/s}^2)(5 \text{ s})^2 = -22.5 \text{ m}$$

Exercise 1 Find (a) the velocity of the stone just before it hits the ground and (b) the total time the stone is in the air.
Answer: (a) -37.1 m/s (b) 5.83 s.

*3.6 KINEMATIC EQUATIONS DERIVED FROM CALCULUS

This is an optional section that assumes that the reader is familiar with the techniques of integral calculus. If you have not studied integration in your calculus course as yet, this section should be skipped or covered at some later time after you become familiar with integration.

The velocity of a particle moving in a straight line can be obtained from a knowledge of its position as a function of time. Mathematically, the velocity equals the derivative of the coordinate with respect to time. It is also possible to find the displacement of a particle if its velocity is known as a function of time. In the calculus, this procedure is referred to as integration, or the antiderivative. Graphically, it is equivalent to finding the area under a curve.

Suppose the velocity versus time plot for a particle moving along the x axis is as shown in Figure 3.12. Let us divide the time interval $t_f - t_i$ into many small intervals of duration Δt_n. From the definition of average velocity, we see that the displacement during any small interval such as the shaded one in Figure

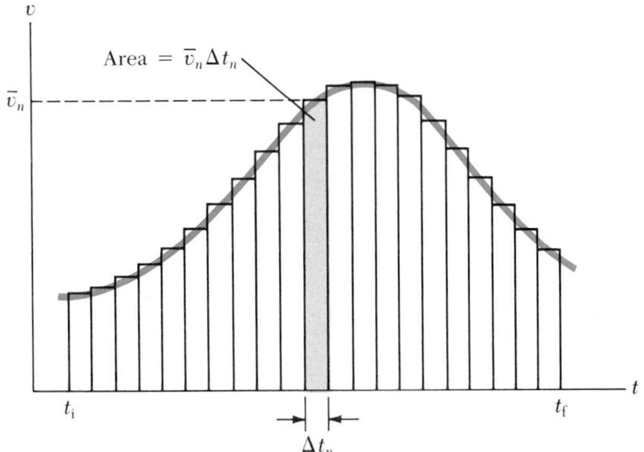

Figure 3.12 Velocity versus time curve for a particle moving along the x axis. The area of the shaded rectangle is equal to the displacement Δx in the time interval Δt_n, while the total area under the curve is the total displacement of the particle.

3.12 is given by $\Delta x_n = \bar{v}_n \, \Delta t_n$, where $\bar{v}_n$ is the average velocity in that interval. Therefore, the displacement during this small interval is simply the area of the shaded rectangle. The total displacement for the interval $t_f - t_i$ is the sum of the areas of all the rectangles:

$$\Delta x = \sum_n \bar{v}_n \, \Delta t_n$$

where the sum is taken over all the rectangles from t_i to t_f. Now, as each interval is made smaller and smaller, the number of terms in the sum increases and the sum approaches a value equal to the area under the velocity-time graph. Therefore, in the limit $n \rightarrow \infty$, or $\Delta t_n \rightarrow 0$, we see that the displacement is given by

$$\Delta x = \lim_{\Delta t_n \rightarrow 0} \sum_n v_n \, \Delta t_n \tag{3.16}$$

or

Displacement = area under the velocity-time graph

Note that we have replaced the average velocity $\bar{v}_n$ by the instantaneous velocity v_n in the sum. As you can see from Figure 3.12, this approximation is clearly valid in the limit of very small intervals. We conclude that if the velocity-time graph for motion along a straight line is known, the displacement during any time interval can be obtained by measuring the area under the curve.

The limit of the sum in Equation 3.16 is called a **definite integral** and is written

$$\lim_{\Delta t_n \rightarrow 0} \sum_n v_n \, \Delta t_n = \int_{t_i}^{t_f} v(t) \, dt \tag{3.17}$$

Definite integral

where $v(t)$ denotes the velocity at any time t. If the explicit functional form of $v(t)$ is known, the specific integral can be evaluated.

If a particle moves with a constant velocity v_0 as in Figure 3.13, its displacement during the time interval Δt is simply the area of the shaded rectangle, that is,

$$\Delta x = v_0 \, \Delta t \qquad (\text{when } v = v_0 = \text{constant})$$

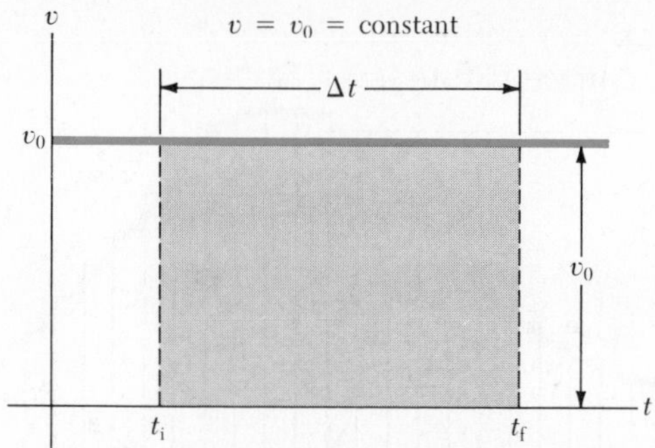

Figure 3.13 The velocity versus time curve for a particle moving with constant velocity v_0.

As another example, consider a particle moving with a velocity that is proportional to t, as in Figure 3.14. Taking $v = at$, where a is the constant of proportionality (the acceleration), we find that the displacement of the particle during the time interval $t = 0$ to $t = t_1$ is the area of the shaded triangle in Figure 3.14:

$$\Delta x = \tfrac{1}{2}(t_1)(at_1) = \tfrac{1}{2}at_1^2$$

Kinematic Equations

We will now make use of the defining equations for acceleration and velocity to derive the kinematic equations.

The defining equation for acceleration,

$$a = \frac{dv}{dt}$$

may also be written in terms of an integral (or antiderivative) as

$$v = \int a\,dt + C_1$$

where C_1 is a constant of integration. For the special case where the acceleration a is a constant, this reduces to

$$v = at + C_1$$

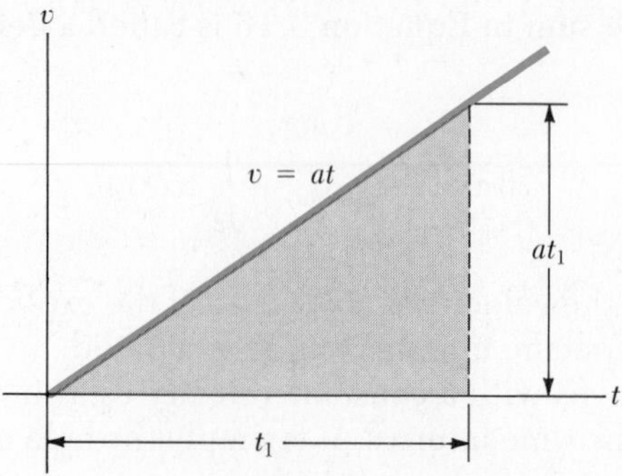

Figure 3.14 The velocity versus time curve for a particle moving with a velocity that is proportional to the time.

The value of C_1 depends on the initial conditions of the motion. If we take $v = v_0$ when $t = 0$, and substitute these into the last equation, we have

$$v_0 = a(0) + C_1$$

or

$$C_1 = v_0$$

Hence, we obtain the first kinematic equation:

$$v = v_0 + at \qquad \text{(for constant } a\text{)}$$

Now let us consider the defining equation for velocity:

$$v = \frac{dx}{dt}$$

We can also write this in integral form as

$$x = \int v\, dt + C_2$$

where C_2 is another constant of integration. Since $v = v_0 + at$, this becomes

$$x = \int (v_0 + at)\, dt + C_2$$

$$x = \int v_0\, dt + \int at\, dt + C_2$$

$$x = v_0 t + \tfrac{1}{2}at^2 + C_2$$

To find C_2, we make use of the initial condition that $x = x_0$ when $t = 0$. This gives $C_2 = x_0$. Therefore, we have

$$x = x_0 + v_0 t + \tfrac{1}{2}at^2 \qquad \text{(for constant } a\text{)}$$

This is the second equation of kinematics. Recall that $x - x_0$ is equal to the displacement of the object, where x_0 is its initial position.

3.7 SUMMARY

The **average velocity** of a particle during some time interval is equal to the ratio of the displacement, Δx, and the time interval, Δt:

$$\bar{v} \equiv \frac{\Delta x}{\Delta t} \qquad (3.1)$$

Average velocity

The **instantaneous velocity** of a particle is defined as the limit of the ratio $\Delta x/\Delta t$ as Δt approaches zero. By definition, this equals the derivative of x with respect to t, or the time rate of change of the position:

$$v \equiv \lim_{\Delta t \to 0} \frac{\Delta x}{\Delta t} = \frac{dx}{dt} \qquad (3.3)$$

Instantaneous velocity

The **speed** of a particle equals the absolute value of the velocity.

Average acceleration

The **average acceleration** of a particle during some time interval is defined as the ratio of the change in its velocity, Δv, and the time interval, Δt:

$$\bar{a} \equiv \frac{\Delta v}{\Delta t} \tag{3.4}$$

Instantaneous acceleration

The **instantaneous acceleration** is equal to the limit of the ratio $\Delta v/\Delta t$ as $\Delta t \to 0$. By definition, this equals the derivative of v with respect to t, or the time rate of change of the velocity:

$$a \equiv \lim_{\Delta t \to 0} \frac{\Delta v}{\Delta t} = \frac{dv}{dt} \tag{3.5}$$

The slope of the tangent to the x versus t curve at any instant equals the instantaneous velocity of the particle.

The slope of the tangent to the v versus t curve equals the instantaneous acceleration of the particle.

The area under the v versus t curve in any time interval equals the displacement of the particle in that interval.

Equations of kinematics

The **equations of kinematics** for a particle moving along the x axis with uniform acceleration a (constant in magnitude and direction) are

$$v = v_0 + at \tag{3.7}$$

$$x - x_0 = \tfrac{1}{2}(v_0 + v)t \tag{3.9}$$

$$x - x_0 = v_0 t + \tfrac{1}{2}at^2 \tag{3.10}$$

$$v^2 = v_0^2 + 2a(x - x_0) \tag{3.11}$$

(constant a only)

Freely falling body

A body falling freely in the presence of the earth's gravity experiences a gravitational acceleration directed toward the center of the earth. If air friction is neglected, and if the altitude of the motion is small compared with the earth's radius, then one can assume that the acceleration of gravity, g, is constant over the range of motion, where g is equal to 9.80 m/s², or 32 ft/s². Assuming y positive upward, the acceleration is given by $-g$, and the equations of kinematics for a body in free fall are the same as those given above, with the substitutions $x \to y$ and $a \to -g$.

QUESTIONS

1. Average velocity and instantaneous velocity are generally different quantities. Can they ever be equal for a specific type of motion? Explain.
2. If the average velocity is nonzero for some time interval, does this mean that the instantaneous velocity is never zero during this interval? Explain.
3. If the average velocity equals zero for some time interval Δt and if $v(t)$ is a continuous function, show that the instantaneous velocity must go to zero some time in this interval. (A sketch of x versus t might be useful in your proof.)
4. Is it possible to have a situation in which the velocity and acceleration have opposite signs? If so, sketch a velocity-time graph to prove your point.
5. If the velocity of a particle is nonzero, can its acceleration ever be zero? Explain.
6. If the velocity of a particle is zero, can its acceleration ever be nonzero? Explain.
7. Can the equations of kinematics (Eqs. 3.7 through 3.11) be used in a situation where the acceleration varies in time? Can they be used when the acceleration is zero?
8. A ball is thrown vertically upward. What are its velocity and acceleration when it reaches its maximum altitude? What is its acceleration just before it strikes the ground?
9. A stone is thrown upward from the top of a building. Does the stone's displacement depend on the location

of the origin of the coordinate system? Does the stone's velocity depend on the origin? (Assume that the coordinate system is stationary with respect to the building.) Explain.

10. A child throws a marble in the air with an initial velocity v_0. Another child drops a ball at the same instant. Compare the accelerations of the two objects while they are in flight.
11. A student at the top of a building of height h throws one ball upward with an initial speed v_0 and then throws a second ball downward with the same initial speed. How do the final velocities of the balls compare when they reach the ground?
12. Can the instantaneous velocity of an object ever be greater in magnitude than the average velocity? Can it ever be less?
13. If a car is traveling eastward, can its acceleration be westward? Explain.
14. If the average velocity of an object is zero in some time interval, what can you say about the displacement of the object for that interval?

PROBLEMS

Section 3.1 Average Velocity

1. A particle moving along the x axis is located initially at $x_i = 2.0$ m. Three minutes later, the particle is at $x_f = -5.0$ m. What is the average velocity of the particle?
2. The displacement over time for a certain particle moving along the x axis is as shown in Figure 3.15. Find the average velocity in the time intervals (a) 0 to 1 s, (b) 0 to 4 s, (c) 1 s to 5 s, (d) 0 to 5 s.

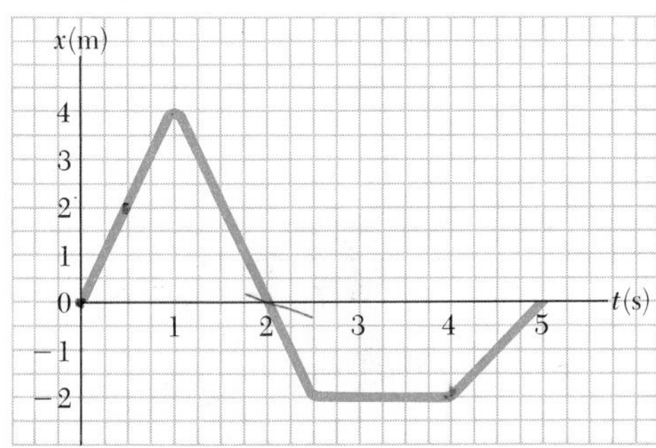

Figure 3.15 (Problems 2 and 5).

3. A jogger runs in a straight line with an average velocity of 5 m/s for 4 min, and then with an average velocity of 4 m/s for 3 min. (a) What is her total displacement? (b) What is her average velocity during this time?
4. An athlete swims the length of a 50-m pool in 20 s and makes the return trip to the starting position in 22 s. Determine his average velocity in (a) the first half of the swim, (b) the second half of the swim, and (c) the round trip.

Section 3.2 Instantaneous Velocity

5. Find the instantaneous velocity of the particle described by Figure 3.15 at the following times: (a) $t = 0.5$ s, (b) $t = 2$ s, (c) $t = 3$ s, (d) $t = 4.5$ s.
6. The position of a particle moving along the x axis varies linearly in time according to the expression $x = At + B$, where A and B are constants. (a) What are the dimensions of A and B? (b) Show by calculus and by graphical arguments that the average velocity equals the instantaneous velocity for this situation.
7. The position-time graph of a particle moving along the x axis is as shown in Figure 3.16. Determine whether the velocity is positive, negative, or zero for the times (a) t_1, (b) t_2, (c) t_3, (d) t_4.

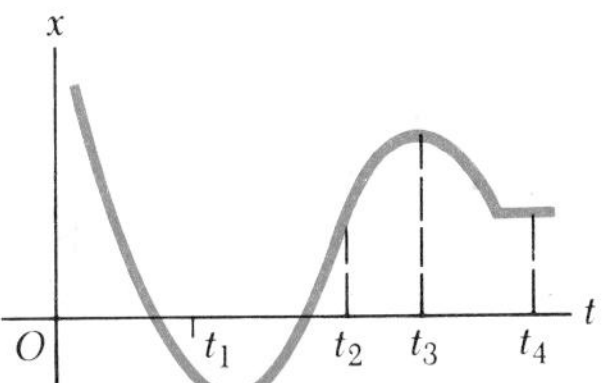

Figure 3.16 (Problem 7).

8. The position-time graph for a particle moving along the x axis is as shown in Figure 3.17. (a) Find the average velocity in the time interval $t = 1.5$ s to $t = 4$ s. (b) Determine the instantaneous velocity at $t = 2$ s by measuring the slope of the tangent line shown in the graph. (c) At what value of t is the velocity zero?

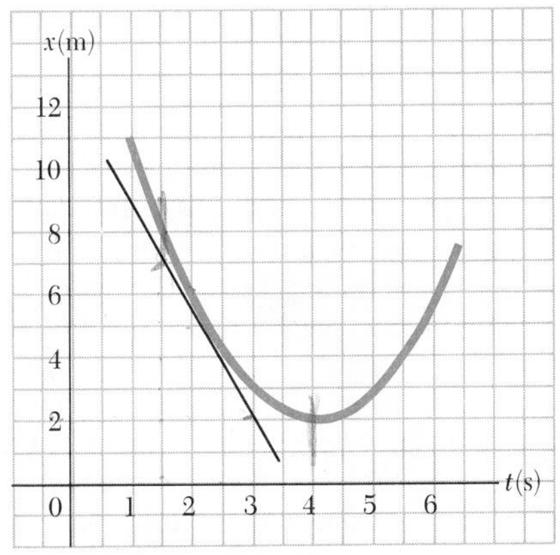

Figure 3.17 (Problem 8).

9. At $t = 1$ s, a particle moving with constant velocity is located at $x = -3$ m, and at $t = 6$ s, the particle is located at $x = 5$ m. (a) From this information, plot the position as a function of time. (b) Determine the velocity of the particle from the slope of this graph.

Section 3.3 Acceleration

10. A particle moves along the x axis according to the equation $x = 2t + 3t^2$, where x is in m and t is in s. Calculate the instantaneous velocity and instantaneous acceleration at $t = 3$ s.
11. A car traveling in a straight line has a velocity of 30 m/s at some instant. Two seconds later its velocity is 25 m/s. What is its average acceleration in this time interval?
12. The position of a particle moving along the y axis is given by $y = At^3 - Bt$, where A and B are constants, y is in m, and t is in s. (a) What are the dimensions of A and B? (b) Find expressions for the velocity and acceleration as functions of time.
13. A particle moving in a straight line has a velocity of 5 m/s at $t = 0$. Its velocity at $t = 4$ s is 21 m/s. (a) What is its average acceleration in this time interval? (b) Can the average velocity be obtained from the information presented? Explain.
14. The velocity-time graph for an object moving along the x axis is as shown in Figure 3.18. (a) Plot a graph of the acceleration versus time. (b) Determine the average acceleration of the object in the time intervals $t = 5$ s to $t = 15$ s and $t = 0$ to $t = 20$ s.

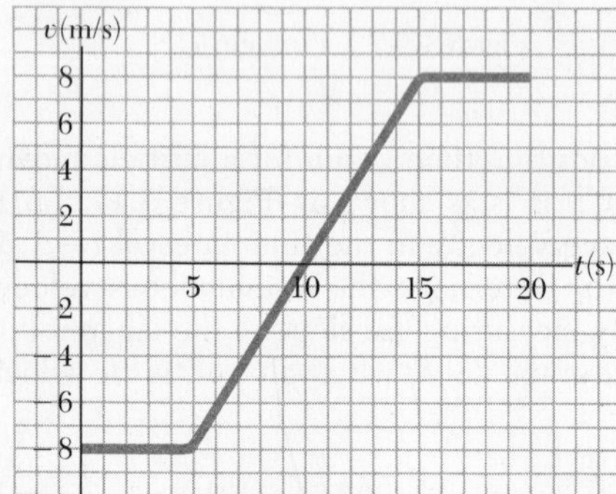

Figure 3.18 (Problem 14).

15. The velocity of a particle as a function of time is shown in Figure 3.19. At $t = 0$, the particle is at $x = 0$. (a) Sketch the acceleration as a function of time. (b) Determine the average acceleration of the particle in the time interval $t = 2$ s to $t = 8$ s. (c) Determine the instantaneous acceleration of the particle at $t = 4$ s.
16. A particle moves along the x axis according to the equation $x = 2 + 3t - t^2$, where x is in m and t is in s. At $t = 3$ s, find (a) the position of the particle, (b) its velocity, and (c) its acceleration.
17. The velocity of a particle moving along the x axis varies in time according to the relation $v = (15 - 8t)$ m/s. Find (a) the acceleration of the particle, (b) its velocity at $t = 3$ s, and (c) its average velocity in the time interval $t = 0$ to $t = 2$ s.

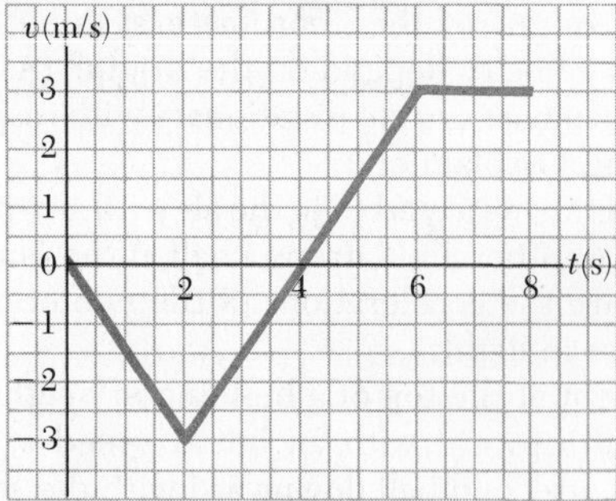

Figure 3.19 (Problem 15).

18. The velocity of a certain particle as a function of time is plotted in Figure 3.20. (a) Sketch the acceleration as a function of time. (b) Does the particle ever travel with constant acceleration? Explain. (c) Estimate the acceleration of the particle at $t = 6$ s.

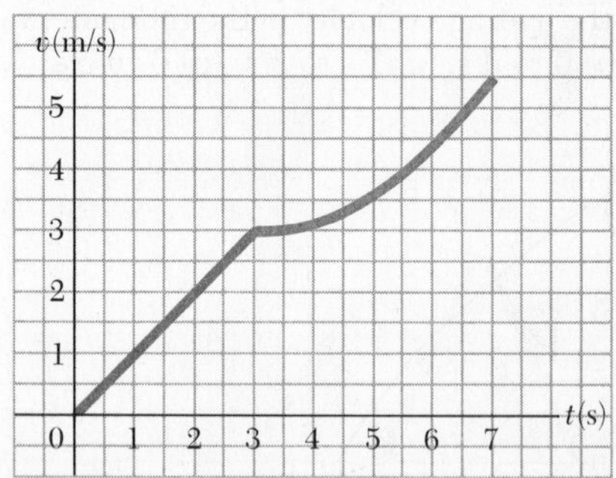

Figure 3.20 (Problem 18).

Section 3.4 One-Dimensional Motion with Constant Acceleration

19. A particle travels in the positive x direction for 10 s at a constant speed of 50 m/s. It then accelerates uniformly to a speed of 80 m/s in the next 5 s. Find (a) the average acceleration of the particle in the first 10 s, (b) its average acceleration in the interval $t = 10$ s to $t = 15$ s, (c) the total displacement of the particle between $t = 0$ and $t = 15$ s, and (d) its average speed in the interval $t = 10$ s to $t = 15$ s.
20. A body has a velocity of 12 cm/s when its x coordinate is 3 cm. If its x coordinate 2 s later is -5 cm, what is the uniform acceleration of the body?
21. The initial velocity of a body moving along the x axis is -6.0 cm/s when it is located at the origin. If it accelerates uniformly at the rate of 8.0 cm/s^2, find (a) its coordinate after 2 s and (b) its velocity after 3 s?
22. A proton has an initial velocity of 2.5×10^5 m/s and undergoes a uniform deceleration of 5.0×10^{10} m/s^2. What is its velocity after moving through a distance of 10 cm?
23. The initial speed of a body is 5.2 m/s. What is its speed after 2.5 s if it (a) accelerates uniformly at 3.0 m/s^2 and

(b) accelerates uniformly at $-3.0\ m/s^2$ (that is, it accelerates in the negative x direction)?

24. The position of a particle moving along the x axis varies in time according to the expression $x = 2 + 8t - 2t^2$, where x is in m and t is in s. Find (a) the displacement of the particle in the first 3 s of motion, (b) its acceleration, (c) its initial velocity, (d) the position where the particle comes momentarily to rest, and (e) its average velocity in the first 3 s of motion.

25. A speedboat increases its speed from 50 ft/s to 80 ft/s in a distance of 200 ft. Find (a) the magnitude of its acceleration and (b) the time it takes the boat to travel this distance.

26. A racing car reaches a speed of 50 m/s. At this instant, it decelerates uniformly using a parachute and braking system and comes to rest 5 s later. (a) Determine the deceleration of the car. (b) How far does the car travel after "turning on the brakes"?

27. The acceleration of gravity on the moon is about one sixth as great as on the earth. A stone is thrown vertically upward on the moon, with an initial speed equal to 20 m/s. (a) How long will the stone remain in motion? (b) What is the maximum height reached by the stone relative to the moon's surface?

28. A particle starts from rest from the top of an inclined plane and slides down with constant acceleration. The inclined plane is 2.0 m long, and it takes 3.0 s for the particle to reach the bottom. Find (a) the acceleration of the particle, (b) its speed at the bottom of the incline, (c) the time it takes the particle to reach the middle of the incline, and (d) its speed at the midpoint.

29. A go-cart travels the first half of a 100-m track with a constant speed of 5 m/s. In the second half of the track, it experiences a mechanical problem and decelerates at $0.2\ m/s^2$. How long does it take the go-cart to travel the 100-m distance?

30. A car moving at a constant speed of 30 m/s suddenly stalls at the bottom of a hill. The car undergoes a constant deceleration of $2\ m/s^2$ while ascending the hill. (a) Write equations for the position and the velocity as functions of time, taking $x = 0$ at the bottom of the hill where $v_0 = 30$ m/s. (b) Determine the maximum distance traveled by the car up the hill after stalling.

31. An electron has an initial velocity of 3.0×10^5 m/s. If it undergoes an acceleration of $8.0 \times 10^{14}\ m/s^2$, (a) how long will it take to reach a velocity of 5.4×10^5 m/s and (b) how far has it traveled in this time?

32. A railroad car is released from a locomotive on an incline. When the car reaches the bottom of the incline, it has a speed of 30 mi/h, at which point it passes through a retarder track that slows it down. If the retarder track is 30 ft long, what deceleration must it produce to bring the car to rest?

33. A bullet is fired through a board, 10 cm thick, in such a way that the bullet's line of motion is perpendicular to the face of the board. If the initial speed of the bullet is 400 m/s and it emerges from the other side of the board with a speed of 300 m/s, find (a) the deceleration of the bullet as it passes through the board and (b) the total time the bullet is in contact with the board.

34. An electron is accelerated from rest with a constant acceleration of $8 \times 10^{12}\ m/s^2$ to the right. It strikes a plate 4 cm from its starting point. (a) Neglecting gravity, find the final speed of the electron and its time of flight. (b) Treating the effect of gravity as a small perturbation on the electron's motion, determine how far the electron falls during its flight. (Treat the motion in the y-direction as independent of the motion in the x-direction.)

35. A hockey player is standing on his skates on a frozen pond when an opposing player skates by with the puck, moving with a uniform speed of 12 m/s. After 3 s, the first player makes up his mind to chase after his opponent. If he accelerates uniformly at $4\ m/s^2$, (a) how long does it take him to catch the opponent and (b) how far has he traveled in this time? (Assume the player with the puck remains in motion at constant speed.)

36. Until recently, the world's land speed record was held by Colonel John P. Stapp, USAF. On March 19, 1954, he rode a rocket-propelled sled that moved down the track at 632 mi/h. He and the sled were safely brought to rest in 1.4 s. Determine (a) the deceleration he experienced and (b) the distance he traveled during this deceleration.

37. A woman is reported to have fallen 144 ft from the 17th floor of a building, landing on a metal ventilator box, which she crushed to a depth of 18 in. She suffered only minor injuries. Neglecting air resistance, calculate (a) the speed of the woman just before she collided with the ventilator, (b) her deceleration while in contact with the box, and (c) the time it took to crush the box.

Section 3.5 Freely Falling Bodies

38. The *Guinness Book of World Records* lists a man who survived a deceleration of 200g, or $1960\ m/s^2$. A person seeking to break this record jumps off a cliff 102 m high onto several mattresses having a total thickness of 2 m. (a) What is the velocity of the "record-breaker" just before hitting the mattresses? (b) If the mattresses are crushed to a depth of 0.5 m, what is the record-breaker's deceleration?

39. A parachutist descending at a speed of 10 m/s drops a camera at an altitude of 50 m. (a) How long does it take the camera to reach the ground? (b) What is the velocity of the camera just before it hits the ground?

40. A ball is thrown vertically upward from the ground with an initial speed of 15 m/s. (a) How long does it take the ball to reach its maximum altitude? (b) What is its maximum altitude? (c) Determine the velocity and acceleration of the ball at $t = 2$ s.

41. A ball thrown vertically upward is caught by the thrower after 3.5 s. Find (a) the initial velocity of the ball and (b) the maximum height it reaches.

42. An object enters the earth's atmosphere with a velocity of 60 mi/h downward when it is 100 miles above the earth. Neglecting the effect of air friction and assuming $g = -32\ \text{ft/s}^2$, find the velocity of the object just before it hits the earth. (Do you think your result is realistic? Describe what you think really happens in this situation.)

43. An object is thrown vertically upward in such a way that it has a speed of 19.6 m/s when it reaches one half its maximum altitude. What are (a) its maximum altitude, (b) its velocity 1 s after it is thrown, and (c) its acceleration when it reaches its maximum altitude?

44. A ball is thrown vertically upward with an initial speed of 10 m/s. One second later, a stone is thrown vertically upward with an initial speed of 25 m/s. Determine (a) the time it takes the stone to reach the same height as the ball, (b) the velocity of the ball and stone when they are at the same height, and (c) the total time each is in motion before returning to the original height.

45. What can be said of the velocity and the acceleration of a particle when the particle's position-time graph, $x(t)$ is found to be (a) a first-degree polynomial (a linear function) and (b) a second-degree polynomial (a quadratic function)?

46. The acceleration-time graph for a particle that starts from rest at the origin is shown in Figure 3.21. (a) Sketch the corresponding velocity-time graph. (b) Sketch the corresponding position-time graph.

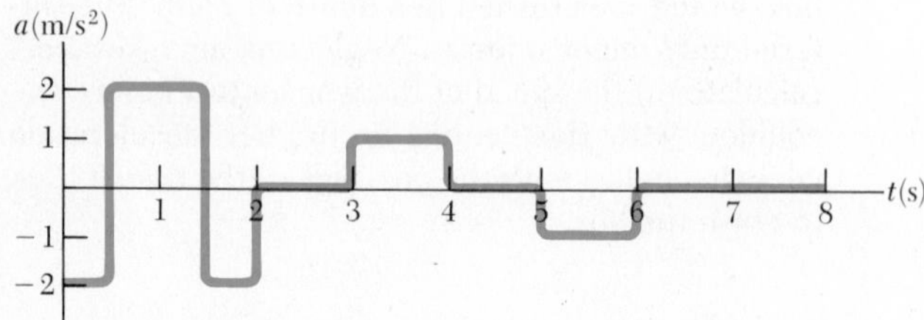

Figure 3.21 (Problem 46).

*3.6 Kinematic Equations Derived from Calculus

47. From the area under the velocity-time graph in Figure 3.19, determine the displacements of the particle for the intervals (a) $t = 0$ to $t = 4$ s, (b) $t = 4$ s to $t = 8$ s, and (c) $t = 0$ to $t = 8$ s. (Note that the areas which fall under the horizontal axis are negative.)

48. Estimate the displacement of the particle described by the velocity-time graph in Figure 3.20 for the interval $t = 0$ to $t = 7$ s.

49. A particle moves along the x axis with an acceleration that is proportional to the time, according to the expression $a = 30t$, where a is in m/s². Initially the particle is at rest at the origin. Find (a) the instantaneous velocity and (b) the instantaneous position as functions of time.

50. A particle is moving along the x axis. Its velocity as a function of time is given by $v = 5 + 10t$, where v is in m/s. The position of the particle at $t = 0$ is 20 m. Find (a) the acceleration as a function of time, (b) the position as a function of time, and (c) the velocity of the particle at $t = 0$.

GENERAL PROBLEMS

51. In a recent California driver's handbook, the following table is given listing data on the distance a typical moving vehicle travels for various initial speeds. The *thinking distance* corresponds to the fact that the driver has a finite reaction time, and the *braking distance* is how far the vehicle travels after the brakes are applied. Inspect these data carefully and determine (a) the thinking time, or reaction time, and (b) the deceleration of the vehicle.

Speed (mi/h)	Thinking Distance (ft)	Braking Distance (ft)	Total Distance (ft)
25	27	34	61
35	38	67	105
45	49	110	159
55	60	165	225
65	71	231	302

52. The following relationships represent the displacement x of a particle as a function of time t. The remaining symbols are constants. Find the velocity and acceleration as a function of time by taking appropriate derivatives and give the correct dimensions for the constants A, b, B, and a: (a) $x = Ae^{-bt}$, (b) $x = B \sin at$.

53. A Shenandoah National Park ranger is driving at the speed limit, 35 mph, when a whitetail deer jumps into the road 200 ft ahead of her position. After a reaction time of t seconds, she applies the brakes and decelerates at $a = -9.0\ \text{ft/s}^2$. What is the *maximum* reaction time allowed if she is to avoid hitting the deer?

54. An inquisitive physics student and mountain climber climbs a 50-m cliff that overhangs a calm pool of water. He throws two stones vertically downward 1 s apart and observes that they cause a single splash. The first stone has an initial velocity of 2 m/s. (a) At what time after release of the first stone will the two stones hit the water? (b) What initial velocity must the second stone have if they are to hit simultaneously? (c) What will the velocity of each stone be at the instant they hit the water?

55. A particle moves along the positive x axis in such a way that its coordinate varies in time according to the expression $x = 4 + 2t - 3t^2$, where x is in m and t is in s. (a) Make a graph of x versus t for the interval $t = 0$ to $t = 2$ s. (b) Determine the initial position and initial velocity of the particle. (c) Determine at what time the particle reaches a *maximum* position coordinate. (*Note:* At this time, $v = 0$.) (d) Calculate the coordinate, velocity, and acceleration at $t = 2$ s.

56. A "superball" is dropped from a height of 2 m above the ground. On the first bounce the ball reaches a height of 1.85 m, where it is caught. Find the velocity

of the ball (a) just as it makes contact with the ground and (b) just as it leaves the ground on the bounce. (c) Neglecting the time the ball spends in contact with the ground, find the total time required for the ball to go from the dropping point to the point where it is caught.

57. The position of a particle traveling along the x axis is given by $x = t^3 - 9t^2 + 6t$, where x is in cm and t is in s. Find (a) the instantaneous velocity of the particle for any time t, (b) the *times* at which the instantaneous velocity is zero, (c) the instantaneous acceleration of the particle at the times found in (b), and (d) the total displacement of the particle in traveling from the first zero to the second zero of the velocity.

58. The coyote, in his relentless attempt to catch the elusive road runner, loses his footing and falls from a sharp cliff, 1500 ft above ground level. After 5 s of free fall the coyote remembers he is wearing his Acme rocket-powered backpack, which he turns on. (a) The coyote comes to the ground with a gentle landing (i.e., zero velocity). Assuming a constant deceleration, find the deceleration of the coyote. (b) Unfortunately for the coyote, he is unable to shut down the rocket as he reaches the ground. Consequently, he is propelled back up into the air. After 5 s the rocket runs out of fuel. Find the maximum height reached by the coyote and his velocity as he reaches the ground for the second time.

59. A young woman named Kathy Kool buys a superdeluxe sports car that can accelerate at the rate of 16 ft/s^2. She decides to test the car by dragging with another speedster, Stan Speedy. Both start from rest, but experienced Stan leaves 1 s before Kathy. If Stan moves with a constant acceleration of 12 ft/s^2 and Kathy maintains an acceleration of 16 ft/s^2, find (a) the time it takes Kathy to overtake Stan, (b) the distance she travels before she catches him, and (c) the velocities of both cars at the instant she overtakes him.

60. A hockey player takes a slap shot at a puck at rest on the ice. The puck glides over the ice for 10 ft without friction, at which point it runs over rough ice. The puck then decelerates at 20 ft/s^2. If the velocity of the puck is 40 ft/s after traveling 100 ft from the point of impact, (a) what is the average acceleration imparted to the puck as it is struck by the hockey stick? (Assume that the time of contact is 0.01 s.) (b) How far in all does the puck travel before coming to rest? (c) What is the total time the puck is in motion, neglecting contact time?

61. A student stands on the edge of a building 100 ft above the ground and throws a baseball upward with some initial velocity. The ball is blown slightly sideways by a crosswind and then falls to the ground, just missing the building. If the total time for the ball to travel from the student's hand to the ground is 6 s, find (a) the initial velocity of the ball, (b) its final velocity as it hits the ground, and (c) its velocity after 3 s.

62. An ice sled powered by a rocket engine starts from rest on a large frozen lake and accelerates at 40 ft/s^2. After some time t_1 the rocket engine is shut down and the sled moves with constant velocity v for a time t_2. If the total distance traveled by the sled is 17 500 ft and the total time is 90 s, find (a) the times t_1 and t_2 and (b) the velocity v. If at the 17 500-ft mark the sled begins to decelerate at 20 ft/s^2, (c) what is the final position of the sled when it comes to rest and (d) how long does it take to come to rest?

63. A person sees a lightning bolt passing close to an airplane flying off in the distance. The person hears thunder 5 s after seeing the bolt and sees the airplane overhead 10 s after hearing the thunder. If the speed of sound in air is 1100 ft/s, (a) find the distance the airplane is from the person at the instant of the bolt. (Neglect the time it takes the light to travel from the bolt to the eye.) (b) Assuming the plane travels with a constant speed toward the person, find the velocity of the airplane. (c) Look up the speed of light in air and defend the approximation used in (a).

•64. The gasoline consumption of a certain car varies with speed according to the relation mi/gal = $1000/(v + 40)$, where v is in mi/h. The car starts from rest and accelerates to a uniform speed of 60 mi/h in 15 s. If it maintains this speed for 5 miles before uniformly decelerating to rest in 2 s, how much gasoline is consumed by the car?

•65. A rocket is fired vertically upward with an initial velocity of 80 m/s. It accelerates upward at 4 m/s^2 until it reaches an altitude of 1000 m. At that point, its engines fail and the rocket goes into free flight, with acceleration -9.80 m/s^2. (a) How long is the rocket in motion? (b) What is its maximum altitude? (c) What is its velocity just before it collides with the earth? (*Hint:* Consider the motion while the engine is operating separate from the free-flight motion.)

•66. A certain trolley car in San Francisco can stop in 10 s when traveling at maximum speed. On one occasion, the driver sees a dog a distance d m in front of the car and slams on the brakes instantly. The car reaches the dog 8 s later, and the dog jumps off the track just in time. If the car travels 4 m beyond the position of the dog before coming to a stop, how far was the car from the dog? (*Hint:* You will need three equations.)

•67. A train travels in time in the following manner. In the first hour, it travels with a speed v, in the next half hour it has a speed $3v$, in the next 90 min it travels with a speed $v/2$, and in the final 2 h it travels with a speed $v/3$, where v is in mi/h. (a) Plot the speed-time graph for this trip. (b) How far does the train travel in this trip? (c) What is the average speed of the train over the entire trip?

68. A commuter train can minimize the time t between two stations by accelerating ($a_1 = 0.1$ m/s^2) for a time t_1 then decelerating ($a_2 = -0.5$ m/s^2) by using his brakes for a time t_2. Since the stations are only 1 km apart, the train never reaches its maximum velocity.

Find the minimum time of travel t, and the time t_1 during which the train accelerates.

•69. In order to protect his food from hungry bears, a boy scout raises his food pack, mass m, with a rope that is thrown over a tree limb of height h above his hands. He walks away from the vertical rope with constant velocity v_0 holding the free end of the rope in his hands (see Fig. 3.22). (a) Show that the velocity v of the food pack is $x(x^2+h^2)^{-1/2}v_0$ where x is the distance he has walked away from the vertical rope. (b) Show that the acceleration a of the food pack is $h^2(x^2+h^2)^{-3/2}v_0{}^2$. (c) What values do the acceleration and velocity v have shortly after the boy scout leaves the vertical rope? (d) What values do the velocity and acceleration approach as the distance x continues to increase?

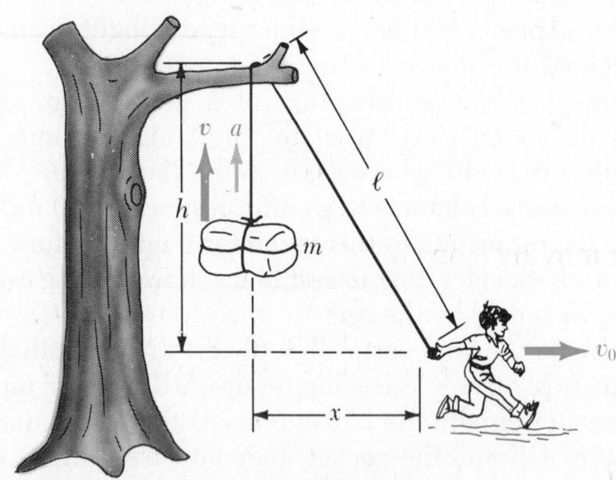

Figure 3.22 (Problems 69 and 70).

CALCULATOR/COMPUTER PROBLEMS

70. In Problem 69 let the height h equal 6 m and the velocity v_0 equal 2 m/s. Assume that the food pack starts from rest on a ledge over a cliff 6 m below the boy scout's hands. (a) Tabulate and graph the velocity-time graph. (b) Tabulate and graph the acceleration-time graph. (Let the range of time be from 0 s to 6 s and the time intervals be 0.5 s.)

71. A particle undergoes a varying acceleration. The velocity is measured at 0.5 s intervals and is tabulated below. (a) Determine the average acceleration in each interval. (b) Use a numerical integration procedure to determine the position of the particle at the end of each time interval. Assume the initial position of the particle is zero.

t(s)	0	0.5	1.0	1.5	2.0	2.5	3.0	3.5	4.0	4.5	5.0
v(m/s)	0	1	3	4.5	7.0	9.5	10.5	12	14	15	17.5

72. The acceleration of a particle moving along the x axis varies with position according to the expression

$$a = a_0 e^{-bx}$$

where $a_0 = 3$ m/s² and $b = 1$ m^{-1}. If the particle starts at rest from the origin, use a numerical integration method to find the position of the particle at $t = 2.37$ s. The accuracy of your calculation should be at least 1%.

73. A particle moving along the x axis undergoes an acceleration given by $a = \sqrt{3+t^3}$ m/s². Use a numerical integration method to find the position and velocity of the particle at $t = 5.7$ s to within 1% accuracy.

4
Motion in Two Dimensions

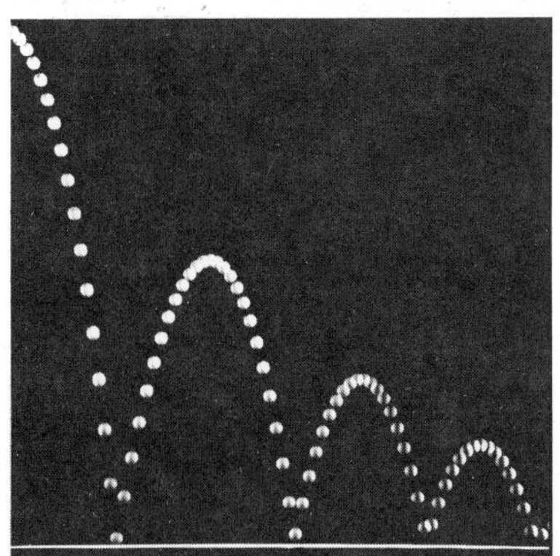

In this chapter we deal with the kinematics of a particle moving in a plane, or two-dimensional motion. Some common examples of motion in a plane are the motion of projectiles and satellites and the motion of charged particles in uniform electric fields. We begin by showing that velocity and acceleration are vector quantities. As in the case of one-dimensional motion, we shall derive the kinematic equations for two-dimensional motion from the fundamental definitions of displacement, velocity, and acceleration. As special cases of motion in two dimensions, we shall treat motion in a plane with constant acceleration and uniform circular motion.

4.1 THE DISPLACEMENT, VELOCITY, AND ACCELERATION VECTORS

In the previous chapter we found that the motion of a particle moving along a straight line is completely known if its coordinate is known as a function of time. Now let us extend this idea to the motion of a particle in the *xy* plane. We begin by describing the position of a particle with a *position vector* $\mathbf{r}$, drawn from the origin of some reference frame to the particle located in the *xy* plane, as in Figure 4.1. At time t_i, the particle is at the point P, and at some later time t_f, the particle is at Q. As the particle moves from P to Q in the time interval $\Delta t = t_f - t_i$, the position vector changes from $\mathbf{r}_i$ to $\mathbf{r}_f$, where the indices i and f refer to initial and final values. Because $\mathbf{r}_f = \mathbf{r}_i + \Delta\mathbf{r}$, the **displacement vector** for the particle is given by

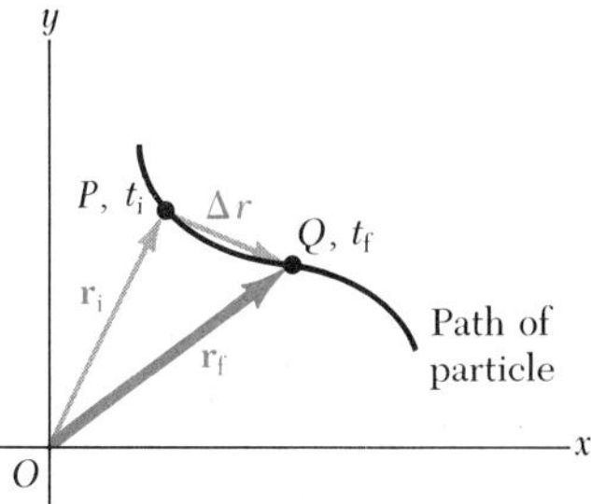

Figure 4.1 A particle moving in the *xy* plane is located with the position vector $\mathbf{r}$ drawn from the origin to the particle. The displacement of the particle as it moves from P to Q in the time interval $\Delta t = t_f - t_i$ is equal to the vector $\Delta\mathbf{r} = \mathbf{r}_f - \mathbf{r}_i$.

Definition of the displacement vector

$$\Delta\mathbf{r} \equiv \mathbf{r}_f - \mathbf{r}_i \tag{4.1}$$

The direction of $\Delta\mathbf{r}$ is indicated in Figure 4.1. Note that the displacement vector equals the difference between the final position vector and the initial position vector. As we see from Figure 4.1, the magnitude of the displacement vector is less than the distance traveled along the curved path.

We now define the **average velocity** of the particle during the time interval Δt as the ratio of the displacement and the time interval for this displacement:

Average velocity

$$\bar{\mathbf{v}} \equiv \frac{\Delta\mathbf{r}}{\Delta t} \tag{4.2}$$

Since the displacement is a vector and the time interval is a scalar, we conclude that the average velocity is a *vector* quantity directed along $\Delta\mathbf{r}$. Note that the average velocity between points P and Q is independent of the path between the two points. This is because the average velocity is proportional to the displacement, which in turn depends only on the initial and final position

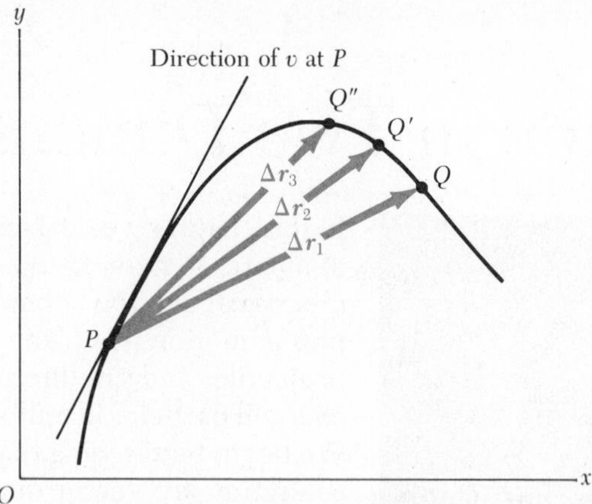

Figure 4.2 As a particle moves between two points, its average velocity is in the direction of the displacement vector $\Delta \mathbf{r}$. As the point Q moves closer to P, the direction of $\Delta \mathbf{r}$ approaches that of the line tangent to the curve at P. By definition, the instantaneous velocity at P is in the direction of this tangent line.

vectors. As we did in the case of one-dimensional motion, we conclude that if a particle starts its motion at some point and returns to this point via any path, its average velocity is zero for this trip since its displacement is zero.

Consider again the motion of a particle between two points in the xy plane, as in Figure 4.2. As the time intervals become smaller and smaller, the displacements, $\Delta \mathbf{r}_1$, $\Delta \mathbf{r}_2$, $\Delta \mathbf{r}_3$, . . . , get progressively smaller and the direction of the displacement approaches that of the line tangent to the path at the point P.

The **instantaneous velocity, v**, is defined as the limit of the average velocity, $\Delta \mathbf{r}/\Delta t$, as Δt approaches zero:

Instantaneous velocity

$$\mathbf{v} \equiv \lim_{\Delta t \to 0} \frac{\Delta \mathbf{r}}{\Delta t} = \frac{d\mathbf{r}}{dt} \tag{4.3}$$

That is, the instantaneous velocity equals the derivative of the position vector with respect to time. The direction of the velocity vector is along a line that is tangent to the path of the particle and in the direction of motion. This is illustrated in Figure 4.3 for two points along the path. The magnitude of the

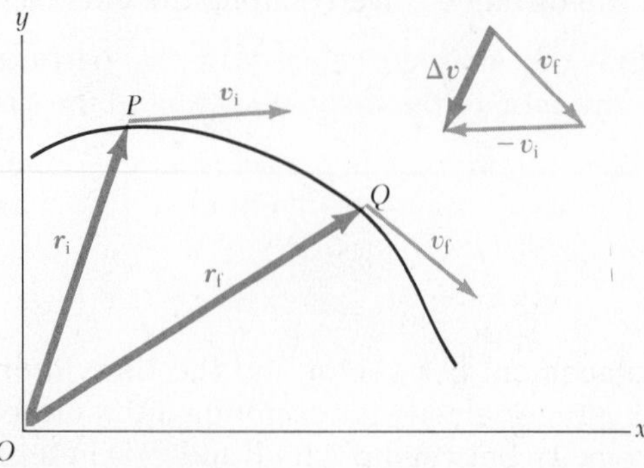

Figure 4.3 The average acceleration vector, $\bar{\mathbf{a}}$, for a particle moving from P to Q is in the direction of the change in the velocity, $\Delta \mathbf{v} = \mathbf{v}_f - \mathbf{v}_i$.

instantaneous velocity vector is called the *speed.* Note that Equation 4.3 is a logical generalization of differentiation as developed in the study of calculus.

As the particle moves from P to Q along some path, its instantaneous velocity vector changes from $\boldsymbol{v}_i$ at time t_i to $\boldsymbol{v}_f$ at time t_f (Figure 4.3).

The **average acceleration** of the particle as it moves from P to Q is defined as the ratio of the change in the instantaneous velocity vector, $\Delta\boldsymbol{v}$, and the elapsed time, Δt:

$$\bar{\boldsymbol{a}} \equiv \frac{\boldsymbol{v}_f - \boldsymbol{v}_i}{t_f - t_i} = \frac{\Delta\boldsymbol{v}}{\Delta t} \quad (4.4)$$

Average acceleration

Since the average acceleration is the ratio of a vector, $\Delta\boldsymbol{v}$, and a scalar, Δt, we conclude that $\bar{\boldsymbol{a}}$ is a vector quantity directed along $\Delta\boldsymbol{v}$. As is indicated in Figure 4.3, the direction of $\Delta\boldsymbol{v}$ is found by adding the vector $-\boldsymbol{v}_i$ (the negative of $\boldsymbol{v}_i$) to the vector $\boldsymbol{v}_f$, since by definition $\Delta\boldsymbol{v} = \boldsymbol{v}_f - \boldsymbol{v}_i$.

The **instantaneous acceleration,** $\boldsymbol{a}$, is defined as the limiting value of the ratio $\Delta\boldsymbol{v}/\Delta t$ as Δt approaches zero:

$$\boldsymbol{a} \equiv \lim_{\Delta t \to 0} \frac{\Delta\boldsymbol{v}}{\Delta t} = \frac{d\boldsymbol{v}}{dt} \quad (4.5)$$

Instantaneous acceleration

In other words, the instantaneous acceleration equals the first derivative of the velocity vector with respect to time.

It is important to recognize that a particle can accelerate for several reasons. First, the magnitude of the velocity vector (the speed) may change with time as in one-dimensional motion. Second, a particle accelerates when the direction of the velocity vector changes with time (a curved path) even though its speed is constant. Finally, the acceleration may be due to a change in both the magnitude and the direction of the velocity vector. In general, the acceleration has both tangential and perpendicular vector components.

4.2 MOTION IN TWO DIMENSIONS WITH CONSTANT ACCELERATION

Let us consider the motion of a particle in two dimensions with constant acceleration. That is, we assume that the magnitude and direction of the acceleration remain unchanged during the motion.

A particle in motion can be described by its position vector $\boldsymbol{r}$. The position vector for a particle moving in the xy plane can be written

$$\boldsymbol{r} = x\boldsymbol{i} + y\boldsymbol{j} \quad (4.6)$$

where x, y, and $\boldsymbol{r}$ change with time as the particle moves. If the position vector is known, the velocity of the particle can be obtained from Equations 4.3 and 4.6, which give

$$\boldsymbol{v} = \frac{d\boldsymbol{r}}{dt} = \frac{dx}{dt}\boldsymbol{i} + \frac{dy}{dt}\boldsymbol{j}$$

$$\boldsymbol{v} = v_x\boldsymbol{i} + v_y\boldsymbol{j} \quad (4.7)$$

Because $\boldsymbol{a}$ is a constant, its components a_x and a_y are also constants. Therefore, we can apply the equations of kinematics to both the x and y components of the velocity vector. Substituting $v_x = v_{x0} + a_x t$ and $v_y = v_{y0} + a_y t$ into Equation 4.7 gives

$$v = (v_{x0} + a_x t)i + (v_{y0} + a_y t)j$$
$$= (v_{x0} i + v_{y0} j) + (a_x i + a_y j)t$$

Velocity vector as a function of time

$$v = v_0 + at \tag{4.8}$$

This result states that the velocity of a particle at some time t equals the vector sum of its initial velocity, v_0, and the additional velocity at acquired in the time t as a result of its constant acceleration.

Similarly, from kinematics we know that the x and y coordinates of a particle moving with constant acceleration are given by

$$x = x_0 + v_{x0}t + \tfrac{1}{2}a_x t^2 \quad \text{and} \quad y = y_0 + v_{y0}t + \tfrac{1}{2}a_y t^2$$

Substituting these expressions into Equation 4.6 gives

$$r = (x_0 + v_{x0}t + \tfrac{1}{2}a_x t^2)i + (y_0 + v_{y0}t + \tfrac{1}{2}a_y t^2)j$$
$$= (x_0 i + y_0 j) + (v_{x0} i + v_{y0} j)t + \tfrac{1}{2}(a_x i + a_y j)t^2$$

or

$$r = r_0 + v_0 t + \tfrac{1}{2}at^2 \tag{4.9}$$

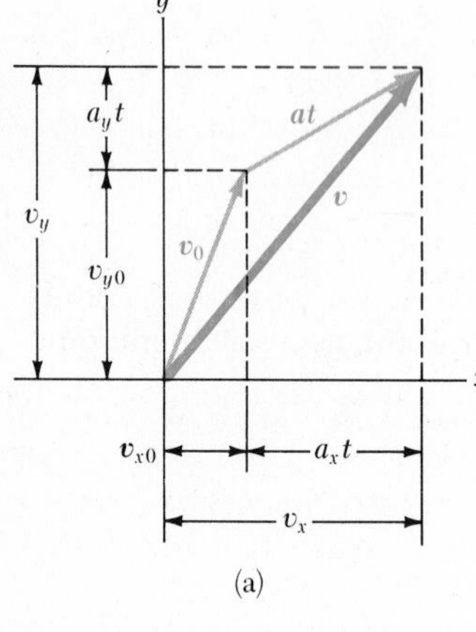

(a)

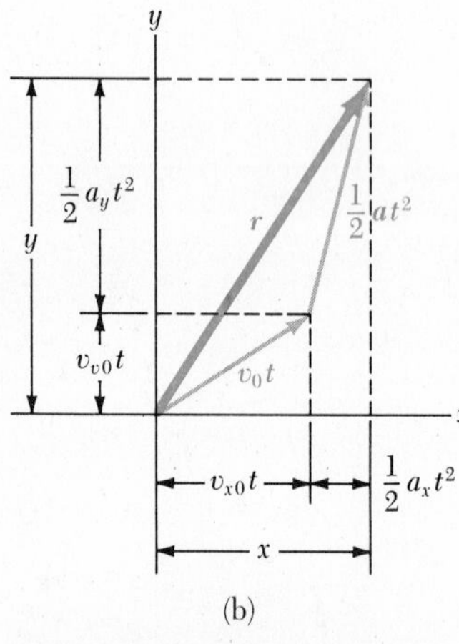

(b)

Figure 4.4 Vector representations and rectangular components of (a) the velocity and (b) the displacement of a particle moving with a uniform acceleration a.

This equation says that the displacement vector $r - r_0$ is the vector sum of a displacement $v_0 t$, arising from the initial velocity of the particle, and a displacement $\frac{1}{2}at^2$, resulting from the uniform acceleration of the particle. Graphical representations of Equations 4.6 and 4.7 are shown in Figures 4.4a and 4.4b. For simplicity in drawing the figure, we have taken $r_0 = 0$ in Figure 4.4b. That is, we assume that the particle is at the origin at $t = 0$. Note from Figure 4.4b that r is generally not along the direction of v_0 or a, since the relation between these quantities is a vector expression. For the same reason, from Figure 4.4a we see that v is generally not along the direction of v_0 or a. Finally, if we compare Figures 4.4a and 4.4b we see that v and r are not in the same direction. This is because v is linear in t, while r is quadratic in t. It is also important to recognize that since Equations 4.8 and 4.9 are *vector* expressions having one or more components (in general, three components), we may write the component forms of these expressions along the x and y axes with $r_0 = 0$

$$v = v_0 + at \qquad \begin{cases} v_x = v_{x0} + a_x t \\ v_y = v_{y0} + a_y t \end{cases}$$

$$r = v_0 t + \tfrac{1}{2}at^2 \qquad \begin{cases} x = v_{x0}t + \tfrac{1}{2}a_x t^2 \\ y = v_{y0}t + \tfrac{1}{2}a_y t^2 \end{cases}$$

These components are illustrated in Figure 4.4. In other words, two-dimensional motion with constant acceleration is equivalent to two independent motions in the x and y directions with constant accelerations a_x and a_y.

EXAMPLE 4.1. Motion in a Plane

A particle moves in the xy plane with an x component of acceleration only, given by $a_x = 4\ \text{m/s}^2$. The particle starts from the origin at $t = 0$ with an initial velocity having an x component of 20 m/s and a y component of -15 m/s. (a) Determine the components of velocity as a function of time and the total velocity vector at any time.

Since $v_{x0} = 20$ m/s and $a_x = 4\ \text{m/s}^2$, the equations of kinematics give

$$v_x = v_{x0} + a_x t = (20 + 4t)\ \text{m/s}$$

Also, since $v_{y0} = -15$ m/s and $a_y = 0$,

$$v_y = v_{y0} = -15\ \text{m/s}$$

Therefore, using the above results and noting that the velocity vector $\boldsymbol{v}$ has two components, we get

$$\boldsymbol{v} = v_x\boldsymbol{i} + v_y\boldsymbol{j} = [(20 + 4t)\boldsymbol{i} - 15\boldsymbol{j}] \text{ m/s}$$

We could also obtain this result using Equation 4.6 directly, noting that $\boldsymbol{a} = 4\boldsymbol{i}$ m/s^2 and $\boldsymbol{v}_0 = (20\boldsymbol{i} - 15\boldsymbol{j})$ m/s. Try it!

(b) Calculate the velocity and speed of the particle at $t = 5$ s.

With $t = 5$ s, the result from (a) gives

$$\boldsymbol{v} = \{[20 + 4(5)]\boldsymbol{i} - 15\boldsymbol{j}\} \text{ m/s} = (40\boldsymbol{i} - 15\boldsymbol{j}) \text{ m/s}$$

That is, at $t = 5$ s, $v_x = 40$ m/s and $v_y = -15$ m/s. The speed is defined as the magnitude of $\boldsymbol{v}$, or

$$v = |\boldsymbol{v}| = \sqrt{v_x^2 + v_y^2} = \sqrt{(40)^2 + (-15)^2} \text{ m/s}$$
$$= 42.7 \text{ m/s}$$

(*Note:* v is larger than v_0. Why?)

The angle θ that $\boldsymbol{v}$ makes with the x axis can be calculated using the fact that $\tan\theta = v_y/v_x$, or

$$\theta = \tan^{-1}\left(\frac{v_y}{v_x}\right) = \tan^{-1}\left(\frac{-15}{40}\right) = -20.6°$$

(c) Determine the x and y coordinates at any time t and the displacement vector at this time.

Since at $t = 0$, $x_0 = y_0 = 0$, the expressions for the x and y coordinates, the equations of kinematics give

$$x = v_{x0}t + \tfrac{1}{2}a_x t^2 = (20t + 2t^2) \text{ m}$$

$$y = v_{y0}t = (-15t) \text{ m}$$

Therefore, the displacement vector at any time t is given by

$$\boldsymbol{r} = x\boldsymbol{i} + y\boldsymbol{j} = [(20t + 2t^2)\boldsymbol{i} - 15t\boldsymbol{j}] \text{ m}$$

Alternatively, we could obtain $\boldsymbol{r}$ by applying Equation 4.7 directly, with $\boldsymbol{v}_0 = (20\boldsymbol{i} - 15\boldsymbol{j})$ m/s and $\boldsymbol{a} = 4\boldsymbol{i}$ m/s^2. Try it!

Thus, for example, at $t = 5$ s, $x = 150$ m and $y = -75$ m, or $\boldsymbol{r} = (150\boldsymbol{i} - 75\boldsymbol{j})$ m. It follows that the distance of the particle from the origin to this point is the magnitude of the displacement, or

$$|\boldsymbol{r}| = r = \sqrt{(150)^2 + (-75)^2} \text{ m} = 168 \text{ m}$$

Note that this is *not* the distance that the particle travels in this time! Can you determine this distance from the available data?

4.3 PROJECTILE MOTION

Anyone who has observed a baseball in motion (or, for that matter, any object thrown in the air) has observed projectile motion. For an arbitrary direction of the initial velocity, the ball moves in a curved path. This very common form of motion is surprisingly simple to analyze if the following three assumptions are made: (1) the acceleration due to gravity, $\boldsymbol{g}$, is constant over the range of motion and is directed downward,[1] (2) the effect of air resistance is negligible,[2] and (3) the rotation of the earth does not affect the motion. With these assumptions, we shall find that the path of a projectile, which we call its *trajectory*, is *always* a parabola. *We shall use these assumptions throughout this chapter.*

Assumptions of projectile motion

If we choose our reference frame such that the y direction is vertical and positive upward, then $a_y = -g$ (as in one-dimensional free fall) and $a_x = 0$ (since air friction is neglected). Furthermore, let us assume that at $t = 0$, the projectile leaves the origin ($x_0 = y_0 = 0$) with a velocity $\boldsymbol{v}_0$, as in Figure 4.5. If the vector $\boldsymbol{v}_0$ makes an angle θ_0 with the horizontal, as in Figure 4.5, then from the definitions of the cosine and sine functions we have

$$\cos\theta_0 = v_{x0}/v_0 \qquad \text{and} \qquad \sin\theta_0 = v_{y0}/v_0$$

Therefore, the initial x and y components of velocity are given by

$$v_{x0} = v_0\cos\theta_0 \qquad \text{and} \qquad v_{y0} = v_0\sin\theta_0$$

Substituting these expressions into Equations 4.8 and 4.9 with $a_x = 0$ and

[1] This approximation is reasonable as long as the range of motion is small compared with the radius of the earth (6.4×10^6 m). In effect, this approximation is equivalent to assuming that the earth is flat over the range of motion considered.

[2] This approximation is generally *not* justified, especially at high velocities. In addition, the spin of a projectile, such as a baseball, can give rise to some very interesting effects associated with aerodynamic forces (for example, a curve thrown by a pitcher).

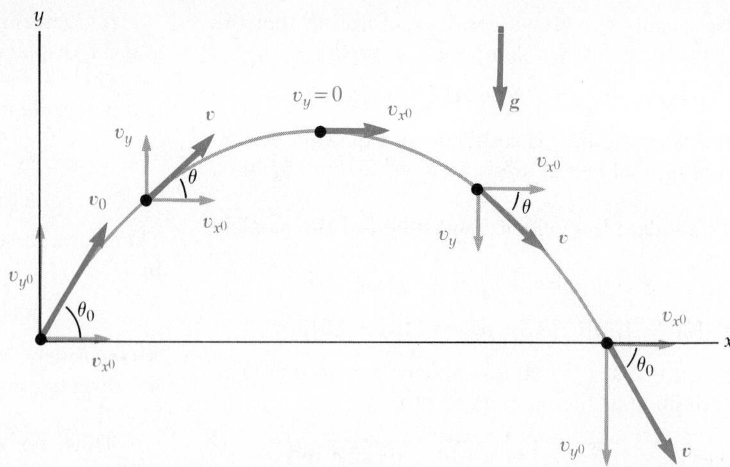

Figure 4.5 The parabolic trajectory of a projectile that leaves the origin with a velocity v_0. Note that the velocity vector, v, changes with time. However, the x component of velocity, v_{x0}, remains constant in time. Also, $v_y = 0$ at the peak.

$a_y = -g$ gives the velocity components and coordinates for the projectile at any time t:

Horizontal velocity component

$$v_x = v_{x0} = v_0 \cos \theta_0 = \text{constant} \tag{4.10}$$

Vertical velocity component

$$v_y = v_{y0} - gt = v_0 \sin \theta_0 - gt \tag{4.11}$$

Horizontal position component

$$x = v_{x0}t = (v_0 \cos \theta_0)t \tag{4.12}$$

Vertical position component

$$y = v_{y0}t - \tfrac{1}{2}gt^2 = (v_0 \sin \theta_0)t - \tfrac{1}{2}gt^2 \tag{4.13}$$

From Equation 4.10 we see that v_x remains constant in time and is equal to the initial x component of velocity, since there is no horizontal component of acceleration. Also, for the y motion we note that v_y and y are identical to the expressions for the freely falling body discussed in Chapter 3. In fact, *all* of the equations of kinematics developed in Chapter 3 are applicable to projectile motion.

If we solve for t in Equation 4.12 and substitute this expression for t into Equation 4.13, we find that

$$y = (\tan \theta_0)x - \left(\frac{g}{2v_0^2 \cos^2 \theta_0}\right) x^2 \tag{4.14}$$

which is valid for the angles in the range $0 < \theta_0 < \pi/2$. This is of the form $y = ax - bx^2$, which is the equation of a parabola that passes through the origin. Thus, we have proved that the trajectory of a projectile is a parabola. Note that the trajectory is *completely* specified if v_0 and θ_0 are known.

One can obtain the speed, v, as a function of time for the projectile by noting that Equations 4.10 and 4.11 give the x and y components of velocity at any instant. Therefore, by definition, since v is equal to the magnitude of $\mathbf{v}$,

$$v = \sqrt{v_x^2 + v_y^2} \tag{4.15}$$

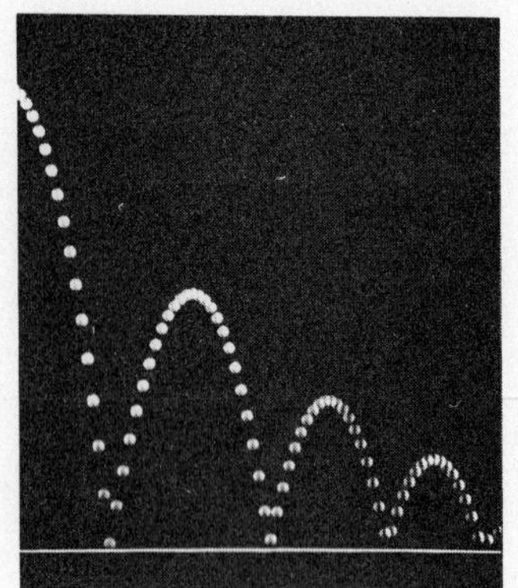

A ball undergoing several bounces off a hard surface. Note the parabolic path of the ball following each bounce. (Photo courtesy of Education Development Center, Newton, MA)

Also, since the velocity vector is tangent to the path at any instant, as shown in

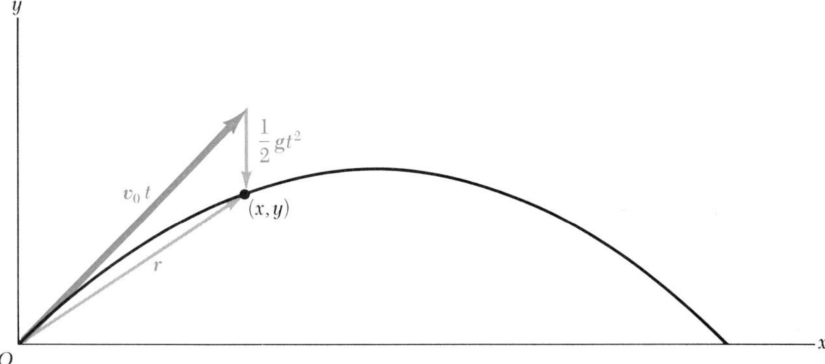

Figure 4.6 The displacement vector, $\boldsymbol{r}$, of a projectile having an initial velocity at the origin of $\boldsymbol{v}_0$. The vector $\boldsymbol{v}_0 t$ would be the displacement of the projectile if gravity were absent, and the vector $\frac{1}{2}\boldsymbol{g}t^2$ is its vertical displacement due to gravity in the time t.

Figure 4.5, the angle θ that $\boldsymbol{v}$ makes with the horizontal can be obtained from v_x and v_y through the expression

Angle of trajectory

$$\tan\theta = \frac{v_y}{v_x} \tag{4.16}$$

The vector expression for the position vector as a function of time for the projectile follows directly from Equation 4.7, with $\boldsymbol{a} = \boldsymbol{g}$:

$$\boldsymbol{r} = \boldsymbol{v}_0 t + \tfrac{1}{2}\boldsymbol{g}t^2$$

This expression is equivalent to Equations 4.12 and 4.13 and is plotted in Figure 4.6. Note that this equation is consistent with Equation 4.13, since the expression for $\boldsymbol{r}$ is a vector equation and $\boldsymbol{a} = \boldsymbol{g} = -g\boldsymbol{j}$ when the upward direction is taken to be positive. It is interesting to note that the motion can be considered the superposition of the term $\boldsymbol{v}_0 t$, which is the displacement if no acceleration were present, and the term $\frac{1}{2}\boldsymbol{g}t^2$, which arises from the acceleration due to gravity. In other words, if there were no gravitational acceleration, the particle would continue to move along a straight path in the direction of $\boldsymbol{v}_0$. Therefore, the vertical distance, $\frac{1}{2}gt^2$, through which the particle "falls" measured from the straight line is that of a freely falling body. *We conclude that projectile motion is the superposition of two motions: (1) the motion of a freely falling body in the vertical direction with constant acceleration and (2) uniform motion in the horizontal direction with constant velocity.*

Horizontal Range and Maximum Height of a Projectile Let us assume that a projectile is fired from the origin at $t = 0$ with a positive v_y component, as in Figure 4.7. There are two special points that are interesting to analyze: the peak with cartesian coordinates labeled $(R/2, h)$ and the point with coordinates $(R, 0)$. The distance R is called the *horizontal range* of the projectile, and h is its *maximum height*. Let us find h and R in terms of v_0, θ_0, and g.

We can determine the maximum height, h, reached by the projectile by noting that at the peak, $v_y = 0$. Therefore, Equation 4.11 can be used to determine the time t_1 it takes to reach the peak:

$$t_1 = \frac{v_0 \sin\theta_0}{g}$$

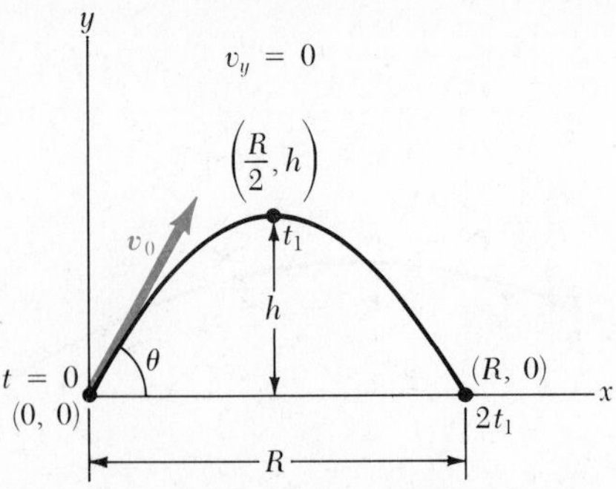

Figure 4.7 A projectile fired from the origin at $t = 0$ with an initial velocity v_0. The maximum height of the projectile is h, and its horizontal range is R.

Substituting this expression for t_1 into Equation 4.13 gives h in terms of v_0 and θ_0:

$$h = (v_0 \sin \theta_0) \frac{v_0 \sin \theta_0}{g} - \tfrac{1}{2} g \left(\frac{v_0 \sin \theta_0}{g} \right)^2$$

Maximum height of projectile

$$h = \frac{v_0^2 \sin^2 \theta_0}{2g} \tag{4.17}$$

The range, R, is the horizontal distance traveled in twice the time it takes to reach the peak, that is, in a time $2t_1$. (This can be seen by setting $y = 0$ in Equation 4.13 and solving the quadratic for t. One solution of this quadratic is $t = 0$, and the second is $t = 2t_1$.) Using Equation 4.12 and noting that $x = R$ at $t = 2t_1$, we find that

$$R = (v_0 \cos \theta_0) 2t_1 = (v_0 \cos \theta_0) \frac{2v_0 \sin \theta_0}{g}$$

$$R = \frac{2v_0^2 \sin \theta_0 \cos \theta_0}{g}$$

Since $\sin 2\theta = 2 \sin \theta \cos \theta$, R can be written in the more compact form

Range of projectile

$$R = \frac{v_0^2 \sin 2\theta_0}{g} \tag{4.18}$$

Keep in mind that Equations 4.17 and 4.18 are useful only for calculating h and R if v_0 and θ_0 are known and only for a symmetric path, as shown in Figure 4.7 (which means that only v_0 has to be specified). The general expressions given by Equations 4.10 through 4.13 are the *most important* results, since they give the coordinates and velocity components of the projectile at *any* time t.

You should note that the maximum value of R from Equation 4.18 is $R_{max} = v_0^2/g$. This result follows from the fact that the maximum value of $\sin 2\theta_0$ is unity, which occurs when $2\theta_0 = 90°$. Therefore, we see that R is a maximum when $\theta_0 = 45°$, as you would expect if air friction is neglected.

Figure 4.8 illustrates various trajectories for a projectile of a given initial speed. As you can see, the range is a maximum for $\theta_0 = 45°$. In addition, for any θ_0 other than 45°, a point with coordinates $(R, 0)$ can be reached with *two* complementary values of θ_0, such as 75° and 15°. Of course, the maximum height and time of flight will be different for these two values of θ_0.

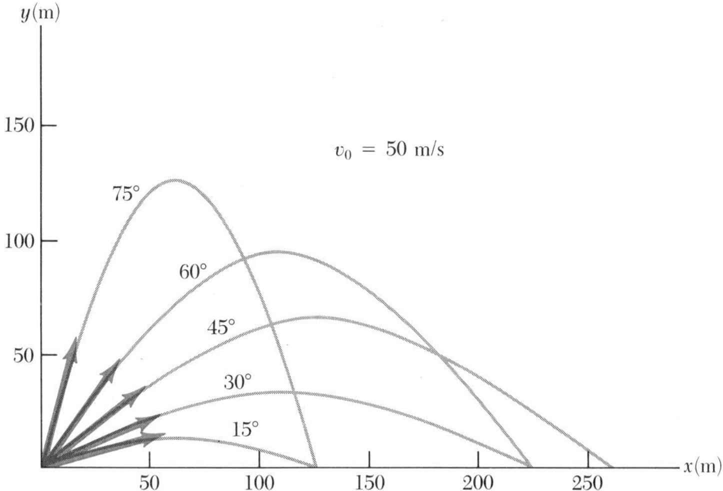

Figure 4.8 A projectile fired from the origin at an initial speed of 50 m/s at various angles of projection. Note that a point along the x axis can be reached at any two complementary values of θ_0.

EXAMPLE 4.2. The Long-jump

A long-jumper leaves the ground at an angle of 20° to the horizontal and at a speed of 11 m/s. (a) How far does he jump? (Assume that the motion of the long-jumper is equivalent to that of a particle.)

Solution: His horizontal motion is described by using Equation 4.12:

$$x = (v_0 \cos \theta_0)t = (11 \text{ m/s})(\cos 20°)t$$

The value of x can be found if t, the total time of the jump, is known. We are able to find t using the expression $v_y = v_0 \sin \theta_0 - gt$ by noting that at the top of the jump the vertical component of velocity goes to zero:

$$v_y = v_0 \sin \theta_0 - gt$$

$$0 = (11 \text{ m/s}) \sin 20° - (9.80 \text{ m/s}^2)t_1$$

$$t_1 = 0.384 \text{ s}$$

Note that t_1 is the time interval to reach the *top* of the jump. Because of the symmetry of the vertical motion, an identical time interval passes before the jumper returns to the ground. Therefore, the *total time* in the air is $t = 2t_1 = 0.768$ s. Substituting this into the expression for x gives

$$x = (11 \text{ m/s})(\cos 20°)(0.768 \text{ s}) = 7.94 \text{ m}$$

(b) What is the maximum height reached?

Solution: The maximum height reached is found using Equation 4.13, with $t = t_1 = 0.384$ s:

$$y_{max} = (v_0 \sin \theta_0)t_1 - \tfrac{1}{2}gt_1^2$$

$$y_{max} = (11 \text{ m/s})(\sin 20°)(0.384 \text{ s}) - \tfrac{1}{2}(9.80 \text{ m/s}^2)(0.384 \text{ s})^2$$

$$= 0.722 \text{ m}$$

The assumption that the motion of the long-jumper is that of a projectile is an oversimplification of the situation. Nevertheless, the values obtained are reasonable. Note that we also could have used Equations 4.17 and 4.18 to find the maximum height and horizontal range. However, the method used in our solution is more instructive.

EXAMPLE 4.3. The Monkey and the Hunter

In a very popular lecture demonstration, a projectile is fired at a falling target in such a way that the projectile leaves the gun at the same time the target is dropped from rest, as in Figure 4.9. Let us show that if the gun is

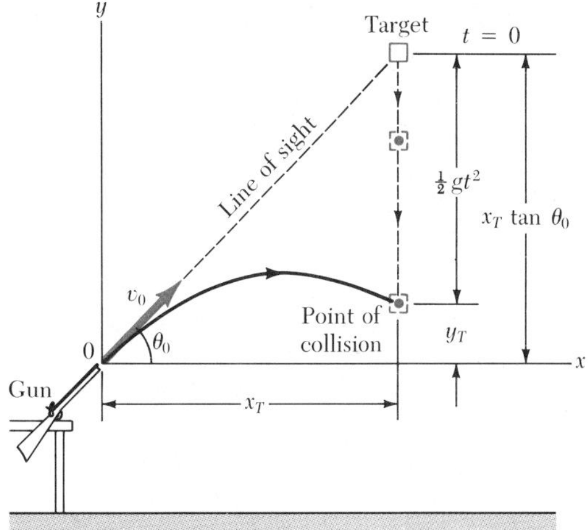

Figure 4.9 (Example 4.3) Schematic diagram of the projectile-and-target demonstration. If the gun is aimed directly at the target and is fired at the same instant the target begins to fall, the projectile will hit the target. Both fall through the same vertical distance in a time t, since both experience the same acceleration, $a_y = -g$.

initially aimed at the target, the projectile will hit the target.[3]

Solution: We can argue that a collision will result under the conditions stated by noting that both the projectile and the target experience the *same* acceleration, $a_y = -g$, as soon as they are released. First, note from Figure 4.9 that the initial y coordinate of the target is $x_T \tan \theta_0$ and that it falls through a distance $\frac{1}{2}gt^2$ in a time t. Therefore, the y coordinate of the target as a function of time is

$$y_T = x_T \tan \theta_0 - \tfrac{1}{2}gt^2$$

Now if we write equations for x and y for the projectile path over time, using Equations 4.12 and 4.13 simultaneously, we get

$$y_P = x_P \tan \theta_0 - \tfrac{1}{2}gt^2$$

Thus, when $x_P = x_T$, we see by comparing the two equations above that $y_P = y_T$ and a collision results.

The result could also be arrived at with vector methods, using expressions for the position vectors for the projectile and target.

You should also note that a collision will *not* always take place. There is the further restriction that a collision will result only when $v_0 \sin \theta_0 \geq \sqrt{gd/2}$, where d is the initial elevation of the target above the *floor*, as in Figure 4.9. If $v_0 \sin \theta_0$ is less than this value, the projectile will strike the floor before reaching the target.

EXAMPLE 4.4. That's Quite an Arm

A stone is thrown from the top of a building upward at an angle of 30° to the horizontal and with an initial speed of 20 m/s, as in Figure 4.10. If the height of the building is 45 m, (a) how long is the stone "in flight"?

Solution: The initial x and y components of the velocity are

$$v_{x0} = v_0 \cos \theta_0 = (20 \text{ m/s})(\cos 30°) = 17.3 \text{ m/s}$$

$$v_{y0} = v_0 \sin \theta_0 = (20 \text{ m/s})(\sin 30°) = 10 \text{ m/s}$$

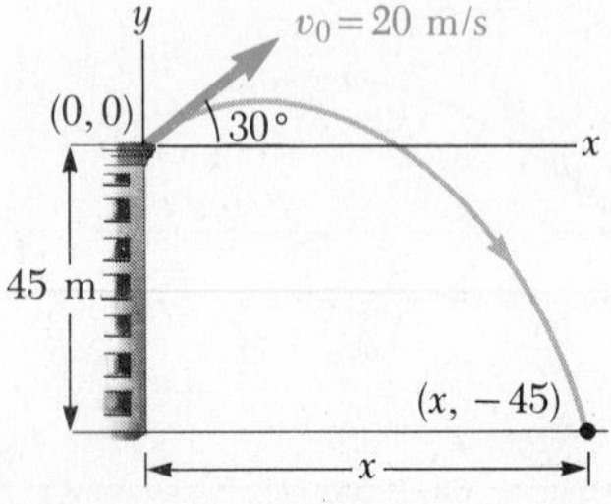

Figure 4.10 (Example 4.4).

To find t, we can use $y = v_{y0}t - \frac{1}{2}gt^2$ (Equation 4.13) with $y = -45$ m and $v_{y0} = 10$ m/s (we have chosen the top of the building as the origin, as shown as in Figure 4.10):

$$-45 \text{ m} = (10 \text{ m/s})t - \tfrac{1}{2}(9.80 \text{ m/s}^2)t^2$$

Solving the quadratic equation for t gives, for the positive root, $t = 4.22$ s. Does the negative root have any physical meaning? (Can you think of another way of finding t from the information given?)

(b) What is the speed of the stone just before it strikes the ground?

Solution: The y component of the velocity just before the stone strikes the ground can be obtained using the equation $v_y = v_{y0} - gt$ (Equation 4.11) with $t = 4.22$ s:

$$v_y = 10 \text{ m/s} - (9.80 \text{ m/s}^2)(4.22 \text{ s}) = 31.4 \text{ m/s}$$

Since $v_x = v_{x0} = 17.3$ m/s, the required speed is given by

$$v = \sqrt{v_x^2 + v_y^2} = \sqrt{(17.3)^2 + (-31.4)^2} \text{ m/s} = 35.9 \text{ m/s}$$

Exercise 1 Where does the stone strike the ground?
Answer: 73 m from the base of the building.

EXAMPLE 4.5. The Stranded Explorers

An Alaskan rescue plane drops a package of emergency rations to a stranded party of explorers, as shown in Figure 4.11. If the plane is traveling horizontally at 40 m/s at a height of 100 m above the ground, where does the package strike the ground relative to the point at which it was released?

Solution: The coordinate system for this problem is selected as shown in Figure 4.11, with the positive x direction to the right and the positive y direction upward.

Consider first the horizontal motion of the package. The only equation available to us is $x = v_{x0}t$

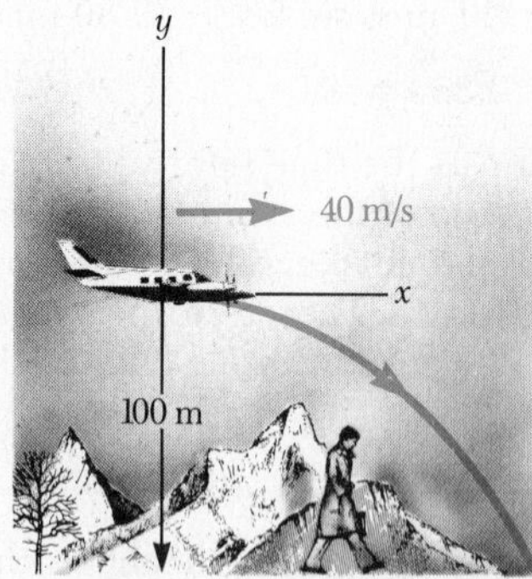

Figure 4.11 (Example 4.5) To an observer on the ground, a package released from the rescue plane travels along the path shown.

[3] In one variation of the demonstration, the target is a tin can held by an electromagnet energized with a small battery. At the instant the projectile leaves the gun, a small switch at the top of the gun is opened by the moving projectile. This opens the circuit containing the electromagnet, allowing the target to fall.

The initial x component of the package velocity is the same as the velocity of the plane when the package was released, 40 m/s. Thus, we have

$$x = (40 \text{ m/s})t$$

If we know t, the length of time the package is in the air, we can determine x, the distance traveled by the package along the horizontal. To find t, we move to the equations for the vertical motion of the package. We know that at the instant the package hits the ground its y coordinate is -100 m. We also know that the initial velocity of the package in the vertical direction, v_{y0}, is zero because the package was released with only a horizontal component of velocity. From Equation 4.13, we have

$$y = -\tfrac{1}{2}gt^2$$

$$-100 \text{ m} = -\tfrac{1}{2}(9.80 \text{ m/s}^2)t^2$$

$$t^2 = 20.4 \text{ s}^2$$

$$t = 4.51 \text{ s}$$

The value for the time of flight substituted into the equation for the x coordinate gives

$$x = (40 \text{ m/s})(4.51 \text{ s}) = 180 \text{ m}$$

Exercise 2 What are the horizontal and vertical components of the velocity of the package just before it hits the ground?
Answer: $v_x = 40$ m/s; $v_y = -44.1$ m/s.

4.4 UNIFORM CIRCULAR MOTION

Figure 4.12a shows an object moving in a circular path with *constant linear speed v*. It is often surprising to students to find that *even though the object moves at a constant speed, it still has an acceleration*. To see why this occurs, consider the defining equation for average acceleration, $\bar{a} = \Delta v/\Delta t$.

Note that the acceleration depends on *the change in the velocity vector*. Because velocity is a vector, there are two ways in which an acceleration can be produced: by a change in the *magnitude* of the velocity and by a change in the *direction* of the velocity. It is the latter situation that is occurring for an object moving in a circular path with constant speed. We shall show that the acceleration vector in this case is perpendicular to the path and always points toward the center of the circle. An acceleration of this nature is called a **centripetal acceleration** (center-seeking) and is given by

$$a_r = \frac{v^2}{r} \qquad (4.19)$$

Centripetal acceleration

To derive Equation 4.19, consider Figure 4.12b. Here an object is seen first at point P with velocity v_i at time t_i and then at point Q with velocity v_f at a later time t_f. Let us also assume here that v_i and v_f differ only in direction; their

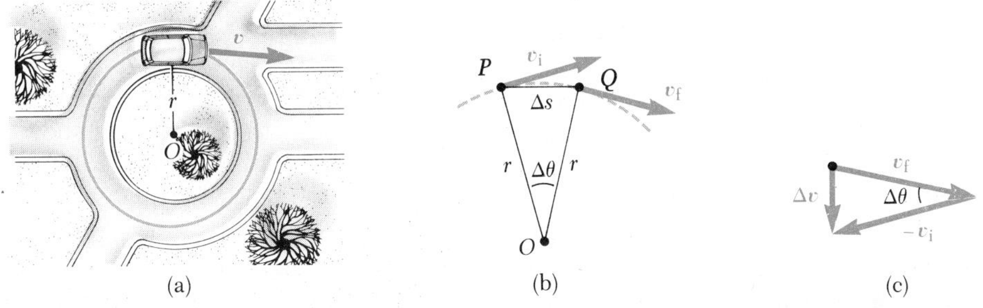

Figure 4.12 (a) Circular motion of an object moving with a constant speed. (b) As the particle moves from P to Q, the direction of its velocity vector changes from v_i to v_f. (c) The construction for determining the direction of the change in velocity, Δv, which is toward the center of the circle.

magnitudes are the same (that is, $v_i = v_f = v$). In order to calculate the acceleration, let us begin with the defining equation for average acceleration:

$$\bar{a} = \frac{\boldsymbol{v}_f - \boldsymbol{v}_i}{t_f - t_i} = \frac{\Delta \boldsymbol{v}}{\Delta t}$$

This equation indicates that we must vectorially subtract $\boldsymbol{v}_i$ from $\boldsymbol{v}_f$, where $\Delta\boldsymbol{v} = \boldsymbol{v}_f - \boldsymbol{v}_i$ is the change in the velocity. That is, $\Delta\boldsymbol{v}$ is obtained by adding to $\boldsymbol{v}_f$ the vector $-\boldsymbol{v}_i$. This can be accomplished graphically as shown by the vector triangle in Figure 4.12c. Note that when Δt is very small, Δs and $\Delta\theta$ are also very small. In this case, $\boldsymbol{v}_f$ will be almost parallel to $\boldsymbol{v}_i$ and the vector $\Delta\boldsymbol{v}$ will be approximately perpendicular to them, pointing toward the center of the circle.

Now consider the triangle in Figure 4.12b, which has sides Δs and r. This triangle and the one with sides Δv and v in Figure 4.12c are similar. (Two triangles are similar if the angle between any two sides is the same for both triangles and if the ratio of lengths of these sides is the same.) This enables us to write a relationship between the lengths of the sides:

$$\frac{\Delta v}{v} = \frac{\Delta s}{r}$$

This equation can be solved for Δv and the expression so obtained can be substituted into $\bar{a} = \Delta v/\Delta t$ to give $\bar{a}\,\Delta t = v\,\Delta s/r$, or

$$\bar{a} = \frac{v}{r}\frac{\Delta s}{\Delta t}$$

Now imagine that points P and Q in Figure 4.12b become extremely close together. In this case $\Delta\boldsymbol{v}$ would point toward the center of the circular path, and because the acceleration is in the direction of $\Delta\boldsymbol{v}$, it too is toward the center. Furthermore, as the two points P and Q approach each other, Δt approaches zero, and the ratio $\Delta s/\Delta t$ approaches the velocity v. Hence, in the limit $\Delta t \rightarrow 0$, the acceleration is

$$a_r = \frac{v^2}{r}$$

Thus we conclude that in uniform circular motion, the acceleration is directed inward toward the center of the circle and has a magnitude given by v^2/r. You should show that the dimensions of a_r are $[L]/[T^2]$, as required because this is a true acceleration. We shall return to the discussion of circular motion in Section 6.1.

4.5 TANGENTIAL AND RADIAL ACCELERATION IN CURVILINEAR MOTION

Let us consider the motion of a particle along a curved path where the velocity changes both in direction and in magnitude, as described in Figure 4.13. In this situation, the velocity of the particle is always tangent to the path; however, the acceleration vector $\boldsymbol{a}$ is now at some angle to the path. As the particle moves along the curved path in Figure 4.13, we see that the direction of the total acceleration vector, $\boldsymbol{a}$, changes from point to point. This vector can be resolved into two component vectors: a radial component vector, $\boldsymbol{a}_r$, and a tangential component vector, $\boldsymbol{a}_t$. That is, the *total* acceleration vector, $\boldsymbol{a}$, can be written as the vector sum of these component vectors:

$$a = a_r + a_t \quad (4.20)$$

Total acceleration

The tangential acceleration arises from the change in the speed of the particle, and its magnitude is given by

$$a_t = \frac{dv}{dt} \quad (4.21)$$

Tangential acceleration

The radial acceleration is due to the time rate of change in direction of the velocity vector and has a magnitude given by

$$a_r = \frac{v^2}{r} \quad (4.22)$$

Centripetal acceleration

where r is the radius of curvature of the path at the point in question. Since $\mathbf{a}_r$ and $\mathbf{a}_t$ are perpendicular component vectors of $\mathbf{a}$, it follows that $a = \sqrt{a_r^2 + a_t^2}$. As in the case of uniform circular motion, $\mathbf{a}_r$ always points toward the center of curvature, as shown in Figure 4.13. Also, at a given speed, a_r is large when the radius of curvature is small (as at points P and Q in Figure 4.13) and small when r is large (such as at point R). The direction of $\mathbf{a}_t$ is either in the same direction as $\mathbf{v}$ (if v is increasing) or opposite $\mathbf{v}$ (if v is decreasing).

Note that in the case of uniform circular motion, where v is constant, $a_t = 0$ and the acceleration is always radial, as we described in Section 4.4. Furthermore, if the direction of $\mathbf{v}$ doesn't change, then there is no radial acceleration and the motion is one-dimensional ($a_r = 0$, $a_t \neq 0$).

It is convenient to write the acceleration of a particle moving in a circular path in terms of unit vectors. We can do this by defining the unit vectors $\hat{\mathbf{r}}$ and $\hat{\boldsymbol{\theta}}$, where $\hat{\mathbf{r}}$ is *a unit vector directed radially outward along the radius vector,* from the center of curvature, and $\hat{\boldsymbol{\theta}}$ is *a unit vector tangent to the circular path,* as in Figure 4.14a. The direction of $\hat{\boldsymbol{\theta}}$ is in the direction of increasing θ, where θ is measured counterclockwise from the positive x axis. Note that both $\hat{\mathbf{r}}$ and $\hat{\boldsymbol{\theta}}$ "move along with the particle" and so vary in time relative to a stationary observer. Using this notation, we can express the total acceleration as

$$\mathbf{a} = \mathbf{a}_t + \mathbf{a}_r = \frac{dv}{dt}\hat{\boldsymbol{\theta}} - \frac{v^2}{r}\hat{\mathbf{r}} \quad (4.23)$$

These vectors are described in Figure 4.14b. The negative sign for $\mathbf{a}_r$ indicates that it is always directed radially inward, *opposite* the unit vector $\hat{\mathbf{r}}$.

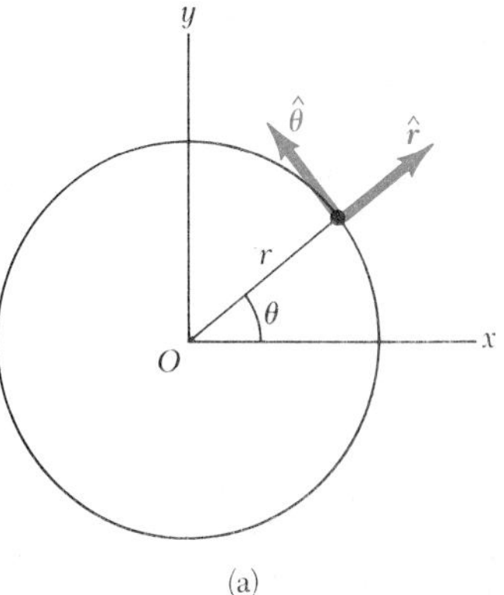

(a)

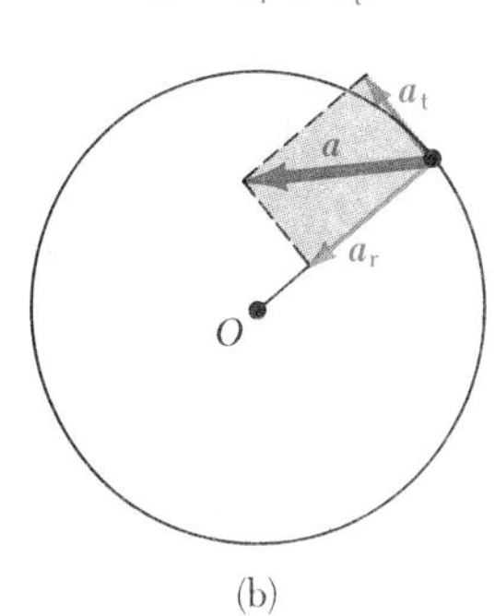

(b)

Figure 4.14 (a) Description of the unit vectors $\hat{\mathbf{r}}$ and $\hat{\boldsymbol{\theta}}$. (b) The total acceleration $\mathbf{a}$ of a particle rotating in a circle consists of a radial component vector, $\mathbf{a}_r$, directed toward the center of rotation, and a tangential component vector, $\mathbf{a}_t$. The component vector $\mathbf{a}_t$ is zero if the speed is constant.

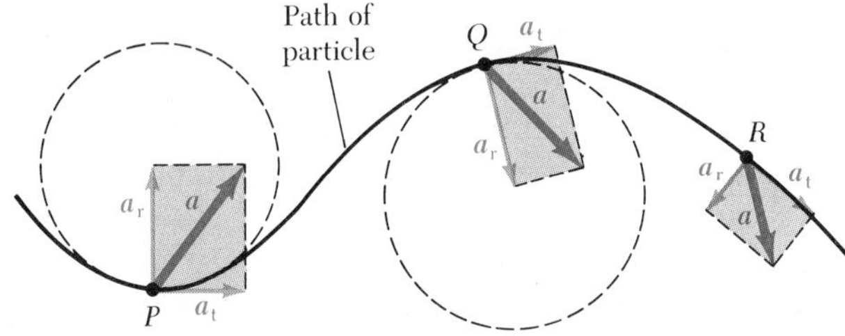

Figure 4.13 The motion of a particle along an arbitrary curved path lying in the xy plane. If the velocity vector $\mathbf{v}$ (always tangent to the path) changes in direction and magnitude, the component vectors of the acceleration of the particle are a tangential vector, $\mathbf{a}_t$, and a radial vector, $\mathbf{a}_r$.

EXAMPLE 4.6. The Rotating Ball
A ball tied to the end of a string 0.5 m in length swings in a vertical circle under the influence of gravity, as in Figure 4.15. When the string makes an angle of $\theta = 20°$ with the vertical, the ball has a speed of 1.5 m/s. (a) Find the radial component of acceleration at this instant.

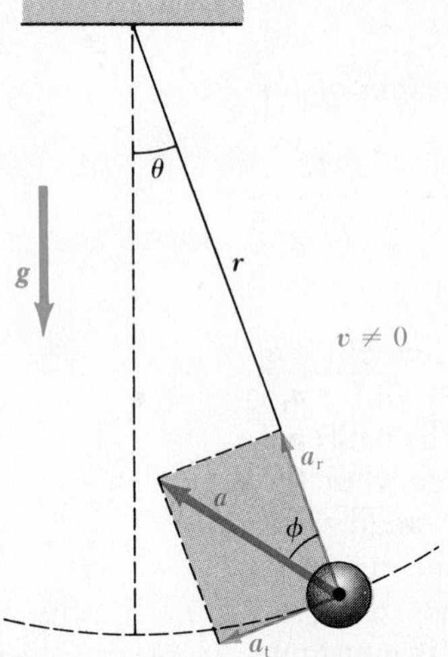

Figure 4.15 (Example 4.6) Circular motion of a ball tied on a string of length r. The ball swings in a vertical plane, and its acceleration, $\boldsymbol{a}$, has a radial component vector, $\boldsymbol{a}_r$, and a tangential component vector, $\boldsymbol{a}_t$.

Since $v = 1.5$ m/s and $r = 0.5$ m, we find that

$$a_r = \frac{v^2}{r} = \frac{(1.5 \text{ m/s})^2}{0.5 \text{ m}} = 4.5 \text{ m/s}^2$$

(b) When the ball is at an angle θ to the vertical, it has a tangential acceleration of magnitude $g \sin \theta$ (the component of g tangent to the circle). Therefore, at $\theta = 20°$, we find that $a_t = g \sin 20° = 3.36$ m/s^2. Find the magnitude and direction of the *total* acceleration at $\theta = 20°$.

Since $\boldsymbol{a} = \boldsymbol{a}_r + \boldsymbol{a}_t$, the magnitude of $\boldsymbol{a}$ at $\theta = 20°$ is given by

$$a = \sqrt{a_r^2 + a_t^2} = \sqrt{(4.5)^2 + (3.36)^2} \text{ m/s}^2 = 5.62 \text{ m/s}^2$$

If ϕ is the angle between $\boldsymbol{a}$ and the string, then

$$\phi = \tan^{-1}\frac{a_t}{a_r} = \tan^{-1}\left(\frac{3.36 \text{ m/s}^2}{4.5 \text{ m/s}^2}\right) = 36.7°$$

Note that all of the vectors—$\boldsymbol{a}$, $\boldsymbol{a}_t$, and $\boldsymbol{a}_r$—change in direction *and* magnitude as the ball swings through the circle. When the ball is at its lowest elevation ($\theta = 0$), $a_t = 0$, since there is no tangential component of g at this angle, and a_r is a *maximum*, since v is a maximum. When the ball is at its highest position ($\theta = 180°$), a_t is again zero but a_r is a minimum, since v is a minimum. Finally, in the two horizontal positions, ($\theta = 90°$ and $270°$), $|a_t| = g$ and a_r is somewhere between its minimum and maximum values.

4.6 RELATIVE VELOCITY AND RELATIVE ACCELERATION

In this section, we describe how observations made by different observers in different frames of reference are related to each other. We shall find that observers in different frames of reference may measure different displacements, velocities, and accelerations for a particle in motion. That is, two observers moving with respect to each other will generally not agree on the outcome of a measurement.

For example, if two cars are moving in the same direction with speeds of 50 mi/h and 60 mi/h, a passenger in the slower car will claim that the speed of the faster car relative to that of the slower car is 10 mi/h. Of course, a stationary observer will measure the speed of the faster car to be 60 mi/h. This simple example demonstrates that velocity measurements differ in different frames of reference.

Next, suppose a person riding on a moving vehicle (observer A) throws a ball straight up in the air according to his frame of reference, as in Figure 4.16a. According to observer A, the ball will move in a vertical path. On the other hand, a stationary observer (B) will see the path of the ball as a parabola, as illustrated in Figure 4.16b.

Another simple example is to imagine a package being dropped from an airplane flying parallel to the earth with a constant velocity. An observer on the airplane would describe the motion of the package as a straight line toward the earth. On the other hand, an observer on the ground would view the trajectory of the package as a parabola. Relative to the ground, the package has

Figure 4.16 (a) Observer A in a moving vehicle throws a ball upward and sees a straight-line path for the ball. (b) A stationary observer B sees a parabolic path for the same ball.

a vertical component of velocity (resulting from the acceleration of gravity and equal to the velocity measured by the observer in the airplane) *and* a horizontal component of velocity (given to it by the airplane's motion). If the airplane continues to move horizontally with the same velocity, the package will hit the ground directly beneath the airplane (assuming that friction is neglected)!

In a more general situation, consider a particle located at the point P in Figure 4.17. Imagine that the motion of this particle is being described by two observers, one in reference frame S, fixed with respect to the earth, and another in reference frame S′, moving to the right relative to S with a constant velocity $\boldsymbol{u}$. (Relative to an observer in S′, S moves to the left with a velocity $-\boldsymbol{u}$.) The location of an observer in his own frame of reference is irrelevant in this discussion, but to be definite the observer can be placed at the origin.

We label the position of the particle with respect to the S frame with the position vector $\boldsymbol{r}$ and label its position relative to the S′ frame with the vector $\boldsymbol{r}'$, at some time t. If the origins of the two reference frames coincide at $t = 0$, then the vectors $\boldsymbol{r}$ and $\boldsymbol{r}'$ are related to each through the expression $\boldsymbol{r} = \boldsymbol{r}' + \boldsymbol{u}t$, or

$$\boldsymbol{r}' = \boldsymbol{r} - \boldsymbol{u}t \tag{4.24}$$

Galilean coordinate transformation

That is, in a time t the S′ frame is displaced to the right by an amount $\boldsymbol{u}t$.

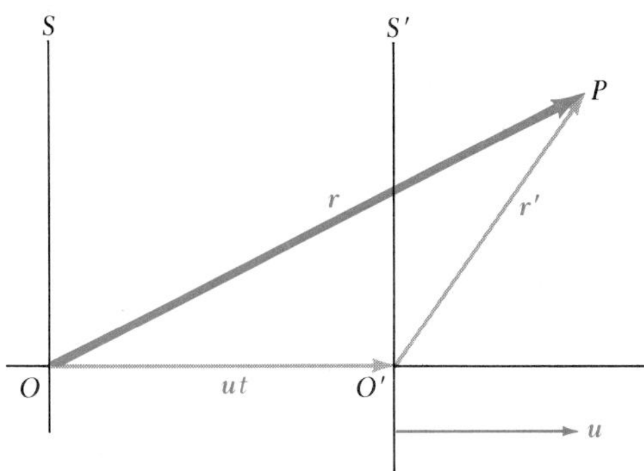

Figure 4.17 A particle located at the point P is described by two observers, one in the fixed frame of reference, S, the other in the frame S′, which moves with a constant velocity $\boldsymbol{u}$ to the right. The vector $\boldsymbol{r}$ is the particle's position vector relative to S, and $\boldsymbol{r}'$ is the position vector relative to S′.

If we differentiate Equation 4.24 with respect to time and note that $\boldsymbol{u}$ is constant, we get

$$\frac{d\boldsymbol{r}'}{dt} = \frac{d\boldsymbol{r}}{dt} - \boldsymbol{u}$$

Galilean velocity transformation

$$\boldsymbol{v}' = \boldsymbol{v} - \boldsymbol{u} \tag{4.25}$$

where $\boldsymbol{v}'$ is the velocity of the particle observed in the S′ frame and $\boldsymbol{v}$ is the velocity observed in the S frame. Equations 4.24 and 4.25 are known as *Galilean transformation equations.* They relate the coordinates and velocity of a particle in the earth's reference frame to those measured in a frame of reference in uniform motion with respect to the earth. However, they are *valid only* at particle speeds (relative to both observers) that are small compared with the speed of light ($\approx 3 \times 10^8$ m/s). When the particle speed according to either observer approaches the speed of light, these transformation equations must be replaced by more exact transformation equations, which are used in the special theory of relativity. As it turns out, the relativity transformation equations reduce to the Galilean transformation equations when the particle speed is small compared with the speed of light. We will discuss this in more detail in Chapter 39.

Although observers in the two different reference frames will measure different velocities for the particles, they will measure the *same acceleration* when $\boldsymbol{u}$ is constant. This can be seen by taking the time derivative of Equation 4.25, which gives

$$\frac{d\boldsymbol{v}'}{dt} = \frac{d\boldsymbol{v}}{dt} - \frac{d\boldsymbol{u}}{dt}$$

But $d\boldsymbol{u}/dt = 0$, since $\boldsymbol{u}$ is constant. Therefore, we conclude that $\boldsymbol{a}' = \boldsymbol{a}$ since $\boldsymbol{a}' = d\boldsymbol{v}'/dt$ and $\boldsymbol{a} = d\boldsymbol{v}/dt$. That is, *the acceleration of the particle measured by an observer in the earth's frame of reference will be the same as that measured by any other observer moving with* constant velocity *with respect to the first observer.*

EXAMPLE 4.7. A Boat Crossing a River

A boat heading due north crosses a wide river with a speed of 10 km/h relative to the water. The river has a uniform speed of 5 km/h due east. Determine the velocity of the boat with respect to a stationary ground observer.

Solution: The moving reference frame, S′, is attached to a cork floating on the river, and the observer is in the stationary reference frame, S (the earth). The vectors $\boldsymbol{u}$, $\boldsymbol{v}$, and $\boldsymbol{v}'$ are defined as follows:

$\boldsymbol{u}$ = velocity of the water with respect to the earth

$\boldsymbol{v}$ = velocity of the boat with respect to the earth

$\boldsymbol{v}'$ = velocity of the boat with respect to the water

In this example, $\boldsymbol{u}$ is to the right, $\boldsymbol{v}'$ is straight up, and $\boldsymbol{v}$ is at the angle θ_1, as defined in Figure 4.18a. Since these

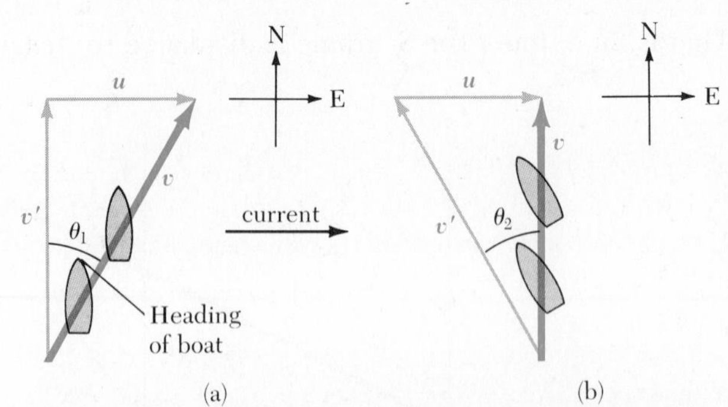

Figure 4.18 (Examples 4.7 and 4.8) (a) If the boat heads north, the motion of the boat relative to the earth is northeast along $\boldsymbol{v}$ when the river flows eastward. (b) If the boat wants to travel north, it must head northwest as shown. In both cases, $\boldsymbol{v} = \boldsymbol{v}' + \boldsymbol{u}$ and the heading of the boat is parallel to $\boldsymbol{v}'$.

three vectors form a right triangle, the speed of the boat with respect to the earth is

$$v = \sqrt{(v')^2 + u^2} = \sqrt{(10)^2 + (5)^2} \text{ km/h} = 11.2 \text{ km/h}$$

and the direction of v is

$$\theta_1 = \tan^{-1}\left(\frac{u}{v'}\right) = \tan^{-1}\left(\frac{5}{10}\right) = 26.6°$$

Therefore, the boat will be traveling 63.4° north of east with respect to the earth.

EXAMPLE 4.8. Which Way Should We Head?
If the boat in Example 4.7 travels with the same speed of 10 km/h relative to the water and wishes to travel due north, as in Figure 4.18b, in what direction should it head?

Solution: Intuitively, we know that the boat must head upstream. For this example, the vectors $\boldsymbol{u}$, $\boldsymbol{v}$, and $\boldsymbol{v}'$ are oriented as shown in Figure 4.18b, where $\boldsymbol{v}'$ is now the hypotenuse of the right triangle. Therefore, the boat's speed relative to the earth is

$$v = \sqrt{(v')^2 - u^2} = \sqrt{(10)^2 - (5)^2} \text{ km/h} = 8.66 \text{ km/h}$$

$$\theta_2 = \tan^{-1}\left(\frac{u}{v}\right) = \tan^{-1}\left(\frac{5}{8.66}\right) = 30°$$

where θ_2 is west of north.

4.7 SUMMARY

If a particle moves with *constant* acceleration $\boldsymbol{a}$ and has a velocity $\boldsymbol{v}_0$ and position $\boldsymbol{r}_0$ at $t = 0$, its velocity and position at some later time t are given by

Velocity vector as a function of time

$$\boldsymbol{v} = \boldsymbol{v}_0 + \boldsymbol{a}t \tag{4.8}$$

Position vector as a function of time

$$\boldsymbol{r} = \boldsymbol{r}_0 + \boldsymbol{v}_0 t + \tfrac{1}{2}\boldsymbol{a}t^2 \tag{4.9}$$

For two-dimensional motion in the xy plane under constant acceleration, these vector expressions are equivalent to two component expressions, one for the motion along x with an acceleration a_x and one for the motion along y with an acceleration a_y.

Projectile motion is two-dimensional motion under constant acceleration, where $a_x = 0$ and $a_y = -g$. In this case, if $x_0 = y_0 = 0$, the components of Equations 4.8 and 4.9 reduce to

Projectile motion equations

$$v_x = v_{x0} = \text{constant} \tag{4.10}$$

$$v_y = v_{y0} - gt \tag{4.11}$$

$$x = v_{x0}t \tag{4.12}$$

$$y = v_{y0}t - \tfrac{1}{2}gt^2 \tag{4.13}$$

where $v_{x0} = v_0 \cos\theta_0$, $v_{y0} = v_0 \sin\theta_0$ is the initial speed of the projectile, and θ_0 is the angle $\boldsymbol{v}_0$ makes with the positive x axis. Note that these expressions give the velocity components (and hence the velocity vector) and the coordinates (and hence the position vector) at *any* time t that the projectile is in motion.

As you can see from Equations 4.10 through 4.13, it is useful to think of projectile motion as the superposition of two motions: (1) uniform motion in the x direction, where v_x remains constant, and (2) motion in the vertical direction, subject to a constant downward acceleration of magnitude $g = 9.80$ m/s^2. Hence, one can analyze the motion in terms of separate horizontal and vertical components of velocity, as in Figure 4.19.

A particle moving in a circle of radius r with constant speed v undergoes a

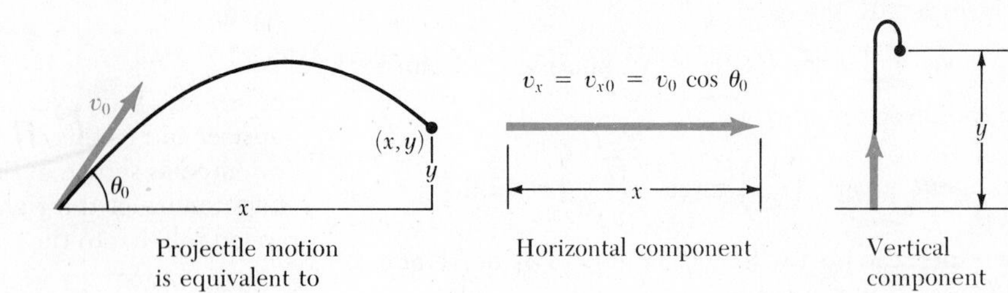

Figure 4.19 Analyzing motion in terms of the horizontal and vertical components of velocity.

centripetal (or radial) acceleration, $\boldsymbol{a}_r$, because the direction of $\boldsymbol{v}$ changes in time. The magnitude of $\boldsymbol{a}_r$ is given by

Centripetal acceleration

$$a_r = \frac{v^2}{r} \tag{4.19}$$

and its direction is always toward the center of the circle.

If a particle moves along a curved path in such a way that the magnitude and direction of $\boldsymbol{v}$ change in time, the particle has an acceleration vector that can be described by two component vectors: (1) a radial component vector, $\boldsymbol{a}_r$, arising from the change in direction of $\boldsymbol{v}$, and (2) a tangential component vector, $\boldsymbol{a}_t$, arising from the change in magnitude of $\boldsymbol{v}$. The magnitude of $\boldsymbol{a}_r$ is v^2/r, and the magnitude of $\boldsymbol{a}_t$ is dv/dt.

The velocity of a particle, $\boldsymbol{v}$, measured in a fixed frame of reference, S, is related to the velocity of the same particle, $\boldsymbol{v}'$, measured in a moving frame of reference, S′, by

Galilean velocity transformation

$$\boldsymbol{v}' = \boldsymbol{v} - \boldsymbol{u} \tag{4.25}$$

where $\boldsymbol{u}$ is the velocity of S′ relative to S.

QUESTIONS

1. If the average velocity of a particle is zero in some time interval, what can you say about the displacement of the particle for that interval?
2. If you know the position vectors of a particle at two points along its path and also know the time it took to get from one point to the other, can you determine the particle's instantaneous velocity? its average velocity? Explain.
3. Describe a situation in which the velocity of a particle is perpendicular to the position vector.
4. Can a particle accelerate if its speed is constant? Can it accelerate if its velocity is constant? Explain.
5. Explain whether or not the following particles have an acceleration: (a) a particle moving in a straight line with constant speed and (b) a particle moving around a curve with constant speed.
6. Correct the following statement: "The racing car rounds the turn at a constant velocity of 90 miles per hour."
7. Determine which of the following moving objects would exhibit an approximate parabolic trajectory: (a) a ball thrown in an arbitrary direction, (b) a jet airplane, (c) a rocket leaving the launching pad, (d) a rocket a few minutes after launch with failed engines, (e) a tossed stone moving to the bottom of a pond.
8. A student argues that as a satellite orbits the earth in a circular path, it moves with a constant velocity and therefore has no acceleration. The professor claims that the student is wrong since the satellite must have a centripetal acceleration as it moves in its circular orbit. What is wrong with the student's argument?

9. What is the fundamental difference between the unit vectors $\hat{r}$ and $\hat{\theta}$ defined in Figure 4.14 and the unit vectors $\boldsymbol{i}$ and $\boldsymbol{j}$?
10. At the end of its arc, the velocity of a pendulum is zero. Is its acceleration also zero at this point?
11. If a rock is dropped from the top of a sailboat's mast, will it hit the deck at the same point whether the boat is at rest or in motion at constant velocity?
12. A stone is thrown upward from the top of the building. Does the stone's displacement depend on the location of the origin of the coordinate system? Does the stone's velocity depend on the location of the origin?
13. Inspect the multiple image photograph of two golf balls released simultaneously under the conditions indicated. Explain why both balls hit the floor simultaneously.

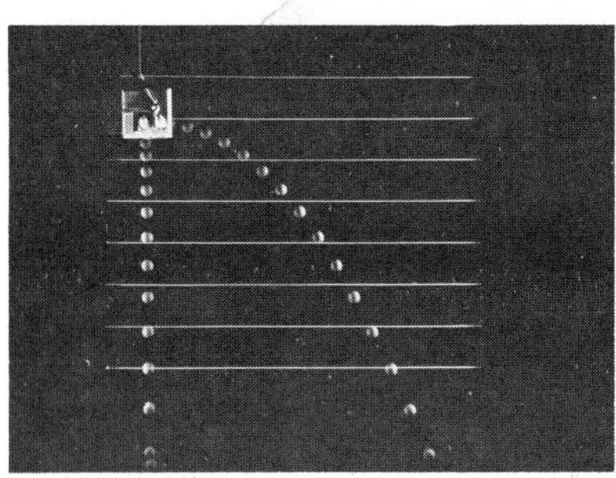

This multiple image photograph of two golf balls released simultaneously illustrates both free fall and projectile motion. The right ball was projected horizontally with an initial velocity of 2 m/s. The light flashes were 1/30 s apart, and the white parallel lines (actually strings) were placed $15\frac{1}{4}$ cm apart.

PROBLEMS

Section 4.2 Motion in Two Dimensions with Constant Acceleration

1. A particle starts from rest at $t = 0$ at the origin and moves in the xy plane with a constant acceleration of $\boldsymbol{a} = (2\boldsymbol{i} + 4\boldsymbol{j})$ m/s^2. After a time t has elapsed, determine (a) the x and y components of velocity, (b) the coordinates of the particle, and (c) the speed of the particle.
2. At $t = 0$, a particle moving in the xy plane with constant acceleration has a velocity of $\boldsymbol{v}_0 = (3\boldsymbol{i} - 2\boldsymbol{j})$ m/s at the origin. At $t = 3$ s, its velocity is $\boldsymbol{v} = (9\boldsymbol{i} + 7\boldsymbol{j})$ m/s. Find (a) the acceleration of the particle and (b) its coordinates at any time t.
3. The vector position of a particle varies in time according to the expression $\boldsymbol{r} = (3\boldsymbol{i} - 6t^2\boldsymbol{j})$ m. (a) Find expressions for the velocity and acceleration as functions of time. (b) Determine the particle's position and velocity at $t = 1$ s.
4. A particle initially located at the origin has an acceleration of $\boldsymbol{a} = 3\boldsymbol{j}$ m/s^2 and an initial velocity of $\boldsymbol{v}_0 = 5\boldsymbol{i}$ m/s. Find (a) the vector position and velocity at any time t and (b) the coordinates and speed of the particle at $t = 2$ s.
5. At $t = 0$ a particle leaves the origin with a velocity of 6 m/s in the positive y direction. Its acceleration is given by $\boldsymbol{a} = (2\boldsymbol{i} - 3\boldsymbol{j})$ m/s^2. When the particle reaches its *maximum* y coordinate, its y component of velocity is zero. At this instant, find (a) the velocity of the particle and (b) its x and y coordinates.

Section 4.3 Projectile Motion (Neglect Air Resistance in All Problems)

6. A football, kicked at an angle of 50° to the horizontal, travels a horizontal distance of 20 m before hitting the ground. Find (a) the initial speed of the football, (b) the time it is in the air, and (c) the maximum height it reaches.
7. An astronaut on a strange planet finds that she can jump a *maximum* horizontal distance of 30 m if her initial speed is 9 m/s. What is the acceleration of gravity on the planet?
8. It has been said that in his youth George Washington threw a silver dollar across a river. Assuming that the river was 300 m wide, (a) what *minimum initial* speed was necessary to get the coin across the river and (b) how long was the coin in flight?
9. A rifle is aimed horizontally through its bore at the center of a large target 150 m away. The initial velocity of the bullet is 450 m/s. (a) Where does the bullet strike the target? (b) To hit the center of the target, the barrel must be at an angle above the line of sight. Find the angle of elevation of the barrel.
10. A ball is thrown horizontally from the top of a building 35 m high. The ball strikes the ground at a point 80 m from the base of the building. Find (a) the time the ball is in flight, (b) its initial velocity, and (c) the x and y components of velocity just before the ball strikes the ground.
11. A projectile is fired in such a way that its horizontal range is equal to three times its maximum height. What is the angle of projection?
12. Show that the horizontal range of a projectile with a fixed initial speed will be the same for any two complementary angles, such as 30° and 60°.
13. The initial speed of a cannon ball is 200 m/s. If it is fired at a target that is at a horizontal distance of 2 km from the cannon, find (a) the two projected angles that will result in a hit and (b) the total time of flight for each of the two trajectories found in (a).

14. The maximum horizontal distance a certain baseball player is able to hit the ball is 150 m. On one pitch, this player hits the ball in such a way that it has the same initial speed as his maximum-distance hit, but makes an angle of 20° with the horizontal. Where will this ball strike the ground with respect to home plate?

15. A student is able to throw a ball vertically to a maximum height of 40 m. What maximum distance (measured horizontally) can the student throw the ball?

Section 4.4 Uniform Circular Motion

16. Find the acceleration of a particle moving with a constant speed of 6 m/s in a circle 3 m in radius.

17. A particle moves in a circular path 0.4 m in radius with constant speed. If the particle makes five revolutions in each second of its motion, find (a) the speed of the particle and (b) its acceleration.

18. The orbit of the moon about the earth is approximately circular, with a mean radius of 3.84×10^8 m. It takes 27.3 days for the moon to complete one revolution about the earth. Find (a) the mean orbital speed of the moon and (b) its centripetal acceleration.

19. A tire 0.5 m in radius rotates at a constant rate of 200 revolutions per minute. Find the speed and acceleration of a small stone lodged in the tread of the tire (on its outer edge).

20. A hunter uses a stone attached to the end of a rope as a crude weapon in attempting to capture an animal running *away* from him at constant velocity. The stone is swung overhead in a horizontal circle 1.6 m in diameter at the rate of 3 revolutions per second. (a) What is the centripetal acceleration of the stone? (b) What minimum speed must the animal have in order to avoid being struck by the stone after it is released?

Section 4.5 Tangential and Radial Acceleration in Curvilinear Motion

21. A student swings a ball attached to the end of a string 0.5 m in length in a vertical circle. The speed of the ball is 4 m/s at its highest point and 6 m/s at its lowest point. Find the acceleration of the ball at (a) its highest point and (b) its lowest point.

22. A pendulum of length 1 m swings in a vertical plane (Figure 4.15). When the pendulum is in the two horizontal positions ($\theta = 90°$ and $\theta = 270°$), its speed is 4 m/s. (a) Find the magnitude of the centripetal acceleration and tangential acceleration for these positions. (b) Draw vector diagrams to determine the direction of the total acceleration for these two positions. (c) Calculate the magnitude and direction of the total acceleration.

23. Figure 4.20 represents the total acceleration of a particle moving clockwise in a circle of radius 3 m at a given instant of time. At this instant of time, find (a) the centripetal acceleration, (b) the speed of the particle, and (c) its tangential acceleration.

24. At some instant of time, a particle moving counterclockwise in a circle of radius 2 m has a speed of 8 m/s

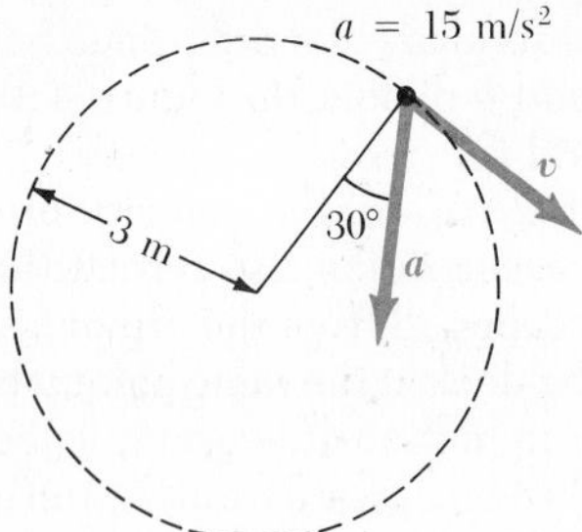

Figure 4.20 (Problem 23).

and its total acceleration is directed as shown in Figure 4.21. At this instant, determine (a) the centripetal acceleration of the particle, (b) the tangential acceleration, and (c) the magnitude of the total acceleration.

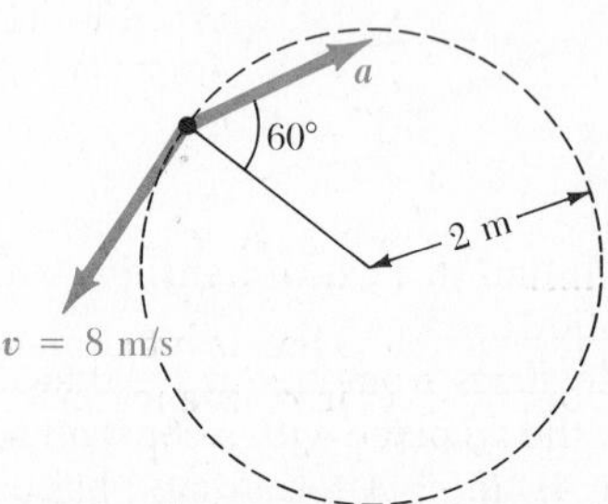

Figure 4.21 (Problem 24).

25. The speed of a particle moving in a circle 2 m in radius increases at the constant rate of 3 m/s². At some instant, the magnitude of the total acceleration is 5 m/s². At this instant, find (a) the centripetal acceleration of the particle and (b) its speed.

Section 4.6 Relative Velocity and Relative Acceleration

26. A car travels north with a speed of 60 km/h on a straight highway. A truck travels in the opposite direction with a speed of 50 km/h. (a) What is the velocity of the car relative to the truck? (b) What is the velocity of the truck relative to the car?

27. The pilot of an aircraft wishes to fly due west in a wind blowing at 50 km/h toward the south. If the speed of the aircraft in the absence of a wind is 200 km/h, (a) in what direction should the aircraft head and (b) what should its speed be relative to the ground?

28. The pilot of an airplane notes that the compass indicates a heading due west. The airplane's speed relative to the air is 150 km/h. If there is a wind of 30 km/h toward the north, find the velocity of the airplane relative to the ground.

29. Car A travels due west with a speed of 40 km/h. Car B travels due north with a speed of 60 km/h. What is the velocity of car B as seen by the driver of car A?

30. A car travels due east with a speed of 50 km/h. Rain is falling vertically with respect to the earth. The traces of the rain on the side windows of the car make an

angle of 60° with the vertical. Find the velocity of the rain with respect to (a) the car and (b) the earth.

31. A river has a steady speed of 0.5 m/s. A student swims upstream a distance of 1 km and returns to the starting point. If the student can swim at a speed of 1.2 m/s in still water, how long does the trip take? Compare this with the time the trip would take if the water were still.

GENERAL PROBLEMS

32. A bomber is flown horizontally with a ground speed of 275 m/s at an altitude of 3000 m over level terrain. Neglect the effects of air resistance. (a) How far from the point vertically under the point of release will a bomb hit the ground? (b) If the plane maintains its original course and speed, where will it be when the bomb hits the ground? (c) For the above conditions, at what angle from the point vertically under the point of release must the telescopic bomb sight be set so that the bomb will hit the target seen in the sight at the time of release?

33. A home run in a baseball game is hit in such a way that the ball just clears a wall 21 m high, located 130 m from home plate. The ball is hit at an angle of 35° to the horizontal, and air resistance is negligible. Find (a) the initial speed of the ball, (b) the time it takes the ball to reach the wall, and (c) the velocity components and the speed of the ball when it reaches the wall. (Assume the ball is hit at a height of 1 m above the ground.)

34. A car is parked on a steep incline overlooking the ocean, where the incline makes an angle of 37° with the horizontal. The negligent driver leaves the car in neutral, and the parking brakes are defective. The car rolls from rest down the incline with a constant acceleration of 4 m/s² and travels 50 m to the edge of the cliff. The cliff is 30 m above the ocean. Find (a) the speed of the car when it reaches the cliff and the time it takes to get there, (b) the velocity of the car when it lands in the ocean, (c) the total time the car is in motion, and (d) the position of the car relative to the base of the cliff when the car lands in the ocean.

35. After delivering his toys in the usual manner, Santa decides to have some fun and slide down an icy roof, as in Figure 4.22. He starts from rest at the top of the roof, which is 8 m in length, and accelerates at the rate of 5 m/s². The edge of the roof is 6 m above a soft snowbank, which Santa lands on. Find (a) Santa's velocity components when he reaches the snowbank, (b) the total time he is in motion, and (c) the distance d between the house and the point where he lands in the snow.

Figure 4.22 (Problem 35).

36. A daredevil is shot out of a cannon at 45° to the horizontal with an initial speed of 25 m/s. A net is located at a horizontal distance of 50 m from the cannon. At what height above the cannon should the net be placed in order to catch the daredevil?

37. The position of a particle moving in the xy plane varies in time according to the equation $r = 3 \cos 2t\mathbf{i} + 3 \sin 2t\mathbf{j}$, where r is in m and t is in s. (a) Show that the path of the particle is a circle 3 m in radius centered at the origin. (*Hint:* Let $\theta = 2t$.) (b) Calculate the velocity and acceleration vectors. (c) Show that the acceleration vector always points toward the origin (opposite r) and has a magnitude of v^2/r.

38. A dart gun is fired while being held horizontally at a height of 1 m above ground level. With the gun at rest relative to the ground, the dart from the gun travels a horizontal distance of 5 m. A child holds the same gun in a horizontal position while sliding down a 45° incline at a constant speed of 2 m/s. How far will the dart travel if the gun is fired when it is 1 m above the ground?

39. A truck is moving due north with a constant velocity of 10 m/s on a horizontal stretch of road. A boy riding on the back of the truck wishes to throw a ball while the truck is moving and to catch the ball after the truck has gone 20 m. (a) Neglecting air resistance, at what angle to the vertical should the ball be thrown? (b) What should be the initial speed of the ball? (c) What is the shape of the path of the ball as seen by the boy? (d) An observer on the ground watches the boy throw the ball up and catch it. In this observer's fixed frame of reference, determine the general shape of the ball's path and the initial velocity of the ball.

40. A student who is able to swim at a speed of 1.5 m/s in still water wishes to cross a river that has a current of velocity 1.2 m/s toward the south. The width of the river is 50 m. (a) If the student starts from the west bank of the river, in what direction should she head in order to swim directly across the river? How long will this trip take? (b) If she heads due east, how long will it take to cross the river? (*Note:* The student travels farther than 50 m in this case.)

41. A rocket is launched at an angle of 53° to the horizontal with an initial speed of 100 m/s. It moves along its initial line of motion with an acceleration of 30 m/s² for 3 s. At this time its engines fail and the rocket proceeds to move as a free body. Find (a) the maximum altitude reached by the rocket, (b) its total time of flight, and (c) its horizontal range.

42. A skier leaves the ramp of a ski jump with a velocity of 10 m/s, 15° above the horizontal, as in Figure 4.23. The slope is inclined at 50°, and air resistance is negligible. Find (a) the distance that the jumper lands down the slope and (b) the velocity components just before landing. (How do you think the results might be affected if air resistance were included? Note that jumpers lean forward in the shape of an airfoil with their hands at their sides to increase their distance. Why does this work?)

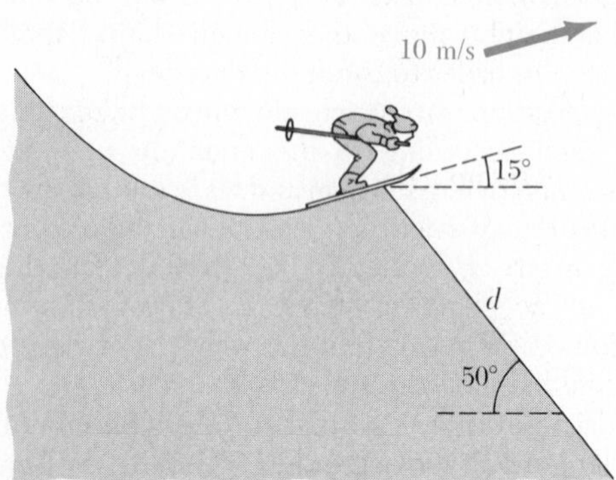

Figure 4.23 (Problem 42).

43. A river flows with a uniform velocity v. A person in a motorboat travels 1 km upstream, at which time a log is seen floating by. The person continues to travel upstream for one more hour at the same speed and then returns downstream to the starting point, where the same log is seen again. Find the velocity of the river. (*Hint:* The time of travel of the boat after it meets the log equals the time of travel of the log.)

44. A man wishes to cross a river 1 km in width in which the current is 5 km/hr toward the north. The man is on the west bank. His boat is propelled with a speed of 4 km/hr relative to the water. (a) In what direction should he head in order to make the crossing in minimum time? (b) How long will the crossing take? (c) Determine the velocity of the boat with respect to a stationary ground observer. (c) Find the final downstream displacement.

45. The man in Problem 44 desires to cross the same river in the same boat starting from the west bank. This time he wants to cross the river such that his downstream displacement is minimized. (a) In what direction should he head? (b) Find the final downstream displacement of the boat. (*Hint:* To minimize the downstream displacement the angle that the boat's velocity vector makes with the west bank must be maximized.)

•46. A truck loaded with cannonball watermelons stops suddenly to avoid running over the edge of a washed-out bridge (see Figure 4.24). The quick stop causes a number of melons to fly off the truck. One melon rolls over the edge with an initial velocity $v_0 = 10$ m/s in the horizontal direction. What are the x and y coordinates of the melon when it splatters on the bank, if a cross-section of the bank has the shape of a parabola ($y^2 = 16x$ where x and y are measured in meters) with its vertex at the edge of the road?

•47. An enemy ship is on the east side of a mountain island as shown in Figure 4.25. The enemy ship can maneuver to within 2500 m of the 1800-m-high mountain peak and can shoot projectiles with an initial speed of 250 m/s. If the western shoreline is horizontally 300 m from the peak, what are the distances from the western shore at which a ship can be safe from the bombardment of the enemy ship?

•48. A flea is at point A on a turntable 10 cm from the center. The turntable is rotating at $33\frac{1}{3}$ rev/min in the clockwise direction. The flea jumps vertically upward to a height of 5 cm and lands on the turntable at point B. Place the coordinate origin at the center of the turntable with the positive x axis fixed in space through the position from which the flea jumped. (a) Find the linear displacement of the flea. (b) Find the position of point A when the flea lands. (c) Find the position of point B when the flea lands.

•49. A football is thrown toward a receiver with an initial speed of 20 m/s at an angle of 30° above the horizontal. At that instant, the receiver is 20 m from the quarterback. In what direction and with what constant speed should the receiver run in order to catch the football at the level at which it was thrown?

•50. The determined coyote is out once more to try to capture the elusive road runner. The coyote wears a pair of Acme jet-powered roller skates, which provide

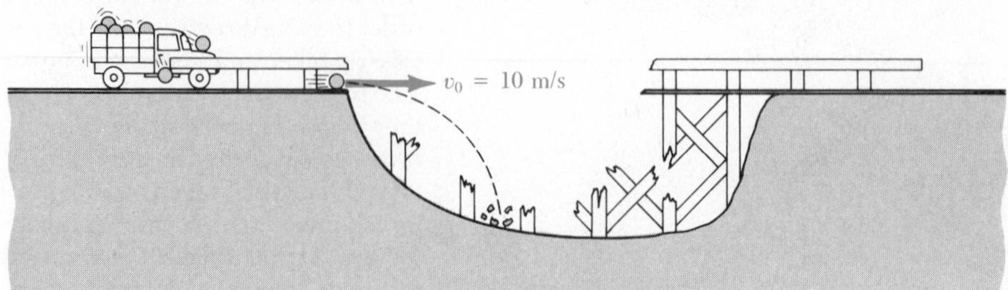

Figure 4.24 (Problem 46).

Figure 4.25 (Problem 47).

a constant horizontal acceleration of 15 m/s^2 (Fig. 4.26). The coyote starts off at rest 70 m from the edge of a cliff at the instant the road runner zips by in the direction of the cliff. (a) If the road runner moves with constant speed, determine the minimum speed he must have in order to reach the cliff before the coyote. (b) If the cliff is 100 m above the base of a canyon, determine where the coyote lands in the canyon (assume his skates are still in operation when he is in "flight"). (c) Determine the coyote's velocity components just before he lands in the canyon. (As usual, the road runner is saved by making a sudden turn at the cliff.)

Figure 4.26 (Problem 50).

•51. A U.S. Olympic decathlon star, who happens to be a bright physics student, is trapped on the roof of a burning building with a pencil, paper, pocket calculator, and his favorite physics textbook. He has about 15 min to decide whether to jump to the next building by either running at top speed horizontally off the edge or by using the long-jump technique. The next building is horizontally 30 ft away and vertically 10 ft below. His 100-m dash time is 10.3 s, and his long-jump distance is 25.5 ft. (Assume he long-jumps at an angle of 45° above the horizontal.) Perform calculations to decide which method (if any) he can use to reach the other building safely.

CALCULATOR/COMPUTER PROBLEM

52. A projectile is fired from the origin with an initial speed of v_0 at an angle θ_0 to the horizontal. Write programs that will enable you to tabulate the projectile's x and y coordinates, displacement, x and y components of velocity, and its speed as functions of time. Tabulate the above values for the following inputs of $v_0 = 50$ m/s, $\theta_0 = 60°$ at time intervals of 0.2 s until a total time of 4.4 s is reached.

5
The Laws of Motion

In the previous two chapters on kinematics, we described the motion of particles based on the definition of displacement, velocity, and acceleration. However, we would like to be able to answer specific questions related to the causes of motion, such as "What mechanism causes motion?" and "Why do some objects accelerate at a higher rate than others?" In this chapter, we shall describe the change in motion of particles using the concepts of force, mass, and momentum. We shall then discuss the three basic laws of motion, which are based on experimental observations and were formulated nearly three centuries ago by Sir Isaac Newton.

5.1 INTRODUCTION TO CLASSICAL MECHANICS

The purpose of classical mechanics is to provide a connection between the acceleration of a body and the forces acting on it. Keep in mind that classical mechanics deals with objects that are large compared with the dimensions of atoms ($\approx 10^{-10}$ m) and move at speeds that are much less than the speed of light (3×10^8 m/s).

We shall see that it is possible to describe the acceleration of an object in terms of the resultant force acting on it and the mass of the object. This force represents the interaction of the object with its environment. The mass of an object is a measure of the object's inertia, that is, the tendency of the object to resist an acceleration when a force acts on it.

We shall also discuss *force laws,* which describe the quantitative method of calculating the force on an object if its environment is known. We shall see that although the force laws are rather simple in form, they successfully explain a wide variety of phenomena and experimental observations. These force laws, together with the laws of motion, are the foundations of classical mechanics.

5.2 THE CONCEPT OF FORCE

Everyone has a basic understanding of the concept of force from everyday experiences. When you push or pull an object, you exert a force on it. You exert a force when you throw or kick a ball. In these examples, the word *force* is associated with the result of muscular activity and some change in the state of motion of an object. Although forces can cause changes in the motion, it does not necessarily follow that forces acting on an object will always cause it to move. For example, as you sit reading this book, the force of gravity acts on your body, and yet you remain stationary. You can push on a block of stone and not move it.

What force (if any) causes a distant star to drift freely through space? Newton answered such questions by stating that the change in velocity of an object is caused by forces. Therefore, if an object moves with uniform motion (constant velocity), no force is required to maintain the motion. Since only a force can cause a change in velocity, we can think of force as that which causes a body to accelerate.

A body accelerates due to an external force

Now consider a situation in which several forces act simultaneously on an object. In this case, the object will accelerate only if the *net force* acting on it is not equal to zero. We shall often refer to the net force as the *resultant force*, or the *unbalanced force. If the net force is zero, the acceleration is zero and the velocity of the object remains constant.* That is, if the net force acting on the object is zero, either the object will be at rest or it will move with constant velocity. *When the velocity of a body is constant or if the body is at rest, it is said to be in equilibrium.*

Definition of equilibrium

When a force acts on an object, its position, velocity, or acceleration can change. Furthermore, the force can change the shape of the object. In this chapter, we shall be concerned with the relation between the force on an object and the acceleration of that object. If you pull on a coiled spring, as in Figure 5.1a, the spring stretches. If the spring is calibrated, the distance that it stretches can be used to measure the strength of the force. If you pull hard enough on a cart to overcome friction, as in Figure 5.1b, it will move. Finally,

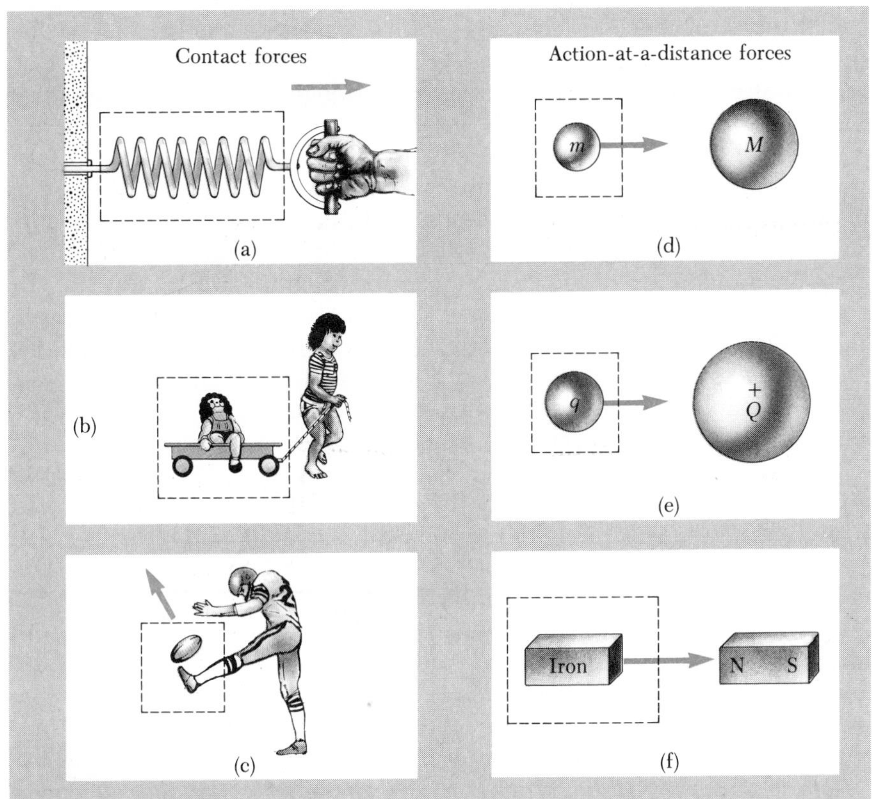

Figure 5.1 Some examples of forces applied to various objects. In each case a force is exerted on the particle or object within the boxed area. The environment external to the boxed area provides the force on the object.

when a football is kicked, as in Figure 5.1c, the football is both deformed and set in motion. These are all examples of a class of forces called *contact forces.* That is, they represent the result of physical contact between two objects. Other examples of contact forces include the force of a gas on the walls of a container (the result of the collisions of molecules with the walls) and the force of our feet on the floor.

Another class of forces, which do not involve physical contact between two objects but act through empty space, are known as *action-at-a-distance forces.* The force of gravitational attraction between two objects is an example of this class of force, illustrated in Figure 5.1d. This force keeps objects bound to the earth and gives rise to what we commonly call the *weight* of an object. The planets of our solar system are bound under the action of gravitational forces. Another common example of an action-at-a-distance force is the electric force that one electric change exerts on another electric charge, as in Figure 5.1e. These might be an electron and proton forming the hydrogen atom. A third example of an action-at-a-distance force is the force that a bar magnet exerts on a piece of iron, as shown in Figure 5.1f. The forces between atomic nuclei are also action-at-a-distance forces but are usually very short-range. They are the dominating interaction for particle separations of the order of 10^{-15} m.

Early scientists, including Newton himself, were uneasy with the concept of a force acting at a distance. To overcome this conceptual problem, Michael Faraday (1791–1867) introduced the concept of a *field.* According to this approach, when a mass m_1 is placed at some point P near a mass m_2, one can say that m_1 interacts with m_2, by virtue of the gravitational field that exists at P. The field at P is produced by mass m_2. In Chapter 23, we shall see that the field concept is also useful in describing electrical interactions between charged particles. We should mention that the distinction between contact forces and action-at-a-distance forces is not as sharp as you may have been led to believe by the above discussion. At the atomic level, the so-called contact forces are actually due to repulsive forces between charges, which themselves are action-at-a-distance forces. Nevertheless, in developing models for macroscopic phenomena, it is convenient to use both classifications of forces. However, the only known *fundamental* forces in nature are (1) gravitational attraction between objects because of their masses, (2) electromagnetic forces between charges at rest or in motion, (3) strong nuclear forces between subatomic particles, and (4) weak nuclear forces (the so-called weak interaction), which arise in certain radioactive decay processes. In classical physics, we shall be concerned only with gravitational and electromagnetic forces.

Fundamental forces in nature

It is convenient to use the deformation of a spring to measure force. Suppose a force is applied vertically to a spring with a fixed upper end, as in Figure 5.2a. We can calibrate the spring by defining the unit force, $\mathbf{F}_1$, as the force that produces an elongation of 1 cm. If a force $\mathbf{F}_2$, applied horizontally as in Figure 5.2b, produces an elongation of 2 cm, the magnitude of $\mathbf{F}_2$ is 2 units. If the two forces $\mathbf{F}_1$ and $\mathbf{F}_2$ are applied simultaneously, as in Figure 5.2c, the elongation of the spring is found to be $\sqrt{5} = 2.24$ cm. The single force, $\mathbf{F}$, that would produce this same elongation is the vector sum of $\mathbf{F}_1$ and $\mathbf{F}_2$, as described in Figure 5.2c. That is, $|\mathbf{F}| = \sqrt{F_1{}^2 + F_2{}^2} = \sqrt{5}$ units, and its direction is $\theta = \arctan(-0.5) = -26.6°$. *Because forces are vectors, you must use the rules of vector addition to get the resultant force on a body.* Springs that elongate in proportion to an applied force are said to obey *Hooke's law.* Such springs can be constructed and calibrated to measure unknown forces.

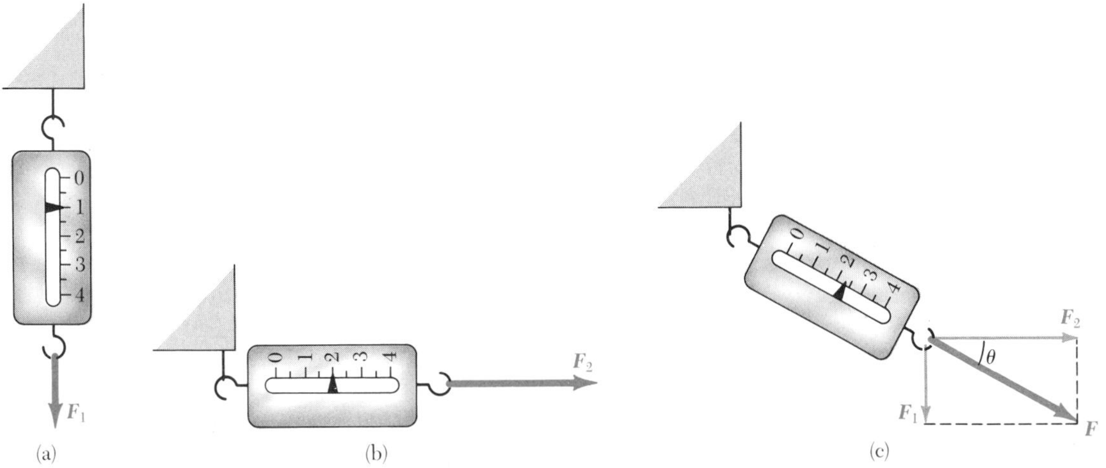

Figure 5.2 The vector nature of a force is tested with a spring scale. (a) The downward vertical force $\boldsymbol{F}_1$ elongates the spring 1 unit. (b) The horizontal force $\boldsymbol{F}_2$ elongates the spring 2 units. (c) The combination $\boldsymbol{F}_1$ and $\boldsymbol{F}_2$ elongates the spring $\sqrt{1^2 + 2^2} = \sqrt{5}$ units.

5.3 NEWTON'S FIRST LAW AND INERTIAL FRAMES

Before we state Newton's first law, consider the following simple experiment. Suppose a book is lying on a table. Obviously, the book will remain at rest in the absence of any influences. Now imagine that you push the book with a horizontal force large enough to overcome the force of friction, which is always present, between the book and table. The book can then be set in motion with constant velocity if the applied force is equal in magnitude to the force of friction and in the direction opposite the friction force. If the applied force exceeds the force of friction, the book accelerates. If the book is released, it stops sliding after moving a short distance since the force of friction retards its motion (or causes a deceleration). Now imagine that the book is pushed across a smooth, highly waxed floor. The book will again come to rest, but not as quickly as before. If you could imagine the possibility of a floor so highly polished that friction is completely absent, the book, once set in motion, will slide until it hits the wall.

Isaac Newton (1642–1727), an English physicist and mathematician, was one of the most brilliant scientists in history. Before the age of 30, he formulated the basic concepts and laws of mechanics, discovered the law of universal gravitation, and invented the mathematical methods of calculus. As a consequence of his theories, Newton was able to explain the motion of the planets, the ebb and flow of the tides, and many special features of the motion of the moon and earth. He also interpreted many fundamental observations concerning the nature of light. His contributions to physical theories dominated scientific thought for two centuries and remain important today. (Courtesy AIP Niels Bohr Library)

Before about 1600, scientists felt that the natural state of matter was the state of rest. Galileo was the first to take quite a different approach to motion and the natural state of matter. He devised thought experiments, such as an object moving on a frictionless surface, and concluded that it is not the nature of an object to stop once set in motion: rather, it is its nature to resist deceleration and acceleration.

This new approach to motion was later formalized by Newton in a form that has come to be known as **Newton's first law of motion:**

> An object at rest will remain at rest and an object in motion will continue in motion with a constant velocity (that is, constant speed in a straight line) unless it experiences a net external force (or resultant force).

In simpler terms, we can say that *when the resultant force on a body is zero, its acceleration is zero*. That is, when $\Sigma \boldsymbol{F} = 0$, then $\boldsymbol{a} = 0$. From the first law, we conclude that an isolated body (a body that does not interact with its environment) is either at rest or moving with constant velocity. Actually, Newton was not the first to state this law. Several decades earlier Galileo wrote, "Any

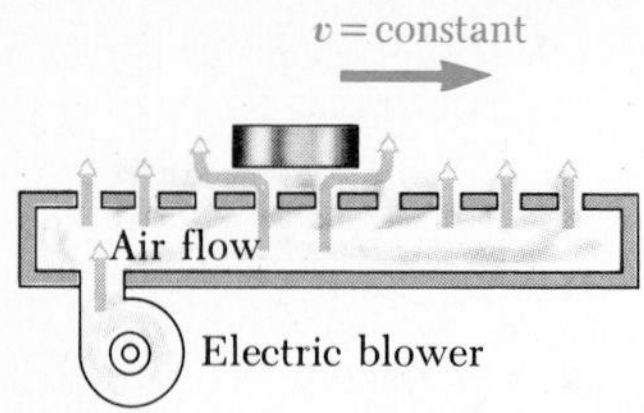

Figure 5.3 A disk moving on a column of air is an example of uniform motion, that is, motion in which the acceleration is zero.

velocity once imparted to a moving body will be rigidly maintained as long as the external causes of retardation are removed."

Another example of uniform motion on a nearly frictionless plane is the motion of a light disk on a column of air (the lubricant), as in Figure 5.3. If the disk is given an initial velocity, it will coast a great distance before coming to rest. This idea is used in the game of air hockey, where the disk makes many collisions with the walls before coming to rest.

Finally, consider a spaceship traveling in space and far removed from any planets or other matter. The spaceship requires some propulsion system to *change* its velocity. However, if the propulsion system is turned off when the spaceship reaches a velocity v, the spaceship will "coast" in space with the same velocity, and the astronauts get a "free ride" (that is, no propulsion system is required to keep them moving at the velocity v).

Inertial Frames

Newton's first law is sometimes called the *law of inertia*, and it applies to objects in an inertial frame of reference.

Inertial frame

> An **inertial frame of reference** is one in which an object, subject to no force, moves with constant velocity. That is, a reference frame in which Newton's first law is valid is called an inertial frame.

In effect, Newton's first law defines an inertial frame of reference. A reference frame that moves with constant velocity relative to the distant stars is the best approximation of an inertial frame. The earth is not an inertial frame because of its orbital motion about the sun and rotational motion about its own axis. As the earth travels in its nearly circular orbit about the sun, it experiences a centripetal acceleration of about 4.4×10^{-3} m/s^2 toward the sun. In addition, since the earth rotates about its own axis once every 24 h, a point on the equator experiences an additional centripetal acceleration of 3.37×10^{-2} m/s^2 toward the center of the earth. However, these are small compared with g and can often be neglected. In most situations *we shall assume that the earth is an inertial frame.*

Thus, if an object is in uniform motion (v = constant) an observer in one inertial frame (say, one at rest with respect to the object) will claim that the acceleration and the resultant force on the object are zero. An observer in *any other* inertial frame will also find that $\boldsymbol{a} = 0$ and $\boldsymbol{F} = 0$ for the object. According to the first law, a body at rest and one moving with constant velocity are equivalent. Unless stated otherwise, we shall usually write the laws of motion with respect to an observer "at rest" in an inertial frame.

5.4 INERTIAL MASS

Inertia

If you attempt to change the state of motion of any body, the body will resist this change. **Inertia** is the property of matter that relates to the tendency of an object to remain at rest or in uniform motion. For instance, consider two large, solid cylinders of equal size, one being balsa wood and the other steel. If you were to push the cylinders along a horizontal, rough surface, it would certainly take more effort to get the steel cylinder rolling. Likewise, once they are in motion, it would require more effort to bring the steel cylinder to rest. Therefore, we say that the steel cylinder has more inertia than the balsa-wood cylinder.

Mass is a term used to measure inertia and the SI unit of mass is the kilogram. The greater the mass of a body, the less it will accelerate (change its state of motion) under the action of an applied force. For example, if a given force acting on a 3-kg mass produces an acceleration of 4 m/s², the same force when applied to a 6-kg mass will produce an acceleration of 2 m/s². This idea will be used to obtain a quantitative description of the concept of mass.

Mass and weight are different quantities

It is important to point out that mass should not be confused with weight. *Mass and weight are two different quantities.* The weight of a body is equal to the force of gravity acting on the body and varies with location. For example, a person who weighs 180 lb on earth weighs only about 30 lb on the moon. On the other hand, the mass of a body is the same everywhere, regardless of location. An object having a mass of 2 kg on earth will also have a mass of 2 kg on the moon.

A quantitative measurement of mass can be made by comparing the accelerations that a given force will produce on different bodies. Suppose a force acting on a body of mass m_1 produces an acceleration $\mathbf{a}_1$, and the *same force* acting on a body of mass m_2 produces an acceleration $\mathbf{a}_2$. The ratio of the two masses is defined as the *inverse* ratio of the magnitudes of the accelerations produced by the same force:

$$\frac{m_1}{m_2} \equiv \frac{a_2}{a_1} \tag{5.1}$$

If one of these is a standard known mass of, say, 1 kg, the mass of an unknown can be obtained from acceleration measurements. For example, if the standard 1-kg mass undergoes an acceleration of 3 m/s² under the influence of some force, a 2-kg mass will undergo an acceleration of 1.5 m/s² under the action of the same force.

Mass is an inherent property of a body and is independent of the body's surroundings and of the method used to measure the mass. It is an experimental fact that *mass is a scalar quantity.* Finally, *mass is a quantity that obeys the rules of ordinary arithmetic.* That is, several masses can be combined in a simple numerical fashion. For example, if you combine a 3-kg mass with a 5-kg mass, their total mass would be 8 kg. This can be verified experimentally by comparing the acceleration of each object produced by a known force with the acceleration of the combined system using the same force.

5.5 NEWTON'S SECOND LAW

Newton's first law explains what happens to an object when the resultant of all external forces on it is zero. In such instances, the object either remains at rest or moves in a straight line with constant speed. Newton's second law answers the question of what happens to an object that has a nonzero resultant force acting on it.

Imagine a situation in which you are pushing a block of ice across a smooth horizontal surface, such that frictional forces can be neglected. When you exert some horizontal force $\mathbf{F}$, the block moves with some acceleration $\mathbf{a}$. If you apply a force twice as large, the acceleration doubles. Likewise, if the applied force is increased to $3\mathbf{F}$, the acceleration is tripled, and so on. From such observations, we can conclude that *the acceleration of an object is directly proportional to the resultant force acting on it.* The acceleration of an object also depends on its mass. This can be understood by considering the following set of experiments. If you apply a force $\mathbf{F}$ to a block of ice on a frictionless surface, it will undergo some acceleration $\mathbf{a}$. If the mass of the block is doubled, the same applied force will produce an acceleration $\mathbf{a}/2$. If the mass is tripled, the same applied force will produce an acceleration $\mathbf{a}/3$, and so on. According to this observation, we conclude that *the acceleration of an object is inversely proportional to its mass.*

These observations are summarized in **Newton's second law,** which states that

the acceleration of an object is directly proportional to the resultant force acting on it and inversely proportional to its mass.

Note that if the resultant force is zero, then $\boldsymbol{a} = 0$, which corresponds to the equilibrium situation where $\boldsymbol{v}$ is equal to a constant. Thus we can relate mass and force through the following mathematical statement of Newton's second law:[1]

Newton's second law

$$\sum \boldsymbol{F} = m\boldsymbol{a} \tag{5.2}$$

You should note that Equation 5.2 is a *vector* expression and hence is equivalent to the following three component equations:

Newton's second law—component form

$$\sum F_x = ma_x \qquad \sum F_y = ma_y \qquad \sum F_z = ma_z \tag{5.3}$$

To state the law in its most general form, we first define the **momentum,** $\boldsymbol{p}$, of a particle as the product of the mass, m, and the velocity, $\boldsymbol{v}$:

Definition of momentum

$$\boldsymbol{p} \equiv m\boldsymbol{v} \tag{5.4}$$

Momentum is a vector quantity that is in the direction of $\boldsymbol{v}$ and has dimensions of ML/T (kg · m/s in SI units).

Newton's second law of motion states that

the time rate of change of momentum of a particle is equal to the resultant external force acting on the particle:

General form of Newton's second law

$$\sum \boldsymbol{F} = \frac{d\boldsymbol{p}}{dt} = \frac{d}{dt}(m\boldsymbol{v}) \tag{5.5}$$

This is the most general form of Newton's second law, which is valid in any inertial frame of reference. The notation $\Sigma \boldsymbol{F}$ represents the *vector sum* of all forces acting on the particle.

If m is treated as a constant, then Equation 5.5 can be expressed

$$\sum \boldsymbol{F} = \frac{d}{dt}(m\boldsymbol{v}) = m\frac{d\boldsymbol{v}}{dt} \tag{5.6}$$

Since acceleration is defined as $\boldsymbol{a} = d\boldsymbol{v}/dt$, Equation 5.6 can be written

$$\sum \boldsymbol{F} = m\boldsymbol{a}$$

which is in agreement with Equation 5.2.

Units of Force and Mass

The SI unit of force is the **newton,** which is defined as the force that, when acting on a 1-kg mass, produces an acceleration of 1 m/s^2.

[1] Equation 5.2 is valid only when the speed of the particle is much less than the speed of light. We will treat the relativistic situation in Chapter 39.

TABLE 5.1 Units of Force, Mass, and Acceleration[a]

System of Units	Mass	Acceleration	Force
SI	kg	m/s^2	$N = kg \cdot m/s^2$
cgs	g	cm/s^2	$dyne = g \cdot cm/s^2$
British engineering (conventional)	slug	ft/s^2	$lb = slug \cdot ft/s^2$

[a] $1\ N = 10^5$ dyne $= 0.225$ lb

From this definition and Newton's second law, we see that the newton can be expressed in terms of the following fundamental units of mass, length, and time:

$$1\ N \equiv 1\ kg \cdot m/s^2 \tag{5.7}$$

Definition of newton

The unit of force in the cgs system is called the **dyne** and is defined as that force that, when acting on a 1-g mass, produces an acceleration of $1\ cm/s^2$:

$$1\ dyne \equiv 1\ g \cdot cm/s^2 \tag{5.8}$$

Definition of dyne

In the British engineering system, the unit of force is the **pound,** defined as the force that, when acting on a 1-slug mass,[2] produces an acceleration of $1\ ft/s^2$:

$$1\ lb = 1\ slug \cdot ft/s^2 \tag{5.9}$$

Definition of pound

Since $1\ kg = 10^3$ g and $1\ m = 10^2$ cm, it follows that $1\ N = 10^5$ dynes. It is left as a problem to show that $1\ N = 0.225$ lb. The units of force, mass, and acceleration are summarized in Table 5.1.

[2] The *slug* is the *unit of mass* in the British engineering system and is that system's counterpart of the SI *kilogram*. When we speak of going on a diet to lose a few pounds, we really mean that we want to lose a few slugs, that is, we want to reduce our mass. When we lose those few slugs, the force of gravity (pounds) on our reduced mass decreases (since $W = mg$) and that is how we "lose a few pounds." Since most of the calculations we shall carry out in our study of classical mechanics will be in SI units, the slug will seldom be used in this text.

EXAMPLE 5.1. An Accelerating Hockey Puck

A hockey puck with a mass of 0.3 kg slides on the horizontal frictionless surface of an ice rink. Two forces act on the puck as shown in Figure 5.4. The force $\boldsymbol{F}_1$ has a magnitude of 5 N, and $\boldsymbol{F}_2$ has a magnitude of 8 N. Determine the acceleration of the puck.

Solution: The resultant force in the x direction is

$$\sum F_x = F_{1x} + F_{2x} = F_1 \cos 20° + F_2 \cos 60°$$
$$= (5\ N)(0.940) + (8\ N)(0.500) = 8.70\ N$$

The resultant force in the y direction is

$$\sum F_y = F_{1y} + F_{2y} = -F_1 \sin 20° + F_2 \sin 60°$$
$$= -(5\ N)(0.342) + (8\ N)(0.866) = 5.22\ N$$

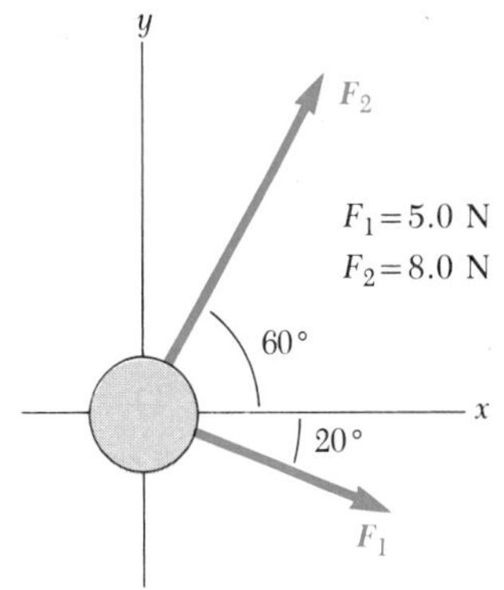

Figure 5.4 (Example 5.1) An object moving on a frictionless surface will accelerate in the direction of the *resultant* force, $\boldsymbol{F}_1 + \boldsymbol{F}_2$.

Now we can use Newton's second law in component form to find the x and y components of acceleration:

$$a_x = \frac{\Sigma F_x}{m} = \frac{8.70 \text{ N}}{0.3 \text{ kg}} = 29.0 \text{ m/s}^2$$

$$a_y = \frac{\Sigma F_y}{m} = \frac{5.22 \text{ N}}{0.3 \text{ kg}} = 17.4 \text{ m/s}^2$$

The acceleration has a magnitude of

$$a = \sqrt{(29.0)^2 + (17.4)^2} \text{ m/s}^2 = 33.8 \text{ m/s}^2$$

and its direction is

$$\theta = \tan^{-1}(a_y/a_x) = \tan^{-1}(17.4/29.0) = 31.0°$$

relative to the positive x axis.

Exercise 1 Determine the components of a third force that when applied to the puck will cause it to be in equilibrium.
Answer: $F_x = -8.70$ N, $F_y = -5.22$ N.

Astronaut Edgar D. Mitchell walking on the moon following the Apollo 14 lunar landing. The weight of this astronaut on the moon is less than on earth, but his mass remains the same. (Courtesy of NASA)

5.6 WEIGHT

We are well aware of the fact that bodies are attracted to the earth. The force exerted by the earth on a body is called the **weight** of the body **W**. This force is directed toward the center of the earth.[3]

We have seen that a freely falling body experiences an acceleration **g** acting toward the center of the earth. Applying Newton's second law to the freely falling body, with $\mathbf{a} = \mathbf{g}$ and $\mathbf{F} = \mathbf{W}$, gives

$$\mathbf{W} = m\mathbf{g} \tag{5.10}$$

Since the weight depends on g, it varies with geographic location. Bodies weigh less at higher altitudes than at sea level. This is because g decreases with increasing distance from the center of the earth. Hence, weight, unlike mass, is not an inherent property of a body. Therefore, you should not confuse mass with weight. For example, if a body has a mass of 70 kg, then the magnitude of its weight in a location where $g = 9.80$ m/s^2 is $mg = 686$ N (about 154 lb). At the top of a mountain where $g = 9.76$ m/s^2, this weight would be 683 N. This corresponds to a decrease in weight of about 0.4 lb. Therefore, if you want to lose weight without going on a diet, climb a mountain or weigh yourself at 30 000 ft during a flight on a jet airplane.

Since $W = mg$, we can compare the masses of two bodies by measuring their weights using a spring scale or a chemical balance. That is, the ratio of the weights of two bodies equals the ratio of their masses at a given location.

5.7 NEWTON'S THIRD LAW

Newton's third law states that if two bodies interact, the force exerted on body 1 by body 2 is equal to and opposite the force exerted on body 2 by body 1. That is,

$$\mathbf{F}_{12} = -\mathbf{F}_{21} \tag{5.11}$$

This law, which is illustrated in Figure 5.5a, is equivalent to stating that *forces always occur in pairs*, or that *a single isolated force cannot exist*. The force that body 1 exerts on body 2 is sometimes called the *action force*, while the force of body 2 on body 1 is called the *reaction force*. Either force can be labeled the

[3] This ignores the fact that the mass distribution of the earth is not perfectly spherical.

action or reaction force. *The action force is equal in magnitude to the reaction force and opposite in direction. In all cases, the action and reaction forces act on different objects.* For example, the force acting on a freely falling projectile is its weight, $\mathbf{W} = m\mathbf{g}$. This equals the force of the earth on the projectile. The reaction to this force is the force of the projectile on the earth, $\mathbf{W}' = -\mathbf{W}$. The reaction force $\mathbf{W}'$, must accelerate the earth toward the projectile just as the action force, $\mathbf{W}$, accelerates the projectile toward the earth. However, since the earth has such a large mass, the acceleration of the earth due to this reaction force is negligibly small.

Another example is shown in Figure 5.5b. The force of the hammer on the nail (the action) is equal to and opposite the force of the nail on the hammer (the reaction). You directly experience Newton's third law if you slam your fist against a wall or if you kick a football with your bare foot. You should be able to identify the action and reaction forces in these cases.

The weight of a body, $\mathbf{W}$, has been defined as the force the earth exerts on the body. If the body is a block at rest on a table, as in Figure 5.6a, the reaction force to $\mathbf{W}$ is the force the block exerts on the earth, $\mathbf{W}'$. The block does not accelerate since it is held up by the table. The table, therefore, exerts an upward action force, $\mathbf{N}$, on the block, called the **normal force.**[4] The normal force is the force that prevents the block from falling through the table, and can have any value needed, up to the point of breaking the table. The normal force balances the weight and provides equilibrium. The reaction to $\mathbf{N}$ is the force of the book on the table, $\mathbf{N}'$. Therefore, we conclude that

$$\mathbf{W} = -\mathbf{W}' \quad \text{and} \quad \mathbf{N} = -\mathbf{N}'$$

Note that the forces acting on the block are $\mathbf{W}$ and $\mathbf{N}$, as in Figure 5.6b. We shall be interested only in such external forces when treating the motion of a body. From the first law, we see that since the book is in equilibrium ($a = 0$), it follows that $W = N = mg$.

[4] The word *normal* is used because in the absence of friction, the direction of $\mathbf{N}$ is always *perpendicular* to the surface.

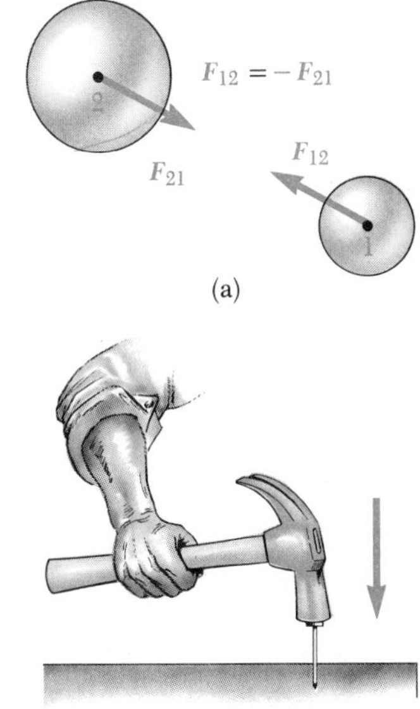

Figure 5.5 Newton's third law. (a) The force of body 1 on body 2 is equal to and opposite the force of body 2 on body 1. (b) The force of the hammer on the nail is equal to and opposite the force of the nail on the hammer.

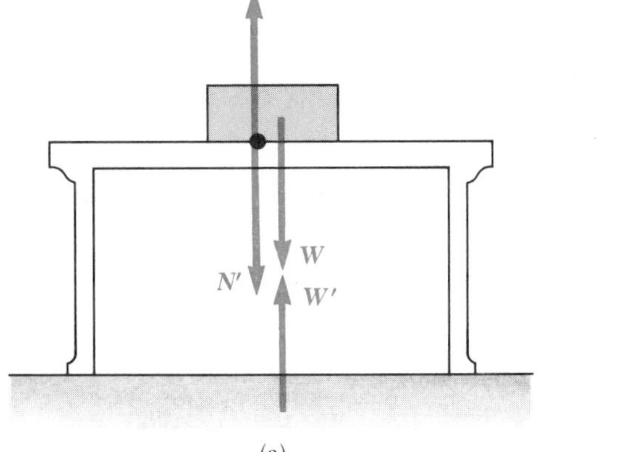

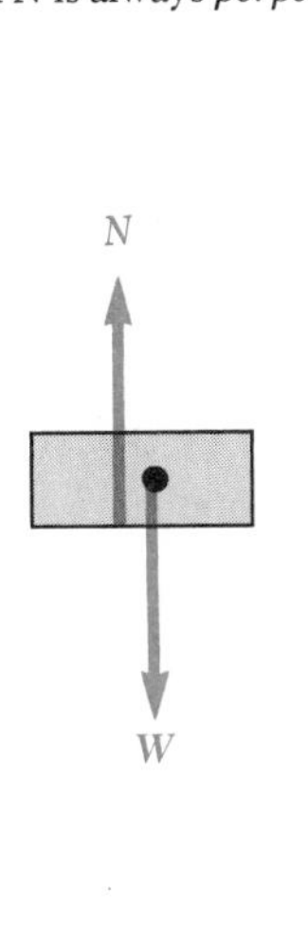

Figure 5.6 When a block is lying on a table, the forces acting on the block are the normal force, $\mathbf{N}$, and the force of gravity, $\mathbf{W}$, as illustrated in (b). The reaction to $\mathbf{N}$ is the force of the block on the table $\mathbf{N}'$. The reaction to $\mathbf{W}$ is the force of the block on the earth, $\mathbf{W}'$.

5.8 SOME APPLICATIONS OF NEWTON'S LAWS

In this section we present some simple applications of Newton's laws to bodies that are either in equilibrium ($a = 0$) or moving linearly under the action of constant external forces. As our model, we shall assume that the bodies behave as particles so that we need not worry about rotational motions. In this section, we shall also neglect the effects of friction for those problems involving motion. This is equivalent to stating that the surfaces are *smooth*. Finally, we shall usually neglect the mass of any ropes involved in a particular problem. In this approximation, the magnitude of the force exerted at any point along the rope is the same at all points along the rope.

Tension

When we apply Newton's laws to a body, we shall be interested only in those external forces that act *on the body*. For example, in Figure 5.6 the only external forces acting on the book are **N** and **W**. The reactions to these forces, **N'** and **W'**, act on the table and on the earth, respectively, and do not appear in Newton's second law as applied to the book.

When an object such as a block is being pulled by a rope attached to it, the rope exerts a force on the object. The **tension** in the rope is defined as the force that the rope exerts on the object. In general, tension in a rope is the force that the rope exerts on what is attached to it.

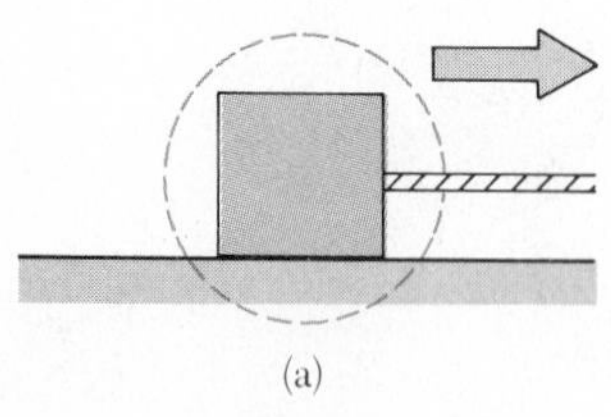

(a)

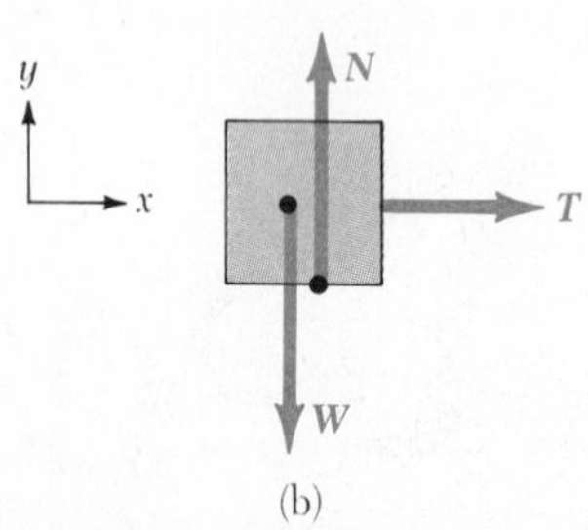

(b)

Figure 5.7 (a) A block being pulled to the right on a smooth surface. (b) The free-body diagram that represents the external forces on the block.

Consider a block being pulled to the right on the smooth, horizontal surface of a table, as in Figure 5.7a. Suppose you are asked to find the acceleration of the block and the force of the table on the block. First, note that the horizontal force being applied to the block acts through the string. The force that the string exerts on the block is denoted by the symbol **T**. The magnitude of **T** is equal to the tension in the string. A dotted circle is drawn around the block in Figure 5.7a to remind you to isolate the block from its surroundings. Since we are interested only in the motion of the block, we must be able to *identify all external forces acting on it*. These are illustrated in Figure 5.7b. In addition to the force **T**, the force diagram for the block includes the weight, **W**, and the normal force, **N**. As before, **W** corresponds to the force of gravity pulling down on the block and **N** is the upward force of the table on the block. Such a force diagram is referred to as a **free-body diagram.** The construction of such a diagram is an important step in applying Newton's laws. The *reactions* to the forces we have listed, namely, the force of the string on the hand, the force of the block on the earth, and the force of the block on the table, are not included in the free-body diagram since they act on *other* bodies and not on the block.

Free-body diagrams are important when applying Newton's laws

We are now in a position to apply Newton's second law in component form to the system. The only force acting in the x direction is **T**. Applying $\Sigma F_x = ma_x$ to the horizontal motion gives

$$\sum F_x = T = ma_x \qquad \text{or} \qquad a_x = \frac{T}{m}$$

In this situation, there is no acceleration in the y direction. Applying $\Sigma F_y = ma_y$ with $a_y = 0$ gives

$$N - W = 0 \qquad \text{or} \qquad N = W$$

That is, the normal force is equal to and opposite the weight.

If T is a *constant* force, then the acceleration, $a_x = T/m$, is also a constant. Hence, the equations of kinematics from Chapter 3 can be used to obtain the displacement, Δx, and velocity, v, as functions of time. Since $a_x = T/m =$ constant, these expressions can be written

$$\Delta x = v_0 t + \tfrac{1}{2}\left(\frac{T}{m}\right)t^2$$

$$v = v_0 + \left(\frac{T}{m}\right)t$$

where v_0 is the velocity at $t = 0$.

The following procedure is recommended for applying Newton's laws:

Procedure for applying Newton's laws

1. Draw a simple, neat diagram of the system.
2. Isolate the object of interest whose motion is being analyzed. Draw a free-body diagram for this object, that is, a diagram showing *all external forces acting on the object.* For systems containing more than one object, draw *separate* diagrams for each object. *Do not* include forces that the object exerts on its surroundings.
3. Establish convenient coordinate axes for each body and find the components of the forces along these axes. Now, apply Newton's second law, $\Sigma F = ma$, in *component* form. Check your dimensions to make sure that all terms have units of force.
4. Solve the component equations for the unknowns. Remember that you must have as many independent equations as you have unknowns in order to obtain a complete solution.
5. It is a good idea to check the predictions of your solutions for extreme values of the variables. You can often detect errors in your results using this procedure.

Consider a lamp of weight W suspended from a chain of negligible weight fastened to the ceiling, as in Figure 5.8a. The free-body diagram for the lamp is shown in Figure 5.8b, where the forces on it are the weight, $\mathbf{W}$, acting downward, and the force of the chain on the lamp, $\mathbf{T}$, acting upward. The force $\mathbf{T}$ is the constraint force in this case. (If we cut the chain, $\mathbf{T} = 0$ and the body executes free fall.)

If we apply the first law to the lamp, noting that $\mathbf{a} = 0$, we see that since there are no forces in the x direction, the equation $\Sigma F_x = 0$ provides no helpful information. The condition $\Sigma F_y = 0$ gives

$$\sum F_y = T - W = 0 \qquad \text{or} \qquad T = W$$

Note that $\mathbf{T}$ and $\mathbf{W}$ are *not* action-reaction pairs. The reaction to $\mathbf{T}$ is $\mathbf{T}'$, the force exerted on the chain by the lamp, as in Figure 5.8c. The force $\mathbf{T}'$ acts downward and is transmitted to the ceiling. That is, the force of the chain on the ceiling, $\mathbf{T}'$, is *downward* and equal to W in magnitude. The ceiling exerts an equal and opposite force, $\mathbf{T}'' = \mathbf{T}$, on the chain, as in Figure 5.8c.

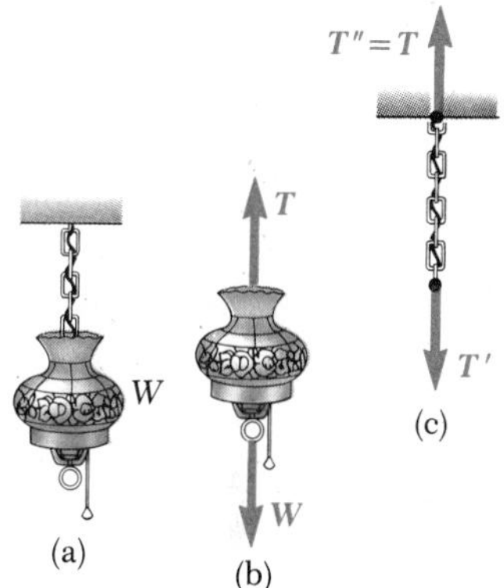

Figure 5.8 (a) A lamp of weight $\mathbf{W}$ suspended by a light chain from a ceiling. (b) The forces acting on the lamp are the force of gravity, $\mathbf{W}$, and the tension in the chain, $\mathbf{T}$. (c) The forces acting on the chain are $\mathbf{T}'$, that exerted by the lamp, and $\mathbf{T}''$, that exerted by the ceiling.

EXAMPLE 5.2. A Traffic Light at Rest

A traffic light weighing 100 N hangs from a cable tied to two other cables fastened to a support, as in Figure 5.9a. The upper cables make angles of 37° and 53° with the horizontal. Find the tension in the three cables.

Solution: First we construct a free-body diagram for the traffic light, as in Figure 5.9b. The tension in the vertical cable, T_3, supports the light, and so we see that $T_3 = W = 100$ N. Now we construct a free-body diagram for the knot that holds the three cables together, as in Figure 5.9c. This is a convenient point to choose because all forces in question act at this point. We choose the coordinate axes as shown in Figure 5.9c and resolve the forces into their x and y components:

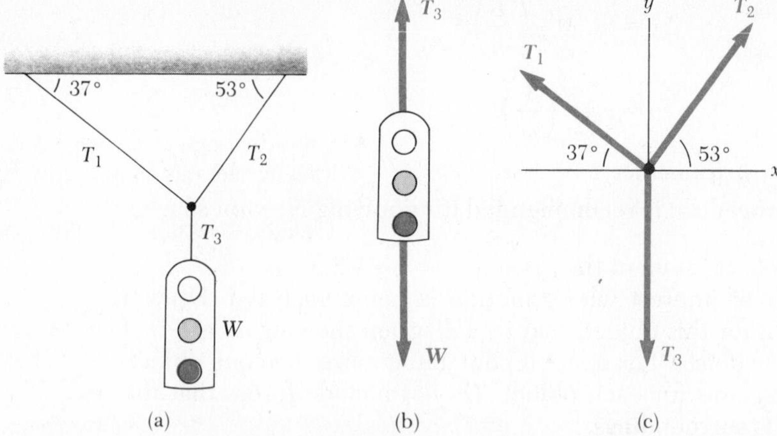

Figure 5.9 (Example 5.2) (a) A traffic light suspended by cables. (b) Free-body diagram for the traffic light. (c) Free-body diagram for the knot.

Force	x component	y component
T_1	$-T_1 \cos 37°$	$T_1 \sin 37°$
T_2	$T_2 \cos 53°$	$T_2 \sin 53°$
T_3	0	-100 N

The first condition for equilibrium gives us the equations

$$(1) \qquad \sum F_x = T_2 \cos 53° - T_1 \cos 37° = 0$$

$$(2) \qquad \sum F_y = T_1 \sin 37° + T_2 \sin 53° - 100\ \text{N} = 0$$

From (1) we see that the horizontal components of T_1 and T_2 must be equal in magnitude, and from (2) we see that the sum of the vertical components of T_1 and T_2 must balance the weight of the light. We can solve (1) for T_2 in terms of T_1 to give

$$T_2 = T_1 \left(\frac{\cos 37°}{\cos 53°} \right) = 1.33 T_1$$

This value for T_2 can be substituted into (2) to give

$$T_1 \sin 37° + (1.33 T_1)(\sin 53°) - 100\ \text{N} = 0$$

$$T_1 = 60.0\ \text{N}$$

$$T_2 = 1.33 T_1 = 79.8\ \text{N}$$

Exercise 2 In what situation will $T_1 = T_2$?
Answer: When the supporting cables make equal angles with the horizontal support.

EXAMPLE 5.3. Block on a Smooth Incline

A block of mass m is placed on a smooth, inclined plane of angle θ, as in Figure 5.10a. (a) Determine the acceleration of the block after it is released.

The free-body diagram for the block is shown in Figure 5.10b. The only forces on the block are the normal force, N, acting perpendicular to the plane and the weight, W, acting vertically downward. *It is convenient to choose the coordinate axes with* x *along the incline and* y *perpendicular to it.* Then, we replace the weight vector by a component of magnitude $mg \sin \theta$ along the *positive* x axis and another of magnitude $mg \cos \theta$ in the *negative y*

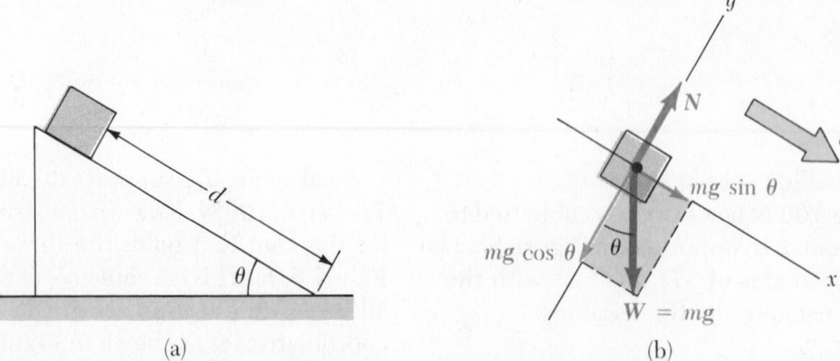

Figure 5.10 (Example 5.3) (a) A block sliding down a smooth incline. (b) The free-body diagram for the block. Note that its acceleration along the incline is $g \sin \theta$.

direction. Applying Newton's second law in component form while noting that $a_y = 0$ gives

$$(1) \qquad \sum F_x = mg \sin\theta = ma_x$$

$$(2) \qquad \sum F_y = N - mg\cos\theta = 0$$

From (1) we see that the acceleration along the incline is provided by the component of weight down the incline:

$$(3) \qquad a_x = g\sin\theta$$

From (2) we conclude that the component of weight perpendicular to the incline is *balanced* by the normal force, or $N = mg\cos\theta$. The acceleration given by (3) is found to be *independent* of the mass of the block! It depends only on the angle of inclination and on g!

Special Cases: We see that when $\theta = 90°$, $a = g$ and $N = 0$. This corresponds to the block in free fall. Also, when $\theta = 0$, $a_x = 0$ and $N = mg$ (its maximum value).

(b) Suppose the block is released from rest at the top, and the distance from the block to the bottom is d. How long does it take the block to reach the bottom, and what is its speed just as it gets there?

Since $a_x =$ constant, we can apply the kinematic equation $x - x_0 = v_{x0}t + \frac{1}{2}a_x t^2$ to the block. Since the displacement $x - x_0 = d$ and $v_{x0} = 0$, we get

$$d = \tfrac{1}{2}a_x t^2$$

or

$$(4) \qquad t = \sqrt{\frac{2d}{a_x}} = \sqrt{\frac{2d}{g\sin\theta}}$$

Also, since $v_x^2 = v_{x0}^2 + 2a_x(x - x_0)$ and $v_{x0} = 0$, we find that

$$v_x^2 = 2a_x d$$

$$(5) \qquad v_x = \sqrt{2a_x d} = \sqrt{2gd\sin\theta}$$

Again, t and v_x are *independent* of the mass of the block. This suggests a simple method of measuring g using an inclined air track or some other smooth incline. Simply measure the angle of inclination, the distance traveled by the block, and the time it takes to reach the bottom. The value of g can then be calculated from (4) and (5).

EXAMPLE 5.4. Atwood's Machine

When two unequal masses are hung vertically over a light, frictionless pulley as in Figure 5.11a, the arrangement is called *Atwood's machine*. The device is sometimes used in the laboratory to measure the acceleration of gravity. Determine the acceleration of the two masses and the tension in the string.

Solution: The free-body diagrams for the two masses are shown in Figure 5.11b, where we assume that $m_2 > m_1$. When Newton's second law is applied to m_1, with a upwards for this mass, we find

$$(1) \qquad \sum F_y = T - m_1 g = m_1 a$$

Similarly, for m_2 we find

$$(2) \qquad \sum F_y = T - m_2 g = -m_2 a$$

The negative sign on the right-hand side of (2) indicates that m_2 accelerates downwards, in the negative y direction.

When (2) is subtracted from (1), T drops out and we get

$$-m_1 g + m_2 g = m_1 a + m_2 a$$

or

$$(3) \qquad a = \left(\frac{m_2 - m_1}{m_1 + m_2}\right) g$$

If (3) is substituted into (1), we get

$$(4) \qquad T = \left(\frac{2m_1 m_2}{m_1 + m_2}\right) g$$

Special Cases: Note that when $m_1 = m_2$, $a = 0$ and $T = m_1 g = m_2 g$, as we would expect for the balanced case. Also, if $m_2 \gg m_1$, $a \approx g$ (a freely falling body) and $T \approx 2m_1 g$.

Exercise 3 Find the acceleration and tension of an Atwood's machine in which $m_1 = 2$ kg and $m_2 = 4$ kg.
Answer: $a = 3.27\ \text{m/s}^2$, $T = 26.1$ N.

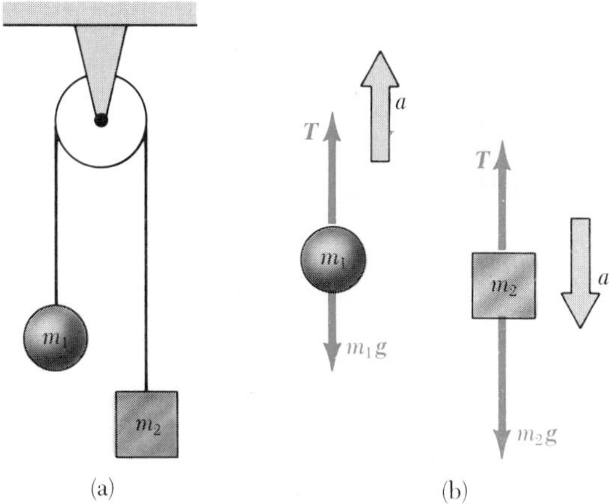

Figure 5.11 (Example 5.4) Atwood's machine. (a) Two masses connected by a light string over a frictionless pulley. (b) Free-body diagrams for m_1 and m_2.

EXAMPLE 5.5. Two Connected Objects

Two unequal masses are attached by a light string that passes over a light, frictionless pulley as in Figure 5.12a. The block of mass m_2 lies on a smooth incline of angle θ. Find the acceleration of the two masses and the tension in the string.

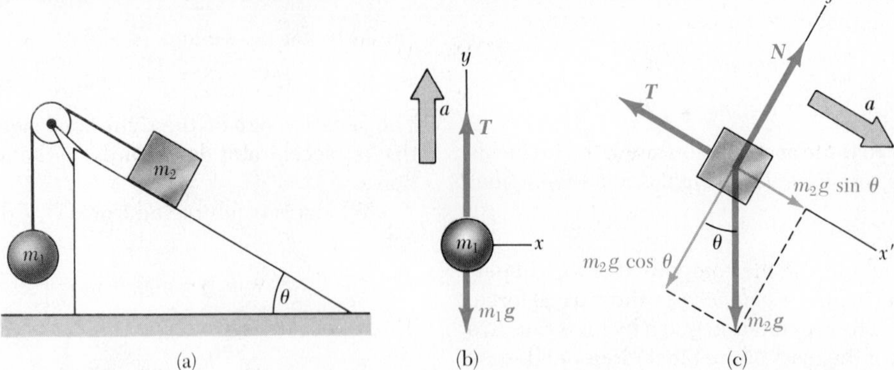

Figure 5.12 (Example 5.5) (a) Two masses connected by a light string over a frictionless pulley. (b) The free-body diagram for m_1. (c) The free-body diagram for m_2 (the incline is smooth).

Solution: Since the two masses are connected by a string (which we assume doesn't stretch), they both have accelerations of the same magnitude. The free-body diagrams for the two masses are shown in Figures 5.12b and 5.12c. Applying Newton's second law in component form to m_1 while *assuming* that a is upward for this mass gives

Equations of motion for m_1:

$$(1) \qquad \sum F_x = 0$$

$$(2) \qquad \sum F_y = T - m_1 g = m_1 a$$

Note that in order for a to be positive, it is necessary that $T > m_1 g$.

Now, for m_2 it is convenient to choose the positive x' axis along the incline as in Figure 5.12c. Applying Newton's second law in component form to m_2 gives

Equations of motion for m_2:

$$(3) \qquad \sum F_{x'} = m_2 g \sin\theta - T = m_2 a$$

$$(4) \qquad \sum F_{y'} = N - m_2 g \cos\theta = 0$$

Expressions (1) and (4) provide no information regarding the acceleration. However, if we solve (2) and (3) simultaneously for the unknowns a and T, we get

$$(5) \qquad a = \frac{m_2 g \sin\theta - m_1 g}{m_1 + m_2}$$

When this is substituted into (2), we find

$$(6) \qquad T = \frac{m_1 m_2 g(1 + \sin\theta)}{m_1 + m_2}$$

Note that m_2 accelerates down the incline if $m_2 \sin\theta$ exceeds m_1 (that is, if a is positive as we assumed). If m_1 exceeds $m_2 \sin\theta$, the acceleration of m_2 is up the incline and downward for m_1. You should also note that the result for the acceleration, (5), can be interpreted as the resultant unbalanced force on the system divided by the total mass of the system.

Exercise 4 If $m_1 = 10$ kg, $m_2 = 5$ kg, and $\theta = 45°$, find the acceleration.
Answer: $a = -4.22 \text{ m/s}^2$, where the negative sign indicates that m_2 accelerates up the incline.

EXAMPLE 5.6. Weighing a Fish in an Elevator
A person weighs a fish on a spring scale attached to the ceiling of an elevator, as shown in Figure 5.13. Show that if the elevator accelerates or decelerates, the spring scale reads a weight different from the true weight of the fish.

Solution: The external forces acting on the fish are its true weight, W, and the upward constraint force, T, exerted on it by the scale. By Newton's third law, T is also the reading of the spring scale. If the elevator is at rest or moving at constant velocity, then the fish is not accelerating and $T = W = mg$ (where $g = 9.80 \text{ m/s}^2$). If the elevator accelerates upward with an acceleration a relative to an observer outside the elevator in an inertial frame (Fig. 5.13a), then the second law applied to the fish of mass m gives the total force F on the fish:

$$(1) \qquad \sum F = T - W = ma \qquad (\text{if } a \text{ is upward})$$

Likewise, if the elevator accelerates downward as in Figure 5.13b, Newton's second law applied to the fish becomes

$$(2) \qquad \sum F = T - W = -ma \qquad (\text{if } a \text{ is downward})$$

Thus, we conclude from (1) that the scale reading, T, is greater than the true weight, W, if a is upward. From (2) we see that T is less than W if a is downward.

For example, if the true weight of the fish is 40 N, and a is 2 m/s² upward, then the scale reading is

$$T = ma + mg = mg\left(\frac{a}{g} + 1\right)$$

$$= W\left(\frac{a}{g} + 1\right) = (40 \text{ N})\left(\frac{2 \text{ m/s}^2}{9.80 \text{ m/s}^2} + 1\right)$$

$$= 48.2 \text{ N}$$

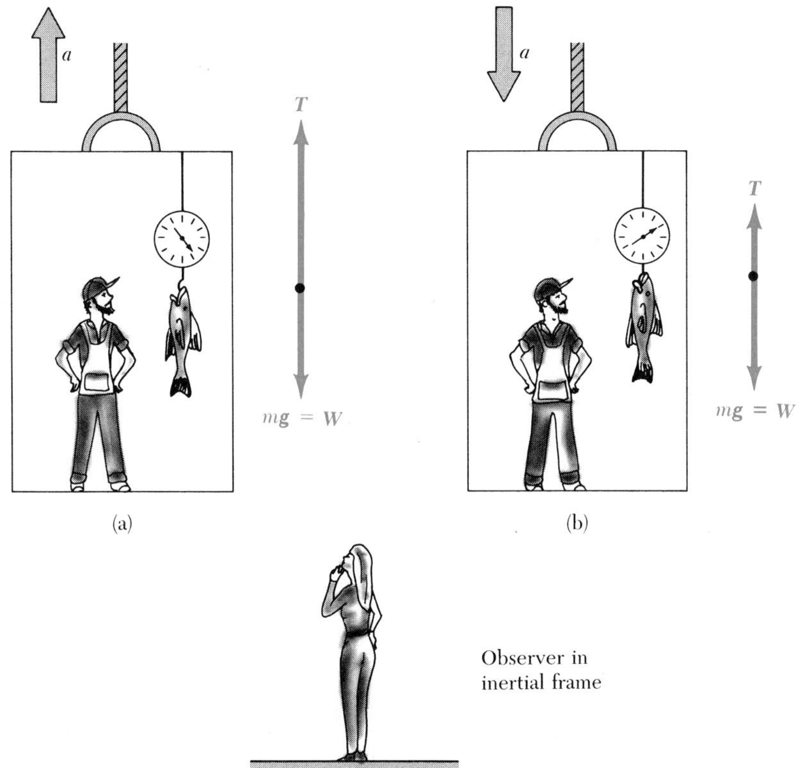

Figure 5.13 (Example 5.6) Apparent weight versus true weight. (a) When the elevator accelerates *upward* the spring scale reads a value *greater* than the true weight. (b) When the elevator accelerates *downward* the spring scale reads a value *less* than the true weight. The spring scale reads the *apparent weight.*

If a is 2 m/s² downward, then

$$T = -ma + mg = mg\left(1 - \frac{a}{g}\right)$$

$$= W\left(1 - \frac{a}{g}\right) = (40 \text{ N})\left(1 - \frac{2 \text{ m/s}^2}{9.80 \text{ m/s}^2}\right)$$

$$= 31.8 \text{ N}$$

Hence, if you buy a fish in an elevator, make sure the fish is weighed while the elevator is at rest or accelerating downward! Furthermore, note that from the information given here, one cannot determine the *direction* of motion of the elevator.

Special Cases: If the elevator cable breaks, then the elevator falls freely and $a = -g$. Since $W = mg$, we see from (1) that the apparent weight, T, is zero, that is, the fish appears to be weightless. If the elevator accelerates *downward* with an acceleration *greater* than g, the fish (along with the person in the elevator) will eventually hit the ceiling since its acceleration will still be that of a freely falling body relative to an outside observer.

5.9 FORCES OF FRICTION

When a body is in motion on a rough surface, or through a viscous medium such as air or water, there is resistance to motion because of the interaction of the body with its surroundings. We call such resistance a **force of friction.** Forces of friction are very important in our everyday lives. Forces of friction allow us to walk or run and are necessary for the motion of wheeled vehicles.

Consider a block on a horizontal table, as in Figure 5.14a. If we apply an external horizontal force $\boldsymbol{F}$ to the block, acting to the right, the block will remain stationary if $\boldsymbol{F}$ is not too large. The force that keeps the block from

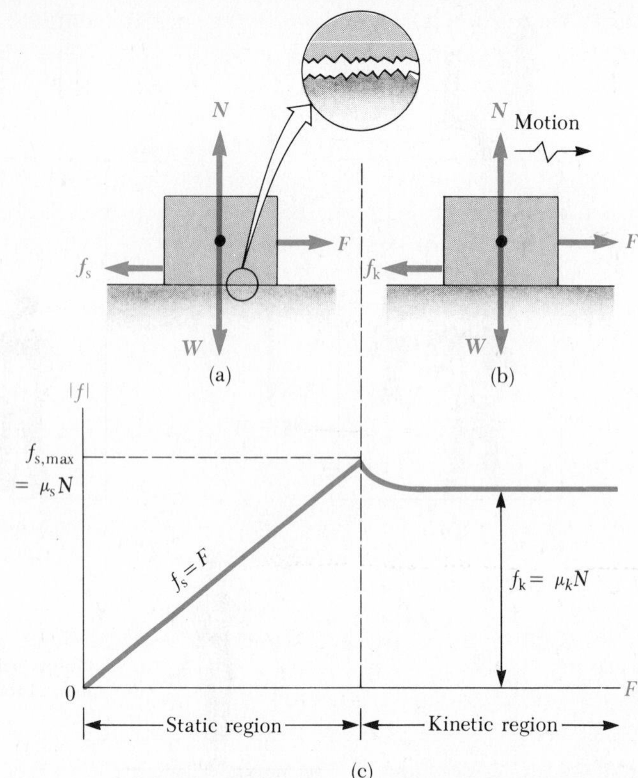

Figure 5.14 The force of friction, f, between a block and a rough surface is opposite the applied force, $\mathbf{F}$. (a) The force of static friction equals the applied force. (b) When the applied force exceeds the force of kinetic friction, the block accelerates to the right. (c) A graph of the magnitude of the frictional force versus the applied force. Note that $f_{s,max} > f_k$.

moving acts to the left and is called the *frictional force, f*. As long as the block is in equilibrium, $f = F$. Since the block is stationary, we call this frictional force the *force of static friction, f_s*. Experiments show that this force arises from the roughness of the two surfaces, so that contact is made only at a few points, as shown in the "magnified" view of the surfaces in Figure 5.14a. Actually, the frictional force is much more complicated than presented here since it ultimately involves the electrostatic forces between atoms or molecules where the surfaces are in contact.

If we increase the magnitude of $\mathbf{F}$, as in Figure 5.14b, the block will eventually slip. When the block is on the verge of slipping, f_s is a maximum. When F *exceeds* $f_{s,max}$ the block moves and accelerates to the right. When the block is in motion, the retarding frictional force becomes *less* than $f_{s,max}$ (Fig. 5.14c). When the block is in motion, we call the retarding force the *force of kinetic friction, f_k*. The unbalanced force in the x direction, $F - f_k$, produces an acceleration to the right. If $F = f_k$ the block moves to the right with constant speed. If the applied force is removed, then the frictional force acting to the left decelerates the block and eventually brings it to rest.

In a simplified model, we can imagine that the force of kinetic friction is less than $f_{s,max}$ because of the reduction in roughness of the two surfaces when the object is in motion.

Experimentally, one finds that both f_s and f_k are *proportional to the normal force acting on the block*. The experimental observations can be summarized by the following laws of friction:

1. The force of static friction between any two surfaces in contact is opposite the applied force and can have values given by

$$f_s \le \mu_s N \qquad (5.12)$$

Force of static friction

where the dimensionless constant μ_s is called the **coefficient of static friction** and N is the normal force. The equality in Equation 5.12 holds when the block is *on the verge* of slipping, that is, when $f_s = f_{s,max} = \mu_s N$. The inequality holds when the applied force is *less* than this value.

2. The force of kinetic friction acting on an object is opposite to the direction of motion of the object and is given by

$$f_k = \mu_k N \qquad (5.13)$$

Force of kinetic friction

where μ_k is the **coefficient of kinetic friction.**

3. The values of μ_k and μ_s depend on the nature of the surfaces, but μ_k is generally less than μ_s. Typical values of μ range from around 0.05 for smooth surfaces to 1.5 for rough surfaces. Table 5.2 lists some reported values.

Finally, the coefficients of friction are nearly independent of the area of contact between the surfaces. Although the coefficient of kinetic friction varies with speed, we shall neglect any such variations.

TABLE 5.2 Coefficients of Friction[a]

	μ_s	μ_k
Steel on steel	0.74	0.57
Aluminum on steel	0.61	0.47
Copper on steel	0.53	0.36
Rubber on concrete	1.0	0.8
Wood on wood	0.25–0.5	0.2
Glass on glass	0.94	0.4
Waxed wood on wet snow	0.14	0.1
Waxed wood on dry snow	—	0.04
Metal on metal (lubricated)	0.15	0.06
Ice on ice	0.1	0.03
Teflon on Teflon	0.04	0.04
Synovial joints in humans	0.01	0.003

[a] All values are approximate.

EXAMPLE 5.7. Experimental Determination of μ_s and μ_k

In this example we describe a simple method of measuring the coefficients of friction between an object and a rough surface. Suppose the object is a small block placed on a surface inclined with respect to the horizontal, as in Figure 5.15. The angle of the inclined plane is increased until the block slips. By measuring the angle θ_c at which this slipping just occurs, we obtain μ_s directly. We note that the only forces acting on the block are its weight, mg, the normal force, N, and the force of static friction, f_s. Taking x parallel to the plane and y perpendicular to the plane, Newton's second law applied to the block gives

Static case:

$$(1) \qquad \sum F_x = mg \sin\theta - f_s = 0$$

$$(2) \qquad \sum F_y = N - mg\cos\theta = 0$$

We can eliminate mg by substituting $mg = N/\cos\theta$ from (2) into (1) to get

$$(3) \qquad f_s = mg\sin\theta = \left(\frac{N}{\cos\theta}\right)\sin\theta = N\tan\theta$$

When the inclined plane is at the critical angle, θ_c,

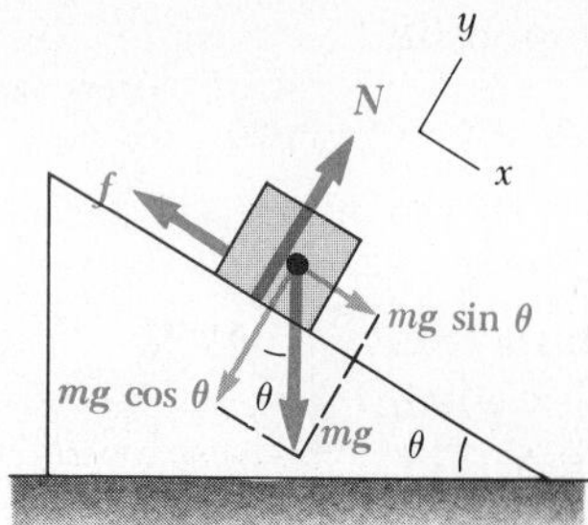

Figure 5.15 (Example 5.7) The external forces acting on a block lying on a rough incline are the weight, mg, the normal force, N, and the force of friction, f. Note that the weight vector is resolved into a component along the incline, $mg \sin \theta$, and a component perpendicular to the incline, $mg \cos \theta$.

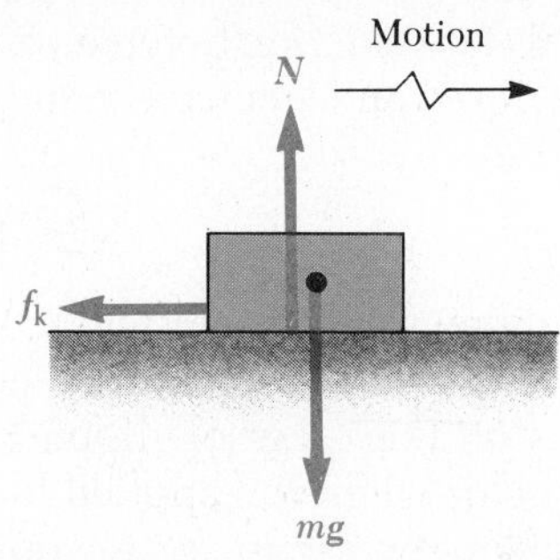

Figure 5.16 (Example 5.8) *After* the puck is given an initial velocity, the external forces acting on it are the weight, mg, the normal force, N, and the force of kinetic friction, f_k.

(called the angle of repose), $f_s = f_{s,\max} = \mu_s N$, and so at this angle, (3) becomes

$$\mu_s N = N \tan \theta_c$$

$$\mu_s = \tan \theta_c$$

For example, if we find that the block just slips at $\theta_c = 20°$, then $\mu_s = \tan 20° = 0.364$. Once the block starts to move at $\theta \geq \theta_c$, it will accelerate down the incline and the force of friction is $f_k = \mu_k N$. However, if θ is reduced below θ_c, an angle θ_c' can be found such that the block moves down the incline with constant speed ($a_x = 0$). In this case, using (1) and (2) with f_s replaced by f_k gives

$$\text{Kinetic case:} \qquad \mu_k = \tan \theta_c'$$

where $\theta_c' < \theta_c$.

You should try this simple experiment using a coin as the block and a notebook as the inclined plane. Also, you can try taping two coins together to prove that you still get the same critical angles as with one coin.

EXAMPLE 5.8. The Sliding Hockey Puck

A hockey puck on a frozen pond is hit and given an initial speed of 20 m/s. If the puck always remains on the ice and slides a distance of 120 m before coming to rest, determine the coefficient of kinetic friction between the puck and the ice.

Solution: The forces acting on the puck after it is in motion are shown in Figure 5.16. If we assume that the force of friction, f_k, remains constant, then this force produces a uniform deceleration of the puck. Applying Newton's second law to the puck in component form gives

$$(1) \qquad \sum F_x = -f_k = ma$$

$$(2) \qquad \sum F_y = N - mg = 0 \qquad (a_y = 0)$$

But $f_k = \mu_k N$, and from (2) we see that $N = mg$. Therefore, (1) becomes

$$-\mu_k N = -\mu_k mg = ma$$

$$a = -\mu_k g$$

The negative sign means that the acceleration is to the left, corresponding to a deceleration of the puck. Also, the acceleration is independent of the mass of the puck and is *constant* since we are assuming that μ_k remains constant.

Since the acceleration is constant, we can use the kinematic equation $v^2 = v_0^2 + 2ax$, with the final speed $v = 0$. This gives

$$v_0^2 + 2ax = v_0^2 - 2\mu_k gx = 0$$

$$\mu_k = \frac{v_0^2}{2gx}$$

In our example, $v_0 = 20$ m/s and $x = 120$ m:

$$\mu_k = \frac{(20 \text{ m/s})^2}{2(9.80 \text{ m/s}^2)(120 \text{ m})} = 0.170$$

Note that μ_k has no dimensions.

EXAMPLE 5.9. Connected Objects with Friction

A block of mass m_1 on a rough, horizontal surface is connected to a second mass m_2 by a light cord over a light, frictionless pulley as in Figure 5.17a. A force of magnitude F is applied to mass m_1 as shown. The coefficient of kinetic friction between m_1 and the surface is μ. Determine the acceleration of the masses and the tension in the cord.

Solution: First we draw the free-body diagrams of m_1 and m_2 as in Figures 5.17b and 5.17c. Note that the force F has components $F_x = F \cos \theta$ and $F_y = F \sin \theta$. Therefore, in this case N is *not* equal to $m_1 g$. Applying Newton's second law to both masses and *assuming* the motion of m_1 is to the right, we get

$$\text{Motion of } m_1: \qquad \Sigma F_x = F \cos \theta - f_k - T = m_1 a$$

$$(1) \qquad \sum F_y = N + F \sin \theta - m_1 g = 0$$

$$\text{Motion of } m_2: \qquad \Sigma F_x = 0$$

$$(2) \qquad \sum F_y = T - m_2 g = m_2 a$$

But $f_k = \mu N$, and from (1), $N = m_1 g - F \sin \theta$; therefore

$$(3) \qquad f_k = \mu(m_1 g - F \sin \theta)$$

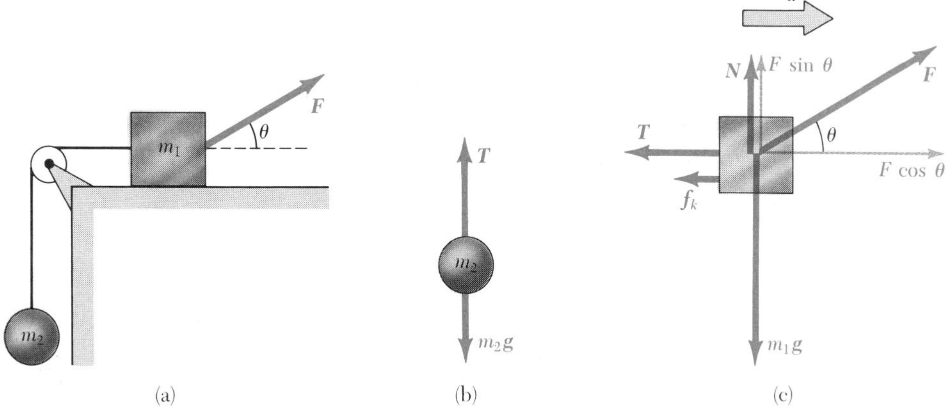

Figure 5.17 (Example 5.9) (a) The external force, F, applied as shown can cause m_1 to accelerate to the right. (b) and (c) The free-body diagrams assuming that m_1 accelerates to the right while m_2 accelerates upward. Note that the force of kinetic friction in this case is given by $f_k = \mu_k N = \mu_k(m_1 g - F\sin\theta)$.

That is, the frictional force is *reduced* because of the positive y component of F. Substituting (3) and the value of T from (2) into (1) gives

$$F\cos\theta - \mu(m_1 g - F\sin\theta) - m_2(a+g) = m_1 a$$

Solving for a, we get

$$(4) \qquad a = \frac{F(\cos\theta + \mu\sin\theta) - g(m_2 + \mu m_1)}{m_1 + m_2}$$

We can find T by substituting this value of a into (2). Note that the acceleration for m_1 can be either to the right or left,[5] depending on the sign of the numerator in (4). If the motion of m_1 is to the *left*, we must reverse the sign of f_k since the frictional force *opposes* the motion. In this case, the value of a is the same as in (4) with μ replaced by $-\mu$.

[5] A close examination of (4) shows that when $\mu m_1 > m_2$, there is a range of values of F for which no motion occurs at a given angle θ.

5.10 SUMMARY

Newton's first law states that a body at rest will remain at rest or a body in uniform motion in a straight line will maintain that motion unless an external resultant force acts on the body. — Newton's first law

Newton's second law states that the time rate of change of momentum of a body is equal to the resultant force acting on the body. If the mass of the body is constant, the net force equals the product of the mass and its acceleration, or $\Sigma \mathbf{F} = m\mathbf{a}$. — Newton's second law

Newton's first and second laws are valid in an inertial frame of reference. An **inertial frame** is one in which an object, subject to no net external force, moves with constant velocity including the special case of $v = 0$. — Inertial frame

Mass is a scalar quantity. The mass that appears in Newton's second law is called **inertial mass.**

The **weight** of a body is equal to the product of its mass and the acceleration of gravity, or $\mathbf{W} = m\mathbf{g}$. — Weight

Newton's third law states that if two bodies interact, the force exerted on body 1 by body 2 is equal to and opposite the force exerted on body 2 by body 1. Thus, an isolated force cannot exist in nature. — Newton's third law

The **maximum force of static friction,** f_s, between a body and a rough surface is proportional to the normal force acting on the body. This maximum force occurs when the body is on the verge of slipping. In general, $f_s \le \mu_s N$, where μ_s is the *coefficient of static friction* and N is the normal force. When a — Forces of friction

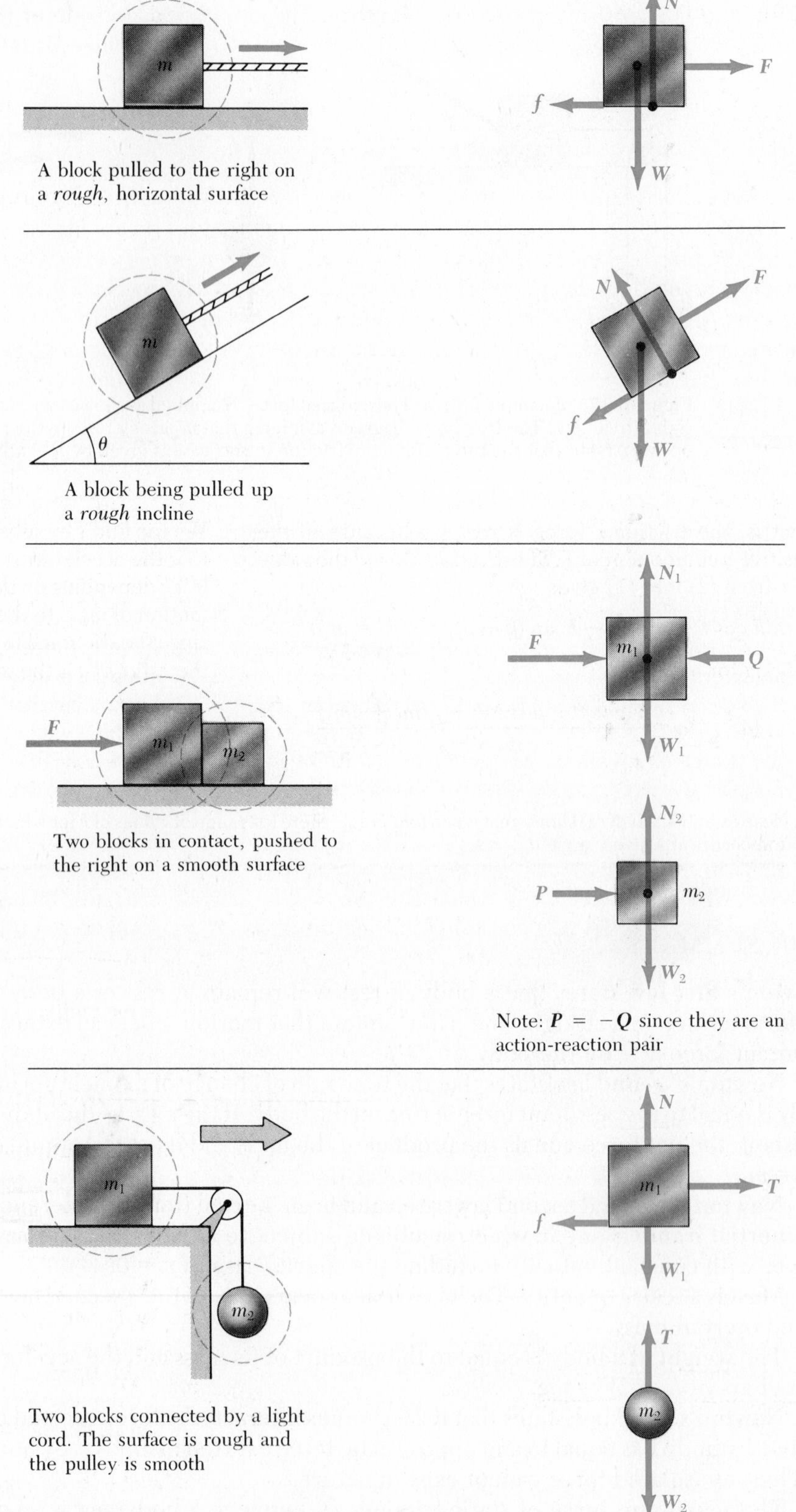

Figure 5.18 Various mechanical configurations (left) and the corresponding free-body diagrams (right).

body slides over a rough surface, the *force of kinetic friction*, f_k, is opposite the motion and is also proportional to the normal force. The magnitude of this force is given by $f_k = \mu_k N$, where μ_k is the *coefficient of kinetic friction*. Usually, $\mu_k < \mu_s$.

More on Free-Body Diagrams

As we have seen throughout this chapter, in order to be successful in applying Newton's second law to a mechanical system you must first be able to recognize all the forces acting on the system. That is, you must be able to construct the correct free-body diagram. The importance of constructing the free-body diagram cannot be overemphasized. In Figure 5.18 a number of mechanical systems are presented together with their corresponding free-body diagrams. You should examine these carefully and then proceed to construct free-body diagrams for other systems described in the problems. When a system contains more than one element, it is important that you construct a free-body diagram for *each* element.

As usual, $\boldsymbol{F}$ denotes some applied force, $\boldsymbol{W} = m\boldsymbol{g}$ is the weight, $\boldsymbol{N}$ denotes a normal force, $\boldsymbol{f}$ is frictional force, and $\boldsymbol{T}$ is the force of tension.

QUESTIONS

1. If an object is at rest, can we conclude that there are no external forces acting on it?
2. If gold were sold by weight, would you rather buy it in Denver or in Death Valley? If sold by mass, at which of the two locations would you prefer to buy it? Why?
3. A passenger sitting in the rear of a bus claims that he was injured when the driver slammed on the brakes, causing a suitcase to come flying toward the passenger from the front of the bus. If you were the judge in this case, what disposition would you make? Why?
4. A space explorer is in a spaceship moving through space far from any planet or star. She notices a large rock, taken as a specimen from an alien planet, floating around the cabin of the spaceship. Should she push it gently toward a storage compartment or kick it toward the compartment? Why?
5. How much does an astronaut weigh out in space, far from any planet?
6. Although the frictional force between two surfaces may decrease as the surfaces are smoothed, the force will again increase if the surfaces are made extremely smooth and flat. How do you explain this?
7. Why is it that the frictional force involved in the rolling of one body over another is less than for a sliding motion?
8. A massive metal object on a rough metal surface may undergo contact welding to that surface. Discuss how this affects the frictional forces between the object and the surface.
9. The observer in the elevator of Example 5.6 would claim that the "weight" of the fish is T, the scale reading. This is obviously wrong. Why does this observation differ from that of a person outside the elevator at rest with respect to the elevator?
10. Identify the action-reaction pairs in the following situations: a man takes a step; a snowball hits a girl in the back; a baseball player catches a ball; a gust of wind strikes a window.
11. While a football is in flight, what forces act on it? What are the action-reaction pairs while the football is being kicked and while it is in flight?
12. A ball is held in a person's hand. (a) Identify all the external forces acting on the ball and the reaction to each. (b) If the ball is dropped, what force is exerted on it while it is falling? Identify the reaction force in this case. (Neglect air resistance.)
13. Identify all the action-reaction pairs that exist for a horse pulling a cart. Include the earth in your examination.
14. If a car is traveling westward with a constant velocity of 20 m/s, what is the resultant force acting on it?
15. A large crate is placed on the bed of a truck without being tied to the truck. (a) As the truck accelerates forward, the crate remains at rest relative to the truck. What force causes the crate to accelerate? (b) If the truck driver slams on the brakes, what could happen to the crate?
16. A child pulls a wagon with some force, causing it to accelerate. Newton's third law says that the wagon exerts an equal and opposite reaction force on the child. How can the wagon accelerate?
17. A rubber ball is dropped onto the floor. What force causes the ball to bounce back into the air?
18. What is wrong with the statement, "Since the car is at rest, there are no forces acting on it"? How would you correct this sentence?

19. Suppose you are driving a car along a highway at a high speed. Why should you avoid slamming on your brakes if you want to stop in the shortest distance?

20. If you have ever taken a ride in an elevator of a high-rise building, you may have experienced the nauseating sensation of "heaviness" and "lightness" depending on the direction of $\boldsymbol{a}$. Explain these sensations. Are we truly weightless in free fall?

PROBLEMS

Section 5.1 through Section 5.7

1. A force, $\boldsymbol{F}$, applied to an object of mass m_1 produces an acceleration of 2 m/s^2. The same force applied to a second object of mass m_2 produces an acceleration of 6 m/s^2. (a) What is the value of the ratio m_1/m_2? (b) If m_1 and m_2 are combined, find their acceleration under the action of the force $\boldsymbol{F}$.
2. An object weighs 25 N at sea level, where $g = 9.8$ m/s^2. What is its weight on planet X, where the acceleration of gravity is 3.5 m/s^2?
3. A person weighs 120 lb. Determine (a) her weight in N and (b) her mass in kg.
4. Verify the following conversions: (a) 1 N $= 10^5$ dynes, (b) 1 N $= 0.225$ lb.
5. An object has a mass of 200 g. Find its weight in dynes and in N.
6. A force of 10 N acts on a body of mass 2 kg. What is (a) the acceleration of the body, (b) its weight in N, and (c) its acceleration if the force is doubled?
7. A 6-kg object undergoes an acceleration of 2 m/s^2. (a) What is the magnitude of the resultant force acting on it? (b) If this same force is applied to a 4-kg object, what acceleration will it produce?
8. A 3-kg mass undergoes an acceleration of $\boldsymbol{a} = (2\boldsymbol{i} + 5\boldsymbol{j})$ m/s^2. Find the resultant force, $\boldsymbol{F}$, and its magnitude.
9. Two forces, $\boldsymbol{F}_1$ and $\boldsymbol{F}_2$, act on a 5-kg mass. If $F_1 = 20$ N and $F_2 = 15$ N, find the acceleration in (a) and (b) of Figure 5.19.

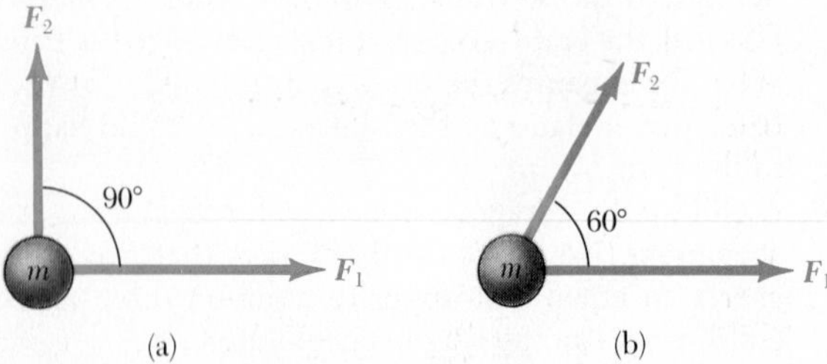

Figure 5.19 (Problem 9).

10. A 3-kg particle starts from rest and moves a distance of 4 m in 2 s under the action of a single, constant force. Find the magnitude of the force.
11. A 3-ton truck provides an acceleration of 3 ft/s^2 to a 10-ton trailer. If the truck exerts the same pull on a 15-ton trailer, what acceleration will result?
12. A ball is held in a person's hand. (a) Identify all the external forces acting on the ball and the reaction to each of these forces. (b) If the ball is dropped, what force is exerted on it while it is in "flight"? Identify the reaction force in this case. (Neglect air resistance.)
13. A 2-kg particle moves along the x axis under the action of a single, constant force. If the particle starts from rest at the origin at $t = 0$ and is observed to have a velocity of $-8.0\boldsymbol{i}$ m/s at $t = 2$ s, what are the magnitude and direction of the force?
14. One or more external forces are exerted on each object shown in Figure 5.1. Clearly identify the reaction to all of these forces. (*Note:* The reaction forces act on other objects.)
15. A bullet of mass 15 g leaves the barrel of a rifle with a speed of 800 m/s. If the length of the barrel is 75 cm, determine the force that accelerates the bullet, assuming the acceleration is constant. (*Note:* The actual force is exerted over a shorter time and is therefore greater than this estimate.)
16. An electron of mass 9.1×10^{-31} kg has an initial speed of 3.0×10^5 m/s. It travels in a straight line, and its speed increases to 7.0×10^5 m/s in a distance of 5.0 cm. Assuming its acceleration is constant, (a) determine the force on the electron and (b) compare this force with the weight of the electron, which we neglected.
17. A 4-kg object has a velocity of $3\boldsymbol{i}$ m/s at one instant. Eight seconds later, its velocity is $(8\boldsymbol{i} + 10\boldsymbol{j})$ m/s. Assuming the object was subject to a constant net force, find (a) the components of the force and (b) its magnitude.
18. A 15-lb block rests on the floor. (a) What force does the floor exert on the block? (b) If a rope is tied to the block and run vertically over a pulley and the other end attached to a free-hanging 10-lb weight, what is the force of the floor on the 15-lb block? (c) If we replace the 10-lb weight in (b) by a 20-lb weight, what is the force of the floor on the 15-lb block?

Section 5.8 Some Applications of Newton's Laws

19. Find the tension in each cord for the systems described in Figure 5.20. (Neglect the mass of the cords.)

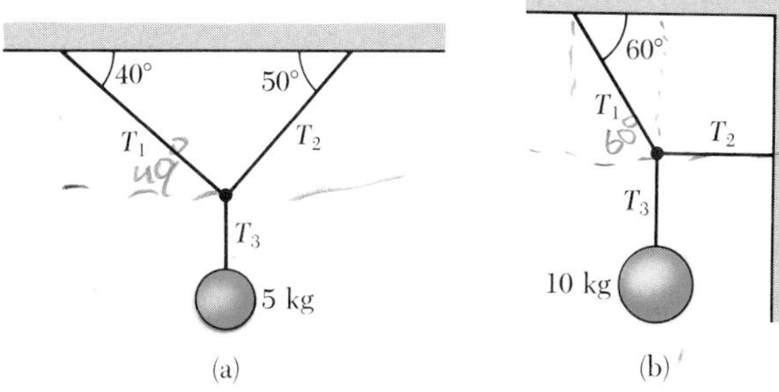

Figure 5.20 (Problem 19).

20. The systems shown in Figure 5.21 are in equilibrium. If the spring scales are calibrated in N, what do they read in each case? (Neglect the mass of the pulleys and strings, and assume the incline is smooth.)

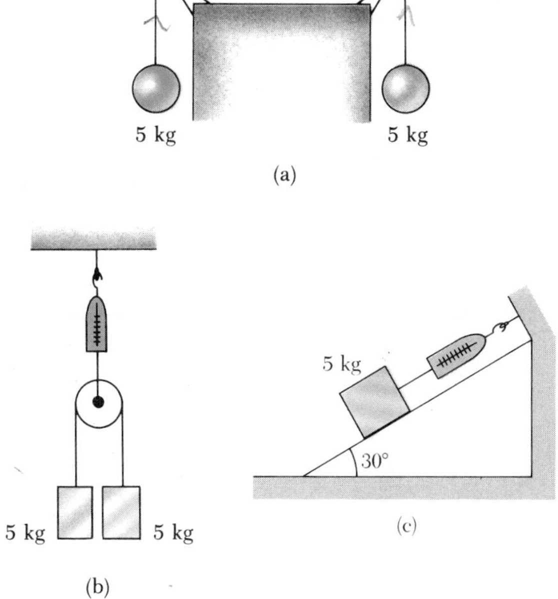

Figure 5.21 (Problem 20).

21. A 200-N weight is tied to the middle of a strong rope, and two people pull at opposite ends of the rope in an attempt to lift the weight. (a) What force $\boldsymbol{F}$ must each person apply to suspend the weight as shown in Figure 5.22? (b) Can they pull in such a way as to make the rope horizontal? Explain.

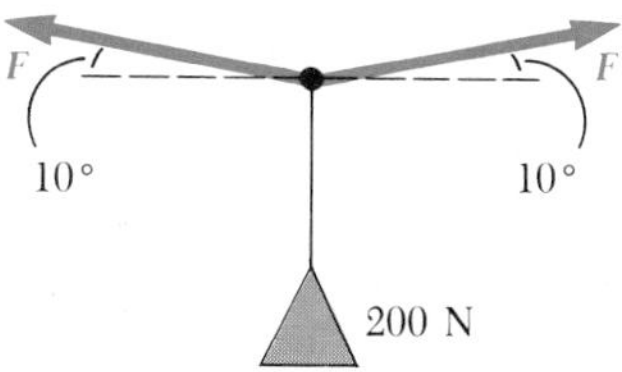

Figure 5.22 (Problem 21).

22. A block slides down a smooth plane having an inclination of $\theta = 15°$ (Fig. 5.23). If the block starts from rest at the top and the length of the incline is 2 m, find (a) the acceleration of the block and (b) its speed when it reaches the bottom of the incline.

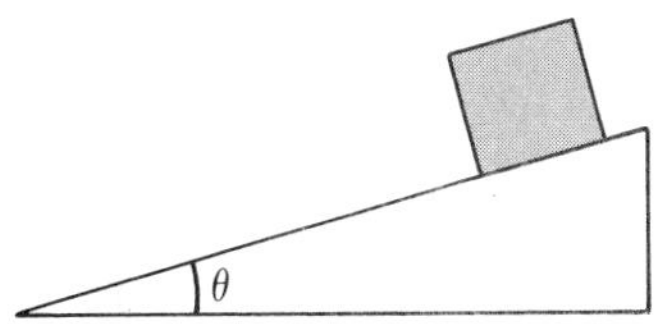

Figure 5.23 (Problems 22 and 23).

23. A block is given an initial velocity of 5 m/s up a smooth 20° incline (Fig. 5.23). How far up the incline does the block slide before coming to rest?

24. A 50-kg mass hangs from a rope 5 m in length, which is fastened to the ceiling. What horizontal force applied to the mass will deflect it 1 m sideways from the vertical and maintain it in that position?

25. Two masses of 3 kg and 5 kg are connected by a light string that passes over a smooth pulley as in Figure 5.11. Determine (a) the tension in the string, (b) the acceleration of each mass, and (c) the distance each mass moves in the first second of motion if they start from rest.

26. Two masses are connected by a light string that passes over a smooth pulley as in Figure 5.12. If the incline is frictionless and if $m_1 = 2$ kg, $m_2 = 6$ kg, and $\theta = 55°$, find (a) the acceleration of the masses, (b) the tension in the string, and (c) the speed of each mass 2 s after they are released from rest.

27. Two masses, m_1 and m_2, situated on a frictionless, horizontal surface are connected by a light string. A force, $\boldsymbol{F}$, is exerted on one of the masses to the right (Fig. 5.24). Determine the acceleration of the system and the tension, T, in the string.

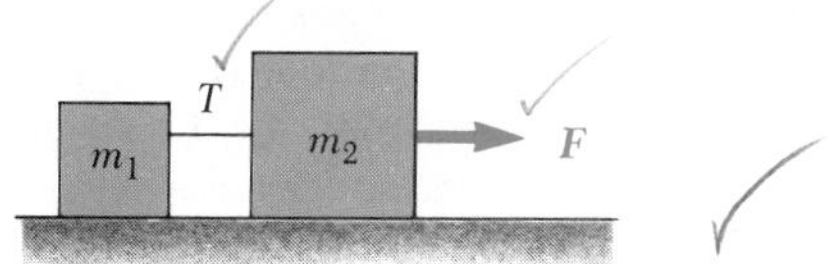

Figure 5.24 (Problems 27 and 35).

28. The parachute on a race car of weight 8820 N opens at the end of a quarter-mile run when the car is traveling at 55 m/s. What is the total retarding force required to stop the car in a distance of 1000 m in the event of a brake failure?

Section 5.9 Forces of Friction

29. A 20-kg block is initially at rest on a rough, horizontal surface. A horizontal force of 75 N is required to set the block in motion. After it is in motion, a horizontal force of 60 N is required to keep the block moving with constant speed. Find the coefficients of static and kinetic friction from this information.

30. The coefficient of static friction between a 4-kg block and a horizontal surface is 0.3. What is the *maximum* horizontal force that can be applied to the block before it slips?

31. A block moves up a 45° incline with constant speed under the action of a force of 15 N applied *parallel* to the incline. If the coefficient of kinetic friction is 0.3, determine (a) the weight of the block and (b) the minimum force required to allow the block to move *down* the incline at constant speed.

32. A car is traveling at 50 mi/h on a horizontal highway. (a) If the coefficient of friction between the road and tires on a rainy day is 0.1, what is the *minimum* distance in which the car will stop? (b) What is the stopping distance when the surface is dry and $\mu = 0.6$? (c) Why should you avoid "slamming on" your brakes if you want to stop in the shortest distance?

33. A racing car accelerates uniformly from 0 to 80 mi/h in 8 s. The external force that accelerates the car is the frictional force between the tires and the road. If the tires do not spin, determine the *minimum* coefficient of friction between the tires and the road.

34. In a game of shuffleboard, a disk is given an initial speed of 5 m/s. It slides a distance of 8 m before coming to rest. What is the coefficient of kinetic friction between the disk and the surface?

35. Two blocks connected by a light rope are being dragged by a horizontal force $\boldsymbol{F}$ (Fig. 5.24). Suppose that $F = 50$ N, $m_1 = 10$ kg, $m_2 = 20$ kg, and the coefficient of kinetic friction between each block and the surface is 0.1. (a) Draw a free-body diagram for each block. (b) Determine the tension, T, and the acceleration of the system.

36. A block slides on a *rough* incline. The coefficient of kinetic friction between the block and the plane is μ_k. (a) If the block accelerates *down* the incline, show that the acceleration of the block is given by $a = g(\sin\theta - \mu_k \cos\theta)$. (b) If the block is projected *up* the incline, show that its deceleration is $a = -g(\sin\theta + \mu_k \cos\theta)$.

37. A 3-kg block starts from rest at the top of a 30° incline and slides a distance of 2 m down the incline in 1.5 s. Find (a) the acceleration of the block, (b) the coefficient of kinetic friction between the block and the plane, (c) the frictional force acting on the block, and (d) the speed of the block after it has slid 2 m.

38. In order to determine the coefficients of friction between rubber and various surfaces, a student uses a rubber eraser and an incline. In one experiment the eraser slips down the incline when the angle of inclination is 36° and then moves down the incline with constant speed when the angle is reduced to 30°. From these data, determine the coefficients of static and kinetic friction for this experiment.

39. Two masses are connected by a light string, which passes over a frictionless pulley as in Figure 5.12. The incline is rough. When $m_1 = 3$ kg, $m_2 = 10$ kg, and $\theta = 60°$, the 10-kg mass accelerates *down* the incline at the rate of 2 m/s². Find (a) the tension in the string and (b) the coefficient of kinetic friction between the 10-kg mass and the plane.

40. A box rests on the back of a truck. The coefficient of static friction between the box and the surface is 0.3. (a) When the truck accelerates, what force accelerates the box? (b) Find the *maximum* acceleration the truck can have before the box slides.

41. A block slides down a 30° incline with *constant* acceleration. The block starts from rest at the top and travels 18 m to the bottom, where its speed is 3 m/s. Find (a) the coefficient of kinetic friction between the block and the incline and (b) the acceleration of the block.

GENERAL PROBLEMS

42. Two masses on a rough horizontal surface are connected by a light, rigid bar. m_1 is to the left of m_2. A horizontal force $\boldsymbol{F}$ is applied to the mass m_1 toward m_2 causing the system to accelerate to the right. The coefficient of kinetic friction between the blocks and the surface is μ. (a) Draw a free-body diagram for *each* mass. Identify all the forces in your diagrams. (b) Write a statement of Newton's second law in the horizontal and vertical direction for each mass in *symbolic* form. (c) Find the contact force between the bar and each block in terms of m_1, m_2, and $\boldsymbol{F}$. (d) Find the acceleration of the system in terms of the given parameters and g.

43. Two blocks of mass 2 kg and 7 kg are connected by a light string that passes over a frictionless pulley (Fig. 5.25). The inclines are smooth. Find (a) the acceleration of each block and (b) the tension in the string.

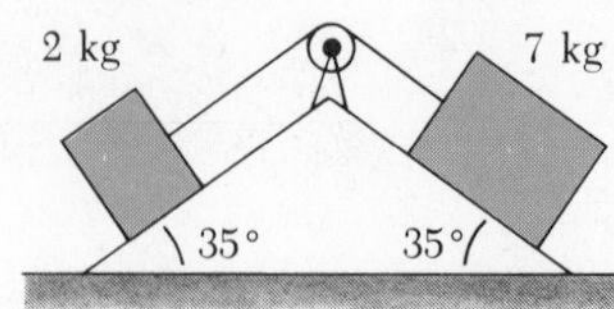

Figure 5.25 (Problems 43 and 44).

44. The system described in Figure 5.25 is observed to have an acceleration of 1.5 m/s² when the inclines are rough. Assume the coefficients of kinetic friction between each block and the inclines are the same. Find (a) the coefficient of kinetic friction and (b) the tension in the string.

*45. A mass M is held in place by an applied force $\boldsymbol{F}_A$ and a pulley system as shown in Figure 5.26. The pulleys are massless and frictionless. Find (a) the tension in each section of rope, T_1, T_2, T_3, T_4, and T_5 and (b) the applied force $\boldsymbol{F}_A$.

46. The force on a particle can be obtained from its momentum as a function of time. Find the force on a particle for each case when the momentum measured

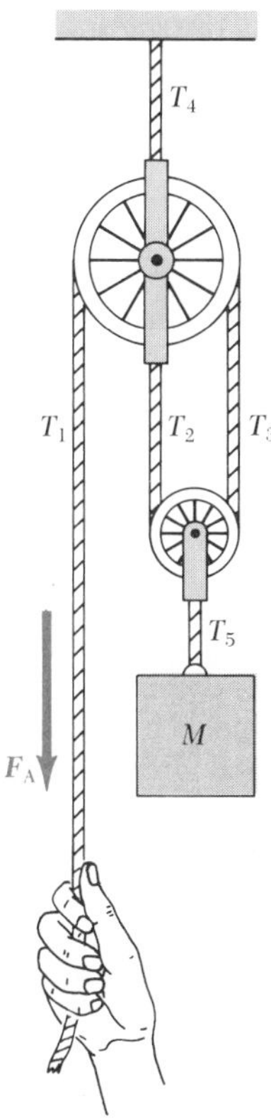

Figure 5.26 (Problem 45).

in kg · m/s varies with time as (a) $\boldsymbol{p} = (4 + 3t)\boldsymbol{j}$, (b) $\boldsymbol{p} = 3t\boldsymbol{i} + 5t^2\boldsymbol{j}$, (c) $\boldsymbol{p} = 4e^{-2t}\boldsymbol{i}$. (d) If the particle has a mass of 2 kg, find its acceleration at $t = 1$ s for cases (a), (b), and (c).

*47. "Big" Al remembered from high school physics that pulleys can be used to aid in lifting heavy objects. Al designed the pulley system, shown in Figure 5.27, to lift a safe to a second-floor office. The safe weighs 400 lb, and Al can pull with a force of 240 lb. (a) Will Big Al be able to raise the safe? (b) What is the maximum weight Big Al can lift using his pulley system? (*Note:* The large pulley is fastened by a yoke to the rope that Big Al is pulling.)

Figure 5.27 (Problem 47).

48. Consider a system consisting of a horse pulling a sled. According to Newton's third law, the force exerted by the horse on the sled is equal to and opposite the force exerted by the sled on the horse. Therefore, one might argue that the system can never move. Explain, using complete force diagrams on the horse and sled, that motion in this system is possible despite Newton's third law. Be sure to identify all of your forces.

49. A 2-kg block is placed on top of a 5-kg block as in Figure 5.28. The coefficient of kinetic friction between the 5-kg block and the surface is 0.2. A horizontal force $\boldsymbol{F}$ is applied to the 5-kg block. (a) Draw a free-body diagram for each block. What force accelerates the 2-kg block? (b) Calculate the force necessary to pull both blocks to the right with an acceleration of 3 m/s². (c) Find the minimum coefficient of static friction between the blocks such that the 2-kg block does not slip under an acceleration of 3 m/s².

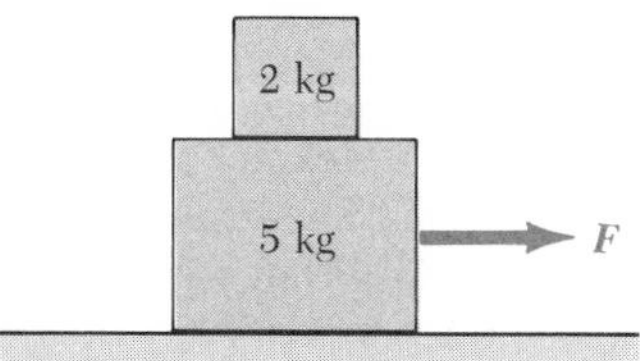

Figure 5.28 (Problem 49).

50. A car moves with a velocity v_0 down a sloped highway having an angle of inclination θ. The coefficient of friction between the car and the road is μ. The driver applies the brakes at some instant. Assuming that the tires do not skid and that the frictional force is a *maximum*, find (a) the deceleration of the car, (b) the distance the car will move before coming to rest after the brakes are applied, and (c) numerical results for the deceleration and the distance traveled if $v_0 = 60$ mi/h, $\theta = 10°$, and $\mu = 0.6$.

51. In Figure 5.29, the coefficient of kinetic friction between the 2-kg and 3-kg blocks is 0.3. The horizontal surface and the pulleys are frictionless. (a) Draw free-body diagrams for each block. (b) Determine the acceleration of each block. (c) Find the tension in the strings.

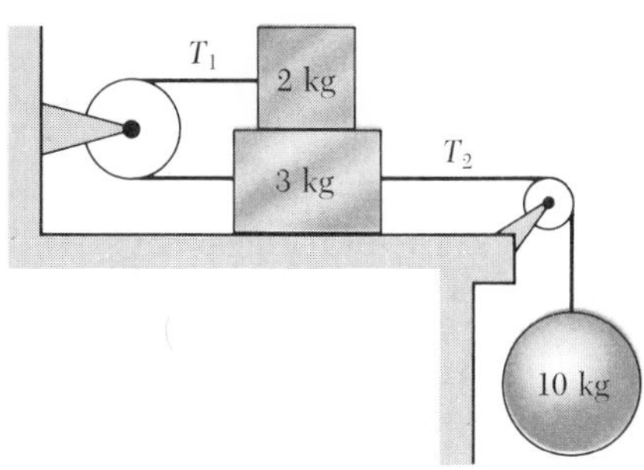

Figure 5.29 (Problem 51).

52. A horizontal force F is applied to a frictionless pulley of mass m_2 as in Figure 5.30. The horizontal surface is smooth. (a) Show that the acceleration of the block of mass m_1 is *twice* the acceleration of the pulley. Find (b) the acceleration of the pulley and the block and (c) the tension in the string. A constant supporting force is applied to the axle of the pulley equal to its weight.

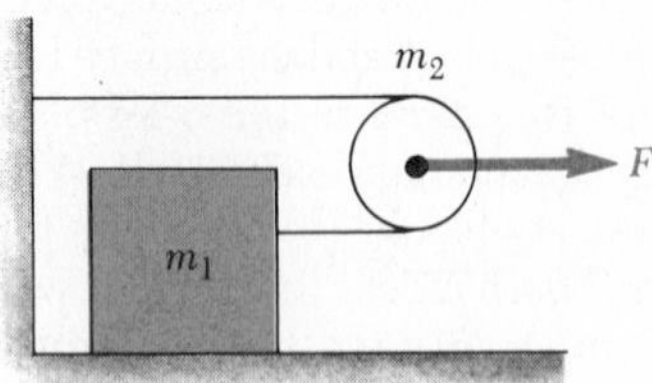

Figure 5.30 (Problem 52).

53. A bowling ball attached to a spring scale is suspended from the ceiling of an elevator as in Figure 5.13. (The ball replaces the fish!) The scale reads 16 lb when the elevator is at rest. (a) What will the scale read if the elevator accelerates *upward* at the rate of 8 ft/s²? (b) What will the scale read if the elevator accelerates *downward* at the rate of 8 ft/s²? (c) If the supporting rope can withstand a maximum tension of 25 lb and the weight of the scale is neglected, what is the maximum acceleration the elevator can have before the rope breaks? (d) If the spring scale weighs 5 lb, which rope breaks first? Why?

54. Three blocks are in contact with each other on a frictionless, horizontal surface as in Figure 5.31. A horizontal force F is applied to m_1. If $m_1 = 2$ kg, $m_2 = 3$ kg, $m_3 = 4$ kg, and $F = 18$ N, find (a) the acceleration of the blocks, (b) the *resultant* force on each block, and (c) the magnitude of the contact forces between the blocks.

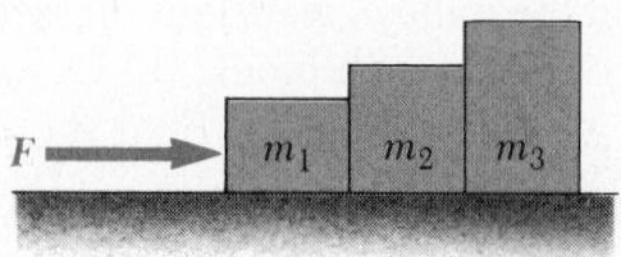

Figure 5.31 (Problems 54 and 55).

55. Repeat Problem 54 given that the coefficient of kinetic friction between the blocks and the surface is 0.1. Use the data given in Problem 54.

56. A 5-kg block is placed on top of a 10-kg block (Fig. 5.32). A horizontal force of 45 N is applied to the 10-kg block, while the 5-kg block is tied to the wall. The coefficient of kinetic friction between the moving surfaces is 0.2. (a) Draw a free-body diagram for each block and identify the action-reaction forces between the blocks. (b) Determine the tension in the string and the acceleration of the 10-kg block.

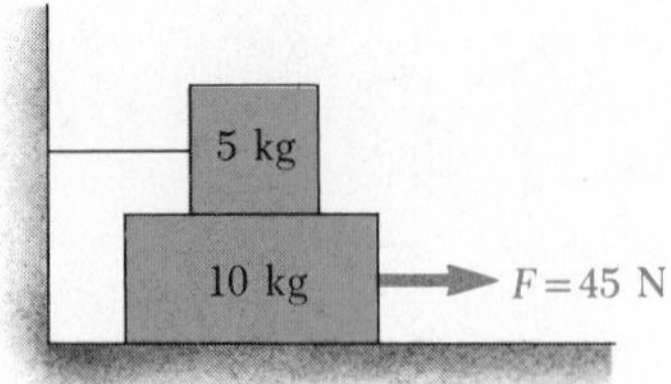

Figure 5.32 (Problem 56).

57. A block of mass m is on a *rough* incline of angle θ. (a) What is the *maximum horizontal* force that can be applied to the block before it slips *up* the plane? (b) What horizontal force will cause the block to move *up* the plane with an acceleration a? Take the coefficients of static and kinetic friction to be μ_s and μ_k, respectively.

58. What horizontal force must be applied to the cart shown in Figure 5.33 in order that the blocks remain *stationary* relative to the cart? Assume all surfaces, wheels, and pulley are frictionless. (*Hint:* Note that the tension in the string accelerates m_1.)

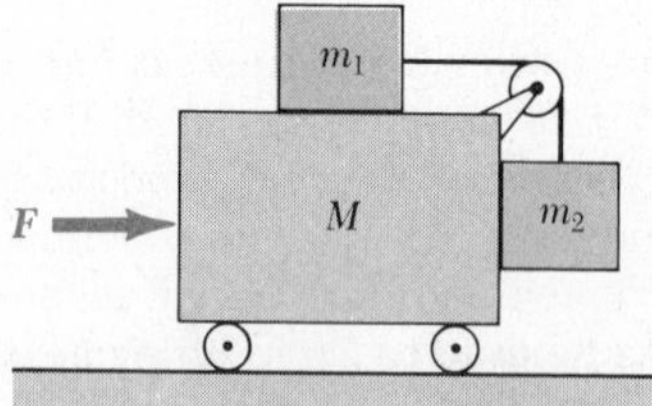

Figure 5.33 (Problems 58 and 69).

59. The three blocks in Figure 5.34 are connected by light strings that pass over frictionless pulleys. The acceleration of the system is 2 m/s² to the left and the surfaces are rough. Find (a) the tensions in the strings and (b) the coefficient of kinetic friction between blocks and surfaces. (Assume the same μ for both blocks.)

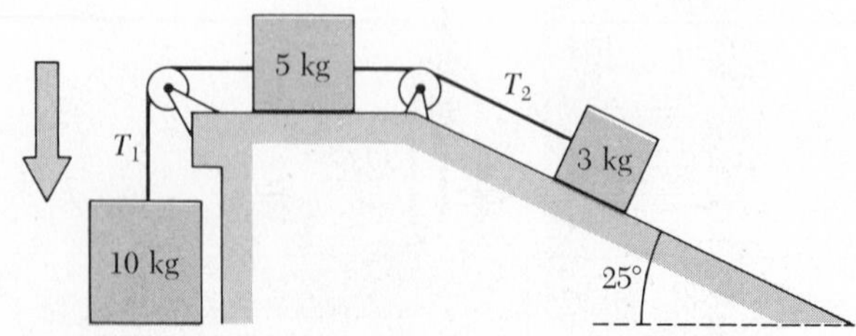

Figure 5.34 (Problem 59).

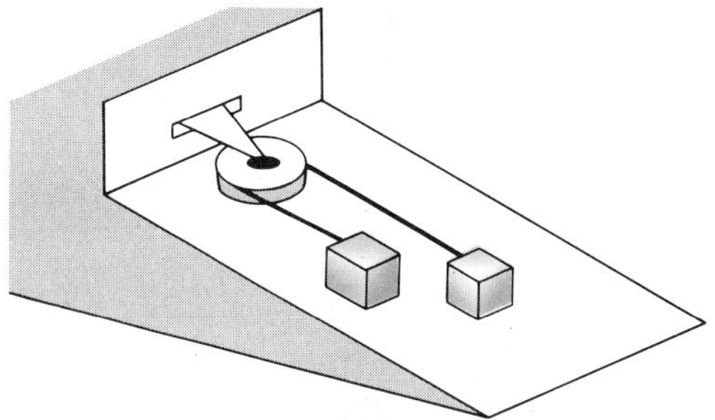

Figure 5.35 (Problem 60).

60. Two blocks on a rough incline are connected by a light string that passes over a frictionless pulley as in Figure 5.35. Assuming $m_1 > m_2$ and taking the coefficient of kinetic friction for each block to be μ, determine expressions for (a) the acceleration of the blocks and (b) the tension in the string. (Assume that the system is in motion.)

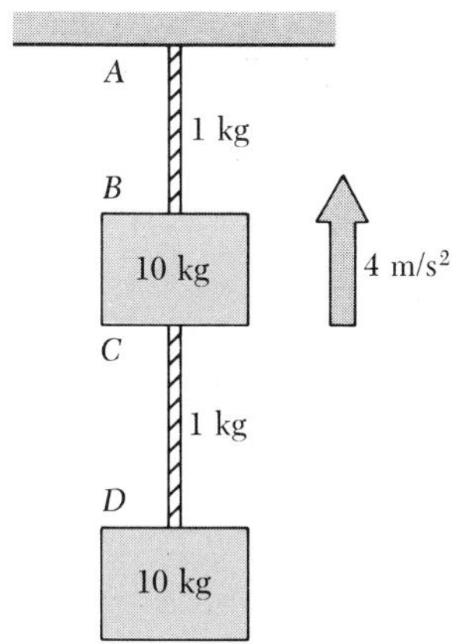

Figure 5.36 (Problem 61).

61. Two blocks are fastened to the top of an elevator as in Figure 5.36. The elevator accelerates upward at 4 m/s^2. Each rope has a mass of 1 kg. Find the tensions in the ropes at points *A*, *B*, *C*, and *D*.

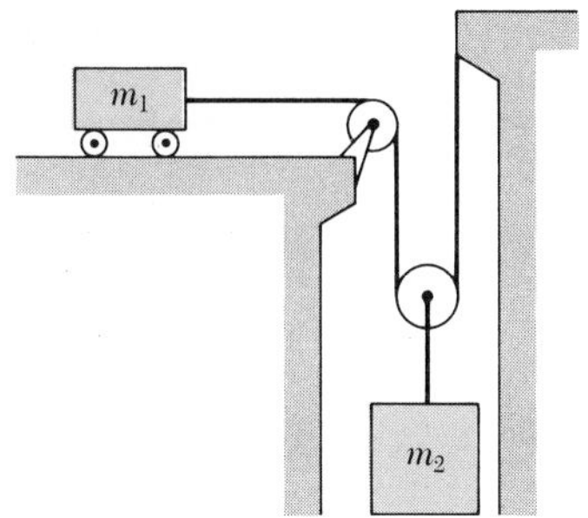

Figure 5.37 (Problem 62).

62. Find the acceleration of the cart and the mass shown in Figure 5.37. The pulleys are light and all surfaces are frictionless. What do these results predict in the limits $m_2 \gg m_1$ and $m_1 \gg m_2$?

•63. An inventive child named Pat wants to reach an apple in a tree without climbing the tree. Sitting in a chair connected to a rope that passes over a frictionless pulley (Fig. 5.38), Pat pulls on the loose end of the rope with such a force that the spring scale reads 60 lb. Pat's true weight is 64 lb and the chair weighs 32 lb. (a) Draw free-body diagrams for Pat and the chair considered as separate systems, and another diagram for Pat and the chair considered as one system. (b) Show that the acceleration of the system is *upward* and find its magnitude. (c) Find the force that Pat exerts on the chair.

Figure 5.38 (Problem 63).

•64. A block of mass m rests on the rough, inclined face of a wedge of mass M as in Figure 5.39. The wedge is free to move on a frictionless, horizontal surface. A horizontal force F is applied to the wedge such that the block is *on the verge* of slipping *up* the incline. If the coefficient of static friction between the block and the wedge is μ, find (a) the acceleration of the system and (b) the horizontal force necessary to produce this acceleration.

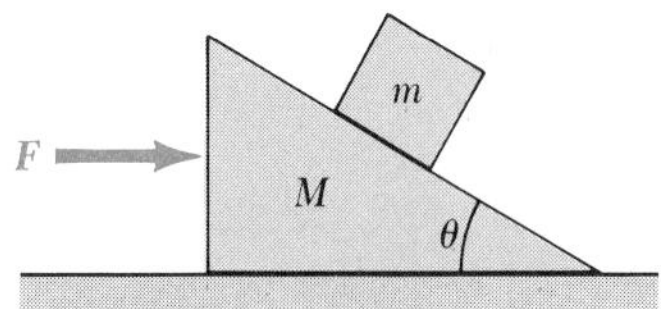

Figure 5.39 (Problem 64).

•**65.** Two masses m and M are attached with strings as shown in Figure 5.40. If the system is in equilibrium, show that $\tan\theta = 1 + \dfrac{2M}{m}$.

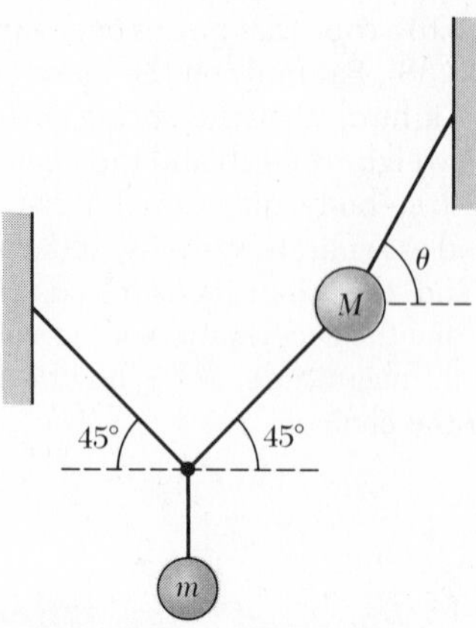

Figure 5.40 (Problem 65).

•**66.** Before 1960 it was commonly believed that the maximum attainable coefficient of static friction of an automobile tire was less than 1. Then about 1962, three companies independently developed racing tires with coefficients of 1.6. Since then, tires have improved, as illustrated in the following problem. According to the 1982 Guinness Book of Records, the fastest $\frac{1}{4}$ mile covered by a wheel-powered car from a standing start is 5.64 s. This record elapsed time was set by Don Garlits in October 1975. (a) Assuming that the torque applied by Garlits' rear wheels nearly lifted his front wheels off the pavement, what is the lowest value of μ_s his tires could have had? (b) Suppose Garlits were able to double his engine power, keeping other things equal. How would this affect his elapsed time?

•**67.** A butterfly mobile is formed by supporting four metal butterflies of equal mass m from a string of length L. The points of support are evenly spaced a distance ℓ apart as shown in Figure 5.41. The string forms an angle θ_1 with the ceiling at each end point. The center section of string is horizontal. (a) Find the tension in each section of string in terms of θ_1, m, and g. (b) Find the angle θ_2, in terms of θ_1, that the sections of string between the outside butterflies and the inside butterflies form with the horizontal. (c) Show that the distance D between the end points of the string is

$$D = \frac{L}{5}\left(2\cos\theta_1 + 2\cos\left[\tan^{-1}\left(\tfrac{1}{2}\tan\theta_1\right)\right] + 1\right).$$

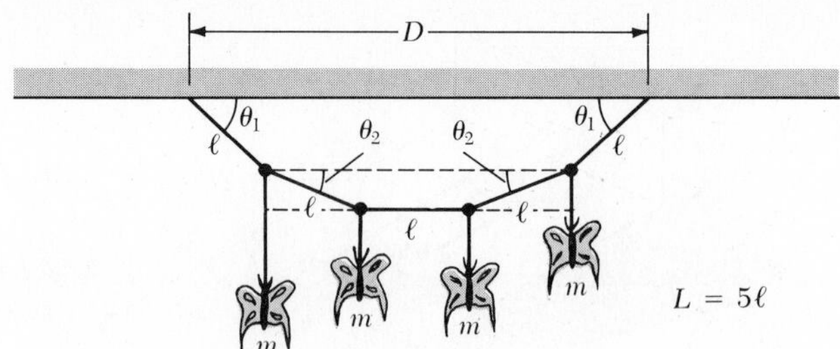

Figure 5.41 (Problem 67).

•**68.** Given a double Atwood's machine composed of two massless pulleys, three masses (40 g, 10 g, and 20 g), and connecting strings as shown in Figure 5.42, determine (a) the tension in the strings, T_1 and T_2 and (b) the magnitude and direction of the acceleration of each mass.

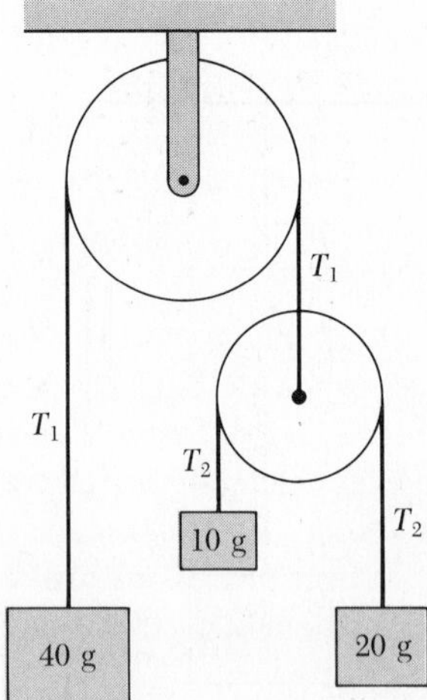

Figure 5.42 (Problem 68).

•**69.** Initially the system of masses shown in Figure 5.33 is held motionless. All surfaces, pulley, and wheels are frictionless. In this case let the force F be zero. At the *instant after* the system of masses is released find: (a) the tension T in the string, (b) the acceleration of m_2, (c) the acceleration of M, and (d) the acceleration of m_1. (*Note:* The pulley accelerates along with the cart.)

6
Circular Motion and Other Applications of Newton's Laws

In the previous chapter we introduced Newton's laws of motion and applied them to situations involving linear motion. In this chapter we shall begin with a brief discussion of the fundamental forces in nature. Then we shall apply Newton's laws of motion to the dynamics of circular motion. We shall also discuss the motion of an object when observed in an accelerated or noninertial frame of reference and the motion of an object through a viscous medium.

6.1 THE FOUR FUNDAMENTAL FORCES

In the previous chapter, we described a variety of forces that are experienced in our everyday activities such as the force of gravity which acts on all bodies at or near the earth's surface, and the force of friction as one surface slides over another. There are many other types of forces which we encounter, including the restoring force in a deformed spring, the electrostatic force between two charged bodies, and the magnetic force between a magnet and a piece of iron.

Forces also act in the microscopic world of atoms and nuclei. For example, atomic forces within the atom are responsible for holding its constituents together, and nuclear forces act on different parts of the nucleus to keep its parts from separating.

In spite of how complex the forces may appear, there are only four fundamental forces in nature: the gravitational force, the electromagnetic force, the strong nuclear force, and the weak nuclear force.

The **gravitational force** is the mutual force of attraction between all masses. We have already encountered the gravitational force when describing the weight of an object. Although gravitational forces can be very significant between macroscopic objects, the gravitational force is the weakest of the four fundamental forces. This statement is based on the relative strengths of the four forces when considering the interaction between elementary particles. For example, the gravitational force between the electron and proton in the hydrogen atom is only about 10^{-47} N, whereas the electrostatic force between the two particles is about 10^{-7} N. Thus, we see that the gravitational force is insignificant in comparison with the electrostatic force.

The **electromagnetic force** is an attraction or repulsion between two charged particles that are in relative motion. Later in the text we shall see that electric and magnetic forces are closely related. In fact, the magnetic force is an additional electric force that acts whenever the interacting charges are in motion. Although the electric force between two charged elementary particles is much stronger than the gravitational force between them, the electric force is of medium strength. It is interesting to note that the forces in our macroscopic world (apart from the gravitational force) are manifestations of the electric force. Friction forces, contact forces, and forces in springs or other deformed bodies are essentially the consequence of electric forces between charged particles in close proximity.

The **strong nuclear force** is responsible for the stability of nuclei. This force represents the "glue" that holds the nuclear constituents (called nucleons) together. It is the strongest of all the fundamental forces. For separations of about 10^{-15} m (a typical nuclear dimension), the strong nuclear force is one to two orders of magnitude stronger than the electric force. However, the strong nuclear force decreases rapidly with increasing separation and is negligible for separations greater than about 10^{-14} m.

Finally, the **weak nuclear force** is a short-range nuclear force that tends to produce instability in certain nuclei. Most radioactive decay reactions are caused by the weak nuclear force. The weak nuclear force is about 12 orders of magnitude weaker than the electrostatic force.

6.2 NEWTON'S SECOND LAW APPLIED TO UNIFORM CIRCULAR MOTION

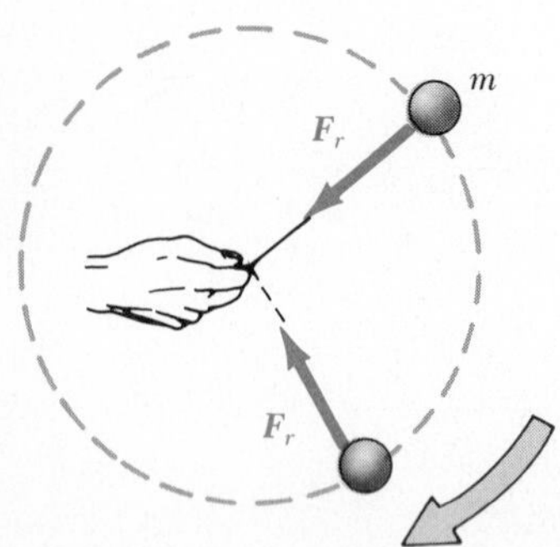

Figure 6.1 A ball moving in a circular path. A force $\boldsymbol{F}_r$ directed toward the center of the circle keeps the ball moving in the circle with a constant speed.

In Section 4.4 we found that a particle moving in a circular path of radius r with uniform speed v experiences an acceleration that has a magnitude

$$a_r = \frac{v^2}{r}$$

Because the velocity vector, $\boldsymbol{v}$, changes its direction continuously during the motion, the acceleration vector, $\boldsymbol{a}_r$, is directed toward the center of the circle and, hence, is called centripetal acceleration. Furthermore, $\boldsymbol{a}_r$ is always perpendicular to the velocity vector, $\boldsymbol{v}$.

Consider a ball of mass m tied to a string of length r and being whirled in a horizontal circular path on a table top as in Figure 6.1. Let us assume that the ball moves with constant speed. The inertia of the ball tends to maintain motion in a straight-line path; however, the string prevents this motion by exerting a force on the ball to make it follow its circular path. This force is directed along the length of the string toward the center of the circle, as shown in Figure 6.1, and is an example of a class of forces called **centripetal forces.** If we apply Newton's second law along the radial direction, we find that the required centripetal force is

Uniform circular motion

$$F_r = ma_r = m\frac{v^2}{r} \tag{6.1}$$

Like the centripetal acceleration, the centripetal force acts toward the center of the circular path followed by the particle. Because they act toward the center of rotation, centripetal forces cause a change in the direction of the velocity. Centripetal forces are no different from any other forces we have encountered. The term *centripetal* is used simply to indicate that *the force is directed toward the center of a circle.* In the case of a ball rotating at the end of a string, the tension force is the centripetal force. For a satellite in a circular orbit around the earth, the force of gravity is the centripetal force. The centripetal force acting on a car rounding a curve on a flat road is the force of friction between the tires and the pavement, and so forth.

Regardless of the example used, if the centripetal force acting on an object should vanish, the object would no longer move in its circular path; instead it would move along a straight-line path tangent to the circle. This idea is illustrated in Figure 6.2 for the case of the ball whirling in a circle at the end of a string. If the string breaks at some instant, the ball will move along the straight-line path tangent to the circle at the point where the string broke.

In general, a body can move in a circular path under the influence of such forces as friction, the gravitational force, or a combination of forces. Let us consider some examples of uniform circular motion. In each case, be sure to recognize the external force (or forces) that causes the body to move in its circular path.

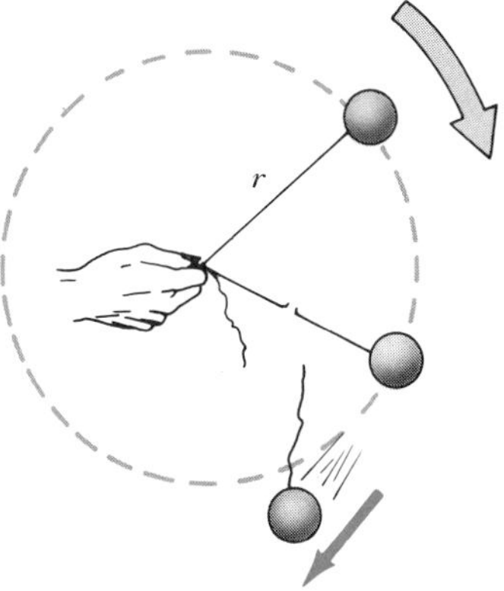

Figure 6.2 When the string breaks, the ball moves in the direction tangent to the circular path.

EXAMPLE 6.1. How Fast Can It Spin?
A ball of mass 0.5 kg is attached to the end of a cord whose length is 1.5 m. The ball is whirled in a horizontal circle as in Figure 6.2. If the cord can withstand a maximum tension of 50 N, what is the maximum speed the ball can have before the cord breaks?

Solution: Because the centripetal force in this case is the tension $\boldsymbol{T}$ in the cord, Equation 6.1 gives

$$T = m\frac{v^2}{r}$$

Solving for v, we have

$$v = \sqrt{\frac{Tr}{m}}$$

The maximum speed that the ball can have will correspond to the maximum value of the tension. Hence, we find

$$v_{\max} = \sqrt{\frac{T_{\max}r}{m}} = \sqrt{\frac{(50\text{ N})(1.5\text{ m})}{0.5\text{ kg}}} = 12.2\text{ m/s}$$

Exercise 1 Calculate the tension in the cord if the speed of the ball is 5 m/s.
Answer: 8.33 N.

EXAMPLE 6.2. The Conical Pendulum
A small body of mass m is suspended from a string of length L. The body revolves in a horizontal circle of radius r with constant speed v, as in Figure 6.3. Since the string sweeps out the surface of a cone, the system is known as a *conical pendulum*. Find the speed of the body and the period of revolution, T_P.

Solution: The free-body diagram for the mass m is shown in Figure 6.3, where the tension, $\boldsymbol{T}$, has been resolved into a vertical component, $T\cos\theta$, and a component $T\sin\theta$ acting toward the center of rotation. Since the body does not accelerate in the vertical direction, the vertical component of the tension must balance the weight. Therefore,

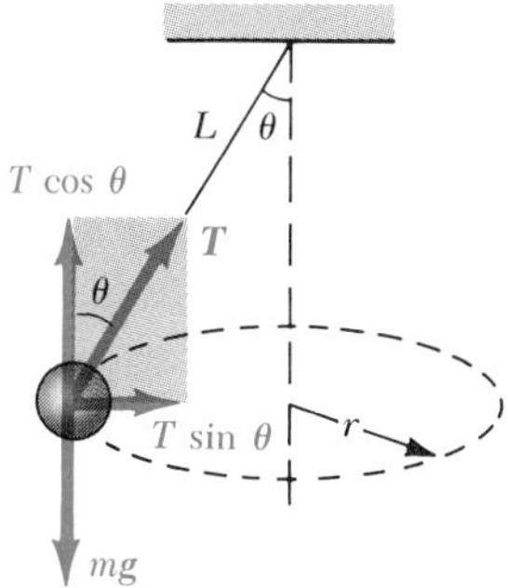

Figure 6.3 (Example 6.2) The conical pendulum and its free-body diagram.

$$(1) \qquad T\cos\theta = mg$$

Since the centripetal force in this example is provided by the component $T\sin\theta$, from Newton's second law we get

$$(2) \qquad T\sin\theta = ma_r = \frac{mv^2}{r}$$

By dividing (2) by (1), we eliminate T and find that

$$\tan\theta = \frac{v^2}{rg}$$

But from the geometry, we note that $r = L\sin\theta$, therefore

$$v = \sqrt{rg\tan\theta} = \sqrt{Lg\sin\theta\tan\theta}$$

The period of revolution, T_P (not to be confused with the tension $\boldsymbol{T}$), is given by

$$(3) \qquad T_P = \frac{2\pi r}{v} = \frac{2\pi r}{\sqrt{rg\tan\theta}} = 2\pi\sqrt{\frac{L\cos\theta}{g}}$$

The intermediate algebraic steps used in obtaining (3) are left to the reader. Note that T_P is independent of m! If we take $L = 1.00$ m and $\theta = 20°$, we find using (3) that

$$T_P = 2\pi \sqrt{\frac{(1.00 \text{ m})(\cos 20°)}{9.80 \text{ m/s}^2}} = 1.95 \text{ s}$$

Is it physically possible to have a conical pendulum with $\theta = 90°$?

EXAMPLE 6.3. What is the Maximum Speed of the Car?
A 1500-kg car moving on a flat road negotiates a curve whose radius is 35 m as in Figure 6.4. If the coefficient of static friction between the tires and the dry pavement is 0.50, find the maximum speed the car can have in order to make the turn successfully.

Solution: In this case, the centripetal force which enables the car to remain in its circular path is the force of static friction. Hence, from Equation 6.1 we have

$$(1) \qquad f_s = m\frac{v^2}{r}$$

The maximum speed that the car can have around the curve corresponds to the speed at which it is on the verge of skidding outwards. At this point, the friction force has its maximum value given by

$$f_{s\max} = \mu N$$

Because the normal force equals the weight in this case, we find

$$f_{s\max} = \mu mg = (0.5)(1500 \text{ kg})(9.80 \text{ m/s}^2) = 7350 \text{ N}$$

Substituting this value into (1), we find that the maximum speed is

$$v_{\max} = \sqrt{\frac{f_{s\max} r}{m}} = \sqrt{\frac{(7350 \text{ N})(35 \text{ m})}{1500 \text{ kg}}} = 13.1 \text{ m/s}$$

Exercise 2 On a wet day, the car described in this example begins to skid on the curve when its speed reaches 8 m/s. What is the coefficient of static friction in this case?
Answer: 0.187.

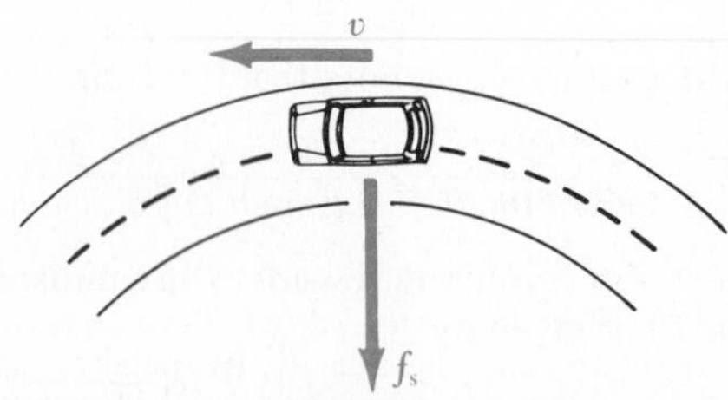

Figure 6.4 (Example 6.3) The force of static friction directed towards the center of the circular arc keeps the car moving in a circle.

EXAMPLE 6.4. The Banked Exit Ramp
An engineer wishes to design a curved exit ramp for a tollroad in such a way that a car will not have to rely on friction to round the curve without skidding. Suppose that a typical car rounds the curve with a speed of 30 mi/h (13.4 m/s) and that the radius of the curve is 50 m. At what angle should the curve be banked?

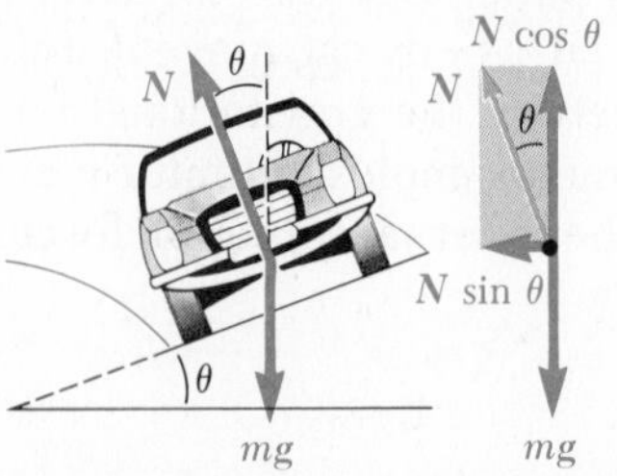

Figure 6.5 (Example 6.4) End view of a car rounding a curve on a road banked at an angle θ to the horizontal. The centripetal force is provided by the horizontal component of the normal force when friction is neglected. Note that N is the *sum* of the forces on the wheels of the car.

Solution: On a level road, the centripetal force must be provided by a force of friction between car and road. However, if the road is banked at an angle θ, as in Figure 6.5, the normal force, N, has a horizontal component $N \sin\theta$ pointing toward the center of the circular path followed by the car. We assume that only the component $N \sin\theta$ furnishes the centripetal force. Therefore, the banking angle we calculate will be one for which *no frictional force is required.* In other words, a car moving at the correct speed (13.4 m/s) can negotiate the curve even on an icy surface. Newton's second law written for the radial direction gives

$$(1) \qquad N \sin\theta = \frac{mv^2}{r}$$

The car is in equilibrium in the vertical direction. Thus, from $\Sigma F_y = 0$, we have

$$(2) \qquad N \cos\theta = mg$$

Dividing (1) by (2) gives

$$\tan\theta = \frac{v^2}{rg}$$

$$\theta = \tan^{-1}\left[\frac{(13.4 \text{ m/s})^2}{(50 \text{ m})(9.80 \text{ m/s}^2)}\right] = 20.1°$$

If a car rounds the curve at a speed lower than 13.4 m/s, the driver will have to rely on friction to keep from sliding down the incline. A driver who attempts to negotiate the curve at a speed higher than 13.4 m/s will have to depend on friction to keep from sliding up the ramp.

Exercise 3 Write Newton's second law applied to the radial direction for the car in a situation in which a fric-

tional force f is directed down the slope of the banked road.
Answer: $N\sin\theta + f\cos\theta = mv^2/r$.

EXAMPLE 6.5. Satellite Motion
This example treats the problem of a satellite moving in a circular orbit about the earth. In order to understand this problem, we must first note that the gravitational force between two particles having masses m_1 and m_2, separated by a distance r, is attractive and has a magnitude given by

$$F = G\frac{m_1 m_2}{r^2} \qquad (6.2)$$

where $G = 6.672 \times 10^{-11}\ \mathrm{N\cdot m^2/kg^2}$. This is Newton's universal law of gravity, which we shall discuss in detail in Chapter 14.

Now consider a satellite of mass m moving in a circular orbit about the earth at a constant speed v and at an altitude h above the earth's surface as in Figure 6.6. (a) Determine the speed of the satellite in terms of G, h, R_e (the radius of the earth), and M_e (the mass of the earth).

Solution: Because the only external force on the satellite is the force of gravity, which acts toward the center of the earth, we have

$$F_r = G\frac{M_e m}{r^2}$$

From Newton's second law, and the fact that $r = R_e + h$, we get

$$G\frac{M_e m}{r^2} = m\frac{v^2}{r}$$

or

$$v = \sqrt{\frac{GM_e}{r}} = \sqrt{\frac{GM_e}{R_e + h}} \qquad (6.3)$$

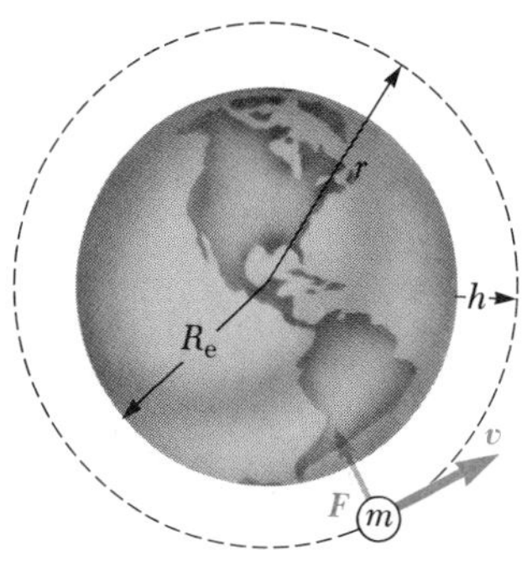

Figure 6.6 (Example 6.5) A satellite of mass m moving in a circular orbit of radius r and with constant speed v around the earth. The centripetal force is provided by the gravitational force between the satellite and the earth.

(b) Determine the satellite's period of revolution, T_p (the time for one revolution about the earth).

Solution: Since the satellite travels a distance of $2\pi r$ (the circumference of the circle) in a time T_p, we find using Equation 6.3 that

$$T_p = \frac{2\pi r}{v} = \frac{2\pi r}{\sqrt{GM_e/r}} = \left(\frac{2\pi}{\sqrt{GM_e}}\right) r^{3/2} \qquad (6.4)$$

The planets move around the sun in approximately circular orbits. The radii of these orbits can be calculated from Equation 6.4 with M_e replaced by the mass of the sun. The fact that the square of the period is proportional to the cube of the radius of the orbit was first recognized as an empirical relation based on planetary data. We shall return to this topic in Chapter 14.

Exercise 4 A satellite is in a circular orbit at an altitude of 1000 km. The radius of the earth is 6.37×10^6 m. Find the speed of the satellite and the period of its orbit.
Answer: 7.35×10^3 m/s; 6.31×10^3 s = 105 min.

6.3 NONUNIFORM CIRCULAR MOTION

In Chapter 4 we found that if a particle moves with varying speed in a circular path, there is, in addition to the centripetal component of acceleration, a tangential component of magnitude dv/dt. Therefore, the force acting on the particle must also have a tangential and a radial component. That is, since the total acceleration is given by $\boldsymbol{a} = \boldsymbol{a}_r + \boldsymbol{a}_t$, the total force is given by $\boldsymbol{F} = \boldsymbol{F}_r + \boldsymbol{F}_t$, as shown in Figure 6.7. The vector component, $\boldsymbol{F}_r$, is directed toward the center of the circle and is responsible for the centripetal acceleration. The vector component, $\boldsymbol{F}_t$, tangent to the circle is responsible for the tangential acceleration, which causes the speed of the particle to change with time. The following example demonstrates this type of motion.

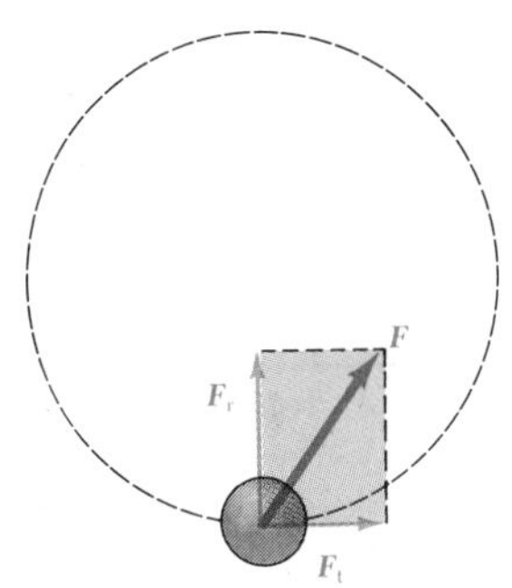

Figure 6.7 When the force acting on a particle has a tangential component F_t, its speed changes. The total force on the particle in this case is the vector sum of the tangential force and the centripetal force. That is, $\boldsymbol{F} = \boldsymbol{F}_t + \boldsymbol{F}_r$.

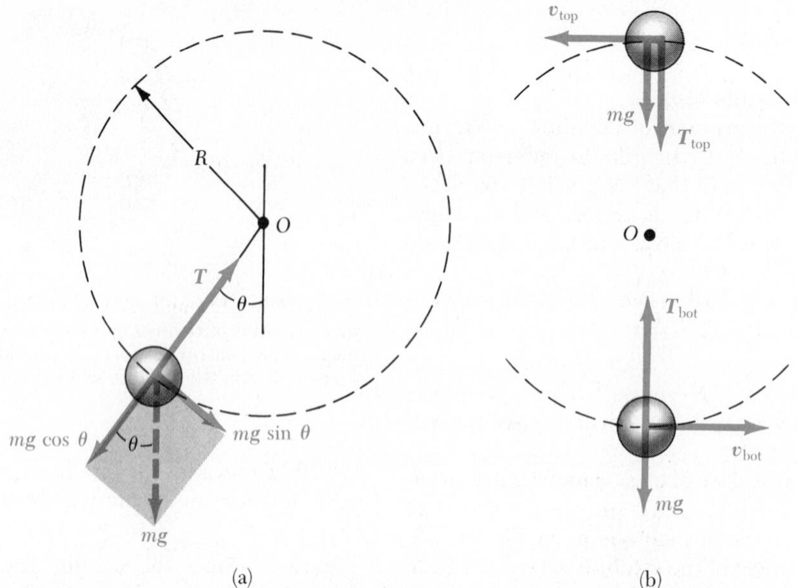

Figure 6.8 (Example 6.6) (a) Forces acting on a mass m connected to a string of length R and rotating in a vertical circle centered at O. (b) Forces acting on m when it is at the top and bottom of the circle. Note that the tension at the bottom is a maximum and the tension at the top is a minimum.

EXAMPLE 6.6. Follow the Rotating Ball

A small sphere of mass m is attached to the end of a cord of length R, which rotates in a *vertical* circle about a fixed point O, as in Figure 6.8a. Let us determine the tension in the cord at any instant that the speed of the sphere is v when the cord makes an angle θ with the vertical.

Solution: First we note that the speed is *not* uniform since there is a tangential component of acceleration arising from the weight of the sphere. From the free-body diagram in Figure 6.8a, we see that the only forces acting on the sphere are the weight, mg, and the constraint force (or tension), $\boldsymbol{T}$. Now we resolve mg into a tangential component, $mg \sin \theta$, and a radial component, $mg \cos \theta$. Applying Newton's second law to the forces in the tangential direction gives

$$\sum F_t = mg \sin \theta = ma_t$$

$$(1) \qquad a_t = g \sin \theta$$

This component causes v to change in time, since $a_t = dv/dt$. Applying Newton's second law to the forces in the radial direction and noting that both $\boldsymbol{T}$ and $\boldsymbol{a}_r$ are directed toward O, we get

$$\sum F_r = T - mg \cos \theta = \frac{mv^2}{R}$$

$$(2) \qquad T = m\left(\frac{v^2}{R} + g \cos \theta\right)$$

Limiting Cases: At the *top* of the path, where $\theta = 180°$, we see from (2) that since $\cos 180° = -1$,

$$T_{top} = m\left(\frac{v_{top}^2}{R} - g\right)$$

This is the *minimum* value of T. Note that at this point $a_t = 0$, and so the acceleration is radial and directed downward, as in Figure 6.8b.

At the *bottom* of the path, where $\theta = 0$, again from (2) we see that since $\cos 0 = 1$,

$$T_{bot} = m\left(\frac{v_{bot}^2}{R} + g\right)$$

This is the *maximum* value of T. Again, at this point $a_t = 0$, and the acceleration is radial and directed upward.

Exercise 5 At what orientation of the system would the cord most likely break if the average speed increased?
Answer: At the bottom of the path, where T has its maximum value.

*6.4 MOTION IN ACCELERATED FRAMES

When Newton's laws of motion were introduced in Chapter 5, we emphasized that the laws are valid when observations are made in an *inertial* frame of reference. In this section, we shall analyze how an observer in a noninertial frame of reference (one that is accelerating) would attempt to apply Newton's second law.

If a particle moves with an acceleration $\boldsymbol{a}$ relative to an observer in an inertial frame, then the inertial observer may use Newton's second law and correctly claim that $\Sigma \boldsymbol{F} = m\boldsymbol{a}$. If an observer in an accelerated frame (the noninertial observer) tries to apply Newton's second law to the motion of the particle, the noninertial observer must introduce *fictitious* forces to make Newton's second law work in that frame. Sometimes, these fictitious forces are referred to as **inertial forces.** These forces "invented" by the noninertial observer *appear* to be real forces in the accelerating frame. However, we emphasize that these fictitious forces *do not* exist when the motion is observed in an inertial frame. The fictitious forces are used only in an accelerating frame but *do not* represent "real" forces on the body. (By "real" forces, we mean the interaction of the body with its environment.) If the fictitious forces are properly defined in the accelerating frame, then the description of motion in this frame will be equivalent to the description by an inertial observer who considers only real forces. Usually, motions are analyzed using inertial reference frames, but there are cases in which an accelerating frame is more convenient.

Fictitious or inertial forces

In order to understand better the motion of a rotating system, consider a car traveling along a highway at a high speed and approaching a curved exit ramp, as in Figure 6.9. As the car takes the sharp left turn onto the ramp, a person sitting in the passenger seat slides to the right across the seat and hits the door. At that point, the force of the door keeps him from being ejected from the car. What causes the passenger to move toward the door? A popular, but *improper,* explanation is that some mysterious force pushes him outward. (This is often called the "centrifugal" force, but we shall not use this term since it often creates confusion.) The passenger invents this fictitious force in order to explain what is going on in his accelerated frame of reference.

The phenomenon is correctly explained as follows. Before the car enters the ramp, the passenger is moving in a straight-line path. As the car enters the ramp and travels a curved path, the passenger, because of inertia, tends to move along the original straight-line path. This is in accordance with Newton's first law: the natural tendency of a body is to continue moving in a straight line. However, if a sufficiently large centripetal force (toward the center of curvature) acts on the passenger, he will move in a curved path along with the car. The origin of this centripetal force is the force of friction between the passenger and the car seat. If this frictional force is not large enough, the passenger will slide across the seat as the car turns under him. Eventually, the passenger encounters the door, which provides a large enough centripetal force to enable the passenger to follow the same curved path as the car. The passenger slides toward the door not because of some mysterious outward force but because *there is no centripetal force large enough to allow him to travel along the circular path followed by the car.*

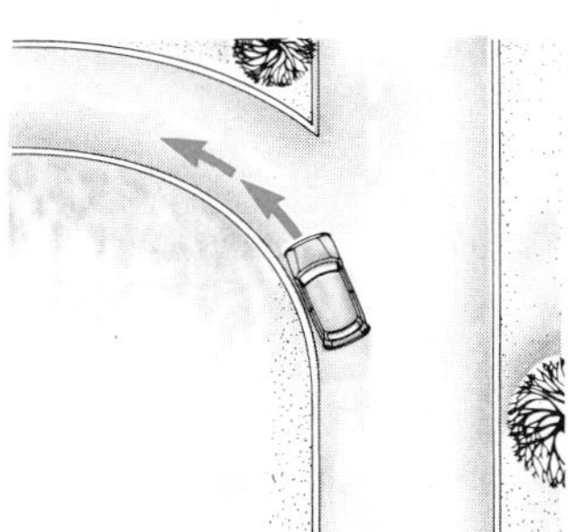

Figure 6.9 A car approaching a curved exit ramp.

In summary, one must be very careful to distinguish real forces from fictitious ones in describing motion in an accelerating frame. An observer in a car rounding a curve is in an accelerating frame and invents a fictitious outward force to explain why he or she is thrown outward. A stationary observer outside the car, however, considers only real forces on the passenger. To this observer, the mysterious outward force *does not exist!* The only real external force on the passenger is the centripetal (inward) force due to friction or the normal force of the door.

EXAMPLE 6.7. Linear Accelerometer
A small sphere of mass m is hung from the ceiling of an accelerating boxcar, as in Figure 6.10. According to the inertial observer at rest (Figure 6.10a), the forces on the sphere are the tension $\mathbf{T}$ and the weight $m\mathbf{g}$. The inertial observer concludes that the acceleration of the sphere of mass m is the same as that of the boxcar and that this acceleration is provided by the horizontal component of $\mathbf{T}$. Also, the vertical component of $\mathbf{T}$ balances the weight. Therefore, the inertial observer writes Newton's second law as $\mathbf{T} + m\mathbf{g} = m\mathbf{a}$, which in component form becomes

$$\text{Inertial observer} \begin{cases} (1) & \sum F_x = T \sin\theta = ma \\ (2) & \sum F_y = T\cos\theta - mg = 0 \end{cases}$$

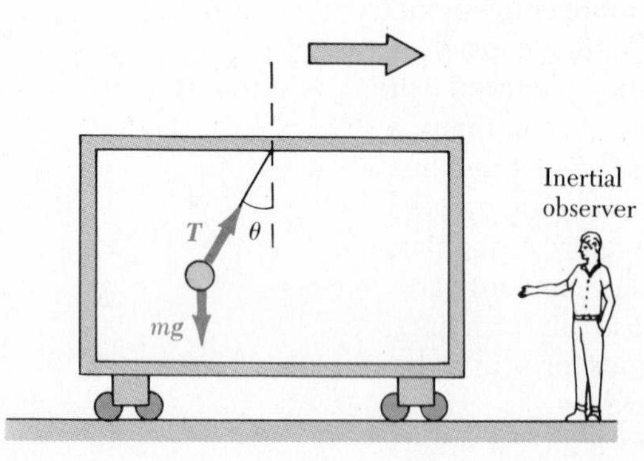

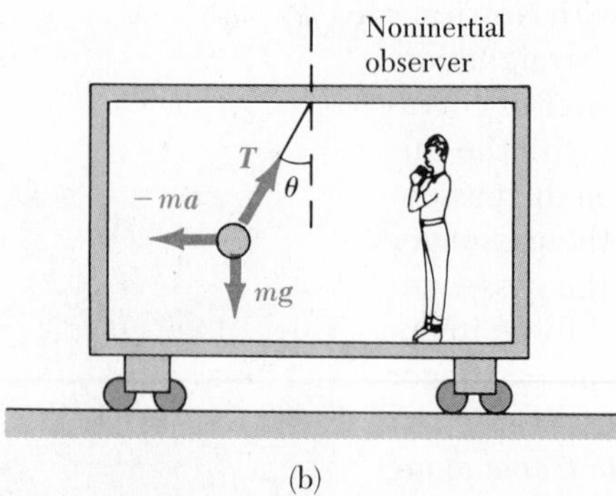

Figure 6.10 (Example 6.7) (a) A ball suspended from the ceiling of a boxcar accelerating to the right is deflected as shown. The inertial observer at rest outside the car claims that the acceleration of the ball is provided by the horizontal component of $\mathbf{T}$. (b) A noninertial observer riding in the car says that the net force on the ball is zero and that the deflection of the string from the vertical is due to a fictitious force, $-m\mathbf{a}$, which balances the horizontal component of $\mathbf{T}$.

Thus, by solving (1) and (2) simultaneously, the inertial observer can determine the acceleration of the car through the relation

$$a = g \tan\theta$$

Therefore, since the deflection of the string from the vertical serves as a measure of the acceleration of the car, *a simple pendulum can be used as an accelerometer.*

According to the noninertial observer riding in the car, described in Figure 6.10b, the sphere is at rest and the acceleration is zero. Therefore, the noninertial observer introduces a *fictitious force*, $-m\mathbf{a}$, to balance the horizontal component of $\mathbf{T}$ and claims that the net force on the sphere is *zero!* In this noninertial frame of reference, Newton's second law in component form gives

$$\text{Noninertial observer} \begin{cases} \sum F_x' = T\sin\theta - ma = 0 \\ \sum F_y' = T\cos\theta - mg = 0 \end{cases}$$

These expressions are equivalent to (1) and (2); therefore the noninertial observer gets the same mathematical results as the inertial observer. However, the physical interpretation of the deflection of the string *differs* in the two frames of reference. Note that even though a pendulum is used, it does not oscillate in this application.

EXAMPLE 6.8. Fictitious Force in a Rotating System
An observer in a rotating system is another example of a noninertial observer. Suppose a block of mass m lying on a horizontal, frictionless turntable is connected to a string as in Figure 6.11. According to an inertial observer, if the block rotates uniformly, it undergoes a centripetal acceleration v^2/r, where v is its tangential speed. The inertial observer concludes that this centripetal acceleration is provided by the force of tension in the string, T, and writes Newton's second law $T = mv^2/r$.

According to a noninertial observer attached to the turntable, the block is at rest. Therefore, in applying Newton's second law, this observer introduces a fictitious *outward* force called the *centrifugal force,* of magnitude mv^2/r. According to the noninertial observer, this "centrifugal" force balances the force of tension and therefore $T - mv^2/r = 0$.

You should be careful when using fictitious forces to describe physical phenomena. Remember that fictitious forces, such as centrifugal force, are used *only* in noninertial, or accelerated, frames of reference. When solving problems, it is generally best to use an inertial frame.

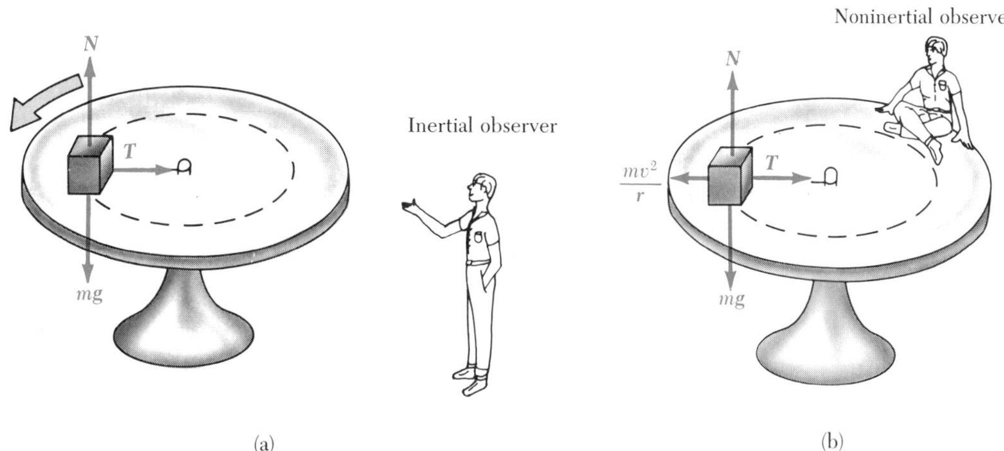

Figure 6.11 (Example 6.8) A block of mass m connected to a string tied to the center of a rotating turntable. (a) The inertial observer claims that the centripetal force is provided by the force of tension, $\boldsymbol{T}$. (b) The noninertial observer claims that the block is not accelerating and therefore introduces a fictitious centrifugal force mv^2/r, which acts outward and balances the tension.

*6.5 MOTION IN THE PRESENCE OF RESISTIVE FORCES

In the previous chapter we discussed the force of sliding friction, that is, the resistive force on an object moving along a rough, solid surface. Such forces are nearly independent of velocity, and matters are simplified by assuming them to be constant in magnitude. Now let us consider what happens when an object moves through a liquid or gas. In such situations, the medium exerts a resistive force $\boldsymbol{R}$ on the object. The magnitude of this force depends on the velocity of the object, and its direction is always opposite the direction of motion of the object relative to the medium. The magnitude of the resistive force is generally found to increase with increasing velocity. Some examples of such resistive forces are the air resistance to flying airplanes and moving cars and the viscous forces on objects moving through a liquid.

In general, the resistive force can have a complicated velocity dependence. In the following discussions, we will consider two situations. First, we will assume that the resistive force is proportional to the velocity. Objects falling through a fluid and very small objects, such as particles of dust moving through air, experience such a force. Second, we will treat situations for which the resistive force is assumed to be proportional to the square of the speed of the object. Large objects, such as a skydiver moving through air in free fall in the presence of gravity, experience such a force.

Resistive Force Proportional to Velocity

When an object moves at low speeds through a viscous medium, it experiences a resistive drag force that is proportional to the velocity of the object. Let us assume that the resistive force, $\boldsymbol{R}$, has the form

$$\boldsymbol{R} = -b\boldsymbol{v} \tag{6.5}$$

where $\boldsymbol{v}$ is the velocity of the object and b is a constant that depends on the properties of the medium and on the shape and dimensions of the object. If the object is a sphere of radius r, then b is found to be proportional to r.

The high cost of fuel has prompted many truck owners to install wind deflectors on their cabs to reduce air drag. (Photo by Lloyd Black)

Consider a sphere of mass m released from rest in a fluid, as in Figure 6.12a. Assuming the only forces acting on the sphere are the resistive force, $-bv$, and the weight, mg, let us describe its motion.[1]

Applying Newton's second law to the vertical motion, choosing the downward direction to be positive, and noting that $\Sigma F_y = mg - bv$, we get

$$mg - bv = m\frac{dv}{dt}$$

where the acceleration is downward. Simplifying the above expression gives

$$\frac{dv}{dt} = g - \frac{b}{m}v \qquad (6.6)$$

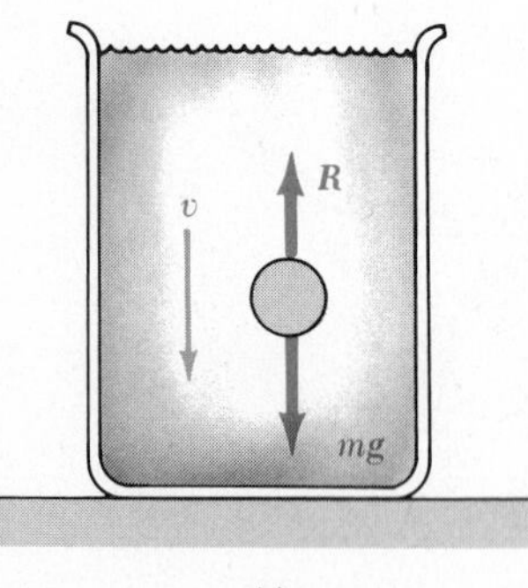

(a)

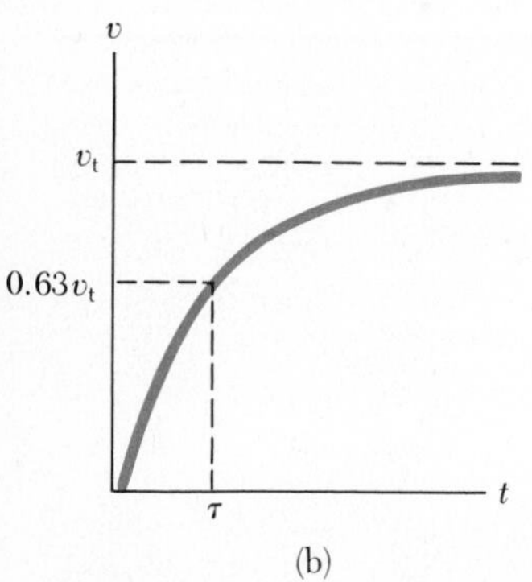

(b)

Figure 6.12 (a) A small sphere falling through a viscous fluid. (b) The velocity-time graph for an object falling through a viscous medium. The object reaches a maximum, or terminal, velocity, v_t, and τ is the time it takes to reach $0.63\, v_t$.

Equation 6.6 is called a *differential equation,* and the methods of solving such an equation may not be familiar to you as yet. However, note that initially, when $v = 0$, the resistive force is zero and the acceleration, dv/dt, is simply g. As t increases, the resistive force increases and the acceleration *decreases.* Eventually, the acceleration becomes zero when the resistive force *equals* the weight. At this point, the body continues to move with zero acceleration, and it reaches its *terminal velocity,* v_t. The terminal velocity can be obtained from Equation 6.6 by setting $a = dv/dt = 0$. This gives

$$mg - bv_t = 0 \qquad \text{or} \qquad v_t = mg/b$$

The expression for v that satisfies Equation 6.6 with $v = 0$ at $t = 0$ is

$$v = \frac{mg}{b}(1 - e^{-bt/m}) = v_t(1 - e^{-t/\tau}) \qquad (6.7)$$

This function is plotted in Figure 6.12b. The time $\tau = m/b$ is the time it takes the object to reach 63% of its terminal velocity. We can check that Equation 6.7 is a solution to Equation 6.6 by direct differentiation:

$$\frac{dv}{dt} = \frac{d}{dt}\left(\frac{mg}{b} - \frac{mg}{b}e^{-bt/m}\right) = -\frac{mg}{b}\frac{d}{dt}e^{-bt/m} = ge^{-bt/m}$$

Substituting this expression and Equation 6.7 into Equation 6.6 shows that our solution satisfies the differential equation.

[1] There is also a *buoyant* force, which is constant and equal to the weight of the displaced fluid. This will only change the weight of the sphere by a constant factor. We shall discuss such buoyant forces in Chapter 15.

EXAMPLE 6.9. Sphere Falling in Oil

A small sphere of mass 2 g is released from rest in a large cylinder filled with oil. The sphere reaches a terminal velocity of 5 cm/s. Determine the constant τ and the speed of the sphere as a function of time.

Solution: Since the terminal velocity is given by $v_t = mg/b$, the constant b is given by

$$b = \frac{mg}{v_t} = \frac{(2\text{ g})(980\text{ cm/s}^2)}{5\text{ cm/s}} = 392\text{ g/s}$$

Therefore, the time τ is given by

$$\tau = \frac{m}{b} = \frac{2\text{ g}}{392\text{ g/s}} = 5.10 \times 10^{-3}\text{ s}$$

The velocity as a function of time can be calculated using Equation 6.7.

$$v(t) = v_t(1 - e^{-t/\tau})$$

Since $v_t = 5$ cm/s and $1/\tau = 196\text{ s}^{-1}$, we have

$$v(t) = 5(1 - e^{-196t})\text{ cm/s}$$

Air Drag

We have seen that an object moving through a fluid experiences a resistive drag force. If the object is small and moves at low speeds, the drag force is proportional to the velocity, as we have already discussed. However, for larger objects moving at high speeds through air, such as airplanes, skydivers, and baseballs, the drag force is approximately proportional to the *square* of the speed. In these situations, the magnitude of the drag force can be expressed as

$$R = \tfrac{1}{2}C\rho Av^2 \tag{6.8}$$

where ρ is the density of air, A is the cross-sectional area of the falling object measured in a plane perpendicular to its motion, and C is a dimensionless empirical quantity called the *drag coefficient.* The drag coefficient has a value of about 0.5 for spherical objects but can be as high as 1 for irregularly shaped objects.

By spreading his arms and legs out from the sides of his body and by keeping the plane of his body parallel to the ground, a skydiver will experience maximum air drag resulting in a specific terminal speed. (U.S. Air Force Photo)

Consider an airplane in flight experiencing such a drag force. Equation 6.8 shows that the drag force is proportional to the density of air and hence decreases with decreasing air density. Since air density decreases with increasing altitude, the drag force on a jet airplane flying at a given speed must also decrease with increasing altitude. Furthermore, if the plane's speed is doubled, the drag force increases by a factor of 4. In order to maintain this increased speed, the propulsive force also increases by a factor of 4 and the power required (force times speed) must increase by a factor of 8.

Now let us analyze the motion of a mass in free fall subject to an upward air drag force given by $R = \tfrac{1}{2}C\rho Av^2$. Suppose a mass m is released from rest from the position $y = 0$ as in Figure 6.13. The mass experiences two external forces: the weight, mg, downward and the drag force, $\mathbf{R}$, upward. There is also an upward buoyant force which we will neglect. Hence, the magnitude of the net force is given by

$$F_{\text{net}} = mg - \tfrac{1}{2}C\rho Av^2 \tag{6.9}$$

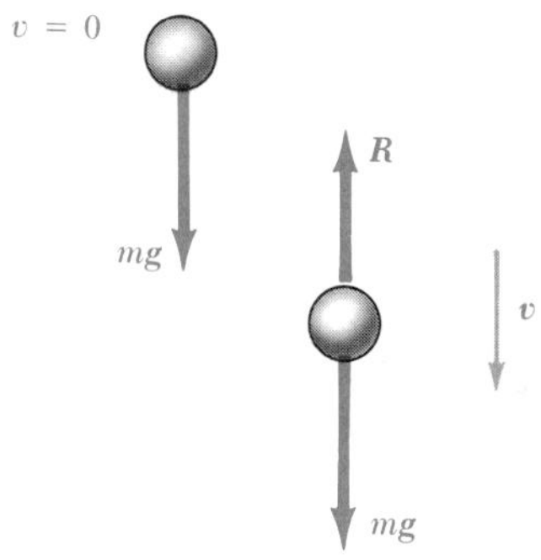

Figure 6.13 An object falling through air experiences a drag force, R, and the force of gravity, mg. The object reaches terminal velocity when the net force is zero, that is, when $R = mg$. Before this occurs, the acceleration varies with speed according to Equation 6.10.

Substituting $F_{\text{net}} = ma$ into Equation 6.9, we find that the mass has a downward acceleration of magnitude

$$a = g - \left(\frac{C\rho A}{2m}\right)v^2 \tag{6.10}$$

Again, we can calculate the terminal velocity, v_t, using the fact that when the weight is balanced by the drag force, the net force is zero and therefore the acceleration is zero. Setting $a = 0$ in Equation 6.10 gives

$$g - \left(\frac{C\rho A}{2m}\right)v_t^2 = 0$$

Terminal velocity

$$v_t = \sqrt{\frac{2mg}{C\rho A}} \tag{6.11}$$

Using this expression, we can determine how the terminal speed depends on the dimensions of the object. Suppose the object is a sphere of radius r. In this case, $A \propto r^2$ and $m \propto r^3$ (since the mass is proportional to the volume). Therefore, $v_t \propto \sqrt{r}$. That is, as r increases, the terminal speed increases with the square root of the radius.

Table 6.1 lists the terminal speeds for several objects falling through air.

TABLE 6.1 Terminal Speed for Various Objects Falling Through Air

Object	Mass (kg)	Area (m^2)	v_t(m/s)[a]
Skydiver	75	0.7	60
Baseball (radius 3.66 cm)	0.145	4.2×10^{-3}	33
Golf ball (radius 2.1 cm)	0.046	1.4×10^{-3}	32
Hailstone (radius 0.5 cm)	4.8×10^{-4}	7.9×10^{-5}	14
Raindrop (radius 0.2 cm)	3.4×10^{-5}	1.3×10^{-5}	9

[a] The drag coefficient, C, is assumed to be 0.5 in each case.

6.6 SUMMARY

There are only four fundamental forces in nature: the gravitational force, the electromagnetic force, the strong nuclear force, and the weak nuclear force.

Newton's second law applied to a particle moving in **uniform circular motion** states that the net force in the radial direction must equal the product of the mass and the centripetal acceleration:

Uniform circular motion

$$F_r = ma_r = \frac{mv^2}{r} \tag{6.1}$$

The force that provides the centripetal acceleration could be, for example, the force of gravity (as in satellite motion), the force of friction, or the force of tension (as in a string). A particle moving in nonuniform circular motion has both a centripetal (or radial) acceleration and a nonzero tangential component of acceleration. In the case of a particle rotating in a vertical circle, the tangential acceleration is provided by gravity.

Fictitious forces

An observer in a noninertial (accelerated) frame of reference must introduce **fictitious forces** when applying Newton's second law in that frame. If these fictitious forces are properly defined, the description of motion in the noninertial frame will be equivalent to that made by an observer in an inertial frame. However, the observers in the two different frames will not agree on the causes of the motion.

A body moving through a liquid or gas experiences a **resistive force** that is velocity dependent. This resistive force, which opposes the motion, generally increases with velocity. The force depends on the shape of the body and the properties of the medium through which the body is moving. In the limiting case for a falling body, when the resistive force equals the weight ($a = 0$), the body reaches its **terminal velocity.**

QUESTIONS

1. Because the earth rotates about its axis and about the sun, it is a noninertial frame of reference. Assuming the earth is a uniform sphere, why would the *apparent weight* of an object be greater at the poles than at the equator?
2. Explain why the earth is not spherical in shape and bulges at the equator.
3. How would you explain the force that pushes a rider toward the side of a car as the car rounds a corner?
4. When an airplane does an inside "loop-the-loop" in a vertical plane, at what point would the pilot appear to be heaviest? What is the constraint force acting on the pilot?

5. A skydiver in free fall reaches terminal velocity. After the parachute is opened, what parameters change to decrease this terminal velocity?
6. Why is it that an astronaut in a space capsule orbiting the earth experiences a feeling of weightlessness?
7. Why does mud fly off a rapidly turning wheel?
8. A pail of water can be whirled in a vertical path such that none is spilled. Why does the water stay in, even when the pail is above your head?
9. Imagine that you attach a heavy object to one end of a spring and then whirl the spring and object in a horizontal circle (by holding the free end of the spring). Does the spring stretch? If so, why? Discuss in terms of centripetal force.
10. It has been suggested that rotating cylinders about 10 mi in length and 5 mi in diameter be placed in space and used as colonies. The purpose of the rotation is to simulate gravity for the inhabitants. Explain this concept for producing an effective gravity.
11. Why does a pilot tend to black out when pulling out of a steep dive?
12. Cite an example of a situation in which an automobile driver can have a centripetal acceleration but no tangential acceleration.
13. Is it possible for a car to move in a circular path in such a way that it has a tangential acceleration but no centripetal acceleration?
14. Analyze the motion of a rock dropped into water in terms of its speed and acceleration as it falls. Assume that there is a resistive force acting on the rock that increases as the velocity increases.

PROBLEMS

Section 6.2 Newton's Second Law Applied to Uniform Circular Motion

1. In a cyclotron (one type of particle accelerator), a deuteron (of atomic mass 2 u) reaches a final velocity of 10% of the speed of light while moving in a circular path of radius 0.48 m. The deuteron is maintained in the circular path by a magnetic force. What magnitude of force is required?
2. What centripetal force is required to keep a 2-kg mass moving in a circle of radius 0.4 m at a speed of 3 m/s?
3. A coin is placed 20 cm from the center of a rotating, horizontal turntable. The coin is observed to slip when its speed is 50 cm/s. (a) What provides the centripetal force when the coin is stationary relative to the turntable? (b) What is the coefficient of static friction between the coin and the turntable?
4. A 3-kg mass attached to a light string rotates in circular motion on a horizontal, frictionless table. The radius of the circle is 0.8 m, and the string can support a mass of 25 kg before breaking. What range of speeds can the mass have before the string breaks?
5. A satellite of mass 300 kg is in a circular orbit about the earth at an altitude equal to the earth's mean diameter (see Example 6.5). Find (a) the satellite's orbital speed, (b) the period of its revolution, and (c) the gravitational force acting on it.
6. A highway curve has a radius of 150 m and is designed for a traffic speed of 40 mi/h (17.9 m/s). (a) If the curve is not banked, determine the minimum coefficient of friction between the car and the road. (b) At what angle should the curve be banked if friction is neglected (Figure 6.4)?
7. In the Bohr model of the hydrogen atom the velocity of the electron is approximately 2.2×10^6 m/s. Find (a) the centripetal force acting on the electron as it revolves in a circular orbit of radius 0.53×10^{-10} m, (b) the centripetal acceleration of the electron, and (c) the number of revolutions per second made by the electron.

Section 6.3 Nonuniform Circular Motion

8. A pail of water is rotated in a vertical circle of radius 1 m. What is the minimum speed of the pail at the top of the circle if no water is to spill out?
9. A roller-coaster vehicle has a mass of 500 kg when fully loaded with passengers (Fig. 6.14). (a) If the vehicle has a speed of 20 m/s at point A, what is the force of the track on the vehicle at this point? (b) What is the maximum speed the vehicle can have at B in order that it remain on the track?

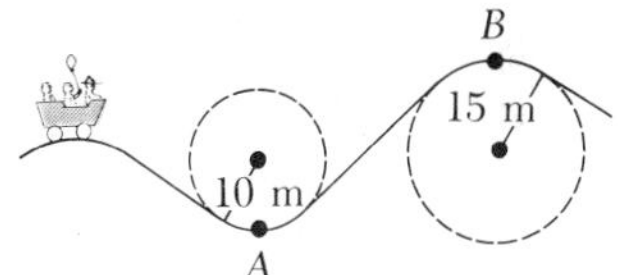

Figure 6.14 (Problem 9).

10. A ball attached to the end of a string 0.8 m in length is rotated in a vertical circle (Fig. 6.8). Determine the minimum speed of the ball at the top of its path if it maintains a circular path. (*Note:* Below this speed, the tension in the string is zero at the top.)
11. A 0.5-kg mass attached to the end of a string swings in a vertical circle of radius $R = 2$ m (Figure 6.8). When $\theta = 25°$, the speed of the mass is 8 m/s. At this instant, find (a) the tension in the string, (b) the tangential and radial components of acceleration, and (c) the magnitude of the total acceleration.
12. A 40-kg child sits in a conventional swing of length 3 m, supported by two chains. If the child's speed is 6 m/s at the lowest point, find (a) the tension in each chain at the lowest point and (b) the force of the seat on the child at the lowest point. (Neglect the mass of the seat.)

*Section 6.4 Motion in Accelerated Frames

13. A ball is suspended from the ceiling of a moving car by a string 25 cm in length. An observer in the car notes that the ball deflects 6 cm from the vertical toward the rear of the car. What is the acceleration of the car?
14. A 0.5-kg object is suspended from the ceiling of an accelerating boxcar as in Figure 6.10. If $a = 3$ m/s^2, find (a) the angle that the string makes with the vertical and (b) the tension in the string.
15. A 5-kg mass attached to a spring scale rests on a smooth, horizontal surface as in Figure 6.15. The spring scale, attached to the front end of a boxcar, reads 18 N when the car is in motion. (a) If the spring scale reads zero when the car is at rest, determine the acceleration of the car. (b) What will the spring scale read if the car moves with constant velocity? (c) Describe the forces on the mass as observed by someone in the car and by someone at rest outside the car.

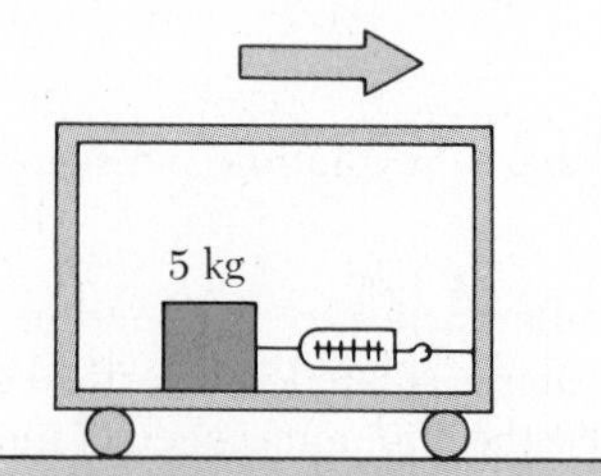

Figure 6.15 (Problem 15).

16. A block is attached to a string, which in turn is connected to a peg at the center of a rotating turntable, as in Figure 6.11. If the turntable is *rough*, describe the forces on the block as observed by (a) someone on the turntable and (b) an observer at rest relative to the turntable. (c) For a given velocity, does the tension in the string increase, decrease, or remain the same as the turntable is made smoother?
17. A mass m rests on spring scales situated on the floor of an elevator. The elevator moves down with a constant downward acceleration a. (a) In the noninertial frame of reference of the elevator, draw the free-body diagram and apply Newton's second law to find the apparent weight indicated by the scale. (b) Repeat part (a) in a fixed (inertial) frame of reference.

*Section 6.5 Motion in the Presence of Resistive Forces

18. A skydiver of mass 80 kg jumps from a slow-moving aircraft and reaches a terminal speed of 50 m/s. (a) What is the acceleration of the skydiver when her speed is 30 m/s? What is the drag force on the diver when her speed is (b) 50 m/s and (c) 30 m/s?
19. A small, spherical bead of mass 3 g is released from rest at $t = 0$ in a bottle of liquid shampoo. The terminal velocity, v_t, is observed to be 2 cm/s. Find (a) the value of the constant b in Equation 6.4, (b) the time, τ, it takes to reach $0.63v_t$, and (c) the value of the retarding force when the bead reaches terminal velocity.

GENERAL PROBLEMS

20. A railroad track has a curve of 400 m radius. The tracks are banked toward the inside at an angle of 6°. For trains of what speed was this track designed? (Assume that the correct speed requires only the normal force to keep the train on the track.)
21. A small turtle, appropriately named "Dizzy," is placed on a horizontal, rotating turntable at a distance of 20 cm from its center. Dizzy's mass is 50 g, and the coefficient of static friction between his feet and the turntable is 0.3. Find (a) the *maximum* number of revolutions per second the turntable can have if Dizzy is to remain stationary relative to the turntable and (b) Dizzy's speed and radial acceleration when he is on the verge of slipping.
22. The pilot of an airplane executes a constant-speed loop-the-loop maneuver in a vertical plane. The speed of the airplane is 300 mi/h, and the radius of the circle is 1200 ft. (a) What is the pilot's apparent weight at the lowest point if his true weight is 160 lb? (b) What is his apparent weight at the highest point? (c) Describe how the pilot could experience weightlessness if both the radius and velocity can be varied. (*Note:* His apparent weight is equal to the force of the seat on his body.)
23. A 4-kg mass is attached to a *horizontal* rod by two strings, as in Figure 6.16. The strings are under tension when the rod rotates about its axis. If the speed of the mass is 4 m/s when observed at the following positions, find the tension in the string when the mass is (a) at its lowest point, (b) in the horizontal position, and (c) at its highest point.

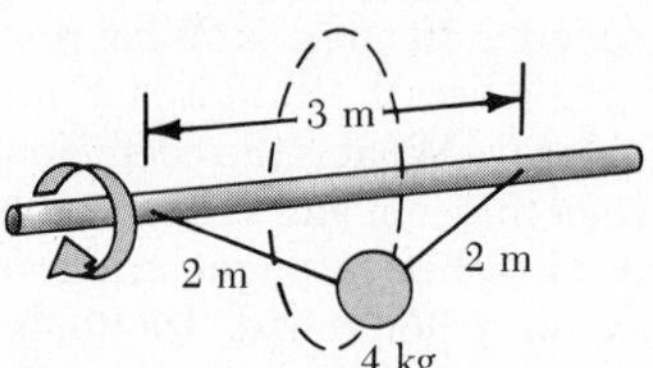

Figure 6.16 (Problems 23 and 24).

24. Suppose the rod in the system shown in Figure 6.16 is made *vertical* and rotates about this axis. If the mass rotates at a constant speed of 6 m/s in a horizontal plane, determine the tensions in the upper and lower strings.
25. A car rounds a banked curve as in Figure 6.5. The radius of curvature of the road is R, the banking angle is θ, and the coefficient of static friction is μ. (a) Determine the *range* of speeds the car can have without slipping up or down the road. (b) Find the minimum value for μ such that the minimum speed is zero. (c) What is the range of speeds possible if $R = 100$ m, $\theta = 10°$, and $\mu = 0.1$ (slippery conditions)?
26. Because of the earth's rotation about its axis, a point on the equator experiences a centripetal acceleration of 0.034 m/s^2, while a point at the poles experiences no

centripetal acceleration. (a) Show that at the equator the gravitational force on an object (the true weight) must *exceed* the object's apparent weight. (b) What is the apparent weight at the equator and at the poles of a person having a mass of 75 kg? (Assume the earth is a uniform sphere and take $g = 9.800$ m/s^2.)

27. An amusement park ride consists of a large vertical cylinder that spins about its axis fast enough that any person inside is held up against the wall when the floor drops away (Figure 6.17). The coefficient of static friction between the person and the wall is μ_s, and the radius of the cylinder is R. (a) Show that the *maximum* period of revolution necessary to keep the person from falling is $T = (4\pi^2 R\mu_s/g)^{1/2}$. (b) Obtain a numerical value for T if $R = 4$ m and $\mu_s = 0.4$. How many revolutions per minute does the cylinder make?

Figure 6.17 (Problem 27).

•28. A bead is threaded on a frictionless vertical wire hoop of radius R. The hoop rotates about a vertical axis through its center, as shown in Figure 6.18. The period of revolution of the hoop is T. The bead has mass m. Answer the following in terms of the given parameters and the acceleration due to gravity g. (a) Find the normal force N exerted on the bead by the wire.

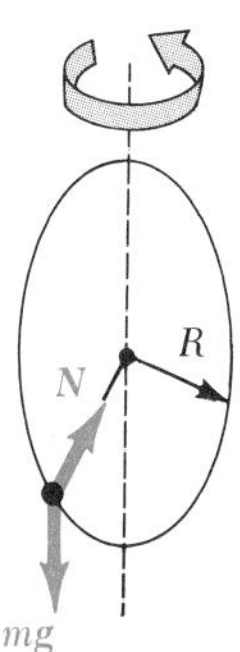

Figure 6.18 (Problem 28).

(b) Find the angle between the normal force and the axis of rotation. (c) Find the centripetal force exerted on the bead.

•29. The following experiment is performed in a spacecraft that is at rest where the net gravitational field is zero. A small sphere is injected into a viscous medium with initial velocity v_0. The sphere experiences a resistive force $\mathbf{R} = -b\mathbf{v}$. Find the velocity of the sphere as a function of time. (*Hint:* Apply Newton's second law, write $\mathbf{a}$ as dr/dt, separate the variables, and integrate the equation.)

CALCULATOR/COMPUTER PROBLEMS

•30. A hailstone of mass 4.8×10^{-4} kg and radius 0.5 cm falls through the atmosphere, and experiences a net force given by Equation 6.9. This expression can be written in the form

$$m\frac{dv}{dt} = mg - Kv^2$$

where $K = \frac{1}{2}C\rho A$, $\rho = 1.29$ kg/m^3 and $C = 0.5$. (a) What is the terminal velocity of the hailstone? (b) Use a method of numerical integration to find the velocity and position of the hailstone at 1-s intervals, taking $v_0 = 0$. Continue your calculation until terminal velocity is reached.

•31. A 0.5-kg block slides down a 30° incline of length 1 m. The coefficient of kinetic friction between the block varies with the block's velocity according to the expression

$$\mu = 0.3 + 1.2\sqrt{v}$$

where v is in m/s. (a) Use a numerical method to find the velocity of the block at intervals of 10 cm during its motion. (b) If the length of the plane is extended to several km, will the block reach terminal velocity? If so, what is its terminal velocity, and at what point does it occur on the incline?

7
Work and Energy

The concept of energy is perhaps one of the most important physical concepts in both contemporary science and engineering practice. In everyday usage, we think of energy in terms of the cost of fuel for transportation and heating, electricity for lights and appliances, and the foods we consume. However, these ideas do not really define energy. They only tell us that fuels are needed to do a job and that those fuels provide us with something we call energy.

Energy is present in various forms, including mechanical energy, electromagnetic energy, chemical energy, thermal (or heat) energy, and nuclear energy. The various forms of energy are related to each other through the fact that when energy is transformed from one form to another, the total amount of energy remains the same. This is the point that makes the energy concept so useful. That is, if an isolated system loses energy in some form, then the law of conservation of energy says that the system will gain an equal amount of energy in other forms. For example, when an electric motor is connected to a battery, chemical energy is converted to electrical energy, which in turn is converted to mechanical energy. The transformation of energy from one form to another is an essential part of the study of physics, chemistry, biology, geology, and astronomy.

7.1 INTRODUCTION

In this chapter, we shall be concerned only with the mechanical form of energy. We shall see that the concepts of work and energy can be applied to the dynamics of a mechanical system without resorting to Newton's laws. However, it is important to note that the work-energy concepts are based upon Newton's laws and therefore do not involve any new physical principles.

Although the approach we shall use provides the same results as Newton's laws in describing the motion of a mechanical system, the general ideas of the work-energy concept can be applied to a wide range of phenomena in the fields of electromagnetism and atomic and nuclear physics. In addition, in a complex situation the "energy approach" can often provide a much simpler analysis than the direct application of Newton's second law.

This alternative method of describing motion is especially useful when the force acting on a particle is not constant. In this case, the acceleration is not constant, and we cannot apply the simple kinematic equations we developed in Chapter 3. Often, a particle in nature is subject to a force that varies with the position of the particle. Such forces include gravitational forces and the force exerted on a body attached to a spring. We shall describe techniques for treating such systems with the help of an extremely important development called the *work-energy theorem*, which is the central topic of this chapter.

We begin by defining work, a concept that provides a link between the concepts of force and energy. In Chapter 8, we shall discuss the law of conservation of energy and apply it to various problems.

7.2 WORK DONE BY A CONSTANT FORCE

Consider an object that undergoes a displacement $\mathbf{s}$ along a straight line under the action of a constant force $\mathbf{F}$, which makes an angle θ with $\mathbf{s}$, as in Figure 7.1.

> The work done by the constant force is defined as the product of the component of the force in the direction of the displacement and the magnitude of the displacement.

Since the component of $\mathbf{F}$ in the direction of $\mathbf{s}$ is $F \cos \theta$, the work W done by $\mathbf{F}$ is given by

$$W \equiv (F \cos \theta)s \tag{7.1}$$

Work done by a constant force

According to this definition, work is done by $\mathbf{F}$ on an object under the following conditions: (1) the object must undergo a displacement and (2) $\mathbf{F}$ must have a nonzero component in the direction of $\mathbf{s}$. From the first condition, we see that a force does no work on an object if the object does not move ($s = 0$). For example, if a person pushes against a brick wall, a force is exerted on the wall but the person does no work since the wall is fixed. However, the person's muscles are contracting (undergoing displacement) in the process so that *internal* energy is being used up.

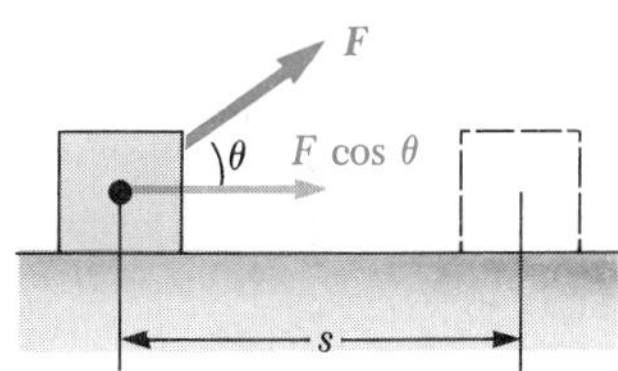

Figure 7.1 If an object undergoes a displacement $\mathbf{s}$, the work done by the force $\mathbf{F}$ is $(F \cos \theta)s$.

From the second condition, note that the work done by a force is also zero when the force is perpendicular to the displacement, since $\theta = 90°$ and $\cos 90° = 0$. For example, in Figure 7.2, both the work done by the normal force and the work done by the force of gravity are zero since both forces are perpendicular to the displacement and have zero components in the direction of $\mathbf{s}$. Likewise, if you hold a weight at arm's length for some period of time, no work is done on the weight (assuming no wiggling or oscillations of the arms). Even though you must exert an upward force to support the weight, the work done by the force is zero since the displacement is zero. After holding the weight for a long period of time, your arms would tire and you would claim that the effort required a considerable amount of "work." Thus, we see that the meaning of work in physics is distinctly different from its meaning in day-to-day affairs.

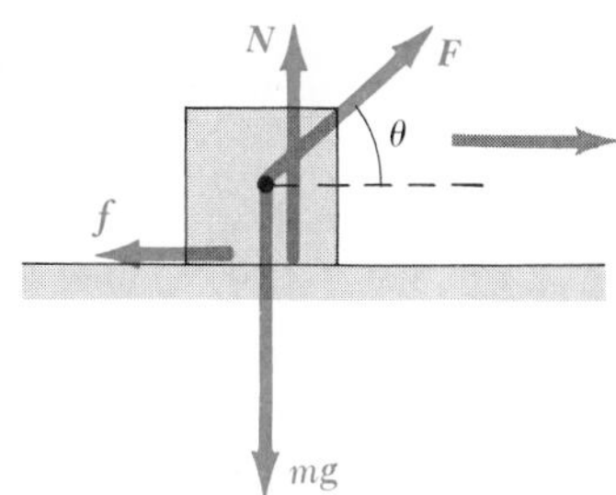

Figure 7.2 When an object is displaced horizontally on a rough surface, the normal force, $\mathbf{N}$, and the weight, $m\mathbf{g}$, do *no* work. The work done by $\mathbf{F}$ is $(F \cos \theta)s$, and the work done by the frictional force is $-fs$.

The sign of the work also depends on the direction of $\mathbf{F}$ relative to $\mathbf{s}$. The work done by the applied force is positive when the vector associated with the component $F \cos \theta$ is in the *same direction* as the displacement. In this situation, the work done by the gravitational force is negative. For example, when an object is lifted, the work done by the applied force is positive since the lifting force is upward, that is, in the same direction as the displacement. In this situation, the work done by the gravitational force is negative. When the vector associated with the component $F \cos \theta$ is in the direction *opposite* the displacement, *W is negative.* A common example in which W is negative is the work done by a frictional force when a body slides over a rough surface. If the force of sliding friction is f, and the body undergoes a linear displacement s, the work done by the frictional force is

$$W_f = -fs \tag{7.2}$$

Work done by a sliding frictional force

TABLE 7.1 Units of Work in the Three Common Systems of Measurement

System	Unit of Work	Name of Combined Unit
SI	newton · meter (N · m)	joule (J)
cgs	dyne · centimeter (dyne · cm)	erg
British engineering (conventional)	pound · foot (lb · ft)	foot · pound (ft · lb)

Does the weight lifter do any work as he holds the weight over his head? Does he do any work as he raises the weight?

where the negative sign comes from the fact that $\theta = 180°$ and $\cos 180° = -1$.

Finally, if an applied force $\boldsymbol{F}$ acts along the direction of the displacement, then $\theta = 0$, and $\cos 0 = 1$. In this case, Equation 7.1 gives

$$W = Fs \tag{7.3}$$

Work is a scalar quantity, and its units are force multiplied by length. Therefore, the SI unit of work is the **newton · meter** (N · m). Another name for the newton-meter is the joule (J). The units of work in the cgs and British engineering systems are **dyne · cm**, which is also called the **erg**, and lb · ft, respectively. These are summarized in Table 7.1. Note that 1 J = 10^7 ergs.

Since work is a scalar quantity we can combine the work done by each of the separate forces to get the total work done. For instance, if there are three forces contributing to the work done, there would be three terms in the sum, each corresponding to the work done by a given force. The following example illustrates this point.

EXAMPLE 7.1. Dragging a Box
A box is dragged across a rough floor by a constant force of magnitude 50 N. The force makes an angle of 37° with the horizontal. A frictional force of 10 N retards the motion, and the box is displaced a distance of 3 m to the right. (a) Calculate the work done by the 50-N force.

Using the definition of work (Equation 7.1) and given that $F = 50$ N, $\theta = 37°$, and $s = 3$ m,

$$W_F = (F \cos \theta)s = (50 \text{ N})(\cos 37°)(3 \text{ m}) = 120 \text{ N} \cdot \text{m} = 120 \text{ J}$$

Note that the vertical component of $\boldsymbol{F}$ does no work.

(b) Calculate the work done by the frictional force.

$$W_f = -fs = (-10 \text{ N})(3 \text{ m}) = -30 \text{ N} \cdot \text{m} = -30 \text{ J}$$

(c) Determine the net work done on the box by all forces acting on it.

Since the normal force, $\boldsymbol{N}$, and the force of gravity, $m\boldsymbol{g}$, are both perpendicular to the displacement, they do no work. Therefore, the net work done on the box is the sum of (a) and (b):

$$W_{net} = W_F + W_f = 120 \text{ J} - 30 \text{ J} = 90 \text{ J}$$

Later we shall show that the net work done on the body equals the change in kinetic energy, which establishes the physical significance of W_{net}.

Exercise 1 Find the net work done on the box if it is pulled a distance of 3 m with a horizontal force of 50 N, assuming the frictional force is 15 N.
Answer: 105 J.

7.3 THE SCALAR PRODUCT OF TWO VECTORS

We have defined work as a *scalar* quantity given by the product of the magnitude of the displacement and the component of the force in the direction of the displacement. It is convenient to express Equation 7.1 in terms of a **scalar product** of the two vectors $\boldsymbol{F}$ and $\boldsymbol{s}$. We write this scalar product $\boldsymbol{F} \cdot \boldsymbol{s}$. Because of the dot symbol used, the scalar product is often called the *dot product.* Thus, we can express Equation 7.1 as a scalar product:

$$W = \mathbf{F} \cdot \mathbf{s} = F s \cos \theta \qquad (7.4)$$

Work expressed as a dot product

In other words, $\mathbf{F} \cdot \mathbf{s}$ (read F dot s) is a shorthand notation for $F s \cos \theta$.

In general, the scalar product of any two vectors **A** and **B** is defined as a scalar quantity equal to the product of the magnitudes of the two vectors and the cosine of the angle θ that is included between the directions of **A** and **B**.

That is, the scalar product (or dot product) of **A** and **B** is defined by the relation

$$\mathbf{A} \cdot \mathbf{B} \equiv AB \cos \theta \qquad (7.5)$$

Scalar product of any two vectors **A** and **B**

where θ is the angle between **A** and **B**, as in Figure 7.3, A is the magnitude of **A**, and B is the magnitude of **B**. Note that **A** and **B** need not have the same units.

Note in Figure 7.3 that $B \cos \theta$ is the projection of **B** onto **A**. Therefore, the definition of $\mathbf{A} \cdot \mathbf{B}$ as given by Equation 7.5 can be considered as the product of the magnitude of **A** and the projection of **B** onto **A**.[1] From Equation 7.5 we also note that the scalar product is *commutative*. That is,

$$\mathbf{A} \cdot \mathbf{B} = \mathbf{B} \cdot \mathbf{A} \qquad (7.6)$$

The order of the dot product can be reversed

Finally, the scalar product obeys the *distributive law of multiplication*, so that

$$\mathbf{A} \cdot (\mathbf{B} + \mathbf{C}) = \mathbf{A} \cdot \mathbf{B} + \mathbf{A} \cdot \mathbf{C} \qquad (7.7)$$

The dot product is simple to evaluate from Equation 7.5 when **A** is either perpendicular or parallel to **B**. If **A** is perpendicular to **B** ($\theta = 90°$), then $\mathbf{A} \cdot \mathbf{B} = 0$. Also, $\mathbf{A} \cdot \mathbf{B} = 0$ in the more trivial case when either **A** or **B** is zero. If **A** and **B** point in the same direction ($\theta = 0°$), then $\mathbf{A} \cdot \mathbf{B} = AB$. If **A** and **B** point in opposite directions ($\theta = 180°$), then $\mathbf{A} \cdot \mathbf{B} = -AB$. Note that the scalar product is negative when $90° < \theta < 180°$.

The unit vectors $\mathbf{i}$, $\mathbf{j}$, and $\mathbf{k}$, which were defined in Chapter 2, lie in the positive x, y, and z directions, respectively, of a right-handed coordinate system. Therefore, it follows from the definition of $\mathbf{A} \cdot \mathbf{B}$ that the scalar products of these unit vectors are given by

$$\mathbf{i} \cdot \mathbf{i} = \mathbf{j} \cdot \mathbf{j} = \mathbf{k} \cdot \mathbf{k} = 1 \qquad (7.8a)$$

$$\mathbf{i} \cdot \mathbf{j} = \mathbf{i} \cdot \mathbf{k} = \mathbf{j} \cdot \mathbf{k} = 0 \qquad (7.8b)$$

Dot products of unit vectors

[1] This is equivalent to stating that $\mathbf{A} \cdot \mathbf{B}$ equals the product of the magnitude of **B** and the projection of **A** onto **B**.

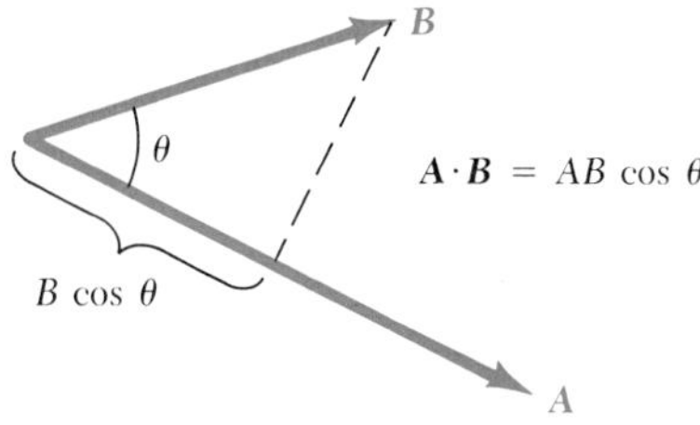

Figure 7.3 The scalar product $\mathbf{A} \cdot \mathbf{B}$ equals the magnitude of **A** multiplied by the projection of **B** onto **A**.

Two vectors $\boldsymbol{A}$ and $\boldsymbol{B}$ can be expressed in component form as

$$\boldsymbol{A} = A_x\boldsymbol{i} + A_y\boldsymbol{j} + A_z\boldsymbol{k}$$

$$\boldsymbol{B} = B_x\boldsymbol{i} + B_y\boldsymbol{j} + B_z\boldsymbol{k}$$

Therefore Equations 7.8a and 7.8b reduces the scalar product of $\boldsymbol{A}$ and $\boldsymbol{B}$ to

$$\boldsymbol{A} \cdot \boldsymbol{B} = A_xB_x + A_yB_y + A_zB_z \tag{7.9}$$

In the special case where $\boldsymbol{A} = \boldsymbol{B}$, we see that

$$\boldsymbol{A} \cdot \boldsymbol{A} = A_x^2 + A_y^2 + A_z^2 = A^2$$

EXAMPLE 7.2. The Scalar Product
The vectors $\boldsymbol{A}$ and $\boldsymbol{B}$ are given by $\boldsymbol{A} = 2\boldsymbol{i} + 3\boldsymbol{j}$ and $\boldsymbol{B} = -\boldsymbol{i} + 2\boldsymbol{j}$. (a) Determine the scalar product $\boldsymbol{A} \cdot \boldsymbol{B}$.

$$\begin{aligned}\boldsymbol{A} \cdot \boldsymbol{B} &= (2\boldsymbol{i} + 3\boldsymbol{j}) \cdot (-\boldsymbol{i} + 2\boldsymbol{j}) \\ &= -2\boldsymbol{i} \cdot \boldsymbol{i} + 2\boldsymbol{i} \cdot 2\boldsymbol{j} - 3\boldsymbol{j} \cdot \boldsymbol{i} + 3\boldsymbol{j} \cdot 2\boldsymbol{j} \\ &= -2 + 6 = 4\end{aligned}$$

where we have used the fact that $\boldsymbol{i} \cdot \boldsymbol{j} = \boldsymbol{j} \cdot \boldsymbol{i} = 0$. The same result is obtained using Equation 7.9 directly, where $A_x = 2$, $A_y = 3$, $B_x = -1$, and $B_y = 2$.

(b) Find the angle θ between $\boldsymbol{A}$ and $\boldsymbol{B}$.

The magnitudes of $\boldsymbol{A}$ and $\boldsymbol{B}$ are given by

$$A = \sqrt{A_x^2 + A_y^2} = \sqrt{(2)^2 + (3)^2} = \sqrt{13}$$

$$B = \sqrt{B_x^2 + B_y^2} = \sqrt{(-1)^2 + (2)^2} = \sqrt{5}$$

Using Equation 7.5 and the result from (a) gives

$$\cos\theta = \frac{\boldsymbol{A} \cdot \boldsymbol{B}}{AB} = \frac{4}{\sqrt{13}\sqrt{5}} = \frac{4}{\sqrt{65}}$$

$$\theta = \cos^{-1}\frac{4}{8.06} = 60.3°$$

EXAMPLE 7.3. Work Done by a Constant Force
A particle moving in the xy plane undergoes a displacement $\boldsymbol{s} = (2\boldsymbol{i} + 3\boldsymbol{j})$ m while a constant force given by $\boldsymbol{F} = (5\boldsymbol{i} + 2\boldsymbol{j})$ N acts on the particle. (a) Calculate the magnitude of the displacement and the force.

$$s = \sqrt{x^2 + y^2} = \sqrt{(2)^2 + (3)^2} = \sqrt{13} \text{ m}$$

$$F = \sqrt{F_x^2 + F_y^2} = \sqrt{(5)^2 + (2)^2} = \sqrt{29} \text{ N}$$

(b) Calculate the work done by the force $\boldsymbol{F}$.

Substituting the expressions for $\boldsymbol{F}$ and $\boldsymbol{s}$ into Equation 7.4 and using Equation 7.8, we get

$$\begin{aligned}W = \boldsymbol{F} \cdot \boldsymbol{s} &= (5\boldsymbol{i} + 2\boldsymbol{j}) \cdot (2\boldsymbol{i} + 3\boldsymbol{j}) \text{ N} \cdot \text{m} \\ &= 5\boldsymbol{i} \cdot 2\boldsymbol{i} + 2\boldsymbol{j} \cdot 3\boldsymbol{j} = 16 \text{ N} \cdot \text{m} = 16 \text{ J}\end{aligned}$$

Exercise 2 Calculate the angle between $\boldsymbol{F}$ and $\boldsymbol{s}$.
Answer: 34.5°.

7.4 WORK DONE BY A VARYING FORCE: THE ONE-DIMENSIONAL CASE

Consider an object being displaced along the x axis under the action of a varying force, as in Figure 7.4. The object is displaced along the x axis from $x = x_i$ to $x = x_f$. In such a situation, we cannot use $W = (F \cos\theta)s$ to calculate the work done by the force, since this relationship applies only when $\boldsymbol{F}$ is constant in magnitude and direction. However, if we imagine that the object undergoes a very small displacement Δx, described in Figure 7.4a, then the x component of the force, F_x, is approximately constant over this interval and we can express the work done by the force for this small displacement as

$$\Delta W = F_x\,\Delta x \tag{7.10}$$

Note that this is just the area of the shaded rectangle in Figure 7.4a. Now, if we imagine that the F_x versus x curve is divided into a large number of such intervals, as in Figure 7.4a, then the total work done for the displacement from x_i to x_f is approximately equal to the sum of a large number of such terms:

$$W \cong \sum_{x_i}^{x_f} F_x\,\Delta x$$

If the displacements are allowed to approach zero, then the number of terms in the sum increases without limit, but the value of the sum approaches a definite value equal to the *true area* under the curve bounded by F_x and the x axis. As you probably have learned in the calculus, this limit of the sum is called an **integral** and is represented by

$$\lim_{\Delta x \to 0} \sum_{x_i}^{x_f} F_x \, \Delta x = \int_{x_i}^{x_f} F_x \, dx$$

The limits on the integral, $x = x_i$ to $x = x_f$, define what is called a **definite integral.** (An *indefinite integral* represents the limit of a sum over a yet-to-be-specified interval. Appendix B.7 gives a brief description of integration.) This definite integral is numerically equal to the area under the F_x versus x curve between x_i and x_f. Therefore, we can express the work done by F_x for the displacement of the object from x_i to x_f as

$$W = \int_{x_i}^{x_f} F_x \, dx \tag{7.11}$$

Note that this equation reduces to Equation 7.1 when $F_x = F \cos \theta$ is constant.

If more than one force acts on the object, the total work done is just the work done by the resultant force. If we express the resultant force in the x direction as ΣF_x (a vector sum), then the *net work* done as the object moves from x_i to x_f is

$$W_{\text{net}} = \int_{x_i}^{x_f} \left(\sum F_x \right) dx \tag{7.12}$$

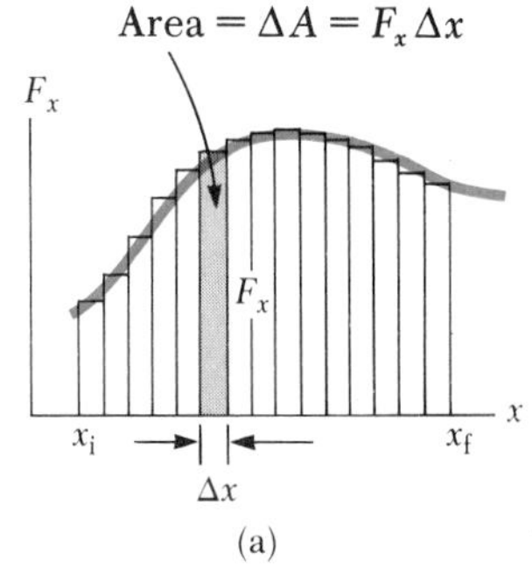

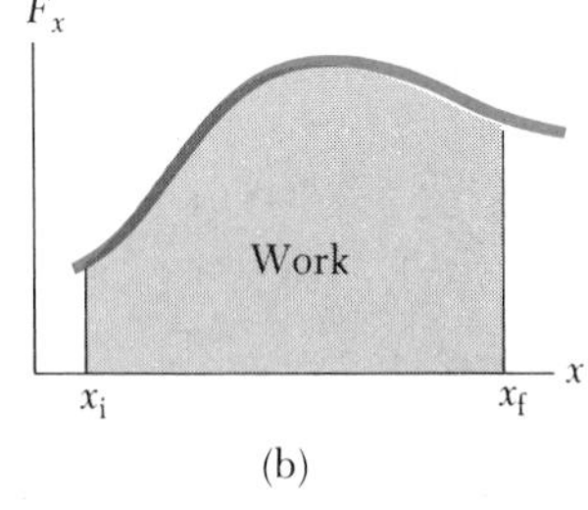

Figure 7.4 (a) The work done by the force F_x for the small displacement Δx is $F_x \, \Delta x$, which equals the area of the shaded rectangle. The total work done for the displacement x_i to x_f is approximately equal to the sum of the areas of all the rectangles. (b) The work done by the variable force F_x as the particle moves from x_i to x_f is *exactly* equal to the area under this curve.

EXAMPLE 7.4

A force acting on an object varies with x as shown in Figure 7.5. Calculate the work done by the force as the object moves from $x = 0$ to $x = 6$ m.

Solution: The work done by the force is equal to the total area under the curve from $x = 0$ to $x = 6$ m. This area is equal to the area of the rectangular section from $x = 0$ to $x = 4$ m plus the area of the triangular section from $x = 4$ m to $x = 6$ m. The area of the rectangle is (4)(5) N · m = 20 J, and the area of the triangle is equal to $\frac{1}{2}$(2)(5) N · m = 5 J. Therefore, the total work done is 25 J.

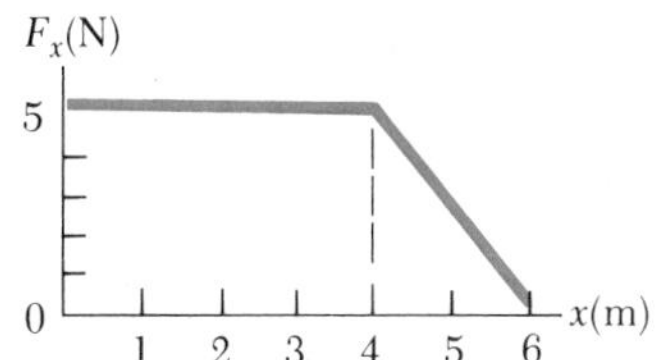

Figure 7.5 (Example 7.4) The force acting on a particle is constant for the first 4 m of motion and then decreases linearly with x from $x = 4$ m to $x = 6$ m. The net work done by this force is the area under this curve.

Work Done by a Spring

A common physical system for which the force varies with position is shown in Figure 7.6. A body on a horizontal, smooth surface is connected to a helical spring. If the spring is stretched or compressed a small distance from its unstretched, or equilibrium, configuration, the spring will exert a force on the body given by

Spring force

$$F_s = -kx \tag{7.13}$$

$F_{spring} = -kx$

where x is the displacement of the body from its unstretched ($x = 0$) position and k is a positive constant called the *force constant* of the spring. As we learned in Chapter 5, this force law for springs is known as **Hooke's law.** Note that Hooke's law is only valid in the limiting case of small displacements. The value of k is a measure of the stiffness of the spring. Stiff springs have large k values, and soft springs have small k values.

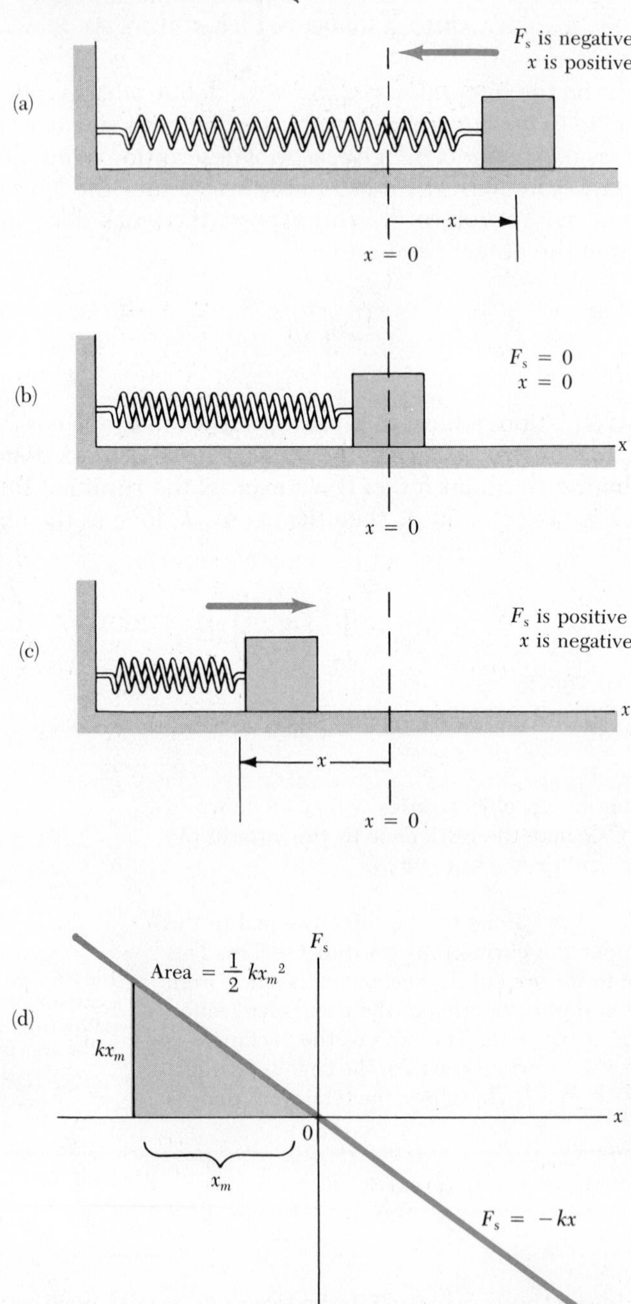

Figure 7.6 The force of a spring on a block varies with the block's displacement from the equilibrium position $x = 0$. (a) When x is positive (stretched spring), the spring force is to the left. (b) When x is zero, the spring force is zero (natural length of the spring). (c) When x is negative (compressed spring), the spring force is to the right. (d) Graph of F_s versus x for systems described above. The work done by the spring force as the block moves from $-x_m$ to 0 is the area of the shaded triangle, $\frac{1}{2}kx_m^2$.

The negative sign in Equation 7.13 signifies that the force exerted by the spring is always directed *opposite* the displacement. For example, when $x > 0$ as in Figure 7.6a, the spring force is to the left, or negative. When $x < 0$ as in Figure 7.6c, the spring force is to the right, or positive. Of course, when $x = 0$ as in Figure 7.6b, the spring is unstretched and $F_s = 0$. Since the spring force always acts toward the equilibrium position, it is sometimes called a *restoring force*. Once the mass is displaced some distance x_m from equilibrium and then released, it will move from $-x_m$ through zero to $+x_m$. The details of the ensuing oscillating motion will be given in Chapter 13.

Suppose that the block is pushed to the left a distance x_m from equilibrium, as in Figure 7.6c, and then released. Let us calculate the *work done by the spring force* as the body moves from $x_i = -x_m$ to $x_f = 0$. Applying Equation 7.11, we get

$$W_s = \int_{x_i}^{x_f} F_s\, dx = \int_{-x_m}^{0} (-kx)\, dx = \tfrac{1}{2}k{x_m}^2 \qquad (7.14a)$$

That is, the work done by the spring force is positive since the spring force is in the same direction as the displacement (both are to the right). However, if we consider the work done by the spring force as the body moves from $x_i = 0$ to $x_f = x_m$, we find that $W_s = -\frac{1}{2}k{x_m}^2$, since for this part of the motion, the displacement is to the right and the spring force is to the left. Therefore, the *net* work done by the spring force as the body moves from $x_i = -x_m$ to $x_f = x_m$ is *zero*.

If we plot F_s versus x as in Figure 7.6d, we arrive at the same results. Note that the work calculated in Equation 7.14a is equivalent to the area of the shaded triangle in Figure 7.6d, with base x_m and height kx_m. The area of this triangle is $\frac{1}{2}k{x_m}^2$, which does equal the work done by the spring, Equation 7.14a.

If the mass undergoes an *arbitrary* displacement from $x = x_i$ to $x = x_f$, the work done by the spring force is given by

Work done by a spring

$$W_s = \int_{x_i}^{x_f} (-kx)\, dx = \tfrac{1}{2}k{x_i}^2 - \tfrac{1}{2}k{x_f}^2 \qquad (7.14b)$$

From this equation, we see that the work done is zero for any motion that ends where it began ($x_i = x_f$). We shall make use of this important result in describing the motion of this system in more detail in the next chapter.

Now let us consider the work done by an *external agent* in *very slowly* stretching a spring from $x_i = 0$ to $x_f = x_m$, as in Figure 7.7. This work can be easily calculated by noting that the *applied force*, $\mathbf{F}_{app}$, is equal to and opposite the spring force, $\mathbf{F}_s$, at any value of the displacement, so that $F_{app} = -(-kx) = kx$. Therefore, the work done by this applied force (the external agent) is given by

$$W_{F_{app}} = \int_0^{x_m} F_{app}\, dx = \int_0^{x_m} kx\, dx = \tfrac{1}{2}k{x_m}^2$$

You should note that this work is equal to the negative of the work done by the spring force for this displacement.

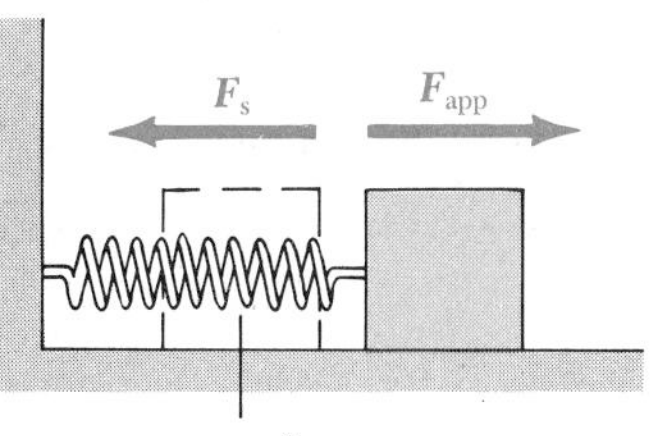

Figure 7.7 A block being pulled to the right on a frictionless surface by a force $\mathbf{F}_{app}$ from $x = 0$ to $x = x_m$. If the process is carried out very slowly, the applied force is equal to and opposite the spring force at all times.

EXAMPLE 7.5. The Spring Force Does Work A block lying on a smooth, horizontal surface is connected to a spring with a force constant of 80 N/m. The spring is compressed a distance of 3.0 cm from equilibrium as in Figure 7.6c. Calculate the work done by the spring force as the block moves from $x_i = -3.0$ cm to its unstretched position, $x_f = 0$.

Solution: Using Equation 7.14a with $x_m = -3.0 \text{ cm} = -3 \times 10^{-2}$ m, we get

$$W_s = \tfrac{1}{2}kx_m^{\,2} = \tfrac{1}{2}\left(80\ \frac{\text{N}}{\text{m}}\right)(-3 \times 10^{-2}\ \text{m})^2$$
$$= 3.6 \times 10^{-2}\ \text{J}$$

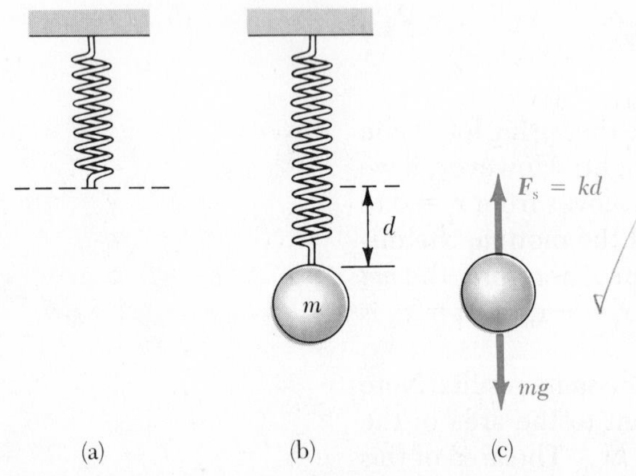

Figure 7.8 (Example 7.6) Determination of the force constant of a helical spring. The elongation d of the spring is due to the weight mg. Since the spring force upward balances the weight, it follows that $k = mg/d$.

EXAMPLE 7.6. Measuring k for a Spring A common technique used to measure the force constant of a spring is described in Figure 7.8. The spring is hung vertically as shown in Figure 7.8a. A body of mass m is then attached to the lower end of the spring as in Figure 7.8b. The spring stretches a distance d from its equilibrium position under the action of the "load" mg. Since the spring force is upward, it must balance the weight mg downward when the system is at rest. In this case, we can apply Hooke's law to give $|F_s| = kd = mg$, or

$$k = mg/d$$

For example, if a spring is stretched a distance of 2.0 cm by a mass of 0.55 kg, the force constant of the spring is

$$k = \frac{mg}{d} = \frac{(0.55\ \text{kg})(9.80\ \text{m/s}^2)}{2.0 \times 10^{-2}\ \text{m}} = 2.7 \times 10^2\ \text{N/m}$$

EXAMPLE 7.7. Work Done in Moving a Car A sports car on a horizontal surface is pushed by a horizontal force that varies with position according to the graph shown in Figure 7.9. Determine an approximate value for the total work done in moving the car from $x = 0$ to $x = 20$ m.

Solution: We can obtain the result from the graph by dividing the total displacement into many small displacements. For simplicity, we choose to divide the total displacement into ten consecutive displacements, each 2 m in length, as shown in Figure 7.9. The work done during each small displacement is *approximately* equal to the area of the dotted rectangle. For example, the work done for the first displacement, from $x = 0$ to $x = 2$ m, is the

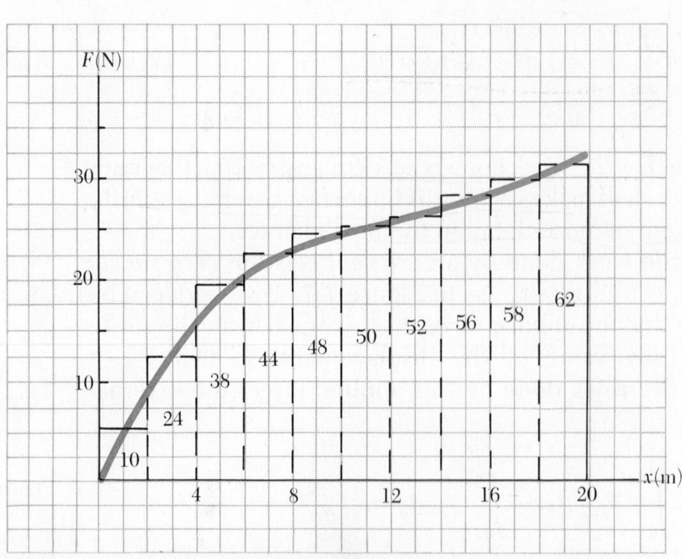

Figure 7.9 (Example 7.7) A graph of force versus position for a car moving along the x axis. The numbers within the rectangles represent the work done (area of rectangle) during that interval.

area of the smallest rectangle, (2 m)(5 N) = 10 J; the work done for the second displacement, from $x = 2$ m to $x = 4$ m, is the area of the second rectangle, (2 m)(12 N) = 24 J. Continuing in this fashion, we get the areas indicated in Figure 7.9, the *sum* of which gives the total work done from $x = 0$ to $x = 20$ m. This result is

$$W_{\text{total}} \approx 442 \text{ J}$$

The accuracy of the result will of course improve as the widths of the intervals are made smaller.

7.5 WORK AND KINETIC ENERGY

In Chapter 5 we found that a particle accelerates when the resultant force on it is not zero. Consider a situation in which a constant force F_x acts on a particle of mass m moving in the x direction. Newton's second law states that $F_x = ma$, where a_x is constant since F_x is constant. If the particle is displaced from $x_i = 0$ to $x_f = s$ the work done by the force F_x is

$$W = F_x s = (ma_x)s \tag{7.15}$$

However, in Chapter 3 we found that the following relationships are valid when a particle undergoes constant acceleration:

$$s = \tfrac{1}{2}(v_i + v_f)t \qquad a_x = \frac{v_f - v_i}{t}$$

where v_i is the velocity at $t = 0$ and v_f is the velocity at time t. Substituting these expressions into Equation 7.15 gives

$$W = m\left(\frac{v_f - v_i}{t}\right)\tfrac{1}{2}(v_i + v_f)t$$

$$W = \tfrac{1}{2}mv_f^2 - \tfrac{1}{2}mv_i^2 \tag{7.16}$$

The product of one half the mass and the square of the speed is defined as the **kinetic energy** of the particle.

That is, the kinetic energy, K, of a particle of mass m and speed v is defined as

Kinetic energy is energy associated with the motion of a body

$$K \equiv \tfrac{1}{2}mv^2 \tag{7.17}$$

This expression is valid in the nonrelativistic limit, that is, when $v \ll c$.

Since the magnitude of the linear momentum of the particle is given by $p = mv$, the kinetic energy is sometimes written

$$K = \tfrac{1}{2}mv^2 = \frac{(mv)^2}{2m} = \frac{p^2}{2m} \tag{7.18}$$

Kinetic energy is a scalar quantity and has the same units as work. For example, a 1-kg mass moving with a speed of 4.0 m/s has a kinetic energy of 8.0 J. We can think of kinetic energy as energy associated with the motion of a body. It is often convenient to write Equation 7.16

$$W = K_f - K_i = \Delta K \tag{7.19}$$

Work-energy theorem

That is,

the work done by the constant force F in displacing a particle equals the change in kinetic energy of the particle.

Work done on a particle equals the change in its kinetic energy

The change here means the final minus the initial value of the kinetic energy.

Equation 7.19 is an important result known as the **work-energy theorem.** This theorem was derived for the case where the force is constant, but we can show that it is valid even when the force is varying: If the resultant force acting on a body in the x direction is ΣF_x, then Newton's second law states that $\Sigma F_x = ma$. Thus, we can use Equation 7.12 and express the net work done as

$$W_{\text{net}} = \int_{x_i}^{x_f} \left(\sum F_x\right) dx = \int_{x_i}^{x_f} ma\, dx$$

Because the resultant force varies with x, the acceleration and velocity also depend on x. We can now use the following chain rule to evaluate W_{net}:

$$a = \frac{dv}{dt} = \frac{dv}{dx}\frac{dx}{dt} = v\frac{dv}{dx}$$

Substituting this into the expression for W gives

$$W_{\text{net}} = \int_{x_i}^{x_f} mv\frac{dv}{dx}\, dx = \int_{v_i}^{v_f} mv\, dv = \tfrac{1}{2}mv_f^2 - \tfrac{1}{2}mv_i^2$$

Note that the limits of the integration were changed because the variable was changed from x to v.

The work-energy theorem given by Equation 7.19 is also valid in the more general case when the force varies in direction and magnitude while the particle moves along an arbitrary curved path in three dimensions. In this situation, we express the work as

General expression for work done by a force F

$$W = \int_{i}^{f} \mathbf{F} \cdot d\mathbf{s} \tag{7.20}$$

where the limits i and f represent the initial and final coordinates of the particle. The integral given by Equation 7.20 is called a *line integral.* Because the infinitesimal displacement vector can by expressed as $d\mathbf{s} = dx\mathbf{i} + dy\mathbf{j} + dz\mathbf{k}$ and because $\mathbf{F} = F_x\mathbf{i} + F_y\mathbf{j} + F_z\mathbf{k}$, Equation 7.20 reduces to

$$W = \int_{x_i}^{x_f} F_x\, dx + \int_{y_i}^{y_f} F_y\, dy + \int_{z_i}^{z_f} F_z\, dz \tag{7.21}$$

This is the general expression that is used to calculate the work done by a force when a particle undergoes a displacement from the point with coordinates (x_i, y_i, z_i) to the point with coordinates (x_f, y_f, z_f).[2]

Work can be positive, negative, or zero

Thus, we conclude that *the net work done on a particle by the resultant force acting on it is equal to the change in the kinetic energy of the particle.* The work-energy theorem also says that the speed of the particle will increase ($K_f > K_i$) if the net work done on it is positive, whereas its speed will decrease ($K_f < K_i$) if the net work done on it is negative. That is, the speed and kinetic energy of a particle will change only if work is done on the particle by some external force. Because of this connection between work and change in kinetic energy, we can also think of the kinetic energy of a body as the work the body can do in coming to rest.

[2] In the general expression, Equation 7.21, the component F_x can depend on y and z as well as on x; similarly for F_y and F_z.

EXAMPLE 7.8. A Block Pulled on a Smooth Surface A 6-kg block initially at rest is pulled to the right along a horizontal smooth surface by a constant, horizontal force of 12 N, as in Figure 7.10a. Find the speed of the block after it moves a distance of 3 m.

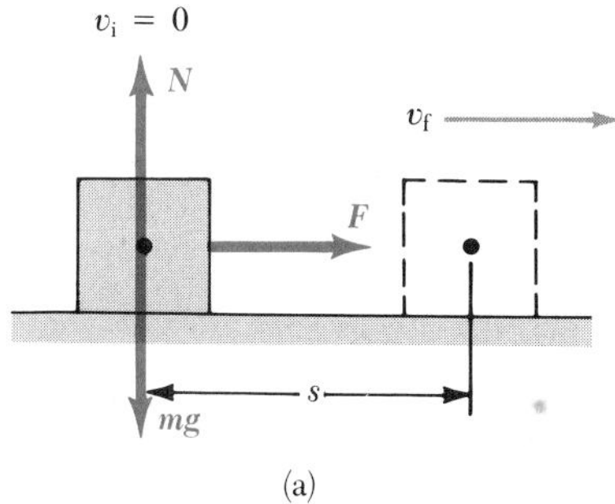

(a)

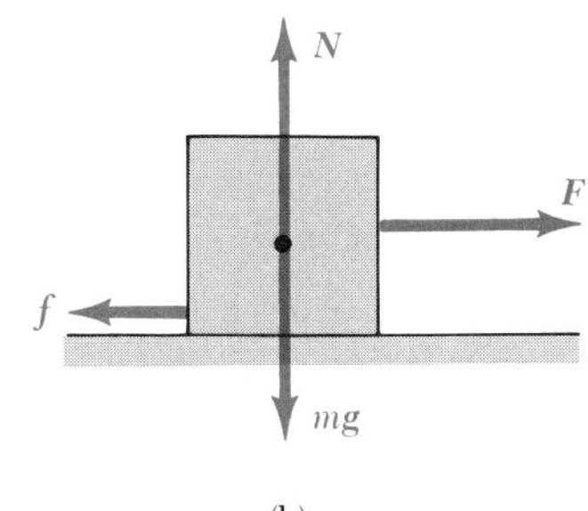

(b)

Figure 7.10 (a) Example 7.8. (b) Example 7.9.

Solution: The weight is balanced by the normal force, and neither of these forces does work since the displacement is horizontal. Since there is no friction, the resultant external force is the 12-N force. The work done by this force is

$$W_F = Fs = (12 \text{ N})(3 \text{ m}) = 36 \text{ N} \cdot \text{m} = 36 \text{ J}$$

Using the work-energy theorem and noting that the initial kinetic energy is zero, we get

$$W_F = K_f - K_i = \tfrac{1}{2}mv_f^2 - 0$$

$$v_f^2 = \frac{2W_F}{m} = \frac{2(36 \text{ J})}{6 \text{ kg}} = 12 \text{ m}^2/\text{s}^2$$

$$v_f = 3.46 \text{ m/s}$$

Exercise 3 Find the acceleration of the block, and determine the final speed of the block using the kinematic equation $v_f^2 = v_i^2 + 2as$.
Answer: $a = 2$ m/s^2; $v_f = 3.46$ m/s.

EXAMPLE 7.9. A Block Pulled on a Rough Surface Find the final speed of the block described in Example 7.8 if the surface is rough and the coefficient of kinetic friction is 0.15.

Solution: In this case, we must calculate the net work done on the block, which equals the sum of the work done by the applied 12-N force and the frictional force f, as in Figure 7.10b. Since the frictional force opposes the displacement, the work this force does is *negative*. The magnitude of the frictional force is given by $f = \mu N = \mu mg$; therefore, the work done by this force is this force multiplied by the displacement (see Eq. 7.2) or

$$W_f = -fs = -\mu mgs = (-0.15)(6)(9.80)(3) = -26.5 \text{ J}$$

Therefore, the net work done on the block is

$$W_{\text{net}} = W_F + W_f = 36.0 \text{ J} - 26.5 \text{ J} = 9.50 \text{ J}$$

Applying the work-energy theorem with $v_i = 0$ gives

$$W_{\text{net}} = \tfrac{1}{2}mv_f^2$$

$$v_f^2 = \frac{2W_{\text{net}}}{m} = \frac{19}{6} \text{ m}^2/\text{s}^2$$

$$v_f = 1.78 \text{ m/s}$$

Exercise 4 Find the acceleration of the block from Newton's second law, and determine the final speed of the block using kinematics.
Answer: $a = 0.530$ m/s^2; $v_f = 1.78$ m/s.

EXAMPLE 7.10. A Mass-Spring System A block of mass 1.6 kg is attached to a spring with a force constant of 10^3 N/m, as in Figure 7.6. The spring is compressed a distance of 2.0 cm and the block is released from rest. (a) Calculate the velocity of the block as it passes through the equilibrium position $x = 0$, if the surface is frictionless.

Following Example 7.5, the work done by the spring with $x_m = -2.0 \text{ cm} = -2 \times 10^{-2}$ m is

$$W_s = \tfrac{1}{2}kx_m^2 = \tfrac{1}{2}\left(10^3 \frac{\text{N}}{\text{m}}\right)(-2 \times 10^{-2} \text{ m})^2 = 0.20 \text{ J}$$

Using the work-energy theorem with $v_i = 0$ gives

$$W_s = \tfrac{1}{2}mv_f^2 - \tfrac{1}{2}mv_i^2$$

$$0.20 \text{ J} = \tfrac{1}{2}(1.6 \text{ kg})v_f^2 - 0$$

$$v_f^2 = \frac{0.4 \text{ J}}{1.6 \text{ kg}} = 0.25 \text{ m}^2/\text{s}^2$$

$$v_f = 0.50 \text{ m/s}$$

(b) Calculate the velocity of the block as it passes through the equilibrium position if a constant frictional force of 4.0 N retards its motion.

The work done by the frictional force for a displacement of 2×10^{-2} m is given by

$$W_f = -fs = -(4 \text{ N})(2 \times 10^{-2} \text{ m}) = -0.08 \text{ J}$$

The net work done on the block is the work done by the

spring plus the work done by friction. In part (a), we found $W_s = 0.20$ J, therefore

$$W_{net} = W_s + W_f = 0.20 \text{ J} - 0.08 \text{ J} = 0.12 \text{ J}$$

Applying the work-energy theorem gives

$$\tfrac{1}{2}mv_f^2 = W_{net}$$

$$\tfrac{1}{2}(1.6 \text{ kg})v_f^2 = 0.12 \text{ J}$$

$$v_f^2 = \frac{0.24 \text{ J}}{1.6 \text{ kg}} = 0.15 \text{ m}^2/\text{s}^2$$

$$v_f = 0.39 \text{ m/s}$$

Note that this value for v_f is less than that obtained in the frictionless case. Is this result sensible?

EXAMPLE 7.11. Block Pushed Along an Incline
A block of mass m is pushed up a rough incline by a constant force $\mathbf{F}$ acting parallel to the incline, as in Figure 7.11a. The block is displaced a distance d up the incline.
(a) Calculate the work done by the force of gravity for this displacement.

The force of gravity is downward but has a component *down* the plane. This is given by $-mg \sin \theta$ if the positive x direction is chosen to be up the plane (Figure 7.11b). Therefore, the work done by gravity for the displacement d is

$$W_g = (-mg \sin \theta)d = -mgh$$

where $h = d \sin \theta$ is the *vertical* displacement. That is, the work done by gravity has a magnitude equal to the force of gravity multiplied by the *upward vertical displacement*. In the next chapter, we shall show that this result is valid in general for any particle displaced between two points. Furthermore, the result is independent of the path taken between these points.

(b) Calculate the work done by the applied force $\mathbf{F}$.

Since $\mathbf{F}$ is in the same direction as the displacement, we get

$$W_F = \mathbf{F} \cdot \mathbf{s} = Fd$$

(c) Find the work done by the force of kinetic friction if the coefficient of friction is μ.

The magnitude of the force of friction is $f = \mu N = \mu mg \cos \theta$. Since the direction of this force is *opposite* the direction of the displacement, we find that

$$W_f = -fd = -\mu mgd \cos \theta$$

(d) Find the net work done on the block for this displacement.

Using the results to (a), (b), and (c), we get

$$W_{net} = W_g + W_F + W_f$$
$$= -mgd \sin \theta + Fd - \mu mgd \cos \theta$$

or

$$W_{net} = Fd - mgd(\sin \theta + \mu \cos \theta)$$

For example, if we take $F = 15$ N, $d = 1.0$ m, $\theta = 25°$, $m = 1.5$ kg, and $\mu = 0.30$, we find that

$$W_g = -(mg \sin \theta)d$$
$$= -(1.5 \text{ kg})\left(9.8 \frac{\text{m}}{\text{s}^2}\right)(\sin 25°)(1.0 \text{ m})$$
$$= -6.2 \text{ J}$$

$$W_F = Fd = (15 \text{ N})(1 \text{ m}) = 15 \text{ J}$$

$$W_f = -\mu mgd \cos \theta$$
$$= -(0.30)(1.5 \text{ kg})\left(9.8 \frac{\text{m}}{\text{s}^2}\right)(1.0 \text{ m})(\cos 25°)$$
$$= -4.0 \text{ J}$$

$$W_{net} = W_g + W_F + W_f = 4.8 \text{ J}$$

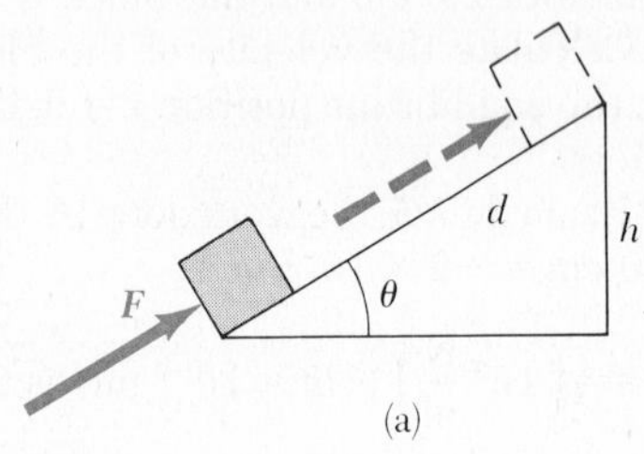

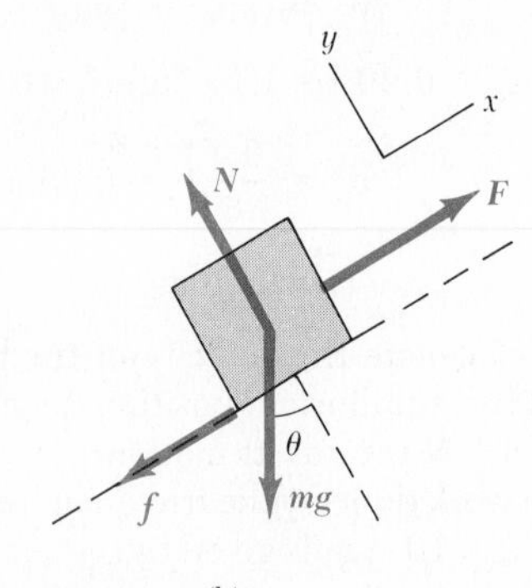

Figure 7.11 (Example 7.11) The block is pushed up the rough incline by a constant force $\mathbf{F}$.

EXAMPLE 7.12. Minimum Stopping Distance
An automobile traveling at 48 km/h can be stopped in a minimum distance of 40 m by applying the brakes. If the same automobile is traveling at 96 km/h, what is the minimum stopping distance?

Solution: We shall assume that when the brakes are applied, the car does not skid. To get the minimum stopping distance, d, we take the frictional force f between the tires and road to be a *maximum*. The work done by this frictional force, $-fd$, must equal the change in kinetic energy of the automobile. Since the kinetic energy has a final value of zero and an initial value of $\frac{1}{2}mv^2$, we get

$$W_f = K_f - K_i$$
$$-fd = 0 - \tfrac{1}{2}mv^2$$

$$d = \frac{mv^2}{2f}$$

If we assume f is the same for the two initial speeds, we can take m and f as constants. Therefore, the ratio of stopping distances is given by

$$\frac{d_2}{d_1} = \left(\frac{v_2}{v_1}\right)^2$$

Taking $v_1 = 48$ km/h, $v_2 = 96$ km/h, and $d_1 = 40$ m gives

$$\frac{d_2}{d_1} = \left(\frac{96}{48}\right)^2 = 4$$

$$d_2 = 4d_1 = 4(40 \text{ m}) = 160 \text{ m}$$

This shows that the *minimum stopping distance varies as the square of the ratio of speeds.* If the speed is doubled, as it is in this example, the distance increases by a factor of 4.

7.6 POWER

From a practical viewpoint, it is interesting to know not only the work done on an object, but also the rate at which the work is being done. **Power** is defined as *the time rate of doing work.*

If an external force is applied to an object, and if the work done by this force is ΔW in the time interval Δt, then the **average power** during this interval is defined as the ratio of the work done to the time interval:

$$\bar{P} \equiv \frac{\Delta W}{\Delta t} \qquad (7.22)$$

Average power

The **instantaneous power,** P, is the limiting value of the average power as Δt approaches zero:

$$P \equiv \lim_{\Delta t \to 0} \frac{\Delta W}{\Delta t} = \frac{dW}{dt} \qquad (7.23)$$

From Equation 7.4, we can express the work done by a force $\mathbf{F}$ for a displacement $d\mathbf{s}$, since $dW = \mathbf{F} \cdot d\mathbf{s}$. Therefore, the instantaneous power can be written

$$P = \frac{dW}{dt} = \mathbf{F} \cdot \frac{d\mathbf{s}}{dt} = \mathbf{F} \cdot \mathbf{v} \qquad (7.24)$$

Instantaneous power

where we have used the fact that $\mathbf{v} = d\mathbf{s}/dt$.

The unit of power in the SI system is J/s, which is also called a *watt,* W (after James Watt):

$$1 \text{ W} = 1 \text{ J/s} = 1 \text{ kg} \cdot \text{m}^2/\text{s}^3$$

The watt

The symbol W for watt should not be confused with the symbol for work.

The unit of power in the British engineering system is the horsepower (hp), where

$$1 \text{ hp} = 550 \text{ ft} \cdot \text{lb/s} = 746 \text{ W}$$

A new unit of energy (or work) can now be defined in terms of the unit of power. One kilowatt-hour (kWh) is the energy converted or consumed in 1 h at the constant rate of 1 kW. The numerical value of 1 kWh is

$$1 \text{ kWh} = (10^3 \text{ W})(3600 \text{ s}) = 3.6 \times 10^6 \text{ J}$$

It is important to note that a kWh is a unit of energy, not power. When you pay your electric bill, you are buying energy, and the amount of electricity used is usually in multiples of kWh. For example, an electric bulb rated at 100 W would "consume" 3.6×10^5 J of energy in 1 h.

Although the W and the kWh are commonly used only in electrical applications, they can be used in other scientific areas. For example, an automobile engine can be rated in kW as well as in hp. Likewise, the power consumption of an electrical appliance can be expressed in hp.

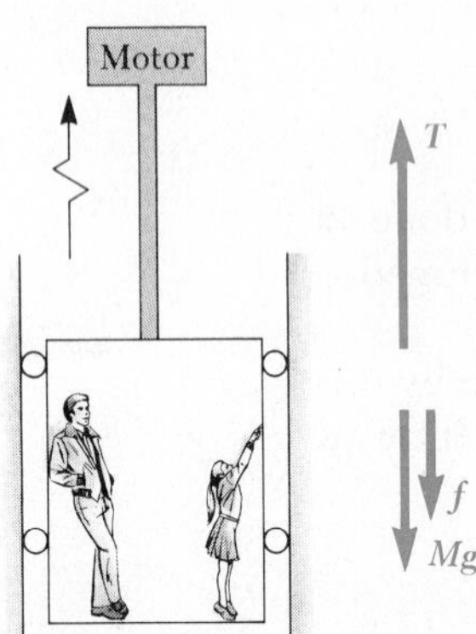

Figure 7.12 (Example 7.13) The motor provides a force T upward on the elevator. A frictional force f and the total weight Mg act downward.

EXAMPLE 7.13. Power Delivered by an Elevator Motor

An elevator has a mass of 1000 kg and carries a maximum load of 800 kg. A constant frictional force of 4000 N retards its motion upward, as in Figure 7.12. (a) What must be the minimum horsepower delivered by the motor to lift the elevator at a constant speed of 3 m/s?

The motor must supply the force T that pulls the elevator upward. From Newton's second law and from the fact that $a = 0$ since v is constant, we get

$$T - f - Mg = 0$$

where M is the *total* mass (elevator plus load), equal to 1800 kg. Therefore,

$$\begin{aligned} T &= f + Mg \\ &= 4 \times 10^3 \text{ N} + (1.8 \times 10^3 \text{ kg})(9.80 \text{ m/s}^2) \\ &= 2.16 \times 10^4 \text{ N} \end{aligned}$$

Using Equation 7.24 and the fact that T is in the same direction as v gives

$$\begin{aligned} P &= \mathbf{T} \cdot \mathbf{v} = Tv \\ &= (2.16 \times 10^4 \text{ N})(3 \text{ m/s}) = 6.48 \times 10^4 \text{ W} \\ &= 64.8 \text{ kW} = 86.9 \text{ hp} \end{aligned}$$

(b) What power must the motor deliver at any instant if it is designed to provide an upward acceleration of 1.0 m/s²?

Applying Newton's second law to the elevator gives

$$T - f - Mg = Ma$$

$$\begin{aligned} T &= M(a + g) + f \\ &= (1.8 \times 10^3 \text{ kg})(1.0 + 9.8) \text{ m/s}^2 + 4 \times 10^3 \text{ N} \\ &= 2.34 \times 10^4 \text{ N} \end{aligned}$$

Therefore, using Equation 7.24 we get for the required power

$$P = Tv = (2.34 \times 10^4\, v) \text{ W}$$

where v is the instantaneous speed of the elevator in m/s. Hence, the power required increases with increasing speed.

*7.7 ENERGY AND THE AUTOMOBILE

Automobiles powered by gasoline engines are known to be very inefficient machines. Even under ideal conditions, less than 15% of the available energy in the fuel is used to power the vehicle. The situation is much worse under stop-and-go driving in the city. The purpose of this section is to use the concepts of energy, power, and forces of friction to analyze some factors that affect automobile fuel consumption.

There are many mechanisms that contribute to the energy losses in a typical automobile.[3] About two thirds of the energy available from the fuel is lost in the engine. Part of this energy ends up in the atmosphere via the exhaust system, and part is used in the engine's cooling system. (The efficiency of

[3] An excellent article on this subject is the one by G. Waring in *The Physics Teacher*, Vol. 18 (1980), p. 494. The data in Tables 7.2 and 7.3 were taken from this article.

TABLE 7.2 Power Losses in a Typical Automobile Assuming a Total Available Power of 136 kW

Mechanism	Power Loss (kW)	Power Loss (%)
Exhaust (heat)	46	33
Cooling system	45	33
Drive train	13	10
Internal friction	8	6
Accessories	5	4
Propulsion of vehicle	19	14

engines will be discussed in Chapter 22.) About 10% of the available energy is lost in the automobile's drive-train mechanism; this loss includes friction in the transmission, drive shaft, wheel and axle bearings, and differential. Friction in other moving parts such as in the motor accounts for about 6% of the energy loss, and 4% of the available energy is used to operate fuel and oil pumps and such accessories as power steering, air conditioning, power brakes, and electrical components. Finally, about 14% of the available energy is used to propel the automobile. This energy is used mainly to overcome road friction and air resistance.

Table 7.2 lists the power losses for an automobile with an available fuel power of 136 kW. These data apply to a typical 1450-kg "gas-guzzler" with a gas consumption rate of 6.4 km/liter (15 mi/gal).

Let us examine the power requirements to overcome road friction and air drag in more detail. The coefficient of rolling friction, μ, between the tires and the road is about 0.016. For a 1450-kg car, the weight is 14 200 N and the force of rolling friction $\mu N = \mu W = 227$ N. As the speed of the car increases, there is a small reduction in the normal force as a result of a reduction in air pressure as air flows over the top of the car. This causes a slight reduction in the force of rolling friction, f_r, with increasing speed, as shown in Table 7.3.

Now let us consider the effect of air friction, that is, the drag force that results from air moving past the various surfaces of the car. The drag force associated with air friction for large objects is proportional to the square of the speed (in m/s) (Section 6.4) and may be written

$$f_a = \tfrac{1}{2} CA\rho v^2 \qquad (7.25)$$

where C is the drag coefficient, A is the cross-sectional area of the moving object, and ρ is the density of air. This expression can be used to calculate the values in Table 7.3 using $C = 0.5$, $\rho = 1.293$ kg/m^3, and $A \approx 2$ m^2.

The magnitude of the total frictional force, f_t, is given by the sum of the rolling friction force and the air drag force:

$$f_t = f_r + f_a \approx \text{constant} + \tfrac{1}{2} CA\rho v^2 \qquad (7.26)$$

TABLE 7.3 Frictional Forces and Power Requirements for a Typical Car

v (km/h)	N (N)	f_r (N)	f_a (N)	f_t (N)	$P = f_t v$ (kW)
0	14 200	227	0	227	0
32	14 100	226	51	277	2.5
64	13 900	222	204	426	7.6
96.5	13 600	218	465	683	18.3
129	13 200	211	830	1041	37.3
161	12 600	202	1293	1495	66.8

In this table, N is the normal force, f_r is road friction, f_a is air friction, f_t is total friction, and P is the power delivered to the wheels.

At low speeds, road resistance and air drag are comparable, but at high speeds air drag is the predominant resistive force, as shown in Table 7.3. Road friction can be reduced by reducing tire flexing (increase the air pressure slightly above recommended values) and using radial tires. Air drag can be reduced by using a smaller cross-sectional area and streamlining the car. Though driving a car with the windows open does create more air drag, resulting in a 3% decrease in mpg, driving with the windows closed and the air conditioner running results in a 12% decrease in mileage.

The total power needed to maintain a constant speed v equals the product $f_t v$. This must equal the power delivered to the wheels. For example, from Table 7.3 we see that at $v = 96.5$ km/h $= 26.8$ m/s, the required power is

$$P = f_t v = (683 \text{ N})\left(26.8 \frac{\text{m}}{\text{s}}\right) = 18.3 \text{ kW}$$

This can be broken into two parts: (1) the power needed to overcome road friction, $f_r v$, and (2) the power needed to overcome air drag, $f_a v$. At $v = 26.8$ m/s, these have the values

$$P_r = f_r v = (218 \text{ N})\left(26.8 \frac{\text{m}}{\text{s}}\right) = 5.8 \text{ kW}$$

$$P_a = f_a v = (465 \text{ N})\left(26.8 \frac{\text{m}}{\text{s}}\right) = 12.5 \text{ kW}$$

Note that $P = P_r + P_a$.

On the other hand, at $v = 161$ km/h $= 44.7$ m/s, we find that $P_r = 9.0$ kW, $P_a = 57.8$ kW, and $P = 66.8$ kW. This shows the importance of air drag at high speeds.

EXAMPLE 7.14. Gas Consumed by Compact Car
A compact car has a mass of 800 kg, and its efficiency is rated at 14%. (That is, 14% of the available fuel energy is delivered to the wheels.) Find the amount of gasoline used to accelerate the car from rest to 60 mi/h (27 m/s). Use the fact that the energy equivalent of one gallon of gasoline is 1.3×10^8 J.

Solution: The energy required to accelerate the car from rest to a speed v is its kinetic energy, $\frac{1}{2}mv^2$. For this case,

$$E = \tfrac{1}{2}mv^2 = \tfrac{1}{2}(800 \text{ kg})\left(27 \frac{\text{m}}{\text{s}}\right)^2 = 2.9 \times 10^5 \text{ J}$$

If the engine were 100% efficient, each gallon of gasoline would supply an energy 1.3×10^8 J. Since the engine is only 14% efficient, each gallon delivers only $(0.14)(1.3 \times 10^8 \text{ J}) = 1.8 \times 10^7$ J. Hence, the number of gallons used to accelerate the car is

$$\text{Number of gal} = \frac{2.9 \times 10^5 \text{ J}}{1.8 \times 10^7 \text{ J/gal}} = 0.016 \text{ gal}$$

At this rate, a gallon of gas would be used after 62 such accelerations. This demonstrates the severe energy requirements for extreme stop-and-start driving.

EXAMPLE 7.15. Power Delivered to Wheels
Suppose the car described in Example 7.14 has a mileage rating of 35 mi/gal when traveling at 60 mi/h. How much power is delivered to the wheels?

Solution: From the given data, we see that the car consumes $60/35 = 1.7$ gal/h. Using the fact that each gallon is equivalent to 1.3×10^8 J, we find that the total power used is

$$P = \frac{(1.7 \text{ gal/h})(1.3 \times 10^8 \text{ J/gal})}{3.6 \times 10^3 \text{ s/h}}$$

$$= \frac{2.2 \times 10^8 \text{ J}}{3.6 \times 10^3 \text{ s}} = 62 \text{ kW}$$

Since 14% of the available power is used to propel the car, we see that the power delivered to the wheels is $(0.14)(62 \text{ kW}) = 8.7$ kW. This is about one half the value obtained for the large 1450-kg car discussed in the text. Size is clearly an important factor in power-loss mechanisms.

EXAMPLE 7.16. Car Accelerating Up a Hill
Consider a car of mass m accelerating up a hill, as in Figure 7.13. Assume that the magnitude of the drag force is given by

$$|f| = (218 + 0.70v^2)\ \text{N}$$

where v is the speed in m/s. Calculate the power that the engine must deliver to the wheels.

Solution: The forces on the car are shown in Figure 7.13, where $\boldsymbol{F}$ is the force that propels the car and the remaining forces have their usual meaning. Newton's second law applied to the motion along the road surface gives

$$\sum F_x = F - |f| - mg \sin \theta = ma$$
$$F = ma + mg \sin \theta + |f|$$
$$= ma + mg \sin \theta + (218 + 0.70v^2)$$

Therefore, the power required for propulsion is

$$P = Fv = mva + mvg \sin \theta + 218v + 0.70v^3$$

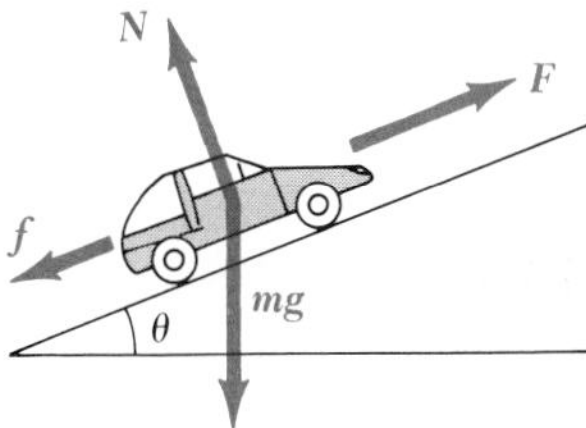

Figure 7.13 (Example 7.16).

In this expression, the term mva represents the power the engine must deliver to accelerate the car. If the car moves at constant speed, this term is zero and the power requirement is reduced. The term $mvg \sin \theta$ is the power required to overcome the force of gravity as the car moves up the incline. This term would be zero for motion on a horizontal surface. The term $218v$ is the power required to overcome road friction. Finally, the term $0.70v^3$ is the power needed to overcome air drag.

If we take $m = 1450$ kg, $v = 27$ m/s (= 60 mi/h), $a = 1\ \text{m/s}^2$, and $\theta = 10°$, the various terms in P are calculated to be

$$mva = (1450\ \text{kg})(27\ \text{m/s})(1\ \text{m/s}^2)$$
$$= 39\ \text{kW} = 52\ \text{hp}$$

$$mvg \sin \theta = (1450\ \text{kg})(27\ \text{m/s})(9.8\ \text{m/s}^2)(\sin 10°)$$
$$= 67\ \text{kW} = 89\ \text{hp}$$

$$218v = 218(27) = 5.9\ \text{kW} = 7.9\ \text{hp}$$

$$0.70v^3 = 0.70(27)^3 = 14\ \text{kW} = 18\ \text{hp}$$

Hence, the total power required is 126 kW, or 167 hp. Note that the power requirements for traveling at *constant* speed on a horizontal surface are only 19.9 kW, or 25.9 hp (the sum of the last two terms). Furthermore, if the mass is halved (as in compact cars), the power required is also reduced by almost the same factor.

7.8 SUMMARY

The **work** done by a *constant* force $\boldsymbol{F}$ acting on a particle is defined as the product of the component of the force in the direction of the particle's displacement and the magnitude of the displacement. If the force makes an angle θ with the displacement $\boldsymbol{s}$, the work done by $\boldsymbol{F}$ is

$$W \equiv Fs \cos \theta \qquad (7.1)$$

Work done by a constant force

The **scalar,** or dot, **product** of any two vectors $\boldsymbol{A}$ and $\boldsymbol{B}$ is defined by the relationship

$$\boldsymbol{A} \cdot \boldsymbol{B} \equiv AB \cos \theta \qquad (7.5)$$

Scalar product

where the result is a scalar quantity and θ is the included angle between the directions of the two vectors. The scalar product obeys the commutative and distributive laws.

The *work* done by a *varying* force acting on a particle moving along the x axis from x_i to x_f is given by

$$W \equiv \int_{x_i}^{x_f} F_x\, dx \qquad (7.11)$$

Work done by a varying force

where F_x is the component of force in the x direction. If there are several forces acting on the particle, the net work done by all forces is the sum of the individual work done by each force.

Kinetic energy

The **kinetic energy** of a particle of mass m moving with a speed v (where v is small compared with the speed of light) is defined as

$$K \equiv \tfrac{1}{2}mv^2 \qquad (7.17)$$

Work-energy theorem

The **work-energy theorem** states that the net work done on a particle by external forces equals the change in kinetic energy of the particle:

$$W = K_f - K_i = \tfrac{1}{2}mv_f^2 - \tfrac{1}{2}mv_i^2 \qquad (7.19)$$

Instantaneous power

The **instantaneous power** is defined as the time rate of doing work. If an agent applies a force $\mathbf{F}$ to an object moving with a velocity $\mathbf{v}$, the power delivered by that agent is given by

$$P \equiv \frac{dW}{dt} = \mathbf{F} \cdot \mathbf{v} \qquad (7.24)$$

QUESTIONS

1. When a particle rotates in a circle, a *centripetal force* acts on it directed toward the center of rotation. Why is it that this force does no work on the particle?
2. Explain why the work done by the force of sliding friction is negative when an object undergoes a displacement on a rough surface.
3. Is there any direction associated with the dot product of two vectors?
4. If the dot product of two vectors is positive, does this imply that the vectors must have positive rectangular components?
5. As the load on a spring hung vertically is increased, one would not expect the F_s versus x curve to always remain linear as in Figure 7.6d. Explain qualitatively what you would expect for this curve as m is increased.
6. Can the kinetic energy of an object have a negative value?
7. If the speed of a particle is doubled, what happens to its kinetic energy?
8. What can be said about the speed of an object if the net work done on that object is zero?
9. Using the work-energy theorem, explain why the force of kinetic friction always has the effect of *reducing* the kinetic energy of a particle.
10. Can the average power ever equal the instantaneous power? Explain.
11. In Example 7.13, does the required power increase or decrease as the force of friction is reduced?
12. An automobile sales representative claims that a "souped-up" 300-hp engine is a necessary option in a compact car (instead of a conventional 15-hp engine). Suppose you intend to drive the car within speed limits ($\leq$55 mi/h) and on flat terrain. How would you counteract this sales pitch?
13. One bullet has twice the mass of a second bullet. If both are fired such that they have the same velocity, which has more kinetic energy? What is the ratio of kinetic energies of the two bullets?
14. When a punter kicks a football, is he doing any work on the ball while his toe is in contact with it? Is he doing any work on the ball after it loses contact with his toe? Are there any forces doing work on the ball while it is in flight?
15. Discuss the work done by a pitcher throwing a baseball. What is the approximate distance through which the force acts as the ball is thrown?
16. Estimate the time it takes you to climb a flight of stairs. Then approximate the power required to perform this task. Express your value in horsepower.
17. Do frictional forces always reduce the kinetic energy of a body? If your answer is no, give examples which illustrate the effect.

PROBLEMS

Section 7.2 Work Done by a Constant Force

1. How much work is done by a person in raising a 20-kg bucket of water from the bottom of a well that is 30 m deep? Assume the speed of the bucket as it is lifted is constant.
2. A tugboat exerts a constant force of 5000 N on a ship moving at constant speed through a harbor. How much work does the tugboat do on the ship in a distance of 3 km?

3. A 15-kg block is dragged over a rough, horizontal surface by a constant force of 70 N acting at an angle of 25° above the horizontal. The block is displaced 5 m, and the coefficient of kinetic friction is 0.3. Find the work done by (a) the 70-N force, (b) the force of friction, (c) the normal force, and (d) the force of gravity. (e) What is the net work done on the block?
4. A horizontal force of 150 N is used to push a 40-kg box on a rough, horizontal surface through a distance of 6 m. If the box moves at constant speed, find (a) the work done by the 150-N force, (b) the work done by friction, and (c) the coefficient of kinetic friction.
5. A 100-kg sled is dragged by a team of dogs a distance of 2 km over a horizontal surface at a constant velocity. If the coefficient of friction between the sled and the snow is 0.15, find the work done by (a) the team of dogs and (b) the force of friction.
6. Verify the following energy unit conversions: (a) 1 J = 10^7 ergs, (b) 1 J = 0.737 ft · lb.

Section 7.3 The Scalar Product of Two Vectors

7. Two vectors are given by $\boldsymbol{A} = 3\boldsymbol{i} + 2\boldsymbol{j}$ and $\boldsymbol{B} = -\boldsymbol{i} + 3\boldsymbol{j}$. Find (a) $\boldsymbol{A} \cdot \boldsymbol{B}$ and (b) the angle between $\boldsymbol{A}$ and $\boldsymbol{B}$.
8. A vector is given by $\boldsymbol{A} = -2\boldsymbol{i} + 3\boldsymbol{j}$. Find (a) the magnitude of $\boldsymbol{A}$ and (b) the angle that $\boldsymbol{A}$ makes with the positive y axis. [In (b), use the definition of the scalar product.]
9. Vector $\boldsymbol{A}$ has a magnitude of 3 units, and $\boldsymbol{B}$ has a magnitude of 8 units. The two vectors make an angle of 40° with each other. Find $\boldsymbol{A} \cdot \boldsymbol{B}$.
10. A force $\boldsymbol{F} = (6\boldsymbol{i} - 2\boldsymbol{j})$ N acts on a particle that undergoes a displacement $\boldsymbol{s} = (3\boldsymbol{i} + \boldsymbol{j})$ m. Find (a) the work done by the force on the particle and (b) the angle between $\boldsymbol{F}$ and $\boldsymbol{s}$.
11. Given two arbitrary vectors $\boldsymbol{A}$ and $\boldsymbol{B}$, show that $\boldsymbol{A} \cdot \boldsymbol{B} = A_xB_x + A_yB_y + A_zB_z$. (*Hint:* Write $\boldsymbol{A}$ and $\boldsymbol{B}$ in unit vector form and use Eq. 7.8.)
12. Vector $\boldsymbol{A}$ is 2 units long and points in the positive y direction. Vector $\boldsymbol{B}$ has a negative x component 5 units long, a positive y component 3 units long, and no z component. Find $\boldsymbol{A} \cdot \boldsymbol{B}$ and the angle between the vectors.
13. Using the definition of the scalar product, find the angles between the following pairs of vectors: (a) $\boldsymbol{A} = 3\boldsymbol{i} - \boldsymbol{j}$ and $\boldsymbol{B} = 2\boldsymbol{i} + 2\boldsymbol{j}$, (b) $\boldsymbol{A} = -\boldsymbol{i} + 4\boldsymbol{j}$ and $\boldsymbol{B} = 2\boldsymbol{i} + \boldsymbol{j} + 2\boldsymbol{k}$, (c) $\boldsymbol{A} = 2\boldsymbol{i} + \boldsymbol{j} + 3\boldsymbol{k}$ and $\boldsymbol{B} = -2\boldsymbol{j} + 2\boldsymbol{k}$.
14. The scalar product of vectors $\boldsymbol{A}$ and $\boldsymbol{B}$ is 6 units. The magnitude of each vector is 4. Find the angle between the vectors.

Section 7.4 Work Done by a Varying Force: The One-Dimensional Case

15. A body is subject to a force F_x that varies with position as in Figure 7.14. Find the work done by the force on the body as it moves (a) from $x = 0$ to $x = 5$ m, (b) from $x = 5$ m to $x = 10$ m, and (c) from $x = 10$ m to $x = 15$ m. (d) What is the total work done by the force over the distance $x = 0$ to $x = 15$ m?

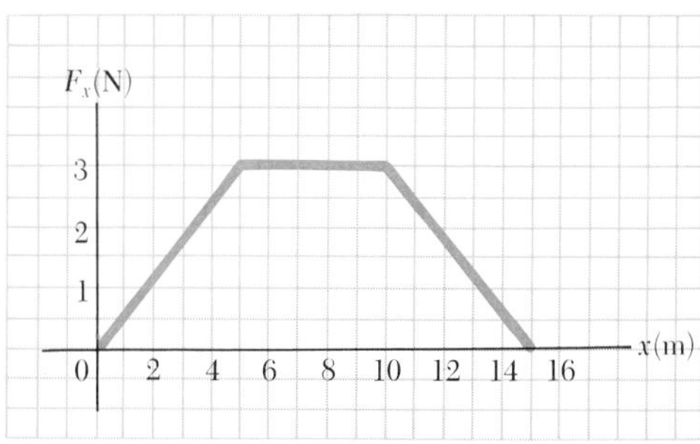

Figure 7.14 (Problems 15 and 25).

16. The force acting on a particle varies as in Figure 7.15. Find the work done by the force as the particle moves (a) from $x = 0$ to $x = 8$ m, (b) from $x = 8$ m to $x = 10$ m, and (c) from $x = 0$ to $x = 10$ m.

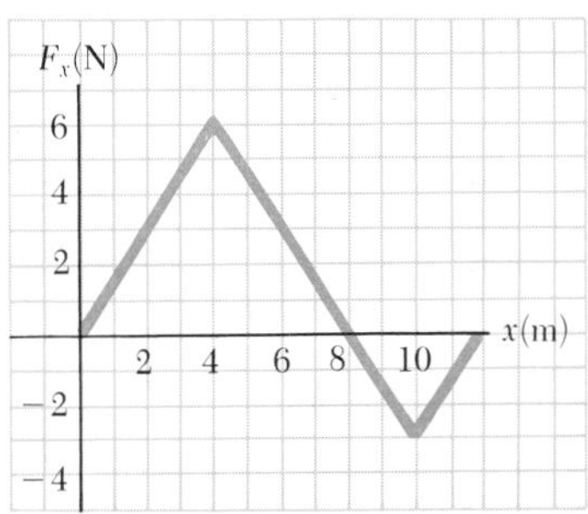

Figure 7.15 (Problems 16 and 26).

17. The force acting on a particle is given by $F_x = (8x - 16)$ N, where x is in m. (a) Make a plot of this force versus x from $x = 0$ to $x = 3$ m. (b) From your graph, find the net work done by this force as the particle moves from $x = 0$ to $x = 3$ m.
18. When a 3-kg mass is hung vertically on a certain light spring that obeys Hooke's law, the spring stretches 1.5 cm. If the 3-kg mass is removed, (a) how far will the spring stretch if a 1-kg mass is hung on it, and (b) how much work must an external agent do to stretch the same spring 4.0 cm from its unstretched position?

Section 7.5 Work and Kinetic Energy

19. A 0.2-kg ball has a speed of 15 m/s. (a) What is its kinetic energy? (b) If its speed is doubled, what is its kinetic energy?
20. Calculate the kinetic energy of a 1000-kg satellite orbiting the earth at a speed of 7×10^3 m/s.
21. A 3-kg mass has an initial velocity $\boldsymbol{v}_0 = (5\boldsymbol{i} - 3\boldsymbol{j})$ m/s. (a) What is its kinetic energy at this time? (b) Find the *change* in its kinetic energy if its *velocity changes* to $(8\boldsymbol{i} + 4\boldsymbol{j})$ m/s. (*Hint:* Remember that $v^2 = \boldsymbol{v} \cdot \boldsymbol{v}$.)
22. A 0.6-kg particle has a speed of 3 m/s at point A and a speed of 5 m/s at point B. What is its kinetic energy (a) at point A and (b) at point B? (c) What is the total work done on the particle as it moves from A to B?

23. A mechanic pushes a 2000-kg car from rest to a speed of 3 m/s with a constant horizontal force. During this time, the car moves a distance of 30 m. Neglecting friction between the car and the road, determine (a) the work done by the mechanic and (b) the horizontal force exerted on the car.

24. A 40-kg box initially at rest is pushed a distance of 5 m along a rough, horizontal floor with a constant applied force of 130 N. If the coefficient of friction between the box and floor is 0.3, find (a) the work done by the applied force, (b) the work done by friction, (c) the change in kinetic energy of the box, and (d) the final speed of the box.

25. A 4-kg particle is subject to a force that varies with position as shown in Figure 7.14. The particle starts from rest at $x = 0$. What is the speed of the particle at (a) $x = 5$ m, (b) $x = 10$ m, (c) $x = 15$ m?

26. The force acting on a 6-kg particle varies with position as shown in Figure 7.15. If its velocity is 2 m/s at $x = 0$, find its speed and kinetic energy at (a) $x = 4$ m, (b) $x = 8$ m, (c) $x = 10$ m.

27. A sled of mass m is given a kick on a frozen pond, imparting to it an initial speed v_0. The coefficient of kinetic friction between the sled and ice is μ_k. (a) Use the work-energy theorem to find the distance the sled moves before coming to rest. (b) Obtain a numerical value for the distance if $v_0 = 5$ m/s and $\mu_k = 0.1$.

28. A 6-kg mass is lifted vertically through a distance of 5 m by a light string with a tension of 80 N. Find (a) the work done by the force of tension, (b) the work done by gravity, and (c) the final speed of the mass if it starts from rest.

29. A 2-kg block is attached to a light spring of force constant 500 N/m as in Figure 7.6. The block is pulled 5 cm to the right of equilibrium and released from rest. Find the speed of the block as it passes through equilibrium if (a) the horizontal surface is frictionless and (b) if the coefficient of friction between the block and surface is 0.35.

30. A 4-kg block is given an initial speed of 8 m/s at the bottom of a 20° incline. The frictional force that retards its motion is 15 N. (a) If the block is directed *up* the incline, how far will it move before it stops? (b) Will it slide back down the incline?

31. A 3-kg block is moved up a 37° incline under the action of a constant *horizontal* force of 40 N. The coefficient of kinetic friction is 0.1, and the block is displaced 2 m up the incline. Calculate (a) the work done by the 40-N force, (b) the work done by gravity, (c) the work done by friction, and (d) the *change* in kinetic energy of the block. (*Note:* The applied force is *not* parallel to the incline.)

32. A 4-kg block attached to a string 2 m in length rotates in a circle on a horizontal surface. (a) If the surface is frictionless, identify all the forces on the block and show that the work done by each force is zero for any displacement of the block. (b) If the coefficient of friction between the block and surface is 0.25, find the work done by the force of friction in each revolution of the block.

Section 7.6 Power

33. A certain automobile engine delivers a power of 30 hp (2.24×10^4 W) to its wheels when moving at a constant speed of 27 m/s ($\approx$60 mi/h). What is the resistive force acting on the automobile at that speed?

34. A speedboat requires 130 hp to move at a constant speed of 15 m/s ($\approx$33 mi/h). Calculate the resistive force due to the water at that speed.

35. A 50-kg student climbs a rope 5 m in length and stops at the top. (a) What must her average speed be in order to match the power output of a 200-W light bulb? (b) How much work does she do?

36. A machine lifts a 300-kg crate through a height of 5 m in 8 s. Calculate its power output.

37. A 200-kg crate is pulled along a level surface by an engine. The coefficient of friction between the crate and surface is 0.4. (a) How much power must the engine deliver to move the crate at a constant speed of 5 m/s? (b) How much work is done by the engine in 3 min?

38. A 65-kg athlete runs a distance of 600 m up a mountain inclined at 20° to the horizontal. He performs this feat in 80 s. Assuming that air resistance is negligible, (a) how much work does he perform and (b) what is his power output during the run?

39. A 1500-kg car accelerates uniformly from rest to a speed of 10 m/s in 3 s. Find (a) the work done on the car in this time, (b) the average power delivered by the engine in the first 3 s, and (c) the instantaneous power delivered by the engine at $t = 2$ s.

40. A single, constant force $\mathbf{F}$ acts on a particle of mass m. The particle starts at rest at $t = 0$. (a) Show that the instantaneous power delivered by the force at any time t is equal to $(F^2/m)t$. (b) If $F = 20$ N and $m = 5$ kg, what is the power delivered at $t = 3$ s?

*Section 7.7 Energy and the Automobile

41. The car described in Table 7.3 travels at a constant speed of 129 km/h. At this speed, determine (a) the power needed to overcome air drag, and (b) the total power delivered to the wheels.

42. A passenger car carrying two people has a fuel economy of 25 mi/gal. It travels a distance of 3000 miles. A jet airplane making the same trip with 150 passengers has a fuel economy of 1 mi/gal. Compare the fuel consumed per passenger for the two modes of transportation.

43. Suppose the car described in Table 7.3 has a fuel economy of 6.4 km/liter (15 mi/gal) when traveling at a speed of 60 mi/h. Assuming an available fuel power of 136 kW and an efficiency of 14%, determine the fuel economy of the car if it carries, in addition to the driver, four passengers, each with an average mass of 70 kg.

44. When an air conditioner is added to the car described in Problem 43, the additional fuel power required to operate the air conditioner is 11 kW. If the fuel economy is 6.4 km/liter without the air conditioner, what is the fuel economy when the air conditioner is operating?

45. A woman raises a 10-kg flag from the ground to the top of a 10-m flagpole at constant velocity, 0.25 m/s. (a) Find the work done by the woman while raising the flag. (b) Find the work done by gravity. (c) What is the power output of the woman while raising the flag?

46. Three vectors that form a closed triangle satisfy the condition $\mathbf{C} = \mathbf{A} - \mathbf{B}$ (Fig. 7.16). Use this fact and the definition of the scalar product to derive the law of cosines in trigonometry,

$$C^2 = A^2 + B^2 - 2AB \cos \theta.$$

(*Hint:* Find the scalar product $\mathbf{C} \cdot \mathbf{C}$ in terms of A and B.)

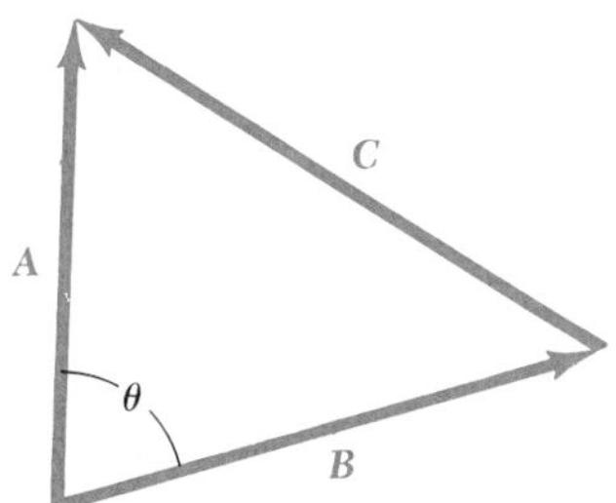

Figure 7.16 (Problem 46).

GENERAL PROBLEMS

47. The direction of an arbitrary vector $\mathbf{A}$ can be completely specified with the angles α, β, and γ that the vector makes with the x, y, and z axes, respectively. If $\mathbf{A} = A_x\mathbf{i} + A_y\mathbf{j} + A_z\mathbf{k}$, (a) find expressions for $\cos \alpha$, $\cos \beta$, and $\cos \gamma$ (these are known as *direction cosines*) and (b) show that these angles satisfy the relation $\cos^2 \alpha + \cos^2 \beta + \cos^2 \gamma = 1$. (*Hint:* Take the scalar product of $\mathbf{A}$ with $\mathbf{i}$, $\mathbf{j}$, and $\mathbf{k}$ separately.)

48. Prove the work-energy theorem, $W = \Delta K$, for a general three-dimensional displacement. (*Note:* $\mathbf{F} = m\, d\mathbf{v}/dt$ and $d\mathbf{s} = \mathbf{v}\, dt$.)

49. The resultant force acting on a 2-kg particle moving along the x axis varies as $F_x = 3x^2 - 4x + 5$, where x is in m and F_x is in N. (a) Find the net work done on the particle as it moves from $x = 1$ m to $x = 3$ m. (b) If the speed of the particle is 5 m/s at $x = 1$ m, what is its speed at $x = 3$ m?

50. A 4-kg particle moves along the x-axis. Its position varies with time according to $x = t + 2t^3$, where x is in m and t is in s. Find (a) the kinetic energy at any time t, (b) the acceleration of the particle and the force acting on it at time t, (c) the power being delivered to the particle at time t, and (d) the work done on the particle in the interval $t = 0$ to $t = 2$ s. (*Note:* $P = dW/dt$.)

51. A block of mass m is attached to a light spring of force constant k as in Figure 7.6. The spring is compressed a distance d from its equilibrium position and released from rest. (a) If the block comes to rest when it first reaches the equilibrium position, what is the coefficient of friction between the block and surface? (b) If the block first comes to rest when the spring is *stretched* a distance of $d/2$ from equilibrium, what is μ?

52. A projectile, mass m, is shot horizontally with initial velocity v_0 from a height h above a flat desert floor. The instant before the projectile hits the desert floor find (a) the work done on the projectile by gravity, (b) the change in kinetic energy since the projectile was fired, and (c) the final kinetic energy of the projectile.

53. Referring to Problem 52, find (a) the instantaneous rate at which work is being done on the projectile and (b) if the mass of the projectile is 10 kg and the initial height is 40 m, the instantaneous rate that work is being done after 1 s, 2 s, and 3 s. (*Note:* Be careful of the elapsed time.)

54. A small sphere of mass m hangs from a string of length L as in Figure 7.17. A variable horizontal force $\mathbf{F}$ is applied to the mass in such a way that it moves slowly from the vertical position until the string makes an angle θ with the vertical. Assuming the sphere is always in equilibrium, (a) show that $F = mg \tan \theta$. (b) Make use of Equation 7.20 to show that the work done by the force $\mathbf{F}$ is equal to $mgL(1 - \cos \theta)$. (*Hint:* Note that $s = L\theta$, and so $ds = L\, d\theta$.)

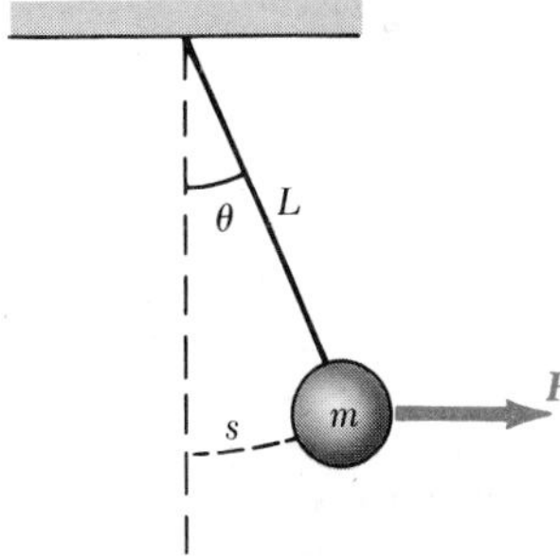

Figure 7.17 (Problem 54).

55. A car of mass m travels with constant speed v on a level road for a distance d. According to actual tests, the drag force is *approximately* given by $f = -Kmv$, where $K = 0.018\ \text{s}^{-1}$. (a) Show that the work done by the engine to overcome the drag force is given by $Kmvd$. (b) Show that the power that the engine must deliver to the wheels to maintain this speed is Kmv^2. (c) Obtain numerical values for the work done and power delivered taking $m = 1500$ kg, $v = 27$ m/s, and $d = 100$ km. (d) If the car has a fuel economy of 15 mi/gal, what is the efficiency of the engine? (In this case, we define efficiency as the work done divided by the energy consumed.)

56. A 0.4-kg particle slides on a horizontal, circular track 1.5 m in radius. It is given an initial speed of 8 m/s. After one revolution, its speed drops to 6 m/s because of friction. (a) Find the work done by the force of friction in one revolution. (b) Calculate the coefficient of kinetic friction. (c) What is the total number of revolutions the particle will make before coming to rest?

57. A passenger car of mass 1500 kg accelerates from rest to 97 km/h in 10 s. (a) Find the acceleration of the car. (b) Show that the coefficient of friction between the rear tires and road must be at least 0.55. (c) Determine the limiting frictional force on the car. (d) Find the average power delivered by the engine. (Assume that the normal force on each tire is $\frac{1}{4}mg$.)

58. Suppose a car is modeled as a cylinder moving with a speed v, as in Figure 7.18. In a time Δt, a column of air of mass Δm must be moved a distance $v\,\Delta t$ and hence must be given a kinetic energy $\frac{1}{2}(\Delta m)v^2$. Using this model, show that the power loss due to air resistance is $\frac{1}{2}\rho Av^3$ and the drag force is $\frac{1}{2}\rho Av^2$, where ρ is the density of air.

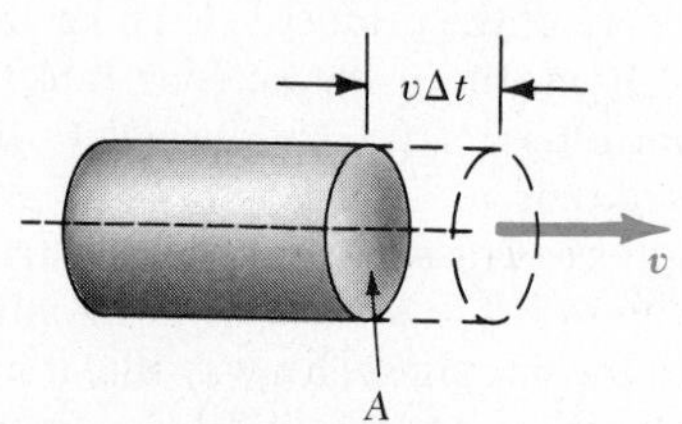

Figure 7.18 (Problem 58).

CALCULATOR/COMPUTER PROBLEMS

•59. A 5-kg particle starts at the origin and moves along the x axis. The net force acting on the particle is measured at intervals of 1 m to be: 27.0, 28.3, 36.9, 34.0, 34.5, 34.5, 46.9, 48.2, 50.0, 63.5, 13.6, 12.2, 32.7, 46.6, 27.9 (in newtons). Determine the total work done on the particle over this interval.

•60. A 0.178-kg particle moves along the x axis from $x = 12.8$ m to $x = 23.7$ m under the influence of a force given by

$$F = \frac{375}{x^3 + 3.75x}$$

where F is in newtons and x is in meters. Use a method of numerical integration to estimate the total work done by this force during this displacement. Your calculations should have an accuracy of at least 2%.

8
Potential Energy and Conservation of Energy

In Chapter 7 we introduced the concept of kinetic energy, which is associated with the motion of an object. We found that the kinetic energy of an object can change only if work is done on the object. In this chapter we introduce another form of mechanical energy associated with the position or configuration of an object, called *potential energy*. We shall find that the potential energy of a system can be thought of as energy stored in the system that can be converted to kinetic energy or do work. The potential energy concept can be used only when dealing with a special class of forces called *conservative forces*. When only internal conservative forces, such as gravitational or spring forces, act on a system, the kinetic energy gained (or lost) by the system as its members change their relative positions is compensated by an equal energy loss (or gain) in the form of potential energy. This is known as the *law of conservation of mechanical energy*. A more general energy conservation law applies to an isolated system when all forms of energy and energy transformations are taken into account.

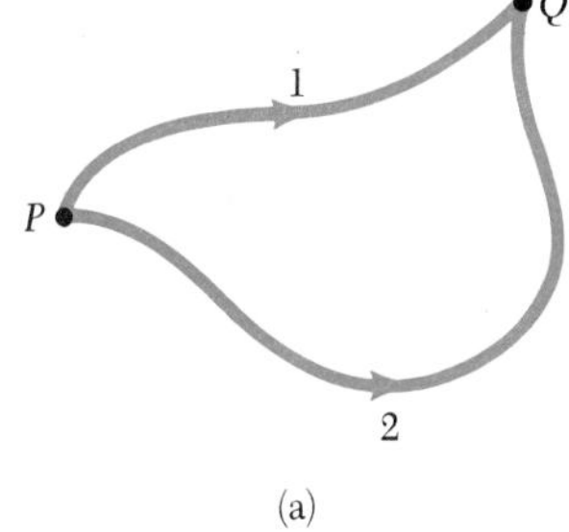

(a)

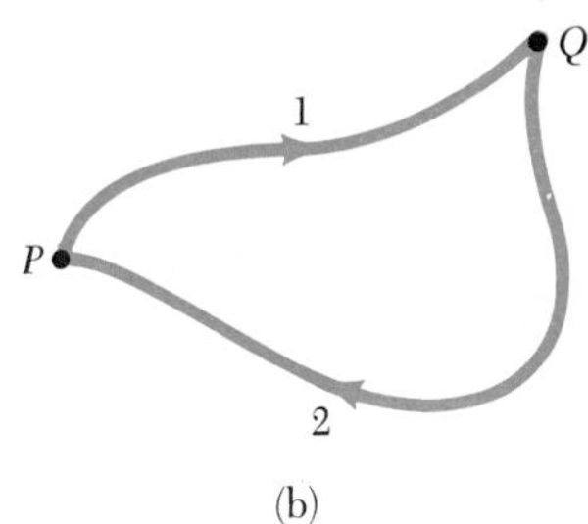

(b)

Figure 8.1 (a) A particle moves from P to Q along two different paths. The work done by a conservative force acting on the particle is the same along each path. If the force is nonconservative, the work done by this force differs along the two paths. (b) A particle moves from P to Q and then from Q back to P along a different path. That is, it moves in a closed path.

8.1 CONSERVATIVE AND NONCONSERVATIVE FORCES

Conservative Forces

In the previous chapter (Example 7.11), we found that the work done by the gravitational force acting on a particle equals the weight of the particle multiplied by its vertical displacement, assuming that g is constant over the range of the displacement. As we shall see in Section 8.2, this result is valid for an arbitrary displacement of the particle. That is, the work done by gravity depends only on the initial and final coordinates and is independent of the path taken between these points. When a force exhibits these properties, it is called a **conservative force.** In addition to the gravitational force, other examples of conservative forces are the electrostatic force and the restoring force in a spring.

> In general, a force is conservative if the work done by that force acting on a particle moving between two points is independent of the path the particle takes between the points.

That is, the work done on a particle by a conservative force depends only on the initial and final coordinates of the particle. With reference to the *arbitrary* paths shown in Figure 8.1a, we can write this condition

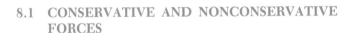

Property of a conservative force

$$W_{PQ} \text{ (along 1)} = W_{PQ} \text{ (along 2)}$$

A conservative force has another property, which can be derived from the above condition. Suppose the particle moves from P to Q along path 1, and then from Q to P along path 2, as in Figure 8.1b. The work done by a conservative force in the reverse path 2 from Q to P is equal to the *negative* of the work done from P to Q along path 2. Therefore, we can write the original condition of a conservative force

$$W_{PQ} \text{ (along 1)} = -W_{QP} \text{ (along 2)}$$

$$W_{PQ} \text{ (along 1)} + W_{QP} \text{ (along 2)} = 0$$

Hence, a conservative force also has the property that

> the total work done by a conservative force on a particle is zero when the particle moves around any closed path and returns to its initial position.

We can interpret this property of a conservative force in the following manner. The work-energy theorem says that the net work done on a particle displaced between two points equals the change in its kinetic energy. Therefore, if all the forces acting on the particle are conservative, then $W = 0$ for a round trip. This means that the particle will return to its starting point with the same kinetic energy it had when it started its motion.

To illustrate that the force of gravity is conservative, recall that the work done by the gravitational force as a particle of mass m moves between two points of elevation y_i and y_f is given by

Work done by the force of gravity

$$W_g = -mg(y_f - y_i)$$

That is, the work done by the gravitational force mg (in the negative y direction) equals the force multiplied by the displacement in the y direction. From this expression, we first note that W_g depends only on the initial and final y coordinates and is *independent* of the path taken. Furthermore, if y_i and y_f are at the same elevation or if the particle makes a round trip, then $y_i = y_f$ and $W_g = 0$. For example, if a ball is thrown vertically upward with an initial speed v_i, and if air resistance is neglected, the ball must return to the thrower's hand with the same speed (and same kinetic energy) it had at the start of its motion.

Another example of a conservative force is the force of a spring on a block attached to the spring, where the restoring force is given by $F_s = -kx$. In the previous chapter, we found that the work done by the spring on the block is

Work done by the spring force

$$W_s = \tfrac{1}{2}kx_i^2 - \tfrac{1}{2}kx_f^2$$

where the initial and final coordinates of the block are measured from the equilibrium position of the block, $x = 0$. We see that W_s again depends only on the initial and final x coordinates. In addition, $W_s = 0$ for a round trip, where $x_i = x_f$.

Nonconservative Forces

> A force is **nonconservative** if the work done by that force on a particle moving between two points depends on the path taken.

That is, the work done by a nonconservative force in taking a particle from P to Q in Figure 8.1a will differ for paths 1 and 2. We can write this

Property of a nonconservative force

$$W_{PQ} \text{ (along 1)} \neq W_{PQ} \text{ (along 2)}$$

Furthermore, from this condition we can show that if a force is nonconservative, the work done by that force on a particle that moves through any closed path is *not necessarily zero*. Since the work done in going from P to Q along path 2 is equal to the negative of the work done in going from Q to P along path 2, it follows from the first condition of a nonconservative force that

$$W_{PQ} \text{ (along 1)} \neq -W_{QP} \text{ (along 2)}$$

$$W_{PQ} \text{ (along 1)} + W_{QP} \text{ (along 2)} \neq 0$$

The force of sliding friction is a good example of a nonconservative force. If an object is moved over a rough, horizontal surface between two points along various paths, the work done by the frictional force certainly depends on the path. The negative work done by the frictional force along any particular path between two points will equal the force of friction multiplied by the length of the path. Paths of different lengths involve different amounts of work. The absolute magnitude of the least work done by the frictional force will correspond to a straight-line path between the two points. Furthermore, for a closed path you should note that the total work done by friction is nonzero since the force of friction opposes the motion along the entire path.

As an instructive example, suppose you were to displace a book between two points on a rough, horizontal surface such as a table. If the book is displaced in a straight line between two points, A and B in Figure 8.2, the work done by friction is simply $-fd$, where d is the distance between the points. However, if the book is moved along *any other* path between the two points, the work done by friction would be *greater* (in absolute magnitude) than $-fd$. For example, the work done by friction along the semicircular path in Figure 8.2 is equal to $-f(\pi d/2)$, where d is the diameter of the circle. Finally, if the book is moved through any closed path (such as a circle), the work done by friction would clearly be nonzero since the frictional force opposes the motion.

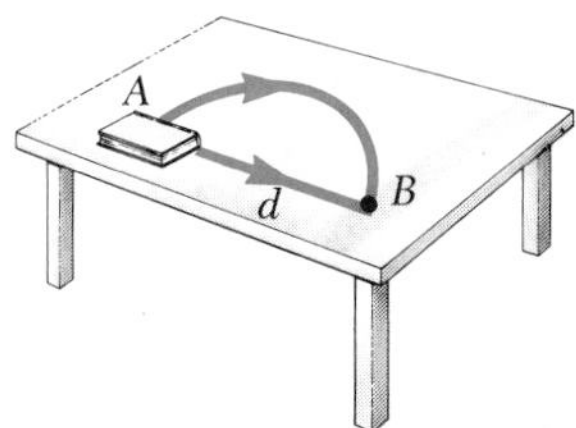

Figure 8.2 The work done by the force of friction depends on the path taken as the book is moved from A to B.

In the example of a ball thrown vertically in the air with an initial speed v_i, careful measurements would show that because of air resistance, the ball would return to the thrower's hand with a speed less than v_i. Consequently, the final kinetic energy is less than the initial kinetic energy. The presence of a nonconservative force has reduced the ability of the system to do work by virtue of its motion. We shall sometimes refer to a nonconservative force as a *dissipative force*. For this reason, frictional forces are often referred to as being dissipative.

8.2 POTENTIAL ENERGY

In the previous section we found that the work done by a conservative force does not depend on the path taken by the particle and is independent of the particle's velocity. The work done is a function only of the particle's initial and final coordinates. For these reasons, we can define a potential energy function U such that the work done equals the decrease in the potential energy. That is, the work done by a conservative force $\boldsymbol{F}$ as the particle moves along the x axis is[1]

$$W_c = \int_{x_i}^{x_f} F_x\, dx = -\Delta U = U_i - U_f \qquad (8.1)$$

[1] For a general displacement, the work done in two or three dimensions also equals $U_i - U_f$, where $U = U(x, y, z)$. We write this formally $W = \int_i^f \boldsymbol{F} \cdot d\boldsymbol{s} = U_i - U_f$.

That is, *the work done by a conservative force equals the negative of the change in the potential energy associated with that force,* where the change in the potential energy is defined as $\Delta U = U_f - U_i$. We can also express Equation 8.1 as

$$\Delta U = U_f - U_i = -\int_{x_i}^{x_f} F_x \, dx \tag{8.2}$$

where F_x is the component of $\boldsymbol{F}$ in the direction of the displacement.

It is often convenient to establish some particular location, x_i, to be a reference point and to then measure all potential energy differences with respect to this point. With this understanding, we can define the potential energy function as

$$U_f(x) = -\int_{x_i}^{x_f} F_x \, dx + U_i \tag{8.3}$$

Furthermore, the value of U_i is often taken to be zero at some arbitrary reference point. It really doesn't matter what value we assign to U_i, since it only shifts $U_f(x)$ by a constant, and it is only the *change* in potential energy that is physically meaningful. (In the next section, we shall see that the change in the particle's potential energy is related to a change in its kinetic energy.) If the conservative force is known as a function of position, we can use Equation 8.3 to calculate the change in potential energy of a body as it moves from x_i to x_f. It is interesting to note that in the one-dimensional case, a force is *always* conservative if it is a function of position only. This is generally not true for motion involving two- or three-dimensional displacements.

The work done by a nonconservative force does depend on the path as a particle moves from one position to another and could also depend on the particle's velocity or on other quantities. Therefore, the work done is not simply a function of the initial and final coordinates of the particle. We conclude that there is no potential energy *function* associated with a nonconservative force.

8.3 CONSERVATION OF MECHANICAL ENERGY

Suppose a particle moves along the x axis under the influence of only *one* conservative force, $\boldsymbol{F}_x$. If this is the only force acting on the particle, then the work-energy theorem tells us that the work done by that force equals the change in kinetic energy of the particle:

$$W_c = \Delta K$$

Since the force is conservative, according to Equation 8.1 we can write $W_c = -\Delta U$. Hence,

$$\Delta K = -\Delta U$$

$$\Delta K + \Delta U = \Delta(K + U) = 0 \tag{8.4}$$

This is the **law of conservation of mechanical energy,** which can be written in the alternative form

Conservation of mechanical energy

$$K_i + U_i = K_f + U_f \tag{8.5}$$

If we now define the total mechanical energy of the system, E, as the sum of the kinetic energy and potential energy, we can express the conservation of mechanical energy as

$$E_i = E_f \tag{8.6a}$$

where

$$E \equiv K + U \tag{8.6b}$$

Total mechanical energy

The law of conservation of mechanical energy states that the total mechanical energy of a system remains constant if the only force that does work is a conservative force. This is equivalent to the statement that if the kinetic energy of a conservative system increases (or decreases) by some amount, the potential energy must decrease (or increase) by the same amount.

If more than one conservative force acts on the system, then there is a potential energy function associated with *each* force. In such a case, we can write the law of conservation of mechanical energy

$$K_i + \sum U_i = K_f + \sum U_f \tag{8.7}$$

Conservation of mechanical energy

where the number of terms in the sums equals the number of conservative forces present. For example, if a mass connected to a spring oscillates vertically, two conservative forces act on it; the spring force and the force of gravity. We will discuss this situation later in a worked example.

8.4 GRAVITATIONAL POTENTIAL ENERGY NEAR THE EARTH'S SURFACE

When an object moves in the presence of the earth's gravity, the gravitational force can do work on that object. In the case of a freely falling object, the work done by gravity is a function of the *vertical* displacement of the object. This result is also valid in the more general case where the object undergoes both a horizontal and vertical displacement, such as in the case of a projectile.

Consider a particle being displaced from P to Q along various paths in the presence of a constant gravitational force[2] (Fig. 8.3). The work done along the path PAQ can be broken into two segments. The work done along PA is $-mgh$ (since mg is opposite to this displacement), and the work done along AQ is zero (since mg is perpendicular to this path). Hence, $W_{PAQ} = -mgh$. Likewise, the work done along PBQ is also $-mgh$, since $W_{PB} = 0$ and $W_{BQ} = -mgh$. Now consider the general path described by the solid line from P to Q. The curve is broken down into a series of horizontal and vertical steps. There is no work done by the force of gravity along the horizontal steps, since mg is perpendicular to these elements of displacement. Work is done by the force of gravity only along the vertical displacements, where the work done in the nth vertical step is $-mg\ \Delta y_n$. Thus, the total work done by the force of gravity as the particle is displaced upward a distance h is the sum of the work done along each vertical displacement. Summation of all such terms gives

$$W_g = -mg \sum_n \Delta y_n = -mgh$$

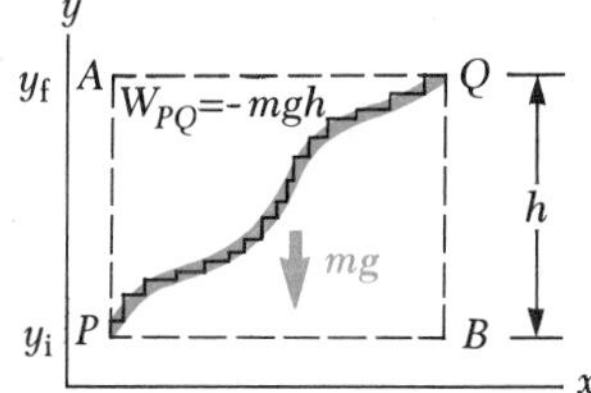

Figure 8.3 A particle that moves between the points P and Q under the influence of gravity can be envisioned as moving along a series of horizontal and vertical steps. The work done by gravity along each horizontal element is zero, and the net work done by gravity is equal to the sum of the works done along the vertical displacements.

[2] The assumption that the force of gravity is constant is a good one as long as the vertical displacement is small compared with the earth's radius.

Since $h = y_f - y_i$, we can express W_g as

Gravitational potential energy

$$W_g = mgy_i - mgy_f \tag{8.8}$$

We conclude that since the work done by the force of gravity is independent of the path, the gravitational force is a conservative force.

Since the force of gravity is conservative, we can define a **gravitational potential energy function** U_g as

Change in potential energy

$$U_g \equiv mgy \tag{8.9}$$

where we have chosen to take $U_g = 0$ at $y = 0$. Note that this function depends on the choice of origin of coordinates and is valid only when the displacement of the object in the vertical direction is small compared with the earth's radius. A general expression for the gravitational potential energy will be developed in Chapter 14.

Estimate the gravitational potential energy of this pole-vaulter at the top of his flight. (Courtesy of Dave Coskey, Villanova University)

Substituting the definition of U_g (Eq. 8.9) into the expression for the work done by the force of gravity (Eq. 8.8) gives

$$W_g = U_i - U_f = -\Delta U_g \tag{8.10}$$

That is, *the work done by the force of gravity is equal to the initial value of the potential energy minus the final value of the potential energy.* We conclude from Equation 8.10 that when the displacement is upward, $y_f > y_i$, and therefore $U_i < U_f$ and the work done by gravity is negative. This corresponds to the case where the force of gravity is *opposite* the displacement. When the object is displaced downward, $y_f < y_i$, and so $U_i > U_f$ and the work done by gravity is positive. In this case, mg is in the *same* direction as the displacement.

The term *potential energy* implies that the object has the potential, or capability, of gaining kinetic energy when released from some point under the influence of gravity. The choice of the origin of coordinates for measuring U_g is completely arbitrary, since only differences in potential energy are important. However, it is often convenient to choose the surface of the earth as the reference position $y_i = 0$.

If the force of gravity is the *only* force acting on a body, then the total mechanical energy of the body is conserved (Eq. 8.5). Therefore, the law of conservation of mechanical energy for a freely falling body can be written

Conservation of mechanical energy for a freely falling body

$$\tfrac{1}{2}mv_i^2 + mgy_i = \tfrac{1}{2}mv_f^2 + mgy_f \tag{8.11}$$

EXAMPLE 8.1. Ball in Free Fall

A ball of mass m is dropped from a height h above the ground as in Figure 8.4. (a) Determine the speed of the ball when it is at a height y above the ground, neglecting air resistance.

Since the ball is in free fall, the only force acting on it is the gravitational force. Therefore, we can use the law of conservation of mechanical energy. When the ball is released from rest at a height h above the ground, its kinetic energy is $K_i = 0$ and its potential energy is $U_i = mgh$, where the y coordinate is measured from ground level. When the ball is at a distance y above the ground, its kinetic energy is $K_f = \frac{1}{2}mv_f^2$ and its potential energy relative to the ground is $U_f = mgy$. Applying Equation 8.11, we get

$$K_i + U_i = K_f + U_f$$

$$0 + mgh = \tfrac{1}{2}mv_f^2 + mgy$$

$$v_f^2 = 2g(h - y)$$

(b) Determine the speed of the ball at y if it is given an initial speed v_i at the initial altitude h.

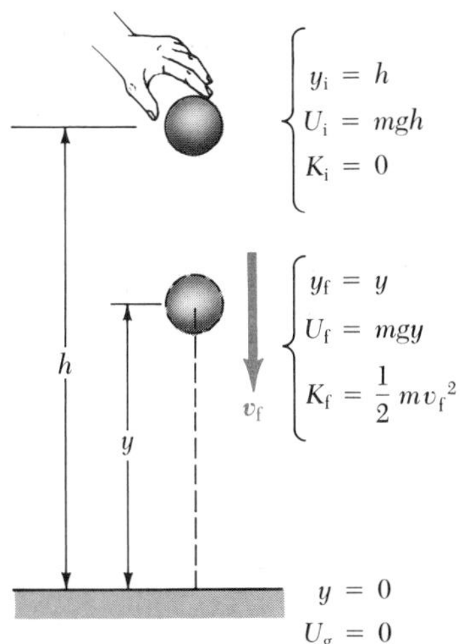

Figure 8.4 (Example 8.1) A ball is dropped from a height h above the floor. Initially, its total energy is its potential energy, equal to mgh relative to the floor. At the elevation y, its energy is the sum of the kinetic and potential energies.

In this case, the initial energy includes kinetic energy equal to $\frac{1}{2}mv_i^2$ and Equation 8.11 gives

$$\tfrac{1}{2}mv_i^2 + mgh = \tfrac{1}{2}mv_f^2 + mgy$$

$$v_f^2 = v_i^2 + 2g(h - y)$$

Note that this result is consistent with an expression from kinematics, $v_y^2 = v_{y0}^2 - 2g(y - y_0)$, where $y_0 = h$. Furthermore, this result is valid even if the initial velocity is at an angle to the horizontal (the projectile situation) if the vertical components of the initial and final velocities are substituted for v_i and v_f respectively.

EXAMPLE 8.2. The Pendulum

A pendulum consists of a sphere of mass m attached to a light cord of length L as in Figure 8.5. The sphere is released from rest when the cord makes an angle θ_0 with the vertical, and the pivot at 0 is frictionless. (a) Find the speed of the sphere when it is at the lowest point, b.

The only force that does work on m is the force of gravity, since the force of tension is always perpendicular to each element of the displacement and hence does no work. Since the force of gravity is a conservative force, the total mechanical energy is conserved. Therefore, as the pendulum swings, there is a continuous transfer between potential and kinetic energy. At the instant the pendulum is released, the energy is entirely

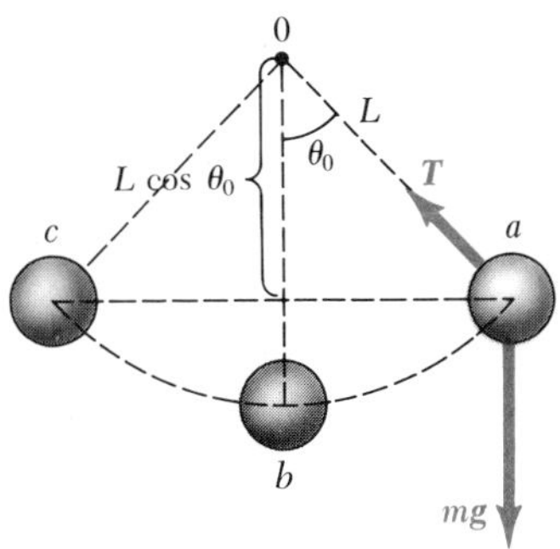

Figure 8.5 (Example 8.2) If the pendulum is released from rest at the angle θ_0, it will never swing above this position during its motion. At the start of the motion, position a, its energy is entirely potential energy. This is transformed into kinetic energy at the lowest elevation, position b.

potential energy. At point b, the pendulum has kinetic energy but has lost some potential energy. At point c, the pendulum has regained its initial potential energy and its kinetic energy is again zero. If we measure the y coordinates from the center of rotation, then $y_a = -L \cos \theta_0$ and $y_b = -L$. Therefore, $U_a = -mgL \cos \theta_0$ and $U_b = -mgL$. Applying the principle of conservation of mechanical energy gives

$$K_a + U_a = K_b + U_b$$

$$0 - mgL \cos \theta_0 = \tfrac{1}{2}mv_b^2 - mgL$$

$$(1) \qquad v_b = \sqrt{2gL(1 - \cos \theta_0)}$$

(b) What is the tension T in the cord at b?

Since the force of tension does no work, it cannot be determined using the energy method. To find T_b, we can apply Newton's second law to the radial direction. First, recall that the centripetal acceleration of a particle moving in a circle is equal to v^2/r directed toward the center of rotation. Since $r = L$ in this example, we get

$$(2) \qquad \sum F_r = T_b - mg = ma_r = mv_b^2/L$$

Substituting (1) into (2) gives for the tension at point b

$$(3) \qquad T_b = mg + 2mg(1 - \cos \theta_0) = mg(3 - 2 \cos \theta_0)$$

Exercise 1 A pendulum of length 2.0 m and mass 0.5 kg is released from rest when the supporting cord makes an angle of 30° with the vertical. Find the speed of the sphere and the tension in the cord when the sphere is at its lowest point.

Answer: 2.3 m/s; 6.21 N.

8.5 NONCONSERVATIVE FORCES AND THE WORK-ENERGY THEOREM

In real physical systems, nonconservative forces, such as friction, are usually present. Therefore, the total mechanical energy is not a constant. However, we can use the work-energy theorem to account for the presence of nonconservative forces. If W_{nc} represents the work done on a particle by all nonconservative forces and W_c is the work done by all conservative forces, we can write the work-energy theorem

$$W_{nc} + W_c = \Delta K$$

Since $W_c = -\Delta U$ (Eq. 8.1), this equation reduces to

$$W_{nc} = \Delta K + \Delta U = (K_f - K_i) + (U_f - U_i) \tag{8.12}$$

That is, the work done by all nonconservative forces equals the change in kinetic energy plus the change in potential energy. Since the total mechanical energy is given by $E = K + U$, we can also express Equation 8.12 as

Work done by nonconservative forces

$$W_{nc} = (K_f + U_f) - (K_i + U_i) = E_f - E_i \tag{8.13}$$

That is, *the work done by all nonconservative forces equals the change in the total mechanical energy of the system.* Of course, when there are no nonconservative forces present, it follows that $W_{nc} = 0$ and $E_i = E_f$; that is, the total mechanical energy is conserved.

EXAMPLE 8.3. Block Moving on Incline

A 3-kg block slides down a rough incline 1 m in length as in Figure 8.6a. The block starts from rest at the top and experiences a constant force of friction of magnitude 5 N; the angle of inclination is 30°. (a) Use energy methods to determine the speed of the block when it reaches the bottom of the incline.

Since $v_i = 0$, the initial kinetic energy is zero. If the y coordinate is measured from the bottom of the incline, they $y_i = 0.50$ m. Therefore, the total mechanical energy of the block at the top is potential energy given by

$$E_i = U_i = mgy_i = (3 \text{ kg})\left(9.80 \frac{\text{m}}{\text{s}^2}\right)(0.50 \text{ m}) = 14.7 \text{ J}$$

When the block reaches the bottom, its kinetic energy is $\frac{1}{2}mv_f^2$, but its potential energy is *zero* since its elevation is $y_f = 0$. Therefore, the total mechanical energy at the bottom is $E_f = \frac{1}{2}mv_f^2$. However, we cannot say that $E_i = E_f$ in this case, because there is a nonconservative force that does work on the block, namely, the force of friction, $W_{nc} = -fs$, where s is the displacement along the plane. (Recall that the forces normal to the plane do no work on the block since they are perpendicular to the displacement.) In this case, $f = 5$ N and $s = 1$ m, therefore,

$$W_{nc} = -fs = (-5 \text{ N})(1 \text{ m}) = -5 \text{ J}$$

That is, some mechanical energy is lost because of the presence of the retarding force. Applying the work-energy theorem in the form of Equation 8.13 gives

$$W_{nc} = E_f - E_i$$

$$-fs = \tfrac{1}{2}mv_f^2 - mgy_i$$

$$\tfrac{1}{2}mv_f^2 = 14.7 \text{ J} - 5 \text{ J} = 9.7 \text{ J}$$

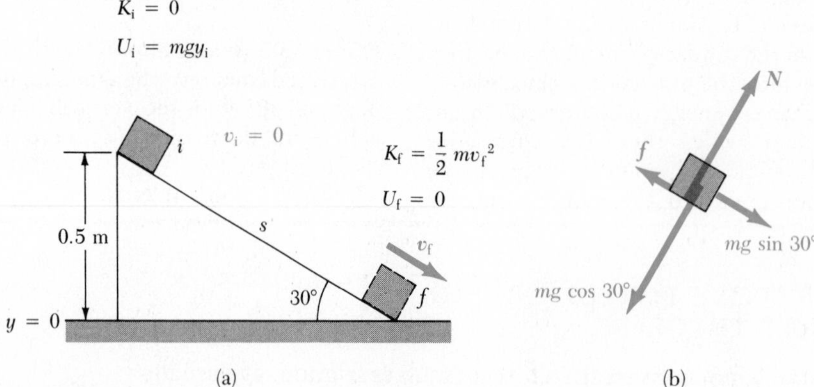

Figure 8.6 (Example 8.3) (a) A block slides down a rough incline under the influence of gravity. Its potential energy decreases while its kinetic energy increases. (b) Free-body diagram for the block.

$$v_f^2 = \frac{19.4 \text{ J}}{3 \text{ kg}} = 6.47 \text{ m}^2/\text{s}^2$$

$$v_f = 2.54 \text{ m/s}$$

(b) Check the answer to (a) using Newton's second law to first find the acceleration.

Summing the forces along the plane gives

$$mg \sin 30° - f = ma$$

$$a = g \sin 30° - \frac{f}{m} = 9.80(0.5) - \frac{5}{3} = 3.23 \text{ m/s}^2$$

Since the acceleration is constant, we can apply the expression $v_f^2 = v_i^2 + 2as$, where $v_i = 0$:

$$v_f^2 = 2as = 2(3.23 \text{ m/s}^2)(1 \text{ m}) = 6.46 \text{ m}^2/\text{s}^2$$

$$v_f = 2.54 \text{ m/s}$$

Exercise 2 If the inclined plane is assumed to be frictionless, find the final speed of the block and its acceleration along the incline.
Answer: 3.13 m/s; 4.90 m/s².

EXAMPLE 8.4. Motion on a Curved Track

A child of mass m takes a ride on an irregularly curved slide of height h, as in Figure 8.7. The child starts from rest at the top. (a) Determine the speed of the child at the bottom, assuming there is no friction present.

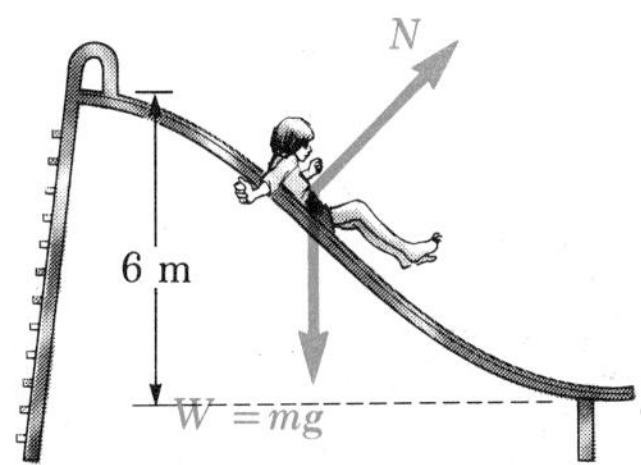

Figure 8.7 (Example 8.4) If the slide is frictionless, the speed of the child at the bottom depends only on the height of the slide and is independent of the shape of the slide.

First, note that the normal force, **N**, does no work on the child since this force is always perpendicular to each element of the displacement. Furthermore, since there is no friction, $W_{nc} = 0$ and we can apply the law of conservation of mechanical energy. If we measure the y coordinate from the bottom of the slide, then $y_i = h$, $y_f = 0$, and we get

$$K_i + U_i = K_f + U_f$$

$$0 + mgh = \tfrac{1}{2}mv_f^2 + 0$$

$$v_f = \sqrt{2gh}$$

Note that this result is the same as if the child had fallen vertically through a distance h! For example, if $h = 6$ m, then

$$v_f = \sqrt{2gh} = \sqrt{2\left(9.80 \frac{\text{m}}{\text{s}^2}\right)(6 \text{ m})} = 10.8 \text{ m/s}$$

(b) If there were a frictional force acting on the child, what would be the work done by this force?

In this case, $W_{nc} \neq 0$ and mechanical energy is *not* conserved. We can use Equation 8.13 to find the work done by friction, assuming the final velocity at the bottom is known:

$$W_{nc} = E_f - E_i = \tfrac{1}{2}mv_f^2 - mgh$$

For example, if $v_f = 8.0$ m/s, $m = 20$ kg, and $h = 6$ m, we find that

$$W_{nc} = \tfrac{1}{2}(20 \text{ kg})(8.0 \text{ m/s})^2 - (20 \text{ kg})\left(9.80 \frac{\text{m}}{\text{s}^2}\right)(6 \text{ m})$$

$$= -536 \text{ J}$$

Again, W_{nc} is negative since the *work done by sliding friction is always negative.* Note, however, that because the slide is curved, the normal force changes in magnitude and direction during the motion. Therefore, the frictional force, which is proportional to **N**, also changes during the motion. Do you think it would be possible to determine μ from these data?

8.6 POTENTIAL ENERGY STORED IN A SPRING

Now let us consider another mechanical system that is conveniently described using the concept of potential, or stored, energy. A block of mass m slides on a frictionless, horizontal surface with constant velocity v_i and collides with a light coiled spring as in Figure 8.8. The description that follows is greatly simplified by assuming that the spring is very light and therefore its kinetic energy is negligible. The spring exerts a force on the block to the left as the spring is compressed, and eventually the block comes to rest (Fig. 8.8c). The initial energy in the system (block + spring) is the initial kinetic energy of the block. When the block comes to rest after colliding with the spring, its kinetic energy is zero. Because the spring force is conservative and because there are no external forces that can do work on the system (including gravity), the total mechanical energy of the system must remain constant. Thus, there is a

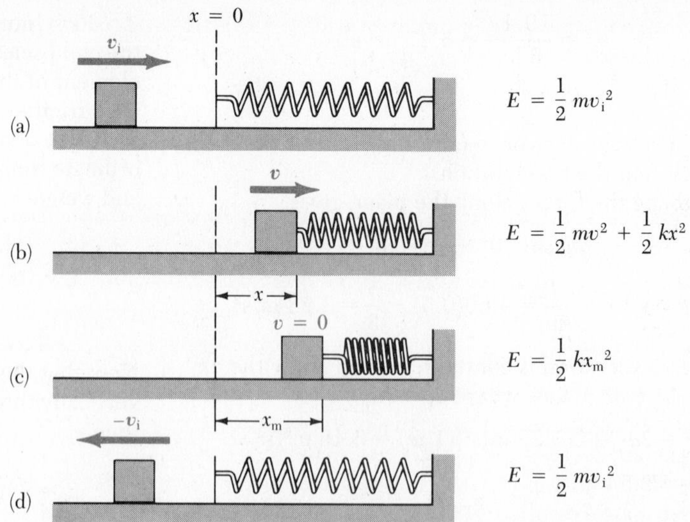

Figure 8.8 A block sliding on a smooth, horizontal surface collides with a light spring. (a) Initially the mechanical energy is all kinetic energy. (b) The mechanical energy is the sum of the kinetic energy of the block and the elastic potential energy stored in the spring. (c) The energy is entirely potential energy. (d) The energy is transformed back to the kinetic energy of the block. Note that the total energy remains constant.

transfer of energy from kinetic energy of the block to potential energy stored in the spring. Eventually, the block moves in the opposite direction and regains all of its initial kinetic energy, as described in Figure 8.8d.

To describe the potential energy stored in the spring, recall from the previous chapter that the work done by the spring on the block as the block moves from $x = x_i$ to $x = x_f$ is

$$W_s = \tfrac{1}{2}kx_i^2 - \tfrac{1}{2}kx_f^2$$

The quantity $\frac{1}{2}kx^2$ is defined as the **elastic potential energy** stored in the spring, denoted by the symbol U_s:

Potential energy stored in a spring

$$U_s = \tfrac{1}{2}kx^2 \tag{8.14}$$

Note that the elastic potential energy stored in the spring is zero when the spring is unstretched, or undeformed, ($x = 0$). Furthermore, U_s is a *maximum* when the spring has reached its maximum compression (Fig. 8.8c). Finally, U_s is *always* positive since it is proportional to x^2.

The total mechanical energy of the block-spring system can be expressed as

$$E = \tfrac{1}{2}mv_i^2 + \tfrac{1}{2}kx_i^2 = \tfrac{1}{2}mv_f^2 + \tfrac{1}{2}kx_f^2 \tag{8.15}$$

Applying this expression to the system described in Figure 8.8 and noting that $x_i = 0$, we get

$$E = \tfrac{1}{2}mv_i^2 = \tfrac{1}{2}mv_f^2 + \tfrac{1}{2}kx_f^2 \tag{8.16}$$

This expression says that for any displacement x_f, when the speed of the block is v_f, the sum of the kinetic and potential energies is equal to a *constant* E, which equals the total energy. In this case, the total energy is the initial kinetic energy of the block.

Now suppose there are nonconservative forces acting on the block-spring system. In this case, we can apply the work-energy theorem in the form of Equation 8.13, which gives

$$W_{nc} = (\tfrac{1}{2}mv_f^2 + \tfrac{1}{2}kx_f^2) - (\tfrac{1}{2}mv_i^2 + \tfrac{1}{2}kx_i^2) \qquad (8.17)$$

That is, the total mechanical energy is not a constant of the motion when nonconservative forces act on the system. Again, if W_{nc} is due to a force of friction, then W_{nc} is *negative* and the final energy is less than the initial energy.

EXAMPLE 8.5. Mass-Spring Collision

A mass of 0.80 kg is given an initial velocity $v_i = 1.2$ m/s to the right and collides with a light spring of force constant $k = 50$ N/m, as in Figure 8.8. (a) If the surface is frictionless, calculate the initial maximum compression of the spring after the collision.

The total mechanical energy is conserved since $W_{nc} = 0$. Applying Equation 8.15 to this system with $v_f = 0$ gives

$$\tfrac{1}{2}mv_i^2 + 0 = 0 + \tfrac{1}{2}kx_f^2$$

$$x_f = \sqrt{\frac{m}{k}}\,v_i = \sqrt{\frac{0.8\text{ kg}}{50\text{ N/m}}}\,(1.2\text{ m/s}) = 0.15\text{ m}$$

(b) If a constant force of friction acts between the block and the surface with $\mu = 0.5$ and if the speed of the block just as it collides with the spring is $v_i = 1.2$ m/s, what is the maximum compression in the spring?

In this case, the mechanical energy of the system is *not* conserved because of the presence of friction, which does negative work on the system. The magnitude of the frictional force is

$$f = \mu N = \mu mg = 0.5(0.80\text{ kg})\left(9.8\ \frac{\text{m}}{\text{s}^2}\right) = 3.9\text{ N}$$

Therefore, the work done by the force of friction as the block is displaced from $x_i = 0$ to $x_f = x$ is

$$W_{nc} = -fx = (-3.9x)\text{ J}$$

Substituting this into Equation 8.17 gives

$$W_{nc} = (0 + \tfrac{1}{2}kx^2) - (\tfrac{1}{2}mv_i^2 + 0)$$

$$-3.9x = \frac{50}{2}\,x^2 - \frac{1}{2}\,(0.80)(1.2)^2$$

$$25x^2 + 3.9x - 0.58 = 0$$

Solving the quadratic equation for x gives $x = 0.093$ m and $x = -0.25$ m. The physically acceptable root is $x = 0.093$ m $= 9.3$ cm. The negative root is unacceptable since the block must be displaced to the right of the origin after coming to rest. Note that 9.3 cm is *less* than the distance obtained in the frictionless case (a). This result is what we should expect, since the force of friction retards the motion of the system.

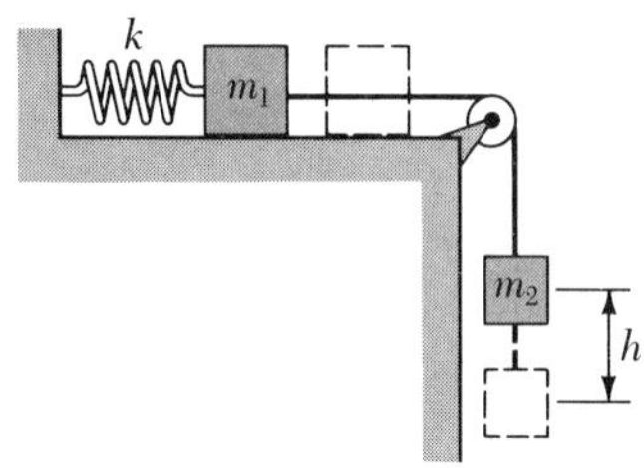

Figure 8.9 (Example 8.6) As the system moves from the highest to the lowest elevation of m_2, the system loses gravitational potential energy but gains elastic potential energy stored in the spring. Some mechanical energy is lost because of the presence of the nonconservative force of friction between m_1 and the surface.

EXAMPLE 8.6. Connected Blocks in Motion

Two blocks are connected by a light string that passes over a frictionless pulley as in Figure 8.9. The block of mass m_1 lies on a rough surface and is connected to a spring of force constant k. The system is released from rest when the spring is unstretched. If m_2 falls a distance h before coming to rest, calculate the coefficient of kinetic friction between m_1 and the surface.

Solution: In this situation there are two forms of potential energy to consider: the gravitational potential energy and the elastic potential energy stored in the spring. We can write the work-energy theorem

$$(1) \qquad W_{nc} = \Delta K + \Delta U_g + \Delta U_s$$

where ΔU_g is the *change* in the gravitational potential energy and ΔU_s is the *change* in the elastic potential energy of the system. In this situation, $\Delta K = 0$ since the initial and final velocities of the system are zero. Also, W_{nc} is the work done by friction, given by

$$(2) \qquad W_{nc} = -fh = -\mu m_1 gh$$

The change in the gravitational potential energy is associated only with m_2 since the vertical coordinate of m_1 does not change. Therefore, we get

$$(3) \qquad \Delta U_g = U_f - U_i = -m_2 gh$$

where the coordinates have been measured from the lowest position of m_2. The change in the elastic potential energy stored in the spring is given by

(4) $\Delta U_s = U_f - U_i = \frac{1}{2}kh^2 - 0$

Substituting (2), (3), and (4) into (1) gives

$$-\mu m_1 gh = -m_2 gh + \tfrac{1}{2}kh^2$$

$$\mu = \frac{m_2 g - \frac{1}{2}kh}{m_1 g}$$

This represents a possible experimental technique for measuring the coefficient of kinetic friction. For example, if $m_1 = 0.50$ kg, $m_2 = 0.30$ kg, $k = 50$ N/m, and $h = 5.0 \times 10^{-2}$ m, we find that

$$\mu = \frac{(0.30\text{ kg})\left(9.8\ \frac{\text{m}}{\text{s}^2}\right) - \frac{1}{2}\left(50\ \frac{\text{N}}{\text{m}}\right)(5.0 \times 10^{-2}\text{ m})}{(0.50\text{ kg})\left(9.8\ \frac{\text{m}}{\text{s}^2}\right)}$$

$$= 0.34$$

8.7 RELATIONSHIP BETWEEN CONSERVATIVE FORCES AND POTENTIAL ENERGY

In the previous sections we saw that the concept of potential energy is related to the configuration, or coordinates, of a system. In a few examples, we showed how to obtain the potential energy from a knowledge of the conservative force. (Remember that one can associate a potential energy function only with a conservative force.)

According to Equation 8.1, the change in the potential energy of a particle under the action of a conservative force equals the negative of the work done by the force. If the system undergoes an infinitesimal displacement, dx, we can express the infinitesimal change in potential energy, dU, as

$$dU = -F_x\, dx$$

Therefore, the conservative force is related to the potential energy function through the relationship

Relation between force and potential energy

$$F_x = -\frac{dU}{dx} \tag{8.18}$$

That is, *the conservative force equals the negative derivative of the potential energy with respect to x.*[3]

We can easily check this relationship for the two examples already discussed. In the case of the deformed spring, $U_s = \frac{1}{2}kx^2$, and therefore

$$F_s = -\frac{dU_s}{dx} = -\frac{d}{dx}\left(\tfrac{1}{2}kx^2\right) = -kx$$

which corresponds to the restoring force in the spring. Since the gravitational potential energy function is given by $U_g = mgy$, it follows from Equation 8.18 that $F_g = -mg$.

We now see that U is an important function, since the conservative force can be derived from it. Furthermore, Equation 8.18 should clarify the fact that adding a constant to the potential energy is unimportant.

*8.8 ENERGY DIAGRAMS AND STABILITY OF EQUILIBRIUM

The qualitative behavior of the motion of a system can often be understood through an analysis of its potential energy curve. Consider the potential en-

[3] In a three-dimensional problem, where U depends on x, y, z, the force is related to U through the expression $\mathbf{F} = -\mathbf{i}\ \partial U/\partial x - \mathbf{j}\ \partial U/\partial y - \mathbf{k}\ \partial U/\partial z$, where $\partial/\partial x$, etc., are partial derivatives. In the language of vector calculus, $\mathbf{F}$ is said to equal the negative of the gradient of the scalar quantity $U(x, y, z)$.

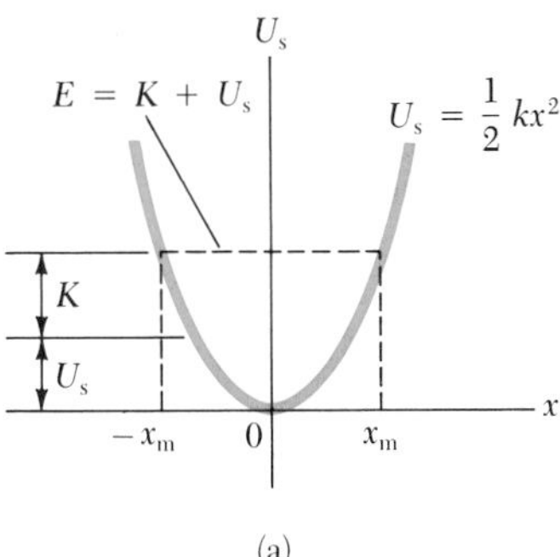

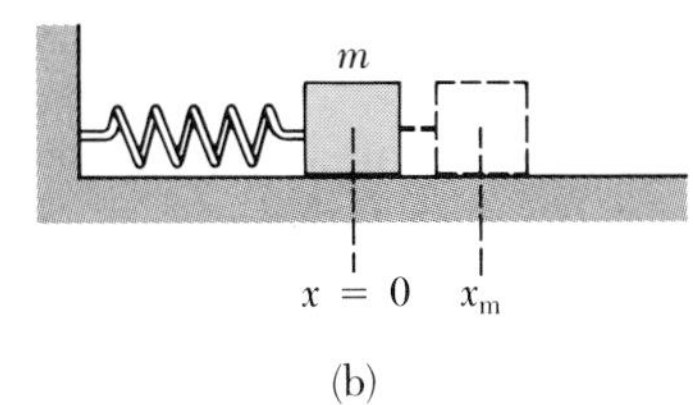

Figure 8.10 (a) The potential energy as a function of x for the block-spring system described in (b). The block oscillates between the turning points, which have the coordinates $x = \pm x_m$. Note that the restoring force of the spring always acts toward $x = 0$, the position of stable equilibrium.

ergy function for the mass-spring system, given by $U_s = \frac{1}{2}kx^2$. This function is plotted versus x in Figure 8.10a. The force is related to U through the expression

$$F_s = -\frac{dU_s}{dx} = -kx$$

That is, the force is equal to the negative of the *slope* of the U versus x curve. When the mass is placed at rest at the equilibrium position ($x = 0$), where $F = 0$, it will remain there unless some external force acts on it. If the spring is stretched from equilibrium, x is positive and the slope dU/dx is positive; therefore F_s is negative and the mass accelerates back toward $x = 0$. If the spring is compressed, x is negative and the slope is negative; therefore F_s is positive and again the mass accelerates toward $x = 0$.

Stable equilibrium

From this analysis, we conclude that the $x = 0$ position is one of **stable equilibrium.** That is, any movement away from this position results in a force that is directed back toward $x = 0$. In general, *positions of stable equilibrium correspond to those points for which U(x) has a minimum value.*

From Fig. 8.10 we see that if the mass is given an initial displacement x_m and released from rest, its total energy initially is the potential energy stored in the spring, given by $\frac{1}{2}kx_m^2$. As motion commences, the system acquires kinetic energy at the expense of losing an equal amount of potential energy. Since the total energy must remain constant, the mass oscillates between the two points $x = \pm x_m$, called the *turning points.* In fact, because there is no energy loss (no friction), the mass will oscillate between $-x_m$ and $+x_m$ forever. (We shall discuss these oscillations further in Chapter 13.) From an energy viewpoint, the energy of the system cannot exceed $\frac{1}{2}kx_m^2$; therefore the mass must stop at these points and, because of the spring force, accelerate toward $x = 0$.

Another simple mechanical system that has a position of stable equilibrium is that of a ball rolling about in the bottom of a spherical bowl. If the ball is displaced from its lowest position, it will always tend to return to that position when released.

Now consider an example where the U versus x curve is as shown in Fig. 8.11. In this case, $F_x = 0$ at $x = 0$, and so the particle is in equilibrium at this point. However, this is a position of **unstable equilibrium** for the following reason. Suppose the particle is displaced to the *right* ($x > 0$). Since the slope is negative for $x > 0$, $F_x = -dU/dx$ is positive and the particle will accelerate away from $x = 0$. Now suppose that the particle is displaced to the left ($x < 0$). In this case, the force is *negative* since the slope is positive for $x < 0$. There-

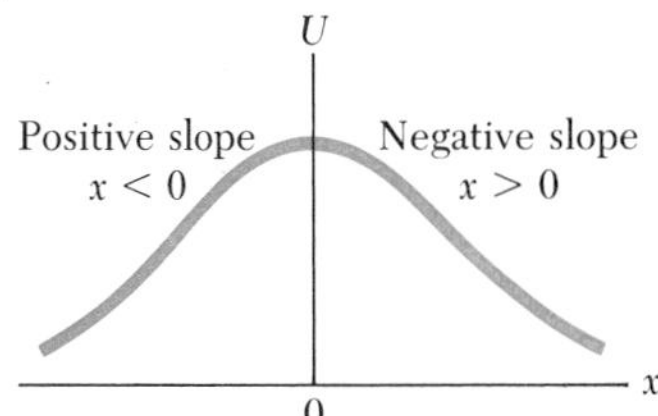

Figure 8.11 A plot of U versus x for a system that has a position of unstable equilibrium, located at $x = 0$. In this case, the force on the system for finite displacements is directed away from $x = 0$.

fore, the particle will again accelerate away from the equilibrium position. Therefore, the $x = 0$ position in this situation is called a position of *unstable equilibrium,* since for any displacement from this point, the force pushes the particle farther away from equilibrium. In fact, the force pushes the particle toward a position of lower potential energy. A ball placed on the top of an inverted spherical bowl is obviously in a position of unstable equilibrium. That is, if the ball is displaced slightly from the top and released, it will surely roll off the bowl. In general, *positions of unstable equilibrium correspond to those points for which* U(x) *has a maximum value.*[4]

Neutral equilibrium

Finally, a situation may arise where U is constant over some region, and hence $F = 0$. This is called a position of **neutral equilibrium.** Small displacements from this position produce neither restoring nor disrupting forces.

8.9 CONSERVATION OF ENERGY IN GENERAL

We have seen that the total mechanical energy of a system is conserved when only conservative forces act on the system. Furthermore, we were able to associate a potential energy function with each conservative force. In other words, mechanical energy is lost when nonconservative forces, such as friction, are present.

We can generalize the energy conservation principle to include all forces acting on the system, both conservative and nonconservative. In the study of thermodynamics we shall find that mechanical energy can be transformed into thermal energy. For example, when a block slides over a rough surface, the mechanical energy lost is transformed into internal energy temporarily stored in the block, as evidenced by a measurable increase in its temperature. On a submicroscopic scale, we shall see that this internal energy is associated with the vibration of atoms about their equilibrium positions. Since this internal atomic motion has kinetic and potential energy, one can say that frictional forces arise fundamentally from conservative atomic forces.[5] Therefore, if we include this increase in the internal energy of the system in our work-energy theorem, the total energy is conserved.

Total energy is always conserved

This is just one example of how you can analyze a system and always find that the total energy of an isolated system does not change, as long as you account for all forms of energy. That is, *energy can never be created or destroyed. Energy may be transformed from one form to another, but the total energy of an isolated system is always constant.* From a universal point of view, we can say that the *total energy of the universe is constant.* Therefore, if one part of the universe gains energy in some form, another part must lose an equal amount of energy. No violation of this principle has been found.

Other examples of energy transformations include the energy carried by sound waves resulting from the collision of two objects, the energy radiated by an accelerating charge in the form of electromagnetic waves (a radio antenna), and the elaborate sequence of energy conversions in a thermonuclear reaction.

In subsequent chapters, we shall see that the energy concept, and especially transformations of energy between various forms, join together the

[4] Mathematically, you can test whether an extreme of U is stable or unstable by examining the sign of d^2U/dx^2.

[5] By introducing the nonconservative force, friction, we are able to limit the system we are studying. We have, in effect, avoided the complex problem of describing the dynamics of 10^{23} molecules and their interactions.

various branches of physics. In other words, one cannot really separate the subjects of mechanics, thermodynamics, and electromagnetism. Finally, from a practical viewpoint, all mechanical and electronic devices rely on some forms of energy transformation.

*8.10 ENERGY FROM THE TIDES

Newton's law of gravity says that the attractive force exerted on one object by another depends inversely on the square of their distance of separation (see Chapter 6). This means that objects on the side of the earth closest to the moon are attracted by the moon more strongly than are objects on the side of the earth opposite the moon. This decrease in gravitational force from one side of the earth to the other is responsible for tides. As shown in Figure 8.12, the attraction of the moon for a mass of water at point *A* is greater than the attraction for a mass of water of the same magnitude centered at *B*. Likewise, the moon attracts a comparable mass of water at *C* even less. The effect of this is to cause a bulge in the water at *A* toward the moon as the water is "pulled" away from the earth. A similar bulge appears at C as the earth is "pulled" away from the water. These bulges are locations at which the water level is higher than average and are called high tides. As the earth rotates on its axis, these bulges occur twice daily. Frictional effects between the ocean floor and the water prevent high tides from occurring at precisely the instant the moon is directly overhead at a particular location.

In certain parts of the world, the tidal variations can be as much as 16 m, largely because of the physical nature of the basins holding the water (as opposed to any extra effect of the moon). When these large surges of water occur in narrow channels, energy can be extracted from the water in the

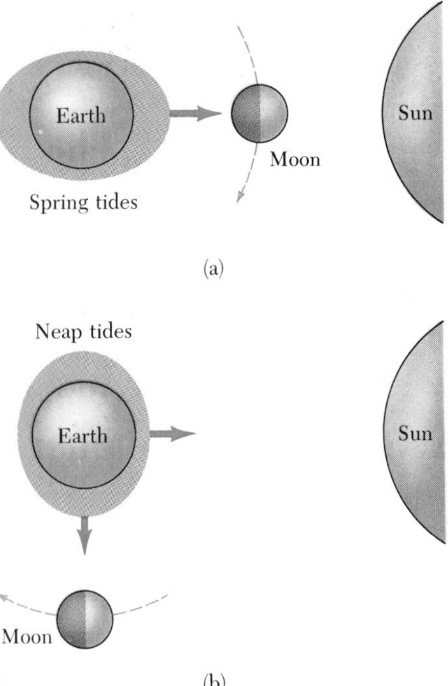

Figure 8.12 Schematic diagram of the tides. (a) The spring tides occur when the earth, moon, and sun are aligned. (b) The neap tides occur when the moon and sun are at right angles relative to the earth. (The figures are not drawn to scale, and the tidal bulges are exaggerated.)

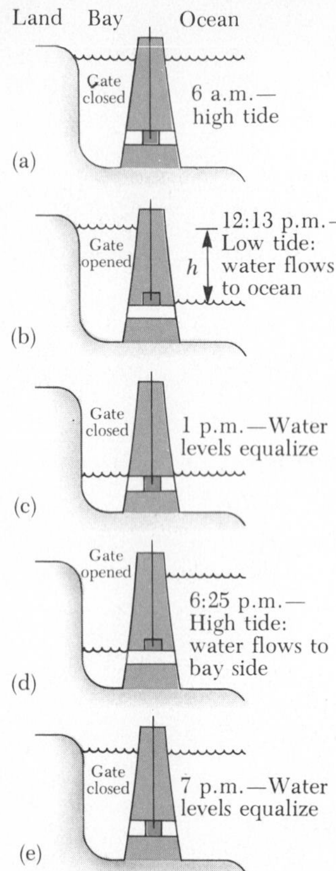

Figure 8.13 Cross-sectional views of a dam used to trap water in a bay during high and low tides. When the gate valve is opened, water flows through the gate and generates electrical power.

following manner. A large dam is constructed in the channel, trapping water on the bay side, as in Figure 8.13. Large gates in the lower portion of the dam allow water to flow when they are opened and trap the water when they are closed. The gates are closed at high tide, when the water levels on the bay and ocean sides are equal. At low tides, the ocean water has dropped a distance h and the gates are opened, allowing water to flow out of the bay. The flowing water is used to drive turbines, which generate electricity. After the water levels are again equal at low tide, the gates are closed. At the next high tide, the gates are again opened, allowing water to flow to the bay side and generating more power. At the next high tide, the gates are closed and the cycle is repeated. In this manner, the water flows through the gates four times per day.

We can estimate the power that can be generated in this manner using the following simple model. Suppose the bay has an area A and the variation between high and low tides is h, as in Figure 8.14. The center of mass of this volume of water must fall through a distance $h/2$; hence the potential energy of the trapped water is

$$U = mg\frac{h}{2}$$

The mass of the trapped water is $m = \rho V = \rho Ah$, where ρ is the density of water; therefore

$$U = \tfrac{1}{2}\rho Agh^2$$

Because the water flows through the gates four times per day, the energy available from the tides each day is four times this value, and the available power is

$$P_{\text{max}} = 4\frac{U}{t} = \frac{2\rho Agh^2}{t}$$

where t is one day.

For example, if we take $A = 5 \times 10^7\ \text{m}^2$ (about 2 square miles) and $h = 2$ m, we find that

$$P_{\text{max}} = \frac{2\left(1000\ \dfrac{\text{kg}}{\text{m}^3}\right)(5 \times 10^7\ \text{m}^2)\left(9.8\ \dfrac{\text{m}}{\text{s}^2}\right)(2\ \text{m})^2}{(24\ \text{h})\left(3600\ \dfrac{\text{s}}{\text{h}}\right)}$$

$$= 45 \times 10^6\ \text{W} = 45\ \text{MW}$$

Because of the inefficiency of the electrical generating facilities and other limiting factors, the actual power is 10 to 25% of this value, or 4.5 to 11 MW.

The most successful operating facility using this principle is located on the Rance River in France, where the tides rise as high as 15 m. This facility has the potential of generating around 240 MW, but the average output is about 62 MW. Clearly, tidal power will be useful only in areas that have large tidal variations and convenient natural bays. Some potential sites include the Cook Inlet in Alaska, the San José Gulf in Argentina, and the Passamaquoddy Bay between Maine and Canada.

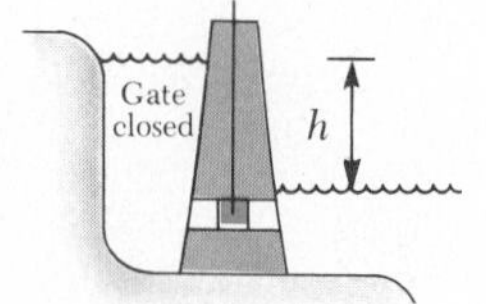

Figure 8.14 The center of gravity of the water trapped in the bay must fall through a distance $h/2$ at low tide before the levels are equalized.

8.11 SUMMARY

A force is **conservative** if the work done by that force acting on a particle is independent of the path the particle takes between two given points. Alternatively, a force is conservative if the work done by that force is zero when the particle moves through an arbitrary closed path and returns to its initial position. A force that does not meet these criteria is said to be **nonconservative.**

A potential energy function U can be associated only with a conservative force. If a conservative force $\mathbf{F}$ acts on a particle that moves along the x axis from x_i to x_f, *the change in the potential energy equals the negative of the work done by that force:*

$$U_f - U_i = -\int_{x_i}^{x_f} F_x \, dx \qquad (8.2)$$

Change in potential energy

The **law of conservation of mechanical energy** states that if the only force acting on a mechanical system is conservative, the total mechanical energy is conserved:

$$K_i + U_i = K_f + U_f \qquad (8.5)$$

Conservation of mechanical energy

The **total mechanical energy of a system** is defined as the sum of the kinetic energy and potential energy:

$$E \equiv K + U \qquad (8.6b)$$

Total mechanical energy

The **gravitational potential energy** of a particle of mass m that is elevated a distance y near the earth's surface is given by

$$U_g \equiv mgy \qquad (8.9)$$

Gravitational potential energy

The **work-energy theorem** states that the work done by all nonconservative forces acting on a system equals the change in the total mechanical energy of the system:

$$W_{nc} = E_f - E_i \qquad (8.13)$$

Work done by nonconservative forces

The **elastic potential energy** stored in a spring of force constant k is

$$U_s \equiv \tfrac{1}{2}kx^2 \qquad (8.14)$$

Potential energy stored in a spring

QUESTIONS

1. A bowling ball is suspended from the ceiling of a lecture hall by a strong cord. The bowling ball is drawn away from its equilibrium position and released from rest at the tip of the demonstrator's nose. If the demonstrator remains stationary, explain why she will not be struck by the ball on its return swing. Would the demonstrator be safe if the ball were given a push from this position?
2. Can the gravitational potential energy of an object ever have a negative value? Explain.
3. A ball is dropped by a person from the top of a building, while another person at the bottom observes its motion. Will these two people agree on the value of the ball's potential energy? on the *change* in potential energy of the ball? on the kinetic energy of the ball?
4. When a person runs in a track event at constant velocity, is any work done? (*Note:* Although the runner may move with constant velocity, the legs and arms undergo acceleration.) How does air resistance enter into the picture?
5. Our body muscles exert forces when we lift, push, run, jump, etc. Are these forces conservative?
6. When nonconservative forces act on a system, does the total mechanical energy remain constant?
7. If three different conservative forces and one nonconservative force act on a system, how many potential energy terms will appear in the work-energy theorem?

8. A block is connected to a spring that is suspended from the ceiling. If the block is set in motion and air resistance is neglected, will the total energy of the system be conserved? How many forms of potential energy are there for this situation?
9. Consider a ball fixed to one end of a rigid rod with the other end pivoted on a horizontal axis so that the rod can rotate in a vertical plane. What are the positions of stable and unstable equilibrium?
10. A ball rolls on a horizontal surface. Is the ball in stable, unstable, or neutral equilibrium?
11. Is it physically possible to have a situation where $E - U < 0$?
12. What will the curve of U versus x look like if a particle is in a region of neutral equilibrium?
13. Explain the energy transformations that occur during the following athletic events: (a) the pole vault, (b) the shotput, (c) the high jump. What is the source of energy in each case?
14. Discuss all the energy transformations that occur during the operation of an automobile.
15. A ball is thrown straight up into the air. At what position is its kinetic energy a maximum? At what position is its gravitational potential energy a maximum?

PROBLEMS

Section 8.1 Conservative and Nonconservative Forces

1. A 3-kg particle moves from the origin to the position having coordinates $x = 5$ m and $y = 5$ m under the influence of gravity acting in the negative y direction (Fig. 8.15.) Using Equation 7.21, calculate the work done by gravity in going from O to C along the following paths: (a) OAC, (b) OBC, (c) OC. Your results should all be identical. Why?

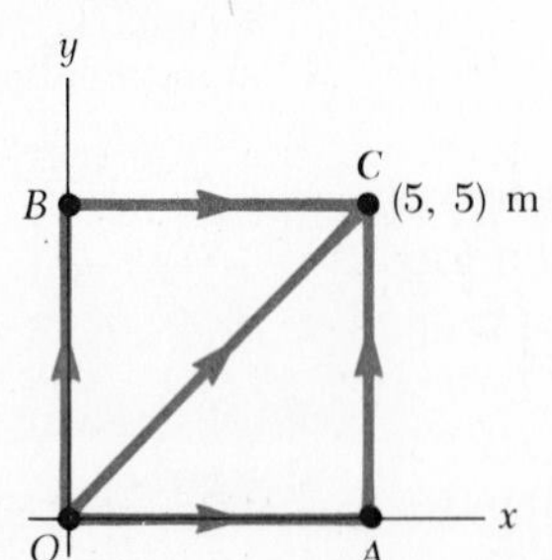

Figure 8.15 (Problems 1, 2, 3, and 5).

2. (a) Starting with Equation 7.20 for the definition of work, show that *any constant force* is conservative. (b) As a special case, suppose a particle of mass m is under the influence of force $\boldsymbol{F} = (2\boldsymbol{i} + 5\boldsymbol{j})$ N and moves from O to C in Figure 8.15. Calculate the work done by $\boldsymbol{F}$ along the three paths OAC, OBC, and OC, and show that they are identical.
3. A particle moves in the xy plane in Figure 8.15 under the influence of a frictional force that opposes its displacement. If the frictional force has a magnitude of 3 N, calculate the total work done by friction along the following *closed* paths: (a) the path OA followed by the return path AO, (b) the path OA followed by AC and the return path CO, and (c) the path OC followed by the return path CO. (d) Your results for the three closed paths should all be different and nonzero. What is the significance of this?
4. A single conservative force acting on a particle varies as $\boldsymbol{F} = (-Ax + Bx^2)\boldsymbol{i}$ N, where A and B are constants and x is in m. (a) Calculate the potential energy associated with this force, taking $U = 0$ at $x = 0$. (b) Find the change in potential energy and change in kinetic energy as the particle moves from $x = 2$ m to $x = 3$ m.
5. A force acting on a particle moving in the xy plane is given by $\boldsymbol{F} = (2y\boldsymbol{i} + x^2\boldsymbol{j})$ N, where x and y are in m. The particle moves from the origin to a final position having coordinates $x = 5$ m and $y = 5$ m, as in Figure 8.15. Calculate the work done by $\boldsymbol{F}$ along (a) OAC, (b) OBC, (c) OC. (d) Is $\boldsymbol{F}$ conservative or nonconservative? Explain.

Section 8.3 Conservation of Mechanical Energy

6. A single conservative force acts on a particle. If its associated potential energy increases by 50 J, find (a) the change in the kinetic energy of the particle, (b) the change in its total energy, and (c) the work done on the particle.
7. A 3-kg particle moves along the x axis under the influence of a single conservative force. If the work done on the particle is 70 J as the particle moves from $x = 2$ m to $x = 5$ m, find (a) the change in the particle's kinetic energy, (b) the change in its potential energy, and (c) its speed at $x = 5$ m if it starts at rest at $x = 2$ m.
8. A single conservative force $F_x = (3x + 5)$ N acts on a 5-kg particle, where x is in m. As the particle moves along the x axis from $x = 1$ m to $x = 4$ m, calculate (a) the work done by this force, (b) the change in the potential energy of the particle, and (c) its kinetic energy at $x = 4$ m if its speed at $x = 1$ m is 3 m/s.
9. At time t_i, the kinetic energy of a particle is 20 J and its potential energy is 10 J. At some later time t_f, its kinetic energy is 15 J. (a) If only conservative forces act on the particle, what is its potential energy at time t_f? What is its total energy? (b) If the potential energy at time t_f is 5 J, are there any nonconservative forces acting on the particle? Explain.
10. A single constant force $\boldsymbol{F} = (3\boldsymbol{i} + 5\boldsymbol{j})$ N acts on a 4-kg particle. (a) Calculate the work done by this force if the particle moves from the origin to the point with vector position $\boldsymbol{r} = (2\boldsymbol{i} - 3\boldsymbol{j})$ m. Does this result depend on the path? Explain. (b) What is the speed of the particle at $\boldsymbol{r}$ if its speed at the origin is 4 m/s? (c) What is the change in the potential energy of the particle?

Section 8.4 Gravitational Potential Energy Near the Earth's Surface

11. A 2-kg ball hangs at the end of a string 1 m in length from the ceiling of a room. The height of the room is 3 m. What is the gravitational potential energy of the ball relative to (a) the ceiling, (b) the floor, and (c) a point at the same elevation as the ball?

12. A rocket is launched at an angle of 37° to the horizontal from an altitude h with a speed v_0. (a) Use energy methods to find the speed of the rocket when its altitude is $h/2$. (b) Find the x and y components of velocity when the rocket's altitude is $h/2$, using the fact that $v_x = v_{x0} =$ constant (since $a_x = 0$) and the results to (a).

13. A 3-kg mass is attached to a light string of length 1.5 m to form a pendulum (Fig. 8.5). The mass is given an initial speed of 4 m/s at its lowest position. When the string makes an angle of 30° with the vertical, find (a) the *change* in the potential energy of the mass, (b) the speed of the mass, and (c) the tension in the string. (d) What is the maximum height reached by the mass above its lowest position?

14. A 0.4-kg ball is thrown vertically upward with an initial speed of 15 m/s. Assuming its initial potential energy is zero, find its kinetic energy, potential energy, and total mechanical energy (a) at its initial position, (b) when its height is 3 m, and (c) when it reaches the top of its flight. (d) Find its maximum height using the law of conservation of energy.

15. A 0.3-kg ball is thrown into the air and reaches a maximum altitude of 50 m. Taking its initial position as the point of zero potential energy and using energy methods, find (a) its initial speed, (b) its total mechanical energy, and (c) the ratio of its kinetic energy to its potential energy when its altitude is 10 m.

16. A 200-g particle is released from rest at point A along the diameter on the inside of a smooth hemispherical bowl of radius $R = 30$ cm (Fig. 8.16). Calculate (a) its gravitational potential energy at point A relative to point B, (b) its kinetic energy at point B, (c) its speed at point B, and (d) its kinetic energy and potential energy at point C.

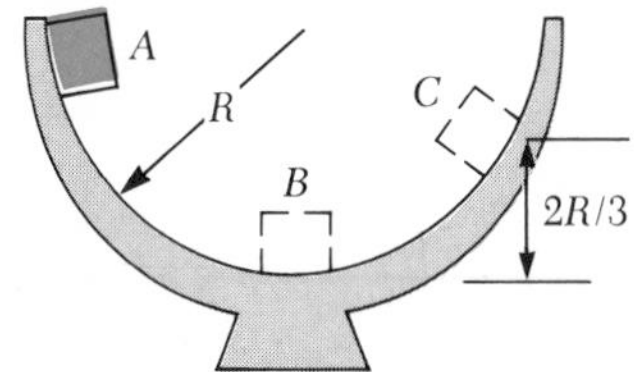

Figure 8.16 (Problems 16 and 17).

Section 8.5 Nonconservative Forces and the Work-Energy Theorem

17. The particle described in Problem 16 (Fig. 8.16) is released from point A at rest. The speed of the particle at point B is 1.5 m/s. (a) What is its kinetic energy at B? (b) How much energy is lost as a result of friction as the particle goes from A to B? (c) Is it possible to determine μ from these results in any simple manner? Explain.

18. A 2-kg block is projected up the incline shown in Fig. 8.6 with an initial speed of 3 m/s at the bottom. The coefficient of friction between the block and the incline is 0.7. Find (a) the distance the block will travel up the incline before coming to rest, (b) the total work done by friction while the block is in motion on the incline, and (c) the change in potential energy and change in kinetic energy when the block has traveled 0.3 m up the incline.

19. The total initial mechanical energy of a particle moving along the x axis is 80 J. A frictional force of 6 N is the *only* force acting on the particle. When the total mechanical energy is 30 J, find (a) the distance the particle has traveled, (b) the change in the particle's kinetic energy, and (c) the change in its potential energy.

20. In a given displacement of a particle, its kinetic energy *decreases* by 25 J while its potential energy *increases* by 10 J. Are there any nonconservative forces acting on the particle? If so, how much work is done by these forces?

21. A child starts from rest at the top of a slide of height $h = 4$ m (Fig. 8.7). (a) What is her speed at the bottom if the incline is frictionless? (b) If she reaches the bottom with a speed of 6 m/s, what percentage of her total energy is lost as a result of friction?

22. A 3-kg particle moving along the x axis has a velocity of $6\mathbf{i}$ m/s when its x coordinate is 3 m. The only force acting on it is a constant retarding force of $-12\mathbf{i}$ N. (a) Find its coordinate when it comes to rest. (b) How much work is done by friction as the particle moves from the origin to the point where it is at rest? (c) What is the change in kinetic energy as the particle moves from the origin to $x = 3$ m?

23. A 0.4-kg bead slides on a curved wire, starting from rest at point A in Figure 8.17. The segment from A to B is frictionless, and the segment from B to C is rough. (a) Find the speed of the bead at B. (b) If the bead comes to rest at C, find the total work done by friction in going from B to C. (c) What is the net work done by nonconservative forces as the bead moves from A to C?

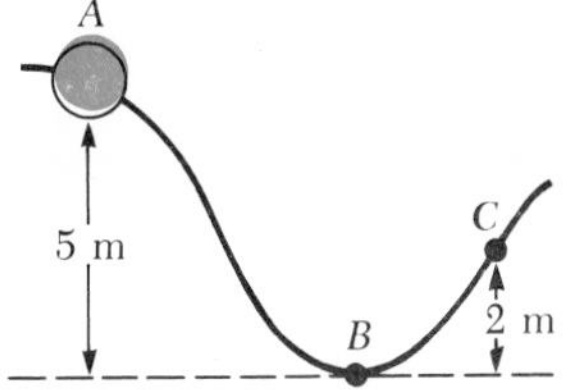

Figure 8.17 (Problem 23).

24. A 25-kg child on a swing 2 m long is released from rest when the swing supports make an angle of 30° with the vertical. (a) Neglecting friction, find the child's speed at the lowest position. (b) If the speed of the child at the lowest position is 2 m/s, what is the energy loss due to friction?

Section 8.6 Potential Energy Stored in a Spring

25. A spring has a force constant of 400 N/m. How much work must be done on the spring to stretch it (a) 3 cm from its equilibrium position and (b) from $x = 2$ cm to $x = 3$ cm, where $x = 0$ is its equilibrium position? (In the unstretched position, the potential energy is defined to be zero.)
26. A spring has a force constant of 500 N/m. What is the elastic potential energy stored in the spring when (a) it is stretched 4 cm from equilibrium, (b) it is compressed 3 cm from equilibrium, and (c) it is unstretched?
27. A block of mass m is released from rest and slides down a frictionless track of height h above a table (Fig. 8.18). At the bottom of the track, where the surface is horizontal, the block strikes and sticks to a light spring. (a) Find the maximum distance the spring is compressed. (b) Obtain a numerical value for this distance if $m = 0.2$ kg, $h = 1$ m, and $k = 490$ N/m.

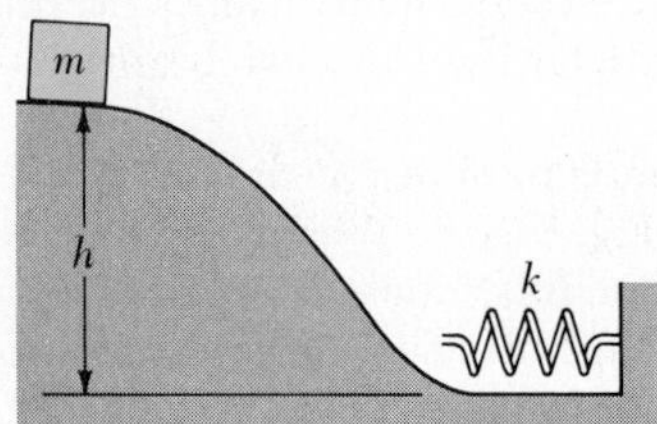

Figure 8.18 (Problem 27).

28. An 8-kg block travels on a rough, horizontal surface and collides with a spring as in Figure 8.8. The speed of the block *just before* the collision is 4 m/s. As the block rebounds to the left with the spring uncompressed, its speed as it leaves the spring is 3 m/s. If the coefficient of kinetic friction between the block and surface is 0.4, determine (a) the work done by friction while the block is in contact with the spring and (b) the maximum distance the spring is compressed.
29. A 3-kg mass is fastened to a light spring that passes over a pulley (Fig. 8.19). The pulley is frictionless, and the mass is released from rest when the spring is unstretched. If the mass drops a distance of 10 cm before coming to rest, find (a) the force constant of the spring, and (b) the speed of the mass when it is 5 cm below its starting point.

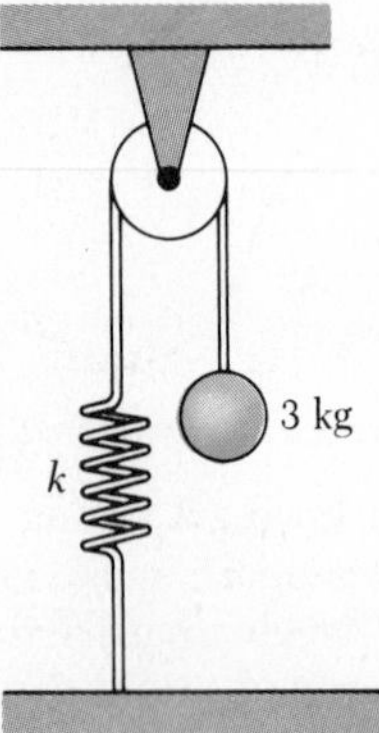

Figure 8.19 (Problem 29).

30. A child's toy consists of a piece of plastic attached to a spring (Fig. 8.20). The spring is compressed against the floor a distance of 2 cm, and the toy is released. If the mass of the toy is 100 g and it rises to a maximum height of 60 cm, estimate the force constant of the spring.

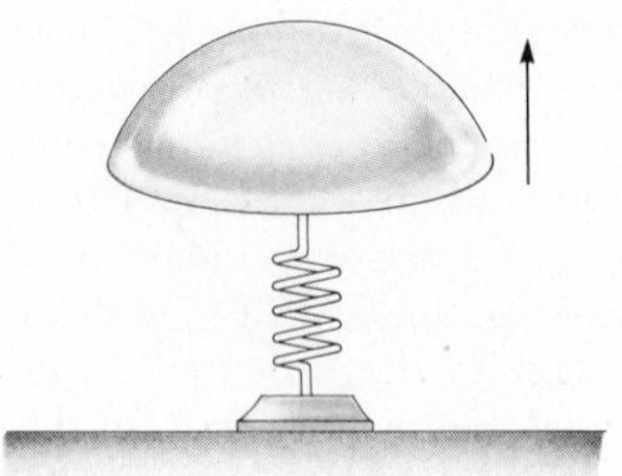
Figure 8.20 (Problem 30).

Section 8.7 Relationship Between Conservative Forces and Potential Energy

31. The potential energy of a two-particle system separated by a distance r is given by $U(r) = A/r$, where A is a constant. Find the radial force F_r.
32. The potential energy function for a system is given by $U = ax^2 - bx$, where a and b are constants. (a) Find the force F_x associated with this potential energy function. (b) At what value of x is the force zero?

*Section 8.8 Energy Diagrams and Stability of Equilibrium

33. Consider the potential energy curve $U(x)$ versus x shown in Figure 8.21. (a) Determine whether the force F_x is positive, negative, or zero at the various points indicated. (b) Indicate points of stable, unstable, or neutral equilibrium.

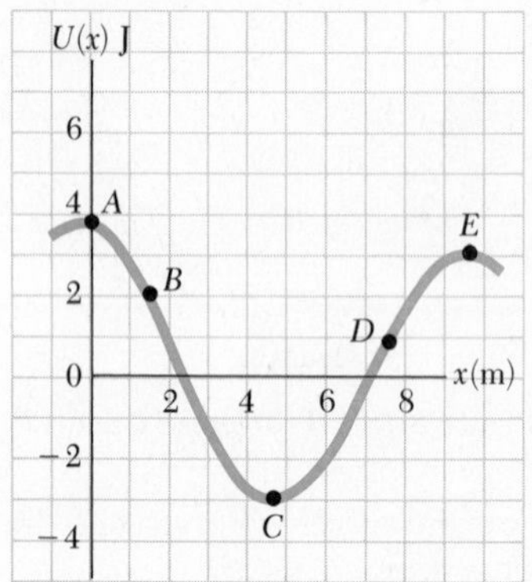

Figure 8.21 (Problems 33 and 34).

34. With reference to the potential energy curve in Figure 8.21, make a rough sketch of the F_x versus x curve from $x = 0$ to $x = 8$ m.

35. A right circular cone can be balanced on a horizontal surface in three different ways. Sketch these three equilibrium configurations, and identify them as being positions of stable, unstable, or neutral equilibrium.

°Section 8.10 Energy from the Tides

36. The Bay of Fundy in Canada has an average tidal range of 8 m and an area of 13 000 km^2. What is the average power available from this supply of water assuming an overall efficiency of 25%?

GENERAL PROBLEMS

37. The masses of the javelin, the discus, and the shot are 0.8 kg, 2.0 kg, and 7.2 kg, respectively, and record throws in the track events using these objects are about 89 m, 69 m, and 21 m, respectively. Neglecting air resistance, (a) calculate the minimum initial kinetic energies that would produce these throws, and (b) estimate the average force exerted on each object during the throw assuming the force acts over a distance of 2 m. (c) Do your results suggest that air resistance is an important factor?

38. An olympic high jumper whose height is 2 m makes a record leap of 2.3 m over a horizontal bar. Estimate the speed with which he must leave the ground to perform this feat. (*Hint:* Estimate the position of his center of gravity before jumping, and assume he is in a horizontal position when he reaches the peak of his jump.)

39. Prove that the following forces are conservative and find the change in potential energy corresponding to these forces taking $x_i = 0$ and $x_f = x$: (a) $F_x = ax + bx^2$, (b) $F_x = Ae^{\alpha x}$. (a, b, A, and α are all constants.)

40. Find the forces corresponding to the following potential energy functions: (a) K/y, (b) bx^3, (c) e^{-ar}/r. (K, b, and a are all constants.)

41. A 2-kg block situated on a rough incline is connected to a light spring having a force constant of 100 N/m (Fig. 8.22). The block is released from rest when the spring is unstretched and the pulley is frictionless. The block moves 20 cm down the incline before coming to rest. Find the coefficient of kinetic friction between the block and the incline.

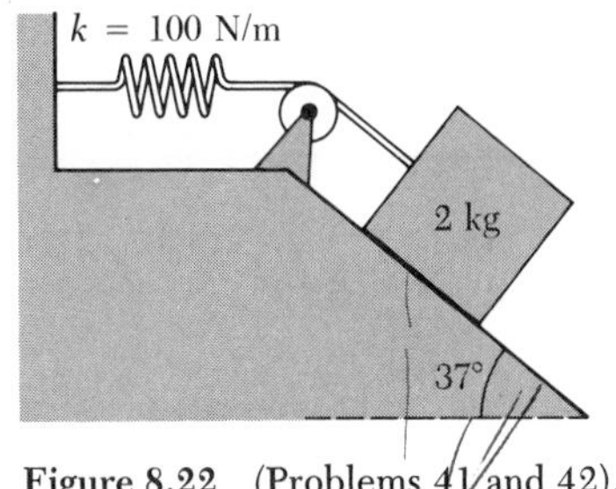

Figure 8.22 (Problems 41 and 42).

42. Suppose the incline is *smooth* for the system described in Problem 7 (Fig. 8.22). The block is released from rest with the spring initially unstretched. (a) How far does it move down the incline before coming to rest? (b) What is the acceleration of the block when it reaches its lowest point? Is the acceleration constant? (c) Describe the energy transformations that occur during the descent of the block.

43. A ball whirls around in a vertical circle at the end of a string. If the ball's total energy remains constant, show that the tension in the string at the bottom is greater than the tension at the top by six times the weight of the ball.

44. A pendulum of length L swings in the vertical plane. The string hits a peg located a distance d below the point of suspension (Fig. 8.23). (a) Show that if the pendulum is released at a height *below* that of the peg, it will return to this height after striking the peg. (b) Show that if the pendulum is released from the horizontal position ($\theta = 90°$) and the pendulum is to swing in a complete circle centered on the peg, then the minimum value of d must be $3L/5$.

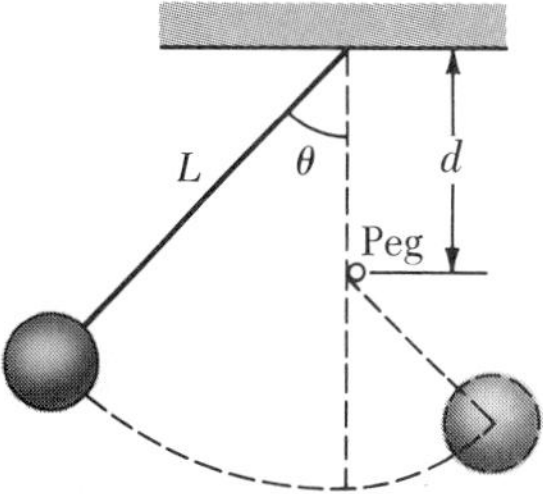

Figure 8.23 (Problem 44).

45. A 25-kg block is connected to a 30-kg block by a light string that passes over a frictionless pulley. The 30-kg block is connected to a light spring of force constant 200 N/m, as in Figure 8.24. The spring is unstretched when the system is as shown in the figure, and the incline is smooth. The 25-kg block is pulled a distance of 20 cm down the incline (so that the 30-kg block is 40 cm above the floor) and is released from rest. Find the speed of each block when the 30-kg block is 20 cm above the floor (that is, when the spring is unstretched).

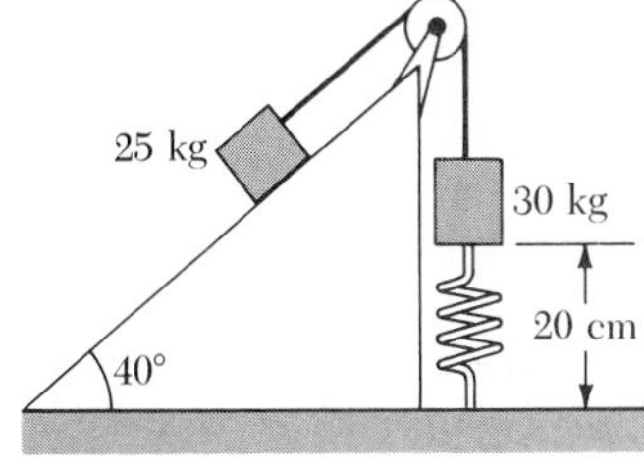

Figure 8.24 (Problem 45).

46. A potential energy function for a system is given by $U(x) = 3x + 4x^2 - x^3$. (a) Determine the force F_x as a function of x. (b) For what values of x is the force equal to zero? (c) Plot $U(x)$ versus x and F_x versus x and indicate points of stable and unstable equilibrium.

47. A 2-kg mass is suspended by means of a light string that passes over a light, frictionless pulley as shown in Figure 8.25. The other end of the string is connected to a 1-kg mass that rests upon a horizontal frictionless surface. The system starts in motion with the string connected to the 1-kg mass making an angle of 30° with the horizontal. When the string makes an angle of 45° with the horizontal, how much work has been done on the 1-kg mass? (The pulley is 2 m above the surface and the surface is frictionless.)

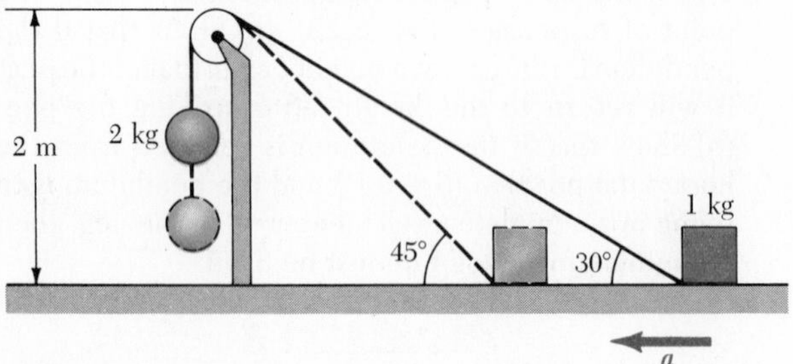

Figure 8.25 (Problem 47).

48. A frictionless roller coaster is given an initial velocity v_0 at a height h, as in Figure 8.26. The radius of curvature of the track at point A is R. (a) Find the *maximum* value of v_0 necessary in order that the roller coaster *not* leave the track at A. (b) Using the value of v_0 calculated in (a), determine the value of h' necessary if the roller coaster is to just make it to point B.

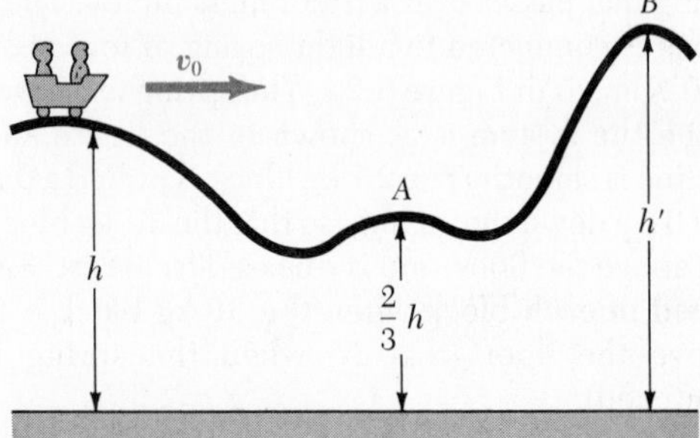

Figure 8.26 (Problem 48).

49. A 1.0-kg mass slides to the right on a surface with coefficient of friction $\mu = 0.25$ (Fig. 8.27). It has a speed of $v_i = 3$ m/s when contact is made with a spring with spring constant $k = 50$ N/m. The mass comes to rest after the spring has been compressed a distance d. The mass is then forced toward the left by the spring and it continues to move in that direction beyond the unstretched position. Finally the mass comes to rest a distance D to the left of the unstretched position. Find the following: (a) the compressed distance d, (b) the velocity v at the unstretched position, and (c) the distance D where the mass will come to rest to the left of the unstretched position.

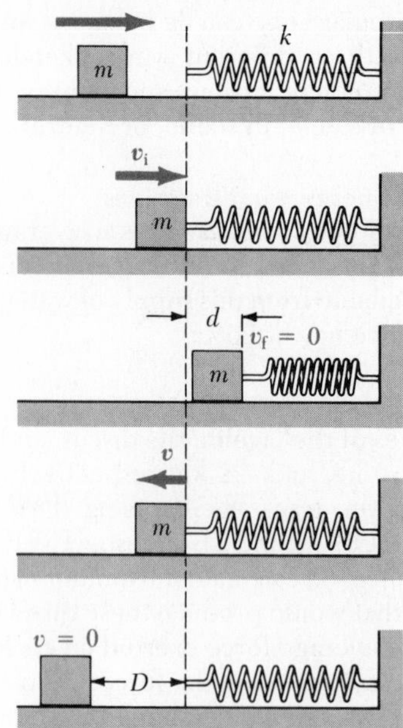

Figure 8.27 (Problem 49).

•50. A skier starts at rest from the top of a large hill that is shaped like a hemisphere (Fig. 8.28). Neglecting friction, show that the skier will leave the hill and become "air-borne" at a distance $h = R/3$ below the top of the hill. (*Hint:* At this point, the normal force goes to zero.)

•51. A block of mass m is dropped from rest at a height h directly above the top of a vertical spring having a force constant k. Find the *maximum* distance the spring will be compressed.

•52. A uniform rope of length L lies on a horizontal smooth table. Part of the rope, of length d, hangs over the table, and the rope is released from rest (Fig. 8.29). Using energy methods, find (a) the velocity of the rope at the instant all of the rope leaves the table and (b) the time it takes this to occur. (*Hint:* Note the motion of the center of gravity of the rope.)

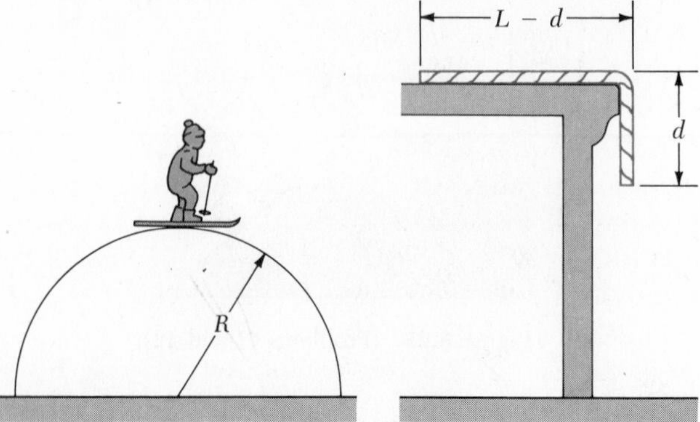

Figure 8.28 (Problem 50). Figure 8.29 (Problem 52).

9

Linear Momentum and Collisions

In this chapter we shall analyze the motion of a system containing many particles. We shall introduce the concept of the linear momentum of the system of particles and show that this momentum is conserved when the system is isolated from its surroundings. The law of momentum conservation is especially useful for treating such problems as the collisions between particles and for analyzing rocket propulsion. The concept of the center of mass of a system of particles will also be introduced. We shall show that the overall motion of a system of interacting particles can be represented by the motion of an equivalent particle located at the center of mass.

9.1 LINEAR MOMENTUM AND IMPULSE

The **linear momentum** of a particle of mass m moving with a velocity $\boldsymbol{v}$ was defined in Chapter 5 to be the product of the mass and velocity:[1]

$$\boldsymbol{p} \equiv m\boldsymbol{v} \tag{9.1}$$

Definition of linear momentum of a particle

Momentum is a vector quantity since it equals the product of a scalar, m, and a vector, $\boldsymbol{v}$. Its direction is along $\boldsymbol{v}$, and it has dimensions of ML/T. In the SI system, momentum has the units kg · m/s.

If a particle is moving in an arbitrary direction, $\boldsymbol{p}$ will have three components and Equation 9.1 is equivalent to the component equations given by

$$p_x = mv_x \qquad p_y = mv_y \qquad p_z = mv_z \tag{9.2}$$

We can relate the linear momentum to the force acting on the particle using Newton's second law of motion: *The time rate of change of the momentum of a particle is equal to the resultant force on the particle.* That is,

$$\boldsymbol{F} = \frac{d\boldsymbol{p}}{dt} \tag{9.3}$$

Newton's second law for a particle

From Equation 9.3 we see that if the resultant force is zero, the momentum of the particle must be constant. In other words, the linear momentum of a particle is conserved when $\boldsymbol{F} = 0$. Of course, if the particle is *isolated* (that is, if it does not interact with its environment), then by necessity, $\boldsymbol{F} = 0$ and $\boldsymbol{p}$ remains unchanged. This result can also be obtained directly through the application of Newton's second law in the form $\boldsymbol{F} = m\, d\boldsymbol{v}/dt$. That is, when the force is zero, the acceleration of the particle is zero and the velocity remains constant.

[1] This expression is nonrelativistic, and so it is valid only when $v \ll c$. For relativistic speeds, $p = mv/(1 - v^2/c^2)^{1/2}$.

Equation 9.3 can be written

$$dp = F\,dt \tag{9.4}$$

We can integrate this expression to find the change in the momentum of a particle. If the momentum of the particle changes from p_i at time t_i to p_f at time t_f, then integrating Equation 9.4 gives

$$\Delta p = p_f - p_i = \int_{t_i}^{t_f} F\,dt \tag{9.5}$$

The quantity on the right side of Equation 9.5 is called the *impulse* of the force F for the time interval $\Delta t = t_f - t_i$. Impulse is a vector defined by

Impulse of a force

$$I = \int_{t_i}^{t_f} F\,dt = \Delta p \tag{9.6}$$

That is,

> the impulse of the force F equals the change in the momentum of the particle.

Impulse-momentum theorem

This statement, known as the **impulse-momentum theorem,** is equivalent to Newton's second law. From this definition, we see that impulse is a vector quantity having a magnitude equal to the area under the force-time curve, as described in Figure 9.1a. In this figure, it is assumed that the force varies in time in the general manner shown and is nonzero in the time interval $\Delta t = t_f - t_i$. The direction of the impulse vector is the same as the direction of the change in momentum. Impulse has the dimensions of momentum, that is, ML/T. Note that impulse is *not* a property of the particle itself, but is a quantity that measures the degree to which an external force changes the momentum of the particle. Therefore, when we say that an impulse is given to a particle, it is implied that momentum is transferred from an external agent to that particle.

Since the force can generally vary in time as in Figure 9.1a, it is convenient to define a time-averaged force $\overline{F}$, given by

$$\overline{F} = \frac{1}{\Delta t}\int_{t_i}^{t_f} F\,dt \tag{9.7}$$

where $\Delta t = t_f - t_i$. Therefore, we can express Equation 9.6 as

$$I = \Delta p = \overline{F}\,\Delta t \tag{9.8}$$

This average force, described in Figure 9.1b, can be thought of as the constant force that would give the same impulse to the particle in the time interval Δt as the actual time-varying force gives over this same interval.

In principle, if F is known as a function of time, the impulse can be calculated from Equation 9.6. The calculation becomes especially simple if the force acting on the particle is constant. In this case, $\overline{F} = F$ and Equation 9.8 becomes

$$I = \Delta p = F\,\Delta t \tag{9.9}$$

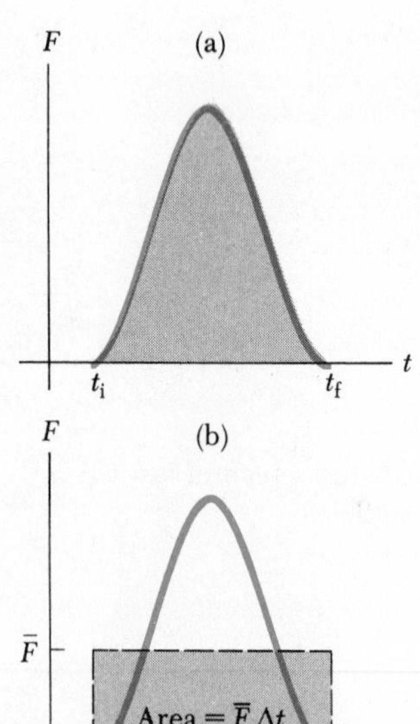

Figure 9.1 (a) A force acting on a particle may vary in time. The impulse is the area under the force versus time curve. (b) The average force (horizontal line) would give the same impulse to the particle in the time Δt as the real time-varying force described in (a).

In many physical situations, we shall use the so-called **impulse approximation.** In this approximation, *we assume that one of the forces exerted on a* *particle acts for a short time but is much larger than any other force present.* This approximation is especially useful in treating collisions, where the duration of the collision is very short. When this approximation is made, we refer to the force as an *impulsive force*. For example, when a baseball is struck with a bat, the time of the collision is about 0.01 s, and the average force the bat exerts on the ball in this time is typically several thousand pounds. This is much greater than the force of gravity, and so the impulse approximation is justified. When we use this approximation, it is important to remember that p_i and p_f represent the momenta *immediately* before and after the collision, respectively. Therefore, in the impulse approximation there is very little motion of the particle during the collision.

EXAMPLE 9.1. Teeing Off

A golf ball of mass 50 g is struck with a club (Fig. 9.2). The force on the ball varies from zero when contact is made up to some maximum value (where the ball is deformed) back to zero when the ball leaves the club. Thus, the force-time curve is qualitatively described by Figure 9.1. Assuming that the ball travels a distance of 200 m, (a) estimate the impulse due to the collision.

Neglecting air resistance, we can use the expression for the range of a projectile (Chapter 4) given by

$$R = \frac{v_0^2}{g} \sin 2\theta_0$$

Let us assume that the launch angle is 45°, which provides the maximum range for any given launch speed. The initial velocity of the ball is then estimated to be

$$v_0 = \sqrt{Rg} = \sqrt{(200 \text{ m})(9.80 \text{ m/s}^2)} = 44.3 \text{ m/s}$$

Since $v_i = 0$ and $v_f = v_0$ for the ball, the magnitude of the impulse imparted to the ball is

$$I = \Delta p = mv_0 = (50 \times 10^{-3} \text{ kg})\left(44 \ \frac{\text{m}}{\text{s}}\right) = 2.2 \text{ kg} \cdot \text{m/s}$$

(b) Estimate the time of the collision.

From Figure 9.2, it appears that a reasonable estimate of the distance the ball travels while in contact with the club is the radius of the ball, about 2 cm. The time it takes the club to move this distance (the contact time) is then

$$\Delta t = \frac{\Delta x}{v_0} = \frac{2 \times 10^{-2} \text{ m}}{44 \text{ m/s}} = 4.5 \times 10^{-4} \text{ s}$$

Figure 9.2 A golf ball being struck by a club.

Exercise 1 Estimate the magnitude of the average force exerted on the ball during the collision with the club.
Answer: 4.91×10^3 N. Note that this force is extremely large compared with the weight (gravity force) of the ball, which is only 0.49 N.

EXAMPLE 9.2. Follow the Bouncing Ball

A ball of mass 100 g is dropped from a height $h = 2$ m above the floor (Fig. 9.3). It rebounds vertically to a height $h' = 1.5$ m after colliding with the floor. (a) Find the momentum of the ball immediately before and after the ball collides with the floor.

Using the energy methods, we can find v_i, the velocity of the ball just before it collides with the floor, through the relationship

$$\tfrac{1}{2}mv_i^2 = mgh$$

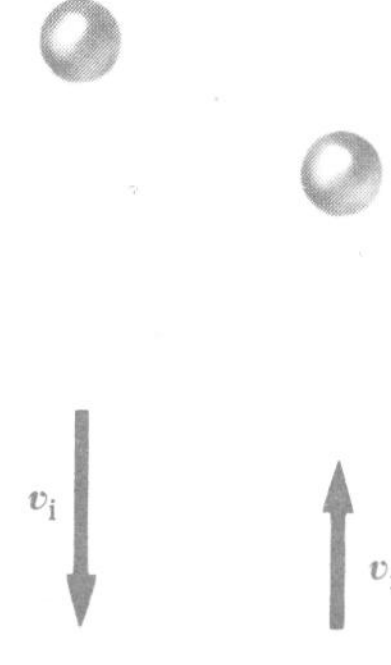

Figure 9.3 (Example 9.2) (a) The ball is dropped from a height h and reaches the floor with a velocity v_i. (b) The ball rebounds from the floor with a velocity v_f and reaches a height h'.

Likewise, v_f, the ball's velocity right after colliding with the floor, is obtained from the energy expression

$$\tfrac{1}{2}mv_f^2 = mgh'$$

Substituting into these expressions the values $h = 2.0$ m and $h' = 1.5$ m gives

$$v_i = \sqrt{2gh} = \sqrt{(2)(9.80)(2)}\ \text{m/s} = 6.26\ \text{m/s}$$

$$v_f = \sqrt{2gh'} = \sqrt{(2)(9.80)(1.5)}\ \text{m/s} = 5.42\ \text{m/s}$$

Since $m = 0.1$ kg, the vector expressions for the initial and final linear momenta are given by

$$\mathbf{p}_i = m\mathbf{v}_i = -0.63\mathbf{j}\ \text{kg}\cdot\text{m/s}$$

$$\mathbf{p}_f = m\mathbf{v}_f = 0.54\mathbf{j}\ \text{kg}\cdot\text{m/s}$$

(b) Determine the average force exerted by the floor on the ball. Assume the time of the collision is 10^{-2} s (a typical value).

Using Equation 9.5 and the definition of $\overline{\mathbf{F}}$, we get

$$\Delta\mathbf{p} = \mathbf{p}_f - \mathbf{p}_i = \overline{\mathbf{F}}\ \Delta t$$

$$\overline{\mathbf{F}} = \frac{[0.54\mathbf{j} - (-0.63\mathbf{j})]\ \text{kg}\cdot\text{m/s}}{10^{-2}\ \text{s}} = 1.2\times 10^2\mathbf{j}\ \text{N}$$

Note that this average force is much greater than the force of gravity ($mg \approx 1.0$ N). That is, the impulsive force due to the collision with the floor overwhelms the gravitational force. In this inelastic collision, the energy lost by the ball is transformed into heat, sound, and distortions of the ball and floor.

9.2 CONSERVATION OF LINEAR MOMENTUM FOR A TWO-PARTICLE SYSTEM

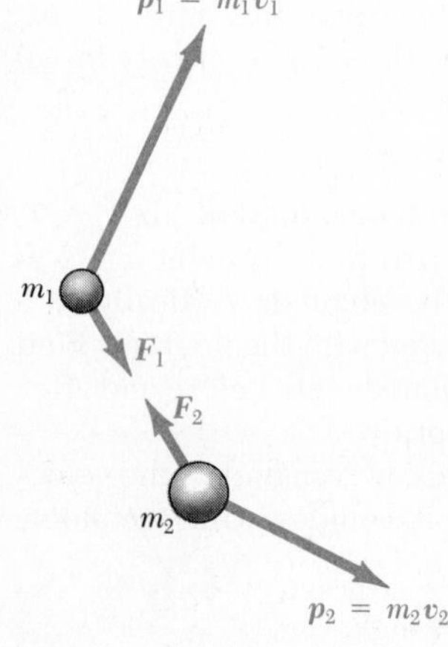

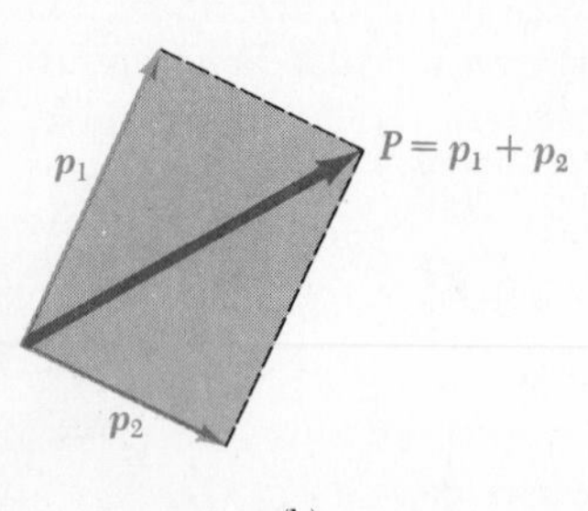

Figure 9.4 (a) At some instant, the momentum of m_1 is $\mathbf{p}_1 = m_1\mathbf{v}_1$ and the momentum of m_2 is $\mathbf{p}_2 = m_2\mathbf{v}_2$. If the particles are isolated, $\mathbf{F}_1 = -\mathbf{F}_2$. (b) The total momentum of the system, $\mathbf{P}$, is equal to the vector sum $\mathbf{p}_1 + \mathbf{p}_2$.

Consider two particles that can interact with each other but are isolated from their surroundings (Fig. 9.4). That is, the particles exert forces on each other, but no external forces are present.[2] Suppose that at some time t, the momentum of particle 1 is $\mathbf{p}_1$ and the momentum of particle 2 is $\mathbf{p}_2$. We can apply Newton's second law to each particle and write

$$\mathbf{F}_{12} = \frac{d\mathbf{p}_1}{dt} \qquad \text{and} \qquad \mathbf{F}_{21} = \frac{d\mathbf{p}_2}{dt}$$

where $\mathbf{F}_{12}$ is the force on particle 1 due to particle 2 and $\mathbf{F}_{21}$ is the force on particle 2 due to particle 1. These forces could be gravitational forces, electrostatic forces, or of some other origin. This really isn't important for the present discussion. However, Newton's third law tells us that $\mathbf{F}_{12}$ and $\mathbf{F}_{21}$ are equal in magnitude and opposite in direction. That is, they form an action-reaction pair and $\mathbf{F}_{12} = -\mathbf{F}_{21}$. We can also express this condition as

$$\mathbf{F}_{12} + \mathbf{F}_{21} = 0$$

or

$$\frac{d\mathbf{p}_1}{dt} + \frac{d\mathbf{p}_2}{dt} = \frac{d}{dt}(\mathbf{p}_1 + \mathbf{p}_2) = 0$$

Since the time derivative of the total momentum, $\mathbf{P} = \mathbf{p}_1 + \mathbf{p}_2$, is *zero*, we conclude that the *total* momentum, $\mathbf{P}$, must remain constant, that is,

$$\mathbf{P} = \mathbf{p}_1 + \mathbf{p}_2 = \text{constant} \tag{9.10}$$

This vector equation is equivalent to three component equations. In other words, Equation 9.10 in component form says that the total momenta in the x, y, and z directions are all independently conserved, or

$$P_{ix} = P_{fx} \qquad P_{iy} = P_{fy} \qquad P_{iz} = P_{fz}$$

[2] A truly isolated system cannot be achieved in the laboratory, since gravitational forces and friction will always be present.

We can state this law, known as **the conservation of linear momentum**, as follows:

> If two particles of masses m_1 and m_2 form an isolated system, then the total momentum of the system is conserved, regardless of the nature of the force between them (provided the force obeys Newton's third law). More simply, whenever two particles collide their total momentum remains constant provided they are isolated.

Suppose $\mathbf{v}_{1i}$ and $\mathbf{v}_{2i}$ are the initial velocities of particles 1 and 2, and $\mathbf{v}_{1f}$ and $\mathbf{v}_{2f}$ are their velocities at some later time. Applying Equation 9.10, we can express the conservation of linear momentum of this isolated system in the form

$$m_1\mathbf{v}_{1i} + m_2\mathbf{v}_{2i} = m_1\mathbf{v}_{1f} + m_2\mathbf{v}_{2f} \tag{9.11}$$

$$\mathbf{p}_{1i} + \mathbf{p}_{2i} = \mathbf{p}_{1f} + \mathbf{p}_{2f} \tag{9.12}$$

Conservation of momentum

That is, *the total momentum of the isolated system at all times equals its initial total momentum*. We can also describe the law of conservation of momentum in another way. Since we require that the system be isolated, the only forces acting must be internal to the system (the action-reaction pair). In other words, if there are no external forces present, the total momentum of the system remains constant. Therefore, momentum conservation for an isolated system is an alternative and equivalent statement of Newton's third law.

The law of conservation of momentum is considered to be one of the most important laws of mechanics. That is, mechanical energy is only conserved for an isolated system when conservative forces alone act on a system. On the other hand, momentum is conserved for an isolated two-particle system *regardless* of the nature of the internal forces. In fact, in Section 9.7 we shall show that the law of conservation of linear momentum also applies to an isolated system of n particles.

EXAMPLE 9.3. The Recoiling Cannon
A 3000-kg cannon rests on a frozen pond as in Figure 9.5. The cannon is loaded with a 30-kg cannon ball and is fired horizontally. If the cannon recoils to the right with a velocity of 1.8 m/s, what is the velocity of the cannon ball just after it leaves the cannon?

Solution: In this example, the system consists of the cannon ball and the cannon. The system is not really isolated because of the force of gravity. However, this external force acts in the vertical direction, while the motion of the system is in the horizontal direction. Therefore, momentum is conserved in the x direction since there are no external forces in this direction (assuming the surface is frictionless).

The total momentum of the system before firing is zero. Therefore, the total momentum after firing must be zero, or

$$m_1v_1 + m_2v_2 = 0$$

With $m_1 = 3000$ kg, $v_1 = 1.8$ m/s, and $m_2 = 30$ kg, solving for v_2, the velocity of the cannon ball, gives

$$v_2 = -\frac{m_1}{m_2}v_1 = -\left(\frac{3000 \text{ kg}}{30 \text{ kg}}\right)(1.8 \text{ m/s}) = -180 \text{ m/s}$$

The negative sign for v_2 indicates that the ball is moving to the left after firing, in the direction opposite the movement of the cannon.

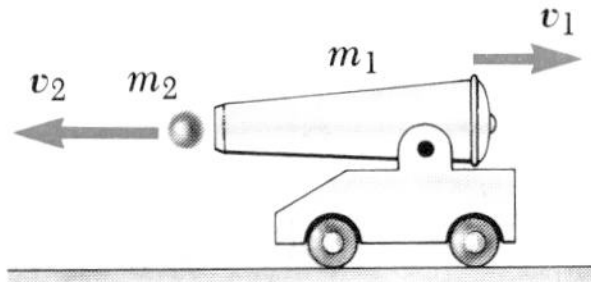

Figure 9.5 (Example 9.3) When the cannon is fired, it recoils to the right.

9.3 COLLISIONS

In this section we shall use the law of conservation of momentum to describe what happens when two particles collide with each other. We shall use the term **collision** to represent the event of two particles coming together for a short time, producing impulsive forces on each other. *The impulsive force due to the collision is assumed to be much larger than any external forces present.*

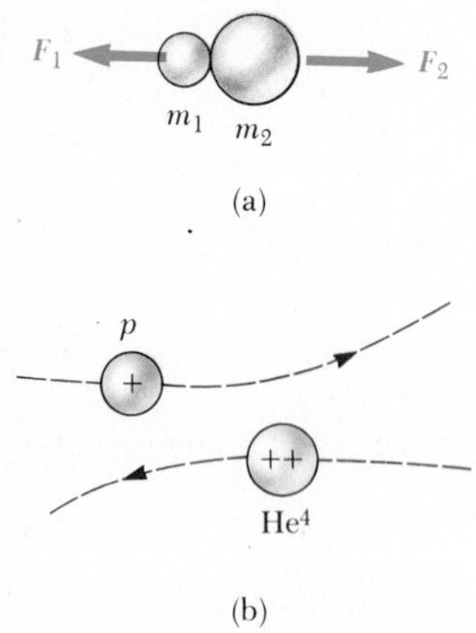

Figure 9.6 (a) The collision between two objects as the result of direct contact. (b) The collision between two charged particles.

The collision process may be the result of physical contact between two objects, as described in Figure 9.6a. This is a common observation when two macroscopic objects, such as two billiard balls or a baseball and a bat, collide. The notion of what we mean by a collision must be generalized since "contact" on a submicroscopic scale is ill-defined and meaningless. More accurately, impulsive forces arise from the electrostatic interaction of the electrons in the surface atoms of the two bodies.

To understand this on a more fundamental basis, consider a collision on an atomic scale (Fig. 9.6b), such as the collision of a proton with an alpha particle (the nucleus of the helium atom). Since the two particles are positively charged, they repel each other because of the strong electrostatic force between them at close separations. Such a process is commonly called a *scattering process.*

When the two particles of masses m_1 and m_2 collide as in Figure 9.6, the impulse forces may vary in time in a complicated way such as described in Figure 9.7. If $\mathbf{F}_{12}$ is the force on m_1 due to m_2, then the change in momentum of m_1 due to the collision is given by

$$\Delta \mathbf{p}_1 = \int_{t_i}^{t_f} \mathbf{F}_{12}\, dt$$

Likewise, if $\mathbf{F}_{21}$ is the force on m_2 due to m_1, the change in momentum of m_2 is given by

$$\Delta \mathbf{p}_2 = \int_{t_i}^{t_f} \mathbf{F}_{21}\, dt$$

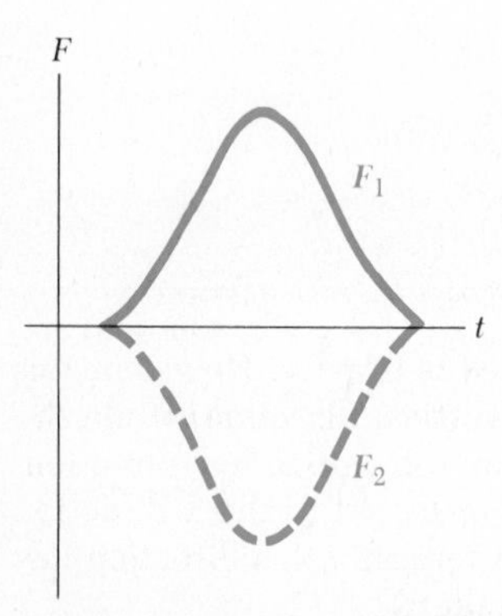

Figure 9.7 The force as a function of time for the two colliding particles described in Figure 9.6a. Note that $\mathbf{F}_1 = -\mathbf{F}_2$.

However, Newton's third law states that the force on m_1 due to m_2 is equal to and opposite the force on m_2 due to m_1, or $\mathbf{F}_{12} = -\mathbf{F}_{21}$. (This is described graphically in Fig. 9.7.) Hence, we conclude that

$$\Delta \mathbf{p}_1 = -\Delta \mathbf{p}_2$$

$$\Delta \mathbf{p}_1 + \Delta \mathbf{p}_2 = 0$$

Since the total momentum of the system is $\mathbf{P} = \mathbf{p}_1 + \mathbf{p}_2$, we conclude that the *change* in the momentum of the system due to the collision is zero, that is,

$$\mathbf{P} = \mathbf{p}_1 + \mathbf{p}_2 = \text{constant}$$

This is precisely what we expect if there are no external forces acting on the system (Section 9.2). However, the result is also valid if we consider the motion just before and just after the collision. Since the impulsive forces due to the collision are internal, they do not affect the total momentum of the system. Therefore, we conclude that for any type of collision, the total momentum of the system just before the collision equals the total momentum of the system just after the collision.

Momentum is conserved for any collision

Whenever a collision occurs between two bodies, we have seen that *the total momentum is always conserved.* However, the total kinetic energy is generally *not* conserved when a collision occurs because some of the kinetic energy is converted into heat and internal elastic potential energy when the bodies are deformed during the collision.

Inelastic collision

We define an **inelastic collision** as a *collision in which momentum is conserved but kinetic energy is not.* For a general inelastic collision, we can apply the law of conservation of momentum in the form given by Equation 9.11. The collision of a rubber ball with a hard surface is inelastic since some of the kinetic energy of the ball is lost when it is deformed while it is in contact with the surface. When two objects collide and stick together after the collision, the collision is called **perfectly inelastic.** This is an extreme case of an inelastic collision. For example, if two pieces of putty collide, they stick together and move with some common velocity after the collision. If a meteorite collides with the earth, it becomes buried in the earth and the collision is considered perfectly inelastic. However, not all of the initial kinetic energy is necessarily lost even in a perfectly inelastic collision.

Elastic collision

An **elastic collision** is defined as a *collision in which both momentum and kinetic energy are conserved.* Billiard ball collisions and the collisions of air molecules with the walls of a container at ordinary temperatures are highly elastic. In reality, collisions in the macroscopic world, such as those between billiard balls, can be only approximately elastic because in such collisions there is always some deformation of the objects; hence there is always some loss of kinetic energy. However, truly elastic collisions do occur between atomic and subatomic particles. Note that elastic and perfectly inelastic collisions are *limiting* cases, and most collisions are cases in between.

Properties of inelastic and elastic collision

1. An **inelastic collision** is one in which momentum is conserved, but kinetic energy is not.
2. A **perfectly inelastic collision** between two objects is an inelastic collision in which the two objects stick together after the collision, so their final velocities are the same.
3. An **elastic collision** is one in which both momentum and kinetic energy are conserved.

9.4 COLLISIONS IN ONE DIMENSION

In this section, we treat collisions in one dimension and consider two extreme types of collisions: (1) perfectly inelastic and (2) elastic. The important distinction between these two types of collisions is the fact that *momentum is conserved in both cases, but kinetic energy is conserved only in the case of an elastic collision.*

Perfectly Inelastic Collisions

Consider two particles of masses m_1 and m_2 moving with initial velocities v_{1i} and v_{2i} along a straight line, as in Figure 9.8. We shall assume that the particles collide "head-on," so that they will be moving along the same line of motion after the collision. If the two particles stick together and move with some common velocity v_f after the collision, then only the linear momentum of the system is conserved. Therefore, we can say that the total momentum before the collision equals the total momentum of the composite system after the collision, that is,

$$m_1v_{1i} + m_2v_{2i} = (m_1 + m_2)v_f \tag{9.13}$$

$$v_f = \frac{m_1v_{1i} + m_2v_{2i}}{m_1 + m_2} \tag{9.14}$$

Before collision

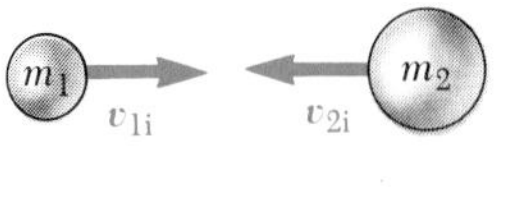

(a)

After collision

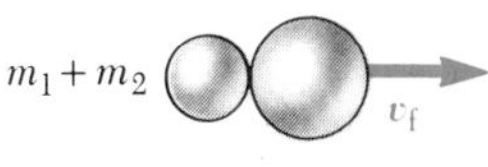

(b)

Figure 9.8 Schematic representation of a perfectly inelastic head-on collision between two particles: (a) before the collision and (b) after the collision.

EXAMPLE 9.4. The Cadillac Versus the "Beetle"
A large luxury car with a mass of 1800 kg stopped at a traffic light is struck from the rear by a compact car with a mass of 900 kg. The two cars become entangled as a result of the collision. (a) If the compact car was moving at 20 m/s before the collision, what is the velocity of the entangled mass after the collision?

Solution: The momentum before the collision is that of the compact car alone because the large car was initially at rest. Thus, we have for the momentum before the collision

$$p_i = m_1 v_i = (900 \text{ kg})(20 \text{ m/s}) = 1.80 \times 10^4 \text{ kg} \cdot \text{m/s}$$

After the collision, the mass that moves is the sum of the masses of the large car plus that of the compact car, and the momentum of the combination is

$$p_f = (m_1 + m_2)v_f = (2700 \text{ kg})(v_f)$$

Equating the momentum before to the momentum after and solving for v_f, the velocity of the wreckage, we have

$$v_f = \frac{p_i}{m_1 + m_2} = \frac{1.80 \times 10^4 \text{ kg} \cdot \text{m/s}}{2700 \text{ kg}} = 6.67 \text{ m/s}$$

(b) How much kinetic energy is lost in the collision?

Solution: Since the luxury car is at rest before the collision, $v_{2i} = 0$, hence the initial kinetic energy (before the collision) is

$$K_i = \tfrac{1}{2}m_1 v_{1i}^2 + \tfrac{1}{2}m_2 v_{2i}^2$$
$$= \tfrac{1}{2}(900 \text{ kg})(20 \text{ m/s})^2 + 0 = 1.80 \times 10^5 \text{ J}$$

Because the vehicles move with a common velocity v_f after the collision, the final kinetic energy (after the collision) is

$$K_f = \tfrac{1}{2}(m_1 + m_2)v_f^2 = \tfrac{1}{2}(900 \text{ kg} + 1800 \text{ kg})(6.67 \text{ m/s})^2$$
$$= 0.60 \times 10^5 \text{ J}$$

Hence, the *loss* in kinetic energy is

$$K_i - K_f = 1.20 \times 10^5 \text{ J}$$

EXAMPLE 9.5. The Ballistic Pendulum
The ballistic pendulum (Fig. 9.9) is a system used to measure the velocity of a fast-moving projectile, such as a bullet. The bullet is fired into a large block of wood suspended from some light wires. The bullet is stopped by the block, and the entire system swings through a height h. Since the collision is perfectly inelastic and momentum is conserved, Equation 9.14 gives the velocity of the system *right after* the collision in the impulse approximation. The kinetic energy *right after* the collision is given by

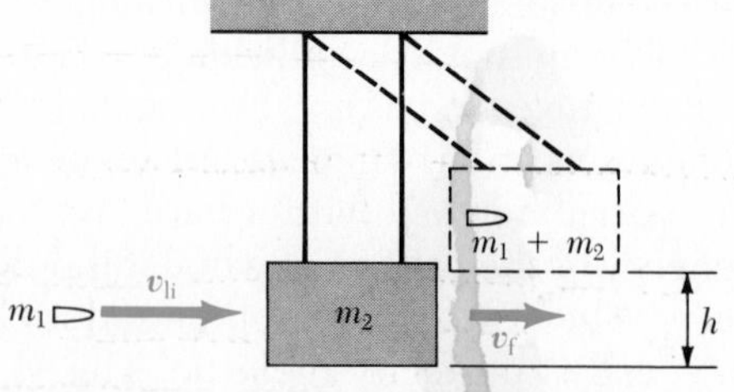

Figure 9.9 (Example 9.5) Diagram of a ballistic pendulum. Note that v_f is the velocity of the system right after the perfectly inelastic collision.

$$(1) \qquad K = \tfrac{1}{2}(m_1 + m_2)v_f^2$$

With $v_{2i} = 0$, Equation 9.14 becomes

$$(2) \qquad v_f = \frac{m_1 v_{1i}}{m_1 + m_2}$$

Substituting this value of v_f into (1) gives

$$K = \frac{m_1^2 v_{1i}^2}{2(m_1 + m_2)}$$

where v_{1i} is the initial velocity of the bullet. Note that this kinetic energy is *less* than the initial kinetic energy of the bullet. However, *after* the collision, energy is conserved and the kinetic energy at the bottom is transformed into potential energy in the bullet and in the block at the height h; that is,

$$\frac{m_1^2 v_{1i}^2}{2(m_1 + m_2)} = (m_1 + m_2)gh$$

$$v_{1i} = \left(\frac{m_1 + m_2}{m_1}\right)\sqrt{2gh}$$

Hence, it is possible to obtain the initial velocity of the bullet by measuring h and the two masses. Why would it be incorrect to equate the initial kinetic energy of the incoming bullet to the final gravitational energy of the bullet-block combination?

Exercise 2 In a ballistic pendulum experiment, suppose that $h = 5$ cm, $m_1 = 5$ g, and $m_2 = 1$ kg. Find (a) the initial speed of the projectile, and (b) the loss in energy due to the collision.
Answer: 199 m/s; 98.5 J.

Elastic Collisions

Now consider two particles that undergo an elastic head-on collision (Fig. 9.10). In this case, both momentum and kinetic energy are conserved; therefore we can write these conditions

$$m_1v_{1i} + m_2v_{2i} = m_1v_{1f} + m_2v_{2f} \tag{9.15}$$

$$\tfrac{1}{2}m_1v_{1i}^2 + \tfrac{1}{2}m_2v_{2i}^2 = \tfrac{1}{2}m_1v_{1f}^2 + \tfrac{1}{2}m_2v_{2f}^2 \tag{9.16}$$

where v is positive if a particle moves to the right and negative if it moves to the left.

In a typical problem involving elastic collisions, there will be two unknown quantities and Equations 9.15 and 9.16 can be solved simultaneously to find these. However, an alternative approach, one that involves a little mathematical manipulation of Equation 9.16, often simplifies this process. To see this, let's cancel the factor of $\frac{1}{2}$ in Equation 9.16 and rewrite it as

$$m_1(v_{1i}^2 - v_{1f}^2) = m_2(v_{2f}^2 - v_{2i}^2)$$

Here we have moved the terms containing m_1 to one side of the equation and those containing m_2 to the other. Next, let us factor both sides of the equation:

$$m_1(v_{1i} - v_{1f})(v_{1i} + v_{1f}) = m_2(v_{2f} - v_{2i})(v_{2f} + v_{2i}) \tag{9.17}$$

We now separate the terms containing m_1 and m_2 in the equation for the conservation of momentum (Eq. 9.15) to get

$$m_1(v_{1i} - v_{1f}) = m_2(v_{2f} - v_{2i}) \tag{9.18}$$

Our final result is obtained by dividing Equation 9.17 by Equation 9.18 to get

$$v_{1i} + v_{1f} = v_{2f} + v_{2i}$$

or

$$v_{1i} - v_{2i} = -(v_{1f} - v_{2f}) \tag{9.19}$$

This equation, in combination with the equation for conservation of momentum, will be used to solve problems dealing with perfectly elastic collisions. Note that Equation 9.19 says that the relative velocity of the two objects before the collision, $v_{1i} - v_{2i}$, equals the negative of the relative velocity of the two objects after the collision, $-(v_{1f} - v_{2f})$.

Before collision

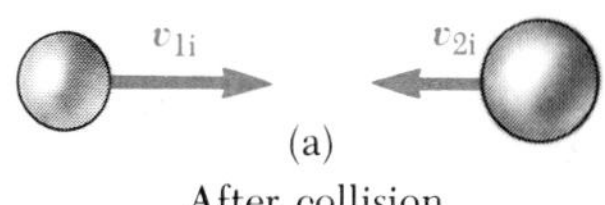

(a)

After collision

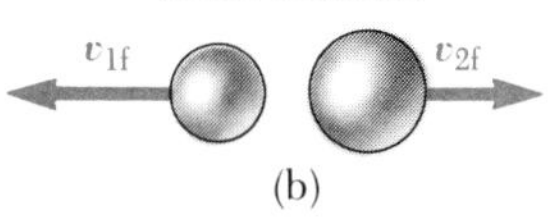

(b)

Figure 9.10 Schematic representation of an elastic head-on collision between two particles; (a) before the collision and (b) after the collision.

Suppose that the masses and the initial velocities of both particles are known. Equations 9.15 and 9.16 can be solved for the final velocities in terms of the initial velocities, since there are two equations and two unknowns. Solving for v_{1f} and v_{2f} gives

Elastic collision: relations between final and initial velocities

$$v_{1f} = \left(\frac{m_1 - m_2}{m_1 + m_2}\right)v_{1i} + \left(\frac{2m_2}{m_1 + m_2}\right)v_{2i} \tag{9.20}$$

$$v_{2f} = \left(\frac{2m_1}{m_1 + m_2}\right)v_{1i} + \left(\frac{m_2 - m_1}{m_1 + m_2}\right)v_{2i} \tag{9.21}$$

Again, note that the appropriate signs for v_{1i} and v_{2i} must be included in Equations 9.20 and 9.21 since they are vectors. For example, if m_2 is moving to the left initially, as in Figure 9.10, then v_{2i} is negative.

Let us consider some special cases: If $m_1 = m_2$, then we see that $v_{1f} = v_{2i}$ and $v_{2f} = v_{1i}$. That is, the particles exchange velocities if they have equal masses. This is what one observes in billiard ball collisions.

If m_2 is initially at rest, $v_{2i} = 0$, and Equations 9.20 and 9.21 become

$$v_{1f} = \left(\frac{m_1 - m_2}{m_1 + m_2}\right)v_{1i} \tag{9.22}$$

$$v_{2f} = \left(\frac{2m_1}{m_1 + m_2}\right) v_{1i} \qquad (9.23)$$

If m_1 is very large compared with m_2, we see from Equations 9.22 and 9.23 that $v_{1f} \approx v_{1i}$ and $v_{2f} \approx 2v_{1i}$. That is, when a very heavy particle collides with a very light one initially at rest, the heavy particle continues its motion unaltered after the collision, while the light particle rebounds with a velocity equal to about twice the initial velocity of the heavy particle. An example of such a collision would be the collision of a moving heavy atom, such as uranium, with a light atom, such as hydrogen.

If m_2 is much larger than m_1, and m_2 is initially at rest, then we note from Equations 9.22 and 9.23 that $v_{1f} \approx -v_{1i}$ and $v_{2f} \approx 0$. That is, when a very light particle collides with a very heavy particle initially at rest, the light particle will have its velocity reversed, while the heavy particle will remain approximately at rest. For example, imagine what happens when a marble is thrown at a stationary bowling ball.

EXAMPLE 9.6. Slowing Down Neutrons by Collisions In a nuclear reactor, neutrons are produced when the isotope $^{235}_{92}U$ undergoes fission. These neutrons are moving at high speeds (typically 10^7 m/s) and must be slowed down to about 10^3 m/s. Once the neutrons have slowed down, they have a high probability of producing another fission event and hence a sustained chain reaction. The high-speed neutrons can be slowed down by passing them through a solid or liquid material called a *moderator*. The slowing-down process involves elastic collisions. Let us show that a neutron can lose most of its kinetic energy if it collides elastically with a moderator containing light nuclei, such as deuterium and carbon. Hence, the moderator material is usually heavy water (D_2O) or graphite (which contains carbon nuclei).

Solution: Let us assume that the moderator nucleus of mass m_2 is at rest initially and that the neutron of mass m_1 has an initial velocity v_{1i}. Since momentum and energy are conserved, Equations 9.22 and 9.23 apply to the head-on collision of a neutron with the moderator nucleus. The initial kinetic energy of the neutron is

$$K_i = \tfrac{1}{2}m_1 v_{1i}^2$$

After the collision, the neutron has a kinetic energy given by $\frac{1}{2}m_1 v_{1f}^2$, where v_{1f} is given by Equation 9.22. We can express this energy as

$$K_1 = \tfrac{1}{2}m_1 v_{1f}^2 = \frac{m_1}{2}\left(\frac{m_1 - m_2}{m_1 + m_2}\right)^2 v_{1i}^2$$

Therefore, the *fraction* of the total kinetic energy possessed by the neutron *after* the collision is given by

$$(1) \qquad f_1 = \frac{K_1}{K_i} = \left(\frac{m_1 - m_2}{m_1 + m_2}\right)^2$$

From this result, we see that the final kinetic energy of the neutron is small when m_2 is close to m_1 and is zero when $m_1 = m_2$.

We can calculate the kinetic energy of the moderator nucleus after the collision using Equation 9.23:

$$K_2 = \tfrac{1}{2}m_2 v_{2f}^2 = \frac{2m_1^2 m_2}{(m_1 + m_2)^2} v_{1i}^2$$

Hence, the fraction of the total kinetic energy transferred to the moderator nucleus is given by

$$(2) \qquad f_2 = \frac{K_2}{K_i} = \frac{4m_1 m_2}{(m_1 + m_2)^2}$$

Note that since the total energy is conserved, (2) can also be obtained from (1) with the condition that $f_1 + f_2 = 1$, so that $f_2 = 1 - f_1$.

Suppose that heavy water is used for the moderator. Collisions of the neutrons with deuterium nuclei in D_2O ($m_2 = 2m_1$) predict that $f_1 = 1/9$ and $f_2 = 8/9$. That is, 89% of the neutron's kinetic energy is transferred to the deuterium nucleus. In practice, the moderator efficiency is reduced because head-on collisions are very unlikely to occur. How would the result differ if graphite were used as the moderator?

9.5 TWO-DIMENSIONAL COLLISIONS

In the previous section and in Section 9.2, it was shown that the total momentum of a system of two particles is conserved when the system is isolated. For a general collision of two particles, this implies that the total momentum in *each* of the directions x, y, and z is conserved (Eq. 9.12). Thus, for a three-dimen-

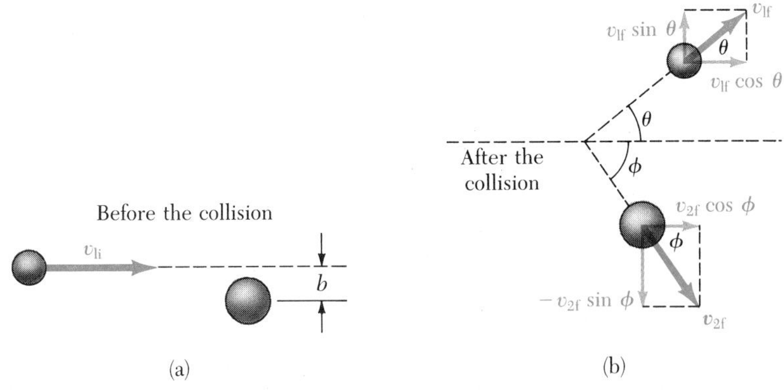

Figure 9.11 Schematic representation of an elastic glancing collision between two particles: (a) before the collision and (b) after the collision. Note that the impact parameter, b, must be greater than zero for a glancing collision.

sional problem we would get three component equations for the conservation of momentum.

Let us consider a two-dimensional problem in which a particle of mass m_1 collides with a particle of mass m_2, where m_2 is initially at rest (Fig. 9.11). The collision is not head-on, but glancing. The parameter b, defined in Figure 9.11, is called the *impact parameter*. As you can see from Figure 9.11, if b is zero, the collision is head-on. After the collision, m_1 moves at an angle θ with respect to the horizontal and m_2 moves at an angle ϕ with respect to the horizontal. Applying the law of conservation of momentum in component form, $P_{xi} = P_{xf}$ and $P_{yi} = P_{yf}$, and noting that $P_{yi} = 0$, we get

$$m_1 v_{1i} = m_1 v_{1f} \cos \theta + m_2 v_{2f} \cos \phi \qquad (9.24a)$$

x component of momentum

$$0 = m_1 v_{1f} \sin \theta - m_2 v_{2f} \sin \phi \qquad (9.24b)$$

y component of momentum

Now let us assume that the collision is elastic, in which case we can also write a third equation for the conservation of kinetic energy, in the form

$$\tfrac{1}{2} m_1 v_{1i}^2 = \tfrac{1}{2} m_1 v_{1f}^2 + \tfrac{1}{2} m_2 v_{2f}^2 \qquad (9.25)$$

Conservation of energy

If we know the initial velocity, v_{1i}, and the masses, we are left with four unknowns. Since we only have three equations, one of the four remaining quantities (v_{1f}, v_{2f}, θ, or ϕ) must be given to determine the motion after the collision from conservation principles alone.

EXAMPLE 9.7. Collision at an Intersection

A 1500-kg car traveling east with a speed of 25 m/s collides at an intersection with a 2500-kg van traveling north at a speed of 20 m/s, as shown in Figure 9.12. Find the direction and magnitude of the velocity of the wreckage after the collision, assuming that the vehicles undergo a perfectly inelastic collision (that is, they stick together).

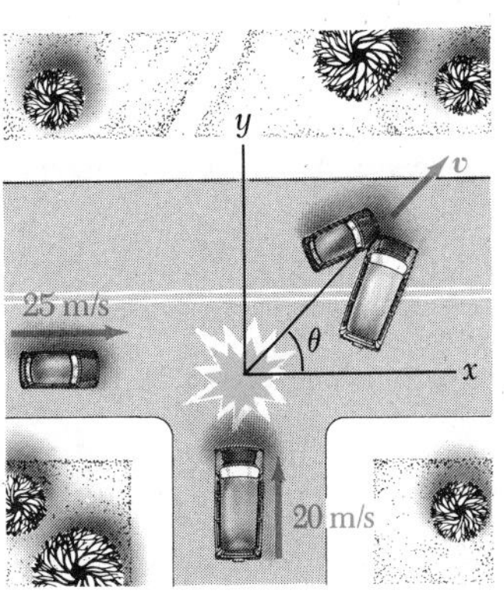

Figure 9.12 Top view of a car colliding with a truck.

Solution: Let us choose east to be along the positive x direction and north to be along the positive y direction, as in Figure 9.12. Before the collision, the only object having momentum in the x direction is the car. Thus, the total initial momentum of the system (car plus van) in the

x direction is

$$\sum p_{xi} = (1500 \text{ kg})(25 \text{ m/s}) = 37\ 500 \text{ kg} \cdot \text{m/s}$$

Now let us assume that the wreckage moves at an angle θ and speed v after the collision, as in Figure 9.12. The total momentum in the x direction after the collision is

$$\sum p_{xf} = (4000 \text{ kg})(v \cos \theta)$$

Because momentum is conserved in the x direction, we can equate these two equations to get

$$(1) \qquad 37\ 500 \text{ kg} \cdot \text{m/s} = (4000 \text{ kg})(v \cos \theta)$$

Similarly, the total initial momentum of the system in the y direction is that of the van, which has the value (2500 kg)(20 m/s). Applying conservation of momentum to the y direction, we have

$$\sum p_{yi} = \sum p_{yf}$$

$$(2500 \text{ kg})(20 \text{ m/s}) = (4000 \text{ kg})(v \sin \theta)$$

$$(2) \qquad 50{,}000 \text{ kg} \cdot \text{m/s} = (4000 \text{ kg})(v \sin \theta)$$

If we divide (2) by (1), we get

$$\tan \theta = \frac{50\ 000}{37\ 500} = 1.33$$

$$\theta = 53.1°$$

When this angle is substituted into (2)—or alternatively into (1)—the value of v is

$$v = \frac{50\ 000 \text{ kg} \cdot \text{m/s}}{(4000 \text{ kg})(\sin 53°)} = 15.6 \text{ m/s}$$

EXAMPLE 9.8. Proton–Proton Collision

A proton collides in a perfectly elastic fashion with another proton initially at rest. The incoming proton has an initial speed of 3.5×10^5 m/s and makes a glancing collision with the second proton, as in Figure 9.11. (At close separations, the protons exert a repulsive electrostatic force on each other.) After the collision, one proton is observed to move at an angle of 37° to the original direction of motion, and the second deflects at an angle ϕ to the same axis. Find the final speeds of the two protons and the angle ϕ.

Solution: Since $m_1 = m_2$, $\theta = 37°$, and we are given $v_{1i} = 3.5 \times 10^5$ m/s, Equations 9.24 and 9.25 become

$$v_{1f} \cos 37° + v_{2f} \cos \phi = 3.5 \times 10^5$$

$$v_{1f} \sin 37° - v_{2f} \sin \phi = 0$$

$$v_{1f}^2 + v_{2f}^2 = (3.5 \times 10^5)^2$$

Solving these three equations with three unknowns simultaneously gives

$$v_{1f} = 2.8 \times 10^5 \text{ m/s} \qquad v_{2f} = 2.1 \times 10^5 \text{ m/s} \qquad \phi = 53°$$

It is interesting to note that $\theta + \phi = 90°$. This result is not accidental. *Whenever two equal masses collide elastically with an impact parameter greater than zero and one of them is initially at rest, their final velocities are* always *at right angles to each other.* The next example illustrates this point in more detail.

EXAMPLE 9.9. Billiard Ball Collision

In a game of billiards, the player wishes to "sink" the target ball in the corner pocket, as shown in Figure 9.13. If the angle to the corner pocket is 35°, at what angle θ is the cue ball deflected? Assume that friction and rotational motion ("English") are unimportant, and assume the collision is elastic.

Solution: Since the target is initially at rest, $v_{2i} = 0$ and conservation of kinetic energy gives

$$\tfrac{1}{2}m_1 v_{1i}^2 = \tfrac{1}{2}m_1 v_{1f}^2 + \tfrac{1}{2}m_2 v_{2f}^2$$

But $m_1 = m_2$, so that

$$(1) \qquad v_{1i}^2 = v_{1f}^2 + v_{2f}^2$$

Applying conservation of momentum to the two-dimensional collision gives

$$(2) \qquad \mathbf{v}_{1i} = \mathbf{v}_{1f} + \mathbf{v}_{2f}$$

If we square both sides of (2), we get

$$v_{1i}^2 = (\mathbf{v}_{1f} + \mathbf{v}_{2f}) \cdot (\mathbf{v}_{1f} + \mathbf{v}_{2f})$$

$$= v_{1f}^2 + v_{2f}^2 + 2\mathbf{v}_{1f} \cdot \mathbf{v}_{2f}$$

But $\mathbf{v}_{1f} \cdot \mathbf{v}_{2f} = v_{1f}v_{2f} \cos(\theta + 35°)$, and so

$$(3) \qquad v_{1i}^2 = v_{1f}^2 + v_{2f}^2 + 2v_{1f}v_{2f} \cos(\theta + 35°)$$

Subtracting (1) from (3) gives

$$2v_{1f}v_{2f} \cos(\theta + 35°) = 0$$

$$\cos(\theta + 35°) = 0$$

$$\theta + 35° = 90° \quad \text{or} \quad \theta = 55°$$

Again, this shows that whenever two equal masses undergo a glancing elastic collision and one of them is initially at rest, they will move at right angles to each other after the collision.

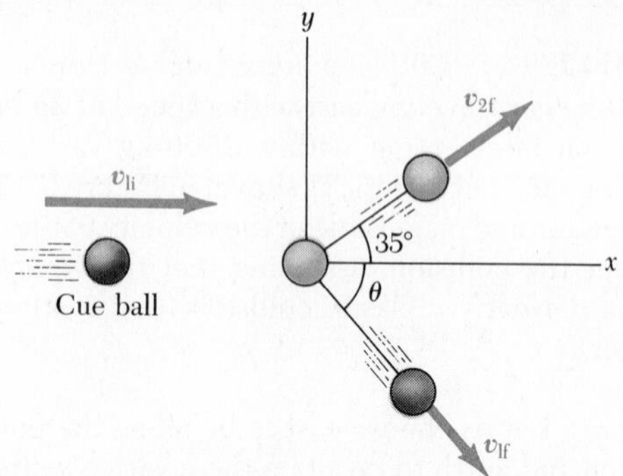

Figure 9.13 (Example 9.9).

9.6 THE CENTER OF MASS

In this section we describe the overall motion of a mechanical system in terms of a very special point called the *center of mass* of the system. The mechanical system can be either a system of particles or an extended object. We shall see that the mechanical system moves as if all its mass were concentrated at the center of mass. Furthermore, if the resultant external force on the system is $\boldsymbol{F}$ and the total mass of the system is M, the center of mass moves with an acceleration given by $\boldsymbol{a} = \boldsymbol{F}/M$. That is, the system moves as if the resultant external force were applied to a single particle of mass M located at the center of mass. This result was implicitly assumed in earlier chapters since nearly all examples referred to the motion of extended objects.

Consider a mechanical system consisting of a pair of particles connected by a light, rigid rod (Fig. 9.14). The center of mass is located somewhere on the line joining the particles and is closer to the larger mass. If a single force is applied at some point on the rod closer to the smaller mass, the system will rotate clockwise (Fig. 9.14a). If the force is applied at a point on the rod closer to the larger mass, the system will rotate in the counterclockwise direction (Fig. 9.14b). If the force is applied at the center of mass, the system will move in the direction of $\boldsymbol{F}$ without rotating (Fig. 9.14c). Thus, the center of mass can be easily located.

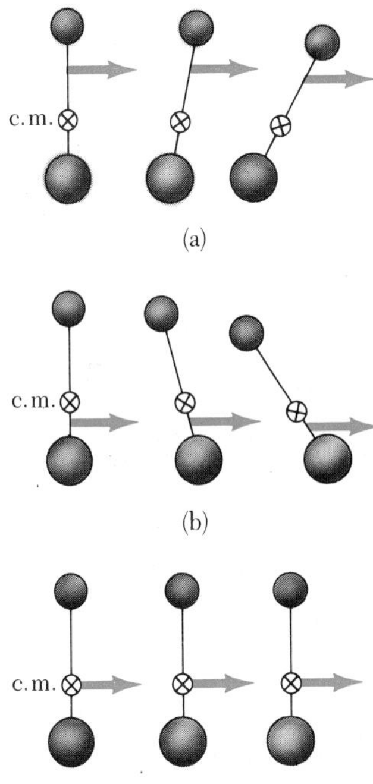

Figure 9.14 Two unequal masses are connected by a light, rigid rod. (a) The system rotates clockwise when a force is applied above the center of mass. (b) The system rotates counterclockwise when a force is applied below the center of mass. (c) The system moves in the direction of F without rotating when a force is applied at the center of mass.

The center of mass of the pair of particles described in Fig. 9.15 is located on the x axis and lies somewhere between the particles. The x coordinate of the center of mass in this case is defined to be

$$x_c \equiv \frac{m_1x_1 + m_2x_2}{m_1 + m_2} \tag{9.26}$$

For example, if $x_1 = 0$, $x_2 = d$, and $m_2 = 2m_1$, we find that $x_c = \frac{2}{3}d$. That is, the center of mass lies closer to the more massive particle. If the two masses are equal, the center of mass lies midway between the particles.

We can extend the center of mass concept to a system of many particles in three dimensions. The x coordinate of the center of mass of n particles is defined to be

$$x_c = \frac{m_1x_1 + m_2x_2 + m_3x_2 + \cdot\cdot\cdot + m_nx_n}{m_1 + m_2 + m_3 + \cdot\cdot\cdot + m_n} = \frac{\Sigma m_ix_i}{\Sigma m_i} \tag{9.27}$$

where x_i is the x coordinate of the ith particle and Σm_i is the *total mass* of the system. For convenience, we shall express the total mass as $M = \Sigma m_i$, where the sum runs over all n particles. The y and z coordinates of the center of mass are similarly defined by the equations

$$y_c = \frac{\Sigma m_iy_i}{M} \quad \text{and} \quad z_c = \frac{\Sigma m_iz_i}{M} \tag{9.28}$$

The center of mass can also be located by its position vector, $\boldsymbol{r}_c$. The rectangular coordinates of this vector are x_c, y_c, and z_c, defined in Equations 9.27 and 9.28. Therefore,

$$\boldsymbol{r}_c = x_c\boldsymbol{i} + y_c\boldsymbol{j} + z_c\boldsymbol{k}$$
$$= \frac{\Sigma m_ix_i\boldsymbol{i} + \Sigma m_iy_i\boldsymbol{j} + \Sigma m_iz_i\boldsymbol{k}}{M} \tag{9.29}$$

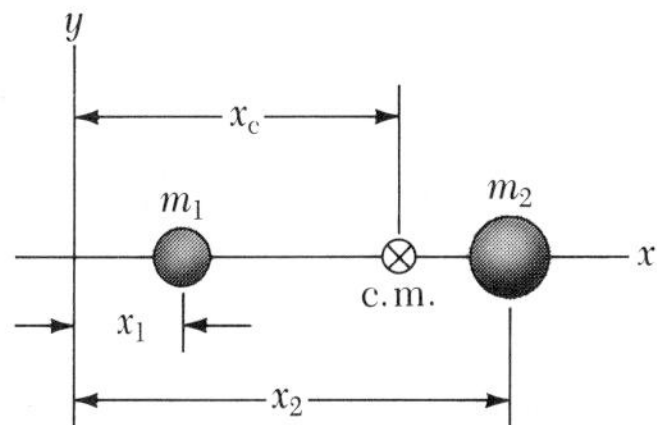

Figure 9.15 The center of mass of two particles on the x axis is located at x_c, a point between the particles, closer to the larger mass.

Vector position of the center of mass for a system of particles

$$r_c = \frac{\Sigma m_i r_i}{M} \tag{9.30}$$

where r_i is the position vector of the ith particle, defined by

$$r_i = x_i i + y_i j + z_i k$$

Although the location of the center of mass for a rigid body is somewhat more cumbersome, the basic ideas we have discussed still apply. We can think of a general rigid body as a system of a large number of particles (Fig. 9.16). The particle separation is very small, and so the body can be considered to have a continuous mass distribution. By dividing the body into elements of mass Δm_i, with coordinates x_i, y_i, z_i, we see that the x coordinate of the center of mass is approximately

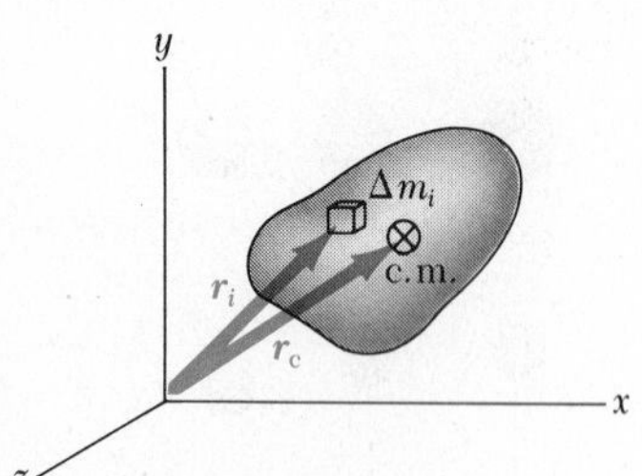

Figure 9.16 A rigid body can be considered a distribution of small elements of mass Δm_i. The center of mass is located at the vector position r_c, which has coordinates x_c, y_c, and z_c.

$$x_c \approx \frac{\Sigma x_i \, \Delta m_i}{M}$$

with similar expressions for y_c and z_c. If we let the number of elements, n, approach infinity, then x_c will be given precisely. In this limit, we replace the sum by an integral and replace Δm_i by the differential element dm, so that

$$x_c = \lim_{\Delta m_i \to 0} \frac{\Sigma x_i \, \Delta m_i}{M} = \frac{1}{M} \int x \, dm \tag{9.31}$$

Likewise, for y_c and z_c we get

$$y_c = \frac{1}{M} \int y \, dm \quad \text{and} \quad z_c = \frac{1}{M} \int z \, dm \tag{9.32}$$

We can express the vector position of the center of mass of a rigid body in the form

$$r_c = \frac{1}{M} \int r \, dm \tag{9.33}$$

where this is equivalent to the three scalar expressions given by Equations 9.31 and 9.32.

The center of mass of various homogeneous, symmetric bodies must lie on an axis of symmetry. For example, the center of mass of a homogeneous rod must lie on the rod, midway between its ends. The center of mass of a homogeneous sphere or a homogeneous cube must lie at its geometric center. One can determine the center of mass of an irregularly shaped planar body experimentally by suspending the body from two different points (Fig. 9.17). The body is first hung from point A, and a vertical line AB is drawn when the body is in equilibrium. The body is then hung from point C, and a second vertical line, CD, is drawn. The center of mass coincides with the intersection of these two lines. In fact, if the body is hung freely from any point, the vertical line through this point must pass through the center of mass.

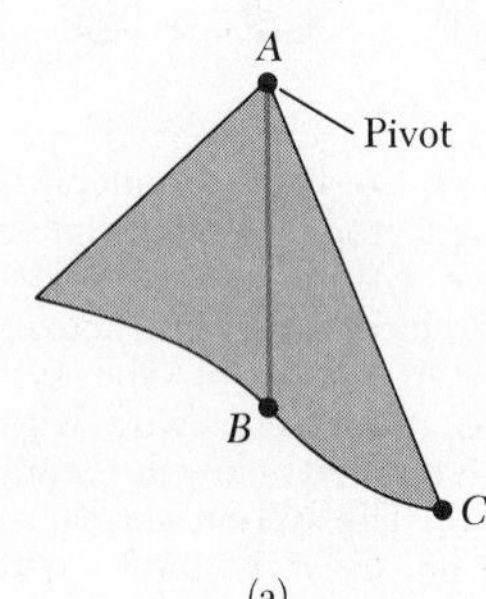

(a)

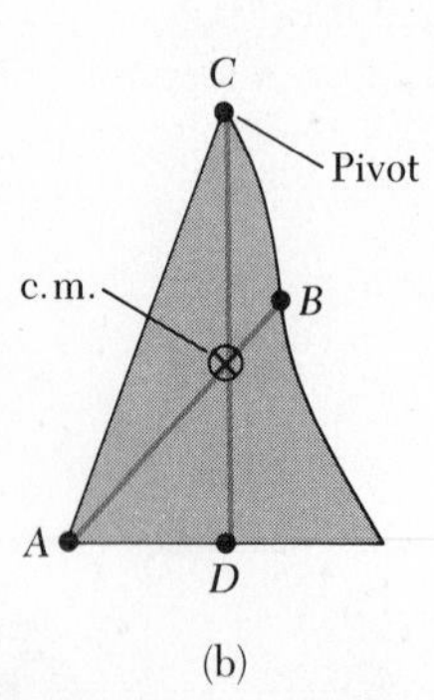

(b)

Figure 9.17 An experimental technique for determining the center of mass of an irregular planar object. The object is hung freely from two different pivots, A and C. The intersection of the two vertical lines AB and CD locates the center of mass.

Since a rigid body is a continuous distribution of mass, each portion is acted upon by the force of gravity. The net effect of all of these forces is equivalent to the effect of a single force, Mg, acting through a special point, called the **center of gravity.** If g is constant over the mass distribution, then the center of gravity coincides with the center of mass. If a rigid body is pivoted at its center of gravity, it will be balanced in any orientation.

EXAMPLE 9.10. The Center of Mass of Three Particles

A system consists of three particles located at the corners of a right triangle as in Figure 9.18. Find the center of mass as measured from the origin.

Solution: Using the basic defining equations for the coordinates of the center of mass, and noting that $z_c = 0$, we get

$$x_c = \frac{\Sigma m_i x_i}{M} = \frac{2md + m(d+b) + 4m(d+b)}{7m} = d + \frac{5}{7}b$$

$$y_c = \frac{\Sigma m_i y_i}{M} = \frac{2m(0) + m(0) + 4mh}{7m} = \frac{4}{7}h$$

Therefore, we can express the position vector to the center of mass as

$$r_c = x_c i + y_c j = (d + \frac{5}{7}b)i + \frac{4}{7}hj$$

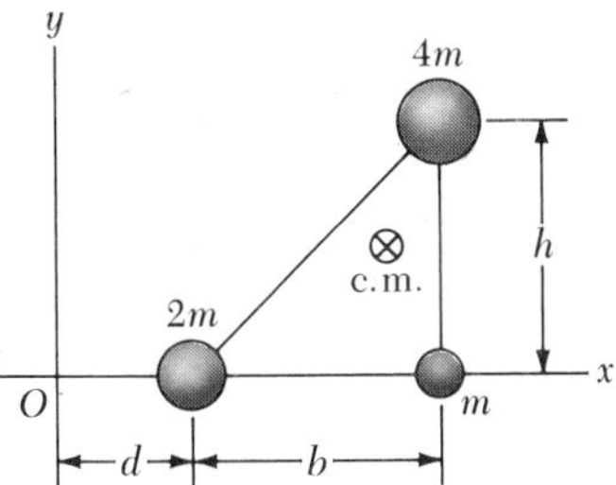

Figure 9.18 (Example 9.10) The center of mass of the three particles is located inside the triangle.

EXAMPLE 9.11. The Center of Mass of a Uniform Rod

(a) Show that the center of mass of a uniform rod of mass M and length L lies midway between its ends (Fig. 9.19).

By symmetry, we see that $y_c = z_c = 0$ if the rod is placed along the x axis. Furthermore, if we call the mass per unit length λ (the linear mass density), then $\lambda = M/L$ for a uniform rod. If we divide the rod into elements of length dx, then the mass of each element is $dm = \lambda\, dx$. Since an arbitrary element is at a distance x from the origin, Equation 9.31 gives

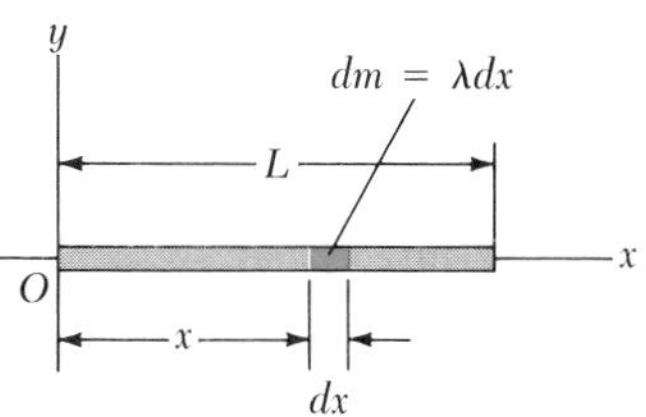

Figure 9.19 (Example 9.11) The center of mass of a uniform rod of length L is located at $x_c = L/2$.

$$x_c = \frac{1}{M}\int_0^L x\, dm = \frac{1}{M}\int_0^L x\lambda\, dx = \frac{\lambda}{M}\frac{x^2}{2}\Big]_0^L = \frac{\lambda L^2}{2M}$$

Because $\lambda = M/L$, this reduces to

$$x_c = \frac{L^2}{2M}\left(\frac{M}{L}\right) = \frac{L}{2}$$

(b) Suppose the rod is *nonuniform* and the mass per unit length varies linearly with x according to the expression $\lambda = \alpha x$, where α is a constant. Find the x coordinate of the center of mass as a fraction of L.

In this case, we replace dm by $\lambda\, dx$, where λ is *not* constant. Therefore, x_c is given by

$$x_c = \frac{1}{M}\int_0^L x\, dm = \frac{1}{M}\int_0^L x\lambda\, dx = \frac{\alpha}{M}\int_0^L x^2\, dx = \frac{\alpha L^3}{3M}$$

We can also eliminate α by noting that the total mass of the rod is related to α through the relation

$$M = \int dm = \int_0^L \lambda\, dx = \int_0^L \alpha x\, dx = \frac{\alpha L^2}{2}$$

Substituting this into the expression for x_c gives

$$x_c = \frac{\alpha L^3}{3\alpha L^2/2} = \frac{2}{3}L$$

9.7 MOTION OF A SYSTEM OF PARTICLES

We can begin to understand the physical significance and utility of the center of mass concept by taking the time derivative of the position vector of the center of mass, r_c, given by Equation 9.30. Assuming that M remains constant, that is, no particles enter or leave the system, we get the following expression for the **velocity of the center of mass:**

$$v_c = \frac{dr_c}{dt} = \frac{1}{M}\sum m_i \frac{dr_i}{dt} = \frac{\Sigma m_i v_i}{M} \qquad (9.34)$$

Velocity of the center of mass

where v_i is the velocity of the ith particle. Rearranging Equation 9.34 gives

Total momentum of a system of particles

$$Mv_c = \sum m_i v_i = \sum p_i = P \qquad (9.35)$$

The right side of Equation 9.35 equals the total momentum of the system. Therefore, we conclude that *the* **total momentum of the system** *equals the total mass multiplied by the velocity of the center of mass.* In other words, the total momentum of the system is equal to that of a single particle of mass M moving with a velocity v_c.

If we now differentiate Equation 9.34 with respect to time, we get the **acceleration of the center of mass:**

Acceleration of the center of mass

$$a_c = \frac{dv_c}{dt} = \frac{1}{M}\sum m_i \frac{dv_i}{dt} = \frac{1}{M}\sum m_i a_i \qquad (9.36)$$

Rearranging this expression and using Newton's second law, we get

$$Ma_c = \sum m_i a_i = \sum F_i \qquad (9.37)$$

where F_i is the force on particle i.

The forces on any particle in the system may include both external forces (from outside the system) and internal forces (from within the system). However, by Newton's third law, the force of particle 1 on particle 2, for example, is equal to and opposite the force of particle 2 on particle 1. Thus, when we sum over all internal forces in Equation 9.37, they cancel in pairs and the net force on the system is due *only* to external forces. Thus, we can write Equation 9.37 in the form

Newton's second law for a system of particles

$$\sum F_{ext} = Ma_c = \frac{dP}{dt} \qquad (9.38)$$

That is, the resultant external force on the system of particles equals the total mass of the system multiplied by the acceleration of the center of mass. If we compare this to Newton's second law for a single particle, we see that the center of mass moves like an imaginary particle of mass M under the influence of the resultant external force on the system.

Finally, we see that if the resultant external force is zero, then from Equation 9.38 it follows that

$$\frac{dP}{dt} = Ma_c = 0$$

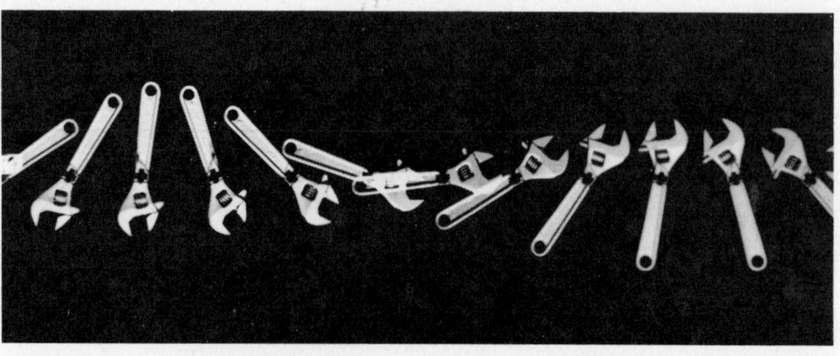

The center of mass of the wrench moves in a straight line as the wrench rotates about this point shown by the black marker. (Education Development Center, Newton, Mass.)

so that

$$P = Mv_c = \text{constant} \qquad (\text{when } \sum F_{ext} = 0) \qquad (9.39)$$

That is, the total linear momentum of a system of particles is conserved if there are no external forces acting on the system. Therefore, it follows that for an *isolated* system of particles, both the total momentum and velocity of the center of mass are constant in time. This is a generalization to a many-particle system of the law of conservation of momentum that was derived in Section 9.2 for a two-particle system.

Suppose an isolated system consisting of two or more members is at rest. The center of mass of such a system will remain at rest unless acted upon by an external force. For example, consider a system made up of a swimmer and a raft, with the system initially at rest. When the swimmer dives off the raft, the center of mass of the system will remain at rest (if we neglect the friction between raft and water). Furthermore, the momentum of the diver will be equal in magnitude to the momentum of the raft, but opposite in direction.

As another example, suppose an unstable atom initially at rest suddenly decays into two fragments of masses M_1 and M_2, with velocities v_1 and v_2, respectively. (An example of such a radioactive decay is that of the uranium-238 nucleus, which decays into an alpha particle—the helium nucleus—and the thorium-234 nucleus.) Since the total momentum of the system before the decay is zero, the total momentum of the system after the decay must also be zero. Therefore, we see that $M_1v_1 + M_2v_2 = 0$. If the velocity of one of the fragments after the decay is known, the recoil velocity of the other fragment can be calculated. Can you explain the origin of the kinetic energy of the fragments?

EXAMPLE 9.12. Exploding Projectile

A projectile is fired into the air and suddenly explodes into several fragments (Fig. 9.20). What can be said about the motion of the fragments after the collision?

Solution: The only external force on the projectile is the force of gravity. Thus, the projectile follows a parabolic path. If the projectile did not explode, it would continue to move along the parabolic path indicated by the broken line in Figure 9.20. Since the forces due to the explosion are internal, they do not affect the motion of the center of mass. Thus, after the explosion the center of mass of the fragments follows the *same* parabolic path the projectile would have followed if there had been no explosion.

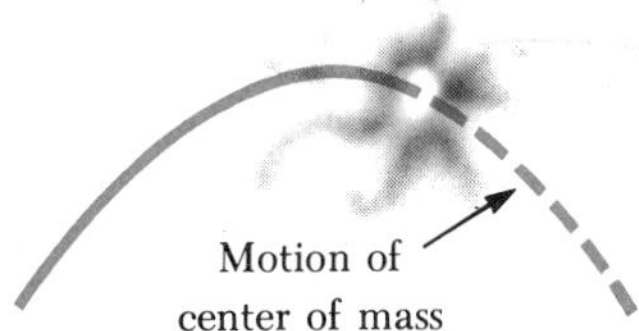

Figure 9.20 (Example 9.12) When a projectile explodes into several fragments, the center of mass of the fragments follows the same parabolic path the projectile would have taken had there been no explosion.

EXAMPLE 9.13. The Exploding Rocket

A rocket is fired vertically upward. It reaches an altitude of 1000 m and a velocity of 300 m/s. At this instant, the rocket explodes into three equal fragments. One fragment continues to move upward with a speed of 450 m/s right after the explosion. The second fragment has a speed of 240 m/s moving in the easterly direction right after the explosion. (a) What is the velocity of the third fragment right after the explosion?

Let us call the total mass of the rocket M; hence the mass of each fragment is $M/3$. The total momentum just before the explosion must equal the total momentum of the fragments right after the explosion since the forces of the explosion are internal to the system and cannot affect the total momentum of the system.

Before the explosion: $\quad P_i = Mv_0 = 300Mj$

After the explosion:

$$P_f = 240\left(\frac{M}{3}\right)i + 450\left(\frac{M}{3}\right)j + \frac{M}{3}v$$

where v is the unknown velocity of the third fragment. Equating these two expressions gives

$$M\frac{v}{3} + 80Mi + 150Mj = 300Mj$$

$$v = (-240i + 450j) \text{ m/s}$$

(b) What is the position of the center of mass relative to the ground 3 s after the explosion? (Assume the rocket engine is nonoperative after the explosion.)

The center of mass of the fragments moves as a freely falling body since the explosion doesn't affect the motion of the center of mass (Example 9.12). If $t = 0$ is the time of the explosion, then $y_0 = 1000$ m and $v_0 = 300$ m/s for the center of mass. Using an expression from kinematics, we get for the y coordinate of the center of mass

$$y_c = y_0 + v_0 t - \tfrac{1}{2}gt^2 = 1000 + 300t - 4.9t^2$$

Thus, at $t = 3$ s,

$$y_c = [1000 + 300\,(3) - 4.9(3)^2] \text{ m} \approx 1856 \text{ m}$$

Note that the x coordinate of the center of mass doesn't change. That is, in a given time interval the second fragment moves to the right by the same distance that the third fragment moves to the left.

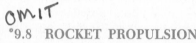

*9.8 ROCKET PROPULSION

Lift-off of the space shuttle Columbia. Massive amounts of thrust are generated by the shuttle's liquid-fueled engines, aided by the two solid fuel boosters. (NASA)

When ordinary vehicles, such as automobiles, boats, and locomotives, are propelled, the driving force for the motion is one of friction. In the case of the automobile, the driving force is the force of the road on the car. A locomotive "pushes" against the tracks; hence the driving force is the force of the tracks on the locomotive. However, a rocket moving in space has no air, tracks, or water to "push" against. Therefore, the source of the propulsion of a rocket must be different. *The operation of a rocket depends upon the law of conservation of momentum as applied to a system of particles, where the system is the rocket plus its ejected fuel.*

The propulsion of a rocket can be understood by first considering the mechanical system consisting of a machine gun mounted on a cart on wheels. As the machine gun is fired, each bullet receives a momentum mv in some direction where v is measured with respect to a stationary earth frame. For each bullet that is fired, the gun and cart must receive a compensating momentum in the opposite direction (as in Example 9.3). That is, the reaction force of the bullet on the gun accelerates the cart and gun. If there are n bullets fired each second, then the average force on the gun is equal to $F_{av} = nmv$.

In a similar manner, as a rocket moves in free space (a vacuum), *its momentum changes when some of its mass is released in the form of ejected gases* (Fig. 9.21). *Since the ejected gases acquire some momentum, the rocket receives a compensating momentum in the opposite direction.* Therefore, *the rocket is accelerated as a result of the "push," or thrust, from the exhaust gases.* In free space, the center of mass of the entire system moves uniformly, independent of the propulsion process.

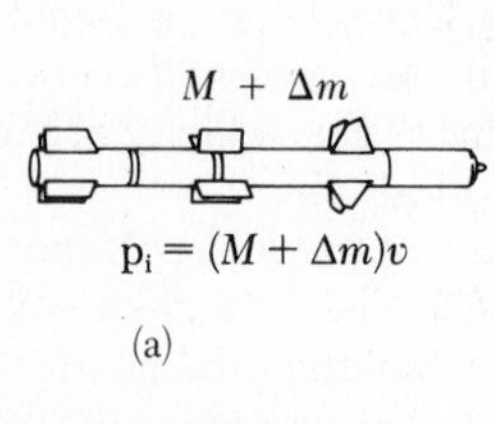

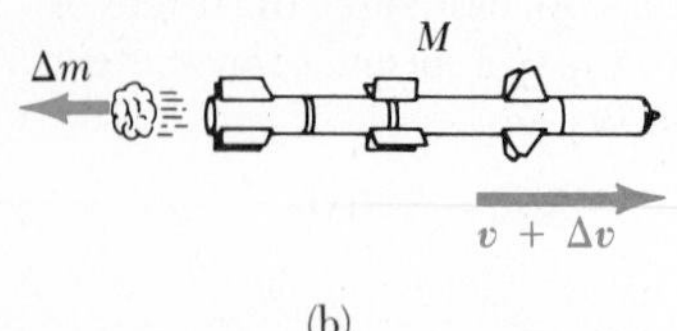

Figure 9.21 Rocket propulsion. (a) The initial mass of the rocket is $M + \Delta m$ at a time t, and its speed is v. (b) At a time $t + \Delta t$, the rocket's mass has reduced to M, and an amount of fuel Δm has been ejected. The rocket's speed increases by an amount Δv.

Suppose that at some time t, the momentum of the rocket plus the fuel is $(M + \Delta m)v$ (Fig. 9.21a). At some short time later, Δt, the rocket ejects some fuel of mass Δm and the rocket's speed therefore increases to $v + \Delta v$ (Fig. 9.21b). If the fuel is ejected with a velocity v_e *relative to the rocket*, then the velocity of the fuel relative to a stationary frame of reference is $v - v_e$. Thus, if we equate the total initial momentum of the system to the total final momentum, we get

$$(M + \Delta m)v = M(v + \Delta v) + \Delta m(v - v_e)$$

Simplifying this expression gives

$$M\,\Delta v = v_e\,\Delta m$$

We also could have arrived at this result by considering the system in the center of mass frame of reference; that is, a frame whose velocity equals the center of mass velocity. In this frame, the total momentum is zero; therefore if the rocket gains a momentum $M\,\Delta v$ by ejecting some fuel, the exhaust gases obtain a momentum $v_e\,\Delta m$ in the *opposite* direction, and so $M\,\Delta v - v_e\,\Delta m = 0$. If we now take the limit as Δt goes to zero, then $\Delta v \rightarrow dv$ and $\Delta m \rightarrow dm$. Furthermore, the increase in the exhaust mass, dm, corresponds to an equal decrease in the rocket mass, so that $dm = -dM$. Note that $dM < 0$. Using this fact, we get

$$M\,dv = -v_e\,dM \tag{9.40}$$

Integrating this equation, and taking the initial mass of the rocket plus fuel to be M_i and the final mass of the rocket plus its remaining fuel to be M_f, we get

$$\int_{v_i}^{v_f} dv = -v_e \int_{M_i}^{M_f} \frac{dM}{M}$$

$$v_f - v_i = v_e \ln\left(\frac{M_i}{M_f}\right) \tag{9.41}$$

Expression for rocket propulsion

This is the basic expression of rocket propulsion. First, it tells us that the increase in velocity is proportional to the exhaust velocity, v_e. Therefore, the exhaust velocity should be very high. Second, the increase in velocity is proportional to the logarithm of the ratio M_i/M_f. Therefore, this ratio should be as large as possible, which means that the rocket should carry as much fuel as possible.

The *thrust* on the rocket is the force exerted on the rocket by the ejected exhaust gases. We can obtain an expression for the thrust from Equation 9.40:

$$\text{Thrust} = M\frac{dv}{dt} = \left|v_e \frac{dM}{dt}\right| \tag{9.42}$$

Here again we see that the thrust increases as the exhaust velocity increases and as the rate of change of mass (burn rate) increases.

EXAMPLE 9.13. A Rocket in Space

A rocket moving in free space has a speed of 3×10^3 m/s. Its engines are turned on, and fuel is ejected in a direction opposite the rocket's motion at a speed of 5×10^3 m/s relative to the rocket. (a) What is the speed of the rocket once its mass is reduced to one half its mass before ignition?

Applying Equation 9.41, we get

$$v_f = v_i + v_e \ln\left(\frac{M_i}{M_f}\right)$$

$$= 3 \times 10^3 + 5 \times 10^3 \ln\left(\frac{M_i}{0.5M_i}\right)$$

$$= 6.47 \times 10^3 \text{ m/s}$$

(b) What is the thrust on the rocket if it burns fuel at the rate of 50 kg/s?

$$\text{Thrust} = \left|v_e \frac{dM}{dt}\right| = \left(5 \times 10^3 \frac{\text{m}}{\text{s}}\right)\left(50 \frac{\text{kg}}{\text{s}}\right)$$

$$= 2.5 \times 10^5 \text{ N}$$

9.9 SUMMARY

The **linear momentum** of a particle of mass m moving with a velocity $\boldsymbol{v}$ is defined to be

$$\boldsymbol{p} \equiv m\boldsymbol{v} \tag{9.1}$$

Impulse

The **impulse** of a force $\mathbf{F}$ on a particle is equal to the change in the momentum of the particle and is given by

$$\mathbf{I} = \Delta \mathbf{p} = \int_{t_i}^{t_f} \mathbf{F}\, dt \tag{9.6}$$

Impulsive forces are forces that are very strong compared with other forces on the system, and usually act for a very short time, as in the case of collisions.

The **law of conservation of momentum** for two interacting particles states that if two particles form an isolated system, their total momentum is conserved regardless of the nature of the force between them. Therefore, the total momentum of the system at all times equals its initial total momentum, or

Conservation of momentum

$$\mathbf{p}_{1i} + \mathbf{p}_{2i} = \mathbf{p}_{1f} + \mathbf{p}_{2f} \tag{9.12}$$

Elastic and inelastic collision

When two particles collide, the total momentum of the system before the collision always equals the total momentum after the collision, regardless of the nature of the collision. An **inelastic collision** is a collision for which the mechanical energy is not conserved, but momentum is conserved. A perfectly inelastic collision corresponds to the situation where the colliding bodies stick together after the collision. An **elastic collision** is one in which both momentum and kinetic energy are conserved.

In a two- or three-dimensional collision, the components of momentum in each of the three directions (x, y, and z) are conserved independently.

The **vector position** *of the center of mass of a system of particles* is defined as

Center of mass for a system of particles

$$\mathbf{r}_c \equiv \frac{\Sigma m_i \mathbf{r}_i}{M} \tag{9.30}$$

where $M = \Sigma m_i$ is the total mass of the system and $\mathbf{r}_i$ is the vector position of the ith particle.

The *vector position of the center of mass of a rigid body* can be obtained from the integral expression

Center of mass for a rigid body

$$\mathbf{r}_c = \frac{1}{M}\int \mathbf{r}\, dm \tag{9.33}$$

The **velocity of the center of mass** *for a system of particles* is given by

Velocity of the center of mass

$$\mathbf{v}_c = \frac{\Sigma m_i \mathbf{v}_i}{M} \tag{9.34}$$

The total momentum of a system of particles equals the total mass multiplied by the velocity of the center of mass, that is, $\mathbf{P} = M\mathbf{v}_c$.

Newton's second law applied to a system of particles is given by

Newton's second law for a system of particles

$$\sum \mathbf{F}_{\text{ext}} = M\mathbf{a}_c = \frac{d\mathbf{P}}{dt} \tag{9.38}$$

where $\mathbf{a}_c$ is the acceleration of the center of mass and the sum is over all external forces. Therefore, the center of mass moves like an imaginary particle of mass M under the influence of the resultant external force on the system. It follows from Equation 9.38 that the total momentum of the system is conserved if there are no external forces acting on it.

QUESTIONS

1. If the kinetic energy of a particle is zero, what is its linear momentum? If the total energy of a particle is zero, is its linear momentum necessarily zero? Explain.
2. If the velocity of a particle is doubled, by what factor is its momentum changed? What happens to its kinetic energy?
3. If two particles have equal kinetic energies, are their momenta necessarily equal? Explain.
4. Does a large force always produce a larger impulse on a body than a smaller force? Explain.
5. An isolated system is initially at rest. Is it possible for parts of the system to be in motion at some later time? If so, explain how this might occur.
6. If two objects collide and one is initially at rest, is it possible for both to be at rest after the collision? Is it possible for one to be at rest after the collision? Explain.
7. Explain why momentum is conserved when a ball bounces from a floor.
8. Is it possible to have a collision in which all of the kinetic energy is lost? If so, cite an example.
9. In a perfectly elastic collision between two particles, do both particles have the same kinetic energy after the collision? Explain.
10. When a ball rolls down an incline, its momentum increases. Does this imply that momentum is not conserved? Explain.
11. Consider an inelastic collision between a car and a large truck. Which vehicle loses more kinetic energy as a result of the collision?
12. Can the center of mass of a body lie outside the body? If so, give examples.
13. A boy stands at one end of a floating raft that is stationary relative to the shore. He then walks to the opposite end of the raft, away from the shore. What happens to the center of mass of the system (boy + raft)? Does the raft move? Explain.
14. Three balls are thrown into the air simultaneously. What is the acceleration of their center of mass while they are in motion?
15. Two isolated particles undergo a head-on collision. What is the acceleration of the center of mass after the collision?
16. A meter stick is balanced in a horizontal position with the index fingers of the right and left hands. If the two fingers are brought together, the stick remains balanced and the two fingers always meet at the 50-cm mark regardless of their original positions (try it!). Carefully explain this observation.
17. A hunter shoots a polar bear on a glacier. How might the hunter, knowing her own weight, be able to *estimate* the weight of the polar bear using a measuring tape and a rope?
18. If the forward momentum of a bullet is the same as the backward momentum of the gun, why isn't it as dangerous to be hit by the gun as by the bullet?
19. A box slides across the frictionless surface of a frozen lake. What happens to the speed of the box as water collects in it from a rain shower? Explain.
20. A piece of mud is thrown against a brick wall and sticks to the wall. What happens to the momentum of the mud? Is momentum conserved? Explain.
21. Early in this century, Robert Goddard proposed sending a rocket to the moon. Critics took the position that in a vacuum, such as exists between the earth and the moon, the gases emitted by the rocket would have nothing to push against to propel the rocket. According to *Scientific American* (January 1975), Goddard placed a gun in a vacuum and fired a blank cartridge from it. (A blank cartridge fires only the hot gases of the burning gunpowder.) What happened when the gun was fired?
22. An astronaut walking in space accidentally severs the safety cord attaching her to the spacecraft. If she happens to have with her a can of aerosol spray deodorant, how could she use this to return safely to her ship?
23. A pole vaulter falls from a height of 15 ft onto a foam rubber pad. Could you calculate his velocity just before he reaches the pad? Would you be able to calculate the force exerted on him due to the collision? Explain.
24. As a ball falls toward the earth, its momentum increases. How would you reconcile this fact with the law of conservation of momentum?
25. A man is at rest sitting at one end of a boat in the middle of a lake. If he walks to the opposite end of the boat toward the east, why does the boat move west? What can you say about the center of mass of the boat-man system?
26. Explain how you would use a balloon to demonstrate the mechanism responsible for rocket propulsion.
27. Explain the maneuver of decelerating a spacecraft. What other maneuvers are possible?
28. Does the center of mass of a rocket in free space accelerate? Explain. Can the speed of a rocket exceed the exhaust velocity of the fuel? Explain.

PROBLEMS

Section 9.1 Linear Momentum and Impulse

1. A 3-kg particle has a velocity of $(2\boldsymbol{i} - 4\boldsymbol{j})$ m/s. Find its x and y components of momentum and the magnitude of its total momentum.
2. The momentum of a 1500-kg car is equal to the momentum of a 5000-kg truck traveling at a speed of 25 mi/h. What is the speed of the car?
3. A 1500-kg automobile travels eastward at a speed of 8 m/s. It makes a 90° turn to the north in a time of 3 s and continues with the same speed. Find (a) the impulse delivered to the car as a result of the turn and (b) the average force exerted on the car during the turn.
4. A 0.3-kg ball moving along a straight line has a velocity of $5\boldsymbol{i}$ m/s. It collides with a wall and rebounds with a velocity of $-3\boldsymbol{i}$ m/s. Find (a) the change in its momentum and (b) the average force exerted on the wall if the ball is in contact with the wall for 5×10^{-3} s.
5. The force F_x acting on a 2-kg particle varies in time as shown in Figure 9.22. Find (a) the impulse of the force, (b) the final velocity of the particle if it is initially at rest, and (c) the final velocity of the particle if it is initially moving along the x axis with a velocity of -2 m/s.

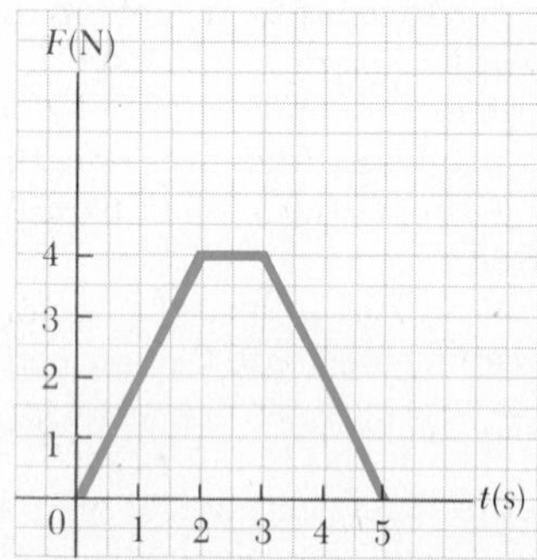

Figure 9.22 (Problems 5 and 6).

6. Find the average force exerted on the particle described in Figure 9.22 for the time interval $t_i = 0$ to $t_f = 5$ s.
7. An estimated force-time curve for a baseball struck by a bat is shown in Figure 9.23. From this curve, determine (a) the impulse delivered to the ball, (b) the average force exerted on the ball, and (c) the peak force exerted on the ball.

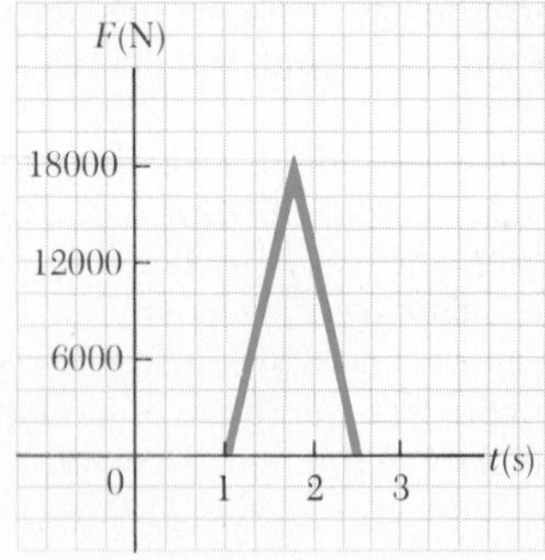

Figure 9.23 (Problem 7).

8. Calculate the magnitude of the linear momentum for the following cases: (a) a proton of mass 1.67×10^{-27} kg moving with a speed of 5×10^6 m/s; (b) a 15-g bullet moving with a speed of 500 m/s; (c) a 75-kg sprinter running at a speed of 12 m/s, and (d) the earth (mass 5.98×10^{24} kg) moving with an orbital speed of 2.98×10^4 m/s.
9. If the momentum of an object is doubled in magnitude, what happens to its kinetic energy? (b) If the kinetic energy of an object is tripled, what happens to its momentum?
10. A 3-kg particle is initially moving along the y axis with a velocity of 5 m/s. After 5 s, it is moving along the x axis with a velocity of 3 m/s. Find (a) the impulse delivered to the particle and (b) the average force exerted on it in the 5-s interval.
11. A 1.5-kg football is thrown with a speed of 15 m/s. A stationary receiver catches the ball and brings it to rest in 0.02 s. (a) What is the impulse delivered to the ball? (b) What is the average force exerted on the receiver?
12. A single constant force of 80 N accelerates a 5-kg object from a speed of 2 m/s to a speed of 8 m/s. Find (a) the impulse acting on the object in this interval and (b) the time interval over which this impulse is delivered.
13. A 0.16-kg baseball is thrown with a speed of 40 m/s. It is hit straight back at the pitcher with a speed of 55 m/s. (a) What is the impulse delivered to the baseball? (b) Find the average force exerted by the bat on the ball if the ball is in contact with the bat for 2×10^{-3} s. Compare this with the weight of the ball and determine whether or not the impulse approximation is valid in the situation.

Section 9.2 Conservation of Linear Momentum for a Two-Particle System

14. A 40-kg child standing on a frozen pond throws a 2-kg stone to the east with a speed of 5 m/s. Neglecting friction between the child and ice, find the recoil velocity of the child.
15. Two blocks of masses M and $3M$ are placed on a horizontal, frictionless surface. A light spring is attached to one of them, and the blocks are pushed together with the spring between them (Fig. 9.24). A string holding them together is burned, after which the block of mass $3M$ moves to the right with a speed of 2 m/s. What is the speed of the block of mass M? (Assume they are initially at rest.)
16. Consider the cannon described in Example 9.3 and Figure 9.5. (a) If the cannon's mass is doubled, what is the velocity of the cannon ball after firing? (b) If the cannon is bolted to the ground, the cannon ball is fired

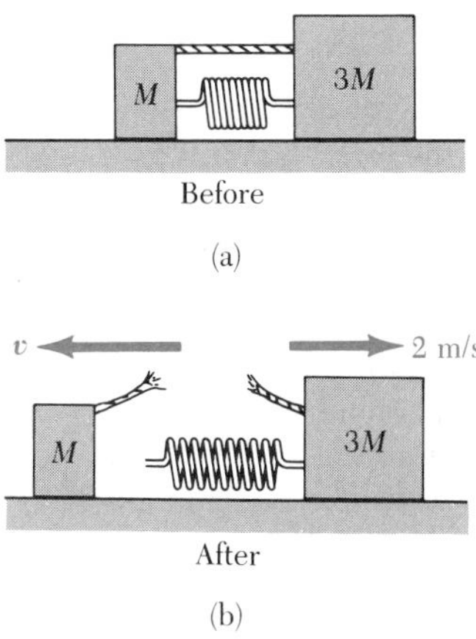

Figure 9.24 (Problem 15).

with some velocity but the cannon apparently doesn't move. Does this mean momentum is not conserved? Explain.

17. A 65-kg boy and a 40-kg girl, both wearing skates, face each other at rest. The boy pushes the girl, sending her eastward with a speed of 4 m/s. Describe the subsequent motion of the boy. (Neglect friction.)

Section 9.3 Collisions and Section 9.4 Examples of Collisions in One Dimension

18. A 3-kg mass moving initially with a speed of 8 m/s makes a perfectly inelastic head-on collision with a 5-kg mass initially at rest. (a) Find the final velocity of the composite particle. (b) How much energy is lost in the collision?
19. A 2000-kg meteorite has a speed of 80 m/s just before colliding head-on with the earth. Determine the recoil speed of the earth (mass 5.98×10^{24} kg).
20. Consider the ballistic pendulum described in Example 9.5 and shown in Figure 9.9. (a) Show that the ratio of the kinetic energy after the collision to the kinetic energy before the collision is given by the ratio $m_1/(m_1 + m_2)$, where m_1 is the mass of the bullet and m_2 is the mass of the block. (b) If $m_1 = 8$ g and $m_2 = 2$ kg, what percentage of the original energy is left after the inelastic collision? What accounts for the missing energy?
21. An 8-g bullet is fired into a 2.5-kg ballistic pendulum and becomes embedded in it. If the pendulum rises a vertical distance of 6 cm, calculate the initial speed of the bullet.
22. A 90-kg halfback running north with a speed of 9 m/s is tackled by a 120-kg opponent running south with a speed of 3 m/s. If the collision is assumed to be perfectly inelastic and head-on, (a) calculate the velocity of the players just after the tackle and (b) determine the energy lost as a result of the collision. Can you account for the missing energy?
23. A 3-kg sphere makes a perfectly inelastic collision with a second sphere initially at rest. The composite system moves with a speed equal to one third the original speed of the 3-kg sphere. What is the mass of the second sphere?
24. A 1200-kg car traveling initially with a speed of 27 m/s in an easterly direction crashes into the rear end of a 9000-kg truck moving in the same direction at 22 m/s (Fig. 9.25). The velocity of the car right after the collision is 20 m/s to the east. (a) What is the velocity of the truck right after the collision? (b) How much mechanical energy is lost in the collision? How do you account for this loss in energy?
25. A railroad car of mass 2×10^4 kg moving with a speed of 5 m/s collides and couples with three other coupled railroad cars each of the same mass as the single car and moving in the same direction with an initial speed of 2 m/s. (a) What is the speed of the four cars after the collision? (b) How much energy is lost in the collision?
26. A 3-kg mass moving with a velocity of $8\mathbf{i}$ m/s makes an elastic head-on collision with a 5-kg mass initially at rest. Find (a) the final velocity of each mass and (b) the final kinetic energy of each mass.
27. A neutron in a reactor makes an elastic head-on collision with the nucleus of a carbon atom initially at rest. (a) What fraction of the neutron's kinetic energy is transferred to the carbon nucleus? (b) If the initial kinetic energy of the neutron is 1 MeV $= 1.6 \times 10^{-13}$ J, find its final kinetic energy and the kinetic energy of the carbon nucleus after the collision. (The mass of the carbon nucleus is about 12 times the mass of the neutron.)
28. A neutron moving with a velocity of $2 \times 10^6\mathbf{i}$ m/s makes a head-on elastic collision with a stationary helium nucleus (the mass of He is 4 u). Find (a) the final velocity of each particle and (b) the fraction of the initial kinetic energy transferred to the helium nucleus.

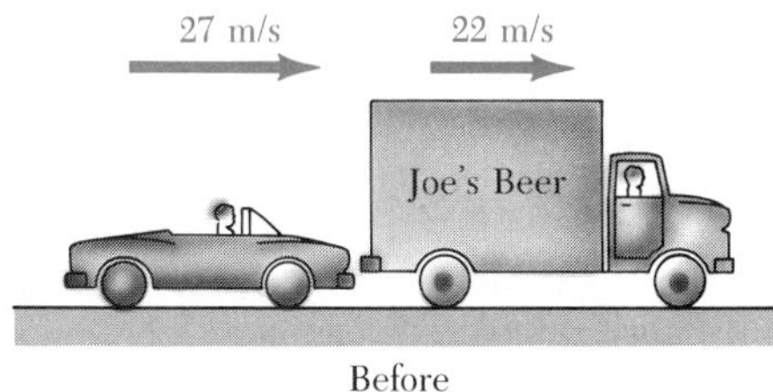

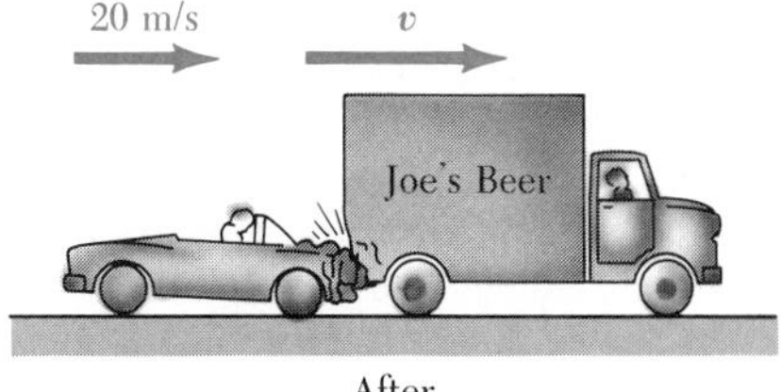

Figure 9.25 (Problem 24).

29. A 5-g particle moving to the right with a speed of 20 cm/s makes an elastic head-on collision with a 10-g particle initially at rest. Find (a) the final velocity of each particle and (b) the fraction of the total energy transferred to the 10-g particle.

30. Two billiard balls have velocities of 1.5 m/s and −0.4 m/s before they meet in an elastic head-on collision. What are their final velocities?

31. Two particles of equal mass m collide head-on as shown in Figure 9.10. Determine which of the following collisions are perfectly elastic for these particles: (a) $v_{1i} = 3$ m/s, $v_{2i} = 0$, $v_{1f} = 0$, $v_{2f} = 2$ m/s; (b) $v_{1i} = 0$, $v_{2i} = -5$ m/s, $v_{1f} = -5$ m/s, $v_{2f} = 0$; (c) $v_{1i} = 4$ m/s, $v_{2i} = -2$ m/s, $v_{1f} = -2$ m/s, $v_{2f} = 4$ m/s.

32. Verify Equations 9.20 and 9.21 for a perfectly elastic head-on collision.

33. A 1-kg mass moving with an initial speed of 5 m/s collides with and sticks to a 6-kg mass initially at rest. The combined mass then proceeds to collide with and stick to a 2-kg mass also at rest initially. If the collisions are all head-on, find (a) the final speed of the system and (b) the amount of kinetic energy lost.

34. A 0.2-kg ball fastened to the end of a string 1.5 m in length to form a pendulum is released in the horizontal position. At the bottom of its swing, the ball collides with a 0.3-kg block initially resting on a frictionless surface (Fig. 9.26). (a) If the collision is elastic, calculate the speed of the ball and of the block just after the collision. (b) If the collision is completely inelastic (they stick), determine the height that the center of mass rises after the collision.

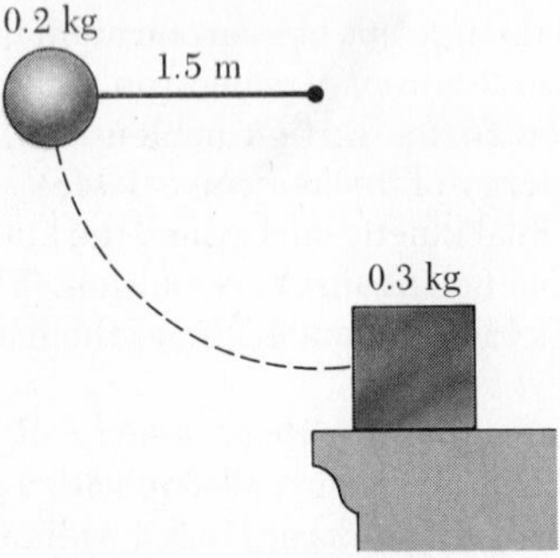

Figure 9.26 (Problem 34).

Section 9.5 Two-Dimensional Collisions

35. A 200-g cart moves on a horizontal, frictionless surface with a constant speed of 30 cm/s. A 50-g piece of modeling clay is dropped vertically onto the cart. (a) If the clay sticks to the cart, find the final speed of the system. (b) After the collision, the clay has no momentum in the vertical direction. Does this mean that the law of conservation of momentum is violated?

36. A bomb initially at rest explodes into three equal fragments. The velocities of two fragments are $(3\boldsymbol{i} + 2\boldsymbol{j})$ m/s and $(-\boldsymbol{i} - 3\boldsymbol{j})$ m/s. Find the velocity of the third fragment.

37. A 2-kg mass with an initial velocity of $5\boldsymbol{i}$ m/s collides with and sticks to a 3-kg mass with an initial velocity of $-3\boldsymbol{j}$ m/s. Find the final velocity of the composite mass.

38. A proton moving with a velocity $v_0\boldsymbol{i}$ collides elastically with another proton initially at rest. If both protons have the same speed after the collision, find (a) the speed of each proton after the collision in terms of v_0 and (b) the direction of the velocity vectors after the collision.

39. An unstable nucleus of mass 17×10^{-27} kg initially at rest disintegrates into three particles. One of the particles, of mass 5.0×10^{-27} kg, moves along the y axis with a velocity of 6×10^6 m/s. Another particle, of mass 8.4×10^{-27} kg, moves along the x axis with a velocity of 4×10^6 m/s. Find (a) the velocity of the third particle and (b) the total energy given off in the process.

40. A 0.3-kg puck, initially at rest on a horizontal, frictionless surface, is struck by a 0.2-kg puck moving initially along the x axis with a velocity of 2 m/s. After the collision, the 0.2-kg puck has a speed of 1 m/s at an angle of $\theta = 53°$ to the positive x axis (Fig. 9.11). (a) Determine the velocity of the 0.3-kg puck after the collision. (b) Find the fraction of kinetic energy lost in the collision.

41. Two shuffleboard disks of equal mass, one orange and the other yellow, are involved in a perfectly elastic glancing collision. The yellow disk is initially at rest and is struck by the orange disk moving with a speed of 4 m/s. After the collision, the orange disk moves along a direction that makes an angle of 30° with its initial direction of motion and the velocity of the yellow disk is perpendicular to that of the orange disk (after the collision). Determine the final speed of each disk.

Section 9.6 The Center of Mass

42. A 3-kg particle is located on the x axis at $x = -4$ m, and a 5-kg particle is on the x axis at $x = 2$ m. Find the center of mass.

43. The mass of the moon is about 0.0123 times the mass of the earth. The earth-moon separation measured from their centers is about 3.84×10^8 m. Determine the location of the center of mass of the earth-moon system as measured from the center of the earth.

44. The separation between the hydrogen and chlorine atoms of the HCl molecule is about 1.30×10^{-10} m. Determine the location of the center of mass of the molecule as measured from the hydrogen atom. (Chlorine is 35 times more massive than hydrogen.)

45. Three masses located in the xy plane have the following coordinates: a 2-kg mass has coordinates given by (3, −2) m; a 3-kg mass has coordinates (−2, 4) m; a 1-kg mass has coordinates (2, 2) m. Find the coordinates of the center of mass.

Section 9.7 Motion of a System of Particles

46. A 5-kg particle moves along the x axis with a velocity of 3 m/s. A 3-kg particle moves along the x axis with a velocity of -2 m/s. Find (a) the velocity of the center of mass and (b) the total momentum of the system.
47. A 2-kg particle has a velocity of $(2\mathbf{i} - \mathbf{j})$ m/s, and a 3-kg particle has a velocity of $(\mathbf{i} + 6\mathbf{j})$ m/s. Find (a) the velocity of the center of mass and (b) the total momentum of the system.
48. A 2-kg particle has a velocity of $\mathbf{v}_1 = -10t\mathbf{j}$ m/s, where t is in s. A 3-kg particle moves with a constant velocity of $\mathbf{v}_2 = 4\mathbf{i}$ m/s. At $t = 0.5$ s, find (a) the velocity of the center of mass, (b) the acceleration of the center of mass, and (c) the total momentum of the system.
49. A particle of mass M has an acceleration of 3 m/s^2 in the x direction. A particle of mass $2M$ has an acceleration of 3 m/s^2 in the y direction. Find the acceleration of the center of mass.
50. Two particles each of mass 0.5 kg move in the xy plane. At some instant, their coordinates, velocity components, and acceleration components are as tabulated below. At this instant, find (a) the vector position of the center of mass, (b) the velocity of the center of mass, and (c) the acceleration of the center of mass.

	x(m)	y(m)	v_x (m/s)	v_y (m/s)	a_x (m/s^2)	a_y (m/s^2)
Particle 1	2	3	5	-4	4	0
Particle 2	-2	3	3	8	2	-2

Section 9.8 Rocket Propulsion

51. A rocket engine consumes 75 kg of fuel per second. If the exhaust velocity is 4×10^3 m/s, calculate the thrust on the rocket.
52. The first stage of a Saturn V space vehicle consumes fuel at the rate of 1.5×10^4 kg/s, with an exhaust velocity of 2.6×10^3 m/s. (These are approximate figures.) (a) Calculate the thrust produced by these engines. (b) If the initial mass of the vehicle is 3×10^6 kg, find its *initial* acceleration on the launch pad. [You must include the force of gravity to solve (b).]
53. A rocket moving in free space with its engines off coasts with a speed of 5×10^3 m/s. Its engines are turned on, and at some later time when its mass is reduced to 90% of its initial mass, the speed of the rocket is 6.5×10^3 m/s. Find the exhaust velocity of the ejected fuel, assuming a uniform burn rate.

GENERAL PROBLEMS

54. Consider a sphere of radius R and mass density ρ that is solid except for a spherical hollow volume of radius $R/2$. The center of the spherical void is located at a distance $R/2$ from the center of the large sphere. Find the center of mass of the body. (*Hint:* Treat the void as a negative mass.)
55. Two children in a 90-kg boat are drifting southward with a constant speed of 1.5 m/s. Each child has a mass of 50 kg. What is the velocity of the boat *immediately* after (a) one of the children falls off the rear of the boat, (b) one of the children dives off the rear in the northerly direction with a speed of 2 m/s relative to a stationary land observer, and (c) one of the children dives eastward (perpendicular to the boat) with a speed of 3 m/s.
56. A 30-06 caliber hunting rifle fires a bullet of mass 0.012 kg with a muzzle velocity of 800 m/s to the right. The rifle has a mass of 4.0 kg. (a) What is the recoil velocity of the rifle as the bullet leaves the rifle? (b) If the rifle is stopped by the hunter's shoulder in a distance of 2 cm, what is the average force exerted on the shoulder by the rifle? (c) If the hunter's shoulder is partially restricted from recoiling, would the force exerted on the shoulder be the same as in part (b)? Explain.
57. A projectile of mass m collides with a flat surface. The angle between the incident path and the surface equals the angle θ between the reflected path and the surface. The speed of the projectile is unchanged by the collision. The time the projectile is in contact with the surface is t. Find (a) the change in momentum of the projectile in the x direction, (b) the change in momentum in the y direction, and (c) the average force exerted on the surface by the projectile during the collision.
58. The vector position of a 1-g particle moving in the xy plane varies in time according to the vector expression $\mathbf{r}_1 = (3\mathbf{i} + 3\mathbf{j})t + 2\mathbf{j}t^2$. At the same time, the vector position of a 2-g particle moving in the xy plane varies as $\mathbf{r}_2 = 3\mathbf{i} - 2\mathbf{i}t^2 - 6\mathbf{j}t$, where t is in s and $\mathbf{r}$ is in cm. At $t = 2$ s, determine (a) the vector position of the center of mass, (b) the linear momentum of the system, (c) the velocity of the center of mass, and (d) the acceleration of the center of mass.
59. A 6-g bullet is fired into a 2-kg block initially at rest at the edge of a table of height 1 m (Fig. 9.27). The bullet remains in the block, and after impact the block lands 2 m from the bottom of the table. Determine the initial speed of the bullet.

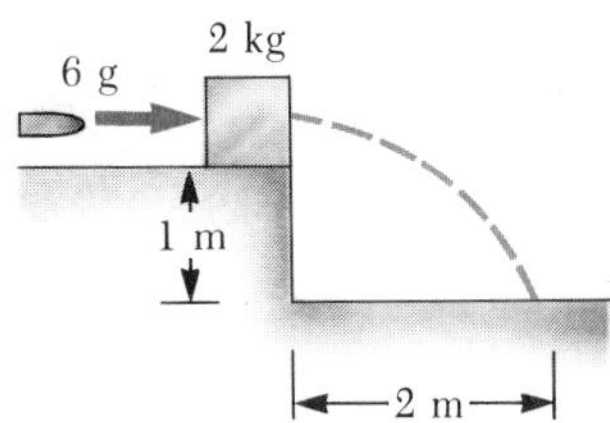

Figure 9.27 (Problem 59).

60. A 40-kg child stands at one end of a 70-kg boat that is 4 m in length (Fig. 9.28). The boat is initially 3 m from the pier. The child notices a turtle on a rock at the far end of the boat and proceeds to walk to that end to catch the turtle. Neglecting friction between the boat and water, (a) describe the subsequent motion of the system (child + boat). (b) Where will the child be *relative to the pier* when he reaches the far end of the boat? (c) Will he catch the turtle? (Assume he can reach out 1 m from the end of the boat.)

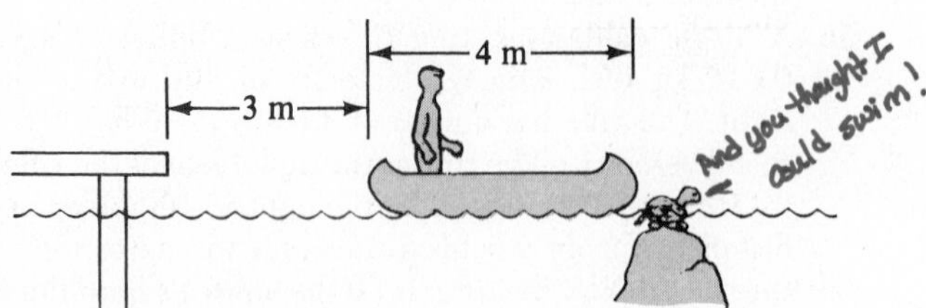

Figure 9.28 (Problem 60).

61. A block of mass M is given an initial velocity v_0 on a rough, horizontal surface. After traveling a distance d, it makes a head-on elastic collision with a block of mass $2M$. How far will the second block move before coming to rest? (Assume the coefficient of friction is the same for each block.)

62. A 7-g bullet is fired into a 1.5-kg ballistic pendulum as in Figure 9.9. The bullet emerges from the block after the collision with a speed of 200 m/s, and the block rises to a maximum height of 12 cm. Find the initial speed of the bullet.

63. A machine gun held by a soldier fires bullets at the rate of three per second. Each bullet has a mass of 30 g and a speed of 1200 m/s. Find the average force exerted on the soldier.

64. An object of mass M is in the shape of a right triangle with dimensions as shown in Figure 9.29. Locate the coordinates of the center of mass, assuming the object has a uniform mass per unit area.

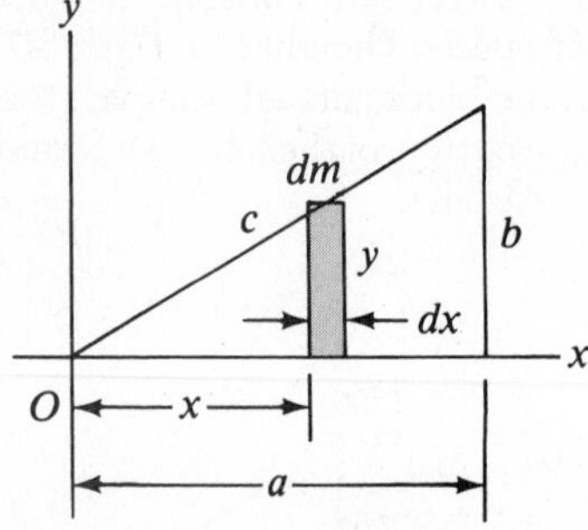

Figure 9.29 (Problem 64).

•65. A 60-kg firefighter slides down a pole while a constant frictional force of 300 N retards his motion. A horizontal 20-kg platform is supported by a spring at the bottom of the pole to cushion the fall. The firefighter starts from rest 5 m above the platform, and the spring constant is 2500 N/m. Find (a) the firefighter's speed just before he collides with the platform and (b) the maximum distance the spring will be compressed. (Assume the frictional force acts during the entire motion.)

•66. A 5-g bullet moving with an initial speed of 400 m/s is fired into and passes through a 1-kg block, as in Figure 9.30. The block, initially at rest on a frictionless, horizontal surface, is connected to a spring of force constant 900 N/m. If the block moves a distance of 5 cm to the right after impact, find (a) the speed at which the bullet emerges from the block and (b) the energy lost in the collision.

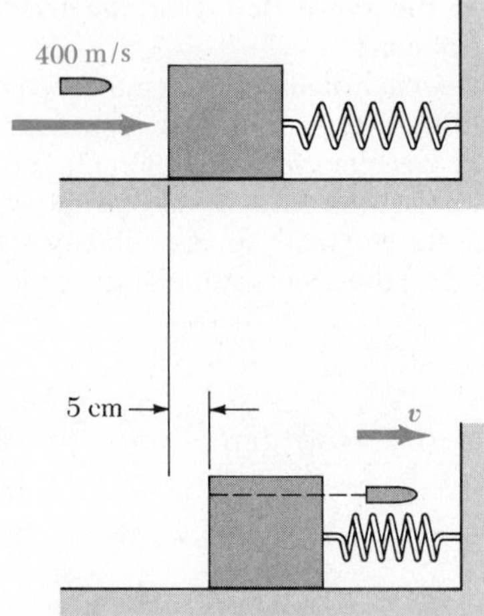

Figure 9.30 (Problem 66).

•67. A chain of length L and total mass M is released from rest with its lower end just touching the top of a table, as in Figure 9.31a. Find the force of the table on the chain after the chain has fallen through a distance x, as in Figure 9.31b. (Assume each link comes to rest the instant it reaches the table.)

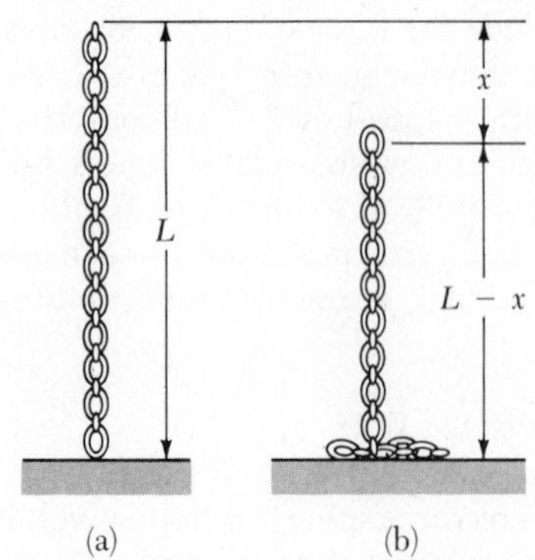

Figure 9.31 (Problem 67).

•68. Two gliders are set in motion on an air track. A spring of force constant k is attached to the near side of one glider. The first glider of mass m_1 has velocity v_1 and the second glider of mass m_2 has velocity v_2 as shown in Figure 9.32 ($v_1 > v_2$). When m_1 collides with the spring attached to m_2 and compresses the spring to its maximum compression x_m, the velocity of the gliders is v. In terms of v_1, v_2, m_1, m_2, and k, find (a) the velocity v at maximum compression, (b) the maximum compression x_m, and (c) the velocities of each glider after the first glider has again lost contact with the spring.

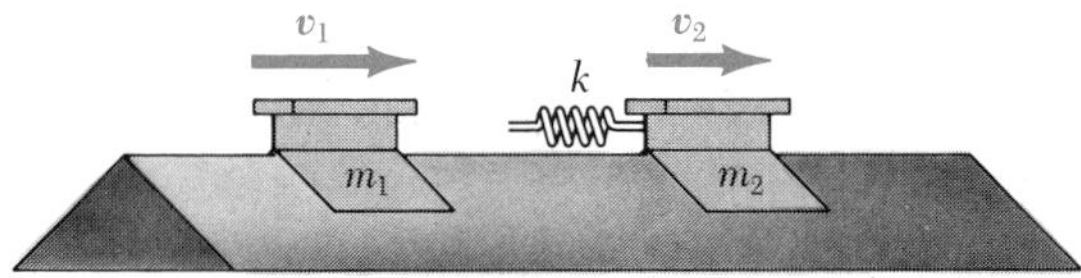

Figure 9.32 (Problem 68).

CALCULATOR/COMPUTER PROBLEM

69. Consider a head-on elastic collision between a moving particle of mass m_1 and an initially stationary particle of mass m_2 (see Example 9.6). (a) Plot f_2, the fraction of energy transferred to m_2, as a function of the ratio m_2/m_1 and show that f_2 reaches a maximum when $m_2/m_1 = 1$. (b) Perform an analytical calculation that verifies that f_2 is a maximum when $m_1 = m_2$.

10
Rotation of a Rigid Body About a Fixed Axis

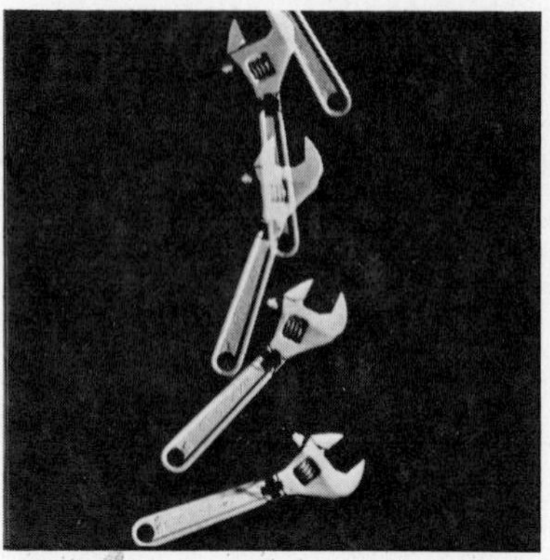

When an extended body, such as a wheel, rotates about its axis, the motion cannot be analyzed by treating the body as a particle, since at any given time different parts of the body will have different velocities and accelerations. For this reason, it is convenient to consider an extended object as a large number of particles, each with its own velocity and acceleration.

Rigid body

In dealing with the rotation of a body, analysis is greatly simplified by assuming the body to be rigid. A **rigid body** is defined as a body that is nondeformable, or one in which the separations between all pairs of particles in the body remain constant. Needless to say, all real bodies in nature are deformable to some extent; however, our rigid-body model is useful in many situations where deformation is negligible. In this chapter, we shall treat the rotation of a rigid body about a fixed axis, commonly referred to as *pure rotational motion.*

The vector nature of angular velocity and angular acceleration rotations in space, and the concept of angular momentum will be presented in detail in Chapter 11.

10.1 ANGULAR VELOCITY AND ANGULAR ACCELERATION

Figure 10.1 illustrates a planar rigid body of arbitrary shape confined to the xy plane and rotating about a fixed axis through O perpendicular to the plane of the figure. A particle on the body at P is at a fixed distance r from the origin and rotates in a circle of radius r about O. In fact, *every* particle on the body undergoes circular motion about O. It is convenient to represent the position of the point P with its polar coordinates (r, θ). In this representation, the only coordinate that changes in time is the angle θ; r remains constant. (In rectangular coordinates, both x and y vary in time.) As the particle moves along the circle from the positive x axis ($\theta = 0$) to the point P, it moves through an arc length s, which is related to the angular position θ through the relation

Figure 10.1 Rotation of a rigid body about a fixed axis through O perpendicular to the plane of the figure (the z axis). Note that a particle at P rotates in a circle of radius r centered at O.

$$s = r\theta \tag{10.1a}$$

$$\theta = s/r \tag{10.1b}$$

It is important to make note of the units of θ as expressed by Equation 10.1b. The angle θ is the ratio of an arc length and the radius of the circle, and hence is a pure number. However, we commonly refer to the unit of θ as a **radian** (rad), where

Radian

one rad is the angle subtended by an arc length equal to the radius of the arc.

Since the circumference of a circle is $2\pi r$, it follows that 360° corresponds to an angle of $2\pi r/r$ rad or 2π rad (one revolution). Hence, 1 rad $= 360°/2\pi \approx 57.3°$. To convert an angle in degrees to an angle in radians, we can use the fact that 2π radians $= 360°$; hence

$$\theta\ (\text{rad}) = \frac{\pi}{180°}\ \theta\ (\text{deg})$$

For example, 60° equals $\pi/3$ rad, and 45° equals $\pi/4$ rad.

As the particle travels from P to Q in Figure 10.2 in a time Δt, the radius vector sweeps out an angle $\Delta\theta = \theta_2 - \theta_1$, which equals the **angular displacement.** We define the **average angular velocity** $\overline{\omega}$ (omega) as the ratio of this angular displacement to the time interval Δt:

$$\overline{\omega} \equiv \frac{\theta_2 - \theta_1}{t_2 - t_1} = \frac{\Delta\theta}{\Delta t} \qquad (10.2)$$

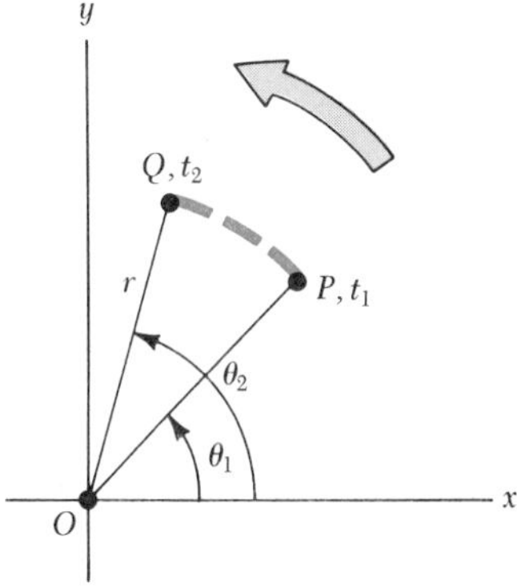

Figure 10.2 A particle on a rotating rigid body moves from P to Q along the arc of a circle. In the time interval $\Delta t = t_2 - t_1$, the radius vector sweeps out an angle $\Delta\theta = \theta_2 - \theta_1$.

In analogy to linear velocity, the **instantaneous angular velocity,** ω, is defined as the limit of the ratio in Equation 10.2 as Δt approaches zero:

$$\omega \equiv \lim_{\Delta t \to 0} \frac{\Delta\theta}{\Delta t} = \frac{d\theta}{dt} \qquad (10.3)$$

Angular velocity has units of rad/s, or s^{-1}, since radians are not dimensional. Let us adopt the convention that the fixed axis of rotation for the rigid body is the z axis, as in Figure 10.1. We shall take ω to be positive when θ is increasing (counterclockwise motion) and negative when θ is decreasing (clockwise motion).

If the instantaneous angular velocity of a body changes from ω_1 to ω_2 in the time interval Δt, the body has an angular acceleration. The **average angular acceleration** $\overline{\alpha}$ (alpha) of a rotating body is defined as the ratio of the change in the angular velocity to the time interval Δt:

Average angular acceleration

$$\overline{\alpha} \equiv \frac{\omega_2 - \omega_1}{t_2 - t_1} = \frac{\Delta\omega}{\Delta t} \qquad (10.4)$$

In analogy to linear acceleration, the **instantaneous angular acceleration** is defined as the limit of the ratio $\Delta\omega/\Delta t$ as Δt approaches zero:

Instantaneous angular acceleration

$$\alpha \equiv \lim_{\Delta t \to 0} \frac{\Delta\omega}{\Delta t} = \frac{d\omega}{dt} \qquad (10.5)$$

Angular acceleration has units of rad/s^2 or s^{-2}. Note that α is positive when ω is increasing in time and negative when ω is decreasing in time.

For rotation about a fixed axis, we see that every *particle on the rigid body has the* same *angular velocity and the* same *angular acceleration.* That is, the quantities ω and α characterize the rotational motion of the entire rigid body. Using these quantities, we can greatly simplify the analysis of rigid-body rotation. Notice that the angular displacement (θ), angular velocity (ω), and angular acceleration (α) are analogous to linear displacement (x), linear velocity (v), and linear acceleration (a), respectively, for the one-dimensional motion discussed in Chapter 3. The variables θ, ω, and α differ dimensionally from the variables x, v, and a, only by a length factor.

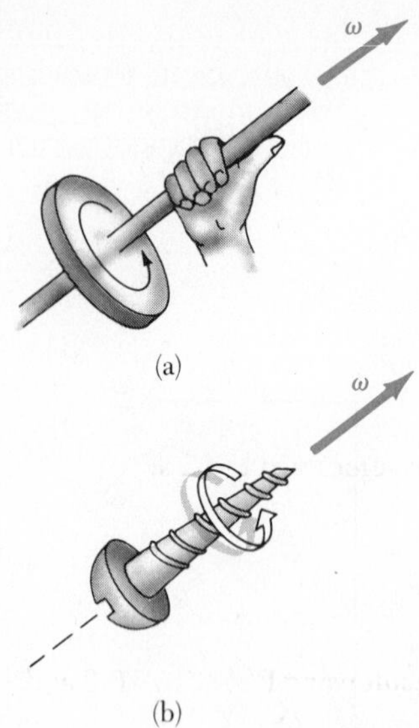

Figure 10.3 (a) The right-hand rule for determining the direction of the angular velocity. (b) The direction of ω is in the direction of advance of a right-handed screw.

We have already indicated how the signs for ω and α are determined; however, we have not specified any direction in space associated with these vector quantities.[1] For rotation about a fixed axis, the only direction in space that uniquely specifies the rotational motion is the direction along the axis of rotation. However, we must also decide on the sense of these quantities, that is, whether they point into or out of the plane of Figure 10.1.

As we have already mentioned, the direction of ω is along the axis of rotation, which is the z axis in Figure 10.1. By convention, we take the direction of ω to be *out* of the plane of the diagram when the rotation is counterclockwise and *into* the plane of the diagram when the rotation is clockwise. To further illustrate this convention, it is convenient to use the *right-hand rule* shown in Figure 10.3a. The four fingers of the right hand are wrapped in the direction of the rotation. The extended right thumb points in the direction of ω. Figure 10.3b illustrates that ω is also in the direction of advance of a similarly rotating right-handed screw. Finally, the sense of α follows from its definition as $d\omega/dt$. It is the same as ω if the angular speed (the magnitude of ω) is increasing in time and antiparallel to ω if the angular speed is decreasing in time.

10.2 ROTATIONAL KINEMATICS: ROTATIONAL MOTION WITH CONSTANT ANGULAR ACCELERATION

In the study of linear motion, we found that the simplest form of accelerated motion to analyze is motion under constant linear acceleration (Chapter 3). Likewise, for rotational motion about a fixed axis the simplest accelerated motion to analyze is motion under constant angular acceleration. Therefore, we shall next develop kinematic relations for rotational motion under constant angular acceleration. If we write Equation 10.5 in the form $d\omega = \alpha\, dt$ and let $\omega = \omega_0$ at $t_0 = 0$, we can integrate this expression directly:

Rotational kinematic equations

$$\omega = \omega_0 + \alpha t \qquad (\alpha = \text{constant}) \tag{10.6}$$

Likewise, substituting Equation 10.6 into Equation 10.3 and integrating once more (with $\theta = \theta_0$ at $t_0 = 0$), we get

$$\theta = \theta_0 + \omega_0 t + \tfrac{1}{2}\alpha t^2 \tag{10.7}$$

If we eliminate t from Equations 10.6 and 10.7, we get

$$\omega^2 = \omega_0{}^2 + 2\alpha(\theta - \theta_0) \tag{10.8}$$

Notice that these kinematic expressions for rotational motion under constant angular acceleration are of the *same form* as those for linear motion under constant linear acceleration with the substitutions $x \to \theta$, $v \to \omega$, and $a \to \alpha$. Table 10.1 gives a comparison of the kinematic equations for rotational and linear motion. Furthermore, the expressions are valid for both rigid-body rotation and particle motion about a *fixed* axis.

[1] Although we do not verify it here, the instantaneous angular velocity and instantaneous acceleration are vector quantities, but the corresponding average values are not. This is because angular displacement is not a vector quantity for finite rotations.

TABLE 10.1 A Comparison of Kinematic Equations for Rotational and Linear Motion Under Constant Acceleration

Rotational Motion About Fixed Axis with α = Constant. Variables: θ and ω	Linear Motion with a = Constant. Variables: x and v
$\omega = \omega_0 + \alpha t$	$v = v_0 + at$
$\theta = \theta_0 + \omega_0 t + \frac{1}{2}\alpha t^2$	$x = x_0 + v_0 t + \frac{1}{2}at^2$
$\omega^2 = \omega_0{}^2 + 2\alpha(\theta - \theta_0)$	$v^2 = v_0{}^2 + 2a(x - x_0)$

EXAMPLE 10.1. Rotating Wheel
A wheel rotates with a constant angular acceleration of 3.5 rad/s². If the angular velocity of the wheel is 2.0 rad/s at $t_0 = 0$, (a) what angle does the wheel rotate through in 2 s?

$$\theta - \theta_0 = \omega_0 t + \tfrac{1}{2}\alpha t^2$$

$$= \left(2.0\ \frac{\text{rad}}{\text{s}}\right)(2\ \text{s}) + \tfrac{1}{2}\left(3.5\ \frac{\text{rad}}{\text{s}^2}\right)(2\ \text{s})^2$$

$$= 11\ \text{rad} = 630° = 1.75\ \text{rev}$$

(b) What is the angular velocity at $t = 2$ s?

$$\omega = \omega_0 + \alpha t = 2.0\ \text{rad/s} + \left(3.5\ \frac{\text{rad}}{\text{s}^2}\right)(2\ \text{s})$$

$$= 9.0\ \text{rad/s}$$

We could also obtain this result using Equation 10.8 and the results of (a). Try it!

Exercise 1 Find the angle that the wheel rotates through between $t = 2$ s and $t = 3$ s.
Answer: 10.8 rad.

10.3 RELATIONSHIPS BETWEEN ANGULAR AND LINEAR QUANTITIES

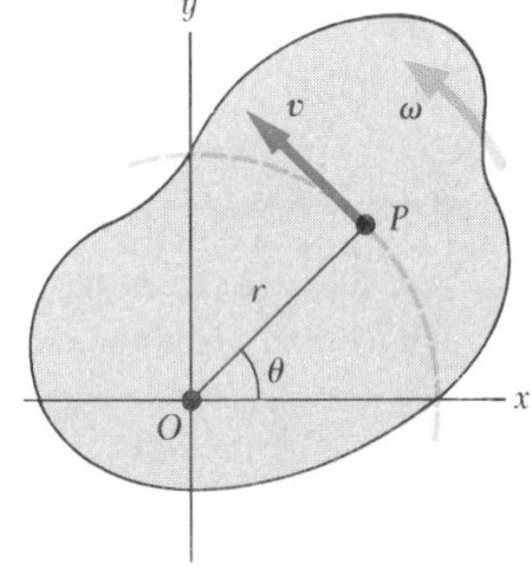

Figure 10.4 As a rigid body rotates about the fixed axis through O, the point P has a linear velocity v, which is always tangent to the circular path of radius r.

In this section we shall derive some useful relationships between the angular velocity and acceleration of a rotating rigid body and the linear velocity and acceleration of an arbitrary point in the body. In order to do so, we should keep in mind that when a rigid body rotates about a fixed axis, *every* particle of the body moves in a circle the center of which is the axis of rotation (Fig. 10.4).

We can first relate the angular velocity of the rotating body to the tangential velocity, v, of a point P on the body. Since P moves in a circle, the linear velocity vector is always tangent to the circular path, and hence the phrase *tangential velocity*. The magnitude of the tangential velocity of the point P is, by definition, ds/dt, where s is the distance traveled by this point measured along the circular path. Recalling that $s = r\theta$ and noting that r is constant, we get

$$v = \frac{ds}{dt} = r\frac{d\theta}{dt}$$

Relationship between linear and angular speed

$$v = r\omega \tag{10.9}$$

That is, the tangential velocity of a point on a rotating rigid body equals the distance of that point from the axis of rotation multiplied by the angular velocity. Therefore, although every point on the rigid body has the same *angular* velocity, not every point has the same *linear* velocity. In fact, Equation 10.9 shows that the linear velocity of a point on the rotating body increases as one moves outward from the center of rotation toward the rim, as you would intuitively expect.

We can relate the angular acceleration of the rotating rigid body to the tangential acceleration of the point P by taking the time derivative of v:

$$a_t = \frac{dv}{dt} = r\frac{d\omega}{dt}$$

$$a_t = r\alpha \tag{10.10}$$

That is, the tangential component of the linear acceleration of a point on a rotating rigid body equals the distance of that point from the axis of rotation multiplied by the angular acceleration.

In Chapter 4 we found that a point rotating in a circular path undergoes a centripetal, or radial, acceleration of magnitude v^2/r and directed toward the center of rotation (Fig. 10.5). Since $v = r\omega$ for the point P on the rotating body, we can express the centripetal acceleration as

$$a_r = \frac{v^2}{r} = r\omega^2 \tag{10.11}$$

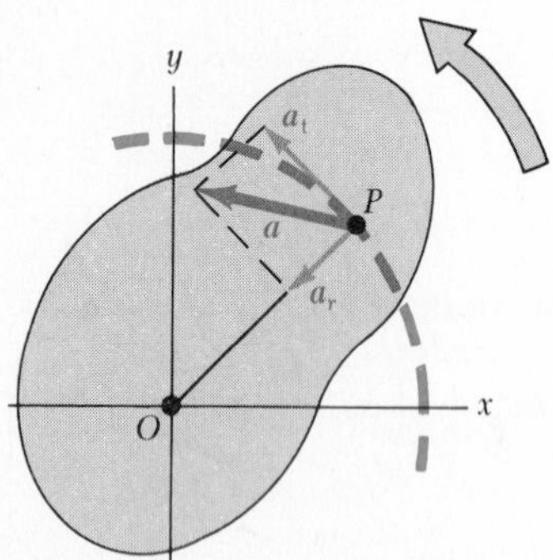

Figure 10.5 As a rigid body rotates about a fixed axis through O, the point P experiences a tangential component of acceleration, a_t, and a centripetal component of acceleration, a_r. The total acceleration of this point is $\mathbf{a} = \mathbf{a}_t + \mathbf{a}_r$.

The *total linear acceleration* of the particle is $\mathbf{a} = \mathbf{a}_t + \mathbf{a}_r$. Therefore, the magnitude of the total linear acceleration of the point P on the rotating rigid body is given by

$$a = \sqrt{a_t^2 + a_r^2} = \sqrt{r^2\alpha^2 + r^2\omega^4} = r\sqrt{\alpha^2 + \omega^4} \tag{10.12}$$

EXAMPLE 10.2. A Rotating Turntable

The turntable of a record player rotates initially at a rate of 33 revolutions/min and takes 20 s to come to rest. (a) What is the angular acceleration of the turntable, assuming the acceleration is uniform?

Recalling that 1 rev = 2π rad, we see that the initial angular velocity is given by

$$\omega_0 = \left(33\,\frac{\text{rev}}{\text{min}}\right)\left(2\pi\,\frac{\text{rad}}{\text{rev}}\right)\left(\frac{1}{60}\,\frac{\text{min}}{\text{s}}\right) = 3.46 \text{ rad/s}$$

Using $\omega = \omega_0 + \alpha t$ and the fact that $\omega = 0$ at $t = 20$ s, we get

$$\alpha = -\frac{\omega_0}{t} = -\frac{3.46 \text{ rad/s}}{20 \text{ s}} = -0.173 \text{ rad/s}^2$$

where the negative sign indicates an angular deceleration (ω is decreasing).

(b) How many rotations does the turntable make before coming to rest?

Using Equation 10.7, we find that the angular displacement in 20 s is

$$\Delta\theta = \theta - \theta_0 = \omega_0 t + \tfrac{1}{2}\alpha t^2$$

$$= [3.46(20) + \tfrac{1}{2}(-0.173)(20)^2] \text{ rad} = 34.6 \text{ rad}$$

This corresponds to $34.6/2\pi$ rev, or 5.51 rev.

(c) What are the magnitudes of the radial and tangential components of the linear acceleration of a point on the rim at $t = 0$?

We can use $a_t = r\alpha$ and $a_r = r\omega^2$, which gives

$$a_t = r\alpha = (14 \text{ cm})\left(0.173\,\frac{\text{rad}}{\text{s}^2}\right) = 2.42 \text{ cm/s}^2$$

$$a_r = r\omega_0^2 = (14 \text{ cm})\left(3.46\,\frac{\text{rad}}{\text{s}}\right)^2 = 168 \text{ cm/s}^2$$

Exercise 2 If the radius of the turntable is 14 cm, what is the initial linear speed of a point on the rim of the turntable?
Answer: 48.4 cm/s.

10.4 ROTATIONAL KINETIC ENERGY

Let us consider a rigid body as a collection of small particles and let us assume that the body rotates about the fixed z axis with an angular velocity ω (Fig. 10.6). Each particle of the body has some kinetic energy, determined by its mass and velocity. If the mass of the ith particle is m_i and its speed is v_i, the kinetic energy of this particle is

$$K_i = \tfrac{1}{2}m_i v_i^2$$

To proceed further, we must recall that although every particle in the rigid body has the same angular velocity, ω, the individual linear velocities depend on the distance r_i from the axis of rotation according to the expression $v_i = r_i\omega$ (Eq. 10.9). The *total* kinetic energy of the rotating rigid body is the sum of the kinetic energies of the individual particles:

$$K = \sum K_i = \sum \tfrac{1}{2}m_i v_i^2 = \tfrac{1}{2}\sum m_i r_i^2 \omega^2$$

$$K = \tfrac{1}{2}\left(\sum m_i r_i^2\right)\omega^2 \qquad (10.13)$$

where we have factored ω^2 from the sum since it is common to every particle. The quantity in parentheses is called the **moment of inertia, I:**

$$I = \sum m_i r_i^2 \qquad (10.14)$$

Using this notation, we can express the kinetic energy of the rotating rigid body (Eq. 10.13) as

$$K = \tfrac{1}{2}I\omega^2 \qquad (10.15)$$

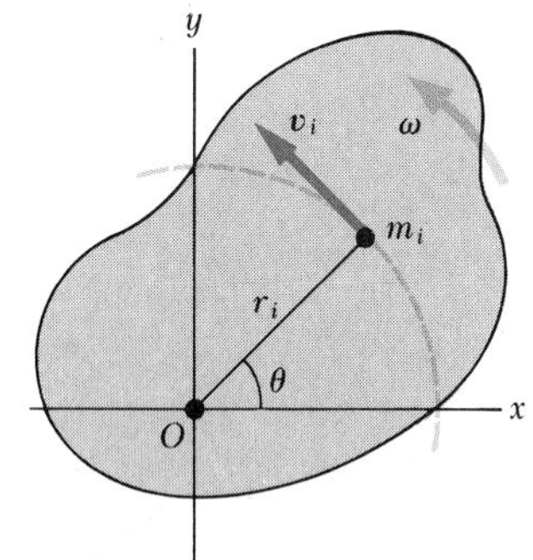

Figure 10.6 A rigid body rotating about the z axis with angular velocity ω. The kinetic energy of the particle of mass m_i is $\frac{1}{2}m_i v_i^2$. The total kinetic energy of the body is $\frac{1}{2}I\omega^2$.

From the definition of moment of inertia, we see that it has dimensions of ML^2 (kg · m² in SI units and g · cm² in cgs units). It plays the role of mass in *all* rotational equations. Although we shall commonly refer to the quantity $\frac{1}{2}I\omega^2$ as the **rotational kinetic energy,** you should note that it is not a new form of energy. It is ordinary kinetic energy, since it was derived from a sum over individual kinetic energies of the particles contained in the rigid body. However, the form of the kinetic energy given by Equation 10.15 is a very convenient one in dealing with rotational motion, providing we know how to calculate I. It is important that you recognize the analogy between kinetic energy associated with linear motion, $\frac{1}{2}mv^2$, and rotational kinetic energy, $\frac{1}{2}I\omega^2$. The quantities I and ω in rotational motion are analogous to m and v in linear motion, respectively. We shall describe how to calculate moments of inertia for rigid bodies in the next section. The following examples illustrate how to calculate moments of inertia and rotational kinetic energy for a distribution of particles.

EXAMPLE 10.3. The Oxygen Molecule

Consider the diatomic molecule oxygen, O_2, which is rotating in the xy plane about the z axis passing through its center, perpendicular to its length. At room temperature, the "average" separation between the two oxygen atoms is 1.21×10^{-10} m (the atoms are treated as point masses). (a) Calculate the moment of inertia of the molecule about the z axis.

Since the mass of an oxygen atom is 2.77×10^{-26} kg and the distance of each atom from the z axis is $d/2$, the moment of inertia about the z axis is

$$I = \sum m_i r_i^2 = m\left(\frac{d}{2}\right)^2 + m\left(\frac{d}{2}\right)^2 = \frac{md^2}{2}$$

$$= \left(\frac{2.77 \times 10^{-26}}{2}\ \text{kg}\right)(1.21 \times 10^{-10}\ \text{m})^2$$

$$= 2.03 \times 10^{-46}\ \text{kg}\cdot\text{m}^2$$

(b) If the angular velocity about the z axis is 2.0×10^{12} rad/s, what is the rotational kinetic energy of the molecule?

$$K = \tfrac{1}{2}I\omega^2$$

$$= \tfrac{1}{2}(2.03 \times 10^{-46}\ \text{kg}\cdot\text{m}^2)\left(2.0 \times 10^{12}\ \frac{\text{rad}}{\text{s}}\right)^2$$

$$= 4.1 \times 10^{-22}\ \text{J}$$

This is about one order of magnitude smaller than the average kinetic energy associated with the translational motion of the molecule at room temperature, which is about 6.2×10^{-21} J.

EXAMPLE 10.4. Four Rotating Particles

Four point masses are fastened to the corners of a frame of negligible mass lying in the xy plane (Fig. 10.7). (a) If the rotation of the system occurs about the y axis with an

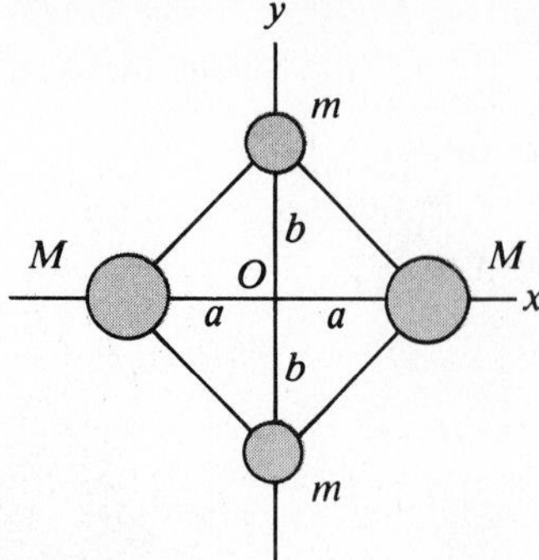

Figure 10.7 (Example 10.4) All particles are at a fixed separation as shown. The moment of inertia depends on the axis about which it is evaluated.

angular velocity ω, find the moment of inertia about the y axis and the rotational kinetic energy about this axis.

First, note that the particles of mass m that lie on the y axis do not contribute to I_y (that is, $r_i = 0$ for these particles about this axis). Applying Equation 10.14, we get

$$I_y = \sum m_i r_i^2 = Ma^2 + Ma^2 = 2Ma^2$$

Therefore, the rotational kinetic energy about the y axis is

$$K = \tfrac{1}{2} I_y \omega^2 = \tfrac{1}{2}(2Ma^2)\omega^2 = Ma^2\omega^2$$

The fact that the masses m do not enter into this result makes sense, since these particles have no motion about the chosen axis of rotation; hence they have no kinetic energy.

(b) Now suppose the system rotates in the xy plane about an axis through O (the z axis). Calculate the moment of inertia about the z axis and the rotational kinetic energy about this axis.

Since r_i in Equation 10.14 is the *perpendicular* distance to the axis of rotation, we get

$$I_z = \sum m_i r_i^2 = Ma^2 + Ma^2 + mb^2 + mb^2$$
$$= 2Ma^2 + 2mb^2$$
$$K = \tfrac{1}{2} I_z \omega^2 = \tfrac{1}{2}(2Ma^2 + 2mb^2)\omega^2 = (Ma^2 + mb^2)\omega^2$$

Comparing the results for (a) and (b), we conclude that the moment of inertia, and therefore the rotational kinetic energy associated with a given angular speed, depend on the axis of rotation. In (b), we would expect the result to include all masses and distances, since all particles are in motion for rotation in the xy plane. Furthermore, the fact that the kinetic energy in (a) is smaller than in (b) indicates that it would take less effort (work) to set the system into rotation about the y axis than about the z axis.

10.5 CALCULATION OF MOMENTS OF INERTIA FOR RIGID BODIES

We can evaluate the moment of inertia of a rigid body by imagining that the body is divided into volume elements, each of mass Δm. Now we can use the definition $I = \Sigma r^2 \, \Delta m$ and take the limit of this sum as $\Delta m \rightarrow 0$. In this limit, the sum becomes an integral over the whole body, where r is the perpendicular distance from the axis of rotation to the element Δm. Hence,

$$I = \lim_{\Delta m \to 0} \sum r^2 \, \Delta m = \int r^2 \, dm \qquad (10.16)$$

To evaluate the moment of inertia using Equation 10.16, it is necessary to express the element of mass dm in terms of its coordinates. It is common to define a mass density in various forms. For a three-dimensional body, it is appropriate to use the *local volume density*, that is, *mass per unit volume*. In this case, we can write

$$\rho = \lim_{\Delta V \to 0} \frac{\Delta m}{\Delta V} = \frac{dm}{dV}$$

$$dm = \rho dV$$

Therefore, the moment of inertia can be expressed in the form

$$I = \int \rho r^2 \, dV$$

If the body is homogeneous, then ρ is constant and the integral can be evaluated for a known geometry. If ρ is not constant, then its variation with position must be specified. When dealing with an object in the form of a sheet of uniform thickness t, it is convenient to define a surface density $\sigma = \rho t$, which signifies *mass per unit area.* Finally, when mass is distributed along a uniform rod of cross-sectional area A, we sometimes use linear density, $\lambda = \rho A$, where λ is defined as *mass per unit length.*

EXAMPLE 10.5. Uniform Hoop
Find the moment of inertia of a uniform hoop of mass M and radius R about an axis perpendicular to the plane of the hoop, through its center (Fig. 10.8).

Solution: All elements of mass are at the same distance $r = R$ from the axis. Therefore, applying Equation 10.16 we get for the moment of inertia about the z axis through O:

$$I_z = \int r^2\,dm = R^2 \int dm = MR^2$$

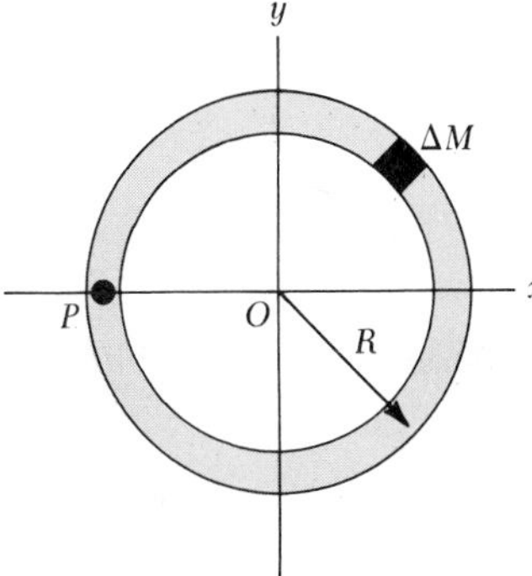

Figure 10.8 (Example 10.5) The mass elements of a uniform hoop are all the same distance from O.

EXAMPLE 10.6. Uniform Rigid Rod
Calculate the moment of inertia of a uniform rigid rod of length L and mass M (Fig. 10.9) about an axis perpendicular to the rod (the y axis) passing through its center of mass.

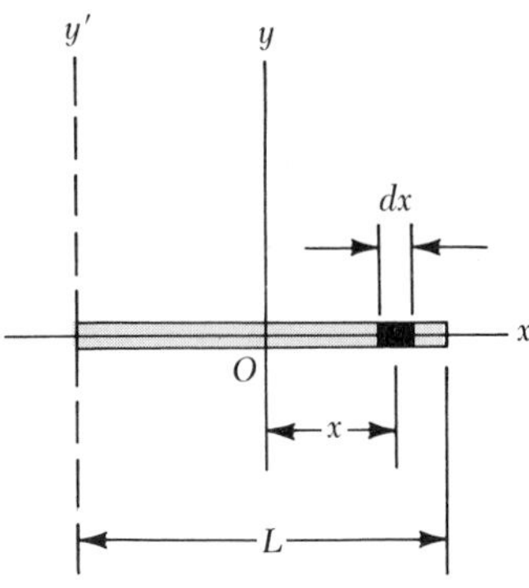

Figure 10.9 (Example 10.6) A uniform rigid rod of length L. The moment of inertia about the y axis is less than that about the y' axis.

Solution: The shaded element of width dx has a mass dm equal to the mass per unit length multiplied by the element of length, dx. That is, $dm = \frac{M}{L}\,dx$. Substituting this into Equation 10.16, with $r = x$, we get

$$I_y = \int r^2\,dm = \int_{-L/2}^{L/2} x^2 \frac{M}{L}\,dx = \frac{M}{L}\int_{-L/2}^{L/2} x^2\,dx$$

$$= \frac{M}{L}\left[\frac{x^3}{3}\right]_{-L/2}^{L/2} = \frac{1}{12}ML^2$$

Exercise 3 Calculate the moment of inertia of a uniform rigid rod about an axis perpendicular to the rod through one end (the y' axis). Note that the calculation requires that the limits of integration be from $x = 0$ to $x = L$.
Answer: $\frac{1}{3}ML^2$.

EXAMPLE 10.7. Uniform Solid Cylinder
A uniform solid cylinder has a radius R, mass M, and length L. Calculate the moment of inertia of the cylinder about an axis through its center, along its length (the z axis in Fig. 10.10).

Solution: In this example, it is convenient to divide the cylinder into cylindrical shells of radius r, thickness dr, and length L, as in Figure 10.10. In this case, cylindrical

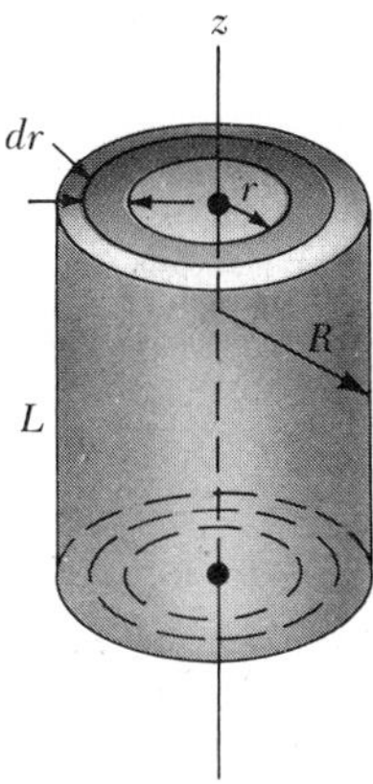

Figure 10.10 (Example 10.7) Calculating I about the z axis for a uniform solid cylinder.

shells are chosen because one wants all mass elements dm to have a single value for r, which makes the calculation of I more straightforward. The volume of each shell is its cross-sectional area multiplied by the length, or $dV = dA \cdot L = (2\pi r\ dr)L$. If the *mass per unit volume* is ρ, then the mass of this differential volume element is $dm = \rho dV = \rho\ 2\pi rL\ dr$. Substituting this into Equation 10.16, we get

$$I_z = \int r^2\, dm = 2\pi\rho L \int_0^R r^3\, dr = \frac{\pi\rho L R^4}{2}$$

However, since the total volume of the cylinder is $\pi R^2 L$, $\rho = M/V = M/\pi R^2 L$. Substituting this into the above result gives

$$I_z = \tfrac{1}{2}MR^2$$

As we saw in the previous examples, the moments of inertia of rigid bodies with simple geometry (high symmetry) are relatively easy to calculate provided the reference axis coincides with an axis of symmetry. Table 10.2 lists, for some common rigid bodies, moments of inertia about an axis through the center of mass and about an axis parallel to this.[2]

The calculation of moments of inertia about an arbitrary axis can be somewhat cumbersome, even for a highly symmetric body, such as a sphere. In this

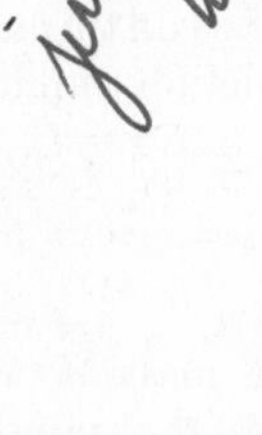

TABLE 10.2 Moments of Inertia of Homogeneous Rigid Bodies with Different Geometries

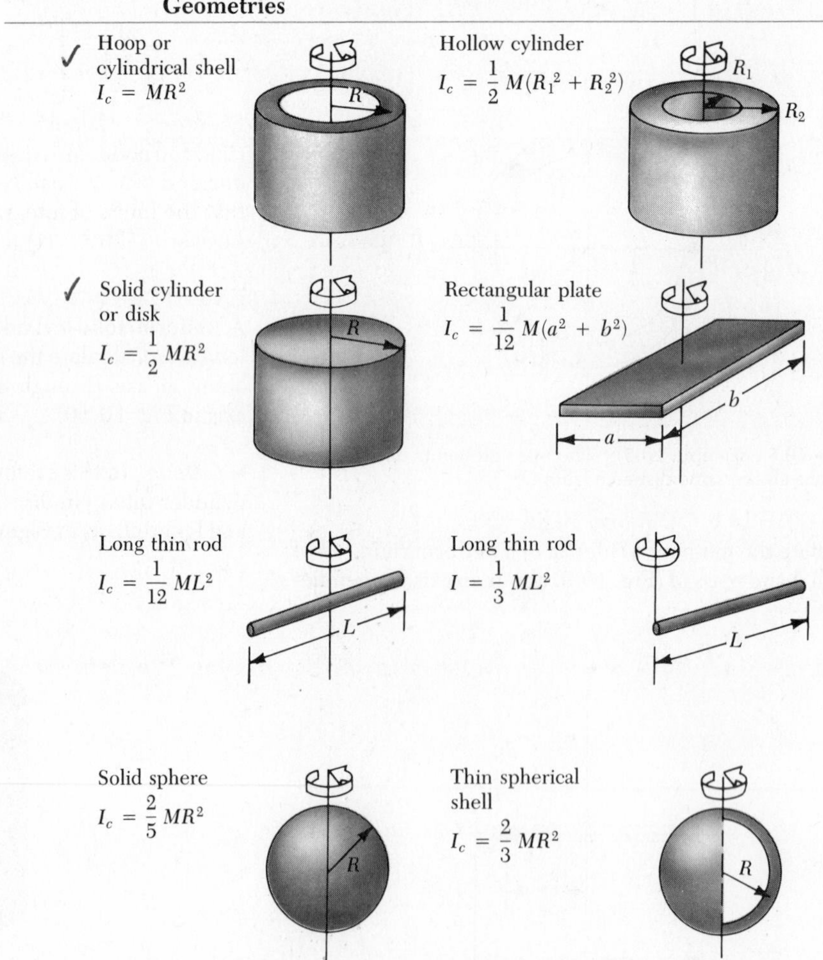

[2] Civil engineers use the moment of inertia concept to characterize the elastic properties (rigidity) of such structures as loaded beams. Hence, it is often useful even in a nonrotational context.

regard, there is an important theorem, called the *parallel-axis theorem*, that often simplifies the calculation of moments of inertia. Suppose the moment of inertia about any axis through the center of mass is I_c. The parallel-axis theorem states that the moment of inertia about any axis that is *parallel* to and a distance d away from the axis that passes through the center of mass is given by

Parallel-axis theorem

$$I = I_c + Md^2 \tag{10.17}$$

For those interested, a discussion of the parallel-axis theorem follows.

Suppose a body rotates in the xy plane about an axis through O as in Figure 10.11 and the coordinates of the center of mass are x_c, y_c. Let the element Δm have coordinates x, y relative to the origin. Since this element is at a distance $r = \sqrt{x^2 + y^2}$ from the z axis, the moment of inertia about the z axis through O is

$$I = \int r^2\, dm = \int (x^2 + y^2) dm$$

However, we can relate the coordinates x, y to the coordinates of the center of mass, x_c, y_c, and the coordinates relative to the center of mass, x', y' through the relations $x = x' + x_c$ and $y = y' + y_c$. Therefore,

$$\begin{aligned} I &= \int [(x' + x_c)^2 + (y' + y_c)^2] dm \\ &= \int [(x')^2 + (y')^2] dm + 2x_c \int x'\, dm + 2y_c \int y'\, dm + (x_c^2 + y_c^2) \int dm \end{aligned}$$

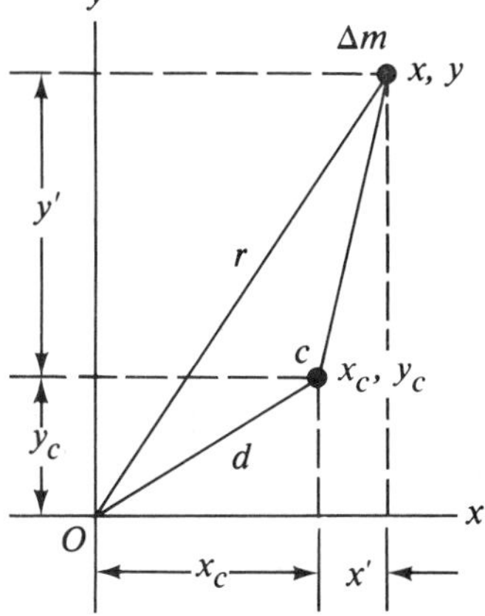

Figure 10.11 The parallel-axis theorem. If the moment of inertia about an axis perpendicular to the figure through the center of mass at c is I_c, then the moment of inertia about the z axis is $I_z = I_c + Md^2$.

The first term on the right is, by definition, the moment of inertia about an axis parallel to the z axis, through the center of mass. The second two terms on the right are zero, since by definition of the center of mass $\int x' dm = \int y'\, dm = 0$ (x', y' are the coordinates of the mass element relative to the center of mass). Finally, the last term on the right is simply Md^2, since $\int dm = M$ and $d^2 = x_c^2 + y_c^2$. Therefore, we conclude that

$$I = I_c + Md^2$$

10.6 TORQUE

When a force is properly exerted on a rigid body pivoted about some axis, the body will tend to rotate about that axis. The tendency of a force to rotate a body about some axis is measured by a quantity called the **torque** (τ). Consider the wrench pivoted about the axis through O in Figure 10.12. The applied force $\boldsymbol{F}$ generally can act at an angle ϕ to the horizontal. We define the

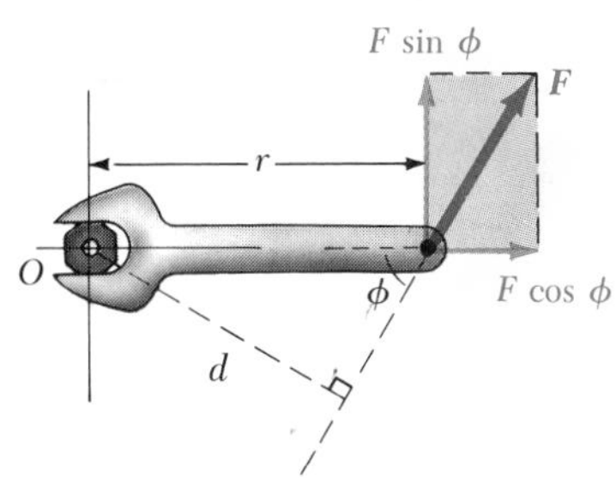

Figure 10.12 The force $\boldsymbol{F}$ has a greater rotating tendency about O as F increases and as the moment arm, d, increases. It is the component $F \sin \phi$ that tends to rotate the system about O.

magnitude of the torque, τ, (Greek letter tau), resulting from the force $\mathbf{F}$ by the expression

Definition of torque

$$\tau \equiv rF \sin \phi = Fd \tag{10.18}$$

Moment arm

It is very important that you recognize that *torque is defined only when a reference axis is specified.* The quantity $d = r \sin \phi$, called the **moment arm** (or *lever arm*) of the force $\mathbf{F}$, represents the perpendicular distance from the rotation axis to the line of action of $\mathbf{F}$. Note that the only component of $\mathbf{F}$ that tends to cause a rotation is $F \sin \phi$, the component perpendicular to r. The horizontal component, $F \cos \phi$, passes through O and has no tendency to produce a rotation. If there are two or more forces acting on a rigid body, as in Figure 10.13, then each has a tendency to produce a rotation about the pivot at O. For example, $\mathbf{F}_2$ has a tendency to rotate the body clockwise, and $\mathbf{F}_1$ has a tendency to rotate the body counterclockwise. We shall use the convention that the sign of the torque resulting from a force is positive if its turning tendency is counterclockwise and negative if its turning tendency is clockwise. For example, in Figure 10.13, the torque resulting from $\mathbf{F}_1$, which has a moment arm of d_1, is *positive* and equal to $+F_1d_1$; the torque from $\mathbf{F}_2$ is *negative* and equal to $-F_2d_2$. Hence, the *net* torque acting on the rigid body about O is

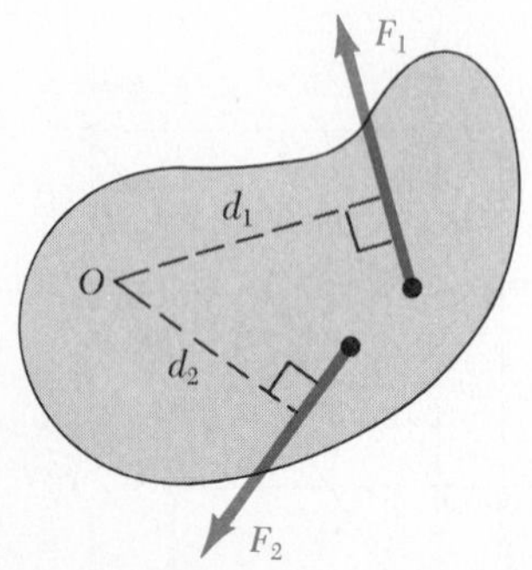

Figure 10.13 The force $\mathbf{F}_1$ tends to rotate the body counterclockwise about O, and $\mathbf{F}_2$ tends to rotate the body clockwise.

$$\tau_{\text{net}} = \tau_1 + \tau_2 = F_1d_1 - F_2d_2$$

From the definition of torque, we see that the rotating tendency increases as F increases and as d increases. For example, it is easier to close a door if we push at the doorknob rather than at a point close to the hinge. *Torque should not be confused with force.* Torque has units of force times length, or N·m in SI units. In Section 10.7 we shall see that the concept of torque is convenient for analyzing the rotational dynamics of a rigid body. The vector nature of torque will be described in detail in the next chapter.

EXAMPLE 10.8. The Net Torque on a Cylinder

A solid cylinder is pivoted about a frictionless axle as in Figure 10.14. A rope wrapped around the outer radius, R_1, exerts a force $\mathbf{F}_1$ to the right on the cylinder. A second rope wrapped around another section of radius R_2 exerts a force $\mathbf{F}_2$ downward on the cylinder. (a) What is the net torque acting on the cylinder about the z axis through O?

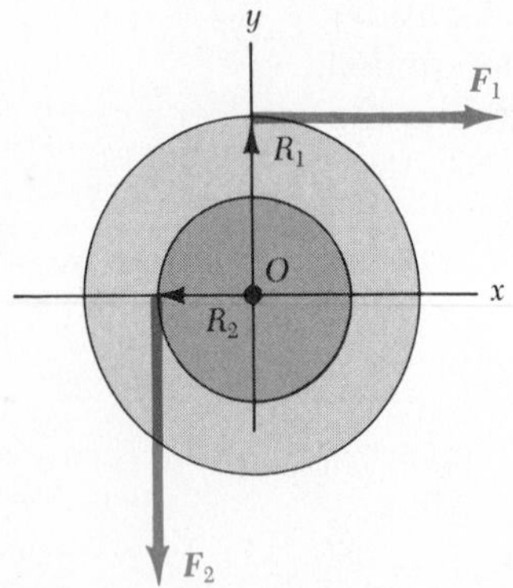

Figure 10.14 (Example 10.8) A solid cylinder pivoted about the z axis through O. The moment arm of $\mathbf{F}_1$ is R_1, and the moment arm of $\mathbf{F}_2$ is R_2.

The torque due to $\mathbf{F}_1$ is $-R_1F_1$ and is negative because it tends to produce a clockwise rotation. The torque due to $\mathbf{F}_2$ is $+R_2F_2$ and is positive because it tends to produce a counterclockwise rotation. Therefore, the net torque is

$$\tau_{\text{net}} = \tau_1 + \tau_2 = R_2F_2 - R_1F_1$$

(b) Suppose $F_1 = 5$ N, $R_1 = 1.0$ m, $F_2 = 6$ N, and $R_2 = 0.5$ m. What is the net torque and which way will the cylinder rotate?

$$\tau_{\text{net}} = (6 \text{ N})(0.5 \text{ m}) - (5 \text{ N})(1.0 \text{ m}) = -2 \text{ N} \cdot \text{m}$$

Since the net torque is negative, the cylinder will rotate in the clockwise direction.

10.7 RELATIONSHIP BETWEEN TORQUE AND ANGULAR ACCELERATION

In this section we shall show that the angular acceleration of a rigid body rotating about a fixed axis is proportional to the net torque acting about that axis. Before discussing the more complex case of rigid-body rotation, it is instructive to first briefly discuss the case of a particle rotating about some fixed point under the influence of an external force. The ideas embodied in this situation will then be extended to the case of a rigid body rotating about a fixed axis.

Consider a particle of mass m rotating in a circle of radius r under the influence of a tangential force $\mathbf{F}_t$ as in Figure 10.15. The tangential force provides a tangential acceleration a_t, and

$$F_t = ma_t$$

The torque about the origin due to the force $\mathbf{F}_t$ is the product of the magnitude of the force, F_t, and the moment arm of the force:

$$\tau = F_t r = (ma_t)r$$

Since the tangential acceleration is related to the angular acceleration through the relation $a_t = r\alpha$, the torque can be expressed

$$\tau = (mr\alpha)r = (mr^2)\alpha$$

Recall that the quantity mr^2 is the moment of inertia of the rotating mass about the z axis passing through the origin, so that

$$\tau = I\alpha \tag{10.19}$$

Relationship between torque and angular acceleration

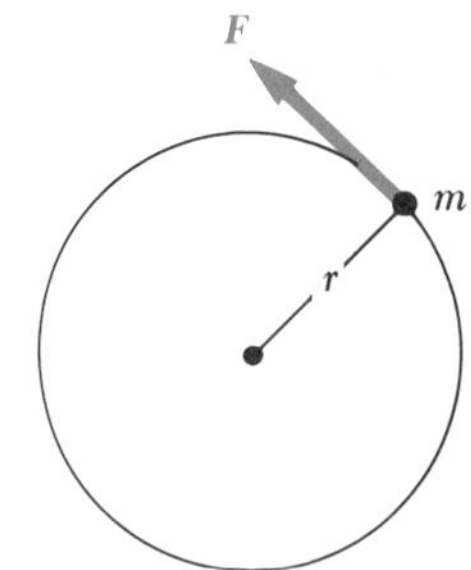

Figure 10.15 A particle rotating in a circle under the influence of a tangential force $\mathbf{F}_t$. A centripetal force $\mathbf{F}_r$ (not shown) must also be present to maintain the circular motion.

That is, *the torque acting on the particle is proportional to its angular acceleration,* and the proportionality constant is the moment of inertia. It is important to note that $\tau = I\alpha$ is the rotational analogue of Newton's second law of motion, $F = ma$.

Now let us extend this discussion to a rigid body of arbitrary shape rotating about a fixed axis as in Figure 10.16. The body can be regarded as an infinite number of mass elements dm of infinitesimal size. Each mass element rotates in a circle about the origin, and each has a tangential acceleration $\mathbf{a}_t$ produced by a tangential force $d\mathbf{F}_t$. For any given element, we know from Newton's second law that

$$dF_t = (dm)a_t$$

The torque $d\tau$ associated with the force $d\mathbf{F}_t$ acting about the origin is given by

$$d\tau = r\ dF_t = (r\ dm)a_t$$

Since $a_t = r\alpha$, the expression for $d\tau$ becomes

$$d\tau = (r\ dm)r\alpha = (r^2\ dm)\alpha$$

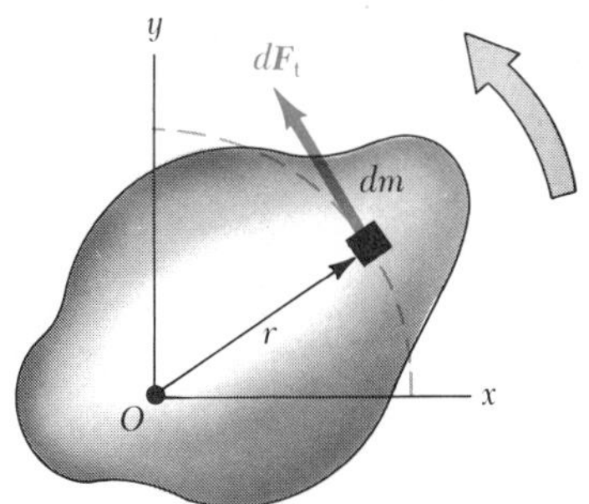

Figure 10.16 A rigid body pivoted about an axis through O. Each mass element dm rotates about O with the same angular acceleration α, and the net torque on the body is proportional to α.

It is important to recognize that although each point of the rigid body may have a different a_t, all mass elements have the *same* angular acceleration, α. With this in mind, the above expression can be integrated to obtain the net torque about O:

$$\tau_{\text{net}} = \int (r^2\ dm)\alpha = \alpha \int r^2\ dm$$

where α can be taken outside the integral since it is common to all mass elements. Since the moment of inertia of the body about the rotation axis through O is defined by $I = \int r^2\,dm$, the expression for τ_{net} becomes

Torque is proportional to angular acceleration

$$\tau_{net} = I\alpha \tag{10.20}$$

Again we see that the net torque about the rotation axis is proportional to the angular acceleration of the body with the proportionality factor being I, which depends upon the axis of rotation and upon the size and shape of the body.

In view of the complex nature of the system, the important result that $\tau_{net} = I\alpha$ is strikingly simple and in complete agreement with experimental observations. The simplicity of the result lies in the manner in which the motion is described.

Every point has the same ω and α

> Although each point on a rigid body rotating about a fixed axis may not experience the same force, linear acceleration, or linear velocity, every point on the body has the same angular acceleration and angular velocity at any instant. Therefore, at any instant the rotating rigid body as a whole is characterized by specific values for angular acceleration, net torque, and angular velocity.

Finally, you should note that the result $\tau_{net} = I\alpha$ would also apply if the forces acting on the mass elements had radial components as well as tangential components. This is because the line of action of all radial components must pass through the axis of rotation, and hence would produce *zero* torque about that axis.

EXAMPLE 10.9. Rotating Rod
A uniform rod of length L and mass M is free to rotate about a frictionless pivot at one end, as in Figure 10.17. The rod is released from rest in the horizontal position. What is the *initial* angular acceleration of the rod and the *initial* linear acceleration of the right end of the rod?

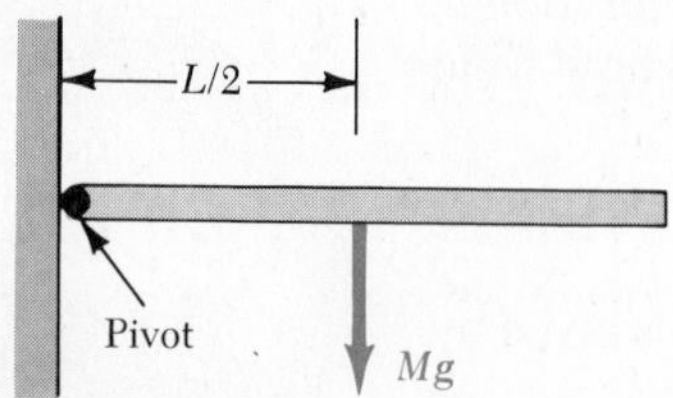

Figure 10.17 (Example 10.9) The uniform rod is pivoted at the left end.

Solution: The weight Mg, located at the geometric center of the rod, acts at its center of mass as shown in Figure 10.17. The magnitude of the torque due to this force about an axis through the pivot is

$$\tau = \frac{MgL}{2}$$

Note that the support force at the hinge has zero torque about an axis through the pivot, because this force passes through the axis (hence $r = 0$). Since $\tau = I\alpha$, where $I = \frac{1}{3}ML^2$ for this axis of rotation (see Table 10.2), we get

$$I\alpha = Mg\,\frac{L}{2}$$

$$\alpha = \frac{Mg(L/2)}{\frac{1}{3}ML^2} = \frac{3g}{2L}$$

This angular acceleration is common to *all* points on the rod.

To find the linear acceleration of the right end of the rod, we use the relation $a_t = R\alpha$, with $R = L$. This gives

$$a_t = L\alpha = \tfrac{3}{2}g$$

This result is rather interesting, since $a_t > g$. That is, the end of the rod has an acceleration *greater* than the acceleration due to gravity. Therefore, if a coin were placed at the end of the rod, the end of the rod would fall faster than the coin when released.

Other points on the rod have a linear acceleration less than $\frac{3}{2}g$. For example, the middle of the rod has an acceleration $\frac{3}{4}g$.

EXAMPLE 10.10. Angular Acceleration of a Wheel
A wheel of radius R, mass M, and moment of inertia I is mounted on a frictionless, horizontal axle as in Figure 10.18. A light cord wrapped around the wheel supports a

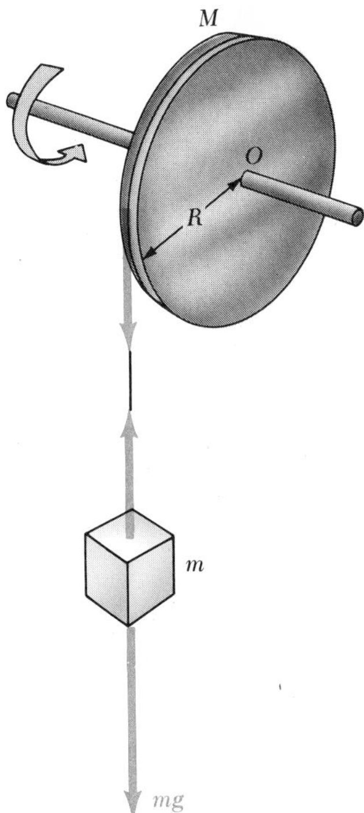

Figure 10.18 (Example 10.10) The cord attached to m is wrapped around the pulley, which produces a torque about the axle through O.

body of mass m. Calculate the linear acceleration of the suspended body, the angular acceleration of the wheel, and the tension in the cord.

Solution: The torque acting on the wheel about its axis of rotation is $\tau = TR$. The weight of the wheel and the normal force of the axle on the wheel pass through the axis of rotation and produce no torque. Since $\tau = I\alpha$, we get

$$\tau = I\alpha = TR$$

$$(1) \qquad \alpha = TR/I$$

Now let us apply Newton's second law to the motion of the suspended mass m, making use of the free-body diagram (Fig. 10.18):

$$\sum F_y = T - mg = -ma$$

$$(2) \qquad a = \frac{mg - T}{m}$$

The linear acceleration of the suspended mass is equal to the tangential acceleration of a point on the rim of the wheel. Therefore, the angular acceleration of the wheel and this linear acceleration are related by $a = R\alpha$. Using this fact together with (1) and (2) gives

$$a = R\alpha = \frac{TR^2}{I} = \frac{mg - T}{m}$$

$$T = \frac{mg}{1 + \dfrac{mR^2}{I}}$$

Likewise, solving for a and α gives

$$a = \frac{g}{1 + I/mR^2}$$

$$\alpha = \frac{a}{R} = \frac{g}{R + I/mR}$$

Exercise 4 The wheel in Figure 10.18 is a solid disk of $M = 2.0$ kg, $R = 30$ cm, and $I = 0.09$ kg · m^2. The suspended object has a mass of $m = 0.5$ kg. Find the tension in the cord and the angular acceleration of the wheel. Answer: 3.27 N; 10.9 rad/s^2.

10.8 WORK AND ENERGY IN ROTATIONAL MOTION

The description of a rotating rigid body would not be complete without a discussion of the rotational kinetic energy and how its change is related to the work done by external forces.

Again, we shall restrict our discussion to rotation about a fixed axis located in an inertial frame. Furthermore, we shall see that the important relationship $\tau_{net} = I\alpha$ derived in the previous section can also be obtained by considering the rate at which energy is changing with time.

Consider a rigid body pivoted at the point O in Figure 10.19. Suppose a single external force $\mathbf{F}$ is applied at the point P. The work done by $\mathbf{F}$ as the body rotates through an infinitesimal distance $ds = r\,d\theta$ in a time dt is

$$dW = \mathbf{F} \cdot d\mathbf{s} = (F \sin \phi) r\, d\theta$$

where $F \sin \phi$ is the tangential component of $\mathbf{F}$, or the component of the force along the displacement. Note from Figure 10.19 that *the radial component of* $\mathbf{F}$ *does no work since it is perpendicular to the displacement.*

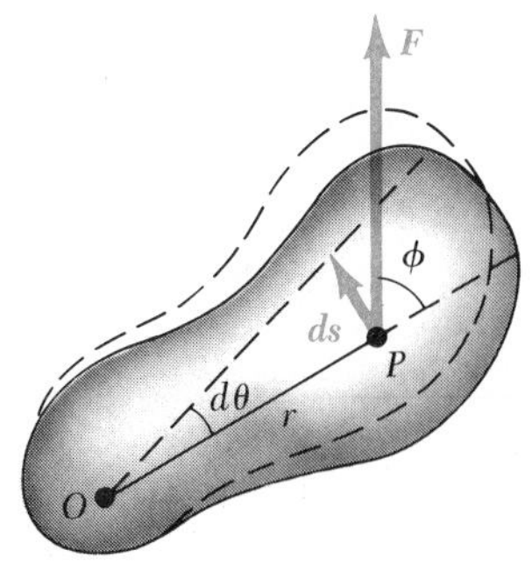

Figure 10.19 A rigid body rotates about an axis through Q under the action of an external force $\mathbf{F}$ applied at P.

Since the magnitude of the torque due to **F** about the origin was defined as $rF \sin \phi$, we can write the work done for the infinitesimal rotation

$$dW = \tau \, d\theta \tag{10.21}$$

The rate at which work is being done by **F** for rotation about the fixed axis is obtained by dividing the left and right sides of Equation 10.21 by dt:

$$\frac{dW}{dt} = \tau \frac{d\theta}{dt} \tag{10.22}$$

But the quantity dW/dt is, by definition, the instantaneous power, P, delivered by the force. Furthermore, since $d\theta/dt = \omega$, Equation 10.22 reduces to

Power delivered to a rigid body

$$P = \frac{dW}{dt} = \tau\omega \tag{10.23}$$

This expression is analogous to $P = Fv$ in the case of linear motion, and the expression $dW = \tau \, d\theta$ is analogous to $dW = F_x \, dx$.

The Work-Energy Theorem in Rotational Motion

In linear motion, we found the energy concept, and in particular the work-energy theorem, to be extremely useful in describing the motion of a system. The energy concept can be equally useful in simplifying the analysis of rotational motion. From what we learned of linear motion, we expect that for rotation of a symmetric object (such as a symmetric wheel) about a fixed axis, the work done by external forces will equal the change in the rotational kinetic energy. To show that this is in fact the case, let us begin with $\tau = I\alpha$. Using the chain rule from the calculus, we can express the torque as

$$\tau = I\alpha = I \frac{d\omega}{dt} = I \frac{d\omega}{d\theta} \frac{d\theta}{dt} = I \frac{d\omega}{d\theta} \omega$$

Rearranging the above expression and noting that $\tau \, d\theta = dW$, we get

$$\tau \, d\theta = dW = I\omega \, d\omega$$

Integrating this expression and noting that I is a constant, we get for the total work done

Work-energy theorem for rotational motion

$$W = \int_{\theta_0}^{\theta} \tau \, d\theta = \int_{\omega_0}^{\omega} I\omega \, d\omega = \tfrac{1}{2}I\omega^2 - \tfrac{1}{2}I\omega_0^{2} \tag{10.24}$$

where the angular velocity changes from ω_0 to ω as the angular displacement changes from θ_0 to θ. Note that this expression is analogous to the expression for the work-energy theorem in linear motion with m replaced by I and v replaced by ω. That is,

> the net work done by external forces in rotating a symmetric rigid body about a fixed axis equals the change in the body's rotational kinetic energy.

Table 10.3 lists the various equations we have discussed pertaining to rotational motion, together with the analogous expressions for linear motion.

pre study sheet

TABLE 10.3 A Comparison of Useful Equations in Rotational and Translational Motion

Rotational Motion About a Fixed Axis	Linear Motion
Angular velocity $\omega = d\theta/dt$	Linear velocity $v = dx/dt$
Angular acceleration $\alpha = d\omega/dt$	Linear acceleration $a = dv/dt$
Resultant torque $\Sigma\tau = I\alpha$	Resultant force $\Sigma F = Ma$
$\alpha = \text{constant}\begin{cases}\omega = \omega_0 + \alpha t \\ \theta - \theta_0 = \omega_0 t + \frac{1}{2}\alpha t^2 \\ \omega^2 = \omega_0{}^2 + 2\alpha(\theta - \theta_0)\end{cases}$	$a = \text{constant}\begin{cases}v = v_0 + at \\ x - x_0 = v_0 t + \frac{1}{2}at^2 \\ v^2 = v_0{}^2 + 2a(x - x_0)\end{cases}$
Work $W = \int_{\theta_0}^{\theta} \tau\, d\theta$	Work $W = \int_{x_0}^{x} F_x\, dx$
Kinetic energy $K = \frac{1}{2}I\omega^2$	Kinetic energy $K = \frac{1}{2}mv^2$
Power $P = \tau\omega$	Power $P = Fv$
Angular momentum $L = I\omega$	Linear momentum $p = mv$
Resultant torque $\tau = dL/dt$	Resultant force $F = dp/dt$

The last two equations, involving the concept of angular momentum L, will be discussed in Chapter 11 and are included only for completeness. In all cases, note the similarity between the equations of rotational motion and those of linear motion.

EXAMPLE 10.11. Rotating Rod—Revisited

A uniform rod of length L and mass M is free to rotate on a frictionless pin through one end (Fig. 10.20). The rod is released from rest in the horizontal position. (a) What is the angular velocity of the rod when it is at its lowest position?

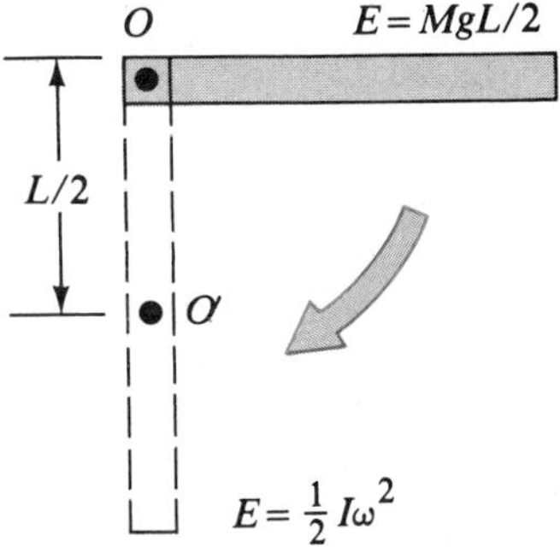

Figure 10.20 (Example 10.11) A uniform rigid rod pivoted at O rotates in a vertical plane under the action of gravity.

The question can be easily answered by considering the mechanical energy of the system. When the rod is in the horizontal position, it has no kinetic energy. Its potential energy relative to the lowest position of its center of mass (O') is $MgL/2$. When it reaches its lowest position, the energy is entirely kinetic energy, $\frac{1}{2}I\omega^2$, where I is the moment of inertia about the pivot. Since $I = \frac{1}{3}ML^2$ (Table 10.2) and since mechanical energy is conserved, we have

$$\tfrac{1}{2}MgL = \tfrac{1}{2}I\omega^2 = \tfrac{1}{2}(\tfrac{1}{3}ML^2)\omega^2$$

$$\omega = \sqrt{\frac{3g}{L}}$$

For example, if the rod is a meter stick, we find that $\omega = 5.42$ rad/s.

(b) Determine the linear velocity of the center of mass and the linear velocity of the lowest point on the rod in the vertical position.

$$v_c = r\omega = \frac{L}{2}\omega = \tfrac{1}{2}\sqrt{3gL}$$

The lowest point on the rod has a velocity equal to $2v_c = \sqrt{3gL}$.

EXAMPLE 10.12. Connected Masses

Consider two masses connected by a string passing over a pulley having a moment of inertia I about its axis of rotation, as in Figure 10.21. The string does not slip on the pulley, and the system is released from rest. Find the linear velocities of the masses after m_2 descends through a distance h, and the angular velocity of the pulley at this time.

Solution: If we neglect friction in the system, then mechanical energy is conserved and we can state that the increase in kinetic energy of the system equals the decrease in potential energy. Since $K_i = 0$ (the system is initially at rest), we have

$$\Delta K = K_f - K_i = \tfrac{1}{2}m_1v^2 + \tfrac{1}{2}m_2v^2 + \tfrac{1}{2}I\omega^2$$

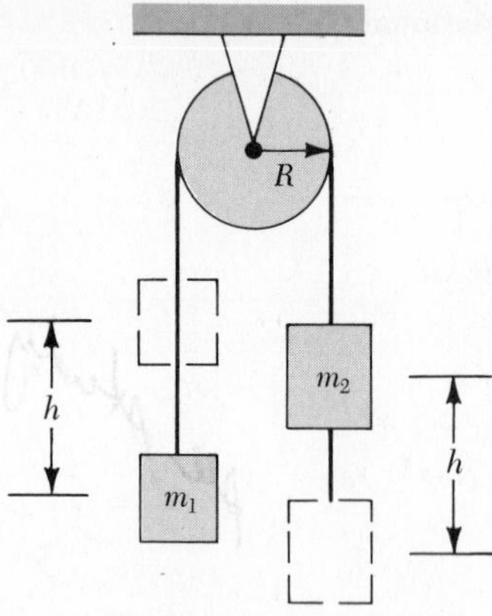

Figure 10.21 (Example 10.12).

where m_1 and m_2 have a common speed. But $v = R\omega$, so that

$$\Delta K = \tfrac{1}{2}\left(m_1 + m_2 + \frac{I}{R^2}\right)v^2$$

From Figure 10.21, we see that m_2 loses potential energy while m_1 gains potential energy. That is, $\Delta U_2 = -m_2gh$ and $\Delta U_1 = m_1gh$. Applying the law of conservation of energy in the form $\Delta K + \Delta U_1 + \Delta U_2 = 0$ gives

$$\tfrac{1}{2}\left(m_1 + m_2 + \frac{I}{R^2}\right)v^2 + m_1gh - m_2gh = 0$$

$$v = \left[\frac{2(m_2 - m_1)gh}{\left(m_1 + m_2 + \dfrac{I}{R^2}\right)}\right]^{1/2}$$

Since $v = R\omega$, the angular velocity of the pulley at this instant is given by $\omega = v/R$.

Exercise 5 Repeat the calculation of v in Example 10.12 using $\tau_{net} = I\alpha$ applied to the pulley and Newton's second law applied to m_1 and m_2. Make use of the procedure presented in Example 10.10.

10.9 SUMMARY

The **instantaneous angular velocity** of a particle rotating in a circle or of a rigid body rotating about a fixed axis is given by

Instantaneous angular velocity

$$\omega = \frac{d\theta}{dt} \tag{10.3}$$

where ω is in rad/s, or s^{-1}.

The **instantaneous angular acceleration** of a rotating body is given by

Instantaneous angular acceleration

$$\alpha = \frac{d\omega}{dt} \tag{10.5}$$

and has units of rad/s^2, or s^{-2}.

When a rigid body rotates about a fixed axis, every part of the body has the same angular velocity and the same angular acceleration. However, different parts of the body, in general, have different linear velocities and linear accelerations.

If a particle or body undergoes rotational motion about a fixed axis under constant angular acceleration α, one can apply equations of kinematics in analogy with kinematic equations for linear motion under constant linear acceleration:

Rotational kinematic equations

$$\omega = \omega_0 + \alpha t \tag{10.6}$$

$$\theta = \theta_0 + \omega_0 t + \tfrac{1}{2}\alpha t^2 \tag{10.7}$$

$$\omega^2 = \omega_0{}^2 + 2\alpha(\theta - \theta_0) \tag{10.8}$$

When a rigid body rotates about a fixed axis, the angular velocity and angular acceleration are related to the linear velocity and tangential linear acceleration through the relationships

$$v = r\omega \qquad (10.9)$$

Relationship between linear and angular speed

$$a_t = r\alpha \qquad (10.10)$$

Relationship between linear and angular acceleration

a system of particles is given by

$$I = \sum m_i r_i^2 \qquad (10.14)$$

Moment of inertia for a system of particles

t a fixed axis with angular velocity ω, its **kinetic**

$$K = \tfrac{1}{2} I\omega^2 \qquad (10.15)$$

Kinetic energy of a rotating rigid body

ia about the axis of rotation.
a rigid body is given by

$$I = \int r^2\, dm \qquad (10.16)$$

Moment of inertia for a rigid body

ie mass element dm to the axis of rotation.
h a force $\boldsymbol{F}$ acting on a body has a magnitude

$$\tau = Fd \qquad (10.18)$$

Torque

the force, which is the perpendicular distance
action of the force. Torque is a measure of the
the body about some axis.
te about a fixed axis has a **net external torque**
ergo an angular acceleration α, where

$$\tau_{\text{net}} = I\alpha \qquad (10.20)$$

Net torque

eing done by external forces in rotating a rigid
power *delivered,* is given by

$$P = \tau\omega \qquad (10.23)$$

Power delivered to a rigid body

rnal forces in rotating a rigid body about a fixed
otational kinetic energy of the body:

$$V = \tfrac{1}{2} I\omega^2 - \tfrac{1}{2} I\omega_0^2 \qquad (10.24)$$

Work-energy theorem for rotational motion

m applied to rotational motion.

QUESTIONS

1. What is the magnitude of the angular velocity, ω, of the second hand of a clock? What is the direction of ω as you view a clock hanging vertically? What is the angular acceleration, ω, of the second hand?
2. A wheel rotates counterclockwise in the xy plane. What is the direction of ω? What is the direction of α if the angular velocity is decreasing in time?
3. Are the kinematic expressions for θ, ω, and α valid when the angular displacement is measured in degrees instead of in radians?
4. A turntable rotates at a constant rate of 45 rotations/min. What is the magnitude of its angular velocity in rad/s? What is its angular acceleration?

5. When a wheel of radius R rotates about a fixed axis, do all points on the wheel have the same angular velocity? Do they all have the same linear velocity? If the angular velocity is constant and equal to ω_0, describe the linear velocities and linear accelerations of the points at $r = 0$, $r = R/2$, and $r = R$.
6. Suppose $a = b$ and $M > m$ in the system of particles described in Figure 10.7. About what axis (x, y, or z) does the moment of inertia have the smallest value? the largest value?
7. A wheel is in the shape of a hoop as in Figure 10.8. In two separate experiments, the wheel is rotated from rest to an angular velocity ω. In one experiment, the rotation occurs about the z axis through O; in the other, the rotation occurs about an axis parallel to z through P. Which rotation requires more work?
8. Suppose the rod in Figure 10.9 has a nonuniform mass distribution. In general, would the moment of inertia about the y axis still equal $\frac{1}{12}ML^2$? If not, could the moment of inertia be calculated without knowledge of the manner in which the mass is distributed?
9. With reference to Figure 10.14, is it possible to have a situation where the resultant torque on the cylinder due to the two applied forces is zero? Explain.
10. Suppose that only two external forces act on a rigid body, and the two forces are equal in magnitude but opposite in direction. Under what condition will the body rotate?
11. Explain how you might use the apparatus described in Example 10.10 to determine the moment of inertia of the wheel. (Note that if the wheel is not a uniform disk the moment of inertia is not necessarily equal to $\frac{1}{2}MR^2$.)
12. Using the results from Example 10.10, how would you calculate the angular velocity of the wheel and the linear velocity of the suspended mass at, say, $t = 2$ s, if the system is released from rest at $t = 0$? Is the relation $v = R\omega$ valid in this situation?
13. If a small sphere of mass M were placed at the end of the rod in Figure 10.20, would the result for ω be greater than, less than, or equal to the value obtained in Example 10.11?
14. Explain why changing the axis of rotation of an object should change its moment of inertia.
15. Is it possible to change the translational kinetic energy of an object without changing its rotational kinetic energy?
16. Two cylinders having the same dimensions are set into rotation about their axes with the same angular velocity. One is hollow, and the other is filled with water. Which cylinder would be easier to stop rotating? Explain.

PROBLEMS

Section 10.1 Angular Velocity and Angular Acceleration

1. A particle moves in a circle 1.5 m in radius. Through what angle in radians does it rotate if it moves through an arc length of 2.5 m? What is this angle in degrees?
2. If a particle moving in a circle makes n rev/min, what is its angular velocity in rad/s?
3. A wheel rotates at a constant rate of 3600 rev/min. (a) What is its angular speed? (b) Through what angle (in radians) does it rotate in 1.5 s?
4. Convert the following to degrees: (a) 3.5 rad, (b) 5π rad, (c) 2.2 rev.

Section 10.2 Rotational Kinematics: Rotational Motion with Constant Angular Acceleration

5. A wheel starts from rest and rotates with constant angular acceleration to an angular velocity of 10 rad/s in a time of 2 s. Find (a) the angular acceleration of the wheel and (b) the angle in radians through which it rotates in this time.
6. The turntable of a record player rotates at the rate of $33\frac{1}{3}$ rev/min and takes 90 s to come to rest when switched off. Calculate (a) its angular acceleration and (b) the number of revolutions it makes before coming to rest.
7. What is the angular speed in rad/s of (a) the earth in its orbit about the sun and (b) the moon in its orbit about the earth?
8. A wheel rotates in such a way that its angular displacement in a time t is given by $\theta = at^2 + bt^3$, where a and b are constants. Determine equations for (a) the angular speed and (b) the angular acceleration, both as functions of time.

Section 10.3 Relationships Between Angular and Linear Quantities

9. A racing car travels on a circular track of radius 200 m. If the car moves with a constant speed of 80 m/s, find (a) the angular speed of the car and (b) the magnitude and direction of the car's acceleration.
10. The racing car described in Problem 9 starts from rest and accelerates uniformly to a speed of 80 m/s in 30 s. Find (a) the average angular speed of the car in this interval, (b) the angular acceleration of the car, (c) the magnitude of the car's linear acceleration at $t = 10$ s, and (d) the total distance traveled in the first 30 s.
11. A wheel 4 m in diameter rotates with a constant angular acceleration of 4 rad/s². The wheel starts at rest at $t = 0$, and the radius vector at point P on the rim makes an angle of 57.3° with the horizontal at this time. At $t = 2$ s, find (a) the angular speed of the wheel, (b) the linear velocity and acceleration of the point P, and (c) the position of the point P.
12. A cylinder of radius 12 cm starts from rest and rotates about its axis with a constant angular acceleration of

5 rad/s². At $t = 3$ s, what is (a) its angular velocity, (b) the linear speed of a point on its rim, and (c) the radial and tangential components of acceleration of a point on its rim?

13. A disk 6 cm in radius rotates at a constant rate of 1200 rev/min about its axis. Determine (a) the angular speed of the disk, (b) the linear speed at a point 2 cm from its center, (c) the radial acceleration of a point on the rim, and (d) the total distance a point on the rim moves in 2 s.

Section 10.4 Rotational Kinetic Energy

14. A tire of moment of inertia 50 kg · m² rotates about a fixed central axis at the rate of 600 rev/min. What is its kinetic energy?
15. The four particles in Figure 10.22 are connected by light, rigid rods. If the system rotates in the xy plane about the z axis with an angular velocity of 8 rad/s, calculate (a) the moment of inertia of the system about the z axis and (b) the kinetic energy of the system.
16. The system of particles described in Problem 15 (Fig. 10.22) rotates about the y axis. Calculate (a) the moment of inertia about the y axis and (b) the work required to take the system from rest to an angular speed of 8 rad/s.
17. Three particles are connected by light, rigid rods lying along the y axis (Fig. 10.23). If the system rotates about the x axis with an angular speed of 2 rad/s, find (a) the moment of inertia about the x axis and the total kinetic energy evaluated from $\frac{1}{2}I\omega^2$ and (b) the linear speed of each particle and the total kinetic energy evaluated from $\Sigma\frac{1}{2}m_iv_i^2$.

Section 10.5 Calculation of Moments of Inertia for Rigid Bodies

18. Following the procedure used in Example 10.6, prove that the moment of inertia about the y' axis of the rigid rod in Figure 10.9 is $\frac{1}{3}ML^2$.
19. Use the parallel-axis theorem and Table 10.2 to find the moments of inertia of (a) a solid cylinder about an axis parallel to the center of mass axis and passing through the edge of the cylinder and (b) a solid sphere about an axis tangent to the surface of the sphere.

Section 10.6 Torque

20. Calculate the net torque (magnitude and direction) on the beam shown in Figure 10.24 about (a) an axis through O, perpendicular to the figure and (b) an axis through C, perpendicular to the figure.

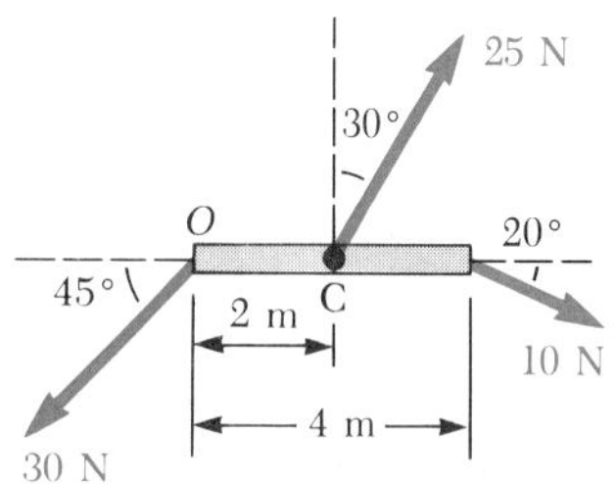

Figure 10.24 (Problem 20).

21. Find the net torque on the wheel in Figure 10.25 about the axle through O if $a = 5$ cm and $b = 20$ cm.

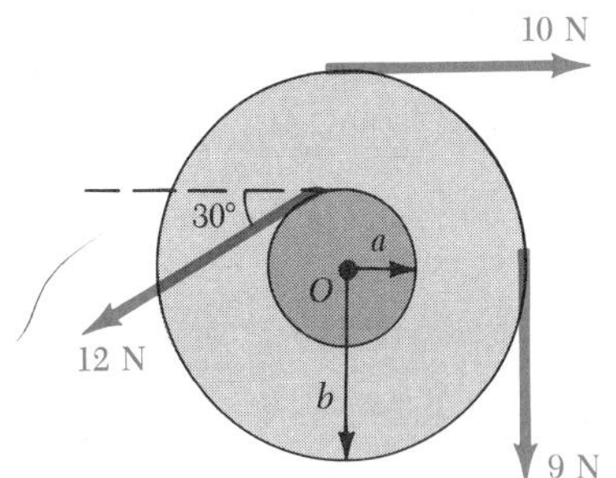

Figure 10.25 (Problem 21).

22. Find the mass m needed to balance the 60-kg cart on the incline shown in Figure 10.26. The angle of inclination θ is 45°. Assume all pulleys are frictionless and massless.

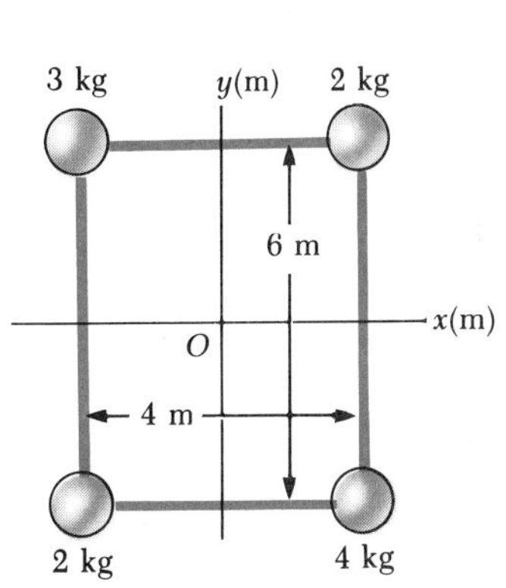

Figure 10.22 (Problems 15 and 16).

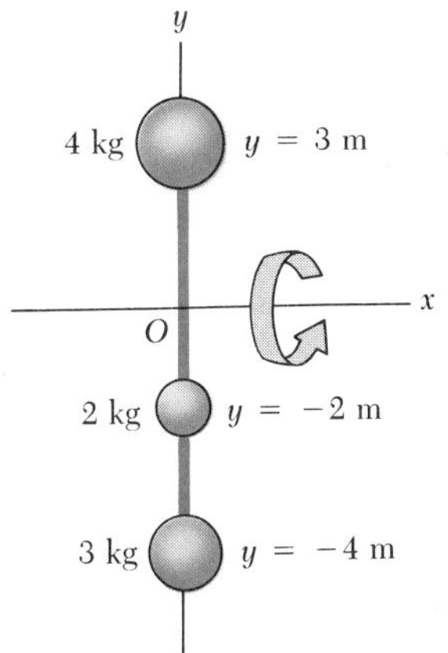

Figure 10.23 (Problem 17).

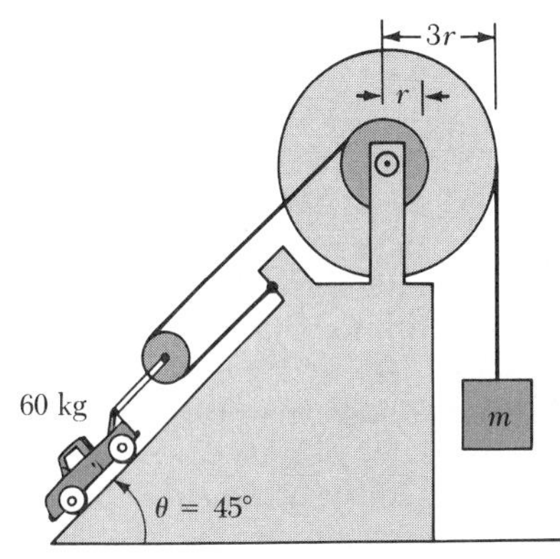

Figure 10.26 (Problem 22).

Section 10.7 Relationship Between Torque and Angular Acceleration

23. The combination of an applied force and a frictional force produces a constant total torque of 24 N · m on a wheel rotating about a fixed axis. The applied force acts for 5 s, during which time the angular speed of the wheel increases from 0 to 10 rad/s. The applied force is then removed, and the wheel comes to rest in 50 s. Find (a) the moment of inertia of the wheel, (b) the magnitude of the frictional torque, and (c) the total number of revolutions of the wheel.
24. If a motor is to produce a torque of 50 N · m on a wheel rotating at 2400 rev/min, how much power must the motor deliver?
25. The system described in Example 10.10 (Fig. 10.18) is released from rest. After the mass m has fallen through a distance h, find (a) the linear velocity of the mass m and (b) the angular speed of the wheel.

Section 10.8 Work and Energy in Rotational Motion

26. A wheel 1 m in diameter rotates on a fixed, frictionless, horizontal axle. Its moment of inertia about this axis is 5 kg · m^2. A constant tension of 20 N is maintained on a rope wrapped around the rim of the wheel, so as to cause the wheel to accelerate. If the wheel starts from rest at $t = 0$, find (a) the angular acceleration of the wheel, (b) the wheel's angular speed at $t = 3$ s, (c) the kinetic energy of the wheel at $t = 3$ s, and (d) the length of rope unwound in the first 3 s.
27. A 12-kg mass is attached to a cord that is wrapped around a wheel of radius $r = 10$ cm (Fig. 10.27). The acceleration of the mass down the frictionless incline is measured to be 2.0 m/s^2. Assuming the axle of the wheel to be frictionless, determine (a) the tension in the rope, (b) the moment of inertia of the wheel, and (c) the angular speed of the wheel 2 s after it begins rotating, starting from rest.

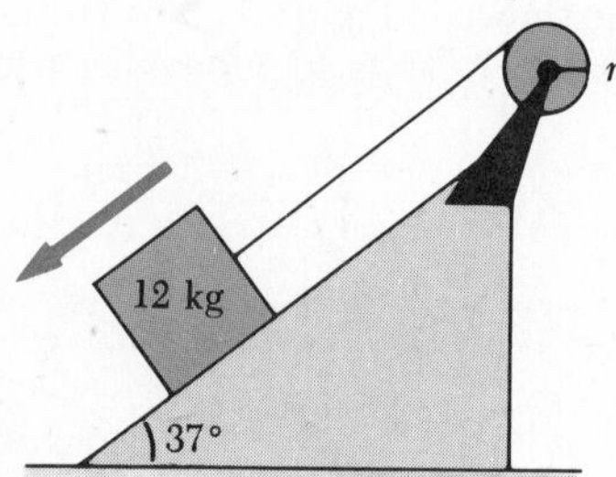

Figure 10.27 (Problem 27).

28. (a) A uniform solid disk of radius R and mass M is free to rotate on a frictionless pivot through a point on its rim (Fig. 10.28). If the disk is released from rest in the position shown by the solid line, what is the velocity of its center of mass when it reaches the position indicated by the broken line? (b) What is the speed of the lowest point on the disk in the dotted position? (c) Repeat part (a) if the object is a uniform hoop.

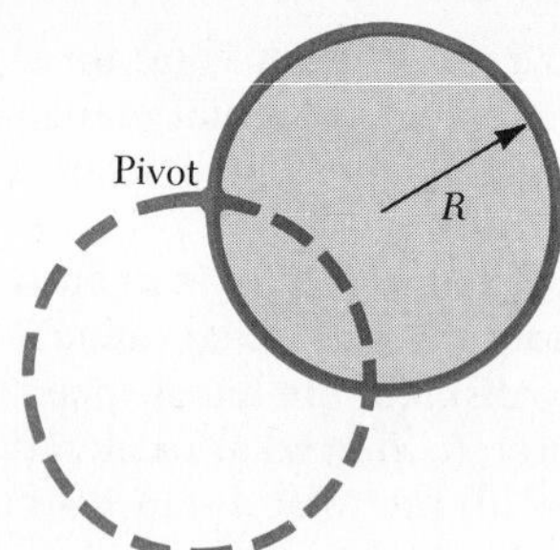

Figure 10.28 (Problem 28).

GENERAL PROBLEMS

29. A string is wound around a uniform disk of radius 0.20 m and mass 5 kg. The disk starts at rest and is free to turn about its axis. The end of the string is pulled with a constant force of 10 N. At time $t = 2$ s after the constant force is applied, determine (a) the torque exerted on the disk, (b) the angular acceleration of the disk, (c) the acceleration of the end of the string, (d) the angular velocity of the disk, (e) the velocity of the end of the string, (f) the kinetic energy of the disk, (g) the work done on the disk, (h) the angle θ through which the disk has turned, and (i) the length of rope pulled from the disk.
30. Calculate the moment of inertia of a uniform solid sphere of mass M and radius R about a diameter (see Table 10.2). (*Hint:* Treat the sphere as a set of disks of various radii, and first obtain an expression for the moment of inertia of one of these disks about the symmetry axis.)
31. A uniform solid cylinder of mass M and radius R rotates on a horizontal, frictionless axle (Fig. 10.29). Two equal masses hang from light cords wrapped around the cylinder. If the system is released from rest, find (a) the tension in each cord, (b) the acceleration of each mass, and (c) the angular velocity of the cylinder after the masses have descended a distance h.

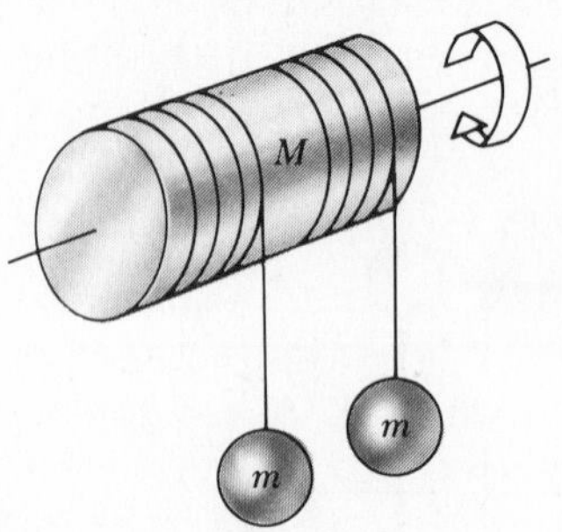

Figure 10.29 (Problem 31).

32. Suppose the pulley in Figure 10.21 has a moment of inertia I and radius R. If the cord supporting m_1 and m_2 does not slip, $m_2 > m_1$, and the axle is frictionless, find (a) the acceleration of the masses, (b) the tension sup-

porting m_1 and the tension supporting m_2 (note they are different), and (c) numerical values for T_1, T_2, and a if $I = 5$ kg · m², $R = 0.5$ m, $m_1 = 2$ kg, and $m_2 = 5$ kg.

33. A mass m_1 is connected by a light cord to a mass m_2, which slides on a smooth surface (Fig. 10.30). The pulley rotates about a frictionless axle and has a moment of inertia I and radius R. Assuming the cord does not slip on the pulley, find (a) the acceleration of the two masses, (b) the tensions T_1 and T_2, and (c) numerical values for a, T_1, and T_2 if $I = 0.5$ kg · m², $R = 0.3$ m, $m_1 = 4$ kg, and $m_2 = 3$ kg. (d) What would your answers be if the inertia of the pulley was neglected?

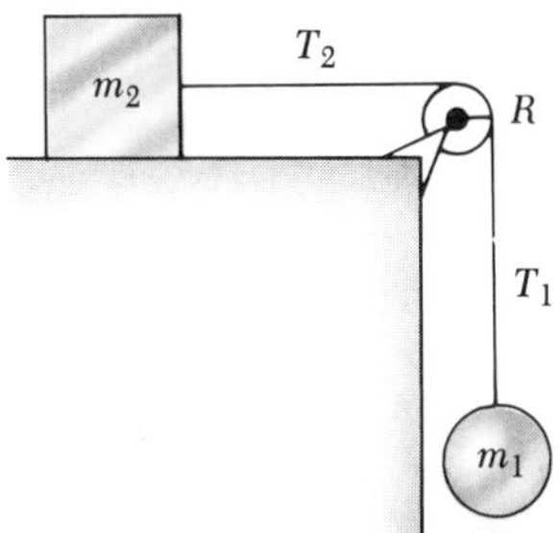

Figure 10.30 (Problem 33).

34. Find by integration the moment of inertia of a hollow cylinder about its symmetry axis. The mass of the cylinder is M, its inner radius is R_1, and its outer radius is R_2. (Check your result against the value given in Table 10.2.)

35. A 3-m length of light nylon cord is wound around a uniform cylindrical spool of radius 0.6 m and 1-kg mass. The spool is mounted on a frictionless axle and is initially at rest. The cord is pulled from the spool with a constant acceleration of 2.5 m/s². (a) How much work has been done on the spool, when it reaches an angular speed, $\omega = 6$ rad/s? (b) Assuming there is enough cord on the spool, how long will it take the spool to reach an angular speed of 6 rad/s? (c) Is there enough cord on the spool to enable the spool to reach this angular speed of 6 rad/s?

36. Many machines make use of heavy circular disks, called flywheels, to help maintain uniformity in rotational motion. The rotational inertia of a flywheel smooths fluctuations in rotational velocity incurred during operation, such as between power strokes in a gasoline engine. A particular flywheel of diameter 0.6 m and mass 200 kg is mounted on a frictionless bearing. A motor connected to the flywheel accelerates the flywheel from rest to 1000 rpm. (a) What is the moment of inertia of the flywheel? (b) How much work is done on the flywheel during this acceleration? (c) After 1000 rpm is achieved the motor is disengaged from the flywheel. A friction brake is used to slow the rotational rate to 500 rpm. How much energy is dissipated as heat within the friction brake?

37. A long uniform rod of length L and mass M is pivoted about a horizontal, frictionless pin through one end. The rod is released from rest in a vertical position as in Fig. 10.31. At the instant the rod is horizontal, find (a) the angular velocity of the rod, (b) its angular acceleration, (c) the x and y components of the acceleration of its center of mass, and (d) the components of the reaction force at the pivot.

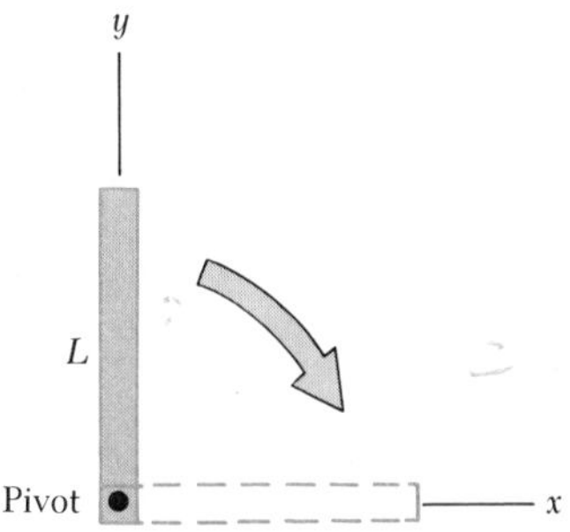

Figure 10.31 (Problem 37).

38. The pulley shown in Figure 10.32 has a radius R and moment of inertia I. One end of the mass m is connected to a spring of force constant k, and the other end is fastened to a cord wrapped around the pulley. The pulley axle and the incline are smooth. If the pulley is wound counterclockwise so as to stretch the spring a distance d from its *unstretched* position and then released from rest, find (a) the angular velocity of the pulley when the spring is again unstretched and (b) a numerical value for the angular velocity at this point if $I =$ kg · m², $R = 0.3$ m, $k = 50$ N/m, $m = 0.5$ kg, $d = 0.2$ m, and $\theta = 37°$.

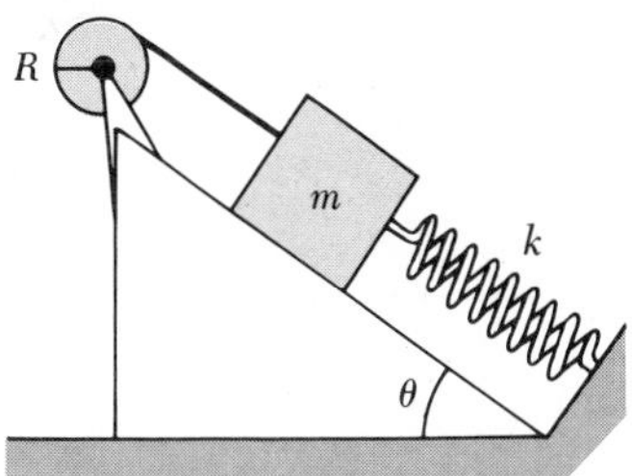

Figure 10.32 (Problem 38).

39. For any given rotational axis, the radius of gyration, K, of a rigid body is defined by the expression $K^2 = I/M$, where M is the total mass of the body and I is the moment of inertia about the given axis. That is, the radius of gyration is equal to the distance of an imaginary point mass M from the axis of rotation such that I for the point mass about that axis is the same as for the rigid body about the same axis. Find the radius of gyration of (a) a solid disk of radius R, (b) a uniform rod of length L, and (c) a solid sphere of radius R, all three rotating about a central axis.

40. A uniform horizontal plank of mass M and length L is supported at each end by vertical ropes. At the instant one of the ropes breaks, show that (a) the angular acceleration of the plank is $3g/2L$, (b) the acceleration of the center of mass is $3g/4$, and (c) the tension in the supporting rope is $Mg/4$.

•41. A mass m is supported by a cord that is wound around a spool of radius r and mass M. A motor accelerates the spool to angular velocity ω and then disengages from the spool. (a) How far will the mass m rise (in terms of m, M, r, ω, and g) after the motor has been disengaged? (b) Show that the acceleration of the mass m is

$$a = -\left(\frac{m}{m + \frac{1}{2}M}\right) g$$

(c) Show that the mass m rises to maximum height in time

$$t = \left(\frac{m + \frac{1}{2}M}{m}\right) \frac{r\omega}{g}$$

•42. A bright physics student purchases a wind vane for her father's garage, consisting of a rooster sitting on top of an arrow. The vane is fixed to a vertical shaft of radius r and mass m that is free to turn in its roof mount as shown in Figure 10.33. The student sets up an experiment to measure the rotational inertia of the rooster and arrow attached to the shaft. String wound about the vertical shaft passes over a pulley and is connected to a mass M hanging over the edge of the garage roof. When the mass M is released, the student determines the time t that the mass takes to fall through a distance h. From these data the student is able to find the rotational inertia I of the rooster and arrow. Find the expression for I in terms of m, M, r, g, h, and t.

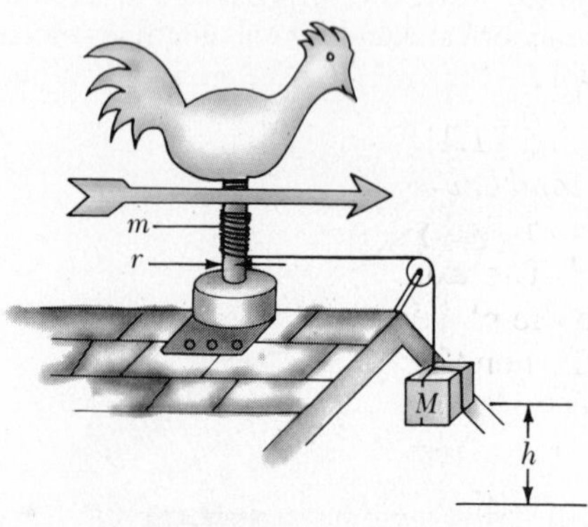

Figure 10.33 (Problem 42).

•43. A uniform hollow cylindrical spool has an inside radius of $R/2$, outside radius R and mass M (see Fig. 10.34). It is mounted so as to rotate on a rough, fixed horizontal axle. A mass m is connected to the end of a string that is wound around the spool. The mass m is observed to fall from rest through a distance y in time t. Show that the torque due to the frictional forces between the spool and the axle is

$$\tau_F = R\left[m\left(g - \frac{2y}{t^2}\right) - \frac{3}{4}M\left(\frac{y}{t^2}\right)\right]$$

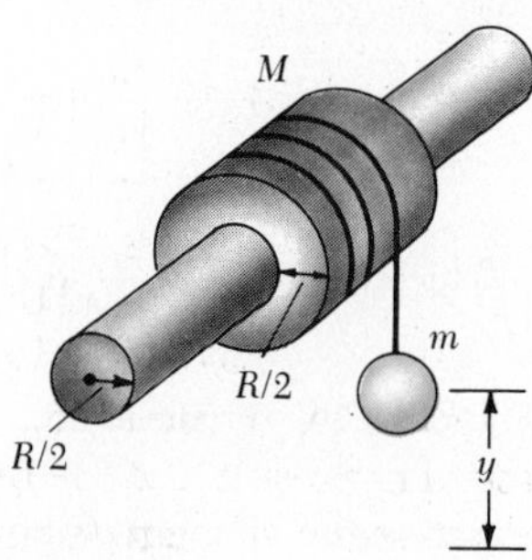

Figure 10.34 (Problem 43).

•44. A cord is wrapped around a pulley of mass m and radius r. The free end of the cord is connected to a block of mass M. The block starts from rest and then slides down a rough incline which makes an angle θ with the horizontal. The coefficient of kinetic friction between the block and the incline is μ. (a) Use the work-energy theorem to show that the block's velocity v as a function of displacement d down the incline is

$$v = \left[4gd\left(\frac{M}{m + 2M}\right)(\sin\theta - \mu\cos\theta)\right]^{1/2}$$

(b) Find the acceleration of the block in terms of μ, m, M, g, and θ.

11
Angular Momentum and Torque as Vector Quantities

In the previous chapter we learned how to treat the rotation of a rigid body about a fixed axis. This chapter deals in part with the more general case, where the axis of rotation is not fixed in space. We begin by defining a vector product. The vector product is a convenient mathematical tool for expressing such quantities as torque and angular momentum. The central point of this chapter is to develop the concept of the angular momentum of a system of particles, a quantity that plays a key role in rotational dynamics. In analogy to the conservation of linear momentum, we shall find that the angular momentum of any isolated system (an isolated rigid body or any other isolated collection of particles) is always conserved. This conservation law is a special case of the result that the time rate of change of the total angular momentum of any system of particles equals the resultant external torque acting on the system.

11.1 THE VECTOR PRODUCT AND TORQUE

Consider a force $\boldsymbol{F}$ acting on a rigid body at the vector position $\boldsymbol{r}$ (Fig. 11.1). *The origin O is assumed to be in an inertial frame, so that Newton's second law is valid.* The *magnitude* of the torque due to this force relative to the origin is, by definition, equal to $rF \sin \phi$, where ϕ is the angle between $\boldsymbol{r}$ and $\boldsymbol{F}$. The axis about which $\boldsymbol{F}$ would tend to produce rotation is perpendicular to the plane formed by $\boldsymbol{r}$ and $\boldsymbol{F}$. If the force lies in the xy plane as in Figure 11.1, then the torque, $\boldsymbol{\tau}$ is represented by a vector parallel to the z axis. The force in Figure 11.1 tends to rotate the body counterclockwise looking down the z axis, so the sense of $\boldsymbol{\tau}$ is toward increasing z, and $\boldsymbol{\tau}$ is in the positive z direction. If we reversed the direction of $\boldsymbol{F}$ in Figure 11.1, $\boldsymbol{\tau}$ would then be in the negative z direction. The torque involves two vectors, $\boldsymbol{r}$ and $\boldsymbol{F}$, and is in fact defined to be equal to the *vector product*, or *cross product*, of $\boldsymbol{r}$ and $\boldsymbol{F}$:

$$\boldsymbol{\tau} \equiv \boldsymbol{r} \times \boldsymbol{F} \tag{11.1}$$

We must now give a formal definition of the vector product. Given any two vectors $\boldsymbol{A}$ and $\boldsymbol{B}$, the vector product $\boldsymbol{A} \times \boldsymbol{B}$ is defined as a third vector $\boldsymbol{C}$, the *magnitude* of which is $AB \sin \theta$, where θ is the angle included between $\boldsymbol{A}$ and $\boldsymbol{B}$. That is, if $\boldsymbol{C}$ is given by

$$\boldsymbol{C} = \boldsymbol{A} \times \boldsymbol{B} \tag{11.2}$$

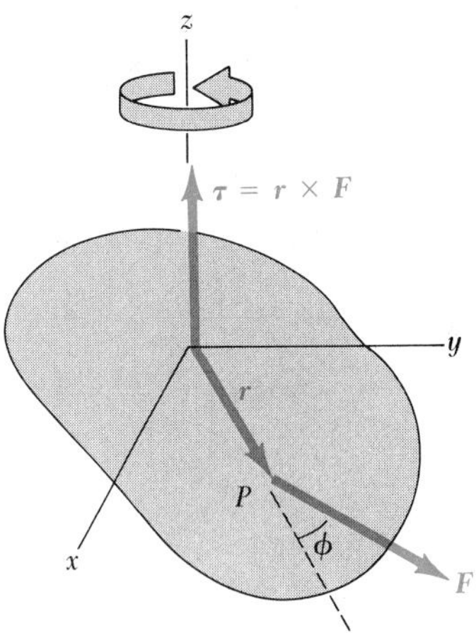

Figure 11.1 The torque vector $\boldsymbol{\tau}$ lies in a direction perpendicular to the plane formed by the position vector $\boldsymbol{r}$ and the applied force $\boldsymbol{F}$.

then its magnitude is

Magnitude of the cross product

$$C \equiv |\mathbf{C}| = |AB \sin\theta| \tag{11.3}$$

Note that the quantity $AB \sin\theta$ is equal to the area of the parallelogram formed by **A** and **B**, as shown in Figure 11.2. The *direction* of $\mathbf{A} \times \mathbf{B}$ is perpendicular to the plane formed by **A** and **B**, as in Figure 11.2, and its sense is determined by the advance of a right-handed screw when turned from **A** to **B** through the angle θ. A more convenient rule to use for the direction of $\mathbf{A} \times \mathbf{B}$ is the right-hand rule illustrated in Figure 11.2. The four fingers of the right hand are pointed along **A** and then "wrapped" into **B** through the angle θ. The direction of the erect right thumb is the direction of $\mathbf{A} \times \mathbf{B}$. Because of the notation, $\mathbf{A} \times \mathbf{B}$ is often read "A cross B"; hence the term *cross product*.

Some properties of the vector product which follow from its definition are as follows:

Properties of the vector product

1. Unlike the scalar product, the order in which the two vectors are multiplied in a cross product is important, that is,

$$\mathbf{A} \times \mathbf{B} = -(\mathbf{B} \times \mathbf{A}) \tag{11.4}$$

   Therefore, if you change the order of the cross product, you must change the sign. You could easily verify this relation with the right-hand rule (Fig. 11.2).
2. If **A** is parallel to **B** ($\theta = 0°$ or $180°$), then $\mathbf{A} \times \mathbf{B} = 0$; therefore, it follows that $\mathbf{A} \times \mathbf{A} = 0$.
3. If **A** is perpendicular to **B**, then $|\mathbf{A} \times \mathbf{B}| = AB$.
4. It is also important to note that the vector product obeys the *distributive law*, that is,

$$\mathbf{A} \times (\mathbf{B} + \mathbf{C}) = \mathbf{A} \times \mathbf{B} + \mathbf{A} \times \mathbf{C} \tag{11.5}$$

5. Finally, the derivative of the cross product with respect to some variable such as t is given by

$$\frac{d}{dt}(\mathbf{A} \times \mathbf{B}) = \mathbf{A} \times \frac{d\mathbf{B}}{dt} + \frac{d\mathbf{A}}{dt} \times \mathbf{B} \tag{11.6}$$

   where it is important to preserve the multiplicative order of **A** and **B**, in view of Equation 11.4.

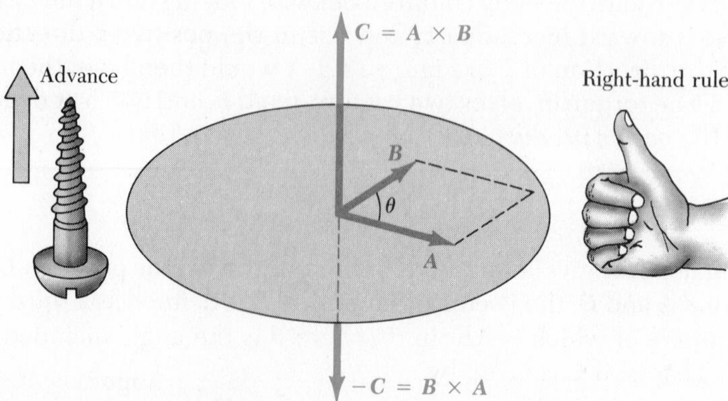

Figure 11.2 The vector product $\mathbf{A} \times \mathbf{B}$ is a third vector **C** having a magnitude $AB \sin\theta$ equal to the area of the parallelogram shown. The direction of **C** is perpendicular to the plane formed by **A** and **B**, and its sense is determined by the right-hand rule.

It is left as an exercise to show from Equations 11.2 and 11.3 and the definition of unit vectors that the cross products of the rectangular unit vectors $\boldsymbol{i}, \boldsymbol{j}$, and $\boldsymbol{k}$ obey the following expressions:

Cross products of unit vectors

$$\boldsymbol{i}\times\boldsymbol{i}=\boldsymbol{j}\times\boldsymbol{j}=\boldsymbol{k}\times\boldsymbol{k}=0 \tag{11.7a}$$

$$\boldsymbol{i}\times\boldsymbol{j}=-\boldsymbol{j}\times\boldsymbol{i}=\boldsymbol{k} \tag{11.7b}$$

$$\boldsymbol{j}\times\boldsymbol{k}=-\boldsymbol{k}\times\boldsymbol{j}=\boldsymbol{i} \tag{11.7c}$$

$$\boldsymbol{k}\times\boldsymbol{i}=-\boldsymbol{i}\times\boldsymbol{k}=\boldsymbol{j} \tag{11.7d}$$

Note that signs are interchangeable. For example, $\boldsymbol{i}\times(-\boldsymbol{j})=-\boldsymbol{i}\times\boldsymbol{j}=-\boldsymbol{k}$.

The cross product of *any* two vectors $\boldsymbol{A}$ and $\boldsymbol{B}$ can be expressed in the following determinant form:

$$\boldsymbol{A}\times\boldsymbol{B}=\begin{vmatrix} \boldsymbol{i} & \boldsymbol{j} & \boldsymbol{k} \\ A_x & A_y & A_z \\ B_x & B_y & B_z \end{vmatrix}$$

Expanding this determinant gives the result

$$\boldsymbol{A}\times\boldsymbol{B}=(A_yB_z-A_zB_y)\boldsymbol{i}+(A_zB_x-A_xB_z)\boldsymbol{j}+(A_xB_y-A_yB_x)\boldsymbol{k} \tag{11.8}$$

EXAMPLE 11.1. The Cross Product

Two vectors lying in the xy plane are given by the equations $\boldsymbol{A}=2\boldsymbol{i}+3\boldsymbol{j}$ and $\boldsymbol{B}=-\boldsymbol{i}+2\boldsymbol{j}$. Find $\boldsymbol{A}\times\boldsymbol{B}$, and verify explicitly that $\boldsymbol{A}\times\boldsymbol{B}=-\boldsymbol{B}\times\boldsymbol{A}$.

Solution: Using Equations 11.7a through 11.7d for the cross product of unit vectors gives

$$\boldsymbol{A}\times\boldsymbol{B}=(2\boldsymbol{i}+3\boldsymbol{j})\times(-\boldsymbol{i}+2\boldsymbol{j})$$
$$=2\boldsymbol{i}\times2\boldsymbol{j}+3\boldsymbol{j}\times(-\boldsymbol{i})=4\boldsymbol{k}+3\boldsymbol{k}=7\boldsymbol{k}$$

(We have omitted the terms in $\boldsymbol{i}\times\boldsymbol{i}$ and $\boldsymbol{j}\times\boldsymbol{j}$, which are zero.)

$$\boldsymbol{B}\times\boldsymbol{A}=(-\boldsymbol{i}+2\boldsymbol{j})\times(2\boldsymbol{i}+3\boldsymbol{j})$$
$$=-\boldsymbol{i}\times3\boldsymbol{j}+2\boldsymbol{j}\times2\boldsymbol{i}=-3\boldsymbol{k}-4\boldsymbol{k}=-7\boldsymbol{k}$$

Therefore, $\boldsymbol{A}\times\boldsymbol{B}=-\boldsymbol{B}\times\boldsymbol{A}$.

As an alternative method for finding $\boldsymbol{A}\times\boldsymbol{B}$, we could use Equation 11.8, with $A_x=2, A_y=3, A_z=0$ and $B_x=-1, B_y=2, B_z=0$. This gives

$$\boldsymbol{A}\times\boldsymbol{B}=(0)\boldsymbol{i}+(0)\boldsymbol{j}+[2\times2-3\times(-1)]\boldsymbol{k}=7\boldsymbol{k}$$

Exercise 1 Use the results to this example and Equation 11.3 to find the angle between $\boldsymbol{A}$ and $\boldsymbol{B}$.
Answer: 60.3°.

11.2 ANGULAR MOMENTUM OF A PARTICLE

A particle of mass m, located at the vector position $\boldsymbol{r}$, moves with a velocity $\boldsymbol{v}$ (Fig. 11.3).

The **instantaneous angular momentum** $\boldsymbol{L}$ of the particle relative to the origin O is defined by the cross product of its instantaneous vector position and the instantaneous linear momentum $\boldsymbol{p}$:

$$\boldsymbol{L}\equiv\boldsymbol{r}\times\boldsymbol{p} \tag{11.9}$$

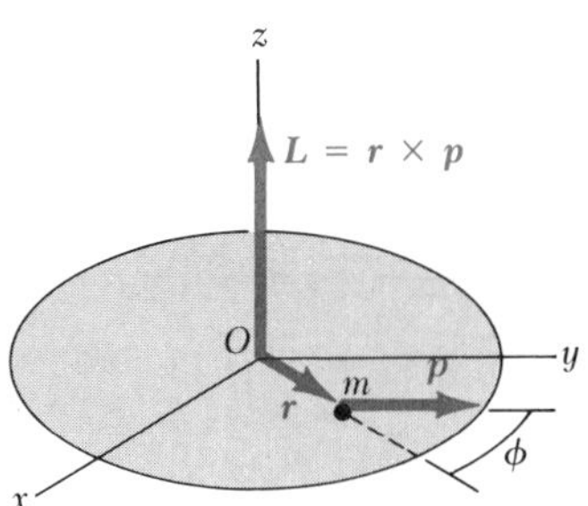

Figure 11.3 The angular momentum $\boldsymbol{L}$ of a particle of mass m and momentum $\boldsymbol{p}$ located at the position $\boldsymbol{r}$ is a vector given by $\boldsymbol{L}=\boldsymbol{r}\times\boldsymbol{p}$. Note that the value of $\boldsymbol{L}$ depends on the origin and is a vector perpendicular to both $\boldsymbol{r}$ and $\boldsymbol{p}$.

The SI units of angular momentum are kg · m²/s. It is important to note that both the magnitude and direction of $\boldsymbol{L}$ depend on the choice of the origin. The direction of $\boldsymbol{L}$ is perpendicular to the plane formed by $\boldsymbol{r}$ and $\boldsymbol{p}$, and its sense is governed by the right-hand rule. For example, in Figure 11.3 $\boldsymbol{r}$ and $\boldsymbol{p}$ are assumed to be in the xy plane, so that $\mathbf{L}$ points in the z direction. Since $\mathbf{p} = m\mathbf{v}$, the magnitude of $\mathbf{L}$ is given by

$$L = mvr \sin \phi \tag{11.10}$$

where ϕ is the angle between $\mathbf{r}$ and $\mathbf{p}$. It follows that L is zero when $\mathbf{r}$ is parallel to $\mathbf{p}$ ($\phi = 0$ or $180°$). In other words, when the particle moves along a line that passes through the origin, it has zero angular momentum with respect to the origin. This is equivalent to stating that it has no tendency to rotate about the origin. On the other hand, if $\mathbf{r}$ is perpendicular to $\mathbf{p}$ ($\phi = 90°$), then L is a maximum and equal to mrv. In this case, the particle has maximum tendency to rotate about the origin. In fact, at that instant the particle moves exactly as though it were on the rim of a wheel rotating about the origin in a plane defined by $\mathbf{r}$ and $\mathbf{p}$.

Alternatively, one may note that a particle has nonzero angular momentum about some point if its position vector measured from that point rotates about the point. On the other hand, if the position vector simply increases or decreases in length, the particle moves along a line passing through the origin and therefore has zero angular momentum with respect to that origin.

In the case of the linear motion of a particle, we found that the resultant force on a particle equals the time rate of change of its linear momentum. We shall now show that Newton's second law implies that the resultant torque acting on a particle equals the time rate of change of its angular momentum. Let us start by writing the torque on the particle in the form

$$\boldsymbol{\tau} = \mathbf{r} \times \mathbf{F} = \mathbf{r} \times \frac{d\mathbf{p}}{dt} \tag{11.11}$$

where we have used the fact that $\mathbf{F} = d\mathbf{p}/dt$. Now let us differentiate Equation 11.9 with respect to time using the rule given by Equation 11.6.

$$\frac{d\mathbf{L}}{dt} = \frac{d}{dt}(\mathbf{r} \times \mathbf{p}) = \mathbf{r} \times \frac{d\mathbf{p}}{dt} + \frac{d\mathbf{r}}{dt} \times \mathbf{p}$$

It is important to adhere to the order of terms since $\mathbf{A} \times \mathbf{B} = -\mathbf{B} \times \mathbf{A}$.

The last term on the right in the above equation is zero, since $\mathbf{v} = d\mathbf{r}/dt$ is parallel to $\mathbf{p}$. Therefore,

$$\frac{d\mathbf{L}}{dt} = \mathbf{r} \times \frac{d\mathbf{p}}{dt} \tag{11.12}$$

Comparing Equations 11.11 and 11.12, we see that

Torque equals time rate of change of angular momentum

$$\boldsymbol{\tau} = \frac{d\mathbf{L}}{dt} \tag{11.13}$$

which is the rotational analog of Newton's second law, $\mathbf{F} = d\mathbf{p}/dt$. This result says that

> the **torque** acting on a particle is equal to the time rate of change of its angular momentum.

It is important to note that Equation 11.13 is valid only if the origins of $\boldsymbol{\tau}$ and $\mathbf{L}$ are the *same*. It is left as an exercise to show that Equation 11.13 is also valid when there are several forces acting on the particle, in which case $\boldsymbol{\tau}$ is the *net* torque on the particle. *Furthermore, the expression is valid for any origin fixed in an inertial frame.* Of course, the same origin must be used in calculating all torques as well as the angular momentum.

A System of Particles

The total angular momentum, $\boldsymbol{L}$, of a system of particles about some point is defined as the vector sum of the angular momenta of the individual particles:

$$\boldsymbol{L} = \boldsymbol{L}_1 + \boldsymbol{L}_2 + \cdots + \boldsymbol{L}_n = \sum \boldsymbol{L}_i$$

where the vector sum is over all of the n particles in the system.

Since the individual momenta of the particles may change in time, the total angular momentum may also vary in time. In fact, from Equations 11.11 through 11.13, we find that the time rate of change of the total angular momentum equals the vector sum of *all* torques, including those associated with internal forces between particles and those associated with external forces. However, the net torque associated with internal forces is zero. To understand this, recall that Newton's third law tells us that the internal forces occur in equal and opposite pairs that lie along the line of separation of each pair of particles. Therefore, the torque due to each action-reaction force pair is zero. By summation, we see that *the net internal torque vanishes.* Finally, we conclude that the total angular momentum can vary with time *only* if there is a net *external* torque on the system, so that we have

$$\sum \boldsymbol{\tau}_{\text{ext}} = \sum \frac{d\boldsymbol{L}_i}{dt} = \frac{d}{dt}\sum \boldsymbol{L}_i = \frac{d\boldsymbol{L}}{dt} \qquad (11.14)$$

That is,

> the time rate of change of the total angular momentum of the system about some origin in an inertial frame equals the net external torque acting on the system about that origin.

Note that Equation 11.14 is the rotational analog of $\boldsymbol{F}_{\text{ext}} = d\boldsymbol{p}/dt$ for a system of particles (Chapter 9).

EXAMPLE 11.2. Linear Motion

A particle of mass m moves in the xy plane with a velocity $\boldsymbol{v}$ along a straight line (Fig. 11.4). What is the magnitude and direction of its angular momentum with respect to the origin O?

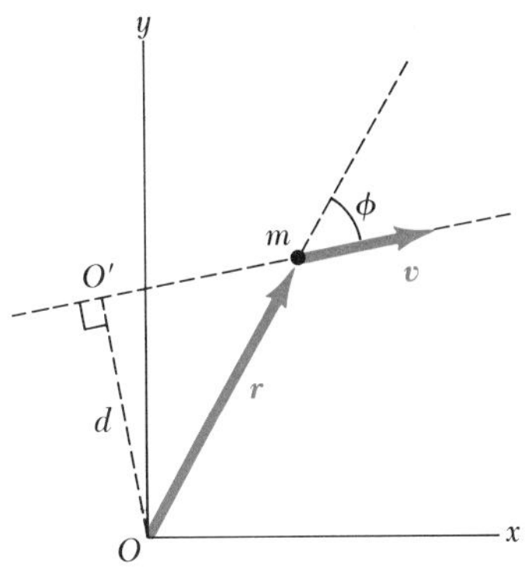

Figure 11.4 (Example 11.2) A particle moving in a straight line with a velocity $\boldsymbol{v}$ has an angular momentum equal in magnitude to mvd relative to O, where $d = r \sin \phi$ is the distance of closest approach to the origin. The vector $\boldsymbol{L} = \boldsymbol{r} \times \boldsymbol{p}$ points *into* the diagram in this case.

Solution: From the definition of angular momentum, $\boldsymbol{L} = \boldsymbol{r} \times \boldsymbol{p} = rmv \sin \phi\,(-\boldsymbol{k})$. Therefore the magnitude of L is given by

$$L = mvr \sin \phi = mvd$$

where $d = r \sin \phi$ is the distance of closest approach of the particle from the origin. The direction of $\boldsymbol{L}$ from the right-hand rule is *into* the diagram, and we can write the vector expression $\boldsymbol{L} = -(mvd)\boldsymbol{k}$. Note that the angular momentum relative to the origin O′ is zero.

EXAMPLE 11.3. Circular Motion

A particle moves in the xy plane in a circular path of radius r, as in Figure 11.5. (a) Find the magnitude and direction of its angular momentum relative to O when its velocity is $\boldsymbol{v}$.

Since $\boldsymbol{r}$ is perpendicular to $\boldsymbol{v}$, $\phi = 90°$ and the magnitude of $\boldsymbol{L}$ is simply

$$L = mvr \sin 90° = mvr \quad \text{(for } r \text{ perpendicular to } v\text{)}$$

The direction of L is perpendicular to the plane of the circle, and its sense depends on the direction of v. If the sense of the rotation is counterclockwise, as in Figure 11.5, then by the right-hand rule, the direction of $L = r \times p$ is *out* of the paper. Hence, we can write the vector expression $L = (mvr)k$. On the other hand, if the particle were to move clockwise, L would point into the paper.

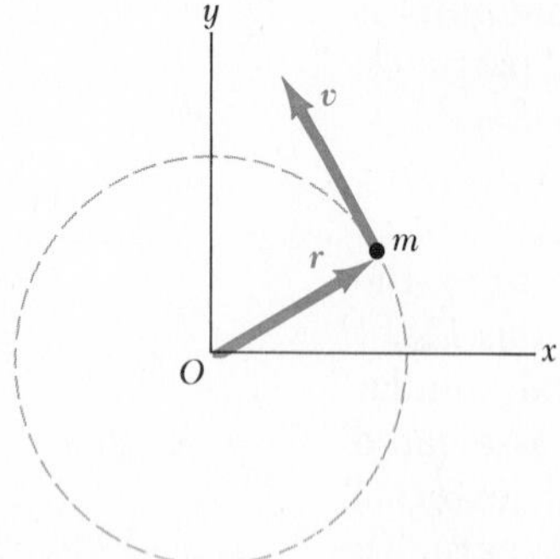

Figure 11.5 (Example 11.3) A particle moving in a circle of radius r has an angular momentum equal in magnitude to mvr relative to the center. The vector $L = r \times p$ points *out* of the diagram.

(b) Find an alternative expression for L in terms of the angular velocity, ω.

Since $v = r\omega$ for a particle rotating in a circle, we can express L as

$$L = mvr = mr^2\omega = I\omega$$

where I is the moment of inertia of the particle about the z axis through O. Furthermore, in this case the angular momentum is in the *same* direction as the angular velocity vector, ω (see Section 10.1), and so we can write $L = I\omega = I\omega k$.

Exercise 2 A car of mass 1500 kg moves in a circular race track of radius 50 m with a speed of 40 m/s. What is the magnitude of its angular momentum relative to the center of the race track?
Answer: 3.00×10^6 kg · m²/s.

11.3 ROTATION OF A RIGID BODY ABOUT A FIXED AXIS

Let us consider a rigid body rotating about an axis that is fixed in direction. We shall assume that the z axis coincides with the axis of rotation, as in Figure 11.6. Each particle of the rigid body rotates in the xy plane about the z axis with an angular velocity ω. The magnitude of the angular momentum of the particle of mass m_i is $m_i v_i r_i$ about the origin O. Because $v_i = r_i\,\omega$, we can express the magnitude of the angular momentum of the ith particle as

$$L_i = m_i r_i^2\,\omega$$

Note that L_i is directed along the z axis, corresponding to the direction of ω.

We can now find the z component of the angular momentum of the rigid body by taking the sum of L_i over all particles of the body:

$$L_z = \sum m_i r_i^2\,\omega = \left(\sum m_i r_i^2\right)\omega$$

or

$$L_z = I\omega \tag{11.15}$$

where L_z is the z component of the angular momentum and I is the moment of inertia of the rigid body about the z axis.

Now let us differentiate Equation 11.15 with respect to time, noting that I is constant for a rigid body:

$$\frac{dL_z}{dt} = I\frac{d\omega}{dt} = I\alpha \tag{11.16}$$

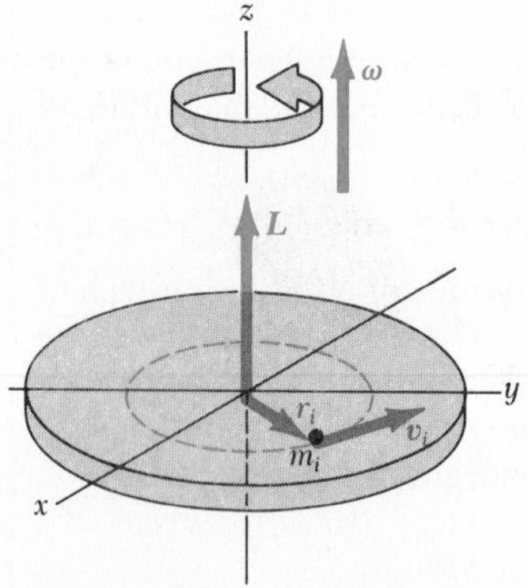

Figure 11.6 When a rigid body rotates about an axis, the angular momentum L is in the same direction as the angular velocity ω, according to the expression $L = I\omega$.

where α is the angular acceleration relative to the axis of rotation. Because the product $I\alpha$ is equal to the net torque (see Eq. 11.14), we can express Equation 11.16 as follows:

$$\sum \tau_{\text{ext}} = \frac{dL_z}{dt} = I\alpha \qquad (11.17)$$

That is, the net external torque acting on a rigid body rotating about a fixed axis equals the moment of inertia about the axis of rotation multiplied by its angular acceleration relative to that axis.

You should note that if a symmetrical rigid body rotates about a fixed axis passing through its center of mass, one can write Equation 11.15 in vector form, $\mathbf{L} = I\boldsymbol{\omega}$, where $\mathbf{L}$ is its total angular momentum measured with respect to the axis of rotation. Furthermore, the expression is valid for any body, regardless of its symmetry, if $\mathbf{L}$ stands for the component of angular momentum along the axis of rotation.[1]

[1] In general, the expression $\mathbf{L} = I\boldsymbol{\omega}$ is not always valid. If a rigid body rotates about an arbitrary axis, $\mathbf{L}$ and $\boldsymbol{\omega}$ may point in different directions. In fact, in this case, the moment of inertia cannot be treated as a scalar. Strictly speaking, $\mathbf{L} = I\boldsymbol{\omega}$ applies only to rigid bodies of any shape that rotate about one of three mutually perpendicular axes (called *principal axes*) through the center of mass. This is discussed in more advanced mechanics texts.

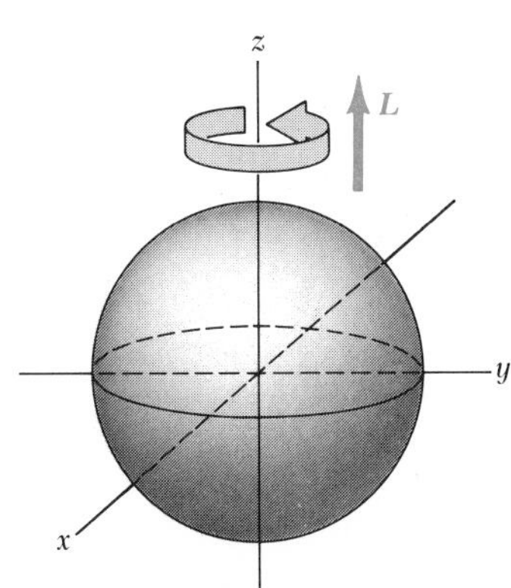

Figure 11.7 (Example 11.4) A sphere that rotates about the z axis in the direction shown has an angular momentum $\mathbf{L}$ in the positive z direction. If the direction of rotation is reversed, $\mathbf{L}$ will point in the negative z direction.

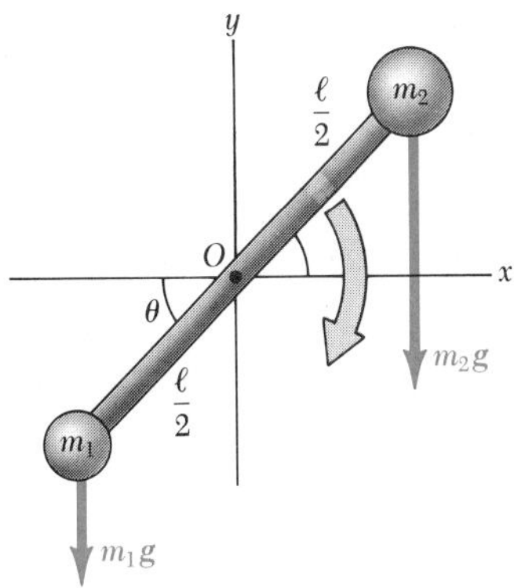

Figure 11.8 (Example 11.5) Since gravitational forces act on the system rotating in a vertical plane, there is in general a net nonzero torque about O when $m_1 \neq m_2$, which in turn produces an angular acceleration according to $\tau_{\text{net}} = I\alpha$.

EXAMPLE 11.4. Rotating Sphere

A uniform solid sphere of radius $R = 0.50$ m and mass 15 kg rotates about the z axis through its center, as in Figure 11.7. Find its angular momentum when the angular velocity is 3 rad/s.

Solution: The moment of inertia of the sphere about an axis through its center is

$$I = \tfrac{2}{5}MR^2 = \tfrac{2}{5}(15 \text{ kg})(0.5 \text{ m})^2 = 1.5 \text{ kg} \cdot \text{m}^2$$

Therefore, the magnitude of the angular momentum is

$$L = I\omega = (1.5 \text{ kg} \cdot \text{m}^2)(3 \text{ rad/s}) = 4.5 \text{ kg} \cdot \text{m}^2/\text{s}$$

EXAMPLE 11.5. Rotating Rod

A rigid rod of mass M and length ℓ rotates in a vertical plane about a frictionless pivot through its center (Fig. 11.8). Particles of masses m_1 and m_2 are attached at the ends of the rod. (a) Determine the angular momentum when the angular velocity is ω.

The moment of inertia of the system equals the sum of the moments of inertia of the three components: the rod, m_1, and m_2. Using Table 10.2 we find that the total moment of inertia about the z axis through O is

$$I = \tfrac{1}{12}M\ell^2 + m_1\left(\frac{\ell}{2}\right)^2 + m_2\left(\frac{\ell}{2}\right)^2 = \frac{\ell^2}{4}\left(\frac{M}{3} + m_1 + m_2\right)$$

Therefore, when the angular velocity is ω, the angular momentum is given by

$$L = I\omega = \frac{\ell^2}{4}\left(\frac{M}{3} + m_1 + m_2\right)\omega$$

(b) Determine the angular acceleration of the system when the rod makes an angle θ with the horizontal.

The torque due to the force m_1g about the pivot is

$$\tau_1 = m_1 g \frac{\ell}{2} \cos\theta \qquad \text{(out of the plane)}$$

The torque due to the force m_2g about the pivot is

$$\tau_2 = -m_2 g \frac{\ell}{2} \cos\theta \qquad \text{(into the plane)}$$

Hence, the net torque about O is

$$\tau_{\text{net}} = \tau_1 + \tau_2 = \tfrac{1}{2}(m_1 - m_2)g\ell \cos\theta$$

You should note that τ_{net} is *out* of the plane if $m_1 > m_2$ and is *into* the plane if $m_1 < m_2$. To find α, we use $\tau_{\text{net}} = I\alpha$, where I was obtained in (a). This gives

$$\alpha = \frac{\tau_{\text{net}}}{I} = \frac{2(m_1 - m_2)g\cos\theta}{\ell\left(\frac{M}{3} + m_1 + m_2\right)}$$

Note that α is zero when θ is $\pi/2$ or $-\pi/2$ (vertical position) and α is a maximum when θ is 0 or π (horizontal position). Furthermore, the angular velocity of the system changes since α varies in time.

Exercise 3 If $m_1 > m_2$, at what value of θ is ω a maximum? Knowing the angular velocity at some instant, how would you calculate the linear velocity of m_1 and m_2?

EXAMPLE 11.6. Two Connected Masses

Two masses, m_1 and m_2, are connected by a light cord that passes over a pulley of radius R and moment of inertia I about its axle as in Figure 11.9. The mass m_2 slides on a frictionless, horizontal surface. Let us determine the acceleration of the two masses using the concepts of angular momentum and torque.

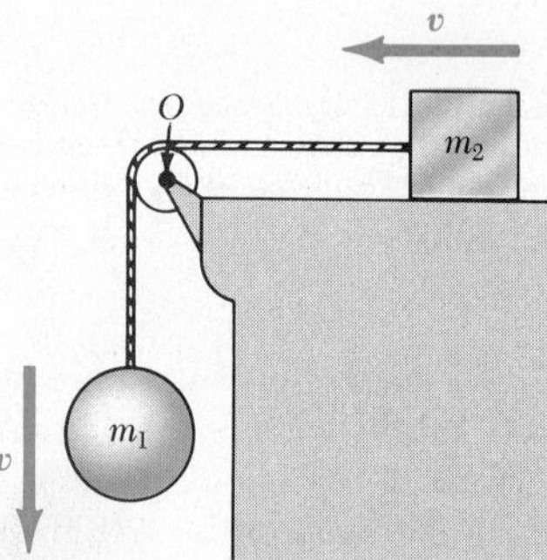

Figure 11.9 (Example 11.6).

Solution: First, let us calculate the angular momentum of the system, which consists of the two masses plus the pulley. We shall calculate the angular momentum about an axis along the axle of the pulley through O. At the instant the masses m_1 and m_2 have a speed v, the angular momentum of m_1 is m_1vR, while the angular momentum of m_2 is m_2vR. At the same instant, the angular momentum of the pulley is $I\omega = Iv/R$. Therefore, the total angular momentum of the system is

$$(1) \quad L = m_1 vR + m_2 vR + I\frac{v}{R}$$

Now let us evaluate the total external torque on the system about the axle. Because the force of the axle on the pulley has zero moment arm, it does not contribute to the torque. Furthermore, the normal force acting on m_2 is balanced by its weight m_2g, hence these forces do not contribute to the torque. The external force m_1g produces a torque about the axle equal in magnitude to m_1gR, where R is the moment arm of the force about the axle. This is the total external torque about O; that is, $\tau_{\text{ext}} = m_1gR$. Using this result, together with (1) and Equation 11.17 gives

$$\tau_{\text{ext}} = \frac{dL}{dt}$$

$$m_1 gR = \frac{d}{dt}\left[(m_1 + m_2)Rv + I\frac{v}{R}\right]$$

or

$$m_1 gR = (m_1 + m_2)R\frac{dv}{dt} + \frac{I}{R}\frac{dv}{dt}$$

Because $dv/dt = a$, we can solve this for a to get

$$a = \frac{m_1 g}{(m_1 + m_2) + I/R^2}$$

You may wonder why we did not include the forces of tension in evaluating the net torque about the axle. The reason is that the forces of tension are *internal* to the system under consideration. It is only the *external* torques that contribute to the change in angular momentum.

11.4 CONSERVATION OF ANGULAR MOMENTUM

In Chapter 9 we found that the total linear momentum of a system of particles remains constant when the resultant external force acting on the system is zero. We have an analogous conservation law in rotational motion which states that *the total angular momentum of a system is constant if the resultant external torque acting on the system is zero.* This follows directly from Equation 11.14, where we see that if

$$\sum \tau_{\text{ext}} = \frac{dL}{dt} = 0 \tag{11.18a}$$

then

$$L = \text{constant} \tag{11.18b}$$

For a system of particles, we write this conservation law as $\Sigma L_i = \text{constant}$. If a body undergoes a redistribution of its mass, then its moment of inertia changes and we express this conservation of angular momentum in the form

$$L_i = L_f = \text{constant}$$

If the system is a body rotating about a *fixed* axis, such as the z axis, then we can write $L_z = I\omega$, where L_z is the component of L along the axis of rotation and I is the moment of inertia about this axis. In this case, we can express the conservation of angular momentum as

Conservation of angular momentum

$$I_i\omega_i = I_f\omega_f = \text{constant} \tag{11.19}$$

This expression is valid for rotations either about a fixed axis or about an axis through the center of mass of the system as long as the axis remains parallel to itself. We only require the net external torque to be zero.

Although we do not prove it here, there is an important theorem concerning the angular momentum relative to the center of mass. This theorem states that

> the resultant torque acting on a body about the center of mass equals the time rate of change of angular momentum regardless of the motion of the center of mass.

This theorem applies even if the center of mass is accelerating, provided τ and L are evaluated relative to the center of mass.

In Equation 11.19 we have a third conservation law to add to our list. Furthermore, we can now state that the energy, linear momentum, and angular momentum of an isolated system all remain constant.

There are many examples that demonstrate conservation of angular momentum, some of which should be familiar to you. You may have observed a figure skater undergoing a spin motion in the finale of an act. The angular velocity of the skater increases upon pulling his or her hands and feet close to the body. Neglecting friction between the skater and the ice, we see that there are no external torques on the skater. The change in angular velocity is due to the fact that since angular momentum is conserved, the product $I\omega$ remains constant and a decrease in the moment of inertia of the skater causes an increase in the angular velocity. Similarly, when divers (or acrobats) wish to make several somersaults, they pull their hands and feet close to their bodies in order to rotate at a higher rate. In these cases, the external force due to gravity acts through the center of mass and hence exerts no torque about this point. Therefore, the angular momentum about the center of mass must be conserved, or $I_i\omega_i = I_f\omega_f$. For example, when divers wish to double their angular velocity, they must reduce their moment of inertia to half its initial value.

The angular velocity of the skater increases when she pulls her arms in close to her body because angular momentum is conserved. (Photo, David Leonardi)

EXAMPLE 11.7. A Projectile–Cylinder Collision

A projectile of mass m and velocity v_0 is fired at a solid cylinder of mass M and radius R (Fig. 11.10). The cylinder is initially at rest and is mounted on a fixed horizontal axle that runs through the center of mass. The line of motion of the projectile is perpendicular to the axle and at a distance $d < R$ from the center. Find the angular speed of the system after the projectile strikes and adheres to the surface of the cylinder.

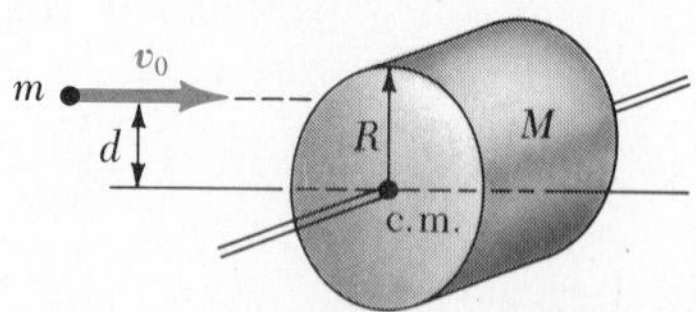

Figure 11.10 (Example 11.7) The angular momentum of the system before the collision equals the angular momentum right after the collision with respect to the center of mass if we neglect the weight of the projectile.

Solution: Let us evaluate the angular momentum of the system (projectile + cylinder) about the axle of the cylinder. About this point, the net external torque on the system is zero if we neglect the force of gravity on the projectile. Hence, the angular momentum of the system is the same before and after the collision.

Before the collision, only the projectile has angular momentum with respect to a point on the axle. The magnitude of this angular momentum is mv_0d, and it is directed along the axle into the paper. After the collision, the total angular momentum of the system is $I\omega$, where I is the total moment of inertia about the axle (projectile + cylinder). Since the total angular momentum is conserved, we get

$$mv_0d = I\omega = (\tfrac{1}{2}MR^2 + mR^2)\omega$$

$$\omega = \frac{mv_0d}{\tfrac{1}{2}MR^2 + mR^2}$$

This suggests another technique for measuring the velocity of a bullet.

Exercise 4 In this example, mechanical energy is *not* conserved, since the collision is inelastic. Show that $\tfrac{1}{2}I\omega^2 < \tfrac{1}{2}mv_0{}^2$. What do you suppose accounts for the energy loss?

EXAMPLE 11.8. The Merry-Go-Round

A horizontal platform in the shape of a circular disk rotates in a horizontal plane about a frictionless vertical axle (Fig. 11.11). The platform has a mass of 100 kg and a radius of 2 m. A student whose mass is 60 kg walks slowly from the rim of the platform toward the center. If the angular velocity of the system is 2 rad/s when the student is at the rim, (a) calculate the angular velocity when the student has reached a point 0.5 m from the center.

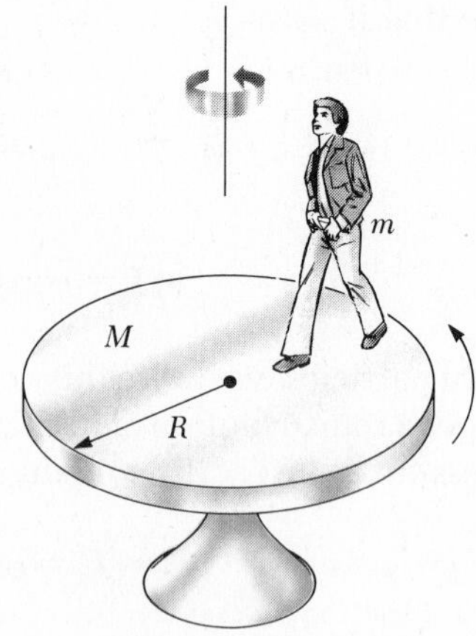

Figure 11.11 (Example 11.8) As the student walks toward the center of the rotating platform, the angular velocity of the system increases since the angular momentum must remain constant.

Let us call the moment of inertia of the platform I_p and the moment of inertia of the student I_s. Treating the student as a point mass m, we can write the *initial* moment of inertia of the system about the axle of rotation

$$I_i = I_p + I_s = \tfrac{1}{2}MR^2 + mR^2$$

where M and R are the mass and radius of the platform, respectively. When the student has walked to the position $r < R$, the moment of inertia of the system *reduces* to

$$I_f = \tfrac{1}{2}MR^2 + mr^2$$

Since there are no external torques on the system (student + platform) about the axis of rotation, we can apply the law of conservation of angular momentum:

$$I_i\omega_i = I_f\omega_f$$

$$(\tfrac{1}{2}MR^2 + mR^2)\omega_i = (\tfrac{1}{2}MR^2 + mr^2)\omega_f$$

$$\omega_f = \left(\frac{\tfrac{1}{2}MR^2 + mR^2}{\tfrac{1}{2}MR^2 + mr^2}\right)\omega_i$$

Substituting the values given for M, R, m, and ω_i we get

$$\omega_f = \left(\frac{200 + 240}{200 + 15}\right)(2 \text{ rad/s}) = 4.1 \text{ rad/s}$$

(b) Calculate the initial and final kinetic energies of the system.

$$K_i = \tfrac{1}{2}I_i\omega_i{}^2 = \tfrac{1}{2}(440 \text{ kg}\cdot\text{m}^2)\left(2\,\frac{\text{rad}}{\text{s}}\right)^2 = 880 \text{ J}$$

$$K_f = \tfrac{1}{2}I_f\omega_f{}^2 = \tfrac{1}{2}(215 \text{ kg}\cdot\text{m}^2)\left(4.1\,\frac{\text{rad}}{\text{s}}\right)^2 = 1800 \text{ J}$$

Note that the kinetic energy of the system *increases!* Although this result may surprise you, it can be explained as follows: In the process of walking toward the center of the platform, the student had to exert some muscular effort and perform positive work, which in turn is transformed into an increase in kinetic energy of the system. In other words, internal forces within the system did work. Since the student is in a rotating, noninertial frame of reference, he senses an outward "centrifugal" force that varies with r. He must exert a counteracting force and hence he must perform work, or exert energy.

Exercise 5 Show that the gain in kinetic energy can be accounted for using the work-energy theorem.

EXAMPLE 11.9. Spinning on a Stool

A student sits on a pivoted stool while holding a pair of weights, as in Figure 11.12. The stool is free to rotate about a vertical axis with negligible friction. The student is set in rotating motion with the weights outstretched. Why does the angular velocity of the system increase as the weights are pulled inward?

Solution: The initial angular momentum of the system is $I_i\omega_i$, where I_i refers to the initial moment of inertia of the entire system (student + weights + stool). After the weights are pulled in, the angular momentum of the system is $I_f\omega_f$. Note that $I_f < I_i$ since the weights are now closer to the axis of rotation, reducing the moment of inertia. Since the net external torque on the system is zero, angular momentum is conserved, so $I_i\omega_i = I_f\omega_f$. Therefore, $\omega_f > \omega_i$, or the angular velocity increases. As in the previous example the kinetic energy of the system increases as the weights are pulled inward. The increase in kinetic energy arises from the fact that the student must do work in pulling the weights toward the axis of rotation.

Exercise 6 Suppose the student were to drop the weights to his side rather than pull them inward horizontally. What would account for the increase in the kinetic energy of the system in this situation?

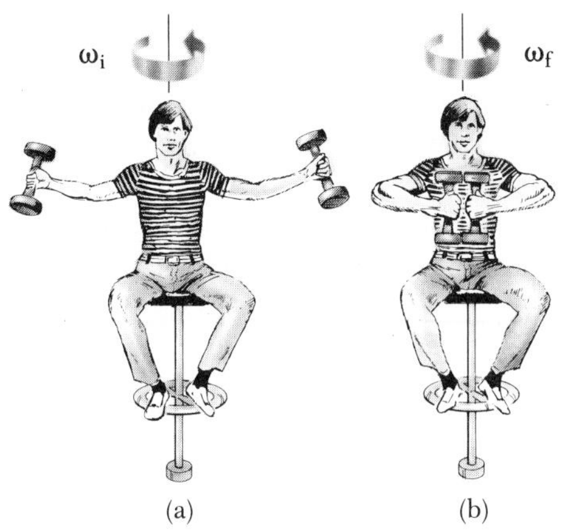

Figure 11.12 (Example 11.9) (a) The student is given an initial angular velocity while holding two weights as shown. (b) When the weights are pulled in close to the body, the angular velocity of the system increases. Why?

EXAMPLE 11.10. The Spinning Bicycle Wheel

In another favorite classroom demonstration, a student holds the axle of a spinning bicycle wheel while seated on a pivoted stool (Fig. 11.13). The student and stool are initially *at rest* while the wheel is spinning in a horizontal plane with an initial angular momentum L_0 pointing upward. Explain what happens if the wheel is inverted about its center by 180°.

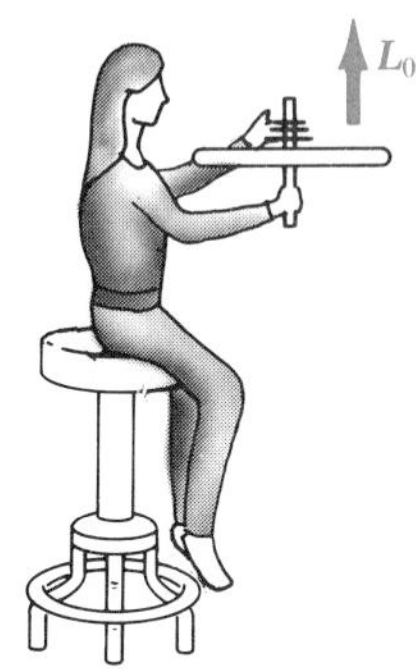

Figure 11.13 (Example 11.10) The wheel is initially spinning when the student is at rest. What happens when the wheel is inverted?

Solution: In this situation, the system consists of the student, wheel, and stool. Initially, the total angular momentum of the system is L_0, corresponding to the contribution from the spinning wheel. As the wheel is inverted, a torque is supplied by the student, but this is *internal* to the system. There is *no* external torque acting on the system about the vertical axis. Therefore, *the angular momentum of the system must be conserved.*

Initially, we have

$$L_{\text{system}} = L_0 \qquad \text{(upward)}$$

After the wheel is inverted,

$$L_{\text{system}} = L_0 = L_{\text{student+stool}} + L_{\text{wheel}}$$

In this case, $L_{\text{wheel}} = -L_0$ since it is now rotating in the opposite sense. Therefore

$$L_0 = L_{\text{student+stool}} - L_0$$

$$L_{\text{student+stool}} = 2L_0$$

This shows that *the student and stool will start to turn, acquiring an angular momentum having a magnitude twice that of the spinning wheel and directed upward.*

Exercise 7 How much angular momentum would the student acquire if the wheel is tilted through an angle θ measured from the vertical axis?
Answer: $L_0(1 - \cos\theta)$.

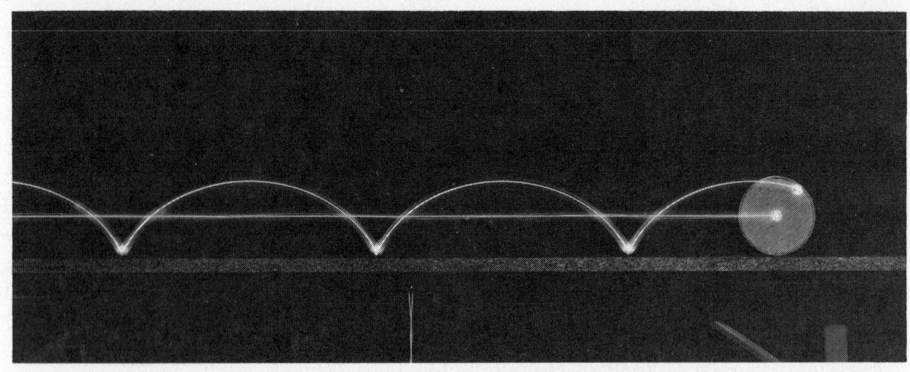

Light sources at the center and rim of a rolling cylinder illustrate the different paths which these points take. The center of mass moves in a straight line, while a point on the rim moves in the path of a cycloid. (Education Development Center, Newton, Mass.)

*11.5 ROLLING MOTION OF A RIGID BODY

In this section we shall treat the motion of a rigid body that is rotating about a moving axis. The general motion of a rigid body in space is very complex. However, we can simplify matters by restricting our discussion to a homogeneous rigid body having a high degree of symmetry, such as a cylinder, sphere, or hoop. Furthermore, we shall assume that the body undergoes rolling motion in a plane.

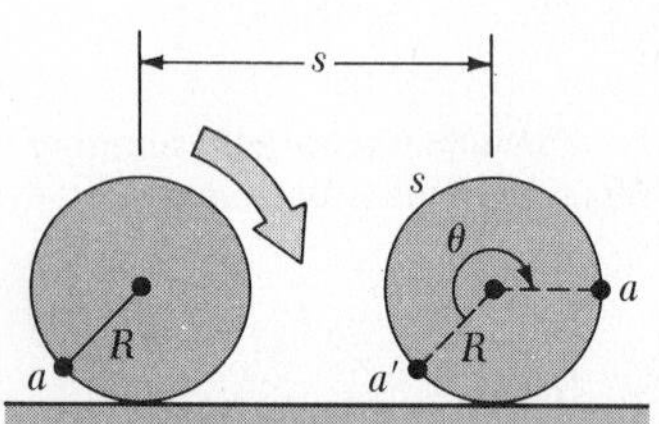

Figure 11.14 For pure rolling motion, as the cylinder rotates through an angle θ, the center of mass moves a distance $s = R\theta$.

Consider a uniform cylinder of radius R rolling on a rough, horizontal surface (Fig. 11.14). As the cylinder rotates through an angle θ, its center of mass moves a distance $s = R\theta$. Therefore, the velocity and acceleration of the center of mass for *pure rolling motion* are given by

$$v_c = \frac{ds}{dt} = R\frac{d\theta}{dt} = R\omega \tag{11.20}$$

$$a_c = \frac{dv_c}{dt} = R\frac{d\omega}{dt} = R\alpha \tag{11.21}$$

The linear velocities of various points on the rolling cylinder are illustrated in Figure 11.15. Note that the linear velocity of any point is in a direction perpendicular to the line from that point to the contact point. At any instant, the point P is at rest relative to the surface since sliding does not occur. For that reason, the axis through P perpendicular to the diagram is called the *instantaneous axis of rotation.*

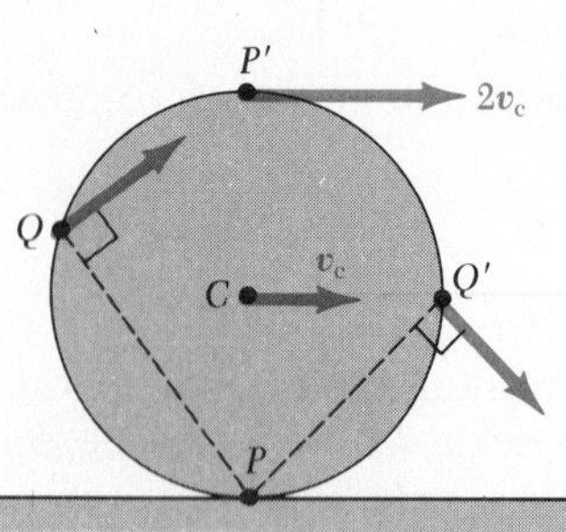

Figure 11.15 All points on a rolling body move in a direction perpendicular to an axis through the contact point P. The center of mass moves with a velocity v_c, while the point P' moves with the velocity $2v_c$.

A general point on the cylinder, such as Q, has both horizontal and vertical components of velocity. However, the points P and P' and the point at the center of mass are unique and of special interest. The center of mass moves with a velocity $v_c = R\omega$, whereas the contact point P has zero velocity. The point P' has a velocity equal to $2v_c = 2R\omega$, since all points on the cylinder have the same angular velocity.

We can express the total kinetic energy of the rolling cylinder as

$$K = \tfrac{1}{2}I_P\omega^2 \tag{11.22}$$

where I_P is the moment of inertia about the axis through P. Applying the parallel-axis theorem, we can substitute $I_P = I_c + MR^2$ into Equation 11.22 to get

$$K = \tfrac{1}{2}I_c\omega^2 + \tfrac{1}{2}MR^2\omega^2$$

$$K = \frac{1}{2}I_c\omega^2 + \frac{1}{2}Mv_c^2 \qquad (11.23)$$

Total kinetic energy of a rolling body

where we have used the fact that $v_c = R\omega$.

We can think of Equation 11.23 as follows: The first term on the right, $\frac{1}{2}I_c\omega^2$, represents the rotational kinetic energy about the center of mass, and the term $\frac{1}{2}Mv_c^2$ represents the kinetic energy the cylinder would have if it were just translating through space without rotating. Thus, we can say that

> the total kinetic energy of an object undergoing pure rolling motion is the sum of a rotational kinetic energy about the center of mass plus the translational kinetic energy of the center of mass.

We can use energy methods to treat a class of problems concerning the rolling motion of a rigid body down a rough incline. We shall assume that the rigid body in Figure 11.16 does not slip and is released from rest at the top of the incline. Note that rolling motion is possible only if a frictional force is present between the object and the incline to produce a net torque about the center of mass. Despite the presence of friction, there is no loss of mechanical energy since the contact point is at rest relative to the surface at any instant. On the other hand, if the rigid body were to slide, mechanical energy would be lost as motion progresses.

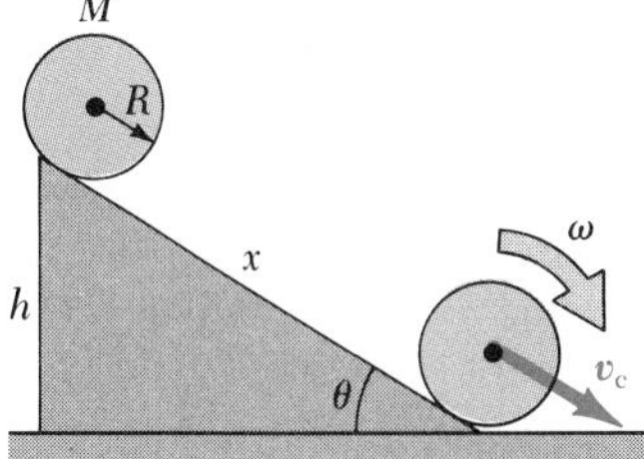

Figure 11.16 A round object rolling down an incline. Mechanical energy is conserved if no slipping occurs.

Using the fact that $v_c = R\omega$ for pure rolling motion, we can express Equation 11.23 as

$$K = \frac{1}{2}I_c\left(\frac{v_c}{R}\right)^2 + \frac{1}{2}Mv_c^2$$

$$K = \frac{1}{2}\left(\frac{I_c}{R^2} + M\right)v_c^2 \qquad (11.24)$$

When the rolling cylinder reaches the bottom of the incline, it has lost potential energy Mgh, where h is the height of the incline. If the body starts from rest at the top, its kinetic energy at the bottom, given by Equation 11.24, must equal its potential energy at the top. Therefore, the velocity of the center of mass at the bottom can be obtained by equating the two quantities:

$$\frac{1}{2}\left(\frac{I_c}{R^2} + M\right)v_c^2 = Mgh$$

$$v_c = \left(\frac{2gh}{1 + I_c/MR^2}\right)^{1/2} \qquad (11.25)$$

EXAMPLE 11.11. Sphere Rolling Down an Incline If the rigid body shown in Fig. 11.16 is a solid sphere, calculate the velocity of its center of mass at the bottom and determine the linear acceleration of the center of mass of the sphere.

Solution: For a uniform solid sphere, $I_c = \frac{2}{5}MR^2$, and therefore Equation 11.25 gives

$$v_c = \left(\frac{2gh}{1 + \frac{2}{5}\frac{MR^2}{MR^2}}\right)^{1/2} = \left(\frac{10}{7}gh\right)^{1/2}$$

The vertical displacement is related to the displacement x along the incline through the relation $h = x \sin\theta$. Hence, after squaring both sides, we can express the equation above as

$$v_c^2 = \frac{10}{7}gx\sin\theta$$

Comparing this with the familiar expression from kinematics, $v_c^2 = 2a_cx$, we see that the acceleration of the center of mass is given by

$$a_c = \frac{5}{7}g\sin\theta$$

The results are quite amazing! Both the velocity and acceleration of the center of mass are *independent* of the mass and radius of the sphere! That is, *all homogeneous solid spheres would experience the same velocity and acceleration on a given incline.* If we repeated the calculations for a hollow sphere, a solid cylinder, or a hoop, we would obtain similar results. The constant factors that appear in the expressions for v_c and a_c depend on the moment of inertia about the center of mass for the specific body. In all cases, the acceleration of the center of mass will be *less* than $g \sin\theta$, the value it would have if the plane were frictionless and no rolling occurred.

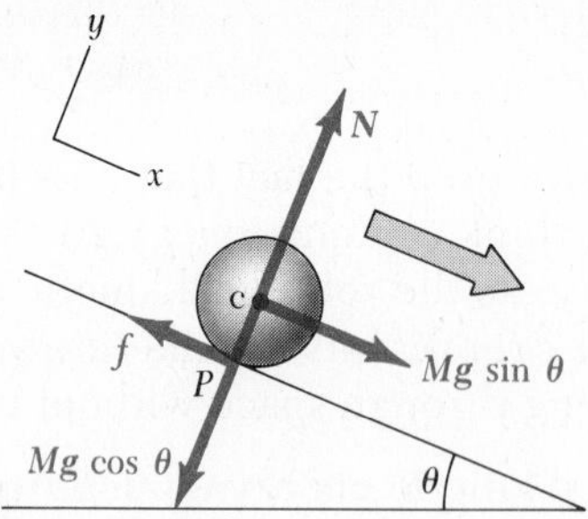

Figure 11.17 (Example 11.12) Free-body diagram for a solid sphere rolling down an incline.

EXAMPLE 11.12. Another Look at the Rolling Sphere
In this example, let us consider the solid sphere rolling down an incline and verify the results of Example 11.11 using dynamic methods. The free-body diagram for the sphere is illustrated in Figure 11.17.

Solution: Newton's second law applied to the center of mass motion gives

$$(1)\ \sum F_x = Mg\sin\theta - f = Ma_c$$

$$\sum F_y = N - Mg\cos\theta = 0$$

where x is measured downward along the inclined plane. Now let us write an expression for the torque acting on the sphere. A convenient axis to choose is an axis through the center of the sphere, perpendicular to the plane of the figure.[2] Since N and Mg go through this origin, they have zero moment arms and do not contribute to the torque. However, the force of friction produces a torque about this axis equal to fR in the clockwise direction; therefore

$$\tau_c = fR = I_c\alpha$$

Since $I_c = \frac{2}{5}MR^2$ and $\alpha = a_c/R$, we get

$$(2)\ f = \frac{I_c\alpha}{R} = \left(\frac{\frac{2}{5}MR^2}{R}\right)\frac{a_c}{R} = \frac{2}{5}Ma_c$$

Substituting (2) into (1) gives

$$a_c = \frac{5}{7}g\sin\theta$$

which agrees with the result of Example 11.11. Note that $a_c < g\sin\theta$ because of the retarding frictional force.

[2] You should note that although the point at the center of mass is not an inertial frame, the expression $\tau_c = I\alpha$ still applies in the center of mass frame.

*11.6 THE MOTION OF GYROSCOPES AND TOPS

Precessional motion

A very unusual and fascinating type of motion that you probably have observed is that of a top spinning about its axis of symmetry as in Fig. 11.18a. If the top spins about its axis very rapidly, the axis will rotate about the vertical direction as indicated, thereby sweeping out a cone. The motion of the axis of the top about the vertical, known as **precessional motion,** is usually slow compared with the spin motion of the top. It is quite natural to wonder why the top doesn't fall over. Since the center of mass is not above the pivot point O, there is clearly a net torque acting on the top about O due to the weight force Mg. From this description, it is easy to see that the top would certainly fall if it were not spinning. However, because the top is spinning, it has an angular momentum $\mathbf{L}$ directed along its axis of symmetry. As we shall show, the motion of the rotation axis about the z axis (the precessional motion) arises from the fact that the torque produces a change in the *direction* of the rotation axis. This is an excellent example of the importance of the directional nature of angular momentum.

The two forces acting on the top are the downward force of gravity, Mg, and the normal force, N, acting upward at the pivot point O. The normal force produces no torque about the pivot since its moment arm is zero. However, the force of gravity produces a torque $\tau = r \times Mg$ about O, where the direction of τ is perpendicular to the plane formed by r and Mg. By necessity, the vector τ lies in a horizontal plane perpendicular to the angular momentum vector. The net torque and angular momentum of the body are related through the expression

$$\tau = \frac{dL}{dt}$$

From this expression, we see that the nonzero torque produces a *change* in angular momentum dL, which is in the same direction as τ. Therefore, like the torque vector, dL must also be at right angles to L. Figure 11.18b illustrates the resulting precessional motion of the axis of the top. In a time Δt, the change in angular momentum $\Delta L = L_f - L_i = \tau\ \Delta t$. Note that because ΔL is perpendicular to L the magnitude of L doesn't change ($|L_i| = |L_f|$). Rather, what is changing is the *direction* of L. Since the change in angular momentum is in the direction of τ, which lies in the xy plane, the top undergoes precessional motion. Thus, the effect of the torque is to deflect the angular momentum of the top in a direction perpendicular to its spin axis.

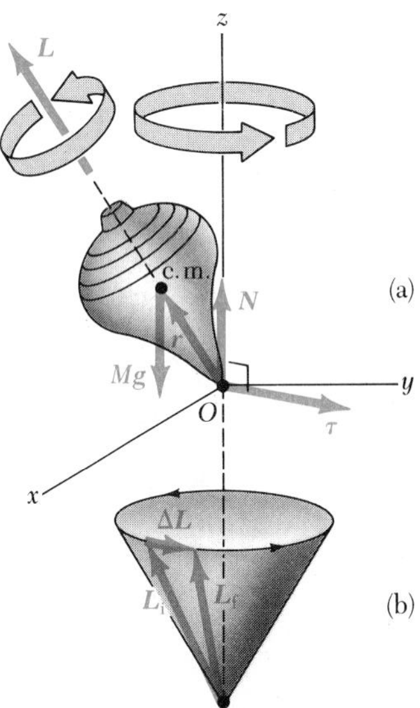

Figure 11.18 Precessional motion of a top spinning about its axis of symmetry. The only external forces acting on the top are the normal force, N, and the force of gravity, Mg. The direction of the angular momentum, L, is along the axis of symmetry.

We have presented a rather qualitative description of the motion of a top. In general, the motion of such an object is very complex. However, the essential features of the motion can be illustrated by considering the simple gyroscope shown in Fig. 11.19. This device consists of a wheel free to spin about an axle that is pivoted at a distance h from the center of mass of the wheel. If the wheel is given an angular velocity ω about its axis, the wheel will have a spin angular momentum $L = I\omega$ directed along the axle as shown. Let us consider the torque acting on the wheel about the pivot O. Again, the force, N, of the support on the axle produces no torque about O. On the other hand, the weight Mg produces a torque of magnitude Mgh about O. The direction of this torque is *perpendicular* to the axle (and perpendicular to L), as described in Figure 11.19. This torque causes the angular momentum to change in the direction perpendicular to the axle. Hence, the axle moves in the direction of the torque, that is, in the horizontal plane. There is an assumption that we must make in order to simplify the description of the system. The *total* angular

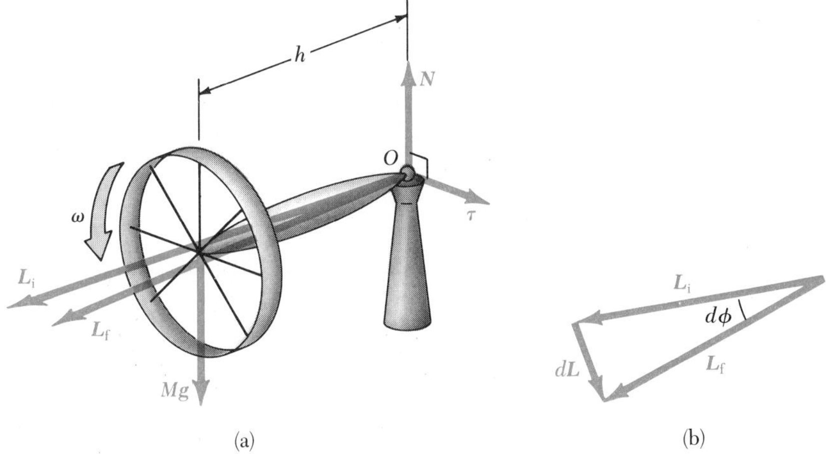

Figure 11.19 The motion of a simple gyroscope pivoted a distance h from its center of gravity. Note that the weight Mg produces a torque about the pivot that is perpendicular to the axle. This results in a change in angular momentum dL in the direction perpendicular to the axle. The axle sweeps out an angle $d\phi$ in a time dt.

momentum of the precessing wheel is actually the sum of the spin angular momentum, $I\omega$, and the angular momentum due to the motion of the center of mass about the pivot. In our treatment, we shall neglect the contribution from the center of mass motion and take the total angular momentum to be just $I\omega$. In practice, this is a good approximation if ω is made very large.

In a time dt the torque due to the weight force adds to the system an *additional* angular momentum equal to $dL = \tau\, dt$ in the direction perpendicular to $\mathbf{L}$. This additional angular momentum, $\tau\, dt$, when added vectorially to the original spin angular momentum, $I\omega$, *causes a shift in the direction of the total angular momentum.* We can express the magnitude of this change in angular momentum as

$$dL = \tau\, dt = (Mgh)\, dt$$

The vector diagram in Fig. 11.19 shows that in the time dt, the angular momentum vector rotates through an angle $d\phi$, which is also the angle through which the axle rotates. From the vector triangle formed by the vectors $\mathbf{L}_i$, $\mathbf{L}_f$, and $d\mathbf{L}$ and from the expression above, we see that

$$d\phi = \frac{dL}{L} = \frac{(Mgh)\, dt}{L}$$

Using $L = I\omega$, we find that the rate at which the axle rotates about the vertical axis is given by

Precessional frequency

$$\omega_p = \frac{d\phi}{dt} = \frac{Mgh}{I\omega} \tag{11.26}$$

The angular frequency ω_p is called the **precessional frequency.** You should note that this result is valid only when $\omega_p \ll \omega$. Otherwise, a much more complicated motion is involved. As you can see from Equation 11.26, the condition that $\omega_p \ll \omega$ is met when $I\omega$ is large compared with Mgh. Furthermore, note that the precessional frequency decreases as ω increases, that is as the wheel spins faster about its axis of symmetry.

OMIT

*11.7 ANGULAR MOMENTUM AS A FUNDAMENTAL QUANTITY

We have seen that the concept of angular momentum is very useful for describing the motion of macroscopic systems. However, the concept is also valid on a submicroscopic scale and has been used extensively in the development of modern theories of atomic, molecular, and nuclear physics. In these developments, it was found that the angular momentum of a system is a *fundamental* quantity. The word *fundamental* in this context implies that angular momentum is an inherent property of atoms, molecules, and their constituents.

In order to explain the results of a variety of experiments on atomic and molecular systems, it is necessary to assign discrete values to the angular momentum. These discrete values are some multiple of a fundamental unit of angular momentum, which equals $\hbar = h/2\pi$, where h is Planck's constant.

$$\text{Fundamental unit of angular momentum} = \hbar = 1.054 \times 10^{-34}\ \frac{\text{kg} \cdot \text{m}^2}{\text{s}^2}$$

Let us accept this postulate for the time being and show how it can be used to estimate the rotational frequency of a diatomic molecule. Consider the O_2

molecule as a rigid rotor, that is, two atoms separated by a fixed distance d and rotating about the center of mass (Fig. 11.20). Equating the rotational angular momentum to the fundamental unit $\hbar$, we can estimate the lowest rotational frequency:

$$I_c\omega \approx \hbar \qquad \text{or} \qquad \omega \approx \frac{\hbar}{I_c}$$

In Example 10.3, we found that the moment of inertia of the O_2 molecule about this axis of rotation is 2.03×10^{-46} kg · m². Therefore,

$$\omega \approx \frac{\hbar}{I_c} = \frac{1.054 \times 10^{-34}\ \text{kg} \cdot \text{m}^2/\text{s}}{2.03 \times 10^{-46}\ \text{kg} \cdot \text{m}^2} = 5.19 \times 10^{11}\ \text{rad/s}$$

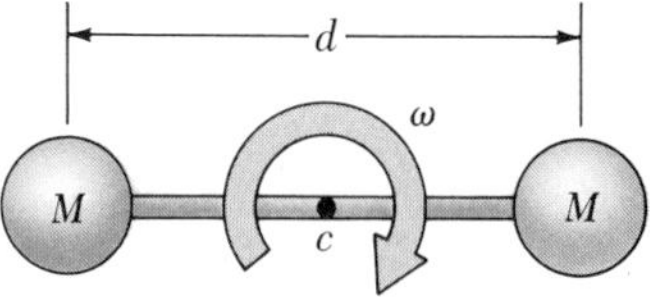

Figure 11.20 The rigid-rotor model of the diatomic molecule. The rotation occurs about the center of mass in the plane of the diagram.

This result is in good agreement with measured rotational frequencies. Furthermore, the rotational frequencies are much lower than the vibrational frequencies of the molecule, which are typically of the order of 10^{13} Hz.

This simple example shows that certain classical concepts and mechanical models might be useful in describing some features of atomic and molecular systems. However, a wide variety of phenomena on the submicroscopic scale can be explained only if one assumes discrete values of the angular momentum associated with a particular type of motion.

Historically, the Danish physicist Niels Bohr (1885–1962) was the first to suggest this radical idea in his theory of the hydrogen atom. Strictly classical models were unsuccessful in describing many properties of the hydrogen atom, such as the fact that the atom absorbs and emits radiation at discrete frequencies. Bohr postulated that the electron could only occupy circular orbits about the proton for which the orbital angular momentum was equal to $n\hbar$, where n is an integer. From this rather simple model, one can estimate the rotational frequencies of the electron in the various orbits (Problem 32).

Although Bohr's theory provided some insight concerning the behavior of matter at the atomic level, it is basically incorrect. Subsequent developments in quantum mechanics from 1924 to 1930 provided models and interpretations that are still accepted. We shall discuss this further in Chapter 40.

Later developments in atomic physics indicated that the electron also possesses another kind of angular momentum, called *spin*, which is also an inherent property of the electron. The spin angular momentum is also restricted to discrete values. We shall return to this important property later in the text and discuss its great impact on modern physical science.

11.8 SUMMARY

The **torque** $\boldsymbol{\tau}$ due to a force $\boldsymbol{F}$ about an origin in an inertial frame is defined to be

$$\boldsymbol{\tau} \equiv \boldsymbol{r} \times \boldsymbol{F} \tag{11.1}$$

Torque

Given two vectors $\boldsymbol{A}$ and $\boldsymbol{B}$, their **cross product** $\boldsymbol{A} \times \boldsymbol{B}$ is a vector $\boldsymbol{C}$ having a magnitude

$$C \equiv |AB \sin \theta| \tag{11.3}$$

Magnitude of the cross product

where θ is the angle included between $\boldsymbol{A}$ and $\boldsymbol{B}$. The direction of the vector $\boldsymbol{C} = \boldsymbol{A} \times \boldsymbol{B}$ is perpendicular to the plane formed by $\boldsymbol{A}$ and $\boldsymbol{B}$, and its sense is determined by the right-hand rule. Some properties of the cross product include the facts that $\boldsymbol{A} \times \boldsymbol{B} = -\boldsymbol{B} \times \boldsymbol{A}$ and $\boldsymbol{A} \times \boldsymbol{A} = 0$.

The **angular momentum** L of a particle of linear momentum $p = mv$ is given by

Angular momentum of a particle

$$L = r \times p = mr \times v \tag{11.9}$$

where r is the vector position of the particle relative to an origin in an inertial frame. If ϕ is the angle between r and p, the magnitude of L is given by

$$L = mvr \sin \phi \tag{11.10}$$

The **net external torque** acting on a particle or rigid body is equal to the time rate of change of its angular momentum:

$$\sum \tau_{ext} = \frac{dL}{dt} \tag{11.14}$$

The *z component of angular momentum* of a rigid body rotating about a fixed axis (the z axis) is given by

Angular momentum of a rigid body about a fixed axis

$$L_z = I\omega \tag{11.15}$$

where I is the moment of inertia about the axis of rotation, and ω is its angular velocity.

The *net external torque* acting on a rigid body equals the product of its moment of inertia about the axis of rotation and its angular acceleration:

$$\sum \tau_{ext} = I\alpha \tag{11.17}$$

If the net external torque acting on a system is zero, the total angular momentum of the system is constant. Applying this **conservation of angular momentum** law to a body whose moment of inertia changes gives

Conservation of angular momentum

$$I_i\omega_i = I_f\omega_f = \text{constant} \tag{11.19}$$

The **total kinetic energy** of a rigid body, such as a cylinder, that is rolling on a rough surface without slipping equals the rotational kinetic energy about its center of mass, $\frac{1}{2}I_c\omega^2$, plus the translational kinetic energy of the center of mass, $\frac{1}{2}Mv_c{}^2$:

Total kinetic energy of a rolling body

$$K = \tfrac{1}{2}I_c\omega^2 + \tfrac{1}{2}Mv_c{}^2 \tag{11.23}$$

In this expression, v_c is the velocity of the center of mass and $v_c = R\omega$ for pure rolling motion.

QUESTIONS

1. Is it possible to calculate the torque acting on a rigid body without specifying the origin? Is the torque independent of the location of the origin?
2. Is the triple product defined by $A \cdot (B \times C)$ equal to a scalar or vector quantity? Note that the operation $(A \cdot B) \times C$ has no meaning. Explain.
3. In the expression for torque, $\tau = r \times F$, is r equal to the moment arm? Explain.
4. If a particle moves in a straight line, is its angular momentum zero with respect to an arbitrary origin? Is its angular momentum zero with respect to any specific origin? Explain.
5. If the linear velocity of a particle is constant in time, can its angular momentum vary in time about an arbitrary origin?

article about an *arbitrary* say about its angular mo-

t line, and you are told zero about some unspe- sarily imply that the net an you conclude that its

tor of a particle is com- ou conclude about the entum vector with re- on?

rigid body is nonzero ny other origin about

otion, is it possible for to be zero about some

y that it does not spin mean that the angular bitrary origin? Explain.

balance on a moving st?

istance from a bellhop . When the unaware ing the suitcase, it sud- or some unknown rea- bellhop dropped the u suppose might have

rizontal surface as in oints on the cylinder ponent of velocity at hey?

es—a solid sphere, a inder—are placed at 1). If they all are re- leased from rest at the same elevation and roll without slipping, which reaches the bottom first? Which reaches last? You should try this at home and note that the result is *independent* of the masses and radii. This is quite amazing!

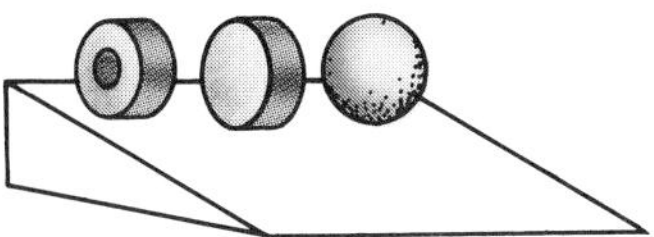

Figure 11.21 (Question 15) Which object wins the race?

16. A mouse is initially at rest on a horizontal turntable mounted on a frictionless vertical axle. If the mouse begins to walk around the perimeter, what happens to the turntable? Explain.
17. Stars originate as large bodies of slowly rotating gas. Because of gravity, these clumps of gas slowly decrease in size. What happens to the angular velocity of a star as it shrinks? Explain.
18. Use the principle of conservation of angular momentum to form a hypothesis that explains how a cat can always land on its feet regardless of the position from which it is dropped.
19. Often when a high diver wants to turn a flip in midair, she will draw her legs up against her chest. Why does this make her rotate faster? What should she do when she wants to come out of her flip?
20. As a tether ball winds around a pole, what happens to its angular velocity? Explain.
21. Space colonies have been proposed that consist of large cylinders placed in space. Gravity would be simulated in these cylinders by setting them into rotation about their long axis. Discuss the difficulties that would be encountered in attempting to set the cylinders into rotation.

---

nd Torque

$3\boldsymbol{i} + \boldsymbol{j}$ and $\boldsymbol{B} = \boldsymbol{i} - 2\boldsymbol{j}$.

between **A** and **B**.

product, verify Equa- he vector product of

$= 3\boldsymbol{j}$ and $\boldsymbol{B} = 2\boldsymbol{i} + 2\boldsymbol{j}$, (c) $\boldsymbol{A} = 3\boldsymbol{j} + \boldsymbol{k}$ and

4. Vector **A** is in the negative y direction, and vector **B** is in the negative x direction. What is the direction of (a) $\boldsymbol{A} \times \boldsymbol{B}$ and (b) $\boldsymbol{B} \times \boldsymbol{A}$?
5. A particle is located at the vector position $\boldsymbol{r} = (2\boldsymbol{i} + 4\boldsymbol{j})$ m, and the force acting on it is $\boldsymbol{F} = (3\boldsymbol{i} + \boldsymbol{j})$ N. What is the torque about (a) the origin and (b) the point having coordinates (0, 6) m?
6. If $|\boldsymbol{A} \times \boldsymbol{B}| = \boldsymbol{A} \cdot \boldsymbol{B}$, what is the angle between **A** and **B**?
7. Verify Equation 11.8 for the cross product of any two vectors **A** and **B**, and show that the cross product may be written in the following determinant form:

$$\boldsymbol{A} \times \boldsymbol{B} = \begin{vmatrix} \boldsymbol{i} & \boldsymbol{j} & \boldsymbol{k} \\ A_x & A_y & A_z \\ B_x & B_y & B_z \end{vmatrix}$$

8. The vectors **A** and **B** form two sides of a parallelogram. (a) Show that the area of the parallelogram is given by $|\boldsymbol{A} \times \boldsymbol{B}|$. (b) If $\boldsymbol{A} = (3\boldsymbol{i} + 3\boldsymbol{j})$ m and $\boldsymbol{B} = (\boldsymbol{i} - 2\boldsymbol{j})$ m, find the area of the parallelogram.

Section 11.2 Angular Momentum of a Particle

9. A light rigid rod 1 m in length rotates in the xy plane about a pivot through the rod's center. Two particles of mass 2 kg and 3 kg are connected to its ends (Fig. 11.22). Determine the angular momentum of the system about the origin at the instant the speed of each particle is 5 m/s.

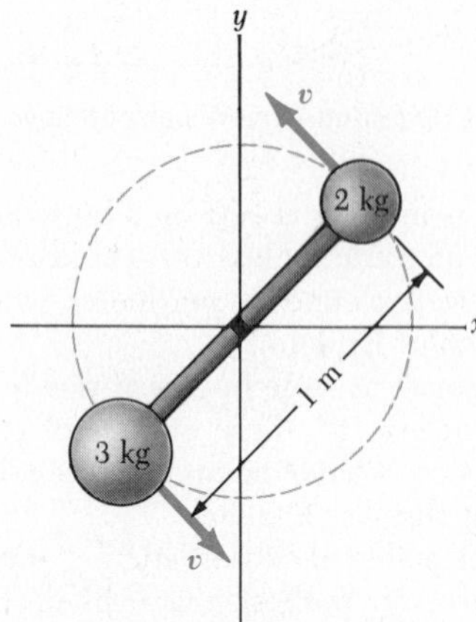

Figure 11.22 (Problem 9).

10. A particle of mass 0.3 kg moves in the xy plane. At the instant its coordinates are (2, 4) m, its velocity is given by $(3\mathbf{i} + 2\mathbf{j})$ m/s. At this instant, determine the angular momentum of the particle relative to the origin.
11. A 4-kg particle moves in the xy plane with a constant speed of 2 m/s in the x direction along the line $y = -3$ m. What is its angular momentum relative to (a) the origin and (b) the point $(0, -5)$ m?
12. Two particles move in opposite directions along a straight line (Fig. 11.23). The particle of mass m moves to the right with a speed v while the particle of mass $3m$ moves to the left with a speed v. What is the *total* angular momentum of the system relative to (a) the point A, (b) the point O, and (c) the point B?

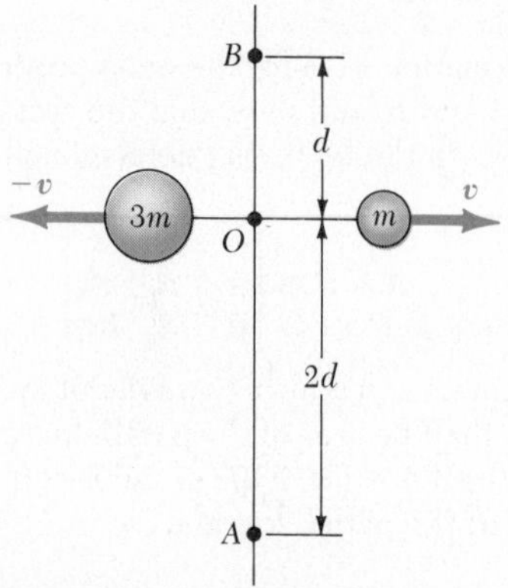

Figure 11.23 (Problem 12).

13. A particle of mass m moves in a straight line with a constant velocity $\mathbf{v} = v\mathbf{j}$ along the positive y axis. Determine the angular momentum of the particle (both magnitude and direction) relative to (a) the point having coordinates $(-d, 0)$, (b) the point having coordinates $(2d, 0)$, and (c) the origin.
14. An airplane of mass 5000 kg flies level to the ground at an altitude of 8 km with a constant speed of 200 m/s relative to the earth. (a) What is the magnitude of the airplane's angular momentum relative to a ground observer who is directly below the airplane? (b) Does this value change as the airplane continues its motion along a straight line?
15. A particle of mass m is given a velocity $-v_0\mathbf{j}$ at the point $(-d, 0)$ and proceeds to accelerate in the presence of earth's gravity (Fig. 11.24). (a) Find an expression for the angular momentum as a function of time with respect to the origin. (b) Calculate the torque acting on the particle at any time relative to the origin. (c) Using your results to (a) and (b), verify that $\tau = dL/dt$.

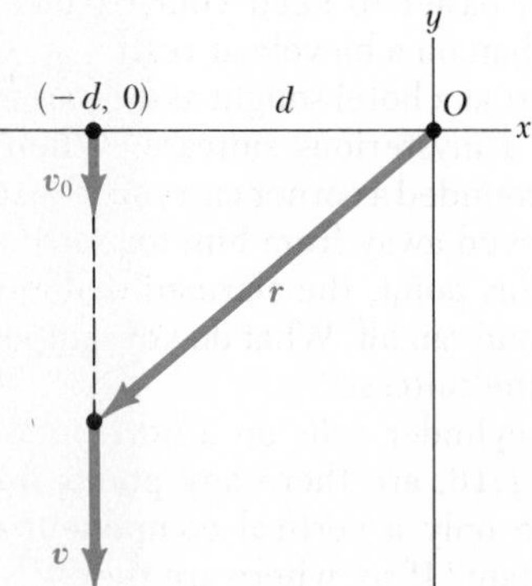

Figure 11.24 (Problem 15).

16. (a) Calculate the angular momentum of the earth due to its spinning motion about its axis. (b) Calculate the angular momentum of the earth due to its orbital motion about the sun and compare this with (a). (Assume that the earth is a homogeneous sphere of radius 6.37×10^6 m and mass 5.98×10^{24} kg. Take the earth-sun distance to be 1.49×10^{11} m.)
17. A 3-kg mass is attached to a light cord, which is wound around a pulley (Fig. 10.19). The pulley is a uniform solid cylinder of radius 8 cm and mass 1 kg. (a) What is the net torque on the system about the point O? (b) When the 3-kg mass has a speed v, the pulley has an angular velocity $\omega = v/R$. Determine the total angular momentum of the system about O. (c) Using the fact that $\tau = dL/dt$ and your result from (b), calculate the acceleration of the 3-kg mass.

Section 11.3 Rotation of a Rigid Body About a Fixed Axis

18. A uniform solid disk of mass 3 kg and radius 0.2 m rotates about a fixed axis perpendicular to its face. If the angular frequency of rotation is 5 rad/s, calculate

the angular momentum of the disk when the axis of rotation (a) passes through its center of mass and (b) passes through a point midway between the center and the rim.

19. A particle of mass 0.3 kg is attached to the 100-cm mark of a meter stick of mass 0.2 kg. The meter stick rotates on a horizontal, smooth table with an angular velocity of 4 rad/s. Calculate the angular momentum of the system if the stick is pivoted about an axis (a) perpendicular to the table through the 50-cm mark and (b) perpendicular to the table through the 0-cm mark.

Section 11.4 Conservation of Angular Momentum

20. A cylinder with moment of inertia I_1 rotates with angular velocity ω_0 about a vertical, frictionless axle. A second cylinder, with moment of inertia I_2 initially not rotating, drops onto the first cylinder (Fig. 11.25). Since the surfaces are rough, the two eventually reach the same angular velocity ω. (a) Calculate ω. (b) Show that energy is lost in this situation and calculate the ratio of the final to the initial kinetic energy.

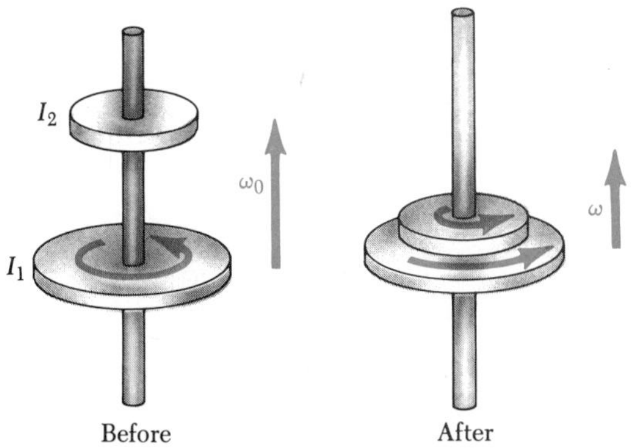

Figure 11.25 (Problem 20).

21. A uniform solid cylinder of mass 1 kg and radius 25 cm rotates about a fixed vertical, frictionless axle with an angular speed of 10 rad/s. A 0.5-kg piece of putty is dropped vertically onto the cylinder at a point 15 cm from the axle. If the putty sticks to the cylinder, calculate the final angular speed of the system. (Assume the putty is a particle.)

22. A uniform rod of mass 100 g and length 50 cm rotates in a horizontal plane about a fixed, vertical, frictionless pin through its center. Two small beads, each of mass 30 g, are mounted on the rod such that they are able to slide without friction along its length. Initially the beads are held by catches at positions 10 cm on each side of center, at which time the system rotates at an angular speed of 20 rad/s. Suddenly, the catches are released and the small beads slide outward along the rod. (a) Find the angular speed of the system at the instant the beads reach the ends of the rod. (b) Find the angular speed of the rod after the beads fly off the ends.

23. A woman whose mass is 70 kg stands at the rim of a horizontal turntable having a moment of inertia of 500 kg · m^2 and a radius of 2 m. The system is initially at rest, and the turntable is free to rotate about a frictionless, vertical axle through its center. The woman then starts walking around the rim in a clockwise direction at a constant speed of 1.5 m/s relative to the earth. (a) In what direction and with what angular speed does the turntable rotate? (b) How much work does the woman do to set the system into motion?

24. A bullet of mass 10 g is shot *through* a door initially at rest. The moment of inertia of the door is 4 kg · m^2 about an axis through its hinges. The bullet is fired perpendicular to the door with an initial velocity of 400 m/s, and the angular speed of the door after the collision is 0.3 rad/s. If the bullet passes through the door 0.4 m from the hinge, find (a) the final speed of the bullet and (b) the loss in mechanical energy.

25. The student in Figure 11.12 holds two weights, each of mass 10 kg. When his arms are extended horizontally, the weights are 1 m from the axis of rotation and he rotates with an angular speed of 3 rad/s. The moment of inertia of the student plus the stool is 8 kg · m^2 and is assumed to be constant. If the student pulls the weights horizontally to 0.3 m from the rotation axis, calculate (a) the final angular speed of the system and (b) the change in the mechanical energy of the system.

26. A particle of mass $m = 10$ g and speed $v_0 = 5$ m/s collides with and sticks to the edge of a uniform solid sphere of mass $M = 1$ kg and radius $R = 20$ cm (Fig. 11.26). If the sphere is initially at rest and is pivoted about a frictionless axle through O perpendicular to the plane, (a) find the angular velocity of the system after the collision and (b) determine how much energy is lost in the collision.

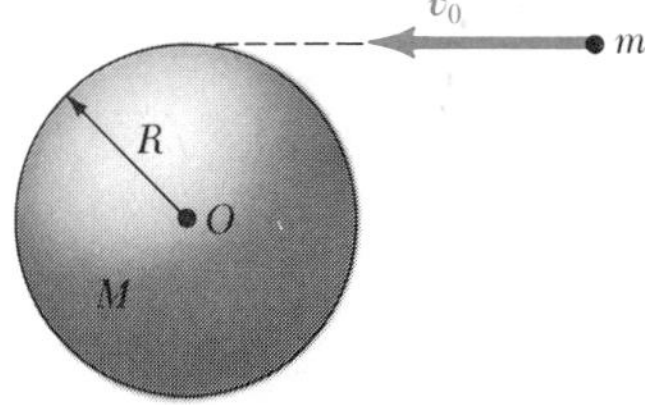

Figure 11.26 (Problem 26).

*Section 11.5 Rolling Motion of a Rigid Body

27. (a) Determine the acceleration of the center of mass of a uniform solid disk rolling down an incline and compare this acceleration with that of a uniform hoop. (b) What is the minimum coefficient of friction required to maintain pure rolling motion for the disk?

28. A solid sphere has a radius of 0.2 m and a mass of 150 kg. How much work is required to get the sphere rolling with an angular speed of 50 rad/s on a horizontal surface? (Assume the sphere starts from rest and rolls without slipping.)

29. A cylinder of mass 10 kg rolls without slipping on a rough surface. At the instant its center of mass has a speed of 10 m/s, determine (a) the translational kinetic energy of its center of mass, (b) the rotational kinetic energy about its center of mass, and (c) its total kinetic energy.

30. A uniform solid disk and a uniform hoop are placed side by side at the top of a rough incline of height h. If they are released from rest and roll without slipping, determine their velocities when they reach the bottom. Which object reaches the bottom first?

31. A spherical shell rolls down a rough incline of height h and angle θ (Fig. 11.16). (a) If the shell is released from rest at the top of the incline, what is the velocity of its center of mass when it reaches the bottom? (b) Calculate the acceleration of its center of mass. Compare your results with those for a solid sphere determined in Example 11.11.

*Section 11.7 Angular Momentum as a Fundamental Quantity

32. In the Bohr model of the hydrogen atom, the electron moves in a circular orbit of radius 0.529×10^{-10} m around the proton. Assuming the orbital angular momentum of the electron is equal to $\hbar$, calculate (a) the orbital speed of the electron, (b) the kinetic energy of the electron, and (c) the angular frequency of the electron's motion.

GENERAL PROBLEMS

33. A uniform solid sphere of radius r is placed on the inside surface of a hemispherical bowl of radius R. The sphere is released from rest at an angle θ to the vertical and rolls without slipping (Fig. 11.27). Determine the angular speed of the sphere when it reaches the bottom of the bowl.

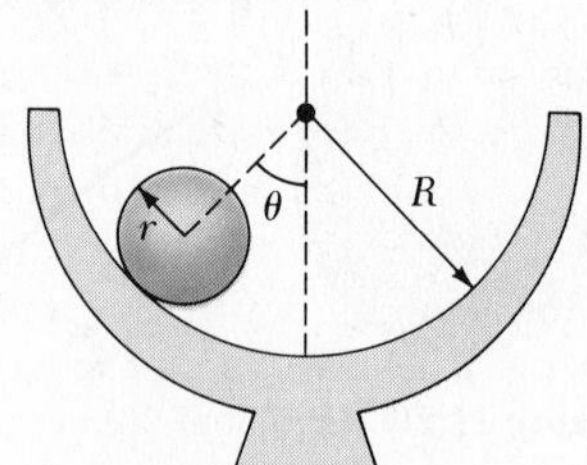

Figure 11.27 (Problem 33).

34. A smooth cube of mass m and side length r slides with speed v on a horizontal surface with negligible friction. The cube then moves up a smooth incline that makes an angle θ with the horizontal. A cylinder of mass m and radius r rolls without slipping with its center of mass moving with speed v and encounters an incline of the same angle of inclination but of nonnegligible friction. (a) Which object will go the greater distance up the incline? (b) Find the difference between the maximum distances the objects travel up the incline. (c) Explain what accounts for this difference in distances travelled.

35. A thin uniform cylindrical turntable of radius 2 m and mass 30 kg rotates in a horizontal plane with an initial angular velocity, 4π rad/s. The turntable bearing is frictionless. A small clump of clay of mass 2.5 kg is dropped onto the turntable and sticks at a point 1.8 m from the center of rotation. (a) Find the final angular velocity of the clay and turntable. (Treat the clay as a point mass.) (b) Is mechanical energy conserved in this collision? Explain and use numerical results to verify your answer.

36. A string is wound around a uniform disk of radius R and mass M. The disk is released from rest with the string vertical and its top end tied to a fixed support (Fig. 11.28). As the disk descends, show that (a) the tension in the string is one third the weight of the disk, (b) the acceleration of the center of mass is $2g/3$, and (c) the velocity of the center of mass is $(4gh/3)^{1/2}$. Verify your result to (c) using the energy approach.

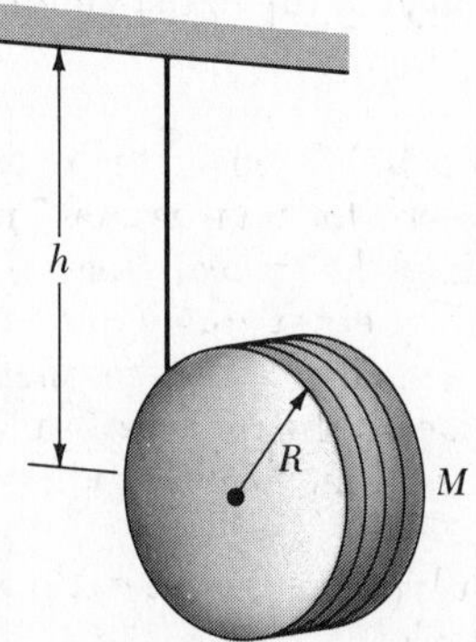

Figure 11.28 (Problem 36).

37. A constant horizontal force F is applied to a lawn roller in the form of a uniform solid cylinder of radius R and mass M (Fig. 11.29). If the roller rolls without slipping on the horizontal surface, show that (a) the acceleration of the center of mass is $2F/3M$ and (b) the minimum coefficient of friction necessary to prevent slipping is $F/3Mg$. (*Hint:* Take the torque with respect to the center of mass.)

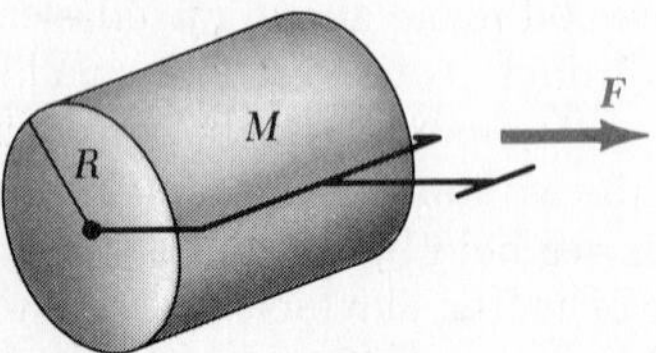

Figure 11.29 (Problem 37).

38. A light rope passes over a light, frictionless pulley. One end is fastened to a bunch of bananas of mass M, and a monkey of mass M clings to the other end of the

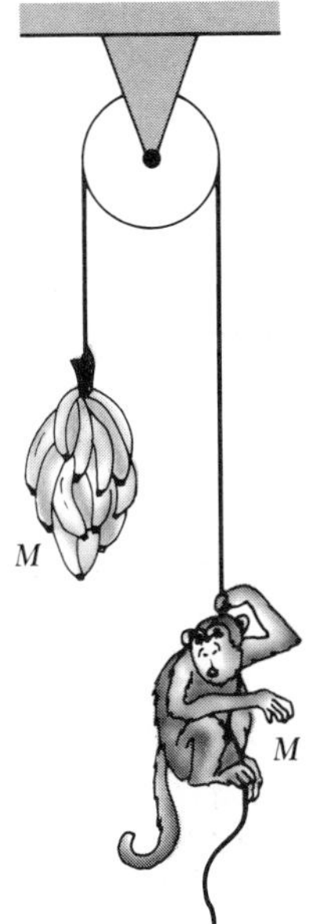

Figure 11.30 (Problem 38).

rope (Fig. 11.30). The monkey climbs the rope in an attempt to reach the bananas. (a) Treating the system as consisting of the monkey, bananas, rope, and pulley, evaluate the net torque about the pulley axis. (b) Using the results to (a), determine the total angular momentum about the pulley axis and describe the motion of the system. Will the monkey reach the bananas?

39. A small, solid sphere of mass m and radius r rolls without slipping along the track shown in Figure 11.31. If it starts from rest at the top of the track, (a) what is the minimum value of h (in terms of the radius of the loop R) such that the sphere completes the loop? (b) What are the force components on the sphere at the point P if $h = 3R$?

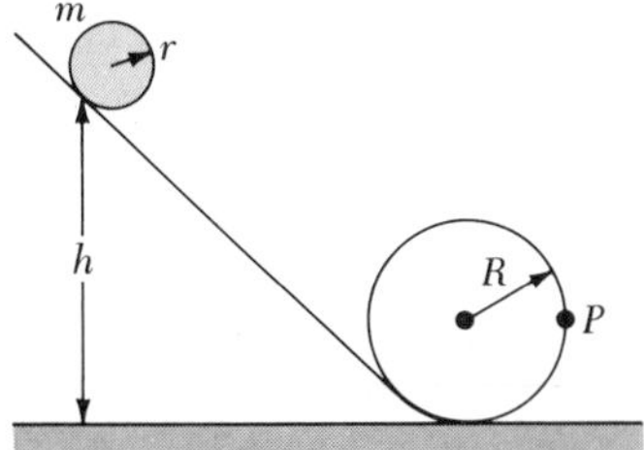

Figure 11.31 (Problem 39).

40. A spool of wire of mass M and radius R is unwound under a constant force $\mathbf{F}$ (Fig. 11.32). Assuming the spool is a uniform solid cylinder that *doesn't slip*, show that (a) the acceleration of the center of mass is $4\mathbf{F}/3M$ and (b) the force of friction is to the *right* and equal to $\mathbf{F}/3$.

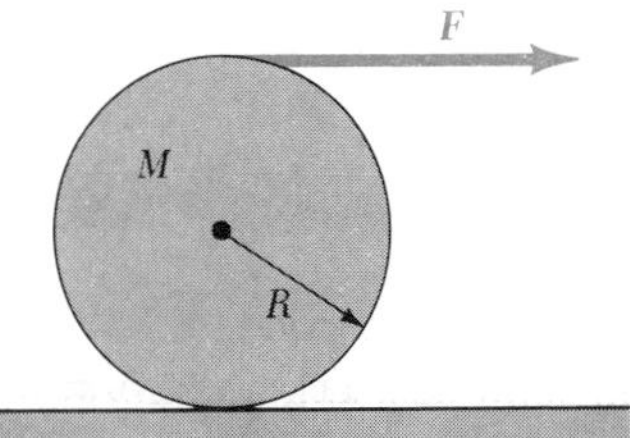

Figure 11.32 (Problems 40 and 41).

41. If the cylinder in Fig. 11.32 starts from rest and rolls without slipping, what is the velocity of its center of mass after it has rolled through a distance d? (Assume the force remains constant.)

42. Consider the problem of the solid sphere rolling down an incline as described in Example 11.11. (a) Choose the axis of the origin for the torque equation as the instantaneous axis through the contact point P and show that the acceleration of the center of mass is given by $a_c = \frac{5}{7}g \sin \theta$. (b) Show that the *minimum* coefficient of friction such that the sphere will roll without slipping is given by $\mu_{\min} = \frac{2}{7}\tan \theta$.

43. A uniform solid disk is set into rotation about an axis through its center with an angular velocity ω_0. The rotating disk is lowered to a *rough*, horizontal surface with this angular velocity and released. (a) What is the angular velocity of the disk once pure rolling takes place? (b) Find the fractional loss in kinetic energy from the time the disk is released until pure rolling occurs. (*Hint:* Angular momentum is conserved about an axis through the point of contact.)

44. A particle of mass m is located at the vector position $\mathbf{r}$ and has a linear momentum $\mathbf{p}$. (a) If $\mathbf{r}$ and $\mathbf{p}$ both have nonzero x, y, and z components, show that the angular momentum of the particle relative to the origin has components given by $L_x = yp_z - zp_y$, $L_y = zp_x - xp_z$, and $L_z = xp_y - yp_x$. (b) If the particle moves only in the xy plane, prove that $L_x = L_y = 0$ and $L_z \neq 0$.

45. A force $\mathbf{F}$ acts on the particle described in Problem 44. (a) Find the components of the torque acting on the particle about the origin when the particle is located at the position $\mathbf{r}$ and the force has three components. (b) From this result, show that if the particle moves in the xy plane and the force has only x and y components, the torque (and angular momentum) must be in the z direction.

•46. A large, cylindrical roll of tissue paper of initial radius R lies on a long, horizontal surface with the open end of the paper nailed to the surface so that it can unroll easily. The roll is given a *slight* shove ($v_0 \approx 0$) and commences to unroll. (a) Determine the speed of the center of mass of the roll when its radius has diminished to r. (b) Calculate a numerical value for this speed at $r = 1$ mm, assuming $R = 6$ m. (c) What happens to the energy of the system when the paper is completely unrolled? (*Hint:* Assume the roll has a uniform density and apply energy methods.)

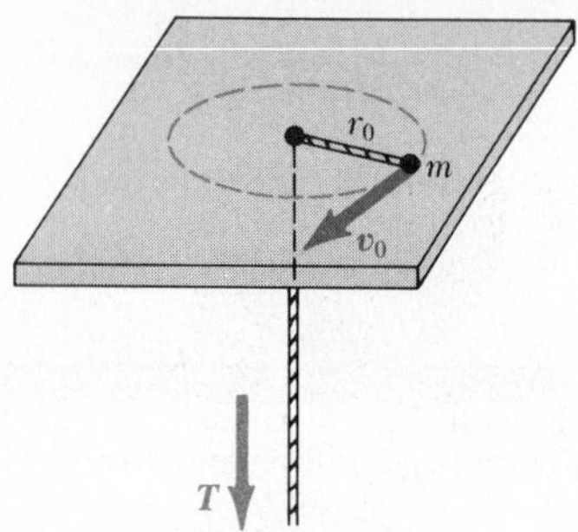

Figure 11.33 (Problem 47).

•**47.** A mass m is attached to a cord passing through a small hole in a frictionless, horizontal surface (Fig. 11.33). The mass is initially orbiting in a circle of radius r_0 with velocity v_0. The cord is then slowly pulled from below, decreasing the radius of the circle to r. (a) What is the velocity of the mass when the radius is r? (b) Find the tension in the cord as a function of r. (c) How much work is done in moving m from r_0 to r? (*Note:* The tension depends on r.) (d) Obtain numerical values for v, T, and W when $r = 0.1$ m, if $m = 50$ g, $r_0 = 0.3$ m, and $v_0 = 1.5$ m/s.

•**48.** A bowling ball is given an initial speed v_0 on an alley such that it *initially slides without rolling.* The coefficient of friction between the ball and the alley is μ. At the time *pure rolling motion occurs,* show that (a) the velocity of the ball's center of mass is $5v_0/7$ and (b) the distance it has traveled is $12v_0{}^2/49\ \mu g$. (*Hint:* When pure rolling motion occurs, $v_c = R\omega$ and $\alpha = a_c/R$. Since the frictional force provides the deceleration, from Newton's second law it follows that $a_c = -\mu g$.)

•**49.** A trailer with loaded weight W is being pulled by a vehicle with a force F, as in Figure 11.34. The trailer is loaded such that its center of gravity is located as shown. Neglect the force of rolling friction and assume the trailer has an acceleration a. (a) Find the vertical component of F in terms of the given parameters. (b) If $a = 2$ m/s^2 and $h = 1.5$ m, what must be the value of d in order that $F_y = 0$ (no vertical load on the vehicle)? (c) Find the values of F_x and F_y given that $W = 1500$ N, $d = 0.8$ m, $L = 3$ m, $h = 1.5$ m, and $a = -2$ m/s^2.

•**50.** Suppose a solid disk of radius R is given an angular velocity ω_0 about an axis through its center and is then lowered to a rough, horizontal surface and released, as in Problem 43. Furthermore, assume that the coefficient of friction between the disk and surface is μ. (a) Show that the *time* it takes pure rolling motion to occur is given by $R\omega_0/3\ \mu g$. (b) Show that the *distance* the disk travels before pure rolling occurs is given by $R^2\omega_{02}/18\ \mu g$. (See hint in Problem 48.)

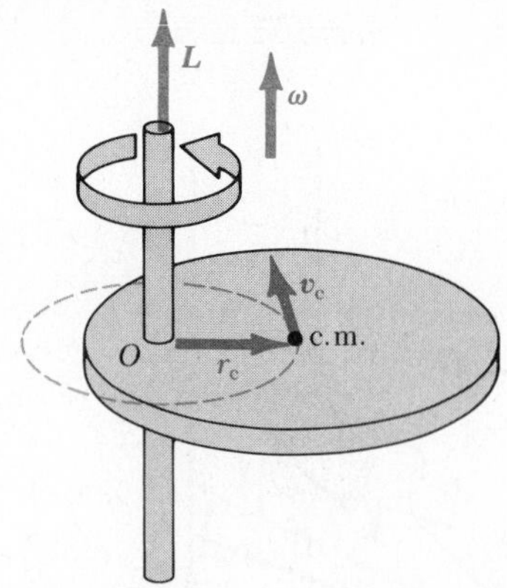

Figure 11.35 (Problem 51).

•**51.** A uniform solid disk of mass M rotates about an axis parallel to the symmetry axis through its center, as in Figure 11.35. Show that the angular momentum of the disk is given by

$$L = I_c\omega + r_c \times Mv_c$$

where I_c is the moment of inertia about the axis through its center of mass, r_c is the vector from O to the center of mass, and v_c is the velocity of the center of mass. The first term on the right side of this expression is called the *spin angular momentum* since it refers to that part of the angular momentum associated with the spin of the system about the center of mass. The second term on the right side is usually referred to as the *orbital angular momentum.* (*Hint:* Use the parallel-axis theorem.)

•**52.** A solid cube of side $2a$ and mass M is sliding on a frictionless surface with uniform velocity v_0 as in Figure 11.36a. It hits a small obstacle at the end of the table, which causes the cube to tilt as in Figure 11.36b. Find the minimum value of v_0 such that the cube falls off the table. Note that the moment of inertia of the cube about an axis along one of its edges is $8Ma^2/3$. (*Hint:* The cube undergoes an *inelastic collision* at the edge.)

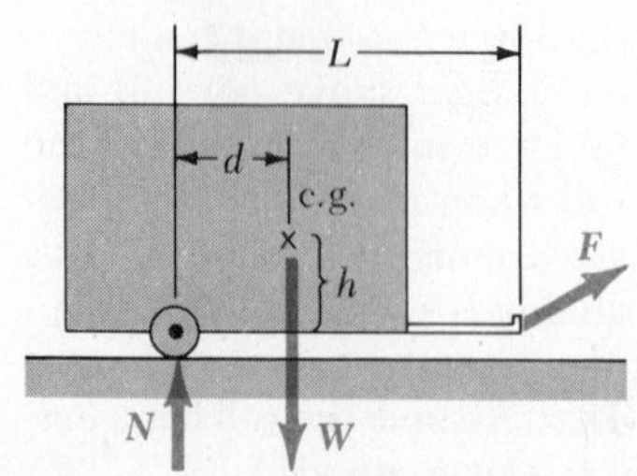

Figure 11.34 (Problem 49).

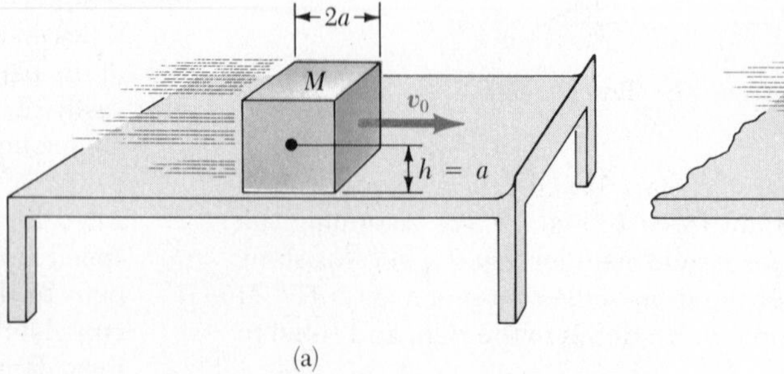

Figure 11.36 (Problem 52).

12
Static Equilibrium of a Rigid Body

This chapter is concerned with the conditions under which a rigid body is in equilibrium. The term *equilibrium* implies that the body is either at rest or that its center of mass moves with constant velocity. We shall deal with bodies at rest, or bodies in *static equilibrium.* This represents a common situation in engineering practice, and the principles involved are of special interest to civil engineers, architects, and mechanical engineers, who deal with various structural designs, such as bridges and buildings. Those of you who are engineering students will undoubtedly take an intensified course in statics in the future.

In Chapter 5 we stated that one necessary condition for equilibrium is that the net force on an object must be zero. If the object is treated as a single particle, this is the *only* condition that must be satisfied in order that the particle be in equilibrium. That is, if the net force on the particle is zero, it will remain at rest (if originally at rest) or move with constant velocity in a straight line (if originally in motion).

The situation with rigid bodies is somewhat more complex because real bodies cannot be treated as particles. A real body has a definite size, shape, and mass distribution. In order that such a body be in static equilibrium, the net force on it must be zero *and* the body must have no tendency to rotate. This second condition of equilibrium requires that *the net torque about any origin must be zero.* In order to establish whether or not a body is in equilibrium, we must know the size and shape of the body, the forces acting on different parts of the body, and the points of application of the various forces.

Rigid body

The bodies that will be treated in this chapter are assumed to be rigid. *A rigid body is defined as a body that does not deform under the application of external forces.* That is, all parts of a rigid body remain at a fixed separation with respect to each other when subjected to external forces. In reality, all bodies will deform to some extent under load conditions. Such deformations are usually small and will not affect the conditions of equilibrium. However, deformation is an important consideration in understanding the mechanics of materials, as we shall see in Chapter 15.

12.1 THE CONDITIONS OF EQUILIBRIUM OF A RIGID BODY

Consider a single force $\mathbf{F}$ acting on a rigid body that is pivoted about an axis through the point O as in Figure 12.1. The effect of the force on the body depends on its point of application, P. If $\mathbf{r}$ is the position vector of this point relative to O, *the torque associated with the force* $\mathbf{F}$ *about* O *is given by*

$$\boldsymbol{\tau} = \mathbf{r} \times \mathbf{F} \tag{12.1}$$

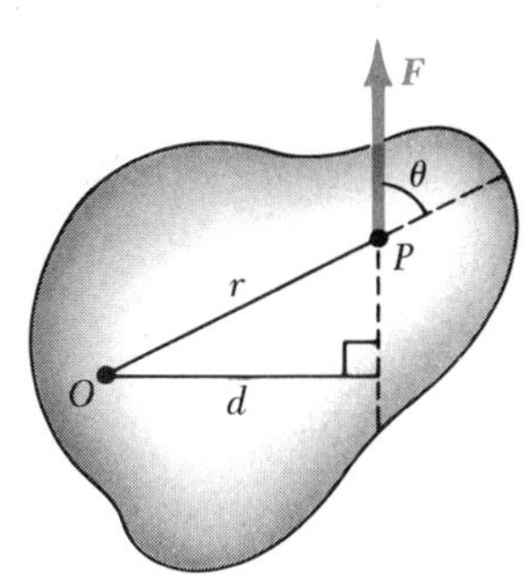

Figure 12.1 A rigid body pivoted about an axis through O. A single force $\mathbf{F}$ acts at the point P. The moment arm of $\mathbf{F}$ is the perpendicular distance d to the line of action of $\mathbf{F}$.

Recall that the vector τ is perpendicular to the plane formed by $\boldsymbol{r}$ and $\boldsymbol{F}$. Furthermore, the sense of τ is determined by the sense of the rotation that $\boldsymbol{F}$ tends to give to the body. The right-hand rule can be used to determine the direction of τ: Close your right hand such that your four fingers wrap in the direction of rotation that $\boldsymbol{F}$ tends to give the body; your thumb will point in the direction of τ. Hence, in Figure 12.1 τ is directed *out* of the paper.

As you can see from Figure 12.1, the tendency of $\boldsymbol{F}$ to make the body rotate about an axis through O depends on the moment arm d (the perpendicular distance to the line of action of the force) as well as on the magnitude of $\boldsymbol{F}$. By definition, the magnitude of τ is given by Fd.

Now suppose two forces, $\boldsymbol{F}_1$ and $\boldsymbol{F}_2$, act on a rigid body. The two forces will have the same effect on the body only if they have the same magnitude, the same direction, and the same line of action. In other words,

Equivalent forces

two forces $\boldsymbol{F}_1$ and $\boldsymbol{F}_2$ are equivalent if and only if $\boldsymbol{F}_1 = \boldsymbol{F}_2$ and if they have the same torque about any given point.

An example of two equal and opposite forces that are *not* equivalent is shown in Figure 12.2. The force directed toward the right tends to rotate the body clockwise about an axis through O, whereas the force directed toward the left tends to rotate it counterclockwise about that axis.

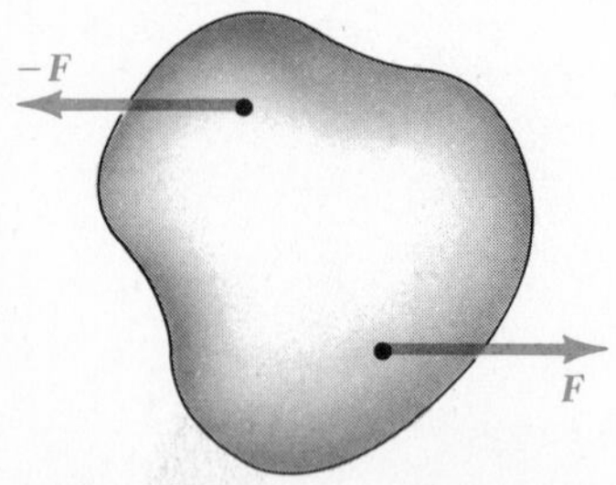

Figure 12.2 The two forces acting on the body are equal in magnitude and opposite in direction, yet the body is not in equilibrium.

When a rigid body is pivoted about an axis through its center of mass, the body will undergo an angular acceleration about this axis if there is a nonzero torque acting on the body. As an example, suppose a rigid body is pivoted about an axis through its center of mass as in Figure 12.3. Two equal and opposite forces act in the directions shown, such that their lines of action do not pass through the center of mass. Such a pair of forces acting in this manner form what is called a **couple.** Since each force produces the same torque, Fd, the net torque has a magnitude given by $2Fd$. Clearly, the body will rotate in a clockwise direction and will undergo an angular acceleration about the axis. This is a nonequilibrium situation as far as the rotational motion is concerned. That is, the "unbalanced," or net, torque on the body gives rise to an angular acceleration α according to the relationship $\tau_{\text{net}} = 2Fd = I\alpha$.

In general, a rigid body will be in rotational equilibrium only if its angular acceleration $\alpha = 0$. Since $\tau_{\text{net}} = I\alpha$ for rotation about a fixed axis, a necessary condition of equilibrium for a rigid body is that *the net torque about any origin must be zero.* We now have *two necessary conditions for equilibrium of a rigid body,* which can be stated as follows:

1. The resultant external force must equal zero.
$$\sum \boldsymbol{F} = 0 \tag{12.2}$$

2. The resultant external torque must be zero about *any* origin.
$$\sum \tau = 0 \tag{12.3}$$

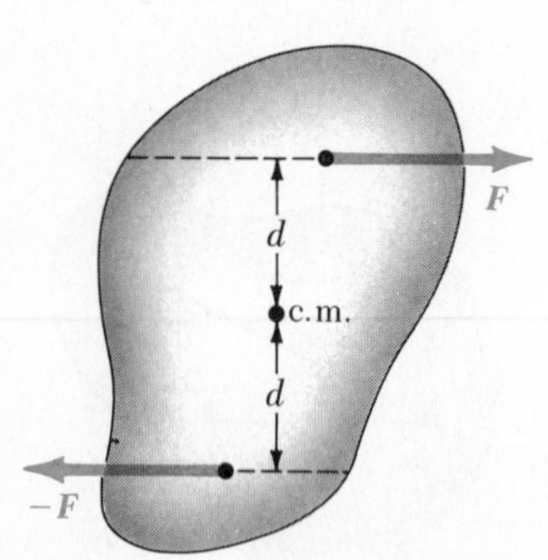

Figure 12.3 Two equal and opposite forces acting on the body form a couple. In this case, the body will rotate clockwise. The net torque on the body about the center of mass is $2Fd$.

The first condition is a statement of **translational equilibrium,** that is, the linear acceleration of the center of mass of the body must be zero when viewed from an inertial reference frame. The second condition is a statement of **rotational equilibrium,** that is, the angular acceleration about any axis must be zero. In the special case of **static equilibrium,** which is the main subject of this chapter, the body is at rest so that is has no linear or angular velocity (that is, $v_c = 0$ and $\omega = 0$).

The two vector expressions given by Equations 12.2 and 12.3 are equivalent, in general, to six scalar equations. Three of these come from the first

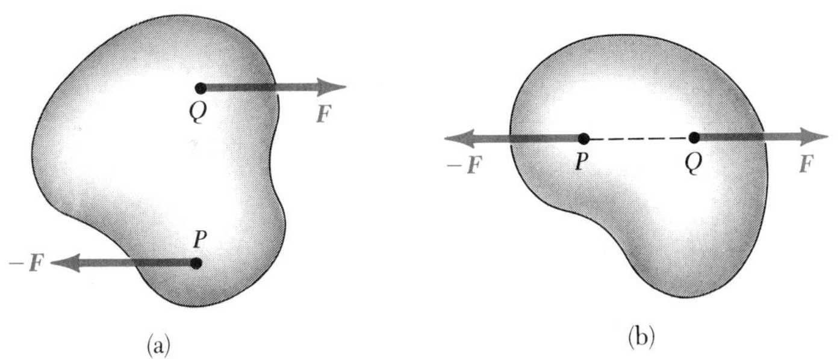

Figure 12.4 (a) The body is not in equilibrium since the two forces do not have the same line of action. (b) The body is in equilibrium since the two forces act along the same line.

condition of equilibrium, and three follow from the second condition (corresponding to x, y, and z components). Hence, in a complex system involving several forces acting in various directions, you would be faced with solving a set of linear equations with many unknowns. We will restrict our discussion to situations in which all the forces lie in a common plane, which we assume to be the xy plane. Forces whose vector representations are in the same plane are said to be *coplanar*. In this case, we shall have to deal with only *three* scalar equations. Two of these come from balancing the forces in the x and y directions. The third comes from the torque equation, namely, that the net torque about *any* point in the xy plane must be zero. Hence, these two conditions of equilibrium provide the equations

$$\sum F_x = 0 \qquad \sum F_y = 0 \qquad \sum \tau_z = 0 \tag{12.4}$$

where the axis of the torque equation is *arbitrary*, as we shall show later.

There are two cases of equilibrium that are often encountered. The first case deals with a rigid body subjected to only two forces, and the second case is concerned with a rigid body subjected to three forces.

Case I. *If a rigid body is subjected to two forces, the body is in equilibrium if and only if the two forces are equal in magnitude and opposite in direction and have the same line of action.* Figure 12.4a shows a situation in which the body is not in equilibrium because the two forces are not along the same line. Note that the torque about any axis, such as one through P, is not zero, which violates the second condition of equilibrium. In Figure 12.4b, the body is in equilibrium because the forces have the same line of action. In this situation, it is easy to see that the net torque about any axis is zero.

Case II. *If a rigid body subjected to three forces is in equilibrium, the lines of action of the three forces must intersect at a common point.* That is, the forces must be *concurrent.* (One exception to this rule is the situation in which none of the lines of action intersect. In this situation, the forces must be parallel.) Figure 12.5 illustrates the general rule. The lines of action of the three forces pass through the point S. The conditions of equilibrium require that $\mathbf{F}_1 + \mathbf{F}_2 + \mathbf{F}_3 = 0$ and that the net torque about any axis be zero. Note that as long as the forces are concurrent, the net torque about an axis through S must be zero.

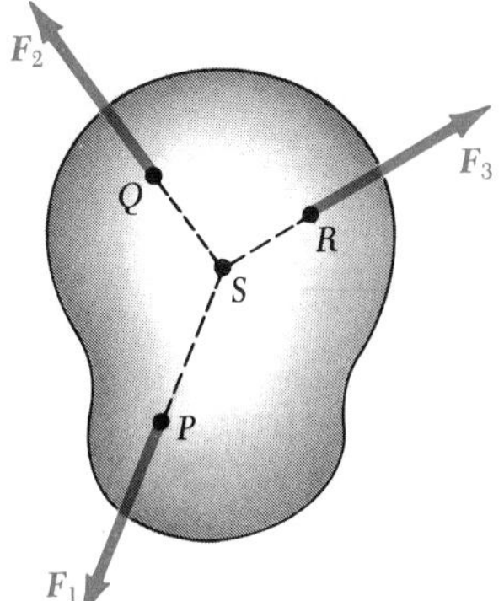

Figure 12.5 If three forces act on a body that is in equilibrium, their lines of action must intersect at a point S.

We can easily show that if a body is in translational equilibrium and the net torque is zero with respect to one point, it must be zero about *any* point. The point can be inside or outside the boundaries of the object. Consider a body under the action of several forces such that the resultant force $\Sigma \mathbf{F} = \mathbf{F}_1 +$

Essay

ARCH STRUCTURES

Gordon Batson
Clarkson University, Potsdam, N.Y.

Of all structures built for various utilitarian purposes, a bridge and its structural components are the most visible. The load-carrying tasks of the principal structural components can be comprehended easily; the supporting cables of a suspension bridge are under tension induced by the weight and loads on the bridge.

The arch is another type of structure whose shape indicates that the loads are carried by compression. The arch can be visualized as an up-side-down suspension cable.

The stone arch is one of the oldest existing structures found in buildings, walls, and bridges. Other materials, such as timber, may have been used prior to stone, but nothing of these remains today most likely because of fires, warfare, and the decay processes of nature. Although stone arches were constructed prior to the Roman Empire, the Romans constructed some of the largest and most enduring stone arches.

Before the development of the arch, the principal method of spanning a space was the simple post-and-beam construction (Fig. 4E.1a), in which a horizontal beam is supported by two columns. This type of construction was used to build the great Greek temples. The columns of these temples are closely spaced because of the limited length of available stones. Much larger spans can now be achieved using steel beams, but the spans are limited because the beams tend to sag under heavy loads.

The corbeled arch (or false arch) shown in Figure 4E.1b is another primitive structure; it is only a slight improvement over post-and-beam construction. The stability of this false arch depends upon the horizontal projection of one stone over another and the downward weight of stones from above.

The semicircular arch (Fig. 4E.2a) developed by the Romans was a great technological achievement in architectural design. The stability of this true (or voussoir) arch depends on the compression between its wedge-shaped stones. (That is, the stones are forced to squeeze against each other.) This results in horizontal outward forces at the springing of the arch (where it starts curving), which must be supported by the foundation (abutments) on the stone wall shown on the sides of the arch (Fig. 4E.2a). It is common to use very heavy walls (buttresses) on either side of the arch to provide the horizontal stability. If the foundation of the arch should move, the compressive forces between the wedge-shaped stones may decrease to the extent that the arch collapses. The surfaces of the stones used in the semicircular arches constructed by the Romans were cut, or "dressed," to make a very tight joint; it is interesting to note that mortar was usually not used in these joints. The resistance to slipping between stones was provided by the compression force and the friction between the stone faces.

Another important architectural innovation was the pointed Gothic arch shown in Figure 4E.2b. This type of structure was first used in Europe beginning in the 12th century, followed by the construction of several magnificent Gothic cathedrals in France in the 13th century. One of the most striking features of these cathedrals is their extreme height. For example, the cathedral at Chartres rises to 118 ft and the one at Reims has a height of 137 ft. It is interesting to note that such magnificent Gothic structures evolved over a very short period of time, without the benefit of any mathematical theory of structures. However, Gothic arches required flying but-

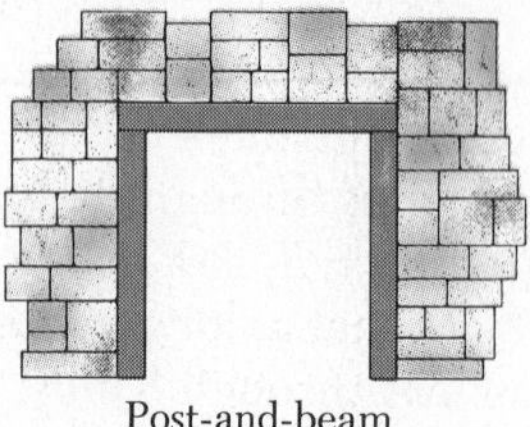

Post-and-beam
(a)

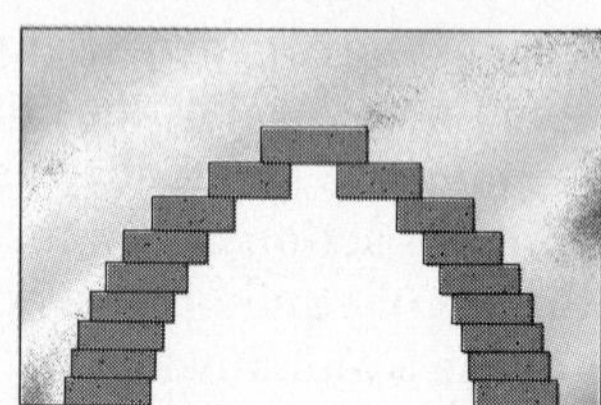

Corbeled (false) arch
(b)

Figure 12E.1 Some methods of spanning space: (a) simple post-and-beam structure and (b) corbeled, or false, arch.

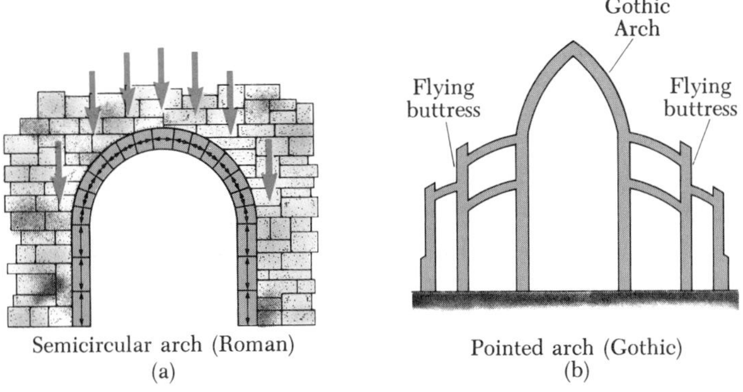

Figure 12E.2 (a) The semicircular arch developed by the Romans. (b) Gothic arch with flying buttresses to provide lateral support. (Typical cross section of a church or cathedral.) The buttresses transfer the spreading forces of the arch by vertical loads to the foundation of the structure.

tresses to prevent the spreading of the arch supported by the tall, narrow columns. The fact that they have been stable for more than 700 years attests to the technical skill of their builders and architects, which was probably acquired through experience and intuition.

Figure 4E.3 shows how the horizontal force at the base of an arch varies with arch height for an arch hinged at the peak. For a given load P, the horizontal force at the base is doubled when the height is reduced by a factor of 2. This explains why the horizontal force required to support a high pointed arch is less than that required for a circular arch. For a given span L, the horizontal force at the base is proportional to the total load P and inversely proportional to the height h. Therefore, in order to minimize the horizontal force at the base, the arch must be made as light and high as possible.

With the advent of more advanced methods of structural analysis, it has become possible to determine the optimum shape of an arch under given load conditions.

One of the most impressive modern arches, the St. Louis Gateway Arch, designed by Eero Saarinen, has a span of 192 m and a height of 192 m. The largest steel-truss arch bridge, the New River Gorge Bridge in Charleston, West Virginia, has a span of 520 m. Beautiful concrete arch bridges were designed and built in the 1920s and 1930s by Robert Maillart in Switzerland. The Sando Bridge in Sweden, a single arch of reinforced concrete, spans 264 m. Today, the arch is still the most common structure used to span large distances.

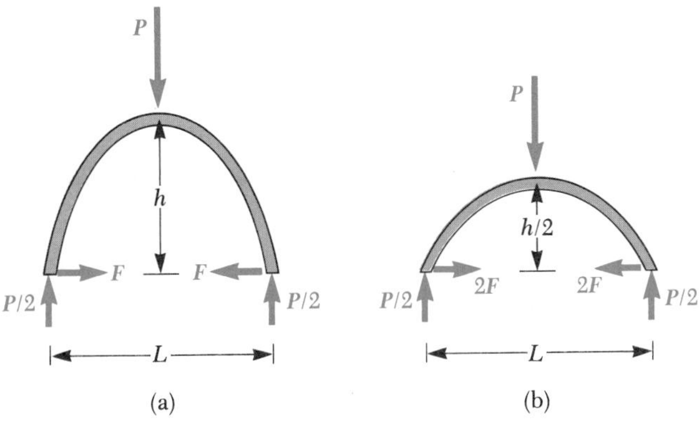

Figure 12E.3 When the height of an arch is reduced by a factor of 2, and the load force P remains the same, the horizontal at the base is doubled.

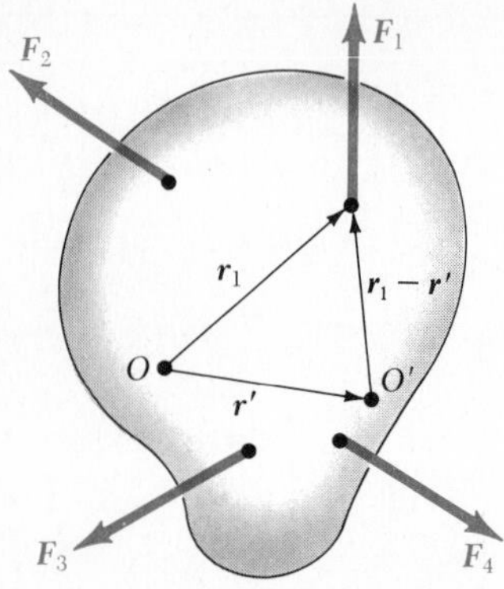

Figure 12.6 Construction for showing that if the net torque about origin O is zero, the net torque about any other origin, such as O', must be zero.

$\boldsymbol{F}_2 + \boldsymbol{F}_3 + \cdots = 0$. Figure 12.6 describes this situation (for clarity, only four forces are shown). The point of application of $\boldsymbol{F}_1$ is specified by the position vector $\boldsymbol{r}_1$. Similarly, the points of application of $\boldsymbol{F}_2, \boldsymbol{F}_3, \ldots$ are specified by $\boldsymbol{r}_2, \boldsymbol{r}_3, \ldots$ (not shown). The net torque about O is

$$\sum \tau_0 = \boldsymbol{r}_1 \times \boldsymbol{F}_1 + \boldsymbol{r}_2 \times \boldsymbol{F}_2 + \boldsymbol{r}_3 \times \boldsymbol{F}_3 + \cdots$$

Now consider another arbitrary point, O', having a position vector $\boldsymbol{r}'$ relative to O. The point of application of $\boldsymbol{F}_1$ relative to this point is identified by the vector $\boldsymbol{r}_1 - \boldsymbol{r}'$. Likewise, the point of application of $\boldsymbol{F}_2$ relative to O' is $\boldsymbol{r}_2 - \boldsymbol{r}'$, and so forth. Therefore, the torque about O' is

$$\sum \tau_{O'} = (\boldsymbol{r}_1 - \boldsymbol{r}') \times \boldsymbol{F}_1 + (\boldsymbol{r}_2 - \boldsymbol{r}') \times \boldsymbol{F}_2 + (\boldsymbol{r}_3 - \boldsymbol{r}') \times \boldsymbol{F}_3 + \cdots$$

$$\sum \tau_{O'} = \boldsymbol{r}_1 \times \boldsymbol{F}_1 + \boldsymbol{r}_2 \times \boldsymbol{F}_2 + \boldsymbol{r}_3 \times \boldsymbol{F}_3 + \cdots - \boldsymbol{r}' \times (\boldsymbol{F}_1 + \boldsymbol{F}_2 + \boldsymbol{F}_3 + \cdots)$$

Since the net force is assumed to be zero, the last term in this last expression vanishes and we see that $\Sigma \tau_{O'} = \Sigma \tau_O$. Hence,

> if a body is in translational equilibrium and the net torque is zero about one point, it must be zero about any other point.

12.2 THE CENTER OF GRAVITY

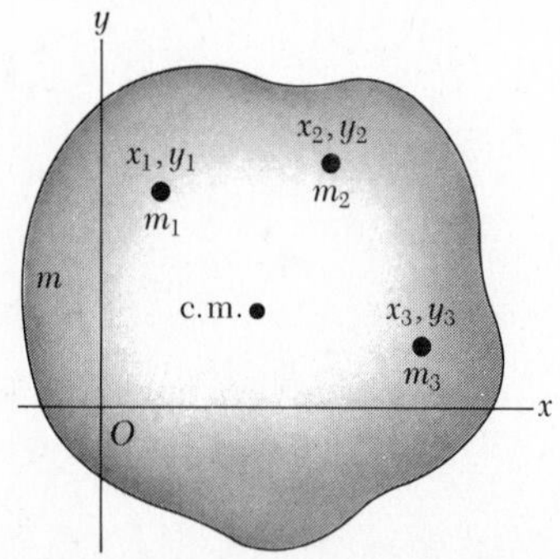

Figure 12.7 A rigid body can be divided into many small particles with specific masses and coordinates. These can be used to locate the center of mass.

Whenever we deal with rigid bodies, one of the forces that must be considered is the weight of the body, that is, the force of gravity acting on the body. In order to compute the torque due to the weight force, all of the weight can be considered as being concentrated at a single point called the *center of gravity*. As we shall see, the center of gravity of a body coincides with its center of mass if the body is in a uniform gravitational field.

Consider a body of arbitrary shape lying in the xy plane, as in Figure 12.7. Suppose the body is divided into a large number of very small particles of masses $m_1, m_2, m_3, \ldots$ having coordinates $(x_1, y_1), (x_2, y_2), (x_3, y_3), \ldots$. In Chapter 9 we defined the x coordinate of the center of mass of such an object to be

$$x_c = \frac{m_1x_1 + m_2x_2 + m_3x_3 + \cdots}{m_1 + m_2 + m_3 + \cdots} = \frac{\Sigma m_i x_i}{\Sigma m_i}$$

The y coordinate of the center of mass is similar to this, with x_i replaced by y_i.

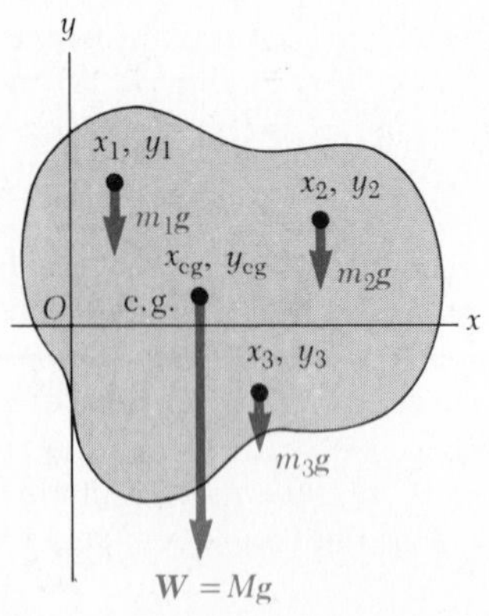

Figure 12.8 The center of gravity of the rigid body is located at the center of mass if the value of g is constant over the body.

Let us now examine the situation from another point of view by considering the weight of each part of the body, as in Figure 12.8. Each particle contributes a torque about the origin equal to its weight multiplied by its moment arm. For example, the torque due to the weight m_1g_1 is $m_1g_1x_1$, and so forth. We wish to locate the one position of the single force $\boldsymbol{W}$ (the total weight of the body) whose effect on the rotation of the body is the same as that of the individual particles. This point is called the *center of gravity* of the body. Equating the torque exerted by $\boldsymbol{W}$ acting at the center of gravity to the sum of the torques acting on the individual particles gives

$$(m_1g_1 + m_2g_2 + m_3g_3 + \cdots)x_{\text{c.g.}} = m_1g_1x_1 + m_2g_2x_2 + m_3g_3x_3 + \cdots$$

where this expression accounts for the fact that the acceleration of gravity can in general vary over the body. If we assume that g is uniform over the body (as is usually the case), then the g terms in the above equation cancel and we get

$$x_{\text{c.g.}} = \frac{m_1x_1 + m_2x_2 + m_3x_3 + \cdots}{m_1 + m_2 + m_3 + \cdots} \tag{12.5}$$

In other words, *the center of gravity is located at the center of mass as long as the body is assumed to be in a uniform gravitational field.*

In several examples that will be presented in the next section, we shall be concerned with homogeneous, symmetric bodies for which the center of gravity coincides with the geometric center of the body. Note that a rigid body in a uniform gravitational field can be balanced by a single force equal in magnitude to the weight of the body, as long as the force is directed upward through the center of gravity.

12.3 EXAMPLES OF RIGID BODIES IN STATIC EQUILIBRIUM

In this section we present several examples of rigid bodies in static equilibrium. In working such problems, it is important to first recognize *all* external forces acting on the body being considered. Failure to do so will result in an incorrect analysis. The following procedure is recommended when analyzing a body in equilibrium under the action of several external forces:

Procedure for analyzing a body in equilibrium

1. Make a sketch of the body under consideration.
2. Draw a free-body diagram and label all external forces acting on the object. Try to guess the correct direction for each force. If you select an incorrect direction that leads to a negative sign in your solution for a force, do not be alarmed; this merely means that the direction of the force is the opposite of what you assumed.
3. Resolve all forces into rectangular components, choosing a convenient coordinate system. Then apply the first condition for equilibrium, which balances forces. Remember to keep track of the signs of the various force components.
4. Choose a convenient axis for calculating the net torque on the rigid body. Remember that the choice of the origin for the torque equation is *arbitrary;* therefore choose an origin that will simplify your calculation as much as possible. Becoming adept at this is a matter of practice.
5. The first and second conditions of equilibrium give a set of linear equations with several unknowns. All that is left is to solve the simultaneous equations for the unknowns in terms of the known quantities.

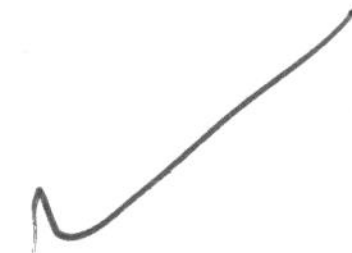

EXAMPLE 12.1. The Seesaw

A uniform board of weight 40 N supports two children weighing 500 N and 350 N, as shown in Figure 12.9. If the support (often called the *fulcrum*) is under the center of gravity of the board and if the 500 N child is 1.5 m from the center, (a) determine the upward force N exerted on the board by the support.

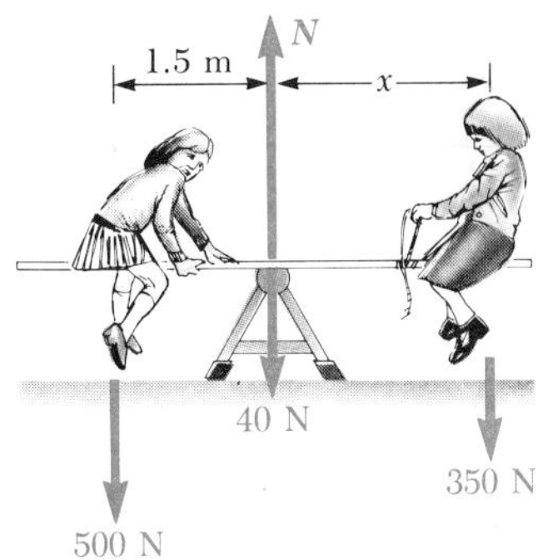

Figure 12.9 (Example 12.1) A balanced system.

Solution: First note that, in addition to N, the external forces acting on the board are the weights of the children and the weight of the board, all of which act downward. We can assume that the center of gravity of the board is at its geometric center because we were told that the board is uniform. Since the system is in equilibrium, the upward force N must balance all the downward forces. From $\Sigma F_y = 0$, we have

$$N - 500 \text{ N} - 350 \text{ N} - 40 \text{ N} = 0 \qquad \text{or} \qquad N = 890 \text{ N}$$

It should be pointed out here that the equation $\Sigma F_x = 0$ also applies to this situation, but it is unnecessary to consider this equation because we have no forces acting horizontally on the board.

(b) Determine where the 350-N child should sit to balance the system.

Solution: To find this position, we must invoke the second condition for equilibrium. Taking the center of gravity of the board as the axis for our torque equation, we see from $\Sigma\tau = 0$ that

$$(500 \text{ N})(1.5 \text{ m}) - (350 \text{ N})(x) = 0$$

$$x = 2.14 \text{ m}$$

Exercise 1 If the fulcrum did not lie under the center of gravity of the board, what other information would you need to solve the problem?

EXAMPLE 12.2. A Weighted Hand

A 50-N weight is held in the hand with the forearm in the horizontal position, as in Figure 12.10a. The biceps muscle is attached 5 cm from the joint, and the weight is 35 cm from the joint. Find the upward force that the biceps exerts on the forearm (made up of the radius and ulna) and the downward force on the upper arm (the humerus) acting at the joint. Neglect the weight of the forearm.

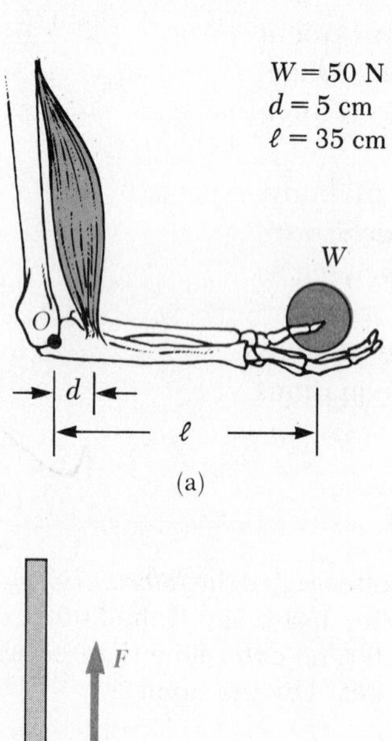

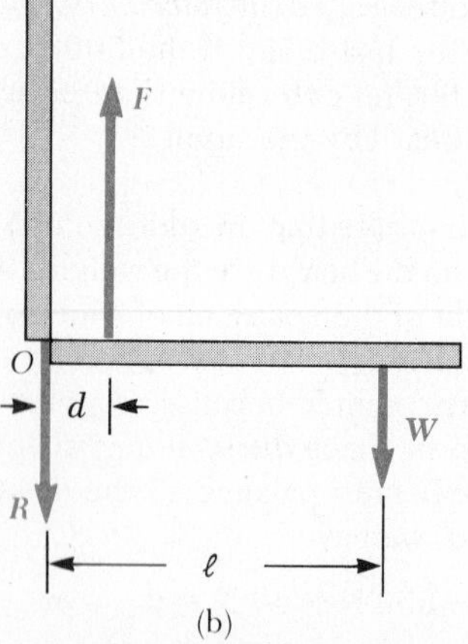

Figure 12.10 (Example 12.2) (a) Forces operative at the elbow joint. (b) The mechanical model for the system described in (a).

Solution: The forces acting on the forearm are equivalent to those acting on a bar of length 35 cm, as shown in Figure 12.10b, where $\boldsymbol{F}$ is the upward force of the biceps and $\boldsymbol{R}$ is the downward force at the joint. From the first condition for equilibrium, we have

$$(1) \qquad \sum F_y = F - R - 50 \text{ N} = 0$$

From the second condition for equilibrium, we know that the sum of the torques about any point must be zero. With the joint O as the axis, we have

$$Fd - WL = 0$$

$$F(5 \text{ cm}) - (50 \text{ N})(35 \text{ cm}) = 0$$

$$F = 350 \text{ N}$$

This value for F can be substituted into (1) to give $R = 300$ N. These values correspond to $F = 79$ lb and $R = 68$ lb. Hence, the forces at joints and in muscles can be extremely large.

Exercise 2 In reality, the biceps makes an angle of 15° with the vertical, so that $\boldsymbol{F}$ has both a vertical and a horizontal component. Find the value of $\boldsymbol{F}$ and the components of $\boldsymbol{R}$ including this fact in your analysis.
Answer: $F = 362$ N, $R_x = 93.8$ N, $R_y = 300$ N.

EXAMPLE 12.3. Standing on a Horizontal Beam

A uniform horizontal beam of length 8 m and weight 200 N is attached to a wall by a pin connection that allows the beam to rotate. Its far end is supported by a cable that makes an angle of 53° with the horizontal (Fig. 12.11a).

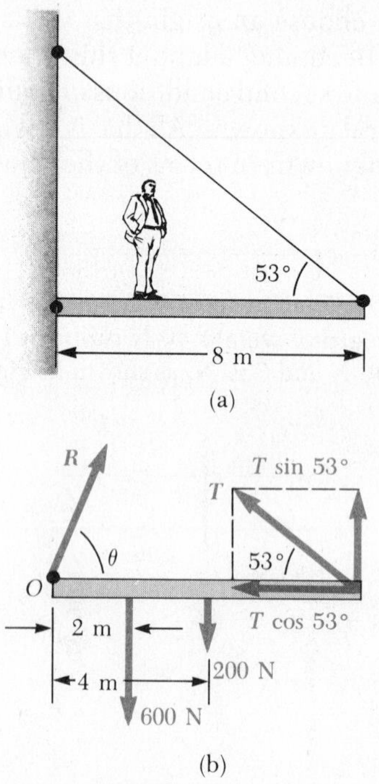

Figure 12.11 (Example 12.3) (a) A uniform beam supported by a cable. (b) The free-body diagram for the beam.

If a 600-N person stands 2 m from the wall, find the tension in the cable and the force exerted on the beam by the wall.

Solution: First we must identify all the external forces acting on the beam. These are its weight, the tension, $\boldsymbol{T}$, in the cable, the force $\boldsymbol{R}$ exerted by the wall at the pivot (the direction of this force is unknown), and the weight of the person on the beam. These are all indicated in the free-body diagram for the beam (Fig. 12.11b). If we resolve $\boldsymbol{T}$ and $\boldsymbol{R}$ into horizontal and vertical components and apply the first condition for equilibrium, we get

(1) $\sum F_x = R\cos\theta - T\cos 53° = 0$

(2) $\sum F_y = R\sin\theta + T\sin 53° - 600\text{ N} - 200\text{ N} = 0$

Because R, T, and θ are all unknown, we cannot obtain a solution from these expressions alone. (The number of simultaneous equations must equal the number of unknowns in order for us to be able to solve for the unknowns.)

Now let us invoke the condition for rotational equilibrium. A convenient axis to choose for our torque equation is the one that passes through the pivot at O. The feature that makes this point so convenient is that the force $\boldsymbol{R}$ and the horizontal component of $\boldsymbol{T}$ both have a lever arm of zero, and hence zero torque, about this pivot. Recalling our convention for the sign of the torque about an axis and noting that the lever arms of the 600-N, 200-N, and $T\sin 53°$ forces are 2 m, 4 m, and 8 m, respectively, we get

(3) $$\sum \tau_O = (T\sin 53°)(8\text{ m}) - (600\text{ N})(2\text{ m}) - (200\text{ N})(4\text{ m}) = 0$$

$$T = 313\text{ N}$$

Thus the torque equation with this axis gives us one of the unknowns directly! This value is substituted into (1) and (2) to give

$$R\cos\theta = 188\text{ N}$$

$$R\sin\theta = 550\text{ N}$$

We divide these two equations and recall the trigonometric identity $\sin\theta/\cos\theta = \tan\theta$ to get

$$\tan\theta = \frac{550\text{ N}}{188\text{ N}} = 2.93$$

$$\theta = 71.1°$$

Finally,

$$R = \frac{188\text{ N}}{\cos\theta} = \frac{188\text{ N}}{\cos 71.1°} = 581\text{ N}$$

If we had selected some other axis for the torque equation, the solution would have been the same. For example, if the axis were to pass through the center of gravity of the beam, the torque equation would involve both T and R. However, this equation, coupled with (1) and (2), could still be solved for the unknowns. Try it!

When many forces are involved in a problem of this nature, it is convenient to "keep the books straight" by setting up a table of forces, their lever arms, and their torques. For instance, in the example just given, we would construct the following table. Setting the sum of the terms in the last column equal to zero represents the condition of rotational equilibrium.

Force Component	Lever Arm Relative to O (m)	Torque About O (N · m)
$T\sin 53°$	8	$8T\sin 53°$
$T\cos 53°$	0	0
200 N	4	$-4(200)$
600 N	2	$-2(600)$
$R\sin\theta$	0	0
$R\cos\theta$	0	0

EXAMPLE 12.4. The Leaning Ladder

A uniform ladder of length ℓ and weight $W = 50$ N rests against a smooth, vertical wall (Fig. 12.12a). If the coefficient of static friction between the ladder and ground is $\mu = 0.40$, find the *minimum* angle θ_{min} such that the ladder will *not* slip.

Solution: The free-body diagram showing all the external forces acting on the ladder is illustrated in Figure 12.12b. Note that the reaction force at the ground, $\boldsymbol{R}$, is the vector sum of a normal force, $\boldsymbol{N}$, and the force of friction, $\boldsymbol{f}$. The reaction force at the wall, $\boldsymbol{P}$, is horizontal, since the wall is smooth. From the first condition for equilibrium applied to the ladder, we have

$$\sum F_x = f - P = 0$$

$$\sum F_y = N - W = 0$$

Since $W = 50$ N, we see from the equation above that $N = W = 50$ N. Furthermore, *when the ladder is on the*

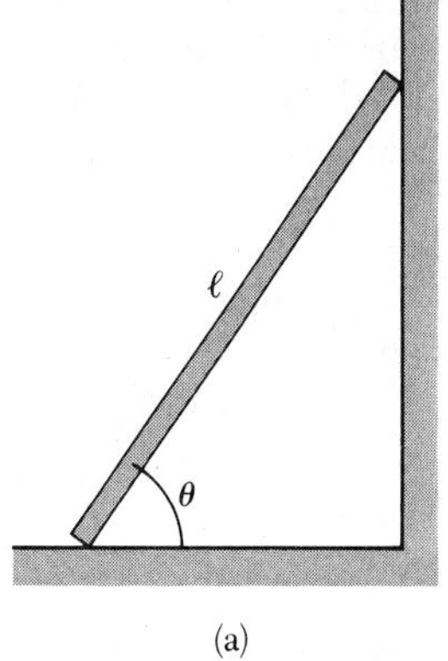

(a)

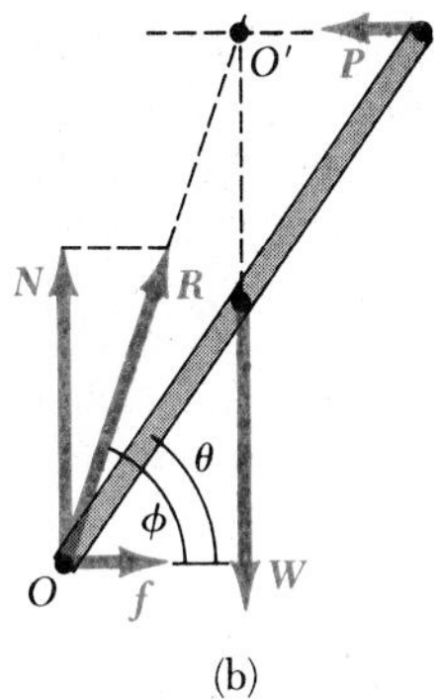

(b)

Figure 12.12 (Example 12.4) (a) A uniform ladder at rest, leaning against a smooth wall. The floor is rough. (b) The free-body diagram for the ladder. Note that the forces $\boldsymbol{R}$, $\boldsymbol{W}$, and $\boldsymbol{P}$ pass through a common point O'.

verge of slipping, the force of friction must be a maximum, given by $f_{max} = \mu N = 0.40(50\ N) = 20\ N$. (Recall that $f_s \le \mu N$.) Thus, at this angle, $P = 20$ N.

To find the value of θ, we must use the second condition of equilibrium. When the torques are taken about the origin O at the bottom of the ladder, we get

$$\sum \tau_O = P\ell \sin\theta - W\frac{\ell}{2}\cos\theta = 0$$

But $P = 20$ N when the ladder is about to slip and $W = 50$ N, so that the expression above gives

$$\tan\theta_{min} = \frac{W}{2P} = \frac{50\ N}{40\ N} = 1.25$$

$$\theta_{min} = 51.3°$$

It is interesting to note that the result does not depend on ℓ!

An alternative approach to analyzing this problem is to consider the intersection O' of the forces **W** and **P**. Since the torque about any origin must be zero, the torque about O' must be zero. This requires that the line of action of **R** (the resultant of **N** and **f**) pass through O'! That is, since this is a three-force body, the forces must be concurrent. With this condition, one could then obtain the angle ϕ that **R** makes with the horizontal (where ϕ is greater than θ), assuming the length of the ladder is known.

Exercise 3 With reference to Figure 12.12, show that $\tan\phi = 2\tan\theta$.

EXAMPLE 12.5. Raising a Cylinder

A cylinder of weight W and radius R is to be raised onto a step of height h as shown in Figure 12.13. A rope is wrapped around the cylinder and pulled horizontally. Assuming the cylinder doesn't slip on the step, find the *minimum* force **F** necessary to raise the cylinder and the reaction force at P.

Solution: When the cylinder is just ready to be raised, the reaction force at Q goes to zero. Hence, at this time there are only three forces on the cylinder, as shown in Figure 12.13b. From the dotted triangle drawn in Figure 12.13a, we see that the moment arm d of the weight relative to the point P is given by

$$d = \sqrt{R^2 - (R-h)^2} = \sqrt{2Rh - h^2}$$

The moment arm of **F** relative to P is $2R - h$. Therefore, the net torque acting on the cylinder about P is

$$Wd - F(2R-h) = 0$$

$$W\sqrt{2Rh - h^2} - F(2R-h) = 0$$

$$F = \frac{W\sqrt{2Rh - h^2}}{2R - h}$$

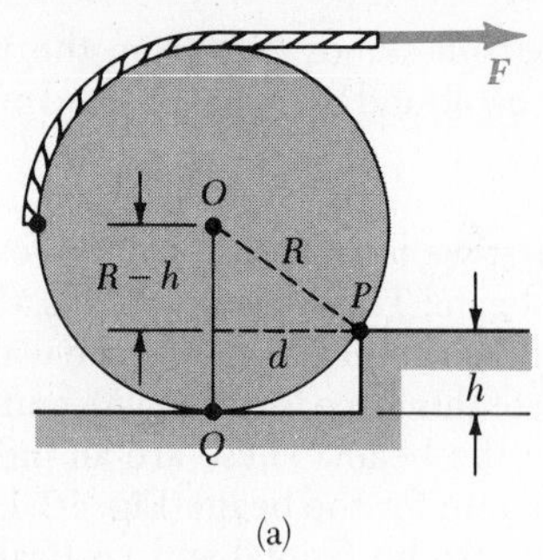

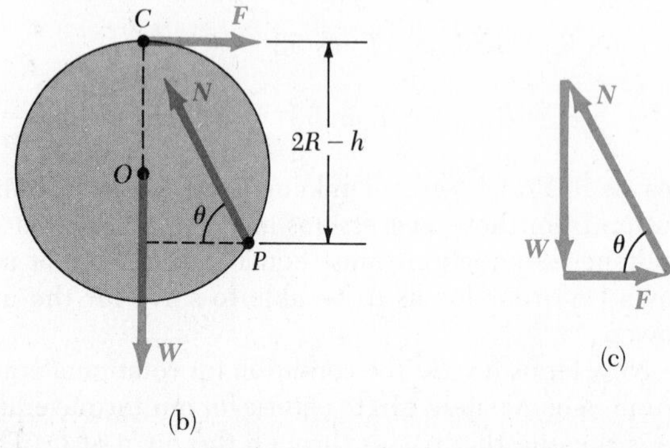

Figure 12.13 (Example 12.5) (a) A cylinder of weight **W** being pulled by a force **F** over a step. (b) The free-body diagram for the cylinder when it is just ready to be raised. (c) The *vector* sum of the three external forces is zero.

Hence, the second condition of equilibrium was sufficient to obtain the magnitude of **F**. We can determine the components of **N** by using the first condition of equilibrium:

$$\sum F_x = F - N\cos\theta = 0$$

$$\sum F_y = N\sin\theta - W = 0$$

Dividing gives

$$(1) \qquad \tan\theta = \frac{W}{F}$$

and solving for N gives

$$(2) \qquad N = \sqrt{W^2 + F^2}$$

For example, if we take $W = 500$ N, $h = 0.3$ m, and $R = 0.8$ m, we find that $F = 385$ N, $\theta = 52.4°$, and $N = 631$ N.

Exercise 4 Solve this problem by noting that the three forces acting on the cylinder are concurrent and must pass through the point C. The three forces form the sides of the triangle shown in Figure 12.13c.

12.4 SUMMARY

A rigid body is in **equilibrium** if and only if the following conditions are satisfied: (1) *the resultant external force must be zero* and (2) *the resultant external torque must be zero about* any *origin*. That is,

Conditions for equilibrium

$$\sum F = 0 \tag{12.2}$$

$$\sum \tau = 0 \tag{12.3}$$

The first condition is the *condition of translational equilibrium*, and the second is the *condition of rotational equilibrium*.

If two forces act on a rigid body, the body is in equilibrium if and only if the forces are equal in magnitude and opposite in direction and have the same line of action.

When three forces act on a rigid body that is in equilibrium, the three forces must be concurrent, that is, their lines of action must intersect at a common point.

The *center of gravity* of a rigid body coincides with the center of mass if the body is in a uniform gravitational field.

QUESTIONS

1. Can a body be in equilibrium if only one external force acts on it? Explain.
2. Can a body be in equilibrium if it is in motion? Explain.
3. Locate the center of gravity for the following uniform objects: (a) a sphere, (b) a cube, (c) a cylinder.
4. The center of gravity of an object may be located outside the object. Give a few examples for which this is the case.
5. You are given an arbitrarily shaped piece of plywood, together with a hammer, nail, and plumb bob. How could you use these items to locate the center of gravity of the plywood? (*Hint:* Use the nail to suspend the plywood.)
6. In order for a chair to be balanced on one leg, where must the center of gravity of the chair be located?
7. Give an example in which the net torque acting on an object is zero and yet the net force is nonzero.
8. Give an example in which the net force acting on an object is zero and yet the net torque is nonzero.
9. Can an object be in equilibrium if the only torques acting on it produce clockwise rotation?
10. A tall crate and a short crate of equal mass are placed side by side on an incline (without touching each other). As the incline angle is increased, which crate will tip first? Explain.
11. A male and a female student are asked to do the following task. Face a wall, step three foot lengths away from the wall, and then lean over and touch the wall with your nose, keeping your hands behind your back. The male usually fails, but the female succeeds. How would you explain this?
12. When lifting a heavy object, why is it recommended to straighten your back as much as possible rather than bend over and lift mainly with the arms?

PROBLEMS

Section 12.1 The Conditions of Equilibrium of a Rigid Body

1. Write the necessary condition of equilibrium for the body shown in Figure 12.14. Take the origin of the torque equation at the point O.
2. Write the necessary conditions of equilibrium for the body shown in Figure 12.15. Take the origin of the torque equation at the point O.
3. A uniform beam of weight W and length ℓ has weights W_1 and W_2 at two positions, as in Figure 12.16. The beam is resting at two points. For what value of x will the beam be balanced at P such that the normal force at O is zero?

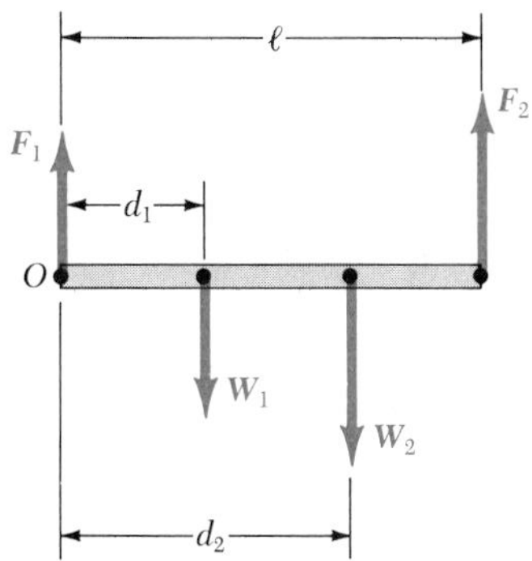

Figure 12.14 (Problem 1).

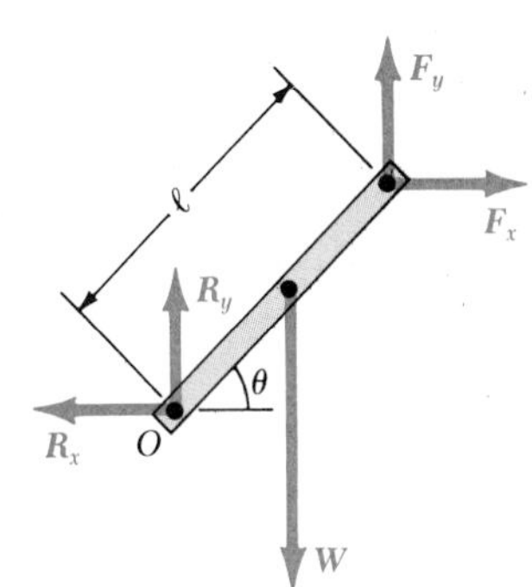

Figure 12.15 (Problem 2).

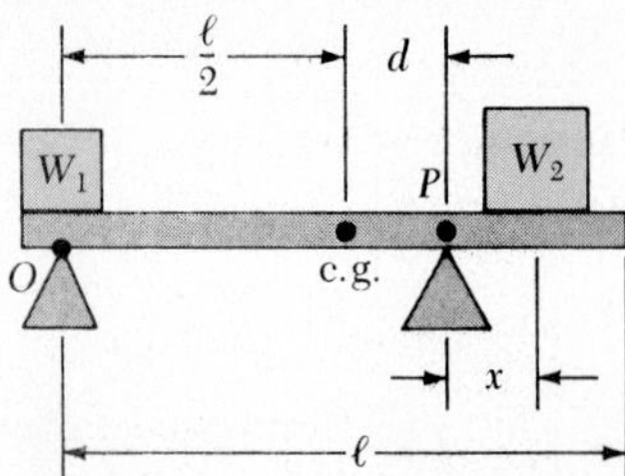

Figure 12.16 (Problems 3 and 4).

4. With reference to Figure 12.16, find x such that the normal force at O will be one half the normal force at P. Neglect the weight of the beam.

Section 12.2 The Center of Gravity

5. A flat plate in the shape of a letter T is cut with the dimensions shown in Figure 12.17. Locate the center of gravity. (*Hint:* Note that the weights of the two rectangular parts are proportional to their volumes.)

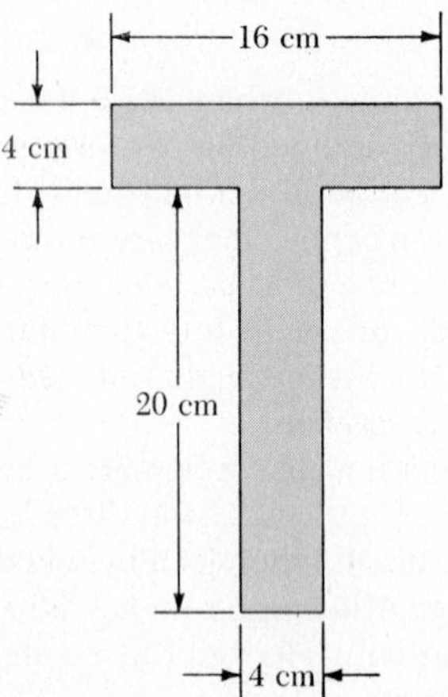

Figure 12.17 (Problem 5).

6. A carpenter's square has the shape of an L, as in Figure 12.18. Locate the center of gravity. (See hint in Problem 5.)

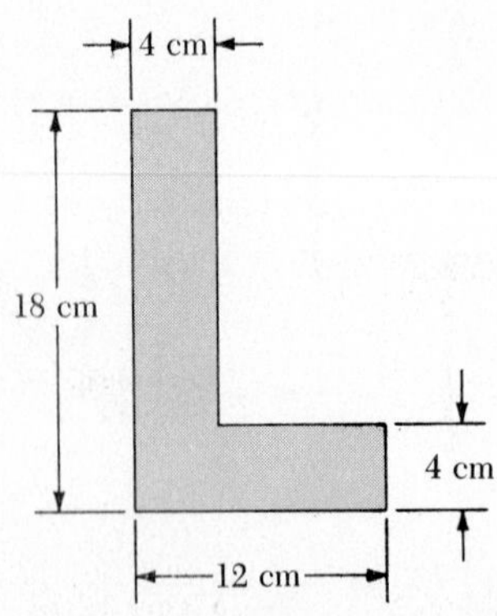

Figure 12.18 (Problem 6).

7. Three masses are located in a rectangular coordinate system as follows: a 2-kg mass is at (3, −2), a 3-kg mass is at (−2, 4), and a 1-kg mass is at (2, 2), where all distances are in meters. Find the coordinates of the center of gravity for the three masses.
8. Four masses are located in a rectangular coordinate system as follows: a 2-kg mass is at (0, 0), a 3-kg mass is at (0, 2), a 4-kg mass is at (2, 2), and a 5-kg mass is at (2, 0), where all distances are in meters. Find the location of the center of gravity for the four masses.

Section 12.3 Examples of Rigid Bodies in Static Equilibrium

9. A meter stick supported at the 50-cm mark has masses of 300 g and 200 g hanging from it at the 10-cm and 60-cm marks, respectively. Determine the position at which one would hang a third, 400-g mass to keep the meter stick balanced.
10. A 48-kg diver stands at the end of a 3-m-long diving board. What torque does the weight of the diver produce about an axis perpendicular to and in the plane of the diving board through its midpoint?
11. Repeat Example 12.3 taking the axis of the torque equation through the center of the beam in Figure 12.11a. Your results for R, T, and θ should be identical with those obtained in Example 12.3.

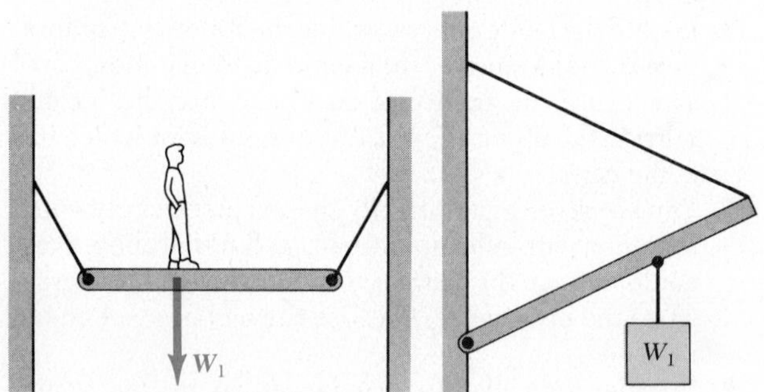

Figure 12.19 (Problem 12).

12. Draw free-body diagrams for each of the rigid beams shown in Figure 12.19. Assume the beams are uniform and have a weight W.
13. An automobile has a mass of 1600 kg. The distance between the front and rear axles is 3 m. If the normal force on the front tires is 20% larger than the normal force on the rear tires, (a) where is the center of gravity relative to the front axle, and (b) what is the normal force on each tire?
14. A uniform plank of length 6 m and mass 30 kg rests horizontally on a scaffold, with 1.5 m of the plank hanging over one end of the scaffold. How far can a painter of mass 70 kg walk on the overhanging part of the plank before it tips?

GENERAL PROBLEMS

15. A bridge of length 50 m and mass 8×10^4 kg is supported at each end as in Figure 12.20. A truck of mass

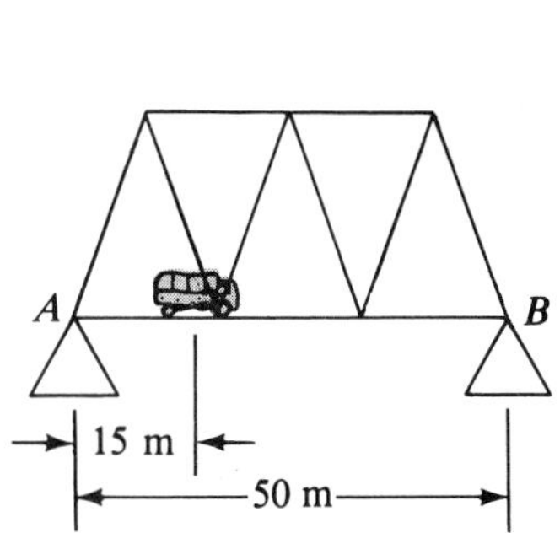

Figure 12.20 (Problem 15).

Figure 12.22 (Problem 17).

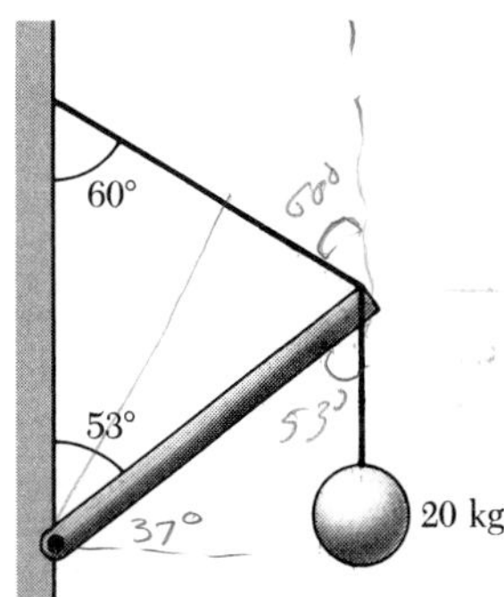

Figure 12.24 (Problem 19).

3×10^4 kg is located 15 m from one end. What are the forces on the bridge at the points of support?

16. A sign of weight W and width 2ℓ hangs from a light, horizontal beam, hinged at the wall and supported by a cable (Fig. 12.21). Determine (a) the tension in the cable and (b) the components of the reaction force at the hinge in terms of W, d, ℓ, and θ.

19. A uniform beam of length 4 m and mass 10 kg supports a 20-kg mass as in Figure 12.24. (a) Draw a free-body diagram for the beam. (b) Determine the tension in the supporting wire and the components of the reaction force at the pivot.

20. A 300-lb uniform boom is supported by a cable as in Figure 12.25. The boom is pivoted at the bottom, and a 500-lb weight hangs from its top. Find the tension in the supporting cable and the components of the reaction force on the boom at the hinge.

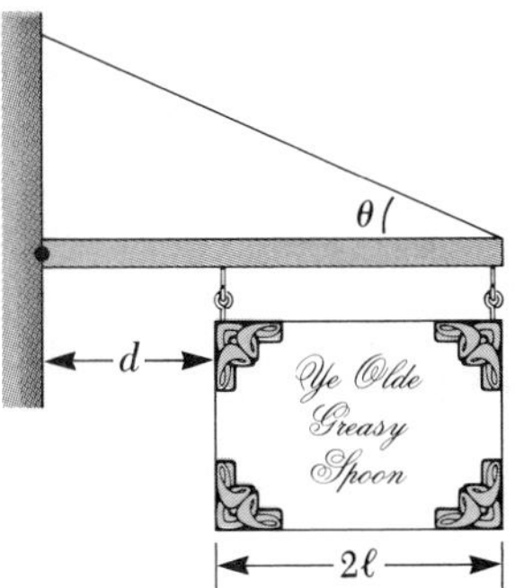

Figure 12.21 (Problem 16).

Figure 12.23 (Problem 18).

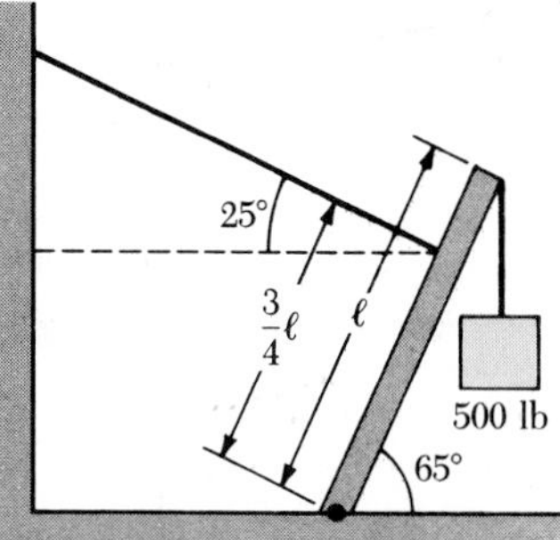

Figure 12.25 (Problem 20).

17. A 24-lb monkey walks up a 30-lb uniform ladder of length ℓ, as in Figure 12.22. The upper and lower ends of the ladder rest on frictionless surfaces. The lower end of the ladder is fastened to the wall by a horizontal rope that can support a maximum tension of 25 lb. (a) Draw a free-body diagram for the ladder. (b) Find the tension in the rope when the monkey is one third the way up the ladder. (c) Find the maximum distance d the monkey can walk up the ladder before the rope breaks, expressing your answer as a fraction of the length ℓ.

18. A hungry bear weighing 160 lb walks out on a beam in an attempt to retrieve some "goodies" hanging at the end of the beam (Fig. 12.23). The beam is uniform, weighs 50 lb, and is 20 ft long; the goodies weigh 20 lb. (a) Draw a free-body diagram for the beam. (b) When the bear is at $x = 3$ ft, find the tension in the wire and the components of the reaction force at the hinge. (c) If the wire can withstand a maximum tension of 200 lb, what is the maximum distance the bear can walk before the wire breaks?

21. A 150-kg mass rests on a 50-kg beam as in Figure 12.26. The weight is also connected to one end of the beam through a rope and pulley. Assuming the system is in equilibrium, (a) draw free-body diagrams for the weight and beam and (b) find the tension in the rope and the components of the reaction force at the pivot O.

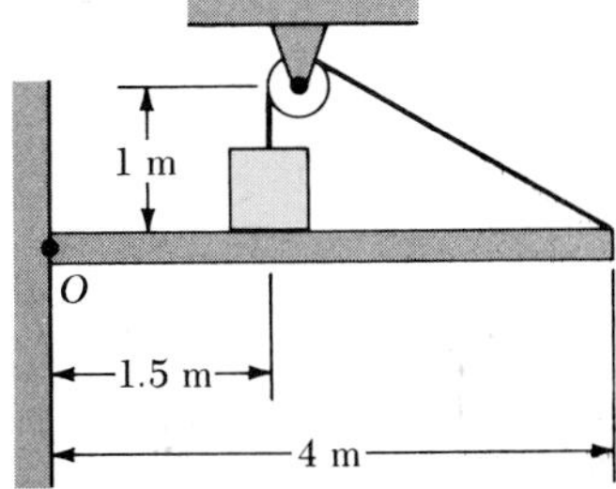

Figure 12.26 (Problem 21).

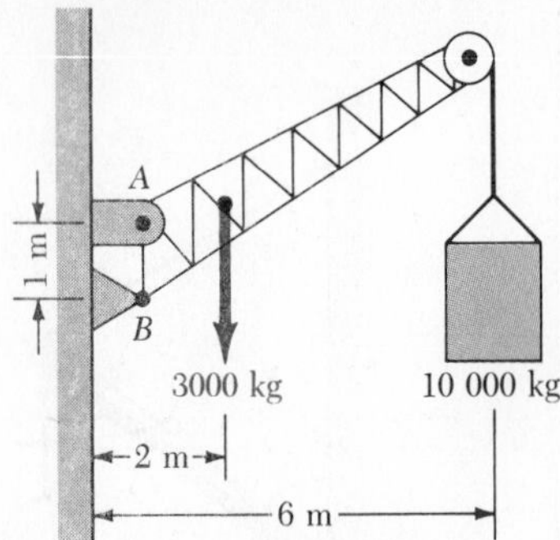

Figure 12.27 (Problem 22).

22. A crane of mass 3000 kg supports a load of 10 000 kg as in Figure 12.27. The crane is pivoted with a smooth pin at *A* and rests against a smooth support at *B*. Find the reaction forces at *A* and *B*.
23. A 15-m uniform ladder weighing 500 N rests against a frictionless wall. The ladder makes a 60° angle with the horizontal. (a) Find the horizontal and vertical forces that the earth exerts on the base of the ladder when an 800-N firefighter is 4 m from the bottom. (b) If the ladder is just on the verge of slipping when the firefighter is 9 m up, what is the coefficient of static friction between ladder and ground?
24. An iron trapdoor 1.25 m wide and 2 m long weighs 360 N. Its center of gravity is at its geometric center and the hinges are on the short side of the door. What force applied at right angles to the door is required to lift it (a) when it is horizontal and (b) when it has been opened so that it makes an angle of 30° with the horizontal? (Assume that the force is applied at the edge of the door opposite the hinges.)
25. A 10 000-N shark is supported by a cable attached to a 4-m rod that can pivot at the base. Calculate the cable tension needed to hold the system in the position shown in Figure 12.28. Find the horizontal and vertical forces exerted on the base of the rod. (Neglect the weight of the rod.)

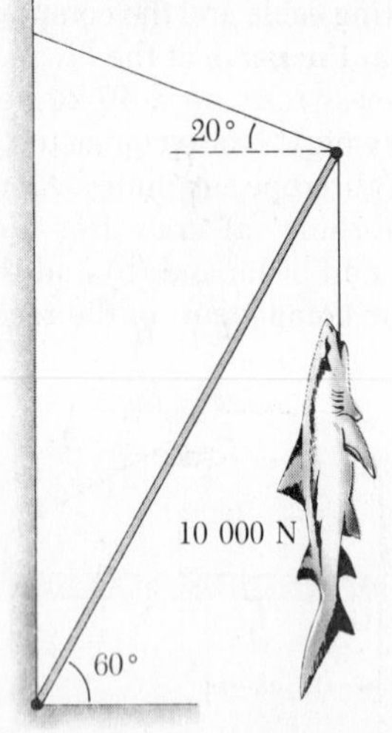

Figure 12.28 (Problem 25).

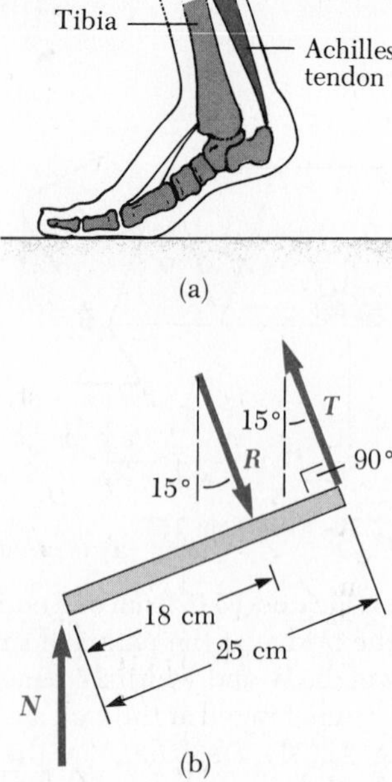

Figure 12.29 (Problem 26).

26. When a person stands on tiptoe (a strenuous position), the position of the foot is as shown in Figure 12.29a. The total weight **W** is supported by the force **N** of the floor on the toe. A mechanical model for the situation is shown in Figure 12.29b, where **T** is the tension in the Achilles tendon and **R** is the force on the foot due to the tibia. Find the values of *T* and *R* using the model and dimensions given, with $W = 700$ N.
27. A person bends over and lifts a 200-N weight as in Figure 12.30a, with the back in the horizontal position. The back muscle attached at a point two thirds up the spine maintains the position of the back, where the angle between the spine and this muscle is 12°. Using the mechanical model shown in Figure 12.30b and taking the weight of the upper body to be 350 N, find the tension in the back muscle and the compressional force in the spine.

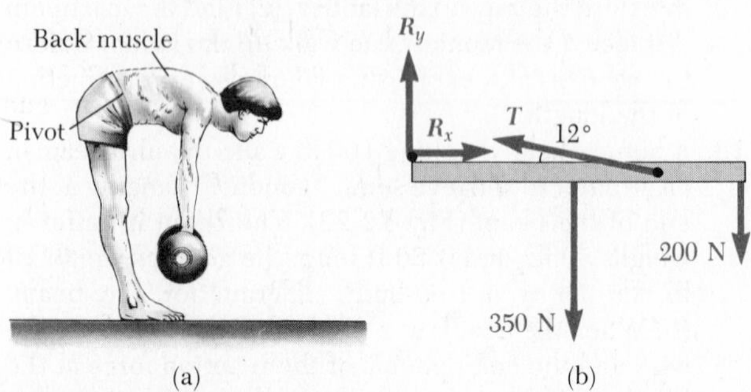

Figure 12.30 (Problem 27).

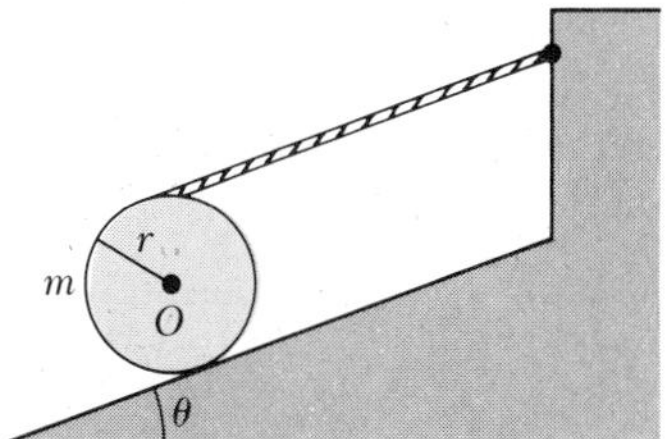

Figure 12.31 (Problem 28).

28. A disk of mass m and of radius r rests on an inclined surface and is supported by a rope that is tangent to the disk and parallel to the inclined surface. The inclined surface makes an angle θ with the horizontal as shown in Figure 12.31. Find (a) the minimum value of the coefficient of static friction, in terms of θ, that will prevent the disk from slipping down the inclined surface and (b) the tension in the rope in terms of m, g, and θ.

•29. A force $\boldsymbol{F}$ acts on a rectangular block weighing 100 lb as in Figure 12.32. (a) If the block slides with constant speed when $F = 50$ lb and $h = 1$ ft, find the coefficient of sliding friction and the position of the resultant normal force. (b) If $F = 75$ lb, find the value of h for which the block will just begin to tip from a vertical position.

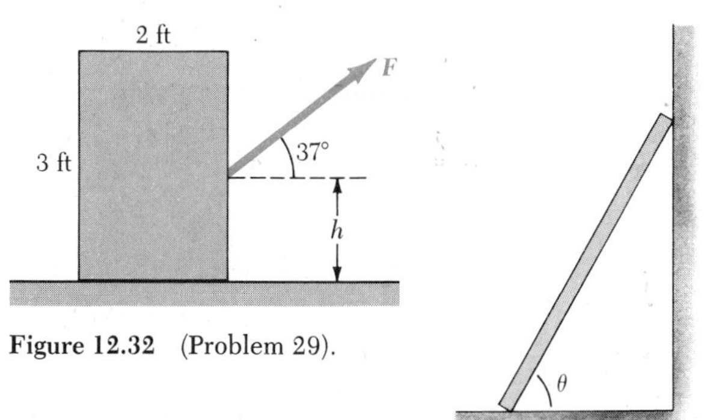

Figure 12.32 (Problem 29).

Figure 12.33 (Problem 30).

•30. A uniform ladder weighing 50 lb is leaning against a wall (Fig. 12.33). The ladder slips when θ is 60°, where θ is the angle between the ladder and the horizontal. Assuming the coefficients of static friction at the wall and the floor *are the same*, obtain a value for μ_s.

•31. A uniform beam of weight w is inclined at an angle θ to the horizontal with its upper end supported by a horizontal rope tied to a wall and its lower end resting on a rough floor (Fig. 12.34). (a) If the coefficient of static friction between the beam and floor is μ_s, determine an expression for the *maximum* weight W that can be suspended from the top before the beam slips. (b) Determine the magnitude of the reaction force at the floor and the magnitude of the force of the beam on the rope at P in terms of w, W, and μ_s.

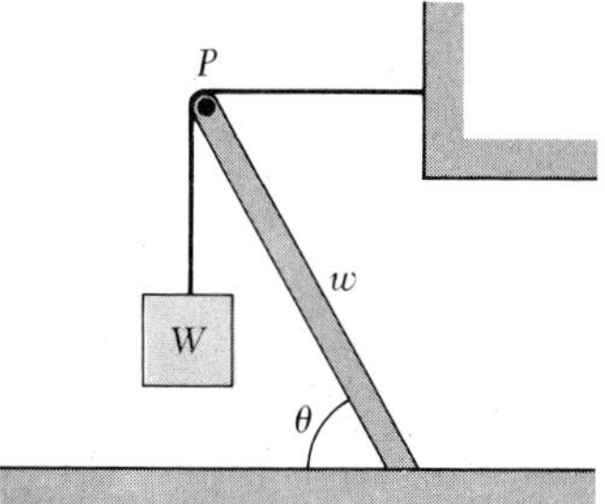

Figure 12.34 (Problem 31).

•32. The cylinder shown in Figure 12.35 is held in position by a rope that supplies a force $\boldsymbol{F}$ and by static friction. What is the *minimum* value of μ_s such that the cylinder will remain in equilibrium when $\boldsymbol{F}$ is at the angle θ with the horizontal?

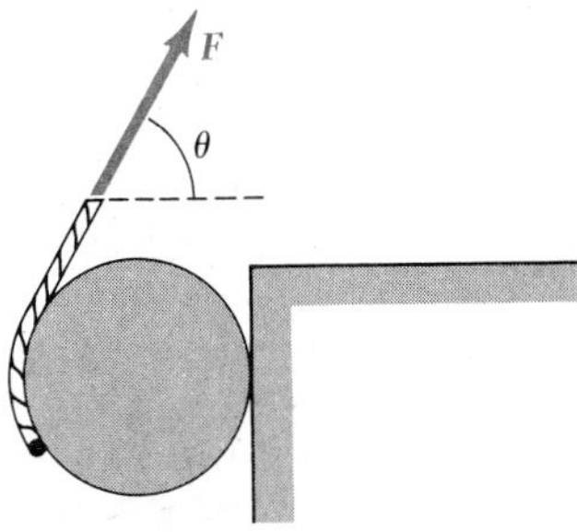

Figure 12.35 (Problem 32).

•33. A stepladder of negligible weight is constructed as shown in Figure 12.36. A painter of mass 70 kg stands on the ladder 3 m from the bottom. Assuming the floor is frictionless, find (a) the tension in the horizontal bar connecting the two halves of the ladder, (b) the normal forces at A and B, and (c) the components of the reaction force at the hinge C that the left half of the ladder exerts on the right half. (*Hint:* Treat each half of the ladder separately.)

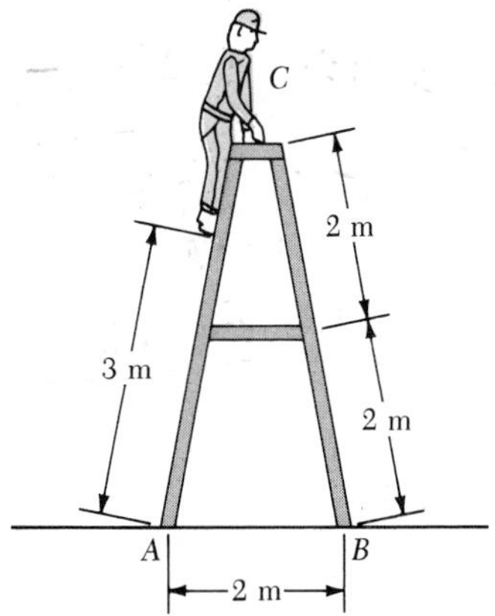

Figure 12.36 (Problem 33).

•**34.** On a hot summer afternoon Kathy Kool is driving in her sports car and enjoying a *thick* vanilla milk shake. She places the milkshake on the carpeted floor to enable her to shift into low gear at a red light. (The height of the cylindrical cup is twice its diameter, and the carpet prevents the cup from sliding.) What is the maximum acceleration the car can have before the cup tips over?

•**35.** A disk of mass m and radius R rests on a surface which makes an angle θ with the horizontal and is supported by a horizontal rope as shown in Figure 12.37. (a) Find the minimum value of the coefficient of static friction, in terms of θ, that will prevent the disk from slipping down the inclined surface. (b) Find the tension in the rope in terms of m, g, and θ.

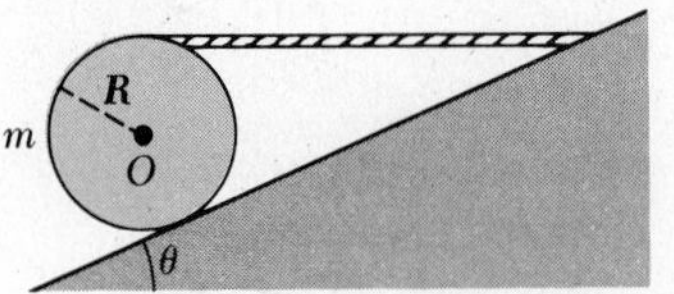

Figure 12.37 (Problem 35).

13
Oscillatory Motion

In previous chapters we pointed out that the motion of a body can be predicted if the external forces acting on it are known. If a force varies in time, the velocity and acceleration of the body will also change with time. A very special kind of motion occurs when the force on a body is proportional to the displacement of the body from equilibrium. If this force always acts toward the equilibrium position of the body, a back-and-forth motion will result about this position. We call such a motion *periodic* or *oscillatory*.

You are most likely familiar with several examples of periodic motion, such as the oscillations of a mass on a spring, the motion of a pendulum, and the vibrations of a stringed musical instrument. However, the number of systems that exhibit oscillatory motion is much more extensive. For example, the molecules in a solid oscillate about their equilibrium positions; electromagnetic waves, such as light waves, radar, and radio waves, are characterized by oscillating electric and magnetic field vectors; and in alternating-current circuits, voltage, current, and electrical charge vary periodically with time.

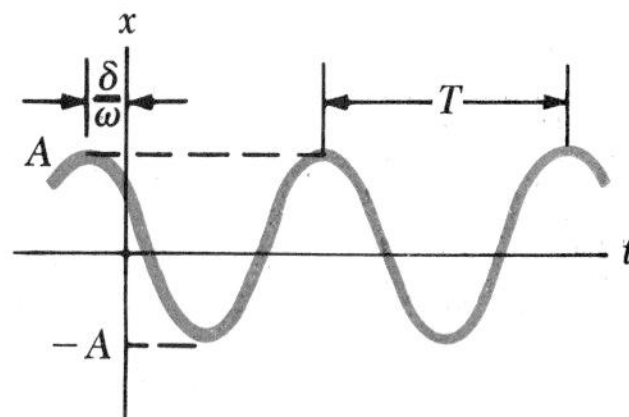

Figure 13.1 Displacement versus time for a particle undergoing simple harmonic motion. The amplitude of the motion is A and the period is T.

Most of the material in this chapter deals with *simple harmonic motion*. For this type of motion, an object oscillates between two spatial positions for an indefinite period of time, with no loss in mechanical energy. In real mechanical systems, retarding (or frictional) forces are always present. Such forces reduce the mechanical energy of the system as motion progresses, and the oscillations are said to be *damped*. If an external driving force is applied such that the energy loss is balanced by the energy input, we call the motion a *forced oscillation*.

13.1 SIMPLE HARMONIC MOTION

A particle moving along the x axis is said to exhibit **simple harmonic motion** when its displacement from equilibrium, x, varies in time according to the relationship

Displacement versus time for simple harmonic motion

$$x = A\cos(\omega t + \delta) \qquad (13.1)$$

where A, ω, and δ are constants of the motion. In order to give physical significance to these constants, it is convenient to plot x as a function of t, as in Figure 13.1. First, we note that A, called the **amplitude** of the motion, is simply the *maximum displacement* of the particle in either the positive or negative x direction. The constant ω is called the *angular frequency* (defined in Eq. 13.4). The constant angle δ is called the **phase constant** (or phase angle) and along with the amplitude A is determined uniquely by the initial displacement and velocity of the particle. The constants δ and A tell us what the displacement was at time $t = 0$. The quantity $(\omega t + \delta)$ is called the **phase** of the motion and is useful in comparing the motions of two systems of particles. Note that the function x is periodic and repeats itself when ωt increases by 2π radians.

The **period,** T, is the time for the particle to go through one full cycle of its motion. That is, the value of x at time t equals the value of x at time $t + T$. We can show that the period of the motion is given by $T = 2\pi/\omega$ by using the fact that the phase increases by 2π radians in a time T:

$$\omega t + \delta + 2\pi = \omega(t + T) + \delta$$

Hence, $\omega T = 2\pi$ or

Period

$$T = \frac{2\pi}{\omega} \tag{13.2}$$

The inverse of the period is called the **frequency** of the motion, f. The frequency represents the *number of oscillations the particle makes per unit time:*

Frequency

$$f = \frac{1}{T} = \frac{\omega}{2\pi} \tag{13.3}$$

The units of f are cycles/s, or hertz (Hz).

Rearranging Equation 13.3 gives

Angular frequency

$$\omega = 2\pi f = \frac{2\pi}{T} \tag{13.4}$$

The constant ω is called the **angular frequency** and has units of rad/s. We shall discuss the geometric significance of ω in Section 13.4.

We can obtain the velocity of a particle undergoing simple harmonic motion by differentiating Equation 13.1 with respect to time:

Velocity in simple harmonic motion

$$v = \frac{dx}{dt} = -\omega A \sin(\omega t + \delta) \tag{13.5}$$

The acceleration of the particle is given by dv/dt:

Acceleration in simple harmonic motion

$$a = \frac{dv}{dt} = -\omega^2 A \cos(\omega t + \delta) \tag{13.6}$$

Since $x = A \cos(\omega t + \delta)$, we can express Equation 13.6 in the form

$$a = -\omega^2 x \tag{13.7}$$

From Equation 13.5 we see that since the sine and cosine functions oscillate between ± 1, the extreme values of v are equal to $\pm\omega A$. Equation 13.6 tells us that the extreme values of the acceleration are $\pm\omega A^2$. Therefore, the *maximum* values of the velocity and acceleration are given by

Maximum values of velocity and acceleration in simple harmonic motion

$$v_{max} = \omega A \tag{13.8}$$

$$a_{max} = \omega^2 A \tag{13.9}$$

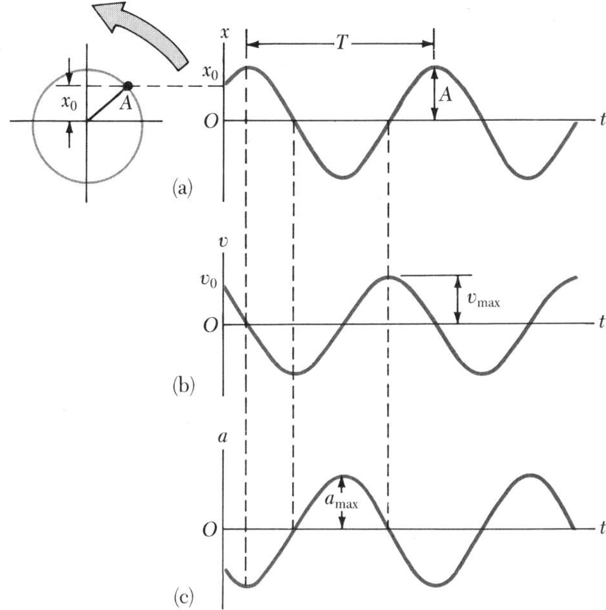

Figure 13.2 Graphical representation of simple harmonic motion: (a) the displacement versus time, (b) the velocity versus time, and (c) the acceleration versus time. Note that the velocity is 90° out of phase with the displacement and the acceleration is 180° out of phase with the displacement.

Figure 13.2a represents the displacement versus time for an arbitrary value of the phase constant. The projection of a point moving with uniform circular motion on a reference circle of radius A also moves in sinusoidal fashion. This will be discussed in more detail in Section 13.5.

The velocity and acceleration versus time curves are illustrated in Figures 13.2b and 13.2c. These curves show that the phase of the velocity differs from the phase of the displacement by $\pi/2$ rad, or 90°. That is, when x is a maximum or a minimum, the velocity is zero. Likewise, when x is zero, the speed is a maximum. Furthermore, note that the phase of the acceleration differs from the phase of the displacement by π radians, or 180°. That is, when x is a maximum, a is a maximum in the opposite direction.

As we stated earlier, the solution $x = A\cos(\omega t + \delta)$ is a general solution of the equation of motion, where the phase constant δ and the amplitude A must be chosen to meet the initial conditions of the motion. The phase constant is important when comparing the motion of two or more oscillating particles. Suppose that the initial position x_0 and initial velocity v_0 of a single oscillator are given, that is, at $t = 0$, $x = x_0$ and $v = v_0$. Under these conditions, the equations $x = A\cos(\omega t + \delta)$ and $v = -\omega A \sin(\omega t + \delta)$ give

$$x_0 = A\cos\delta \qquad \text{and} \qquad v_0 = -\omega A \sin\delta$$

Dividing these two equations eliminates A, giving

$$\frac{v_0}{x_0} = -\omega \tan\delta$$

The phase angle δ and amplitude A can be obtained from the initial conditions

$$\tan\delta = -\frac{v_0}{\omega x_0} \tag{13.10a}$$

Furthermore, if we take the sum $x_0{}^2 + \left(\frac{v_0}{\omega}\right)^2 = A^2\cos^2\delta + A^2\sin^2\delta$ and solve for A, we find that

$$A = \sqrt{x_0^2 + \left(\frac{v_0}{\omega}\right)^2} \tag{13.10b}$$

Thus, we see that δ and A are known if x_0, ω, and v_0 are specified. We shall treat a few specific cases in the next section.

We conclude this section by pointing out the following important properties of a particle moving in simple harmonic motion:

Properties of simple harmonic motion

1. The displacement, velocity, and acceleration all vary sinusoidally with time but are not in phase.
2. The acceleration of the particle is proportional to the displacement, but in the opposite direction.
3. The frequency and the period of motion are independent of the amplitude.

EXAMPLE 13.1. An Oscillating Body

A body oscillates with simple harmonic motion along the x axis. Its displacement varies with time according to the equation

$$x = 4.0 \cos\left(\pi t + \frac{\pi}{4}\right)$$

where x is in m, t is in s, and the angles in the parentheses are in radians. (a) Determine the amplitude, frequency, and period of the motion.

By comparing this equation with the general relation for simple harmonic motion, $x = A\cos(\omega t + \delta)$, we see that $A = 4.0$ m and $\omega = \pi$ rad/s; therefore we find $f = \omega/2\pi = \pi/2\pi = 0.50\ \text{s}^{-1}$ and $T = 1/f = 2.0$ s.

(b) Calculate the velocity and acceleration of the body at any time t.

$$v = \frac{dx}{dt} = -4.0 \sin\left(\pi t + \frac{\pi}{4}\right)\frac{d}{dt}(\pi t)$$

$$= -4\pi \sin\left(\pi t + \frac{\pi}{4}\right) \text{ m/s}$$

$$a = \frac{dv}{dt} = -4\pi \cos\left(\pi t + \frac{\pi}{4}\right)\frac{d}{dt}(\pi t)$$

$$= -4\pi^2 \cos\left(\pi t + \frac{\pi}{4}\right) \text{ m/s}^2$$

(c) Using the results to (b), determine the position, velocity, and acceleration of the body at $t = 1$ s.

Noting that the angles in the trigonometric functions are in radians, we get at $t = 1$ s

$$x = 4.0 \cos\left(\pi + \frac{\pi}{4}\right) = 4.0 \cos\left(\frac{5\pi}{4}\right)$$

$$= 4.0(-0.71) = -2.8 \text{ m}$$

$$v = -4\pi \sin\left(\frac{5\pi}{4}\right) = -4\pi(-0.71) = 8.9 \text{ m/s}$$

$$a = -4\pi^2 \cos\left(\frac{5\pi}{4}\right) = -4\pi^2(-0.71) = 28 \text{ m/s}^2$$

(d) Determine the maximum speed and maximum acceleration of the body.

By analyzing the general relations for v and a in (b), we note that the maximum values of the sine and cosine functions are unity. Therefore, v varies between $\pm 4\pi$ m/s, and a varies between $\pm 4\pi^2$ m/s^2. Thus, $v_{max} = 4\pi$ m/s and $a_{max} = 4\pi^2$ m/s^2. The same results are obtained using $v_{max} = \omega A$ and $a_{max} = \omega^2 A$, where $A = 4.0$ and $\omega = \pi$ rad/s.

(e) Find the displacement of the body between $t = 0$ and $t = 1$ s.

The x coordinate at $t = 0$ is given by

$$x_0 = 4.0 \cos\left(0 + \frac{\pi}{4}\right) = 4.0(0.71) = 2.8 \text{ m}$$

In (c), we found that the coordinate at $t = 1$ s was -2.8 m; therefore the displacement between $t = 0$ and $t = 1$ s is

$$\Delta x = x - x_0 = -2.8 \text{ m} - 2.8 \text{ m} = -5.6 \text{ m}$$

Because the particle's velocity changes sign during the first second, the magnitude of Δx is *not* the same as the distance traveled in the first second.

(f) What is the phase of the motion at $t = 2$ s?

The phase is defined as $\omega t + \delta$, where in this case $\omega = \pi$ and $\delta = \pi/4$. Therefore, at $t = 2$ s, we get

$$\text{Phase} = (\omega t + \delta)_{t=2} = \pi(2) + \pi/4 = 9\pi/4 \text{ rad}$$

13.2 MASS ATTACHED TO A SPRING

In Chapter 7 we introduced the physical system consisting of a mass attached to the end of a spring, where the mass is free to move on a horizontal, frictionless surface (Fig. 13.3). We know from experience that such a system will oscillate back and forth if disturbed from the equilibrium position $x = 0$, where the spring is unstretched. If the surface is frictionless, the mass will exhibit simple harmonic motion. One possible experimental arrangement that clearly demonstrates that such a system exhibits simple harmonic motion is illustrated in Figure 13.4, in which a mass oscillating vertically on a spring has a marking pen attached to it. While the mass is in motion, a sheet of paper is moved horizontally as shown, and the marking pen traces out a sinusoidal pattern. We can understand this qualitatively by first recalling that when the mass is displaced a small distance x from equilibrium, the spring exerts a force on m given by **Hooke's law,**

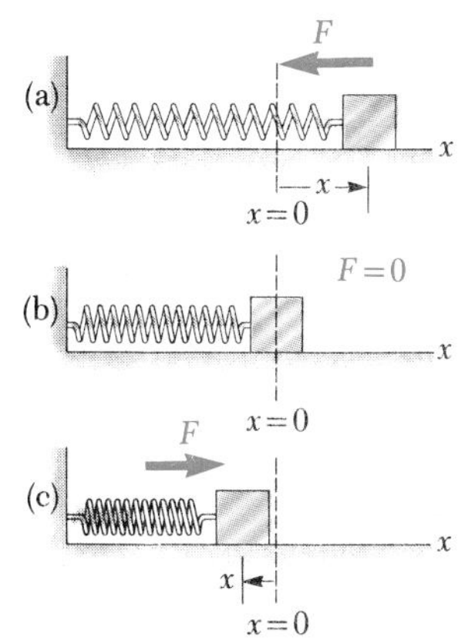

Figure 13.3 A mass attached to a spring on a frictionless surface exhibits simple harmonic motion. (a) When the mass is displaced to the right of equilibrium, the displacement is positive and the acceleration is negative. (b) At the equilibrium position, $x = 0$, the acceleration is zero but the speed is a maximum. (c) When the displacement is negative, the acceleration is positive.

$$F = -kx \tag{13.11}$$

where k is the force constant of the spring. We call this a **linear restoring force** since it is linearly proportional to the displacement and is always directed toward the equilibrium position, *opposite* the displacement. That is, when the mass is displaced to the right in Figure 13.3, x is positive and the restoring force is to the left. When the mass is displaced to the left of $x = 0$, then x is negative and $\mathbf{F}$ is to the right. If we now apply Newton's second law to the motion of m in the x direction, we get

$$F = -kx = ma$$

$$a = -\frac{k}{m}x \tag{13.12}$$

The acceleration of a mass-spring system is proportional to the displacement

that is, *the acceleration is proportional to the displacement of the mass from equilibrium and is in the opposite direction.* If the mass is displaced a maximum distance $x = A$ at some initial time and released from rest, its *initial* acceleration will be $-kA/m$ (that is, it has a maximum negative value). When it passes through the equilibrium position, $x = 0$ and its acceleration is zero. At this instant, its velocity is a maximum. It will then travel to the left of equilibrium and finally reach $x = -A$, at which time its acceleration is kA/m (maximum positive) and its velocity is again zero. Thus, we see that the mass will oscillate between the turning points $x = \pm A$. In one full cycle of its motion, the mass travels a distance $4A$.

We shall now describe the motion in a quantitative fashion. This can be accomplished by recalling that $a = dv/dt = d^2x/dt^2$. Thus, we can express Equation 13.12 as

$$\frac{d^2x}{dt^2} = -\frac{k}{m}x \tag{13.13}$$

If we denote the ratio k/m by the symbol ω^2,

$$\omega^2 = k/m \tag{13.14}$$

then Equation 13.13 can be written in the form

$$\frac{d^2x}{dt^2} = -\omega^2 x \tag{13.15}$$

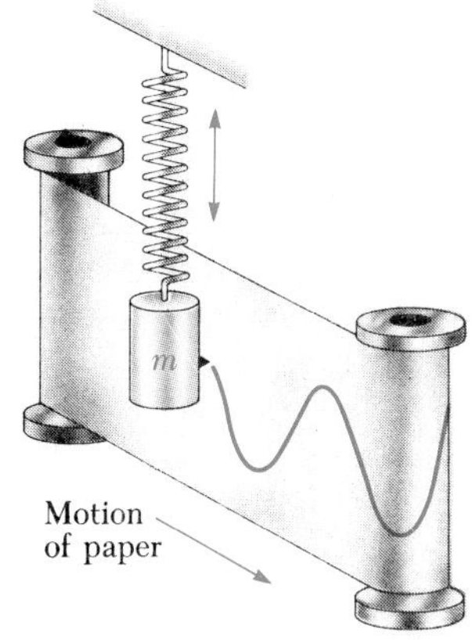

Figure 13.4 An experimental apparatus for demonstrating simple harmonic motion. A pen attached to the oscillating mass traces out a sine wave on the moving chart paper.

What we now require is a solution to Equation 13.15, that is, a function $x(t)$ that satisfies this second-order differential equation. The nature of such a solution $x(t)$ as an algebraic relationship is that it reduces the differential equation to an identity. However, by noting that Equations 13.15 and 13.7 are equivalent, we see that the solution must be that of simple harmonic motion:

$$x(t) = A\cos(\omega t + \delta)$$

To see this explicitly, note that if

$$x = A\cos(\omega t + \delta)$$

then

$$\frac{dx}{dt} = A\frac{d}{dt}\cos(\omega t + \delta) = -\omega A\sin(\omega t + \delta)$$

$$\frac{d^2x}{dt^2} = -\omega A\frac{d}{dt}\sin(\omega t + \delta) = -\omega^2 A\cos(\omega t + \delta)$$

Comparing the expressions for x and d^2x/dt^2, we see that $d^2x/dt^2 = -\omega^2 x$ and Equation 13.15 is satisfied.

The following general statement can be made based on the above discussion:

> Whenever the force acting on a particle is linearly proportional to the displacement and in the opposite direction, the particle will exhibit simple harmonic motion.

We shall give additional physical examples in subsequent sections.

Since the period is given by $T = 2\pi/\omega$ and the frequency is the inverse of the period, we can express the period and frequency of the motion for this system as

Period and frequency for mass-spring system

$$T = \frac{2\pi}{\omega} = 2\pi\sqrt{\frac{m}{k}} \tag{13.16}$$

$$f = \frac{1}{T} = \frac{1}{2\pi}\sqrt{\frac{k}{m}} \tag{13.17}$$

That is, the period and frequency depend *only* on the mass and on the force constant of the spring. As we might expect, the frequency is larger for a stiffer spring and decreases with increasing mass.

It is interesting to note that a mass suspended from a vertical spring attached to a fixed support will also exhibit simple harmonic motion. Although there is a gravitational force to consider in this case, the equation of motion still reduces to Equation 13.15, where the displacement is measured from the equilibrium position of the suspended mass. The proof of this is left as a problem (Problem 5).

Special Case I In order to better understand the physical significance of our solution of the equation of motion, let us consider the following special case. Suppose we extend the mass from equilibrium by a distance A and release it from rest from this stretched position, as in Figure 13.5. We must then require that our solution for $x(t)$ obey the *initial conditions* that at $t = 0$, $x_0 = A$ and $v_0 = 0$. These conditions will be met if we choose $\delta = 0$, giving $x = A\cos\omega t$ as our solution. Note that this is consistent with $x = A\cos(\omega t + \delta)$, where $x_0 = A$

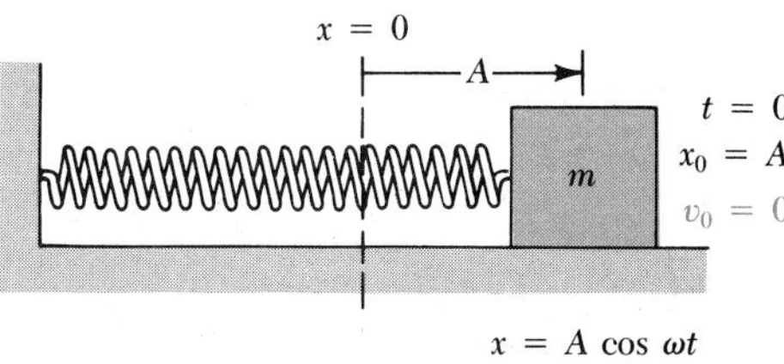

Figure 13.5 A mass-spring system that starts from rest at $x_0 = A$. In this case, $\delta = 0$, and so $x = A \cos \omega t$.

and $\delta = 0$. To check this, we see that the solution $x = A \cos \omega t$ satisfies the condition that $x_0 = A$ at $t = 0$, since $\cos 0 = 1$. Thus, we see that A and δ contain the information on initial conditions. Now let us investigate the behavior of the velocity and acceleration for this special case. Since $x = A \cos \omega t$

$$v = \frac{dx}{dt} = -\omega A \sin \omega t$$

and

$$a = \frac{dv}{dt} = -\omega^2 A \cos \omega t$$

From the velocity expression $v = -\omega A \sin \omega t$, we see that at $t = 0$, $v_0 = 0$, as we require. The expression for the acceleration tells us that at $t = 0$, $a = -\omega^2 A$. Physically this makes sense, since the force on the mass is to the left when the displacement is positive. In fact, at this position $F = -kA$ (to the left), and the initial acceleration is $-kA/m$.

We could also use a more formal approach to show that $x = A \cos \omega t$ is the correct solution by using the relation $\tan \delta = -v_0/\omega x_0$ (Eq. 13.10a). Since $v_0 = 0$ at $t = 0$, $\tan \delta = 0$ and so $\delta = 0$.

The displacement, velocity, and acceleration versus time are plotted in Figure 13.6 for this special case. Note that the acceleration reaches extreme values of $\pm\omega^2 A$ when the displacement has extreme values of $\pm A$. Further-

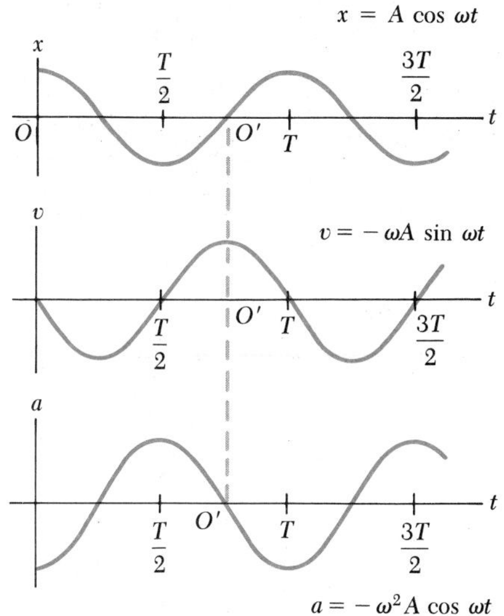

Figure 13.6 Displacement, velocity, and acceleration versus time for a particle undergoing simple harmonic motion under the initial conditions that at $t = 0$, $x_0 = A$ and $v_0 = 0$.

more, the velocity has extreme values of $\pm\omega A$, which both occur at $x = 0$. Hence, the quantitative solution agrees with our qualitative description of this system.

Special Case II Now suppose that the mass starts from the unstretched position moving to the right so that at $t = 0$, $x_0 = 0$ and $v = v_0$ (Fig. 13.7). Our particular solution must now satisfy these initial conditions. Since the mass is moving toward positive x values at $t = 0$, and $x_0 = 0$ at $t = 0$, the solution has the form $x = A \sin \omega t$.

Applying $\tan \delta = -v_0/\omega x_0$ and the initial condition that $x_0 = 0$ at $t = 0$ gives $\tan \delta = -\infty$ or $\delta = -\pi/2$. Hence, the solution is $x = A \cos(\omega t - \pi/2)$, which can be written $x = A \sin \omega t$. Furthermore, from Equation 13.10b we see that $A = v_0/\omega$; therefore we can express our solution as

$$x = \frac{v_0}{\omega} \sin \omega t$$

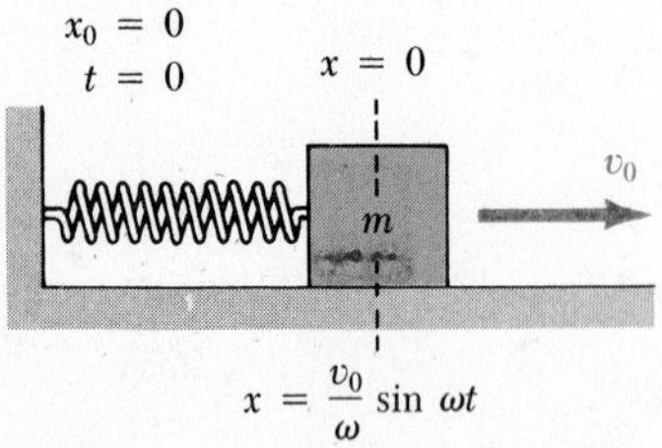

Figure 13.7 The mass-spring system starts its motion at the equilibrium position, $x_0 = 0$ at $t = 0$. If its initial velocity is v_0 to the right, its x coordinate varies as $x = \frac{v_0}{\omega} \sin \omega t$.

The velocity and acceleration in this case are given by

$$v = \frac{dx}{dt} = v_0 \cos \omega t$$

$$a = \frac{dv}{dt} = \omega v_0 \sin \omega t$$

This is consistent with the fact that the mass always has a maximum speed at $x = 0$, while the force and acceleration are zero at this position. The graphs of these functions versus time in Figure 13.6 correspond to the origin at O'. What would be the solution for x if the mass is initially moving to the left in Figure 13.7?

EXAMPLE 13.2. That Car Needs a New Set of Shocks
A car of mass 1300 kg is constructed using a frame supported by four springs. Each spring has a force constant of 20 000 N/m. If two people riding in the car have a combined mass of 160 kg, find the frequency of vibration of the car when it is driven over a pot hole in the road.

Solution: We shall assume that the weight is evenly distributed. Thus, each spring supports one fourth of the load. The total mass supported by the springs is 1460 kg, and therefore each spring supports 365 kg. Hence, the frequency of vibration is

$$f = \frac{1}{2\pi}\sqrt{\frac{k}{m}} = \frac{1}{2\pi}\sqrt{\frac{20\ 000\ \text{N/m}}{365\ \text{kg}}} = 1.18\ \text{Hz}$$

Exercise 1 How long does it take the car to execute two complete vibrations?
Answer: 1.69 s.

EXAMPLE 13.3. A Mass-Spring System
A mass of 200 g is connected to a light spring of force constant 5 N/m and is free to oscillate on a horizontal, frictionless surface. If the mass is displaced 5 cm from equilibrium and released from rest, as in Figure 13.5, (a) find the period of its motion.

First, note that this situation corresponds to Case I, where $x = A \cos \omega t$ and $A = 5 \times 10^{-2}$ m. Therefore,

$$\omega = \sqrt{\frac{k}{m}} = \sqrt{\frac{5\ \text{N/m}}{200 \times 10^{-3}\ \text{kg}}} = 5\ \text{rad/s}$$

Therefore

$$T = \frac{2\pi}{\omega} = \frac{2\pi}{5} = 1.26\ \text{s}$$

(b) Determine the maximum speed of the mass.

$$v_{max} = \omega A = (5\ \text{rad/s})(5 \times 10^{-2}\ \text{m}) = 0.25\ \text{m/s}$$

(c) What is the maximum acceleration of the mass?

$$a_{max} = \omega^2 A = (5\ \text{rad/s})^2(5 \times 10^{-2}\ \text{m}) = 1.25\ \text{m/s}^2$$

(d) Express the displacement, speed, and acceleration as functions of time.

The expression $x = A \cos \omega t$ is our special solution for Case I, and so we can use the results from (a), (b), and (c) to get

$$x = A \cos \omega t = (5 \times 10^2 \cos 5t)\ \text{m}$$

$$v = -\omega A \sin \omega t = (-0.25 \sin 5t)\ \text{m/s}$$

$$a = -\omega^2 A \cos \omega t = (-1.25 \cos 5t)\ \text{m/s}^2$$

13.3 ENERGY OF THE SIMPLE HARMONIC OSCILLATOR

Let us examine the mechanical energy of the mass-spring system described in Figure 13.6. Since the surface is frictionless, we expect that the total mechanical energy is conserved, as was shown in Chapter 8. We can use Equation 13.5 to express the kinetic energy as

Kinetic energy of a simple harmonic oscillator

$$K = \tfrac{1}{2}mv^2 = \tfrac{1}{2}m\omega^2A^2 \sin^2(\omega t + \delta) \qquad (13.18)$$

The elastic potential energy stored in the spring for any elongation x is given by $\tfrac{1}{2}kx^2$. Using Equation 13.1, we get

Potential energy of a simple harmonic oscillator

$$U = \tfrac{1}{2}kx^2 = \tfrac{1}{2}kA^2 \cos^2(\omega t + \delta) \qquad (13.19)$$

We see that K and U are *always* positive quantities. Since $\omega^2 = k/m$, we can express the *total energy* of the simple harmonic oscillator as

$$E = K + U = \tfrac{1}{2}kA^2[\sin^2(\omega t + \delta) + \cos^2(\omega t + \delta)]$$

But $\sin^2 \theta + \cos^2 \theta = 1$, where $\theta = \omega t + \delta$, therefore this equation reduces to

Total energy of a simple harmonic oscillator

$$E = \tfrac{1}{2}kA^2 \qquad (13.20)$$

That is,

the energy of a simple harmonic oscillator is a constant of the motion and proportional to the square of the amplitude.

In fact, the total mechanical energy is just equal to the maximum potential energy stored in the spring when $x = \pm A$. At these points, $v = 0$ and there is no kinetic energy. At the equilibrium position, $x = 0$ and $U = 0$, so that the total energy is all in the form of kinetic energy. That is, at $x = 0$, $E = \tfrac{1}{2}mv_{\text{max}}^2 = \tfrac{1}{2}m\omega^2A^2$.

Plots of the kinetic and potential energies versus time are shown in Figure 13.8a, where we have taken $\delta = 0$. Note that both K and U are always positive and their sum at all times is a constant equal to $\tfrac{1}{2}kA^2$, the total energy of the system. The variations of K and U with displacement are plotted in Figure

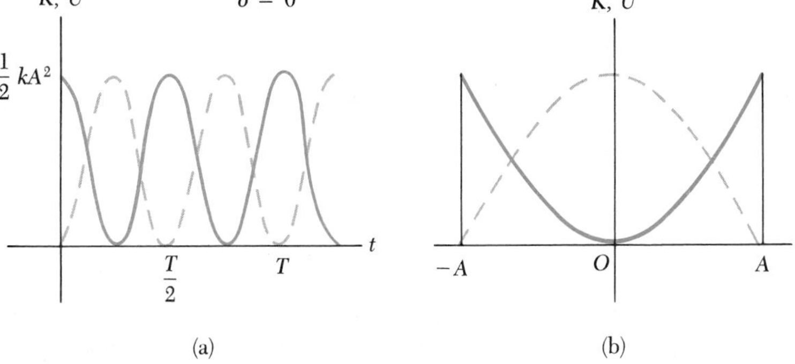

Figure 13.8 (a) Kinetic energy and potential energy versus time for a simple harmonic oscillator with $\delta = 0$. (b) Kinetic energy and potential energy versus displacement for a simple harmonic oscillator. In either plot, note that $K + U =$ constant.

13.8b. You should note that energy is continuously being transferred between potential energy stored in the spring and the kinetic energy of the mass. Figure 13.9 illustrates the position, velocity, acceleration, kinetic energy, and potential energy of the mass-spring system for one full period of the motion. Most of the ideas discussed so far are incorporated in this important figure. We suggest that you study this figure carefully.

Finally, we can use energy conservation to obtain the velocity for an arbitrary displacement x by expressing the total energy at some arbitrary position as

$$E = K + U = \tfrac{1}{2}mv^2 + \tfrac{1}{2}kx^2 = \tfrac{1}{2}kA^2$$

Velocity as a function of position for a simple harmonic oscillator

$$v = \pm\sqrt{\frac{k}{m}(A^2 - x^2)} \qquad (13.21)$$

Again, this expression substantiates the fact that the speed is a maximum at $x = 0$ and is zero at the turning points, $x = \pm A$.

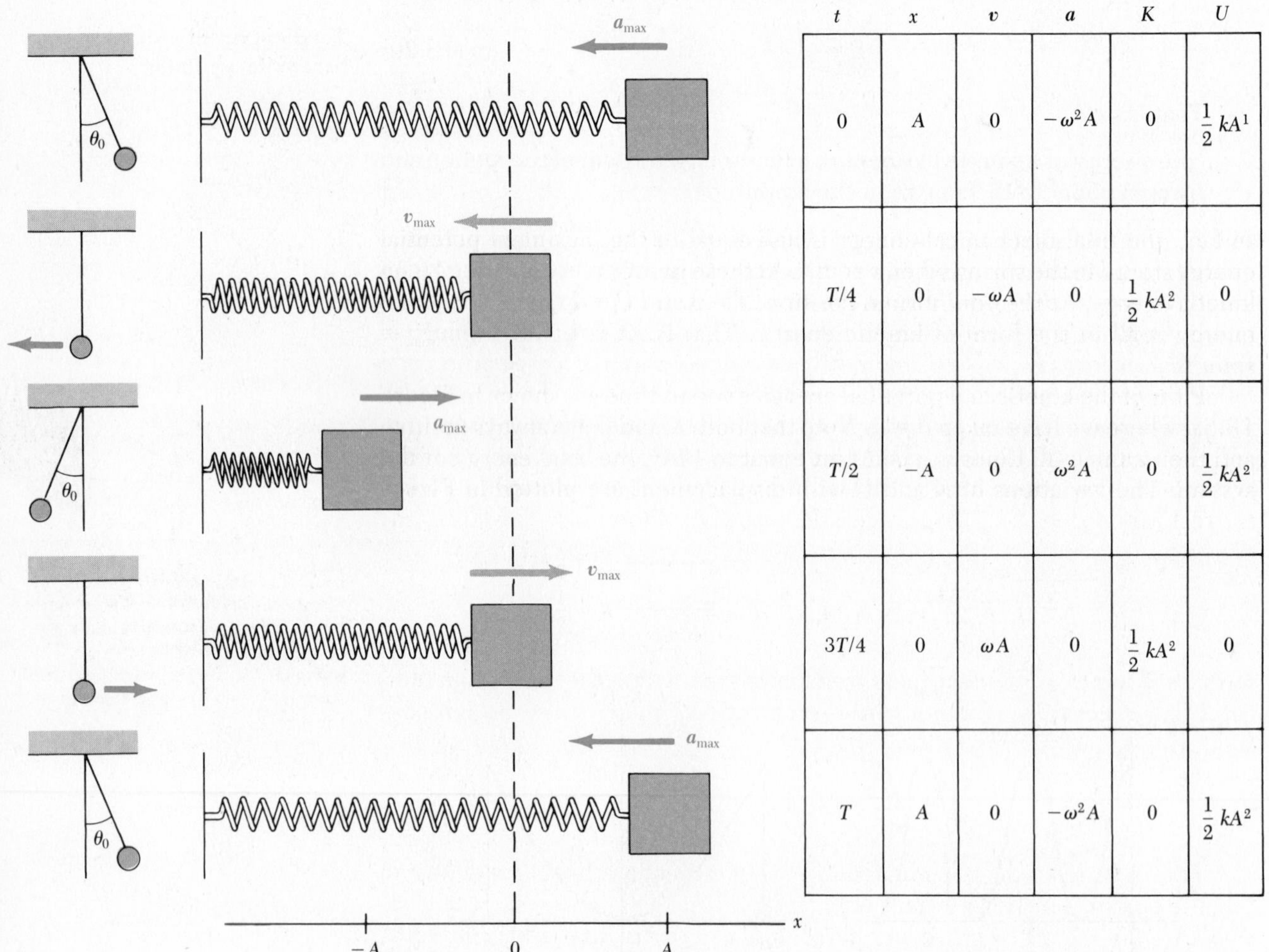

t	x	v	a	K	U
0	A	0	$-\omega^2 A$	0	$\frac{1}{2}kA^1$
$T/4$	0	$-\omega A$	0	$\frac{1}{2}kA^2$	0
$T/2$	$-A$	0	$\omega^2 A$	0	$\frac{1}{2}kA^2$
$3T/4$	0	ωA	0	$\frac{1}{2}kA^2$	0
T	A	0	$-\omega^2 A$	0	$\frac{1}{2}kA^2$

Figure 13.9 Simple harmonic motion for a mass-spring system and its analogy to the motion of a simple pendulum. The parameters in the table at the right refer to the mass-spring system, assuming that at $t = 0$, $x = A$ so that $x = A \cos \omega t$ (Case I).

EXAMPLE 13.4. Oscillations on a Horizontal Surface A mass of 0.5 kg connected to a light spring of force constant 20 N/m oscillates on a horizontal, frictionless surface. (a) Calculate the total energy of the system and the maximum speed of the mass if the amplitude of the motion is 3 cm.

Using Equation 13.20, we get

$$E = \tfrac{1}{2}kA^2 = \tfrac{1}{2}\left(20\ \frac{\text{N}}{\text{m}}\right)(3 \times 10^{-2}\ \text{m})^2$$
$$= 9.0 \times 10^{-3}\ \text{J}$$

When the mass is at $x = 0$, $U = 0$ and $E = \frac{1}{2}mv_{\text{max}}^2$; therefore

$$\tfrac{1}{2}mv_{\text{max}}^2 = 9 \times 10^{-3}\ \text{J}$$

$$v_{\text{max}} = \sqrt{\frac{18 \times 10^{-3}\ \text{J}}{0.5\ \text{kg}}} = 0.19\ \text{m/s}$$

(b) What is the velocity of the mass when the displacement is equal to 2 cm?

We can apply Equation 13.21 directly:

$$v = \pm\sqrt{\frac{k}{m}(A^2 - x^2)} = \pm\sqrt{\frac{20}{0.5}(3^2 - 2^2) \times 10^4}$$
$$= \pm 0.14\ \text{m/s}$$

The positive and negative signs indicate that the mass could be moving to the right or left at this instant.

(c) Compute the kinetic and potential energies of the system when the displacement equals 2 cm.

Using the result to (b), we get

$$K = \tfrac{1}{2}mv^2 = \tfrac{1}{2}(0.5\ \text{kg})(0.14\ \text{m/s})^2 = 5.0 \times 10^{-3}\ \text{J}$$

$$U = \tfrac{1}{2}kx^2 = \tfrac{1}{2}\left(20\ \frac{\text{N}}{\text{m}}\right)(2 \times 10^{-2}\ \text{m})^2 = 4.0 \times 10^{-3}\ \text{J}$$

Note that the sum $K + U$ equals the total energy, E.

Exercise 2 For what values of x does the speed of the mass equal 0.10 m/s?
Answer: ± 2.6 m.

13.4 THE PENDULUM

The Simple Pendulum

The simple pendulum is another mechanical system that exhibits periodic, oscillatory motion. It consists of a point mass m suspended by a light string of length L, where the upper end of the string is fixed as in Figure 13.10. The motion occurs in a vertical plane and is driven by the force of gravity. We shall show that the motion is that of a simple harmonic oscillator, provided the angle θ that the pendulum makes with the vertical is small.

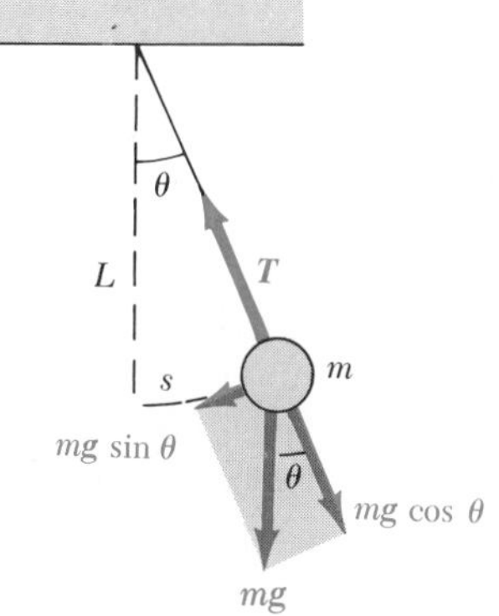

Figure 13.10 When θ is small, the simple pendulum oscillates with simple harmonic motion about the equilibrium position ($\theta = 0$). The restoring force is $mg \sin \theta$, the component of weight tangent to the circle.

The forces acting on the mass are the tension, $\mathbf{T}$, acting along the string, and the weight $m\mathbf{g}$. The tangential component of the weight, $mg \sin \theta$, always acts toward $\theta = 0$, opposite the displacement. Therefore, the tangential force is a restoring force, and we can write the equation of motion in the tangential direction

$$F_t = -mg \sin \theta = m\frac{d^2s}{dt^2}$$

where s is the displacement measured along the arc and the minus sign indicates that F_t acts toward the equilibrium position. Since $s = L\theta$ and L is constant, this equation reduces to

$$\frac{d^2\theta}{dt^2} = -\frac{g}{L}\sin \theta$$

The right side is proportional to $\sin \theta$, rather than to θ; hence we conclude that the motion is not simple harmonic motion. However, if we assume that θ is *small*, we can use the approximation $\sin \theta \approx \theta$, where θ is measured in *radians*.[1]

[1] This approximation can be understood by examining the series expansion for $\sin \theta$, which is $\sin \theta = \theta - \theta^3/3! + \cdots$. For small values of θ, we see that $\sin \theta \approx \theta$. The difference between θ and $\sin \theta$ for $\theta = 15°$ is only about 1%.

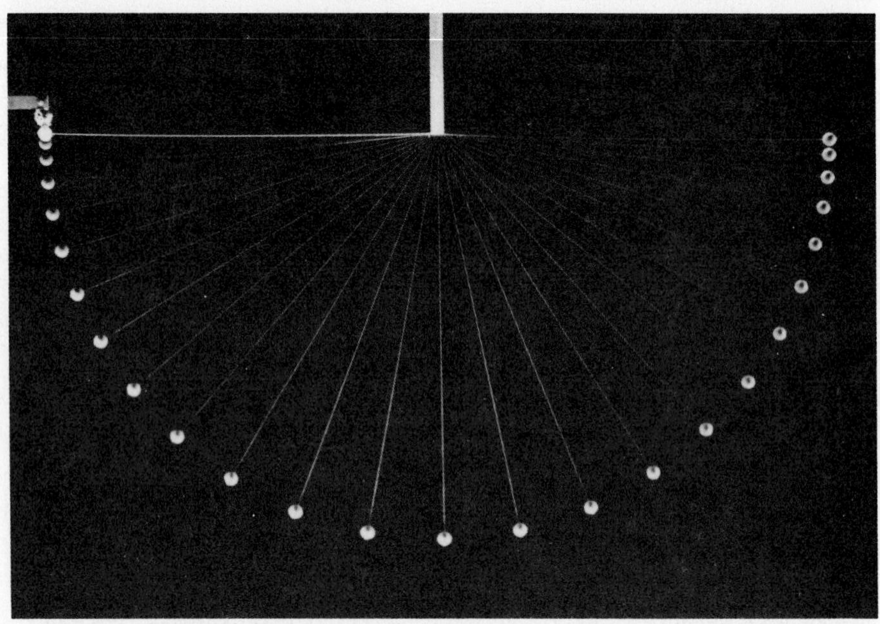

The motion of a simple pendulum captured with multiflash photography. Is the motion simple harmonic in this case? (Photograph © Bernice Abbott, 1963)

Therefore, the equation of motion becomes

Equation of motion for the simple pendulum (small θ)

$$\frac{d^2\theta}{dt^2} = -\frac{g}{L}\,\theta \tag{13.22}$$

Now we have an expression that is of exactly the same form as Equation 13.15, and so we conclude that the motion is simple harmonic motion. Therefore, θ can be written as $\theta = \theta_0 \cos(\omega t + \delta)$, where θ_0 is the *maximum angular displacement* and the angular velocity ω is given by

Angular velocity of motion for the simple pendulum

$$\omega = \sqrt{\frac{g}{L}} \tag{13.23}$$

The period of the motion is

Period of motion for the simple pendulum

$$T = \frac{2\pi}{\omega} = 2\pi\sqrt{\frac{L}{g}} \tag{13.24}$$

In other words, *the period and frequency of a simple pendulum depend only on the length of the string and the acceleration of gravity.* Since the period is *independent* of the mass, we conclude that *all* simple pendula of equal length at the same location oscillate with equal periods.[2] The analogy between the motion of a simple pendulum and the mass-spring system is illustrated in Figure 13.9.

[2] The period of oscillation for the simple pendulum with arbitrary amplitude is

$$T = 2\pi\sqrt{\frac{L}{g}}\left(1 + \frac{1}{4}\sin^2\frac{\theta_0}{2} + \frac{9}{64}\sin^4\frac{\theta_0}{2} + \cdots\right)$$

where θ_0 is the maximum angular displacement in radians.

The simple pendulum can be used as a timekeeper. It is also a convenient device for making precise measurements of the acceleration of gravity. Such measurements are important since variations in local values of **g** can provide information on the location of oil and other valuable underground resources.

EXAMPLE 13.5. What is the Height of That Tower?
A man enters a tall tower. He needs to know the height of the tower. He notes that a long pendulum extends from the ceiling almost to the floor and that its period is 12 s. How tall is the tower?

Solution: If we use $T = 2\pi\sqrt{L/g}$ and solve for L, we get

$$L = \frac{gT^2}{4\pi^2} = \frac{(9.80\ \text{m/s}^2)(12\ \text{s})^2}{4\pi^2} = 35.7\ \text{m}$$

Exercise 3 If the pendulum described in this example is taken to the moon, where the acceleration due to gravity is 1.67 m/s^2, what would its period be there?
Answer: 29.1 s.

The Physical Pendulum

A physical, or compound, pendulum consists of any rigid body suspended from a fixed axis that does not pass through the body's center of mass. The system will oscillate when displaced from its equilibrium position. Consider a rigid body pivoted at a point O that is a distance d from the center of mass (Fig. 13.11). The torque about O is provided by the force of gravity, and its magnitude is $mgd \sin\theta$. Using the fact that $\tau = I\alpha$, where I is the moment of inertia about the axis through O, we get

$$-mgd \sin\theta = I\frac{d^2\theta}{dt^2}$$

The minus sign on the left indicates that the torque about O tends to decrease θ. That is, the force of gravity produces a restoring torque.

If we again assume that θ is small, then the approximation $\sin\theta \approx \theta$ is valid and the equation of motion reduces to

$$\frac{d^2\theta}{dt^2} = -\left(\frac{mgd}{I}\right)\theta = -\omega^2\theta \qquad (13.25)$$

Thus, we note that the equation is of the same form as Equation 13.15, and so the motion is simple harmonic motion. That is, the solution of Equation 13.25 is $\theta = \theta_0 \cos(\omega t + \delta)$, where θ_0 is the maximum angular displacement and

$$\omega = \sqrt{\frac{mgd}{I}}$$

The period is given by

$$T = \frac{2\pi}{\omega} = 2\pi\sqrt{\frac{I}{mgd}} \qquad (13.26)$$

One can use this result to measure the moment of inertia of a planar rigid body. If the location of the center of mass, and hence of d, are known, the moment of inertia can be obtained through a measurement of the period. Finally, note that Equation 13.26 reduces to the period of a simple pendulum (Eq. 13.24) when $I = md^2$, that is, when all the mass is concentrated at the center of mass.

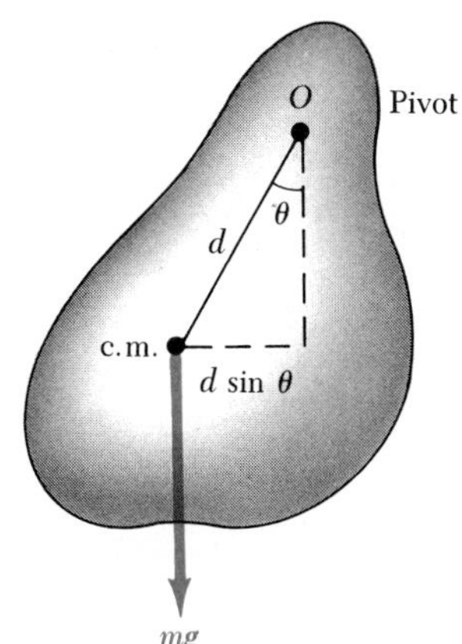

Figure 13.11 The physical pendulum consists of a rigid body pivoted at the point O, and not through the center of mass. At equilibrium, the weight vector passes through O, corresponding to $\theta = 0$. The restoring torque about O when the system is displaced through an angle θ is $mgd \sin\theta$.

EXAMPLE 13.6. A Swinging Rod

A uniform rod of mass M and length L is pivoted about one end and oscillates in a vertical plane (Fig. 13.12). Find the period of oscillation if the amplitude of the motion is small.

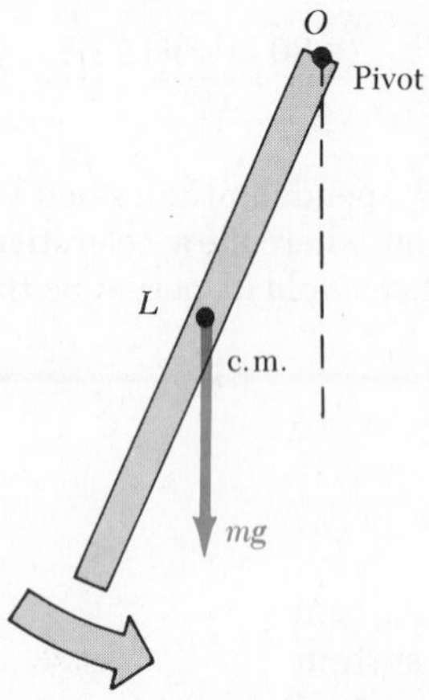

Figure 13.12 (Example 13.6) A rigid rod oscillating about a pivot through one end is a physical pendulum with $d = L/2$ and $I_0 = \frac{1}{3}ML^2$.

Solution: In Chapter 10 we found that the moment of inertia of a uniform rod about an axis through one end is $\frac{1}{3}ML^2$. The distance d from the pivot to the center of mass is $L/2$. Substituting these quantities into Equation 13.26 gives

$$T = 2\pi \sqrt{\frac{\frac{1}{3}ML^2}{Mg\frac{L}{2}}} = 2\pi \sqrt{\frac{2L}{3g}}$$

Comment: In one of the early moon landings, an astronaut walking on the moon's surface had a belt hanging from his spacesuit, and the belt oscillated as a compound pendulum. A scientist on earth observed this motion on TV and was able to estimate the acceleration of gravity on the moon from this observation. How do you suppose this calculation was done?

Exercise 4 Calculate the period of a meter stick pivoted about one end and oscillating in a vertical plane as in Figure 13.12.
Answer: 1.64 s.

Torsional Pendulum

Figure 13.13 shows a rigid body suspended by a wire attached at the top to a fixed support. When the body is twisted through some small angle θ, the twisted wire exerts a restoring torque on the body proportional to the angular displacement. That is,

$$\tau = -\kappa\theta$$

where κ (the Greek letter kappa) is called the *torsion constant* of the support wire. The value of κ can be obtained by applying a known torque to twist the wire through a measurable angle θ. Applying Newton's second law for rotational motion gives

$$\tau = -\kappa\theta = I\frac{d^2\theta}{dt^2}$$

$$\frac{d^2\theta}{dt^2} = -\frac{\kappa}{I}\theta \tag{13.27}$$

Again, this is the equation of motion for a simple harmonic oscillator, with $\omega = \sqrt{\kappa/I}$ and a period

$$T = 2\pi\sqrt{\frac{I}{\kappa}} \tag{13.28}$$

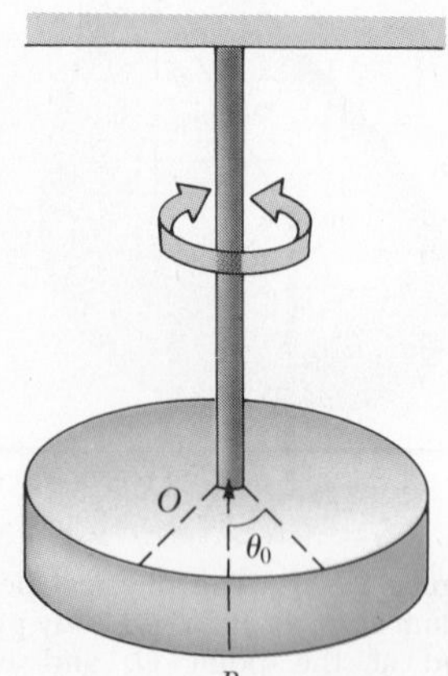

Figure 13.13 A torsional pendulum consists of a rigid body suspended by a wire attached to a rigid support. The body oscillates about the line OP with an amplitude θ_0.

This system is called a *torsional pendulum.* Note that there is no small angle restriction in this situation, as long as the elastic limit of the wire is not exceeded. The balance wheel of a watch oscillates as a torsional pendulum, energized by the mainspring. Torsional pendulums are also used in laboratory galvanometers and the Cavendish torsional balance.

OMIT

*13.5 COMPARING SIMPLE HARMONIC MOTION WITH UNIFORM CIRCULAR MOTION

We can better understand and visualize many aspects of simple harmonic motion along a straight line by looking at its relationship to uniform circular motion. Figure 13.14 shows an experimental arrangement useful for developing this concept. This figure represents a top view of a ball attached to the rim of a phonograph turntable of radius A, illuminated from the side by a lamp. Rather than concentrating on the ball, let us focus our attention on the shadow that the ball casts on the screen. We find that *as the turntable rotates with constant angular velocity, the shadow of the ball moves back and forth with simple harmonic motion.*

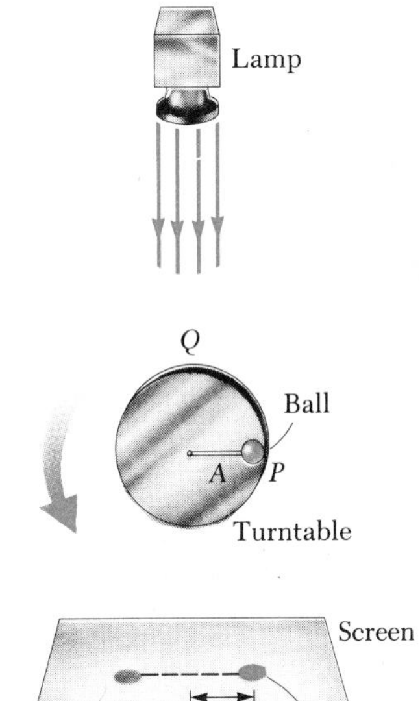

Figure 13.14 Experimental setup for demonstrating the connection between simple harmonic motion and uniform circular motion. As the ball rotates on the turntable with constant angular velocity, its shadow on the screen moves back and forth with simple harmonic motion.

Consider a particle at point P moving in a circle of radius A with constant angular velocity ω (Fig. 13.15a). We shall refer to this circle as the *reference circle* for the motion. As the particle rotates, the position vector of the particle rotates about the origin, O. At some instant of time, t, the angle between OP and the x axis is $\omega t + \delta$, where δ is the angle that OP makes with the x axis at $t = 0$. We take this as our reference point for measuring the angular displacement. As the particle rotates on the reference circle, the angle that OP makes with the x axis *changes* with time. Furthermore, the projection of P onto the x axis, labeled point Q, moves back and forth along a line parallel to the diameter of the reference circle, between the limits $x = \pm A$.

Note that points P and Q have the *same* x coordinate. From the right triangle OPQ, we see that the x coordinate of P and Q is given by

$$x = A\cos(\omega t + \delta) \tag{13.29}$$

This expression shows that the point Q moves with simple harmonic motion along the x axis. Therefore, we conclude that

> simple harmonic motion along a straight line can be represented by the projection of uniform circular motion along a diameter.

By a similar argument, you can see from Figure 13.15a that the projection of P along the y axis also exhibits simple harmonic motion. Therefore, *uniform circular motion can be considered a combination of two simple harmonic motions*, one along x and one along y, where the two differ in phase by 90°.

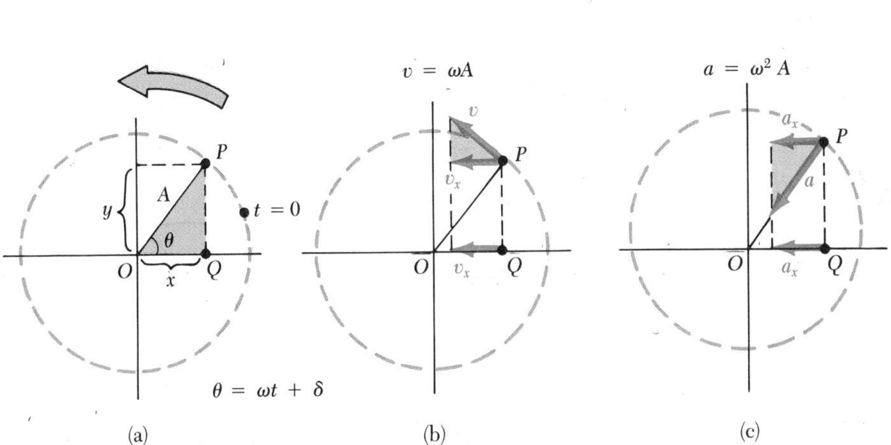

Figure 13.15 Relationship between the uniform circular motion of a point P and the simple harmonic motion of the point Q. A particle at P moves in a circle of radius A with constant angular velocity ω. (a) The x components of the points P and Q are equal and vary in time as $x = A\cos(\omega t + \delta)$. (b) The x component of velocity of P equals the velocity of Q. (c) The x component of the acceleration of P equals the acceleration of Q.

The geometric interpretation we have presented shows that the time for one complete revolution of the point P on the reference circle is equal to the period of motion, T, for simple harmonic motion between $x = \pm A$. That is, the angular speed of the point P is the same as the angular frequency, ω, of simple harmonic motion along the x axis. The phase constant δ for simple harmonic motion corresponds to the initial angle that OP makes with the x axis. The radius of the reference circle, A, equals the amplitude of the simple harmonic motion.

Since the relationship between linear and angular velocity for circular motion is $v = r\omega$, the particle moving on the reference circle of radius A has a velocity of magnitude ωA. From the geometry in Figure 13.15b, we see that the x component of this velocity is given by $-\omega A \sin(\omega t + \delta)$. By definition, the point Q has a velocity given by dx/dt. Differentiating Equation 13.29 with respect to time, we find that the velocity of Q is the same as the x component of velocity of P.

The acceleration of the point P on the reference circle is directed radially inward toward O and has a magnitude given by $v^2/A = \omega^2 A$. From the geometry in Figure 13.15c, we see that the x component of this acceleration is equal to $-\omega^2 A \cos(\omega t + \delta)$. This also coincides with the acceleration of the projected point Q along the x axis, as you can easily verify from Equation 13.29.

EXAMPLE 13.7. Circular Motion With Constant Speed

A particle rotates counterclockwise in a circle of radius 3.0 m with a constant angular speed of 8 rad/s, as in Figure 13.15. At $t = 0$, the particle has an x coordinate of 2.0 m. (a) Determine the x coordinate as a function of time.

Since the amplitude of the particle's motion equals the radius of the circle and $\omega = 8$ rad/s, we have

$$x = A\cos(\omega t + \delta) = 3.0 \cos(8t + \delta)$$

We can evaluate δ using the initial condition that $x =$ 2.0 m at $t = 0$:

$$2.0 = 3.0 \cos(0 + \delta)$$

$$\delta = \cos^{-1}(\tfrac{2}{3}) = 48^\circ = 0.84 \text{ rad}$$

Therefore, the x coordinate versus time is of the form

$$x = 3.0 \cos(8t + 0.84) \text{ m}$$

Note that the angles in the cosine function are in radians.

(b) Find the x components of the particle's velocity and acceleration at any time t.

$$v_x = \frac{dx}{dt} = (-3.0)(8) \sin(8t + 0.84)$$

$$= -24 \sin(8t + 0.84) \text{ m/s}$$

$$a_x = \frac{dv_x}{dt} = (-24)(8) \cos(8t + 0.84)$$

$$= -192 \cos(8t + 0.84) \text{ m/s}^2$$

From these results, we conclude that $v_{max} = 24$ m/s and $a_{max} = 192$ m/s^2. Note that these values also equal the tangential velocity, ωA, and centripetal acceleration, $\omega^2 A$.

*13.6 DAMPED OSCILLATIONS

The oscillatory motions we have considered so far have dealt with an ideal system, that is, one that oscillates indefinitely under the action of a linear restoring force. In realistic systems, dissipative forces, such as friction, are present and retard the motion of the system. Consequently, the mechanical energy of the system will diminish in time, and the motion is said to be *damped.*

One common type of retarding force, which we discussed in Chapter 6, is proportional to the velocity and acts in the direction opposite the motion. This is often observed for the motion of an object through a liquid. Because the retarding force can be expressed as $\mathbf{R} = -b\mathbf{v}$, where b is a constant, and the restoring force is $-kx$, we can write Newton's second law

$$\sum F_x = -kx - bv = ma_x$$

$$-kx - b\frac{dx}{dt} = m\frac{d^2x}{dt^2} \qquad (13.30)$$

The solution of this equation requires mathematics that may not be familiar to you as yet, and so it will simply be stated without proof. When the retarding force is small compared with kx, that is, when b is small, the solution to Equation 13.30 is

$$x = A\,e^{-\frac{b}{2m}t}\cos(\omega t + \delta) \qquad (13.31)$$

where the frequency of motion is

$$\omega = \sqrt{\frac{k}{m} - \left(\frac{b}{2m}\right)^2} \qquad (13.32)$$

This can be verified by substitution of the solution into Equation 13.30. Figure 13.16a shows the displacement as a function of time in this case. We see that *when the dissipative force is small compared with the restoring force, the oscillatory character of the motion is preserved but the amplitude of vibration decreases in time,* and the motion will ultimately cease. This is known as an **underdamped oscillator.** The dotted line in Figure 13.16a, which is the *envelope* of the oscillatory curve, represents the exponential factor that appears in Equation 13.31. This shows that *the amplitude decays exponentially with time.* For motion with a given spring constant and particle mass, the oscillations dampen more rapidly as the maximum value of the dissipative force approaches the maximum value of the restoring force. One example of a damped harmonic oscillator is a mass immersed in a fluid as in Figure 13.16b.

It is convenient to express the frequency of vibration in the form

$$\omega = \sqrt{\omega_0^2 - \left(\frac{b}{2m}\right)^2}$$

where $\omega_0 = \sqrt{k/m}$ represents the frequency of oscillation in the absence of a resistive force (the undamped oscillator). In other words, when $b = 0$, the resistive force is zero and the system oscillates with its natural frequency, ω_0. As the magnitude of the resistive force approaches the value of the restoring

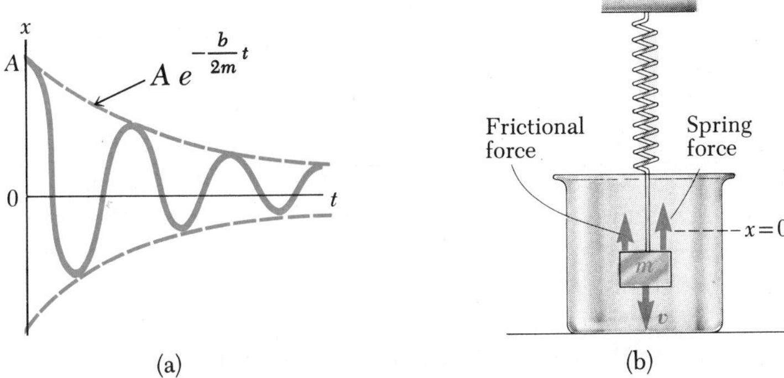

Figure 13.16 (a) Graph of the displacement versus time for an underdamped oscillator. Note the decrease in amplitude with time. (b) One example of a damped oscillator is a mass submersed in a liquid.

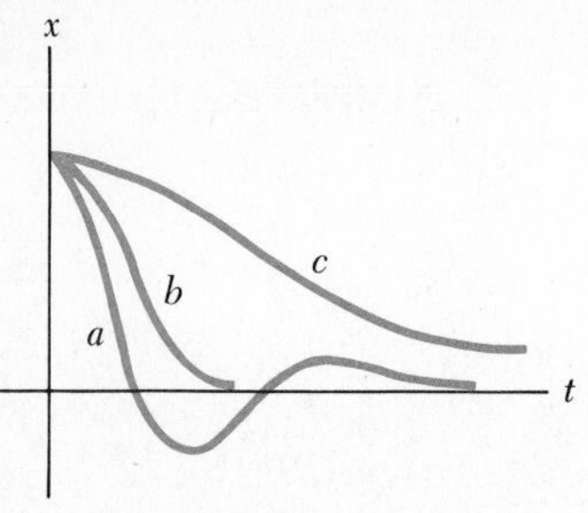

Figure 13.17 Plots of displacement versus time for (*a*) a critically damped oscillator and (*b*) an overdamped oscillator.

force in the spring, the oscillations dampen more rapidly. When b reaches a critical value b_c such that $b_c/2m = \omega_0$, the system does not oscillate and is said to be **critically damped.** In this case, the system returns to equilibrium in an exponential manner with time, as in Figure 13.17.

If the medium is so viscous that the resistive force is greater than the restoring force, that is, if $b/2m > \omega_0$, the system will be **overdamped.** Again, the displaced system does not oscillate but simply returns to its equilibrium position. As the damping increases, the time it takes the displacement to reach equilibrium also increases, as indicated in Figure 13.17. In any case, when friction is present, the energy of the oscillator will eventually fall to zero. The loss in mechanical energy dissipates into heat energy in the resistive medium.

*13.7 FORCED OSCILLATIONS

We have seen that the energy of a damped oscillator decreases in time as a result of the dissipative force. It is possible to compensate for this energy loss by applying an external force that does positive work on the system. At any instant, energy can be put into the system by an applied force that acts in the direction of motion of the oscillator. For example, a child on a swing can be kept in motion by appropriately timed "pushes." The amplitude of motion will remain constant if the energy input per cycle of motion exactly equals the energy lost as a result of friction.

A common example of a forced oscillator is a damped oscillator driven by an external force that varies harmonically, such as $F = F_0 \cos \omega t$, where ω is the angular frequency of the force and F_0 is a constant. Adding this driving force to the left side of Equation 13.30 gives

$$F_0 \cos \omega t - b\frac{dx}{dt} - kx = m\frac{d^2x}{dt^2} \tag{13.33}$$

Again, the solution of this equation is rather lengthy and will not be presented. However, after a sufficiently long period of time, when the energy input per cycle equals the energy lost per cycle, a *steady-state* condition is reached in which the oscillations proceed with constant amplitude. At this time, when the system is in steady state, Equation 13.33 has the following solution:

$$x = A\cos(\omega t + \delta) \tag{13.34}$$

where

$$A = \frac{F_0/m}{\sqrt{(\omega^2 - \omega_0{}^2)^2 + \left(\frac{b\omega}{m}\right)^2}} \tag{13.35}$$

and where $\omega_0 = \sqrt{k/m}$ is the frequency of the undamped oscillator ($b = 0$).

Equation 13.35 shows that the motion of the forced oscillator is not damped since it is being driven by an external force. That is, the external agent provides the necessary energy to overcome the losses due to the resistive force. Note that the mass oscillates at the frequency of the driving force, ω. For small damping, the amplitude becomes large when the frequency of the driving force is near the natural frequency of oscillation, or when $\omega \approx \omega_0$. The dramatic increase in amplitude near the natural frequency is called **resonance,** and the frequency ω_0 is called the **resonance frequency** of the system.

Physically, the reason for large-amplitude oscillations at the resonance frequency is that energy is being transferred to the system under the most favorable conditions. This can be better understood by taking the first time derivative of x, which gives an expression of the velocity of the oscillator. In doing so, one finds that v is proportional to $\sin(\omega t + \delta)$. When the applied force is in phase with $\boldsymbol{v}$, the rate at which work is done on the oscillator by the force $\boldsymbol{F}$ (or the power) equals Fv. Since the quantity Fv is always positive when $\boldsymbol{F}$ and $\boldsymbol{v}$ are in phase, we conclude that *at resonance the applied force is in phase with the velocity and the power transferred to the oscillator is a maximum.*

A graph of the amplitude as a function of frequency for the forced oscillator with and without a resistive force is shown in Figure 13.18. Note that the amplitude increases with decreasing damping ($b \rightarrow 0$). Furthermore, the resonance curve is broadened as the damping increases. Under steady-state conditions, and at any driving frequency, the energy transferred into the system equals the energy lost because of the damping force; hence the average total energy of the oscillator remains constant. In the absence of a damping force ($b = 0$), we see from Equation 13.35 that the steady-state amplitude approaches infinity as $\omega \rightarrow \omega_0$. In other words, if there are no losses in the system, and we continue to drive an initially motionless oscillator with a sinusoidal force that is in phase with the velocity, the amplitude of motion will build up without limit (Fig. 13.18). This does not occur in practice since some damping will always be present. That is, at resonance the amplitude will be large but finite for small damping.

One experiment that demonstrates a resonance phenomenon is illustrated in Figure 13.19. Several pendula of different lengths are suspended from a common beam. If one of them, such as P, is set in motion, the other will begin to oscillate, since they are coupled by the beam. Of those that are forced into oscillation by this coupling, pendulum Q, whose length is the same as that of P (and hence the two pendula have the same natural frequency), will oscillate with the greatest amplitude.

Later in the text we shall see that the phenomenon of resonance appears in other areas of physics. For example, certain electrical circuits have natural (or resonant) frequencies. A structure such as a bridge has natural frequencies, which can be set into resonance by an appropriate driving force. A striking example of such a structural resonance occurred in 1940, when the Tacoma

These photographs show the collapse of the Tacoma Narrows suspension bridge in 1940 and provide a vivid demonstration of mechanical resonance. High winds set up standing waves in the bridge, causing it to oscillate at a frequency near to one of the natural frequencies of the bridge structure. Once established, this resonance condition led to the bridge's collapse. (United Press International Photo)

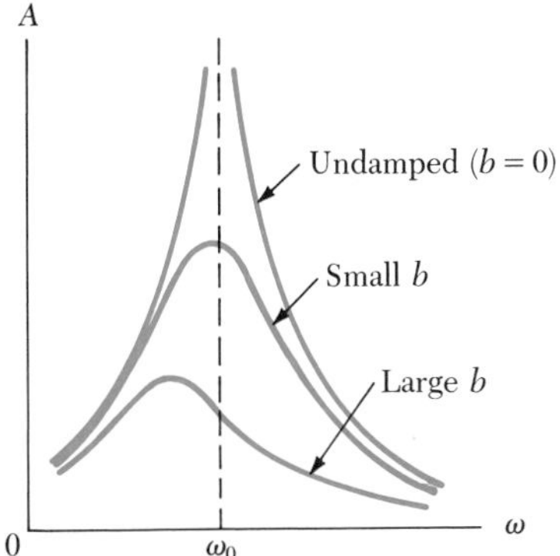

Figure 13.18 Graph of the amplitude versus frequency for a damped oscillator when a periodic driving force is present. When the frequency of the driving force equals the natural frequency, ω_0, resonance occurs. Note that the shape of the resonance curve depends on the size of the damping coefficient, b.

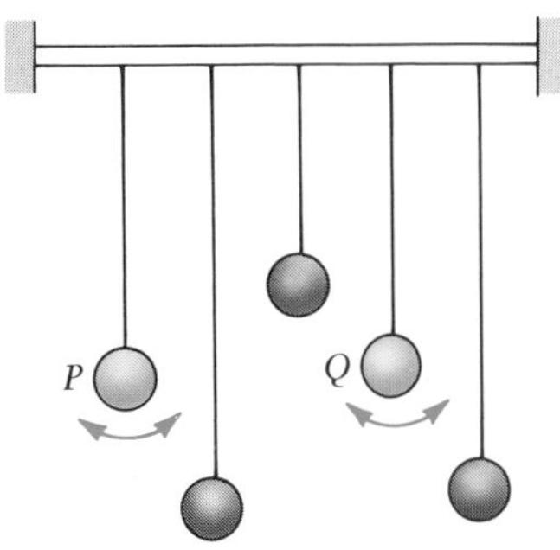

Figure 13.19 If pendulum P is set into oscillation, pendulum Q will eventually oscillate because of the coupling between them and the fact that they have the same natural frequency of vibration.

Narrows bridge in Washington was set into torsional oscillation by heavy winds. The amplitude of these oscillations increased steadily until the bridge ultimately was destroyed.

13.8 SUMMARY

The position of a simple harmonic oscillator varies periodically in time according to the relation

Displacement versus time for simple harmonic motion

$$x = A\cos(\omega t + \delta) \tag{13.1}$$

where A is the amplitude of the motion, ω is the angular frequency, and δ is the phase constant. The value of δ depends on the initial position and velocity of the oscillator.

The time for one complete vibration is called the **period** of the motion, defined by

Period

$$T = \frac{2\pi}{\omega} \tag{13.2}$$

The inverse of the period is the **frequency** of the motion, which equals the number of oscillations per second.

The **velocity** and **acceleration** of a simple harmonic oscillator are given by

Velocity in simple harmonic motion

$$v = \frac{dx}{dt} = -\omega A\sin(\omega t + \delta) \tag{13.5}$$

Acceleration in simple harmonic motion

$$a = \frac{dv}{dt} = -\omega^2 A\cos(\omega t + \delta) \tag{13.6}$$

Thus, the maximum velocity is ωA, and the maximum acceleration is $\omega^2 A$. The velocity is zero when the oscillator is at its turning points, $x = \pm A$, and the speed is a maximum at the equilibrium position, $x = 0$. The magnitude of the acceleration is a maximum at the turning points and is zero at the equilibrium position.

A mass-spring system exhibits simple harmonic motion on a frictionless surface, with a period given by

Period of motion for mass-spring system

$$T = \frac{2\pi}{\omega} = 2\pi\sqrt{\frac{m}{k}} \tag{13.16}$$

where k is the force constant of the spring and m is the mass attached to the spring.

The kinetic energy and potential energy for a simple harmonic oscillator vary with time and are given by

Kinetic and potential energy of a simple harmonic oscillator

$$K = \tfrac{1}{2}mv^2 = \tfrac{1}{2}m\omega^2 A^2\sin^2(\omega t + \delta) \tag{13.18}$$

$$U = \tfrac{1}{2}kx^2 = \tfrac{1}{2}kA^2\cos^2(\omega t + \delta) \tag{13.19}$$

The **total energy** of a simple harmonic oscillator is a constant of the motion and is given by

$$E = \tfrac{1}{2}kA^2 \tag{13.20}$$

Total energy of a simple harmonic oscillator

The potential energy of a simple harmonic oscillator is a maximum when the particle is at its turning points (maximum displacement from equilibrium) and is zero at the equilibrium position. The kinetic energy is zero at the turning points and is a maximum at the equilibrium position.

A simple pendulum of length L exhibits simple harmonic motion for small angular displacements from the vertical, with a *period* given by

$$T = 2\pi \sqrt{\frac{L}{g}} \tag{13.24}$$

Period of motion for a simple pendulum

That is, the period is *independent* of the suspended mass.

A physical pendulum exhibits simple harmonic motion about a pivot that does not go through the center of mass. The period of this motion is

$$T = 2\pi \sqrt{\frac{I}{mgd}} \tag{13.26}$$

Period of motion for a compound pendulum

where I is the moment of inertia about an axis through the pivot and d is the distance from the pivot to the center of mass.

Damped oscillations occur in a system in which a dissipative force opposes the linear restoring force. If such a system is set into motion and then left to itself, the mechanical energy decreases in time because of the presence of the nonconservative damping force. It is possible to compensate for this loss in energy by driving the system with an external periodic force that is in phase with the motion of the system. When the frequency of the driving force matches the natural frequency of the undamped oscillator that starts its motion from rest, energy is continuously transferred to the oscillator and its amplitude increases without limit.

QUESTIONS

1. What is the total distance traveled by a body executing simple harmonic motion in a time equal to its period if its amplitude is A?
2. If the coordinate of a particle varies as $x = -A \cos \omega t$, what is the phase constant δ in Equation 13.1? At what position does the particle begin its motion?
3. Does the displacement of an oscillating particle between $t = 0$ and a later time t necessarily equal the position of the particle at time t? Explain.
4. Determine whether or not the following quantities can be in the same direction for a simple harmonic oscillator: (a) displacement and velocity, (b) velocity and acceleration, (c) displacement and acceleration.
5. Can the amplitude A and phase constant δ be determined for an oscillator if only the position is specified at $t = 0$? Explain.
6. Describe qualitatively the motion of a mass-spring system if the mass of the spring is not neglected.
7. If a mass-spring system is hung vertically and set into oscillation, why does the motion eventually stop?
8. Explain why the kinetic and potential energies of a mass-spring system can never be negative.
9. A mass-spring system undergoes simple harmonic motion with an amplitude A. Does the total energy change if the mass is doubled but the amplitude is not changed? Do the kinetic and potential energies depend on the mass? Explain.
10. What happens to the period of a simple pendulum if its length is doubled? What happens if the mass that is suspended is doubled?
11. A simple pendulum is suspended from the ceiling of a stationary elevator, and the period is determined. Describe the changes, if any, in the period if the elevator (a) accelerates upward, (b) accelerates downward, and (c) moves with constant velocity.
12. A simple pendulum undergoes simple harmonic motion when θ is small. Will the motion be *periodic* if θ is large? How does the period of motion change as θ increases?

13. Give a few examples of damped oscillations that are commonly observed.
14. Will damped oscillations occur for any values of b and k? Explain.
15. Is it possible to have damped oscillations when a system is at resonance? Explain.
16. At resonance, what does the phase constant δ equal in Equation 13.34? (*Hint:* Compare this with the expression for the driving force, and note that the force must be in phase with the velocity at resonance.)
17. A platoon of soldiers marches in step along a road. Why are they ordered to break step when crossing a bridge?

PROBLEMS

Section 13.1 Simple Harmonic Motion

1. The displacement of a particle is given by the expression $x = 4\cos(3\pi t + \pi)$, where x is in m and t is in s. Determine (a) the frequency and period of the motion, (b) the amplitude of the motion, (c) the phase constant, and (d) the position of the particle at $t = 0$.
2. For the particle described in Problem 1, determine (a) the velocity at any time t, (b) the acceleration at any time, (c) the maximum velocity and maximum acceleration, and (d) the velocity and acceleration at $t = 0$.
3. A particle oscillates with simple harmonic motion such that its displacement varies according to the expression as $x = 5\cos(2t + \pi/6)$, where x is in cm and t is in s. At $t = 0$, find (a) the displacement of the particle, (b) its velocity, and (c) its acceleration. (d) Find the period and amplitude of the motion.
4. A particle moving with simple harmonic motion travels a total distance of 20 cm in each cycle of its motion, and its maximum acceleration is 50 m/s^2. Find (a) the angular frequency of the motion and (b) the maximum speed of the particle.
5. The displacement of a body is given by the expression $x = 8.0\cos(2t + \pi/3)$, where x is in cm and t is in s. Calculate (a) the velocity and acceleration at $t = \pi/2$ s, (b) the maximum speed and the earliest time ($t > 0$) at which the particle has this speed, and (c) the maximum acceleration and the earliest time ($t > 0$) at which the particle has this acceleration.
6. At $t = 0$ a particle moving with simple harmonic motion is at $x_0 = 2$ cm, where its velocity is given by $v_0 = -24$ cm/s. If the period of its motion is 0.5 s and the frequency is 2 Hz, find (a) the phase constant; (b) the amplitude; (c) the displacement, velocity, and acceleration as functions of time; and (d) the maximum speed and maximum acceleration.
7. A particle moving along the x axis with simple harmonic motion starts from the origin at $t = 0$ and moves toward the right. If the amplitude of its motion is 2 cm and the frequency is 1.5 Hz, (a) show that its displacement is given by $x = 2\sin 3\pi t$ cm. Determine (b) the maximum speed and the earliest time ($t > 0$) at which the particle has this speed, (c) the maximum acceleration and the earliest time ($t > 0$) at which the particle has this acceleration, and (d) the total *distance* traveled between $t = 0$ and $t = 1$ s.

Section 13.2 Mass Attached to a Spring (neglect spring masses)

8. A spring stretches by 3.9 cm when a 10-g mass is hung from it. If a total mass of 25 g attached to this spring oscillates in simple harmonic motion, calculate the period of motion.
9. The frequency of vibration of a mass-spring system is 5 Hz when a 4-g mass is attached to the spring. What is the force constant of the spring?
10. A 1-kg mass attached to a spring of force constant 25 N/m oscillates on a horizontal, frictionless surface. At $t = 0$, the mass is released from rest at $x = -3$ cm. (That is, the spring is compressed by 3 cm.) Find (a) the period of its motion, (b) the maximum values of its speed and acceleration, and (c) the displacement, velocity, and acceleration as functions of time.
11. A simple harmonic oscillator takes 12 s to undergo 5 complete vibrations. Find (a) the period of its motion, (b) the frequency in Hz, and (c) the angular frequency in rad/s.
12. A mass-spring system oscillates such that the displacement is given by $x = 0.25\cos 2\pi t$ m. (a) Find the speed and acceleration of the mass when $x = 0.10$ m. (b) Determine the maximum speed and maximum acceleration.
13. A 0.5-kg mass attached to a spring of force constant 8 N/m vibrates with simple harmonic motion with an amplitude of 10 cm. Calculate (a) the maximum value of its speed and acceleration, (b) the speed and acceleration when the mass is at $x = 6$ cm from the equilibrium position, and (c) the time it takes the mass to move from $x = 0$ to $x = 8$ cm.

Section 13.3 Energy of the Simple Harmonic Oscillator (neglect spring masses)

14. A 200-g mass is attached to a spring and executes simple harmonic motion with a period of 0.25 s. If the total energy of the system is 2 J, find (a) the force constant of the spring and (b) the amplitude of the motion.
15. A mass-spring system oscillates with an amplitude of 3.5 cm. If the spring constant is 250 N/m and the mass is 0.5 kg, determine (a) the mechanical energy of the system, (b) the maximum speed of the mass, and (c) the maximum acceleration.

16. A simple harmonic oscillator has a total energy E. (a) Determine the kinetic and potential energies when the displacement equals one half the amplitude. (b) For what value of the displacement does the kinetic energy equal the potential energy?
17. The amplitude of a system moving with simple harmonic motion is doubled. Determine the change in (a) the total energy, (b) the maximum velocity, (c) the maximum acceleration, and (d) the period.
18. A mass-spring system of force constant 50 N/m undergoes simple harmonic motion on a horizontal surface with an amplitude of 12 cm. (a) What is the total energy of the system? (b) What is the kinetic energy of the system when the mass is 9 cm from equilibrium? (c) What is its potential energy when $x = 9$ cm?
19. A particle executes simple harmonic motion with an amplitude of 3.0 cm. At what displacement from the midpoint of its motion will its speed equal one half of its maximum speed?

Section 13.4 The Pendulum

20. A simple pendulum has a period of 2.50 s. (a) What is its length? (b) What would its period be on the moon where $g_m = 1.67$ m/s^2?
21. Calculate the frequency and period of a simple pendulum of length 10 m.
22. If the length of a simple pendulum is quadrupled, what happens to (a) its frequency and (b) its period?
23. A simple pendulum 2.00 m in length oscillates in a location where $g = 9.80$ m/s^2. How many complete oscillations will it make in 5 min?
24. A uniform rod is pivoted at one end as in Figure 13.12. If the rod swings with simple harmonic motion, what must its length be in order that its period be equal to that of a simple pendulum 1 m long?
25. A simple pendulum has a length of 3.00 m. Determine the *change* in its period if it is taken from a point where $g = 9.80$ m/s^2 to a higher elevation, where $g = 9.79$ m/s^2.
26. A circular hoop of radius R is hung over a knife edge. Show that its period of oscillation is equal to that of a simple pendulum of length $2R$.
27. A physical pendulum in the form of a planar body exhibits simple harmonic motion with a frequency of 1.5 Hz. If the pendulum has a mass of 2.2 kg and the pivot is located 0.35 m from the center of mass, determine the moment of inertia of the pendulum.

Section 13.6 Damped Oscillations

28. Show that the damping constant, b, has units of kg/s.
29. Show that Equation 13.31 is a solution of Equation 13.30 provided that $b^2 < 4mk$.
30. Show that the time rate of change of mechanical energy for a damped, undriven oscillator is given by $dE/dt = -bv^2$ and hence is *always negative.* (*Hint:* Differentiate the expression for the mechanical energy of an oscillator, $E = \frac{1}{2}mv^2 + \frac{1}{2}kx^2$, and make use of Eq. 13.30.)

Section 13.7 Forced Oscillations

31. A 2-kg mass attached to a spring is driven by an external force $F = 3 \cos 2\pi t$ N. If the force constant of the spring is 20 N/m, determine (a) the period and (b) the amplitude of the motion. (*Hint:* Assume that there is *no* damping, that is, $b = 0$, and make use of Eq. 13.35.)
32. Calculate the resonant frequencies of the following systems: (a) a 3-kg mass attached to a spring of force constant 240 N/m, (b) a simple pendulum 1.5 m in length.
33. Consider an *undamped* forced oscillator ($b = 0$), and show that Equation 13.34 is a solution of Equation 13.33, with an amplitude given by Equation 13.35.

GENERAL PROBLEMS

34. A car with bad shock absorbers bounces up and down with a period of 1.5 s after hitting a bump. The car has a mass of 1500 kg and is supported by four springs of equal force constant k. Determine a value for k.
35. When the simple pendulum illustrated in Figure 13.20 makes an angle θ with the vertical, its speed is v. (a) Calculate the total mechanical energy of the pendulum as a function of v and θ. (b) Show that when θ is small, the potential energy can be expressed as $\frac{1}{2}mgL\theta^2 = \frac{1}{2}m\omega^2 s^2$. (*Hint:* In part (b), approximate $\cos \theta$ by $\cos \theta \approx 1 - \theta^2/2$.)

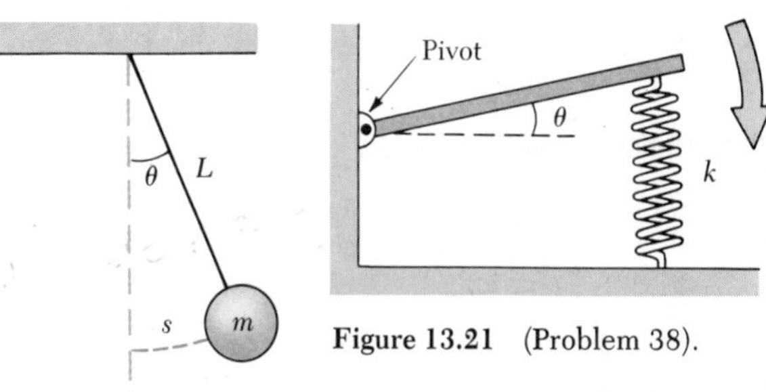

Figure 13.21 (Problem 38).

Figure 13.20 (Problem 35).

36. A horizontal platform vibrates with simple harmonic motion in the horizontal direction with a period of 2 s. A body on the platform starts to slide when the amplitude of vibration reaches 0.3 m. Find the coefficient of static friction between the body and the platform.
37. A particle of mass m slides inside a hemispherical bowl of radius R. Show that for small displacements from equilibrium, the particle exhibits simple harmonic motion with an angular frequency equal to that of a simple pendulum of length R. That is, $\omega = \sqrt{g/R}$.
38. A horizontal plank of mass m and length L is pivoted at one end, and the opposite end is attached to a spring of force constant k (Fig. 13.21). The moment of inertia of the plank about the pivot is $\frac{1}{3}mL^2$. If the plank is displaced a *small* angle θ from the horizontal and released, show that it will move with simple harmonic motion with an angular frequency given by $\omega = \sqrt{3k/m}$.

39. A mass M is attached to the end of a uniform rod of mass m and length L, which is pivoted at the top (Fig. 13.22). (a) Determine the tensions in the rod at the pivot and at the point P when the system is stationary. (b) Calculate the period of oscillation for small displacements from equilibrium, and determine this period for $L = 2$ m. (*Hint:* Assume the mass M is a point mass, and make use of Eq. 13.26.)

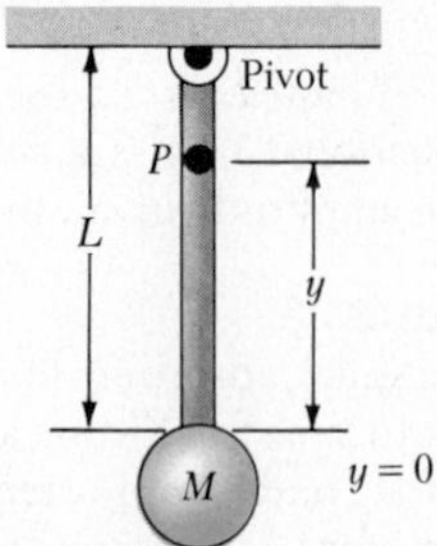

Figure 13.22 (Problem 39).

40. A mass M is connected to a spring of mass m and oscillates in simple harmonic motion on a horizontal, smooth surface (Fig. 13.23). The force constant of the spring is k and the equilibrium length is ℓ. Find (a) the kinetic energy of the system when the mass has a speed v and (b) the period of oscillation. (*Hint:* Assume that all portions of the spring oscillate in phase and that the velocity of a segment dx is proportional to the distance from the fixed end; that is, $v_x = \frac{x}{\ell} v$. Also, note that the mass of a segment of the spring is $dm = \frac{m}{\ell} dx$.)

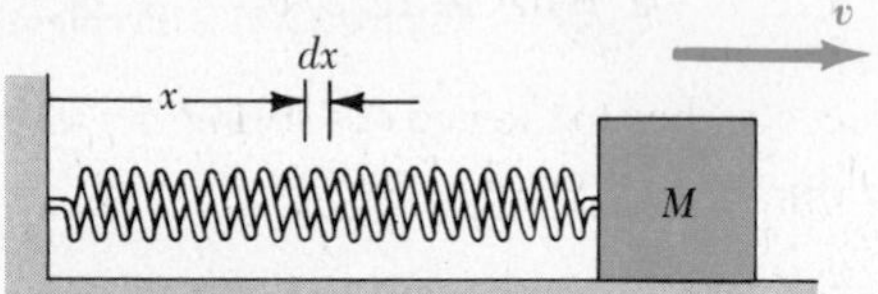

Figure 13.23 (Problem 40).

41. A small thin disk of radius r and mass m is attached rigidly to the face of a second thin disk of radius R and mass M as shown in Figure 13.24. The center of the small disk is located at the edge of the large disk. The large disk is mounted at its center on a frictionless axle. The assembly is rotated through an angle θ from its equilibrium position and released. (a) Show that the magnitude of the velocity of the center of the small disk as it passes through the equilibrium position is

$$v = 2\left[\frac{Rg(1 - \cos\theta)}{(M/m) + (r^2/R)^2 + 2}\right]^{1/2}.$$

(b) Show that the period of the motion is

$$T = 2\pi\left[\frac{(M + 2m)R^2 + mr^2}{2\,mgR}\right]^{1/2}.$$

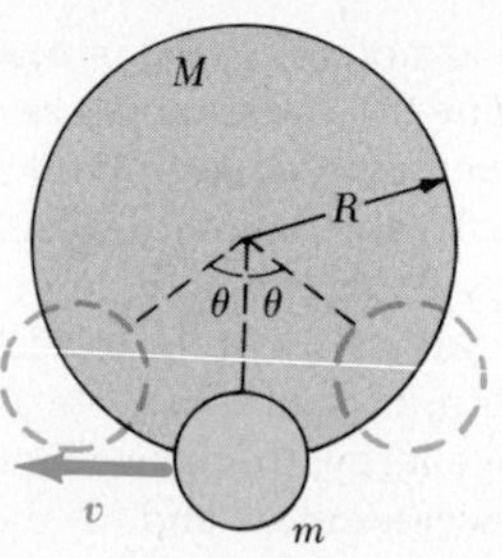

Figure 13.24 (Problem 41).

42. A mass m is connected to two springs of force constants k_1 and k_2 as in Figures 13.25a and 13.25b. Show that in each case the mass exhibits simple harmonic motion with periods (a) $T = 2\pi\sqrt{\frac{m(k_1 + k_2)}{k_1 k_2}}$ and (b) $T = 2\pi\sqrt{\frac{m}{k_1 + k_2}}$.

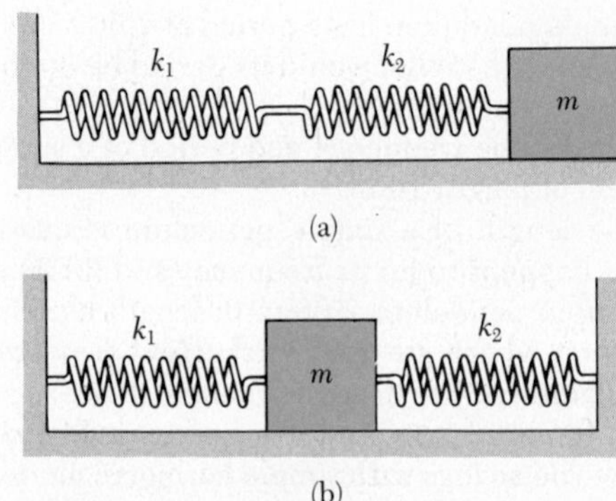

Figure 13.25 (Problem 42).

43. A pendulum of length L and mass M has a spring of force constant k connected to it at a distance h below its point of suspension (Fig. 13.26). Find the frequency of vibration of the system for small values of the amplitude (small θ). (Assume the vertical suspension of length L is rigid, but neglect its mass.)

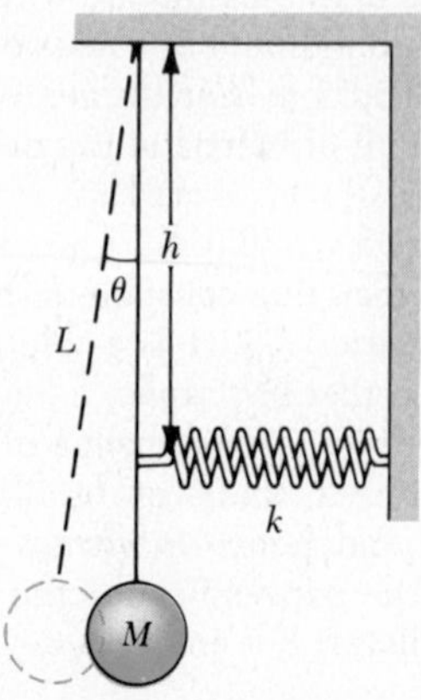

Figure 13.26 (Problem 43).

44. A mass m is attached to a spring of force constant k hanging vertically (Fig. 13.27). If the mass is released when the spring is *unstretched*, show that (a) the system exhibits simple harmonic motion with displacement measured from the unstretched position given by $y = \frac{mg}{k}(1 - \cos \omega t)$, where $\omega = \sqrt{k/m}$, and (b) the maximum tension in the spring is $2mg$.

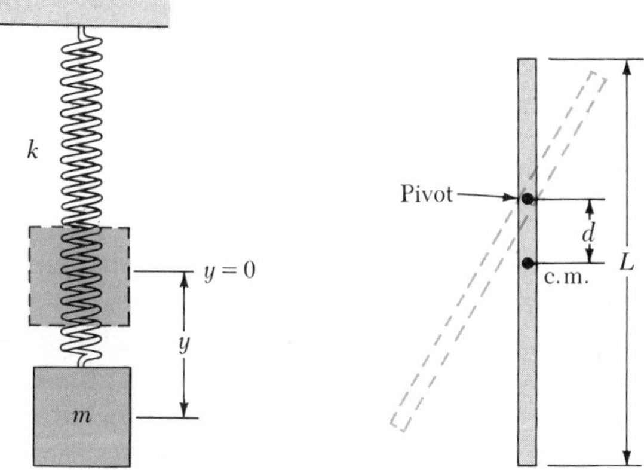

Figure 13.27 (Problem 44). **Figure 13.28** (Problem 45).

45. A homogeneous rod of length L is pivoted at a distance d above its center of mass (Fig. 13.28). For a small displacement from the vertical equilibrium position, the rod exhibits simple harmonic motion. (a) Find the angular frequency of this motion. (b) If the rod is a meter stick pivoted at the 75-cm mark, what is its period? (The lower end is at 0 cm.)

46. A 50-g mass attached to a spring moves on a horizontal, frictionless surface in simple harmonic motion. Its amplitude is 16 cm, and its period is 4 s. At $t = 0$, the mass is released from rest at $x = 16$ cm, as in Figure 13.5. Find (a) the displacement as a function of time and its value at $t = 0.5$ s, (b) the magnitude and direction of the force acting on the mass at $t = 0.5$ s, (c) the minimum time required for the mass to reach the position $x = 8$ cm, (d) the velocity at any time t and the speed at $x = 8$ cm, and (e) the total mechanical energy and force constant of the spring.

•47. A spherical mass m of radius R is suspended from a light string of length $L - R$ (Fig. 13.29). (a) Determine the moment of inertia for this physical pendulum about the point O using the parallel-axis theorem. (b) Calculate the period for small displacements from equilibrium. (c) If $R \ll L$, show that the period is that of a simple pendulum.

•48. A mass m is connected to two rubber bands of length L, each under tension T, as in Figure 13.30. The mass is displaced by a *small* distance y vertically. Assuming the tension does not change appreciably, show that (a) the restoring force is $-(2T/L)y$ and (b) the system exhibits simple harmonic motion with an angular frequency given by $\omega = \sqrt{2T/mL}$.

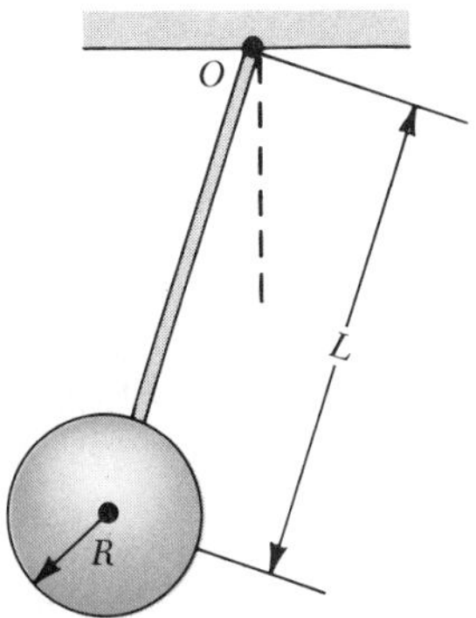

Figure 13.29 (Problem 47).

•49. A light cubical container of volume a^3 is initially filled with a fluid of mass density ρ. The cube is initially supported by a light string to form a pendulum of length L_0 measured from the center of mass of the filled container. The fluid is allowed to flow from the bottom of the cube at a constant rate (dM/dt). At any time t the level of the fluid in the container is ℓ and the length of the pendulum is L (measured relative to the instantaneous center of mass). (a) Sketch the apparatus and label the dimensions a, ℓ, L_0, and L. (b) Find the time rate of change of the period as a function of time t. (c) Find the period T as a function of time.

CALCULATOR/COMPUTER PROBLEMS

50. Using Equations 13.18 and 13.19, plot (a) the kinetic energy versus time and (b) the potential energy versus time for a simple harmonic oscillator. For convenience, take $\delta = 0$. What features do these graphs illustrate?

51. An object attached to the end of a spring vibrates with an amplitude of 20 cm. Find the position of the object at these times: 0, $T/8$, $T/4$, $3T/8$, $T/2$, $5T/8$, $3T/4$, $7T/8$, and T, where T is the period of vibration. Plot your results (position along the vertical axis and time along the horizontal axis).

52. A body oscillates with simple harmonic motion according to the equation $x = -7 \cos 2\pi t$ cm. (a) Determine the velocity and acceleration as functions of time. (b) Make a table of x, v, and a versus t for the interval $t = 0$ to $t = 1$ s in steps of 0.1 s. (c) Plot x, v, and a versus time for this interval.

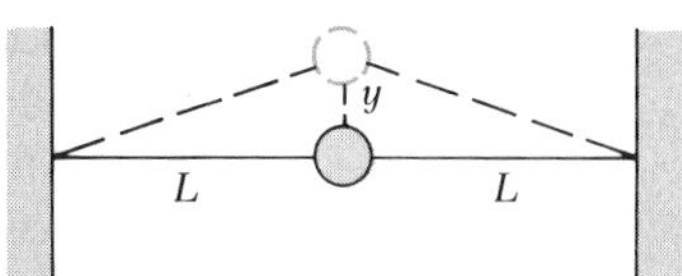

Figure 13.30 (Problem 48).

14
The Law of Universal Gravitation

Prior to 1686, a great mass of data had been collected on the motions of the moon and the planets but a clear understanding of the forces that caused these celestial bodies to move the way they did was not available. In that year, however, Isaac Newton provided the key that unlocked the secrets of the heavens. He knew, from the first law, that a net force had to be acting on the moon. If not, it would move in a straight-line path rather than in its almost circular orbit. Newton reasoned that this force arose as a result of a gravitational attraction that the earth exerted on the moon. He also concluded that there could be nothing special about the earth-moon system or the sun and its planets that would cause gravitational forces to act on them alone. In other words, he saw that the same force of attraction that causes the moon to follow its path also causes an apple to fall to earth from a tree. He wrote, "I deduced that the forces which keep the planets in their orbs must be reciprocally as the squares of their distances from the centers about which they revolve; and thereby compared the force requisite to keep the Moon in her orb with force of gravity at the surface of the Earth; and found them answer pretty nearly."

In this chapter we shall study the law of universal gravitation. Emphasis will be placed on describing the motion of the planets, since astronomical data provide an important test of the validity of the law of universal gravitation. We shall show that the laws of planetary motion developed by Johannes Kepler (1571 – 1630) follow from the law of universal gravitation and the concept of the conservation of angular momentum. A general expression for the gravitational potential energy will be derived, and the energetics of planetary and satellite motion will be treated. The law of universal gravitation will also be used to determine the force between a particle and an extended body.

14.1 NEWTON'S UNIVERSAL LAW OF GRAVITY

It has been said that Newton was struck on the head by a falling apple while napping under a tree (or some variation of this legend). This supposedly prompted Newton to imagine that perhaps all bodies in the universe are attracted to each other in the same way the apple was attracted to the earth. Newton proceeded to analyze astronomical data on the motion of the moon around the earth. From the analysis of such data, Newton made the bold statement that the law of force governing the motion of planets has the *same* mathematical form as the force law that attracts a falling apple to the earth.

In 1687 Newton published his work on the universal law of gravity in his *Mathematical Principles of Natural Philosophy.* **Newton's law of gravitation** states that

> every particle in the universe attracts every other particle with a force that is directly proportional to the product of their masses and inversely proportional to the square of the distance between them.

If the particles have masses m_1 and m_2 and are separated by a distance r, the magnitude of this gravitational force is

Universal law of gravity

$$F = G\frac{m_1 m_2}{r^2} \tag{14.1}$$

where G is a universal constant called the *gravitational constant*, which has been measured experimentally. Its value in SI units is

$$G = 6.672 \times 10^{-11} \frac{\text{N} \cdot \text{m}^2}{\text{kg}^2} \tag{14.2}$$

The force law given by Equation 14.1 is often referred to as an **inverse-square law**, since the magnitude of the force varies as the inverse square of the separation of the particles. We can express this force in vector form by defining a unit vector $\hat{r}_{12}$ (Fig. 14.1). Because this unit vector is in the direction of the displacement vector r_{12} directed from m_1 to m_2, the force on m_2 due to m_1 is given by

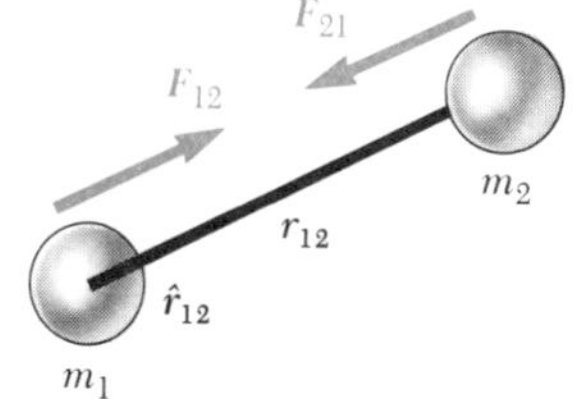

Figure 14.1 The gravitational force between two particles is attractive. The unit vector $\hat{r}_{12}$ is directed from m_1 to m_2. Note that $F_{12} = -F_{21}$.

$$\boldsymbol{F}_{21} = -G\frac{m_1 m_2}{{r_{12}}^2}\hat{r}_{12} \tag{14.3}$$

The minus sign in Equation 14.3 indicates that m_2 is attracted to m_1, and so the force must be directed toward m_1. Likewise, by Newton's third law the force on m_1 due to m_2, designated $\boldsymbol{F}_{12}$, is equal in magnitude to $\boldsymbol{F}_{21}$ and in the opposite direction. That is, these forces form an action-reaction pair, and $\boldsymbol{F}_{12} = -\boldsymbol{F}_{21}$.

Properties of the gravitational force

There are several features of the inverse-square law that deserve some attention. The gravitational force is an action-at-a-distance force, which always exists between two particles, regardless of the medium that separates them. The force varies as the inverse square of the distance between the particles and therefore decreases rapidly with increasing separation. The force is proportional to the mass of each particle, as one might intuitively expect.

Another important fact is that *the gravitational force exerted by a finite-size, spherically symmetric mass distribution on a particle outside the sphere is the same as if the entire mass of the sphere were concentrated at its center.* (The proof of this involves the use of integral calculus and is presented in Section 14.11.) For example, the force on a particle of mass m at the earth's surface has the magnitude

$$F = G\frac{M_e m}{{R_e}^2}$$

where M_e is the earth's mass and R_e is the earth's radius. This force is directed toward the center of the earth.

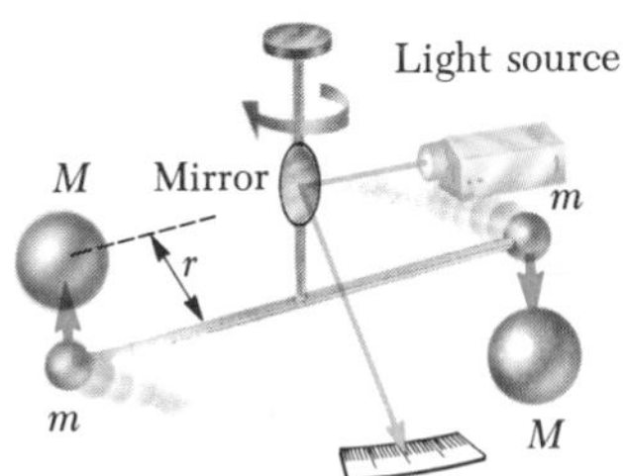

Figure 14.2 Schematic diagram of the Cavendish apparatus for measuring G. The smaller spheres of mass m are attracted to the large spheres of mass M, and the bar rotates through a small angle. A light beam reflected from a mirror on the rotating apparatus measures the angle of rotation.

14.2 MEASUREMENT OF THE GRAVITATIONAL CONSTANT

The gravitational constant, G, was first measured in an important experiment by Sir Henry Cavendish in 1798. The Cavendish apparatus consists of two small spheres each of mass m fixed to the ends of a light horizontal rod suspended by a fine fiber or thin metal wire, as in Figure 14.2. Two large spheres each of mass M are then placed near the smaller spheres. The attractive force between the smaller and larger spheres causes the rod to rotate and twist the wire suspension. If the system is oriented as shown in Figure 14.2, the rod rotates clockwise when viewed from the top. The angle through which the suspended rod rotates is measured by the deflection of a light beam reflected from a mirror attached to the vertical suspension. The deflected spot of light is an effective technique for amplifying the motion. The experiment is carefully repeated with different masses at various separations. In addition to providing a value for G, the results show that the force is attractive, proportional to the product mM, and inversely proportional to the square of the distance r.

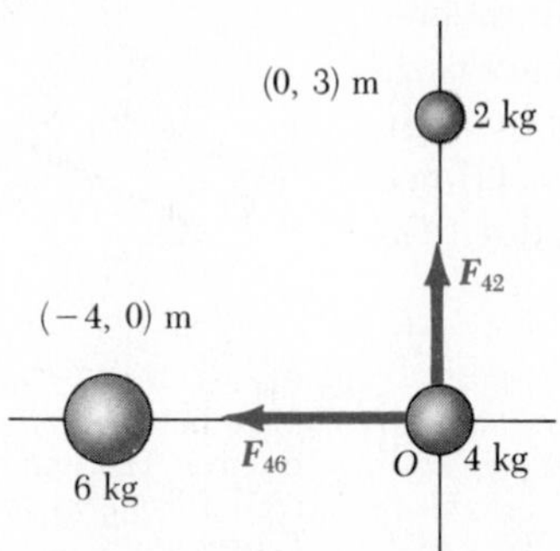

Figure 14.3 (Example 14.1) The *resultant* force on the 4-kg mass is the vector sum $\mathbf{F}_{46} + \mathbf{F}_{42}$.

EXAMPLE 14.1. Three Interacting Masses
Three uniform spheres of mass 2 kg, 4 kg, and 6 kg are placed at the corners of a right triangle as in Figure 14.3, where the coordinates are in m. Calculate the resultant gravitational force on the 4-kg mass, assuming the spheres are isolated from the rest of the universe.

Solution: First we calculate the individual forces on the 4-kg mass due to the 2-kg and 6-kg masses separately, and then we find the vector sum to get the resultant force on the 4-kg mass.

The force on the 4-kg mass due to the 2-kg mass is upward and given by

$$\mathbf{F}_{42} = G\frac{m_4 m_2}{{r_{42}}^2}\mathbf{j} = \left(6.67 \times 10^{-11}\,\frac{\text{N}\cdot\text{m}^2}{\text{kg}^2}\right)\frac{(4\text{ kg})(2\text{ kg})}{(3\text{ m})^2}\,\mathbf{j}$$

$$= 5.93 \times 10^{-11}\,\mathbf{j}\text{ N}$$

The force on the 4-kg mass due to the 6-kg mass is to the left and given by

$$\mathbf{F}_{46} = G\frac{m_4 m_6}{{r_{46}}^2}(-\mathbf{i})$$

$$= \left(-6.67 \times 10^{-11}\,\frac{\text{N}\cdot\text{m}^2}{\text{kg}^2}\right)\frac{(4\text{ kg})(6\text{ kg})}{(4\text{ m})^2}\,\mathbf{i}$$

$$= -10.0 \times 10^{-11}\,\mathbf{i}\text{ N}$$

Therefore, the resultant force on the 4-kg mass is the vector sum of $\mathbf{F}_{42}$ and $\mathbf{F}_{46}$:

$$\mathbf{F}_4 = \mathbf{F}_{42} + \mathbf{F}_{46} = (-10.0\mathbf{i} + 5.93\mathbf{j}) \times 10^{-11}\text{ N}$$

The magnitude of this force is 11.6×10^{-11} N, which is only 2.61×10^{-11} lb! The force makes an angle of 149° with the positive x axis.

14.3 WEIGHT AND GRAVITATIONAL FORCE

In Chapter 5 we defined the weight of a body of mass m as simply mg, where g is the magnitude of the acceleration due to gravity. Now, we are in a position to obtain a more fundamental description of g. Since the force on a freely falling body of mass m near the surface of the earth is given by Equation 14.1, we can equate mg to this expression to give

$$mg = G\frac{M_e m}{{R_e}^2}$$

Acceleration due to gravity

$$g = G\frac{M_e}{{R_e}^2} \tag{14.4}$$

where M_e is the mass of the earth and R_e is the earth's radius. Using the facts that $g = 9.80$ m/s² at the earth's surface and the radius of the earth is approximately 6.38×10^6 m, we find from Equation 14.4 that $M_e = 5.98 \times 10^{24}$ kg. From this result, the average density of the earth is calculated to be

$$\rho_e = \frac{M_e}{V_e} = \frac{M_e}{\frac{4}{3}\pi R_e^3} = \frac{5.98 \times 10^{24} \text{ kg}}{\frac{4}{3}\pi(6.38 \times 10^6 \text{ m})^3} = 5.50 \times 10^3 \text{ kg/m}^3$$

Since this value is about twice the density of most rocks at the earth's surface, we conclude that the inner core of the earth has a much higher density.

Now consider a body of mass m a distance h above the earth's surface, or a distance r from the earth's center, where $r = R_e + h$. The magnitude of the gravitational force acting on this mass is given by

$$F = G\frac{M_e m}{r^2} = G\frac{M_e m}{(R_e + h)^2}$$

If the body is in free fall, then $F = mg'$ and we see that g', the acceleration of gravity at the altitude h, is given by

$$g' = \frac{GM_e}{r^2} = \frac{GM_e}{(R_e + h)^2} \qquad (14.5)$$

Variation of g with altitude

Thus, it follows that g' *decreases* with *increasing altitude.* Since the true weight of a body is mg', we see that as $r \to \infty$, the true weight approaches zero.

EXAMPLE 14.2. Variation of g with Altitude h
Determine the magnitude of the acceleration of gravity at an altitude of 500 km. By what percentage is the weight of a body reduced at this altitude?

Solution: Using Equation 14.5 with $h = 500$ km, $R_e = 6.38 \times 10^6$ m, and $M_e = 5.98 \times 10^{24}$ kg gives

$$g' = \frac{GM_e}{(R_e + h)^2} = \frac{(6.67 \times 10^{-11} \text{ N} \cdot \text{m}^2/\text{kg}^2)(5.98 \times 10^{24} \text{ kg})}{(6.38 \times 10^6 + 0.5 \times 10^6)^2 \text{ m}^2} = 8.43 \text{ m/s}^2$$

Since $g'/g = 8.43/9.8 = 0.86$, we conclude that the weight of a body is reduced by about 14% at an altitude of 500 km. Values of g' at other altitudes are listed in Table 14.1.

TABLE 14.1 Acceleration Due to Gravity, g', at Various Altitudes

Altitude h (km)[a]	g' (m/s^2)
1000	7.33
2000	5.68
3000	4.53
4000	3.70
5000	3.08
6000	2.60
7000	2.23
8000	1.93
9000	1.69
10 000	1.49
50 000	0.13
∞	0

[a] All values are distances above the earth's surface.

14.4 KEPLER'S LAWS

The movements of the planets, stars, and other celestial bodies have been observed by people for thousands of years. In early history, scientists regarded the earth as the center of the universe. This so-called geocentric model was proposed by the Greek astronomer Claudius Ptolemy in the second century A.D. and was accepted for the next 1400 years. In 1543, the Polish astronomer Nicolaus Copernicus (1473–1543) suggested that the earth and the other planets revolve in circular orbits about the sun (the heliocentric hypothesis).

The Danish astronomer Tycho Brahe (1546–1601) made accurate astronomical measurements over a period of 20 years and provided the basis for the currently accepted model of the solar system. It is interesting to note that these precise observations, made on the planets and 777 stars visible to the naked eye, were carried out with a large sextant and compass because the telescope had not yet been invented.

The German astronomer Johannes Kepler, who was Brahe's student, acquired Brahe's astronomical data and spent about 16 years trying to deduce a mathematical model for the motion of the planets. After many laborious calculations, he found that Brahe's precise data on the revolution of Mars about the sun provided the answer. Such data are difficult to sort out because the earth is also in motion about the sun. Kepler's analysis first showed that the concept of circular orbits about the sun had to be abandoned. He eventually discovered that the orbit of Mars could be accurately described by an ellipse with the sun at one focus. He then generalized this analysis to include the motion of all planets. The complete analysis is summarized in three statements, known as **Kepler's laws.** These empirical laws applied to the solar system are:

Kepler's laws

1. All planets move in elliptical orbits with the sun at one of the focal points.
2. The radius vector drawn from the sun to any planet sweeps out equal areas in equal time intervals.
3. The square of the orbital period of any planet is proportional to the cube of the semi-major axis of the elliptical orbit.

About 100 years later, Newton demonstrated that these laws were the consequence of a simple force that exists between any two masses. Newton's law of universal gravitation, together with his development of the laws of motion, provides the basis for a full mathematical solution to the motion of planets and satellites. More important, Newton's universal law of gravity correctly describes the gravitational attractive force between *any* two masses.

14.5 THE LAW OF UNIVERSAL GRAVITATION AND THE MOTION OF PLANETS

In formulating his law of universal gravitation, Newton used the following observation, which suggests that the gravitational force is proportional to the inverse square of the separation. Let us compare the acceleration of the moon in its orbit with the acceleration of an object falling near the earth's surface, such as the legendary apple (Fig. 14.4). Assume that both accelerations have the same cause, namely, the gravitational attraction of the earth. From the inverse-square law, Newton found that the acceleration of the moon toward the earth (centripetal acceleration) should be proportional to $1/r_m{}^2$, where r_m is the earth-moon separation. Furthermore, the acceleration of the apple toward the earth should vary as $1/R_e{}^2$, where R_e is the radius of the earth. Using the values $r_m = 3.84 \times 10^8$ m and $R_e = 6.37 \times 10^6$ m, the ratio of the moon's acceleration, a_m, to the apple's acceleration, g, is predicted to be

$$\frac{a_m}{g} = \frac{(1/r_m)^2}{(1/R_e)^2} = \left(\frac{R_e}{r_m}\right)^2 = \left(\frac{6.37 \times 10^6 \text{ m}}{3.84 \times 10^8 \text{ m}}\right)^2 = 2.75 \times 10^{-4}$$

Therefore

The acceleration of the moon

$$a_m = (2.75 \times 10^{-4})(9.80 \text{ m/s}^2) = 2.70 \times 10^{-3} \text{ m/s}^2$$

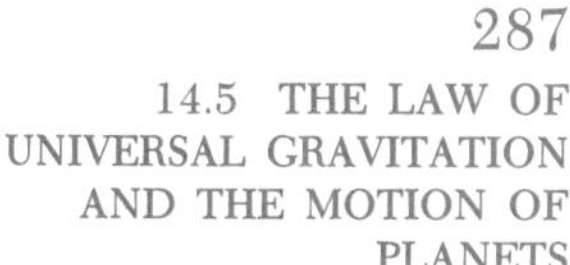

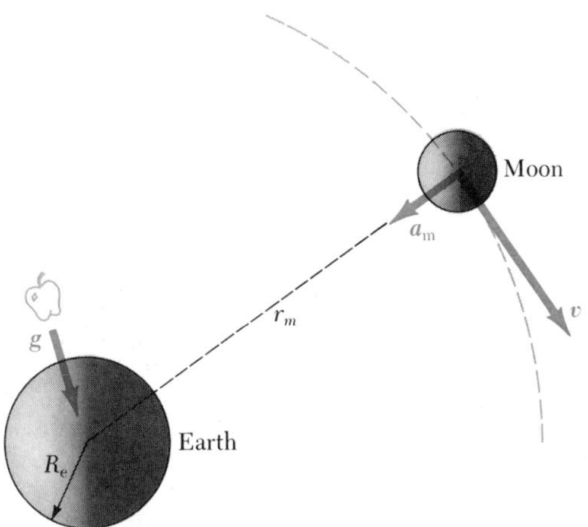

Figure 14.4 As the moon revolves about the earth, the moon experiences a centripetal acceleration a_m directed toward the earth. An object near the earth's surface experiences an acceleration equal to g. (Dimensions are not to scale.)

The centripetal acceleration of the moon can also be calculated kinematically from a knowledge of its orbital period, T, where $T = 27.32$ days $= 2.36 \times 10^6$ s, and its mean distance from the earth, r_m. In a time T, the moon travels a distance $2\pi r_m$, which equals the circumference of its orbit. Therefore, its orbital speed is $2\pi r_m/T$, and its centripetal acceleration is

$$a_m = \frac{v^2}{r_m} = \frac{(2\pi r_m/T)^2}{r_m} = \frac{4\pi^2 r_m}{T^2} = \frac{4\pi^2(3.84 \times 10^8 \text{ m})}{(2.36 \times 10^6 \text{ s})^2} = 2.72 \times 10^{-3} \text{ m/s}^2$$

This agreement provides strong evidence that the inverse-square law of force is correct.

Although these results must have been very encouraging to Newton, he was deeply troubled by an assumption made in the analysis. In order to evaluate the acceleration of an object at the earth's surface, the earth was treated as if its mass were all concentrated at its center. That is, Newton assumed that the earth acts as a particle as far as its influence on an exterior object is concerned. Several years later, and based on his pioneering work in the development of the calculus, Newton proved this point. (The details of the derivation are given in Section 14.11.) For this reason, and because of Newton's inherent shyness, the publication of the theory of gravitation was delayed for about 20 years.

Kepler's Third Law

It is informative to show that Kepler's third law can be predicted from the inverse-square law for circular orbits.[1] Consider a planet of mass M_p which is assumed to be moving about the sun of mass M_s in a circular orbit, as in Figure 14.5. Since the gravitational force on the planet is equal to the centripetal force needed to keep it moving in a circle,

$$\frac{GM_sM_p}{r^2} = \frac{M_pv^2}{r}$$

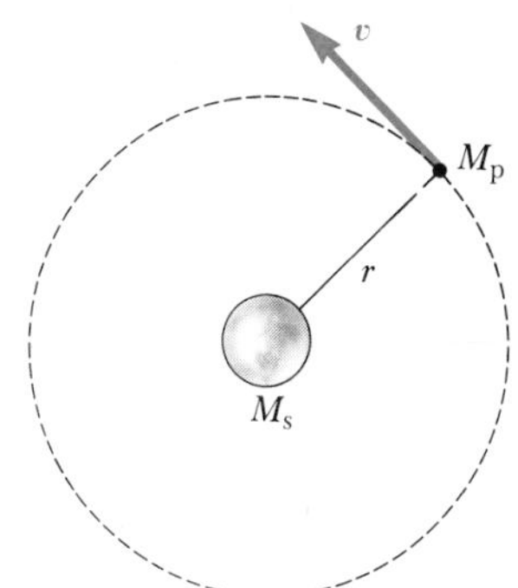

Figure 14.5 A planet of mass M_p moving in a circular orbit about the sun. The orbits of all planets except Mars, Mercury, and Pluto are nearly circular.

[1] The orbits of all planets except Mars, Mercury, and Pluto are very close to being circular. For example, the ratio of the semi-minor to the semi-major axis for the earth is $b/a = 0.99986$.

But the orbital velocity of the planet is simply $2\pi r/T$, where T is its period; therefore the above expression becomes

$$\frac{GM_s}{r^2} = \frac{(2\pi r/T)^2}{r}$$

Kepler's third law

$$T^2 = \left(\frac{4\pi^2}{GM_s}\right)r^3 = K_s r^3 \tag{14.6}$$

where K_s is a constant given by

$$K_s = \frac{4\pi^2}{GM_s} = 2.97 \times 10^{-19}\ \mathrm{s^2/m^3}$$

Equation 14.6 is Kepler's third law. The law is also valid for elliptical orbits if we replace r by the length of the semi-major axis, a (Fig. 14.6). Note that the constant of proportionality, K_s, is independent of the mass of the planet. Therefore, Equation 14.6 is valid for *any* planet. If we were to consider the orbit of a satellite about the earth, such as the moon, then the constant would have a different value, with the sun's mass replaced by the earth's mass. In this case, the proportionality constant equals $4\pi^2/GM_e$.

A collection of useful planetary data is given in Table 14.2. The last column of this table verifies that T^2/r^3 is a constant whose value is given by $K_s = 4\pi^2/GM_s = 2.97 \times 10^{-19}\ \mathrm{s^2/m^3}$.

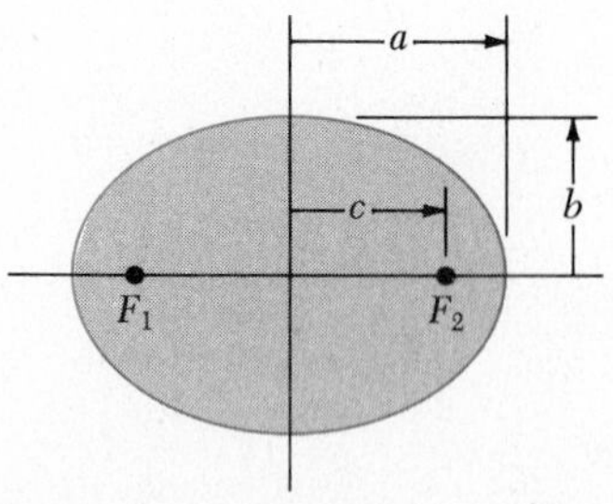

Figure 14.6 Plot of an ellipse. The semi-major axis has a length a, and the semi-minor axis has a length b. The focal points are located at a distance c from the center, and the eccentricity is defined as $e = c/a$.

EXAMPLE 14.3. The Mass of the Sun
Calculate the mass of the sun using the fact that the period of the earth is 3.156×10^7 s and its distance from the sun is 1.496×10^{11} m.

Solution: Using Equation 14.6, we get

$$M_s = \frac{4\pi^2 r^3}{GT^2} = \frac{4\pi^2(1.496 \times 10^{11}\ \mathrm{m})^3}{\left(6.67 \times 10^{-11}\ \frac{\mathrm{N \cdot m^2}}{\mathrm{kg^2}}\right)(3.156 \times 10^7\ \mathrm{s})^2}$$

$$= 1.99 \times 10^{30}\ \mathrm{kg}$$

Note that the sun is 333 000 times as massive as the earth!

TABLE 14.2 Useful Planetary Data

Body	Mass (kg)	Mean Radius (m)	Period (s)	Distance from Sun (m)	$\frac{T^2}{r^3}\left[10^{-19}\left(\frac{\mathrm{s^2}}{\mathrm{m^3}}\right)\right]$
Mercury	3.18×10^{23}	2.43×10^6	7.60×10^6	5.79×10^{10}	2.97
Venus	4.88×10^{24}	6.06×10^6	1.94×10^7	1.08×10^{11}	2.99
Earth	5.98×10^{24}	6.37×10^6	3.156×10^7	1.496×10^{11}	2.97
Mars	6.42×10^{23}	3.37×10^6	5.94×10^7	2.28×10^{11}	2.98
Jupiter	1.90×10^{27}	6.99×10^7	3.74×10^8	7.78×10^{11}	2.97
Saturn	5.68×10^{26}	5.85×10^7	9.35×10^8	1.43×10^{12}	2.99
Uranus	8.68×10^{25}	2.33×10^7	2.64×10^9	2.87×10^{12}	2.95
Neptune	1.03×10^{26}	2.21×10^7	5.22×10^9	4.50×10^{12}	2.99
Pluto	$\approx 1 \times 10^{23}$	$\approx 3 \times 10^6$	7.82×10^9	5.91×10^{12}	2.96
Moon	7.36×10^{22}	1.74×10^6	—	—	—
Sun	1.991×10^{30}	6.96×10^8	—	—	—

For a more complete set of data, see, for example, the *Handbook of Chemistry and Physics,* Boca Raton, Florida, The Chemical Rubber Publishing Co.

Kepler's Second Law and Conservation of Angular Momentum

Consider a planet (or comet) of mass m moving about the sun in an elliptical orbit (Fig. 14.7). The gravitational force acting on the planet is always along the radius vector, directed toward the sun. Such a force directed toward or away from a fixed point (that is, one that is a function of r only) is called a **central force.**[2] The torque acting on the planet due to this central force is clearly zero since $\mathbf{F}$ is parallel to $\mathbf{r}$. That is,

$$\boldsymbol{\tau} = \mathbf{r} \times \mathbf{F} = \mathbf{r} \times F(r)\hat{\mathbf{r}} = 0$$

But recall that the torque equals the time rate of change of angular momentum, or $\boldsymbol{\tau} = d\mathbf{L}/dt$. Therefore,

> because $\boldsymbol{\tau} = 0$, the angular momentum $\mathbf{L}$ of the planet is a constant of the motion:

$$\mathbf{L} = \mathbf{r} \times \mathbf{p} = m\mathbf{r} \times \mathbf{v} = \text{constant}$$

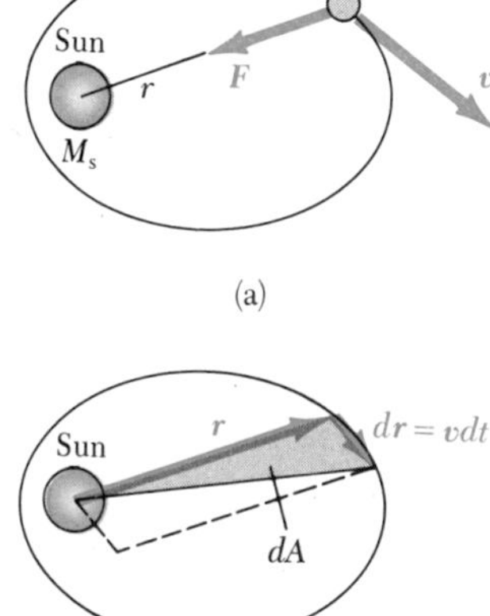

Figure 14.7 (a) The force acting on a planet acts towad the sun, along the radius vector. (b) As a planet orbits the sun, the area swept out by the radius vector in a time dt is equal to one half the area of the parallelogram formed by the vectors $\mathbf{r}$ and $d\mathbf{r} = \mathbf{v}\,dt$.

Since $\mathbf{L}$ is a constant of the motion, we see that the planet's motion at any instant is restricted to the plane formed by $\mathbf{r}$ and $\mathbf{v}$.

We can relate this result to the following geometric consideration. The radius vector $\mathbf{r}$ in Figure 14.7b sweeps out an area dA in a time dt. This area equals one half the area $|\mathbf{r} \times d\mathbf{r}|$ of the parallelogram formed by the vectors $\mathbf{r}$ and $d\mathbf{r}$. Since the displacement of the planet in a time dt is given by $d\mathbf{r} = \mathbf{v}\,dt$, we get

$$dA = \tfrac{1}{2}|\mathbf{r} \times d\mathbf{r}| = \tfrac{1}{2}|\mathbf{r} \times \mathbf{v}\,dt| = \frac{L}{2m}\,dt$$

$$\frac{dA}{dt} = \frac{L}{2m} = \text{constant} \qquad (14.7)$$

Kepler's second law

where L and m are both constants of the motion. Thus, we conclude that

> the radius vector from the sun to any planet sweeps out equal areas in equal times.

It is important to recognize that this result is a consequence of the fact that the force of gravity is a central force, which in turn implies conservation of angular momentum. Therefore, the law applies to *any* situation that involves a central force, whether inverse-square or not.

The inverse-square nature of the force of gravity is not revealed by Kepler's second law. Although we do not prove it here, Kepler's first law is a direct consequence of the fact that the gravitational force varies as $1/r^2$. That is, under an inverse-square force law, the orbits of the planets can be shown to be ellipses with the sun at one focus.

[2] Another example of a central force is the electrostatic force between two charged particles.

EXAMPLE 14.4. Motion in an Elliptical Orbit
A planet of mass m moves in an elliptical orbit about the sun (Fig. 14.8). The minimum and maximum distances of the planet from the sun are called the *perihelion* (indicated by p in Fig. 14.8) and *aphelion* (indicated by a), respectively. If the speed of the planet at p is v_p, what is its speed at a? Assume the distances r_a and r_p are known.

Solution: The angular momentum of the planet relative to the sun is $m\mathbf{r} \times \mathbf{v}$. At the points a and p, $\mathbf{v}$ is perpendic-

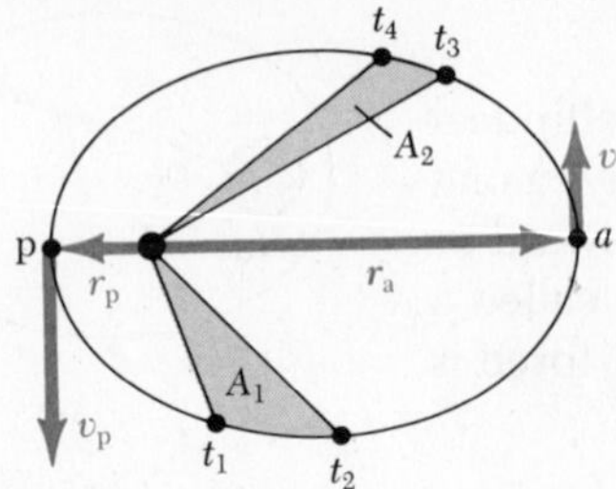

Figure 14.8 (Example 14.4) As a planet moves about the sun in an elliptical orbit, its angular momentum is conserved. Therefore, $mv_ar_a = mv_pr_p$, where the subscripts a and p represent aphelion and perihelion, respectively.

ular to r. Therefore, the magnitude of the angular momentum at these positions is $L_a = mv_ar_a$ and $L_p = mv_pr_p$. The direction of the angular momentum is out of the plane of the paper. Since angular momentum is conserved, we see that

$$mv_ar_a = mv_pr_p$$

$$v_a = \frac{r_p}{r_a} v_p$$

14.6 THE GRAVITATIONAL FIELD

The gravitational force between two masses is an action-at-a-distance type of interaction. That is, the two masses interact even though they are not in contact with each other. An alternative approach in describing the gravitational interaction is to introduce the concept of a **gravitational field, g**, at every point in space. When a particle of mass m is placed at a point where the field is g, the particle experiences a force $F = mg$. In other words, the field g exerts a force on the particle. Hence, the gravitational field is defined by

Gravitational field

$$g \equiv \frac{F}{m} \tag{14.8}$$

That is, the gravitational field at any point equals the gravitational force that a test mass experiences divided by that test mass. Consequently, if g is known at some point in space, a test particle of mass m experiences a gravitational force mg when placed at that point.

As an example, consider an object of mass m near the earth's surface. The gravitational force on the object is directed toward the center of the earth and has a magnitude mg. Thus we see that the gravitational field that the object experiences at some point has a magnitude equal to the acceleration of gravity at that point. Since the gravitational force on the object has a magnitude GM_em/r^2 (where M_e is the mass of the earth), the field g at a distance r from the center of the earth is given by

$$g = \frac{F}{m} = -\frac{GM_e}{r^2}\hat{r}$$

This expression is valid at all points *outside* the earth's surface, assuming the earth is spherical. At the earth's surface, where $r = R_e$, g has a magnitude of 9.80 m/s^2.

The field concept is used in many other areas of physics. In fact, the field concept was first introduced by Michael Faraday (1791–1867) in the field of electromagnetism. Later in the text we shall use the field concept to describe electromagnetic interactions. Gravitational, electrical, and magnetic fields are all examples of *vector fields* since a vector is associated with each point in space. On the other hand, a *scalar field* is one in which a scalar quantity is used to describe each point in space. For example, the variation in temperature over a given region can be described by a scalar temperature field.

In Chapter 8 we introduced the concept of gravitational potential energy, that is, the energy associated with the position of a particle. We emphasized the fact that the gravitational potential energy function, $U = mgy$, is valid only when the particle is near the earth's surface. Since the gravitational force between two particles varies as $1/r^2$, we expect that the correct potential energy function will depend on the amount of separation between the particles.

Before we calculate the specific form for the gravitational potential energy function, we shall first verify that *the gravitational force is conservative.* In order to establish the conservative nature of the gravitational force, we first note that it is a central force. By definition, a central force is one that depends only on the polar coordinate r, and hence can be represented by $F(r)\hat{r}$, where $\hat{r}$ is a unit vector directed from the origin to the particle under consideration. Such a force acts from some origin and is directed parallel to the radius vector.

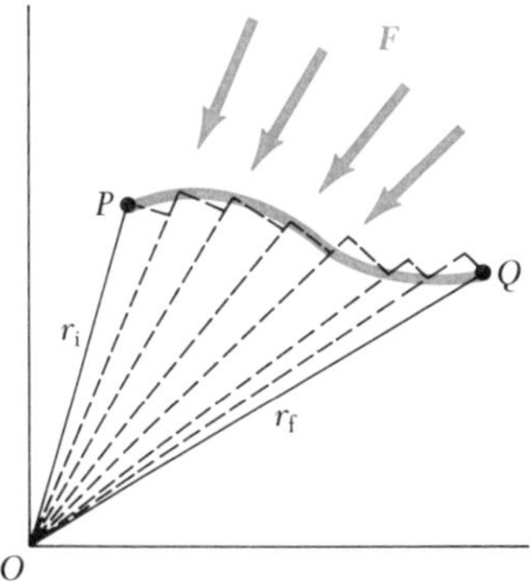

Figure 14.9 A particle moves from P to Q while under the action of a central force $\boldsymbol{F}$, which is in the radial direction. The path is broken into a series of radial and circular segments. Since the work done along the circular segments is zero, the work done is independent of the path.

Consider a central force acting on a particle moving along the general path P to Q in Figure 14.9. The central force acts from the point O. This path can be approximated by a series of radial and circular segments. By definition, a central force is always directed along one of the radial segments; therefore the work done along any *radial segment* is given by

$$dW = \boldsymbol{F} \cdot d\boldsymbol{r} = F(r)\, dr$$

You should recall that by definition the work done by a force that is perpendicular to the displacement is zero. Hence, the work done along any circular segment is *zero* because $\boldsymbol{F}$ is perpendicular to the displacement along these segments. Therefore, the total work done by $\boldsymbol{F}$ is the sum of the contributions along the radial segments:

Work done by a central force

$$W = \int_{r_i}^{r_f} F(r)\, dr$$

where the subscripts i and f refer to the initial and final positions. This result applies to *any* path from P to Q. Therefore, we conclude that *any central force is conservative.* We are now assured that a potential energy function can be obtained once the form of the central force is specified. You should recall from Chapter 8 that the change in the gravitational potential energy associated with a given displacement is defined as the negative of the work done by the gravitational force during that displacement, or

$$\Delta U = U_f - U_i = -\int_{r_i}^{r_f} F(r)\, dr \qquad (14.9)$$

We can use this result to evaluate the gravitational potential energy function. Consider a particle of mass m moving between two points P and Q above the earth's surface (Fig. 14.10). The particle is subject to the gravitational force given by Equation 14.1. We can express the force on m in vector form as

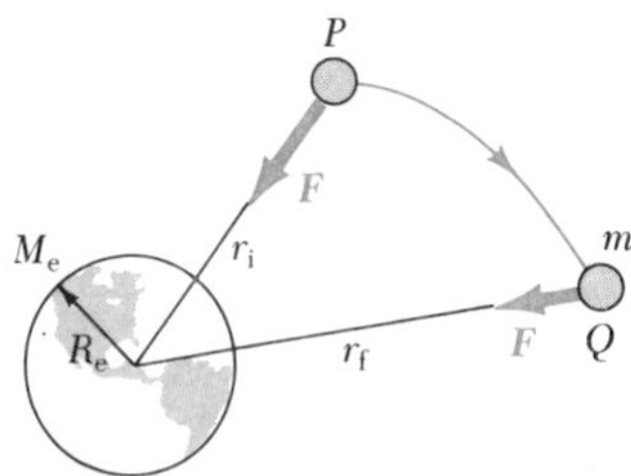

Figure 14.10 As a particle of mass m moves from P to Q above the earth's surface, the potential energy changes according to Equation 14.10.

$$\boldsymbol{F} = -\frac{GM_e m}{r^2}\hat{r}$$

where $\hat{r}$ is a unit vector directed from the earth to the particle and the negative sign indicates that the force is attractive. Substituting this into Equation 14.9, we can compute the change in the gravitational potential energy function:

$$U_f - U_i = GM_em \int_{r_i}^{r_f} \frac{dr}{r^2} = GM_em \left[-\frac{1}{r}\right]_{r_i}^{r_f}$$

$$U_f - U_i = -GM_em \left(\frac{1}{r_f} - \frac{1}{r_i}\right) \tag{14.10}$$

The choice of a reference point for the potential energy is completely arbitrary. It is customary to choose the reference point where the force is zero. Taking $U_i = 0$ at $r_i = \infty$, we obtain the important result

Gravitational potential energy $r > R_e$

$$U(r) = -\frac{GM_em}{r} \tag{14.11}$$

This expression applies to the earth-particle system separated by a distance r, provided that $r > R_e$. The result is not valid for particles moving inside the earth, where $r < R_e$. We shall treat this situation in Section 14.10. Because of our choice of U_i, the function $U(r)$ is always negative (Fig. 14.11).

Although Equation 14.11 was derived for the particle-earth system, it can be applied to *any* two particles. That is, the gravitational potential energy associated with *any pair* of particles of masses m_1 and m_2 separated by a distance r is given by

Gravitational potential energy for a pair of particles

$$U = -\frac{Gm_1m_2}{r} \tag{14.12}$$

This expression shows that the gravitational potential energy for any pair of particles varies as $1/r$, whereas the force between them varies as $1/r^2$. Furthermore, the potential energy is *negative* since the force is attractive and we have taken the potential energy as zero when the particle separation is infinity. Since the force between the particles is attractive, we know that an external agent must do positive work to increase the separation between the two particles. The work done by the external agent produces an increase in the potential energy as the two particles are separated. That is, U becomes less negative as r increases. (Note that part of the work done can also produce a change in kinetic energy of the system. That is, if the work done in separating the particles exceeds the increase in potential energy, the excess energy is accounted for by the increase in kinetic energy of the system.) When the two particles are separated by a distance r, an external agent would have to supply an energy *at least* equal to $+Gm_1m_2/r$ in order to separate the particles by an infinite distance. It is convenient to think of the absolute value of the potential energy as the *binding energy* of the system. If the external agent supplies an energy *greater than* the binding energy, Gm_1m_2/r, the additional energy of the system will be in the form of kinetic energy when the particles are at an infinite separation.

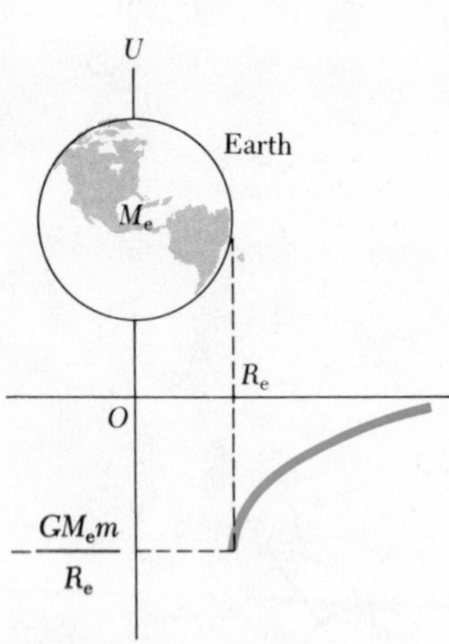

Figure 14.11 Graph of the gravitational potential energy, U, versus r for a particle above the earth's surface. The potential energy goes to zero as r approaches ∞.

We can extend this concept to three or more particles. In this case, the total potential energy of the system is the sum over all *pairs* of particles.[3] Each pair contributes a term of the form given by Equation 14.12. For example, if the system contains three particles as in Figure 14.12, we find that

[3] The fact that one can add potential energy terms for all pairs of particles stems from the experimental fact that gravitational forces obey the superposition principle. That is, if $\Sigma \mathbf{F} = \mathbf{F}_{12} + \mathbf{F}_{13} + \mathbf{F}_{23} + \cdots$ then there exists a potential energy term for each interaction $\mathbf{F}_{ij}$.

$$U_{\text{total}} = U_{12} + U_{13} + U_{23} = -G\left(\frac{m_1m_2}{r_{12}} + \frac{m_1m_3}{r_{13}} + \frac{m_2m_3}{r_{23}}\right) \quad (14.13)$$

This represents the total work done by an external agent against the gravitational force in assembling the system from an infinite separation. If the system consists of four particles, there are six terms in the sum, corresponding to the six distinct pairs of interaction forces.

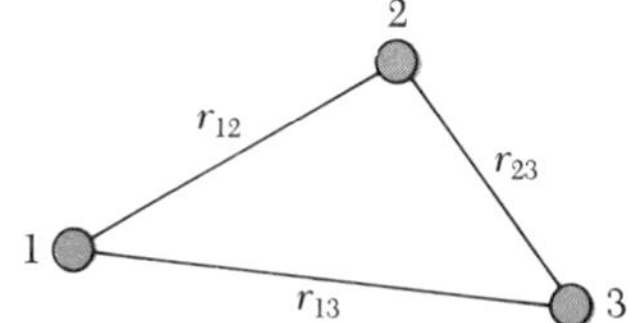

Figure 14.12 Diagram of three interacting particles.

EXAMPLE 14.5. The Change in Potential Energy
A particle of mass m is displaced through a small vertical distance Δy near the earth's surface. Let us show that the general expression for the change in gravitational potential energy given by Equation 14.10 reduces to the familiar relationship $\Delta U = mg\,\Delta y$.

Solution: We can express Equation 14.10 in the form

$$\Delta U = -GM_em\left(\frac{1}{r_f} - \frac{1}{r_i}\right) = GM_em\left(\frac{r_f - r_i}{r_ir_f}\right)$$

If both the initial and the final position of the particle are close to the earth's surface, then $r_f - r_i = \Delta y$ and $r_ir_f \approx R_e^2$. (Recall that r is measured from the center of the earth.) Therefore, the *change* in potential energy becomes

$$\Delta U \approx \frac{GM_em}{R_e^2}\,\Delta y = mg\,\Delta y$$

where we have used the fact that $g = GM_e/R_e^2$. Keep in mind that the reference point is arbitrary, since it is the *change* in potential energy that is meaningful.

14.8 ENERGY CONSIDERATIONS IN PLANETARY AND SATELLITE MOTION

Consider a body of mass m moving with a speed v in the vicinity of a massive body of mass M, where $M \gg m$. The system might be a planet moving around the sun or a satellite in orbit around the earth. If we assume that M is at rest in an inertial reference frame, then the total energy E of the two-body system when the bodies are separated by a distance r is the sum of the kinetic energy of the mass m and the potential energy of the system, given by Equation 14.12.[4] That is,

$$E = K + U$$

$$E = \tfrac{1}{2}mv^2 - \frac{GMm}{r} \quad (14.14)$$

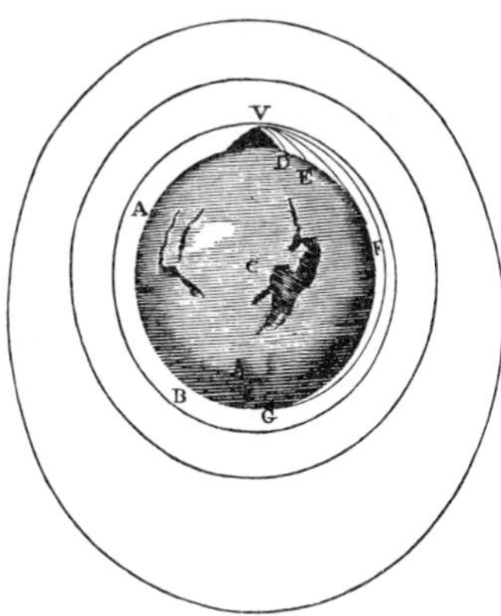

". . . the greater the velocity . . . with which (a stone) is projected, the farther it goes before it falls to the earth. We may therefore suppose the velocity to be so increased, that it would describe an arc of 1, 2, 5, 10, 100, 1000 miles before it arrived at the earth, till at last, exceeding the limits of the earth, it should pass into space without touching."— Newton, *System of the World.*

Furthermore, the total energy is conserved if we assume the system is isolated. Therefore as the mass m moves from P to Q in Figure 14.10, the total energy remains constant and Equation 14.14 gives

$$E = \tfrac{1}{2}mv_i^2 - \frac{GMm}{r_i} = \tfrac{1}{2}mv_f^2 - \frac{GMm}{r_f} \quad (14.15)$$

[4] You might recognize that we have ignored the acceleration and kinetic energy of the larger mass. To see that this is reasonable, consider an object of mass m falling toward the earth. Since the center of mass of the object-earth system is stationary, it follows that $mv = M_ev_e$. Thus, the earth acquires a kinetic energy equal to

$$\tfrac{1}{2}M_ev_e^2 = \tfrac{1}{2}\frac{m^2}{M_e}v^2 = \frac{m}{M_e}K,$$

where K is the kinetic energy of the object. Since $M_e \gg m$, the kinetic energy of the earth is negligible.

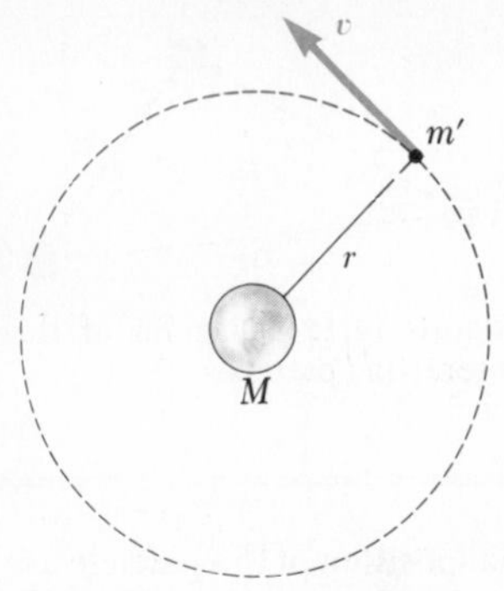

Figure 14.13 A body of mass m moving in a circular orbit about a body of mass M.

This result shows that E may be positive, negative, or zero, depending on the value of the velocity of the mass m. However, for a bound system, such as the earth and sun, E is necessarily less than zero. We can easily establish that $E < 0$ for the system consisting of a mass m moving in a circular orbit about a body of mass M, where $M \gg m$ (Fig. 14.13). Newton's second law applied to the body of mass m gives

$$\frac{GMm}{r^2} = \frac{mv^2}{r}$$

Multiplying both sides by r and dividing by 2 gives

$$\tfrac{1}{2}mv^2 = \frac{GMm}{2r} \tag{14.16}$$

Substituting this into Equation 14.14, we obtain

$$E = \frac{GMm}{2r} - \frac{GMm}{r}$$

Total energy for circular orbits

$$E = -\frac{GMm}{2r} \tag{14.17}$$

This clearly shows that *the total energy must be negative in the case of circular orbits.* Note that *the kinetic energy is positive and equal to one half the magnitude of the potential energy.* The absolute value of E is also equal to the binding energy of the system.

The total mechanical energy is also negative in the case of elliptical orbits.[5] The expression for E for elliptical orbits is the same as Equation 14.17 with r replaced by the semi-major axis length, a.

Both the total energy and the total angular momentum of a planet-sun system are constants of the motion.

EXAMPLE 14.6. Changing the Orbit of a Satellite
Calculate the work required to move an earth satellite of mass m from a circular orbit of radius $2R_e$ to one of radius $3R_e$.

Solution: Applying Equation 14.17, we get for the total initial and final energies

$$E_i = -\frac{GM_em}{4R_e} \qquad E_f = -\frac{GM_em}{6R_e}$$

Therefore, the work required to increase the energy of the system is

$$W = E_f - E_i = -\frac{GM_em}{6R_e} - \left(-\frac{GM_em}{4R_e}\right) = \frac{GM_em}{12R_e}$$

For example, if we take $m = 10^3$ kg, we find that the work required is $W = 5.2 \times 10^9$ J, which is the energy equivalent of 39 gal of gasoline.

If we wish to determine how the energy is distributed after doing work on the system, we find from Equation 14.16 that the change in kinetic energy is $\Delta K = -GM_em/12R_e$ (it decreases), while the corresponding change in potential energy is $\Delta U = GM_em/6R_e$ (it increases). Thus, the work done on the system is given by $W = \Delta K + \Delta U = GM_em/12R_e$, as we calculated above. In other words, part of the work done goes into increasing the potential energy and part goes into decreasing the kinetic energy.

[5] This is shown in more advanced mechanics texts. One can also show that if $E = 0$, the mass would move in a parabolic path, whereas if $E > 0$, its path would be a hyperbola. Nothing in Equation 14.14 precludes a particle with $E \geq 0$ from reaching infinitely great distances from the gravitating center (that is, the particle's orbit is unbound). Infinitely great distances are energetically forbidden to a particle with $E < 0$, and so its orbit must be bound.

Escape Velocity

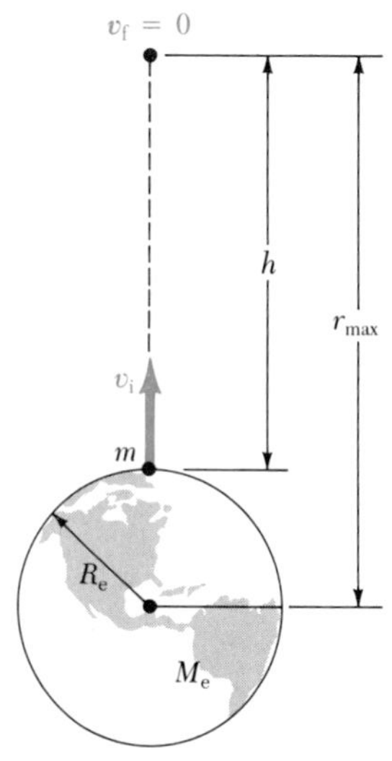

Figure 14.14 An object of mass m projected upward from the earth's surface with an initial speed v_i reaches a maximum altitude h (where $M_e \gg m$).

Suppose an object of mass m is projected vertically upward from the earth's surface with an initial speed v_i, as in Figure 14.14. We can use energy considerations to find the minimum value of the initial speed such that the object will escape the earth's gravitational field. Equation 14.15 gives the total energy of the object at any point when its velocity and distance from the center of the earth are known. At the surface of the earth, where $v_i = v$, $r_i = R_e$. When the object reaches its maximum altitude, $v_f = 0$ and $r_f = r_{max}$. Because the total energy of the system is conserved, substitution of these conditions into Equation 14.15 gives

$$\tfrac{1}{2}mv_i^2 - \frac{GM_e m}{R_e} = -\frac{GM_e m}{r_{max}}$$

Solving for v_i^2 gives

$$v_i^2 = 2GM_e\left(\frac{1}{R_e} - \frac{1}{r_{max}}\right) \tag{14.18}$$

Therefore, if the initial speed is known, this expression can be used to calculate the maximum altitude h, since we know that $h = r_{max} - R_e$.

We are now in a position to calculate the minimum speed the object must have at the earth's surface in order to escape from the influence of the earth's gravitational field. This corresponds to the situation where the object can *just* reach infinity with a final speed of *zero*. Setting $r_{max} = \infty$ in Equation 14.18 and taking $v_i = v_{esc}$ (the escape velocity), we get

$$v_{esc} = \sqrt{\frac{2GM_e}{R_e}} \tag{14.19}$$

Escape velocity

Note that this expression for v_{esc} is independent of the mass of the object projected from the earth. For example, a spacecraft has the same escape velocity as a molecule. If the object is given an initial speed equal to v_{esc}, its *total* energy is equal to zero. This can be seen by noting that when $r = \infty$, the object's kinetic energy and its potential energy are both zero. If v_i is greater than v_{esc}, the *total* energy will be greater than zero and the object will have some residual kinetic energy at $r = \infty$.

EXAMPLE 14.7. Escape Velocity of a Rocket
Calculate the escape velocity from the earth for a 5000-kg spacecraft, and determine the kinetic energy it must have at the earth's surface in order to escape the earth's field.

Solution: Using Equation 14.19 with $M_e = 5.98 \times 10^{24}$ kg and $R_e = 6.37 \times 10^6$ m gives

$$v_{esc} = \sqrt{\frac{2GM_e}{R_e}} = \sqrt{\frac{2(6.67 \times 10^{-11})(5.98 \times 10^{24})}{6.37 \times 10^6}}$$
$$= 1.12 \times 10^4 \text{ m/s}$$

This corresponds to about 25 000 mi/h.

The kinetic energy of the spacecraft is given by

$$K = \tfrac{1}{2}mv_{esc}^2 = \tfrac{1}{2}(5000)(1.12 \times 10^4)^2 = 3.14 \times 10^{11} \text{ J}$$

Finally, you should note that Equations 14.18 and 14.19 can be applied to objects projected vertically from *any* planet. That is, in general, the escape velocity from any planet of mass M and radius R is given by

$$v_{esc} = \sqrt{\frac{2GM}{R}}$$

Essay

A RADIO VIEW OF THE UNIVERSE

George A. Seielstad
National Radio Astronomy Observatory
Green Bank, West Virginia

Until half a century ago, astronomers' vision of the universe was "filtered" through the narrow slice of the electronic spectrum encompassing optical wavelengths. The technology did not exist to receive signals through other portions of the spectrum. The universe deduced with this severely restricted vision was regarded as a tranquil, quiescent, unchanging environment.

This vision has merit. Certainly many physical structures have "relaxed" to configurations that change much too slowly to be noticed by any human within his or her lifetime, or even within the whole of human history. The galaxy in Figure 14E.1, for instance, shows evidence of permanence. Its form indicates rotation, but a typical timescale for a complete revolution about the center by a star near the periphery is a few hundred million years. So whether the photograph was taken recently, or one, ten, or fifty million years from now, its appearance would be the same.

Changes in the universe were, of course, detected by optical astronomers. But these changes were often rhythmic, cyclic, repetitive, and as a consequence, absolutely predictable. Examples are the return of Halley's comet every 76 years, or the seasons on the planets, or the phases of the moon and some of the planets. Since none of these is in the category of one-shot, transient, or short-term events, they reinforce the concept of permanence.

The notion of a tranquil and unchanging universe is not so much wrong as it is incomplete. It illustrates the extent to which physics is dependent upon the instruments used to explore physical phenomena. Once extraterrestrial radio waves were detected in 1932 (accidentally, as it turned out), a whole new window on the electromagnetic spectrum was opened for exploration. Through it a more restless and violent universe was discovered.

Contrast the image of a galaxy emitting powerful radio waves, Cygnus A in Figure 14E.2, with the peaceful system of Figure 14E.1. The Cygnus A images reveal that radio galaxies (1) exhibit *explosive violence on a massive scale,* (2) host *active galactic nuclei,* and (3) exist at *enormous distances.*

Violence can be deduced because the radio emission has a chaotic, turbulent structure. Moreover, it appears to originate in the *center* of the corresponding optical galaxy, both because the bright outer edges of the distant radio lobes signal matter plowing into a surrounding medium, and more directly, a faint channel of radiating material connects one of the lobes to the central galaxy. The optical image of that galaxy is clearly disturbed. Have two galaxies collided? Is one splitting into two? Does a dust lane split the image? Whatever the case, the system is not what, prior to radio astronomy, had been considered "normal."

The large distance of the object follows from the feeble optical image, obtained only after a long photographic exposure, and its small angular diameter. Surprisingly, however, as a radio source, Cygnus A is the second brightest object in the sky, and consequently one of the earliest discovered. Its discoverers realized that the weaker radio signals they were detecting might originate from similar objects located even farther away, at distances theretofore regarded as beyond measurement.

The total radio energy contained in a galaxy like Cygnus A is immense, some 10^{53} joules. This amount is equivalent to the mass contained in some millions of stars the size of the sun if all their matter could be converted with perfect efficiency to energy via physics' most famous relation, $E = mc^2$. The energy is radiated by particles moving at nearly the speed of light in magnetic fields about a thousandth the strength of that at the earth's surface; in other words, Cygnus A is a giant **cosmic synchrotron.** From end to end it measures 10^{22} m. But remember that we only see the longitudinal dimension as projected onto the plane of the sky; its true length is probably greater. Neither energy nor matter could have reached Cygnus A's extremities from this center at any speed faster than that of light. Consequently the flow of matter/energy has persisted for at least a million years.

Radio astronomy did more than shatter prevailing notions of tranquility. It also struck a blow against the notion of permanence. Data collected over the last two decades on extragalactic sources show that the strength of the radio signals received

Figure 14E.1 NGC628, a spiral galaxy. (Palomar Observatory Photograph)

varied over this time. We now know that significant changes occur on timescales of years or even less, not the eons to which astronomers had been accustomed. In similar objects comparable changes have occurred in mere months. Evidently *transient phenomena* punctuate extragalactic history. Their discovery introduced yet another new concept to astronomy, that of *compactness*.

When an object doubles its brightness, a major fraction of its radiating volume must be involved. If it brightens in, say, a year, the radiating volume can scarcely be larger than the distance over which light can travel in one year, called a lightyear and equal to approximately 10^{16} m (about 6 trillion miles). To satisfy yourself that this is so, imagine that an object one lightyear in diameter is turned on instantaneously, like a lightbulb. At first, one will see light arriving from the closest surface. A little later, because it comes from farther away, light from an inner layer will arrive. Only at the end of the year will light from the far side be detected and the brightening cease. And if, instead of a general brightening of the entire radiating volume, individual portions of it had flared at random, the result would have been an insignificant flickering, not a dramatic outburst. Now, while a lightyear seems immense from a human perspective, it is small in a cosmic context, about, for example, the size of the solar system.

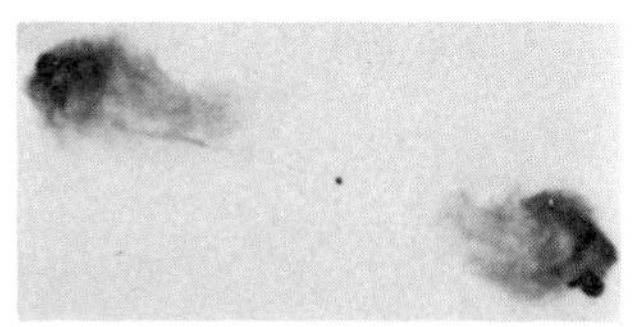

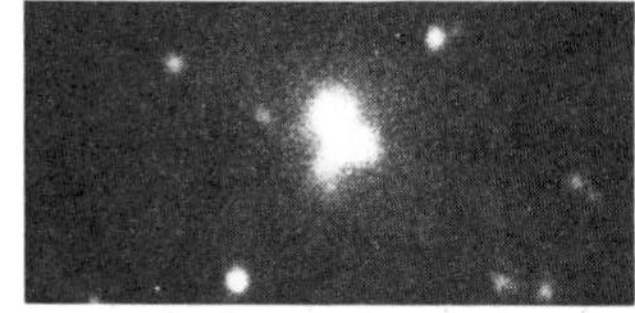

Figure 14E.2 Cygnus A. (a) A radio map of the galaxy. (b) The optical galaxy, which is coincident with the small central dot of the radio map. (NRAO Photograph)

How could objects so bizarre have escaped detection by optical astronomers? Optically 3C273—a seeming intragalactic source of radio noise—is a pinpoint of light, indistinguishable from the innumerable stars seen on every deep photographic exposure of the heavens. Hence the name **quasi-stellar radio source** or **quasar** for short, for objects in 3C273's class. Of course, the intense scrutiny quasars received after radio astronomers called attention to them revealed such peculiarities as the "jet" of light streaking from 3C273's stellar image. And, their attention aroused, optical astronomers proceeded to locate them at distances well beyond the known galaxies. Although quasars are the most distant objects known, both the radio and optical signals received from them are comparable to those from much closer objects, including Cygnus A. Accordingly, their intrinsic powers must be orders of magnitude greater. Quasars appear to be extreme examples of active galactic nuclei.

Objects like 3C273 begin to strain physicists' understanding of their world. Their energies are equivalent to the summed contributions of hundreds or thousands of galaxies. Yet all of it emanates from regions with solar-system dimensions (in length at best a percent of a galaxy's breadth). The resulting high-energy densities suggest correspondingly high matter densities, and together they establish intense gravitational fields. Gravity is implicated as the force responsible for the immense energy production, and astrophysicists suspect that it lurks within these prominent galactic nuclei in its most spectacular manifestation, as **black holes.**

The electromagnetic signals astronomers receive all travel at the same speed, 3×10^8 m/s—enormously fast but still finite. The delays between transmission and reception of signals from very distant objects such as galaxies or quasars can reach hundreds of millions or even billions of years. This means then that as we look farther out into the universe we actually look farther back in time.

Using this fact, astronomers have probed cosmic history almost to the moment when it began. To see how radio signals emerge from the birth of the universe, consider one whose density rises continuously as time is transversed in reverse. This is not an implausible hypothesis, since the universe is known to be expanding at present. At some epoch, a backward-in-time time traveler will encounter a fireball of heat, the consequence of friction when all matter and energy interact at ever closer range. This heat, once having filled the entire universe, will do so forever, since the universe by definition is all of space and time; simply put, the heat has no other place to which to escape.

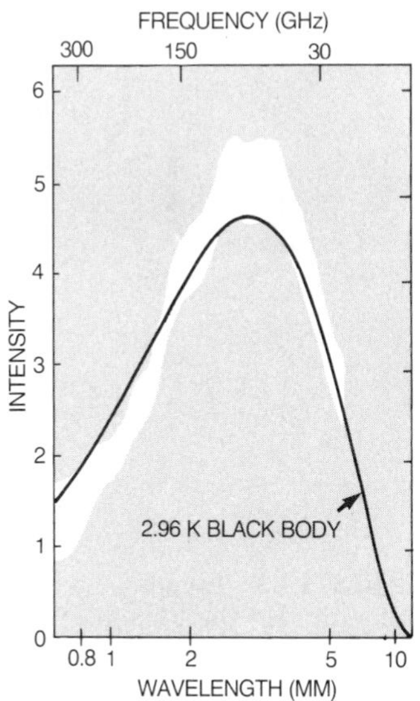

Figure 14E.3 The spectrum of the cosmic background radiation. Observed data (*white, unshaded region*) agrees well with the theoretical curve for a black body (*solid line*). (Pasachoff, J. M., *Contemporary Astronomy,* 3rd ed., Saunders College Publishing, 1985, p. 303)

Radio astronomers have found the expected fossil radiation from a cosmic fireball. Where? Everywhere, with nearly equal strength from every direction. We are evidently immersed within it. What is its signature? A Planck blackbody curve (Figure 14E.3) illustrates that the radiation arose when thermal equilibrium prevailed, as indeed it must have when the universe was so small that everything mixed completely with everything else. The present temperature of the radiation is only 3° Kelvin, or −270° Celsius, with its nearness to absolute zero indicating just how much and for how long the universe has expanded. What radio astronomers have found is no less than an echo of creation, decisive proof that the universe began in a hot, high-density state. This fiery origin has come to be called the **Big Bang.** Its detectable remnant, cool and faint, a whisper of its former glory, is called **cosmic background radiation.**

To this catalog of major discoveries via radio astronomy must be added **pulsars,** whose pulsed signals—brief bursts of radio energy—repeat with precise timing that is typically a few times per second. The pulses are believed to originate when directional beams radiated from rapidly spinning objects are swept across the earth, in lighthouse fashion, by their rotation. Since some of these objects are centered within the remains of stars that have exploded, so-called supernovae, pulsars are believed to be the remnant cores of now-dead stars.

Death presumably accompanied exhaustion of the supply of light elements (mainly hydrogen) whose fusion into heavier ones (helium) had powered the star during its lifetime. Once fusion ceased, no energy rose from the star's interior to balance the weight of all the matter trying to collapse upon itself under the grip of gravity. The stellar core therefore imploded. In order to conserve angular momentum, its rate of rotation increased. Likewise, the magnetic flux density of whatever field it possessed increased as the shrinking core pulled field lines closer together. The resultant dense, rotating, magnetized object is known as a **neutron star,** since compression first crushed (figuratively) the electrons present into the protons, making neutrons, then squeezed those neutrons until only quantum mechanical forces among them halted indefinite collapse. Although neutron stars contain the mass of a sun, they could fit within the borders of a city. Their densities are those of atomic nuclear matter, a teaspoonful weighing billions of tons. Their surface magnetic fields can exceed the earth's by factors of a hundred million.

One final contribution radio astronomy has made is to the field called **astrochemistry.** Many molecules naturally generate microwave signals. These occur at discrete frequencies, characteristic of each molecule. A radio astronomer can tune the receiver on his telescope to the appropriate frequencies, much as you tune your home radio to different commercial stations. The frequency at which a signal is detected identifies the radiating molecule. Over 60 species have been discovered, among them ammonia (NH_3), water (H_2O), hydrogen cyanide (HCN), methyl alcohol (CH_3OH), formic acid (HCOOH), methane (CH_4), and ethyl cyanide (CH_3CH_2CN).

Satellites have permitted the universe to be viewed through all portions of the electromagnetic spectrum. Astronomy, once regarded as the oldest science, is really the youngest: its full exploitation has only recently become possible. Now that our "vision" is broadened to its limit, discoveries will continue to alter our understanding of the universe we live in and our place within it.

TABLE 14.3 Escape Velocities for the Planets and the Moon

Planet	v_{esc} (km/s)
Mercury	4.3
Venus	10.3
Earth	11.2
Moon	2.3
Mars	5.0
Jupiter	60
Saturn	36
Uranus	22
Neptune	24

A list of escape velocities for the planets and the moon is given in Table 14.3. Note that the values vary from 2.3 km/s for the moon to about 60 km/s for Jupiter. These results, together with some ideas from the kinetic theory of gases (Chapter 21), explain why some planets have atmospheres and others do not. As we shall see later, a gas molecule at a given temperature has an average velocity that depends on its mass and the temperature. Lighter atoms, such as hydrogen and helium, have a higher average velocity than the heavier species. When the velocity of the lighter atoms is much greater than this average velocity, a significant fraction of the molecules have a chance to escape from the planet. This mechanism also explains why the earth does not retain hydrogen and helium molecules in its atmosphere while much heavier molecules, such as oxygen and nitrogen, do not escape. On the other hand, Jupiter has a very large escape velocity (60 km/s), which enables it to retain hydrogen, the primary constituent of its atmosphere.

*14.9 THE GRAVITATIONAL FORCE BETWEEN AN EXTENDED BODY AND A PARTICLE

We have emphasized that the law of universal gravitation given by Equation 14.3 is valid only if the interacting objects are considered as particles. In view of this, how can we calculate the force between a particle and an object having finite dimensions? This is accomplished by treating the *extended* object as a collection of particles and making use of integral calculus. We shall take the approach of first evaluating the potential energy function, from which the force can be calculated.

The potential energy associated with a system consisting of a point mass m and an extended body of mass M is obtained by dividing the body into segments of mass ΔM_i (Fig. 14.15). The potential energy associated with this element and with the particle of mass m is $-Gm\,\Delta M_i/r_i$, where r_i is the distance from the particle to the element ΔM_i. The total potential energy of the system is obtained by taking the sum over all segments as $\Delta M_i \rightarrow 0$. In this limit, we can express U in integral form as

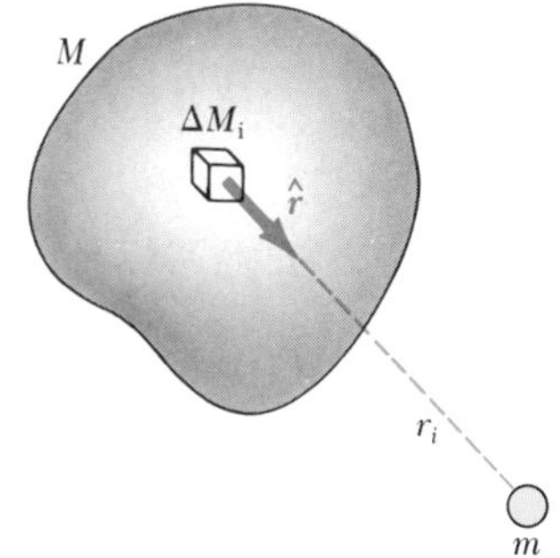

Figure 14.15 A particle of mass m interacting with an extended body of mass M. The potential energy of the system is given by Equation 14.20. The total force on a particle at P due to an extended body can be obtained by taking a vector sum over all forces due to each segment of the body.

$$U = -Gm \int \frac{dM}{r} \tag{14.20}$$

Once U has been evaluated, the force can be obtained by taking the negative derivative of this scalar function (see Section 8.7). If the extended body has spherical symmetry, the function U depends only on r and the force is given by $-dU/dr$. We shall treat this situation in Section 14.10. In principle, one can evaluate U for any specified geometry; however, the integration can be cumbersome.

An alternative approach to evaluating the force between a particle and an extended body is to perform a vector sum over all segments of the body. Using the procedure outlined in evaluating U and the law of universal gravitation (Eq. 14.3), the total force on the particle is given by

Total force between a particle and an extended body

$$\mathbf{F} = -Gm \int \frac{dM}{r^2}\,\hat{\mathbf{r}} \tag{14.21}$$

where $\hat{\mathbf{r}}$ is a unit vector directed from the element dM toward the particle (Fig. 14.15). This procedure is not always recommended, since working with a vector function is more difficult than working with the scalar potential energy function. However, if the geometry is simple, as in the following example, the evaluation of $\mathbf{F}$ can be straightforward.

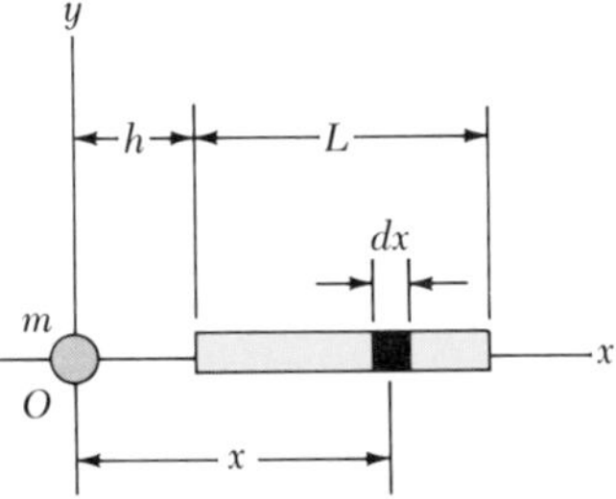

Figure 14.16 (Example 14.8) The force on a particle at the origin due to the bar is to the right. Note that the bar is *not* equivalent to a particle of mass M located at its center of mass.

EXAMPLE 14.8. Force Between a Mass and a Bar

A homogeneous bar of length L and mass M is at a distance h from a point mass m (Fig. 14.16). Calculate the force on m.

Solution: The segment of the bar that has a length dx has a mass dM. Since the mass per unit length is a constant, it then follows that the ratio of masses, dM/M, is equal to the ratio of lengths, dx/L, and so $dM = \frac{M}{L}\,dx$. The variable r in Equation 14.21 is x in our case, and the force on m is to the right; therefore we get

$$F = Gm \int_h^{L+h} \frac{M}{L}\frac{dx}{x^2}\, i$$

$$F = \frac{GmM}{L}\left[-\frac{1}{x}\right]_h^{L+h} i = \frac{GmM}{h(L+h)}\, i$$

We see that the force on m is in the positive x direction, as expected, since the gravitational force is attractive.

Note that in the limit $L \rightarrow 0$, the force varies as $1/h^2$, which is what is expected for the force between two point masses. Furthermore, if $h \gg L$, the force also varies as $1/h^2$. This can be seen by noting that the denominator of the expression for F can be expressed in the form $h^2\left(1+\frac{L}{h}\right)$, which is approximately equal to h^2. Thus, when bodies are separated by distances that are large compared with their characteristic dimensions, they behave like particles.

*14.10 GRAVITATIONAL FORCE BETWEEN A PARTICLE AND A SPHERICAL MASS

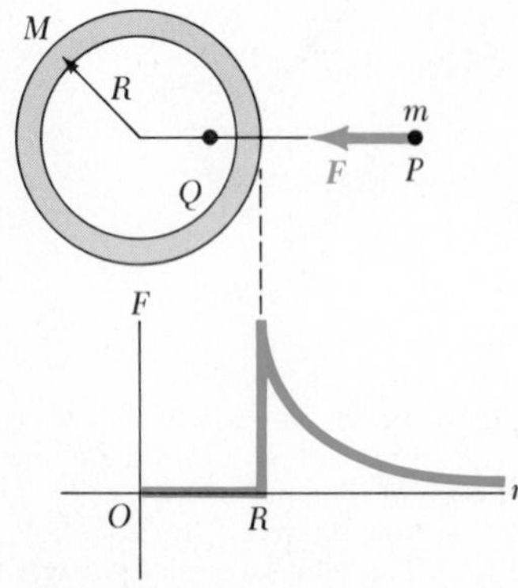

Figure 14.17 The force on a particle when it is outside the spherical shell is given by GMm/r^2 and acts toward the center. The force on the particle is zero everywhere inside the shell.

In this section we shall describe the gravitational force between a particle and a spherically symmetric mass distribution. We have already stated that a large sphere attracts a particle outside it as if the total mass of the sphere were concentrated at its center. This, and other properties of the spherical mass distribution, are proved formally in Section 14.11. Let us describe the nature of the force on a particle when the extended body is either a spherical shell or a solid sphere, and then apply these facts to some interesting systems.

Spherical Shell

1. If a particle of mass m is located *outside* a spherical shell of mass M (say, point P in Fig. 14.17), the spherical shell attracts the particle as though the mass of the shell were concentrated at its center.

2. If the particle is located *inside* the spherical shell (point Q in Fig. 14.17), the force on it is zero. We can express these two important results in the following way:

Force on a particle due to a spherical shell

$$F = -\frac{GMm}{r^2}\hat{r} \qquad \text{for } r > R \qquad (14.22a)$$

$$F = 0 \qquad \text{for } r < R \qquad (14.22b)$$

The force as a function of the distance r is plotted in Figure 14.17.

Solid Sphere

1. If a particle of mass m is located *outside* a homogeneous solid sphere of mass M (point P in Fig. 14.18), the sphere attracts the particle as though the mass of the sphere were concentrated at its center. That is, Equation 14.22a applies in this situation. This follows from case 1 above, since a solid sphere can be considered a collection of concentric spherical shells.

2. If a particle of mass m is located *inside* a homogeneous solid sphere of mass M (point Q in Fig. 14.18), the force on m is due *only* to the mass M' contained within the sphere of radius $r < R$, represented by the dotted line in Figure 14.18. In other words,

Force on a particle due to a solid sphere

$$F = -\frac{GmM}{r^2}\hat{r} \qquad \text{for } r > R \qquad (14.23a)$$

$$F = -\frac{GmM'}{r^2}\hat{r} \qquad \text{for } r < R \qquad (14.23b)$$

Since the sphere is assumed to have a uniform density, it follows that the ratio of masses M'/M is equal to the ratio of volumes V'/V, where V is the total volume of the sphere and V' is the volume within the dotted surface. That is,

$$\frac{M'}{M} = \frac{V'}{V} = \frac{\frac{4}{3}\pi r^3}{\frac{4}{3}\pi R^3} = \frac{r^3}{R^3}$$

Solving this equation for M' and substituting the value obtained into Equation 14.23b, we get

$$F = -\frac{GmM}{R^3} r \hat{r} \qquad \text{for } r < R \tag{14.24}$$

That is, the force goes to zero at the center of the sphere, as we would intuitively expect. The force as a function of r is plotted in Figure 14.18.

3. If a particle is located *inside* a solid sphere having a density ρ that is spherically symmetric but *not* uniform, then M' in Equation 14.23 is given by an integral of the form $M' = \int \rho \, dV$, where the integration is taken over the volume contained *within* the dotted surface. This integral can be evaluated if the radial variation of ρ is given. The integral is easily evaluated if the mass distribution has spherical symmetry, that is, if ρ is a function of r only. In this case, we take the volume element dV as the volume of a spherical shell of radius r and thickness dr, so that $dV = 4\pi r^2 \, dr$. For example, if $\rho(r) = Ar$, where A is a constant, it is left as a problem (Problem 44) to show that $M' = \pi A r^4$. Hence we see from Equation 14.23b that F is proportional to r^2 in this case and is zero at the center.

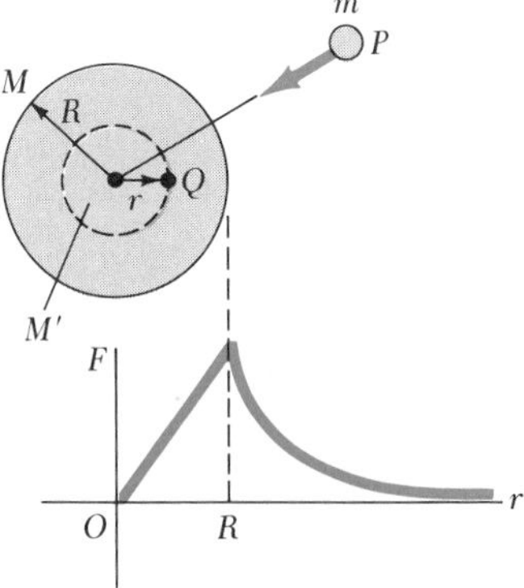

Figure 14.18 The force on a particle when it is outside a uniform solid sphere is given by GMm/r^2 and is directed toward the center. The force on the particle when it is inside such a sphere is proportional to r and goes to zero at the center.

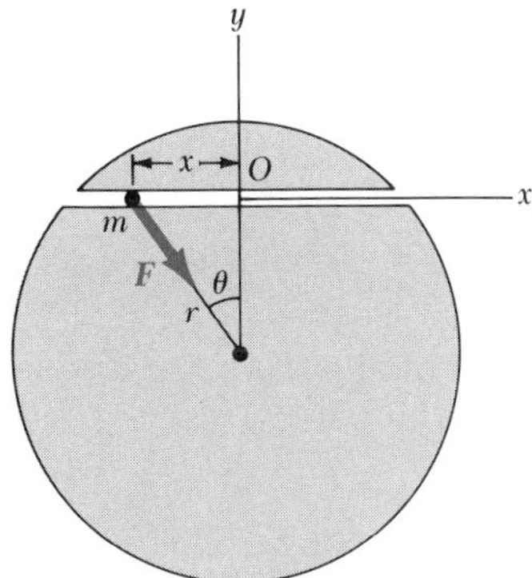

Figure 14.19 A particle moves along a tunnel dug through the earth. The component of the gravitational force F along the x axis is the driving force for the motion. Note that this component always acts toward the origin O.

EXAMPLE 14.9. A Free Ride

An object moves in a smooth, straight tunnel dug between two points on the earth's surface (Fig. 14.19). Show that the object moves with simple harmonic motion and find the period of its motion.

Solution: When the object is in the tunnel, the gravitational force on it acts toward the earth's center and is given by Equation 14.24:

$$F = -\frac{GmM_e}{R_e^3} r$$

The y component of this force is balanced by the normal force exerted by the tunnel wall, and the x component of the force is given by

$$F_x = -\frac{GmM_e}{R_e^3} r \sin \theta$$

Since the x coordinate of the object is given by $x = r \sin \theta$, we can write F_x in the form

$$F_x = -\frac{GmM_e}{R_e^3} x$$

Applying Newton's second law to the motion along x gives

$$F_x = -\frac{GmM_e}{R_e^3} x = ma$$

$$a = -\frac{GM_e}{R_e^3} x = -\omega^2 x$$

But this is the equation of simple harmonic motion with angular velocity ω (Chapter 13), where

$$\omega = \sqrt{\frac{GM_e}{R_e^3}}$$

The period is calculated using the data in Table 14.2 and the above result:

$$T = \frac{2\pi}{\omega} = 2\pi\sqrt{\frac{R_e^{\,3}}{GM_e}}$$

$$= 2\pi\sqrt{\frac{(6.37\times10^6)^3}{(6.67\times10^{-11})(5.98\times10^{24})}}$$

$$= 5.06\times10^6 \text{ s} = 84.3 \text{ min}$$

This period is the same as that of a satellite in a circular orbit just above the earth's surface. Note that the result is *independent* of the length of the tunnel.

It has been proposed to operate a mass-transit system between any two cities using this principle. A one-way trip would take about 42 min. A more precise calculation of the motion must account for the fact that the earth's density is not uniform as we have assumed. More important, there are many practical problems to consider. For instance, it would be impossible to achieve a frictionless tunnel, and so some auxiliary power source would be required. Can you think of other problems?

*14.11 DERIVATION OF THE GRAVITATIONAL EFFECT OF A SPHERICAL MASS DISTRIBUTION

The purpose of this section is to prove Equations 14.22 and 14.23 using integral calculus. Consider a spherical shell of mass M and radius R with a thickness that is small compared with R (Fig. 14.20). A particle of mass m is placed at a point P, some distance r from the center of the shell. We could calculate the force on m directly, but since this is a vector quantity, a vector sum over all parts of the shell would be required. It is easier to first calculate the potential energy associated with the system (a scalar quantity). Since the mass distribution is spherically symmetric, the potential energy, U, is a function only of the radial distance r, that is, $U = U(r)$. The force on m can then be obtained from the relation $F_r = -dU/dr$. This is the approach we shall take.

First, let us calculate the mass of a zone of the shell, where the zone is taken perpendicular to the axis OP (the shaded section in Fig. 14.20). Since the width of this zone is $R\,d\theta$ and its radius is $R \sin\theta$, we see that the outer surface area of the zone is $dA = 2\pi R^2 \sin\theta\,d\theta$. The total surface area of the shell is $4\pi R^2$; hence it follows that the mass of the zone is given by

$$dM = \frac{\text{area of zone}}{\text{area of shell}} \times M = \frac{2\pi R^2 \sin\theta\,d\theta}{4\pi R^2} \times M = \tfrac{1}{2}M \sin\theta\,d\theta$$

Since all parts of the zone are at essentially the same distance s from the point P, from Equation 14.20 we see that the potential energy associated with this zone and the particle is

$$dU = -\frac{Gm\,dM}{s} = -\frac{GmM}{2}\frac{\sin\theta\,d\theta}{s}$$

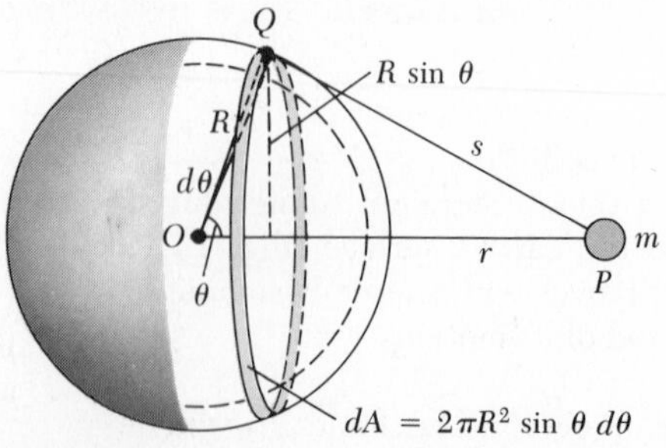

Figure 14.20 Diagram for calculating the gravitational potential energy of a particle interacting with a spherical shell. The shell is divided into circular zones (shaded) for convenience.

The total potential energy of the system is

$$U = -\frac{GmM}{2}\int\frac{\sin\theta\,d\theta}{s} \tag{14.25}$$

We cannot evaluate this integral directly, since it involves two variables, θ and s. However, we can eliminate one of the variables by applying the law of cosines to the triangle OPQ in Figure 14.20:

$$s^2 = r^2 + R^2 - 2rR\cos\theta$$

Differentiating both sides of this equation with respect to θ and noting that r and R are constants for a particular point P, we get

$$2s\frac{ds}{d\theta} = -2rR(-\sin\theta)$$

$$\sin\theta\,d\theta = \frac{s\,ds}{rR}$$

Substituting this into the integrand of Equation 14.25 gives

$$U = -\frac{GmM}{2rR}\int_{s_1}^{s_2} ds \tag{14.26}$$

To evaluate U from Equation 14.26 we must specify the limits of integration. We shall first consider a point P outside the shell, as in Figure 14.20, and then a point inside the shell.

Outside the Shell

When the particle of mass m is outside the shell, where $r > R$, the limits of integration in Equation 14.26 are $s_1 = r - R$ to $s_2 = r + R$. Therefore,

$$U = -\frac{GMm}{2rR}\int_{r-R}^{r+R} ds = -\frac{GMm}{r} \qquad \text{for } r > R$$

Hence the force on m when it is outside the shell is

$$F_r = -\frac{dU}{dr} = -\frac{d}{dr}\left(-\frac{GMm}{r}\right) = -\frac{GMm}{r^2} \qquad \text{for } r > R$$

This verifies Equation 14.22a.

Inside the Shell

When the particle of mass m is inside the shell, where $r < R$, the limits of integration in Equation 14.26 are $s_1 = R - r$ to $s_2 = R + r$. Therefore,

$$U = -\frac{GMm}{2rR}\int_{R-r}^{R+r} ds = -\frac{GMm}{R} \qquad \text{for } r < R$$

Since R is a constant, we see that the potential energy is constant for all points within the sphere. Therefore,

$$F_r = -\frac{dU}{dr} = 0 \qquad \text{for } r < R$$

This verifies Equation 14.22b.

The extension of these ideas to a solid sphere is straightforward, since we can regard a solid sphere as a collection of concentric spherical shells.

14.12 SUMMARY

Kepler's laws

Kepler's laws *of planetary motion* state that

1. All planets move in elliptical orbits with the sun at one of the focal points.
2. The radius vector drawn from the sun to any planet sweeps out equal areas in equal time intervals.
3. The square of the orbital period of any planet is proportional to the cube of the semi-major axis for the elliptical orbit.

Newton's law of universal gravitation states that the gravitational force of attraction between any two particles of masses m_1 and m_2 separated by a distance r has the magnitude

Universal law of gravity

$$F = G\frac{m_1 m_2}{r^2} \tag{14.1}$$

where G is the universal gravitational constant, which has the value $6.672 \times 10^{-11}\ \mathrm{N \cdot m^2/kg^2}$.

Kepler's second law is a consequence of the fact that the force of gravity is a *central force*, that is, one that is directed toward a fixed point. This in turn implies that the angular momentum of the planet-sun system is a constant of the motion.

Kepler's third law is consistent with the inverse-square nature of the law of universal gravitation. Newton's second law, together with the force law given by Equation 14.1, verifies that the period T and radius r of the orbit of a planet about the sun are related by

Kepler's third law

$$T^2 = \left(\frac{4\pi^2}{GM_s}\right)r^3 \tag{14.6}$$

where M_s is the mass of the sun. Most planets have nearly circular orbits about the sun. For elliptical orbits, Equation 14.6 is valid if r is replaced by the semi-major axis, a.

The gravitational force is a conservative force, and therefore a potential energy function can be defined. The **gravitational potential energy** associated with two particles separated by a distance r is given by

Gravitational potential energy for a pair of particles

$$U = -\frac{Gm_1 m_2}{r} \tag{14.12}$$

where U is taken to be zero at $r = \infty$. The total potential energy for a system of particles is the sum of energies for all pairs of particles, with each pair represented by a term of the form given by Equation 14.12.

If an isolated system consists of a particle of mass m moving with a speed v in the vicinity of a massive body of mass M, the *total energy* of the system is given by

$$E = \tfrac{1}{2}mv^2 - \frac{GMm}{r} \tag{14.14}$$

That is, the energy is the sum of the kinetic and potential energies. The total energy is a constant of the motion.

If m moves in a circular orbit of radius r about M, where $M \gg m$, *the total energy of the system is*

$$E = -\frac{GMm}{2r} \tag{14.17}$$

Total energy for circular orbits

The total energy is negative for any bound system, that is, one in which the orbit is closed, such as an elliptical orbit.

The *potential energy* of gravitational attraction between a particle of mass m and an extended body of mass M is given by

$$U = -Gm \int \frac{dM}{r} \tag{14.20}$$

Total potential energy for a particle-extended body system

where the integral is over the extended body, dM is an infinitesimal mass element of the body, and r is the distance from the particle to the element.

If a particle is outside a uniform spherical shell or solid sphere with a spherically symmetric internal mass distribution, the sphere attracts the particle as though the mass of the sphere were concentrated at the center of the sphere.

If a particle is inside a uniform spherical shell, the gravitational force on the particle is zero.

If a particle is inside a homogeneous solid sphere, the force on the particle acts toward the center of the sphere and is linearly proportional to the distance from the center to the particle.

QUESTIONS

1. Estimate the gravitational force between you and a person 2 m away from you.
2. Use Kepler's second law to convince yourself that the earth must move faster in its orbit during the winter, when it is closest to the sun, than it does during the summer, when it is farthest from the sun.
3. How would you explain the fact that planets such as Saturn and Jupiter have periods much greater than one year?
4. If a system consists of five distinct particles, how many terms appear in the expression for the total potential energy?
5. Is it possible to calculate the potential energy function associated with a particle and an extended body without knowing the geometry or mass distribution of the extended body?
6. Does the escape velocity of a rocket depend on its mass? Explain.
7. Compare the energies required to reach the moon for a 10^5-kg spacecraft and a 10^3-kg satellite.
8. Explain why it takes more fuel for a spacecraft to travel from the earth to the moon than for the return trip. Estimate the difference.
9. Is the magnitude of the potential energy associated with the earth-moon system greater than, less than, or equal to the kinetic energy of the moon relative to the earth?
10. Explain carefully why there is no work done on a planet as it moves in a circular orbit around the sun, even though a gravitational force is acting on the planet. What is the *net* work done on a planet during each revolution as it moves around the sun in an elliptical orbit?
11. A particle is projected through a small hole into the interior of a large spherical shell. Describe the motion of the particle in the interior of the shell.
12. Explain why the force on a particle due to a uniform sphere must be directed toward the center of the sphere. Would this be the case if the mass distribution of the sphere were not spherically symmetric?
13. Neglecting the density variation of the earth, what would be the period of a particle moving in a smooth hole dug through the earth's center?
14. With reference to Figure 14.8, consider the area swept out by the radius vector in the time intervals $t_2 - t_1$ and $t_4 - t_3$. Under what condition is A_1 equal to A_2?

15. If A_1 equals A_2 in Figure 14.8, is the average speed of the planet in the time interval $t_2 - t_1$ less than, equal to, or greater than its average speed in the time interval $t_4 - t_3$?
16. At what position in its elliptical orbit is the speed of a planet a maximum? At what position is the speed a minimum?
17. If you are given the mass and radius of planet X, how would you calculate the acceleration of gravity on the surface of this planet?
18. If a hole could be dug to the center of the earth, do you think that the force on a mass m would still obey Equation 14.1 there? What do you think the force on m would be at the center of the earth?

PROBLEMS

Section 14.1 through Section 14.3

1. Two identical, isolated particles, each of mass 2 kg, are separated by a distance of 30 cm. What is the magnitude of the gravitational force of one particle on the other?
2. A 200-kg mass and a 500-kg mass are separated by a distance of 0.40 m. (a) Find the net gravitational force due to these masses acting on a 50-kg mass placed midway between them. (b) At what position (other than infinitely remote ones) would the 50-kg mass experience a net force of zero?
3. Three 5-kg masses are located at the corners of an equilateral triangle having sides 0.25 m in length. Determine the magnitude and direction of the resultant gravitational force on one of the masses due to the other two masses.
4. Two stars of masses M and $4M$ are separated by a distance d. Determine the location of a point measured from M at which the net force on a third mass would be zero.
5. Four particles are located at the corners of a rectangle as in Figure 14.21. Determine the x and y components of the resultant force acting on the particle of mass m.

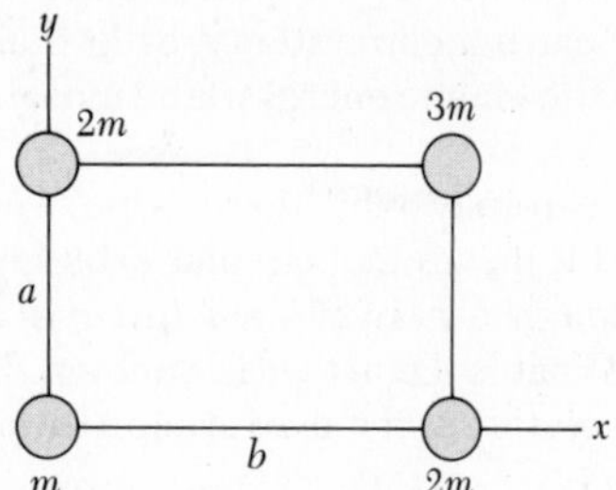

Figure 14.21 (Problem 5).

6. Calculate the acceleration of gravity at a point that is a distance R_e above the surface of the earth, where R_e is the radius of the earth.
7. Using the data given in Figure 14.3, determine a vector expression for the resultant force on the 6-kg mass. What is the magnitude of this force?

Section 4.4 Kepler's Laws
Section 14.5 The Law of Universal Gravitation and the Motion of Planets

8. Given that the moon's period about the earth is 27.32 days and the earth-moon distance is 3.84×10^8 m, estimate the mass of the earth. Assume the orbit is circular. Why do you suppose your estimate is high?
9. A satellite is in a circular orbit about the earth. (a) Evaluate the constant K that appears in Kepler's third law as applied to this situation. (b) What is the period of the orbit if the satellite is at an altitude of 2×10^6 m?
10. The planet Jupiter has at least 14 satellites. One of them, named Callisto, has a period of 16.75 days and a mean orbital radius of 1.883×10^9 m. From this information, calculate the mass of Jupiter.
11. A satellite of Mars has a period of 459 min. The mass of Mars is 6.42×10^{23} kg. From this information, determine the radius of the satellite's orbit.
12. At its aphelion, the planet Mercury is 6.99×10^{10} km from the sun, and at its perihelion, it is 4.60×10^{10} km from the sun. If its orbital speed is 3.88×10^4 m/s at the aphelion, what is its orbital speed at the perihelion?
13. A satellite is to be sent into orbit about the earth in an equatorial plane such that it will always appear to be stationary relative to an observer on earth. Find the radius of its orbit. (*Hint:* The satellite must have the same angular velocity as the earth.)

Section 14.7 Gravitational Potential Energy (Assume $U = 0$ at $r = \infty$)

14. A satellite of the earth has a mass of 100 kg and is at an altitude of 2×10^6 m. (a) What is the potential energy of the satellite-earth system? (b) What is the magnitude of the force on the satellite?
15. A system consists of three particles, each of mass 5 g, located at the corners of an equilateral triangle with sides of 30 cm. (a) Calculate the potential energy of the system. (b) If the particles are released simultaneously, where will they collide?
16. How much energy is required to move a 1000-kg mass from the earth's surface to an altitude equal to twice the earth's radius?
17. Four particles are positioned at the corners of a square as in Figure 14.22. Calculate the total potential energy of the system.

Section 14.8 Energy Considerations in Planetary and Satellite Motion

18. Calculate the escape velocity from the moon using the data in Table 14.2.

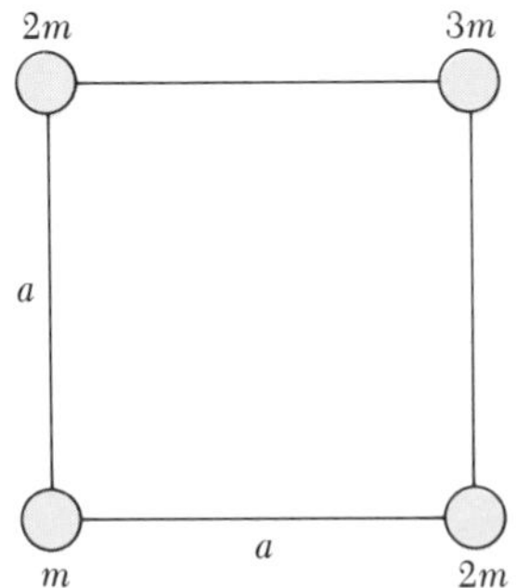

Figure 14.22 (Problem 17).

19. Calculate the escape velocity from Mars using the data in Table 14.2.

20. A spaceship is fired from the earth's surface with an initial speed of 2.0×10^4 m/s. What will its speed be when it is very far from the earth? (Neglect friction.)

21. A 500-kg spaceship is in a circular orbit of radius $2R_e$ about the earth. (a) How much energy is required to transfer the spaceship to a circular orbit of radius $4R_e$? (b) Discuss the change in the potential energy, kinetic energy, and total energy.

22. Two identical spacecrafts, each of mass 1000 kg, travel in free space along the same path. At some instant when their separation is 20 m and each has the *same* velocity, the power is turned off in each vehicle. What are their speeds when they are 2 m apart? (Treat the spacecrafts as particles.)

23. (a) Calculate the minimum energy required to send a 3000-kg spacecraft from the earth to a distant point in space where earth's gravity is negligible. (b) If the journey is to take three weeks, what *average* power would the engines have to supply?

24. A rocket is fired vertically from the earth's surface and reaches a maximum altitude equal to three earth radii. What was the initial speed of the rocket? (Neglect friction, the earth's rotation, and the earth's orbital motion.)

25. A satellite moves in a circular orbit around a planet. Show that the orbital velocity v and escape velocity of the satellite are related by the expression $v_{esc} = \sqrt{2}v$.

°Section 14.9 The Gravitational Force Between an Extended Body and a Particle

26. A uniform rod of mass M is in the shape of a semicircle of radius R (Fig. 14.23). Calculate the force on a point mass m placed at the center of the semicircle.

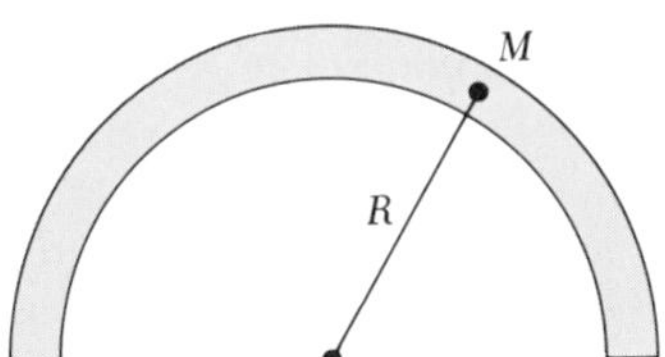

Figure 14.23 (Problem 26).

27. A *nonuniform* rod of length L is placed along the x axis at a distance h from the origin, as in Figure 14.16. The mass per unit length, λ, varies according to the expression $\lambda = \lambda_0 + Ax^2$, where λ_0 and A are constants. Find the force on a particle of mass m placed at the origin. (*Hint:* An element of the rod has a mass $dM = \lambda\, dx$.)

°Section 14.10 Gravitational Force Between a Particle and a Spherical Mass

28. A spherical shell has a radius of 0.5 m and mass of 80 kg. Find the force on a particle of mass 50 g placed (a) 0.3 m from the center of the shell and (b) outside the shell 1 m from its center.

29. A uniform solid sphere has a radius of 0.4 m and a mass of 500 kg. Find the magnitude of the force on a particle of mass 50 g located (a) 1.5 m from the center of the sphere, (b) at the surface of the sphere, and (c) 0.2 m from the center of the sphere.

30. A uniform solid sphere of mass m_1 and radius R_1 is inside and concentric with a spherical shell of mass m_2 and radius R_2 (Fig. 14.24). Find the force on a particle of mass m located at (a) $r = a$, (b) $r = b$, (c) $r = c$, where r is measured from the center of the spheres.

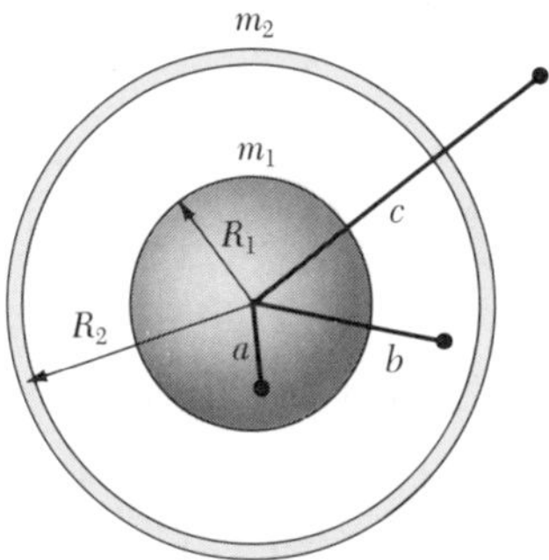

Figure 14.24 (Problem 30).

GENERAL PROBLEMS

31. Two astronauts, each of the same mass M, are seated opposite each other in a space station drifting in free space. The room they are in is a cylinder of radius R that rotates about its symmetry axis (Fig. 14.25). (a) What is the minimum angular speed of the cylinder that will keep the astronauts from moving toward

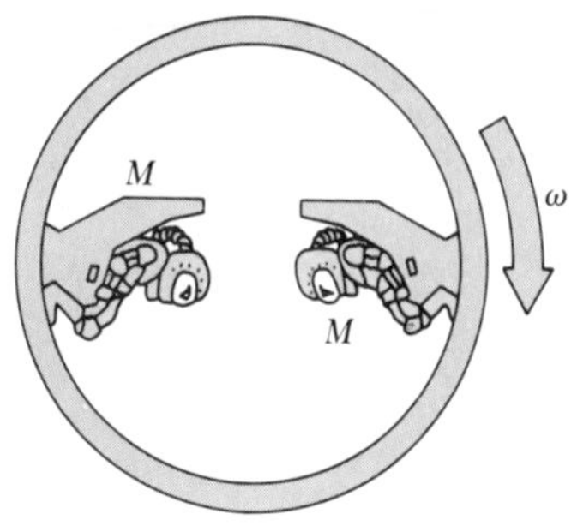

Figure 14.25 (Problem 31).

each other if they are not strapped in their seats? (b) What angular speed must the cylinder have in order to produce a gravitational force equivalent to that experienced on earth? Obtain a numerical value if $R = 4$ m.

32. Arranged in a rectangular coordinate system are a 2-kg mass at the origin, a 3-kg mass at the position (0, 2), and a 4-kg mass at (4, 0), where all distances are in m. Find the resultant gravitational force exerted on the mass at the origin by the other two masses.

33. Three masses are aligned along the x axis of a rectangular coordinate system such that a 2-kg mass is at the origin, a 3-kg mass is at (2, 0) m, and a 4-kg mass is at (4, 0) m. (a) Find the gravitational force exerted on the 4-kg mass by the other two masses. (b) Find the magnitude and direction of the gravitational force exerted on the 3 kg mass by the other two.

34. Four objects are located at the corners of a rectangle, as in Figure 14.26. Determine the magnitude and direction of the resultant force acting on the 2-kg mass at the origin.

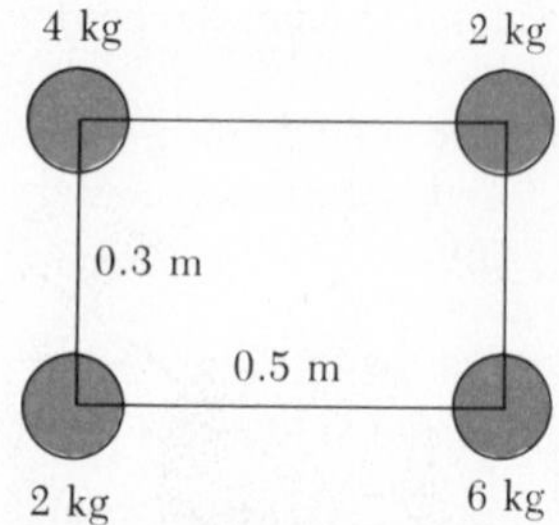

Figure 14.26 (Problem 34).

35. An object of mass m moves in a smooth straight tunnel of length L dug through a chord of the earth as discussed in Example 14.9 (Fig. 14.19). (a) Determine the effective force constant of the harmonic motion and the amplitude of the motion. (b) Using energy considerations, find the maximum speed of the object. Where does this occur? (c) Obtain a numerical value for the maximum speed if $L = 500$ km.

36. The maximum distance from the earth to the sun (at the aphelion) is 1.521×10^{11} m, and the distance of closest approach (at the perihelion) is equal to 1.471×10^{11} m. If the earth's orbital speed at the perihelion is 3.027×10^{4} m/s, determine (a) the earth's orbital speed at the aphelion, (b) the kinetic and potential energy at the perihelion, and (c) the kinetic and potential energy at the aphelion. Is the total energy conserved? (Neglect the effect of the moon and other planets.)

37. Two hypothetical planets of masses m_1 and m_2 and radii r_1 and r_2, respectively, are at rest when they are an infinite distance apart. Because of their gravitational attraction, they head toward each other on a collision course. (a) When their center-to-center separation is d, find the speed of each planet and their *relative* velocity. (b) Find the kinetic energy of each planet *just* before they collide if $m_1 = 2 \times 10^{24}$ kg, $m_2 = 8 \times 10^{24}$ kg, $r_1 = 3 \times 10^{6}$ m, and $r_2 = 5 \times 10^{6}$ m. (*Hint:* Note that both energy and momentum are conserved.)

38. Use the equation $F = mv^2/r$ to calculate the centripetal force needed to make the earth follow its path about the sun. Compare this value for F with the value found by using Newton's universal law of gravity.

39. When the Apollo 11 spacecraft orbited the moon, its mass was 9.979×10^{3} kg, its period was 119 min, and its mean distance from the moon's center was 1.849×10^{6} m. Assuming its orbit was circular and assuming the moon to be a uniform sphere, find (a) the mass of the moon, (b) the orbital speed of the spacecraft, and (c) the minimum energy required for the craft to leave the orbit and escape the moon's gravity.

40. Using the data in Table 14.1, calculate the total potential energy of the sun-moon-earth system. Assume that the moon and earth are at the same distance from the sun.

41. A hypothetical planet of mass M has three moons of equal mass m, each moving in the same circular orbit of radius R (Fig. 14.27). The moons are equally spaced and thus form an equilateral triangle. Find (a) the total potential energy of the system and (b) the orbital speed of each moon such that they maintain this configuration.

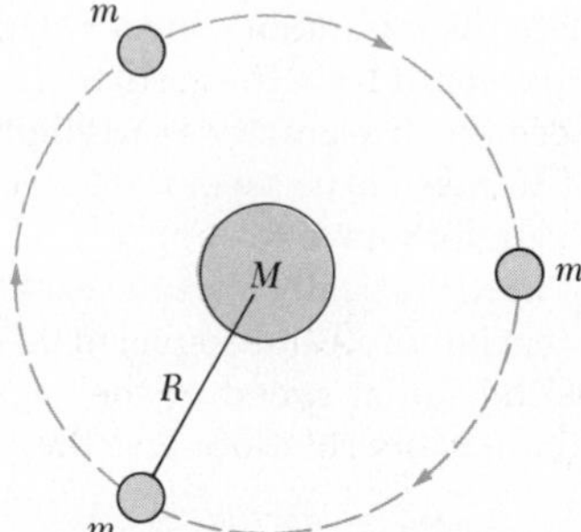

Figure 14.27 (Problem 41).

42. A satellite of mass 600 kg is in a circular orbit about the earth at a height above the earth equal to the earth's mean radius. Find (a) the satellite's orbital speed, (b) the period of its revolution, and (c) the gravitational force acting on it.

•43. A particle of mass m lies along the symmetry axis of a uniform circular ring of mass M and radius R (Fig. 14.28). (a) Find the force on m if it is at a distance d from the plane of the ring. (b) Show that your result to (a) reduces to what you would intuitively expect (1) when m is at the center of the ring ($d = 0$) and (2) when m is distant from the ring ($d \gg R$).

•44. A sphere of mass M and radius R has a *nonuniform* density that varies with r, the distance from its center,

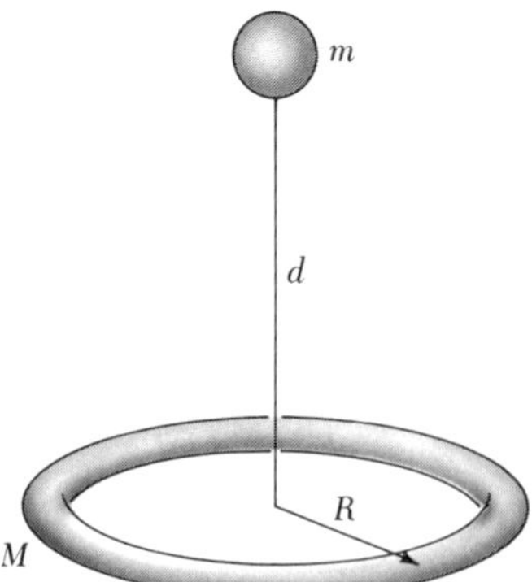

Figure 14.28 (Problem 43).

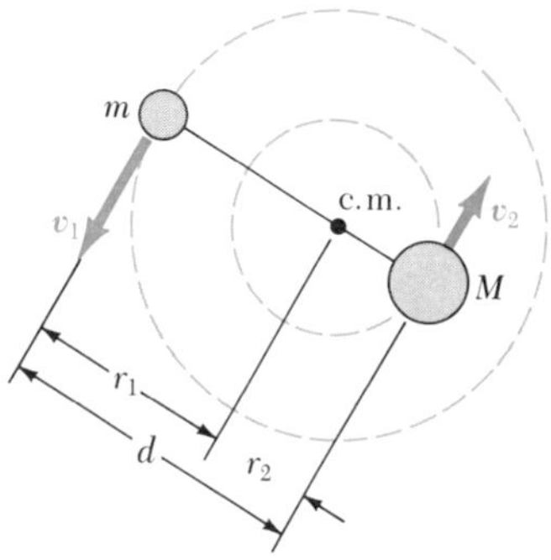

Figure 14.29 (Problem 45).

according to the expression $\rho = Ar$, for $0 \leq r \leq R$. (a) What is the constant A in terms of M and R? (b) Determine the force on a particle of mass m placed *outside* the sphere. (c) Determine the force on the particle if it is *inside* the sphere. (*Hint:* See Section 14.10.)

•**45.** Two stars of masses M and m, separated by a distance d, revolve in circular orbits about their center of mass (Fig. 14.29). Show that each star has a period given by

$$T^2 = \frac{4\pi^2}{G(M+m)} d^3$$

(*Hint:* Apply Newton's second law to each star, and note that the center of mass condition requires that $Mr_2 = mr_1$, where $r_1 + r_2 = d$.)

•**46.** A particle of mass m is located *inside* a uniform solid sphere of radius R and mass M. If the particle is at a distance r from the center of the sphere, (a) show that the gravitational potential energy of the system is given by $U = (GmM/2R^3)r^2 - 3GmM/2R$. (b) How much work is done by the gravitational force in bringing the particle from the surface of the sphere to its center?

•**47.** A satellite is in a circular orbit about a planet of radius R. If the altitude of the satellite is h and its period is T, (a) show that the density of the planet is given by

$$\rho = \frac{3\pi}{GT^2}\left(1 + \frac{h}{R}\right)^3.$$

(b) Calculate the average density of the planet if the period is 200 min and the satellite's orbit is close to the planet's surface.

•**48.** The acceleration of gravity at an altitude h is given by Equation 14.5. If $h \ll R_e$, show that the acceleration of gravity at h is given *approximately* by

$$g' \approx g\left(1 - 2\frac{h}{R_e}\right)$$

(*Hint:* Start with Equation 14.5, and use the binomial expansion for the denominator.)

15
Mechanics of Solids and Fluids

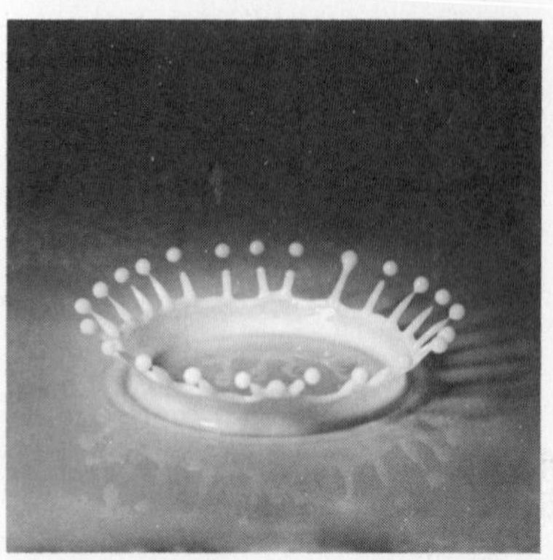

This chapter deals with the mechanical properties of solids and fluids. This subject, often called *continuum mechanics*, treats the macroscopic behavior of a large number of particles. In reality, all bodies will deform to some extent under load conditions. The deformation of a body under external forces will be used to classify matter as being solid, liquid, or gaseous. Deformation is an important consideration in understanding the mechanics of materials and structural designs. Such deformations are usually small and will not affect the conditions of equilibrium. By *small* we mean that when the deforming forces are removed, the body returns to its original shape. Several elastic constants will be defined, each corresponding to a different type of deformation.

In our treatment of the mechanics of fluids, we shall see that no new physical principles are needed to explain such effects as the buoyant force on a submerged object and the dynamic lift on an airplane wing. First, we shall present a discussion of the various states of matter. Next, we shall consider a fluid at rest and derive an expression for the pressure as a function of its density and depth. We shall then treat fluids in motion, or fluid dynamics. A fluid in motion can be described by a model in which certain simplifying assumptions are made. We shall use this model to analyze some situations of practical importance. An underlying principle known as the *Bernoulli effect* will enable us to determine relations between the pressure, density, and velocity at every point in a fluid. As we shall see, the Bernoulli effect is a result of conservation of energy applied to an ideal fluid. We conclude the chapter with a brief discussion of internal friction in a fluid and turbulent motion.

15.1 ELASTIC PROPERTIES OF SOLIDS

In our study of mechanics thus far, we assumed that objects remain undeformed when external forces act on them. In reality, all objects are deformable. That is, it is possible to change the shape or size of a body (or both) through the application of external forces. Although these changes are observed as large-scale deformations, the internal forces that resist the deformation are due to short-range forces between atoms.

We shall discuss the elastic properties of solids in terms of the concepts of stress and strain. **Stress** is a quantity that is proportional to the force causing a deformation; **strain** is a measure of the degree of deformation. It is found that, for sufficiently small stresses, the stress is proportional to the strain and the constant of proportionality depends on the material being deformed and on the nature of the deformation. We call this proportionality constant the **elastic modulus.** The elastic modulus is therefore the ratio of stress to strain:

$$\text{Elastic modulus} = \frac{\text{stress}}{\text{strain}} \qquad (15.1)$$

We shall consider three types of deformation and define an elastic modulus for each:

1. **Young's modulus,** which measures the resistance of a solid to a change in its length
2. **Shear modulus,** which measures the resistance to motion of the planes of a solid sliding past each other
3. **Bulk modulus,** which measures the resistance that solids or liquids offer to changes in their volume

Young's Modulus: Elasticity in Length

Consider a long bar of cross-sectional area A and length L_0 that is clamped at one end (Fig. 15.1). When an external force $\mathbf{F}$ is applied along the bar and perpendicular to the cross section, internal forces in the bar resist distortion ("stretching"), but the bar attains an equilibrium in which its length is greater and in which the external force is exactly balanced by internal forces. In such a situation, the bar is said to be stressed. We define the **tensile stress** as the ratio of the magnitude of the external force F to the cross-sectional area A. The **tensile strain** in this case is defined as the ratio of the change in length, ΔL, to the original length, L_0, and is therefore a dimensionless quantity. Thus, we can use Equation 15.1 to define **Young's modulus,** Y:

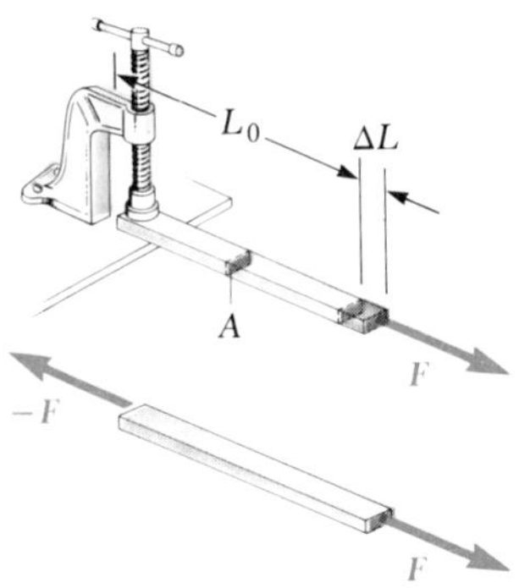

Figure 15.1 A long bar clamped at one end is stretched by an amount ΔL under the action of a force $\mathbf{F}$.

Young's modulus

$$Y \equiv \frac{\text{tensile stress}}{\text{tensile strain}} = \frac{F/A}{\Delta L/L_0} \tag{15.2}$$

This quantity is typically used to characterize a rod or wire stressed under either tension or compression. Note that because the strain is a dimensionless quantity, Y has units of force per unit area. Typical values are given in Table 15.1. Experiments show that (a) the change in length for a fixed applied force is proportional to the original length and (b) the force necessary to produce a given strain is proportional to the cross-sectional area. Both of these observations are in accord with Equation 15.2.

It is possible to exceed the *elastic limit* of a substance by applying a sufficiently large stress (Fig. 15.2). At the *yield point*, the stress-strain curve departs from a straight line. A material subjected to a stress beyond the yield point will ordinarily not return to its original length when the external force is removed. Hence, its shape is permanently changed. As the stress is increased even further, the material will ultimately break.

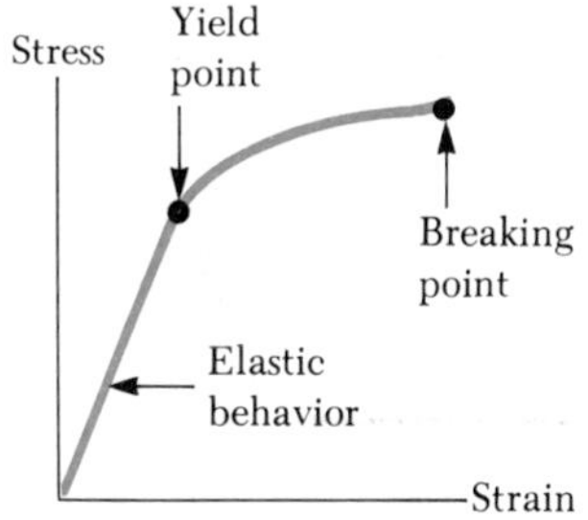

Figure 15.2 Stress versus strain curve for an elastic solid.

TABLE 15.1 Typical Values for Elastic Modulus

Substance	Young's Modulus (N/m²)	Shear Modulus (N/m²)	Bulk Modulus (N/m²)
Aluminum	7.0×10^{10}	2.5×10^{10}	7.0×10^{10}
Brass	9.1×10^{10}	3.5×10^{10}	6.1×10^{10}
Copper	11×10^{10}	4.2×10^{10}	14×10^{10}
Steel	20×10^{10}	8.4×10^{10}	16×10^{10}
Tungsten	35×10^{10}	14×10^{10}	20×10^{10}
Glass	$6.5-7.8 \times 10^{10}$	$2.6-3.2 \times 10^{10}$	$5.0-5.5 \times 10^{10}$
Quartz	5.6×10^{10}	2.6×10^{10}	2.7×10^{10}
Water	—	—	0.21×10^{10}
Mercury	—	—	2.8×10^{10}

Shear Modulus: Elasticity of Shape

Another type of deformation occurs when a body is subjected to a force **F** tangential to one of its faces while the opposite face is held in a fixed position by a force of friction, f_s (Fig. 15.3a). If the object is originally a rectangular block, a shear stress results in a shape whose cross-section is a parallelogram. For this situation, the stress is called a shear stress. A book pushed sideways as in Figure 15.3b is an example of an object under a shear stress. There is no change in volume under this deformation. We define the **shear stress** as F/A, the ratio of the tangential force to the area, A, of the face being sheared. The **shear strain** is defined as the ratio $\Delta x/h$, where Δx is horizontal distance the sheared face moves and h is the height of the object. In terms of these quantities, the **shear modulus,** S, is

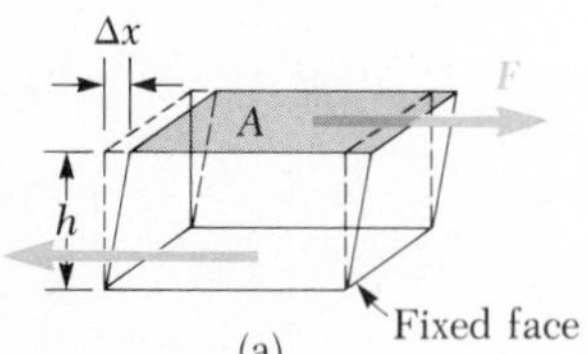

(a)

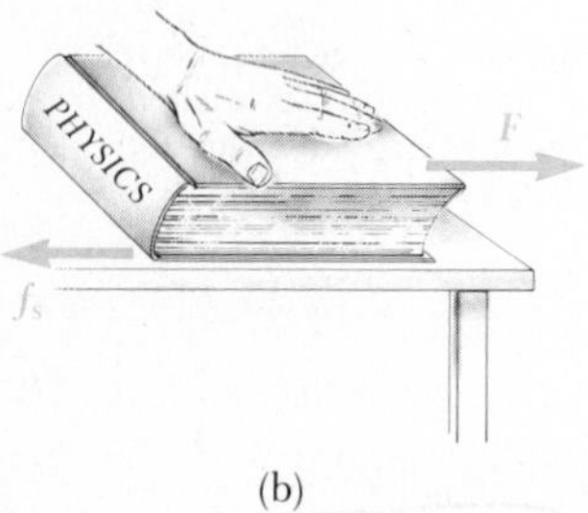

(b)

Figure 15.3 (a) A shear deformation in which a rectangular block is distorted by a force applied tangent to one of its faces. (b) A book under shear stress.

$$S \equiv \frac{\text{shear stress}}{\text{shear strain}} = \frac{F/A}{\Delta x/h} \tag{15.3}$$

2 dimensions

Values of the shear modulus for some representative materials are given in Table 15.1. Note that the units of shear modulus are force per unit area.

Bulk Modulus: Volume Elasticity

Bulk modulus

Finally, we define the bulk modulus of a substance, which characterizes the response of the substance to uniform squeezing. Suppose that the external forces acting on an object are at right angles to all of its faces (Fig. 15.4) and distributed uniformly over all the faces. As we shall see later, this occurs when an object is immersed in a fluid. A body subject to this type of deformation undergoes a change in volume but no change in shape. The **volume stress,** ΔP, is defined as the ratio of the magnitude of the normal force, F, to the area, A. When dealing with fluids, we shall refer to this quantity $\Delta P = F/A$ as the **pressure.** The volume strain is equal to the change in volume, ΔV, divided by the original volume, V. Thus, from Equation 15.1 we can characterize a volume compression in terms of the **bulk modulus,** B, defined as

$$B \equiv \frac{\text{volume stress}}{\text{volume strain}} = -\frac{F/A}{\Delta V/V} = -\frac{\Delta P}{\Delta V/V} \tag{15.4}$$

3 dimensions

Note that a negative sign is inserted in this defining equation so that B will always be a positive number. This is because an increase in pressure (positive ΔP) causes a decrease in volume (negative ΔV) and vice versa.

Table 15.1 lists bulk modulus values for some materials. If you look up such values in a different source, you will often find that the reciprocal of the bulk modulus is listed. The reciprocal of the bulk modulus is called the **compressibility** of the material. You should note from Table 15.1 that both solids and liquids have a bulk modulus. However, there is no shear modulus and no Young's modulus for liquids because a liquid will not sustain a shearing stress or a tensile stress (it will flow instead).

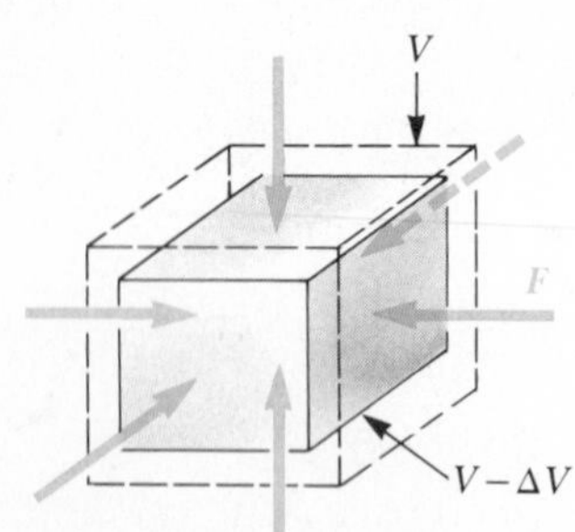

Figure 15.4 When a solid is under uniform pressure, it undergoes a change in volume but no change in shape. This cube is compressed on all sides by forces normal to its surfaces.

Prestressed Concrete

If the stress on a solid object exceeds a certain value, the object will break or fracture. The maximum stress that can be applied before fracture occurs depends on the nature of the material and the type of stress that is applied. For example, concrete has a tensile strength of about 2×10^6 N/m^2, a compressive strength of 20×10^6 N/m^2, and a shear strength of 2×10^6 N/m^2. If the actual

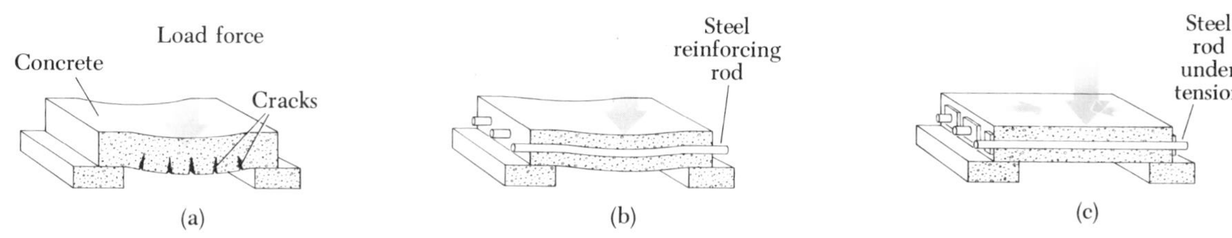

Figure 15.5 (a) A concrete slab with no reinforcement tends to crack under a heavy load. (b) The strength of the concrete slab is increased by using steel reinforcement rods. (c) The slab is further strengthened by prestressing the concrete with steel rods under tension.

stress exceeds these values, the concrete fractures. It is common practice to use large safety factors to prevent failure in concrete structures.

Concrete is normally very brittle when cast in thin sections. Thus, concrete slabs tend to sag and crack at unsupported areas, as in Figure 15.5a. The slab can be strengthened by using steel rods to reinforce the concrete at specific depths, as in Figure 15.5b. Recall that concrete is much stronger under compression than under tension. For this reason, vertical columns of concrete that are under compression can support very heavy loads, whereas horizontal beams of concrete will tend to sag and crack because of their smaller shear strength. A significant increase in shear strength is achieved, however, by prestressing the reinforced concrete, as in Figure 15.5c. As the concrete is being poured, the steel rods are held under tension. The tension is released after the concrete cures, which provides a compressive stress on the concrete. This enables the concrete slab to support a much heavier load.

Another method that has been successful in strengthening concrete is the use of fibers mixed in the cement and aggregate. Problems such as cracking can be controlled by fibrous materials, such as glass, steel, nylon, polypropylene, and, more recently, glass fibers.

EXAMPLE 15.1. Measuring Young's Modulus
A load of 102 kg is supported by a wire of length 2 m and cross-sectional area 0.1 cm². The wire is stretched by 0.22 cm. Find the tensile stress, tensile strain, and Young's modulus for the wire from this information.

Solution

$$\text{Tensile stress} = \frac{F}{A} = \frac{Mg}{A} = \frac{(102\ \text{kg})(9.80\ \text{m/s}^2)}{0.1 \times 10^{-4}\ \text{m}^2}$$

$$= 1.0 \times 10^8\ \text{N/m}^2$$

$$\text{Tensile strain} = \frac{\Delta L}{L_0} = \frac{0.22 \times 10^{-2}\ \text{m}}{2\ \text{m}} = 0.11 \times 10^{-2}$$

$$Y = \frac{\text{tensile stress}}{\text{tensile strain}} = \frac{1.0 \times 10^8\ \text{N/m}^2}{0.11 \times 10^{-2}}$$

$$= 9.1 \times 10^{10}\ \text{N/m}^2$$

Comparing this value for Y with the values in Table 15.1, we conclude that the wire is probably brass.

EXAMPLE 15.2. Squeezing a Lead Sphere
A solid lead sphere of volume 0.5 m³ is lowered to a depth in the ocean where the water pressure is equal to 2×10^7 N/m². The bulk modulus of lead is equal to 7.7×10^9 N/m². What is the change in volume of the sphere?

Solution: From the definition of bulk modulus, we have

$$B = -\frac{\Delta P}{\Delta V/V}$$

or

$$\Delta V = -\frac{V\Delta P}{B}$$

In this case, the change in pressure, ΔP, has the value 2×10^7 N/m². (This is large relative to atmospheric pressure, 1.01×10^5 N/m².) Taking $V = 0.5$ m³ and $B = 7.7 \times 10^9$ N/m², we get

$$\Delta V = -\frac{(0.5\ \text{m}^3)(2 \times 10^7\ \text{N/m}^2)}{7.7 \times 10^9\ \text{N/m}^2} = -1.3 \times 10^{-3}\ \text{m}^3$$

The negative sign indicates a *decrease* in volume.

15.2 STATES OF MATTER

Matter is normally classified as being in one of three states: solid, liquid, or gaseous. Often, this classification is extended to include a fourth state referred to as a plasma.

Everyday experience tells us that a solid has a definite volume and shape. A brick maintains its familiar shape and size day in and day out. We also know that a liquid has a definite volume but no definite shape. For example, when you fill the tank on a lawn mower, the gasoline assumes the shape of the tank on the mower, but if you have a gallon of gasoline before you pour, you will have a gallon after. Finally, a gas has neither definite volume nor definite shape. These definitions help us to picture the states of matter, but they are somewhat artificial. For example, asphalt and plastics are normally considered solids, but over long periods of time they tend to flow like liquids. Likewise, water can be a solid, liquid, or gas (or combinations of these), depending on the temperature and pressure. The response time of the change in shape to an *external* force or pressure determines if we treat the substance as a solid, a very viscous fluid, or another state.

The fourth state of matter can occur when matter is heated to very high temperatures. Under these conditions, one or more electrons surrounding each atom are freed from the nucleus. The resulting substance is a collection of free electrically charged particles: the negatively charged electrons and the positively charged ions. Such a highly ionized gas with equal amounts of positive and negative charges is called a **plasma.** The plasma state exists inside stars, for example. If we were to take a grand tour of our universe, we would find that there is far more matter in the plasma state than in the more familiar forms of solid, liquid, and gas because there are far more stars around than any other form of celestial matter. However, in this chapter we shall ignore this plasma state and concentrate instead on the more familiar solid, liquid, and gaseous forms that make up the environment on our planet.

All matter consists of some distribution of atoms and molecules. The atoms in a solid are held at specific positions with respect to one another by forces that are mainly electrical in origin. The atoms of a solid vibrate about these equilibrium positions because of thermal agitation. However, at low temperatures, this vibrating motion is slight and the atoms can be considered to be almost fixed. As thermal energy (heat) is added to the material, the amplitude of these vibrations increases. One can view the vibrating motion of the atom as that which would occur if the atom were bound in its equilibrium position by springs attached to neighboring atoms. Such a vibrating collection of atoms and imaginary springs is shown in Figure 15.6. If a solid is compressed by external forces, we can picture these external forces as compressing these tiny internal springs. When the external forces are removed, the solid tends to return to its original shape and size. For this reason, a solid is said to have elasticity.

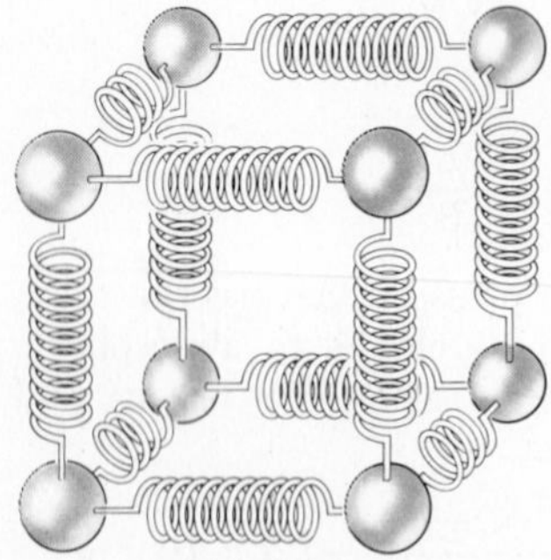

Figure 15.6 A model of a solid. The atoms (spheres) are imagined as being attached to each other by springs, which represent the elastic nature of the interatomic forces.

Solids can be classified as being either crystalline or amorphous. A **crystalline solid** is one in which the atoms have an ordered, periodic structure. For example, in the sodium chloride crystal (common table salt), sodium and chlorine atoms occupy alternate corners of a cube face, as in Figure 15.7a. In an **amorphous solid,** such as glass, the atoms are arranged in a disordered fashion, as in Figure 15.7b.

In any given substance, the liquid state exists at a higher temperature than the solid state. Thermal agitation is greater in the liquid state than in the solid state. As a result, the molecular forces in a liquid are not strong enough to keep

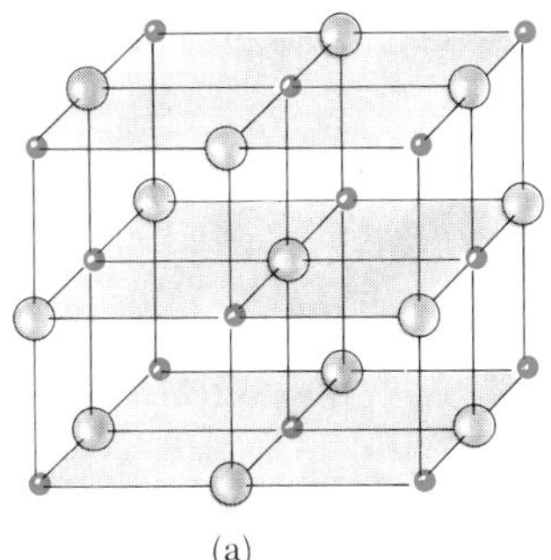

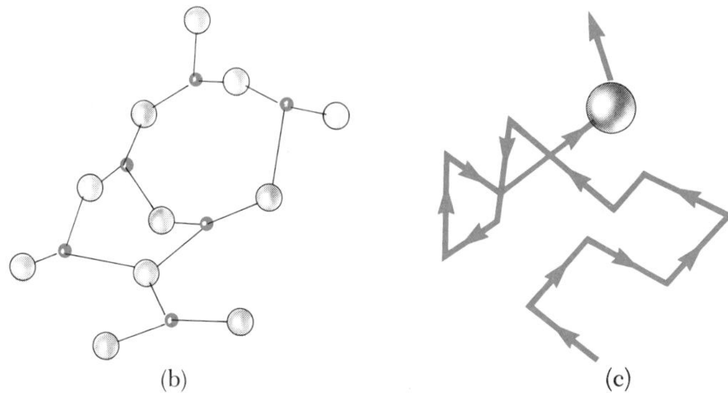

Figure 15.7 (a) The NaCl structure, with Na^+ and Cl^- ions at alternate corners of a cube. The large spheres represent Na^+ ions, and the smaller spheres represent Cl^- ions. (b) In an amorphous solid, the atoms are arranged in a random fashion. (c) Erratic motion of a molecule in a liquid.

the molecules in fixed positions, and the molecules wander through the liquid in a random fashion (Fig. 15.7c). Solids and liquids have the following property in common. When one tries to compress a liquid or a solid, strong repulsive atomic forces act internally to resist the deformation.

In the gaseous state, the molecules are in constant random motion and exert only weak forces on each other. The average separation distances between the molecules of a gas are quite large compared with the dimensions of the molecules. Occasionally, the molecules collide with each other; however, most of the time they move as nearly free, noninteracting particles. We shall have more to say about the properties of gases in subsequent chapters.

15.3 DENSITY AND PRESSURE

The **density** of a homogeneous substance is defined as its mass per unit volume. That is, a substance of mass m and volume V has a density ρ given by

$$\rho \equiv \frac{m}{V} \tag{15.5}$$

Definition of density

The units of density are kg/m^3 in the SI system and g/cm^3 in the cgs system. Table 15.2 lists the densities of various substances. These values vary slightly with temperature, since the volume of a substance is temperature dependent (as we shall see in Chapter 16). Note that under normal conditions the densities of gases are about 1/1000 the densities of solids and liquids. This implies

TABLE 15.2 Density of Some Common Substances

Substance	ρ (kg/m³)[a]	Substance	ρ (kg/m³)[a]
Ice	0.917×10^3	Water	1.00×10^3
Aluminum	2.70×10^3	Glycerin	1.26×10^3
Iron	7.86×10^3	Ethyl alcohol	0.806×10^3
Copper	8.92×10^3	Benzene	0.879×10^3
Silver	10.5×10^3	Mercury	13.6×10^3
Lead	11.3×10^3	Air	1.29
Gold	19.3×10^3	Oxygen	1.43
Platinum	21.4×10^3	Hydrogen	8.99×10^{-2}
		Helium	1.79×10^{-1}

[a] All values are at standard atmospheric pressure and temperature (STP). To convert to grams per cubic centimeter, multiply by 10^{-3}.

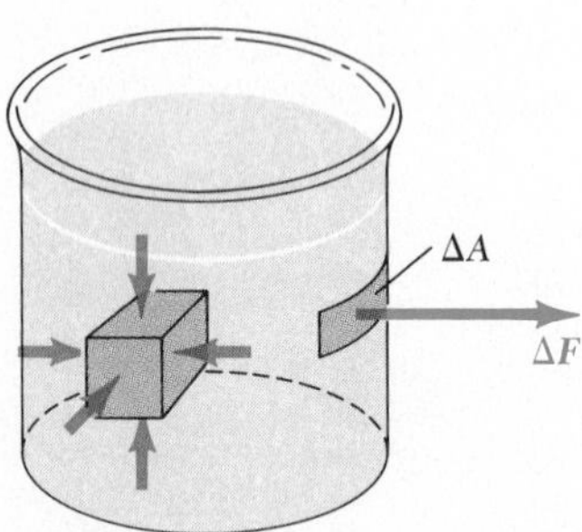

Figure 15.8 The force of the fluid on a submerged object at any point is perpendicular to the surface of the object. The force of the fluid on the walls of the container is perpendicular to the walls at all points.

that the average molecular spacing in a gas under these conditions is about ten times greater than in a solid or liquid.

The **specific gravity** of a substance is defined as the ratio of its density to the density of water at 4°C, which is 1.0×10^3 kg/m³. By definition, specific gravity is a dimensionless quantity. For example, if the specific gravity of a substance is 3, its density is $3(1.0 \times 10^3 \text{ kg/m}^3) = 3.0 \times 10^3 \text{ kg/m}^3$.

We have seen that fluids do not sustain shearing stresses, and thus the only stress that can exist on an object submerged in a fluid is one that tends to compress the object. The force exerted by the fluid on the object is always perpendicular to the surfaces of the object, as shown in Figure 15.8.

The pressure at a specific point in a fluid can be measured with the device pictured in Figure 15.9. The device consists of an evacuated cylinder enclosing a light piston connected to a spring. As the device is submerged in a fluid, the fluid presses down on the top of the piston and compresses the spring until the inward force of the fluid is balanced by the outward force of the spring. The fluid pressure can be measured directly if the spring is calibrated in advance. This is accomplished by applying a known force to the spring to compress it a given distance.

Definition of average pressure

If F is the magnitude of the force on the piston and A is the area of the piston, then the **average pressure,** P, of the fluid at the level to which the device has been submerged is defined as the ratio of force to area:

$$P \equiv \frac{F}{A} \tag{15.6}$$

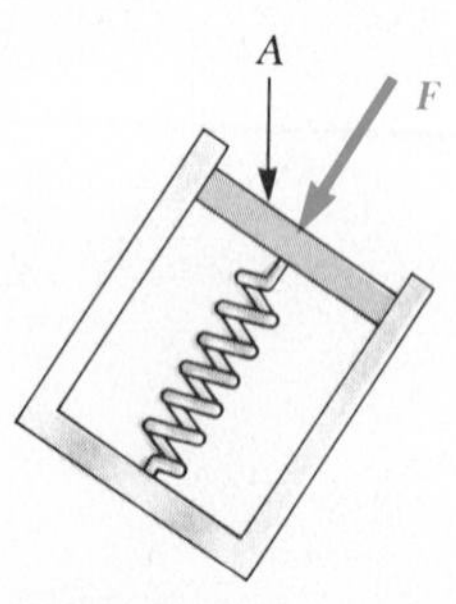

Figure 15.9 A simple device for measuring pressure.

The pressure in a fluid is not the same at all points. To define the pressure at a specific point, consider a fluid enclosed as in Figure 15.8. If the force exerted by the fluid is ΔF over a surface element of area ΔA, then the **pressure** at that point is

$$P = \lim_{\Delta A \to 0} \frac{\Delta F}{\Delta A} = \frac{dF}{dA} \tag{15.7}$$

As we shall see in the next section, the pressure in a fluid in the presence of the force of gravity varies with depth. Therefore, to get the total force on the wall of a container, we would have to integrate Equation 15.7 over the wall surface.

Since pressure is force per unit area, it has units of N/m^2 in the **SI** system. Another name for the SI unit of pressure is **pascal** (Pa).

$$1 \text{ Pa} \equiv 1 \text{ N/m}^2 \tag{15.8}$$

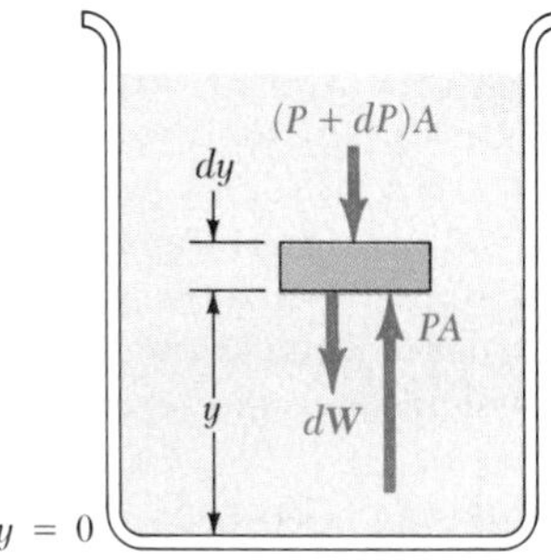

Figure 15.10 The variation of pressure with depth in a fluid. The volume element is at rest, and the forces on it are shown.

15.4 VARIATION OF PRESSURE WITH DEPTH

Consider a fluid at rest in a container (Fig. 15.10). We first note that *all points at the same depth must be at the same pressure.* If this were not the case, a given element of the fluid would not be in equilibrium. Now let us select a portion of the fluid contained within an imaginary cylinder of cross-sectional area A and height dy. The upward force on the bottom of the cylinder is PA, and the downward force on the top is $(P + dP)A$. The weight of the cylinder, the volume of which is dV, is given by $dW = \rho g\, dV = \rho g A\, dy$, where ρ is the density of the fluid. Since the cylinder is in equilibrium, the forces must add to zero, and so we get

$$\sum F_y = PA - (P + dP)A - \rho g A\, dy = 0$$

$$\frac{dP}{dy} = -\rho g \tag{15.9}$$

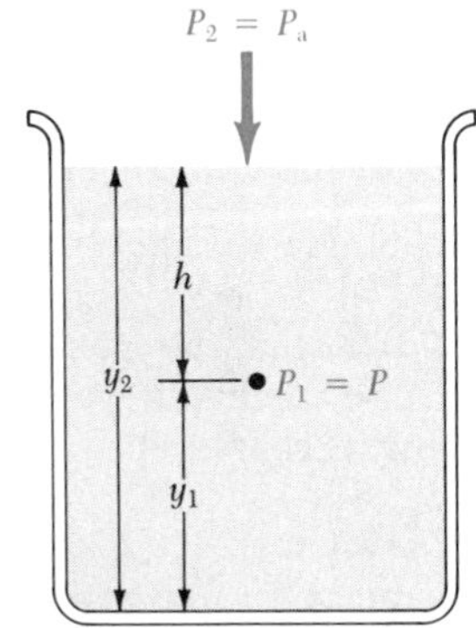

Figure 15.11 The pressure P at a depth h below the surface of a liquid open to the atmosphere is given by $P = P_a + \rho g h$.

From this result, we see that an increase in elevation (positive dy) corresponds to a decrease in pressure (negative dP). If P_1 and P_2 are the pressures at the elevations y_1 and y_2 above the reference level, then integrating Equation 15.9 gives

$$P_2 - P_1 = -\rho g(y_2 - y_1) \tag{15.10}$$

If the vessel is open at the top (Fig. 15.11), then the pressure at the depth h can be obtained from Equation 15.10. Taking atmospheric pressure to be $P_a = P_2$, and noting that the depth $h = y_2 - y_1$, we find that

$$P = P_a + \rho g h \tag{15.11}$$

Pressure at any depth h

where we usually take $P_a \approx 1.01 \times 10^5$ N/m^2 (14.7 lb/in.^{2}). In other words,

> the **absolute pressure** P at a depth h below the surface of a liquid open to the atmosphere is *greater* than atmospheric pressure by an amount $\rho g h$.

This result also verifies that the *pressure is the same at all points having the same elevation.* Furthermore, *the pressure is not affected by the shape of the vessel.*

In view of the fact that the pressure in a fluid depends only upon depth, any increase in pressure at the surface must be transmitted to every point in the fluid. This was first recognized by the French scientist Blaise Pascal (1623–1662) and is called **Pascal's law:**

> A change in pressure applied to an enclosed fluid is transmitted undiminished to every point of the fluid and the walls of the containing vessel.

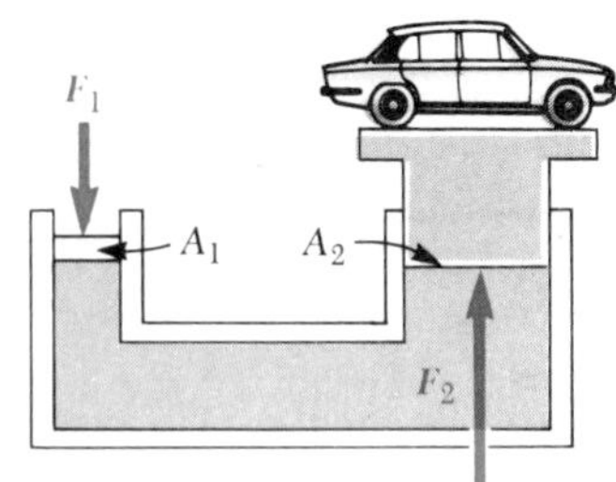

Figure 15.12 Schematic diagram of a hydraulic press. Since the increase in pressure is the same at the left and right sides, a small force $\mathbf{F}_1$ at the left produces a much larger force $\mathbf{F}_2$ at the right.

An important application of Pascal's law is the hydraulic press illustrated by Figure 15.12. A force F_1 is applied to a small piston of area A_1. The pressure is transmitted through a fluid to a larger piston of area A_2. Since the pressure is the same on both sides, we see that $P = F_1/A_1 = F_2/A_2$. Therefore, the force F_2 is larger than F_1 by the multiplying factor A_2/A_1. Hydraulic brakes, car lifts, hydraulic jacks, fork lifts, and so on make use of this principle.

EXAMPLE 15.3. The Car Lift

In a car lift used in a service station, compressed air exerts a force on a small piston having a radius of 5 cm. This pressure is transmitted to a second piston of radius 15 cm. What force must the compressed air exert in order to lift a car weighing 13 300 N? What air pressure will produce this force?

Solution: Because the pressure exerted by the compressed air is transmitted undiminished throughout the fluid, we have

$$F_1 = \left(\frac{A_1}{A_2}\right) F_2 = \frac{\pi(5 \times 10^{-2}\ \text{m})^2}{\pi(15 \times 10^{-2}\ \text{m})^2} (13\ 300\ \text{N})$$
$$= 1.48 \times 10^3\ \text{N}$$

The air pressure that will produce this force is given by

$$P = \frac{F_1}{A_1} = \frac{1.48 \times 10^3\ \text{N}}{\pi(5 \times 10^{-2}\ \text{m})^2} = 1.88 \times 10^5\ \text{N/m}^2$$

This pressure is approximately twice atmospheric pressure.

Note that the input work (the work done by $\boldsymbol{F}_1$) is equal to the output work (the work done by $\boldsymbol{F}_2$), so that energy is conserved.

EXAMPLE 15.4. The Water Bed

A water bed is 2 m on a side and 30 cm deep. (a) Find its weight.

Solution: Since the density of water is 1000 kg/m³, the mass of the bed is

$$M = \rho V = (1000\ \text{kg/m}^3)(1.2\ \text{m}^3) = 1.20 \times 10^3\ \text{kg}$$

and its weight is

$$W = Mg = (1.20 \times 10^3\ \text{kg})(9.80\ \text{m/s}^2) = 1.18 \times 10^4\ \text{N}$$

This is equivalent to approximately 2640 lb. In order to support such a heavy load, you would be well advised to keep your water bed in the basement or on a sturdy, well-supported floor.

(b) Find the pressure that the water bed exerts on the floor when the bed rests in its normal position. Assume that the entire lower surface of the bed makes contact with the floor.

Solution: The weight of the water bed is 1.18×10^4 N. The cross-sectional area is 4 m² when the bed is in its normal position. This gives a pressure exerted on the floor of

$$P = \frac{1.18 \times 10^4\ \text{N}}{4\ \text{m}^2} = 2.95 \times 10^3\ \text{N/m}^2$$

Exercise 1 Calculate the pressure that would be exerted on the floor if the bed rests on its side.
Answer: Since the area of its side is 0.6 m², the pressure is 1.96×10^4 N/m².

EXAMPLE 15.5. Pressure In the Ocean

Calculate the pressure at an ocean depth of 1000 m. Assume the density of water is 1.0×10^3 kg/m³ and $P_a = 1.01 \times 10^5$ N/m².

Solution:

$$P = P_a + \rho g h$$
$$= 1.01 \times 10^5\ \text{N/m}^2 + (1.0 \times 10^3\ \text{kg/m}^3)(9.80\ \text{m/s}^2)(10^3\ \text{m})$$
$$P \approx 9.9 \times 10^6\ \text{N/m}^2$$

This is approximately 100 times greater than atmospheric pressure! Obviously, the design and construction of vessels that will withstand such enormous pressures are not a trivial matter.

Exercise 2 Calculate the total force exerted on the outside of a circular submarine window of diameter 30 cm at this depth.
Answer: 7.0×10^5 N.

EXAMPLE 15.6. The Force on a Dam

Water is filled to a height H behind a dam of width w (Fig. 15.13). Determine the resultant force on the dam.

Solution: The pressure at the depth h beneath the surface at the shaded portion is

$$P = \rho g h = \rho g(H - y)$$

(We have left out atmospheric pressure since it acts on both sides of the dam.) Using Equation 15.7, we find the force on the shaded strip to be

$$dF = P\,dA = \rho g(H - y) w\,dy$$

Therefore, the total force on the dam is

$$F = \int P dA = \int_0^H \rho g(H - y) w\,dy = \tfrac{1}{2}\rho g w H^2$$

For example, if $H = 30$ m and $w = 100$ m, we find that $F = 4.4 \times 10^8\ \text{N} = 9.9 \times 10^7$ lb!

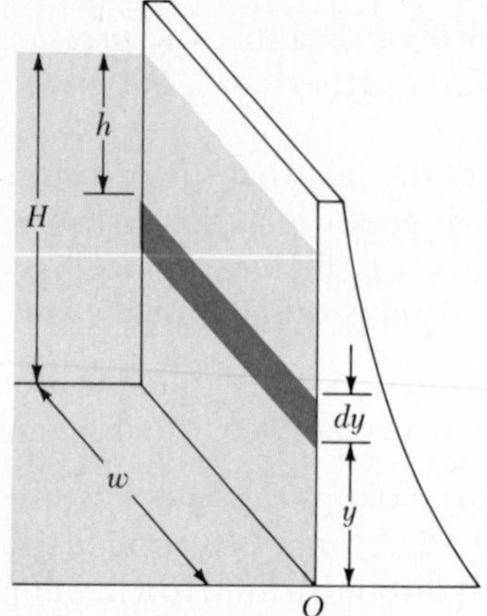

Figure 15.13 (Example 15.6) The total force on a dam must be obtained from the expression $F = \int P\,dA$, where dA is the area of the dark strip.

One simple device for measuring pressure is the open-tube manometer illustrated in Figure 15.14a. One end of a U-shaped tube containing a liquid is open to the atmosphere, and the other end is connected to a system of unknown pressure P. The pressure at point B equals $P_a + \rho g h$, where ρ is the density of the fluid. But the pressure at B equals the pressure at A, which is also the unknown pressure P. Therefore, we conclude that

$$P = P_a + \rho g h$$

The pressure P is called the **absolute pressure,** while $P - P_a$ is called the **gauge pressure.** Thus, if the pressure in the system is greater than atmospheric pressure, h is positive. If the pressure is less than atmospheric pressure (a partial vacuum), h is negative.

Another instrument used to measure pressure is the common barometer, invented by Evangelista Torricelli (1608–1647). A long tube closed at one end is filled with mercury and then inverted into a dish of mercury (Fig. 15.14b). The closed end of the tube is nearly a vacuum, and so its pressure can be taken as zero. Therefore, it follows that $P_a = \rho g h$, where ρ is the density of the mercury and h is the height of the mercury column. One atmosphere of pressure is defined to be the pressure equivalent of a column of mercury that is exactly 0.76 m in height at 0°C, with $g = 9.80665 \text{ m/s}^2$. At this temperature, mercury has a density of $13.595 \times 10^3 \text{ kg/m}^3$; therefore

$$\begin{aligned} P_a = \rho g h &= (13.595 \times 10^3 \text{ kg/m}^3)(9.80665 \text{ m/s}^2)(0.7600 \text{ m}) \\ &= 1.013 \times 10^5 \text{ N/m}^2 \end{aligned}$$

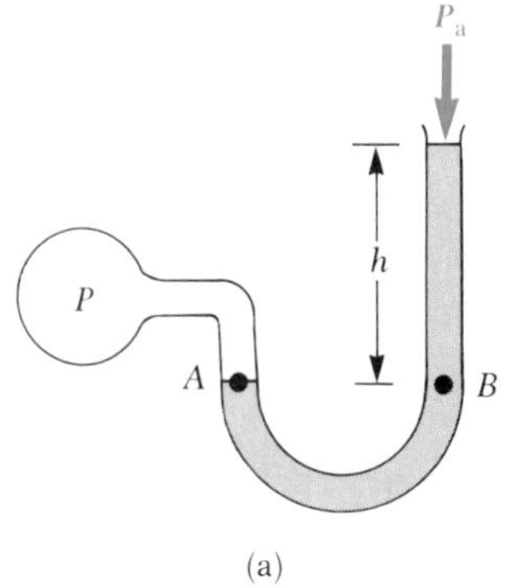

(a)

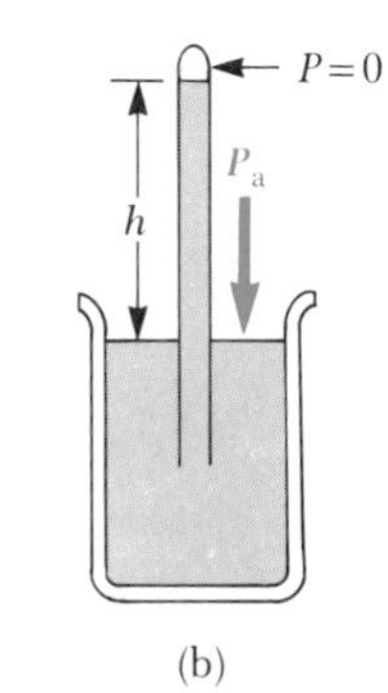

(b)

Figure 15.14 Two devices for measuring pressure: (a) the open-tube manometer; (b) the mercury barometer.

15.6 BUOYANT FORCES AND ARCHIMEDES' PRINCIPLE

Archimedes (287–212 B.C.), a Greek mathematician, physicist, and engineer, was perhaps the greatest scientist of antiquity. He is well known for discovering the nature of the buoyant force acting on objects and was also a gifted inventor. One of his practical inventions, still in use today, is the Archimedes' screw, an inclined rotating coiled tube used originally to lift water from the holds of ships. He also invented the catapult and devised systems of levers, pulleys, and weights for raising heavy loads. Such inventions were successfully used by the soldiers to defend his native city, Syracuse, during a two-year siege by the Romans.

According to legend, Archimedes was asked by King Hieron to determine whether the king's crown was made of pure gold or had been alloyed with some other metal. The task was to be performed without damaging the crown. Archimedes presumably arrived at a solution while taking a bath, noting a partial loss of weight after submerging his arms and legs in the water. As the story goes, he was so excited about his great discovery that he ran through the streets of Syracuse naked shouting, "Eureka!" which is Greek for "I have found it." **Archimedes' principle** can be stated as follows:

Archimedes' principle

> Any body completely or partially submerged in a fluid is buoyed up by a force equal to the weight of the fluid displaced by the body.

Everyone has experienced Archimedes' principle. As an example of a common experience, recall that it is relatively easy to lift someone if the person is in a swimming pool whereas lifting that same individual on dry land may be a very difficult task. Evidently, water provides partial support to any

Archimedes (287 – 212 B.C.), a Greek mathematician, physicist, and engineer, was perhaps the greatest scientist of antiquity. He was the first to accurately compute the ratio of a circle's circumference to its diameter and also showed how to calculate the volume and surface area of spheres, cylinders, and other geometric shapes. He is well known for discovering the nature of the buoyant force acting on floating objects and was also a gifted inventor. One of his practical inventions, still in use today, is the Archimedes screw, a rotating coiled tube used originally to lift water from the holds of ships. He also invented the catapult and devised systems of levers, pulleys, and weights for raising heavy loads. Such inventions were successfully used by the soldiers of his native city, Syracuse, during a two-year siege by the Romans.

object placed in it. We say that an object placed in a fluid is buoyed up by the fluid, and we call this upward force the **buoyant force.** According to Archimedes' principle,

> the magnitude of the buoyant force always equals the weight of the fluid displaced by the object.

The buoyant force acts vertically upward through what was the center of gravity of the displaced fluid.

Archimedes' principle can be verified in the following manner. Suppose we focus our attention on the indicated cube of water in the container of Figure 15.15. This cube of water is in equilibrium under the action of the forces on it. One of these forces is the weight of the cube of water. What cancels this downward force? Apparently, the rest of the water inside the container is buoying up the cube and holding it in equilibrium. Thus, the buoyant force, B, on the cube of water is exactly equal in magnitude to the weight of the water inside the cube:

$$B = W$$

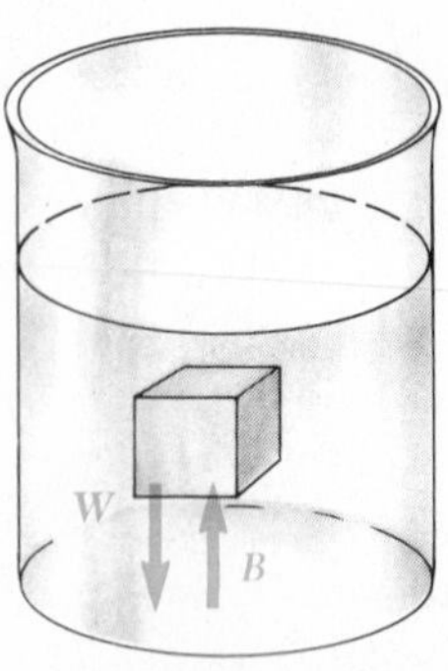

Figure 15.15 The external forces on the cube of water are its weight W and the buoyancy force B. Under equilibrium conditions, $B = W$.

Now, imagine that the cube of water is replaced by a cube of steel of the same dimensions. What is the buoyant force on the steel? The water surrounding a cube will behave in the same way whether a cube of water or a cube of steel is being buoyed up; therefore, *the buoyant force acting on the steel is the same as the buoyant force acting on a cube of water of the same dimensions.* This result applies for a submerged object of any shape, size, or density.

Let us show explicitly that the buoyant force is equal in magnitude to the weight of the displaced fluid. The pressure at the bottom of the cube in Figure 15.15 is greater than the pressure at the top by an amount $\rho_f gh$, where ρ_f is the density of the fluid and h is the height of the cube. Since the pressure difference, ΔP, is equal to the buoyant force per unit area, that is, $\Delta P = B/A$, we see that $B = (\Delta P)(A) = (\rho_f gh)(A) = \rho_f gV$, where V is the volume of the cube. Since the mass of the water in the cube is $M = \rho_f V$, we see that

$$B = W = \rho_f Vg = Mg \qquad (15.12)$$

Buoyant force

where W is the weight of the displaced fluid.

Note that the weight of the submerged object is $\rho_0 Vg$, where ρ_0 is the density of the object. Therefore, if the density of the object is greater than the density of the fluid, the unsupported object will sink. If the density of the object is less than that of the fluid, the unsupported submerged object will accelerate upward and will ultimately float. When a floating object is in equilibrium, part of it is submerged. In this case, the buoyant force equals the weight of the object.

Under normal conditions, the average density of a fish is slightly greater than the density of water. This being the case, a fish would sink if it did not have some mechanism for adjusting its density. This mechanism is supplied by an internal gas bag. If a fish desires to move higher in the water, it causes this gas bag to expand. Likewise, in order to move lower in the water, the fish contracts the bag by compressing the gas.

EXAMPLE 15.7. A Submerged Object
A piece of aluminum is suspended from a string and then completely immersed in a container of water (Fig. 15.16). The mass of the aluminum is 1 kg, and its density is 2.7×10^3 kg/m^3. Calculate the tension in the string before and after the aluminum is immersed.

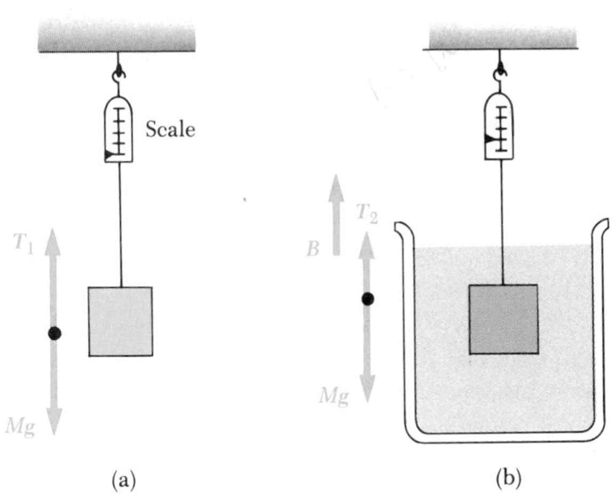

Figure 15.16 (Example 15.7) (a) When the aluminum is suspended in air, the scale reads the true weight, Mg (neglecting the buoyancy of air). (b) When the aluminum is immersed in water, the buoyant force, B, reduces the scale reading to $T_2 = Mg - B$.

Solution: When the piece of aluminum is suspended in air, as in Figure 15.16a, the tension in the string, T_1 (the reading on the scale), is equal to the weight, Mg, of the aluminum, assuming that the buoyant force of air can be neglected:

$$T_1 = Mg = (1 \text{ kg})(9.80 \text{ m/s}^2) = 9.80 \text{ N}$$

When immersed in water, the aluminum experiences an upward buoyant force B, as in Figure 15.16b, which reduces the tension in the string. Since the system is in equilibrium,

$$T_2 + B - Mg = 0$$

$$T_2 = Mg - B = 9.80 \text{ N} - B$$

In order to calculate B, we must first calculate the volume of the aluminum:

$$V_{Al} = \frac{M}{\rho_{Al}} = \frac{1 \text{ kg}}{2.7 \times 10^3 \text{ kg/m}^3} = 3.7 \times 10^{-4} \text{ m}^3$$

Since the buoyant force equals the weight of the water displaced, we have

$$B = M_w g = \rho_w V_{Al} g$$
$$= (1 \times 10^3 \text{ kg/m}^3)(3.7 \times 10^{-4} \text{ m}^3)(9.80 \text{ m/s}^2) = 3.6 \text{ N}$$

Therefore,

$$T_2 = 9.80 \text{ N} - B = 9.80 \text{ N} - 3.6 \text{ N} = 6.2 \text{ N}$$

15.7 LAMINAR AND TURBULENT FLOW

When a fluid is in motion, its flow can be characterized as being one of two types. The flow is said to be **streamline, or laminar,** if every particle that passes a particular point moves along the exact path followed by particles that passed that point earlier. In this case, each particle of the fluid moves along a smooth path called a *streamline*, as shown in Figure 15.17a. The various streamlines cannot cross each other under this steady-flow condition, and the streamline at any point coincides with the direction of fluid velocity at that point. In contrast, the flow of a fluid becomes irregular, or **turbulent**, above a certain velocity or under conditions near its boundaries that can cause abrupt changes in velocity. Irregular motion of the fluid, called *eddy currents*, are characteristic in turbulent flow, as shown in Figure 15.17b.

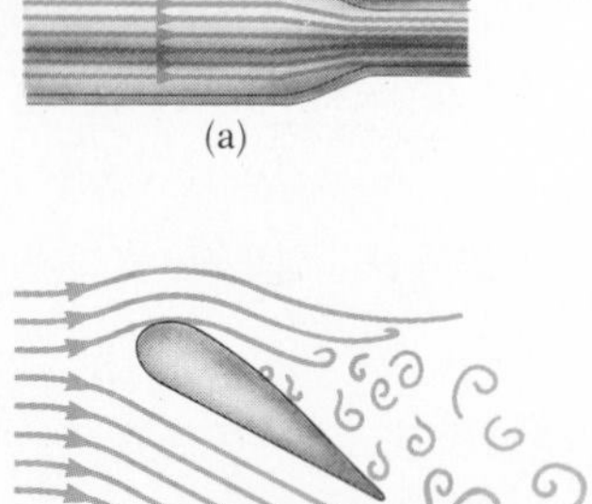

Figure 15.17 (a) Streamline flow through a tube containing a constriction. The lines followed by particles are called streamlines. (b) The flow around the wing of an airplane is streamline below the wing and turbulent above and behind it.

In the discussion of fluid flow, the term **viscosity** is used to characterize the degree of internal friction in the fluid. This internal friction is associated with the resistance of two adjacent layers of the fluid to move relative to each other. A fluid such as kerosene has a lower viscosity than crude oil or molasses.

15.8 FLUID DYNAMICS AND BERNOULLI'S EQUATION

Many features of fluid motion can be understood by considering the behavior of an **ideal fluid**, which satisfies the following conditions:

1. *The fluid is nonviscous,* that is, there is no internal frictional force between adjacent fluid layers.
2. *The fluid is incompressible,* which means that its density is constant.
3. *The fluid motion is steady,* meaning that the velocity, density, and pressure at each point in the fluid do not change in time.
4. *The fluid moves without turbulence.* This implies that each element of the fluid has zero angular velocity about its center, that is, there can be no eddy currents present in the moving fluid.

Equation of Continuity

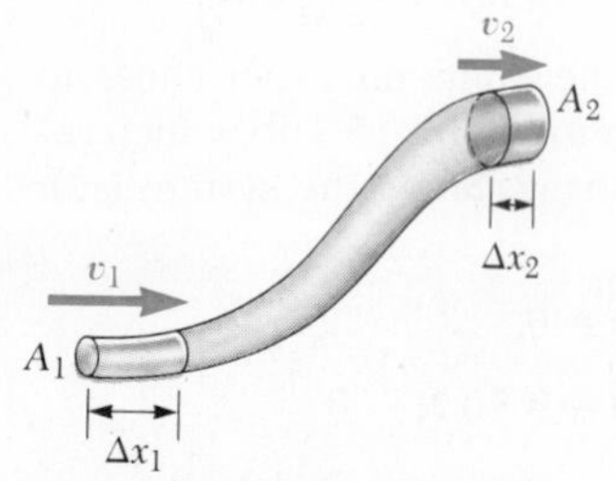

Figure 15.18 A fluid moving with streamline flow through a pipe of varying cross-sectional area. The volume of fluid flowing through A_1 in a time interval Δt must equal the volume flowing through A_2 in the same time interval. Therefore, $A_1v_1 = A_2v_2$.

Figure 15.18 represents a fluid flowing through a pipe of nonuniform size. The particles in the fluid move along the streamlines in steady-state flow. At all points the velocity of the particle is tangent to the streamline along which it moves.

In a small time interval Δt, the fluid at the bottom end of the pipe moves a distance $\Delta x_1 = v_1 \Delta t$. If A_1 is the cross-sectional area in this region, then the mass contained in the shaded region is $\Delta M_1 = \rho_1 A_1 \Delta x_1 = \rho_1 A_1 v_1 \Delta t$. Similarly, the fluid that moves through the upper end of the pipe in the time Δt has a mass $\Delta M_2 = \rho_2 A_2 v_2 \Delta t$. However, since *mass is conserved* and because the flow is steady, the mass that crosses A_1 in a time Δt must equal the mass that crosses A_2 in a time Δt. Therefore $\Delta M_1 = \Delta M_2$, or

$$\rho_1 A_1 v_1 = \rho_2 A_2 v_2 \tag{15.13}$$

This expression is called the **equation of continuity.**

Since ρ is constant for the steady flow of an *incompressible* fluid, Equation 15.13 reduces to

The product Av is constant for an incompressible fluid in steady flow

$$A_1 v_1 = A_2 v_2 = \text{constant} \tag{15.14}$$

That is,

> the product of the area and the fluid speed at all points along the pipe is a constant.

Therefore, as one would expect, the speed is high where the tube is constricted and low where the tube is wide. The product Av, which has the dimensions of volume/time, is called the *volume flux,* or flow rate. The conditions $Av =$ constant is equivalent to the fact that the amount of fluid which enters one end of the tube in a given time interval equals the amount of fluid leaving the tube in the same time interval, assuming no leaks.

EXAMPLE 15.8. Filling a Water Bucket
A water hose 2 cm in diameter is used to fill a 20-liter bucket. If it takes 1 min to fill the bucket, what is the speed v at which the water leaves the hose? (Note that 1 liter $= 10^3$ cm³.)

Solution: The cross-sectional area of the hose is

$$A = \pi \frac{d^2}{4} = \pi \left(\frac{2^2}{4}\right) \text{cm}^2 = \pi \text{ cm}^2$$

According to the data given, the flow rate is equal to 20 liters/min. Equating this to the product Av gives

$$Av = 20 \frac{\text{liters}}{\text{min}} = \frac{20 \times 10^3 \text{ cm}^3}{60 \text{ s}}$$

$$v = \frac{20 \times 10^3 \text{ cm}^3}{(\pi \text{ cm}^2)(60 \text{ s})} = 106 \text{ cm/s}$$

Exercise 3 If the diameter of the hose is reduced to 1 cm, what will the speed of the water be as it leaves the hose, assuming the same flow rate?
Answer: 424 cm/s.

Bernoulli's Equation

As a fluid moves through a pipe of varying cross section and elevation, the pressure will change along the pipe. In 1738 the Swiss physicist Daniel Bernoulli (1700–1782) first derived a most fundamental expression that relates the pressure to fluid speed and elevation. As we shall see, this result is a consequence of energy conservation as applied to our ideal fluid.

Again, *we shall assume that the fluid is incompressible and nonviscous and flows in an irrotational, steady-state manner* as described in the previous section. Consider the flow through a nonuniform pipe in a time Δt, as illustrated in Figure 15.19. The force on the lower end of the fluid is P_1A_1, where P_1 is the pressure at point 1. The work done by this force is $W_1 = F_1\,\Delta x_1 = P_1A_1\,\Delta x_1 = P_1\,\Delta V$, where ΔV is the volume of the lower shaded region. In a similar manner, the work done on the fluid on the upper portion in the time Δt is given by $W_2 = -P_2A_2\,\Delta x_2 = -P_2\,\Delta V$. (Note that the volume that passes through 1 in a time Δt equals the volume that passes through 2 in the same time interval.) This work is negative since the fluid force opposes the displacement. Thus the net work done by these forces in the time Δt is

$$W = (P_1 - P_2)\,\Delta V$$

Part of this work goes into changing the kinetic energy of the fluid, and part goes into changing the gravitational potential energy. If Δm is the mass passing through the pipe in the time Δt, then the change in kinetic energy is

$$\Delta K = \tfrac{1}{2}(\Delta m)v_2^2 - \tfrac{1}{2}(\Delta m)v_1^2$$

The change in potential energy is

$$\Delta U = \Delta m g y_2 - \Delta m g y_1$$

We can apply the work-energy theorem in the form $W = \Delta K + \Delta U$ (Chapter 8) to this volume of fluid to give

$$(P_1 - P_2)\,\Delta V = \tfrac{1}{2}(\Delta m)v_2^2 - \tfrac{1}{2}(\Delta m)v_1^2 + \Delta m g y_2 - \Delta m g y_1$$

If we divide each term by ΔV, and recall that $\rho = \Delta m/\Delta V$, the above expression reduces to

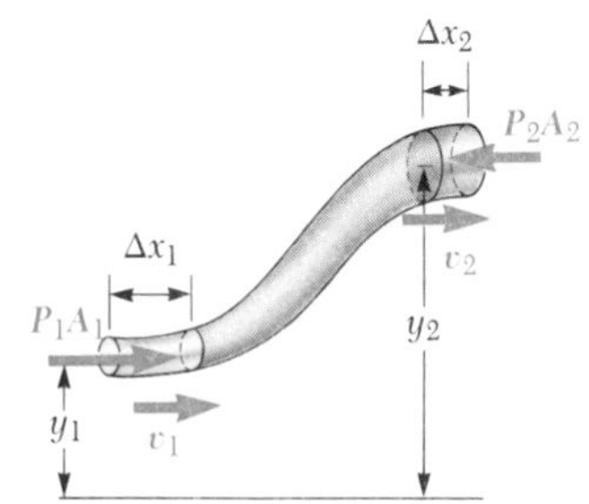

Figure 15.19 A fluid flowing through a constricted pipe with streamline flow. The fluid in the section of length Δx_1 moves to the section of length Δx_2. The volumes of fluid in the two sections are equal.

$$P_1 - P_2 = \tfrac{1}{2}\rho v_2^2 - \tfrac{1}{2}\rho v_1^2 + \rho g y_2 - \rho g y_1$$

Rearranging terms, we get

$$P_1 + \tfrac{1}{2}\rho v_1^2 + \rho g y_1 = P_2 + \tfrac{1}{2}\rho v_2^2 + \rho g y_2 \qquad (15.15)$$

This is **Bernoulli's equation** as applied to a nonviscous, incompressible fluid in steady flow. It is often expressed as

Bernoulli's equation

$$P + \tfrac{1}{2}\rho v^2 + \rho g y = \text{constant} \qquad (15.16)$$

Bernoulli's equation says that the sum of the pressure, (P), the kinetic energy per unit volume ($\tfrac{1}{2}\rho v^2$), and potential energy per unit volume ($\rho g y$) has the same value at all points along a streamline.

When the fluid is at *rest*, $v_1 = v_2 = 0$ and Equation 15.15 becomes

$$P_1 - P_2 = \rho g(y_2 - y_1) = \rho g h$$

which agrees with Equation 15.6.

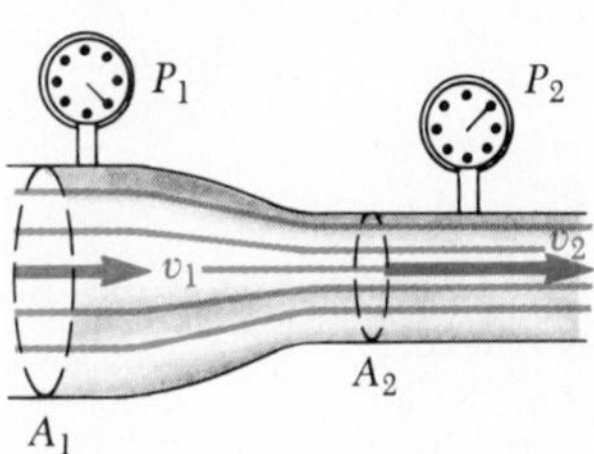

Figure 15.20 (Example 15.9) Schematic diagram of a Venturi tube. The pressure P_1 is greater than the pressure P_2, since $v_1 < v_2$. This device can be used to measure the speed of fluid flow.

EXAMPLE 15.9. The Venturi Tube

The horizontal constricted pipe illustrated in Figure 15.20, known as a *Venturi tube*, can be used to measure flow velocities in an incompressible fluid. Let us determine the flow velocity at point 2 if the pressure difference $P_1 - P_2$ is known.

Solution: Since the pipe is horizontal, $y_1 = y_2$ and Equation 15.15 applied to points 1 and 2 gives

$$P_1 + \tfrac{1}{2}\rho v_1^2 = P_2 + \tfrac{1}{2}\rho v_2^2$$

From the equation of continuity (Eq. 15.14), we see that $A_1 v_1 = A_2 v_2$ or

$$v_1 = \frac{A_2}{A_1} v_2$$

Substituting this expression into the previous equation gives

$$P_1 + \tfrac{1}{2}\rho \left(\frac{A_2}{A_1}\right)^2 v_2^2 = P_2 + \tfrac{1}{2}\rho v_2^2$$

$$v_2 = A_1 \sqrt{\frac{2(P_1 - P_2)}{\rho(A_1^2 - A_2^2)}} \qquad (15.17)$$

We can also obtain an expression for v_1 using this result and the continuity equation. Note that since $A_2 < A_1$, it follows that P_1 is *greater* than P_2. In other words, the pressure is *reduced* in the constricted part of the pipe. This result is somewhat analogous to the following situation: Consider a very crowded room, where people are squeezed together. As soon as a door is opened and people begin to exit, the squeezing (pressure) is least near the door where the motion (flow) is greatest.

EXAMPLE 15.10. Torricelli's law (speed of efflux)

A tank containing a liquid of density ρ has a small hole in its side at a distance y_1 from the bottom (Fig. 15.21). The air above the liquid is maintained at a pressure P. Determine the speed at which the fluid leaves the hole when the liquid level is a distance h above the hole.

Solution: If we assume the tank is large in cross section compared to the hole ($A_2 \gg A_1$), then the fluid will be approximately at rest at the top, point 2. Applying Bernoulli's equation to points 1 and 2, and noting that at the hole $P_1 = P_a$, we get

$$P_a + \tfrac{1}{2}\rho v_1^2 + \rho g y_1 = P + \rho g y_2$$

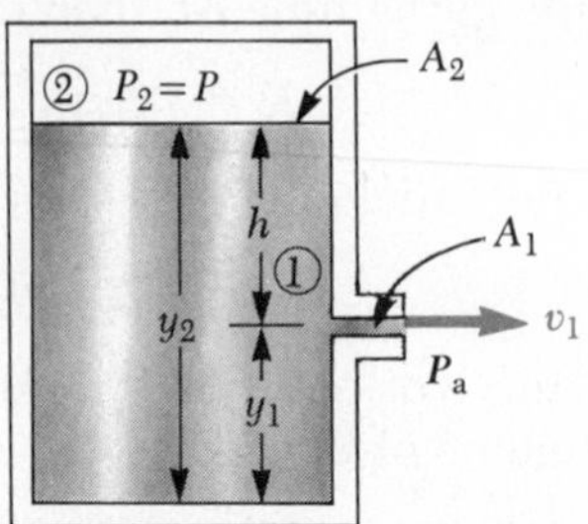

Figure 15.21 (Example 15.10) The speed of efflux, v_1, from the hole in the side of the container is given by $v_1 = \sqrt{2gh}$.

But $y_2 - y_1 = h$, and so this reduces to

Speed of efflux
$$v_1 = \sqrt{\frac{2(P - P_a)}{\rho} + 2gh} \qquad (15.18)$$

If A_1 is the cross-sectional area of the hole, then the flow rate from the hole is given by $A_1 v_1$. When P is large compared with atmospheric pressure (and therefore the term $2gh$ can be neglected), the speed of efflux is mainly a function of P. Finally, if the tank is open to the atmosphere, then $P = P_a$ and $v_1 = \sqrt{2gh}$. In other words, the speed of efflux for an open tank is equal to that acquired by a body falling freely through a vertical distance h. This is known as **Torricelli's law.**

°15.9 OTHER APPLICATIONS OF BERNOULLI'S EQUATION

In this section we shall give a qualitative description of some common phenomena that can be explained at least in part by Bernoulli's equation.

Let us examine the "lift" of an airplane wing (Fig. 15.22). We shall assume that the shape of the wing is such that streamline flow is maintained. The air in the region above the wing moves faster than the air below the wing. (Note the difference in the density of streamlines.) As a result, the air pressure above is less than the air pressure below and there is a net upward force, or "lift," on the wing. Of course, the lift depends on several factors, such as the speed of the airplane and the angle between the wing and the horizontal. As this angle increases, turbulent flow above the wing reduces the lift predicted by the Bernoulli effect.

The curve of a spinning baseball is one example in which Bernoulli's equation arises. The ball in Figure 15.23 is moving toward the right and is rotating counterclockwise. From the point of view of the baseball, the air is streaming by it toward the left. However, because the ball is spinning, some air is "dragged" along with the ball because of its rough surface, and raised laces. The air in region A is held back while the air in region B is helped along because of the direction in which the ball spins. Because the speed of the air is less at A than at B, it follows from Bernoulli's principle that the pressure at B is less than at A. This pressure difference causes the ball to follow the curved path shown by the dashed line in Figure 15.23.

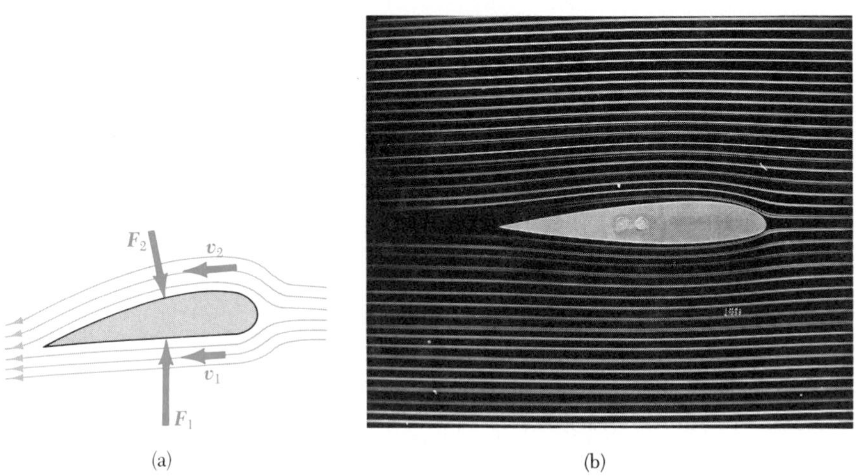

Figure 15.22 (a) Streamline flow around an airplane wing. By Bernoulli's principle, since v_2 is greater than v_1, the pressure at the top is less than the pressure at the bottom, and so there is a net force upward. (b) Streamline flow around an airfoil, made visible by smoke particles moving along the streamlines. The flow is from right to left, simulating motion of the airfoil from left to right. (Courtesy of NASA)

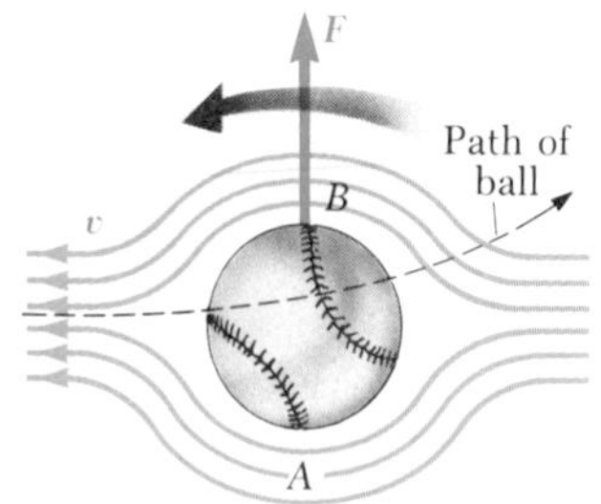

Figure 15.23 Streamline flow around a spinning ball. The ball will curve as shown because of a deflecting force F, which arises from the Bernoulli effect.

Essay

HIGH PRESSURE PHYSICS

A. Jayaraman
AT&T Bell Laboratories
Murray Hill, NJ

Knowledge about the behavior of matter under extreme pressures is of great interest to several scientific disciplines. Scientists working in the field of high pressure research have continually strived to reach higher and higher pressures to expand this knowledge. In recent years a novel pressure-generating device called the *diamond anvil cell* has evolved. With this device, static pressures (sustained application of pressure as opposed to dynamic pressures generated in shock waves, lasting only for microseconds) of over 2 Mbar can be reached under laboratory conditions, and a variety of sophisticated measurements can be performed to probe the behavior of matter under high pressures. [Pressure is force per unit area and is expressed in bars or Pascals; 1 bar = 10^5 Pascals; 1 kbar or kilobar = 10^3 bars; 10 kbar = 1 GPa (gigapascals); 1 Mbar or megabar = 10^6 bars or 100 GPa.]

The late Professor P. W. Bridgman of Harvard University almost single-handedly pioneered high pressure research for over half a century, until his death in 1961. He not only studied an unbelievable number of elements and compounds under pressure, but also invented every technique he used to reach pressures of up to 100 kbar. Bridgman was awarded the Nobel prize in 1946 for the development of pressure-generation techniques and for the discovery of new phenomena under pressure. Generation of very high pressure involves the use of the principle of force multiplication: A force $\boldsymbol{F}$ applied to a larger area A_1 is multiplied by the ratio A_1/A_2 at the delivery end having a smaller area A_2. (See Section 15.4.) This ratio can vary from 10 to 1000 in any practical pressure-generating device. Bridgman used this principle to construct high pressure apparatus.

Diamond has two properties that qualify it as the best material for the containment of high pressure; (1) it is the hardest substance known to science and (2) it is transparent to optical radiation as well as to x-rays. Although these properties of the diamond were well known for a long time, it was only in 1959 that the diamond anvil device was born. A diamond anvil cell capable of generating a pressure of a half a million atmospheres fits easily in the palm of a hand (see Fig. 15E.1 (a) and (b)). One of the greatest advantages of the diamond anvil cell is that pressurized samples can be directly viewed with a microscope. The basic principle of the diamond anvil cell is simple. A sample placed between flat parallel faces of two diamond anvils is subjected to very high pressure by means of a force pushing one anvil against the other. Since the flat faces have a very small area (≈ 0.1 mm^2) there is a large pressure multiplication. A prerequisite to using the diamond anvil cell for pressure generation is that the anvil flats be perfectly aligned axially and set parallel to each other as well. The first

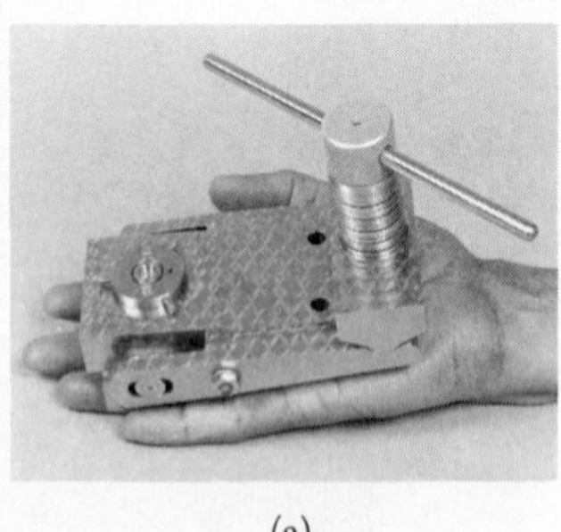

(a)

Figure 15E.1 (a) Photograph of a diamond anvil cell for generating pressure of $\frac{1}{2}$ Mbar.

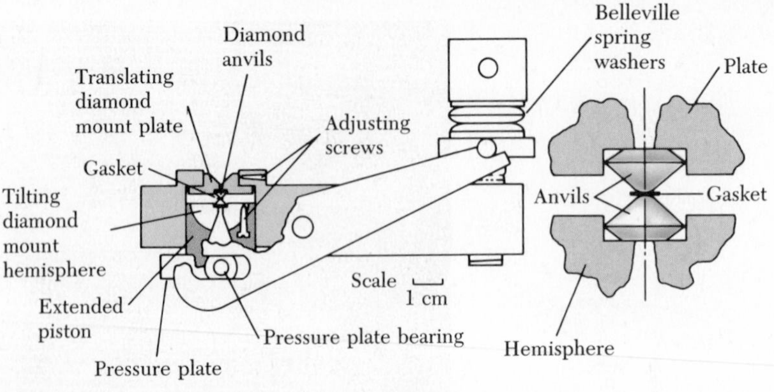

(b)

Figure 15E.1 (b) Diagramatic sketch of a diamond anvil cell according to the National Bureau of Standards design. On the extreme right, the diamond anvil region is shown magnified. To generate hydrostatic pressure, a thin metal gasket with a 200-μm diameter hole is introduced between the anvil faces, and the hole filled with a pressure medium, usually a mixture of methanol and ethanol for hydrostatic pressures up to 10 GPa. Rare gases like argon and xenon can be condensed into the hole for use as a hydrostatic pressure medium to much higher pressures.

objective is easily realized with the help of centering screws, and the second with the help of a tiltable hemispherical mount or a set of two half-cylindrical rocker-mounts. Force on the diamond is applied by compressing the Belleville spring washers with the screws (Fig. 15.1) or other types of lever-arm arrangement. The diamonds used in the diamond anvil cell must be of high quality and flawless. Usually $\frac{1}{3}$ to $\frac{1}{2}$ carat gem-cut diamonds whose culet has been ground to a small flat surface are used. The pressure is measured by the Ruby fluorescence technique, which is in wide use at the present time. Ruby crystals exhibit a strong red fluorescence when excited by light, such as the blue line of a helium-cadmium laser. This red fluorescent emission consists of two well-defined peaks at wavelengths of 694.2 nm (R_1 line) and 692.7 nm (R_2 line). These shift to higher wavelengths with increasing pressure; the R_1 shift has been calibrated, using the equation of state of standard substances, to construct a pressure-scale. The ruby scale is almost linear up to 300 GPa, the shift being 0.365 nm/GPa. At higher pressures the scale begins to bend somewhat towards the pressure axis.

Diamond cells have been adapted for low-temperature studies down to liquid helium temperatures. For high-temperature investigations heating is accomplished with an yttrium aluminum garnet (YAG) laser. If the sample is absorbing at the wavelength of the laser line, it can be very quickly heated to about 4000°C, without heating the cell. The temperature is usually measured by optical pyrometry or spectroscopy. The diamond anvil cell is an excellent tool for high pressure spectroscopy and is well suited for high pressure x-ray diffraction studies. With modern synchrotron x-ray sources, diffraction data can be collected in a fraction of a second.

One of the most useful basic data for understanding the fundamental properties of a solid is the application of the pressure-volume relationship. This information can be obtained from high pressure x-ray measurements. By fitting the pressure-volume data to a theoretical equation of state, the bulk modulus B (inverse of the compressibility) of a solid ($B = V(\partial P/\partial V)T$ and its pressure derivative B' can be evaluated. One of the equations commonly used in this connection is the Murnaghan equation of state given by

$$P = \frac{B}{B'}\left[\left(\frac{V_0}{V}\right)^{B'} - 1\right]$$

The above equation of state assumes that dB/dP is a constant. However, this assumption may not be true for highly compressible solids; in that case nonlinear terms should be incorporated into the equation of state. Many such equations are in use. Figure 15E.2 shows the pressure-volume data for silver, fitted to the Murnaghan equation of state.

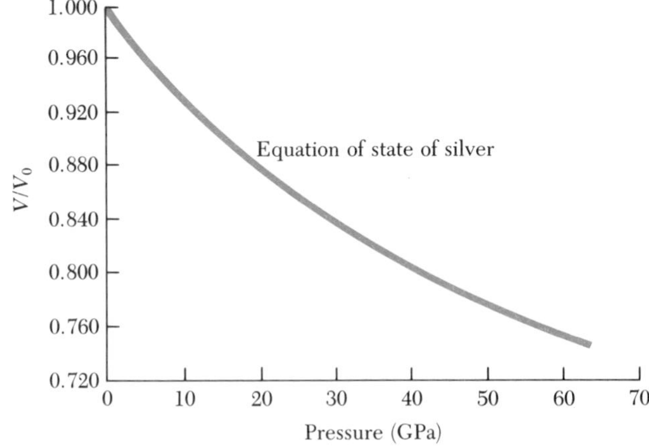

Figure 15E.2 The pressure-volume relationship for silver generated using the Murnaghan equation of state with $B_\beta = 118$ GPa and $B' = 3.8$. The experimental data fit this curve.

To a high pressure researcher the most rewarding experience is discovering a pressure-induced phase transition. The fundamental philosophy behind pressure-induced phase transitions is as follows: The total energy of a system in thermodynamic equilibrium must be at its minimum value. (A system in thermodynamic equilibrium is one in which no changes in pressure, volume or temperature can occur without external influences.) This minimum energy state is called the *stable state* of the system. When the volume is decreased by compression, the total energies of the possible states change (see Fig. 15E.3). The system adjusts itself through either structural transformations or changes in its electronic structure, or both, to minimize the total energy. Structural transformations involve geometrical rearrangement of the atoms, while the electronic structure changes are connected with the electronic energy bands in a solid. In many instances these changes profoundly influence the physical properties of solids. Figure 15E.3 shows the structural stability of silicon, a well-known semiconductor, which has the diamond structure at ambient pressure. Upon compression the energy of the diamond phase increases and crosses the β-tin at

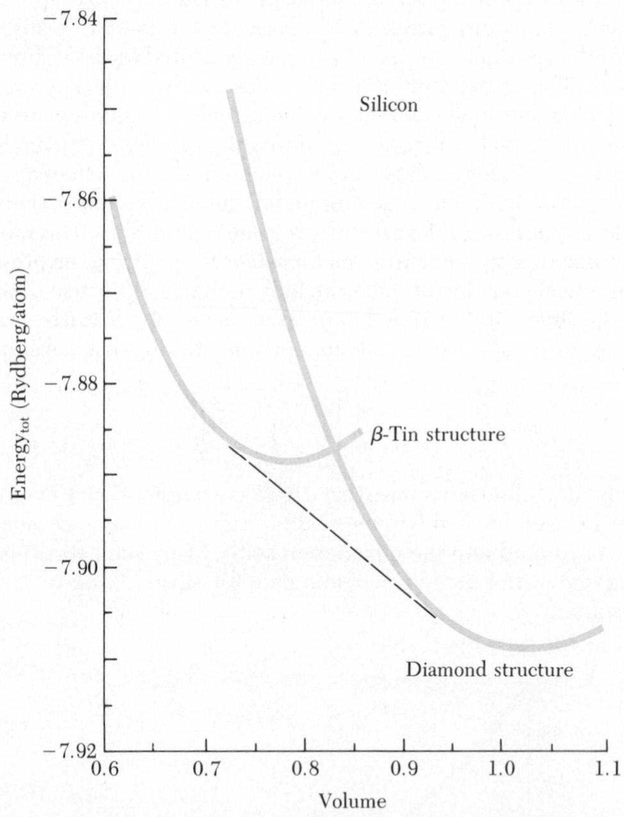

Figure 15E.3 Total energy curves for the diamond and β-tin phases of silicon plotted against relative volume, calculated by Yin and Cohen of the University of California, Berkeley. The dashed line is the common tangent of the energy curves for the two phases. The slope of this line gives the transformation pressure ($dE/dV = P$). The diamond phase of Si transforms to β-tin phase around 12 GPa.

A number of devices operate in the manner described in Figure 15.24 (p. 330). A stream of air passing over an open tube reduces the pressure above the tube. This reduction in pressure causes the liquid to rise into the air stream. The liquid is then dispersed into a fine spray of droplets. You might recognize that this so-called atomizer is used in perfume bottles and paint sprayers. The same principle is used in the carburetor of a gasoline engine. In

V/V_0 (0.82). At a pressure of about 10 GPa, Si transforms to the metallic tin structure called the β-tin, with a large increase in density. The β-tin phase is metallic and superconducting, a spectacular change indeed. Such behavior under the influence of pressure is shown by germanium and many compound semiconductors as well. One of the exciting predictions of solid state theory is that hydrogen becomes metallic at a pressure of about 3 Mbar. Hydrogen solidifies to a clear transparent solid near a pressure of 57 kbar at 298 K. It is a very good insulator but under the influence of pressure, the localized electron orbitals can overlap, leading to a metallic state. According to solid state theory, metallic hydrogen is likely to be a high temperature superconductor. The high pressure behavior of hydrogen is therefore not only of fundamental interest, but it has implications for the internal structure of the giant planets, which are believed to be largely made up of hydrogen. A rather spectacular pressure-induced semiconductor-to-metal transition occurs at a moderate pressure in a compound known as samarium monosulfide (SmS). This transition is a classic case of electronic structure change. The transition looks like "alchemy," for the dull, black-looking substance glitters like gold when the applied pressure exceeds 7 kbar. In the high pressure phase, one bound electron from the $4f$ level per Sm is squeezed out into the $5d$ conduction band, without causing any change in structure.

The most spectacular application of high pressure research to technology is the synthesis of diamond from graphite. If carbon is subjected to a pressure of 65 kbar at $\sim 1400\,^{\circ}\mathrm{C}$ in the presence of nickel or iron, it is spontaneously converted into diamond. Millions of carats of diamonds are synthesized in this manner annually for industrial applications. Perhaps the best scientific application of high pressure research is in the realm of geophysics. Inside the earth both pressure and temperature increase with depth; at the center, pressure is estimated to be 3.5 Mbar and temperature, 4000 °C. Furthermore, the interior of the earth is stratified into a crust, mantle and core, with some minor features superimposed. Experiments on silicate minerals (the major components of the earth) have revealed that the major and minor features of the crust and the mantle region are related to phase transitions induced by high pressure in these minerals. A major phase transition in silicate minerals, recently discovered with the diamond anvil cell, is to the so-called "perovskite" phase, which seems to dominate the entire lower mantle region. Thus, the diamond anvil cell is serving as a transparent window right up to the core-mantle boundary, located at a depth of 2900 km.

Since the advent of the diamond anvil cell, research regarding the properties of materials under very high pressures has made a quantum jump. A new level of understanding of the physics and chemistry of solids at high pressure is emerging. The possibility of reaching megabar pressures has stimulated solid state theoretical calculations. These calculations, facilitated by modern computing, are able to predict pressure-induced transitions and the electronic band structure at high pressure to an astonishing degree of precision.

What is the limit of pressure attainable with the diamond anvils? One thing that would limit the maximum pressure would be a phase transition in diamond. A recent theoretical calculation shows that the diamond structure would be stable up to at least 10 Mbar. However, the ultimate pressure that can be reached in a diamond anvil cell would be determined by the yield strength of diamond. From plasticity theory this is estimated to be about 5 Mbar. If such pressures could be attained, high pressure physics would leap to even greater heights.

this case, the low-pressure region in the carburetor is produced by air drawn in by the piston through the air filter. The gasoline vaporizes, mixes with the air, and enters the cylinder of the engine for combustion.

If a person has advanced arteriosclerosis, the Bernoulli effect produces a sign called vascular flutter. In this situation, the artery is constricted as a result of an accumulation of plaque on its inner walls. In order to maintain a constant flow rate through such a constricted artery, the driving pressure must increase. Such an increase in pressure requires a greater demand on the heart muscle. If the blood velocity is sufficiently high in the constricted region, the artery may collapse under external pressure, causing a momentary interruption in blood flow. At this point, there is no Bernoulli effect and the vessel reopens under arterial pressure. As the blood rushes through the constricted artery, the internal pressure drops and again the artery closes. Such variations in blood flow can be heard with a stethoscope. If the plaque becomes dislodged and ends up in a smaller vessel that delivers blood to the heart, the person can suffer a heart attack.

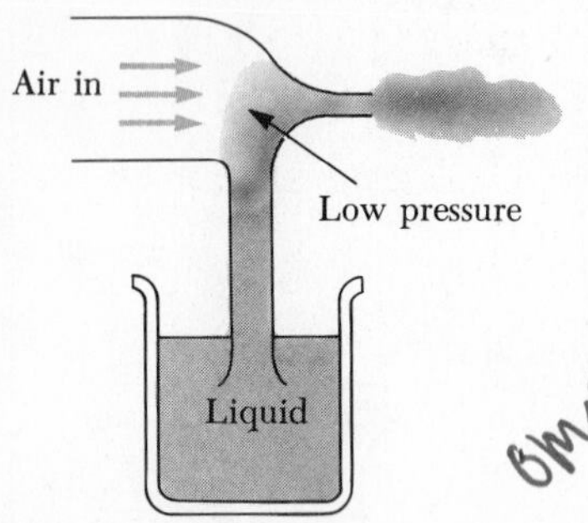

Figure 15.24 A stream of air passing over a tube dipped into a liquid will cause the liquid to rise in the tube as shown.

15.10 ENERGY FROM THE WIND

The wind as a source of energy is not a new concept. In fact, there is some evidence that windmills were used in Babylon and in China as early as 2000 B.C. The kinetic energy carried by the winds originates from solar energy.

Although the wind is a large potential source of energy (about 5 kW per acre in the United States), it has been harnessed only on a small scale. It has been estimated that, on a global scale, the winds account for a total available power of 2×10^{13} W (three times the world energy consumption in 1972). Therefore, if only a small percentage of the available power could be harnessed, wind power would represent a significant fraction of our energy needs. As with all indirect energy resources, wind power systems have some disadvantages, which in this case arise mainly from the variability of wind velocities.

The largest windmill built in the United States was a 1.25-MW generator installed on "Grandpa's Knob" near Rutland, Vermont. The machine's blades were 175 ft in diameter, and the facility operated intermittently between 1941 and 1945. Unfortunately, one of its two main blades broke off as a result of material fatigue and was never repaired. Despite this failure, the windmill was considered a technological success, since wartime needs limited the quality of available materials. Nevertheless, the project was abandoned because costs were not competitive with hydroelectric power. The U.S. Department of Energy is currently planning to develop wind machines capable of generating 1 MW.

We can use some of the ideas developed in this chapter to estimate wind power. Any wind energy machine involves the conversion of the kinetic energy of moving air to mechanical energy, usually a rotating shaft. The kinetic energy per unit volume of a moving column of air is given by

$$\frac{KE}{\text{volume}} = \tfrac{1}{2}\rho v^2$$

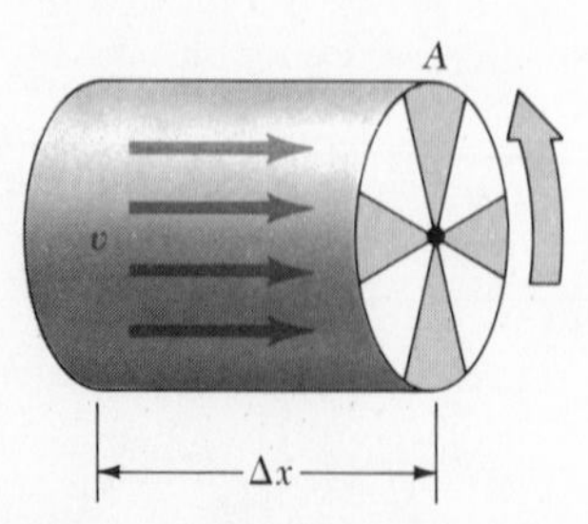

Figure 15.25 Wind moving through a cylindrical column of cross-sectional area A with a speed v.

where ρ is the density of air and v is its speed. The rate of flow of air through a column of cross-sectional area A is Av (Fig. 15.25). This can be considered as the volume of air crossing the area each second. In the working machine, A is the cross-sectional area of the wind-collecting system, such as a set of rotating propeller blades. Multiplying the kinetic energy per unit volume by the flow rate gives the rate at which energy is transferred, or, in other words, the power:

$$\text{Power} = \frac{KE}{\text{volume}} \times \frac{\text{volume}}{\text{time}} = (\tfrac{1}{2}\rho v^2)(Av) = \tfrac{1}{2}\rho v^3 A \qquad (15.19)$$

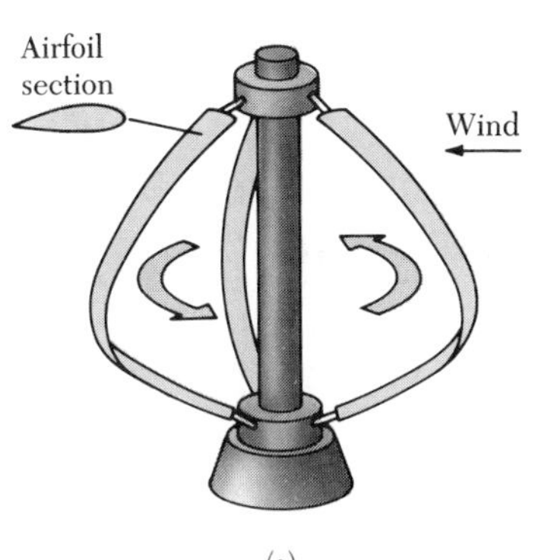

(a)

(b)

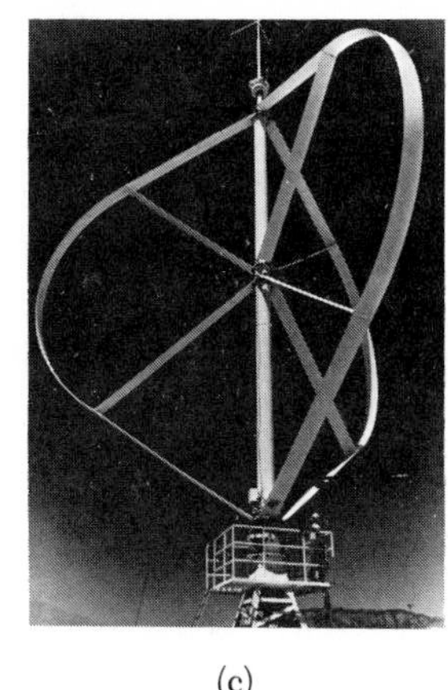

(c)

Figure 15.26 (a) A vertical-axis wind generator. (b) A horizontal-axis wind generator. (c) Photograph of a vertical-axis wind generator. (Courtesy of DOE)

Therefore, the available power per unit area is given by

$$\frac{P}{A} = \tfrac{1}{2}\rho v^3 \tag{15.20}$$

According to this result, if the moving air column could be brought to rest, a power of $\frac{1}{2}\rho v^3$ would be available for each square meter that is intercepted. For example, if we assume a moderate speed of 12 m/s (27 mi/h) and take $\rho = 1.3\ \text{kg/m}^3$, we find that

$$\frac{P}{A} = \tfrac{1}{2}\left(1.3\ \frac{\text{kg}}{\text{m}^3}\right)\left(12\frac{\text{m}}{\text{s}}\right)^3 \approx 1100\ \frac{\text{W}}{\text{m}^2} = 1.1\ \frac{\text{kW}}{\text{m}^2}$$

Since the power per unit area varies as the cube of the velocity, its value doubles if v increases by only 26%. Conversely, the power output would be halved if the velocity decreased by 26%.

This calculation is based on ideal conditions and assumes that all of the kinetic energy is available for power. In reality, the air stream emerges from the wind generator with some residual velocity, and more refined calculations show that, at best, one can extract only 59.3% of this quantity.[1] The expression for the maximum available power per unit area for the ideal wind generator is found to be

$$\frac{P_{max}}{A} = \frac{8}{27}\rho v^3 \tag{15.21}$$

In a real wind machine, further losses resulting from the nonideal nature of the propeller, gearing, and generator reduce the total available power to around 15% of the value predicted by Equation 15.20. Sketches of two types of wind turbines are shown in Figure 15.26.

[1] For more details, see J. H. Krenz, *Energy Conversion and Utilization,* Boston, Allyn and Bacon, 1976, Chapter 8.

EXAMPLE 15.11. Power Output of a Windmill
Calculate the power output of a wind generator having a blade diameter of 80 m, assuming a wind speed of 10 m/s and an overall efficiency of 15%.

Solution: Since the radius of the blade is 40 m, the cross-sectional area of the propellers is given by

$$A = \pi r^2 = \pi(40\text{ m})^2 = 5.0 \times 10^3\text{ m}^2$$

If 100% of the available wind energy could be extracted, the maximum available power would be

$$P_{max} = \tfrac{1}{2}\rho A v^3 = \tfrac{1}{2}\left(1.2\ \frac{\text{kg}}{\text{m}^3}\right)(5.0 \times 10^3\text{ m}^2)\left(10\ \frac{\text{m}}{\text{s}}\right)^3$$
$$= 3.0 \times 10^6\text{ W} = 3.0\text{ MW}$$

Since the overall efficiency is 15%, the output power is

$$P = 0.15P_{max} = 0.45\text{ MW}$$

In comparison, a large steam-turbine plant has a power output of about 1 GW. Hence, one would require 2200 such wind generators to equal this output under these conditions. The large number of generators required for reasonable output power is clearly a major disadvantage of wind power. (See Problem 37.)

omit

*15.11 VISCOSITY

We have seen that a fluid does not support a shearing stress. However, fluids do offer some degree of resistance to shearing motion. This resistance to shearing motion is a form of internal friction which is called *viscosity.* In the case of liquids, the viscosity arises because of a frictional force between adjacent layers of the fluid as they slide past one another. The degree of viscosity of a fluid can be understood with the following example. If two plates of glass are separated by a layer of fluid such as oil, with one plate fixed in position, it is easy to slide one plate over the other (Fig. 15.27). However, if the fluid separating the plates is tar, the task of sliding one plate over the other becomes much more difficult. Thus, we would conclude that tar has a higher viscosity than oil. In Figure 15.27, note that the velocity of successive layers increases linearly from 0 to v as one moves from a layer adjacent to the fixed plate to a layer adjacent to the moving plate.

Recall that in a solid a shearing stress gives rise to a relative displacement of adjacent layers (Section 15.1). In an analogous fashion, adjacent layers of a fluid under shear stress are set into relative motion. Again, consider two parallel layers, one fixed and one moving to the right under the action of an external force F as in Figure 15.27. Because of this motion, a portion of the liquid is distorted from its original shape, *ABCD*, at one instant to the shape *AEFD* after a short time interval. If you refer to Section 15.1, you will recognize that the liquid has undergone a shear strain. (Note that previous sections in this chapter have assumed ideal fluids that do not support shear strains. Viscous fluids can support such strains.) By definition, the shear stress on the liquid is equal to the ratio F/A, while the shear strain is defined by the ratio $\Delta x/\ell$. That is,

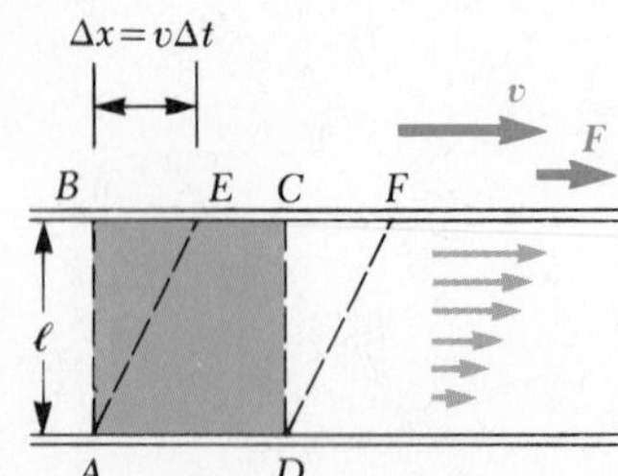

Figure 15.27 A layer of liquid between two solid surfaces in which the lower surface is fixed and the upper surface moves to the right with a velocity v.

$$\text{Shear stress} = \frac{F}{A} \qquad \text{Shear strain} = \frac{\Delta x}{\ell}$$

The upper plate moves with a speed v, and the fluid adjacent to this plate has the same speed. Thus, in a time Δt, the fluid at the upper plate moves a distance $\Delta x = v\ \Delta t$, and we can express the shear strain per unit time as

$$\frac{\text{Shear strain}}{\Delta t} = \frac{\Delta x/\ell}{\Delta t} = \frac{v}{\ell}$$

This equation states that the rate of change of shearing strain is v/ℓ.

TABLE 15.3 The Viscosities of Various Fluids

Fluid	T(°C)	Viscosity η ($N \cdot s/m^2$)
Water	20	1.0×10^{-3}
Water	100	0.3×10^{-3}
Whole blood	37	2.7×10^{-3}
Glycerine	20	830×10^{-3}
10-wt motor oil	30	250×10^{-3}

The **coefficient of viscosity,** η, for the fluid is defined as the ratio of the shearing stress to the rate of change of the shear strain:

$$\eta \equiv \frac{F/A}{v/\ell} = \frac{F\ell}{Av} \qquad (15.22)$$

Coefficient of viscosity

The SI unit of viscosity is $N \cdot s/m^2$. The coefficients of viscosity for some substances are given in Table 15.3. The cgs unit of viscosity is dyne $\cdot$ s/cm^2, which is called the **poise.**

The expression for η given by Equation 15.22 is only valid if the fluid velocity varies linearly with position. In this case, it is common to say that the velocity gradient, v/ℓ, is uniform. If the velocity gradient is *not* uniform, we must express η in the general form

$$\eta \equiv \frac{F/A}{dv/dy} \qquad (15.23)$$

where the velocity gradient dv/dy is the change in velocity with position as measured perpendicular to the direction of velocity.

*15.12 TURBULENCE

If adjacent layers of a viscous fluid flow smoothly over each other, the stable streamline flow is called *laminar flow*. However, at sufficiently high velocities, the fluid flow changes from laminar flow to a highly irregular and random motion of the fluid called *turbulent flow*. The velocity at which turbulence occurs depends on the geometry of the medium surrounding the fluid and the fluid viscosity.

There are many examples of turbulent flow that can be cited. Water flowing in a rock-filled stream or river, and smoke rising from a chimney on a windy day are turbulent in nature. Likewise, the water in the wake of a speedboat and the air in the wakes left by airplanes and other moving vehicles represent turbulent flow.

Experimentally, it is found that the onset of turbulence is determined by a dimensionless parameter called the **Reynolds number,** *RN*, given by

$$RN = \frac{\rho v d}{\eta} \qquad (15.24)$$

Reynolds number

where ρ is the fluid density, v is the fluid speed, η is the viscosity, and d is a geometrical length associated with the flow. For flow through a tube, d would be the diameter of the tube. In the case of flow around a sphere, d would be the diameter of the sphere.

Experiments show that if the Reynolds number is below about 2000, the flow of fluid through a tube is laminar; turbulence occurs if the Reynolds number is above about 3000.

EXAMPLE 15.12. Measuring the Coefficient of Viscosity

A metal plate of area 0.15 m² is connected to an 8-g mass via a string that passes over an ideal pulley, as in Figure 15.28. A lubricant with a film thickness of 0.3 mm is placed between the plate and surface. When released, the plate is observed to move to the right with a constant speed of 0.085 m/s. Find the coefficient of viscosity of the lubricant.

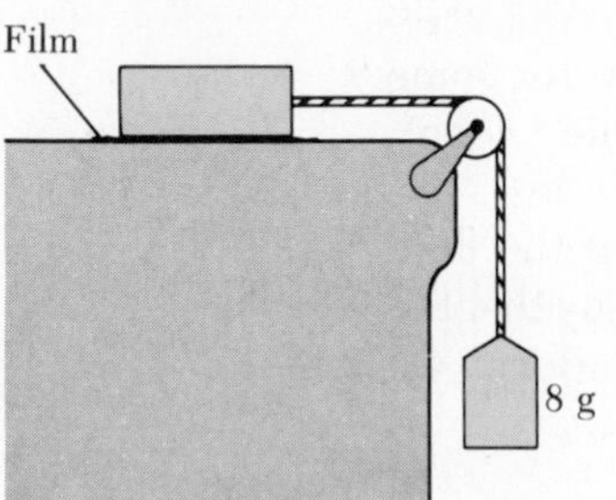

Figure 15.28 (Example 15.12).

Solution: Because the plate moves with constant speed, its acceleration is zero. The plate moves to the right under the action of the tension force, T, and the frictional force, f, associated with the viscous fluid. In this case, the tension is equal in magnitude to the suspended weight, therefore,

$$f = T = mg = (8 \times 10^{-3}\ \text{kg})(9.80\ \text{m/s}^2) = 7.84 \times 10^{-2}\ \text{N}$$

The lubricant in contact with the horizontal surface is at rest, while the layer in contact with the plate moves at the speed of the plate. Assuming the velocity gradient is uniform, we have

$$\eta = \frac{F\ell}{Av} = \frac{(7.84 \times 10^{-2}\ \text{N})(0.3 \times 10^{-3}\ \text{m})}{(0.05\ \text{m}^2)(0.085\ \text{m/s})} = 5.53 \times 10^{-3}\ \text{N} \cdot \text{s/m}^2$$

15.13 SUMMARY

The elastic properties of a solid can be described using the concepts of stress and strain. **Stress** is a quantity proportional to the force producing a deformation; **strain** is a measure of the degree of deformation. Stress is proportional to strain, and the constant of proportionality is the **elastic modulus:**

$$\text{Elastic modulus} \equiv \frac{\text{stress}}{\text{strain}} \tag{15.1}$$

Three common types of deformation are: (1) the resistance of a solid to elongation under a load, characterized by **Young's modulus,** Y; (2) the resistance of a solid to the motion of planes in the solid sliding past each other, characterized by the **shear modulus,** S; (3) the resistance of a solid (or a liquid) to a volume change, characterized by the **bulk modulus,** B.

The **density,** ρ, of a homogeneous substance is defined as its mass per unit volume and has units of kg/m³ in the SI system:

Density

$$\rho \equiv \frac{m}{V} \tag{15.5}$$

The **pressure,** P, in a fluid is the force per unit area that the fluid exerts on an object immersed in the fluid:

$$P \equiv \frac{F}{A} \qquad (15.6)$$

Average pressure

In the SI system, pressure has units of N/m^2, and $1\ N/m^2 = 1$ pascal (Pa).

The pressure in a fluid varies with depth h according to the expression

$$P = P_a + \rho g h \qquad (15.11)$$

Pressure at any depth h

where P_a is atmospheric pressure ($= 1.01 \times 10^5\ N/m^2$) and ρ is the density of the fluid.

Pascal's law states that when pressure is applied to an enclosed fluid, the pressure is transmitted undiminished to every point of the fluid and of the walls of the container.

When an object is partially or fully submerged in a fluid, the fluid exerts an upward force on the object called the **buoyant force.** According to **Archimedes' principle,** the buoyant force is equal to the weight of the fluid displaced by the body.

Various aspects of fluid dynamics (fluids in motion) can be understood by assuming that the fluid is nonviscous and incompressible and that the fluid motion is in a steady state with no turbulence.

Using these assumptions, one obtains two important results regarding fluid flow through a pipe of nonuniform size:

1. The flow rate through the pipe is a constant, which is equivalent to stating that the product of the cross-sectional area, A, and the speed, v, at any point is a constant. That is,

$$A_1 v_1 = A_2 v_2 = \text{constant} \qquad (15.14)$$

Equation of continuity

2. The sum of the pressure, kinetic energy per unit volume, and potential energy per unit volume has the same value at all points along a streamline. That is,

$$P + \tfrac{1}{2}\rho v^2 + \rho g y = \text{constant} \qquad (15.16)$$

Bernoulli's equation

This is known as **Bernoulli's equation** and is fundamental in the study of fluid dynamics.

The viscosity of a fluid is a measure of its resistance to shearing motion. The **coefficient of viscosity** for a fluid is defined as the ratio of the shearing stress to the rate of change of the shear strain.

The onset of turbulence in fluid flow is determined by a parameter called the **Reynolds number,** whose value depends on the fluid density, the fluid speed, the viscosity, and a geometrical factor.

QUESTIONS

1. Two glass tumblers that weigh the same but have different shapes and different cross-sectional areas are filled to the same level with water. According to the expression $P = P_a + \rho g h$, the pressure is the same at the bottom of both tumblers. In view of this, why does one tumbler weigh more than the other?
2. How much force does the atmosphere exert on 1 mi^2 of land?
3. When you drink a liquid through a straw, you reduce the pressure in your mouth and let the atmosphere move the liquid. Explain how this works. Could you use a straw to sip a drink on the moon?

4. Indian fakirs stretch out for a nap on a bed of nails. How is this possible?
5. Pascal used a barometer with water as the working fluid. Why is it impractical to use water for a typical barometer?
6. A person sitting in a boat floating in a small pond throws a heavy anchor overboard. Does the level of the pond rise, fall, or remain the same?
7. Steel is much denser than water. How, then, do boats made of steel float?
8. A helium-filled balloon will rise until its density becomes the same as that of the air. If a sealed submarine begins to sink, will it go all the way to the bottom of the ocean or will it stop when its density becomes the same as that of the surrounding water?
9. A fish rests on the bottom of a bucket of water while the bucket is being weighed. When the fish begins to swim around, does the weight change?
10. Will a ship ride higher in the water of an island lake or in the ocean? Why?
11. If 1 000 000 N of weight was placed on the deck of the World War II battleship North Carolina, it would sink only 2.5 cm lower in the water. What is the cross-sectional area of the ship at water level?
12. Lead has a greater density than iron, and both are denser than water. Is the buoyant force on a lead object greater than, less, than, or equal to the buoyant force on an iron object of the same dimensions?
13. An ice cube is placed in a glass of water. What happens to the level of the water as the ice melts?
14. A woman wearing high-heeled shoes is invited into a home in which the kitchen has a newly installed vinyl floor covering. Why should the homeowner be concerned?
15. A typical silo on a farm has many bands wrapped around its perimeter, as shown in the photograph.

(Question 15).

Why is the spacing between successive bands smaller at the lower regions of the silo?
16. The water supply for a city is often provided from reservoirs built on high ground. Water flows from the reservoir, through pipes, and into your home when you turn the tap on your faucet. Why is the water flow more rapid out of a faucet on the first floor of a building than in an apartment on a higher floor?
17. Smoke rises in a chimney faster when a breeze is blowing. Use Bernoulli's principle to explain this phenomenon.
18. Why do many trailer trucks use wind deflectors on the top of their cabs? (See photograph.) How do such devices reduce fuel consumption?

(Question 18) The high cost of fuel has prompted many truck owners to install wind deflectors on their cabs to reduce air drag. (Photo by Lloyd Black)

19. Consider the cross section of the wing on an airplane. The wing is designed such that the air travels faster over the top than under the bottom. Explain why there is a net upward force (lift) on the wing due to the Bernoulli effect.
20. When a fast-moving train passes a train at rest, the two tend to be drawn together. How does the Bernoulli effect explain this phenomenon?
21. A baseball moves past an observer from left to right spinning counterclockwise. In which direction will the ball tend to deflect?
22. A tornado or hurricane will often lift the roof of a house. Use the Bernoulli effect to explain why this occurs. Why should you keep your windows open during these conditions?
23. If you suddenly turn on your shower water at full speed, why is the shower curtain pushed inward?
24. If you hold a sheet of paper and blow across the top surface, the paper rises. Explain.
25. If air from a hair dryer is blown over the top of a Ping-Pong ball, the ball can be suspended in air. Explain how the ball can remain in equilibrium.
26. Two ships passing near each other in a harbor tend to be drawn together and run the risk of a sideways collision. How does the Bernoulli effect explain this?
27. When ski-jumpers are air-borne, why do they bend their bodies forward and keep their hands at their sides?

28. When an object is immersed in a fluid at rest, why is the net force on it in the horizontal direction equal to zero?
29. Explain why a sealed bottle partially filled with a liquid can float.
30. When is the buoyant force on a swimmer the greatest —when the swimmer is exhaling or inhaling?
31. A piece of wood is partially submerged in a container filled with water. If the container is sealed and pressurized above atmospheric pressure, does the wood rise, fall, or remain at the same level? (*Hint:* Wood is porous.)
32. A flat plate is immersed in a fluid at rest. For what orientation of the plate will the pressure on its flat surface be uniform?
33. Because atmospheric pressure is 14.7 lb/in^2 and the area of a person's chest is about 200 in^2, the force of the atmosphere on one's chest is around 3000 lb! In view of this enormous force, why don't our bodies collapse?
34. Why do you suppose the increase in length of a wire under a given load is proportional to its length? (Use a microscopic model in your argument.)
35. What kind of deformation does a cube of Jello exhibit when it "jiggles"?

PROBLEMS

Section 15.1 Elastic Properties of Solids

1. A steel wire has a length of 3 m and a cross-sectional area of 0.2 cm^2. Under what load will its length increase by 0.05 cm?
2. A mass of 2 kg is supported by a copper wire of length 4 m and diameter 4 mm. Determine (a) the stress in the wire and (b) the elongation of the wire.
3. A cube of steel 5 cm on an edge is subjected to a shearing force of 2000 N while one face is clamped. Find the shearing strain in the cube.
4. The *elastic limit* of a material is defined as the maximum stress that can be applied to the material before it becomes permanently deformed. If the elastic limit of copper is 1.5×10^8 N/m^2, determine the *minimum* diameter a copper wire can have under a load of 10 kg if its elastic limit is not to be exceeded.
5. What increase in pressure is necessary to decrease the volume of a 6-cm-diameter sphere of mercury by 0.05%?
6. Determine the decrease in volume of a cube of copper 10 cm on an edge if it is subjected to a bulk stress (pressure) of 10^8 N/m^2.
7. If the shear stress in steel exceeds about 4.0×10^8 N/m^2, it ruptures. Determine the shearing force necessary to (a) shear a steel bolt 1 cm in diameter and (b) punch a 1-cm-diameter hole in a steel plate that is 0.5 cm thick.
8. Two wires are made of the same metal, but have different dimensions. Wire 1 is four times longer and twice the diameter of wire 2. If they are both under the same load, compare (a) the stresses in the two wires and (b) the elongations of the two wires.

Section 15.3 Density and Pressure

9. Calculate the mass of a solid iron sphere that has a diameter of 3.0 cm.
10. A solid cube of material 5.0 cm on an edge has a mass of 1.31 kg. What is the material made of, assuming it consists of only one element? (Consult Table 15.2.)
11. Estimate the density of the *nucleus* of an atom. What does this result suggest concerning the structure of matter? (Use the fact that the mass of a proton is 1.67×10^{-27} kg and its radius is about 10^{-15} m.)
12. A king orders a gold crown having a mass of 0.5 kg. When it arrives from the metalsmith, the volume of the crown is found to be 185 cm^3. Is the crown made of solid gold?
13. A 50-kg woman balances on one heel of a pair of high-heel shoes. If the heel is circular with radius 0.5 cm, what pressure does she exert on the floor?

Section 15.4 Variation of Pressure with Depth

14. Determine the absolute pressure at the bottom of a lake that is 30 m deep.
15. At what depth in a lake is the absolute pressure equal to three times atmospheric pressure?
16. The small piston of a hydraulic lift has a cross-sectional area of 3 cm^2, and the large piston has an area of 200 cm^2 (Fig. 15.12). What force must be applied to the small piston to raise a load of 15 000 N? (In service stations this is usually accomplished with compressed air.)
17. The spring of the pressure gauge shown in Figure 15.9 has a force constant of 1000 N/m, and the piston has a diameter of 2 cm. Find the depth in water for which the spring compresses by 0.5 cm.
18. A rectangular swimming pool has dimensions $\ell = 10$ m, $w = 5$ m, and $h = 2$ m, where h is the depth. If the pool is completely filled with water, calculate the force exerted *by the water* against (a) the bottom of the pool, (b) the 10-m sides, and (c) the 5-m sides.

Section 15.5 Pressure Measurements

19. The U-shaped tube in Figure 15.14a contains mercury. What is the absolute pressure, P, on the left if $h = 20$ cm? What is the gauge pressure?

20. If the fluid in the barometer illustrated in Figure 15.14b is water, what will be the height of the water column in the vertical tube at atmospheric pressure?

21. The open vertical tube in Figure 15.29 contains two fluids of densities ρ_1 and ρ_2, which do not mix. Show that the pressure at the depth $h_1 + h_2$ is given by $P = P_a + \rho_1 g h_1 + \rho_2 g h_2$.

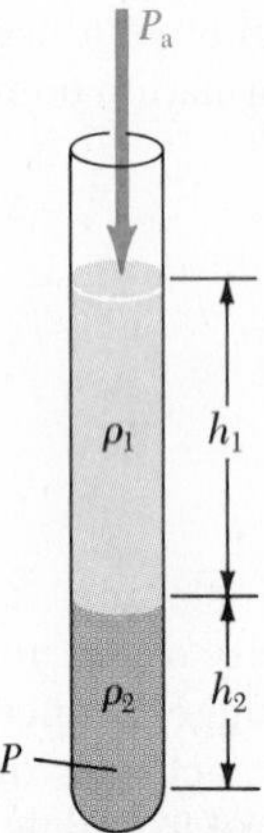

Figure 15.29 (Problem 21).

Section 15.6 Buoyant Forces and Archimedes' Principle

22. Calculate the buoyant force on a solid object made of copper and having a volume of 0.2 m^3 if it is submerged in water. What is the result if the object is made of steel?

23. Show that only 11% of the total volume of an iceberg is above the water level. (Note that sea water has a density of 1.03×10^3 kg/m^3, and ice has a density of 0.92×10^3 kg/m^3.)

24. A solid object has a weight of 5.0 N. When it is suspended from a spring scale and submerged in water, the scale reads 3.5 N (Fig. 15.16b). What is the density of the object?

25. A cube of wood 20 cm on a side and having a density of 0.65×10^3 kg/m^3 floats on water. (a) What is the distance from the top of the cube to the water level? (b) How much lead weight has to be placed on top of the cube so that its top is just level with the water?

26. A balloon filled with helium at atmospheric pressure is designed to support a mass M (payload + empty balloon). (a) Show that the volume of the balloon must be *at least* $V = M/(\rho_a - \rho_{He})$, where ρ_a is the density of air and ρ_{He} is the density of helium. (Ignore the volume of the payload.) (b) If $M = 2000$ kg, what radius should the balloon have?

27. A hollow plastic ball has a radius of 5 cm and a mass of 100 g. The ball has a tiny hole at the top through which lead shot can be inserted. How many grams of lead can be inserted into the ball before it sinks in water? (Assume the ball does not leak.)

28. A 10-kg block of metal measuring 12 cm × 10 cm × 10 cm is suspended from a scale and immersed in water as in Figure 15.16b. The 12-cm dimension is vertical, and the top of the block is 5 cm from the surface of the water. (a) What are the forces on the top and bottom of the block? (Take $P_a = 1.0130 \times 10^5$ N/m^2.) (b) What is the reading of the spring scale? (c) Show that the buoyant force equals the difference between the forces at the top and bottom of the block.

Section 15.8 Fluid Dynamics and Bernoulli's Equation

29. The rate of flow of water through a horizontal pipe is 2 m^3/min. Determine the velocity of flow at a point where the diameter of the pipe is (a) 10 cm, (b) 5 cm.

30. A large storage tank filled with water develops a small hole in its side at a point 16 m below the water level. If the rate of flow from the leak is 2.5×10^{-3} m^3/min, determine (a) the speed at which the water leaves the hole and (b) the diameter of the hole.

31. Water flows through a constricted pipe at a uniform rate (Fig. 15.19). At one point, where the pressure is 2.5×10^4 Pa, the diameter is 8.0 cm; at another point 0.5 m higher, the pressure is 1.5×10^4 Pa and the diameter is 4.0 cm. (a) Find the speed of flow in the lower and upper sections. (b) Determine the rate of flow through the pipe.

32. Water flows through a horizontal constricted pipe. The pressure is 4.5×10^4 Pa at a point where the speed is 2 m/s and the area is A. Find the speed and pressure at a point where the area is $A/4$.

33. The water supply of a building is fed through a main 6-cm-diameter pipe. A 2-cm-diameter faucet tap located 2 m above the main pipe is observed to fill a 25-liter container in 30 s. (a) What is the speed at which the water leaves the faucet? (b) What is the *gauge pressure* in the 6-cm main pipe? (Assume the faucet is the only "leak" in the building.)

°Section 15.9 Other Applications of Bernoulli's Equation

34. An airplane has a mass of 16 000 kg, and each wing has an area of 40 m^2. During level flight, the pressure on the lower wing surface is 7.0×10^4 Pa. Determine the pressure on the upper wing surface.

35. Each wing of an airplane has an area of 25 m^2. If the speed of the air is 50 m/s over the lower wing surface and 65 m/s over the upper wing surface, determine the weight of the airplane. (Assume the plane travels in level flight at constant speed at an elevation where the density of air is 1 kg/m^3. Also assume that all of the lift is provided by the wings.)

°Section 15.10 Energy from the Wind

36. Calculate the power output of a windmill having blades 10 m in diameter if the wind speed is 8 m/s. Assume that the efficiency of the system is 20%.

37. According to one rather ambitious plan, it would take 50 000 windmills, each 800 ft in diameter, to obtain

an average output of 200 GW. These would be strategically located through the Great Plains, along the Aleutian Islands, and on floating platforms along the Atlantic and Gulf coasts and on the Great Lakes. The annual energy consumption in the United States in 1985 is projected to be 1.3×10^{20} J. What fraction of this could be supplied by the array of windmills?

GENERAL PROBLEMS

38. The distortion of the earth's crustal plates is an example of shear on a large scale. A particular type of crustal rock is determined to have a shear modulus of $1.5 \times 10^{10}\ \text{N/m}^2$. What shear stress is involved when a 10 km layer of rock is sheared through a distance of 5 m?
39. A sample of copper is to be subjected to a hydrostatic pressure that will increase its density by 0.1 percent. What pressure is required?
40. One side of the U-shaped tube in Figure 15.30 is filled with a liquid of density ρ_1 while the other side contains a liquid of density ρ_2. If the liquids do not mix, show that $\rho_2 = (h_1/h_2)\rho_1$.

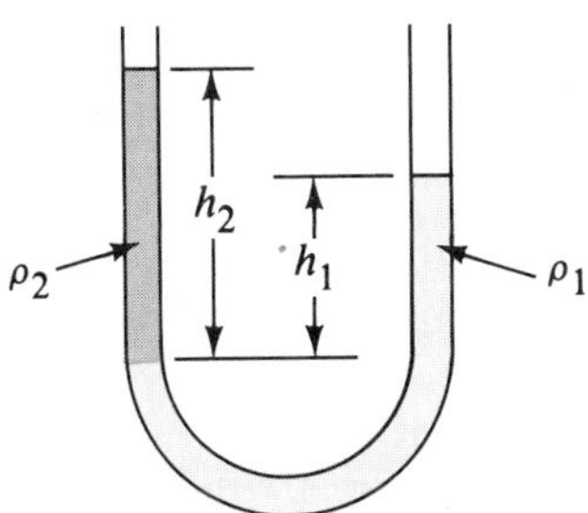

Figure 15.30 (Problem 40).

41. One method of measuring the density of a liquid is illustrated in Figure 15.31. One side of the U-shaped tube is in the liquid being tested; the other side is in water of density ρ_w. When the air is partially removed at the upper part of the tube, show that the density of the liquid on the left is given by $\rho = (h_w/h)\rho_w$.
42. A tank that has a flat bottom of area A and vertical sides is filled to a depth h in the water. There is 1 atm of pressure at the top surface. (a) What is the absolute pressure at the bottom of the tank? (b) Suppose that an object of total mass M (and average density less than the density of water) is placed into the tank and floats there. What is the resulting increase in the absolute pressure at the bottom of the tank? (c) Evaluate your results for a backyard swimming pool ($h = 1.50$ m; a circular tank with 6-m diameter). If two persons with a combined total mass of 150 kg get into the pool and float quietly there, what is the pressure increase at the bottom of the pool?
43. Consider a windmill with blades of cross-sectional area A, as in Figure 15.32, and assume the mill is facing directly into the wind. (a) If the wind speed is v, show that the kinetic energy of the air that passes through the blades in a time Δt is given by the expression $K = \frac{1}{2}\rho A v^3 \Delta t$. (b) What is the maximum available power according to this model? Compare your result with Equation 15.19.
44. A girl weighing 100 lb sits on a 1 m $\times$ 1 m $\times$ 0.06 m raft made of solid Styrofoam. If the raft *just* supports the girl (that is, the raft is totally submerged), determine the density of the Styrofoam.
45. A wire of length L, Young's modulus Y, and cross-sectional area A is stretched elastically by an amount ΔL. By Hooke's law, the restoring force is given by $-k\,\Delta L$. (a) Show that the constant k is given by $k = YA/L$. (b) Show that the work done in stretching the wire by an amount ΔL is given by

$$\text{Work} = \frac{1}{2}\frac{YA}{L}(\Delta L)^2.$$

46. The true weight of a body is its weight when measured in a vacuum where there are no buoyant forces. A body of volume V is weighed in air on a balance using weights of density ρ. If the density of air is ρ_a and the balance reads W', show that the true weight W is given by

$$W = W' + \left(V - \frac{W'}{\rho g}\right)\rho_a g$$

47. A block of cross-sectional area A, height ℓ, and density ρ is in equilibrium between two fluids of densities ρ_1 and ρ_2 (Fig. 15.33), where $\rho_1 < \rho < \rho_2$. The fluids do not mix. (a) Show that the buoyant force on the block is given by $B = [\rho_1 g y + \rho_2 g(\ell - y)]A$. (b) Show that the density of the block is equal to $\rho = [\rho_1 y + \rho_2(\ell - y)]/\ell$.

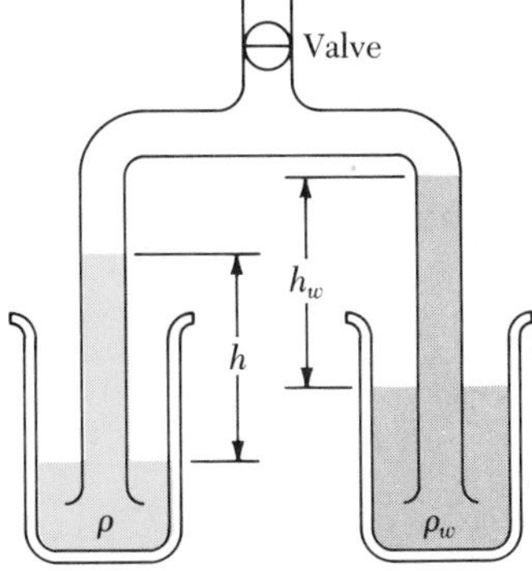

Figure 15.31 (Problem 41).

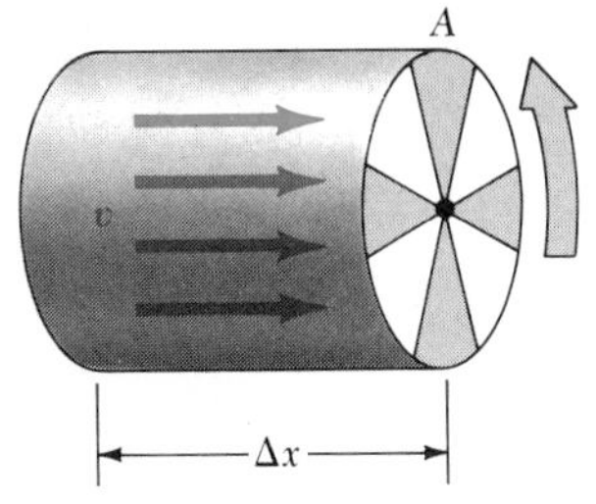

Figure 15.32 (Problem 43).

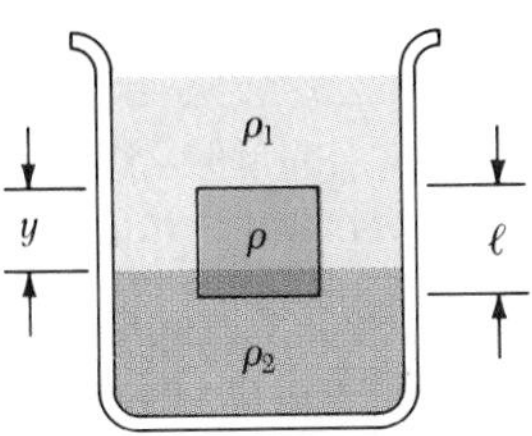

Figure 15.33 (Problem 47).

•**48.** A cube of ice whose edge is 20 mm is floating in a glass of ice-cold water with one of its faces parallel to the water surface. (a) How far below the water surface is the bottom face of the block? (b) Ice-cold ethyl alcohol is quietly poured onto the water surface to form a layer 5 mm thick above the water. When the ice cube attains hydrostatic equilibrium again, what will be the distance from the top of the *water* to the bottom face of the block? (c) Additional cold ethyl alcohol is poured onto the water surface until the top surface of the alcohol coincides with the top surface of the ice cube (in hydrostatic equilibrium). How thick is the required layer of ethyl alcohol?

49. A tank *open to the atmosphere* is filled with liquid, and a leak develops at a distance h below the surface of the liquid (Fig. 15.21). (a) Show that the liquid strikes the floor at a distance $x = 2\sqrt{hy_1}$ from the bottom of the tank. (b) Show that the horizontal distance x is a maximum when the hole is located at $y_1 = h$.

50. As a first approximation, the earth's continents may be thought of as granite blocks floating in a denser rock (called peridotite) in the same way that ice floats in water. (a) Show that a formula describing this phenomenon is

$$\rho_g t = \rho_p d$$

where ρ_g is the density of granite (2800 kg/m^3), ρ_p is the density of peridotite (3300 kg/m^3), t is the thickness of a continent, and d is the depth to which a continent floats in the peridotite. (b) If a continent rises 5 km above the surface of the peridotite (this surface may be thought of as the ocean floor), what is its thickness?

•**51.** A water tank has a conical top surface which slopes upward at an angle α (to the horizontal). The tank is full, but there is a small hole at the apex of the cone, so that the pressure there is 1 atm. If a small hole is opened in the tank wall at a distance s from the apex, the resulting water stream falls back onto the sloping surface a distance s' down from the leak. (The total distance from the apex to the splash point is thus $s + s'$.) (a) Find s' in terms of s and α. (You may assume that the initial velocity vector of the stream is normal to the tank surface.) (b) Evaluate the ratio s'/s for $\alpha = 45°$. (c) Find the value of α for which $s' = s$. (Give your answer in both radians and in degrees.)

52. (a) Find the diameter of the largest helium-filled balloon that can be held down using string that snaps when the tension exceeds 40 lb (178 newtons). Use 1.29 kg/m^3 for the density of air and 0.200 kg/m^3 for the density of helium. Assume that the balloon is spherical and ignore the mass of its skin. (*Note:* The density value given here for helium exceeds the value given in Table 15.2. The pressure within the helium balloon has been assumed to be about 10% above atmospheric pressure.) (b) What is the maximum diameter if the balloon is to be tethered under water?

53. Consider a composite "raft" consisting of two square slabs, each of side s, attached face to face. One slab has density ρ_1 and thickness h_1, while the other has density $\rho_2 > \rho_1$ and thickness h_2. (a) Find the average density $\bar{\rho}$ of the raft. (b) Assume that $\bar{\rho} < \rho_w$, so that the raft floats in water. The raft is placed in water with the denser slab on the bottom. Find d, the depth of the bottom surface of the raft. (c) If the raft is placed in water with the denser slab on the *top* find d', the depth of the bottom surface of the raft. Comment on your answer. (d) For which of the orientations described in (b) and (c) is the gravitational potential energy of the entire system (consisting of the raft and the body of water in which it is floating) greater? Find the potential energy difference.

54. A *siphon* is a device that allows a fluid to seemingly defy gravity (Fig. 15.34). The flow must be initiated by a partial vacuum in the tube, as in a drinking straw. (a) Show that the speed of efflux is given by $v = \sqrt{2gh}$. (b) For what values of y will the siphon work? (Incidentally, it has been told that gasoline tastes terrible!)

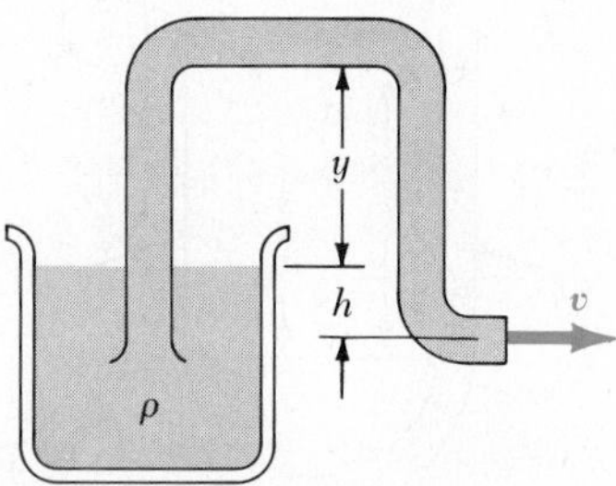

Figure 15.34 (Problem 54).

55. With reference to Figure 15.13, show that the total torque exerted by the water behind the dam about an axis through O is $\frac{1}{6}\rho g w H^3$. Show that the effective line of action of the total force exerted by the water is at a distance $\frac{1}{3}H$ above O.

56. In 1654 Otto von Guericke, inventor of the air pump, evacuated a sphere made of two brass hemispheres. Two teams of eight horses each *could not pull the hemispheres apart* (Fig. 15.35). (a) Show that the force F required to pull the evacuated hemispheres apart is $\pi R^2(P_a - P)$, where R is the radius of the hemispheres and P is the pressure inside the hemispheres, which is much less than P_a. (b) Determine the force if $P = 0.1P_a$ and $R = 0.3$ m.

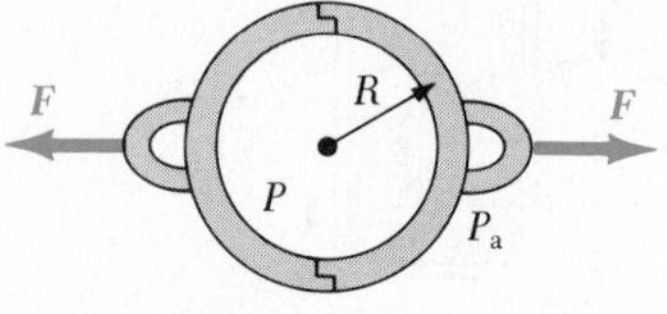

Figure 15.35 (Problem 56).

•**57.** A cable of mass density ρ_c and diameter d extends vertically downward a distance h through water, and a block of mass M_b and density ρ_b is hung from the bottom end of the cable. Both ρ_c and ρ_b exceed ρ_w, the density of water. Find (a) the tension T_ℓ at the lower end of the cable, (b) the tension T_u at the upper end of the cable, and (c) the tensions T_ℓ' and T_u' that would exist at the lower and upper ends of the cable if the entire assembly were in air rather than water. (Neglect the buoyant force provided by the air.) (d) Evaluate T_ℓ, T_u, T_ℓ', and T_u' for the case of a 100-meter steel cable supporting a prefabricated concrete object of mass 2.00 metric tons: $\rho_c = 7.86 \times 10^3$ kg/m^3, $d = 2 \times 10^{-2}$ m, $h = 100$ m, $M_b = 2.00 \times 10^3$ kg, and $\rho_b = 2.38 \times 10^3$ kg/m^3.

•**58.** Show that the variation of atmospheric pressure with altitude is given by $P = P_0 e^{-\alpha h}$, where $\alpha = \rho_0 g/P_0$, P_0 is atmospheric pressure at some reference level, and ρ_0 is the atmospheric density at this level. Assume that the decrease in atmospheric pressure with increasing altitude is given by Equation 15.9 and that the density of air is proportional to the pressure.

PART II
Vibrations and Wave Motion

The impetus is much quicker than the water, for it often happens that the wave flees the place of its creation, while the water does not; like the waves made in a field of grain by the wind, where we see the waves running across the field while the grain remains in place.

LEONARDO DA VINCI

As we look around us, we find many examples of objects that vibrate or oscillate: a pendulum, the strings of a guitar, an object suspended on a spring, the piston of an engine, the head of a drum, the reed of a saxophone. Most elastic objects will vibrate when an impulse is applied to them. That is, once they are distorted, their shape tends to be restored to some equilibrium configuration. Even at the atomic level, the atoms in a solid vibrate about some position as if they were connected to their neighbors by some imaginary springs.

Wave motion is closely related to the phenomenon of vibration. Sound waves, earthquake waves, waves on stretched strings, and water waves are all produced by some source of vibration. As a sound wave travels through some medium, such as air, the molecules of the medium vibrate back and forth; as a water wave travels across a pond, the water molecules vibrate up and down. As waves travel through a medium, the particles of the medium move in repetitive cycles. Therefore, the motion of the particles bears a strong resemblance to the periodic motion of a vibrating pendulum or a mass attached to a spring.

There are many other phenomena in nature whose explanation requires us to first understand the concepts of vibrations and waves. Although many large structures, such as skyscrapers and bridges, appear to be rigid, they actually vibrate, a fact that must be taken into account by the architects and engineers who design and build them. To understand how radio and television work, we must understand the origin and nature of electromagnetic waves and how they propagate through space. Finally, much of what scientists have learned about atomic structure has come from information carried by waves. Therefore, we must first study waves and vibrations in order to understand the concepts and theories of atomic physics.

16
Wave Motion

Most of us experienced waves as children when we dropped a pebble into a pond. The disturbance created by the pebble excites ripple waves, which move outward, finally reaching the shore of the pond. If you were to examine carefully the motion of a leaf floating near the disturbance, you would see that it moves up and down and sideways about its original position, but does not undergo any net displacement away or toward the source of the disturbance. That is, the water wave (or disturbance) moves from one place to another, *yet the water is not carried with it.*

An excerpt from a book by Einstein and Infeld gives the following remarks concerning wave phenomena.[1]

> A bit of gossip starting in Washington reaches New York very quickly, even though not a single individual who takes part in spreading it travels between these two cities. There are two quite different motions involved, that of the rumor, Washington to New York, and that of the persons who spread the rumor. The wind, passing over a field of grain, sets up a wave which spreads across the whole field. Here again we must distinguish between the motion of the wave and the motion of the separate plants, which undergo only small oscillations. . . . The particles constituting the medium perform only small vibrations, but the whole motion is that of a progressive wave. The essential new thing here is that for the first time we consider the motion of something which is not matter, but energy propagated through matter.

Water waves represent only one example of a wide variety of physical phenomena that have wavelike characteristics. The world is full of waves: sound waves; mechanical waves, such as a wave on a string; earthquake waves; shock waves generated by supersonic aircraft; and electromagnetic waves, such as visible light, radio waves, television signals, and x-rays. In the present chapter, we shall confine our attention to mechanical waves, that is, waves that travel only in a material substance.

The wave concept is rather abstract. When we observe what we call a water wave, what we see is a rearrangement of the water's surface. Without the water, there would be no wave. A wave traveling on a string would not exist without the string. Sound waves travel through air as a result of pressure variations from point to point. In such cases, what we interpret as a wave corresponds to the disturbance of a body or medium. Therefore, we can consider a wave to be the *motion of a disturbance.* The motion of the disturbance (that is, the wave itself, or the state of the medium) is not to be confused with the motion of the particles. The mathematics used to describe wave phenomena is common to all waves. In general, we shall find that mechanical

[1] Albert Einstein and Leopold Infeld, *The Evolution of Physics,* New York, Simon and Schuster, 1961. Excerpt from *What is a Wave?*

wave motion is described by specifying the positions of all points of the disturbed medium as a function of time.

16.1 INTRODUCTION

The mechanical waves discussed in this chapter require (1) some source of disturbance, (2) a medium that can be disturbed, and (3) some physical connection or mechanism through which adjacent portions of the medium can influence each other. We shall find that all waves carry energy and momentum. The amount of energy transmitted through a medium and the mechanism responsible for the transport of energy will differ from case to case. For instance, the power carried by ocean waves during a storm is much greater than the power of sound waves generated by a single human voice.

Three physical characteristics are important in characterizing waves: the wavelength, the frequency, and the wave velocity. One **wavelength** is the *distance between any two points on a wave that behave identically.* For example, in the case of water waves, the wavelength is the distance between adjacent crests or between adjacent troughs.

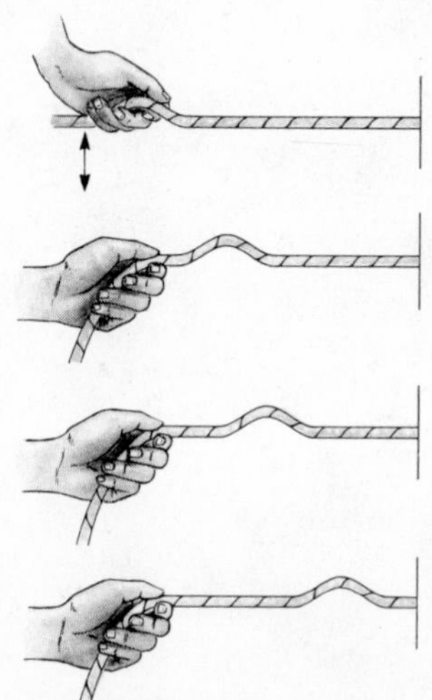

Figure 16.1 A wave pulse traveling down a stretched rope. The shape of the pulse is assumed to remain unchanged as it travels along the rope.

Most waves are periodic in nature. The **frequency** of such periodic waves is *the rate at which the disturbance repeats itself.*

Waves travel, or *propagate*, with a specific velocity, which depends on the properties of the medium being disturbed. For instance, sound waves travel through air at 20°C with a speed of about 344 m/s (781 mi/h), whereas the speed of sound through solids is higher than 344 m/s. A special class of waves that do not require a medium in order to propagate are electromagnetic waves, which travel very swiftly through a vacuum with a speed of about 3×10^8 m/s (186 000 mi/s). We shall discuss electromagnetic waves further in Chapter 34.

16.2 TYPES OF WAVES

One way to demonstrate wave motion is to flip one end of a long rope that is under tension and has its opposite end fixed, as in Figure 16.1. Only a portion of the wave is produced in this manner. It consists of a bump (called a pulse) in the rope that travels (to the right in Fig. 16.1) with a definite speed. This type of disturbance is called a **traveling wave.** Figure 16.1 represents four consecutive "snapshots" of the traveling wave. As we shall see later, the speed of the wave depends on the tension in the rope and on the properties of the rope. The rope is the *medium* through which the wave travels. We shall assume that the shape of the wave pulse does not change as it travels along the rope.[2]

Note that, as the wave pulse travels along the rope, *each segment of the rope that is disturbed moves in a direction perpendicular to the wave motion.* Figure 16.2 illustrates this point for one particular segment, labeled P. Note that there is no motion of any part of the rope in the direction of the wave.

> A traveling wave such as this, in which the particles of the disturbed medium move perpendicular to the wave velocity, is called a **transverse wave.**[3]

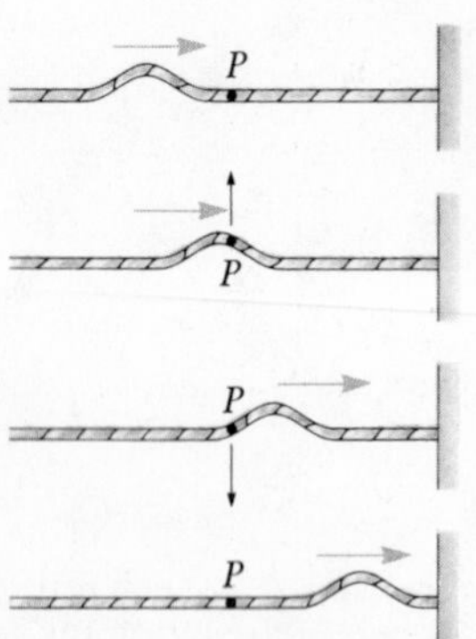

Figure 16.2 A pulse traveling on a stretched rope is a transverse wave. That is, any element P on the rope moves in a direction *perpendicular* to the wave motion.

[2] Strictly speaking, the pulse will change its shape and gradually spread out during the motion. This effect is called *dispersion* and is common to many mechanical waves.

[3] Other examples of transverse waves are electromagnetic waves, such as light, radio, and television waves. At a given point in space, the electric and magnetic fields of an electromagnetic wave are perpendicular to the direction of the wave and to each other, and vary in time as the wave passes. As we shall see later, electromagnetic waves are produced by accelerating charges.

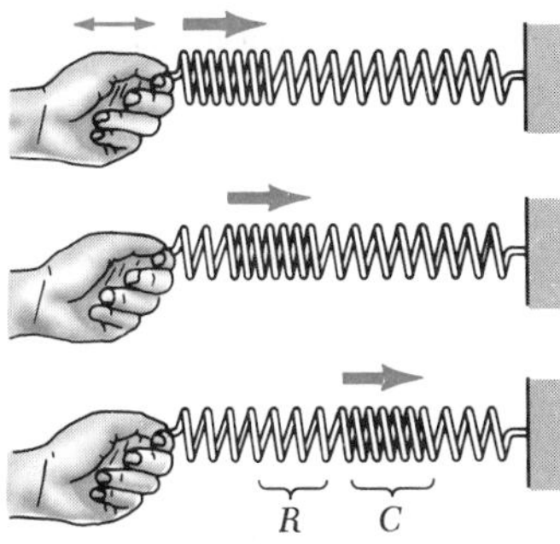

Figure 16.3 A longitudinal pulse along a stretched spring. The disturbance of the medium (the displacement of the coils) is in the direction of the wave motion. For the starting motion described in the text, the compressed region C is followed by an extended region R.

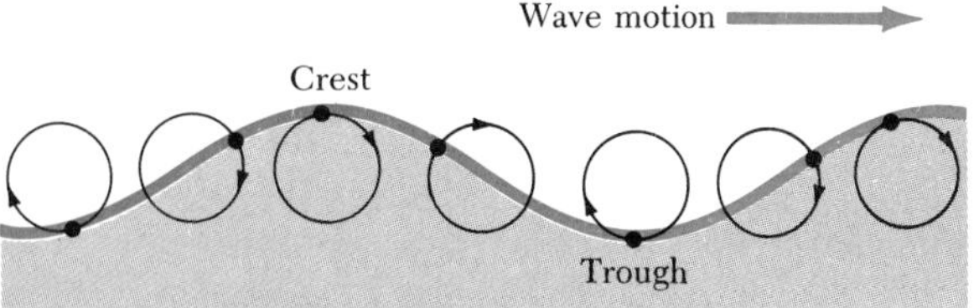

Figure 16.4 Wave motion on the surface of water. The particles at the water's surface move in nearly circular paths. Each particle is displaced horizontally and vertically from its equilibrium position, represented by circles.

In another class of waves, called **longitudinal waves,** the particles of the medium undergo displacements in a direction *parallel* to the direction of wave motion.

Sound waves, which we shall discuss in Chapter 17, are longitudinal waves that result from the disturbance of the medium. The disturbance corresponds to a series of high- and low-pressure regions that travel through air or through any material medium with a certain velocity. A longitudinal pulse can be easily produced in a stretched spring, as in Figure 16.3. The left end of the spring is given a sudden jerk (consisting of a brief push to the right and equally brief pull to the left) along the length of the spring; this creates a sudden compression of the coils. The compressed region C (pulse) travels along the spring, and so we see that the disturbance is parallel to the wave motion. Region C is followed by a region R, where the coils are extended.[4]

Some waves in nature are neither transverse nor longitudinal, but a combination of the two. Surface water waves are a good example. When a water wave travels on the surface of deep water, water molecules at the surface move in nearly circular paths, as shown in Figure 16.4, where the water surface is drawn as a series of crests and troughs. Note that the disturbance has both transverse and longitudinal components. As the wave passes, water molecules at the crests move in the direction of the wave, and molecules at the troughs move in the opposite direction. Hence, there is no *net* displacement of a water molecule after the passage of any number of complete wavelengths.

16.3 ONE-DIMENSIONAL TRAVELING WAVES

So far we have given only a verbal and graphical description of a traveling wave. Let us now give a mathematical description of a one-dimensional traveling wave. Consider again a wave pulse traveling to the right on a long stretched string with constant speed v, as in Figure 16.5. The pulse moves along the x axis (the axis of the string), and the transverse displacement of the string is measured with the coordinate y.

Figure 16.5a represents the shape and position of the pulse at time $t = 0$. At this time, the shape of the pulse, whatever it may be, can be represented as

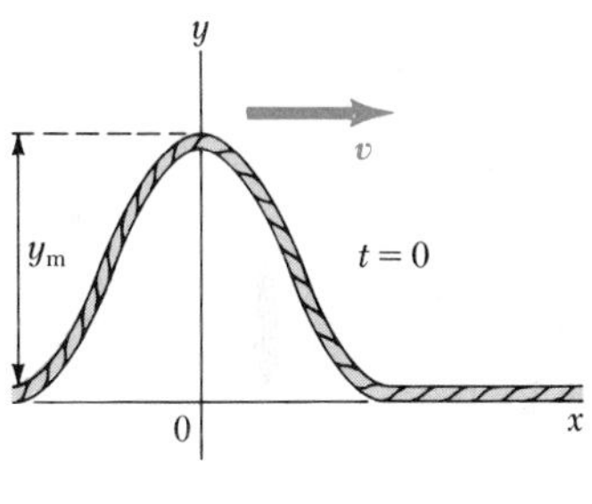

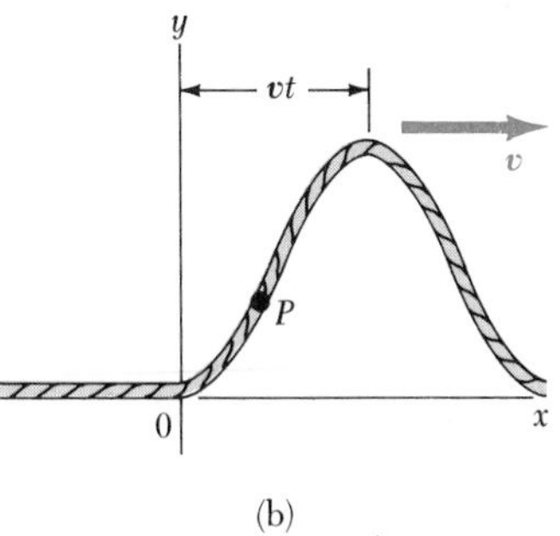

Figure 16.5 A one-dimensional wave pulse traveling to the right with a speed v. (a) At $t = 0$, the shape of the pulse is given by $y = f(x)$. (b) At some later time t, the shape remains unchanged and the vertical displacement is given by $y = f(x - vt)$.

[4] In the case of longitudinal pressure waves in a gas, each compressed area is a region of higher-than-average pressure and density, and each extended region is a region of lower-than-average pressure and density.

$y = f(x)$. That is, y is some definite function of x. The *maximum displacement*, y_m, is called the **amplitude** of the wave. Since the speed of the wave pulse is v, it travels to the right a distance vt in a time t (Fig. 16.5b).

If the shape of the wave pulse doesn't change with time, we can represent the displacement y for all later times measured in a stationary frame with the origin at 0 as

Wave traveling to the right

$$y = f(x - vt) \tag{16.1}$$

Similarly, if the wave pulse travels to the *left*, its displacement is given by

Wave traveling to the left

$$y = f(x + vt) \tag{16.2}$$

The displacement y, sometimes called the *wave function*, depends on the two variables x and t. For this reason, it is often written $y(x, t)$, which is read "y as a function of x and t." It is important to understand the meaning of y.

Consider a particular point P on the string, identified by a particular value of its coordinates. As the wave passes the point P, the y coordinate of this point will increase, reach a maximum, and then decrease to zero. Therefore, the **wave function** *y represents the y coordinate of any point P at any time t.* Furthermore, if t is fixed, then the wave function y as a function of x *defines a curve representing the actual shape of the pulse at this time.* This is equivalent to a "snapshot" of the wave at this time.

For a pulse that moves without changing its shape, the velocity of a wave pulse is the same as the motion of any feature along the pulse profile, such as the crest. To find the velocity of the pulse, we can calculate how far the crest moves in a short time and then divide this distance by the time interval. The crest of the pulse corresponds to that point for which y has its maximum value. In order to follow the motion of the crest, some particular value, say x_0, must be substituted for $x - vt$. (This value x_0 is called the *argument* of the function y.) Regardless of how x and t change individually, we must require that $x - vt = x_0$ in order to stay with the crest. This, therefore, represents the equation of motion of the crest. At $t = 0$, the crest is at $x = x_0$; at a time dt later, the crest is at $x = x_0 + v\,dt$. Therefore, the crest has moved a distance $dx = (x_0 + v\,dt) - x_0 = v\,dt$ in a time dt. Clearly, the wave speed, often called the **phase velocity,** is given by

Phase velocity

$$v = dx/dt \tag{16.3}$$

The wave speed, or phase velocity, must not be confused with the transverse velocity (which is in the y direction) of a particle in the medium.

The following example illustrates how a specific wave function is used to describe the motion of a traveling wave pulse.

EXAMPLE 16.1. A Pulse Moving to the Right
A traveling wave pulse moving to the right along the x axis is represented by the wave function

$$y(x, t) = \frac{2}{(x - 3t)^2 + 1}$$

where x and y are measured in cm and t is in s. Let us plot the waveform at $t = 0$, $t = 1$ s, and $t = 2$ s.

Solution: First, note that this function is of the form $y = f(x - vt)$. By inspection, we see that the speed of the wave is $v = 3$ cm/s. Furthermore, the wave amplitude (the maximum value of y) is given by $y_m = 2$ cm. At times $t = 0$, $t = 1$ s, and $t = 2$ s, the wave function expressions are

$$y(x, 0) = \frac{2}{x^2 + 1} \qquad \text{at } t = 0$$

$$y(x, 1) = \frac{2}{(x-3)^2 + 1} \qquad \text{at } t = 1 \text{ s}$$

$$y(x, 2) = \frac{2}{(x-6)^2 + 1} \qquad \text{at } t = 2 \text{ s}$$

We can now use these expressions to plot the wave function versus x at these times. For example, let us evaluate $y(x, 0)$ at $x = 0.5$ cm:

$$y(0.5, 0) = \frac{2}{(0.5)^2 + 1} = 1.60 \text{ cm}$$

Likewise, $y(1, 0) = 1.0$ cm, $y(2, 0) = 0.40$ cm, etc. A continuation of this procedure for other values of x yields the waveform shown in Figure 16.6a. In a similar manner, one obtains the graphs of $y(x, 1)$ and $y(x, 2)$, shown in Figures 16.6b and 16.6c, respectively. These snapshots show that the wave pulse moves to the right without changing its shape and has a constant speed of 3 cm/s.

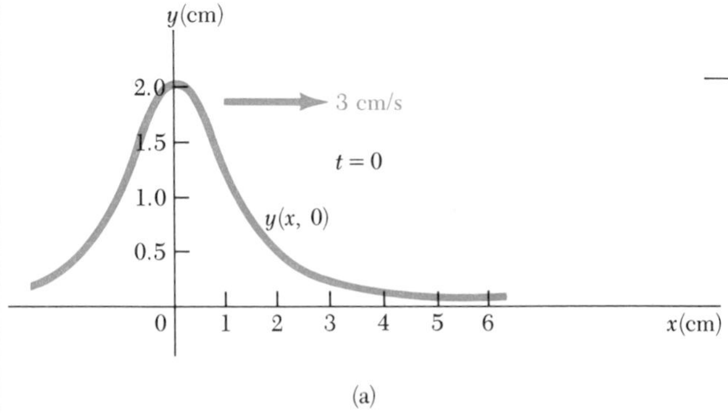

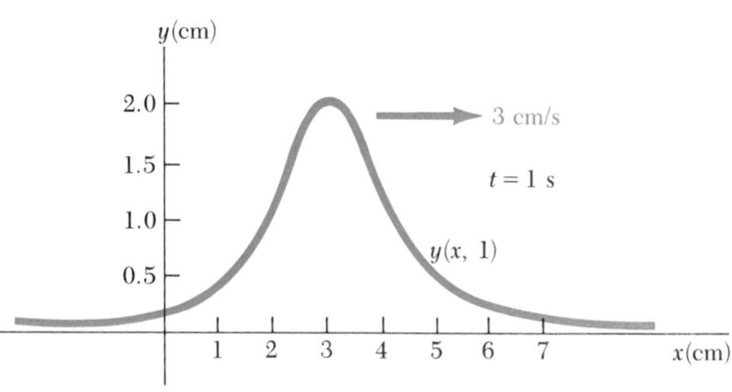

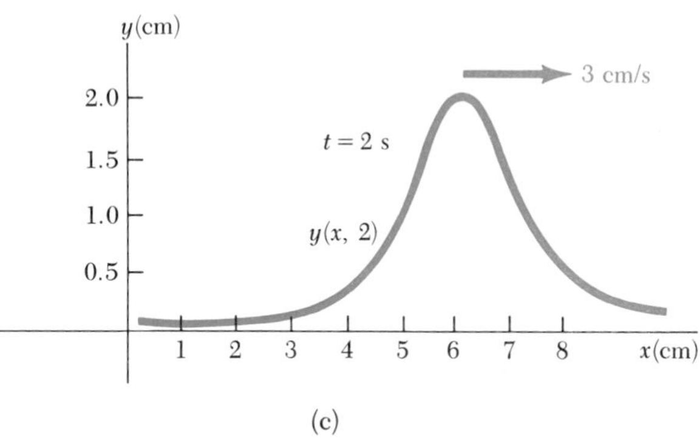

Figure 16.6 (Example 16.1) Graphs of the function $y(x, t) = 2/[(x-3t)^2 + 1]$. (a) $t = 0$, (b) $t = 1$ s, and (c) $t = 2$ s.

16.4 SUPERPOSITION AND INTERFERENCE OF WAVES

Many interesting wave phenomena in nature cannot be described by a single moving pulse. Instead, one must analyze complex waveforms in terms of a combination of many traveling waves. To analyze such wave combinations, one can make use of the **superposition principle:**

> If two or more traveling waves are moving through a medium, the resultant wave function at any point is the algebraic sum of the wave functions of the individual waves.

Linear waves obey the superposition principle

This rather striking property is exhibited by many waves in nature. Waves that obey this principle are called *linear waves*, and they are generally characterized by small wave amplitudes. Waves that violate the superposition principle are called *nonlinear waves* and are often characterized by large amplitudes. In this book, we shall deal only with linear waves.

One consequence of the superposition principle is the observation that *two traveling waves can pass through each other without being destroyed or even altered.* For instance, when two pebbles are thrown into a pond, the expanding circular surface waves do not destroy each other. In fact, the ripples pass through each other. The complex pattern that is observed can be viewed

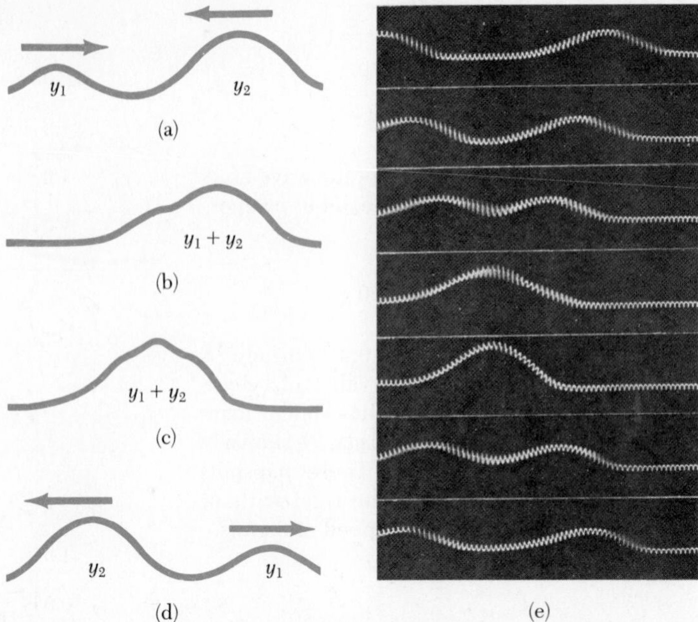

Figure 16.7 (Left) Two wave pulses traveling on a stretched string in opposite directions pass through each other. When the pulses overlap, as in (b) and (c), the net displacement of the string equals the sum of the displacements of each pulse. Since the pulses both have positive displacements, we refer to their superposition as *constructive interference*. (Right) Photograph of superposition of two equal and symmetric pulses traveling in opposite directions on a stretched string. (Photo, Education Development Center, Newton, Mass.)

as two independent sets of expanding circles. Likewise, when sound waves from two sources move through air, they also can pass through each other. The resulting sound one hears at a given point is the resultant of both disturbances.

A simple pictorial representation of the superposition principle is obtained by considering two pulses traveling in opposite directions on a stretched string, as in Figure 16.7. The wave function for the pulse moving to the right is y_1, and the wave function for the pulse moving to the left is y_2. The pulses have the same speed, but different shapes. Each pulse is assumed to be symmetric, and both displacements are taken to be positive. When the waves begin to overlap (Fig. 16.7b), the resulting complex waveform is given by $y_1 + y_2$. When the crests of the pulses exactly coincide (Fig. 16.7c), the resulting waveform $y_1 + y_2$ is symmetric. The two pulses finally separate and continue moving in their original directions (Fig. 16.7d). Note that the final waveforms remain unchanged, as if the two pulses never met! The combination of separate waves in the same region of space to produce a resultant wave is called *interference*. For the two pulses shown in Figure 16.7, the displacements of the individual pulses are in the same direction, and the resultant waveform (when the pulses overlap) exhibits a displacement greater than those of the individual pulses. This type of interference is called **constructive interference.**

Now consider two identical pulses traveling in opposite directions on an infinitely long string, where one is inverted relative to the other, as in Figure 16.8. In this case, when the pulses begin to overlap, the resultant waveform is the *arithmetic difference* between the two separate displacements. Again, the

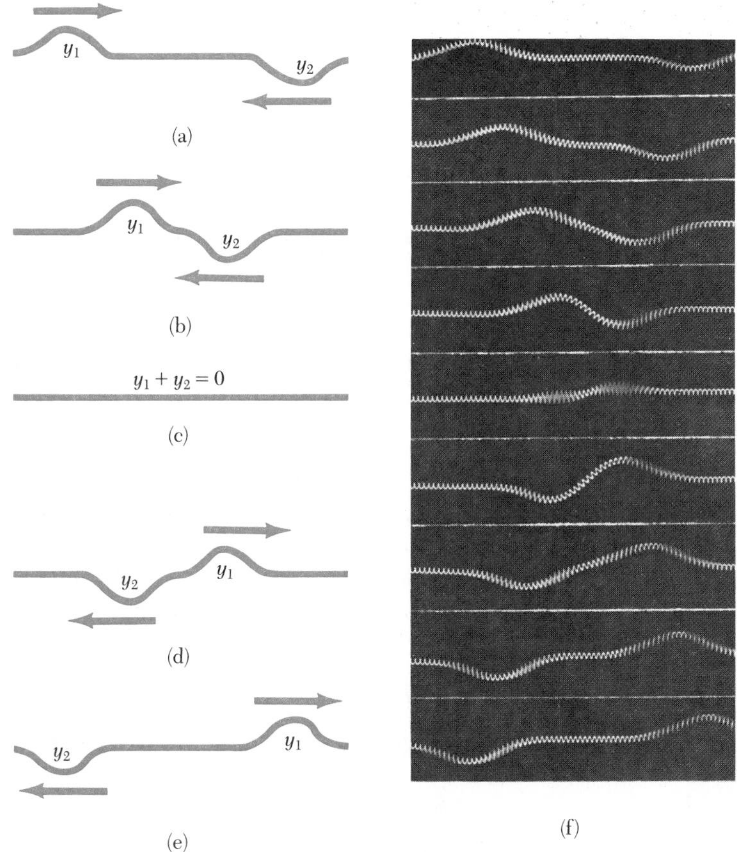

Figure 16.8 (Left) Two wave pulses traveling in opposite directions with equal but opposite displacements. When the two overlap, their displacements subtract from each other, corresponding to *destructive interference.* Note that in (c), the displacement is zero for all values of x. (Right) Photograph of superposition of two symmetric pulses traveling in opposite directions, where one is inverted relative to the other. (Photo, Education Development Center, Newton, Mass.)

two pulses pass through each other as indicated. When the two pulses exactly overlap, they *cancel* each other (assuming the upper positive displacement of the pulse y_1 is equal in magnitude to that of the inverted pulse y_2). At this time, the string is horizontal and the energy associated with the disturbance is contained in the kinetic energy of the string, where the string segments move *vertically.* That is, when the two pulses exactly overlap, the segments of the string on either side of the crossover point are moving vertically, but in opposite directions. When traveling waves cancel each other in this manner, the phenomenon is called **destructive interference.**

16.5 THE VELOCITY OF WAVES ON STRINGS

For linear waves, *the velocity of mechanical waves depends only on the properties of the medium through which the disturbance travels.* In this section, we shall focus our attention on determining the speed of a transverse pulse traveling on a stretched string. If the *tension* in the string is F and its *mass per unit length* is μ, then the wave speed v is given by

Speed of a wave on a stretched string

$$v = \sqrt{F/\mu} \tag{16.4}$$

First, we verify that this expression is dimensionally correct. The dimensions of F are MLT^{-2}, and the dimensions of μ are ML^{-1}. Therefore, the dimensions of F/μ are L^2/T^2; hence the dimensions of $\sqrt{F/\mu}$ are L/T, which are indeed the dimensions of velocity. No other combination of F and μ is dimensionally correct, assuming they are the only variables relevant to the situation.

Now let us use a mechanical analysis to derive the above expression for the speed of a pulse traveling on a stretched string. Consider a pulse moving to the right with a uniform speed v, measured relative to a stationary frame of reference. It is more convenient to choose as our reference frame one that moves along with the pulse with the same speed, so that the pulse appears to be at rest in this frame, as in Figure 16.9a. This is permitted since Newton's laws are valid in either a stationary frame or one that moves with constant velocity. A *small* segment of the string of length Δs forms the arc of a circle of radius R, as shown in Figure 16.9a and magnified in Figure 16.9b. This small segment has a centripetal acceleration equal to v^2/R, which is supplied by the force of tension F in the string. The force F acts on each side of the segment, tangent to the arc, as in Figure 16.9b. The horizontal components of F cancel, and each vertical component $F \sin\theta$ acts radially inward toward the center of the arc. Hence, the total radial force is $2F\sin\theta$. Since the segment is small, θ is small and we can use the small-angle approximation $\sin\theta \approx \theta$. Therefore, the total radial force can be expressed as

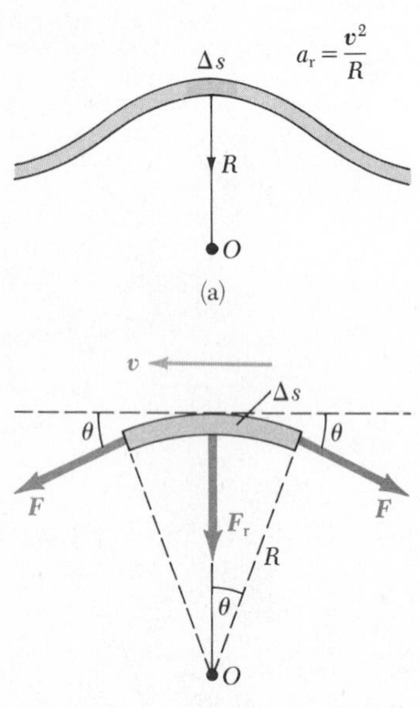

Figure 16.9 (a) To obtain the speed v of a wave on a stretched string, it is convenient to describe the motion of a small segment of the string in a moving frame of reference. (b) The net force on a small segment of length Δs is in the radial direction. The horizontal components of the tension force cancel.

$$F_r = 2F\sin\theta \approx 2F\theta$$

The small segment has a mass given by $m = \mu\Delta s$, where μ is the mass per unit length of the string. Since the segment forms part of a circle and subtends an angle 2θ at the center, $\Delta s = R(2\theta)$, and hence

$$m = \mu\,\Delta s = 2\mu R\theta$$

If we apply Newton's second law to this segment, the radial component of motion gives

$$F_r = mv^2/R \qquad \text{or} \qquad 2F\theta = 2\mu R\theta v^2/R$$

where F_r is the force which supplies the centripetal acceleration of the segment and maintains the curvature at this point.

Solving for v gives

$$v = \sqrt{F/\mu}$$

Notice that this derivation is based on the assumption that the pulse height is small relative to the length of the string. Using this assumption, we were able to use the approximation that $\sin\theta \approx \theta$. Furthermore, the model assumes that the tension F is not affected by the presence of the pulse, so that F is the same at all points on the string. Finally, note that this proof does *not* assume any particular shape for the pulse. Therefore, we conclude that a pulse of *any shape* will travel on the string with speed $v = \sqrt{F/\mu}$ without changing its shape.

EXAMPLE 16.2. The Speed of a Pulse on a Cord
A uniform cord has a mass of 0.3 kg and a length of 6 m. Tension is maintained in the cord by suspending a 2-kg mass from one end (Fig. 16.10). Find the speed of a pulse on this cord.

The tension F in the cord is equal to the weight of the suspended 2-kg mass multiplied by the gravitational acceleration:

$$F = mg = (2\text{ kg})(9.80\text{ m/s}^2) = 19.6\text{ N}$$

(This calculation of the tension neglects the small mass of the cord. Strictly speaking, the cord can never be exactly horizontal, and therefore the tension is not uniform.)

The mass per unit length μ is

$$\mu = \frac{m}{\ell} = \frac{0.3\text{ kg}}{6\text{ m}} = 0.05\text{ kg/m}$$

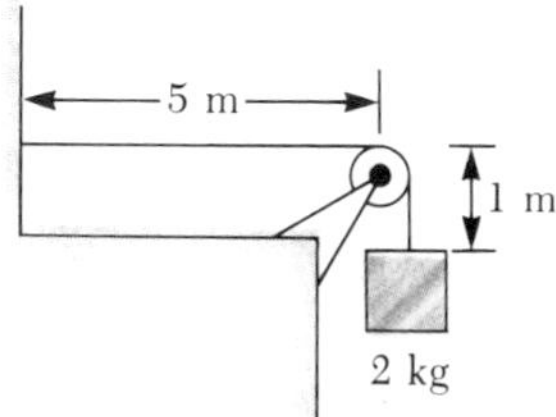

Figure 16.10 (Example 16.2) The tension F in the cord is maintained by the suspended mass. The wave speed is calculated using the expression $v = \sqrt{F/\mu}$.

Therefore, the wave speed is

$$v = \sqrt{F/\mu} = \sqrt{19.6\text{ N}/0.05\text{ kg/m}} = 19.8\text{ m/s}$$

Exercise 1 Find the time it takes the pulse to travel from the wall to the pulley.
Answer: 0.253 s.

16.6 REFLECTION AND TRANSMISSION OF WAVES

Whenever a traveling wave reaches a boundary, part or all of the wave will be reflected. For example, consider a pulse traveling on a string fixed at one end (Fig. 16.11). When the pulse reaches the fixed wall, it will be reflected. Since the support attaching the string to the wall is assumed to be rigid, it does not transmit any part of the disturbance to the wall.

Note that the reflected pulse is inverted. This can be explained as follows. When the pulse meets the end of the string that is fixed at the support, the string produces an upward force on the support. By Newton's third law, the support must then exert an equal and opposite (downward) reaction force on the string. This downward force causes the pulse to invert upon reflection.

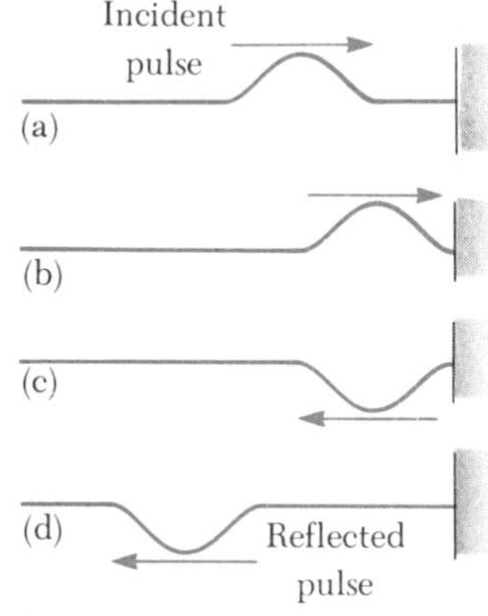

Figure 16.11 The reflection of a traveling wave at the fixed end of a stretched string. Note that the reflected pulse is inverted, but its shape remains the same.

Now consider another case where the pulse arrives at the end of a string that is free to move vertically, as in Figure 16.12. The tension at the free end is maintained by tying the string to a ring of negligible mass that is free to slide vertically on a smooth post. Again, the pulse will be reflected, but this time its displacement is not inverted. As the pulse reaches the post, it exerts a force on the free end, causing the ring to accelerate upward. In the process, the ring "overshoots" the height of the incoming pulse and is then returned to its original position by the downward component of the tension. This produces a reflected pulse that is not inverted, whose amplitude is the same as that of the incoming pulse.

Finally, we may have a situation in which the boundary is intermediate between these two extreme cases, that is, one in which the boundary is neither rigid nor free. In this case, part of the incident energy is transmitted and part is reflected. For instance, suppose a light string is attached to a heavier string as in Figure 16.13. When a pulse traveling on the light string reaches the knot,

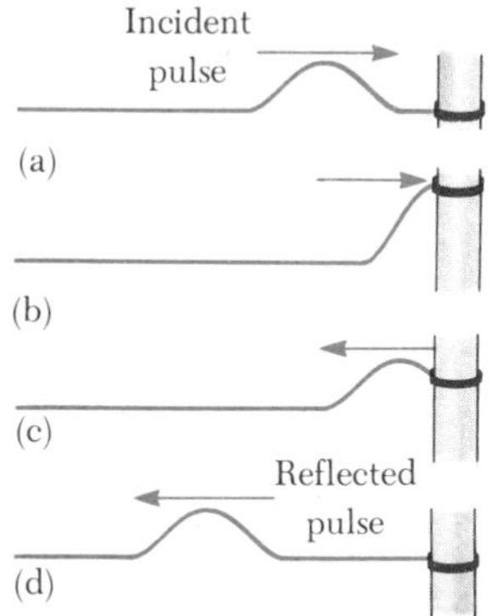

Figure 16.12 The reflection of a traveling wave at the free end of a stretched string. In this case, the reflected pulse is not inverted.

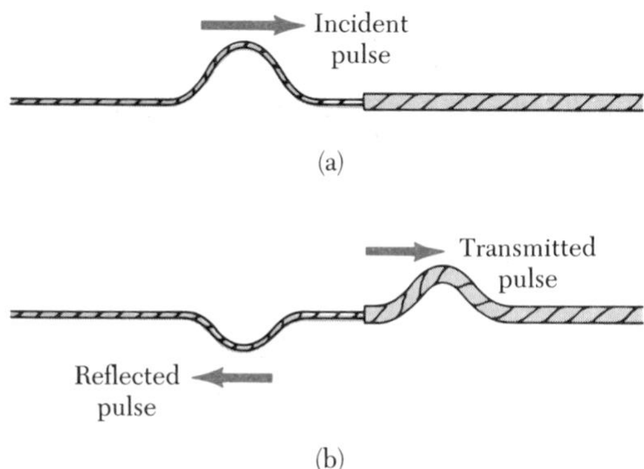

Figure 16.13 (a) A pulse traveling to the right on a light string tied to a heavier string. (b) Part of the incident pulse is reflected (and inverted), and part is transmitted to the heavier string.

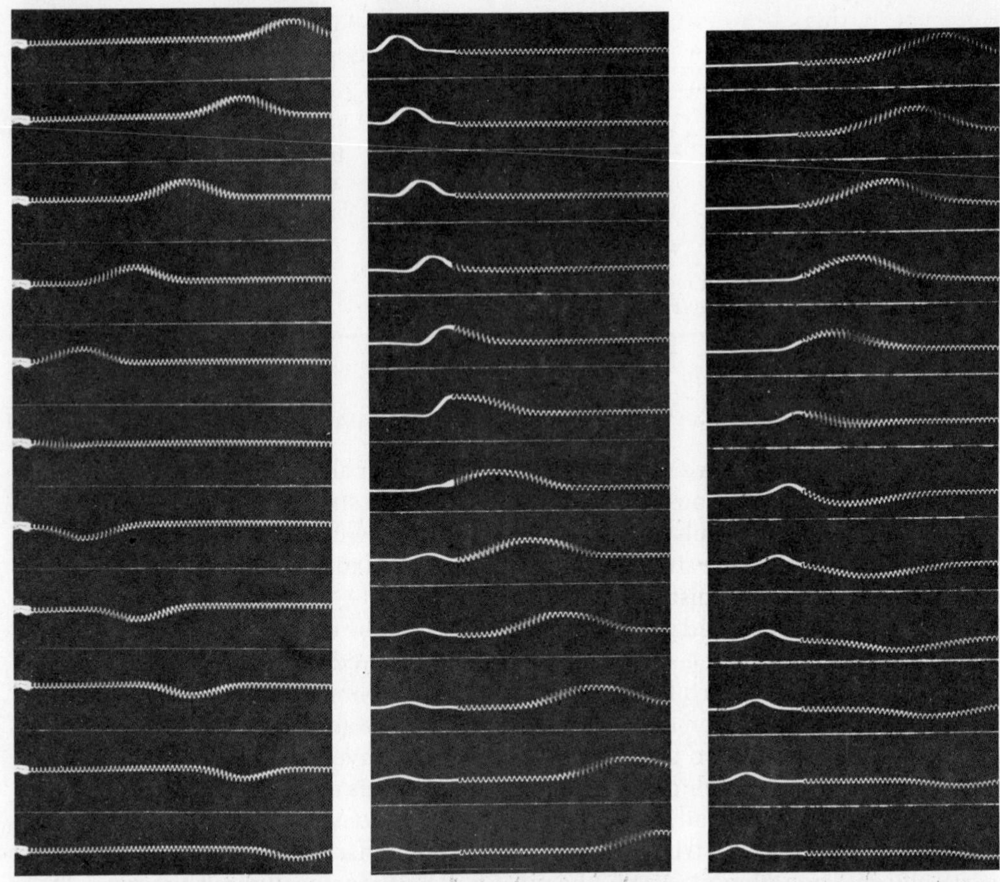

Photographs showing: (Left) Reflection of a pulse from a fixed end. The reflected pulse is inverted. (Center) A pulse passing from a heavy spring to a light spring. At the junction the pulse is partially transmitted and partially reflected. The reflected pulse is not inverted. (Right) A pulse passing from a light spring to a heavy spring. At the junction the pulse is partially transmitted and partially reflected. Note that the reflected pulse is inverted. (Photos, Education Development Center, Newton, Mass.)

part of it is reflected and inverted and part of it is transmitted to the heavier string. As one would expect, the reflected pulse has a smaller amplitude than the incident pulse, since part of the incident energy is transferred to the pulse in the heavier string. The inversion in the reflected wave is similar to the behavior of a pulse meeting a rigid boundary, where it is totally reflected.

When a pulse traveling on a heavy string strikes the boundary of a lighter string, as in Figure 16.14, again part is reflected and part is transmitted.

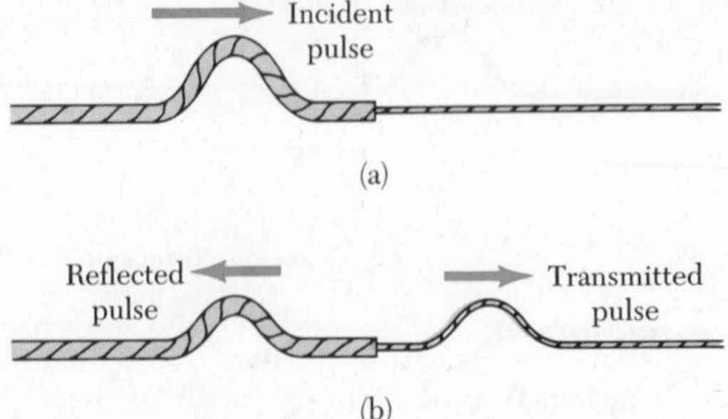

Figure 16.14 (a) A pulse traveling to the right on a heavy string tied to a lighter string. (b) The incident pulse is partially reflected and partially transmitted. In this case, the reflected pulse is not inverted.

However, in this case, the reflected pulse is not inverted. In either case, the relative heights of the reflected and transmitted pulses depend on the relative densities of the two strings.

If the strings are identical, there is no discontinuity at the boundary, and hence no reflection takes place.

In the previous section, we found that the speed of a wave on a string increases as the density of the string decreases. A pulse travels more slowly on a heavy string than on a light string if both are under the same tension. The following general rules apply to reflected waves: *When a wave pulse travels from medium A to medium B and* $v_A > v_B$ *(that is, when B is denser than A), the pulse will be inverted upon reflection. When a wave pulse travels from medium A to medium B and* $v_A < v_B$ *(A is denser than B), it will not be inverted upon reflection.* Similar rules apply to other kinds of waves.

16.7 HARMONIC WAVES

In this section, we introduce an important waveform known as a **harmonic wave.** The shape of a harmonic wave is a sinusoidal curve, as shown in Figure 16.15. The solid curve represents a snapshot of the traveling harmonic wave at $t = 0$, and the dashed curve represents a snapshot of the wave at some later time t. At $t = 0$, the displacement of the curve can be written

$$y = A \sin\left(\frac{2\pi}{\lambda} x\right) \tag{16.5}$$

The constant A, called the **amplitude** of the wave, represents the *maximum* value of the displacement. The constant λ, called the **wavelength** of the wave, equals the distance between two successive maxima, which we shall refer to as *crests*, or between any two adjacent points that have the same phase. Thus, we see that the displacement repeats itself when x is increased by any integral multiple of λ. If the wave moves to the right with a phase velocity v, the wave function at some later time t is given by

$$y = A \sin\left[\frac{2\pi}{\lambda}(x - vt)\right] \tag{16.6}$$

That is, the harmonic wave moves to the right a distance vt in the time t, as in Figure 16.15. Note that the wave function has the form $f(x - vt)$ and represents a wave traveling to the right. If the wave were traveling to the left, the quantity $x - vt$ would be replaced by $x + vt$.

The time it takes the wave to travel a distance of one wavelength is called the **period,** T. Therefore, the phase velocity, wavelength, and period are related by

$$v = \lambda/T \qquad \text{or} \qquad \lambda = vT \tag{16.7}$$

Substituting this into Equation 16.6, we find that

$$y = A \sin\left[2\pi\left(\frac{x}{\lambda} - \frac{t}{T}\right)\right] \tag{16.8}$$

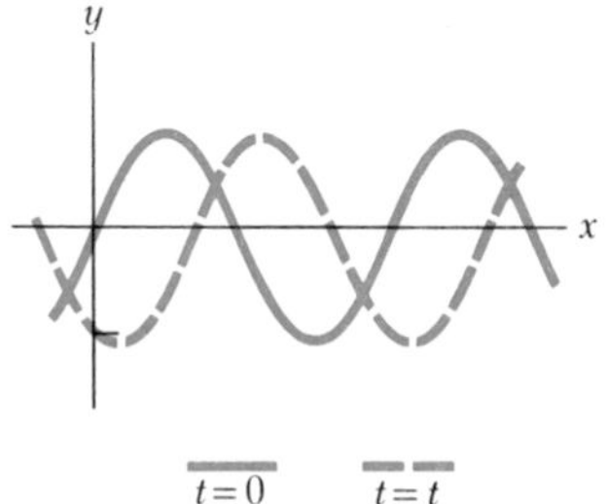

Figure 16.15 A one-dimensional harmonic wave traveling to the right with a speed v. The solid curve represents a snapshot of the wave at $t = 0$, and the dashed curve is a snapshot at some later time t.

This form of the wave function clearly shows the *periodic* nature of y. That is, at any given time t (a snapshot of the wave), y has the *same* value at the positions $x, x + \lambda, x + 2\lambda$, etc. Furthermore, at any given position x, y has the *same* value at times $t, t + T, t + 2T$, etc.

We can express the harmonic wave function in a convenient form by defining two other quantities, called the **wave number** k and the **angular frequency** ω:

Wave number

$$k = 2\pi/\lambda \tag{16.9}$$

Angular frequency

$$\omega = 2\pi/T \tag{16.10}$$

Using these definitions, we see that Equation 16.8 can be written in the more compact form

Wave function for a harmonic wave

$$y = A\sin(kx - \omega t) \tag{16.11}$$

We shall use this form most frequently.

The **frequency** of a harmonic wave equals the number of times a crest (or any other point on the wave) passes a *fixed* point each second. The frequency is related to the period by the relationship

Frequency

$$f = 1/T \tag{16.12}$$

The most common unit for f is s^{-1}, or hertz (Hz). The corresponding unit for T is s.

Using Equations 16.9, 16.10, and 16.12, we can express the phase velocity v in the alternative forms

Velocity of a harmonic wave

$$v = \frac{\omega}{k} \tag{16.13}$$

$$v = \lambda f \tag{16.14}$$

The wave function given by Equation 16.11 assumes that the displacement y is zero at $x = 0$ and $t = 0$. This need not be the case. If the transverse displacement is not zero at $x = 0$ and $t = 0$, we generally express the wave function in the form

General relation for a harmonic wave

$$y = A\sin(kx - \omega t - \phi) \tag{16.15}$$

where ϕ is called the **phase constant.** This constant can be determined from the initial conditions.

EXAMPLE 16.3. A Traveling Sinusoidal Wave

A sinusoidal wave traveling in the positive x direction has an amplitude of 15 cm, a wavelength of 40 cm, and a frequency of 8 Hz. The displacement of the wave at $t = 0$ and $x = 0$ is also 15 cm, as shown in Figure 16.16. (a) Find the wave number, period, angular frequency, and phase velocity of the wave.

Using Equations 16.9, 16.10, 16.12, and 16.14 and given the information that $\lambda = 40$ cm and $f = 8$ Hz, we find the following:

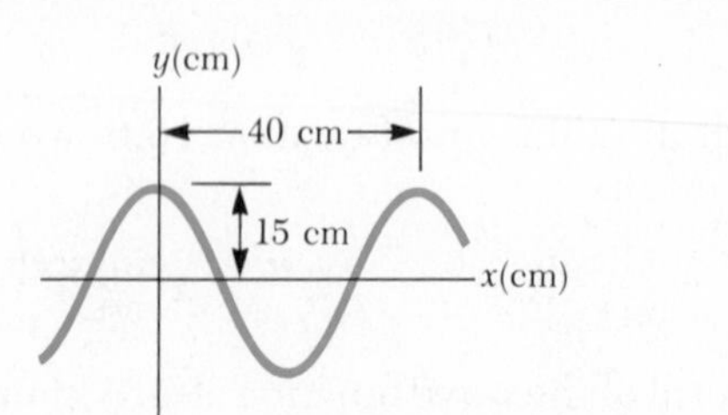

Figure 16.16 (Example 16.3) A harmonic wave of wavelength $\lambda = 40$ cm and amplitude $A = 15$ cm. The wave function can be written in the form $y = A\cos(kx - \omega t)$.

$$k = 2\pi/\lambda = 2\pi/40 \text{ cm} = 0.157 \text{ cm}^{-1}$$

$$T = 1/f = 1/8 \text{ s}^{-1} = 0.125 \text{ s}$$

$$\omega = 2\pi f = 2\pi(8 \text{ s}^{-1}) = 50.3 \text{ rad/s}$$

$$v = f\lambda = (8 \text{ s}^{-1})(40 \text{ cm}) = 320 \text{ cm/s}$$

(b) Determine the phase constant ϕ, and write a general expression for the wave function.

Since the amplitude $A = 15$ cm and since it is given that $y = 15$ cm at $x = 0$ and $t = 0$, substitution into Equation 16.15 gives

$$15 = 15 \sin(-\phi) \quad \text{or} \quad \sin(-\phi) = 1$$

Since $\sin(-\phi) = -\sin\phi$, we see that $\phi = -\pi/2$ rad (or $-90°$). Hence, the wave function is of the form

$$y = A \sin\left(kx - \omega t + \frac{\pi}{2}\right) = A \cos(kx - \omega t)$$

This can be seen by inspection, noting that the cosine function is displaced by 90° from the sine function. Substituting the values for A, k, and ω into this expression gives

$$y = 15 \cos(0.157x - 50.3t) \text{ cm}$$

Harmonic Waves on Strings

One method of producing a wave on a very long string is shown in Figure 16.17. One end of the string is connected to a blade that is set into vibration. As the blade oscillates vertically with simple harmonic motion, a traveling wave moving to the right is set up on the string. Figure 16.17 represents snapshots of the wave at intervals of one quarter of a period. Note that *each particle of the string, such as P, oscillates vertically in the y direction with simple harmonic motion.* This must be the case because each particle follows the simple harmonic motion of the blade. Therefore, every segment of the string can be treated as a simple harmonic oscillator vibrating with a frequency equal to the frequency of vibration of the blade that drives the string.[5] Note that although each segment oscillates in the y direction, the wave (or disturbance) travels in the x direction with a speed v. Of course, this is the definition of a transverse wave. In this case, the energy carried by the traveling wave is supplied by the vibrating blade. (In reality, the oscillations would gradually decrease in amplitude because of air resistance and the energy delivered to the string.)

[5] In this arrangement, we are assuming that the mass always oscillates in a vertical line. The tension in the string would vary if the mass were allowed to move sideways. Such a motion would make the analysis very complex.

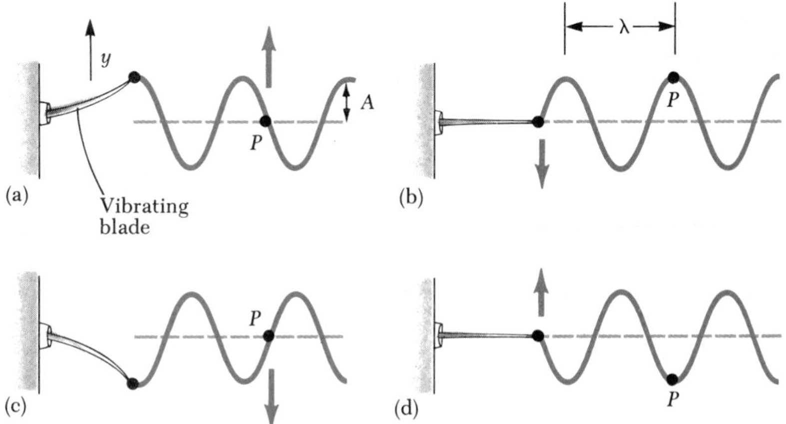

Figure 16.17 One method for producing harmonic waves on a continuous string. The left end of the string is connected to a blade that is set into vibration. Note that every segment, such as P, oscillates with simple harmonic motion in the vertical direction.

If the waveform at $t = 0$ is as described in Figure 16.17a, then the wave function can be written

$$y = A \sin (kx - \omega t)$$

We can use this expression to describe the motion of any point on the string. The point P (or any other point on the string) moves vertically, and so *its x coordinate remains constant.* Therefore, the *transverse velocity,* v_y, (not to be confused with the wave velocity v) and *transverse acceleration,* a_y, are given by

$$v_y = dy_y/dt]_{x=\text{constant}} = \partial y/\partial t = -\omega A \cos(kx - \omega t) \tag{16.16}$$

$$a_y = dv_y/dt]_{x=\text{constant}} = \partial v_y/\partial t = -\omega^2 A \sin(kx - \omega t) \tag{16.17}$$

The *maximum* values of these quantities are simply the absolute values of the coefficients of the cosine and sine functions:

$$(v_y)_{\text{max}} = \omega A \tag{16.18}$$

$$(a_y)_{\text{max}} = \omega^2 A \tag{16.19}$$

You should recognize that the transverse velocity and transverse acceleration do not reach their maximum values simultaneously. In fact, the transverse velocity reaches its maximum value (ωA) when the displacement $y = 0$, whereas the transverse acceleration reaches its maximum value ($\omega^2 A$) when $y = -A$.

EXAMPLE 16.4. A Harmonically Driven String
The string shown in Figure 16.17 is driven at one end at a frequency of 5 Hz. The amplitude of the motion is 12 cm, and the wave speed is 20 m/s. Determine the angular frequency and wave number for this wave, and write an expression for the wave function.

Using Equations 16.10, 16.12, and 16.13 gives

$$\omega = 2\pi/T = 2\pi f = 2\pi(5 \text{ Hz}) = 31.4 \text{ rad/s}$$

$$k = \omega/v = \frac{31.4 \text{ rad/s}}{20 \text{ m/s}} = 1.57 \text{ m}^{-1}$$

Since $A = 12$ cm $= 0.12$ m, we have

$$y = A \sin(kx - \omega t) = 0.12 \sin(1.57x - 31.4t)\text{m}$$

Exercise 2 Calculate the maximum values for the transverse velocity and transverse acceleration of any point on the string.
Answer: 3.77 m/s; 118 m/s^2.

16.8 ENERGY TRANSMITTED BY HARMONIC WAVES ON STRINGS

As waves propagate through a medium, they transport energy and momentum. This is easily demonstrated by hanging a weight on a stretched string and then sending a pulse down the string, as in Figure 16.18. When the pulse meets the weight, the weight will be momentarily displaced, as in Figure 16.18b. In the process, energy is transferred to the weight since work must be done in moving it upward.

In this section, we describe the rate at which energy is transported along a string. We shall assume a sinusoidal wave when we calculate the power transferred for this one-dimensional wave. Later, we shall extend these ideas to three-dimensional waves.

Consider a harmonic wave traveling on a string (Fig. 16.19). The source of the energy is some external agent at the left end of the string, which does work in producing the oscillations. Let us focus our attention on an element of the

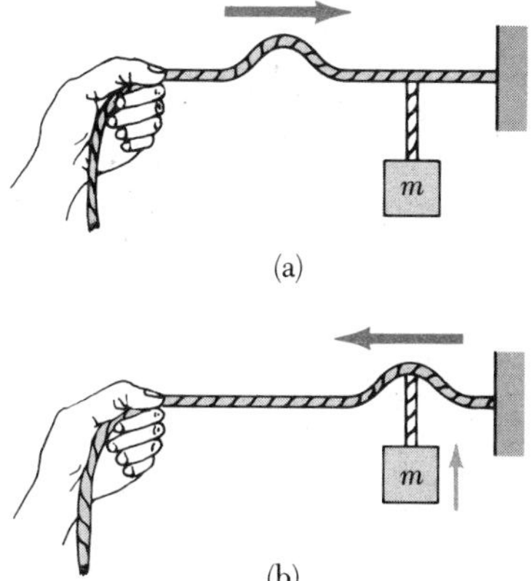

Figure 16.18 A pulse traveling to the right on a stretched string on which a mass has been suspended. (b) Energy and momentum are transmitted to the suspended mass when the pulse arrives.

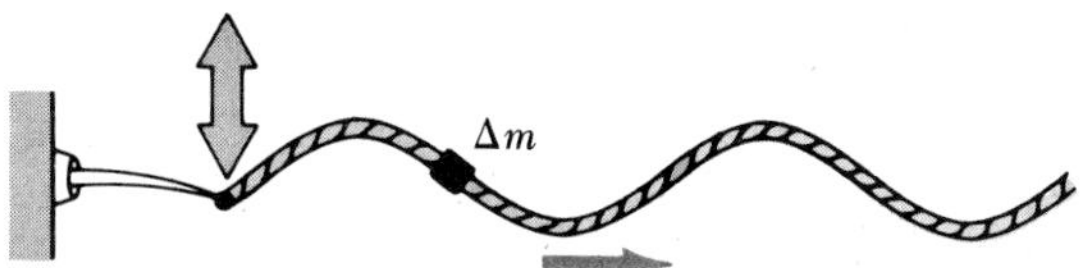

Figure 16.19 A harmonic wave traveling along the x axis on a stretched string. Every segment moves vertically, and each has the same total energy. The power transmitted by the wave equals the energy contained in one wavelength divided by the period of the wave.

string of length Δx and mass Δm. Each such segment moves vertically with simple harmonic motion. Furthermore, each segment has the same frequency, ω, and the same amplitude, A. As we found in Chapter 13, the total energy E associated with a particle moving with simple harmonic motion is $\frac{1}{2}kA^2 = \frac{1}{2}m\,\omega^2A^2$, where k is the equivalent force constant of the restoring force. If we apply this to the element of length Δx, we see that the total energy of this element is

$$\Delta E = \tfrac{1}{2}(\Delta m)\omega^2A^2$$

If μ is the mass per unit length of the string, then the element of length Δx has a mass Δm that is equal to $\mu\,\Delta x$. Hence, we can express the energy ΔE as

$$\Delta E = \tfrac{1}{2}(\mu\,\Delta x)\omega^2A^2 \tag{16.20}$$

If the wave travels from left to right as in Figure 16.19, the energy ΔE arises from the work done on the element Δm by the string element to the left of Δm. Similarly, the element Δm does work on the element to its right, so we see that energy is transmitted to the right. The rate at which energy is transmitted along the string, or the power, is given by dE/dt. If we let Δx approach 0, Equation 16.20 gives

$$\text{Power} = \frac{dE}{dt} = \tfrac{1}{2}\left(\mu\,\frac{dx}{dt}\right)\omega^2A^2$$

Since dx/dt is equal to the wave speed, v, we have

$$\text{Power} = \tfrac{1}{2}\mu\,\omega^2A^2v \tag{16.21}$$

Power

This shows that the power transmitted by a harmonic wave on a string is proportional to (a) the wave speed, (b) the square of the frequency, and (c) the square of the amplitude. In fact, *all* harmonic waves have the following general property: *The power transmitted by any harmonic wave is proportional to the square of the frequency and to the square of the amplitude.*

Thus, we see that a wave traveling through a medium corresponds to energy transport through the medium, with no net transfer of matter. An oscillating source provides the energy and produces a harmonic disturbance of the medium. The disturbance is able to propagate through the medium as the result of the interaction between adjacent particles. In order to verify Equation 16.20 by direct experiment, one would have to design some device at the far end of the string to extract the energy of the wave without producing any reflections.

EXAMPLE 16.5. Power Supplied to a Vibrating Rope A stretched rope having mass per unit length of $\mu = 5 \times 10^{-2}$ kg/m is under a tension of 80 N. How much power must be supplied to the rope to generate harmonic waves at a frequency of 60 Hz and an amplitude of 6 cm?

Solution: The wave speed on the stretched rope is given by

$$v = \sqrt{T/\mu} = \left(\frac{80 \text{ N}}{5 \times 10^{-2} \text{ kg/m}}\right)^{1/2} = 40 \text{ m/s}$$

Since $f = 60$ Hz, the angular frequency ω of the harmonic waves on the string has the value

$$\omega = 2\pi f = 2\pi(60 \text{ Hz}) = 377 \text{ s}^{-1}$$

Using these values in Equation 16.21 for the power, with $A = 6 \times 10^{-2}$ m, gives

$$\begin{aligned} P &= \tfrac{1}{2}\mu\omega^2 A^2 v \\ &= \tfrac{1}{2}(5 \times 10^{-2} \text{ kg/m})(377 \text{ s}^{-1})^2(6 \times 10^{-2} \text{ m})^2 (40 \text{ m/s}) \\ &= 512 \text{ W} \end{aligned}$$

*16.9 THE LINEAR WAVE EQUATION

Earlier in this chapter, we introduced the concept of the wave function to represent waves traveling on a string. All wave functions $y(x, t)$ represent solutions of an equation called the *linear wave equation.* This equation gives a complete description of the wave motion, and from it one can derive an expression for the wave velocity. Furthermore, the wave equation is basic to many forms of wave motion. In this section, we shall derive the wave equation as applied to waves on strings.

Consider a small segment of a string of length Δx and tension F, on which a traveling wave is propagating (Fig. 16.20). Let us assume that the ends of the segment make small angles θ_1 and θ_2 with the x axis. This is equivalent to the assumption that the vertical displacement of the segment is very small compared with its length.

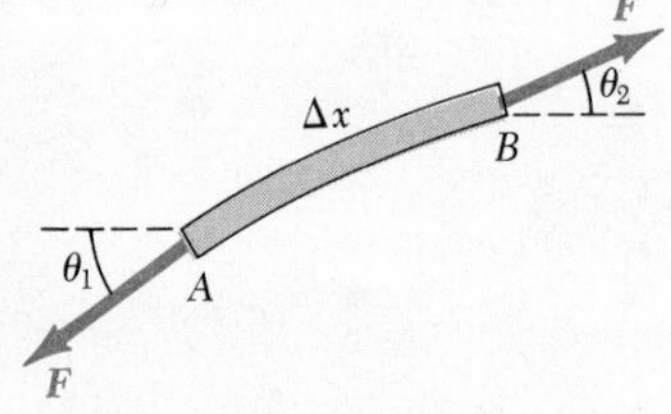

Figure 16.20 A segment of a string under tension F. Note that the slope at points A and B is given by $\tan \theta_1$ and $\tan \theta_2$, respectively.

The net force on the segment in the vertical direction is given by

$$\sum F_y = F \sin \theta_2 - F \sin \theta_1 = F(\sin \theta_2 - \sin \theta_1)$$

Since we have assumed that the angles are small, we can use the small-angle approximation $\sin \theta \approx \tan \theta$ and express the net force as

$$\sum F_y \approx F(\tan \theta_2 - \tan \theta_1)$$

However, the tangents of the angles at A and B are defined as the slope of the curve at these points. Since the slope of a curve is given by $\partial y/\partial x$, we have[6]

$$\sum F_y \approx F[(\partial y/\partial x)_B - (\partial y/\partial x)_A] \qquad (16.22)$$

We now apply Newton's second law, $\Sigma F_y = ma_y$, to the segment, where m is the mass of the segment, given by $m = \mu \, \Delta x$. This gives

$$\sum F_y = ma_y = \mu \, \Delta x(\partial^2 y/\partial t^2) \qquad (16.23)$$

[6] It is necessary to use partial derivatives because y depends on both x and t.

where we have used the fact that $a_y = \partial^2 y/\partial t^2$. Equating Equation 16.23 to Equation 16.22 gives

$$\mu\, \Delta x(\partial^2 y/\partial t^2) = F[(\partial y/\partial x)_B - (\partial y/\partial x)_A]$$

$$\frac{\mu}{F}\frac{\partial^2 y}{\partial t^2} = \frac{[(\partial y/\partial x)_B - (\partial y/\partial x)_A]}{\Delta x} \qquad (16.24)$$

The right side of Equation 16.24 can be expressed in a different form if we note that the derivative of any function is defined as

$$\frac{\partial f}{\partial x} = \lim_{\Delta x \to 0} \frac{f(x + \Delta x) - f(x)}{\Delta x}$$

If we associate $f(x + \Delta x)$ with $(\partial y/\partial x)_B$ and $f(x)$ with $(\partial y/\partial x)_A$, we see that in the limit $\Delta x \to 0$, Equation 16.24 becomes

$$\frac{\mu}{F}\frac{\partial^2 y}{\partial t^2} = \frac{\partial^2 y}{\partial x^2} \qquad (16.25)$$

Linear wave equation

This is the linear wave equation as it applies to waves on a string.

We shall now show that the harmonic wave function represents a solution of this wave equation. If we take the harmonic wave function to be of the form $y(x, t) = A \sin(kx - \omega t)$, the appropriate derivatives are

$$\partial^2 y/\partial t^2 = -\omega^2 A \sin(kx - \omega t)$$

$$\partial^2 y/\partial x^2 = -k^2 A \sin(kx - \omega t)$$

Substituting these expressions into Equation 16.25 gives

$$k^2 = (\mu/F)\omega^2$$

Using the relation $v = \omega/k$ in the above expression, we see that

$$v^2 = \omega^2/k^2 = F/\mu$$

$$v = \sqrt{F/\mu}$$

This represents another proof of the expression for the wave velocity on a stretched string.

The linear wave equation given by Equation 16.25 is often written in the form

$$\frac{\partial^2 y}{\partial x^2} = \frac{1}{v^2}\frac{\partial^2 y}{\partial t^2} \qquad (16.26)$$

Linear wave equation in general

This expression applies in general to various types of waves moving through nondispersive media. For waves on strings, y represents the vertical displacement. For sound waves, y corresponds to variations in the pressure or density of a gas. In the case of electromagnetic waves, y corresponds to electric or magnetic field components.

We have shown that the harmonic wave function is one solution of the linear wave equation. Although we do not prove it here, the linear wave equation is satisfied by *any* wave function having the form $y = f(x \pm vt)$. Furthermore, we have seen that the wave equation is a direct consequence of Newton's second law applied to any segment of the string. Similarly, the wave equation in electromagnetism can be derived from the fundamental laws of electricity and magnetism. This will be discussed further in Chapter 34.

16.10 SUMMARY

Transverse wave

A **transverse wave** is a wave in which the particles of the medium move in a direction *perpendicular* to the direction of the wave velocity. An example is a wave on a stretched string.

Longitudinal wave

Longitudinal waves are waves for which the particles of the medium move in a direction *parallel* to the direction of the wave velocity. Sound waves are longitudinal.

Any one-dimensional wave traveling with a speed v in the positive x direction can be represented by a wave function of the form $y = f(x - vt)$. Likewise, a wave traveling in the negative x direction has the form $y = f(x + vt)$. The shape of the wave at any instant (a snapshot of the wave) is obtained by holding t constant.

Superposition principle

The **superposition principle** says that when two or more linear waves move through a medium, the resultant wave function equals the algebraic sum of the individual wave functions. Waves that obey this principle are said to be *linear*. When two waves combine in space, they interfere to produce a resultant wave. The *interference* may be *constructive* (when the individual displacements are in the same direction) or *destructive* (when the displacements are in opposite directions).

The **speed** of a wave traveling on a stretched string of mass per unit length μ and tension F is

Speed of a wave on a stretched string

$$v = \sqrt{F/\mu} \tag{16.4}$$

When a pulse traveling on a string meets a fixed end, the pulse is reflected and inverted. If the pulse reaches a free end, it is reflected but not inverted.

The **wave function** for a one-dimensional harmonic wave traveling to the right can be expressed as

Wave function for a harmonic wave

$$y = A \sin[(2\pi/\lambda)(x - vt)] = A \sin(kx - \omega t) \qquad (16.6, 16.11)$$

where A is the amplitude, λ is the wavelength, k is the wave number, and ω is the angular frequency. If T is the period (the time it takes the wave to travel a distance equal to one wavelength) and f is the frequency, then v, k and ω can be written

$$v = \lambda/T = \lambda f \qquad (16.7, 16.14)$$

Wave number

$$k = 2\pi/\lambda \tag{16.9}$$

Angular frequency

$$\omega = 2\pi/T = 2\pi f \qquad (16.10, 16.12)$$

The **power** transmitted by a harmonic wave on a stretched string is given by

Power

$$P = \tfrac{1}{2}\mu\omega^2 A^2 v \tag{16.21}$$

The wave function $y(x, t)$ for many kinds of waves satisfies the following **linear wave equation:**

Linear wave equation in general

$$\frac{\partial^2 y}{\partial x^2} = \frac{1}{v^2}\frac{\partial^2 y}{\partial t^2} \tag{16.26}$$

QUESTIONS

1. Why is a wave pulse traveling on a string considered a transverse wave?
2. How would you set up a longitudinal wave in a stretched spring? Would it be possible to set up a transverse wave in a spring?
3. By what factor would you have to increase the tension in a stretched string in order to double the wave speed?
4. When a wave pulse travels on a stretched string, does it always invert upon reflection? Explain.
5. Can two pulses traveling in opposite directions on the same string reflect from one another? Explain.
6. Does the transverse velocity of a segment on a stretched string depend on the wave velocity?
7. If you were to periodically shake the end of a stretched rope three times each second, what would be the period of the harmonic waves set up in the string?
8. Harmonic waves are generated on a string under constant tension by a vibrating source. If the power delivered to the string is doubled, by what factor does the amplitude change? Does the wave velocity change under these circumstances?
9. Consider a wave traveling on a stretched rope. What is the difference, if any, between the speed of the wave and the speed of a small section of the rope?
10. If a long rope is hung from a ceiling and waves are sent up the rope from its lower end, the waves do not ascend with constant speed. Explain.
11. What happens to the wavelength of a wave on a string when the frequency is doubled? Assume the tension in the string remains the same.
12. What happens to the velocity of a wave on a string when the frequency is doubled? Assume the tension in the string remains the same.
13. How do transverse waves differ from longitudinal waves?
14. When all the strings on a guitar are stretched to the same tension, will the velocity of a wave along the more massive bass strings be faster or slower than the velocity of a wave on the lighter strings?

PROBLEMS

Section 16.3 One-Dimensional Traveling Waves

1. At $t = 0$, a transverse wave pulse in a wire is described by the function

$$y = \frac{6}{x^2 - 3}$$

where x and y are in m. Write the function $y(x, t)$ that describes this wave if it is traveling in the positive x direction with a speed of 4.5 m/s.

2. Two wave pulses A and B are moving in *opposite* directions along a stretched string with a speed of 2 cm/s. The amplitude of A is twice the amplitude of B. The pulses are shown in Figure 16.21 at $t = 0$. Sketch the shape of the string at $t = 1$, 1.5, 2, 2.5, and 3 s.

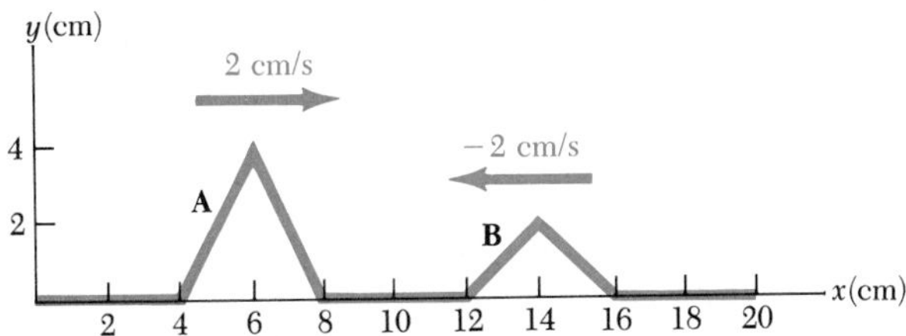

Figure 16.21 (Problem 2).

3. A traveling wave pulse moving to the right along the x axis is represented by the following wave function:

$$y(x, t) = \frac{4}{2 + (x - 4t)^2}$$

where x and y are measured in cm and t is in s. Plot the shape of the waveform at $t = 0$, 1, and 2 s.

Section 16.5 The Velocity of Waves on Strings

4. Transverse waves with a speed of 50 m/s are to be produced in a stretched string. A 5-m length of string with a total mass of 0.06 kg is used. What is the required tension in the string?
5. Calculate the wave speed in the string described in Problem 4 if the tension in the string is 8 N.
6. The tension in a cord 15 m in length is 20 N. The measured transverse wave speed in the cord is 60 m/s. Calculate the total mass of the cord.
7. Transverse waves travel with a speed of 20 m/s in a string under a tension of 6 N. What tension is required for a wave speed of 30 m/s in the same string?
8. Tension is maintained in a horizontal string as shown in Figure 16.10. The observed wave speed is 24 m/s when the suspended mass is 3 kg. What is the linear density of the string?

Section 16.7 Harmonic Waves

9. (a) Plot y versus t at $x = 0$ for a harmonic wave of the form $y = 15 \cos(0.157x - 50.3t)$, where x and y are in cm and t is in s. (b) Determine the period of vibration from this plot and compare your result with the value found in Example 16.3.
10. For a certain transverse wave, it is observed that the distance between two successive maxima is 1.2 m. It is also noted that eight crests, or maxima, pass a given point along the direction of travel every 12 s. Calculate the wave speed.

11. A harmonic wave is traveling along a rope. It is observed that the oscillator that generates the wave completes 40 vibrations in 30 s. Also, a given maximum travels 425 cm along the rope in 10 s. What is the wavelength?

12. When a particular wire is vibrating with a frequency of 4 Hz, a transverse wave of wavelength 60 cm is produced. Determine the speed of wave pulses along the wire.

13. One form of the wave function for a harmonic wave is given by Equation 16.11. The displacement y is expressed as a function of x and t in terms of the wave number k and the angular frequency ω. Write equivalent equations in which y is shown as a function of x and t in terms of (a) k and v, (b) λ and v, (c) λ and f, and (d) f and v.

14. A harmonic wave train is described by

$$y = 0.15 \sin(0.2x - 30t),$$

where x and y are in m and t is in s. Determine for this wave the (a) amplitude, (b) angular frequency, (c) wave number, (d) wavelength, (e) wave speed, and (f) direction of motion.

15. Determine the quantities (a) through (f) of Problem 10 when the wave train is described by $y = 0.2 \sin 4\pi(0.4x + t)$. Again x and y are in m and t is in s.

16. In Example 16.3 the harmonic wave was found to be described by $y = 15 \cos(0.157x - 50.3t)$, where x and y are in cm and t is in s. (a) Plot y versus x at $t = 0$ and $t = 0.125$ s. (b) Determine the wave speed from this plot and compare your result with the value found in Example 16.3.

17. (a) Write the expression for y as a function of x and t for a sinusoidal wave traveling along a rope in the *negative* x direction with the following characteristics: $y_{max} = 8$ cm, $\lambda = 80$ cm, $f = 3$ Hz, and $y(0, t) = 0$ at $t = 0$. (b) Write the expression for y as a function of x for the wave in (a) assuming that $y(x, 0) = 0$ at the point $x = 10$ cm.

18. Consider a wave in a string described by

$$y = 15 \sin[(\pi/16)(2x - 64t)],$$

where x and y are in cm and t is in s. (a) Calculate the maximum transverse velocity of a point on the string. (b) Calculate the transverse velocity of the point at $x = 6$ cm when $t = 0.25$ s.

19. For the wave described in Problem 18, calculate (a) the maximum transverse acceleration and (b) the transverse acceleration for the point located at $x = 6$ cm when $t = 0.25$ s.

20. (a) Write the expression for y as a function of x and t for a sinusoidal wave traveling along a rope in the positive x direction with the following characteristics: $y_{max} = 8$ cm, $\lambda = 80$ cm, $f = 3$ Hz, and $y(0, t) = y_{max}$ at $t = 0$. (b) Determine the speed and wave number for the wave described in (a).

Section 16.8 Energy Transmitted by Harmonic Waves on Strings

21. A stretched rope has a mass of 0.18 kg and a length of 3.6 m. What power must be supplied in order to generate harmonic waves having an amplitude of 0.1 m and a wavelength of 0.5 m and traveling with a speed of 30 m/s?

22. A wire of mass 0.24 kg is 48 m long and under a tension of 60 N. An electric vibrator operating at an angular frequency of 80π rad/s is generating harmonic waves in the wire. The vibrator can supply energy to the wire at a maximum rate of 400 J/s. What is the maximum amplitude of the wave pulses?

23. Transverse waves are being generated on a rope under *constant tension*. By what factor will the required power be increased or decreased if (a) the length of the rope is doubled and the angular frequency remains constant, (b) the amplitude is doubled and the angular frequency is halved, (c) both the wavelength and the amplitude are doubled, and (d) both the length of the rope and the wavelength are halved?

24. Harmonic waves 5 cm in amplitude are to be transmitted along a string that has a linear density of 4×10^{-2} kg/m. If the maximum power delivered by the source is 300 W and the string is under a tension of 100 N, what is the highest vibrational frequency at which the source can operate?

Section 16.9 The Linear Wave Equation

25. Show that the wave function $y = \ln[A(x - vt)]$ is a solution to Equation 16.25, where A is a constant.

26. In Section 16.9 it is verified that $y_1 = A \sin(kx - \omega t)$ is a solution to the wave equation. The wave function $y_2 = B \cos(kx - \omega t)$ describes a wave $\pi/2$ radians out of phase with the first. (a) Determine whether or not $y = A \sin(kx - \omega t) + B \cos(kx - \omega t)$ is a solution to the wave equation. (b) Determine if $y = A(\sin kx)B(\cos \omega t)$ is a solution to the wave equation.

GENERAL PROBLEMS

27. A traveling wave propagates according to the expression $y = 4.0 \sin(2.0x - 3.0t)$ cm where x is in cm. Determine (a) the amplitude, (b) the wavelength, (c) the frequency, (d) the period, and (e) the direction of travel of the wave.

28. A traveling wave on a string is harmonic, and its transverse displacement is given by

$$y = 3.0 \cos(\pi x - 4\pi t) \text{ cm},$$

where x is in cm. (a) Determine the wavelength and period of the wave. (b) Find the transverse velocity and transverse acceleration at any time t. (c) Calculate the transverse velocity and transverse acceleration at $t = 0$ for a point located at $x = 0.25$ cm. (d) What are the maximum values of the transverse velocity and transverse acceleration?

29. (a) Determine the speed of transverse waves on a stretched string that is under a tension of 80 N if the string has a length of 2 m and a mass of 5 g. (b) Calculate the power required to generate these waves if they have a wavelength of 16 cm and are 4 cm in amplitude.

30. A harmonic wave in a rope is described by the wave function $y = 0.2 \sin[\pi(0.75x - 18t)]$ where x and y are in m and t is in s. This wave is traveling in a rope that has a linear mass density of 0.25 kg/m. If the tension in the rope is provided by an arrangement like the one illustrated in Figure 16.10, what is the value of the suspended mass?

31. Consider the sinusoidal wave of Example 16.3, for which it was determined that

$$y = 15 \cos(0.157x - 50.3t) \text{ cm.}$$

At a given instant, let point A be at the origin and point B be the first point along x that is 60° out of phase with point A. What is the coordinate of point B?

32. A harmonic traveling wave moving in the positive x direction has an amplitude of 2.0 cm, a wavelength of 4.0 cm, and a frequency of 5 Hz. (a) Determine the speed of the wave and (b) write an expression for the transverse displacement as a function of x and t.

33. A transverse wave propagating along the positive x axis has the following properties: $y_{max} = 6$ cm, $\lambda = 8\pi$ cm, $v = 48$ cm/s, and the displacement of the wave at $t = 0$ and $x = 0$ is -2 cm. Determine the (a) wave number, (b) angular frequency, and (c) phase constant for the wave. (d) What is the first value of t for which the displacement at $x = 0$ will be $+2$ cm? (e) For this initial condition, find the coordinate of the particle on the positive x axis closest to the origin for which $y = 0$.

34. A rope of total mass m and length L is suspended vertically. Show that a transverse wave pulse will travel the length of the rope in a time $t = 2\sqrt{L/g}$. (*Hint:* First find an expression for the velocity at any point a distance x from the lower end of the rope, by considering the tension in the rope as resulting from the weight of the segment below that point.)

35. An aluminum wire is clamped at each end under zero tension at room temperature (22°C). The tension in the wire is increased by reducing the temperature, which results in a decrease in the wire's length. What fractional decrease in length ($\Delta L/L$) will result in a transverse wave speed of 100 m/s? Take the cross-sectional area of the wire to be 5×10^{-6} m², and use the following properties of aluminum: density, $\rho = 2.7 \times 10^3$ kg/m³; and Young's modulus, $Y = 6.8 \times 10^{11}$ N/m².)

36. (a) Show that the speed of longitudinal waves along a spring of force constant k is $v = \sqrt{kL/\mu}$, where L is the unstretched length of the spring and μ is the mass per unit length. (b) A spring of mass 0.4 kg has an unstretched length of 2 m and a force constant of 100 N/m. Using the results to (a), determine the speed of longitudinal waves along this spring.

•37. It is stated in Problem 34 that a wave pulse will travel from the bottom to the top of a rope of length L in a time $t = 2\sqrt{L/g}$. Use this result to answer the following questions. (It is *not* necessary to set up any new integrations.) (a) How long does it take for a wave pulse to travel halfway up the rope of length L? (Give your answer as a fraction of the quantity $(2\sqrt{L/g}$.) (b) A pulse starts traveling up the rope. How far has the pulse traveled after a time $\sqrt{L/g}$?

•38. A string of length L consists of two distinct sections. The left half has mass density $\mu \equiv \mu_0/2$, while the right half has mass per unit length $\mu' = 3\mu = 3\mu_0/2$. Tension in the string is F_0. Notice that this string has the same total mass as a uniform string of length L and mass per unit length μ_0. (a) Find the speeds v and v' at which transverse wave pulses travel in the two sections of the string. Express the speeds in terms of F_0 and μ_0, and also as multiples of the speed $v_0 \equiv \sqrt{F_0/\mu_0}$. (b) Find the time required for a wave to travel from one end of the string to the other. Give your result as a multiple of $T_0 \equiv L/v_0$.

CALCULATOR/COMPUTER PROBLEM

39. Two transverse wave pulses traveling in opposite directions along the x axis are represented by the following wave functions:

$$y_1(x, t) = \frac{6}{(x - 3t)^2}; \quad y_2(x, t) = -\frac{3}{(x + 3t)^2}$$

where x and y are measured in cm and t is in s. Write a program which will enable you to obtain the shape of the composite waveform $y_1 + y_2$ as a function of time. Use your program and make plots of the waveform at $t = 0$, 0.5, 1, 1.5, 2, 2.5, and 3 s.

17
Sound Waves

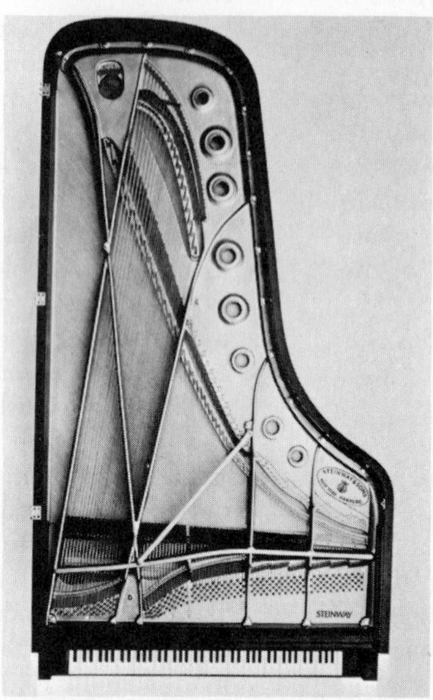

This chapter deals with the properties of longitudinal waves traveling through various media. Sound waves are the most important example of longitudinal waves. They can travel through any material medium (that is, gases, solids, or liquids) with a speed that depends on the properties of the medium. As sound waves travel through a medium, the particles in the medium vibrate to produce density and pressure changes along the direction of motion of the wave. This is in contrast to a transverse wave, where the particle motion is perpendicular to the direction of wave motion. The displacements that occur as a result of sound waves involve the longitudinal displacements of individual molecules from their equilibrium positions. This results in a series of high- and low-pressure regions called *condensations* and *rarefactions,* respectively. If the source of the sound waves, such as the diaphragm of a loudspeaker, vibrates sinusoidally, the pressure variations will also be sinusoidal. We shall find that the mathematical description of harmonic sound waves is identical to that of harmonic string waves discussed in the previous chapter.

There are three categories of longitudinal mechanical waves that cover different ranges of frequency: (1) *Audible waves* are sound waves that lie within the range of sensitivity of the human ear, typically, 20 Hz to 20 000 Hz. They can be generated in a variety of ways, such as by musical instruments, human vocal cords, and loudspeakers. (2) *Infrasonic waves* are longitudinal waves with frequencies below the audible range. Earthquake waves are an example. (3) *Ultrasonic waves* are longitudinal waves with frequencies above the audible range. For example, they can be generated by inducing vibrations in a quartz crystal with an applied alternating electric field. Any device that transforms one form of power into another is called a *transducer.* In addition to the loudspeaker and the quartz crystal, ceramic and magnetic phonograph pickups are common examples of sound transducers. Some transducers can generate ultrasonic waves. Such devices are used in the construction of ultrasonic cleaners and for underwater navigation.

17.1 VELOCITY OF SOUND WAVES

Sound waves are compressional waves traveling through a compressible medium, such as air. The compressed region of air which propagates corresponds to a variation in the normal value of the air pressure. The speed of such compressional waves depends on the compressibility of the medium and on the inertia of the medium. If the compressible medium has a bulk modulus B and an equilibrium density ρ, the speed of sound in that medium is

$$v = \sqrt{B/\rho} \tag{17.1}$$

Speed of sound

Recall that the **bulk modulus** (Section 12.4) is defined as the ratio of the change in pressure, ΔP, to the resulting fractional change in volume, $-\Delta V/V$:

$$B = -\frac{\Delta P}{\Delta V/V} \tag{17.2}$$

Bulk modulus

Note that B is always positive, since an increase in pressure (positive ΔP) results in a decrease in volume. Hence, the ratio $\Delta P/\Delta V$ is always negative.

It is interesting to compare Equation 17.1 with the expression for the speed of transverse waves on a string, $v = \sqrt{F/\mu}$, discussed in the previous chapter. In both cases, the wave speed depends on an elastic property of the medium (B or F) and on an inertial property of the medium (ρ or μ). In fact, the speed of *all mechanical waves* follows an expression of the general form

$$v = \sqrt{\text{elastic property/inertial property}}$$

In order to understand the origin of Equation 17.1, let us first describe pictorially the motion of a longitudinal pulse moving through a long tube containing a compressible gas or liquid (Fig. 17.1). A piston at the left end can be moved to the right to compress the fluid and create the longitudinal pulse. This is a convenient arrangement, since the wave motion is one-dimensional. Before the piston is moved, the medium is undisturbed and of uniform density, as described by the uniformly spaced vertical lines in Figure 17.1a. When the piston is suddenly pushed to the right (Fig. 17.1b), the medium just in front of it is compressed (represented by the shaded region). The pressure and density in this shaded region are higher than normal. When the piston comes to rest (Fig. 17.1c), the compressed region continues to move to the right, corresponding to a longitudinal pulse traveling down the tube with a speed v. Note that the piston speed does *not* equal v. Furthermore, the compressed region does not "stay with" the piston until it stops.

Let us assume that the equilibrium values of the pressure and density of the medium are P and ρ, respectively, as in Figure 17.2a. If the piston is pushed to the right with a constant speed u, the distance it moves in a time Δt is equal to $u\,\Delta t$. Let us assume that the boundary of the compressed region (the leading edge of the longitudinal pulse) moves with a velocity v, which corresponds to the velocity of the disturbance. In the time interval Δt, the wavefront advances a distance $v\,\Delta t$. Furthermore, let us assume that all the fluid in the shaded region moves with the velocity u of the piston. We can now apply the impulse-momentum theorem to this shaded region.

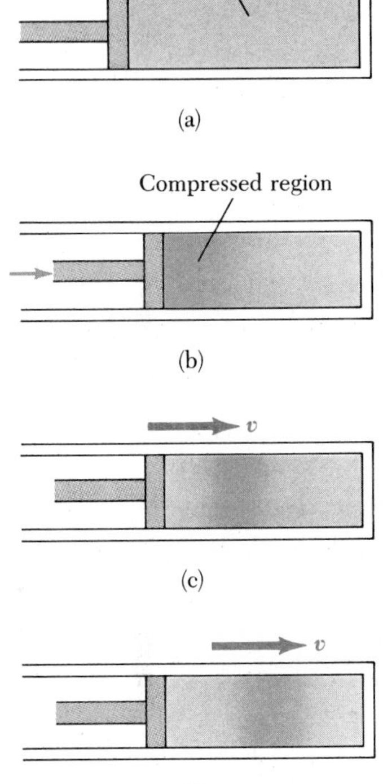

Figure 17.1 Motion of a longitudinal pulse through a compressible medium. The compression (dark region) is produced by the moving piston.

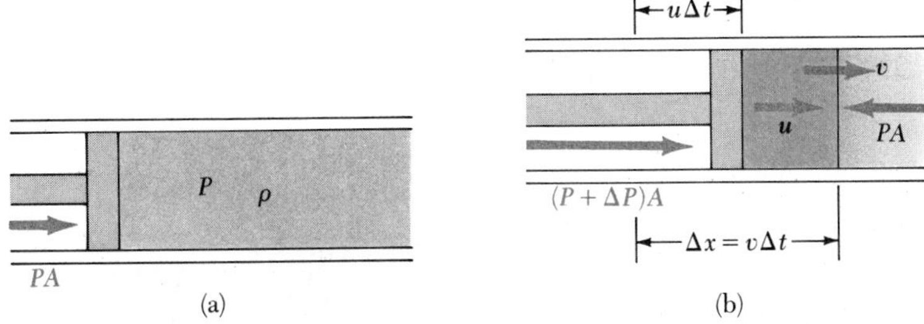

Figure 17.2 A longitudinal wave pulse produced by a piston that is suddenly moved to the right with a speed u. In a time Δt, the wave moves a distance $v\,\Delta t$, while the piston moves a distance $u\,\Delta t$.

The net force on the compressed region is $A\,\Delta P$, where ΔP is the *increase* in pressure necessary to compress the fluid and A is the cross-sectional area of the piston. Hence, the impulse imparted to the shaded region in a time Δt is given by

$$\text{Impulse} = F\,\Delta t = (A\,\Delta P)\,\Delta t$$

Now let us calculate the change in momentum of the mass of fluid set in motion. The mass Δm that is compressed and set in motion equals the density ρ multiplied by the volume $\Delta V = A\,\Delta x$:

$$\Delta m = \rho\,\Delta V = \rho A\,\Delta x = \rho A v\,\Delta t$$

where $\Delta x = v\,\Delta t$ is the length of the fluid set in motion. Since the initial speed of the fluid is zero and the final speed is u, the *change* in momentum is

$$\text{Change in momentum} = (\Delta m)u = (\rho A v\,\Delta t)u$$

Since the impulse acting on a body equals its change in momentum, we see that

$$(A\,\Delta P)\,\Delta t = (\rho A v\,\Delta t)u$$

$$\Delta P = \rho v u \tag{17.3}$$

We can obtain another expression for ΔP by using the definition of bulk modulus, given by Equation 17.2:

$$\Delta P = -B(\Delta V/V) \tag{17.4}$$

where $\Delta V/V$ is the *fractional* change in volume of the compressed fluid. Since the *original* volume of the compressed fluid is $V = Av\,\Delta t$ and since the change in volume ΔV equals the volume displaced by the piston, where $\Delta V = -Au\,\Delta t$ (the minus sign means the volume has decreased), we see that

$$\frac{\Delta V}{V} = -\frac{Au\,\Delta t}{Av\,\Delta t} = -\frac{u}{v}$$

Substituting this result into Equation 17.4 gives

$$\Delta P = B\frac{u}{v}$$

Finally, equating this expression to Equation 17.3, we find that

$$\rho v u = B\frac{u}{v}$$

Rearranging this expression, we arrive at Equation 17.1:

$$v = \sqrt{B/\rho}$$

Let us now determine the speed of sound waves in various media.

EXAMPLE 17.1. Sound Waves in a Solid Bar
If a solid bar is struck at one end with a hammer, a longitudinal pulse will propagate down the bar with a speed

$$v = \sqrt{Y/\rho} \tag{17.5}$$

where Y is the Young's modulus for the material, defined as the longitudinal stress divided by the longitudinal strain (Chapter 15). Find the speed of sound in an aluminum bar.

Solution: Using Equation 17.5 and the available data for aluminum, $Y = 7.0 \times 10^{10}\ \text{N/m}^2$ and having a density $\rho = 2.7 \times 10^3\ \text{kg/m}^3$, we find that

$$v_{\text{Al}} = \sqrt{\frac{7.0 \times 10^{10}\ \text{N/m}^2}{2.7 \times 10^3\ \text{kg/m}^3}} \approx 5100\ \text{m/s}$$

This is a typical value for the speed of sound in solids. Note that the result is much larger than the speed of sound in gases. This makes sense since the molecules of a solid are close together (in comparison to the molecules of a gas) and hence respond more rapidly to a disturbance.

EXAMPLE 17.2. Speed of Sound in a Liquid
Find the speed of sound in water, which has a bulk modulus of about 2.1×10^9 N/m^2 and a density of about 10^3 kg/m^3.

Solution: Using Equation 17.1, we find that

$$v_{\text{water}} = \sqrt{B/\rho} \approx \sqrt{\frac{2.1 \times 10^9 \text{ N/m}^2}{1 \times 10^3 \text{ kg/m}^3}} = 1500 \text{ m/s}$$

This result is much smaller than that for the speed of sound in aluminum, calculated in the previous example. In general, sound waves travel more slowly in liquids than in solids. This is because liquids are more compressible than solids and hence have a smaller bulk modulus.

The speed of sound in various media is given in Table 17.1.

TABLE 17.1 Speed of Sound in Various Media

Medium	v(m/s)
Gases	
Air (0°C)	331
Air (20°C)	343
Hydrogen (0°C)	1286
Oxygen (0°C)	317
Helium (0°C)	972
Liquids at 25°C	
Water	1493
Methyl alcohol	1143
Sea water	1533
Solids	
Aluminum	5100
Copper	3560
Iron	5130
Lead	1322
Vulcanized rubber	54

17.2 HARMONIC SOUND WAVES

If the source of a longitudinal wave, such as a vibrating diaphragm, oscillates with simple harmonic motion, the resulting disturbance will also be harmonic. One can produce a one-dimensional harmonic sound wave in a long, narrow tube containing a gas by means of a vibrating piston at one end, as in Figure 17.3. The darker regions in this figure represent regions where the gas is compressed, and so the density and pressure are *above* their equilibrium values.

A compressed layer is formed at times when the piston is being pushed into the tube. This compressed region, called a **condensation,** moves down the tube as a pulse, continuously compressing the layers in front of it. When the piston is withdrawn from the tube, the gas in front of it expands and the pressure and density in this region fall below their equilibrium values (represented by the lighter regions in Figure 17.3). These low-pressure regions, called **rarefactions,** also propagate along the tube, following the condensations. Both regions move with a speed equal to the speed of sound in that medium (about 343 m/s in air at 20°C).

As the piston oscillates back and forth in a sinusoidal fashion, regions of condensation and rarefaction are continuously set up. The distance between two successive condensations (or two successive rarefactions) equals the wavelength, λ. As these regions travel down the tube, any small volume of the medium moves with simple harmonic motion parallel to the direction of the wave. If $s(x, t)$ is the displacement of a small volume element measured from its equilibrium position, we can express this harmonic displacement function as

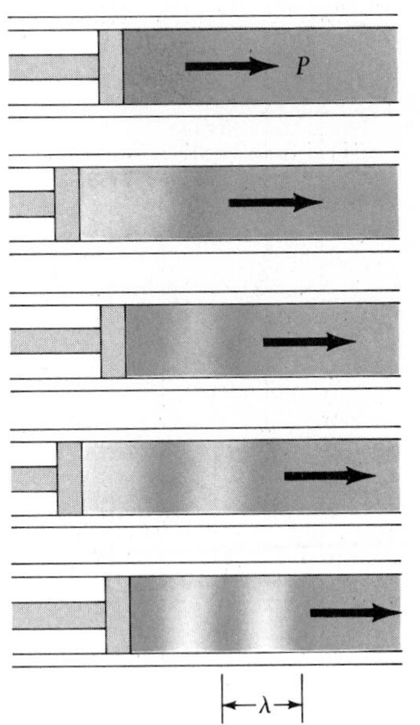

Figure 17.3 A harmonic longitudinal wave propagating down a tube filled with a compressible gas. The source of the wave is a vibrating piston at the left. The high- and low-pressure regions are dark and light, respectively.

$$s(x, t) = s_m \cos(kx - \omega t) \tag{17.6}$$

where s_m is the *maximum displacement from equilibrium* (the displacement amplitude), k is the wave number, and ω is the angular frequency of the piston. Note that the displacement is along x, the direction of motion of the sound wave, which of course means we are describing a longitudinal wave. The variation in the pressure of the gas, ΔP, measured from its equilibrium value is also harmonic and given by

Pressure variation

$$\Delta P = \Delta P_m \sin(kx - \omega t) \tag{17.7}$$

The derivation of this expression will be given below.

The **pressure amplitude** ΔP_m is the *maximum change in pressure from the equilibrium value.* As we shall show later, the pressure amplitude is proportional to the displacement amplitude, s_m, and is given by

Pressure amplitude

$$\Delta P_m = \rho v \omega s_m \tag{17.8}$$

where ωs_m is the maximum longitudinal velocity of the medium in front of the piston.

Thus, we see that a sound wave may be considered as either a displacement wave or a pressure wave. A comparison of Equations 17.6 and 17.7 shows that *the pressure wave is 90° out of phase with the displacement wave.* Graphs of these functions are shown in Figure 17.4. Note that the pressure variation is a maximum when the displacement is zero, whereas the displacement is a maximum when the pressure variation is zero. Since the pressure is proportional to the density, the variation in density from the equilibrium value follows an expression similar to Equation 17.7.

We shall now give a derivation of Equations 17.7 and 17.8. From Equation 17.4, we see that the pressure variation in a gas is given by

$$\Delta P = -B(\Delta V/V)$$

The volume of a layer of thickness Δx and cross-sectioned area A is $V = A\,\Delta x$. The change in the volume ΔV accompanying the pressure change is equal to $A\,\Delta s$, where Δs is the difference in s between x and $x + \Delta x$. That is, $\Delta s = s(x + \Delta x) - s(x)$. Hence, we can express ΔP as

$$\Delta P = -B\frac{\Delta V}{V} = -B\frac{A\,\Delta s}{A\,\Delta x} = -B\frac{\Delta s}{\Delta x}$$

As Δx approaches zero, the ratio $\Delta s/\Delta x$ becomes $\partial s/\partial x$. (The partial derivative is used here to indicate that we are interested in the variation of s with position at a *fixed* time.) Therefore,

$$\Delta P = -B(\partial s/\partial x)$$

If the displacement is the simple harmonic function given by Equation 17.6, we find that

$$\Delta P = -B\frac{\partial}{\partial x}[s_m \cos(kx - \omega t)] = Bs_m k \sin(kx - \omega t)$$

Since the bulk modulus is given by $B = \rho v^2$ (Eq. 17.1), the pressure variation reduces to

$$\Delta P = \rho v^2 s_m k \sin(kx - \omega t)$$

Furthermore, from Equation 16.13, we can write $\omega = kv$, hence ΔP can be expressed as

$$\Delta P = \rho \omega s_m v \sin(kx - \omega t) = \Delta P_m \sin(kx - \omega t)$$

where ΔP_m is the maximum pressure variation, given by Equation 17.8.

$$\Delta P_m = \rho v \omega s_m$$

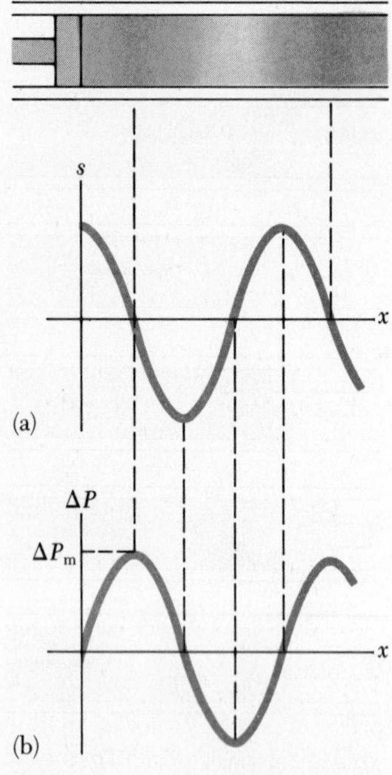

Figure 17.4 (a) Displacement amplitude versus position and (b) pressure amplitude versus position for a harmonic longitudinal wave. Note that the displacement wave is 90° out of phase with the pressure wave.

17.3 ENERGY AND INTENSITY OF HARMONIC SOUND WAVES

In the previous chapter, we showed that waves traveling on stretched strings transport energy. The same concepts are now applied to sound waves. Consider a layer of air of mass Δm and width Δx in front of a piston oscillating with a frequency ω, as in Figure 17.5. The piston transmits energy to the layer of air.[1] Since the average kinetic energy equals the average potential energy in simple harmonic motion (as was shown in Chapter 13), the average total energy of the mass Δm equals its maximum kinetic energy. Therefore, we can express the average energy of the moving layer of gas as

$$\Delta E = \tfrac{1}{2}\Delta m(\omega s_{\mathrm{m}})^2 = \tfrac{1}{2}(\rho A\;\Delta x)(\omega s_{\mathrm{m}})^2$$

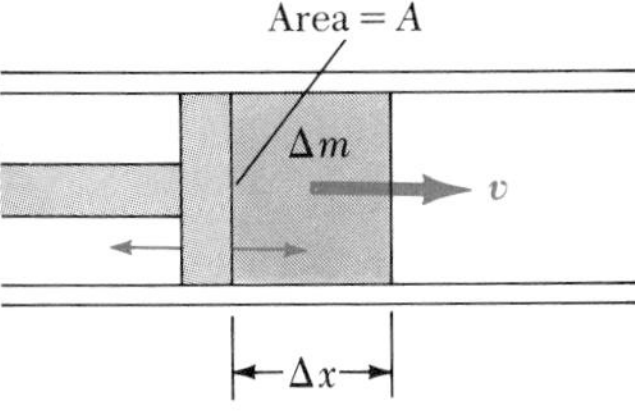

Figure 17.5 An oscillating piston transfers energy to the gas in the tube, causing the layer of width Δx and mass Δm to oscillate with an amplitude s_{m}.

where $A\;\Delta x$ is the volume of the layer. The time rate at which energy is transferred to each layer (or the power) is given by

$$\text{Power} = \frac{\Delta E}{\Delta t} = \tfrac{1}{2}\rho A\left(\frac{\Delta x}{\Delta t}\right)(\omega s_{\mathrm{m}})^2 = \tfrac{1}{2}\rho A v(\omega s_{\mathrm{m}})^2$$

where $v = \Delta x/\Delta t$ is the velocity of the disturbance to the right.

We define the **intensity** I of a wave to be the rate at which sound energy flows through a unit area A perpendicular to the direction of travel of the wave, or the power per unit area.

In this case, the intensity is given by

$$I = \frac{\text{power}}{\text{area}} = \tfrac{1}{2}\rho(\omega s_{\mathrm{m}})^2 v \qquad (17.9)$$

Intensity of a sound wave

Thus, we see that the intensity of the harmonic sound wave is proportional to the square of the amplitude and the square of the frequency (as in the case of a harmonic string wave). This can also be written in terms of the pressure amplitude ΔP_{m}, using Equation 17.8, which gives

$$I = \frac{\Delta P_{\mathrm{m}}{}^2}{2\rho v} \qquad (17.10)$$

[1] Although it is not proved here, the work done by the piston equals the energy carried away by the wave. For a detailed mathematical treatment of this concept, see Frank S. Crawford, Jr., *Waves,* New York, McGraw-Hill, 1968, Berkeley Physics Course, Volume 3, Chapter 4.

EXAMPLE 17.3. Hearing Limitations

The faintest sounds the human ear can detect at a frequency of 1000 Hz correspond to an intensity of about 10^{-12} W/m² (the so-called *threshold of hearing*). Likewise, the loudest sounds that the ear can tolerate correspond to an intensity of about 1 W/m² (the *threshold of pain*). Determine the pressure amplitudes and maximum displacements associated with these two limits.

Solution: First, consider the faintest sounds. Using Equation 17.10 and taking $v = 343$ m/s and the density of air to be $\rho = 1.2$ kg/m³, we get

$$\begin{aligned}\Delta P_{\mathrm{m}} &= (2\rho v I)^{1/2}\\ &= [2(1.2\text{ kg/m}^3)(343\text{ m/s})(10^{-12}\text{ W/m}^2)]^{1/2}\\ &= 2.9\times10^{-5}\text{ N/m}^2\end{aligned}$$

Since atmospheric pressure is about 10^5 N/m², this means the ear can discern pressure fluctuations as small as 3 parts in 10^{10}! The corresponding maximum displacement can be calculated using Equation 17.8, recalling that $\omega = 2\pi f$:

$$\begin{aligned}s_{\mathrm{m}} &= \frac{\Delta P_{\mathrm{m}}}{\rho\omega v} = \frac{2.9\times10^{-5}\text{ N/m}^2}{(1.2\text{ kg/m}^3)(2\pi\times10^3\text{ s}^{-1})(343\text{ m/s})}\\ &= 1.1\times10^{-11}\text{ m}\end{aligned}$$

This is a remarkably small number! If we compare this result for s_m with the diameter of a molecule (about 10^{-10} m), we see that the ear is an extremely sensitive detector of sound waves.

In a similar manner, one finds that the loudest sounds the human ear can tolerate correspond to a pressure amplitude of about 29 N/m^2 and a maximum displacement of 1.1×10^{-5} m. Note that the small pressure amplitudes, called acoustic pressure, correspond to fluctuations taking place above and below atmospheric pressure.

TABLE 17.2 Decibel Scale Intensity for Some Sources

Source of Sound	β (dB)
Nearby jet airplane	150
Jackhammer; machine gun	130
Siren; rock concert	120
Subway; power mower	100
Busy traffic	80
Vacuum cleaner	70
Normal conversation	50
Mosquito buzzing	40
Whisper	30
Rustling leaves	10
Threshold of hearing	0

Intensity in Decibels

The previous example illustrates the wide range of intensities that the human ear can detect. For this reason, it is convenient to use a logarithmic intensity scale, where the **intensity level** β is defined by the equation

Intensity in decibels

$$\beta \equiv 10 \log(I/I_0) \qquad (17.11)$$

The constant I_0 is the *reference intensity,* taken to be at the threshold of hearing ($I_0 = 10^{-12}$ W/m^2), and I is the intensity in W/m^2 at the level β, where β is measured in decibels (dB).[2] On this scale, the threshold of pain ($I = 1$ W/m^2) corresponds to an intensity level of $\beta = 10 \log(1/10^{-12}) = 10 \log(10^{12}) =$ 120 dB. Likewise, the threshold of hearing corresponds to $\beta = 10 \log(1/1) =$ 0 dB. Nearby jet airplanes can create intensity levels of 150 dB, and subways and riveting machines have levels of 90 to 100 dB. The electronically amplified sounds heard at rock concerts can be at levels of up to 120 dB, the threshold of pain. Prolonged exposure to such high intensity levels may produce serious damage to the ear. Ear plugs are recommended whenever intensity levels exceed 90 dB. Recent evidence also suggests that "noise pollution" may be a contributing factor to high blood pressure, anxiety, and nervousness. Table 17.2 gives some typical values of the sound intensities of various sources.

17.4 SPHERICAL AND PLANE WAVES

If a spherical body pulsates or oscillates periodically such that its radius varies harmonically with time, a sound wave with spherical wave fronts will be produced (Fig. 17.6). The wave moves outward from the source at a constant speed if the medium is uniform.

[2] The "bel" is named after the inventor of the telephone, Alexander Graham Bell (1847–1922). The prefix deci- is the metric system scale factor that stands for 10^{-1}.

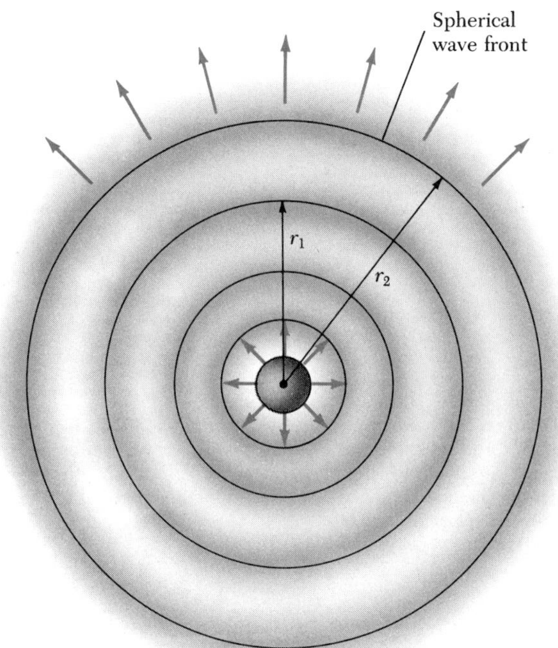

Figure 17.6 A spherical wave propagating radially outward from an oscillating spherical body. The intensity of the spherical wave varies as $1/r^2$.

Since all points on the sphere behave in the same way, we conclude that the energy in a spherical wave will propagate equally in all directions. That is, no one direction is preferred over any other. If P_{av} is the average power emitted by the source, then this power at any distance r from the source must be distributed over a spherical surface of area $4\pi r^2$. Hence, the wave intensity at a distance r from the source is

$$I = P_{av}/A = P_{av}/4\pi r^2 \tag{17.12}$$

Since P_{av} is the same through any spherical surface centered at the source, we see that the intensities at distances r_1 and r_2 are given by

$$I_1 = P_{av}/4\pi r_1^{\,2} \qquad \text{and} \qquad I_2 = P_{av}/4\pi r_2^{\,2}$$

Therefore, the ratio of intensities on these two spherical surfaces is

$$\frac{I_1}{I_2} = \frac{r_2^{\,2}}{r_1^{\,2}}$$

In Equation 17.9 we found that the intensity was also proportional to $s_m^{\,2}$, the square of the wave amplitude. Comparing this result with Equation 17.12, we conclude that the wave amplitude of a spherical wave must vary as $1/r$. Therefore, we can write the wave function ψ (Greek letter "psi") for an outgoing spherical wave in the form

$$\psi(r, t) = (s_0/r)\sin(kr - \omega t) \tag{17.13}$$

where s_0 is a constant.

It is useful to represent spherical waves by a series of circular arcs concentric with the source, as in Figure 17.7. Each arc represents a surface over which the phase of the wave is constant. We call such a surface of constant

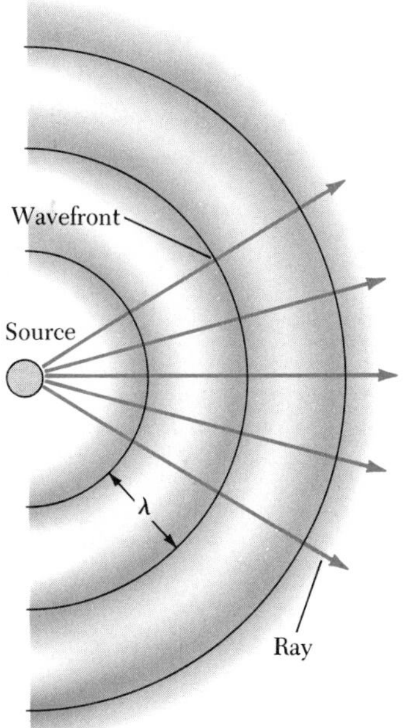

Figure 17.7 Spherical waves emitted by a point source. The circular arcs represent the spherical wavefronts concentric with the source. The rays are radial lines pointing outward from the source perpendicular to the wavefronts.

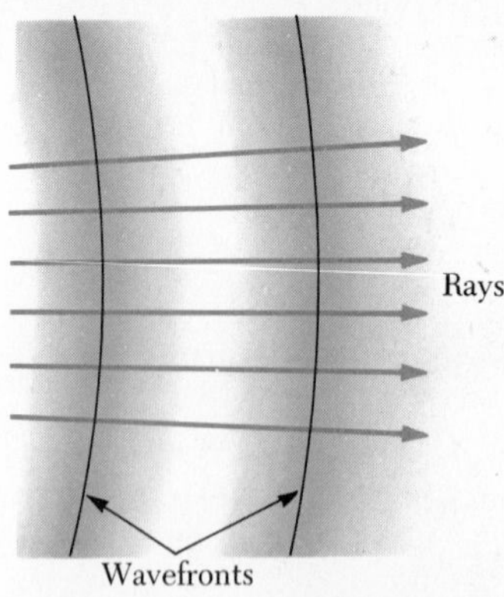

Figure 17.8 At large distances from a point source, the wavefronts are nearly parallel planes and the rays are nearly parallel lines perpendicular to the planes. Hence, a small segment of a spherical wavefront is approximately a planar wave.

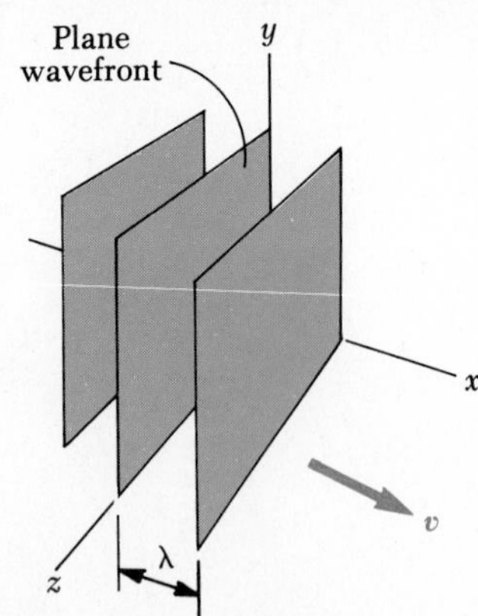

Figure 17.9 Representation of a planar wave moving in the positive x direction. The wavefronts are planes parallel to the yz plane.

Wavefront

phase a **wavefront.** The distance between adjacent wavefronts equals the wavelength, λ. The radial lines pointing outward from the source are called **rays.**

Now consider a small portion of the wavefronts at *large* distances (large relative to λ) from the source, as in Figure 17.8. In this case, the rays are nearly parallel and the wavefronts are very close to being planar. Therefore, at distances from the source that are large compared with the wavelength, we can approximate the wavefronts by parallel planes. We call such a wave a **plane wave.** Any small portion of a spherical wave that is far from the source can be considered a plane wave.

Figure 17.9 illustrates a plane wave propagating along the x axis. If x is taken to be the direction of the wave motion (or rays) in Figure 17.9, then the wavefronts are parallel to the yz plane. In this case, the wave function depends only on x and t and has the form

Plane wave representation

$$\psi(x, t) = s_0 \sin(kx - \omega t) \tag{17.14}$$

That is, the wave function for a plane wave is identical in form to that of a one-dimensional traveling wave. Note that the intensity is the same on successive wavefronts of the plane wave.

EXAMPLE 17.4. Intensity Variations of a Point Source

A source emits sound waves with a power output of 80 W. Assume the source is a point source. (a) Find the intensity at a distance 3 m from the source.

A point source emits energy in the form of spherical waves (Fig. 17.6). Let P_{av} be the average power output of the source. At a distance r from the source, the power is distributed over the surface area of a sphere, $4\pi r^2$. Therefore, the intensity at a distance r from the source, is given by Equation 17.12. Since $P_{av} = 80$ W and $r = 3$ m, we find that

$$I = \frac{P_{av}}{4\pi r^2} = \frac{80\text{ W}}{4\pi(3\text{ m})^2} = 0.71\text{ W/m}^2$$

which is close to the threshold of pain.

(b) Find the distance at which the sound reduces to a level of 40 dB.

We can find the intensity at the 40-dB level by using Equation 17.11 with $I_0 = 10^{-12}$ W/m². This gives

$\log(I/I_0) = 4$

$$I = 10^4 I_0 = 10^{-8}\ \text{W/m}^2$$

Using this value for I in Equation 17.12 and solving for r, we get

$$r = (P_{av}/4\pi I)^{1/2} = \frac{(80\ \text{W})^{1/2}}{(4\pi \times 10^{-8}\ \text{W/m}^2)^{1/2}}$$

$$= 2.5 \times 10^4\ \text{m}$$

which equals about 15 miles!

*17.5 THE DOPPLER EFFECT

When a car or truck is moving while its horn is blowing, the frequency of the sound you hear is higher as the vehicle approaches you and lower as it moves away from you. This is one example of the Doppler effect.[3]

In general, a Doppler effect is experienced whenever there is *relative* motion between the source and the observer. When the source and observer are moving toward each other, the frequency heard by the observer is higher than the frequency of the source. When the source and observer move away from each other, the observer hears a frequency which is lower than the source frequency.

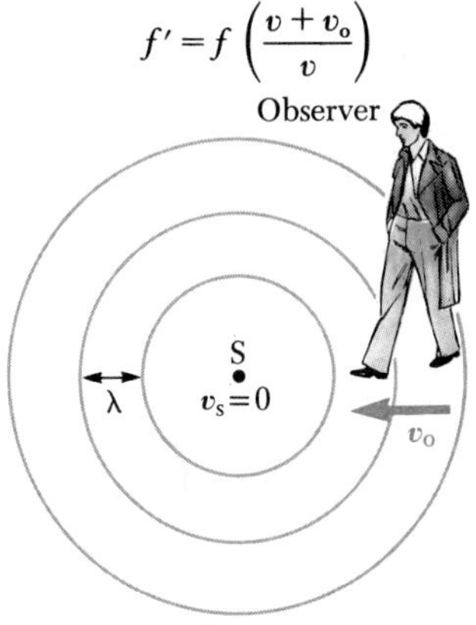

Figure 17.10 An observer O moving with a speed v_o *toward* a stationary point source S hears a frequency f' that is *greater* than the source frequency.

Although the Doppler effect is most commonly experienced with sound waves, it is a phenomenon common to all harmonic waves. For example, there is a shift in frequencies of light waves (electromagnetic waves) produced by the relative motion of source and observer.

First, let us consider the case where the observer O is moving and the sound source S is stationary. For simplicity, we shall assume that the air is also stationary. Figure 17.10 describes the situation when the observer moves with a speed v_O toward the source (considered as a point source), which is at rest ($v_S = 0$). In general, "at rest" means at rest with respect to the medium, air.

We shall take the frequency of the source to be f, the wavelength to be λ, and the velocity of sound to be v. If the observer were also stationary, clearly he or she would detect f wavefronts per second. (That is, when $v_O = 0$ and $v_S = 0$, the observed frequency equals the source frequency.) When the observer travels toward the source, he or she moves a distance $v_O t$ in time of t seconds and in this time detects an *additional* $v_O t/\lambda$ wavefronts. Furthermore, the speed of the waves relative to the observer is $v + v_O$. Since the additional number of wavefronts detected *per second* is v_O/λ, the frequency f' heard by the observer is *increased* and given by

$$f' = f + \Delta f = f + v_O/\lambda$$

Using the fact that $\lambda = v/f$, we see that $v_O/\lambda = (v_O/v)f$, hence f' can be expressed as

$$f' = f\left(\frac{v + v_O}{v}\right) \tag{17.15}$$

Moving *away* from the source, as in Figure 17.11, an observer detects *fewer* wavefronts per second. In this case, the speed of the wave relative to the observer is $v - v_O$. Thus, from Equation 17.15, it follows that the frequency heard by the observer in this case is lowered and given by

$$f' = f\left(\frac{v - v_O}{v}\right) \tag{17.16}$$

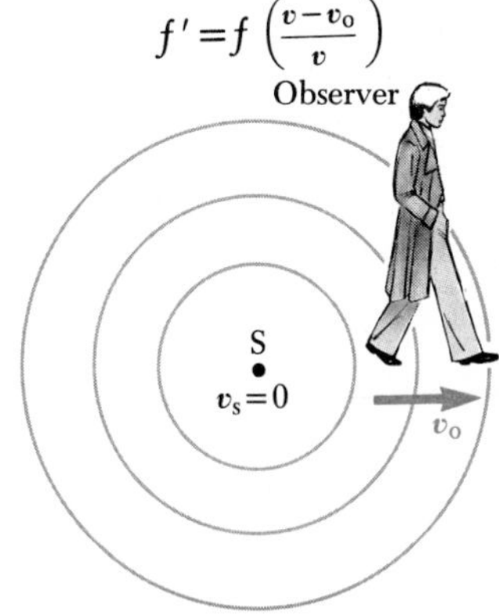

Figure 17.11 An oberserver O moving with a speed v_o *away* from a stationary source S hears a frequency f' that is *lower* than the source frequency.

[3] Named after the Austrian physicist Christian Johann Doppler (1803–1853), who discovered the effect for light waves.

Essay

ON THE PERCEPTION OF MUSICAL SOUND IN A CONCERT HALL

Donald H. White
Western Oregon State College

If a tree falls in a forest and nobody is around to hear its fall, is sound produced? This often-quoted question illustrates the fact that we use the word *sound* in two different ways: the mechanical waves in a conducting medium and a perceived aural stimulation. If we use the perceptual definition, the answer is no.

The crash of a tree in a forest would not be considered musical, however. To create a musical sound, one or more audio frequencies must persist long enough to produce a discernible pitch. It is fairly easy to create a tone, say middle C, by using a sinusoidal oscillator tuned to 132 Hz, an audio amplifier, and a pair of earphones; however, most listeners find such a pure tone to be dull. The same note played in a concert hall by the string section of a symphony orchestra sounds much more pleasant.

There are several factors that contribute to the perceived sound of a musical note, all of which serve to destroy the purity of the sound wave. First, a bowed violin string produces a set of overtones, the frequencies of which are almost, but not quite, integral multiples of the fundamental frequency. The violinist also produces *vibrato* (frequency modulation) by rapidly shaking the finger that holds the string against the fingerboard. Violinists in an orchestra, all playing the same note, will actually each be playing at slightly different frequencies, causing the phases of the sound waves from the different instruments to shift continuously. The interference of the waves shifting in and out of phase produces *tremolo* (amplitude modulation). The resultant *chorus effect* adds to the richness of the sound.

In an auditorium or concert hall, the presence of walls and other surfaces further modifies the sound by causing closely spaced multiple reflections of the sound waves. Partial absorption by the surfaces progressively dampens successive reflections, causing each note to persist audibly for typically 1 to 3 s. This reverberation blends the sound both spatially and temporally, causing the listener to feel bathed in sound. The

In general, when an observer moves with a speed v_O relative to a stationary source, the frequency heard by the observer is

Frequency heard with an observer in motion

$$f' = f\left(\frac{v \pm v_O}{v}\right) \tag{17.17}$$

where the *positive* sign is used when the observer moves *toward* the source and the *negative* sign holds when the observer moves *away* from the source.

Now consider the situation in which the source is in motion and the observer is at rest. If the source moves directly toward observer A in Figure 17.12a, the wavefronts seen by the observer are closer together as a result of the motion of the source in the direction of the outgoing wave. As a result, the wavelength λ' measured by observer A is shorter than the wavelength λ of the source. During each vibration, which lasts for a time T (the period), the source moves a distance $v_S T = v_S/f$ and the wavelength is *shortened* by this amount. Therefore, the observed wavelength λ' is given by

$$\lambda' = \lambda - \Delta\lambda = \lambda - (v_S/f)$$

Since $\lambda = v/f$, the frequency heard by observer A is

$$f' = \frac{v}{\lambda'} = \frac{v}{\lambda - \dfrac{v_S}{f}} = \frac{v}{\dfrac{v}{f} - \dfrac{v_S}{f}}$$

$$f' = f\left(\frac{v}{v - v_S}\right) \tag{17.18}$$

reverberation time, T, was first shown by W. C. Sabine to be approximately proportional to the volume, V, of the room and inversely proportional to the effection absorbing area, A:

$$T(\mathrm{s}) = 0.16\,\frac{V\,(\mathrm{m}^3)}{A\,(\mathrm{m}^2)}$$

Therefore, the desired reverberation time of a concert hall with a specific volume can be engineered by the proper selection and distribution of absorbing surfaces. Many modern concert halls are built with movable absorbing surfaces so that the reverberation time can be tuned to match the type of musical performance.

It is the quality of the reverberant sound, however, that distinguishes acoustically good halls from poor ones. This quality depends on the design of the hall, the structural materials and their distribution, the placement of the musicians and the listeners, and many other factors. For example, walls made of thick solid wood are good reflectors at low frequencies, producing what is called a *warm* sound. Panels suspended from the ceiling cause early first reflection of the direct sound to the listener, producing a sense of *intimacy* even in a large hall. If the space is broken up with many irregularities, such as niches, columns, and chandeliers, the reverberation will be broken into a large number of closely spaced sounds, resulting in a smooth, reverberant *texture.* The use of electronic sound amplification also helps to (1) increase the overall sound intensity, (2) compensate for spatial nonuniformity in loudness, and (3) compensate for nonuniformity in the frequency response of the hall by selectively tuning the amplification in its various frequency ranges.

With good acoustical design, a high-quality musical environment can in principle be created. Ultimately, however, the quality of a musical sound will be determined by the individual listener.

That is, the observed frequency is *increased* when the source moves toward the observer.

In a similar manner, when the source moves away from an observer B at rest (where observer B is to the left of the source, as in Fig. 17.12a), observer B measures a wavelength λ' that is *greater* than λ and hears a *decreased* frequency given by

$$f' = f\left(\frac{v}{v + v_S}\right) \tag{17.19}$$

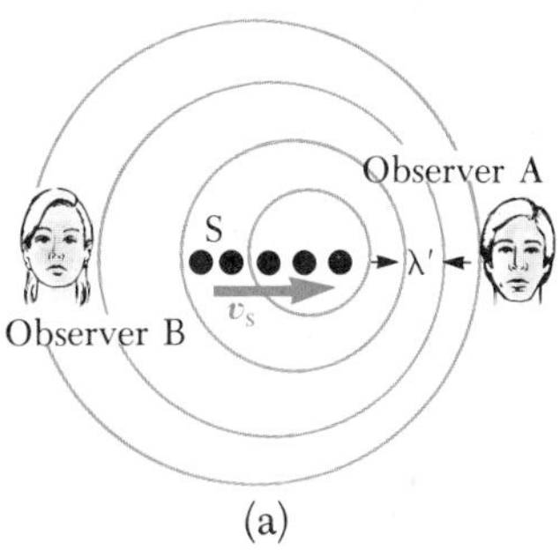

(a)

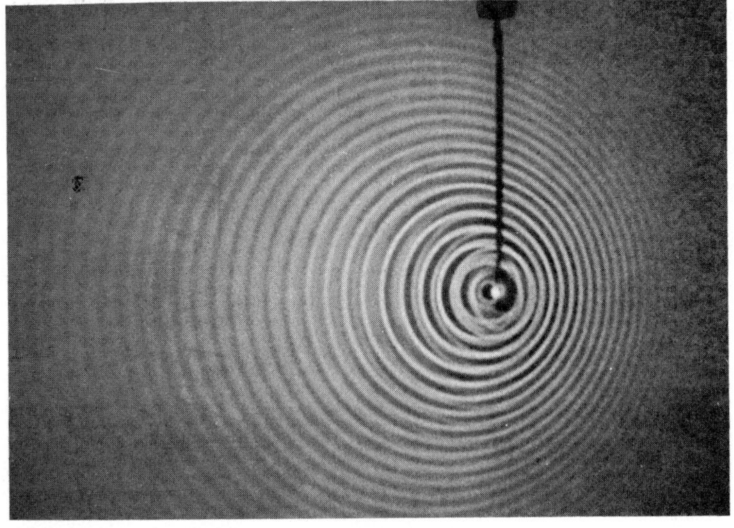

(b)

Figure 17.12 (a) A source S moving with a speed v_S toward a stationary observer A and away from a stationary observer B. Observer A hears an *increased* frequency, and observer B hears a *decreased* frequency. (b) The Doppler effect in water observed in a ripple tank. (Courtesy Educational Development Center, Newton, Mass.)

Combining Equations 17.18 and 17.19, we can express the general relationship for the observed frequency when the source is moving and the observer is at rest as

Frequency heard with source in motion

$$f' = f\left(\frac{v}{v \mp v_S}\right) \tag{17.20}$$

Finally, if both the source and the observer are in motion, one finds the following general relationship for the observed frequency:

Frequency heard with observer and source in motion

$$f' = f\left(\frac{v \pm v_O}{v \mp v_S}\right) \tag{17.21}$$

In this expression, the *upper* signs ($+v_O$ and $-v_S$) refer to motion of one *toward* the other, and the lower signs ($-v_O$ and $+v_S$) refer to motion of one *away* from the other.

EXAMPLE 17.5. The Moving Train Whistle
A train moving at a speed of 40 m/s sounds its whistle, which has a frequency of 500 Hz. Determine the frequencies heard by a stationary observer as the train approaches and recedes from the observer.

Solution: We can use Equation 17.18 to get the apparent frequency as the train approaches the observer. Taking $v = 343$ m/s for the speed of sound in air gives

$$f' = f\left(\frac{v}{v - v_S}\right) = (500 \text{ Hz})\left(\frac{343 \text{ m/s}}{343 \text{ m/s} - 40 \text{ m/s}}\right)$$

$$= 566 \text{ Hz}$$

Likewise, Equation 17.19 can be used to obtain the frequency heard as the train recedes from the observer:

$$f' = f\left(\frac{v}{v + v_S}\right) = (500 \text{ Hz})\left(\frac{343 \text{ m/s}}{343 \text{ m/s} + 40 \text{ m/s}}\right)$$

$$= 448 \text{ Hz}$$

EXAMPLE 17.6. The Noisy Siren
An ambulance travels down a highway at a speed of 75 mi/h. Its siren emits sound at a frequency of 400 Hz. What is the frequency heard by a passenger in a car traveling at 55 mi/h in the opposite direction as the car approaches the ambulance and as the car moves away from the ambulance?

Solution: Let us take the velocity of sound in air to be $v = 343$ m/s and note that 1 mi/h = 0.447 m/s. Therefore, $v_S = 75$ mi/h = 33.5 m/s and $v_O = 55$ mi/h = 24.6 m/s. We can use Equation 17.21 in both cases. As the ambulance and car approach each other, the observed apparent frequency is

$$f' = f\left(\frac{v + v_O}{v - v_S}\right) = (400 \text{ Hz})\left(\frac{343 \text{ m/s} + 24.6 \text{ m/s}}{343 \text{ m/s} - 33.5 \text{ m/s}}\right)$$

$$= 475 \text{ Hz}$$

Likewise, as they recede from each other, a passenger in the car hears a frequency

$$f' = f\left(\frac{v - v_O}{v + v_S}\right) = (400 \text{ Hz})\left(\frac{343 \text{ m/s} - 24.6 \text{ m/s}}{343 \text{ m/s} + 33.5 \text{ m/s}}\right)$$

$$= 338 \text{ Hz}$$

Note that the *change* in frequency as detected from the car is 475 − 338 = 137 Hz, which is more than 30% of the actual frequency emitted.

Shock Waves

Now let us consider what happens when the source velocity v_S *exceeds* the wave velocity v. This situation is described graphically in Figure 17.13. The circles represent spherical wavefronts emitted by the source at various times during its motion. At $t = 0$, the source is at S_0, and at some later time t, the source is at S_n. In the time t, the wavefront centered at S_0 reaches a radius of vt. In this same interval, the source travels a distance v_St to S_n. At the instant the source is at S_n, waves are just beginning to be generated and so the waterfront has zero radius at this point. The line drawn from S_n to the wavefront centered

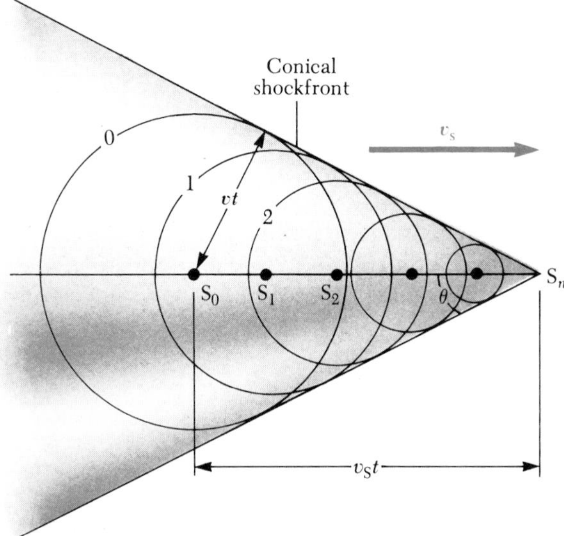

Figure 17.13 Representation of a shock wave produced when a source moves from S_0 to S_n with a speed v_S, which is *greater* than the wave speed v in that medium. The envelope of the wavefronts forms a cone whose apex angle is given by $\sin\theta = v/v_S$.

on S_0 is tangent to all other wavefronts generated at intermediate times. Thus, we see that the envelope of these waves is a cone whose apex angle θ is given by

$$\sin\theta = v/v_S$$

The ratio v_S/v is referred to as the *mach number.* The conical wavefront produced when $v_S > v$ (supersonic speeds) is known as a *shock wave.* An interest-

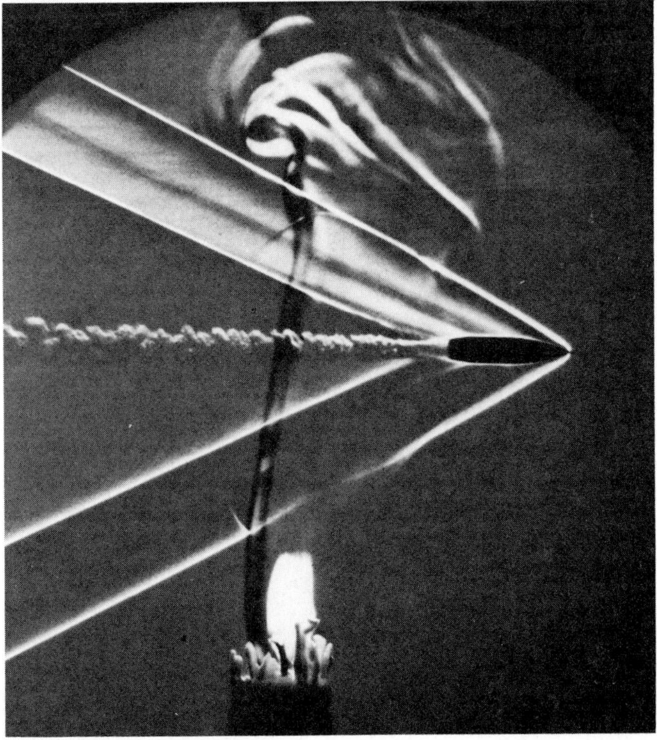

A bullet traveling in air faster than the speed of sound. The shock waves accompanying the bullet are made visible by smoke from a candle.

ing analogy to shock waves is the *V*-shaped wavefronts produced by a boat (the bow wave) when the boat's speed exceeds the speed of the surface water waves.

Jet airplanes traveling at supersonic speeds produce shock waves, which are responsible for the loud explosion, or "sonic boom," one hears. The shock wave carries a great deal of energy concentrated on the surface of the cone, with correspondingly large pressure variations. Such shock waves are unpleasant to hear and can cause damage to buildings when aircraft fly supersonically at low altitudes. In fact, an airplane flying at supersonic speeds produces a double boom because two shock fronts are formed, one from the nose of the plane and one from the tail (Fig. 17.14).

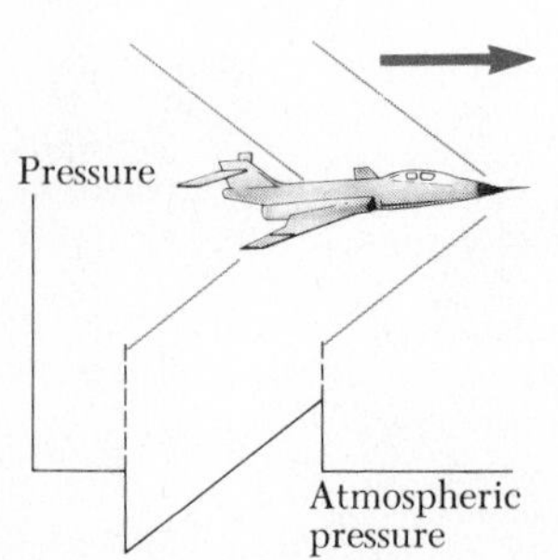

Figure 17.14 Two shock waves produced by the nose and tail of a jet airplane traveling at supersonic speeds.

17.6 SUMMARY

Sound waves are longitudinal in nature and travel through a compressible medium with a speed that depends on the compressibility and inertia of that medium. The **speed of sound** in a medium of compressibility B and density ρ is

Speed of sound in a compressible medium

$$v = \sqrt{B/\rho} \tag{17.1}$$

In the case of harmonic sound waves, the **variation in pressure** from the equilibrium value is given by

Pressure variation

$$\Delta P = \Delta P_{\mathrm{m}} \sin(kx - \omega t) \tag{17.7}$$

where ΔP_{m} is the **pressure amplitude.** The pressure wave is 90° out of phase with the displacement wave. If the displacement amplitude is s_{m}, ΔP_{m} has the value

Pressure amplitude

$$\Delta P_{\mathrm{m}} = \rho v \omega s_{\mathrm{m}} \tag{17.8}$$

The **intensity of a harmonic sound wave,** which is the power per unit area, is given by

Intensity of a sound wave

$$I = \tfrac{1}{2}\rho(\omega s_{\mathrm{m}})^2 v = \frac{\Delta P_{\mathrm{m}}^{\ 2}}{2\rho v} \tag{17.10}$$

The **intensity of a spherical wave** produced by a point source is proportional to the average power emitted and inversely proportional to the square of the distance from the source.

The change in frequency heard by an observer whenever there is relative motion between the source and observer is called the **Doppler effect.** If the observer moves with a speed v_{O} and the source is at rest, the observed frequency f' is

Frequency heard by an observer in motion

$$f' = f\left(\frac{v \pm v_{\mathrm{O}}}{v}\right) \tag{17.17}$$

where the positive sign is used when the observer moves toward the source and the negative sign refers to motion away from the source.

If the *source moves* with a speed v_{S} and the observer is at rest, the observed frequency is

$$f' = f\left(\frac{v}{v \mp v_S}\right) \quad (17.20)$$

Frequency heard with source in motion

where $-v_S$ refers to motion *toward* the observer and $+v_S$ refers to motion *away* from the observer.

When the *observer and source are both moving*, the observed frequency is

$$f' = f\left(\frac{v \pm v_O}{v \mp v_S}\right) \quad (17.21)$$

Frequency heard with observer and source in motion

QUESTIONS

1. Why are sound waves characterized as being longitudinal?
2. As a result of a distant explosion, an observer senses a ground tremor and then hears the explosion. Explain.
3. If an alarm clock is placed in a good vacuum and then activated, no sound will be heard. Explain.
4. Some sound waves are harmonic, whereas others are not. Give an example of each.
5. In Example 17.4, we found that a point source with a power output of 80 W reduces to an intensity level of 40 dB at a distance of about 15 miles. Why do you suppose you cannot normally hear a rock concert going on 15 miles away?
6. If the distance from a point source is tripled, by what factor does the intensity decrease?
7. Explain how the Doppler effect is used with microwaves to determine the speed of an automobile.
8. If you are in a moving vehicle, explain what happens to the frequency of your echo as you move *toward* a canyon wall. What happens to the frequency as you move *away* from the wall?
9. Suppose an observer and a source of sound are both at rest, and a strong wind blows toward the observer. Describe the effect of the wind (if any) on (a) the observed wavelength, (b) the observed frequency, and (c) the wave velocity.
10. Of the following sounds, which is most likely to have an intensity level of 60 dB: a rock concert, the turning of a page in this text, normal conversation, a cheering crowd at a football game, or background noise at a church?
11. Estimate the decibel level of each of the sounds in Question 10.
12. A binary star system consists of two stars revolving about each other. If we observe the light reaching us from one of these stars as it makes one complete revolution about the other, what does the Doppler effect predict will happen to this light?
13. How could an object move with respect to an observer such that the sound from it is not shifted in frequency? Why is the intensity of an echo less than that of the original sound?

PROBLEMS

Section 17.1 Velocity of Sound Waves

1. Find the velocity of sound in mercury, which has a bulk modulus of about 2.8×10^{19} N/m^2 and a density of about 13.7 gm/cm^3.
2. The ocean floor is underlain by a layer of basalt that constitutes the crust, or uppermost layer of the earth in this region. Below this crust is found more dense peridotite rock, which forms the earth's mantle. The boundary between these two layers is called the Mohorovičić discontinuity ("Moho" for short). If an explosive charge is set off at the surface of the basalt, it generates a seismic wave that is reflected back at the Moho. If the velocity of this wave in basalt is 6.5 km/s, and the two-way travel time is 1.85 s, what is the thickness of this oceanic crust?
3. (a) What are the SI units of bulk modulus as expressed in Equation 17.2? (b) Show that the SI units of $\sqrt{B/\rho}$ are m/s, as required by Equation 17.1.
4. The density of aluminum is 2.7×10^3 kg/m^3. Use the value for the speed of sound in aluminum given in Table 17.1 to calculate Young's modulus for this material.

Section 17.2 Harmonic Sound Waves

(In this section, use the following values as needed unless otherwise specified: the equilibrium density of air, $\rho = 1.2$ kg/m^3; the velocity of sound in air, $v =$ 343 m/s. Also, pressure variations ΔP are measured relative to atmospheric pressure.)

5. Calculate the pressure amplitude of a 2000-Hz sound wave in air if the displacement amplitude is equal to 2×10^{-8} m.
6. A sound wave in air has a pressure amplitude equal to 4×10^{-3} N/m^2. Calculate the displacement amplitude of the wave at a frequency of 10 kHz.
7. The pressure amplitude corresponding to the threshold of hearing is 2.9×10^{-5} N/m^2. At what frequency will a sound wave in air have this pressure amplitude if the displacement amplitude is 2.8×10^{-10} m?

8. An experimenter wishes to generate in air a sound wave that has a displacement amplitude equal to 5.5×10^{-6} m. The pressure amplitude is to be limited to 8.4×10^{-1} N/m^2. What is the minimum wavelength the sound wave can have?
9. A sound wave in air has a pressure amplitude of 4 N/m^2 and a frequency of 5000 Hz. $\Delta P = 0$ at the point $x = 0$ when $t = 0$. (a) What is ΔP at $x = 0$ when $t = 2 \times 10^{-4}$ s and (b) what is ΔP at $x = 0.02$ m when $t = 0$?
10. The harmonic displacement of a sound wave is described by $s(x, t) = 0.006 \cos[\pi(5.834x - 2000t)]$, where x is in m and t is in s. (a) What are the values of frequency, wavelength, and speed of the wave? (b) What is the displacement at the point $x = 0.05$ m when $t = 0$? (c) What is the displacement at $x = 0$ when $t = 3.75 \times 10^{-4}$ s?
11. Consider the sound wave whose harmonic displacement is described in Problem 10. What is the pressure variation at $x = 0$ when $t = \pi/2\omega$?
12. Write an expression that describes the pressure variation as a function of position and time for a harmonic sound wave in air if $\lambda = 0.1$ m and $\Delta P_m = 0.2$ N/m^2.
13. Write the function that describes the displacement wave corresponding to the pressure wave in Problem 12.

Section 17.3 Energy and Intensity of Harmonic Sound Waves

14. Calculate the intensity level in dB of a sound wave that has an intensity of 4 μW/m^2.
15. A vacuum cleaner has a measured sound level of 70 dB. What is the intensity of this sound in W/m^2?
16. (a) Calculate the intensity in W/m^2 of the wave described in Problem 9. (b) Express this intensity in dB.
17. The intensity of a sound wave at a fixed distance from a speaker vibrating at 1000 Hz is 0.6 W/m^2. (a) Determine the intensity if the frequency is increased to 2500 Hz while a *constant* displacement amplitude is maintained. (b) Calculate the intensity if the frequency is reduced to 500 Hz and the displacement amplitude is doubled.
18. Calculate the pressure amplitude corresponding to a sound intensity of 120 dB (a rock concert).

Section 17.4 Spherical and Plane Waves

19. An experiment requires a sound intensity of 1.2 W/m^2 at a distance of 4 m from a speaker. What power output is required?
20. A source emits sound waves with a uniform power of 100 W. At what distance will the intensity be just below the threshold of pain, which is 1 W/m^2?
21. The sound level at a distance of 3 m from a source is 120 dB. At what distance will the sound level be (a) 100 dB and (b) 10 dB?
22. Spherical waves of wavelength 25 cm are propagating outward from a point source. (a) Compare the wave amplitude at $r = 50$ cm and $r = 200$ cm. (b) Compare the intensity at $r = 50$ cm with the intensity at $r = 100$ cm. (c) Compare the phase of the wave function at a specific time at $r = 50$ cm and $r = 75$ cm.

*Section 17.5 The Doppler Effect

23. At what speed should a supersonic aircraft fly so that the conical wavefront will have an apex half-angle of 50°?
24. The Concorde flies at mach 1.5. What is the angle between the direction of propagation of the shock wave and the direction of the plane's velocity?
25. A commuter train passes a passenger platform at a constant speed of 40 m/s. The train horn is sounded at its characteristic frequency of 320 Hz. (a) What change in frequency is observed by a person on the platform as the train passes? (b) What wavelength does a person on the platform observe as the train approaches?
26. Standing at a crosswalk, you hear a frequency of 510 Hz from the siren on an approaching police car. After the police car passes, the observed frequency of the siren is 430 Hz. Determine the car's speed from these observations.
27. A projectile has a velocity of 725 m/s in air. (a) What is the apex angle of the shock wave associated with the projectile? (b) What is the mach number of the projectile?
28. A train is moving parallel to a highway with a constant speed of 20 m/s. A car is traveling in the same direction as the train with a speed of 40 m/s. As the auto overtakes and passes the train, the car horn sounds at a frequency of 510 Hz and the train horn sounds at a frequency of 320 Hz. (a) What frequency does an occupant of the car observe for the train horn just before passing? (b) What frequency does a train passenger observe for the car horn just after passing?
29. A train passenger hears a frequency of 520 Hz as the train approaches a bell on a trackside safety gate; the bell is actually emitting a signal of 500 Hz. What frequency will the passenger hear just after passing the bell?
30. When high-energy, charged particles move through a transparent medium with a velocity greater than the velocity of light in that medium, a shock wave, or bow wave, of light is produced. This phenomenon is called the *Cerenkov effect* and can be observed in the vicinity of the core of a swimming pool reactor due to high-speed electrons moving through the water. In a particular case, the Cerenkov radiation produces a wavefront with a cone angle of 53°. Calculate the velocity of the electrons in the water. (Use 2.25×10^8 m/s as the velocity of light in water.)

GENERAL PROBLEMS

31. (a) The sound level of a jackhammer is measured as 130 dB and that of a siren as 120 dB. Find the ratio of the intensities of the two sound sources. (b) Two sources have measured intensities of $I_1 = 100$ μW/m^2 and $I_2 = 200$ μW/m^2. By how many dB is source 1 lower than source 2?

32. The measured speed of sound in copper is 3560 m/s, and the density of copper is 8.89 g/cm^3. Based on this information, by what percent would you expect a block of copper to decrease in volume when subjected to a uniform external (gauge) pressure of 2 atm?

33. Two ships are moving along a line due east. The trailing vessel has a speed relative to a land-based observation point of 64 km/h, and the leading ship has a speed of 45 km/h relative to that station. The two ships are in a region of the ocean where the current is moving uniformly due west at 10 km/h. The trailing ship transmits a sonar signal at a frequency of 1200 Hz. What frequency is monitored by the leading ship? (Use 1520 m/s as the speed of sound in ocean water.)

•34. In order to be able to determine his speed, a skydiver carries a tone generator with him. A friend on the ground at the landing site has equipment for receiving and analyzing sound waves. While the skydiver is falling at terminal speed, his tone generator emits a steady tone of frequency 500 Hz. (Assume that the air is calm and the sound speed is 343 m/s, independent of altitude.) (a) If his friend on the ground (directly beneath the skydiver) receives waves of frequency 610 Hz, what is the skydiver's speed of descent? (b) If the skydiver were also carrying sound-receiving equipment sensitive enough to detect waves reflecting from the ground, what frequency would he receive?

•35. A high-tech model airplane equipped with a sonar range and speed finder is headed straight for a brick wall at constant speed. At $t = 0$, it emits a short burst of waves of frequency f_e. At $t = T$, it receives the echo; the received frequency is f_r. Let v represent the speed of sound in air, and let v_p represent the speed of the airplane. (a) Obtain an equation for v_p in terms of f_e, f_r, and v. (b) Let d_e represent the distance between the model and the wall at $t = 0$. Obtain an equation for d_e in terms of v, v_p, and T, and then use the result of part (a) to write d_e in terms of f_e, f_r, v_p, and T. (c) Use the results of parts (a) and (b) to find d_r, the distance between the model and the wall at $t = T$. Express d_r in terms of f_e, f_r, v, and T. (d) Evaluate v_p, d_e, and d_r for the following case: $v = 343$ m/s, $f_e = 4000$ Hz, $f_r = 4240$ Hz, and $T = 0.295$ s.

36. Consider a longitudinal (compressional) wave of wavelength λ traveling with speed v along the x direction through a medium of density ρ. The *displacement* of the molecules of the medium from their equilibrium position is given by

$$s = s_m \sin(kx - \omega t)$$

Show that the pressure variation in the medium is given by

$$P = -\left(\frac{2\pi\rho v^2}{\lambda} s_m\right)\cos(kx - \omega t)$$

37. A meteoroid the size of a truck enters the earth's atmosphere at a speed of 20 km/s and is not significantly slowed before entering the ocean. (a) What is the mach angle of the shock wave from the meteoroid in the atmosphere? (Use 331 m/s as the sound speed.) (b) Assuming that the meteoroid survives the impact with the ocean surface, what is the (initial) mach angle of the shock wave that the meteoroid produces in the water? (Use the wave speed for sea water given in Table 17.1.)

38. (a) Use values from Table 17.2 to determine the resultant intensity in dB when a vacuum cleaner and a power mower are operated against a background of busy traffic. (b) In Table 17.2, a buzzing mosquito is rated at 40 dB and normal conversation at 50 dB. How many buzzing mosquitos are required to equal normal conversation in sound intensity?

39. By proper excitation, it is possible to produce both longitudinal and transverse waves in a long metal rod. A particular metal rod is 150 cm long and has a radius of 0.2 cm and a mass of 50.9 g. Young's modulus for the material is 6.8×10^{11} dynes/cm^2. What must the tension in the rod be if the ratio of the speed of longitudinal waves to the speed of transverse waves is 8?

•40. Three metal rods are located relative to each other as shown in Figure 17.15, where $L_1 + L_2 = L_3$. Values of density and Young's modulus for the three materials are $\rho_1 = 2.7 \times 10^3$ kg/m^3, $Y_1 = 7 \times 10^{10}$ N/m^2, $\rho_2 = 11.3 \times 10^3$ kg/m^3, $Y_2 = 1.6 \times 10^{10}$ N/m^2, $\rho_3 = 8.8 \times 10^3$ kg/m^3, and $Y_3 = 11 \times 10^{10}$ N/m^2. (a) If $L_3 = 1.5$ m, what must the ratio L_1/L_2 be if a sound wave is to travel the length of rods 1 *and* 2 in the same time required to travel the length of rod 3? (b) If the frequency of the source is 4000 Hz, determine the phase difference between the wave traveling along rods 1 and 2 and the one traveling along rod 3.

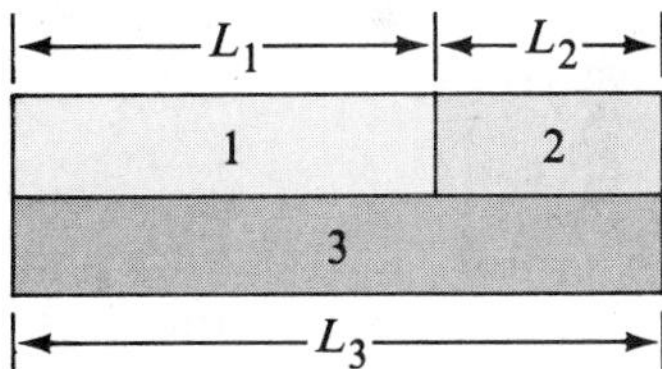

Figure 17.15 (Problem 40).

41. The gas filling the tube shown in Figure 17.3 is air at 20°C and at a pressure of 1.5×10^5 N/m^2. The piston shown is driven at a frequency of 600 Hz. The diameter of the piston is 10 cm, and the amplitude of its motion is 0.1 cm. What power must be supplied to maintain the oscillation of the piston?

42. Consider plane harmonic sound waves propagating in three different media at 0°C: air, water, and iron. Each wave has the same intensity (I_0) and the same angular frequency (ω_0). (a) Compare the values of λ (the wavelength) in the three media. (b) Compare the values of s_m (the displacement amplitude) in the three media. (*Hint:* Refer to Tables 15.2 and 17.1.) (c) Compare the values of ΔP_m (the pressure amplitude) in the three media. (d) For $\omega_0 = 2000\pi$ rad/s and $I_0 = 10^{-6}$ W/m^2 (60 dB), evaluate λ, s_m, and ΔP_m for each of three media.

18
Superposition and Standing Waves

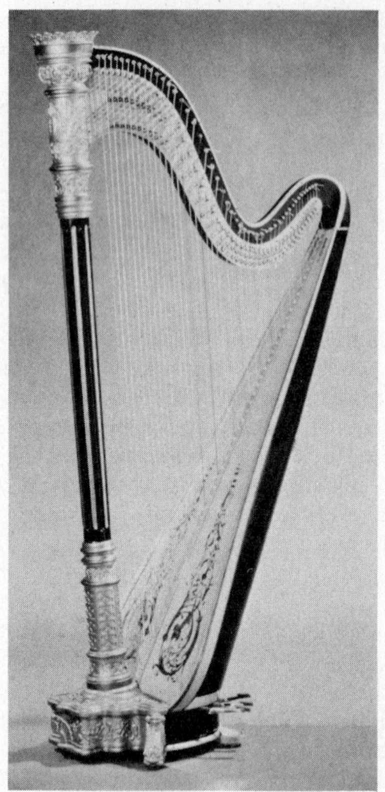

An important aspect of waves is the combined effect of two or more waves traveling in the same medium. For instance, what happens to a string when a wave traveling toward its fixed end is reflected back on itself? What is the pressure variation in the air when the instruments of an orchestra sound together?

In a linear medium, that is, one in which the restoring force of the medium is proportional to the displacement of the medium, one can apply the *principle of superposition* to obtain the resultant disturbance. This principle can be applied to many types of waves, including waves on strings, sound waves, surface water waves, and electromagnetic waves. The superposition principle states that the actual displacement of any part of the disturbed medium equals the algebraic sum of the displacements caused by the individual waves. We discussed this principle as it applied to wave pulses in Chapter 16. The term *interference* was used to describe the effect produced by combining two waves moving simultaneously through a medium.

This chapter is concerned with the superposition principle as it applies to harmonic waves. If the harmonic waves that combine in a given medium have the same frequency and wavelength, one finds that a stationary pattern, called a *standing wave,* can be produced at certain frequencies under certain circumstances. For example, a stretched string fixed at both ends has a discrete set of oscillation patterns, called *modes of vibration,* which depend upon the tension and mass per unit length of the string. These modes of vibration are found in stringed musical instruments. Other musical instruments, such as the organ and flute, make use of the natural frequencies of sound waves in hollow pipes. Such frequencies depend upon the length of the pipe, its shape, and upon whether one end is open or closed.

We also consider the superposition and interference of waves with different frequencies and wavelengths. When two sound waves with nearly the same frequency interfere, one hears variations in the loudness called *beats.* The beat frequency corresponds to the rate of alternation between constructive and destructive interference. Finally, we describe how any complex periodic waveform can, in general, be described by a sum of sine and cosine functions.

18.1 SUPERPOSITION AND INTERFERENCE OF HARMONIC WAVES

The superposition principle tells us that when two or more waves move in the same linear medium, the net displacement of the medium (the resultant wave) at any point equals the algebraic sum of the displacements of all the waves. Let us apply this superposition principle to two harmonic waves traveling in the same direction in a medium. If the two waves are traveling to the right and have the same frequency, wavelength, and amplitude but differ in phase, we can express their individual wave functions as

$$y_1 = A_0 \sin(kx - \omega t) \qquad \text{and} \qquad y_2 = A_0 \sin(kx - \omega t - \phi)$$

Hence, the resultant wave function y is given by

$$y = y_1 + y_2 = A_0 \, [\sin(kx - \omega t) + \sin(kx - \omega t - \phi)]$$

In order to simplify this expression, it is convenient to make use of the following trigonometric identity:

$$\sin a + \sin b = 2 \cos\left(\frac{a-b}{2}\right) \sin\left(\frac{a+b}{2}\right)$$

If we let $a = kx - \omega t$ and $b = kx - \omega t - \phi$, we find that the resultant wave y reduces to

Resultant of two traveling harmonic waves

$$y = \left(2A_0 \cos \frac{\phi}{2}\right) \sin\left(kx - \omega t - \frac{\phi}{2}\right) \tag{18.1}$$

There are several important features of this result. The resultant wave function y is also harmonic and has the *same* frequency and wavelength as the individual waves. The amplitude of the resultant wave is $2A_0 \cos(\phi/2)$, and its phase is equal to $\phi/2$. If the phase constant ϕ equals 0, then $\cos(\phi/2) = \cos 0 = 1$ and the amplitude of the resultant wave is $2A_0$. In other words, the amplitude of the resultant wave is twice as large as the amplitude of either individual wave. In this case, the waves are said to be everywhere *in phase* and thus *interfere constructively*. That is, the crests and troughs of the individual waves occur at the same positions, as is shown by the broken lines in Figure 18.1a. In general, constructive interference occurs when $\cos(\phi/2) = \pm 1$, or when $\phi = 0, 2\pi, 4\pi, \ldots$. On the other hand, if ϕ is equal to π radians (or any *odd* multiple of π) then $\cos(\phi/2) = \cos(\pi/2) = 0$ and the resultant wave has *zero* amplitude everywhere. In this case, the two waves *interfere destructively*. That is, the crest of one wave coincides with the trough of the second (Fig. 18.1b) and their displacements cancel at every point. Finally, when the phase constant has an arbitrary value between 0 and π, as in Figure 18.1c, the resultant wave has an amplitude whose value is somewhere between 0 and $2A_0$.

Constructive interference

Destructive interference

Interference of Sound Waves

One simple device for demonstrating interference of sound waves is illustrated in Figure 18.2. Sound from a loudspeaker S is sent into a tube at P, where there is a T-shaped junction. Half the sound intensity travels in one direction and half in the opposite direction. Thus, the sound waves that reach the receiver R at the other side can travel along two different paths. The receiver may be a microphone whose output is amplified and fed into ear-

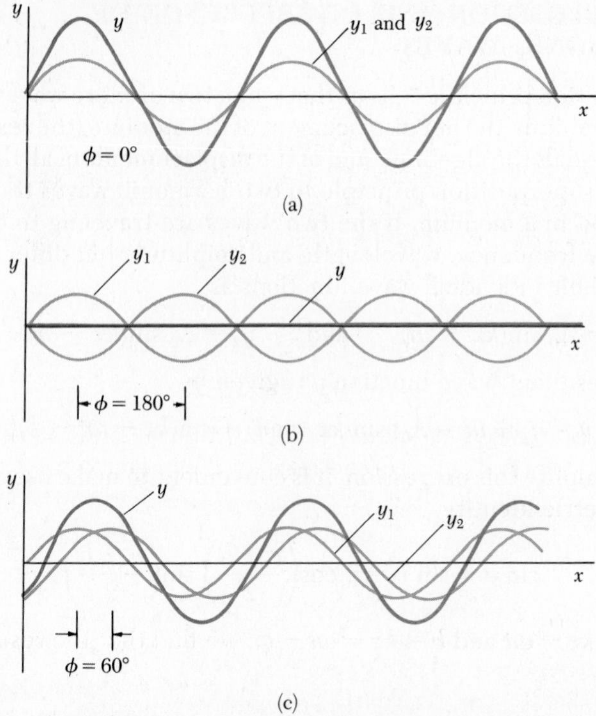

Figure 18.1 The superposition of two waves with amplitudes y_1 and y_2. (a) When the two waves are in phase, the result is constructive interference. (b) When the two waves are 180° out of phase, the result is destructive interference. (c) When the phase angle lies in the range $0 < \phi < 180°$, the resultant y fails somewhere between that shown in (a) and that shown in (b).

phones or an oscilloscope. The total distance from the speaker to the receiver is called the *path length, r*. The path length for the lower path is fixed at r_1. Along the upper path, the path length r_2 can be varied by sliding the U-shaped tube, similar to that on a slide trombone. When the difference in the path lengths $\Delta r = |r_2 - r_1|$ is either zero or some integral multiple of the wavelength λ, the two waves reaching the receiver will be in phase and will interfere constructively, as in Figure 18.1a. For this case, a maximum in the sound intensity will be detected at the receiver. If the path length r_2 is adjusted such that the path difference Δr is $\lambda/2$, $3\lambda/2$, . . . , $n\lambda/2$ (for n odd), the two waves will be exactly 180° out of phase at the receiver and hence will cancel each

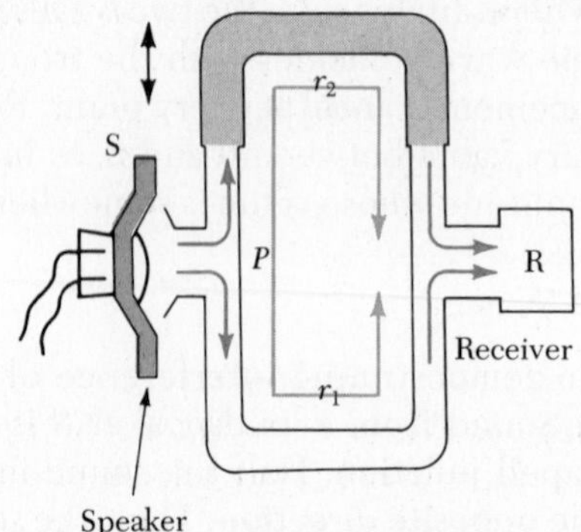

Figure 18.2 An acoustical system for demonstrating interference of sound waves. Sound from the speaker propagates into a tube and splits into two parts at P. The two waves, which superimpose at the opposite side, are detected at R. Note that the upper path length, r_2, can be varied by the sliding section.

other. In this case of completely destructive interference, no sound will be detected at the receiver. This simple experiment is a striking illustration of the phenomenon of interference. In addition, it demonstrates the fact that a phase difference may arise between two waves generated by the same source when they travel along paths of unequal lengths.

It is often useful to express the path difference in terms of the phase difference ϕ between the two waves. Since a path difference of one wavelength corresponds to a phase difference of 2π radians, we obtain the ratio $\lambda/2\pi = \Delta r/\phi$, or

$$\Delta r = \frac{\lambda}{2\pi}\phi \tag{18.2}$$

Relationship between path difference and phase angle

There are many other examples of interference phenomena in nature. Later, in Chapter 37, we shall describe several interesting interference effects involving light waves.

EXAMPLE 18.1. Two Speakers Driven by the Same Source

Two speakers are driven by the same oscillator at a frequency of 2000 Hz. The speakers are separated by a distance of 3 m, as in Figure 18.3. A listener is originally at a point O located 8 m away along the center line. How far must the listener walk, perpendicular to the center line, before reaching the first minimum in the sound intensity?

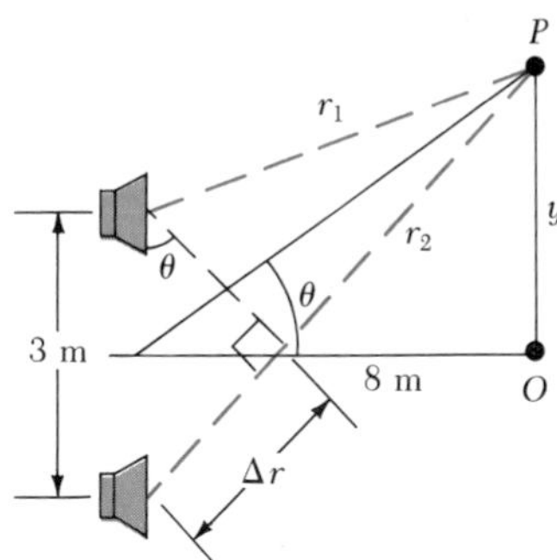

Figure 18.3 (Example 18.1).

Solution: Since the speed of sound in air is 330 m/s and since $f = 2000$ Hz, the wavelength is given by

$$\lambda = \frac{v}{f} = \frac{330 \text{ m/s}}{2000 \text{ Hz}} = 0.165 \text{ m}$$

The first minimum occurs when the two waves reaching P are 180° out of phase, or when their path difference, $r_2 - r_1$, equals $\lambda/2$. Therefore, the path difference is given by

$$\Delta r = r_2 - r_1 = \frac{\lambda}{2} = \frac{0.165 \text{ m}}{2} = 0.0825 \text{ m}$$

From the small right triangle in Figure 18.3, we see that, to a good approximation, $\sin\theta = \Delta r/3$ for small values of θ, or

$$\sin\theta = \frac{\Delta r}{3} = \frac{0.0825 \text{ m}}{3 \text{ m}} = 0.0275$$

$$\theta = 1.58^\circ$$

From the large right triangle in Figure 18.3, we find that $\tan\theta = y/8$, or

$$y = 8\tan\theta = 8\tan 1.58^\circ = 0.22 \text{ m}$$

That is, the listener will hear minima in the resultant sound intensity 22 cm to either side of the center line. If the listener remains at these positions, at what other frequencies will minima be heard?

18.2 STANDING WAVES

If a stretched string is clamped at both ends, traveling waves will reflect from the fixed ends, creating waves traveling in both directions. The incident and reflected waves will combine according to the superposition principle.

Consider two sinusoidal waves in the same medium with the same amplitude, frequency, and wavelength, but traveling in *opposite* directions. Their wave functions can be written

$$y_1 = A_0\sin(kx - \omega t) \qquad \text{and} \qquad y_2 = A_0\sin(kx + \omega t)$$

where y_1 represents a wave traveling to the right and y_2 represents a wave traveling to the left. Adding these two functions gives the resultant wave function y:

$$y = y_1 + y_2 = A_0 \sin(kx - \omega t) + A_0 \sin(kx + \omega t)$$

where $k = 2\pi/\lambda$ and $\omega = 2\pi f$, as usual. Using the trigonometric identity $\sin(a \pm b) = \sin a \cos b \pm \cos a \sin b$, this reduces to

Wave function for a standing wave

$$y = (2A_0 \sin kx) \cos \omega t \tag{18.3}$$

This expression represents the wave function of a **standing wave.** From this result, we see that a standing wave has an angular frequency ω and an amplitude given by $2A_0 \sin kx$ (the quantity in the parentheses of Eq. 18.3). That is, every particle of the string vibrates in simple harmonic motion with the same frequency. However, the amplitude of motion of a given particle depends on x. This is in contrast to the situation involving a traveling harmonic wave, in which all particles oscillate with both the same amplitude and the same frequency.

Because the amplitude of the standing wave at any value of x is equal to $2A_0 \sin kx$, we see that the *maximum* amplitude has the value $2A_0$. This occurs when the coordinate x satisfies the condition $\sin kx = 1$, or when

$$kx = \frac{\pi}{2}, \frac{3\pi}{2}, \frac{5\pi}{2}, \ldots$$

Since $k = 2\pi/\lambda$, the positions of maximum amplitude, called **antinodes,** are given by

Position of antinodes

$$x = \frac{\lambda}{4}, \frac{3\lambda}{4}, \frac{5\lambda}{4}, \ldots = \frac{n\lambda}{4} \tag{18.4}$$

where $n = 1, 3, 5, \ldots$. Note that *adjacent antinodes are separated by a distance of* $\lambda/2$. Similarly, the standing wave has a *minimum* amplitude of zero when x satisfies the condition $\sin kx = 0$, or when

$$kx = \pi, 2\pi, 3\pi, \ldots$$

giving

Position of nodes

$$x = \frac{\lambda}{2}, \lambda, \frac{3\lambda}{2}, \ldots = \frac{n\lambda}{2} \tag{18.5}$$

where $n = 1, 2, 3, \ldots$. These points of zero amplitude, called **nodes,** *are also spaced by* $\lambda/2$. The distance between a node and an adjacent antinode is $\lambda/4$.

A graphical description of the standing wave patterns produced at various times by two waves traveling in opposite directions is shown in Figure 18.4. The upper part of each figure represents the individual traveling waves, and the lower part represents the standing wave patterns. The nodes of the standing wave are labeled N, and the antinodes are labeled A. At $t = 0$ (Fig. 18.4a), the two waves are identical spatially, giving a standing wave of maximum amplitude, $2A_0$. One quarter of a period later, at $t = T/4$ (Fig. 18.4b), the individual waves have moved one quarter of a wavelength (one to the right and the other to the left). At this time, the individual amplitudes are equal and opposite for all values of x, and hence the resultant wave has zero amplitude everywhere. At $t = T/2$ (Fig. 18.4c), the individual waves are again identical

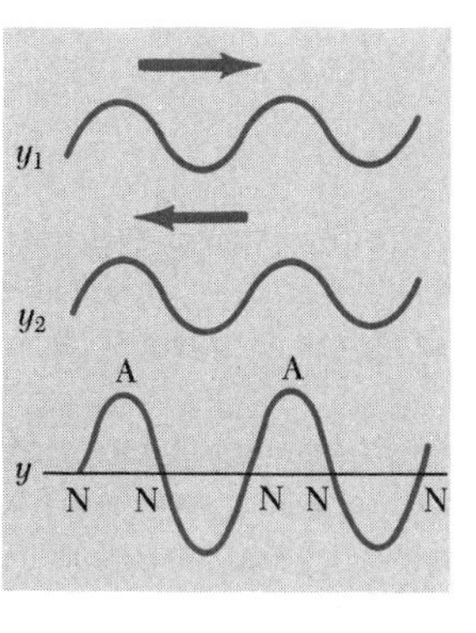

(a) $t = 0$

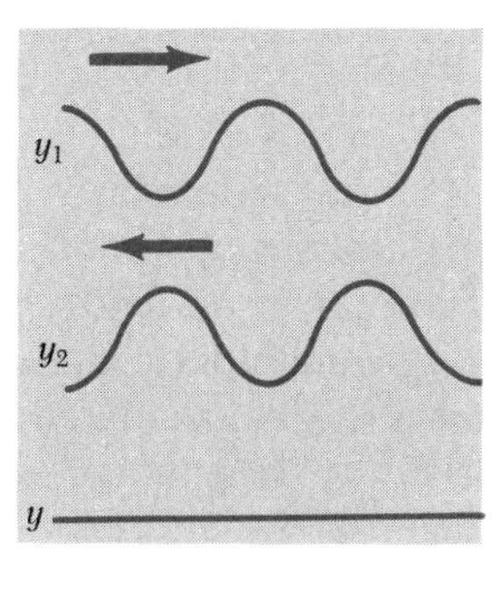

(b) $t = T/4$

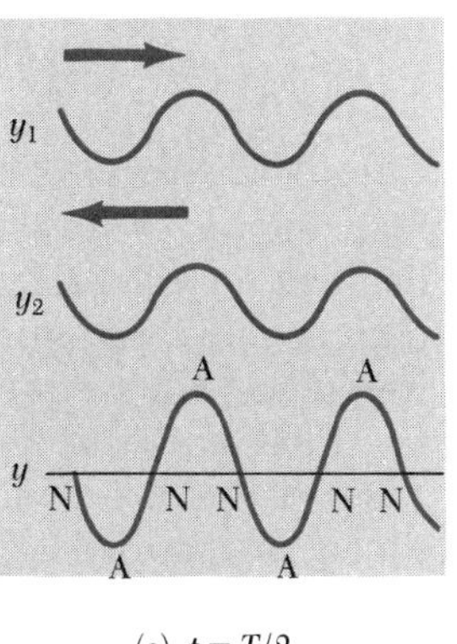

(c) $t = T/2$

Figure 18.4 Standing wave patterns at various times produced by two waves of equal amplitude traveling in *opposite* directions. For the resultant wave y, the nodes (N) are points of zero displacement, and the antinodes (A) are points of maximum amplitude.

spatially, producing a standing wave pattern that is inverted relative to the $t = 0$ pattern.

It is instructive to describe the energy associated with the motion of a standing wave. To illustrate this point, consider a standing wave formed on a stretched string fixed at each end, as in Figure 18.5. All points on the string oscillate vertically with the same frequency except for the nodes, which are stationary. Furthermore, the various points have different amplitudes of motion. Figure 18.5 represents snapshots of the standing wave at various times over one half of a cycle. Note that since the nodal points are stationary, no energy is transmitted along the string across the center nodal point. For this reason, standing waves are often called **stationary waves.** Each point on the string executes simple harmonic motion in the vertical direction. That is, one can view the standing wave as a large number of oscillators vibrating parallel to each other. The energy of the vibrating string continuously alternates between elastic potential energy, at which time the string is momentarily stationary (Fig. 18.5a), and kinetic energy, at which time the string is horizontal and the particles have their maximum speed (Fig. 18.5c). The string particles have both potential energy and kinetic energy at intermediate times (Figs. 18.5b and 18.5d).

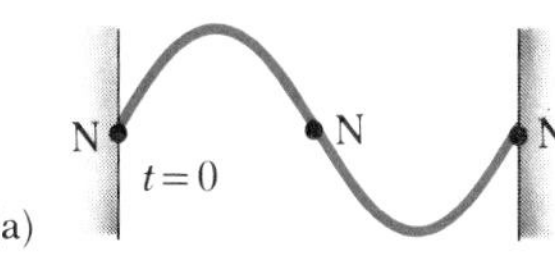

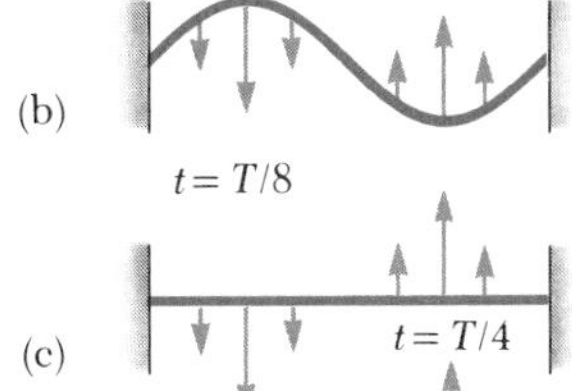

(c) $t = T/4$

(d) $t = 3T/8$

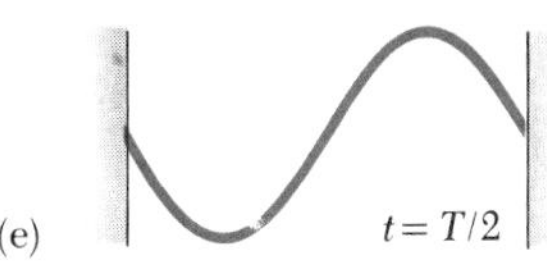

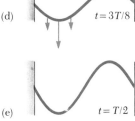

Figure 18.5 A standing wave pattern in a stretched string showing snapshots during one half of a cycle. (a) At $t = 0$, the string is momentarily at rest, and so $K = 0$ and all of the energy is potential energy U associated with the vertical displacements of the string segments (deformation energy). (b) At $t = T/8$, the string is in motion, and the energy is half kinetic and half potential, (c) At $t = T/4$, the string is horizontal (undeformed) and therefore $U = 0$ and all of the energy is kinetic. The motion continues as indicated, and ultimately the initial configuration (a) is repeated.

EXAMPLE 18.2. Formation of a Standing Wave

Two waves traveling in opposite directions produce a standing wave. The individual wave functions are given by

$$y_1 = 4 \sin(3x - 2t) \text{ cm}$$

$$y_2 = 4 \sin(3x + 2t) \text{ cm}$$

where x and y are in cm. (a) Find the maximum displacement of the motion at $x = 2.3$ cm.

When the two waves are summed up, the result is a standing wave whose function is given by Equation 18.3, with $A_0 = 4$ cm and $k = 3 \text{ cm}^{-1}$:

$$y = (2A_0 \sin kx) \cos \omega t = (8 \sin 3x) \cos \omega t \text{ cm}$$

Thus, the *maximum* displacement of the motion at the position $x = 2.3$ cm is given by

$$y_{\max} = 8 \sin 3x]_{x=2.3} = 8 \sin(6.9 \text{ rad}) = 4.63 \text{ cm}$$

(b) Find the positions of the nodes and antinodes.

Since $k = 2\pi/\lambda = 3 \text{ cm}^{-1}$, we see that $\lambda = 2\pi/3$ cm. Therefore, from Equation 18.4 we find that the *antinodes* are located at

$$x = n\left(\frac{\pi}{6}\right) \text{ cm} \qquad (n = 1,3,5, \ldots)$$

and from Equation 18.5 we find that the *nodes* are located at

$$x = n\frac{\lambda}{2} = n\left(\frac{\pi}{3}\right) \text{ cm} \qquad (n = 1,2,3, \ldots)$$

18.3 STANDING WAVES IN A STRING FIXED AT BOTH ENDS

Consider a string of length L that is fixed at both ends, as in Figure 18.6. Standing waves are set up in the string by a continuous superposition of waves incident on and reflected from the ends. The string has a number of natural patterns of vibration, called **normal modes.** Each of these has a characteristic frequency; the frequencies are easily calculated.

First, note that the ends of the string must be nodes since these points are *fixed.* If the string is displaced at its midpoint and released, the vibration shown in Figure 18.6b is produced, in which the center of the string is an antinode. For this normal mode, the length of the string equals $\lambda/2$ (the distance between nodes):

$$L = \lambda_1/2 \qquad \text{or} \qquad \lambda_1 = 2L$$

The next normal mode, of wavelength λ_2 (Fig. 18.6c), occurs when the length of the string equals one wavelength, that is, when $\lambda_2 = L$. The third normal mode (Fig. 18.6d) corresponds to the case where the length equals $3\lambda/2$; therefore, $\lambda_3 = 2L/3$. In general, the wavelengths of the various normal modes can be conveniently expressed as

Wavelengths of normal modes

$$\lambda_n = 2L/n \qquad (n = 1,2,3, \ldots) \tag{18.6}$$

where the index n refers to the nth mode of vibration. The natural frequencies associated with these modes are obtained from the relationship $f = v/\lambda$, where the *wave speed v is the same for all frequencies.* Using Equation 18.6, we find that the frequencies of the normal modes are given by

Frequencies of normal modes

$$f_n = \frac{v}{\lambda_n} = \frac{n}{2L}\, v \qquad (n = 1,2,3, \ldots) \tag{18.7}$$

Because $v = \sqrt{F/\mu}$, where F is the tension in the string and μ is its mass per unit length, we can also express the natural frequencies of a stretched string as[1]

Normal modes of a stretched string

$$f_n = \frac{n}{2L}\sqrt{F/\mu} \qquad (n = 1,2,3, \ldots) \tag{18.8}$$

The lowest frequency, corresponding to $n = 1$, is called the *fundamental* or the **fundamental frequency,** f_1, and is given by

$$f_1 = \frac{1}{2L}\sqrt{F/\mu} \tag{18.9}$$

Clearly, the frequencies of the remaining modes (sometimes called *harmonics*) are integral multiples of the fundamental frequency, that is, $2f_1$, $3f_1$, $4f_1$, and so on. These higher natural frequencies, together with the fundamental frequency, are seen to form a **harmonic series.** The fundamental, f_1, is the first harmonic; the frequency $f_2 = 2f_1$ is the second harmonic; the frequency f_n is the nth harmonic. In musical terms, the various allowed frequencies are called *overtones.* For example, if all harmonics are present, the second harmonic is the first overtone, the third harmonic is the second overtone, and so on.

[1] The laws governing the sound produced by a vibrating string were first published in 1636 by a Franciscan friar, Pére Mersenne, in a treatise entitled "Harmonie Universelle."

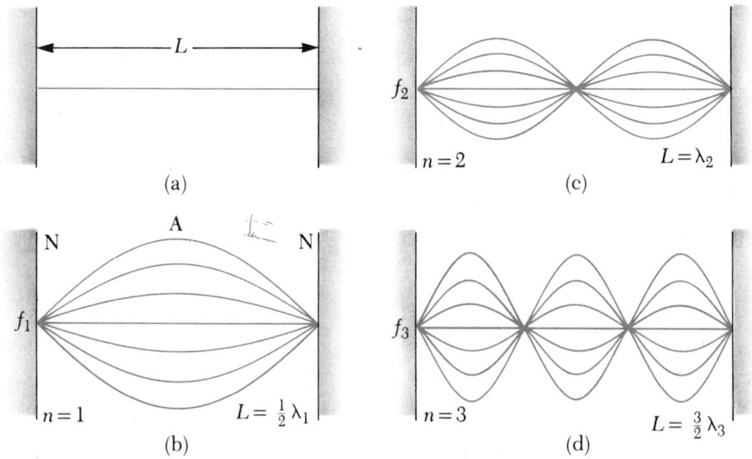

Figure 18.6 (a) Standing waves in a string of length L fixed at both ends. The normal modes of vibration shown as multiple exposures form a harmonic series: (b) the fundamental frequency, or first harmonic; (c) the second harmonic; and (d) the third harmonic.

We can obtain the above results in an alternative manner. Since we require that the string be fixed at $x = 0$ and $x = L$, the wave function $y(x, t)$ given by Equation 18.3 must be *zero* at these points for *all* times. That is, the boundary conditions require that $y(0, t) = 0$ and $y(L, t) = 0$ for all values of t. Since $y = (2A_0 \sin kx) \cos \omega t$, the first condition, $y(0, t) = 0$, is automatically satisfied because $\sin kx = 0$ at $x = 0$. To meet the second condition, $y(L, t) = 0$, we require that $\sin kL = 0$. This condition is satisfied when the angle kL equals an integral multiple of π (180°). Therefore, the allowed values of k are[2]

$$k_nL = n\pi \qquad (n = 1,2,3, \ldots) \tag{18.10}$$

Since $k_n = 2\pi/\lambda_n$, we find that

$$(2\pi/\lambda_n)L = n\pi \qquad \text{or} \qquad \lambda_n = 2L/n$$

which is identical to Equation 18.6.

When a stretched string is distorted such that its initial shape corresponds to any one of its harmonics, after being released it will vibrate at the frequency of that harmonic. However, if the string is struck or bowed, the resulting vibration will include frequencies of various harmonics, including the fundamental. Waves of the "wrong" frequency destroy each other in traveling on a string fixed at both ends. In effect, the string "selects" the normal-mode frequencies when disturbed by a nonharmonic disturbance (which happens, for example, when a guitar string is plucked).

Figure 18.7 shows a stretched string vibrating with its first and second harmonics simultaneously. In this figure, the combined vibration is the superposition of the two vibrations shown in Figures 18.6b and 18.6c. The large loop corresponds to the fundamental frequency of vibration, f_1, and the smaller loops correspond to the second harmonic, f_2. In general, the resulting motion, or displacement, can be described by a superposition of the various harmonic wave functions, with different frequencies and amplitudes. Hence, the sound that one hears corresponds to a complex waveform associated with these various modes of vibration. We shall return to this point in Section 18.8.

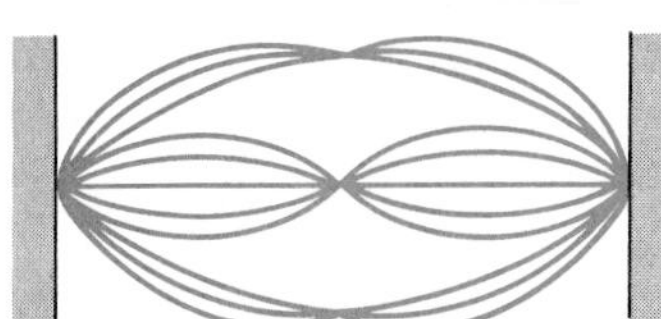

Figure 18.7 Multiple exposures of a stretched string vibrating in its first harmonic and second harmonic simultaneously.

The frequency and pitch of a stringed instrument can be changed either by varying the tension F or by changing the length L. For example, the tension

[2] We exclude $n = 0$ since this corresponds to the trivial case where no wave exists ($k = 0$).

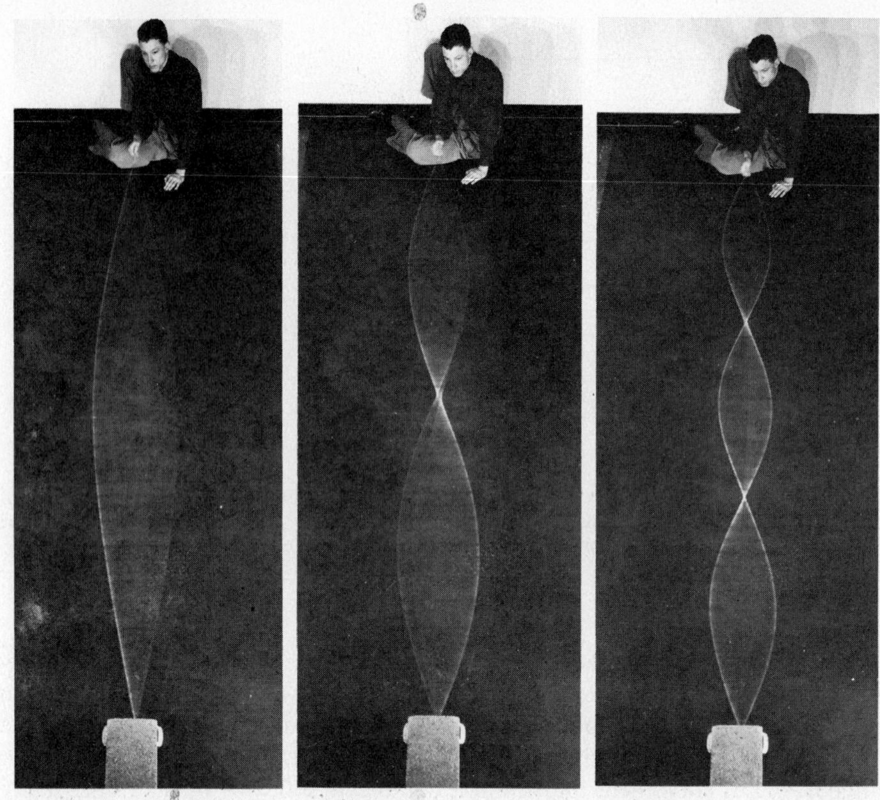

Photographs of standing waves. As one end of the tube is moved from side to side with increasing frequency, patterns with more and more loops are formed; only certain definite frequencies will produce fixed patterns. (Photos, Education Development Center, Newton, Mass.)

in the strings of guitars and violins is varied by a screw adjustment mechanism or by turning pegs located on the neck of the instrument. As the tension is increased, the frequency of the normal modes increases according to Equation 18.8. Once the instrument is "tuned," the player varies the pitch by moving his or her fingers along the neck, thereby changing the length of the vibrating portion of the string. As the length is shortened, the pitch increases, since the normal-mode frequencies are inversely proportional to string length.

EXAMPLE 18.3. Give Me a C Note
The C note of the C-major scale on a piano has a fundamental frequency of 264 Hz, and the A note has a fundamental frequency of 440 Hz. (a) Calculate the frequencies of the first two overtones of the C note.

Since $f_1 = 264$ Hz, we can use Equations 18.8 and 18.9 to find the frequencies f_2 and f_3:

$$f_2 = 2f_1 = 528 \text{ Hz}$$

$$f_3 = 3f_1 = 792 \text{ Hz}$$

(b) If the two piano strings for the A and C notes are assumed to have the same mass per unit length and the same length, determine the ratio of tensions in the two strings.

Using Equation 18.8 for the two strings vibrating at their fundamental frequencies gives

$$f_{1A} = \frac{1}{2L}\sqrt{F_A/\mu} \quad \text{and} \quad f_{1C} = \frac{1}{2L}\sqrt{F_C/\mu}$$

$$f_{1A}/f_{1C} = \sqrt{F_A/F_C}$$

$$F_A/F_C = (f_{1A}/f_{1C})^2 = (440/264)^2 = 2.78$$

(c) While the string densities are, in fact, equal, the A string is 64% as long as the C string. What is the ratio of their tensions?

$$f_{1A}/f_{1C} = (L_C/L_A)\sqrt{F_A/F_C} = (100/64)\sqrt{F_A/F_C}$$

$$F_A/F_C = (0.64)^2(440/264)^2 = 1.14$$

18.4 RESONANCE

We have seen that a system such as a stretched string is capable of oscillating in one or more natural modes of vibration. *If a periodic force is applied to such a system, the resulting amplitude of motion of the system will be larger when the frequency of the applied force is equal or nearly equal to one of the natural frequencies of the system* than when the driving force is applied at some other frequency. We have already discussed this phenomena, known as *resonance,* for mechanical systems.

The corresponding natural frequencies of oscillation of the system are often referred to as **resonant frequencies.** The resonance phenomenon is of great importance in the production of musical sounds. At the atomic level, the electrons and nuclei of atoms and molecules exhibit resonant behavior when exposed to certain frequencies of electromagnetic radiation and applied magnetic fields.

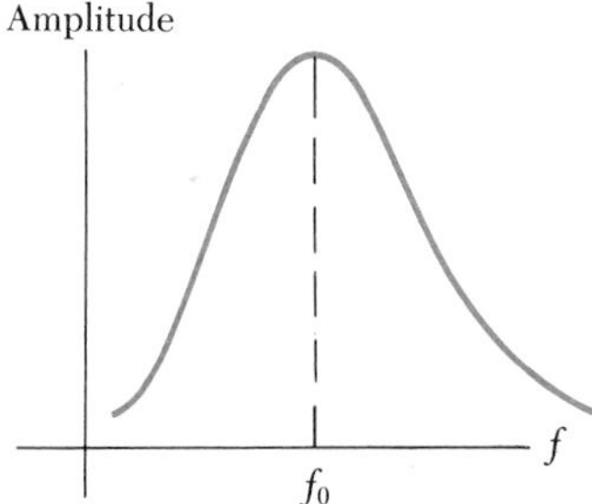

Figure 18.8 Amplitude (response) versus driving frequency for an oscillating system. The amplitude is a maximum at the resonance frequency, f_0.

Whenever a system capable of oscillating is driven by a periodic force, or a regular series of impulses, the resulting amplitude of motion will be large only when the frequency of the driving force is nearly equal to one of the resonant frequencies of the system. Figure 18.8 shows the response of a system to various frequencies, where the peak of the curve represents the resonant frequency, f_0. Note that the amplitude is largest when the frequency of the driving force equals the resonant frequency. When the frequency of the driving force exactly matches one of the resonant frequencies, the amplitude of the motion will be limited by friction in the system. Once maximum amplitude is reached, the work done by the periodic force is used to overcome friction. A system is said to be *weakly damped* when the amount of friction is small. Such a system undergoes a large amplitude of motion when driven at one of its resonant frequencies. The oscillations in such a system will persist for a long time after the driving force is removed. On the other hand, a system with considerable friction, that is, one that is *strongly damped,* will undergo small amplitude oscillations which will decrease rapidly with time once the driving force is removed.

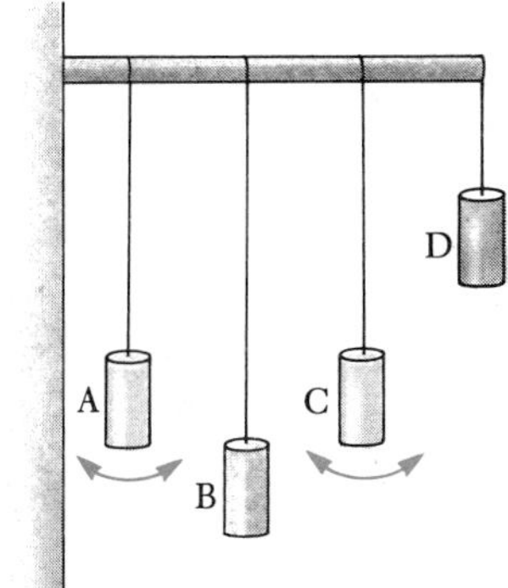

Figure 18.9 If pendulum A is set into oscillation, only pendulum C, whose length is close to the length of A, will eventually oscillate with large amplitude, or resonate.

Examples of Resonance

A playground swing is a pendulum with a natural frequency that depends on its length. Whenever we push a child in a swing with a series of regular impulses, the swing will go higher if the frequency of the periodic force equals the natural frequency of the swing. One can demonstrate a similar effect by suspending several pendula of different lengths from a horizontal support, as in Figure 18.9. If pendulum A is set into oscillation, the other pendula will soon begin to oscillate as a result of the longitudinal waves transmitted along the beam. However, you will find that those pendula, such as C, whose length is close to the length of A will oscillate with a much larger amplitude than those whose length is much different from the length of A, such as B and D. This is because the natural frequency of C is nearly the same as the driving frequency associated with A.

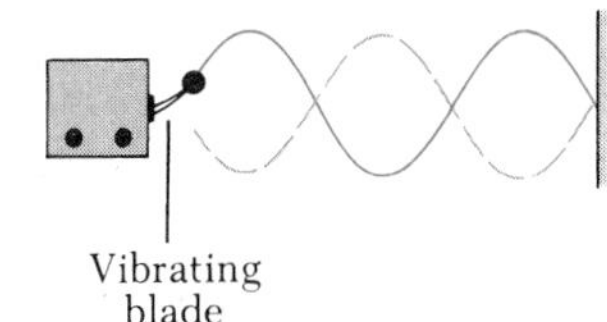

Figure 18.10 Standing waves are set up in a stretched string having one end connected to a vibrating blade when the natural frequencies of the string are nearly the same as those of the vibrating blade.

Next, consider a stretched string fixed at one end and connected at the opposite end to a vibrating blade as in Figure 18.10. As the blade oscillates, transverse waves sent down the string are reflected from the fixed end. As we found in Section 18.3, the string has natural frequencies of vibration that are determined by its length, tension, and mass per unit length (Eq. 18.8). When the frequency of the vibrating blade equals one of the natural frequencies of the string, standing waves will be produced and the string will vibrate with a large amplitude. In this case, the wave being generated by the vibrating blade is *in phase* with the wave that has been reflected at the fixed end, and so the string absorbs energy from the blade at resonance. Once the amplitude of the standing-wave oscillations reaches a maximum, the energy delivered by the blade and absorbed by the system is lost because of the damping forces. Note that the fixed end is a node, and the point P, which is near the end connected to the vibrating blade, is very nearly a node, since the amplitude of the blade's motion is small compared with that of the string.

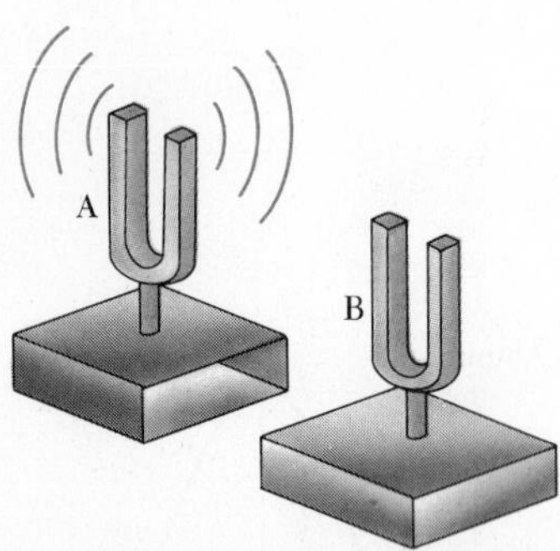

Figure 18.11 If tuning fork A is set into vibration, tuning fork B will eventually vibrate at the same frequency, or resonate, if the two forks are identical.

As a final example of resonance, consider two identical tuning forks mounted on separate hollow boxes (Fig. 18.11). The hollow boxes augment the sound wave intensity generated by the vibrating tuning forks. If tuning fork A is set into vibration (by someone's striking it, say), tuning fork B will be set into vibration as longitudinal sound waves are received from A. The frequencies of vibration of A and B will be the same, assuming the tuning forks are identical. The energy exchange, or resonance behavior, will not occur if the two have different natural frequencies of vibration. One can test this by changing the natural frequency of the receiving fork B by placing a bit of wax on its tip.

18.5 STANDING WAVES IN AIR COLUMNS

Standing longitudinal waves can be set up in a tube of air, such as an organ pipe, as the result of interference between longitudinal waves traveling in opposite directions. The phase relationship between the incident wave and the wave reflected from one end depends on whether that end is open or closed. This is analogous to the phase relationships between incident and reflected transverse waves at the ends of a string. *The closed end of an air column is a displacement node,* just as the fixed end of a vibrating string is a displacement node. As a result, at a closed end of a tube of air, the reflected wave is 180° out of phase with the incident wave. Furthermore, since the pressure wave is 90° out of phase with the displacement wave (Section 17.2), *the closed end of an air column corresponds to a pressure antinode* (that is, a point of maximum pressure variation).

If the end of an air column is open to the atmosphere, the air molecules have complete freedom of motion. Therefore, the wave reflected from an open end is nearly in phase with the incident wave when the tube's diameter is small relative to the wavelength of the sound. Consequently, *the open end of an air column is approximately a displacement antinode and a pressure node.*

Strictly speaking, the open end of an air column is not exactly an antinode. When a condensation reaches an open end, it does not reach full expansion until it passes somewhat beyond the end. For a thin-walled tube of circular cross section, this end correction is about $0.6R$, where R is the tube's radius. Hence, the effective length of the tube is somewhat longer than the true length L.

The first three modes of vibration of a pipe open at both ends are shown in Figure 18.12a. By directing air against an edge at the left, longitudinal standing waves are formed and the pipe resonates at its natural frequencies. All modes of vibration are excited simultaneously (although not with the same amplitude). Note that the ends are displacement antinodes (approximately). In the fundamental mode, the wavelength is twice the length of the pipe, and hence the frequency of the fundamental, f_1, is given by $v/2L$. Similarly, one finds that the frequencies of the overtones are $2f_1$, $3f_1$, ... Thus, in a pipe open at both ends, the natural frequencies of vibration form a harmonic series, that is, the overtones are integral multiples of the fundamental frequency.

Since all harmonics are present, we can express the natural frequencies of vibration as

Natural frequencies of a pipe open at both ends

$$f_n = n\frac{v}{2L} \qquad (n = 1,2,3, \ldots) \tag{18.11}$$

where v is the speed of sound in air.

If a pipe is closed at one end and open at the other, the closed end is a displacement node (Fig. 18.12b). In this case, the wavelength for the fundamental mode is four times the length of the tube. Hence, the fundamental, f_1, is equal to $v/4L$, and the frequencies of the overtones are equal to $3f_1$, $5f_1$, ... That is,

in a pipe closed at one end, only odd harmonics are present, and these are given by

Natural frequencies of a pipe closed at one end

$$f_n = n\frac{v}{4L} \qquad (n = 1,3,5, \ldots) \tag{18.12}$$

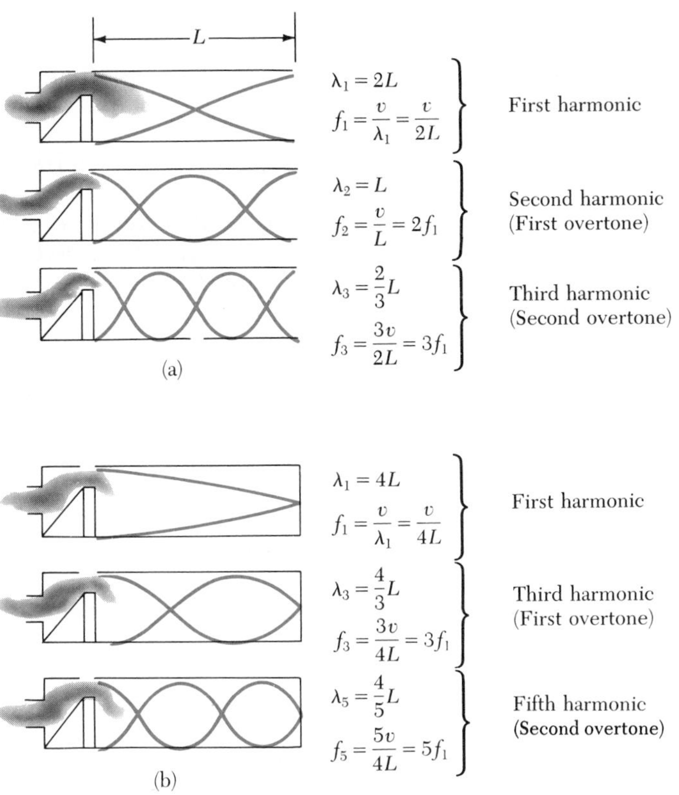

Figure 18.12 (a) Standing longitudinal waves in an organ pipe open at both ends. The natural frequencies which form a harmonic series are f_1, $2f_1$, $3f_1$, ... (b) Standing longitudinal waves in an organ pipe closed at one end. Note that only the *odd* harmonics are present, and so the natural frequencies are f_1, $3f_1$, $5f_1$, ...

EXAMPLE 18.4. Resonance in a Pipe
A pipe has a length of 1.23 m. (a) Determine the frequencies of the fundamental and the first two overtones if the pipe is open at each end. Take $v = 344$ m/s as the speed of sound in air.

The fundamental frequency of an open pipe is

$$f_1 = \frac{v}{2L} = \frac{344 \text{ m/s}}{2(1.23 \text{ m})} = 140 \text{ Hz}$$

Since all harmonics are present, the first and second overtones are given by $f_2 = 2f_1 = 280$ Hz and $f_3 = 3f_1 = 420$ Hz.

(b) What are the three frequencies determined in (a) if the pipe is closed at one end?

The fundamental frequency of a pipe closed at one end is

$$f_1 = \frac{v}{4L} = \frac{344 \text{ m/s}}{4(1.23 \text{ m})} = 70 \text{ Hz}$$

In this case, only odd harmonics are present, and so the first and second overtones have frequencies given by $f_3 = 3f_1 = 210$ Hz and $f_5 = 5f_1 = 350$ Hz.

(c) For the case of the open pipe, how many harmonics are present in the normal human hearing range (20 to 20 000 Hz)?

Since all harmonics are present, $f_n = nf_1$. Hence, the highest frequency corresponds to $n = 20\,000/140 = 142$, so that 142 harmonics are present. Actually, only the first few harmonics will have sufficient amplitude to be heard.

EXAMPLE 18.5. Measuring the Frequency of a Tuning Fork
A simple apparatus for demonstrating resonance in a tube is described in Figure 18.13a. A long, vertical, open tube is partially submerged in a beaker of water, and a vibrating tuning fork of unknown frequency is placed near the top. The length of the air column, L, is adjusted by moving the tube vertically. The sound waves generated by the fork are reinforced when the length of the column corresponds to one of the resonant frequencies of the tube. The smallest value of L for which a peak occurs in the sound intensity is 9 cm. From this measurement, determine the frequency of the tuning fork and the value of L for the next two resonant modes.

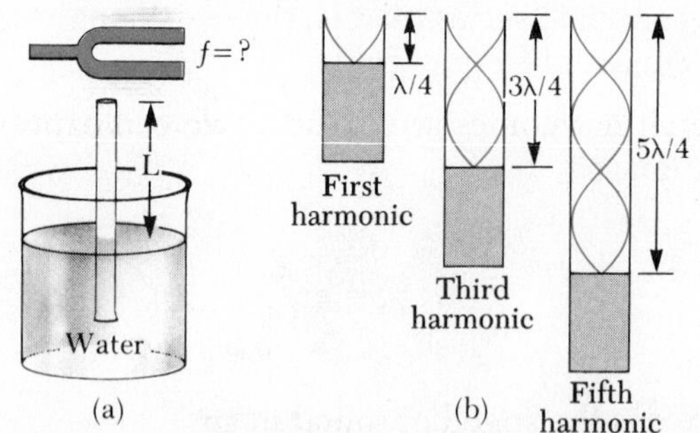

Figure 18.13 (a) Apparatus for demonstrating the resonance of sound waves in a tube closed at one end. The length L of the air column is varied by moving the tube vertically while it is partially submerged in water. (b) The first three normal modes of the system shown in (a).

Solution: Since this setup represents a pipe closed at one end, the fundamental has a frequency of $v/4L$ (Fig. 18.13b). Taking $v = 344$ m/s for the speed of sound in air and $L = 0.09$ m, we get

$$f_1 = \frac{v}{4L} = \frac{344 \text{ m/s}}{4(0.09 \text{ m})} = 956 \text{ Hz}$$

From this information about the fundamental mode, we see that the wavelength is given by $\lambda = 4L = 0.36$ m. Since the frequency of the source is constant, we see that the next two resonance modes (Fig. 18.13b) correspond to lengths of $3\lambda/4 = 0.27$ m and $5\lambda/4 = 0.45$ m.

*18.6 STANDING WAVES IN RODS AND PLATES

Standing wave vibrations can also be set up in rods and plates. If a rod is clamped in the middle and stroked at one end, it will undergo longitudinal vibrations as described in Figure 18.14a. Note that the broken lines in Figure 18.14 represent *longitudinal* displacements of various parts of the rod. The midpoint is a displacement node since it is fixed by the clamp, whereas the ends are displacement antinodes since they are free to vibrate. This is analogous to vibrations set up in a pipe open at each end. The broken lines in Figure 18.14a represent the fundamental mode for which the wavelength is $2L$ and the frequency is $v/2L$, where v is the speed of longitudinal waves in the rod. Other modes may be excited by clamping the rod at different points. For example, the second harmonic (Fig. 18.14b) is excited by clamping the rod at a point that is a distance $\lambda/4$ away from one end.

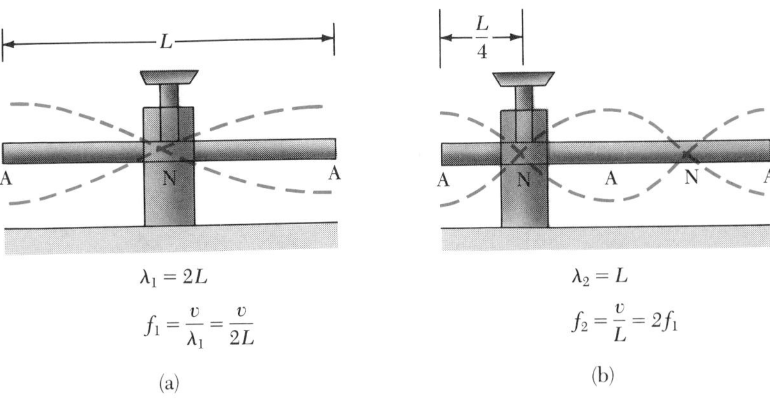

$\lambda_1 = 2L$

$f_1 = \frac{v}{\lambda_1} = \frac{v}{2L}$

(a)

$\lambda_2 = L$

$f_2 = \frac{v}{L} = 2f_1$

(b)

Figure 18.14 Normal longitudinal vibrations of a rod of length L (a) clamped at the middle and (b) clamped at an approximate distance of $L/4$ from one end.

Two-dimensional vibrations can be set up in a flexible membrane stretched over a circular hoop, such as a drumhead. As the membrane is struck at some point, wave pulses traveling toward the fixed boundary are reflected many times. The resulting sound is not melodious, but rather explosive in nature. This is because the vibrating drumhead and the drum's hollow interior produce a disorganized set of waves, which create an unrecognizable note when they reach a listener's ear. This is in contrast to wind and stringed instruments, which produce melodious, recognizable notes.

Some possible normal modes of oscillation of a vibrating, two-dimensional, circular membrane are shown in Figure 18.15. Note that the nodes are *lines* rather than points, which was the case for a vibrating string. The fixed outer perimeter is one such nodal line. Some other nodal lines are indicated with arrows. The lowest mode of vibration with frequency f_1 (the fundamental) is a symmetric mode with one nodal line, the circumference of the membrane. Note that the other possible modes of vibration are *not* integral multiples of f_1; hence the normal frequencies *do not* form a harmonic series. When a drum is struck, many of these modes are excited simultaneously. However, the higher-frequency modes dampen out more rapidly. With this information, one can understand why the drum is a nonmelodious instrument.

*18.7 BEATS: INTERFERENCE IN TIME

The interference phenomena we have been dealing with so far involve the superposition of two or more waves with the same frequency traveling in opposite directions. Since the resultant waveform in this case depends on the coordinates of the disturbed medium, we can refer to the phenomenon as *spatial interference.* Standing waves in strings and pipes are common examples of spatial interference.

We now consider another type of interference effect, one that results from the superposition of two waves with slightly *different frequencies* traveling in the *same direction.* In this case, when the two waves are observed at a given point, they are periodically in and out of phase. That is, there is an alternation in time between constructive and destructive interference. Thus, we refer to this phenomenon as *interference in time* or *temporal interference.* For example, if two tuning forks of slightly different frequencies are struck, one hears a sound of pulsating intensity, called **beats.**

Figure 18.15 (a) Six normal modes of vibration of a circular membrane (drumhead) fixed at its perimeter. Arrows indicate the nodal lines. (From P. M. Morse, *Vibration and Sound,* 2nd ed., New York, McGraw-Hill, 1948, with permission of the publishers.) (b) Representation of some natural modes of vibration on a circular membrane fixed at its perimeter. Note that the frequencies of vibration *do not* form a harmonic series. (From M. L. Warren, *Introductory Physics,* New York, W. H. Freeman, 1979, with permission.)

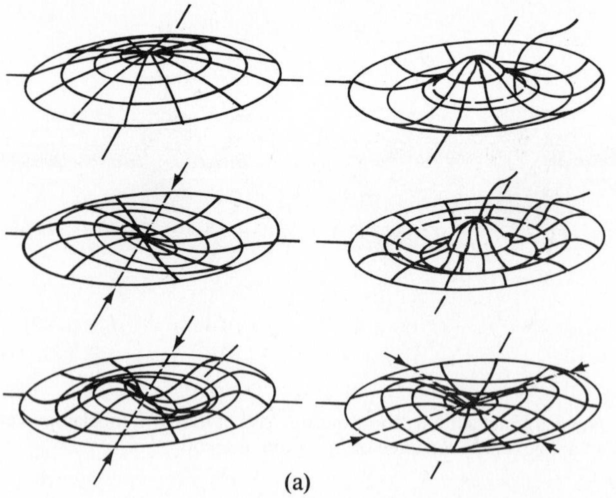

(a)

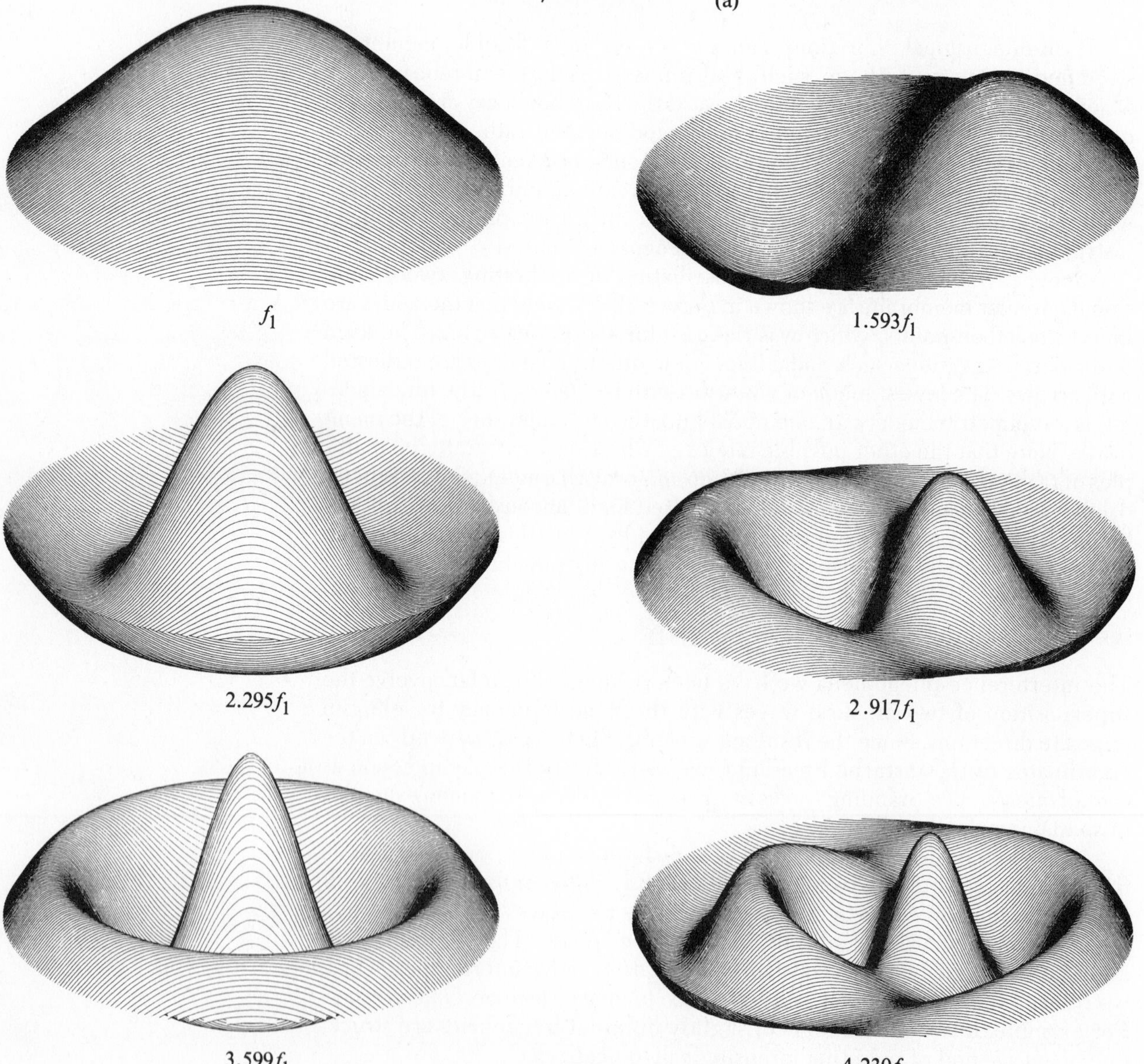

f_1

$1.593f_1$

$2.295f_1$

$2.917f_1$

$3.599f_1$

$4.230f_1$

(b)

Beats can therefore be defined as the periodic variation in intensity at a given point due to the superposition of two waves having slightly different frequencies.

Definition of beats

The number of beats one hears per second, or *beat frequency*, equals the difference in frequency between the two sources. The maximum beat frequency that the human ear can detect is about 20 beats/s.

When the beat frequency exceeds this value, it blends indistinguishably with the compound sounds producing the beats. One can use beats to tune a stringed instrument, such as a piano, by beating a note against a reference tone of known frequency. The string can then be adjusted to equal the frequency of the reference by tightening or loosening it until no beats are heard.

Consider two waves with equal amplitudes traveling through a medium in the *same* direction, but with slightly different frequencies, f_1 and f_2. We can represent the displacement that each wave would produce at a point as

$$y_1 = A_0 \cos 2\pi f_1 t \quad \text{and} \quad y_2 = A_0 \cos 2\pi f_2 t$$

Using the superposition principle, we find that the resultant displacement at that point is given by

$$y = y_1 + y_2 = A_0 (\cos 2\pi f_1 t + \cos 2\pi f_2 t)$$

It is convenient to write this in a form that uses the trigonometric identity

$$\cos a + \cos b = 2 \cos \left(\frac{a-b}{2}\right) \cos \left(\frac{a+b}{2}\right)$$

Letting $a = 2\pi f_1 t$ and $b = 2\pi f_2 t$, we find that

$$y = 2A_0 \cos 2\pi \left(\frac{f_1 - f_2}{2}\right) t \cos 2\pi \left(\frac{f_1 + f_2}{2}\right) t \qquad (18.13)$$

Resultant of two waves of different frequencies but equal amplitude

Graphs demonstrating the individual waveforms as well as the resultant wave are shown in Figure 18.16. From the factors in Equation 18.13, we see that the resultant vibration at a point has an effective frequency equal to the average frequency, $(f_1 + f_2)/2$, and an amplitude given by

$$A = 2A_0 \cos 2\pi \left(\frac{f_1 - f_2}{2}\right) t \qquad (18.14)$$

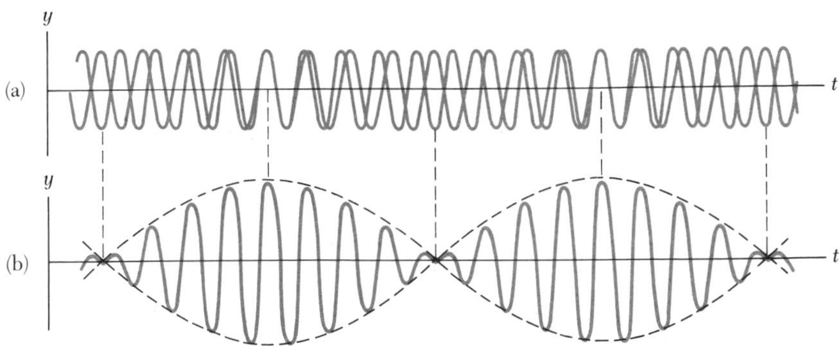

Figure 18.16 Beats are formed by the combination of two waves of slightly different frequencies traveling in the same direction. (a) The individual waves. (b) The combined wave has an amplitude (broken line) that oscillates in time. (From R. Resnick and D. Halliday, *Physics*, New York, Wiley, 1977; by permission of the publisher)

That is, the *amplitude varies in time* with a frequency given by $(f_1 - f_2)/2$. When f_1 is close to f_2, this amplitude variation is slow, as illustrated by the envelope (broken line) of the resultant waveform in Figure 18.16b.

Note that a beat, or a maximum in amplitude, will be detected whenever

$$\cos 2\pi\left(\frac{f_1 - f_2}{2}\right)t = \pm 1$$

That is, there will be *two* maxima in each cycle. Since the amplitude varies with frequency as $(f_1 - f_2)/2$, the number of beats per second, or the beat frequency f_b, is twice this value. That is,

Beat frequency

$$f_b = f_1 - f_2 \tag{18.15}$$

For instance, if two tuning forks vibrate individually at frequencies of 438 Hz and 442 Hz, the resultant sound wave of the combination would have a frequency of 440 Hz (the fundamental of a piano's A note) and a beat frequency of 4 Hz. That is, the listener would hear the 440-Hz sound wave go through an intensity maximum four times every second.

(a)

Tuning fork

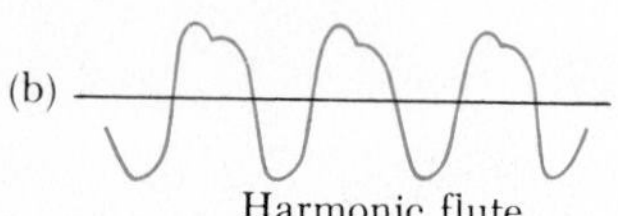

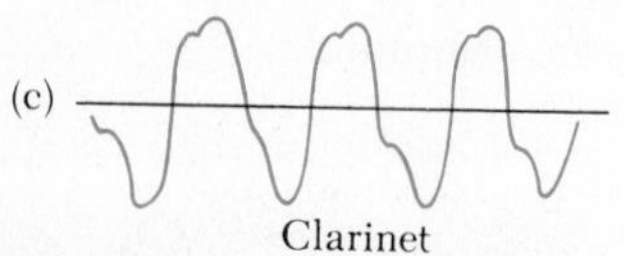

Figure 18.17 Waveform produced by (a) a tuning fork, (b) harmonic flute, and (c) a clarinet, each at approximately the same frequency. (Adapted from C. A. Culver, *Musical Acoustics*, 4th ed., New York, McGraw-Hill, 1956, p. 128.)

*18.8 COMPLEX WAVES

The sound wave patterns produced by most instruments are very complex. Some characteristic waveforms produced by a tuning fork, a harmonic flute, and a clarinet, each playing the same pitch are shown in Figure 18.17. Although each instrument has its own characteristic pattern, Figure 18.17 shows that each of the waveforms is periodic in nature. Furthermore, note that a struck tuning fork produces only one harmonic (the fundamental frequency), whereas the flute and clarinet produce many frequencies, which include the fundamental and various harmonics. Thus, the complex waveforms produced by a violin or clarinet, and the corresponding richness of musical tones, are the result of the superposition of various harmonics. This is in contrast to the drum, in which the overtones do not form a harmonic series.

The problem of analyzing complex waveforms appears at first sight to be a rather formidable task. However, if the waveform is periodic, it can be represented with arbitrary precision by the combination of a sufficiently large number of sinusoidal waves that form a harmonic series. In fact, one can represent any periodic function or any finite function as a series of sine and cosine terms by using a mathematical technique based on **Fourier's theorem.**[3] The corresponding sum of terms that represents the periodic waveform is called a **Fourier series.**

Let $y(t)$ be any function that is periodic in time with period T, such that $y(t + T) = y(t)$. Fourier's theorem states that this function can be written

Fourier's theorem

$$y(t) = \sum_n (A_n \sin 2\pi f_n t + B_n \cos 2\pi f_n t) \tag{18.16}$$

where the lowest frequency $f_1 = 1/T$.

The higher frequencies are integral multiples of the fundamental, so that $f_n = nf_1$. The coefficients A_n and B_n represent the amplitudes of the various waves. The amplitude of the nth harmonic is proportional to $\sqrt{A_n^2 + B_n^2}$, and its intensity is proportional to $A_n^2 + B_n^2$.

[3] Developed by Jean Baptiste Joseph Fourier (1786–1830).

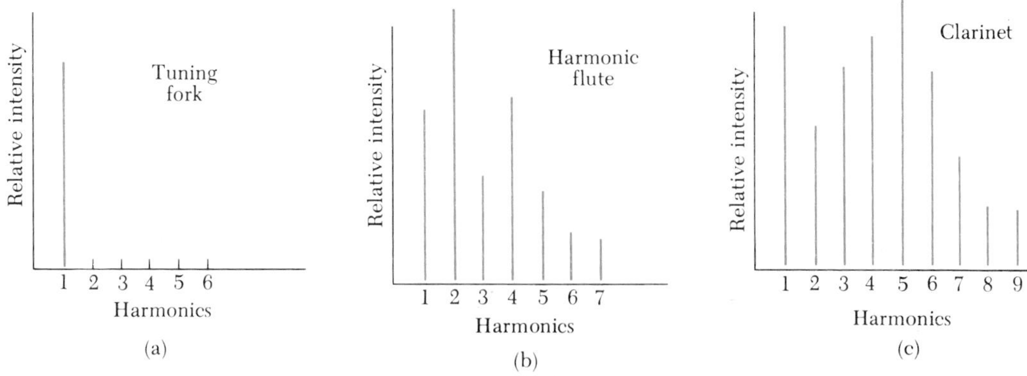

Figure 18.18 Harmonics of the waveforms shown in Figure 18.17. Note the variations in intensities of the various harmonics. (Adapted from C. A. Culver, *Musical Acoustics*, 4th ed., New York, McGraw-Hill, 1956.)

Figure 18.18 represents a harmonic analysis of the waveforms shown in Figure 18.17. Note the variation of relative intensity with harmonic content for the flute and clarinet. In general, any pleasing periodic musical sound (that is, one with good tone quality) contains components that are members of a harmonic set with varying relative intensities.

As an example of *Fourier synthesis*, consider the periodic square wave shown in Figure 18.19. Note that the square wave is synthesized by a series of *odd* harmonics of the fundamental. The series contains only sine functions (that is, $B_n = 0$ for all n). Only the first four odd harmonics and their respective amplitudes are shown. One obtains a better fit to the true waveform by adding more harmonics.

Using modern technology, one can generate musical sounds electronically by mixing any number of harmonics with varying amplitudes. These

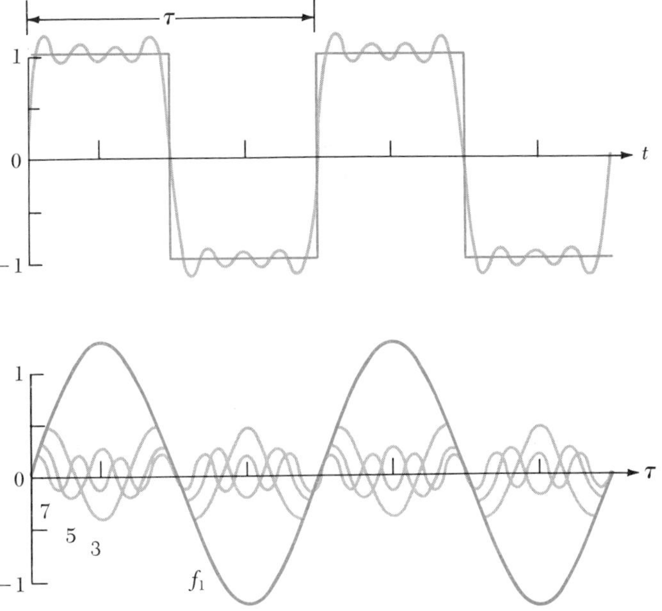

Figure 18.19 Harmonic synthesis of a square wave, which can be represented by the sum of odd harmonics of the fundamental. (From M. L. Warren, *Introductory Physics*, San Francisco, W. H. Freeman, 1979, p. 178; by permission of the publisher)

widely used electronic music synthesizers are able to produce an infinite variety of musical tones and repetitive sequences.

18.9 SUMMARY

When two waves with equal amplitudes and frequencies superimpose, the resultant wave has an amplitude that depends on the phase angle ϕ between the two waves. **Constructive interference** occurs when the two waves are *in phase* everywhere, corresponding to $\phi = 0, 2\pi, 4\pi, \ldots$. **Destructive interference** occurs when the two waves are 180° out of phase everywhere, corresponding to $\phi = \pi, 3\pi, 5\pi, \ldots$.

Standing waves are formed from the superposition of two harmonic waves having the same frequency, amplitude, and wavelength, but traveling in *opposite* directions. The resultant standing wave is described by the wave function

Wave function for a standing wave

$$y = (2A_0 \sin kx) \cos \omega t \tag{18.3}$$

Hence, its amplitude varies as sin kx. The maximum amplitude points (called **antinodes**) occur at $x = n\pi/2k = n\lambda/4$ (for odd n). The points of zero amplitude (called **nodes**) occur at $x = n\pi/k = n\lambda/2$ (for integral values of n).

One can set up standing waves with specific frequencies in such systems as stretched strings, hollow pipes, rods, and drumheads. The natural frequencies of vibration of a stretched string of length L, fixed at both ends, have frequencies given by

Normal modes of a stretched string

$$f_n = \frac{n}{2L}\sqrt{F/\mu} \qquad (n = 1,2,3, \ldots) \tag{18.8}$$

where F is the tension in the string and μ is its mass per unit length. The natural frequencies of vibration form a **harmonic series,** that is, $f_1, 2f_1, 3f_1, \ldots$.

The standing wave patterns for longitudinal waves in a hollow pipe depend on whether the ends of the pipe are open or closed. If the pipe is open at both ends, the natural frequencies of vibration form a harmonic series. If one end is closed, only odd harmonics of the fundamental are present.

A system capable of oscillating is said to be in **resonance** with some driving force whenever the frequency of the driving force matches one of the natural frequencies of the system. When the system is resonating, it responds by oscillating with a relatively large amplitude.

The phenomenon of **beats** occurs from the superposition of two waves of slightly different frequencies, traveling in the same direction. For sound waves at a given point, one would hear an alternation in sound intensity with time. Thus, beats correspond to *interference in time.*

Any periodic waveform can be represented by the combination of the sinusoidal waves that form a harmonic series. The process is called *Fourier synthesis* and is based upon *Fourier's theorem.*

QUESTIONS

1. For certain positions of the movable section in Figure 18.2, there is no sound detected at the receiver, corresponding to destructive interference. This suggests that perhaps energy is somehow lost! What happens to the energy transmitted by the receiver?
2. Does the phenomenon of wave interference apply only to harmonic waves?
3. When two waves interfere constructively or destructively, is there any gain or loss in energy? Explain.

4. A standing wave is set up on a string as in Figure 18.5. Explain why no energy is transmitted along the string.
5. What is common to *all* points (other than the nodes) on a string supporting a standing wave?
6. Some singers claim to be able to shatter a wine glass by maintaining a certain pitch in their voice over a period of several seconds (see photo). What mechanism causes the glass to break? (The glass must be very clean in order for it to break.)

(Question 6) A wine glass shattered by the amplified sound of a human voice. (Courtesy Memorex Corporation)

7. What limits the amplitude of motion of a real vibrating system that is driven at one of its resonant frequencies?
8. If the temperature of the air in an organ pipe increases, what happens to the resonance frequencies?
9. Explain why your voice seems to sound better than usual when you sing in the shower.
10. What is the purpose of the slide on a trombone or the valves on a trumpet?
11. Explain why all harmonics are present in an organ pipe open at both ends, but only the odd harmonics are present in a pipe closed at one end.
12. Explain how a musical instrument such as a piano may be tuned using the phenomenon of beats.
13. An airplane mechanic notices that the sound from a twin-engine aircraft rapidly varies in loudness when both engines are running. What could be causing this variation from loud to soft?
14. At certain speeds, an automobile driven on a washboard road will vibrate disastrously and lose traction and braking effectiveness. At other speeds, either lesser or greater, the vibration is more manageable. Explain. Why are "rumble strips," which work on this same principle, often used just before stop signs?
15. Why does a vibrating guitar string sound louder when placed on the instrument than it would if allowed to vibrate in the air while off the instrument?

PROBLEMS

18.1 Superposition and Interference of Harmonic Waves

1. Two harmonic waves are described by

$$y_1 = 5 \sin[\pi(4x - 1200t)]$$

$$y_2 = 5 \sin[\pi(4x - 1200t - 0.25)]$$

where x, y_1, and y_2 are in m and t is in s. (a) What is the amplitude of the resultant wave? (b) What is the frequency of the resultant wave?

2. Two harmonic waves are described by

$$y_1 = 6 \sin\left(\frac{\pi}{15} x - \frac{\pi}{0.005} t\right)$$

$$y_2 = 6 \sin\left(\frac{\pi}{15} x - \frac{\pi}{0.005} t - \phi\right)$$

where x, y_1, and y_2 are in m and t is in s. (a) What is the amplitude of the resultant wave when $\phi = (\pi/6)$ rad? (b) For what value of ϕ will the amplitude of the resultant wave have its maximum value?

3. A harmonic wave is described by

$$y_1 = 8 \sin[2\pi(0.1x - 80t)]$$

where y_1 and x are in m and t is in s. Write an expression for a wave that has the same frequency, amplitude, and wavelength as y_1, but when added to y_1 will give a resultant with an amplitude of $8\sqrt{3}$ m.

4. Two speakers are arranged similar to those shown in Figure 18.3. The distance between the two speakers is 2 m, and they are driven at a frequency of 1500 Hz. An observer is initially at a point 6 m along the perpendicular bisector of the line joining the two speakers. (a) What distance must the observer move along a line parallel to the line joining the two speakers before reaching the first minimum in intensity? (Use $v = 330$ m/s.) (b) At what distance from the perpendicular bisector will the observer find the first relative maximum in intensity?
5. Two identical sound sources are located along the y axis. Source S_1 is located at (0, 0.1) m and source S_2 is located at (0, −0.1) m. The two sources radiate isotropically at a frequency of 1650 Hz and the amplitude of each wave separately is A. A listener is located along the y axis a distance of 5 m from source S_1. (a) What is the phase difference between the sound waves at the position of the listener? (b) What is the amplitude of the resultant wave at the location of the listener? (Use $v = 330$ m/s.)
6. Two identical sound sources are located as described in Problem 5. The frequency of each source is variable. An observer is located at the point (1, 0.5) m. (a) What is the lowest frequency that will produce a relative maximum at the location of the observer? (b) What is the lowest frequency that will produce a relative minimum at the observer's location?

7. Two speakers are driven by a common oscillator at 800 Hz and face each other at a distance of 1.25 m. Locate the two points along a line joining the two speakers where relative minima would be expected. (Use $v = 330$ m/s.)
8. For the arrangement shown in Figure 18.2, let the path length $r_1 = 1.20$ m and the path length $r_2 = 0.80$ m. (a) Calculate the three lowest speaker frequencies that will result in intensity maxima at the receiver. (b) What is the highest frequency within the audible range (20–20 000 Hz) that will result in a minimum at the receiver?

18.2 Standing Waves

9. Two harmonic waves are described by

$$y_1 = 3 \sin\pi(x + 0.6t) \text{ cm}$$

$$y_2 = 3 \sin\pi(x - 0.6t) \text{ cm}$$

Determine the *maximum* displacement of the motion at (a) $x = 0.25$ cm, (b) $x = 0.5$ cm, and (c) $x = 1.5$ cm. (d) Find the three smallest values of x corresponding to antinodes.

10. Use the trigonometric identity

$$\sin(a \pm b) = \sin a \cos b \pm \cos a \sin b$$

to show that the resultant of two wave functions each of amplitude A_0, angular frequency ω, and propagation number k and traveling in opposite directions can be written

$$y = (2A_0 \sin kx) \cos \omega t$$

11. The wave function for a standing wave in a string is given by

$$y = 0.3 \sin(0.25x) \cos(120\pi t)$$

where x is in m and t is in s. Determine the wavelength and frequency of the interfering traveling waves.

12. Two harmonic waves traveling in opposite directions interfere to produce a standing wave described by

$$y = 1.5 \sin(0.4x) \cos(200t)$$

where x is in m and t is in s. Determine the wavelength, frequency, and speed of the interfering waves.

13. A standing wave is formed by the interference of the following two traveling waves, each of which has an amplitude $A = \pi$ cm, propagation number $k = (\pi/2)$ cm^{-1}, and angular frequency $\omega = 10\pi$ rad/s. (a) Calculate the distance between the first two antinodes. (b) What is the amplitude of the standing wave at $x = 0.25$ cm?
14. Verify by direct substitution that the wave function for a standing wave given in Equation 18.3,

$$y = 2A_0 \sin kx \cos \omega t,$$

is a solution of the general linear wave equation, Equation 16.26.

$$\frac{\partial^2 y}{\partial x^2} = \frac{1}{v^2}\frac{\partial^2 y}{\partial t^2}$$

18.3 Standing Waves in a String Fixed at Both Ends

15. A standing wave is established in a 120-cm-long string fixed at both ends. The string vibrates in four segments when driven at 120 Hz. (a) Determine the wavelength. (b) What is the fundamental frequency?
16. A stretched string is 160 cm long and has a linear density of 0.015 g/cm. What tension in the string will result in a second harmonic of 460 Hz?
17. Consider a tuned guitar string of length L. At what point along the string (fraction of length from one end) should the string be plucked and at what point should the finger be held lightly against the string in order that the first overtone be the most prominent mode of vibration?
18. A string 50 cm long has a mass per unit length of 20×10^{-5} kg/m. To what tension should this string be stretched if its fundamental frequency is to be (a) 20 Hz and (b) 4500 Hz?
19. Find the fundamental frequency and the next three frequencies that could cause a standing wave pattern on a string that is 30 m long, has a mass per unit length 9×10^{-3} kg/m, and is stretched to a tension of 20 N.
20. A stretched string of length L is observed to vibrate in five equal length segments when driven by a 630-Hz oscillator. What oscillator frequency will set up a standing wave such that the string vibrates in three segments?
21. A string with $L = 16$ m and $\mu = 0.015$ g/cm is stretched with a tension of 557 N ($\approx$125 lb). What is the highest harmonic of this string that is within the typical human's audible range (up to 20 000 Hz)?
22. Two pieces of steel wire having identical cross sections have lengths of L and $2L$. The wires are each fixed at both ends and stretched such that the tension in the longer wire is four times greater than that in the shorter wire. If the fundamental frequency in the shorter wire is 60 Hz, what is the frequency of the second harmonic in the longer wire?
23. A stretched string fixed at each end has a mass of 40 g and a length of 8 m. The tension in the string is 49 N. Determine the position of the nodes and antinodes for the third harmonic.
24. A string of length L, mass per unit length μ, and tension F is vibrating at its fundamental frequency. What effect will the following have on the fundamental frequency? (a) The length of the string is doubled with all other factors held constant. (b) The mass per unit length is doubled with all other factors held constant. (c) The tension is doubled with all other factors held constant.
25. A 60-cm guitar string under a tension of 25 N has a mass per unit length of 0.1 g/cm. What is the highest resonant frequency that can be heard by a person capable of hearing frequencies up to 20 000 Hz?
26. Write an expression for the harmonic number of the nth overtone for a sounding device that has (a) all harmonics present and (b) only odd harmonics present. (The fundamental corresponds to the 0th overtone.)

18.5 Standing Waves in Air Columns

(In this section, unless otherwise indicated, assume that the velocity of sound in air is 344 m/s.)

27. A resonance condition is set up in a pipe with a tuning fork whose frequency is f. Write an expression for the length of the pipe that will cause it to resonate in its nth overtone if the pipe is (a) open at both ends and (b) closed at one end. (Assume that the speed of sound is v.)
28. If an organ pipe is to resonate at 20 Hz, what is its required length if it is (a) open at both ends and (b) closed at one end?
29. A tuning fork of frequency 512 Hz is placed near the top of the tube shown in Figure 18.13a. The water level is lowered so that the length L slowly increases from an initial value of 20 cm. Determine the next two values of L that correspond to resonant modes.
30. A pipe open at each end has a fundamental frequency of 300 Hz when the velocity of sound in air is 333 m/s. (a) What is the length of the pipe? (b) What is the frequency of the second harmonic when the temperature of the air is increased so that the velocity of sound in the pipe is 344 m/s?
31. Calculate the minimum length for a pipe that has a fundamental frequency of 240 Hz if the pipe is (a) closed at one end and (b) open at both ends.
32. A tunnel beneath a river is approximately 2 km long. At what frequencies can this tunnel resonate? What does your answer say about the instructions often given at the mouth of such a tunnel that you should not blow your car horn in the tunnel?
33. An organ pipe open at both ends is vibrating in its third harmonic with a frequency of 748 Hz. The length of the pipe is 0.7 m. Determine the speed of sound in air in the pipe.
34. Determine the frequency corresponding to the first three harmonics of a 30-cm pipe when it is (a) open at both ends and (b) closed at one end.
35. An air column 2 m in length is open at both ends. The frequency of a certain harmonic is 410 Hz, and the frequency of the next higher harmonic is 492 Hz. Determine the speed of sound in the air column.
36. At a particular instant, the tube in Figure 18.13a is adjusted so that L, the length above the water surface, is 40 cm. The tuning fork in the figure is replaced by a variable-frequency oscillator that has a frequency range between 20 and 2000 Hz. What are the (a) lowest and (b) highest frequencies within this range that will excite resonant modes in the air column?

°18.6 Standing Waves in Rods and Plates

37. An aluminum rod is clamped at the one-quarter position and set into longitudinal vibration by a variable-frequency driving source. The lowest frequency that produces resonance is 4400 Hz. The speed of sound in aluminum is 5100 m/s. Determine the length of the rod.
38. A 60-cm metal bar that is clamped at one end is struck with a hammer. If the speed of longitudinal (compressional) waves in the bar is 4500 m/s, what is the lowest frequency with which the struck bar will resonate?
39. Longitudinal waves move with a speed v in a bar of length L. Write an expression for the frequencies of the nth overtone of a metal bar that is (a) clamped at its center, as shown in Figure 18.14a, and (b) clamped at one-fourth the length of the bar from one end, as shown in Figure 18.14b.

°18.7 Beats: Interference in Time

40. The tension in the strings of certain instruments normally decreases in time, which results in frequencies lower than intended. It is noted that when a string tuned to 256 Hz is plucked and an "out of tune" string of an identical instrument is plucked simultaneously, beats occur at a rate of 5 beats per second. What is the frequency of the "out of tune" string?
41. Two waves with equal amplitude but with slightly different frequencies are traveling in the same direction through a medium. At a given point the separate displacements are described by

$$y_1 = A_0 \cos \omega_1 t \qquad \text{and} \qquad y_2 = A_0 \cos \omega_2 t$$

Use the trigonometric identity

$$\cos a + \cos b = 2 \cos\left(\frac{a-b}{2}\right)\cos\left(\frac{a+b}{2}\right)$$

to show that the resultant displacement due to the two waves is given by

$$y = 2A_0 \left[\cos\left(\frac{\omega_1 - \omega_2}{2}\right)t\right]\left[\cos\left(\frac{\omega_1 + \omega_2}{2}\right)t\right]$$

GENERAL PROBLEMS

42. A variable-length air column as shown in Figure 18.13a is placed just below a vibrating wire fixed at both ends. The length of the air column is gradually increased from zero until the first position of resonance is observed at $L = 34$ cm. The wire is 120 cm in length and is vibrating in its third harmonic. If the speed of sound in air is 340 m/s, what is the speed of transverse waves in the wire?
43. Two speakers are arranged as shown in Figure 18.3. For this problem, assume that point O is 12 m along the center line and the speakers are separated by a distance of 1.5 m. As the listener moves toward point P from point O, a series of alternating minima and maxima are encountered. The distance between the first minimum and the next maximum is 0.4 m. Using 340 m/s as the speed of sound in air, determine the frequency of the speakers. (Use the approximation $\sin \theta \approx \tan \theta$.)
44. Two pipes are each open at one end and are of adjustable length. Each has a fundamental frequency of 480 Hz at 300 K. The air temperature is increased in one pipe to 305 K. (a) If the two pipes are sounded together, what beat frequency will result? (b) By what percent should the length of the 300 K pipe be increased to again match the frequencies? (Use $v = 331(T/273)^{1/2}$ m/s as the speed of sound in air, where T is the air temperature in K.)

45. The frequency of the second overtone of an organ pipe open at both ends is equal to the frequency of the second overtone of an organ pipe closed at one end. (a) Find the ratio of the length of the "closed" pipe to the length of the "open" pipe. (b) If the fundamental frequency of the open pipe is 256 Hz, what is the length of each pipe? (Use $v = 340$ m/s.)

46. A speaker at the front of a room and an identical speaker at the rear of the room are being driven by the same oscillator at 456 Hz. A student walks at a uniform rate of 1.5 m/s along the length of the room. How many beats does the student hear per second?

47. To maintain a string 1.25 m long under tension in a horizontal position, one end of the string is connected to a vibrating blade and the other end is passed over a pulley and attached to a mass. The mass of the string is 10 g. (a) When the suspended mass is 10 kg, the string vibrates in three equal length segments. Determine the vibration frequency of the blade. (Assume that the point where the string passes over the pulley and the point where it is attached to the blade are both nodes. Also, ignore the contribution to the tension due to the string's mass.) (b) What mass should be attached to the string if it is to vibrate in four equal segments?

48. While waiting for Stan Speedy to arrive on a late passenger train, Kathy Kool notices beats occurring as a result of two trains blowing their whistles simultaneously. One train is at rest and the other is approaching her at a speed of 20 km/hr. Assume that both whistles have the same frequency and that the speed of sound is 330 m/s. If Kathy hears 4 beats per second, what is the frequency of the whistles?

49. A light rope 1.5 m in length lies along the x axis. It is set into vibration with *one* end fixed at $x = 0$. (a) What is the wavelength of the standing wave corresponding to the fundamental mode? (b) If the rope resonates in its third overtone at a frequency of 320 Hz, what is the speed of transverse waves in the rope? (c) Write an expression for the wave function of the standing wave if the displacement at $x = \lambda/2$ is 4 cm.

50. In an arrangement like the one shown in Figure 18.2, paths r_1 and r_2 are each 1.75 m in length. The top portion of the tube (corresponding to r_2) is filled with air at 0°C (273 K). Air in the lower portion is quickly heated to 200°C (473 K). What is the lowest speaker frequency that will produce an intensity maximum at the receiver? (You may determine the speed of sound in air in different temperatures by using the expression $v = 331(T/273)^{1/2}$ m/s, where T is in K.

51. An air column 2 m in length is open at both ends. The frequency of its nth harmonic is 410 Hz and the $(n + 1)$ harmonic frequency is 492 Hz. Determine the speed of sound in air under these conditions.

52. A student located several meters in front of a smooth reflecting wall is holding a board on which a wire is fixed at each end. The wire, vibrating in its third harmonic, is 75 cm long, has a mass of 2.25 g and is under a tension of 400 N. A second student located between the vibrating wire and the wall is moving toward the wall and hears 8.3 beats per second. At what speed does the moving student approach the wall? Use 340 m/s as the speed of sound in air.

53. Two identical steel wires each fixed at both ends are under equal tension and are vibrating in their third harmonic at 963 Hz. The tension in one wire is increased by 3%. Determine the beat frequency when the two wires now vibrate in their *fundamental* modes.

•54. A string of length L consists of two distinct sections. The left half has mass per unit length μ; the right half has mass per unit length $\mu' > \mu$. It can be shown that the nth normal mode of this composite string has a profile that consists of portions of two sinusoids (one in each section) that join smoothly (the slopes match). (a) What is the ratio k_n'/k_n of the wave numbers that characterize the two sinusoids? (*Hint:* Use the relationship among wave number, wave speed, and frequency. Remember that the two sections of the string vibrate with the same frequency.) (b) In the fundamental mode of a *uniform* string, the profile of the string is one half-cycle of a sine curve, so there are no displacement nodes (excluding those at the ends). The profile of the fundamental mode of the composite string is also devoid of nodes. Sketch the profile of the fundamental mode in the composite string, keeping in mind that $\mu' > \mu$. In which section of the string does the maximum of the profile occur?

CALCULATOR/COMPUTER PROBLEMS

55. Sketch the resultant waveform due to the interference of the two waves y_1 and y_2 in Problem 2 at $t = 0$ s for (a) $\phi = 0$, (b) $\phi = 90°$, and (c) $\phi = 270°$. Let x range over the interval 0 to 30 m.

56. A standing wave is described by the function

$$y = 6 \sin(\pi x/2) \cos(100\pi t)$$

where x and y are in m and t is in s. (a) Plot $y(x)$ versus t for $t = 0$, 0.0005 s, 0.001 s, 0.0015 s, and 0.002 s. (b) What is the frequency of the wave? (c) What is the wavelength λ?

PART III
Thermodynamics

When dining, I had often observed that some particular dishes retained their Heat much longer than others; and that apple pies, and apples and almonds mixed (a dish in great repute in England) remained hot a surprising length of time. Much struck with this extraordinary quality of retaining Heat, which apples appeared to possess, it frequently occurred to my recollection; and I never burnt my mouth with them, or saw others meet with the same misfortune, without endeavouring, but in vain, to find out some way of accounting, in a satisfactory manner, for this surprising phenomenon.

BENJAMIN THOMPSON
(Count Rumford)

As we saw in the first part of this textbook, newtonian mechanics explains a wide range of phenomena on a macroscopic scale, such as the motion of baseballs, rockets, and the planets of our solar system. We now turn to the study of thermodynamics, which is concerned with the concepts of heat and temperature. As we shall see, thermodynamics is very successful in explaining the bulk properties of matter and the correlation between these properties and the mechanics of atoms and molecules.

Historically, the development of thermodynamics paralleled the development of the atomic theory of matter. By the middle of the 19th century, chemical experiments provided solid evidence for the existence of atoms. At that time, scientists recognized that there must be a connection between the theory of heat and temperature, and the structure of matter. In 1827, the botanist Robert Brown reported that grains of pollen suspended in a liquid move erratically from one place to another, as if under constant agitation. In 1905, Albert Einstein developed a theory in which he used thermodynamics to explain the cause of this erratic motion, today called brownian motion. Einstein explained this phenomenon by assuming that the grains of pollen are under constant bombardment by "invisible" molecules in the liquid, which themselves undergo an erratic motion. This important experiment and Einstein's insight gave scientists a means of discovering vital information concerning molecular motion.

Have you ever wondered how a refrigerator is able to cool its contents or what types of transformations occur in a power plant or in the engine of your automobile or what happens to the kinetic energy of an object when it falls to the ground and comes to rest? The laws of thermodynamics and the concepts of heat and temperature will enable us to answer such practical questions.

Many things can happen to an object when it is heated. Its size will change slightly, but it may also melt, boil, ignite, or even explode. The outcome depends upon the composition of the object and the degree to which it is heated. In general, thermodynamics must concern itself with the physical and chemical transformations of matter in all of its forms: solid, liquid, and gas.

19
Temperature, Thermal Expansion, and Ideal Gases

The subject of thermal physics deals with phenomena involving energy transfer between bodies at different temperatures. In the study of mechanics such concepts as mass, force, and kinetic energy were carefully defined in order to make the subject quantitative. Likewise, a quantitative description of thermal phenomena requires a careful definition of the concepts of temperature, heat, and internal energy. The science of thermodynamics is concerned with the study of heat flow from a *macroscopic* viewpoint. The laws of thermodynamics provide us with a relationship between heat flow, work, and internal energy of a system. In practice, suitable observable quantities must be selected to describe the overall behavior of a system. For example, the macroscopic quantities, pressure, volume, and temperature are used to characterize the properties of a gas. Thermal phenomena can also be understood using a *microscopic* approach, which describes what is happening on a microscopic scale. For example, the temperature of a gas is a measure of the average kinetic energy of the gas molecules.

The composition of a body is an important factor when dealing with thermal phenomena. For example, liquids and solids will expand only slightly when heated. On the other hand, a gas will tend to undergo appreciable expansion when heated. If the gas is not free to expand, its pressure will rise when heated. Certain substances may melt, boil, burn, or explode, depending on their composition and structure. Thus, the thermal behavior of a substance is closely related to its structure.

It would be far beyond the scope of this book to attempt to present applications of thermodynamics to a wide variety of substances. Instead, we shall examine some rather simple systems, such as a dilute gas and a homogeneous solid. Emphasis will be placed on understanding the key principles of thermodynamics and on providing a basis upon which the thermal behavior of all matter can be understood.

19.1 TEMPERATURE AND THE ZEROTH LAW OF THERMODYNAMICS

When we speak of the temperature of an object, we often associate this concept with the degree of "hotness" or "coldness" of the object when we touch it. Thus, our senses provide us with a qualitative indication of temperature. However, our senses are unreliable and often misleading. For example, if we remove an ice tray and a package of frozen vegetables from the freezer, the ice tray feels colder to the hand even though both are at the same temperature. This is because metal is a better conductor of heat than cardboard. What we need is a reliable and reproducible method for establishing the relative "hotness" or "coldness" of bodies. Scientists have developed various types of thermometers for making such quantitative measurements. Some typical thermometers will be described in Section 19.2.

We are familiar with the fact that two objects at different initial temperatures will eventually reach some intermediate temperature when placed in contact with each other. For example, a piece of meat placed on a block of ice in a well-insulated container will eventually reach a temperature near 0°C. Likewise, if an ice cube is dropped into a container of warm water, the ice cube will eventually melt and the water's temperature will decrease. If the process takes place in a thermos bottle, the system (water + ice) is approximately isolated from its surroundings.

In order to understand the concept of temperature, it is useful to first define two often used phrases, *thermal contact* and *thermal equilibrium.* Two objects are in **thermal contact** with each other if energy exchange can occur between them in the absence of macroscopic work done by one on the other. **Thermal equilibrium** is a situation in which two objects in thermal contact with each other cease to have any net energy exchange due to a difference in their temperatures. The time it takes the two objects to reach thermal equilibrium depends on the properties of the objects and on the pathways available for energy exchange.

Now consider two objects, A and B, which are not in thermal contact, and a third object, C, which will be our thermometer. We wish to determine whether or not A and B are in thermal equilibrium with each other. The thermometer (object C) is first placed in thermal contact with A until thermal equilibrium is reached. At that point, the thermometer's reading will remain constant. The thermometer is then placed in thermal contact with B, and its reading is recorded after thermal equilibrium is reached. If the readings after contact with A and B are the same, then A and B are in thermal equilibrium with each other. We can summarize these results in a statement known as the **zeroth law of thermodynamics** *(the law of equilibrium):*

> If objects A and B are separately in thermal equilibrium with a third object, C, then A and B are in thermal equilibrium with each other.

This statement, although it may seem obvious, is most fundamental in the field of thermodynamics since it can be used to define temperature. We can think of temperature as the property that determines whether or not an object is in thermal equilibrium with other objects. That is, *two objects in thermal equilibrium with each other are at the same temperature.* Conversely, if two objects have different temperatures, they cannot be in thermal equilibrium with each other at that time.

19.2 THERMOMETERS AND TEMPERATURE SCALES

Thermometers are devices used to define and measure the temperature of a system. A thermometer in thermal equilibrium with a system measures both the temperature of the system and its own temperature. All thermometers make use of the change in some physical property with temperature. Some of these physical properties are (1) the change in volume of a liquid, (2) the change in length of a solid, (3) the change in pressure of a gas at constant volume, (4) the change in volume of a gas at constant pressure, (5) the change in electric resistance of a conductor, and (6) the change in color of a very hot body. A temperature scale can be established for a given substance using one of these physical quantities.

The most common thermometer in everyday use consists of a glass bulb connected to a glass capillary tube. The glass bulb is filled with a volume of mercury that expands into the capillary tube when heated (Fig. 19.1). Thus, the physical property in this case is the thermal expansion of the mercury. One can now define any temperature change to be proportional to the change in length of the mercury column. The thermometer can be calibrated by placing it in thermal contact with some natural systems that remain at constant temperature (called a *fixed-point temperature*). One of the fixed-point temperatures normally chosen is that of a mixture of water and ice at atmospheric pressure, which is defined to be zero degrees Celsius, written 0°C. (This was formerly called *degrees centigrade.*) Another convenient fixed point is the temperature of a mixture of water and water vapor (steam) in equilibrium at atmospheric pressure. The temperature of this *steam point* is 100°C. Once the mercury levels have been established at these fixed points, the column is divided into 100 equal segments, each denoting a change in temperature of one Celsius degree.

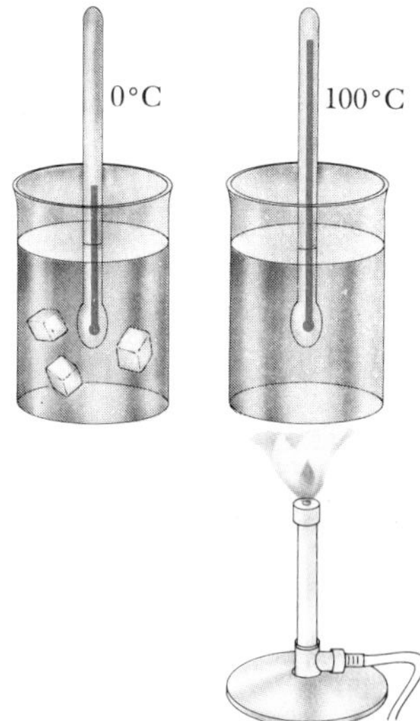

Figure 19.1 Schematic diagram of a mercury thermometer. As a result of thermal expansion, the level of the mercury rises as the mercury is heated from 0°C (the ice point) to 100°C (the steam point).

Thermometers calibrated in this way do present problems, however, when extremely accurate readings are needed. For instance, an alcohol thermometer calibrated at the ice and steam points of water might agree with a mercury thermometer only at the calibration points. Because mercury and alcohol have different thermal expansion properties, when one reads a temperature of 50°C, say, the other may indicate a slightly different value. The discrepancies between thermometers are especially large when the temperatures to be measured are far from the calibration points.[1] An additional practical problem of any thermometer is its limited temperature range. A mercury thermometer, for example, cannot be used below the freezing point of mercury, which is −39°C. What we need is a universal thermometer whose readings are independent of the substance used. The gas thermometer meets this requirement.

19.3 THE CONSTANT-VOLUME GAS THERMOMETER AND THE KELVIN SCALE

In a gas thermometer, the temperature readings are nearly independent of the substance used in the thermometer. One version of this is the constant-volume gas thermometer shown in Figure 19.2. The physical property in this device is the pressure variation with temperature of a fixed volume of gas. As the gas is heated, its pressure increases and the height of the mercury column shown in Figure 19.2 increases. When the gas is cooled, its pressure decreases, hence the column height decreases. Thus, we can define temperature in terms of the concept of pressure discussed in Chapter 15. If the variation of temperature, T, with pressure is assumed to be linear, then

$$T = aP + b \tag{19.1}$$

where a and b are constants. These constants can be determined from two fixed points, such as the ice and steam points described in Section 19.2.

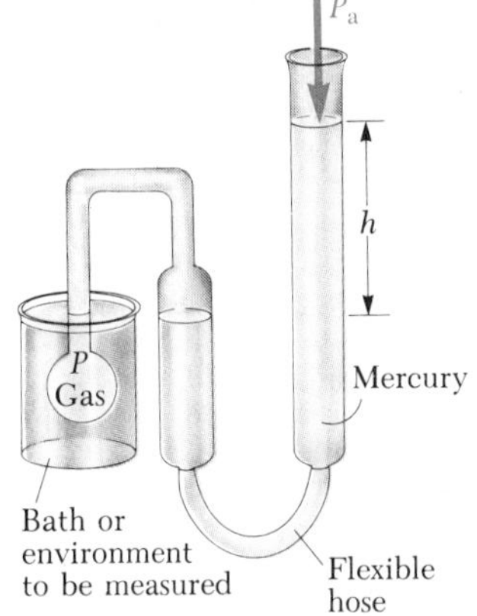

Figure 19.2 A constant-volume gas thermometer measures the pressure of the gas contained in the flask on the left. The volume of gas in the flask is kept constant by raising or lowering the column on the right such that the mercury level on the left remains constant.

Now suppose that temperatures are measured with various gas thermometers containing different gases. Experiments show that the thermometer readings are nearly independent of the type of gas used, so long as the gas

[1] Thermometers that use the same material may also give different readings. This is due in part to difficulties in constructing uniform-bore glass capillary tubes.

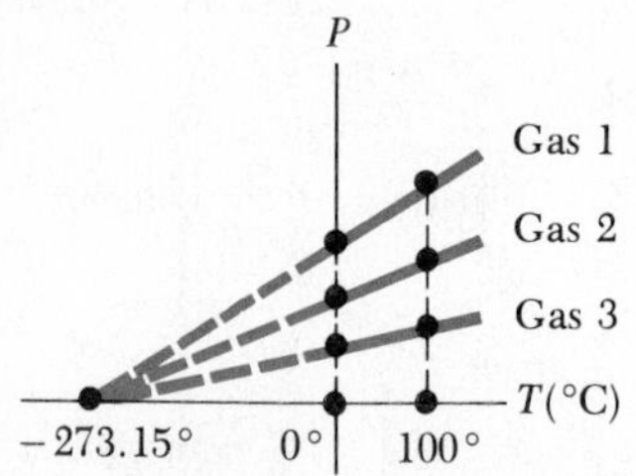

Figure 19.3 Pressure versus temperature for dilute gases. Note that, for all gases, the pressure extrapolates to zero at the unique temperature of −273.15°C.

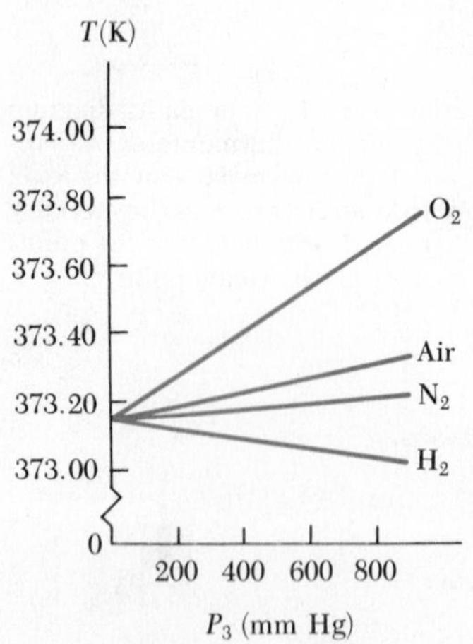

Figure 19.4 The temperature read with a constant-volume gas thermometer versus P_3, the pressure at the triple point of water, for various gases. Note that as the pressure is reduced, the steam-point temperature of water approaches a common value of 373.15 K regardless of which gas is used in the thermometer. Furthermore, the data for helium are nearly independent of pressure, which suggests it behaves like an ideal gas over this range.

pressure is low and the temperature is well above the liquefaction point. The agreement among thermometers using various gases improves as the pressure is reduced. This agreement of all gas thermometers at low pressure and high temperature implies that the intercept b appearing in Equation 19.1 is the same for *all* gases. This fact is illustrated in Figure 19.3. When the pressure versus temperature curve is extrapolated to very low temperatures, one finds that the pressure is zero when the temperature is −273.15°C. This temperature corresponds to the constant b in Equation 19.1. An extrapolation is necessary since all gases liquefy before reaching this temperature.

Early gas thermometers made use of the ice point and steam point as standard temperatures. However, these points are experimentally difficult to duplicate since they are very sensitive to dissolved impurities in the water. For this reason, a new temperature scale based on a single fixed point with b equal to zero was adopted in 1954 by the International Committee on Weights and Measures. The *triple point of water*, which corresponds to the single temperature and pressure at which water, water vapor, and ice can coexist in equilibrium, was chosen as a convenient and reproducible reference temperature for this new scale. The triple point of water occurs at a temperature of about 0.01°C and a pressure of 4.58 mm Hg. The temperature at the triple point of water on the new scale was set at 273.16 kelvin, abbreviated 273.16 K.[2] This choice was made so that the old temperature scale based on the ice- and steam-points would agree closely with the new scale based on the triple point. This new scale is called the **thermodynamic temperature scale** and the SI unit of thermodynamic temperature,

> the **kelvin**, is defined as the fraction 1/273.16 of the temperature of the triple point of water.

If we take $b = 0$ in Equation 19.1 and call P_3 the pressure at the triple-point temperature, $T_3 = 273.16$ K, then we see that $a = (273.16\text{ K})/P_3$. Therefore, the temperature at a measured gas pressure P for a constant-volume gas thermometer is defined to be

$$T = \left(\frac{273.16\text{ K}}{P_3}\right)P \qquad \text{(constant } V\text{)} \qquad (19.2)$$

As mentioned earlier, one finds experimentally that as the pressure P_3 decreases, the measured value of the temperature approaches the same value for all gases. An example of such a measurement is illustrated in Figure 19.4, which shows the steam-point temperature measured with a constant-volume gas thermometer using various gases. As P_3 approaches zero, all measurements approach a common value of 373.15 K. Similarly, one finds that the ice-point temperature is 273.15 K.

In the limit of low gas pressures and high temperatures, real gases behave as what is known as an **ideal gas**, which will be discussed in detail in Section 19.6 and Chapter 20. The temperature scale defined in this limit of low gas pressures is called the **ideal gas temperature,** T, given by

Definition of ideal gas temperature

$$T \equiv 273.16\text{ K} \lim_{P_3 \to 0} \frac{P}{P_3} \qquad \text{(constant } V\text{)} \qquad (19.3)$$

[2] A second fixed point at 0 K is implied by Equation 19.1. We shall describe the meaning of this point in Chapter 22 when we discuss the second law of thermodynamics.

TABLE 19.1 Fixed-Point Temperatures

Fixed Point	Temperature (°C)	Temperature (K)
Triple point of hydrogen	−259.34	13.81
Boiling point of hydrogen at 33360.6 N/m² pressure	−256.108	17.042
Boiling point of hydrogen	−252.87	20.28
Triple point of neon	−246.048	27.102
Triple point of oxygen	−218.789	54.361
Boiling point of oxygen	−182.962	90.188
Triple point of water	0.01	273.16
Boiling point of water	100.00	373.15
Freezing point of tin	231.9681	505.1181
Freezing point of zinc	419.58	692.73
Freezing point of silver	961.93	1235.08
Freezing point of gold	1064.43	1337.58

All values from National Bureau of Standards Special Publication 420, U. S. Department of Commerce, May 1975.

Thus the constant-volume gas thermometer defines a temperature scale that can be reproduced in laboratories throughout the world. Although the scale depends on the properties of a gas, it is independent of which gas is used. In practice, one can use a gas thermometer down to around 1 K using low-pressure helium gas. Helium liquefies below this temperature; other gases liquefy at even higher temperatures.

It would be convenient to have a temperature scale that is independent of the property of any substance. Such a scale is called an **absolute temperature scale,** or **kelvin scale.** Later we shall find that the ideal gas scale is identical with the absolute temperature scale for temperatures above 1 K, where gas thermometers can be used. In anticipation of this, we shall also use the symbol T to denote absolute temperature. The absolute temperature scale will be properly defined when we study the second law of thermodynamics in Chapter 22.

Other methods of thermometry calibrated against gas thermometers have been used to provide various other fixed-point temperatures. The "International Practical Temperature Scale of 1968," which was established by international agreement, is based on measurements in various national standard laboratories. The assigned temperatures of particular fixed points associated with various substances are given in Table 19.1. The platinum resistance thermometer was used to establish all but the last two points in this table. Note that the scale is not defined below 13.81 K.

19.4 THE CELSIUS, FAHRENHEIT, AND RANKINE TEMPERATURE SCALES[3]

The Celsius temperature, T_C, is shifted from the absolute (or kelvin) temperature T by 273.15°, since by definition the triple point of water (273.16 K) corresponds to 0.01°C. Therefore,

$$T_C = T - 273.15 \tag{19.4}$$

[3] Named after Anders Celsius (1701–1744), Gabriel Fahrenheit (1686–1736), and William MacQuorn Rankine (1820–1872).

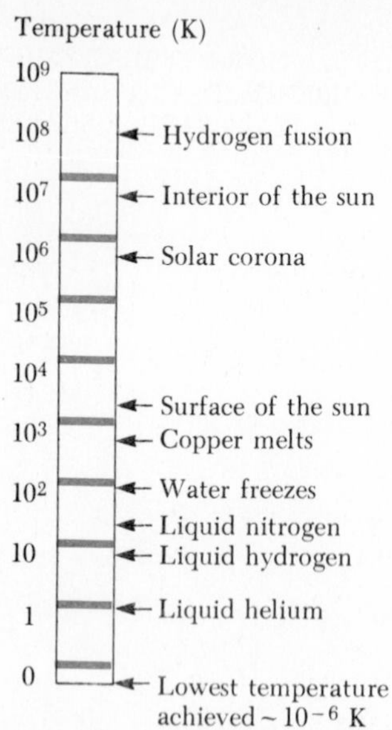

Figure 19.5 Absolute temperatures at which various physical processes take place. Note that the scale is logarithmic.

From this we see that the size of a degree on the kelvin scale is the same as on the Celsius scale. In other words, a temperature difference of 5 Celsius degrees, written 5 C°, is equal to a temperature difference of 5 K. The two scales differ only in the choice of the zero point. Furthermore, the ice point (273.15 K) corresponds to 0.00°C, and the steam pont (373.15 K) is equivalent to 100.00°C.

Two other scales used in the United States and in Great Britain are the *Rankine scale* and the *Fahrenheit scale.* The Rankine temperature, T_R, (written °R) is related to the kelvin temperature through the relation

$$T_R = \tfrac{9}{5} T \tag{19.5}$$

The Fahrenheit temperature, T_F, (written °F) is shifted from the Rankine temperature according to the relation

$$T_F = T_R - 459.67 \tag{19.6}$$

Substituting Equations 19.4 and 19.5 into Equation 19.6 we get

$$T_F = \tfrac{9}{5} T_C + 32 \text{ F}^\circ \tag{19.7}$$

From this expression it follows that the ice point (0.00°C) equals 32°F and the steam point (100.00°C) equals 212°F. Figure 19.5 shows on a logarithmic scale the absolute temperatures for various physical processes and structures.

EXAMPLE 19.1. Converting Temperatures
An object has a temperature of 50°F. What is its temperature in degrees Celsius and in kelvins?

Solution: Substituting $T_F = 50°\text{F}$ into Equation 19.7, we get

$$T_C = \tfrac{5}{9}(T_F - 32) = \tfrac{5}{9}(50 - 32) = 10°\text{C}$$

From Equation 19.4, we find that

$$T = T_C + 273.15 = 283.15 \text{ K}$$

EXAMPLE 19.2. Heating a Pan of Water
A pan of water is heated from 25°C to 80°C. What is the *change* in its temperature on the kelvin scale and on the Fahrenheit scale?

Solution: From Equation 19.4, we see that the change in temperature on the Celsius scale equals the change on the kelvin scale. Therefore,

$$\Delta T = \Delta T_C = 80 - 25 = 55 \text{ C}^\circ = 55 \text{ K}$$

From Equation 19.7, we find that the change in temperature on the Fahrenheit scale is greater than the change on the Celsius scale by the factor 9/5. That is,

$$\Delta T_F = \tfrac{9}{5}\Delta T_C = \tfrac{9}{5}(80 - 25) = 99 \text{ F}^\circ$$

In other words, 55 C° = 99 F°, where the notations C° and F° refer to temperature *differences*, not to be confused with actual temperatures, which are written °C and °F.

Other Thermometers

A technique that is often used as a temperature standard in thermometry makes use of a pure platinum wire because its electrical resistance changes with temperature. The **platinum resistance thermometer** is essentially a coil of platinum wire mounted in a strain-free glass capsule. The platinum resistance changes by about 0.3% for a temperature change of 1 K. It is commonly used for temperatures ranging from about 14 K to 900 K and can be calibrated to within ±0.0003 K at the triple point of water.

Thermocouple

One of the most useful thermometers for scientific and engineering applications is a device called a **thermocouple.** The thermocouple is essentially a junction formed by two different metals or alloys, labeled A and B in Figure 19.6. The test junction is placed in the material whose temperature is to be

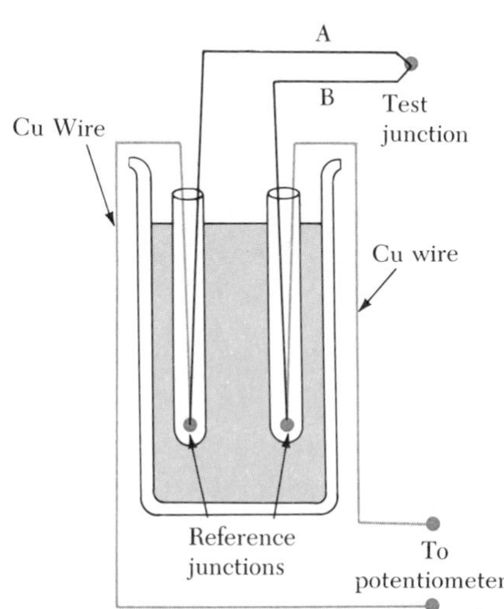

Figure 19.6 Schematic diagram of a thermocouple, which consists of two dissimilar metals, A and B. The reference junctions usually kept at 0°C.

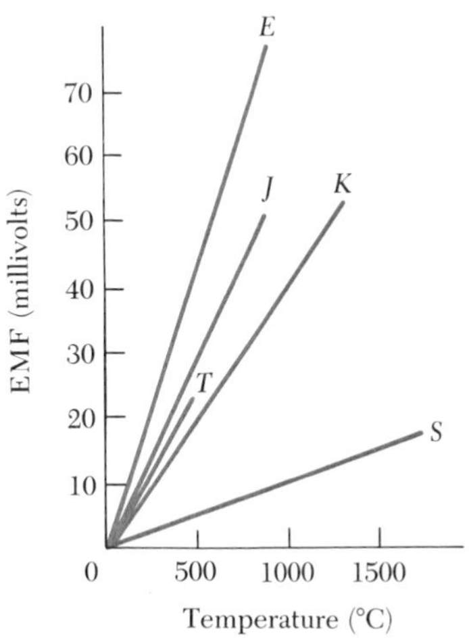

Figure 19.7 Plot of emf (junction voltage) versus temperature for various thermocouples: E—chromel/constantan; J—iron/constantan; T—copper/constantan; K—chromel/alumel; S—platinum/platinum-10% rhodium.

measured, while the opposite ends of the thermocouple wires are maintained at some constant reference temperature (usually in a water-ice mixture) to form two junctions. When the reference temperature is different from the temperature of the test junction, a voltage called the *electromotive force* (emf) is set up in the circuit. The value of this emf is proportional to the temperature difference and therefore can be used to measure an unknown temperature. An instrument called a *potentiometer* is used to measure the emf. In practice, one usually uses junctions for which calibration curves are available.

One advantage of the thermocouple is its small mass, which enables it to quickly reach thermal equilibrium with the material being probed. Some common examples of thermocouple junction materials are copper/constantan (an alloy), which is useful over the temperature range of about −180°C to 400°C, and platinum/platinum-10% rhodium, which is useful over the range from about 0°C to 1500°C. Some typical outputs for various thermocouples are given in Figure 19.7. where the reference junction is at 0°C.

Another thermometer that has extremely high sensitivity is a device called a **thermistor.** This device consists of a small piece of semiconductor material whose electrical resistance changes with temperature. Thermistors are usually fabricated from oxides of various metals, such as nickel, manganese, iron, cobalt, and copper, and can be encapsulated in an epoxy. A careful measurement of the resistance serves as an indicator of temperature, with a typical accuracy of ± 0.1 C°. Temperature changes as small as about 10^{-3} C° can be detected with these devices. Most thermistors operate reliably over the temperature range from about −50°C to 100°C. They are often used as clinical thermometers (with digital readout) and in various biological applications.

19.5 THERMAL EXPANSIONS OF SOLIDS AND LIQUIDS

Most bodies expand as their temperature increases. This phenomenon plays an important role in numerous engineering applications. For example, thermal expansion joints must be included in buildings, concrete highways, railroad tracks, and bridges to compensate for changes in dimensions with temperature variations.

The overall thermal expansion of a body is a consequence of the change in the average separation between its constituent atoms or molecules. To understand this, consider a crystalline solid, which consists of a regular array of atoms held together by electrical forces. We can obtain a mechanical model of these forces by imagining that the atoms are connected by a set of stiff springs as in Figure 19.8. The interatomic forces are taken to be elastic in nature. At ordinary temperatures, the atoms vibrate about their equilibrium positions with an amplitude of about 10^{-11} m and a frequency of about 10^{13} Hz. The average spacing between the atoms is of the order of 10^{-10} m. As the temperature of the solid increases, the atoms vibrate with larger amplitudes and the average separation between them increases.[4] Consequently, the solid as a whole expands with increasing temperature. If the expansion of an object is sufficiently small compared with its initial dimensions, then the change in any dimension (length, width, or thickness) is, to a good approximation, a linear function of the temperature.

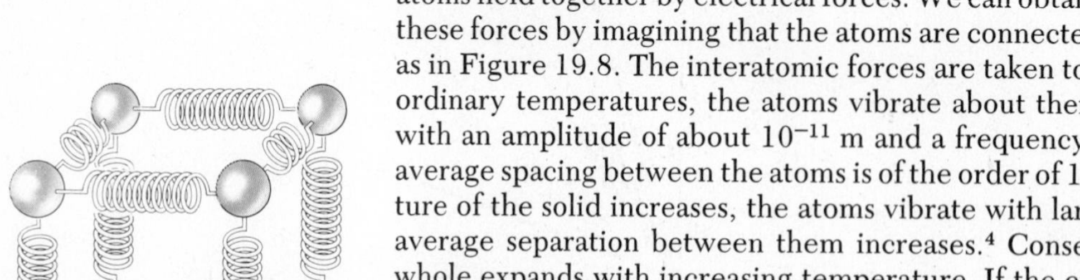

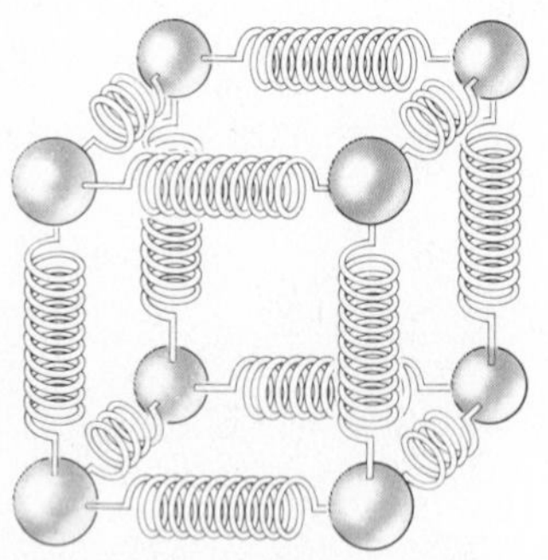

Figure 19.8 A mechanical model of a crystalline solid. The atoms (solid spheres) are imagined to be attached to each other by springs, which reflect the elastic nature of the interatomic forces.

Suppose the linear dimension of a body along some direction is ℓ at some temperature. The length increases by an amount $\Delta\ell$ for a change in temperature ΔT. Experiments show that the change in length is proportional to the temperature change and to the original length when ΔT is small enough. Thus the basic equation for the expansion of a solid is

$$\Delta\ell = \alpha\ell\Delta T \tag{19.8}$$

where the proportionality constant α is called the **average coefficient of linear expansion** for a given material. From this expression, we see that

$$\alpha = \frac{1}{\ell}\frac{\Delta\ell}{\Delta T} \tag{19.9}$$

In other words, the average coefficient of linear expansion of a solid is the fractional change in length ($\Delta\ell/\ell$) per degree change in temperature. The unit of α is deg^{-1}. For example, an α value of 11×10^{-6} (C°)$^{-1}$ means that the length of an object changes by 11 parts per million of its original length for every Celsius degree change in temperature. It may be helpful to think of thermal expansion as an effective magnification or as a photographic enlargement of an object when it is heated. For example, as a metal washer is heated (Fig. 19.9) all dimensions increase, including the radius of the hole.

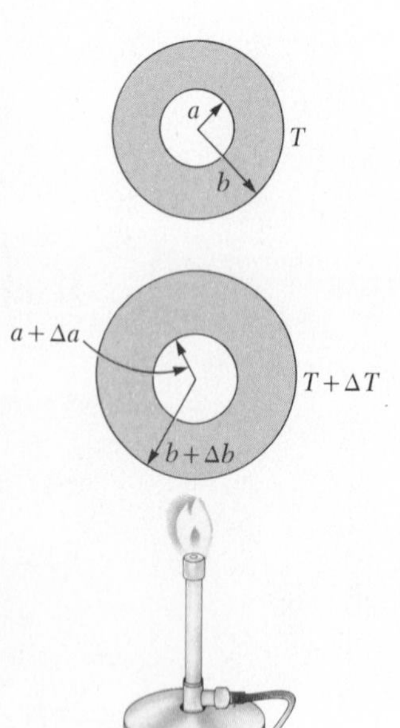

Figure 19.9 Thermal expansion of a homogeneous metal washer. Note that as the washer is heated, all dimensions increase. The expansion is exaggerated.

The coefficient of linear expansion generally varies with temperature. Usually this temperature variation is negligible over the temperature range of most everyday measurements. Table 19.2 lists the average coefficient of linear expansion for various materials. Note that α is positive for these materials, indicating an increase in length with increasing temperature. This is not always the case. For example, some single anisotropic crystalline substances, such as calcite ($CaCO_3$), expand along one dimension (positive α) and contract along another (negative α) with increasing temperature.

Because the linear dimensions of a body change with temperature, it follows that the area and volume of a body also change with temperature. The

[4] Strictly speaking, thermal expansion arises from the *asymmetric* nature of the potential energy curve for the atoms in a solid. If the oscillators were truly harmonic, the average atomic separations would not change regardless of the amplitude of vibration.

TABLE 19.2 Expansion Coefficients for Some Materials Near Room Temperature

Material	Linear Expansion Coefficient α (C°)$^{-1}$	Material	Volume Expansion Coefficient β (C°)$^{-1}$
Aluminum	24×10^{-6}	Alcohol, ethyl	1.12×10^{-4}
Brass and bronze	19×10^{-6}	Benzene	1.24×10^{-4}
Copper	17×10^{-6}	Acetone	1.5×10^{-4}
Glass (ordinary)	9×10^{-6}	Glycerin	4.85×10^{-4}
Glass (pyrex)	3.2×10^{-6}	Mercury	1.82×10^{-4}
Lead	29×10^{-6}	Turpentine	9.0×10^{-4}
Steel	11×10^{-6}	Gasoline	9.6×10^{-4}
Invar (Ni-Fe alloy)	0.9×10^{-6}	Air	3.67×10^{-3}
Concrete	12×10^{-6}	Helium	3.665×10^{-3}

change in volume at constant pressure is proportional to the original volume V and to the change in temperature according to the relation

$$\Delta V = \beta V \, \Delta T \tag{19.10}$$

where β is the **average coefficient of volume expansion.** *For an isotropic solid, the coefficient of volume expansion is approximately three times the linear expansion coefficient,* or $\beta = 3\alpha$ (An **isotropic solid** is one in which the coefficient of linear expansion is the same in all directions.) Therefore, Equation 19.10 can be written

$$\Delta V = 3\alpha V \, \Delta T \tag{19.11}$$

Change in volume of an isotropic solid at constant pressure

The derivation of $\beta = 3\alpha$ is given below.

To show that $\beta = 3\alpha$ for an isotropic solid, consider an object in the shape of a box of dimensions ℓ, w, and h. Its volume at some temperature T is $V = \ell wh$. If the temperature changes to $T + \Delta T$, its volume changes to $V + \Delta V$, where each dimension changes according to Equation 19.8. Therefore,

$$\begin{aligned} V + \Delta V &= (\ell + \Delta\ell)(w + \Delta w)(h + \Delta h) \\ &= (\ell + \alpha\ell \, \Delta T)(w + \alpha w \, \Delta T)(h + \alpha h \, \Delta T) \\ &= \ell wh(1 + \alpha \, \Delta T)^3 \\ &= V[1 + 3\alpha \, \Delta T + 3(\alpha \, \Delta T)^2 + (\alpha \, \Delta T)^3] \end{aligned}$$

Hence the fractional change in volume is

$$\frac{\Delta V}{V} = 3\alpha \, \Delta T + 3(\alpha \, \Delta T)^2 + (\alpha \, \Delta T)^3$$

Since the product $\alpha \, \Delta T$ is small compared with unity for typical values of ΔT (less than ≈ 100°C), we can neglect the terms $3(\alpha \, \Delta T)^2$ and $(\alpha \, \Delta T)^3$. In this approximation, we see that

$$\beta = \frac{1}{V}\frac{\Delta V}{\Delta T} = 3\alpha$$

Expansion joints allow bridges to expand and contract safely.

A sheet or flat plate can be described by its area. You should show (Problem 33) that the change in the area of an isotropic plate is given by

$$\Delta A = 2\alpha A \, \Delta T \tag{19.12}$$

Change in area of an isotropic solid

EXAMPLE 19.3. Expansion of a Railroad Track
A steel railroad track has a length of 30 m when the temperature is 0°C. (a) What is its length on a hot day when the temperature is 40°C?

Making use of Table 19.2 and noting that the change in temperature is 40 C°, we find that the increase in length is

$$\Delta \ell = \alpha \ell \, \Delta T = [11 \times 10^{-6} (\text{C}^\circ)^{-1}](30 \text{ m})(40 \text{ C}^\circ) = 0.013 \text{ m}$$

Therefore, its length at 40°C is 30.013 m.

(b) Suppose the ends of the rail are rigidly clamped at 0°C so as to prevent expansion. Calculate the thermal stress set up in the rail if its temperature is raised to 40°C.

From the definition of Young's modulus for a solid (Chapter 12), we have

$$\text{Tensile stress} = \frac{F}{A} = Y \frac{\Delta \ell}{\ell}$$

Since Y for steel is 20×10^{10} N/m² we have

$$\frac{F}{A} = \left(20 \times 10^{10} \frac{\text{N}}{\text{m}^2}\right)\left(\frac{0.013 \text{ m}}{30 \text{ m}}\right) = 8.67 \times 10^{7} \text{ N/m}^2$$

Exercise 1 If the rail has a cross-sectional area of 30 cm², calculate the force of compression in the rail.
Answer: 2.60×10^5 N or 58 500 lb!

Liquids generally increase in volume with increasing temperature, and have volume expansion coefficients about ten times greater than those of solids (Table 19.2). Water is an exception to this rule, as we can see from its density versus temperature curve, shown in Figure 19.10. As the temperature increases from 0°C to 4°C, the water contracts, and thus its density increases. Above 4°C, the water expands with increasing temperature. The density of water reaches a *maximum* value of 1000 kg/m³ at 4°C.

We can explain why a pond or lake freezes at the surface from this unusual thermal expansion behavior of water. As the pond cools, the cooler, denser water at the surface initially flows to the bottom. When the temperature at the surface reaches 4°C, this flow ceases. Consequently, when the surface of the pond is below 4°C, equilibrium is reached when the coldest water is at the surface. As the water freezes at the surface, it remains there since ice is less dense than water. The ice continues to build up at the surface, while water near the bottom remains at 4°C. If this did not happen, fish and other forms of marine life would not survive. In fact, if it were not for this peculiarity of water, among others, life as we now know it wouldn't exist!

19.6 MACROSCOPIC DESCRIPTION OF AN IDEAL GAS

In this section we shall be concerned with the properties of a gas of mass m confined to a container of volume V at a pressure P and temperature T. It

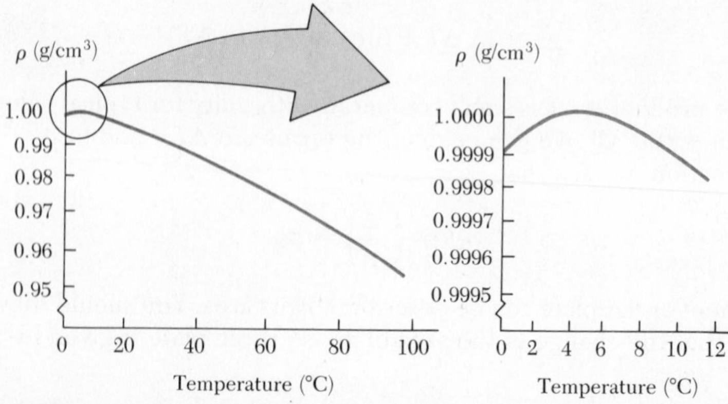

Figure 19.10 The variation of density with temperature for water at atmospheric pressure. The maximum density occurs at 4°C as can be seen in the magnified graph at the right.

would be useful to know how these quantities are related. In general, the equation that interrelates these quantities, called the *equation of state,* is very complicated. However, if the gas is maintained at a very low pressure (or low density), the equation of state is experimentally found to be quite simple. Such a low-density gas is commonly referred to as an **ideal gas.**[5] Most gases at room temperature and atmospheric pressure behave as ideal gases.

It is convenient to express the amount of gas in a given volume in terms of the number of moles, n. By definition, **one mole** of any substance is that mass of the substance that contains a specific number of molecules called Avogadro's number, N_A. The value of N_A is approximately 6.022×10^{23} molecules/mole. Avogadro's number is defined to be the number of carbon atoms in 12 g of the isotope carbon-12. The number of moles of a substance is related to its mass m through the expression

$$n = \frac{m}{M} \qquad (19.13)$$

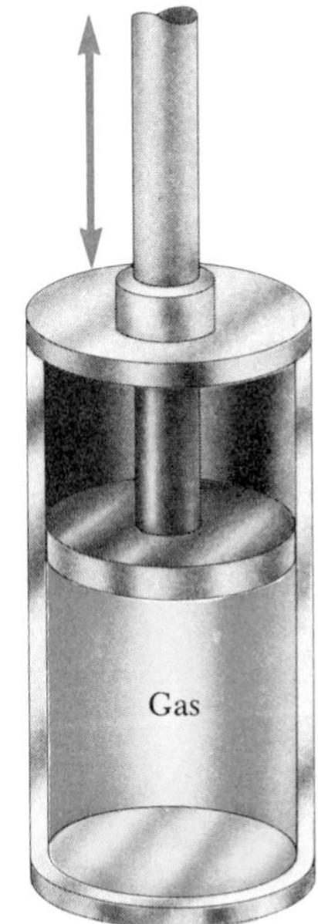

Figure 19.11 An ideal gas contained in a cylinder with a movable piston that allows the volume to be varied. The state of the gas is defined by its pressure, volume, and temperature.

where M is a quantity called the **molecular weight** of the substance, usually expressed in g/mole. For example, the molecular weight of oxygen, O_2, is 32.0 g/mol. Therefore, the mass of one mole of oxygen is 32.0 g.

Now suppose an ideal gas is confined to a cylindrical container whose volume can be varied by means of a movable piston, as in Figure 19.11. We shall assume that the cylinder does not leak, and hence the mass (or the number of moles) remains constant. For such a system, experiments provide the following information. First, when the gas is kept at a constant temperature, its pressure is inversely proportional to the volume (Boyle's law). Second, when the pressure of the gas is kept constant, the volume is directly proportional to the temperature (the law of Charles and Gay-Lussac). These observations can be summarized by the following **equation of state for an ideal gas:**

Equation of state for an ideal gas

$$PV = nRT \qquad (19.14)$$

In this expression, R is a constant for a specific gas, which can be determined from experiments, and T is the absolute temperature in kelvin. Experiments on several gases show that as the pressure approaches zero, the quantity PV/nT approaches the same value of R for all gases. For this reason, R is called the **universal gas constant.** In metric units, where pressure is expressed in N/m^2 and volume in m^3, the product PV has units of N · m, or J, and R has the value

The universal gas constant

$$R = 8.31 \text{ J/mole} \cdot \text{K} \qquad (19.15)$$

If the pressure is expressed in atmospheres and the volume in liters (1 liter = 10^3 cm^3 = 10^{-3} m^3), then R has the value

$$R = 0.0821 \text{ liter} \cdot \text{atm/mole} \cdot \text{K}$$

Using this value of R and Equation 19.14, one finds that the volume occupied by 1 mole of any gas at atmospheric pressure and 0°C (273 K) is 22.4 liters.

The ideal gas law is often expressed in terms of the total number of molecules, N. Since the total number of molecules equals the product of the number of moles and Avogadro's number, we can write Equation 19.14

[5] To be more specific, the assumption here is that the temperature of the gas is sufficiently high and its pressure sufficiently low that it does not condense into a liquid.

$$PV = nRT = \frac{N}{N_A} RT$$

$$PV = NkT \tag{19.16}$$

where k is called **Boltzmann's constant,** which has the value

Boltzmann's constant

$$k = \frac{R}{N_A} = 1.38 \times 10^{-23}\ \mathrm{J/K} \tag{19.17}$$

We have defined an ideal gas as one that obeys the equation of state, $PV = nRT$, under all conditions. In reality, an ideal gas does not exist. However, the concept of an ideal gas is very useful in view of the fact that real gases behave as ideal gases at low pressures. It is common to call quantities such as P, V, and T the **thermodynamic variables** of the system. We note that if the equation of state is known, then one of the variables can always be expressed as some function of the other two thermodynamic variables. That is, given two of the variables, the third can be determined from the equation of state. Other thermodynamic systems are often described with different thermodynamic variables. For example, a wire under tension at constant pressure is described by its length, the tension in the wire, and the temperature.

EXAMPLE 19.4. How Many Gas Molecules Are in the Container?
An ideal gas occupies a volume of 100 cm³ at 20°C and a pressure of 10^{-3} atm. Determine the number of moles of gas in the container.

Solution: The quantities given are volume, pressure, and temperature: $V = 100\ \mathrm{cm^3} = 0.1$ liter, $P = 10^{-3}$ atm, and $T = 20°\mathrm{C} = 293$ K. Using Equation 19.14, we get

$$n = \frac{PV}{RT} = \frac{(10^{-3}\ \mathrm{atm})(0.1\ \mathrm{liter})}{(0.0821\ \mathrm{liter \cdot atm/mole \cdot K})(293\ \mathrm{K})}$$
$$= 4.16 \times 10^{-6}\ \mathrm{moles}$$

Note that you must express T as an absolute temperature (K) when using the ideal gas law.

Exercise 2 Calculate the number of molecules in the container, using the fact that Avogadro's number is 6.02×10^{23} molecules/mole.
Answer: 2.50×10^{18} molecules.

EXAMPLE 19.5. Squeezing a Tank of Gas
Pure helium gas is admitted into a tank containing a movable piston. The initial volume, pressure, and temperature of the gas are 15 liters, 2 atm, and 300 K. If the volume is decreased to 12 liters and the pressure increased to 3.5 atm, find the final temperature of the gas. (Assume that helium behaves like an ideal gas.)

Solution: If no gas escapes from the tank, the number of moles remains constant; therefore using $PV = nRT$ at the initial and final points gives

$$\frac{P_i V_i}{T_i} = \frac{P_f V_f}{T_f}$$

where i and f refer to the initial and final values. Solving for T_f, we get

$$T_f = \left(\frac{P_f V_f}{P_i V_i}\right) T_i = \frac{(3.5\ \mathrm{atm})(12\ \mathrm{liters})}{(2\ \mathrm{atm})(15\ \mathrm{liters})}(300\ \mathrm{K}) = 420\ \mathrm{K}$$

EXAMPLE 19.6. Heating a Bottle of Air
A sealed glass bottle containing air at atmospheric pressure and having a volume of 30 cm³ is at 23°C. It is then tossed into an open fire. When the temperature of the air in the bottle reaches 200°C, what is the pressure inside the bottle? Assume any volume changes of the bottle are small enough to be negligible.

Solution: This example is approached in the same fashion as that used in Example 19.5. We start with the expression

$$\frac{P_i V_i}{T_i} = \frac{P_f V_f}{T_f}$$

Since the initial and final volumes of the gas are assumed equal, this expression reduces to

$$\frac{P_i}{T_i} = \frac{P_f}{T_f}$$

This gives

$$P_f = \left(\frac{T_f}{T_i}\right)(P_i) = \left(\frac{473\ \text{K}}{300\ \text{K}}\right)(1\ \text{atm}) = 1.58\ \text{atm}$$

Obviously, the higher the temperature, the higher the pressure exerted by the trapped air. Of course, if the pressure rises high enough, the bottle will shatter.

Exercise 3 In this example, we neglected the change in volume of the bottle. If the coefficient of volume expansion for glass is $27 \times 10^{-6}\ (\text{C}°)^{-1}$, find the magnitude of this volume change.
Answer: $0.14\ \text{cm}^3$.

19.7 SUMMARY

Two bodies are in **thermal equilibrium** with each other if they have the same temperature.

The **zeroth law of thermodynamics** states that if bodies A and B are separately in thermal equilibrium with a third body, C, then A and B are in thermal equilibrium with each other.

The SI unit of thermodynamic temperature is the **kelvin,** which is defined to be the fraction 1/273.16 of the temperature of the triple point of water.

When a substance is heated, it generally expands. The linear expansion of an object is characterized by an **average expansion coefficient,** α, defined by

$$\alpha = \frac{1}{\ell}\frac{\Delta \ell}{\Delta T} \qquad (19.9)$$

Average coefficient of linear expansion

where ℓ is the initial length of the object and $\Delta\ell$ is the change in length for a temperature change ΔT. The **average volume expansion coefficient,** β, for a homogeneous substance is equal to 3α.

An **ideal gas** is one that obeys the *equation of state,*

$$PV = nRT \qquad (19.14)$$

Equation of state for an ideal gas

where n equals the number of moles of gas, V is its volume, R is the universal gas constant (8.31 J/mole · K), and T is the absolute temperature in kelvins. A real gas behaves approximately as an ideal gas at very low pressures. An ideal gas is used as the working substance in a constant-volume gas thermometer, which defines the absolute temperature scale in kelvins. This absolute temperature T is related to temperatures on the Celsius scale by $T = T_C + 273.15$.

QUESTIONS

1. Is it possible for two objects to be in thermal equilibrium if they are not in contact with each other? Explain.
2. A piece of copper is dropped into a beaker of water. If the water's temperature rises, what happens to the temperature of the copper? When will the water and copper be in thermal equilibrium?
3. In principle, any gas can be used in a gas thermometer. Why is it not possible to use oxygen for temperatures as low as 15 K? What gas would you use? (Look at the data in Table 19.1.)
4. Explain why a column of mercury in a thermometer first descends slightly and then rises when placed in hot water.
5. Explain why the thermal expansion of a spherical shell made of an isotropic solid is equivalent to that of a solid sphere of the same material.
6. A steel wheel bearing is 1 mm smaller in diameter than an axle. How can it be fit onto the axle without removing any material?
7. Determine the number of grams in one mole of the following gases: (a) hydrogen, (b) helium, and (c) carbon monoxide.
8. Two identical cylinders at the same temperature each contain the same kind of gas. If cylinder A contains three times more gas than cylinder B, what can you say about the relative pressures in the cylinders?

9. Why is it necessary to use absolute temperature when using the ideal gas law?
10. An inflated rubber balloon filled with air is immersed in a flask of liquid nitrogen that is at 77 K. Describe what happens to the balloon.
11. Markings to indicate length are placed on a steel tape in a room that has a temperature of 22°C. Are measurements made with the tape on a day when the temperature is 27°C too long, too short, or accurate? Defend your answer.
12. What would happen if the glass of a thermometer expanded more upon heating than did the liquid inside?

PROBLEMS

Section 19.3 The Constant-Volume Gas Thermometer and the Kelvin Scale

1. The gas thermometer shown in Figure 19.2 reads a pressure of 50 mm Hg at the triple-point temperature. What pressure will it read at (a) the boiling point of water, and (b) the boiling point of sulfur (444.6°C)?
2. A constant-volume gas thermometer registers a pressure of 40 mm Hg when it is at a temperature of 350 K. (a) What is the pressure at the triple point of water? (b) What is the temperature when the pressure reads 2 mm Hg?
3. The pressure in a constant-volume gas thermometer is 0.600 atm at 100°C and 0.439 atm at 0°C. (a) What is the temperature when the pressure is 0.400 atm? (b) What is the pressure at 450°C?

Section 19.4 The Celsius, Fahrenheit, and Rankine Temperature Scales

4. Liquid hydrogen has a boiling point of −252.87°C at atmospheric pressure. Express this temperature in (a) degrees Fahrenheit, (b) degrees Rankine, and (c) kelvins.
5. The boiling point of sulfur is 444.6°C. Express this in (a) degrees Fahrenheit and (b) kelvins.
6. The temperature of one northeastern state varies from 95°F in the summer to −30°F in winter. Express this range of temperatures in degrees Celsius.
7. The normal human body temperature is 98.6°F. A person with a fever reaches a temperature of 103°F. Express these temperatures in degrees Celsius.
8. A substance is heated from 70°F to 195°F. What is its change in temperature on (a) the Celsius scale and (b) the kelvin scale?
9. Two thermometers are calibrated, one in degrees Celsius, and the other in degrees Fahrenheit. At what temperature are their readings the same?

Section 19.5 Thermal Expansion of Solids and Liquids (use Table 19.2)

10. A copper pipe is 2 m long at 25°C. What is its length at (a) 100°C and (b) 0°C?
11. A structural steel I-beam is 20 m long when installed at 20°C. How much will its length change over the temperature extremes −25°C to 40°C?
12. Calculate the *fractional* change in the volume of an aluminum bar that undergoes a change in temperature of 30 C°. (Note that $\beta = 3\alpha$ for an isotropic substance.)
13. The concrete sections of a certain superhighway are designed to have a length of 30 m. The sections are poured and cured at 10°C. What minimum spacing should the engineer leave *between the sections* to eliminate "buckling" if the concrete is to reach a temperature of 45°C?
14. A steel washer has an inner diameter of 2.000 cm and an outer diameter of 2.500 cm at 20°C. To what temperature must the washer be heated to just fit over a rod that is 2.005 cm in diameter?
15. An automobile fuel tank is filled to the brim with 22 gal of gasoline at −20°C. Immediately afterwards, the vehicle is parked in a garage at 25°C. How much gasoline overflows from the tank as a result of expansion? (Neglect the expansion of the tank.)
16. A metal rod made of some alloy is to be used as a thermometer. At 0°C its length is 30.000 cm and at 100°C its length is 30.050 cm. (a) What is the linear expansion coefficient of the alloy? (b) When the rod is 30.015 cm long, what is the temperature?
17. The active element of a certain laser (light amplifier) is made of a glass rod 20 cm long and 1 cm in diameter. If the temperature of the rod increases by 75 C°, find the increase in (a) its length, (b) its diameter, and (c) its volume. [Take $\alpha = 9 \times 10^{-6} (\mathrm{C}°)^{-1}$]

Section 19.6 Macroscopic Description of an Ideal Gas

18. An ideal gas is held in a container at constant volume. Initially, its temperature is 20°C and its pressure is 3 atm. Find the pressure when its temperature is 50°C.
19. A cylinder with a movable piston contains gas at a temperature of 27°C, a pressure of 0.2×10^5 Pa, and a volume of 1.5 m^3. What will be its final temperature if the gas is compressed to 0.7 m^3 and the pressure increases to 0.8×10^5 Pa?
20. A gas is heated from 27°C to 127°C while maintained at constant pressure in a vessel whose volume increases. By what factor does the volume change?
21. One mole of oxygen gas is at a pressure of 5 atm and a temperature of 27°C. (a) If the gas is heated at constant volume until the pressure is doubled, what is the final temperature? (b) If the gas is heated such that both the pressure and volume are doubled, what is the final temperature?
22. A cylinder of volume 12 liters contains helium gas at a pressure of 136 atm. How many balloons can be filled

with this cylinder at atmospheric pressure if each balloon has a volume of 1 liter?

23. The tire of a bicycle is filled with air to a gauge pressure of 50 lb/in.^{2} at 20°C. What is the gauge pressure in the tire on a day when the temperature rises to 35°C? (Assume the volume does not change, and recall that gauge pressure means absolute pressure in the tire minus atmospheric pressure. Furthermore, assume that the atmospheric pressure remains constant.)

24. Gas is contained in a 3-liter vessel at a temperature of 25°C and a pressure of 5 atm. (a) Determine the number of moles of gas in the vessel. (b) How many molecules are there in the vessel?

25. In modern vacuum systems, pressures as low as 10^{-9} mm Hg are common. Calculate the number of molecules in a 1-m^3 vessel at this pressure if the temperature is 20°C. (*Note:* One atm of pressure corresponds to 760 mm Hg.)

26. Show that one mole of any gas at atmospheric pressure (1.01×10^5 N/m^2) and standard temperature (273 K) occupies a volume of 22.4 liters.

GENERAL PROBLEMS

27. Precise temperature measurements are often made using the change in the electrical resistance of a metal or semiconductor with temperature. The resistance varies approximately according to the expression $R = R_0(1 + AT_C)$, where R_0 and A are constants and T_C is the temperature in degrees Celsius. A certain element has a resistance of 50.0 ohms at 0°C and 82.5 ohms at the freezing point of zinc (419.58°C). (a) Determine the constants A and R_0. (b) At what temperature is the resistance equal to 65.5 ohms?

28. A fluid has a density ρ. (a) Show that the *fractional* change in density for a change in temperature ΔT is given by $\Delta\rho/\rho = -\beta\,\Delta T$. What does the negative sign signify? (b) Water has a maximum density of 1.000 g/cm^3 at 4°C. At 10°C, its density is 0.9997 g/cm^3. What is β for water over this temperature interval?

29. A mercury thermometer is constructed as in Figure 19.12. The capillary tube has a diameter of 0.005 cm, and the bulb has a diameter of 0.30 cm. Neglecting the expansion of the glass, find the change in height of the mercury column for a temperature change of 25 C°.

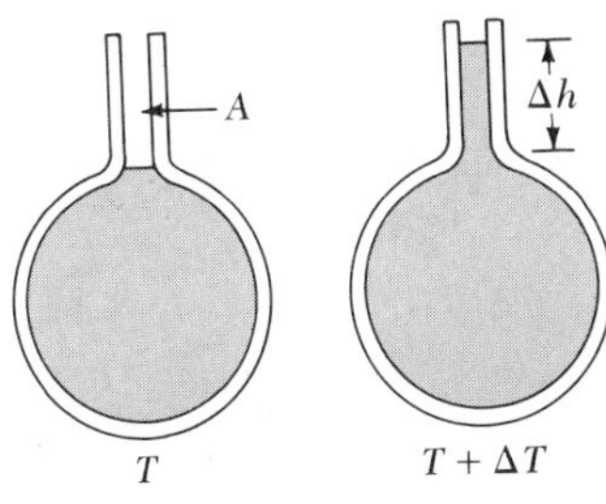

Figure 19.12 (Problems 29 and 32).

30. A steel ball bearing is 2.000 cm in diameter at 20°C. An aluminum plate has a hole in it that is 1.995 cm in diameter at 20°C. What common temperature must they have in order that the ball just squeeze through the hole?

31. A pendulum clock with a steel suspension system has a period of 1 s at 20°C. If the temperature increases to 25°C, (a) by how much will its period change, and (b) how much time will the clock gain or lose in one week?

32. A liquid with a coefficient of volume expansion β just fills a spherical shell of volume V at a temperature T (Fig. 19.12). The shell is made of a material that has a coefficient of linear expansion of α. The liquid is free to expand into a capillary of cross-sectional area A at the top. (a) If the temperature increases by ΔT, show that the liquid rises in the capillary by an amount Δh given by $\Delta h = \dfrac{V}{A}(\beta - 3\alpha)\,\Delta T$. (b) For a typical system, such as a mercury thermometer, why is it a good approximation to neglect the expansion of the shell?

33. The rectangular plate shown in Figure 19.13 has an area A equal to ℓw. If the temperature increases by ΔT, show that the increase in area is given by $\Delta A = 2\alpha A\,\Delta T$, where α is the coefficient of linear expansion. What approximation does this expression assume? (*Hint:* Note that each dimension increases according to $\Delta\ell = \alpha\ell\,\Delta T$.)

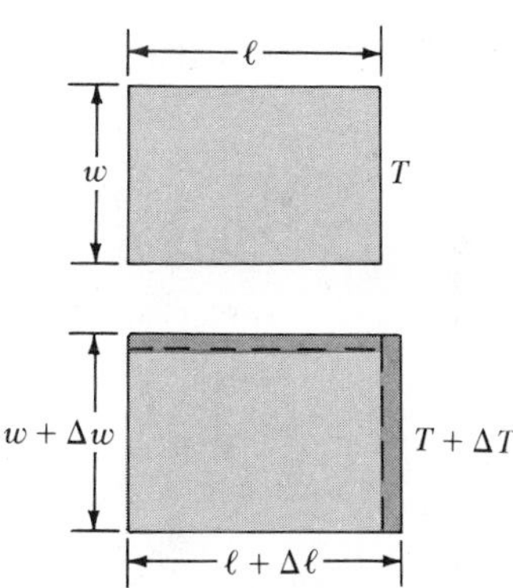

Figure 19.13 (Problem 33).

34. At $T = 0$°C, each one of three metal bars (two of aluminum and one of invar) is drilled with two holes a distance ℓ apart. Pins are put through the holes to create an equilateral triangle. If the bars are then heated to 100°C, what will be the angle between the two aluminum bars?

35. At $T = 0$°C, a container is completely full of liquid mercury. If the container does not expand when heated (and is pushed out to a negligible extent when the mercury begins to exert an outward pressure), what will be the internal pressure when the temperature is raised to 20°C? Express your answer in N/m^2 and in atmospheres. (Refer to Tables 15.2 and 19.2.)

36. (a) Show that the volume coefficient of thermal expansion for an ideal gas at constant pressure is given by $\beta = 1/T$, where T is the kelvin temperature. Start with the definition of β and use the equation of state, $PV = nRT$. (b) What value does this expression predict for β at 0°C? Compare this with the experimental values for helium and air in Table 19.2.

37. An air bubble originating from a deep-sea diver has a radius of 2 mm at some depth h. When the bubble reaches the surface of the water, it has a radius of 3 mm. Assuming the temperature of the air in the bubble remains constant, determine (a) the depth h of the diver, and (b) the absolute pressure at this depth.

38. Starting with Equation 19.14, show that the total pressure P in a container filled with a mixture of several different ideal gases is given by $P = P_1 + P_2 + P_3 + \cdots$, where P_1, P_2, etc., are the pressures that each gas would exert if it alone filled the container (or the *partial pressures* of the respective gases). This is known as *Dalton's law of partial pressures.*

39. (a) Show that the density of n moles of a gas occupying a volume V is given by $\rho = nM/V$, where M is the molecular weight. (b) Determine the density of one mole of nitrogen gas at atmospheric pressure and 0°C.

40. A vertical cylinder of cross-sectional area A is fitted with a tight-fitting, frictionless piston of mass m (Fig. 19.14). (a) If there are n moles of an ideal gas in the cylinder at a temperature T, determine the height h at which the piston will be in equilibrium under its own weight. (b) What is the value for h if $n = 3$ moles, $T = 500$ K, $A = 0.05$ m^2, and $m = 5$ kg?

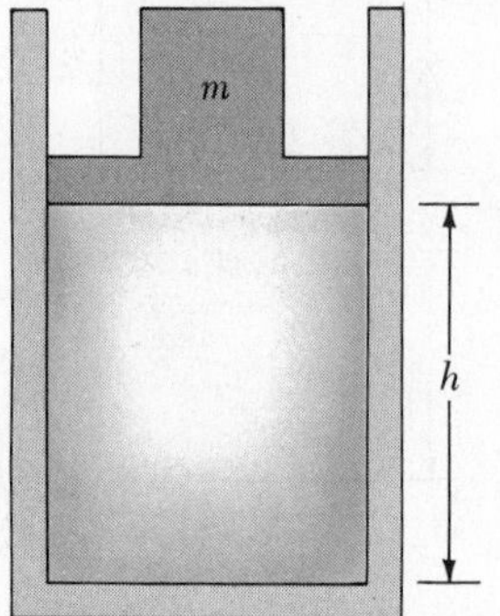

Figure 19.14 (Problem 40).

41. Consider an object with any one of the shapes displayed in Table 10.2. What is the percentage increase in the moment of inertia of the object when it is heated from 0°C to 100°C, if it is composed of (a) lead, (b) invar? (See Table 19.2. Assume that the linear expansion coefficients do not vary between 0°C and 100°C.)

•42. An aluminum pot has the shape of a right circular cylinder. The pot is initially at 4°C, at which temperature it has an inside diameter of 28.00 cm. The pot contains 3.000 gallons of water at 4°C. (a) What is the depth of water in the pot? (1 gallon = 3785 cm^3) (b) The pot and the water in it are heated to a final temperature of 90°C. Allowing for the expansion of the water, but ignoring the expansion of the pot, what is the change in depth of the water? Express the change as percentage of the original depth and also in millimeters. (The density of water is 1.000 g/cm^3 at 4°C and 0.965 g/cm^3 at 90°C.) (c) Modify your solution for part (b) to allow for the expansion of the pot. (Refer to Table 19.2.)

•43. An aluminum wire and copper wire, each of diameter 1.00 mm, are joined end to end. At 40°C, each has an unstretched length of 1.000 m; they are connected between fixed supports 2 m apart on a tabletop, so that the aluminum wire extends from $x = -1.000$ m to $x = 0$, the copper wire extends from $x = 0$ to $x = 1.000$ m, and the tension is negligible. The temperature is then lowered to 0°C, while the supports are held fixed at a separation of 2.000 m. At this lower temperature, find the tension in the wire and the location (x coordinate) of the junction between the aluminum and the copper. (Refer to Tables 15.2 and 19.2.)

•44. A steel guitar string with a diameter of 1.00 mm is stretched between supports 80 cm apart. The temperature is 0°C. (a) Find the mass per unit length of this string. (Use the value 7.86×10^3 kg/m^3 for the density.) (b) The fundamental frequency of transverse oscillations of the string is 200 Hz. What is the tension in the string? (c) If the temperature is raised to 30°C, find the resulting values of the tension and the fundamental frequency. [Assume that both the Young's modulus (Table 15.2) and the coefficient of thermal expansion (Table 19.2) of steel have constant values between 0°C and 30°C.]

20
Heat and the First Law of Thermodynamics

It is well known that when two objects at different temperatures are placed in thermal contact with each other, the temperature of the warmer body decreases while the temperature of the cooler body increases. If they are left in contact for some time, they eventually reach a common equilibrium temperature somewhere between the two initial temperatures. When such processes occur, we say that heat is transferred from the warmer to the cooler body. But what is the nature of this heat transfer? Early investigators believed that heat was an invisible, material substance called *caloric,* which was transferred from one body to another. According to this theory, caloric could neither be created nor destroyed. Although the caloric theory was successful in describing heat transfer, it eventually was abandoned when various experiments showed that caloric was in fact not conserved.

The first experimental observation suggesting that caloric was not conserved was made by Benjamin Thompson (1753 – 1814) at the end of the 18th century. Thompson, an American-born scientist, emigrated to Europe during the Revolutionary War because of his Tory sympathies. Following his appointment as director of the Bavarian Arsenal, he was given the title Count Rumford. While supervising the boring of an artillery cannon in Munich, Thompson noticed the great amount of heat generated by the boring tool. The water being used for cooling had to be replaced continuously as it boiled away during the boring process. On the basis of the caloric theory, he reasoned that the ability of the metal filings to retain caloric should decrease as the size of the filings decreased. These heated filings, in turn, presumably transfer caloric to the cooling water, causing it to boil. To his surprise, Thompson discovered that the amount of water boiled away by a blunt boring tool was comparable to the quantity boiled away by a sharper tool for a given turning rate. He then reasoned that if the tool were turned long enough, an almost infinite amount of caloric could be produced from a finite amount of metal filings. For this reason, Thompson rejected the caloric theory and suggested that heat is not a substance, but some form of motion that is transferred from the boring tool to the water. In another experiment, he showed that the heat generated by friction was proportional to the mechanical work done by the boring tool.

Benjamin Thompson (1753 – 1814). "Being engaged, lately, in superintending the boring of cannon, in the workshops of the military arsenal at Munich, I was struck with the very considerable degree of Heat which a brass gun acquires, in a short time, in being bored; and with the still more intense Heat (much greater than that of boiling water, as I found by experiment) of the metallic chips separated from it by the borer."

There are many other experiments that are at odds with the caloric theory. For example, if you rub two blocks of ice together on a day when the temperature is below 0°C, the blocks will melt. This experiment was first conducted by Sir Humphry Davy (1778 – 1829). To properly account for this "creation of caloric," we note that mechanical work is done on the system. Thus, we see that the effects of doing mechanical work on a system and of adding heat to it directly, as with a flame, are equivalent. That is, heat and work are both forms of energy.

Although Thompson's observations provided evidence that heat energy is not conserved, it was not until the middle of the 19th century that the modern mechanical model of heat was developed. Before this period, the subjects of heat and mechanics were considered to be two distinct branches of science, and the law of conservation of energy seemed to be a rather specialized result used to describe certain kinds of mechanical systems. After the two disciplines were shown to be intimately related, the law of conservation of energy emerged as a universal law of nature. In this new view, heat is treated as another form of energy that can be transformed into mechanical energy. Experiments performed by James Joule (1818–1889) and others in this period showed that whenever heat is gained or lost by a system during some process, the gain or loss could be accounted for by an equivalent quantity of mechanical work done on the system. Thus, by broadening the concept of energy to include heat as a form of energy, the law of energy conservation was extended.

20.1 HEAT AND THERMAL ENERGY

Definition of heat

The concepts of heat and the internal energy of a substance appear to be synonymous, but there is a major distinction between them. The word *heat* should be used only when describing energy transferred from one place to another. That is, *heat flow is an energy transfer that takes place as a consequence of temperature differences only.* On the other hand, *internal energy* is the energy a substance has at some temperature. In the next chapter, we shall show that the energy of an ideal gas is associated with the internal motion of its atoms and molecules. In other words, the internal energy of a gas is essentially its kinetic energy on a microscopic scale; the higher the temperature of the gas, the greater its internal energy. As an analogy, consider the distinction between work and energy that we discussed in Chapter 7. The work done on (or by) a system is a measure of energy transfer, whereas the mechanical energy (kinetic and/or potential) is a consequence of the motion and coordinates of the system. Thus, when you do work on a system, energy is transferred from you to the system. It makes no sense to talk about the work *of* a system—one can refer only to the *work done on or by a system* when some process has occurred in which the system has changed in some way. Likewise, it makes no sense to use the term *heat* unless the thermodynamic variables of the system have undergone a change during some process.

It is also important to note that energy can be transferred between two systems even when there is no heat flow. For example, when two objects are rubbed together, their internal energy increases since mechanical work is done on them. When an object slides across a surface and comes to rest as a result of friction, its kinetic energy is transformed into internal energy contained in the block and surface. In such cases, the work done on the system adds energy to the system. The changes in internal energy are measured by corresponding changes in temperature.

20.2 HEAT CAPACITY AND SPECIFIC HEAT

It is useful to define a quantity of heat Q in terms of a specific process. The heat unit that is commonly used is the **calorie** (cal).

The calorie

> The calorie is defined as the amount of heat necessary to raise the temperature of 1 g of water from 14.5°C to 15.5°C.[1]

[1] Originally, the calorie was defined as the heat necessary to raise the temperature of 1 g of water by 1 C°. However, careful measurements showed that energy depends somewhat on temperature; hence, a more precise definition evolved.

The **kilocalorie** (kcal) is the heat necessary to raise the temperature of 1 kg of water from 14.5°C to 15.5°C (1 kcal = 10^3 cal). (Note that the "Calorie," which is used in describing the energy equivalent of foods, is actually a kilocalorie.) The unit of heat in the British engineering system is the **British thermal unit** (Btu), defined as the heat required to raise the temperature of 1 lb of water from 63°F to 64°F. Of course, since we have already recognized that heat is a form of energy, it can be expressed in whatever units happen to be convenient, such as joules, electron-volts, ergs, or foot-pounds. The relationship between the calorie and the mechanical energy unit, the joule, is found from experiment to be

Mechanical equivalent of heat

$$1 \text{ cal} = 4.186 \text{ J}$$

This result for the so-called **mechanical equivalent of heat** was first established by Joule using an apparatus that will be described in Section 20.5.

The quantity of heat energy required to raise the temperature of a given mass of a substance by some amount varies from one substance to another. For example, the heat required to raise the temperature of 1 g of water by 1 C° is 1 cal, whereas the heat needed to change the temperature of 1 g of carbon by 1 C° is only 0.12 cal.

Heat capacity

> The **heat capacity,** C', of a particular sample of a substance is defined as the amount of heat energy needed to raise the temperature of that sample by one Celsius degree.

Therefore, by definition, the heat capacity of 5 g of water is 5 cal/C°, and the heat capacity of 5 g of carbon is 0.60 cal/C°. We shall often refer to a *heat reservoir,* which is considered to be a massive system with a very large heat capacity, such as a lake. The temperature of a heat reservoir is assumed to remain constant during a process. That is, a heat reservoir can exchange heat with another system without itself undergoing any appreciable temperature change.

In practice, it is often more useful to work with the **specific heat,** c, defined as the heat capacity per unit mass:

Specific heat

$$c = \frac{\text{heat capacity}}{\text{mass}} = \frac{C'}{m} \tag{20.1}$$

The **molar heat capacity,** C, of a substance is defined as the heat capacity per mole:

$$C = \frac{C'}{n} \tag{20.2}$$

Since the number of moles, n, equals the mass, m, divided by the molecular weight, M, we can express the molar heat capacity in the form

Molar heat capacity

$$C = \frac{C'}{n} = \frac{mc}{m/M} = Mc \tag{20.3}$$

Tables found in handbooks usually give the specific heats or the molar heat capacities of substances.

From the definition of heat capacity, we can express the heat energy Q transferred between a system of mass m and its surroundings for a temperature change ΔT as

Heat required to raise the temperature of a substance

$$Q = C' \, \Delta T = mc \, \Delta T \tag{20.4}$$

For example, the heat energy required to raise the temperature of 500 g of water by 3 C° is equal to (500 g)(1 cal/g · C°)(3 C°) = 1500 cal. If the number of moles of the system is specified, we can write Q in the form

$$Q = nC \, \Delta T \tag{20.5}$$

Note that when the temperature increases, ΔT and Q are both positive, corresponding to heat flowing into the system. Likewise, when the temperature decreases, ΔT and Q are negative and heat flows out of the system.

Heat capacities of all materials vary somewhat with temperature. If the temperature intervals are not too great, the temperature variation can be ignored and c can be treated as a constant.[2] For example, the specific heat of water (1.00 cal/g·C°) varies by only about 1% from 0°C to 100°C at atmospheric pressure. Unless stated otherwise, we shall neglect such variations. When specific heats are measured, one also finds that the amount of heat needed to raise the temperature of a substance depends on conditions of the measurement. In general, measurements made at constant pressure are different from those made at constant volume. Specific heats measured under conditions of constant pressure are designated c_{p}, and those measured at constant volume are designated c_{v}. The difference between the two specific heats for liquids and solids is usually no more than a few percent and is often neglected. Since experimental measurements on solids and liquids are easier to perform under constant-pressure conditions, it is usually c_{p} that is measured. Table 20.1 gives the specific heat and molar heat capacity of several solid elements. Note that these values are valid at room temperature and atmospheric pressure. Furthermore, these values are considerably less than that of water. Therefore, it takes more heat to raise the temperature of a given mass of water than for most other substances. Large bodies of water will therefore tend to stabilize temperatures in their vicinity, since large heat flows are required to produce significant temperature changes.

Measuring specific heat

One technique for measuring the specific heat of solids or liquids is simply to heat the substance to some known temperature, place it in a vessel containing water of known mass and temperature, and measure the temperature after equilibrium is reached. Since a negligible amount of mechanical work is done in this process, the law of conservation of energy implies that the heat that leaves the warmer body (of unknown c) must equal the heat that enters the water.[3] Suppose that m_x is the mass of the substance whose specific heat we wish to determine, c_x its specific heat, and T_x its initial temperature. Likewise, let m_{w}, c_{w}, and T_{w} represent the corresponding values for the water. If T is the final equilibrium temperature after everything is mixed, then from Equation 20.4, we find that the heat gained by the water is $m_{\text{w}}c_{\text{w}}(T - T_{\text{w}})$, and the heat

[2] The definitions given by Equations 20.4 and 20.5 assume that the specific heat does not vary with temperature over the interval ΔT. In general, if c and C vary with temperature over the range T_{i} to T_{f}, the correct expression for Q is

$$Q = m \int_{T_{\text{i}}}^{T_{\text{f}}} c \, dT = n \int_{T_{\text{i}}}^{T_{\text{f}}} C \, dT$$

[3] For precise measurements, the container for the water should be included in our calculations, since it also gains heat. This would require a knowledge of its mass and composition. However, if the mass of the water is large compared with that of the container, we can neglect the heat gained by the container. Furthermore, precautions must be taken in such measurements to minimize heat transfer between the system and the surroundings.

TABLE 20.1 Specific Heat and Molar Heat Capacity for Some Solids at 25°C and Atmospheric Pressure

Substance	Specific Heat, c_p (cal/g · C°)	(J/g · C°)	Molar Heat Capacity (cal/mol · C°)	(J/mol · C°)
Aluminum	0.215	0.900	5.81	24.3
Beryllium	0.436	1.83	3.93	16.5
Cadmium	0.055	0.230	6.18	25.9
Copper	0.0924	0.387	5.86	24.5
Germanium	0.077	0.322	5.59	23.4
Gold	0.0308	0.129	6.07	25.4
Iron	0.107	0.448	5.98	25.0
Lead	0.0305	1.28	6.31	26.4
Silicon	0.168	0.703	4.72	19.8
Silver	0.056	0.234	6.06	25.4

lost by the substance of unknown c is $-m_xc_x(T-T_x)$. Assuming that the system (water + unknown) does not lose or gain any heat, it follows that the heat gained by the water must equal the heat lost by the unknown (conservation of energy):

$$m_wc_w(T-T_w) = -m_xc_x(T-T_x)$$

Solving for c_x gives

$$c_x = \frac{m_wc_w(T-T_w)}{m_x(T_x-T)} \tag{20.6}$$

EXAMPLE 20.1. Cooling a Hot Ingot
A 50-g chunk of metal is heated to 200°C and then dropped into a beaker containing 400 g of water initially at 20°C. If the final equilibrium temperature of the mixed system is 22.4°C, find the specific heat of the metal.

Solution: Because the heat lost by the metal equals the heat gained by the water, we can use Equation 20.6 directly. In our case, $T_x = 200°\text{C}$, $T = 22.4°\text{C}$, $T_w = 20°\text{C}$, $m_x = 50$ g, $m_w = 400$ g, and $c_w = 1$ cal/g · C°. Substituting these values into Equation 20.6 gives

$$c_x = \frac{(400\text{ g})(1\text{ cal/g}\cdot\text{C}°)(2.4\text{ C}°)}{(50\text{ g})(177.6\text{ C}°)} = 0.108\text{ cal/g}\cdot\text{C}°$$

The metal is most likely iron, as can be seen by comparing this result with the data in Table 20.1.

Exercise 1 What is the total heat transferred to the water in cooling the ingot?
Answer: 960 cal.

20.3 LATENT HEAT

A substance usually undergoes a change in temperature when heat is transferred between the substance and its surroundings. There are situations, however, where the flow of heat does not result in a change in temperature. This occurs whenever the physical characteristics of the substance change from one form to another, commonly referred to as a **phase change.** Some common phase changes are solid to liquid (melting), liquid to gas (boiling), and the change in crystalline structure of a solid. All such phase changes involve a change in internal energy. The energy required is called the **heat of transformation.**

The heat required to change the phase of a given mass m of a pure substance is given by

$$Q = mL \tag{20.7}$$

where L is called the **latent heat** (hidden heat) of the substance[4] and depends on the nature of the phase change as well as on the properties of the substance. The **latent heat of fusion, L_f**, is used when the phase change is from a solid to a liquid, and the **latent heat of vaporization, L_v**, is the latent heat corresponding to the liquid-to-gas phase change.[5] The latent heat of fusion for water at atmospheric pressure is 79.7 cal/g, and the latent heat of vaporization of water is 540 cal/g. The latent heats of various substances vary considerably, as is seen in Table 20.2.

Consider, for example, the heat required to convert a 1-g block of ice at −30°C to steam (water vapor) at 120°C. Figure 20.1 indicates the experimental results obtained when heat is gradually added to the ice. Let us examine each portion of the curve separately.

Part A During this portion of the curve, we are changing the temperature of the ice from −30°C to 0°C. Since the specific heat of ice is 0.5 cal/g · C°, we can calculate the amount of heat added as follows:

$$Q = m_i c_i \, \Delta T = (1\text{ g})(0.5\text{ cal/g}\cdot\text{C}^\circ)(30\text{ C}^\circ) = 15\text{ cal}$$

Part B When the ice reaches 0°C, it remains at this temperature—even though heat is being added—until all the ice melts. Because the latent heat of fusion for water at atmospheric pressure is 79.7 cal/g, the heat required to melt 1 g of ice at 0°C is

$$Q = mL_f = (1\text{ g})(79.7\text{ cal/g}) = 79.7\text{ cal}$$

Part C Between 0°C and 100°C, nothing surprising happens. No phase change occurs in this region. The heat added to the water is being used to increase its temperature. The amount of heat necessary to increase the temperature from 0°C to 100°C is

$$Q = m_w c_w \, \Delta T = (1\text{ g})(1\text{ cal/g}\cdot\text{C}^\circ)(100\text{ C}^\circ) = 100\text{ cal}$$

TABLE 20.2 Latent Heats of Fusion and Vaporization

Substance	Melting Point (°C)	Latent Heat of Fusion (cal/g)	Latent Heat of Fusion (J/g)	Boiling Point (°C)	Latent Heat of Vaporization (cal/g)	Latent Heat of Vaporization (J/g)
Helium	−269.65	1.25	5.23	−268.93	4.99	20.9
Nitrogen	−209.97	6.09	25.5	−195.81	48.0	201
Oxygen	−218.79	3.30	13.8	−182.97	50.9	213
Ethyl alcohol	−114	24.9	104	78	204	854
Water	0.00	79.7	334	100.00	540	2260
Sulfur	119	9.10	38.1	444.60	77.9	326
Lead	327.3	5.85	24.5	1750	208	871
Aluminum	660	21.5	90.0	2450	2720	11386
Silver	960.80	21.1	88.3	2193	558	2336
Gold	1063.00	15.4	64.5	2660	377	1578
Copper	1083	32.0	134	1187	1210	5065

[4] The word *latent* is from the Latin *latere*, meaning *hidden or concealed*.

[5] When a gas cools, it eventually returns to the liquid phase, or *condenses*. The heat per unit mass given up is called the *latent heat of condensation*, which equals the latent heat of vaporization. Likewise, when a liquid cools it eventually solidifies, and the *latent heat of solidification* equals the latent heat of fusion.

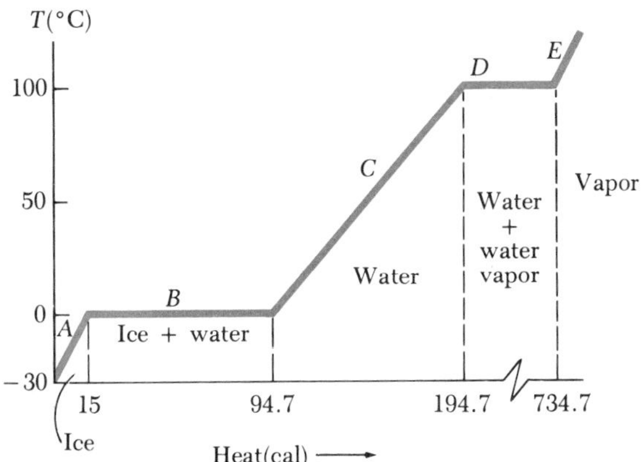

Figure 20.1 A plot of temperature versus heat added when 1 g of ice initially at −30°C is converted to steam.

Part D At 100°C, another phase change occurs as the water changes from water at 100°C to steam at 100°C. We can find the amount of heat required to produce this phase change by using Equation 20.7. In this case, we must set $L = L_v$, the **latent heat of vaporization.** Since the latent heat of vaporization for water is 540 cal/g, the amount of heat we must add to convert 1 g of water to steam at 100°C is

$$Q = mL_v = (1\text{ g})(540\text{ cal/g}) = 540\text{ cal}$$

Part E On this portion of the curve, heat is being added to the steam with no phase change occurring. Using 0.48 cal/g · C° for the specific heat of steam, we find that the heat we must add to raise the temperature of the steam to 120°C is

$$Q = m_s c_s \,\Delta T = (1\text{ g})(0.48\text{ cal/g}\cdot\text{C}^\circ)(20\text{ C}^\circ) = 9.6\text{ cal}$$

The *total amount of heat* that must be added to change one gram of ice at −30°C to steam at 120°C is about 744 cal. It should be noted that this process is *reversible.* That is, if we cool steam at 120°C down to the point at which we have ice at −30°C, we must remove 744 cal of heat.

The fact that a substance such as water gives off heat as it cools is often used by farmers to protect fruits and vegetables stored in a cellar on nights when the temperature is expected to fall below 0°C. Large vats of water are placed in the cellar, and as the water freezes at 0°C, each gram of ice formed liberates about 80 cal of heat to the surroundings. This helps to keep the cellar temperature high enough to prevent damage to the stored food.

Phase changes can be described in terms of a rearrangement of molecules when heat is added to or removed from a substance. Consider first the liquid-to-gas phase change. The molecules in the liquid phase are close together, and the forces between them are stronger than in the gas phase, where the molecules are far apart. Therefore, work must be done on the liquid against these attractive molecular forces in order to separate the molecules. The latent heat of vaporization is the amount of energy that must be added to the liquid to accomplish this.

Similarly, at the melting point of a solid, we imagine that the amplitude of vibration of the atoms about their equilibrium position becomes large enough to overcome the attractive forces binding them together. The thermal energy required to totally melt a given mass of solid is equal to the work required to break the intermolecular bonds and transform the mass from the ordered solid

James Prescott Joule (1818–1889). "First: That the quantity of heat produced by the friction of bodies, whether solid or liquid, is always proportional to the quantity of energy expended. And second: That the quantity of heat capable of increasing the temperature of a pound of water . . . by 1°Fahr. requires for its evolution the expenditure of a mechanical energy represented by the fall of 772 lb through the distance of one foot."

phase to the disordered liquid phase. Because the mean distance between atoms in the gas phase is much larger than in either the liquid or solid phase, we could expect that more work is required to vaporize a given mass of a substance than to melt it. Therefore, it is not surprising that the latent heat of vaporization is much larger than the latent heat of fusion for a given substance (Table 20.2).

EXAMPLE 20.2. Boiling Liquid Helium
Liquid helium has a very low boiling point, 4.2 K, and a very low heat of vaporization, 4.99 cal/g (Table 20.2). A constant power of 10 W (1 W = 1 J/s) is transferred to some helium using an immersed electrical heating element. At this rate, how long does it take to boil away 1 kg of liquid helium? (Liquid helium has a density of 0.125 g/cm³, so that 1 kg corresponds to 8×10^3 cm³, or 8 liters of liquid.)

Solution: Since $L_v = 4.99$ cal/g for liquid helium, we must supply 4.99×10^3 cal of energy to boil away 1 kg of liquid. The mechanical equivalent of 4.99×10^3 cal is

$$4.99 \times 10^3 \text{ cal} = (4.99 \times 10^3 \text{ cal})(4.186 \text{ J/cal}) = 2.09 \times 10^4 \text{ J}$$

The power supplied to the helium is 10 W = 10 J/s. That is, in 1 s, 10 J of energy are transferred to the helium. Therefore, the time it takes to transfer 2.09×10^4 J is

$$t = \frac{2.09 \times 10^4 \text{ J}}{10 \text{ J/s}} = 2.09 \times 10^3 \text{ s} \approx 35 \text{ min}$$

Since 1 kg of helium corresponds to 8 liters of liquid, this means a "boil off" rate of about 0.23 liters/min. In contrast, 1 kg of liquid nitrogen would boil away in about 3.4 h at the rate of 10 J/s.

Exercise 2 If 10 W of power is supplied to 1 kg of water at 100°C, how long will it take the water to boil away completely?
Answer: 38 h.

20.4 HEAT TRANSFER

In practice, it is important to understand the rate at which heat is transferred between a system and its surroundings and the mechanisms responsible for the heat transfer. You may have used a Thermos bottle or some other thermally insulated vessel to store hot coffee for a length of time. The vessel reduces heat transfer between the outside air and the hot coffee. Ultimately, of course, the coffee will reach air temperature since the vessel is not a perfect insulator. In general, there will be no heat transfer between a system and its surroundings when they are at the same temperature.

Melted snow pattern on a parking lot indicates the presence of underground steam pipes used to aid snow removal. Heat from the steam is conducted to the pavement from the pipes, causing the snow to melt. (Courtesy of Dr. Albert A. Bartlett, University of Colorado, Boulder)

Heat Conduction

The easiest heat transfer process to describe quantitatively is called *heat conduction.* In this process, the heat transfer can be viewed on an atomic scale as an exchange of kinetic energy between molecules, where the less energetic particles gain energy by colliding with the more energetic particles. For example, if you insert a copper rod into a flame while holding one end, you will find that the temperature of the metal in your hand increases rapidly. The heat reaches your hand through conduction. The manner in which heat is transferred from the flame, through the copper rod, and to your hand can be understood by examining what is happening to the atoms of the metal. Initially, before the rod is inserted into the flame, the copper atoms are vibrating about their equilibrium positions. As the flame heats the rod, those copper atoms near the flame begin to vibrate with larger and larger amplitudes. These atoms vibrating with large amplitudes collide with their neighbors and transfer some of their energy in the collisions. Slowly, copper atoms farther down the rod increase their amplitude of vibration, until the large amplitude vibrations arrive at the end being held. The effect of this increased vibration results in an increase in temperature of the metal, and possibly a burned hand.

Although the transfer of heat through a metal can be partially explained by atomic vibrations, the rate of heat conduction also depends on the properties of the substance being heated. For example, it is possible to hold a piece of asbestos in a flame indefinitely. This implies that very little heat is being conducted through the asbestos. In general, metals are good conductors of heat and materials such as asbestos, cork, paper, and fiber glass are poor conductors. Gases also are poor heat conductors because of their dilute nature. Metals are good conductors of heat because they contain large numbers of electrons that are relatively free to move through the metal and transport energy from one region to another. Thus, in a good conductor, such as copper, heat conduction takes place via the vibration of atoms and via the motion of free electrons.

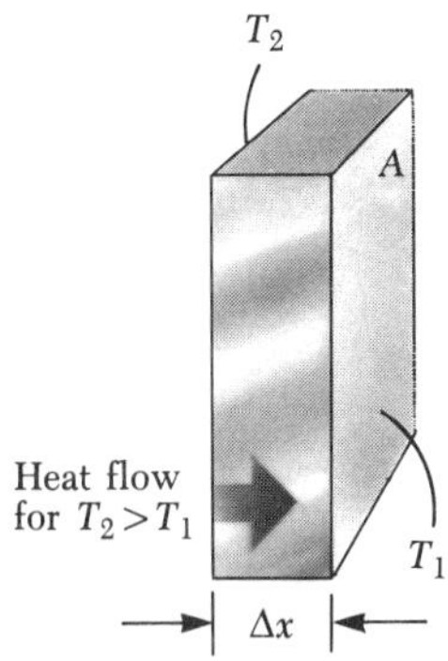

Figure 20.2 Heat transfer through a conducting slab of cross-sectional area A and thickness Δx. The opposite faces are at different temperatures, T_1 and T_2.

Consider a slab of material of thickness Δx and cross-sectional area A with its opposite faces at different temperatures T_1 and T_2, where $T_2 > T_1$ (Fig. 20.2). One finds from experiment that the heat ΔQ transferred in a time Δt flows from the hotter end to the colder end. The rate at which heat flows, $\Delta Q/\Delta t$, is found to be proportional to the cross-sectional area, the temperature difference, and inversely proportional to the thickness. That is,

$$\frac{\Delta Q}{\Delta t} \propto A\frac{\Delta T}{\Delta x}$$

For a slab of infinitesimal thickness dx and temperature difference dT, we can write the **law of heat conduction**

Law of heat conduction

$$\frac{dQ}{dt} = -kA\frac{dT}{dx} \tag{20.8}$$

where the proportionality constant k is called the **thermal conductivity** of the material and dT/dx is known as the **temperature gradient** (the variation of temperature with position). The minus sign in Equation 20.8 denotes that heat flows in the direction of decreasing temperature.

Suppose a substance is in the shape of a long uniform rod of length L, as in Figure 20.3, and is insulated so that no heat can escape from its surface except at the ends, which are in thermal contact with heat reservoirs having temperatures T_1 and T_2. When a steady state has been reached, the temperature at each point along the rod is constant in time. In this case, the temperature gradient is the same everywhere along the rod and is given by $\dfrac{dT}{dx} = \dfrac{T_1 - T_2}{L}$.

Thus the heat transferred in a time Δt is

$$\frac{\Delta Q}{\Delta t} = kA\frac{(T_2 - T_1)}{L} \tag{20.9}$$

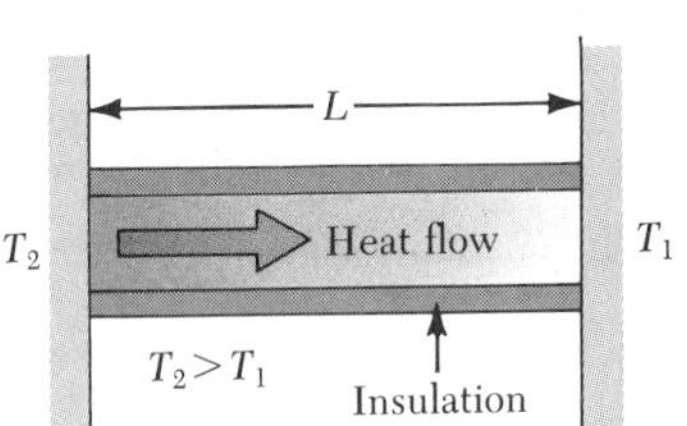

Figure 20.3 Conduction of heat through a uniform, insulated rod of length L. The opposite ends are in thermal contact with heat reservoirs at two different temperatures.

Substances that are good heat conductors have large thermal conductivity values, whereas good thermal insulators have low thermal conductivity values. Table 20.3 lists thermal conductivities for various substances. We see that metals are generally better thermal conductors than nonmetals.

For a compound slab containing several materials of thicknesses L_1,

TABLE 20.3 Thermal Conductivities

Substance	Thermal Conductivity k (cal/s · cm · C°)	(J/s · cm · C°)
	Metals (at 25°C)	
Aluminum	0.57	2.4
Copper	0.95	4.0
Gold	0.75	3.1
Iron	0.19	0.80
Lead	0.083	0.35
Silver	1.02	4.27
	Gases (at 20°C)	
Air	5.6×10^{-4}	2.3×10^{-3}
Helium	3.3×10^{-4}	1.4×10^{-3}
Hydrogen	4.1×10^{-4}	1.7×10^{-3}
Nitrogen	5.6×10^{-5}	2.3×10^{-4}
Oxygen	5.7×10^{-5}	2.4×10^{-4}
	Nonmetals (approximate values)	
Glass	2×10^{-3}	8×10^{-3}
Wood	2×10^{-4}	8×10^{-4}
Asbestos	2×10^{-4}	8×10^{-4}
Concrete	2×10^{-3}	8×10^{-3}
Ice	4×10^{-3}	2×10^{-2}
Rubber	5×10^{-4}	2×10^{-3}

$L_2, \ldots$ and thermal conductivities $k_1, k_2, \ldots$, the rate of heat transfer through the slab at steady state is given by

$$\frac{\Delta Q}{\Delta t} = \frac{A(T_2 - T_1)}{\sum_i (L_i/k_i)} \tag{20.10}$$

where T_1 and T_2 are the temperatures of the outer extremities of the slab (which are held constant) and the summation is over all slabs.

EXAMPLE 20.3. Heat Transfer Through Two Slabs Two slabs of thickness L_1 and L_2 and thermal conductivities k_1 and k_2 are in thermal contact with each other as in Figure 20.4. The temperatures of their outer surfaces are T_1 and T_2, respectively, and $T_2 > T_1$. Determine the temperature at the interface and the rate of heat transfer through the slabs in the steady-state condition.

Solution: If T is the temperature at the interface, then the rate at which heat is transferred through slab 1 is given by

$$(1) \qquad \frac{\Delta Q_1}{\Delta t} = \frac{k_1 A(T - T_1)}{L_1}$$

Likewise, the rate at which heat is transferred through slab 2 is

$$(2) \qquad \frac{\Delta Q_2}{\Delta t} = \frac{k_2 A(T_2 - T)}{L_2}$$

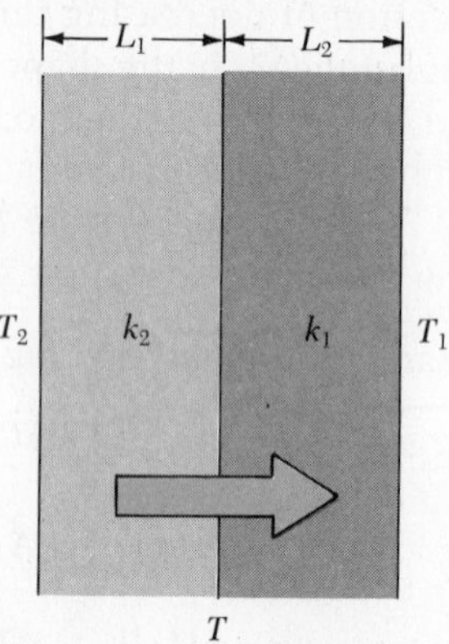

Figure 20.4 Heat transfer by conduction through two slabs in thermal contact with each other. At steady state, the rate of heat transfer through slab 1 equals the rate of heat transfer through slab 2.

When a steady state is reached, these two rates must be equal; hence

$$\frac{k_1A(T - T_1)}{L_1} = \frac{k_2A(T_2 - T)}{L_2}$$

Solving for T gives

$$(3) \qquad T = \frac{k_1L_2T_1 + k_2L_1T_2}{k_1L_2 + k_2L_1}$$

Substituting (3) into either (1) or (2), we get

$$\frac{\Delta Q}{\Delta t} = \frac{A(T_2 - T_1)}{(L_1/k_1) + (L_2/k_2)}$$

An extension of this model to several slabs leads to Equation 20.10.

*Home Insulation

If you would like to do some calculating to determine whether or not to add insulation to a ceiling or to some other portion of a building, what you have just learned about conduction needs to be modified slightly, for two reasons. (1) The insulating properties of materials used in buildings are usually expressed in engineering rather than SI units. For example, measurements stamped on a package of fiber glass insulating board will be in units such as British thermal units, feet, and degrees Fahrenheit. (2) In dealing with the insulation of a building, we must consider heat conduction through a compound slab, with each portion of the slab having a different thickness and a different thermal conductivity. For example, a typical wall in a house will consist of an array of materials, such as wood paneling, dry wall, insulation, sheathing, and wood siding.

In Example 20.3, we showed how to deal with heat conduction through a two-layered slab. A general formula for heat transfer through a compound slab is given by Equation 20.10. For example, if the slab consists of three different materials, the denominator of Equation 20.10 will consist of the sum of three terms. In engineering practice, the term L/k for a particular substance is referred to as the R value of the material. Thus, Equation 20.10 reduces to

$$\frac{\Delta Q}{\Delta t} = \frac{A(T_2 - T_1)}{\sum_i R_i} \qquad (20.11)$$

The R values for a few common building materials are given in Table 20.4 (note the units).

TABLE 20.4 *R* Values for Some Common Building Materials

Material	*R* value ($ft^2 \cdot F^\circ \cdot h/BTU$)
Hardwood siding (1 in. thick)	0.91
Wood shingles (lapped)	0.87
Brick (4 in. thick)	4.00
Concrete block (filled cores)	1.93
Fiber glass batting (3.5 in. thick)	10.90
Fiber glass batting (6 in. thick)	18.80
Fiber glass board (1 in. thick)	4.35
Cellulose fiber (1 in. thick)	3.70
Flat glass (0.125 in. thick)	0.89
Insulating glass (0.25-in. space)	1.54
Vertical air space (3.5 in. thick)	1.01
Air film	0.17
Dry wall (0.5 in. thick)	0.45
Sheathing (0.5 in. thick)	1.32

Also, it should be noted that near any vertical surface there is a very thin, stagnant layer of air that must be considered when finding the total R value for a wall. The thickness of this stagnant layer on an outside wall depends on the velocity of the wind. As a result, heat loss from a house on a day when the wind is blowing hard is greater than heat loss on a day when the wind velocity is zero. A representative R value for this stagnant layer of air is given in Table 20.4.

EXAMPLE 20.4. The *R* Value of a Typical Wall
Calculate the total R value for a wall constructed as shown in Figure 20.5a. Starting outside the house (to the left in Fig. 20.5a) and moving inward, the wall consists of brick, 0.5 in. of sheathing, a vertical air space 3.5 in. thick, and 0.5 in. of dry wall. Do not forget the dead-air layers inside and outside the house.

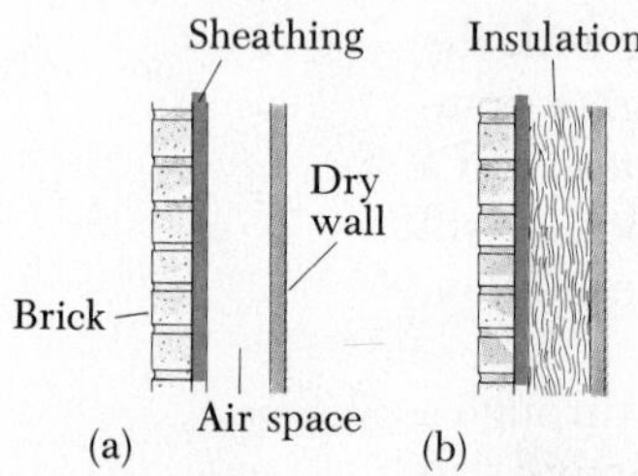

Figure 20.5 (Example 20.4) Cross-sectional view of an exterior wall containing (a) an air space and (b) insulation.

Solution: Referring to Table 20.4, we find the total R value for the wall as follows:

R_1 (outside air film) $= 0.17\ \text{ft}^2 \cdot \text{F}^\circ \cdot \text{h/BTU}$
R_2 (brick) $= 4.00$
R_3 (sheathing) $= 1.32$
R_4 (air space) $= 1.01$
R_5 (dry wall) $= 0.45$
R_6 (inside air film) $= 0.17$

$R_{\text{total}} = 7.12\ \text{ft}^2 \cdot \text{F}^\circ \cdot \text{h/BTU}$

Exercise 3 If a layer of fiberglass insulation 3.5 in. thick is placed inside the wall to replace the air space as in Figure 20.5b, what is the total R value of the wall? By what factor is the heat loss reduced?
Answer: $R = 17\ \text{ft}^2 \cdot \text{F}^\circ \cdot \text{h/BTU}$; a factor of 2.5.

Convection and Radiation

Two other important heat transfer processes, which we shall discuss only briefly, are convection and radiation. *Convection* is heat transfer as the result of the actual movement of a heated substance from one place to another. In some cases, such as in hot-air and hot-water heating systems, the heated substance is forced to move by a fan or pump. This is known as *forced convection.* In *natural*, or *free, convection* the motion is produced as the result of the differences in density between hot and cold regions. Since warmer fluids are generally less dense than cooler fluids, the heated portions will rise according to Archimedes' principle. Convection is the mechanism for the mixing of warm and cool air masses in the atmosphere and hence is a key factor in weather conditions.

The third mechanism of heat transfer is *radiation.* All bodies radiate energy continuously in the form of electromagnetic waves, which we shall discuss in Chapter 34. For example, when we see that the heating element on an electric range is "red hot," we are observing electromagnetic radiation emitted by the hot element. Likewise, the tungsten wire in an incandescent lamp and the surface of the sun also emit radiant energy. Techniques for converting this free solar radiation into useful forms of energy are of current interest.

The rate at which a body emits radiant energy is proportional to the fourth power of its absolute temperature. This is known as **Stefan's law,** often written in the form

Stefan's law

$$P = \sigma A e T^4 \tag{20.12}$$

where P is the power radiated by the body in W (or J/s), σ is a universal constant equal to 5.6696×10^{-8} W/m$^2 \cdot$ K^4, A is the surface area of the body in m^2, e is a constant called the **emissivity,** and T is the absolute temperature. The value of e can vary between 0 and 1 depending on the properties of the surface.

An object radiates energy at a rate given by Equation 20.12. At the same time, the object also absorbs electromagnetic radiation. If the latter process did not occur, an object would eventually radiate all of its energy and its temperature would reach absolute zero. The energy that a body absorbs comes from its surroundings, which consists of other objects which radiate energy. If an object is at a temperature T and its surroundings are at a temperature T_0, the net energy gained or lost each second by the object as a result of radiation is given by

$$P_{\text{net}} = \sigma A e(T^4 - T_0^{\,4}) \tag{20.13}$$

When an object is in *equilibrium* with its surroundings, *it radiates and absorbs energy at the same rate, and so its temperature remains constant.* When an object is hotter than its surroundings, it radiates more energy than it absorbs and so it cools. An *ideal absorber* is defined as an object that absorbs all of the energy incident on it. The emissivity of an ideal absorber is equal to unity. Such an object is often referred to as a **black body.** An ideal absorber is also an ideal radiator of energy. In contrast, an object with an emissivity equal to zero absorbs none of the energy incident on it. Such an object reflects all the incident energy and so is a perfect reflector.

The Dewar

The Thermos bottle, called a *Dewar flask*[6] in the scientific community, is a practical example of a container designed to minimize heat losses by conduction, convection, and radiation. Such a container is used to store either cold or hot liquids for long periods of time. The standard construction (Fig. 20.6) consists of a double-walled pyrex vessel with silvered inner walls. The space between the walls is evacuated to minimize heat transfer by conduction and convection. The silvered surfaces minimize heat transfer by radiation by reflecting most of the radiant heat. Very little heat is lost over the neck of the flask since glass is a poor heat conductor. A further reduction in heat loss is obtained by reducing the size of the neck. Dewar flasks are commonly used to store liquid nitrogen (boiling point 77 K) and liquid oxygen (boiling point 90 K).

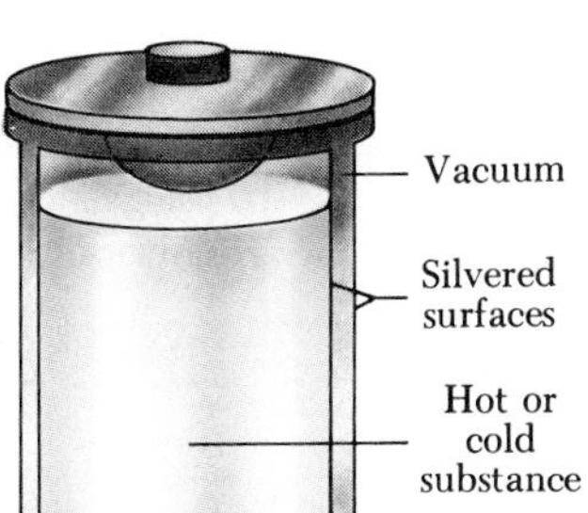

Figure 20.6 Cross-sectional view of a Dewar vessel, used to store hot or cold liquids or other substances.

For other cryogenic liquids, such as liquid helium, which has a very low specific heat (boiling point 4.2 K), it is often necessary to use a double Dewar system in which the Dewar flask containing the liquid is surrounded by a second Dewar flask. The space between the two flasks is filled with liquid nitrogen.

20.5 THE MECHANICAL EQUIVALENT OF HEAT

When the concept of mechanical energy was introduced in Chapters 7 and 8, we found that whenever friction is present in a mechanical system, some mechanical energy is lost, or is not conserved. Experiments of various sorts show that this lost mechanical energy does not simply disappear, but is trans-

[6] Invented by Sir James Dewar (1842–1923).

formed into thermal energy. Although this connection between mechanical and thermal energy was first suggested by Thompson's crude cannon boring experiment, it was Joule who first established the equivalence of the two forms of energy.

A schematic diagram of Joule's most famous experiment is shown in Figure 20.7. The system of interest is the water in a thermally insulated container. Work is done on the water by a rotating paddle wheel, which is driven by weights falling at a constant speed. The water, which is stirred by the paddles, is warmed due to the friction between it and the paddles. If the energy lost in the bearings and through the walls is neglected, then the loss in potential energy of the weights equals the work done by the paddle wheel on the water. If the two weights fall through a distance h, the loss in potential energy is $2mgh$, and it is this energy that is used to heat the water. By varying the conditions of the experiment, Joule found that the loss in mechanical energy, $2mgh$, is proportional to the increase in temperature of the water, ΔT. The proportionality constant (the specific heat of water) was found to be equal to 4.18 J/g · C°. Hence, 4.18 J of mechanical energy will raise the temperature of 1 g of water from 14.5°C to 15.5°C. One calorie is now defined to be *exactly* 4.186 J:

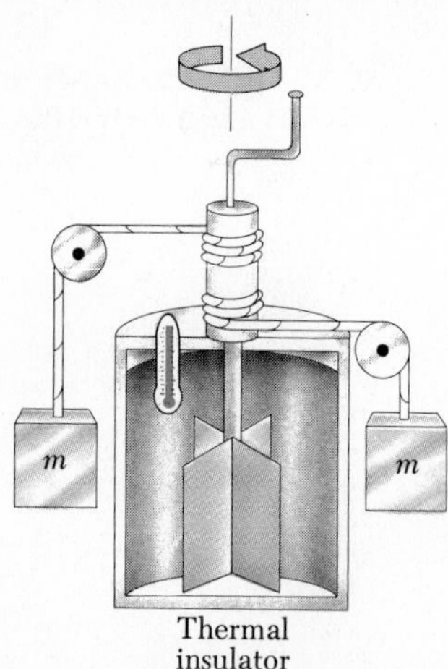

Figure 20.7 An illustration of Joule's experiment for measuring the mechanical equivalent of heat. The falling weights rotate the paddles, causing the temperature of the water to increase.

Mechanical equivalent of heat

$$1 \text{ cal} = 4.186 \text{ J} \tag{20.14}$$

EXAMPLE 20.5. Losing Weight the Hard Way

A student eats a dinner rated at 2000 (food) Calories. He wishes to do an equivalent amount of work in the gymnasium by lifting a 50-kg mass. How many times must he raise the weight to expend this much energy? Assume he raises the weight a distance of 2 m each time and that no work is done when the weight is dropped to the floor.

Solution: Since 1 (food) Calorie = 10^3 cal, the work required is 2×10^6 cal. Converting this to J, we have for the total work required

$$W = (2 \times 10^6 \text{ cal})\ (4.186 \text{ J/cal}) = 8.37 \times 10^6 \text{ J}$$

The work done in lifting the weight once through a distance h is equal to mgh, and the work done in lifting the weight n times is $nmgh$. Equating this to the total work required gives

$$W = nmgh = 8.37 \times 10^6 \text{ J}$$

Since $m = 50$ kg and $h = 2$ m, we get

$$n = \frac{8.37 \times 10^6 \text{ J}}{(50 \text{ kg})(9.80 \text{ m/s}^2)(2 \text{ m})} = 8.54 \times 10^3 \text{ times}$$

If the student is in good shape and lifts the weight, say, once every 5 s, it would take him about 30 h to perform this feat. Clearly, it is much easier to lose weight by dieting.

20.6 WORK AND HEAT IN THERMODYNAMIC PROCESSES

In the macroscopic approach to thermodynamics we describe the *state* of a system with such variables as pressure, volume, temperature, and internal energy. The number of macroscopic variables needed to characterize a system depends on the nature of the system. For a homogeneous system, such as a gas containing only one type of molecule, usually only two variables are needed, such as pressure and volume. However, it is important to note that a *macroscopic state* of an isolated system can be specified only if the system is in thermal equilibrium internally. In the case of a gas in a container, internal thermal equilibrium requires that every part of the container be at the same pressure and temperature.

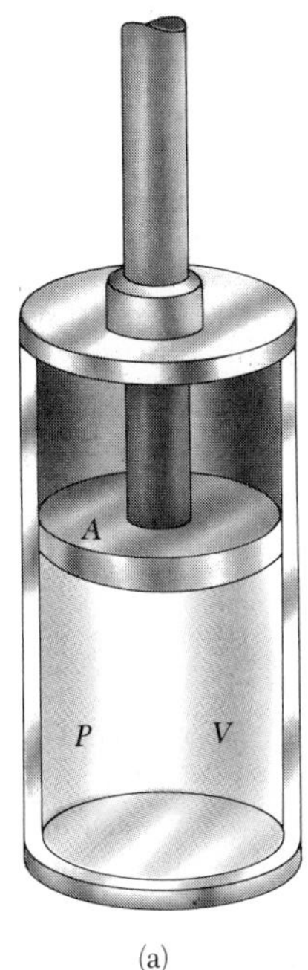

(a)

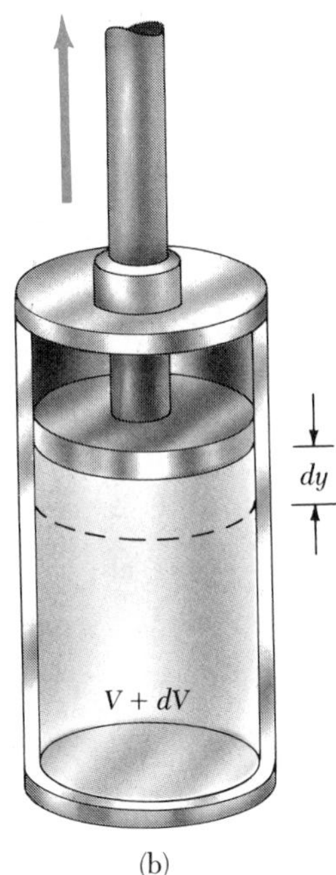

(b)

Figure 20.8 Gas contained in a cylinder at a pressure P does work on a moving piston as the system expands from a volume V to a volume $V + dV$.

Consider gas contained in a cylinder fitted with a movable piston (Fig. 20.8). In equilibrium, the gas occupies a volume V and exerts a uniform pressure P on the cylinder walls and piston. If the piston has a cross-sectional area A, the force exerted by the gas on the piston is $F = PA$. Now let us assume that the gas expands **quasi-statically,** that is, slowly enough to allow the system to move through an (infinite) series of equilibrium states. As the piston moves up a distance dy, the work done by the gas on the piston is

$$dW = F\,dy = PA\,dy$$

Since $A\,dy$ is the increase in volume of the gas, dV, we can express the work done as

$$dW = P\,dV \tag{20.15}$$

If the gas expands, as in Figure 20.8b, then dV is positive and the work done by the gas is positive, whereas if the gas is compressed, dV is negative, indicating that the work done by the gas is negative. (In the latter case, negative work can be interpreted as being work done *on* the system.) Clearly, the work done by the system is zero when the volume remains constant. The total work done by the gas as its volume changes from V_i to V_f is given by the integral of Equation 20.15:

$$W = \int_{V_i}^{V_f} P\,dV \tag{20.16}$$

Work done by a gas

To evaluate this integral, one must know how the pressure varies during the process. (Note that a *process* is *not* specified merely by giving the initial and final states. That is, a process is a *fully specified* change in state of the system.) In general, the pressure is not constant, but depends on the volume and temperature. If the pressure and volume are known at each step of the process, the states of the gas can then be represented as a curve on a *PV* diagram, as in Figure 20.9.

Work equals area under the curve in a *PV* diagram

The work done in the expansion from the initial state to the final state is the area under the curve in a *PV* diagram.

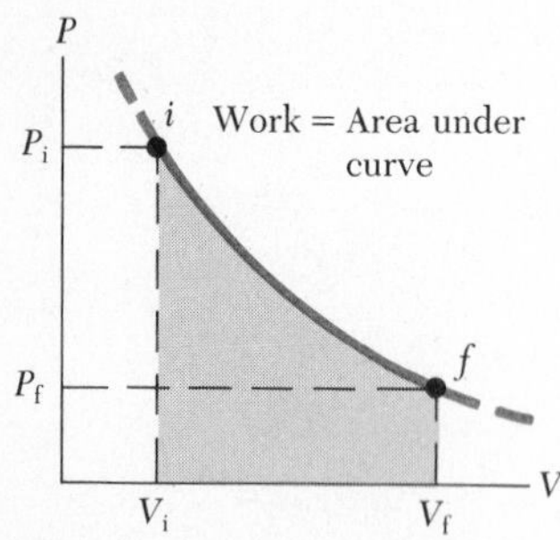

Figure 20.9 A gas expands reversibly (slowly) from state *i* to state *f*. The work done by the gas equals the area under the *PV* curve.

As one can see from Figure 20.9, the work done in the expansion from the initial state, i, to the final state, f, will depend on the specific path taken between these two states. To illustrate this important point, consider several different paths connecting i and f (Fig. 20.10). In the process described in Figure 20.10a, the pressure of the gas is first reduced from P_i to P_f by cooling at constant volume V_i, and the gas then expands from V_i to V_f at constant pressure P_f. The work done along this path is $P_f(V_f - V_i)$. In Figure 20.10b, the gas first expands from V_i to V_f at constant pressure P_i, and then its pressure is reduced to P_f at constant volume V_f. The work done along this path is $P_i(V_f - V_i)$, which is greater than that for the process described in Figure 20.10a. Finally, for the process described in Figure 20.10c, where both *P* and *V* change continuously, the work done has some value intermediate between the values obtained in the first two processes. To evaluate the work in this case, the shape of the *PV* curve must be known. Therefore, we see that

Work done depends on the path between the initial and final states

the work done by a system depends on the process by which the system goes from the initial to the final state. In other words, the work done depends on the initial, final, and intermediate states of the system.

In a similar manner, the heat transferred into or out of the system is also found to depend on the process. This can be demonstrated by considering the situations described in Figure 20.11. In each case, the gas has the same initial volume, temperature, and pressure and is assumed to be ideal. In Figure 20.11a, the gas is in thermal contact with a heat reservoir. If the pressure of the gas is infinitesimally greater than atmospheric pressure, the gas will expand and cause the piston to rise. During this expansion to some final volume V_f, sufficient heat to maintain a constant temperature T_i will be transferred from the reservoir to the gas.

Now consider the thermally insulated system shown in Figure 20.11b. When the membrane is broken, the gas expands rapidly into the vacuum until it occupies a volume V_f. In this case, the gas does no work since there is no movable piston. Furthermore, no heat is transferred through the thermally insulated wall, which we call an *adiabatic wall.* This process is often referred to as **adiabatic free expansion,** or simply *free expansion.* In general, an adiabatic process is one in which no heat is transferred between the system and its surroundings.

Free expansion of a gas

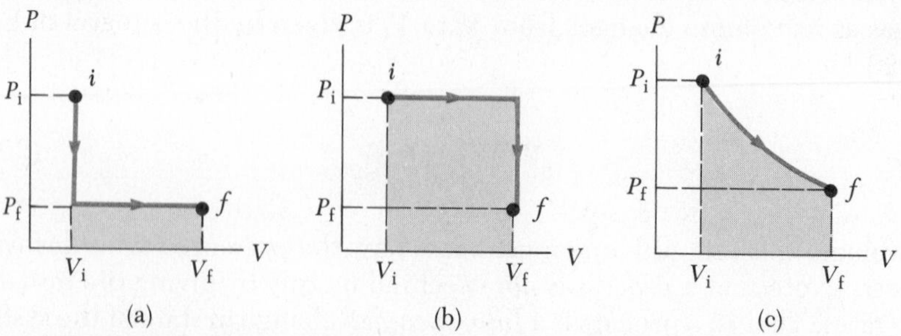

Figure 20.10 The work done by a gas as it is taken from an initial state to a final state depends on the intermediate path between these states.

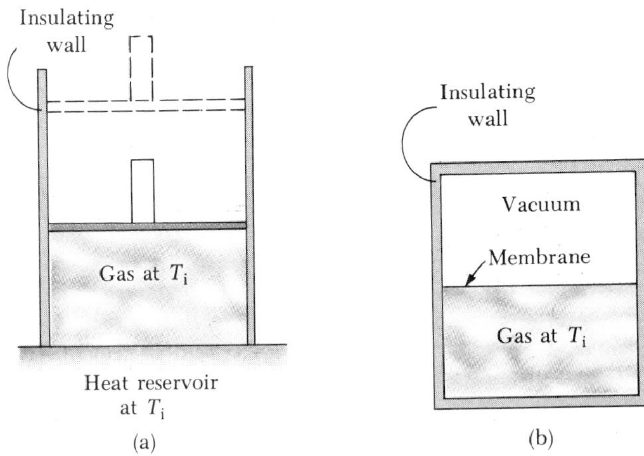

Figure 20.11 (a) A gas at temperature T_i expands slowly by absorbing heat from a reservoir at the same temperature. (b) A gas expands rapidly into an evacuated region by breaking a membrane.

The initial and final states of the ideal gas in Figure 20.11a are identical to the initial and final states in Figure 20.11b, but the paths are different. In the first case, heat is transferred slowly to the gas, and the gas does work on the piston. In the second case, no heat is transferred and the work done is zero. Therefore, we conclude that *heat, like work, depends on the initial, final, and intermediate states of the system*. Furthermore, since heat and work depend on the path, neither quantity is independently conserved during a thermodynamic process.

20.7 THE FIRST LAW OF THERMODYNAMICS

When the law of conservation of energy was first introduced in Chaper 8, it was stated that the mechanical energy of a system is conserved in the absence of nonconservative forces, such as friction. That is, the changes in the internal energy of the system were not included in this mechanical model.

> The **first law of thermodynamics** is a generalization of the law of conservation of energy that includes possible changes in internal energy.

It is a universally valid law that can be applied to all kinds of processes. Furthermore, it provides us with a connection between the microscopic and macroscopic worlds.

We have seen that energy can be transferred between a system and its surroundings in two ways. One is work done by (or on) the system. This mode of energy exchange results in measurable changes in the macroscopic variables of the system, such as the pressure, temperature, and volume of a gas. The other is heat transfer, which takes place at the microscopic level.

Change in internal energy

To put these ideas on a more quantitative basis, suppose a thermodynamic system undergoes a change from an initial state to a final state in which Q units of heat are absorbed (or removed) and W is the work done by (or on) the system.[7] For example, the system may be a gas whose pressure and volume change from P_i, V_i to P_f, V_f. If the quantity $Q - W$ is measured for various paths connecting the initial and final equilibrium states (that is, for various *pro-*

[7] We use the convention that Q is positive if the system absorbs heat and negative if it loses heat. Likewise, the work done is positive if the system does work on the surroundings and negative if work is done on the system.

cesses), one finds that $Q - W$ is the same for *all* paths connecting the initial and final states. We conclude that the quantity $Q - W$ is determined completely by the initial and final states of the system, and we call the quantity $Q - W$ the *change in the internal energy of the system.* Although Q and W both depend on the path, the quantity $Q - W$, that is, *the change in internal energy, is independent of the path.* If we represent the internal energy function by the letter U, then the *change* in internal energy, $\Delta U = U_f - U_i$, can be expressed as

First law of thermodynamics

$$\Delta U = U_f - U_i = Q - W \tag{20.17}$$

where all quantities must have the same energy units. Equation 20.17 is known as the **first law of thermodynamics.** When it is used in this form, we must note that Q is positive when heat *enters* the system and W is positive when work is done *by* the system.

When a system undergoes an infinitesimal change in state, where a small amount of heat, $đQ$, is transferred and a small amount of work, $đW$, is done, the internal energy also changes by a small amount, dU. Thus, for infinitesimal processes we can express the first law as[8]

First law of thermodynamics for infinitesimal changes

$$dU = đQ - đW \tag{20.18}$$

On a microscopic level, the internal energy of a system includes the kinetic and potential energies of the molecules making up the system. In thermodynamics, we do not concern ourselves with the specific form of the internal energy. We simply use Equation 20.17 as a definition of the change in internal energy. One can make an analogy here between the potential energy function associated with a body moving under the influence of gravity without friction. The potential energy function is independent of the path, and it is only its change that is of concern. Likewise, the change in internal energy of a thermodynamic system is what matters, since only differences are defined. Any reference state can be chosen for the internal energy since absolute values are not defined.

Now let us look at some special cases. First consider an *isolated system,* that is, one that does not interact with its surroundings. In this case, there is no heat flow and the work done is zero; hence the internal energy remains constant. That is, since $Q = W = 0$, $\Delta U = 0$, and so $U_i = U_f$. We conclude that

Isolated systems

the internal energy of an isolated system remains constant.

Next consider a process in which the system is taken through a **cyclic process,** that is, one that originates and ends at the same state. In this case, the change in the internal energy is *zero* and the heat added to the system must equal the work done during the cycle. That is, in a cyclic process,

Cyclic process

$$\Delta U = 0 \quad \text{and} \quad Q = W$$

Note that *the net work done per cycle equals the area enclosed by the path representing the process on a PV diagram.* As we shall see in Chapter 22, cyclic processes are very important in describing the thermodynamics of *heat engines,* which are devices in which some part of the heat energy input is extracted as mechanical work.

If a process occurs in which the work done is zero, then the change in internal energy equals the heat entering or leaving the system. If heat enters

[8] Note that $đQ$ and $đW$ are not true differential quantities, although dU is a true differential. For further details on this point, see an advanced text in thermodynamics, such as M. W. Zemansky and R. H. Dittman, *Heat and Thermodynamics,* New York, McGraw-Hill, 1981.

the system, Q is positive and the internal energy increases. For a gas, we can associate this increase in internal energy with an increase in the kinetic energy of the molecules. On the other hand, if a process occurs in which the heat transferred is zero and work is done by the system, then the magnitude of the change in internal energy equals the negative of the work done by the system. That is, the internal energy of the system decreases. For example, if a gas is compressed with no heat transferred (by a moving piston, say), the work done is negative and the internal energy again increases. This is because kinetic energy is transferred from the moving piston to the gas molecules.

We have seen that there is really no distinction between heat and work on a microscopic scale. Both can produce a change in the internal energy of a system. Although the macroscopic quantities Q and W are *not* properties of a system, they are related to the internal energy of the system through the first law of thermodynamics. Once a process or path is defined, Q and W can be calculated or measured, and the change in internal energy can be found from the first law. One of the important consequences of the first law is that there is a quantity called *internal energy*, the value of which is determined by the state of the system. The internal energy function is therefore called a *state function*.

20.8 SOME APPLICATIONS OF THE FIRST LAW OF THERMODYNAMICS

In order to apply the first law of thermodynamics to specific systems, it is useful to first define some common thermodynamic processes.

Adiabatic process

An **adiabatic process** is defined as a process for which no heat enters or leaves the system, that is, $Q = 0$.

Applying the first law of thermodynamics in this case, we see that

First law for an adiabatic process

$$\Delta U = -W \qquad (20.19)$$

An adiabatic process can be achieved either by thermally insulating the system from its surroundings (say, with Styrofoam or an evacuated wall) or by performing the process rapidly. From this result, we see that if a gas expands adiabatically, W is positive and so ΔU is negative and the gas cools. Conversely, a gas is heated when it is compressed adiabatically.

Adiabatic processes are very important in engineering practice. Some common examples of adiabatic processes include the expansion of hot gases in an internal combustion engine, the liquefaction of gases in a cooling system, and the compression stroke in a diesel engine.

The *free expansion process* described in Figure 20.11b is an adiabatic process in which no work is done on or by the gas. Since $Q = 0$ and $W = 0$, we see from the first law that $\Delta U = 0$ for this process. That is, *the initial and final internal energies of a gas are equal in an adiabatic free expansion.* As we shall see in the next chapter, the internal energy of an ideal gas depends only on its temperature. Thus, we would expect no change in temperature during an adiabatic free expansion. This is in accord with experiments performed at low pressures. Careful experiments at high pressures for real gases show a slight decrease in temperature after the expansion.

Isobaric process

A process that occurs at constant pressure is called an **isobaric process.** When such a process occurs, the heat transferred and the work done are both nonzero. The work done is simply the pressure multiplied by the change in volume, or $P(V_f - V_i)$.

A process that takes place at constant volume is called an **isovolumetric process.** In such a process, the work done is clearly zero. Hence from the first law we see that

First law for a constant-volume process

$$\Delta U = Q \tag{20.20}$$

This tells us that *if heat is added to a system kept at constant volume, all of the heat goes into increasing the internal energy of the system.* When a mixture of gasoline vapor and air explodes in the cylinder of an engine, the temperature and pressure rise suddenly because the cylinder volume doesn't change appreciably during the short duration of the explosion.

Isothermal process

A process that occurs at constant temperature is called an **isothermal process,** and a plot of P versus V at constant temperature for an ideal gas yields a hyperbolic curve called an *isotherm.* The internal energy of an ideal gas is a function of temperature only. Hence, in an isothermal process of an ideal gas, $\Delta U = 0$.

Isothermal Expansion of an Ideal Gas

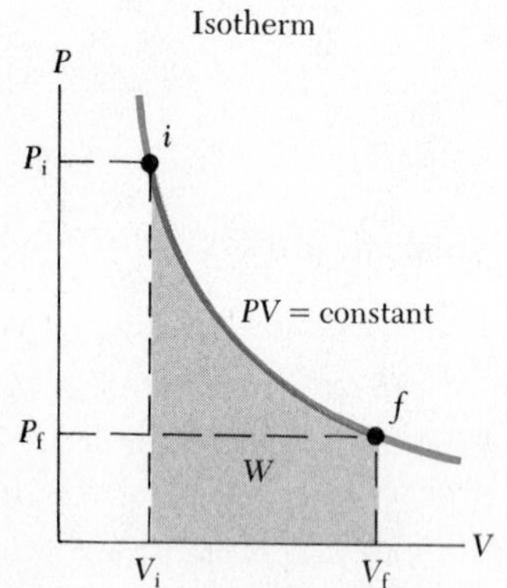

Figure 20.12 The PV diagram for an isothermal expansion of an ideal gas from an initial state to a final state. The curve is a hyperbola.

Suppose an ideal gas is allowed to expand quasi-statically at constant temperature as described by the PV diagram in Figure 20.12. The curve is a hyperbola that obeys the equation $PV =$ constant. Let us calculate the work done by the gas in the expansion from state i to state f.

The isothermal expansion of the gas can be achieved by placing the gas in good thermal contact with a heat reservoir at the same temperature, as in Figure 20.11a.

The work done by the gas is given by Equation 20.16. Since the gas is ideal and the process is quasi-static, we can apply $PV = nRT$ for each point on the path. Therefore, we have

$$W = \int_{V_i}^{V_f} P\,dV = \int_{V_i}^{V_f} \frac{nRT}{V}\,dV$$

But T is constant in this case; therefore it can be removed from the integral. This gives

$$W = nRT \int_{V_i}^{V_f} \frac{dV}{V}$$

To evaluate this integral, we use the fact that $\int \frac{dx}{x} = \ln x$ (Table B.5 in Appendix B), which gives

Work done in an isothermal process

$$W = nRT \ln V \Big]_{V_i}^{V_f} = nRT \ln\left(\frac{V_f}{V_i}\right) \tag{20.21}$$

Numerically, this work equals the shaded area under the PV curve in Figure 20.12. If the gas expands isothermally, then $V_f > V_i$ and we see that the work done by the gas is positive, as we would expect. If the gas is compressed isothermally, then $V_f < V_i$ and the work done by the gas is negative. (Negative work here implies that positive work must be done *on* the gas by some external agent to compress it.) In the next chapter we shall find that the internal energy of an ideal gas depends only on temperature. Hence, for an isothermal process

$\Delta U = 0$, and from the first law we conclude that the heat given up by the reservoir (and transferred to the gas) equals the work done by the gas, or $Q = W$.

EXAMPLE 20.6. Work Done During an Isothermal Expansion
Calculate the work done by 1 mole of an ideal gas that is kept at 0°C in an expansion from 3 liters to 10 liters.

Solution: Substituting these values into Equation 20.21 gives

$$W = nRT \ln\left(\frac{V_f}{V_i}\right)$$

$$= (1 \text{ mole})(8.31 \text{ J/mole}\cdot\text{K})(273 \text{ K}) \ln\left(\frac{10}{3}\right)$$

$$= 2.73 \times 10^3 \text{ J}$$

The heat that must be supplied to the gas from the reservoir to keep T constant is also 2.73×10^3 J.

The Boiling Process

Suppose that a liquid of mass m vaporizes at constant pressure P. Its volume in the liquid state is V_ℓ, and its volume in the vapor state is V_v. Let us find the work done in the expansion and the change in internal energy of the system.

Since the expansion takes place at constant pressure, the work done by the system is

$$W = \int_{V_\ell}^{V_v} P\,dV = P\int_{V_\ell}^{V_v} dV = P(V_v - V_\ell)$$

The heat that must be transferred to the liquid to vaporize all of it is equal to $Q = mL_v$, where L_v is the latent heat of vaporization of the liquid. Using the first law and the result above, we get

$$\Delta U = Q - W = mL_v - P(V_v - V_\ell) \tag{20.22}$$

EXAMPLE 20.7. Boiling Water
One gram of water occupies a volume of 1 cm³ at atmospheric pressure. When this amount of water is boiled, it becomes 1671 cm³ of steam. Calculate the change in internal energy for this process.

Solution: Since the heat of vaporization of water is 540 cal/g at atmospheric pressure, the heat required to boil 1 g of water is

$$Q = mL_v = 540 \text{ cal} = 2259 \text{ J}$$

The work done by the system is positive and equal to

$$W = P(V_v - V_\ell)$$
$$= (1.013 \times 10^5 \text{ N/m}^2)[(1671 - 1) \times 10^{-6} \text{ m}^3]$$
$$= 169 \text{ J}$$

Hence, the change in internal energy is given by

$$\Delta U = Q - W = 2259 \text{ J} - 169 \text{ J} = 2090 \text{ J}$$

The internal energy of the system *increases* since ΔU is positive. We see that most of the heat (93%) that is transferred to the liquid goes into increasing the internal energy. Only a small fraction of the heat (7%) goes into external work.

EXAMPLE 20.8. Heat Transferred to a Solid
The internal energy of a solid also increases when heat is transferred to it from its surroundings.

A 1-kg bar of copper is heated at atmospheric pressure. If its temperature increases from 20°C to 50°C, (a) find the work done by the copper.

The change in volume of the copper can be calculated using Equation 19.11 and the volume expansion coefficient for copper taken from Table 19.2 (remembering that $\beta = 3\alpha$):

$$\Delta V = \beta V\,\Delta T = [5.1 \times 10^{-5}\,(\text{C}^\circ)^{-1}](50^\circ\text{C} - 20^\circ\text{C})V$$
$$= 1.5 \times 10^{-3}\,V$$

But the volume is equal to m/ρ, and the density of copper is 8.92×10^3 kg/m³. Hence,

$$\Delta V = (1.5 \times 10^{-3})\left(\frac{1\text{ kg}}{8.92 \times 10^3\text{ kg/m}^3}\right)$$
$$= 1.7 \times 10^{-7}\text{ m}^3$$

Since the expansion takes place at constant pressure, the work done is given by

$$W = P\,\Delta V = (1.013 \times 10^5\text{ N/m}^2)(1.7 \times 10^{-7}\text{ m}^3)$$
$$= 1.9 \times 10^{-2}\text{ J}$$

(b) What quantity of heat is transferred to the copper?

Taking the specific heat of copper from Table 20.1 and using Equation 20.4, we find that the heat transferred is

$$Q = mc\,\Delta T = (1 \times 10^3\text{ g})(0.0924\text{ cal/g}\cdot\text{C°})(30\text{ C°})$$
$$= 2.77 \times 10^3\text{ cal} = 1.16 \times 10^4\text{ J}$$

(c) What is the increase in internal energy of the copper?

From the first law of thermodynamics, the increase in internal energy is found to be

$$\Delta U = Q - W = 1.16 \times 10^4\text{ J}$$

Note that almost *all* of the heat transferred goes into increasing the internal energy. The fraction of heat energy that is used to do work against the atmosphere is only about 10^{-6}! Hence, in the thermal expansion of a solid or a liquid, the small amount of work done is usually ignored.

20.9 SUMMARY

Heat flow is a form of energy transfer that takes place as a consequence of a temperature difference only. The **internal energy** of a substance is a function of its temperature and generally increases with increasing temperature.

The **calorie** is the amount of heat necessary to raise the temperature of 1 g of water from 14.5°C to 15.5°C. The **mechanical equivalent of heat** is found from experiment to be 1 cal = 4.186 J.

The **heat capacity,** C', of any substance is defined as the amount of heat energy needed to raise the temperature of the substance by one Celsius degree. The heat required to change the temperature of a substance by ΔT is

Heat required to raise the temperature of a substance

$$Q = C'\,\Delta T = mc\,\Delta T \tag{20.4}$$

where m is the mass of the substance and c is its **specific heat,** or heat capacity per unit mass.

The heat required to change the phase of a pure substance of mass m is given by

Latent heat

$$Q = mL \tag{20.7}$$

The parameter L is called the **latent** (hidden) **heat** of the substance and depends on the nature of the phase change and the properties of the substance.

Heat may be transferred by three fundamentally distinct mechanisms: conduction, convection, and radiation. The *conduction* process can be viewed as an exchange of kinetic energy between colliding molecules. The rate at which heat flows by conduction through a slab of area A is given by

Law of heat conduction

$$\frac{dQ}{dt} = -kA\frac{dT}{dx} \tag{20.8}$$

where k is the **thermal conductivity** and $\frac{dT}{dx}$ is the **temperature gradient.**

Convection is a heat transfer process in which the heated substance moves from one place to another.

All bodies radiate and absorb energy in the form of electromagnetic waves. A body that is hotter than its surroundings radiates more energy than it absorbs, whereas a body that is cooler than its surroundings absorbs more energy than it radiates. An **ideal radiator,** or black body, is one that absorbs all energy incident on it; an ideal radiator is also a good emitter of radiation.

A **quasi-static process** is one that proceeds slowly enough to allow the system to always be in a state of equilibrium.

The **work done** by a gas as its volume changes from some initial value V_i to some final value V_f is

$$W = \int_{V_i}^{V_f} P\, dV \tag{20.16}$$

Work done by a gas

where P is the pressure, which may vary during the process. In order to evaluate W, the nature of the process must be specified—that is, P and V must be known during each step of the process. Since the work done depends on the initial, final, and intermediate states, it therefore depends on the path taken between the initial and final states.

From the **first law of thermodynamics** we see that when a system undergoes a change from one state to another, the change in its internal energy, ΔU, is given by

$$\Delta U = Q - W \tag{20.17}$$

First law of thermodynamics

where Q is the heat transferred into (or out of) the system and W is the work done by the system. Although Q and W both depend on the path taken from the initial state to the final state, the quantity ΔU is path-independent.

In a **cyclic process** (one that originates and terminates at the same state), $\Delta U = 0$, and therefore $Q = W$. That is, the heat transferred into the system equals the work done during the cycle.

An **adiabatic process** is one in which no heat is transferred between the system and its surroundings ($Q = 0$). In this case, the first law gives $\Delta U = -W$. That is, the internal energy changes as a consequence of work being done by (or on) the system.

In an **adiabatic free expansion** of a gas, $Q = 0$ and $W = 0$, and so $\Delta U = 0$. That is, the internal energy of the gas does not change in such a process.

An **isobaric process** is one that occurs at constant pressure. The work done in such a process is simply $P\,\Delta V$.

An **isothermal process** is one that occurs at constant temperature. The work done by an ideal gas during an isothermal process is

$$W = nRT \ln\left(\frac{V_f}{V_i}\right) \tag{20.21}$$

Work done in an isothermal process

QUESTIONS

1. Ethyl alcohol has about one half the specific heat of water. If equal masses of alcohol and water in separate beakers are supplied with the same amount of heat, compare the temperature increases of the two liquids.
2. Give one reason why coastal regions tend to have a more moderate climate than inland regions.
3. A small crucible is taken from a 200°C oven and immersed in a tub full of water at room temperature (often referred to as *quenching*). What is the approximate final equilibrium temperature?
4. In a daring lecture demonstration, an instructor dips his wetted fingers into molten lead (327°C) and withdraws them quickly, without getting burned. How is this possible? (Note that this is a dangerous experiment, which you should not attempt.)

5. In the winter you might notice that some roofs are uniformly covered with snow, while others have regions where the snow has melted. Which houses would you say are better insulated as in the photograph?

(Question 5) Alternating pattern of snow-covered and exposed roof. (Courtesy of Dr. Albert A. Bartlett, University of Colorado, Boulder)

6. Why is it possible to hold a lighted match, even when it is burned to within a few millimeters of your fingertips?
7. If you wish to cook a piece of meat thoroughly on an open fire, why should you not use a high flame? (Note that carbon is a good thermal insulator.)
8. When insulating a wood-frame house, is it better to place the insulation against the cooler outside wall or against the warmer inside wall? (In either case, there is an air barrier to consider.)
9. Why is it necessary to store liquid nitrogen or liquid oxygen in vessels equipped with either Styrofoam insulation or a double-evacuated wall?
10. A Thermos bottle is constructed with double silvered-glass walls, with the space between them evacuated. Give reasons for the silvered walls and the vacuum jacket.
11. When a sealed Thermos bottle full of hot coffee is shaken, what are the changes, if any, in (a) the temperature of the coffee and (b) the internal energy of the coffee?
12. Using the first law of thermodynamics, explain why the *total* energy of an isolated system is always conserved.
13. Is it possible to convert internal energy to mechanical energy? Explain with examples.
14. Concrete has a higher specific heat than does soil. Use this fact to explain (partially) why cities have a higher average temperature than the surrounding countryside. If a city is hotter than the surrounding countryside, would you expect breezes to blow from city to country or from country to city? Explain.
15. Pioneers stored fruits and vegetables in underground cellars. Discuss as fully as possible this choice for a storage site.
16. Why can you get a more severe burn from steam at 100°C than from water at 100°C?
17. A piece of paper is wrapped around a rod made half of wood and half of copper. When held over a flame, the paper in contact with the wood burns but the half in contact with the metal does not. Explain.
18. If water is a poor conductor of heat, why can it be heated quickly when placed over a flame?
19. Why does a piece of metal feel colder than a piece of wood when they are at the same temperature?
20. Updrafts of air are familiar to all pilots. What causes these currents?
21. A tile floor in a bathroom may feel uncomfortably cold to your bare feet, but a carpeted floor in an adjoining room at the same temperature will feel warm. Why?
22. Why can potatoes be baked more quickly when a piece of metal has been inserted through them?
23. The U.S. penny is now made of copper-coated zinc. Can a calorimetric experiment be devised to test for the metal content in a collection of pennies? If so, describe the procedure you would use.
24. If you hold water in a paper cup over a flame, you can bring the water to a boil without burning the cup. How is this possible?
25. A 500-g brass object is heated to the boiling point of water and is then placed into 500 g of water in a beaker at room temperature. Calculate the final equilibrium temperature.

PROBLEMS

Section 20.2 Heat Capacity and Specific Heat

1. How many calories of heat are required to raise the temperature of 5 kg of aluminum from 25°C to 50°C?
2. A 50-g piece of copper is at 25°C. If 300 cal of heat is added to the copper, what is its final temperature?
3. What is the final equilibrium temperature when 20 g of milk at 10°C is added to 150 g of coffee at 90°C? (Assume the heat capacities of the two liquids are the same as that of water, and neglect the heat capacity of the container.)
4. It takes 3.5×10^3 cal to heat 400 g of an unknown substance from 20°C to 35°C. What is the specific heat of the substance?
5. A 2-kg iron horseshoe initially at 500°C is dropped into a bucket containing 30 kg of water at 20°C. What is the final equilibrium temperature? (Neglect the heat capacity of the container.)
6. Lead pellets, each of mass 1 g, are heated to 200°C. How many pellets must be added to 500 g of water initially at 20°C to make the final equilibrium temperature 25°C? (Neglect the heat capacity of the container.)
7. If 100 g of water is contained in a 300-g aluminum vessel at 20°C and an additional 200 g of water at 100°C is poured into the container, what is the final equilibrium temperature of the system?

8. A 250-g chunk of aluminum is heated in a furnace and then dropped into a 500-g copper vessel containing 300 g of water. If the temperature of the water rises from 20°C to 35°C, what was the initial temperature of the aluminum?
9. A 50-g ice cube at 0°C is heated until 45 g has become water at 100°C and 5 g has been converted to steam. How much heat was added to do this?
10. How much heat must be added to 10 g of copper at 20°C to completely melt it?
11. One liter of water at 25°C is used to make iced tea. How much ice at 0°C must be added to lower the temperature of the tea to 10°C? (Ice has a specific heat of .50 cal/g°C.)
12. In an insulated vessel, 300 g of ice at 0°C is added to 550 g of water at 16°C. (a) What is the final temperature of the system? (b) How much ice remains? (Ice has a specific heat of 0.50 cal/g°C.)
13. A 1-kg block of aluminum, initially at 20°C, is dropped into a large vessel of liquid nitrogen, which is boiling at 77 K. Assuming the vessel is thermally insulated from its surroundings, calculate the number of liters of nitrogen that boils away by the time the aluminum reached 77 K. (*Note:* Nitrogen has a specific heat of 0.21 cal/g · C°, a heat of vaporization of 48 cal/g, and a density of 0.8 g/cm^3.)
14. If 90 g of molten lead at 327.3°C is poured into a 300-g casting made of iron and initially at 20°C, what is the final temperature of the system? (Assume there are no heat losses.)

Section 20.4 Heat Transfer

15. A glass windowpane has an area of 2 m^2 and a thickness of 0.4 cm. If the temperature difference between its faces is 25 C°, how much heat flows through the window per hour?
16. The earth's thermal gradient, as measured at the surface, is 30°C/km, and the earth's radius is 6400 km. Assume that this gradient remains the same all the way to the center of the earth. What is the temperature of the earth at its center if we take the surface temperature to be 0°C? Do you think this is a reasonable answer or is it necessary to refine our assumption of the nature of the thermal gradient with depth?
17. The surface of the sun has a temperature of about 5800 K. Taking the radius of the sun to be equal to 6.96×10^8 m, calculate the total power radiated by the sun. (Assume $e = 1$.)
18. The rod shown in Figure 20.3 is made of aluminum and has a length of 50 cm and a cross-sectional area of 2 cm^2. One end is maintained at 80°C, and the other end is at 0°C. At steady state, find (a) the temperature gradient, (b) the rate of heat transfer, and (c) the temperature in the rod 15 cm from the cold end.
19. A bar of copper is in thermal contact with a bar of aluminum of the same length and area (Fig. 20.13). One end of the compound bar is maintained at 90°C while the opposite end is at 20°C. When the heat flow reaches steady state, find the temperature.

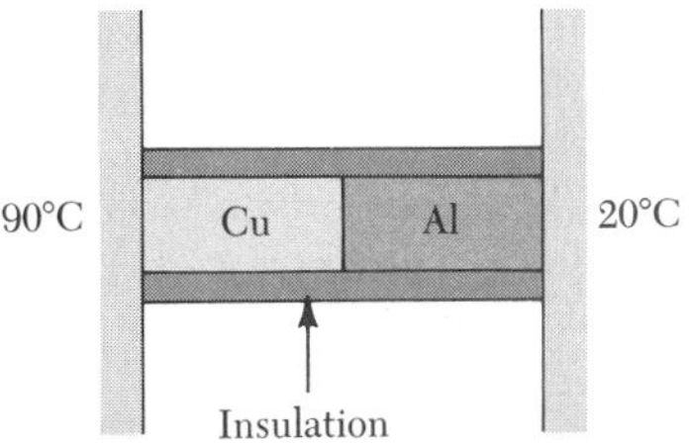

Figure 20.13 (Problem 19).

20. A Styrofoam container in the shape of a box has a surface area of 0.8 m^2 and a thickness of 2 cm. The inside is at 5°C and the outside is at 25°C. If it takes 8 h for 5 kg of ice to melt in the container, determine the thermal conductivity of the Styrofoam.
21. A Thermopane window 5 m^2 in area is constructed of two layers of glass, each 3 mm thick, separated by an air space of 5 mm. If the inside is at 20°C and the outside is at −30°C, what is the heat loss through the window?

Section 20.5 The Mechanical Equivalent of Heat

22. Consider Joule's apparatus described in Figure 20.7. The two masses are 2 kg each, and the tank is filled with 150 kg of water. What is the increase in the temperature of the water after the masses fall through a distance of 1 m?
23. A 75-kg weight-watcher wishes to climb a mountain to work off the equivalent of a large piece of chocolate cake rated at 500 (food) Calories. How high must the person climb?
24. A 5-g lead bullet traveling with a speed of 275 m/s is stopped by a large tree. If all of its initial kinetic energy is converted to heat in the bullet, find the increase in the temperature of the bullet.
25. A 3-g copper penny at 20°C drops a distance of 30 m to the ground. (a) If 75% of its initial potential energy goes into increasing the internal energy of the penny, determine its final temperature. (b) Does the result depend on the mass of the penny? Explain.
26. Water at the top of Niagara Falls has a temperature of 10°C. If it falls through a distance of 50 m and all of its potential energy goes into heating the water, calculate the temperature of the water at the bottom of the falls.
27. A 1.5-kg copper block is given an initial speed of 3 m/s on a rough, horizontal surface. Because of friction, it finally comes to rest. (a) If 85% of its initial kinetic energy is absorbed by the block in the form of heat, calculate the increase in temperature of the block. (b) What happens to the remaining energy?

Section 20.6 Work and Heat in Thermodynamic Processes

28. Using the fact that 1 atm = 1.013×10^5 N/m^2, verify the conversion 1 liter · atm = 101.3 J = 24.2 cal.

29. Gas in a container is at a pressure of 2 atm and a volume of 3 m^3. What is the work done by the gas if (a) it expands at constant pressure to twice its initial volume and (b) it is compressed at constant pressure to one third its initial volume?

30. A gas expands from I to F along three possible paths as indicated in Figure 20.14. Calculate the work done by the gas along paths IAF, IF, and IBF.

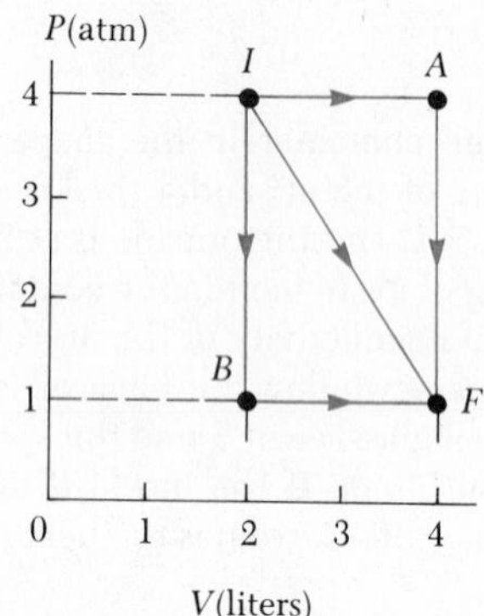

Figure 20.14 (Problems 30 and 31).

Section 20.7 The First Law of Thermodynamics

31. A gas expands from I to F as in Figure 20.14. The heat added to the gas is 100 cal when the gas goes from I to F along the diagonal path. (a) What is the change in internal energy of the gas? (b) How much heat must be added to the gas for the indirect path IAF to give the same change in internal energy?

32. A gas is compressed at a constant pressure of 0.3 atm from a volume of 8 liters to a volume of 3 liters. In the process, 400 J of heat energy flows out of the gas. (a) What is the work done by the gas? (b) What is the change in internal energy of the gas?

33. A thermodynamic system undergoes a process in which its internal energy decreases by 300 J. If at the same time, 120 J of work is done on the system, find the heat transferred to or from the system.

34. A gas is taken through the cyclic process described in Figure 20.15. (a) Find the net heat transferred to the system during one complete cycle. (b) If the cycle is reversed, that is, the process goes along $ACBA$, what is the net heat transferred per cycle?

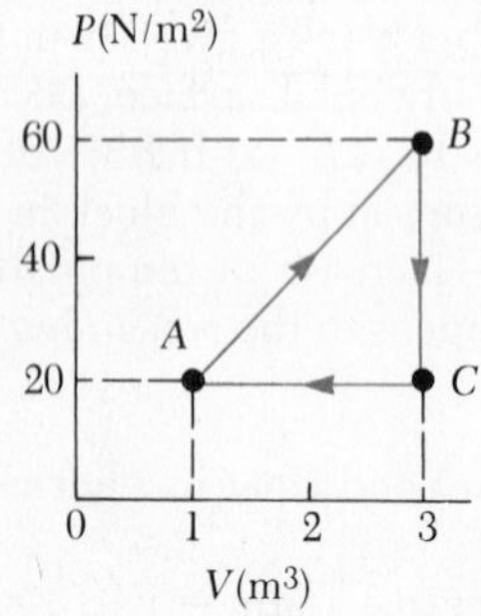

Figure 20.15 (Problems 34 and 35).

35. Consider the cyclic process described by Figure 20.15. If Q is negative for the process $B \rightarrow C$ and ΔU is negative for the process $C \rightarrow A$, determine the signs of Q, W, and ΔU associated with each process.

Section 20.8 Some Applications of the First Law of Thermodynamics

36. A 15-g silicon wafer used in a solar cell is heated from 20°C to 150°C at atmospheric pressure. What is the change in its internal energy?

37. Two moles of an ideal gas expands isothermally at 27°C to three times its initial volume. Find (a) the work done by the gas and (b) the heat flow into the system.

38. An ideal gas initially at 300 K undergoes an isobaric expansion at a pressure of 25 N/m^2. If the volume increases from 1 m^3 to 3 m^3 and 80 J of heat is added to the gas, find (a) the change in internal energy of the gas and (b) its final temperature.

39. One mole of helium gas initially at a temperature of 300 K and pressure of 0.2 atm is compressed isothermally to a pressure of 0.8 atm. Find (a) the final volume of the gas, (b) the work done by the gas, and (c) the heat transferred.

40. One mole of gas initially at a pressure of 2 atm and a volume of 0.3 liters has an internal energy equal to 91 J. In its final state, the pressure is 1.5 atm, the volume is 0.8 liters, and the internal energy equals 182 J. For the three paths IAF, IBF, and IF in Figure 20.16, calculate (a) the work done by the gas and (b) the net heat transferred in the process.

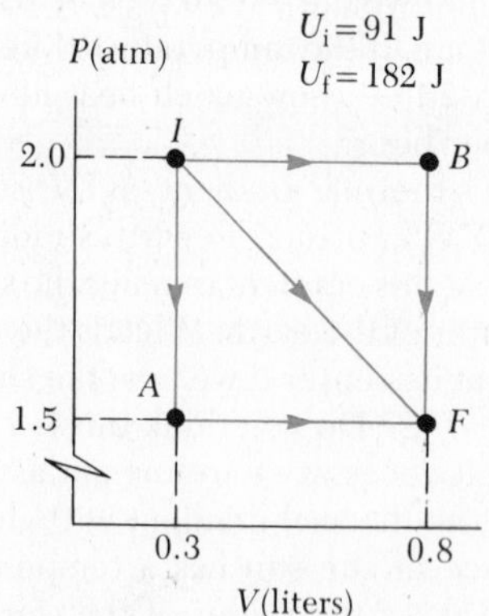

Figure 20.16 (Problem 40).

GENERAL PROBLEMS

41. One mole of an ideal gas is contained in a cylinder with a movable piston. The initial pressure, temperature, and volume are P_0, V_0, and T_0, respectively. Find the work done by the gas for the following processes and show the processes in a PV diagram: (a) an isobaric compression in which the final volume is one third the initial volume, (b) an isothermal compression in which the final pressure is twice the initial pressure, (c) an isovolumetric process in which the final pressure is twice the initial pressure.

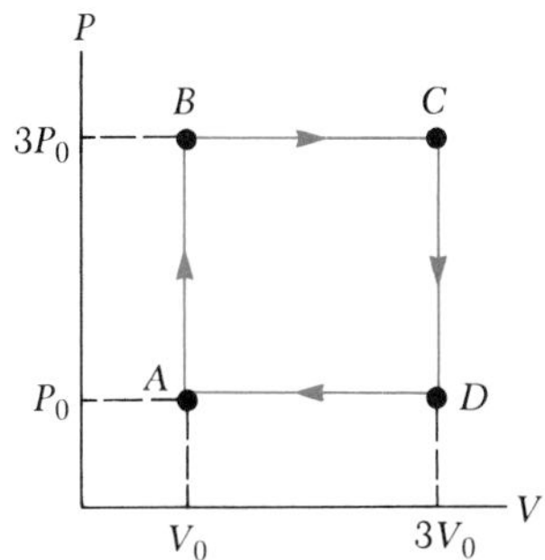

Figure 20.17 (Problem 42).

42. An ideal gas initially at pressure P_0, volume V_0, and temperature T_0 is taken through a cycle as described in Figure 20.17. (a) Find the net work done by the gas per cycle. (b) What is the net heat added to the system per cycle? (c) Obtain a numerical value for the net work done per cycle for one mole of gas initially at 0°C.
43. A would-be alchemist places 10 kg of molten lead at 327.3°C and 1 kg of ice at 0°C into an insulated chamber where they reach a common final temperature. (Assume that the specific heats of lead and water are constant throughout the temperature ranges encountered in this problem.) Find (a) the final temperature and (b) the heat transferred during the equilibration process.
44. An iron plate is held against an iron wheel so that there is a sliding frictional force of 50 N acting between the two pieces of metal. The relative speed at which the two surfaces slide over each other is 40 m/s. (a) Calculate the rate at which mechanical energy is converted to heat. (b) The plate and the wheel have a mass of 5 kg each, and each receives 50% of the frictional heat. If the system is run as described for 10 s and each object is then allowed to reach a uniform internal temperature, what is the resultant temperature increase?
45. The density of water is 999.17 kg/m³ at 14.5°C and 999.02 kg/m³ at 15.5°C. (a) Calculate the work done against the surrounding atmosphere when 1 kg of water expands as its temperature is increased from 14.5°C to 15.5°C. (b) Compare this work with the required heat input of 4186 J.
46. An automobile has a mass of 1500 kg, and its aluminum brakes have an overall mass of 60 kg. (a) Assuming that all of the frictional heat produced when the car stops is deposited in the brakes, and neglecting heat transfer, how many times could the car be braked to rest from 25 m/s (56 mph) before the brakes would begin to melt? (Assume an initial temperature of 20°C.) (b) Identify some effects that are neglected in part (a) but are likely to be important in a more realistic assessment of the heating of brakes.
47. An aluminum kettle has a circular cross section and is 9 cm in radius and 0.2 cm thick. It is placed on a hotplate and filled with 1 kg of water. If the bottom of the kettle is maintained at 101°C and the inside at 100°C, find (a) the rate of heat flow into the water and (b) the time it takes for all of the water to boil away. (Neglect heat transferred from the sides.)
48. A gas expands from a volume of 2 m³ to a volume of 6 m³ along two different paths as described in Figure 20.18. The heat added to the gas along the path *IAF* is equal to 4×10^5 cal. Find (a) the work done by the gas along the path *IAF*, (b) the work done along the path *IF*, (c) the change in internal energy of the gas, and (d) the heat transferred in the process along the path *IF*.

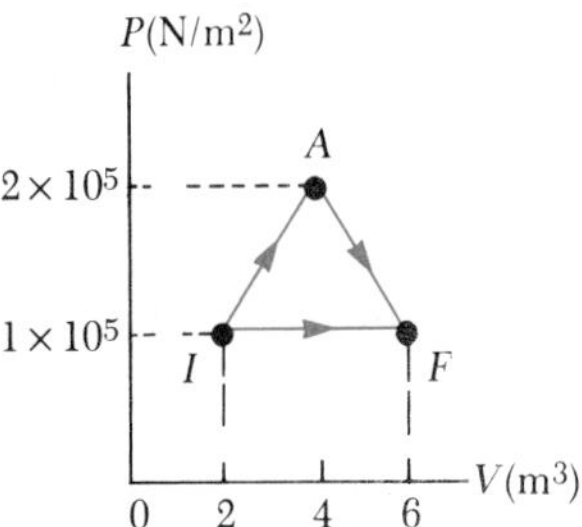

Figure 20.18 (Problem 48).

49. Using the data in Example 20.2 and Table 20.2, calculate the change in internal energy when 1 cm³ of liquid helium at 4.2 K is converted to helium gas at 273.15 K and atmospheric pressure. (Assume that the molar heat capacity of helium gas is 24.9 J/mole · K, and note that 1 cm³ of liquid helium is equivalent to 3.1×10^{-2} moles.)
50. The inside of a hollow cylinder is maintained at a temperature T_a while the outside is at a lower temperature, T_b (Fig. 20.19). The wall of the cylinder has a thermal conductivity k. Neglecting end effects, show that the rate of heat flow from the inner to the outer wall in the radial direction is given by

$$\frac{dQ}{dt} = 2\pi Lk \left[\frac{T_a - T_b}{\ln(b/a)}\right]$$

(*Hint:* The temperature gradient is given by dT/dr. Note that a radial heat current passes through a concentric cylinder of area $2\pi rL$.)

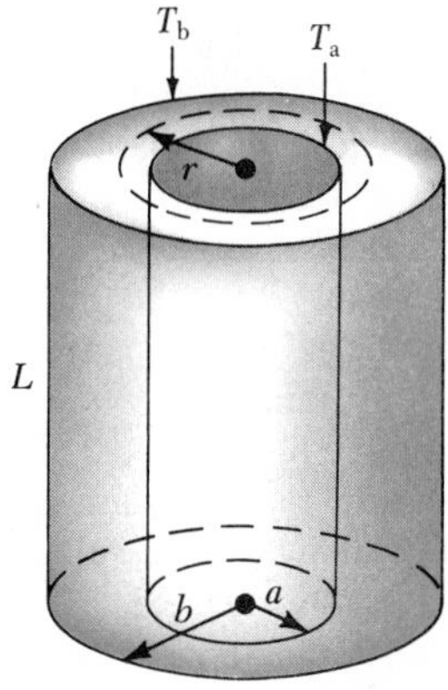

Figure 20.19 (Problem 50).

51. The passenger section of a jet airliner is in the shape of a cylindrical tube of length 30 m and inner radius 2 m. Its walls are lined with a 5-cm thickness of insulating material of thermal conductivity 3×10^{-5} cal/s · cm · C°. The inside is to be maintained at 20°C while the outside is at −40°C. What heating rate is required to maintain this temperature difference? (Use the result from Problem 50.)

52. Estimate the minimum heat required to transform 300 g of lead at 20°C to a gas at atmospheric pressure? Assume that C_p for lead is constant over the temperature range 20°C to 1750°C. (Use Tables 20.1 and 20.2.)

53. A *flow calorimeter* is an apparatus used to measure the specific heat of a liquid. The technique is to measure the temperature difference between the input and output points of a flowing stream of the liquid while adding heat at a known rate. In one particular experiment, a liquid of density 0.72 g/cm³ flows through the calorimeter at the rate of 3.5 cm³/s. At steady state, a temperature difference of 5.8 C° is established between the input and output points when heat is supplied at the rate of 40 J/s. What is the specific heat of the liquid?

54. A vessel in the shape of a spherical shell has an inner radius a and outer radius b. The wall has a thermal conductivity k. If the inside is maintained at a temperature T_1 and the outside is at a temperature T_2, show that the rate of heat flow between the surfaces is given by

$$\frac{dQ}{dt} = \left(\frac{4\pi kab}{b-a}\right)\left(T_1 - T_2\right)$$

55. An aluminum rod, 1 m in length and of cross-sectional area 2 cm², is inserted vertically into a thermally insulated vessel containing liquid helium at 4.2 K. The rod is initially at 300 K. (a) If one half of the rod is inserted into the helium, how many liters of helium boil off by the time the inserted half cools to 4.2 K? (Assume the upper half does not cool.) (b) If the upper portion of the rod is maintained at 300 K, what is the *approximate* boil-off rate of liquid helium *after* the lower half has reached 4.2 K? (Note that aluminum has a thermal conductivity of 31 J/s · cm · K at 4.2 K, a specific heat of 0.21 cal/g · C°, and a density of 2.7 g/cm³. See Example 20.2 for data on helium.)

56. A Thermos bottle in the shape of a cylinder has an inner radius of 4 cm, outer radius of 4.5 cm, and length of 30 cm. The insulating walls have a thermal conductivity equal to 2×10^{-5} cal/s · cm · C°. One liter of hot coffee at 90°C is poured into the bottle. If the outside wall remains at 20°C, how long does it take for the coffee to cool to 50°C? (Neglect end effects and losses by radiation and convection. Use the result from Problem 50 and assume that coffee has the same properties as water.)

57. A "solar cooker" consists of a curved reflecting mirror that focuses sunlight onto the object to be heated (Fig. 20.20). The solar power per unit area reaching the earth at some location is 600 W/m², and a small solar cooker has a diameter of 0.5 m. Assuming that 50% of the incident energy is converted into heat energy, how long would it take to evaporate 1 liter of water initially at 20°C? (Neglect the heat capacity of the container.)

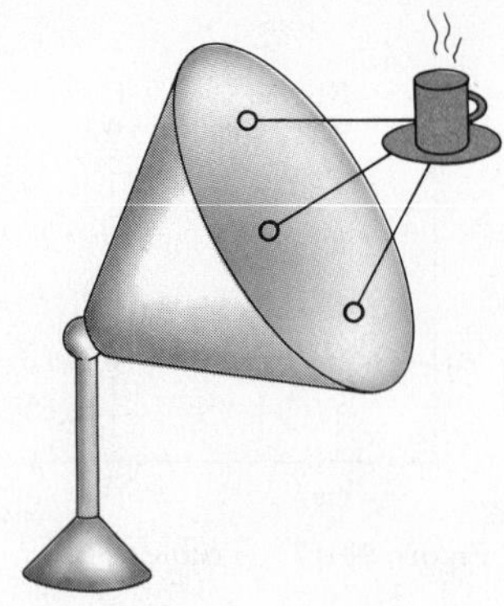

Figure 20.20 (Problem 57).

58. A one-person research submarine has a spherical iron hull 1.50 m in outer radius and 2 cm thick, lined with an equal thickness of rubber. If the submarine is used in arctic waters (temperature 4.0°C) and the total rate of heat release within the sub (including the occupant's metabolic heat) is 1500 W, find the equilibrium temperature of the interior.

•59. Consider a mass M of liquid that partially fills a cylindrical container of cross-section A. The container has a negligible coefficient of thermal expansion, but the liquid has a volume coefficient of expansion β. (a) Show that the fractional increase $\Delta h/h$ in the depth of the liquid in response to a temperature increase ΔT is given by:

$$\frac{\Delta h}{h} = \beta\,\Delta T$$

(b) Show that the corresponding increase in the potential energy of the liquid (in the gravitational field of the earth) is equal to

$$\frac{Mg\,\Delta h}{2} \quad \text{or} \quad \frac{Mgh\beta\,\Delta T}{2}$$

(c) The nominal heat requirement of 4186 J to raise the temperature of 1 kg of water by 1°C does not include any allowance for energy invested in increased gravitational potential energy. Use the expression given in part (b) to assess this additional energy requirement. Specifically consider water of mass $M = 1$ kg heated from 14.5°C to 15.5°C in a container of cross-section $A = 50$ cm², so that $h \simeq 20$ cm. At 15°C, $\beta = 1.5 \times 10^{-4}(°\text{C})^{-1}$. Evaluate $(Mg\,\Delta h)/2$ in joules, and also express that energy as a fraction of 4186 J. (d) Does your result suggest that the increase in gravitational potential energy is a significant additional energy requirement when a container of water is heated?

21
The Kinetic Theory of Gases

In the previous chapter we discussed the properties of an ideal gas using such macroscopic variables as pressure, volume, and temperature. We shall now show that such large-scale properties can be described on a microscopic scale, where matter is treated as a collection of molecules. Newton's laws of motion applied to a collection of particles in a statistical manner provide a reasonable description of thermodynamic processes. In order to keep the mathematics relatively simple, we shall consider only the molecular behavior of gases, where the interactions between molecules are much weaker than in liquids or solids. In the current view of gas behavior, called the *kinetic theory*, gas molecules move about in a random fashion, colliding with the walls of their container and with each other. Perhaps the most important consequence of this theory is that it shows the equivalence between the kinetic energy of molecular motion and the internal energy of the system. Furthermore, the kinetic theory provides us with a physical basis upon which the concept of temperature can be understood.

In the simplest model of a gas, each molecule is considered to be a hard sphere that collides elastically with other molecules or with the container wall. The hard-sphere model assumes that the molecules do not interact with each other except during collisions and that they are not deformed by collisions. This description is adequate only for monatomic gases, where the energy is entirely translational kinetic energy. One must modify the theory for more complex molecules, such as O_2 and CO_2, to include the internal energy associated with rotations and vibrations of the molecules.

21.1 MOLECULAR MODEL FOR THE PRESSURE OF AN IDEAL GAS

We begin this chapter by developing a microscopic model of an ideal gas which shows that the pressure that a gas exerts on the walls of its container is a consequence of the collisions of the gas molecules with the walls. As we shall see, the model is consistent with the macroscopic description of the preceding chapter. The following assumptions will be made:

Assumptions of the molecular model of an ideal gas

1. *The number of molecules is large, and the average separation between them is large* compared with their dimensions. Therefore, the molecules occupy a negligible volume compared with the volume of the container.
2. *The molecules obey Newton's laws of motion, but the individual molecules move in a random fashion.* By random fashion, we mean that the molecules move in all directions with equal probability and with various speeds. This distribution of velocities does not change in time, despite the collisions between molecules.
3. *The molecules undergo elastic collisions with each other and with the walls of the container.* Thus, the molecules are considered to be structureless (that

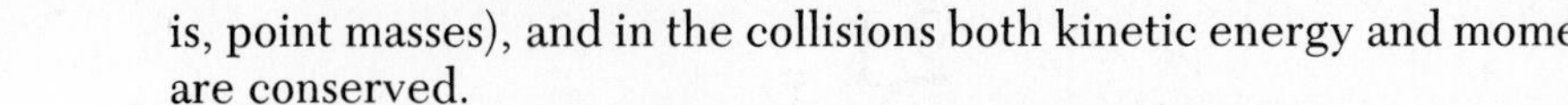

is, point masses), and in the collisions both kinetic energy and momentum are conserved.

4. *The forces between molecules are negligible except during a collision.* The forces between molecules are short-range, so that the only time the molecules interact with each other is during a collision.
5. *The gas under consideration is a pure gas.* That is, all molecules are identical.

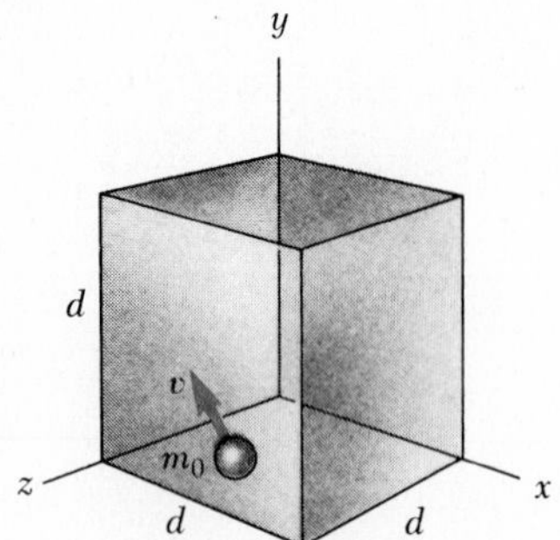

Figure 21.1 A cubical box of sides d containing an ideal gas. The molecule shown moves with velocity v.

Now let us derive an expression for the pressure of an ideal gas consisting of N molecules in a container of volume V. The container is assumed to be in the shape of a cube with edges of length d (Fig. 21.1). Consider the collision of one molecule moving with a velocity v toward the right-hand face of the box. The molecule has velocity components v_x, v_y, and v_z. As it collides with the wall elastically, its x component of velocity is reversed, while its y and z components of velocity remained unaltered (Fig. 21.2). Since the x component of momentum of the molecule is mv_x before the collision, and $-mv_x$ afterward, the *change* in momentum of the molecule is given by

$$\Delta p_x = -mv_x - (mv_x) = -2mv_x$$

The momentum delivered to the wall for each collision is $2mv_x$, since the momentum of the system (molecule + container) is conserved. In order that a molecule makes two successive collisions with the same wall, it must travel a distance $2d$ in the x direction in a time Δt. But in a time Δt, the molecule moves a distance $v_x \, \Delta t$ in the x direction; therefore the time between two successive collisions is $\Delta t = 2d/v_x$. If F is the magnitude of the average force exerted by a molecule on the wall in the time Δt, then from the definition of impulse (which equals change in momentum) we have

$$F \, \Delta t = \Delta p = 2mv_x$$

$$F = \frac{2mv_x}{\Delta t} = \frac{2mv_x}{2d/v_x} = \frac{mv_x^2}{d} \qquad (21.1)$$

The total force on the wall is the sum of all such terms for all particles. To get the total pressure on the wall, we divide the total force by the area, d^2:

$$P = \frac{\Sigma F}{A} = \frac{m}{d^3}(v_{x1}^2 + v_{x2}^2 + \cdots)$$

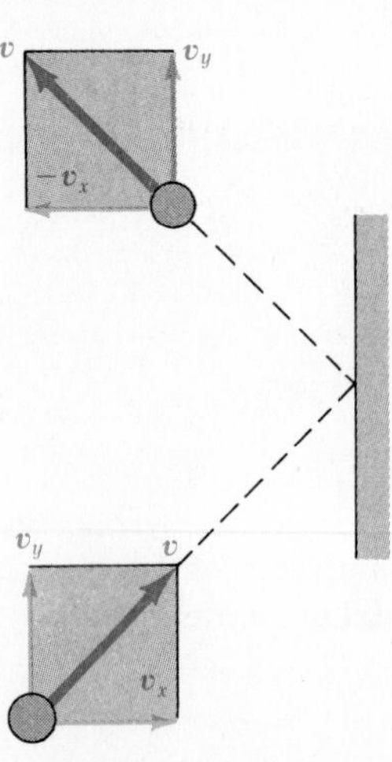

Figure 21.2 A molecule makes an elastic collision with the wall of the container. Its x component of momentum is reversed, thereby imparting momentum to the wall, while its y component remains unchanged.

where $v_{x1}, v_{x2}, \ldots$ refer to the x components of velocity for particles 1, 2, etc. Since the average value of v_x^2 is given by

$$\overline{v_x^2} = \frac{v_{x1}^2 + v_{x2}^2 + \cdots}{N}$$

and the volume is given by $V = d^3$, we can express the pressure in the form

$$P = \frac{Nm}{V}\overline{v_x^2} \qquad (21.2)$$

The square of the speed for any one particle is given by

$$v^2 = v_x^2 + v_y^2 + v_z^2$$

Since there is no preferred direction for the molecules, the average values $\overline{v_x^2}$, $\overline{v_y^2}$, and $\overline{v_z^2}$ are equal to each other. Using this fact and the above result, we find that

$$\overline{v_x^2} = \overline{v_y^2} = \overline{v_z^2} = \tfrac{1}{3}\overline{v^2}$$

Hence, the pressure from Equation 21.2 can be expressed as

$$P = \frac{1}{3}\frac{Nm}{V}\overline{v^2} \tag{21.3}$$

Pressure and molecular speed

The quantity Nm is the total mass of the molecules, which is equal to nM, where n is the number of moles of the gas and M is its molecular weight in g/mole. Therefore, the pressure can also be expressed in the alternate form

$$P = \frac{1}{3}\frac{nM}{V}\overline{v^2} \tag{21.4}$$

By rearranging Equation 21.3, we can also express the pressure as

$$P = \frac{2}{3}\frac{N}{V}\left(\tfrac{1}{2}m\overline{v^2}\right) \tag{21.5}$$

Pressure and molecular kinetic energy

This equation tells us that the pressure is proportional to the number of molecules per unit volume and to the average translational kinetic energy per molecule.

With this simplified model of an ideal gas, we have arrived at an important result that relates the macroscopic quantities of pressure and volume to a microscopic quantity, average molecular speed. Thus we have a key link between the microscopic world of the gas molecules and the macroscopic world as measured, in this case, with a pressure gauge and meter stick.

In the derivation of this result, note that we have not accounted for collisions between gas molecules. When these collisions are considered, the results do not change since collisions will only affect the momenta of the particles, with no net effect on the walls. This is consistent with one of our initial assumptions, namely, that the distribution of velocities does not change in time. In addition, although our result was derived for a cubical container, it is valid for a container of any shape.

21.2 MOLECULAR INTERPRETATION OF TEMPERATURE

We can obtain some insight into the meaning of temperature by first writing Equation 21.5 in the more familiar form

$$PV = \frac{2}{3}N\left(\tfrac{1}{2}m\overline{v^2}\right)$$

Let us now compare this with the empirical equation of state for an ideal gas (Eq. 19.14):

$$PV = NkT$$

Recall that the equation of state is based on experimental facts concerning the macroscopic behavior of gases. Equating the right sides of these expressions, we find that

$$T = \frac{2}{3k}\left(\tfrac{1}{2}m\overline{v^2}\right) \tag{21.6}$$

Temperature is proportional to average kinetic energy

That is, the absolute temperature of an ideal gas is a measure of the average of the square of the speed of its molecular constituents. Furthermore, since $\frac{1}{2}m\overline{v^2}$

is the average translational kinetic energy per molecule, we see that *temperature is a direct measure of the average molecular kinetic energy.*

By rearranging Equation 21.6, we can relate the translational molecular kinetic energy to the temperature:

Average kinetic energy per molecule

$$\tfrac{1}{2}m\overline{v^2} = \tfrac{3}{2}kT \qquad (21.7)$$

That is, the average translational kinetic energy per molecule is $\frac{3}{2}kT$. Since $\overline{v_x^2} = \frac{1}{3}\overline{v^2}$, it follows that

Equipartition of energy

$$\tfrac{1}{2}m\overline{v_x^2} = \tfrac{1}{2}kT \qquad (21.8)$$

That is, the average translational kinetic energy per molecule associated with motion in the x direction is $\frac{1}{2}kT$. In a similar manner, for the y and z motions it follows that

$$\tfrac{1}{2}m\overline{v_y^2} = \tfrac{1}{2}kT \qquad \text{and} \qquad \tfrac{1}{2}m\overline{v_z^2} = \tfrac{1}{2}kT$$

Thus, each translational degree of freedom contributes an equal amount of energy to the gas, namely, $\frac{1}{2}kT$. (In general, the degrees of freedom refers to the number of independent means by which a molecule can possess energy.)

> A generalization of this result, known as **the theorem of equipartition of energy**, says that the energy of a system in thermal equilibrium is equally divided among all degrees of freedom.

We shall return to this important point in Section 21.5.

The total translational kinetic energy of N molecules of gas is simply N times the average energy per molecule, which is given by Equation 21.7:

Total kinetic energy of N molecules

$$E = N\left(\tfrac{1}{2}m\overline{v^2}\right) = \tfrac{3}{2}NkT = \tfrac{3}{2}nRT \qquad (21.9)$$

where we have used $k = R/N_A$ for Boltzmann's constant and $n = N/N_A$ for the number of moles of gas.

The square root of $\overline{v^2}$ is called the *root mean square* (rms) *speed* of the molecules. From Equation 21.7 we get for the rms speed

Root mean square speed

$$v_{rms} = \sqrt{\overline{v^2}} = \sqrt{\frac{3kT}{m}} = \sqrt{\frac{3RT}{M}} \qquad (21.10)$$

The expression for the rms speed shows that at a given temperature, lighter molecules move faster, on the average, than heavier molecules. For example, hydrogen, with a molecular weight of 2 g/mole, moves four times as fast as oxygen, whose molecular weight is 32 g/mole. Note that the rms speed is not the speed at which a gas molecule will move across a room, since it undergoes several billion collisions per second with other molecules under standard conditions. We shall describe this in more detail in Section 21.6.

Table 21.1 lists the rms speeds for various molecules at 20°C.

EXAMPLE 21.1. A Tank of Helium

A tank of volume 0.3 m³ contains 2 moles of helium gas at 20°C. Assuming the helium behaves like an ideal gas, (a) find the total internal energy of the system.

Using Equation 21.9 with $n = 2$ and $T = 293$ K, we get

$$E = \tfrac{3}{2}nRT = \tfrac{3}{2}(2 \text{ moles})(8.31 \text{ J/mole}\cdot\text{K})(293 \text{ K})$$
$$= 7.30 \times 10^3 \text{ J}$$

(b) What is the average kinetic energy per molecule?

From Equation 21.7, we see that the average kinetic energy per molecule is equal to

$$\tfrac{1}{2}m\overline{v^2} = \tfrac{3}{2}kT = \tfrac{3}{2}(1.38 \times 10^{-23} \text{ J/K})(293 \text{ K})$$
$$= 6.07 \times 10^{-21} \text{ J}$$

Exercise 1 Using the fact that the molecular weight of helium is 4 g/mole, determine the rms speed of the atoms at 20°C.

Answer: 1.35×10^3 m/s.

TABLE 21.1 Some rms Speeds

Gas	Molecular Weight (g/mole)	v_{rms} at 20°C (m/s)*
H_2	2.02	1902
He	4.0	1352
H_2O	18	637
Ne	20.1	603
N_2 and CO	28	511
NO	30	494
CO_2	44	408
SO_2	48	390

* All values calculated using Equation 21.10.

21.3 HEAT CAPACITY OF AN IDEAL GAS

We have found that the temperature of a gas is a measure of the average translational kinetic energy of the gas molecules. It is important to note that this kinetic energy is associated with the motion of the center of mass of each molecule. It does not include the energy associated with the internal motion of the molecule, namely, vibrations and rotations about the center of mass. This should not be surprising, since the simple kinetic theory model assumes a structureless molecule.

In view of this, let us first consider the simplest case of an ideal monatomic gas, that is, a gas containing one atom per molecule, such as helium, neon, and argon. Essentially, all of the kinetic energy of such molecules is associated with the motion of their centers of mass. When energy is added to a monatomic gas in a container of fixed volume (by heating, say) all of the added energy goes into increasing the translational kinetic energy of the molecules.[1] There is no other way to store the energy in a monatomic gas. Therefore, from Equation 21.9 we see that the total internal energy U of N molecules (or n moles) of an ideal monatomic gas is given by

Internal energy of an ideal monatomic gas

$$U = \tfrac{3}{2}NkT = \tfrac{3}{2}nRT \qquad (21.11)$$

If heat is transferred to the system at *constant volume*, the work done by the system is zero. That is, since $\Delta V = 0$, $W = \int P\,dV = 0$. Hence, from the first law of thermodynamics we see that

$$Q = \Delta U = \tfrac{3}{2}nR\,\Delta T \qquad (21.12)$$

In other words, all of the heat transferred goes into increasing the internal energy (and temperature) of the system. The constant-volume process from i to f is described in Figure 21.3, where ΔT is the temperature difference between the two isotherms. Substituting the value for Q given by Equation 20.5 into Equation 21.12, we get

$$nC_v\,\Delta T = \tfrac{3}{2}nR\,\Delta T$$

$$C_v = \tfrac{3}{2}R \qquad (21.13)$$

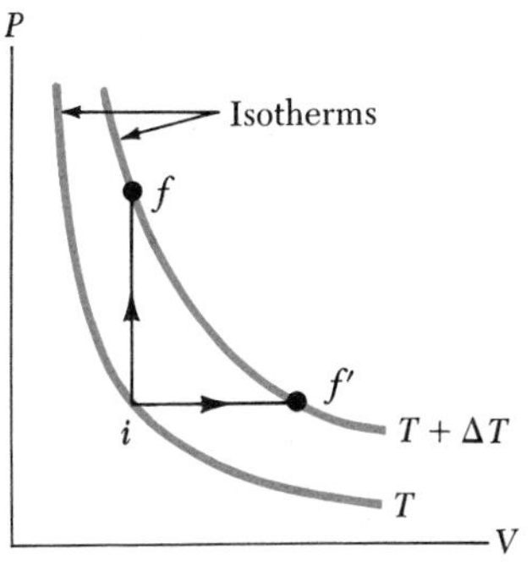

Figure 21.3 Heat is added to an ideal gas in two ways. For the constant-volume path *if*, the heat added goes into increasing the internal energy of the gas since no work is done. Along the constant-pressure path *if′*, part of the heat added goes into work done by the gas. Note that the internal energy is constant along any isotherm.

[1] If the gas is raised to sufficiently high temperatures, the atom can also be excited or even ionized.

TABLE 21.2 Molar Heat Capacities of Various Gases

	Molar Heat Capacity (cal/mole · K)			
	C_p	C_v	$C_p - C_v$	$\gamma = C_p/C_v$
		Monatomic Gases		
He	4.97	2.98	1.99	1.67
A	4.97	2.98	1.99	1.67
		Diatomic Gases		
H_2	6.87	4.88	1.99	1.41
N_2	6.95	4.96	1.99	1.40
O_2	7.03	5.04	1.99	1.40
CO	7.01	5.02	1.99	1.40
Cl_2	8.29	6.15	2.14	1.35
		Polyatomic Gases		
CO_2	8.83	6.80	2.03	1.30
SO_2	9.65	7.50	2.15	1.29
H_2O	8.46	6.46	2.00	1.30
CH_4	8.49	6.48	2.01	1.31

Note: All values obtained at 300 K.

In this notation, C_v is the molar heat capacity of the gas at constant volume. Note that this expression predicts a value of $\frac{3}{2}R = 2.98$ cal/mole · K for all monatomic gases. This is in excellent agreement with measured values of molar heat capacities for such gases as helium and argon over a wide range of temperatures (Table 21.2).

In the limit of differential changes, we can use Equation 21.12 and the first law of thermodynamics to express the molar heat capacity in the form

$$C_v = \frac{1}{n}\frac{dU}{dT} \qquad (21.14)$$

For an ideal monatomic gas, where $U = \frac{3}{2}nRT$, Equation 21.14 gives $C_v = \frac{3}{2}R$ in agreement with Equation 21.13.

Now suppose that the gas is taken along the constant-pressure path $i \rightarrow f'$ in Figure 21.3. Along this path, the temperature again increases by ΔT. The heat that must be transferred to the gas in this process is given by $Q = nC_p\,\Delta T$, where C_p is the molar heat capacity at constant pressure. Since the volume increases in this process, we see that the work done by the gas is $W = P\,\Delta V$. Applying the first law to this process gives

$$\Delta U = Q - W = nC_p\,\Delta T - P\,\Delta V \qquad (21.15)$$

In this case, the heat added to the gas is transferred in two forms. Part of it is used to do external work by moving a piston, and the remainder increases the internal energy of the gas. But the change in internal energy for the process $i \rightarrow f'$ is equal to the change for the process $i \rightarrow f$, since U depends only on temperature for an ideal gas and ΔT is the same for each process. In addition, since $PV = nRT$, we note that for a constant-pressure process $P\,\Delta V = nR\,\Delta T$. Substituting this into Equation 21.15 with $\Delta U = nC_v\,\Delta T$ gives

$$nC_v\,\Delta T = nC_p\,\Delta T - nR\,\Delta T$$

or

$$C_p - C_v = R \qquad (21.16)$$

This expression applies to *any* ideal gas. It shows that the molar heat capacity of an ideal gas at constant pressure is greater than the molar heat capacity at constant volume by an amount R, the universal gas constant (which has the value 1.99 cal/mole · K). This is in good agreement with real gases under standard conditions (Table 21.2).

Since $C_v = \frac{3}{2}R$ for a monatomic ideal gas, Equation 21.16 predicts a value $C_p = \frac{5}{2}R = 4.98$ cal/mole · K for the molar heat capacity of a monatomic gas at constant pressure. The ratio of these heat capacities is a dimensionless quantity γ given by

$$\gamma = \frac{C_p}{C_v} = \frac{\frac{5}{2}R}{\frac{3}{2}R} = \frac{5}{3} = 1.67 \qquad (21.17)$$

Ratio of heat capacities for an ideal gas.

The values of C_p and γ are in excellent agreement with experimental values for monatomic gases, but in serious disagreement with the values for the more complex gases (Table 21.2). This is not surprising since the value $C_v = \frac{3}{2}R$ was derived for a monatomic ideal gas, and we expect some additional contribution to the specific heat from the internal structure of the more complex molecules. In Section 21.5, we describe the effect of molecular structure on the specific heat of a gas. We shall find that the internal energy and hence the specific heat of a complex gas must include contributions from the rotational and vibrational motions of the molecule.

We have seen that the heat capacities of gases at constant pressure are greater than the heat capacities at constant volume. This difference is a consequence of the fact that in a constant-volume process, no work is done and all of the heat goes into increasing the internal energy (and temperature) of the gas, whereas in a constant-pressure process some of the heat energy is transformed into work done by the gas. In the case of solids and liquids heated at constant pressure, very little work is done since the thermal expansion is small. Consequently C_p and C_v are approximately equal for solids and liquids.

EXAMPLE 21.2. Heating a Cylinder of Helium
A cylinder contains 3 moles of helium gas at a temperature of 300 K. (a) How much heat must be transferred to the gas to increase its temperature to 500 K if the gas is heated at constant volume?

For the constant-volume process, the work done is zero. Therefore from Equation 21.12, we get

$$Q_1 = \tfrac{3}{2}nR\,\Delta T = nC_v\,\Delta T$$

But $C_v = 2.98$ cal/mole · K for He and $\Delta T = 200$ K; therefore

$$Q_1 = (3\text{ moles})(2.98\text{ cal/mole}\cdot\text{K})(200\text{ K})$$
$$= 1.79 \times 10^3\text{ cal} = 7.49 \times 10^3\text{ J}$$

(b) How much heat must be transferred to the gas at constant pressure to raise the temperature to 500 K?

Making use of Table 21.2, we get

$$Q_2 = nC_p\,\Delta T = (3\text{ moles})(4.97\text{ cal/mole}\cdot\text{K})(200\text{ K})$$
$$= 2.98 \times 10^3\text{ cal} = 12.5 \times 10^3\text{ J}$$

Exercise 2 What is the work done by the gas in this process?
Answer: $W = Q_2 - Q_1 = 5.01 \times 10^3$ J.

21.4 ADIABATIC PROCESS FOR AN IDEAL GAS

An **adiabatic process** is one in which there is no heat transfer between the system and its surroundings.

Definition of an adiabatic process

In reality, true adiabatic processes cannot occur since a perfect heat insulator between a system and its surroundings does not exist. However, there are processes that are nearly adiabatic. For example, if a gas is compressed (or expands) very rapidly, very little heat flows into (or out of) the system, and so the process is nearly adiabatic. Such processes occur in the cycle of a gasoline engine, which we shall discuss in detail in the next chapter.

It is also possible for a process to be both quasi-static and adiabatic. For example, if a gas that is thermally insulated from its surroundings is allowed to expand slowly against a piston, the process is a quasi-static, adiabatic expansion. In general,

> a **quasi-static, adiabatic process** is one that is slow enough to allow the system to always be near equilibrium, but fast compared with the time it takes the system to exchange heat with its surroundings.

Suppose that an ideal gas undergoes a *quasi-static, adiabatic* expansion. *At any time during the process, we assume that the gas is in an equilibrium state, so that the equation of state, $PV = nRT$, is valid.* In addition, we shall show that the pressure and volume at any time during the adiabatic process are related by the expression

Relation between P and V for an adiabatic process involving an ideal gas

$$PV^{\gamma} = \text{constant} \tag{21.18}$$

where $\gamma = C_p/C_v$ is assumed to be constant during the process. Thus, we see that all the thermodynamic variables, P, V, and T, change during an adiabatic process.

When a gas expands adiabatically in a thermally insulated cylinder, there is no heat transferred between the gas and its surroundings, and so $Q = 0$. Let us take the change in volume to be ΔV and the change in temperature to be ΔT. The work done by the gas is $W = P\,\Delta V$. Since the internal energy of an ideal gas depends only on temperature, the change in internal energy is given by $\Delta U = nC_v\,\Delta T$. Hence, the first law of thermodynamics gives

$$\Delta U = nC_v\,\Delta T = -P\,\Delta V$$

From the equation of state of an ideal gas, $PV = nRT$, we see that

$$P\,\Delta V + V\,\Delta P = nR\,\Delta T$$

Eliminating ΔT from these two equations we find that

$$P\,\Delta V + V\,\Delta P = \frac{R}{C_v}\,P\,\Delta V$$

Substituting $R = C_p - C_v$ and dividing by PV, we get

$$\frac{\Delta V}{V} + \frac{\Delta P}{P} = -\left(\frac{C_p - C_v}{C_v}\right)\frac{\Delta V}{V} = (1-\gamma)\frac{\Delta V}{V}$$

$$\frac{\Delta P}{P} + \gamma\frac{\Delta V}{V} = 0$$

Taking the limits of differential changes ($\Delta P \to dP$ and $\Delta V \to dV$) and integrating, we get

$$\ln P + \gamma \ln V = \ln\,(\text{constant})$$

which is equivalent to Equation 21.18:

$$PV^{\gamma} = \text{constant}$$

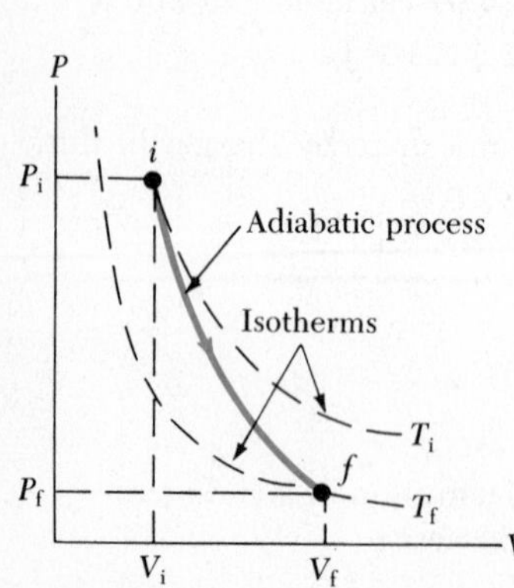

Figure 21.4 The PV diagram for an adiabatic expansion. Note that $T_f < T_i$ in this process.

The PV diagram for an adiabatic expansion is shown in Figure 21.4. Because $\gamma > 1$, the PV curve for the adiabatic expansion is steeper than that for an isothermal expansion. As the gas expands adiabatically, no heat is transferred in or out of the system. Hence, from the first law, we see that ΔU is negative so that ΔT is also negative. Thus, we see that the gas cools ($T_f < T_i$) during an adiabatic expansion. Conversely, the temperature increases if the gas is compressed adiabatically. Applying Equation 21.18 to the initial and final states, we see that

$$P_i V_i^{\gamma} = P_f P_f^{\gamma} \qquad (21.19)$$

Adiabatic process

Using $PV = nRT$, it is left as a problem (Problem 19) to show that Equation 21.19 can also be expressed as

$$T_i V_i^{\gamma-1} = T_f V_f^{\gamma-1} \qquad (21.20)$$

Note that the above analysis is valid only in processes that are slow enough to allow the system to always remain near equilibrium, but fast enough to prevent the system from exchanging heat with its surroundings.

EXAMPLE 21.3. A Diesel Engine Cylinder
Air in the cylinder of a diesel engine at 20°C is compressed from an initial pressure of 1 atm and volume of 200 cm³ to a volume of 15 cm³. Assuming that air behaves as an ideal gas ($\gamma = 1.40$) and that the compression is adiabatic, find the final pressure and temperature.

Solution: Using Equation 21.19, we find that

$$P_f = P_i(V_i/V_f)^{\gamma} = 1 \text{ atm } (200 \text{ cm}^3/15 \text{ cm}^3)^{1.4}$$
$$= 37.6 \text{ atm}$$

Since $PV = nRT$ is always valid during the process and since no gas escapes from the cylinder,

$$\frac{P_i V_i}{T_i} = \frac{P_f V_f}{T_f}$$

$$T_f = \frac{P_f V_f}{P_i V_i} T_i = \frac{(37.6 \text{ atm})(15 \text{ cm}^3)}{(1 \text{ atm})(200 \text{ cm}^3)} (293 \text{ K})$$
$$= 826 \text{ K} = 553°\text{C}$$

21.5 SOUND WAVES IN A GAS

Most gases are poor heat conductors. Therefore, when a sound wave propagates through a gas, very little heat is transferred between regions of high and low densities. To a good approximation, we can assume that the variations of pressure and volume occur adiabatically, corresponding to no heat transfer between portions of the gas. This is equivalent to assuming that all of the work done in compressing the gas goes into increasing the internal energy of the gas. We shall use this fact to determine an expression for the speed of sound in a gas. First, recall from Chapter 17 that the speed of a longitudinal wave is given by

$$v = \sqrt{\frac{B}{\rho}}$$

where ρ is the density of the medium and B is its bulk modulus, given by Equation 17.2, $B = -B(\Delta P/\Delta V)$. If ΔP and ΔV are replaced by dP and dV, respectively, we find

$$B = -\frac{1}{V}\left(\frac{dP}{dV}\right)$$

In Section 21.4, we found that if the gas is ideal, the pressure and volume during an adiabatic process are related by the expression $PV^{\gamma} =$ constant. Differentiating this with respect to V gives

$$\gamma P V^{\gamma-1} + V^{\gamma}\left(\frac{dP}{dV}\right) = 0$$

or

$$\frac{dP}{dV} = -\gamma \frac{P}{V}$$

Substituting this into the expression for B gives

$$B_{\text{adiabatic}} = -\frac{1}{V}\left(-\gamma \frac{P}{V}\right) = \gamma P$$

Therefore the speed of sound in a gas is

Speed of sound in a gas

$$v = \sqrt{\frac{\gamma P}{\rho}} \qquad (21.21)$$

Equation 21.21 can be expressed in another useful form, which uses the equation of state of an ideal gas, $PV = nRT$, or

$$P = nRT/V = \rho RT/M$$

where R is the gas constant, M is the molecular weight, and n is the number of moles of gas. Substituting this expression for P into Equation 21.21 gives

Speed of sound in a gas

$$v = \sqrt{\frac{\gamma RT}{M}} \qquad (21.22)$$

It is interesting to compare this result with the rms speed of molecules in a gas (Eq. 21.10), where $v_{\text{rms}} = \sqrt{3RT/M}$. The two results differ only by the factors γ and 3. It is known that γ lies between 1 and 1.67; hence the two speeds are nearly the same! Since sound waves propagate through air as a result of collisions between gas molecules, one would expect the wave speed to increase as the temperature (and molecular speed) increase.

EXAMPLE 21.4. The Speed of Sound in Air
Calculate the speed of sound in air at atmospheric pressure and at 0°C, taking $P = 1.01 \times 10^5$ N/m², $\gamma = 1.40$, and $\rho = 1.29$ kg/m³.

Solution: Using Equation 21.21, we find that

$$v_{\text{air}} = \sqrt{(1.4)(1.01) \times 10^5 \text{ N/m}^2)/1.29 \text{ kg/m}^3} = 331 \text{ m/s}$$

This is in excellent agreement with the measured speed of sound in air.

It is interesting to note that the speed of sound in helium is much greater than this because of the lower density of helium. An amusing demonstration of this fact is the variation in the human voice when the vocal cavities are partially filled with helium. The demonstrator talks before and after taking a deep breath of helium, an inert gas. The result is a high-pitched voice sounding a bit like that of Donald Duck. The increase in frequency corresponds to an increase in the speed of sound in helium, since frequency is proportional to velocity.

21.6 THE EQUIPARTITION OF ENERGY

We have found that model predictions based on specific heat agree quite well with the behavior of monatomic gases, but not with the behavior of complex gases (Table 21.2). Furthermore, the value predicted by the model for the quantity $C_p - C_v = R$ is the same for all gases. This is not surprising, since this difference is the result of the work done by the gas, which is independent of its molecular structure.

In order to explain the variations in C_v and C_p in going from monatomic gases to the more complex gases, let us explain the origin of the specific heat. So far, we have assumed that the sole contribution to the internal energy of a gas is the translational kinetic energy of the molecules. However, the internal energy of a gas actually includes contributions from the translational, vibrational, and rotational motion of the molecules. The rotational and vibrational motions of molecules with structure can be activated by collisions and therefore are "coupled" to the translational motion of the molecules. The branch of physics known as *statistical mechanics* has shown that for a large number of particles obeying newtonian mechanics, the available energy is, on the average, shared equally by each independent degree of freedom. Recall that the **equipartition theorem** states that at equilibrium each degree of freedom contributes, on the average, $\frac{1}{2}kT$ of energy per molecule.

Let us consider a diatomic gas, which we can visualize as a dumbbell-shaped molecule (Fig. 21.5). In this model, the center of mass of the molecule can translate in the x, y, and z directions (Fig. 21.5a). In addition, the molecule can rotate about three mutually perpendicular axes (Fig. 21.5b). We can neglect the rotation about the y axis since the moment of inertia and the rotational energy, $\frac{1}{2}I\omega^2$, about this axis are negligible compared with those associated with the x and z axes. If the two atoms of the molecule are taken to be point masses, then I_y is identically zero. Thus there are five degrees of freedom: three associated with the translational motion and two associated with the rotational motion. Since *each degree of freedom contributes, on the average, $\frac{1}{2}kT$ of energy per molecule,* the total energy for N molecules is

$$U = 3N(\tfrac{1}{2}kT) + 2N(\tfrac{1}{2}kT) = \tfrac{5}{2}NkT = \tfrac{5}{2}nRT$$

We can use this result and Equation 21.14 to get the molar heat capacity at constant volume:

$$C_v = \frac{1}{n}\frac{dU}{dT} = \frac{1}{n}\frac{d}{dT}(\tfrac{5}{2}nRT) = \tfrac{5}{2}R$$

From Equations 21.16 and 21.17 we find that

$$C_p = C_v + R = \tfrac{7}{2}R$$

$$\gamma = \frac{C_p}{C_v} = \frac{\frac{7}{2}R}{\frac{5}{2}R} = \frac{7}{5} = 1.40$$

These results agree quite well with most of the data given in Table 21.2 for diatomic molecules. This is rather surprising since we have not yet accounted for the possible vibrations of the molecule. In the vibratory model, the two atoms are joined by an imaginary spring. The vibratory motion adds two more degrees of freedom, corresponding to the kinetic and potential energies associated with vibrations along the length of the molecule. Hence, the equipartition theorem predicts an internal energy of $\frac{7}{2}nRT$ and a higher heat capacity than what is observed. Examination of the experimental data (Table 21.2) suggests that some diatomic molecules, such as H_2 and N_2, do not vibrate at room temperature, and others, such as Cl_2, do. For molecules with more than two atoms, the number of degrees of freedom is even larger and the vibrations are more complex. This results in an even higher predicted heat capacity, which is in qualitative agreement with experiment.

We have seen that the equipartition theorem is successful in explaining some features of the heat capacity of molecules with structure. However, the equipartition theorem does not explain the observed temperature variation in heat capacities. As an example of such a temperature variation, C_v for the

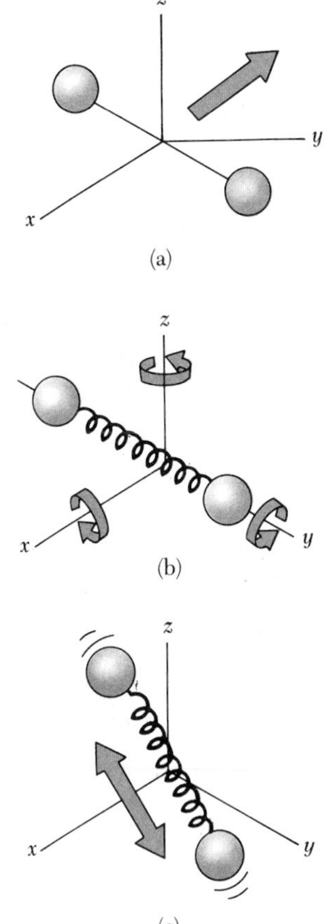

Figure 21.5 Possible motions of a diatomic molecule: (a) translational motion of the center of mass, (b) rotational motion about the various axes, and (c) vibrational motion along the molecular axis.

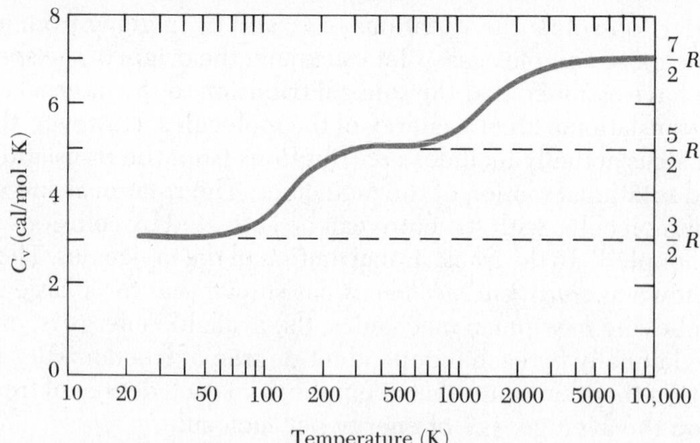

Figure 21.6 The molar heat capacity, C_v, of hydrogen as a function of temperature. The horizontal scale is logarithmic. Note that hydrogen liquefies at 20 K.

hydrogen molecule is $\frac{5}{2}R$ from about 250 K to 750 K and then increases steadily to about $\frac{7}{2}R$ well above 750 K (Fig. 21.6). This suggests that vibrations occur at very high temperatures. At temperatures well below 259 K, C_v has a value of about $\frac{3}{2}R$, suggesting that the molecule has only translational energy at low temperatures.

The failure of the equipartition theorem to explain such phenomena is due to the inadequacy of classical mechanics when applied to molecular systems. For a more satisfactory description, it is necessary to use a quantum-mechanical model in which the energy of an individual molecule is quantized. The magnitude of the energy separation between adjacent vibrational energy levels for a molecule such as H_2 is about ten times as great as the average kinetic energy of the molecule at room temperature. Consequently, collisions between molecules at low temperatures do not provide enough energy to change the vibrational state of the molecule. It is often stated that such degrees of freedom are "frozen out." This explains why the vibrational energy does not contribute to the heat capacities of molecules at low temperatures.

The rotational energy levels are also quantized, but their spacing at ordinary temperatures is small compared witk kT. Since the spacing between rotational levels is so small compared with kT, the system behaves classically. However, at sufficiently low temperatures (typically less than 50 K), where kT is small compared with the spacing between rotational levels, intermolecular collisions may not be energetic enough to alter the rotational states. This explains why C_v reduces to $\frac{3}{2}R$ for H_2 in the range from 20 K to about 100 K.

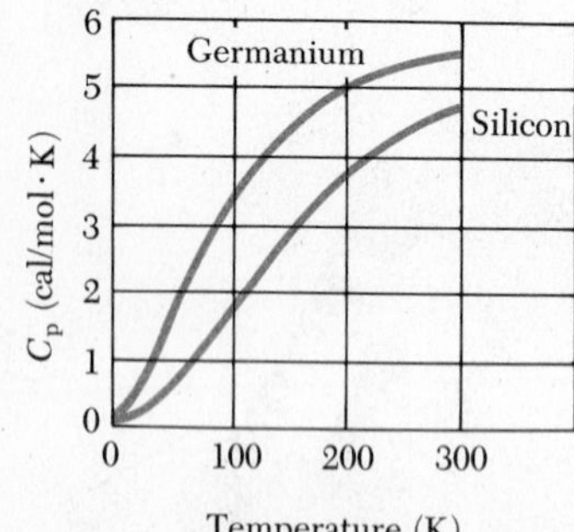

Figure 21.7 Molar heat capacity, C_p, of silicon and germanium. As T approaches zero, the heat capacity also approaches zero. (From C. Kittel, *Introduction to Solid State Physics*, New York, John Wiley, 1971.)

Heat Capacities of Solids

Measurements of heat capacities of solids also show a marked temperature dependence. The heat capacities of solids generally decrease in a nonlinear manner with decreasing temperature and approach zero as the absolute temperature approaches zero. At high temperatures (usually above 500 K), the heat capacities of solids approach the value of about $3R \approx 6$ cal/mole · K, a result known as the *DuLong-Petit law*. The typical data shown in Figure 21.7 demonstrate the temperature dependence of the heat capacity for two semiconducting solids, silicon and germanium.

The heat capacity of a solid at high temperatures can be explained using the equipartition theorem. For small displacements of an atom from its equilibrium position, each atom executes simple harmonic motion in the x, y, and z directions. The energy associated with vibrational motion in the x direction is

$$E_x = \tfrac{1}{2}mv_x^2 + \tfrac{1}{2}kx^2$$

There are analogous expressions for E_y and E_z. Therefore, each atom of the solid has six degrees of freedom. According to the equipartition theorem, this corresponds to an average vibrational energy of $6(\tfrac{1}{2}kT) = 3\ kT$ per atom. Therefore, the total internal energy of a solid consisting of N atoms is given by

Total internal energy of a solid

$$U = 3NkT = 3nRT$$

From this result, we find that the molar heat capacity

$$C_v = \frac{1}{n}\frac{dU}{dT} = 3R,$$

which agrees with the empirical law of DuLong and Petit. The discrepancies between this model and the experimental data at low temperatures are again due to the inadequacy of classical physics in the microscopic world. One can attribute the decrease in heat capacity with decreasing temperature to a "freezing out" of various vibrational excitations.

*21.7 DISTRIBUTION OF MOLECULAR SPEEDS

Thus far we have not concerned ourselves with the fact that not all molecules in a gas have the same speed and energy. Their motion is extremely chaotic. Any individual molecule is colliding with others at the enormous rate of typically a billion times per second. Each collision results in a change in the speed and direction of motion of each of the participant molecules. From Equation 21.10, we see that average molecular speeds increase with increasing temperature. What we would like to know now is the distribution of molecular speeds. For example, how many molecules of a gas have a speed in the range of, say, 400 to 410 m/s? Intuitively, we expect that the speed distribution depends on temperature. Furthermore, we expect that the distribution peaks in the vicinity of v_{rms}. That is, few molecules are expected to have speeds much less than or much greater than v_{rms}, since these extreme speeds will result only from an unlikely chain of collisions.

The development of a reliable theory for the speed distribution of a large number of particles appears, at first, to be an almost impossible task. However, in 1860 James Clerk Maxwell (1831–1879) derived an expression that describes the distribution of molecular speeds in a very definite manner. His work, and developments by other scientists shortly thereafter, were highly controversial, since experiments at that time were not capable of directly detecting molecules. However, about 60 years later experiments were devised which confirmed Maxwell's predictions.

One experimental arrangement for observing the speed distribution of molecules is illustrated in Figure 21.8. A substance is vaporized in an oven and forms gas molecules, which are permitted to escape through a hole. The molecules enter an evacuated region and pass through a series of slits to form a collimated beam. The beam is incident on two slotted rotating disks separated by a distance s and displaced from each other by an angle θ. A molecule passing through the first slotted disk will pass through the second slotted disk only if its speed is $v = s\omega/\theta$, where ω is the angular velocity of the disks. Molecules with

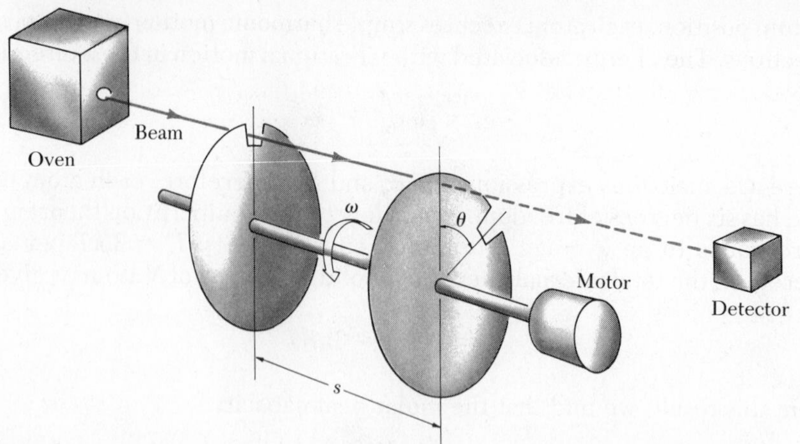

Figure 21.8 A schematic diagram of one apparatus used to measure the speed distribution of gas molecules.

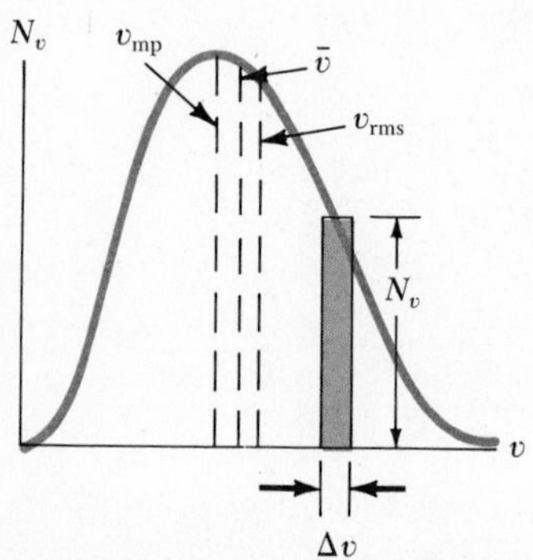

Figure 21.9 The speed distribution of gas molecules at some temperature. The number of molecules in the range Δv is equal to the area of the shaded rectangle, $N_v\,\Delta v$. The function N_v approaches zero as v approaches infinity.

other speeds will necessarily collide with the second disk and hence will not reach the detector. By varying ω and θ, one can measure the number of molecules in a given range of speeds.

The observed speed distribution of gas molecules in thermal equilibrium is shown in Figure 21.9. The quantity N_v (which is called the *distribution function*) represents the number of molecules per unit interval of speed.

The number of molecules having a speed in the range from v to $v + \Delta v$ is equal to $N_v\,\Delta v$, represented in Figure 21.9 by the area of the shaded rectangle. If N is the total number of molecules, then the fraction of molecules with speeds between v and $v + \Delta v$ is equal to $N_v\,\Delta v/N$. This fraction is also equal to the probability that any given molecule has a speed in the range from v to $v + \Delta v$.

The total number of molecules numerically equals the total area under the speed distribution curve. Since the abscissa ranges from $v = 0$ to $v = \infty$ (classically, all molecular speeds are possible), we can express the total number of particles as the sum of the areas of all shaded rectangles. In the limit $\Delta v \to 0$, this sum is replaced by an integral:

$$N = \lim_{\Delta v \to 0} \left(\sum_{v=0}^{\infty} N_v\,\Delta v \right) = \int_0^{\infty} N_v\,dv \qquad (21.23)$$

The fundamental expression (derived by Maxwell) that describes the most probable distribution of speeds of N gas molecules is given by

Maxwell speed distribution function

$$N_v = 4\pi N \left(\frac{m}{2\pi kT} \right)^{3/2} v^2 e^{-mv^2/2kT} \qquad (21.24)$$

where m is mass of a gas molecule, k is Boltzmann's constant, and T is the absolute temperature.[2] The function given by Equation 21.24 satisfies Equation 21.23. Furthermore, N_v approaches zero in the low- and high-speed limits, as expected. We also note that the speed distribution for a given gas depends only on temperature.

[2] For the derivation of this expression, see any text on thermodynamics, such as M. W. Zemansky and R. H. Dittman, *Heat and Thermodynamics*, New York, McGraw-Hill, 1981.

As indicated in Figure 21.9, the average speed, $\bar{v}$, is somewhat lower than the rms speed. The most probable speed, $v_{\rm mp}$, is the speed at which the distribution curve reaches a peak. Using Equation 21.24, one finds that

rms speed

$$v_{\rm rms} = \sqrt{\overline{v^2}} = \sqrt{3kT/m} = 1.73\sqrt{kT/m} \quad (21.25)$$

Average speed

$$\bar{v} = \sqrt{8kT/\pi m} = 1.60\sqrt{kT/m} \quad (21.26)$$

Most probable speed

$$v_{\rm mp} = \sqrt{2kT/m} = 1.41\sqrt{kT/m} \quad (21.27)$$

The details of these calculations are left for the student (Problems 36 and 50), but from these equations we see that $v_{\rm rms} > \bar{v} > v_{\rm mp}$.

Figure 21.10 represents specific speed distribution curves for nitrogen molecules. The curves were obtained by using Equation 21.24 to evaluate the distribution function, N_v, at various speeds and at two temperatures (300 K and 900 K). Note that the curve shifts to the right as T increases, indicating that the average speed increases with increasing temperature, as expected. The asymmetric shape of the curves is due to the fact that the lowest speed possible is zero while the upper classical limit of the speed is infinity. Furthermore, as temperature increases the distribution curve broadens and the range of speeds also increases.

Equation 21.24 shows that the distribution of molecular speeds in a gas depends on mass as well as temperature. At a given temperature, the fraction of particles with speeds exceeding a fixed value increases as the mass decreases. This explains why lighter molecules, such as hydrogen and helium, escape more readily from the earth's atmosphere than heavier molecules, such as nitrogen and oxygen. (See the discussion of escape velocity in Chapter 14. Notice that gas molecules escape even more readily from the moon's surface because its escape velocity is lower.)

The evaporation process

The speed distribution of molecules in a liquid is similar to that shown in Figure 21.10. The phenomenon of evaporation of a liquid can be understood from this distribution in speeds using the fact that some molecules in the liquid are more energetic than others. Some of the faster-moving molecules in the liquid penetrate the surface and leave the liquid even at temperatures well below the boiling point. The molecules that escape the liquid by evaporation are those that have sufficient energy to overcome the attractive forces of the

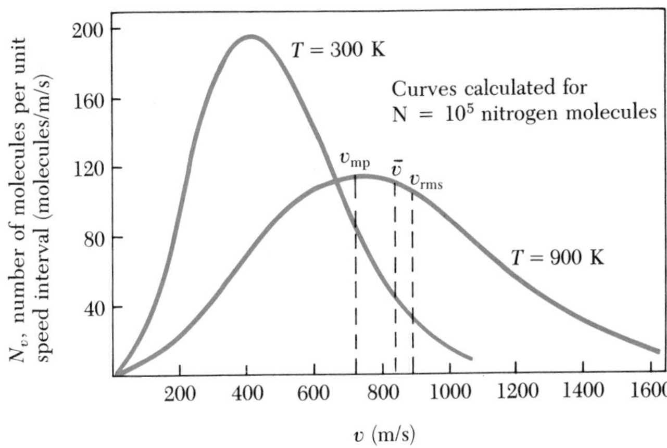

Figure 21.10 The Maxwell speed distribution function for 10^5 nitrogen molecules at temperatures of 300 K and 900 K. The total area under either curve is equal to the total number of molecules, which, in this case, equals 10^5. Note that $v_{\rm rms} > \bar{v} > v_{\rm mp}$.

molecules in the liquid phase. Consequently, the molecules left behind in the liquid phase have a lower average kinetic energy, causing the temperature of the liquid to decrease. Hence evaporation is a cooling process.

EXAMPLE 21.5. A System of Nine Particles
Nine particles have speeds of 5, 8, 12, 12, 12, 14, 14, 17, and 20 m/s. (a) Find the average speed.

The average speed is the sum of the speeds divided by the total number of particles:

$$\bar{v} = \frac{5 + 8 + 12 + 12 + 12 + 14 + 14 + 17 + 20}{9}$$

$$= 12.7 \text{ m/s}$$

(b) What is the rms speed?

The average value of the square of the speed is given by

$$\overline{v^2} = \frac{5^2 + 8^2 + 12^2 + 12^2 + 12^2 + 14^2 + 14^2 + 17^2 + 20^2}{9}$$

$$= 178 \text{ m}^2/\text{s}^2$$

Hence, the rms speed is

$$v_{\text{rms}} = \sqrt{\overline{v^2}} = \sqrt{178 \text{ m}^2/\text{s}^2} = 13.3 \text{ m/s}$$

(c) What is the most probable speed of the particles?

Three of the particles have a speed of 12 m/s, two have a speed of 14 m/s, and the remaining have different speeds. Hence, we see that the most probable speed, v_{mp}, is 12 m/s.

*21.8 MEAN FREE PATH

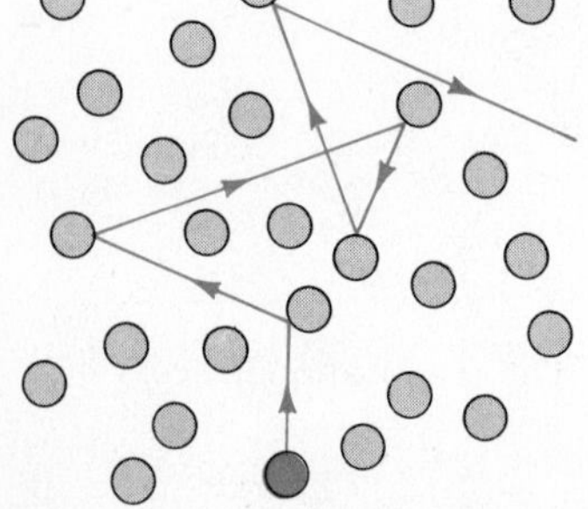

Figure 21.11 A molecule moving through a gas collides with other molecules in a random fashion. This behavior is sometimes referred to as a *random-walk process*. The mean free path increases as the number of molecules per unit volume decreases. Note that the motion is *not* limited to the plane of the paper.

Most of us are familiar with the fact that the strong odor associated with a gas such as ammonia may take several minutes to diffuse through a room. However, since average molecular speeds are typically several hundred meters per second at room temperature, we might expect a time much less than one second. To understand this apparent contradiction, we note that molecules collide with each other, since they are not geometrical points. Therefore, they do not travel from one side of a room to the other in a straight line. Between collisions, the molecules move with constant speed along straight lines.[3] The average distance between collisions is called the **mean free path.** The path of individual molecules is random and resembles that shown in Figure 21.11. As we would expect from this description, the mean free path is related to the diameter of the molecules and the density of the gas.

We shall now describe how to estimate the mean free path for a gas molecule. For this calculation we shall assume that the molecules are spheres of diameter d. We see from Figure 21.12a that no two molecules will collide

[3] Actually, there is a small curvature in the path because of the force of gravity at the earth's surface. However, this effect is small and can be neglected.

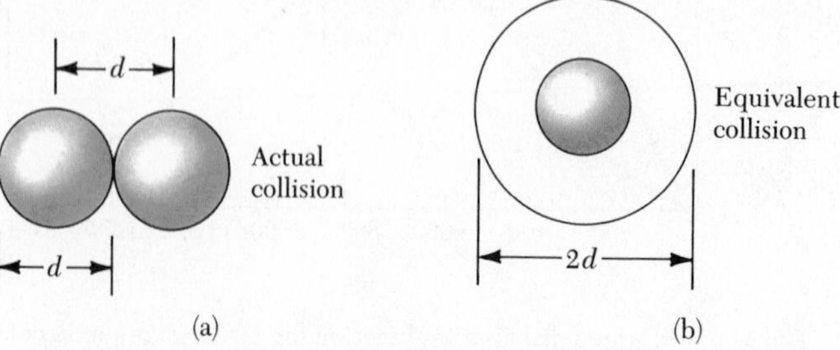

Figure 21.12 (a) Two molecules, each of diameter d, collide if their centers are within a distance d of each other. (b) The collision between the two molecules is equivalent to a point mass colliding with a molecule having an effective diameter of $2d$.

unless their centers are less than a distance d apart as they approach each other. An equivalent description of the collisions is to imagine that one of the molecules has a diameter $2d$ and the rest are geometrical points (Fig. 21.12b). In a time t, the molecule having the speed that we shall take to be the average speed, $\bar{v}$, will travel a distance $\bar{v}t$. In this same time interval, our molecule with equivalent diameter $2d$ will sweep out a cylinder having a cross-sectional area of πd^2 and a length of $\bar{v}t$ (Fig. 21.13). Hence the volume of the cylinder is $\pi d^2\bar{v}t$. If n_v is the number of particles per unit volume, then the number of particles in the cylinder is $(\pi d^2\bar{v}t)n_v$. The molecule of equivalent diameter $2d$ will collide with every particle in this cylinder in the time t. Hence, the number of collisions in the time t is equal to the number of particles in the cylinder, which we found was $(\pi d^2\bar{v}t)n_v$.

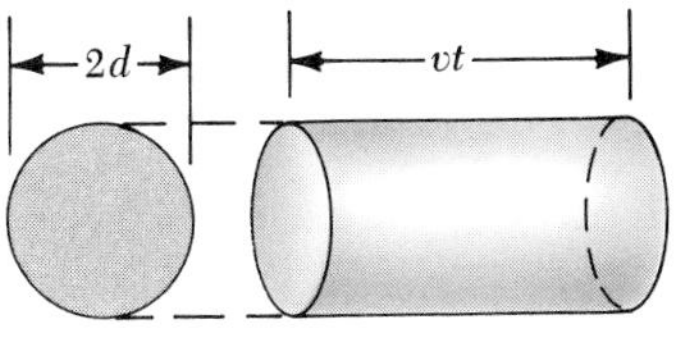

Figure 21.13 In a time t, a molecule of effective diameter $2d$ will sweep out a cylinder of length $\bar{v}t$, where $\bar{v}$ is its average speed. In this time, it will collide with every molecule within this cylinder.

The **mean free path,** ℓ, which is the mean distance between collisions, equals the average distance $\bar{v}t$ traveled in a time t divided by the number of collisions that occurs in the time:

$$\ell = \frac{\bar{v}t}{(\pi d^2\bar{v}t)n_v} = \frac{1}{\pi d^2 n_v}$$

Since the number of collisions in a time t is $(\pi d^2\bar{v}t)n_v$, the number of collisions per unit time, or **collision frequency** f, is given by

$$f = \pi d^2\bar{v}n_v$$

The inverse of the collision frequency is the average time between collisions, called the **mean free time.**

Our analysis has assumed that particles in the cylinder are stationary. When the motion of the particles is included in the calculation, the correct results are

$$\ell = \frac{1}{\sqrt{2}\,\pi d^2 n_v} \qquad (21.28)$$

Mean free path

$$f = \sqrt{2}\,\pi d^2\bar{v}n_v = \frac{\bar{v}}{\ell} \qquad (21.29)$$

Collision frequency

EXAMPLE 21.6. A Collection of Nitrogen Molecules
Calculate the mean free path and collision frequency for nitrogen molecules at a temperature of 20°C and a pressure of 1 atm. Assume a molecular diameter of 2×10^{-10} m.

Solution: Assuming the gas is ideal, we can use the equation $PV = NkT$ to obtain the number of molecules per unit volume under these conditions:

$$n_v = \frac{N}{V} = \frac{P}{kT} = \frac{1.01 \times 10^5\ \text{N/m}^2}{(1.38 \times 10^{-23}\ \text{J/K})(293\ \text{K})}$$
$$= 2.50 \times 10^{25}\ \frac{\text{molecules}}{\text{m}^3}$$

Hence, the mean free path is

$$\ell = \frac{1}{\sqrt{2}\,\pi d^2 n_v}$$
$$= \frac{1}{\sqrt{2}\,\pi(2 \times 10^{-10}\ \text{m})^2\left(2.50 \times 10^{25}\ \frac{\text{molecules}}{\text{m}^3}\right)}$$
$$= 2.25 \times 10^{-7}\ \text{m}$$

This is about 10^3 times greater than the molecular diameter. Since the average speed of a nitrogen molecule at 20°C is about 511 m/s (Table 21.2), the collision frequency is

$$f = \frac{\bar{v}}{\ell} = \frac{511\ \text{m/s}}{2.25 \times 10^{-7}\ \text{m}} = 2.27 \times 10^9/\text{s}$$

The molecule collides with other molecules at the average rate of about two billion times each second!

You should note that the mean free path, ℓ, is *not* the same as the average separation between particles. In fact, the average separation, d, between particles is given approximately by $n_v^{-1/3}$. In this example, the average molecular separation is

$$d = \frac{1}{n_v^{1/3}} = \frac{1}{(2.5 \times 10^{25})^{1/3}} = 3.4 \times 10^{-9} \text{ m}$$

*21.9 VAN DER WAALS' EQUATION OF STATE

Thus far we have assumed all gases to be ideal, that is, to obey the equation of state, $PV = nRT$. To a very good approximation, real gases behave as ideal gases at ordinary temperatures and pressures. In the kinetic theory derivation of the ideal-gas law, we neglected the volume occupied by the molecules and assumed that intermolecular forces were negligible. Now let us investigate the qualitative behavior of real gases and the conditions under which deviations from ideal-gas behavior are expected.

Consider a gas contained in a cylinder fitted with a movable piston. As noted in Chapter 20, if the temperature is kept constant while the pressure is measured at various volumes, a plot of P versus V yields a hyperbolic curve (an *isotherm*) as predicted by the ideal-gas law (Fig. 21.14).

Now let us describe what happens to a real gas. Figure 21.15 gives some typical experimental curves taken on a gas at various temperatures. At the higher temperatures, the curves are approximately hyperbolic and the gas behavior is close to ideal. However, as the temperature is lowered, the deviations from the hyperbolic shape are very pronounced.

There are two major reasons for this behavior. First, we must account for the volume occupied by the gas molecules. If V is the volume of the container and b is the volume occupied by the molecules, then $V - b$ is the empty volume available to the gas. The constant b is equal to the number of molecules of gas multiplied by the volume per molecule. As V decreases for a given quantity of gas, the fraction of the volume occupied by the molecules increases.

The second important effect concerns the intermolecular forces when the molecules are close together. At close separations, the molecules attract each other, as we might expect, since gases condense to form liquids. This attractive force reduces the pressure that the molecules exert on the container walls. In other words, a molecule that is on the verge of colliding with the walls is under the influence of attractive forces directed toward the body of the gas. Consequently, the average energy of the molecules colliding with the walls is reduced and the resulting pressure is decreased from that of an ideal gas. The net inward force on a molecule near the wall is proportional to the density of molecules, or inversely proportional to the volume. In addition, the pressure at the wall is proportional to the density of molecules. The net pressure is reduced by a factor proportional to the square of the density, which varies as $1/V^2$. Hence, the pressure P is replaced by an effective pressure $P + a/V^2$, where a is a constant.

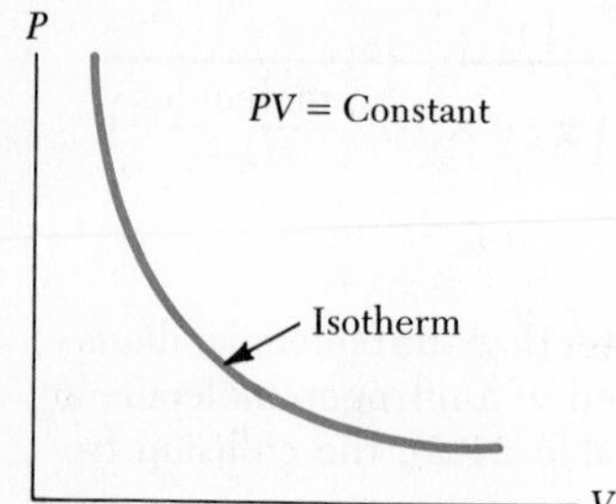

Figure 21.14 The PV diagram of an isothermal process for an ideal gas. In this case, the pressure and volume are related by $PV =$ constant.

The two effects just described can now be incorporated into a modified equation of state proposed by J. D. van der Waals (1837–1923) in 1873. For one mole of gas, **van der Waals' equation of state** is given by

$$\left(P + \frac{a}{V^2}\right)(V - b) = RT \qquad (21.30)$$

The constants a and b are empirical and are chosen to provide the best fit to the experimental data for a particular gas.

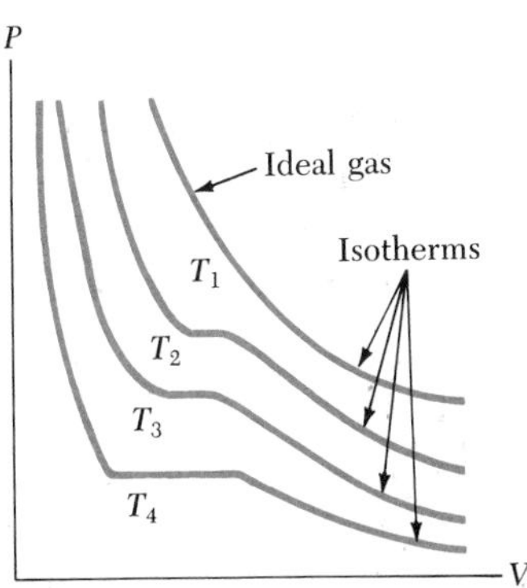

Figure 21.15 Isotherms for a real gas at various temperatures. At higher temperatures, such as T_2, the behavior is nearly ideal. The behavior is not ideal at the lower temperatures.

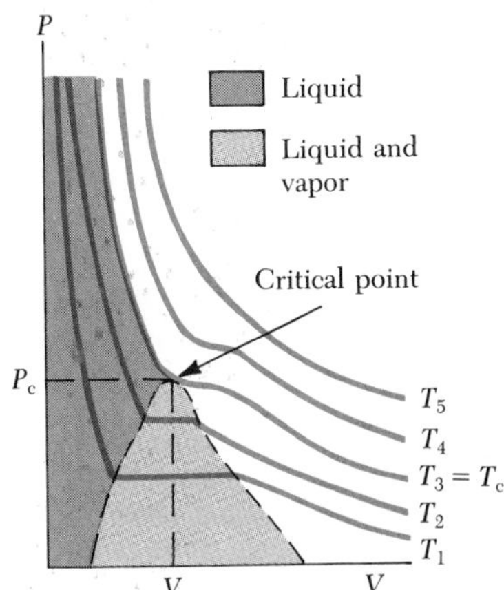

Figure 21.16 Isotherms for CO_2 at various temperatures. Below the critical temperature, T_c, the substance could be in the liquid state, the liquid-vapor equilibrium state, or the gaseous state, depending on the pressure and volume. (Adapted from K. Mendelssohn, *The Quest for Absolute Zero*, New York, McGraw-Hill, World University Library, 1966.)

The experimental curves in Figure 21.16 for CO_2 are described quite accurately by van der Waals' equation at the higher temperatures (T_3, T_4, and T_5) and outside the shaded regions. Within the shaded region there are major discrepancies. If the van der Waals equation of state is used to predict the PV relationship at a temperature such as T_1, then a nonlinear curve is obtained that is unlike the observed flat portion of the curve in the figure.

The departure from the predictions of van der Waals equation at the lower temperatures and higher densities is due to the onset of liquefaction. That is, the gas begins to liquefy at the pressure P_c, called the **critical pressure.** In the region within the dotted line below P_c the gas is partially liquefied and the gas vapor and liquid coexist. In the flat portions of the low-temperature isotherms, as the volume is decreased more gas liquefies and the pressure remains constant. At even lower volumes, the gas is completely liquefied. Any further decrease in volume leads to large increases in pressure because liquids are not easily compressed.

It is now realized that a real gas cannot be rigorously described by any simple equation of state, such as Equation 21.30, because of the complex nature of the intermolecular forces. Nevertheless, the basic concepts involved in Equation 21.30 are correct. At very low temperatures, the low-energy molecules attract each other and the gas tends to liquefy, or condense. A further pressure increase will accelerate the rate of liquefaction. At the higher temperatures, the average kinetic energy is large enough to overcome the attractive intermolecular forces; hence the molecules do not bind together at the higher temperatures and the gas phase is maintained.

21.10 SUMMARY

The **pressure** of N molecules of an ideal gas contained in a volume V is given by

Pressure and molecular kinetic energy

$$P = \frac{2}{3}\frac{N}{V}(\tfrac{1}{2}m\overline{v^2}) \qquad (21.5)$$

where $\frac{1}{2}m\overline{v^2}$ is the average kinetic energy per molecule.

Temperature is proportional to kinetic energy

The **temperature** of an ideal gas is related to the average kinetic energy per molecule through the expression

$$T = \frac{2}{3k}\left(\tfrac{1}{2}m\overline{v^2}\right) \qquad (21.6)$$

where k is Boltzmann's constant.

Average kinetic energy per molecule

The **average translational kinetic energy per molecule** of a gas is given by

$$\tfrac{1}{2}m\overline{v^2} = \tfrac{3}{2}kT \qquad (21.7)$$

Each translational degree of freedom (x, y, or z) has $\frac{1}{2}kT$ of energy associated with it.

The **equipartition of energy theorem** states that the energy of a system in thermal equilibrium is equally divided among all degrees of freedom.

The **total energy** of N molecules (or n moles) of an ideal monatomic gas is given by

Internal energy of an ideal monatomic gas

$$U = \tfrac{3}{2}NkT = \tfrac{3}{2}nRT \qquad (21.11)$$

The **molar heat capacity** of an ideal monatomic gas at constant volume is $C_v = \frac{3}{2}R$; the molar heat capacity at constant pressure is $C_p = \frac{5}{2}R$. The ratio of heat capacities is $\gamma = C_p/C_v = 5/3$.

An **adiabatic process** is one in which there is no heat transfer between the system and its surroundings.

If an ideal gas undergoes an **adiabatic expansion or compression,** the first law of thermodynamics together with the equation of state, $PV = nRT$, shows that

Adiabatic process

$$PV^{\gamma} = \text{constant} \qquad (21.18)$$

The speed of sound in a gas of density ρ and at a pressure P is

Speed of sound in gas

$$v = \sqrt{\frac{\gamma P}{\rho}} \qquad (21.21)$$

The *most probable speed distribution* of N gas molecules at a temperature T is given by **Maxwell's speed distribution function:**

Maxwell speed distribution function

$$N_v = 4\pi N\left(\frac{m}{2\pi kT}\right)^{3/2} v^2 e^{-mv^2/2kT} \qquad (21.24)$$

Using this expression, one can find the rms speed, v_{rms}, the average speed, $\bar{v}$, and the most probable speed, v_{mp}:

$$v_{\text{rms}} = \sqrt{\frac{3kT}{m}} \quad (21.25) \qquad \bar{v} = \sqrt{\frac{8kT}{\pi m}} \quad (21.26) \qquad v_{\text{mp}} = \sqrt{\frac{2kT}{m}} \quad (21.27)$$

The molecules of a gas undergo collisions with each other billions of times each second under standard conditions. If the gas has a volume density n_v and each molecule is assumed to have a diameter d, the average distance between collisions, or **mean free path**, ℓ, is found to be

$$\ell = \frac{1}{\sqrt{2}\,\pi d^2 n_v} \qquad (21.28)$$

Mean free path

Furthermore, the number of collisions per second, or **collision frequency,** *f*, is given by

$$f = \sqrt{2}\,\pi d^2 \bar{v} n_v = \frac{\bar{v}}{\ell} \qquad (21.29)$$

Collision frequency

QUESTIONS

1. Dalton's law of partial pressures states: *The total pressure of a mixture of gases is equal to the sum of the partial pressures of gases making up the mixture.* Give a convincing argument of this law based on the kinetic theory of gases.
2. One container is filled with helium gas and another with argon gas. If both containers are at the same temperature, which molecules have the higher rms speed?
3. If you wished to manufacture an after-shave lotion with a scent that is less "likely to get there before you do," would you use a high- or low-molecular-weight lotion?
4. A gas consists of a mixture of He and N_2 molecules. Do the lighter He molecules travel faster than the N_2 molecules? Explain.
5. Although the average speed of gas molecules in thermal equilibrium at some temperature is greater than zero, the average velocity is zero. Explain.
6. Why does a fan make you feel cooler on a hot day?
7. Alcohol taken internally makes you feel warmer. Yet when it is rubbed on your body, it lowers body temperature. Explain the latter effect.
8. A liquid partially fills a container. Explain why the temperature of the liquid decreases when the container is partially evacuated. (Using this technique, it is possible to freeze water at temperatures above 0°C.)
9. A vessel containing a fixed volume of gas is cooled. Does the mean free path increase, decrease, or remain constant in the cooling process? What about the collision frequency?
10. A gas is compressed at a constant temperature. What happens to the mean free path of the molecules in this process?

PROBLEMS

Section 21.1 Molecular Model for the Pressure of an Ideal Gas

1. Find the average square speed of nitrogen molecules under standard conditions. Recall that 1 mole of any gas occupies a volume of 22.4 liters at standard temperature and 1 atm pressure.
2. Two moles of oxygen gas are confined to a 5-liter vessel at a pressure of 8 atm. Find the average kinetic energy of an oxygen molecule under these conditions. (The mass of an O_2 molecule is 5.34×10^{-26} kg.)
3. In a 1-min interval, a machine gun fires 150 bullets, each of mass 8 g and speed 400 m/s. The bullets strike and become imbedded in a stationary target. If the target has an area of 5 m², find the average force and pressure exerted on the target. (*Note:* These are inelastic collisions.)
4. In a period of 1 s, 5×10^{23} nitrogen molecules strike a wall of area 8 cm². If the molecules move with a speed of 300 m/s and strike at an angle of 45° to the normal to the wall, find the pressure exerted on the wall. (The mass of an N_2 molecule is 4.68×10^{-26} kg.)
5. In a 30-s interval, 500 hailstones strike a glass window of area 0.6 m² at an angle of 45° to the window surface. Each hailstone has a mass of 5 g and a speed of 8 m/s. If the collisions are assumed to be elastic, find the average force and pressure on the window.

Section 21.2 Molecular Interpretation of Temperature

6. A cylinder contains a mixture of helium and argon gas in equilibrium at a temperature of 150°C. What is the average kinetic energy of each gas molecule?
7. Calculate the root mean square speed of a H_2 molecule at a temperature of 250°C.
8. (a) Determine the temperature at which the rms speed of a He atom equals 500 m/s. (b) What is the rms speed of He on the surface of the sun, where the temperature is 5800 K?

9. Nitrogen molecules have a rms speed of 517 m/s at 300 K. (a) What is the rms speed of nitrogen at 600 K? at 200 K? (b) Construct a graph of v_{rms} versus temperature for helium in intervals of 200 K, over the temperature range 200 K to 2000 K.
10. What is the temperature at which the rms speed of nitrogen molecules equals the rms speed of helium at 20°C?
11. A 5-liter vessel contains nitrogen gas at a temperature of 27°C and a pressure of 3 atm. Find (a) the total translational kinetic energy of the gas molecules and (b) the average kinetic energy per molecule.

Section 21.3 Heat Capacity of an Ideal Gas (use data in Table 21.2)

12. Calculate the change in internal energy of 3 moles of helium gas when its temperature is increased by 2 K.
13. Two moles of oxygen gas are heated from 300 K to 320 K. How much heat is transferred to the gas if the process occurs at (a) constant volume and (b) constant pressure?
14. The total heat capacity, C', of a monatomic gas measured at constant pressure is 14.9 cal/K. Find (a) the number of moles of gas, (b) the total heat capacity at constant volume, and (c) the internal energy of the gas at 350 K. (Recall that $C' = nC$.)
15. Consider *three* moles of an ideal gas. (a) If the gas is monatomic, find the *total* heat capacity at constant volume and at constant pressure. (b) Repeat (a) for a diatomic gas in which the molecules rotate but do not vibrate.
16. One mole of hydrogen gas is heated at constant pressure from 300 K to 420 K. Calculate (a) the heat transferred to the gas, (b) the increase in internal energy of the gas, and (c) the work done by the gas.
17. In a constant-volume process, 50 cal of heat is transferred to 1 mole of an ideal monatomic gas initially at 300 K. Find (a) the increase in internal energy of the gas, (b) the work done by the gas, and (c) the final temperature of the gas.

Section 21.4 Adiabatic Process for an Ideal Gas

18. Two moles of an ideal gas ($\gamma = 1.40$) expands quasi-statically and adiabatically from a pressure of 5 atm and a volume of 12 liters to a final volume of 30 liters. (a) What is the final pressure of the gas? (b) What are the initial and final temperatures?
19. Show that Equation 21.20 follows from Equation 21.19 for a quasi-static, adiabatic process. (*Note:* $PV = nRT$ applies during the process.)
20. An ideal gas ($\gamma = 2$) expands quasi-statically and adiabatically. If the final temperature is one third the initial temperature, (a) by what factor does its volume change? (b) by what factor does its pressure change?
21. One mole of an ideal monatomic gas ($\gamma = 1.67$) initially at 300 K and 1 atm is compressed quasi-statically and adiabatically to one fourth its initial volume. Find its final pressure and temperature.
22. During the compression stroke of a certain gasoline engine, the pressure increases from 1 atm to 20 atm. Assuming that the process is adiabatic and the gas is ideal with $\gamma = 1.40$, (a) by what factor does the volume change and (b) by what factor does the temperature change?

Section 21.5 Sound Waves in a Gas

23. Calculate the speed of sound in methane (CH_4) at 288 K, using the values $\gamma = 1.31$, the molecular weight of CH_4, $M = 16$ kg/kmole, and $R = 8.314$ J/kmole · K.
24. At what temperature will the speed of sound in methane equal the speed of sound in helium at 288 K? For helium, $\gamma = 1.66$ and $M = 4$ kg/kmole.
25. A worker is at one end of a mile-long section (1.61) km) of iron pipeline when an accidental blast occurs at the other end of the section. The worker receives two sound signals from the blast, one transmitted through the pipe and one through the surrounding air. Use values from Table 17.1 to calculate the elapsed time between the two signals. (*Note:* First find the speed of sound in air at 300 K and take the speed of sound in iron at that temperature to be 5200 m/s.)
26. Xenon has a density of 5.9 kg/m^3 at 0°C and 1 atm pressure. Since it is monatomic, $\gamma = 1.67$. (a) Calculate the speed of sound in xenon at 0°C. (b) What is the bulk modulus of xenon?
27. A sound wave propagating in air has a frequency of 4000 Hz. Calculate the percent change in wavelength when the wavefront, initially in a region where $T =$ 27°C, enters a region where the air temperature decreases to 10°C.
28. A spelunker attempts to determine the depth of a pit in the floor of a cave by dropping a stone into the pit and measuring the time interval between release and the sound of the stone's hitting bottom. If the measured time interval is 10 s, what is the depth of the pit? (Assume a temperature of 15°C.)

Section 21.6 The Equipartition of Energy

29. If a molecule has f degrees of freedom, show that a gas consisting of such molecules has the following properties: (1) its total internal energy is $fnRT/2$; (2) its molar heat capacity at constant volume is $fR/2$; (3) its molar heat capacity at constant pressure is $(f+2)R/2$; (4) the ratio $\gamma = C_p/C_v = (f+2)/f$.
30. Examine the data for polyatomic gases in Table 21.2 and explain why SO_2 has a higher C_v than the other polyatomic gases at 300 K.
31. Inspecting the magnitudes of C_v and C_p for the diatomic and polyatomic gases in Table 21.2, we find that the values increase with increasing molecular

mass. Give a qualitative explanation of this observation.

32. Consider 2 moles of an ideal diatomic gas. Find the *total* heat capacity at constant volume and at constant pressure if (a) the molecules rotate but do not vibrate and (b) the molecules rotate and vibrate.

*Section 21.7 Distribution of Molecular Speeds

33. A vessel containing oxygen gas is at a temperature of 400 K. Find (a) the rms speed, (b) the average speed, and (c) the most probable speed of the gas molecules. (The mass of O_2 is 5.31×10^{-26} kg.)
34. Fifteen identical particles have the following speeds: one has speed 2 m/s; two have speed 3 m/s; three have speed 5 m/s; four have speed 7 m/s; three have speed 9 m/s; two have speed 12 m/s. Find (a) the average speed, (b) the rms speed, and (c) the most probable speed of these particles.
35. Calculate the most probable speed, average speed, and rms speed for nitrogen gas molecules at 900 K. Compare your results with the values obtained from Figure 21.10.
36. Show that the most probable speed of a gas molecule is given by Equation 21.27. Note that the most probable speed corresponds to the point where the slope of the speed distribution curve, dN_v/dv, is zero.
37. Use Figure 21.10 to *estimate* the number of nitrogen molecules with speeds between 400 m/s and 600 m/s at (a) 300 K and (b) 900 K.
38. Show that the Maxwell speed distribution function given by Equation 21.24 satisfies Equation 21.23. (Such a function is said to be *normalized*.)
39. At what temperature would the average velocity of helium atoms equal (a) the escape velocity from earth, 1.12×10^4 m/s, and (b) the escape velocity from the moon, 2.37×10^3 m/s? (See Chapter 14 for a discussion of escape velocity, and note that the mass of helium is 6.66×10^{-27} kg.)
40. Using the data in Figure 21.10, estimate the *fraction* of N_2 molecules that have speeds in the range 1000 m/s to 1200 m/s at 900 K. Note that the total number of molecules is 10^5.

*Section 21.8 Mean Free Path

41. In an ultrahigh vacuum system, the pressure is measured to be 10^{-10} torr (where 1 torr = 133 N/m²). If the gas molecules have a molecular diameter of 3 Å = 3×10^{-10} m and the temperature is 300 K, find (a) the number of molecules in a volume of 1 m³, (b) the mean free path of the molecules, and (c) the collision frequency, assuming an average speed of 500 m/s.
42. Show that the mean free path for the molecules of an ideal gas is given by

$$\ell = \frac{kT}{\sqrt{2}\pi d^2 P}$$

where d is the molecular diameter.

43. A cylinder contains 5 moles of oxygen gas at a pressure of 80 atm and temperature of 300 K. Assuming a molecular diameter of 2.5×10^{-10} m, find (a) the number of molecules per unit volume, (b) the mean free path, and (c) the collision frequency.

*Section 21.9 Van der Waals' Equation of State

44. The constant b that appears in van der Waals' equation of state for oxygen is measured to be equal to 31.8 cm³/mole. Assuming a spherical shape, estimate the diameter of the molecule.
45. Use Equation 21.22 to compute the speed of sound in a mixture of 60% oxygen and 40% nitrogen at 40°C.
46. Consider again the situation described in Problem 25. Show that, in general, the elapsed time interval between arrival of the sound signal through the pipe and through the surrounding air is

$$\Delta t = \frac{\ell(v_m - v_a)}{v_m v_a}$$

where v_m is the speed of sound in the metal, v_a is the speed of sound in air, and ℓ is the length of the pipe.

GENERAL PROBLEMS

47. A mixture of two gases will diffuse through a filter at rates proportional to their rms speeds. If the molecules of the two gases have masses m_1 and m_2, show that the ratio of their rms speeds (or the ratio of diffusion rates) is given by

$$\frac{(v_1)_{rms}}{(v_2)_{rms}} = \sqrt{\frac{m_2}{m_1}}$$

This process is used to obtain uranium enriched with the isotope ^{235}U, which is used in nuclear reactors.

48. A cylinder containing n moles of an ideal gas undergoes a quasi-static, adiabatic process. (a) Starting with the expression $W = \int P\,dV$ and using PV^γ = constant, show that the work done is given by

$$W = \left(\frac{1}{\gamma - 1}\right)(P_i V_i - P_f V_f)$$

(b) Starting with the first law in differential form, prove that the work done is also equal to $nC_v(T_i - T_f)$. Show that this result is consistent with the equation in (a).

49. Twenty particles, each of mass m and confined to a volume V, have the following speeds: two have speed v; three have speed $2v$; five have speed $3v$; four have speed $4v$; three have speed $5v$; two have speed $6v$; one has speed $7v$. Find (a) the average speed, (b) the rms speed, (c) the most probable speed, (d) the pressure they exert on the walls of the vessel, and (e) the average kinetic energy per particle.

50. Verify Equations 21.25 and 21.26 for the rms and average speed of the molecules of a gas at a temperature T. Note that the average value of v^n is given by

$$\overline{v^n} = \frac{1}{N}\int_0^\infty v^n N_v \, dv$$

and make use of the integrals

$$\int_0^\infty x^3 e^{-ax^2}\, dx = \frac{1}{2a^2}$$

and

$$\int_0^\infty x^4 e^{-ax^2}\, dx = \frac{3}{8a^2}\sqrt{\frac{\pi}{a}}$$

51. The internal energy of a gas consisting of n moles of CO_2 at 300 K is given by $U = anRT + b$, where a and b are constants. (a) From this expression, derive the molar heat capacity at constant volume, C_V. (b) What is C_p for this gas? (c) Use Table 21.2 to obtain a value for the constant a. (d) How many degrees of freedom does the molecule have at this temperature?

52. A vessel contains 10^4 oxygen molecules at 500 K. (a) Make an accurate graph of the Maxwell speed distribution function, N_v, versus speed with points at speed intervals of 100 m/s. (b) Determine the most probable speed from this graph. (c) Calculate the average and rms speeds for the molecules and label these points on your graph. (d) From the graph, estimate the fraction of molecules with speeds in the range 300 m/s to 600 m/s.

53. A vessel contains 1 mole of helium gas at a temperature of 300 K. Calculate the approximate number of molecules having speeds in the range from 400 m/s to 410 m/s. (*Hint:* This number is approximately equal to $N_v\,\Delta v$, where Δv is the range of speeds.)

54. The compressibility, κ, of a substance is defined as the fractional change in volume of that substance for a given change in pressure:

$$\kappa = -\frac{1}{V}\frac{dV}{dP}$$

(a) Explain why the negative sign in this expression ensures that κ will always be positive. (b) Show that if an ideal gas is compressed *isothermally*, its compressibility is given by $\kappa_1 = 1/P$. (c) Show that if an ideal gas is compressed *adiabatically*, its compressibility is given by $\kappa_2 = 1/\gamma P$. (d) Determine values for κ_1 and κ_2 for a monatomic ideal gas at a pressure of 2 atm.

55. One mole of a gas obeying van der Waals' equation of state is compressed isothermally. At some critical temperature, T_c, the isotherm has a point of zero slope and zero inflection, as in Figure 21.16. That is, at $T = T_c$,

$$\frac{\partial P}{\partial V} = 0 \quad \text{and} \quad \frac{\partial^2 P}{\partial V^2} = 0.$$

Using Equation 21.30 and these conditions, show that at the critical point, the pressure, volume, and temperature are given by $P_c = a/27b^2$; $V_c = 3b$, and $T_c = 8a/27Rb$.

56. Solve for the molar heat capacities C_v and C_p of an ideal gas in terms of the gas constant R and the adiabatic exponent γ.

57. In Equation 21.22 the temperature T must be in degrees kelvin. (a) Starting with this equation, show that the speed of sound in a gas can be expressed in the form $v = [v_0 + (v_0/546)]t$, where t is the temperature in °C and v_0 is the speed of sound in the gas at 0°C. (*Hint:* Assume that $t \ll 273$°C and use the expansion $(1+x)^{1/2} = 1 + \frac{1}{2}x - \frac{1}{8}x^2 + \ldots$.) (b) In the case of air, show that this result leads to

$$v = (331 + 0.61t)\ \text{m/s}$$

•58. An ideal monatomic gas undergoes an adiabatic expansion for which the final pressure P_f is related to the initial pressure P_i by $P_f = P_i/10$. Find (a) the ratio of the final volume V_f to the initial volume V_i, (b) the ratio of the final temperature T_f to the initial temperature T_i, (c) the ratio of ℓ_f (the mean free path in the final state) to ℓ_i (the mean free path in the initial state), and (d) the ratio of the collision frequency in the final state to the collision frequency in the initial state.

•59. An ideal gas whose constant-volume molar heat capacity is $C_v = \frac{5}{2}R$ undergoes an adiabatic expansion for which $P_f = P_i/10$. Find the ratios requested in parts (a) through (d) of Problem 58.

•60. Consider an ideal gas of triatomic molecules. (a) If the molecule is a linear one (such as CO_2) and the gas temperature is low enough that there is negligible vibrational motion, what will be the value of C_v, the molar heat capacity at constant volume? (Give your result as a multiple of R.) (b) At temperatures high enough that vibrations along the length of the molecule are "fully engaged" what will the value of C_v be? (*Note:* A linear triatomic molecule has two distinct patterns or "modes" of vibrational motion along the axis of the molecule. At sufficiently high temperatures *each* of these two modes contributes R ($\frac{1}{2}R$ potential and $\frac{1}{2}R$ kinetic) to the molar heat capacity.) (c) If the molecule is nonlinear (such as H_2O) and the gas temperature is low enough that there is negligible vibrational motion, what will the value of C_v be? (d) Based on your results for parts (a) and (c), how could specific-heat data be used to determine whether a triatomic molecule (of known molecular weight) is linear or nonlinear?

•61. An ideal gas mixture consists of n_1 moles of a pure gas whose molar heat capacity at constant volume is $R/(\gamma_1 - 1)$ and n_2 moles of a pure gas whose molar heat capacity at constant volume is $R/(\gamma_2 - 1)$. Find (a) the total heat capacity (at constant volume) of the sample, and (b) the total heat capacity at constant pressure.

(*Hint:* The pressure P of the sample obeys the equation of state $PV = (n_1 + n_2) RT$.)

CALCULATOR/COMPUTER PROBLEMS

62. For a Maxwellian gas, find the numerical value of the ratio $\{N_v(v)/N_v(v_{mp})\}$ for the following values of v: $v = (v_{mp}/50)$, $(v_{mp}/10)$, $(v_{mp}/2)$, $2v_{mp}$, $10v_{mp}$, $50v_{mp}$. Give your results to three significant figures.
63. Consider a system of 10^4 oxygen molecules at a temperature T. Write a program that will enable you to calculate the Maxwell distribution function N_v as a function of the speed of the molecules and the temperature. Use your program to evaluate N_v for speeds ranging from $v = 0$ to $v = 2000$ m/s (in intervals of 100 m/s) at temperatures of (a) 300 K and (b) 1000 K. (c) Make graphs of your results (N_v versus v) and use the graph at $T = 1000$ K to calculate the number of molecules having speeds between 800 m/s and 1000 m/s at $T = 1000$ K.

22
Heat Engines, Entropy, and the Second Law of Thermodynamics

The first law of thermodynamics is merely the law of conservation of energy generalized to include heat as a form of energy. This law tells us only that an increase in one form of energy must be accompanied by a decrease in some other form of energy. The first law places no restrictions on the types of energy conversions that can occur. Furthermore, it makes no distinction between heat and work. According to the first law, the internal energy of a body may be increased by either adding heat to it or doing work on it. But there is an important difference between heat and work that is not evident from the first law. For example, it is possible to convert work completely into heat but, in practice, it is impossible to convert heat completely into work without changing the surroundings.

The *second law of thermodynamics* establishes which processes in nature do or do not occur. Of all processes permitted by the first law, only certain types of energy conversions can take place. The following are some examples of processes that are consistent with the first law of thermodynamics but proceed in an order governed by the second law of thermodynamics. (1) When two objects at different temperatures are placed in thermal contact with each other, heat flows from the warmer to the cooler object, but never from the cooler to the warmer. (2) Salt dissolves spontaneously in water, but extracting salt from salt water requires some external influence. (3) When a rubber ball is dropped to the ground, it bounces several times and eventually comes to rest. The opposite process does not occur. (4) The oscillations of a pendulum will slowly decrease in amplitude because of collisions with air molecules and friction at the point of suspension. Eventually the pendulum will come to rest. Thus, the initial mechanical energy of the pendulum is converted into thermal energy. The reverse transformation of energy does not occur.

Lord Kelvin (1824–1907)

These are all examples of *irreversible* processes, that is, processes that occur naturally in only one direction. None of these processes occur in the opposite temporal order; if they did, they would violate the second law of thermodynamics.[1] That is, the one-way nature of thermodynamic processes in fact *establishes* a direction of time.[2] You may have witnessed the humor of an action film running in reverse, which demonstrates the improbable order of events in a time-reversed world.

[1] To be more precise, we should say that the set of events in the time-reversed sense is highly improbable. From this viewpoint, events occur with a vastly higher probability in one direction than in the opposite direction.

[2] See, for example, D. Layzer, "The Arrow of Time," *Scientific American*, December 1975.

The second law of thermodynamics, which can be stated in many equivalent ways, has some very practical applications. From an engineering viewpoint, perhaps the most important application is the limited efficiency of heat engines. Simply stated, the second law says that a machine capable of continuously converting thermal energy completely into other forms of energy cannot be constructed.

22.1 HEAT ENGINES AND THE SECOND LAW OF THERMODYNAMICS

The field of thermodynamics developed from a study of heat engines, an application of great importance today. A **heat engine** is a device that converts thermal energy to other useful forms of energy, such as mechanical and electrical energy. More specifically, a heat engine is a device that carries a substance through a cycle during which (1) heat is absorbed from a source at a high temperature, (2) work is done by the engine, and (3) heat is expelled by the engine to a source at a lower temperature. In a typical process for producing electricity in a power plant, coal or some other fuel is burned and the heat produced is used to convert water to steam. This steam is then directed at the blades of a turbine, setting it into rotation. Finally, the mechanical energy associated with this rotation is used to drive an electric generator. The internal combustion engine in your automobile extracts heat from a burning fuel and converts a fraction of this energy to mechanical energy.

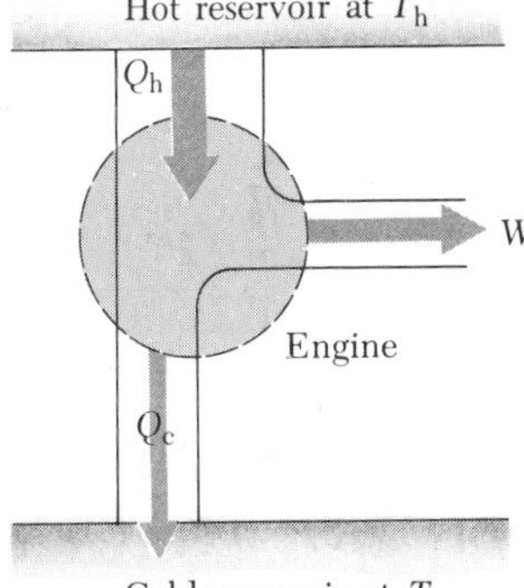

Figure 22.1 Schematic representation of a heat engine. The engine (in the circular area) receives heat Q_h from the hot reservoir, expels heat Q_c to the cold reservoir, and does work W.

As was mentioned above, a heat engine carries some working substance through a cyclic process, defined as one in which the substance eventually returns to its initial state. As an example of a cyclic process, consider the operation of a steam engine in which the working substance is water. The water is carried through a cycle in which it first evaporates into steam in a boiler and then expands against a piston. After the steam is condensed with cooling water, it is returned to the boiler and the process is repeated.

In the operation of any heat engine, a quantity of heat is extracted from a high-temperature source, some mechanical work is done, and some heat is expelled to a low-temperature reservoir. It is useful to represent a heat engine schematically as in Fig. 22.1. The engine (represented by the circle at the center of the diagram) absorbs a quantity of heat Q_h from the high-temperature reservoir. It does work W and gives up heat Q_c to a lower-temperature heat reservoir. Because the working substance goes through a cycle, its initial and final internal energies are equal, so $\Delta U = 0$. Hence, from the first law of thermodynamics we see that the

> net work W done by the engine equals the net heat flowing into the engine.

As we can see from Figure 19.1, $Q_{net} = Q_h - Q_c$; therefore

$$W = Q_h - Q_c \tag{22.1}$$

where Q_h and Q_c are taken to be positive quantities. If the working substance is a gas, *the net work done for a cyclic process is the area enclosed by the curve representing the process on a PV diagram.* This is shown for an arbitrary cyclic process in Figure 22.2.

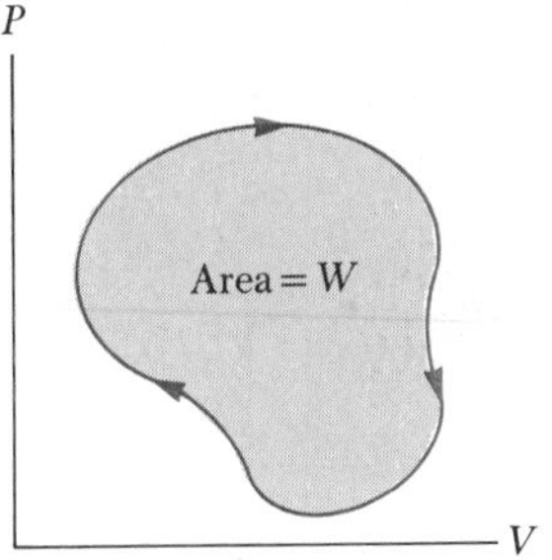

Figure 22.2 The PV diagram for an arbitrary cyclic process. The net work done equals the area enclosed by the curve.

> The **thermal efficiency**, e, of a heat engine is defined as the ratio of the net work done to the heat absorbed during one cycle:

Thermal efficiency

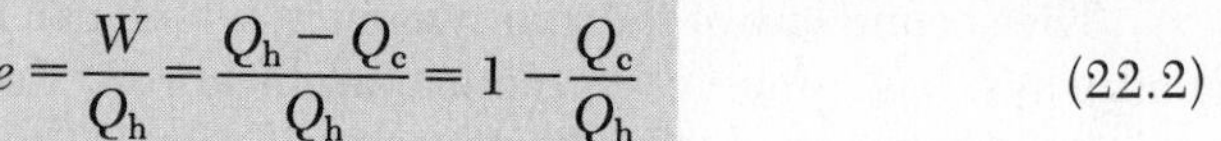

$$e = \frac{W}{Q_h} = \frac{Q_h - Q_c}{Q_h} = 1 - \frac{Q_c}{Q_h} \tag{22.2}$$

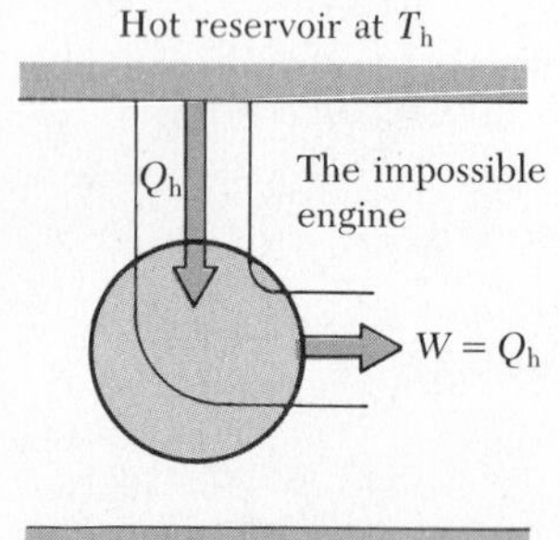

Figure 22.3 Schematic diagram of a heat engine that receives heat Q_h from a hot reservoir and does an equivalent amount of work. This perfect engine is impossible to construct.

We can think of the efficiency as the ratio of "what you get" (mechanical work) to "what you pay for" (energy). This result shows that a heat engine has 100% efficiency ($e = 1$) only if $Q_c = 0$, that is, if no heat is expelled to the cold reservoir. In other words, a heat engine with perfect efficiency would have to convert all of the absorbed heat energy Q_h into mechanical work. The second law of thermodynamics says that this is impossible.

In practice, it is found that all heat engines convert only a fraction of the absorbed heat into mechanical work. For example, a good automobile engine has an efficiency of about 20%, and diesel engines have efficiencies ranging from 35% to 40%. On the basis of this fact, the **Kelvin-Planck** form of the **second law of thermodynamics** states the following:

> It is impossible to construct a heat engine that, operating in a cycle, produces no other effect than the absorption of thermal energy from a reservoir and the performance of an equal amount of work.

This is equivalent to stating that *it is impossible to construct a perpetual-motion machine of the second kind,* that is, a machine that would violate the second law.[3] Figure 22.3 is a schematic diagram of the impossible "perfect" heat engine.

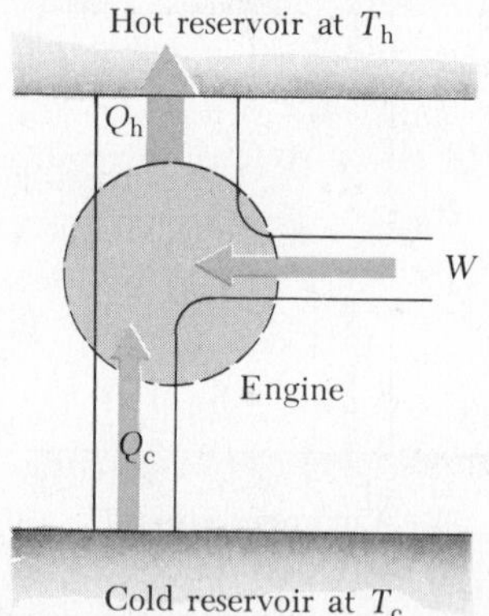

Figure 22.4 Schematic diagram for a refrigerator, which absorbs heat Q_c from the cold reservoir and expels heat Q_h to the hot reservoir. Work W is done *on* the refrigerator.

A refrigerator (or heat pump) is a heat engine running in reverse. This is shown schematically in Figure 22.4, in which the engine absorbs heat Q_c from the cold reservoir and expels heat Q_h to the hot reservoir. This can be accomplished only if work is done *on* the refrigerator. From the first law, we see that the heat given up to the hot reservoir must equal the sum of the work done and the heat absorbed from the cold reservoir. Therefore, we see that the refrigerator transfers heat from a colder body (the contents of the refrigerator) to a hotter body (the room). In practice, it is desirable to carry out this process with a minimum of work. If it could be accomplished without doing any work, we would have a "perfect" refrigerator (Fig. 22.5). Again, this is in violation of the second law of thermodynamics, which in the form of the **Clausius statement**[4] says the following:

> It is impossible to construct a cyclical machine that produces no other effect than to transfer heat continuously from one body to another body at a higher temperature.

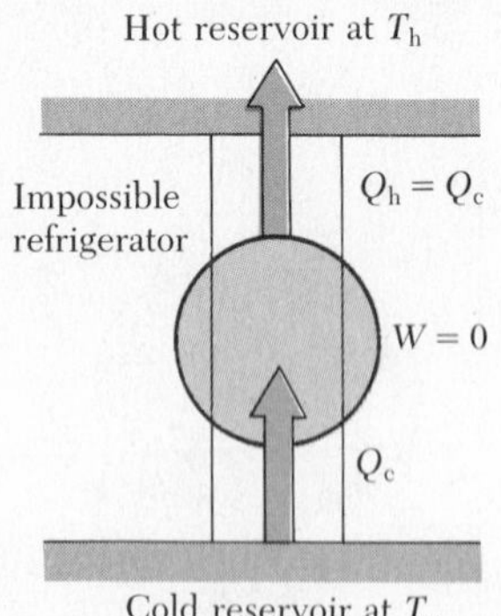

Figure 22.5 Schematic diagram of the impossible refrigerator, that is, one that absorbs heat Q_c from a cold reservoir and expels an equivalent amount of heat to the hot reservoir with $W = 0$.

In effect, this statement of the second law governs the direction of heat flow between two bodies at different temperatures. Heat will flow from the colder to the hotter body only if work is done on the system. For example, homes are cooled in summer by pumping heat out; the work done on the air conditioner is supplied by the power company.

The Clausius and Kelvin-Planck statements of the second law appear, at first sight, to be unrelated. They are, in fact, equivalent in all respects. Although we do not prove it here, one can show that if either statement is false, so is the other.[5]

[3] A perpetual-motion machine of the first kind is one that would violate the first law of thermodynamics (energy conservation). This type of machine is also impossible to construct.

[4] First expressed by Rudolf Clausius (1822–1888).

[5] See, for example, F.W. Sears, *Thermodynamics, The Kinetic Theory of Gases, and Statistical Mechanics,* Reading, Mass., Addison-Wesley, 1953, Chapter 7.

22.2 REVERSIBLE AND IRREVERSIBLE PROCESSES

In our introductory remarks we mentioned that real processes have a preferred direction. Heat flows spontaneously from a hot to a cold body when the two are placed in contact, but the reverse is accomplished only with some external influence. When a block slides on a rough surface, it eventually comes to rest. The mechanical energy of the block is converted into internal energy of the block and table. Such unidirectional processes are called **irreversible** processes. After any irreversible process occurs, it is impossible to return the system to its original state without affecting its surroundings.

Irreversible process

In general, a process is **irreversible** if the system and its surroundings cannot be returned to their initial states. Processes that involve the conversion of mechanical energy to internal energy, such as the block sliding on a rough surface, are irreversible. Once the block has come to rest, the internal energy of the block and table cannot be completely converted back into mechanical energy. The process can only be reversed by doing external work, that is, by changing the surroundings.

Reversible process

A process is **reversible** if the system passes from the initial state to the final state through a succession of equilibrium states. If a real process occurs quasi-statically, that is, slowly enough so that each state departs only infinitesimally from equilibrium, it can be considered reversible. For example, we can imagine compressing a gas quasi-statically by dropping some grains of sand onto a frictionless piston (Fig. 22.6). The pressure, volume, and temperature of the gas are well defined during the isothermal compression. The process is made isothermal by placing the gas in thermal contact with a heat reservoir. Some heat is transferred from the gas to the reservoir during the process. Each time a grain of sand is added to the piston, the volume decreases slightly while the pressure increases slightly. Each added grain of sand represents a change to a new equilibrium state. The process can be reversed by slowly removing grains of sand from the piston.

Since a reversible process is defined by a succession of equilibrium states, it can be represented by a line on a *PV* diagram, which establishes the path for the process (Fig. 22.7). Each point on this line represents one of the intermediate equilibrium states. On the other hand, an irreversible process is one that passes from the initial state to the final state through a series of nonequilibrium states. In this case, only the initial and final equilibrium states can be represented on the *PV* diagram. The intermediate, nonequilibrium states may have

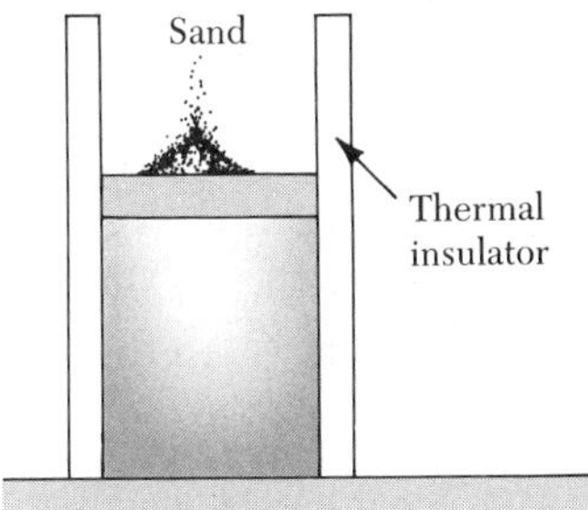

Figure 22.6 A gas in thermal contact with a heat reservoir is compressed slowly by dropping grains of sand onto the piston. The compression is isothermal and reversible.

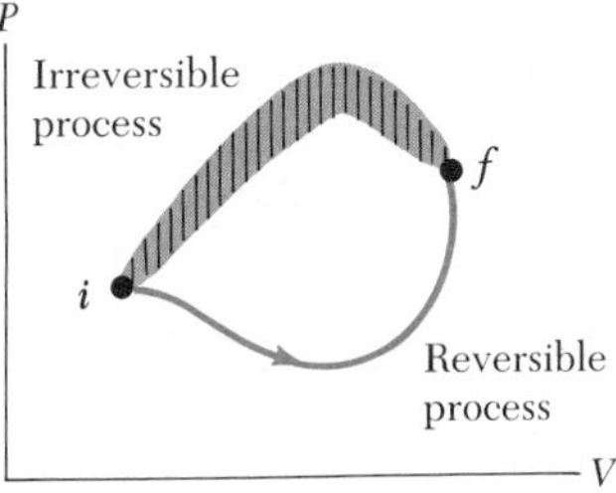

Figure 22.7 A reversible process between the two equilibrium states i and f can be represented by a line on the *PV* diagram. Each point on this line represents an equilibrium state. An irreversible process passes through a series of nonequilibrium states and cannot be represented by a line on this diagram.

well-defined volumes, but these states are not characterized by a unique pressure for the entire system. Instead, there are variations in pressure (and temperature) throughout the volume range, and these variations will not persist if left to themselves (i.e., nonequilibrium conditions). For this reason, an irreversible process cannot be represented by a line on a *PV* diagram.

We have stated that a reversible process must take place quasi-statically. In addition, in a reversible process there can be no dissipative effects that produce heat. Other effects that tend to disrupt equilibrium, such as heat conduction resulting from a temperature difference, must not be present. In reality, such effects are impossible to eliminate completely, and so it is not surprising that processes in nature are irreversible. Nevertheless, it is possible to approximate reversible processes through carefully controlled procedures. As we shall see in the next section, the concept of reversible processes is especially important in establishing the theoretical limit on the efficiency of a heat engine.

22.3 THE CARNOT ENGINE

In 1824 a French engineer named Sadi Carnot (1796–1832) described a working cycle, now called a **Carnot cycle,** that is of great importance from both a practical and a theoretical viewpoint. Using the second law of thermodynamics, he showed that a heat engine operating in this ideal, reversible cycle between two heat reservoirs would be the most efficient engine possible. Such an ideal engine, called a **Carnot engine,** establishes an upper limit on the efficiencies of all engines. That is, the net work done by a working substance taken through the Carnot cycle is the largest possible for a given amount of heat supplied to the working substance.

To describe the Carnot cycle, we shall assume that the substance working between temperatures T_c and T_h is an ideal gas contained in a cylinder with a movable piston at one end. The cylinder walls and the piston are thermally nonconducting. Four stages of the Carnot cycle are shown in Figure 22.8, and the *PV* diagram for the cycle is shown in Figure 22.9. The Carnot cycle consists of two adiabatic and two isothermal processes, all reversible.

Sadi Carnot (1796–1832). "The steam engine works our mines, impels our ships, excavates our ports and our rivers, forges iron. . . . Notwithstanding the work of all kinds done by steam engines, notwithstanding the satisfactory condition to which they have bought today, their theory is very little understood."

1. The process $A \rightarrow B$ is an isothermal expansion at temperature T_h, in which the gas is placed in thermal contact with a heat reservoir at temperature T_h (Fig. 22.8a). During the process, the gas absorbs heat Q_h from the reservoir through the base of the cylinder and does work W_{AB} in raising the piston.
2. In the process $B \rightarrow C$, the base of the cylinder is replaced by a thermally nonconducting wall and the gas expands adiabatically, that is, no heat enters or leaves the system (Fig. 22.8b). During the process, the temperature falls from T_h to T_c and the gas does work W_{BC} in raising the piston.
3. In the process $C \rightarrow D$, the gas is placed in thermal contact with a heat reservoir at temperature T_c (Fig. 22.8c) and is compressed isothermally at temperature T_c. During this time, the gas expels heat Q_c to the reservoir and the work done on the gas by an external agent is W_{CD}.
4. In the final stage, $D \rightarrow A$, the base of the cylinder is replaced by a nonconducting wall (Fig. 22.8d) and the gas is compressed adiabatically. The temperature of the gas increases to T_h and the work done on the gas by an external agent is W_{DA}.

The net work done in this reversible, cyclic process is equal to the area enclosed by the path *ABCDA* of the *PV* diagram (Fig. 22.9). As we showed in Section 22.1, the net work done in one cycle equals the net heat transferred

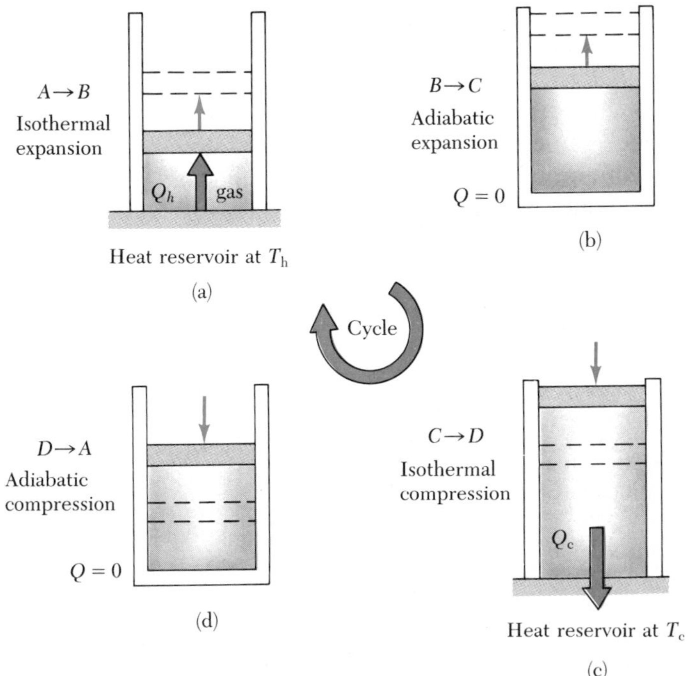

Figure 22.8 The Carnot cycle. In process $A \rightarrow B$, the gas expands isothermally while in contact with a reservoir at T_h. In process $B \rightarrow C$, the gas expands adiabatically ($Q = 0$). In process $C \rightarrow D$, the gas is compressed isothermally while in contact with a reservoir at $T_c < T_h$. In process $D \rightarrow A$, the gas is compressed adiabatically. The upward arrows on the piston indicate sand being removed during the expansions, and the downward arrows indicate the addition of sand during the compressions.

into the system, $Q_h - Q_c$, since the change in internal energy is zero. Hence, the thermal efficiency of the engine is given by Equation 22.2:

Efficiency of a heat engine

$$e = \frac{W}{Q_h} = 1 - \frac{Q_c}{Q_h}$$

Equation 22.2 gives the efficiency of *any* ideal heat engine operating between two reservoirs. In Example 22.1, we show that for a Carnot cycle, the ratio of heats Q_c/Q_h is given by

Ratio of heats for a Carnot cycle

$$\frac{Q_c}{Q_h} = \frac{T_c}{T_h} \qquad (22.3)$$

Hence, the thermal efficiency of a Carnot engine is given by

$$e_c = 1 - \frac{T_c}{T_h} \qquad (22.4)$$

Hence, we see that

all Carnot engines operating between the same two temperatures in a reversible manner have the same efficiency. It can also be shown that the efficiency of any reversible engine operating in a cycle between two temperatures is greater than the efficiency of any irreversible (real) engine operating between the same two temperatures.[6]

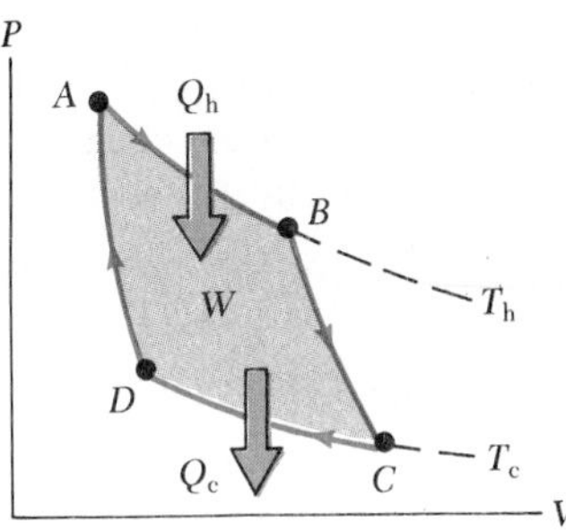

Figure 22.9 The *PV* diagram for the Carnot cycle. The net work done, *W*, equals the net heat received in one cycle, $Q_h - Q_c$. Note that $\Delta U = 0$ for the cycle.

[6] See, for example, F.W. Sears, *Thermodynamics, The Kinetic Theory of Gases, and Statistical Mechanics,* Reading, Mass., Addison-Wesley, 1953, Chapter 7.

Equation 22.4 can be applied to any working substance operating in a Carnot cycle between two heat reservoirs. According to this result, the efficiency is zero if $T_c = T_h$, as one would expect. The efficiency increases as T_c is lowered and as T_h increases. However, the efficiency can only be unity (100%) if $T_c = 0$ K. Such reservoirs are not available, and so the maximum efficiency is always less than unity. In most practical cases, the cold reservoir is near room temperature, about 300 K. Therefore, one usually strives to increase the efficiency by raising the temperature of the hot reservoir. *All real engines are less efficient than the Carnot engine since they are subject to such practical difficulties as friction and heat losses by conduction.*

EXAMPLE 22.1. Efficiency of the Carnot Engine
Show that the efficiency of a heat engine operating in a Carnot cycle using an ideal gas is given by Equation 22.4.

Solution: During the isothermal expansion, $A \rightarrow B$ (Fig. 22.8a), the temperature does not change and so the internal energy remains constant. The work done by the gas is given by Equation 20.21. According to the first law the heat absorbed, Q_h, equals the work done, so that

$$Q_h = W_{AB} = nRT_h \ln \frac{V_B}{V_A}$$

In a similar manner, the heat rejected to the cold reservoir during the isothermal compression $C \rightarrow D$ is given by

$$Q_c = W_{CD} = nRT_c \ln \frac{V_C}{V_D}$$

Dividing these expressions, we find that

$$(1) \qquad \frac{Q_c}{Q_h} = \frac{T_c}{T_h} \frac{\ln(V_C/V_D)}{\ln(V_B/V_A)}$$

We now show that the ratio of the logarithmic quantities is unity by obtaining a relation between the ratio of volumes.

For any quasi-static, adiabatic process, the pressure and volume are related by Equation 21.18:

$$PV^{\gamma} = \text{constant}$$

During any reversible, quasi-static process, the ideal gas must also obey the equation of state, $PV = nRT$. Substituting this into the above expression to eliminate the pressure, we find that

$$TV^{\gamma-1} = \text{constant}$$

Applying this result to the adiabatic processes $B \rightarrow C$ and $D \rightarrow A$, we find that

$$T_h V_B^{\gamma-1} = T_c V_C^{\gamma-1}$$

$$T_h V_A^{\gamma-1} = T_c V_D^{\gamma-1}$$

Dividing these equations, we obtain

$$(V_B/V_A)^{\gamma-1} = (V_C/V_D)^{\gamma-1}$$

$$(2) \qquad \frac{V_B}{V_A} = \frac{V_C}{V_D}$$

Substituting (2) into (1), we see that the logarithmic terms cancel and we obtain the relation

$$\frac{Q_c}{Q_h} = \frac{T_c}{T_h}$$

Using this result and Equation 22.2, the thermal efficiency of the Carnot engine is

$$e_c = 1 - \frac{Q_c}{Q_h} = 1 - \frac{T_c}{T_h} = \frac{T_h - T_c}{T_h}$$

EXAMPLE 22.2. The Steam Engine
A steam engine has a boiler that operates at 500 K. The heat changes water to steam, which drives the piston. The exhaust temperature is that of the outside air, about 300 K. What is the maximum thermal efficiency of this steam engine?

Solution: From the expression for the efficiency of a Carnot engine, we find the maximum thermal efficiency for any engine operating between these temperatures:

$$e_c = 1 - \frac{T_c}{T_h} = 1 - \frac{300\text{ K}}{500\text{ K}} = 0.4, \text{ or } 40\%$$

You should note that this is the highest theoretical efficiency of the engine. In practice, the efficiency will be considerably lower.

Exercise 1 Determine the maximum work the engine can perform in each cycle of operation if it absorbs 200 J of heat from the hot reservoir during each cycle.
Answer: 80 J.

EXAMPLE 22.3. The Carnot Efficiency
The highest theoretical efficiency of a gasoline engine, based on the Carnot cycle, is 30%. If this engine expels its gases into the atmosphere, which has a temperature of 300 K, what is the temperature in the cylinder immediately after combustion?

Solution: The Carnot efficiency is used to find T_h:

$$e_c = 1 - \frac{T_c}{T_h}$$

$$T_h = \frac{T_c}{1 - e_c} = \frac{300\text{ K}}{1 - 0.3} = 429\text{ K}$$

Exercise 2 If the heat engine absorbs 400 J of heat from the hot reservoir during each cycle, how much work can it perform in each cycle?
Answer: 120 J.

22.4 THE ABSOLUTE TEMPERATURE SCALE

In Chapter 19 we defined temperature scales in terms of observed changes in certain physical properties of materials with temperature. It is desirable to define a temperature scale that is independent of material properties. The Carnot cycle provides us with the basis for such a temperature scale. Equation 22.3 tells us that the ratio Q_c/Q_h depends *only* on the temperatures of the two heat reservoirs. The ratio of the two temperatures, T_c/T_h, can be obtained by operating a reversible heat engine in a Carnot cycle between these two temperatures and carefully measuring the heats Q_c and Q_h. A temperature scale can be determined with reference to some fixed-point temperatures. The *absolute*, or *kelvin*, temperature scale is defined by choosing 273.16 K as the absolute temperature of the triple point of water.

The temperature of any substance can be obtained in the following manner: (1) take the substance through a Carnot cycle; (2) measure the heat Q absorbed or expelled by the system at some temperature T; (3) measure the heat Q_3 absorbed or expelled by the system when it is at the temperature of the triple point of water. From Equation 22.3 and this procedure we find that the unknown temperature is given by

$$T = (273.16\text{ K})\frac{Q}{Q_3}$$

The absolute temperature scale is identical to the ideal-gas temperature scale and is independent of the property of the working substance. Therefore it can be applied even at very low temperatures.

In the previous section, we found that the thermal efficiency of any Carnot engine is given by $e_c = 1 - (T_c/T_h)$. This result shows that a 100% efficient engine is possible only if a temperature of absolute zero is maintained for T_c. If this were possible, any Carnot engine operating between T_h and $T_c = 0$ K would convert all of the absorbed heat into work.[7] Using this idea, Lord Kelvin defined absolute zero as follows: *Absolute zero is the temperature of a reservoir at which a Carnot engine will expel no heat.*

22.5 THE GASOLINE ENGINE

In this section we shall discuss the efficiency of the common gasoline engine. Four successive processes occur in each cycle, as illustrated in Figure 22.10. During the *intake stroke* of the piston, air that has been mixed with gasoline vapor in the carburetor is drawn into the cylinder. During the *compression stroke*, the intake valve is closed and the air-fuel mixture is compressed approximately adiabatically. At this point a spark ignites the air-fuel mixture, causing a rapid increase in pressure and temperature at nearly constant vol-

[7] Experimentally, it is not possible to reach absolute zero. Temperatures as low as about 10^{-5} K have been achieved with enormous difficulties using a technique called *nuclear demagnetization*. The fact that absolute zero may be approached but never reached is a law of nature known as the *third law of thermodynamics*.

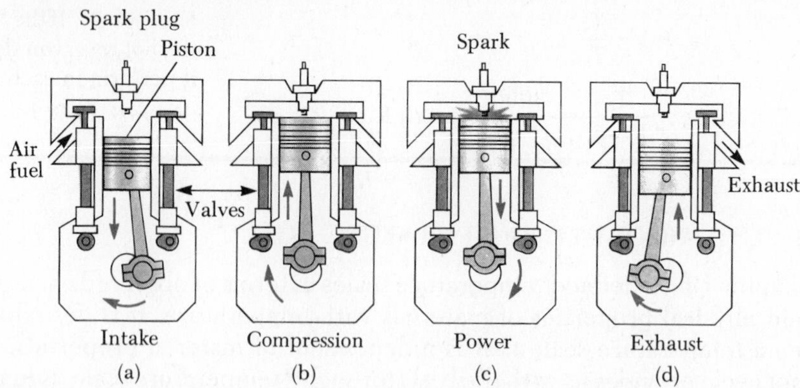

Figure 22.10 The four-stroke cycle of a conventional internal combustion engine. In the intake stroke, air is mixed with fuel. The intake valve is then closed, and the air-fuel mixture is compressed by the piston. The mixture is ignited by the spark plug in the power stroke. Finally, the residual gases are exhausted.

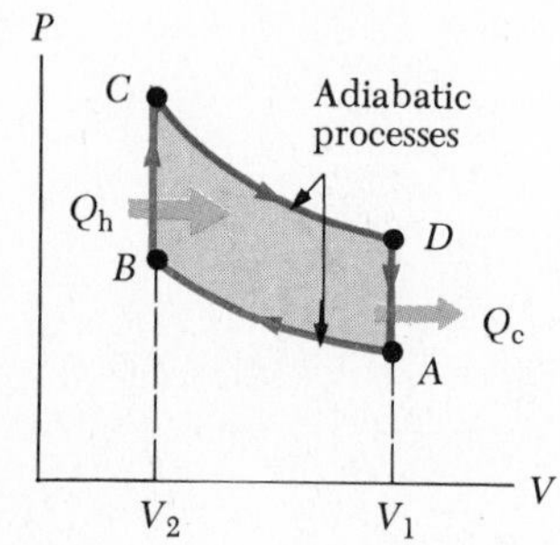

Figure 22.11 The *PV* diagram for the Otto cycle, which approximately represents the processes in the internal combustion engine. No heat is transferred during the adiabatic processes $A \to B$ and $C \to D$.

ume. The burning gases expand and force the piston back, which produces the *power stroke.* Finally, during the *exhaust stroke,* the exhaust valve is opened and the rising piston forces most of the remaining gas out of the cylinder. The cycle is repeated after the exhaust valve is closed and the intake valve is opened.

These four processes can be approximated by the **Otto cycle,** a *PV* diagram of which is illustrated in Figure 22.11.

1. In the process $A \to B$ (compression stroke), the air-fuel mixture is compressed adiabatically from volume V_1 to volume V_2, and the temperature increases from T_A to T_B. The work done on the gas is the area under the curve *AB*.
2. In the process $B \to C$, combustion occurs and heat Q_h is added to the gas. This is not an inflow of heat but rather a release of heat from the combustion process. During this time the pressure and temperature rise rapidly, but the volume remains approximately constant. No work is done on the gas.
3. In the process $C \to D$ (power stroke), the gas expands adiabatically from V_2 to V_1, causing the temperature to drop from T_C to T_D. The work done by the gas equals the area under the curve *CD*.
4. In the final process, $D \to A$ (exhaust stroke), heat Q_c is extracted from the gas as its pressure decreases at constant volume. (Hot gas is replaced by cool gas.) No work is done during this process.

If the air-fuel mixture is assumed to be an ideal gas, the efficiency of the Otto cycle is shown in Example 22.4 to be

Efficiency of the Otto cycle

$$e = 1 - \frac{1}{(V_1/V_2)^{\gamma-1}} \tag{22.5}$$

where γ is the ratio of the molar heat capacities C_p/C_v and V_1/V_2 is called the **compression ratio.** This expression shows that the efficiency increases with increasing compression ratios. For a typical compression ratio of 8 and $\gamma = 1.4$, a theoretical efficiency of 56% is predicted for an engine operating in the idealized Otto cycle. This is much higher than what is achieved in real engines (15% to 20%) because of such effects as friction, heat loss to the cylinder walls, and incomplete combustion of the air-fuel mixture. Diesel engines have higher efficiencies than gasoline engines because of their higher compression ratios (about 16) and higher combustion temperatures.

EXAMPLE 22.4. Efficiency of the Otto Cycle
Show that the thermal efficiency of an engine operating in an idealized Otto cycle (Fig. 22.11) is given by Equation 22.5. Treat the working substance as an ideal gas.

Solution: First, let us calculate the work done by the gas during each cycle. No work is done during the processes $B \rightarrow C$ and $D \rightarrow A$. Work is done on the gas during the adiabatic compression $A \rightarrow B$, and work is done by the gas during the adiabatic expansion $C \rightarrow D$. The net work done equals the area bounded by the closed curve in Figure 22.11. Since the change in internal energy is zero for one cycle, we see from the first law that the net work done for each cycle equals the net heat into the system:

$$W = Q_h - Q_c$$

Since the processes $B \rightarrow C$ and $D \rightarrow A$ take place at constant volume and since the gas is ideal, we find from the definition of heat capacity that

$$Q_h = nC_v(T_C - T_B) \quad \text{and} \quad Q_c = nC_v(T_D - T_A)$$

Using these expressions together with Equation 22.2, we obtain for the thermal efficiency

$$(1) \qquad e = \frac{W}{Q_h} = 1 - \frac{Q_c}{Q_h} = 1 - \frac{T_D - T_A}{T_C - T_B}$$

We can simplify this expression by noting that the processes $A \rightarrow B$ and $C \rightarrow D$ are adiabatic and hence obey the relation $TV^{\gamma-1} =$ constant. Using this condition, and the facts that $V_A = V_D = V_1$ and $V_B = V_C = V_2$, we find that

$$(2) \qquad \frac{T_D - T_A}{T_C - T_B} = \left[\frac{V_2}{V_1}\right]^{\gamma-1}$$

Substituting (2) into (1) gives for the thermal efficiency

$$(3) \qquad e = 1 - \frac{1}{(V_1/V_2)^{\gamma-1}}$$

This can also be expressed in terms of a ratio of temperatures by noting that since $T_A V_1^{\gamma-1} = T_B V_2^{\gamma-1}$, it follows that

$$\left[\frac{V_2}{V_1}\right]^{\gamma-1} = \frac{T_A}{T_B} = \frac{T_D}{T_C}$$

Therefore (3) becomes

$$(4) \qquad e = 1 - \frac{T_A}{T_B} = 1 - \frac{T_D}{T_C}$$

During this cycle, the lowest temperature is T_A and the highest temperature is T_C. Therefore the efficiency of a Carnot engine operating between reservoirs at these two extreme temperatures $\left(\text{which is given by } e_c = 1 - \frac{T_A}{T_C}\right)$ would be *greater* than the efficiency of the Otto cycle, which is given by (4).

22.6 HEAT PUMPS AND REFRIGERATORS

A **heat pump** is a mechanical device that transfers heat from one location to another. For example, when a heat pump is used to warm a building, it extracts heat from the outside air and releases it to the interior of the building. When a heat pump is operated in reverse, it acts as an air conditioner. That is, during the summer, the heat pump is used to cool a building by extracting heat from the interior and releasing it to the outside air.

Figure 22.12 is a schematic representation of a heat pump used to heat a home. Heat is removed from the outside by a fluid, such as Freon, circulating through the heat pump. The heat absorbed by the pump is Q_c and the heat transferred from the pump is Q_h. In order to circulate and compress the fluid in the heat pump, a motor or some other source of energy does work W on the pump.

The effectiveness of a heat pump is described in terms of a number called the **coefficient of performance, COP**. This is defined as the ratio of the heat transferred into the hot reservoir and the work required to transfer that heat:

$$\text{COP(heat pump)} \equiv \frac{\text{heat transferred}}{\text{work done by pump}} = \frac{Q_h}{W} \qquad (22.6)$$

If the outside temperature is 25°F or higher, the COP for a heat pump is about 4. That is, the heat transferred into the house is about four times greater than the work done by the motor in the heat pump. However, as the outside temperature decreases, it becomes more difficult for the heat pump to extract

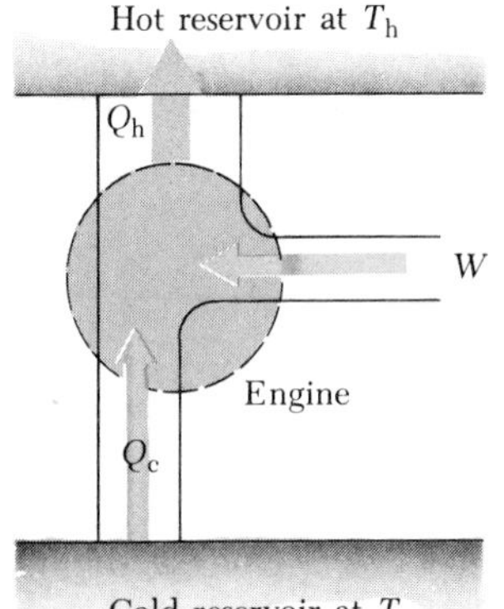

Figure 22.12 Schematic diagram of a heat pump, which absorbs heat Q_c from the cold reservoir and expels heat Q_h to the hot reservoir.

sufficient heat from the air and the COP drops. In fact, the COP can fall below unity for temperatures below the midteens.

Although heat pumps used in buildings are relatively new products in the heating and air conditioning field, the refrigerator has been a standard appliance in homes for years. The refrigerator works much like a heat pump, except that it cools its interior by pumping heat from the food storage compartments into the warmer air outside. During its operation, a refrigerator removes a quantity of heat Q_c from the interior of the refrigerator, and in the process its motor does work W. The coefficient of performance of a refrigerator is given by

$$\text{COP(refrigerator)} = \frac{Q_c}{W} \tag{22.7}$$

An efficient refrigerator is one that removes the greatest amount of heat from the cold reservoir for the least amount of work. Thus, a good refrigerator should have a high coefficient of performance, typically 5 or 6. The impossible (perfect) refrigerator would have an infinite coefficient of performance.

22.7 DEGRADATION OF ENERGY

The first law of thermodynamics is a general statement of the conservation of energy. It makes no distinction between the different forms of energy. The second law of thermodynamics says that thermal energy is different from all other forms of energy. Various forms of energy can be completely converted into thermal energy spontaneously, whereas the reverse transformation is never complete. For example, when a block slides on a table, the force of friction causes the block's kinetic energy to be converted into thermal energy and the block ultimately comes to rest. The reverse energy conversion does not occur. In general, if two kinds of energy, A and B, can be completely converted into each other, we can say that they are *of the same grade.* On the other hand, if form A can be completely converted into form B, but the reverse is never complete, then form A is a higher grade of energy than form B. For example, the kinetic energy of the sliding block is of higher grade than the thermal energy contained in the block and table. Therefore, when high-grade energy is converted into thermal energy, it can never be fully recovered as high-grade energy. This conversion of high-grade energy into thermal energy is referred to as the *degradation of energy.* The energy is said to be degraded because it takes on a form that is less useful for doing work. In other words, in all real processes where heat transfer occurs, the energy available for doing work decreases.

High and low grades of energy

To understand more clearly what we mean by *high-grade* and *low-grade* energy, recall from the previous chapter that thermal energy is actually a measure of the random kinetic energy of the molecules making up a substance. Since the motion of the large number of molecules is chaotic, or disordered, we regard this as a low-grade form of energy. In contrast, the kinetic energy of a macroscopic object, such as a ball, is a high-grade form of energy. It is the result of a highly ordered form of motion since all molecules have a common velocity (apart from their random thermal motions).

In real processes, the disorder in the system increases

When real (irreversible) processes occur, the degree of disorder or chaos in the system increases. For example, consider the isothermal expansion of an ideal gas in a container with a movable piston. As the gas absorbs heat and gradually expands, it maintains a constant temperature by doing work (pushing on the piston). After the expansion, the gas occupies a greater volume than it did originally. The gas molecules become more disordered in that they are not as localized as they were originally. Left by itself, the gas will not become ordered again by giving up its thermal energy to a reservoir. Thus, the flow of heat is in the direction that increases the amount of disorder. In view of these considerations, we can state the second law of thermodynamics as follows: *When an isolated system undergoes a change, the disorder in the system increases.* Furthermore, we can say that the *changes occurring in an isolated system result in a degradation of energy.* Ordered energy is converted into disordered energy.

The measure of the disorder in a system is made quantitative by introducing a quantity called **entropy.** For the moment we can think of entropy as being synonymous with the "degree of disorder" in a system. Simply stated, *an increase in disorder is equivalent to an increase in entropy.* For example, the highly ordered arrangement of atoms in a crystal of sodium chloride has lower entropy than the disordered arrangement of atoms in molten sodium chloride. The vapor phase has even more disorder, and consequently higher entropy. Entropy should not be confused with energy. The total energy of a closed system remains constant, whereas the entropy generally increases, never decreases. In fact, the entropy of a closed system tends to increase toward a maximum value. Entropy is rather abstract and must be defined carefully for every situation. In the next section we shall give a purely thermodynamic definition of entropy.

22.8 ENTROPY

The concept of temperature is involved in the zeroth law of thermodynamics, and the concept of internal energy is involved in the first law. Temperature and internal energy are both state functions. That is, they can be used to describe the thermodynamic state of a system. Another state function related to the second law of thermodynamics is the **entropy,** S. In this section we define entropy on a macroscopic scale as it was first expressed by Clausius in 1865.

Consider a quasi-static, reversible process between two equilibrium states. If $đQ_r$ is the heat absorbed or expelled by the system during some small interval of the path,

the **change in entropy,** dS, between two equilibrium states is given by the heat transferred, $đQ_r$, divided by the absolute temperature, T, of the system in this interval. That is,

$$dS = \frac{đQ_r}{T} \tag{22.8}$$

Clausius definition of change in entropy

The subscript r on the $đQ_r$ is used to emphasize that the definition applies only to *reversible* processes. When heat is absorbed by the system, $đQ_r$ is positive and hence the entropy increases. When heat is expelled by the system, $đQ_r$ is negative and the entropy decreases. Note that Equation 22.8 does not define entropy, but the *change* in entropy. This is consistent with the fact that a change in state always accompanies heat transfer. Hence, the meaningful quantity in describing a process is the *change* in entropy.

Entropy originally found its place in thermodynamics, but its importance grew tremendously as the field of statistical mechanics developed because this method of analysis provided an alternative way of interpreting the concept of entropy. In statistical mechanics, the behavior of a substance is described in terms of the statistical behavior of the atoms and molecules contained in the substance. One of the main results of this treatment is that

isolated systems tend toward disorder and entropy is a measure of this disorder.

For example, consider the molecules of a gas in the air in your room. If all the gas molecules moved together like soldiers marching in step, this would be a very ordered state. It is also an unlikely state. If you could see the molecules, you would see that they move haphazardly in all directions, bumping into one another, changing speed upon collision, some going fast, some slow. This is a highly disordered state, and it is also the most likely state.

All physical processes tend toward the most likely state, and that state is always one in which the disorder increases. Because entropy is a measure of disorder, an alternative way of saying this is

the entropy of the universe increases in all natural processes.

This statement is yet another way of stating the second law of thermodynamics.

To calculate the change in entropy for a finite process, we must recognize that T is generally not constant. If $đQ_r$ is the heat transferred when the system is at a temperature T, then the change in entropy in an arbitrary reversible process between an initial state and a final state is

Changes in entropy for a finite process

$$\Delta S = \int_i^f dS = \int_i^f \frac{đQ_r}{T} \quad \text{(reversible path)} \tag{22.9}$$

Although we do not prove it here, the change in entropy of a system in going from one state to another has the same value for *all* reversible paths connecting the two states.[8] That is,

the change in entropy of a system depends only on the properties of the initial and final equilibrium states.

In the case of a *reversible, adiabatic* process, no heat is transferred between the system and its surroundings, and therefore $\Delta S = 0$ in this case. Since there is no change in entropy, such a process is often referred to as an **isentropic process.**

Consider the changes in entropy that occur in a Carnot heat engine operating between the temperatures T_c and T_h. In one cycle, the engine absorbs heat Q_h from a hot reservoir at a temperature T_h and rejects heat Q_c to a cold reservoir at a temperature T_c. Thus, the total change in entropy for one cycle is

$$\Delta S = \frac{Q_h}{T_h} - \frac{Q_c}{T_c}$$

where the negative sign in the second term represents the fact that heat Q_c is expelled by the system. In Example 22.1 we showed that for a Carnot cycle,

$$\frac{Q_c}{Q_h} = \frac{T_c}{T_h}$$

[8] Note that the quantity $đQ_r$ is called an *inexact differential quantity,* whereas $đQ_r/T = dS$ is a perfect differential. This is because heat is not a property of the system, and hence Q is not a state function. Mathematically, we call $1/T$ the *integrating factor* in this case, since the perfect differential $đQ_r/T$ can be integrated.

Using this result in the previous expression for ΔS, we find that the total change in entropy for a Carnot engine operating in a cycle is *zero*. That is,

Change in entropy for a Carnot cycle is zero

$$\Delta S = 0$$

Now consider a system taken through an arbitrary reversible cycle. Since the entropy function is a state function and hence depends only on the properties of a given equilibrium state, we conclude that $\Delta S = 0$ for *any* reversible cycle. In general, we can write this condition in the mathematical form

$\Delta S = 0$ for any reversible cycle

$$\oint \frac{đQ_r}{T} = 0 \qquad (22.10)$$

where the symbol $\oint$ indicates that the integration is over a *closed* path.

Another important property of entropy is the fact that

the entropy of the universe remains constant in a reversible process.

This can be understood by noting that two bodies A and B that interact with each other reversibly must always be in thermal equilibrium with each other. That is, their temperatures must always be equal. Therefore, when a small amount of heat $đQ$ is transferred from A to B, the increase in entropy of B is $đQ/T$, while the corresponding change in entropy of A is $-đQ/T$. Thus the total change in entropy of the system $(A + B)$ is zero, and the entropy of the universe is unaffected by the reversible process.[9]

As a special case, we next show how to calculate the change in entropy for an ideal gas that undergoes a quasi-static, reversible process in which heat is absorbed from a reservoir.

Quasi-static, Reversible Process for an Ideal Gas

An ideal gas undergoes a quasi-static, reversible process from an initial state T_i, V_i to a final state T_f, V_f. Let us calculate the change in entropy for this process.

According to the first law, $đQ_r = dU + đW$, where $đW = P\,dV$. For an ideal gas, recall that $dU = nC_v\,dT$ and $P = nRT/V$. Therefore, we can express the heat transferred as

$$đQ_r = dU + P\,dV = nC_v\,dT + nRT\frac{dV}{V} \qquad (22.11)$$

We cannot integrate this expression as it stands since the last term contains two variables, T and V. However, if we divide each term by T, we can integrate both terms on the right-hand side:

$$\frac{đQ_r}{T} = nC_v\frac{dT}{T} + nR\frac{dV}{V} \qquad (22.12)$$

Assuming that C_v is constant over the interval in question, and integrating Equation 22.12 from T_i, V_i to T_f, V_f, we get

$$\Delta S = \int_i^f \frac{đQ_r}{T} = nC_v \ln\frac{T_f}{T_i} + nR\ln\frac{V_f}{V_i} \qquad (22.13)$$

Rudolph Clausius (1822–1888). "I propose . . . to call S the entropy of a body, after the Greek word 'transformation.' I have designedly coined the work 'entropy' to be similar to energy, for these two quantities are analogous in their physical significance, that an analogy of denominations seems to be helpful." (AIP Niels Bohr Library, Lande Collection)

[9] Alternatively, we can say that since the universe is, by definition, an isolated system, it never gains or loses heat; hence the change in entropy of the universe is zero for a reversible process.

This expression shows that ΔS depends *only on the initial and final states and is independent of the reversible path.* Furthermore, ΔS can be positive or negative depending on whether the gas absorbs or expels heat during the process. Finally, for a cyclic process ($T_i = T_f$ and $V_i = V_f$), we see that $\Delta S = 0$.

EXAMPLE 22.5. Change in Entropy—Melting Process
A solid substance with a latent heat of fusion L melts at a temperature T_m. Calculate the change in entropy that occurs when m grams of this substance is melted.

Solution: Let us assume that the melting process occurs so slowly that it can be considered a reversible process. In that case the temperature can be considered to be constant and equal to T_m. Making use of Equations 22.8 and 20.7, we find that

$$\Delta S = \int \frac{đQ_r}{T} = \frac{1}{T_m}\int đQ = \frac{Q}{T_m} = \frac{mL}{T_m} \qquad (22.14)$$

Note that we were able to remove T_m from the integral in this case since the process is isothermal. Also, the quantity Q is the total heat required to melt the substance and is equal to mL (Section 20.3).

Exercise 4 Calculate the change in entropy when 300 g of lead melts at 327°C. Lead has a latent heat of fusion equal to 24.5 J/g.
Answer: $\Delta S = 12.3$ J/K.

22.9 ENTROPY CHANGES IN IRREVERSIBLE PROCESSES

By definition, the change in entropy for a system can be calculated only for reversible paths connecting the initial and final equilibrium states. In order to calculate changes in entropy for real (irreversible) processes, we must first recognize that the entropy function (like internal energy) depends only on the *state* of the system. That is, entropy is a state function. Hence, the change in entropy of a system between any two equilibrium states depends only on the initial and final states. Experimentally one finds that the entropy change is the same for all processes between the initial and final states.[10]

In view of the fact that the entropy of a system depends only on the state of the system, we can now calculate entropy changes for irreversible processes between two equilibrium states. This can be accomplished by devising a reversible process (or series of reversible processes) between the same two equilibrium states and computing $\int đQ_r/T$ for the reversible process. The entropy change for the irreversible process is the same as that of the reversible process between the same two equilibrium states. Let us demonstrate this procedure with a few specific cases.

Heat Conduction

Consider the transfer of heat Q from a hot reservoir at temperature T_h to a cold reservoir at temperature T_c. Since the cold reservoir absorbs heat Q, its entropy increases by Q/T_c. At the same time, the hot reservoir loses heat Q and its entropy decreases by Q/T_h. The increase in entropy of the cold reservoir is greater than the decrease in entropy of the hot reservoir since T_c is less than T_h. Therefore, the total change in entropy of the system (universe) is greater than zero:

$$\Delta S_u = \frac{Q}{T_c} - \frac{Q}{T_h} > 0$$

[10] It is also possible to show that if this were not the case, the second law of thermodynamics would be violated.

Free Expansion

An ideal gas in an insulated container initially occupies a volume V_i (Fig. 22.13). A partition separating the gas from another evacuated region is suddenly broken so that the gas expands (irreversibly) to a volume V_f. Let us find the change in entropy of the gas and the universe.

The process is clearly neither reversible nor quasi-static. The work done by the gas against the vacuum is zero, and since the walls are insulating, no heat is transferred during the expansion. That is, $W = 0$ and $Q = 0$. Using the first law, we see that the change in internal energy is zero, therefore $U_i = U_f$, where i and f indicate the initial and final equilibrium states. Since the gas is ideal, U depends on temperature only, and so we conclude that $T_i = T_f$.

We cannot use Equation 22.9 directly to calculate the change in entropy since that equation applies only to reversible processes. In fact, at first sight one might *wrongfully* conclude that $\Delta S = 0$ since there is no heat transferred. To calculate the change in entropy, let us imagine a reversible process between the same initial and final equilibrium states. A simple one to choose is an isothermal, reversible expansion in which the gas pushes slowly against a piston. Since T is constant in this process, Equation 22.8 gives

$$\Delta S = \int \frac{đQ_r}{T} = \frac{1}{T} \int_i^f đQ_r$$

But $\int đQ_r$ is simply the work done by the gas during the isothermal expansion from V_i to V_f, which is given by Equation 20.16. Using this result, we find that

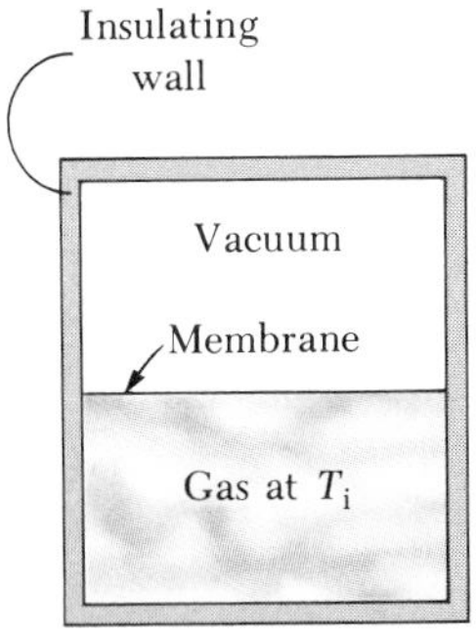

Insulating wall

Vacuum

Membrane

Gas at T_i

Figure 22.13 Free expansion of a gas. When the partition separating the gas from the evacuated region is ruptured, the gas expands freely and irreversibly so that it occupies a greater final volume. The container is thermally insulated from its surroundings, and so $Q = 0$.

Change in entropy during a free expansion

$$\Delta S = nR \ln \frac{V_f}{V_i} \tag{22.15}$$

Since $V_f > V_i$, we conclude that ΔS is positive, and so both the entropy and disorder of the gas (and universe) increase as a result of the irreversible, adiabatic expansion. This result can also be obtained from Equation 22.13, noting that $T_i = T_f$, and so $dT = 0$.

EXAMPLE 22.6. Free Expansion of a Gas
Calculate the change in entropy of 2 moles of an ideal gas that undergoes a free expansion to three times its initial volume.

Solution: Using Equation 22.15 with $n = 2$ and $V_f = 3V_i$, we find that

$$\Delta S = nR \ln \frac{V_f}{V_i} = (3 \text{ moles})(8.31 \text{ J/mole} \cdot \text{K}) \ln 3$$

$$= 27.4 \text{ J/K}$$

Entropy of Mixing

A substance of mass m_1, specific heat c_1, and initial temperature T_1 is mixed with a second substance of mass m_2, specific heat c_2, and initial temperature T_2, where $T_2 > T_1$. (For example, they could both be liquids.) The mixed system is allowed to reach thermal equilibrium. What is the total entropy change for the system?

First, let us calculate the final equilibrium temperature, T_f. Energy conservation requires that the heat lost by one substance equal the heat gained by the other. Since by definition, $Q = mc\,\Delta T$ for each substance, we get $Q_1 = -Q_2$, or

$$m_1c_1\,\Delta T = m_2c_2\,\Delta T$$

$$m_1c_1(T_f - T_1) = m_2c_2(T_f - T_2)$$

Solving for T_f gives

$$T_f = \frac{m_1c_1T_1 + m_2c_2T_2}{m_1c_1 + m_2c_2} \tag{22.16}$$

Note that $T_1 < T_f < T_2$, as would be expected.

The mixing process is irreversible since the system goes through a series of nonequilibrium states. During such a transformation, the temperature at any time is not well defined. However, we can imagine that the hot body at the initial temperature T_i is slowly cooled to the temperature T_f by placing it in contact with a series of reservoirs differing infinitesimally in temperature, where the first reservoir is at the initial temperature T_i and the last is at T_f. Such a series of very small changes in temperature would approximate a reversible process. Applying Equation 22.8 and noting that $dQ = mc\,dT$ for an infinitesimal change, we get

$$\Delta S = \int_1 \frac{đQ_1}{T} + \int_2 \frac{đQ_2}{T} = m_1c_1 \int_{T_1}^{T_f} \frac{dT}{T} + m_2c_2 \int_{T_2}^{T_f} \frac{dT}{T}$$

where we have assumed that the specific heats remain constant. Integrating, we find

Change in entropy for a mixing process

$$\Delta S = m_1c_1 \ln\frac{T_f}{T_1} + m_2c_2 \ln\frac{T_f}{T_2} \tag{22.17}$$

where T_f is given by Equation 22.16. If Equation 22.16 is substituted into Equation 22.17, you can show that one of the terms in Equation 22.17 will always be positive and the other negative. (You may want to verify this for yourself). However, the positive term will always be larger than the negative term, resulting in a positive value for ΔS. Thus, we conclude that the entropy of the universe (system) increases in this irreversible process.

EXAMPLE 22.7. Calculating ΔS for a Mixing Process
One kg of water at 0°C is mixed with an equal mass of water at 100°C. After equilibrium is reached, the mixture has a uniform temperature of 50°C. What is the change in entropy of the system?

Solution: The change in entropy can be calculated from Equation 22.17 using the values $m_1 = m_2 = 1$ kg, $c_1 = c_2 = 1$ cal/g · K = 4.19 J/g · K, $T_1 = 0$°C (= 273 K), $T_2 = 100$°C (= 373 K), and $T_f = 50$°C (= 323 K). Note that you must use absolute temperatures in this calculation.

$$\Delta S = m_1c_1 \ln\frac{T_f}{T_1} + m_2c_2 \ln\frac{T_f}{T_2}$$

$$= (10^3\text{ g})(4.19\text{ J/g}\cdot\text{K}) \ln\frac{323}{273}$$

$$+ (10^3\text{ g})(4.19\text{ J/g}\cdot\text{K}) \ln\frac{323}{373}$$

$$= 705\text{ J/K} - 603\text{ J/K} = 102\text{ J/K}$$

That is, the increase in entropy of the cold water is greater than the decrease in entropy of the warm water as a result of this irreversible mixing process. Consequently, the increase in entropy of the system is 102 J/K.

The cases just described show that the change in entropy of a system is always positive for an irreversible process. In general, the total entropy (and disorder) always increases in irreversible processes. From these considerations, the second law of thermodynamics can be stated as follows: *The total entropy of an isolated system that undergoes a change cannot decrease.* Furthermore, if the process is *irreversible,* the total entropy of an isolated system always *increases.* On the other hand, in a reversible process, the total entropy of an isolated system remains constant. When dealing with interacting bodies that are not isolated, one must be careful to note that the system refers to the bodies *and* their surroundings. When two bodies interact in an irreversible process, the increase in entropy of one part of the system is greater than the decrease in entropy of the other part. Hence, we conclude that

> the change in entropy of the universe must be greater than zero for an irreversible process and equal to zero for a reversible process.

Ultimately, the entropy of the universe should reach a maximum value. At this point, the universe will be in a state of uniform temperature and density. All physical, chemical, and biological processes will cease, since a state of perfect disorder implies no energy available for doing work. This gloomy state of affairs is sometimes referred to as an ultimate "heat death" of the universe.

An illustration from Flammarion's novel *La Fin du Monde,* depicting the "heat-death" of the universe.

*22.10 ENERGY CONVERSION AND THERMAL POLLUTION

The main source of thermal pollution is waste heat from electrical power plants. In the United States, about 85% of the electric power is produced by steam engines, which burn either fossil fuels (coal, oil, or natural gas) or nuclear fuels (uranium-235). The remaining 15% of the electric power is generated by water in hydroelectric plants. The overall thermal efficiency of a modern fossil-fuel plant is about 40%. The actual efficiencies of any power plant must be lower than the theoretical efficiencies derived from the second law of thermodynamics. One always seeks the highest efficiency possible for two reasons. First, higher efficiency results in lower fuel costs. Second, thermal pollution of the environment is reduced since there is less waste energy in a highly efficient power plant. Since any power plant involves several steps of energy conversion, the inefficiency will accumulate in steps.

The burning of fossil fuels in an electrical power plant involves three energy-conversion processes: (1) chemical to thermal energy, (2) thermal to mechanical energy, and (3) mechanical to electrical energy. These are indicated schematically in Figure 22.14.

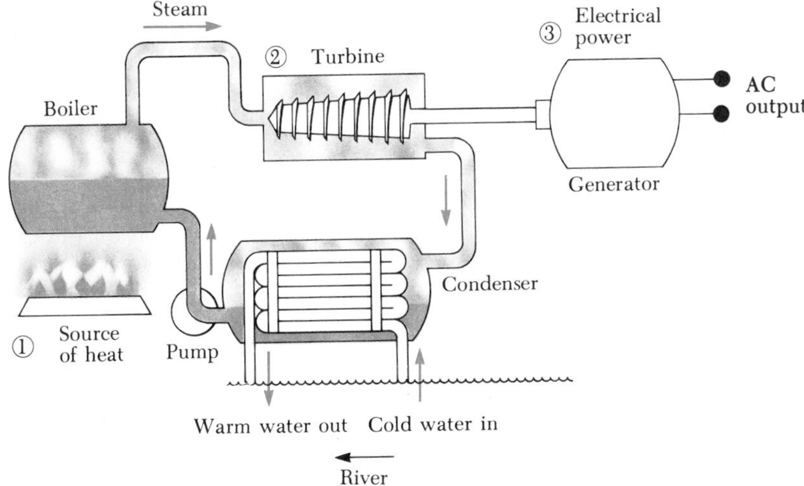

Figure 22.14 Schematic diagram of an electrical power plant.

Essay

SUPERCONDUCTIVITY

David Markowitz
University of Connecticut

As temperature is lowered to a few degrees above absolute zero, many metals undergo a change of phase into what is called a superconducting state. A superconductor has many unusual thermal, electrical, magnetic, and optical properties. The most striking property of a superconductor is its electrical resistance. (The electrical resistance of materials is discussed in more detail in Chapter 27.) Above a temperature T_c, called the critical temperature, the metal, perhaps in the form of a wire, is an ordinary good conductor and has some electrical resistance. At or below the critical temperature, however, the resistance of the wire falls abruptly to zero, as shown in Figure 22E.1.

The critical temperature is a property of a particular metallic substance, just as the freezing point of fresh water is 0°C, whether the water is in a tray in your freezer or in a pond in winter. For example, the critical temperature for mercury is about 4.2 K, which is the boiling point of liquid helium. Above T_c the metal is said to be in the normal (N) state (or phase), and below T_c it is said to be in the superconducting (S) state. This is analogous to the fact that a substance is in the liquid state above its freezing temperature and in the solid state below that temperature.

There are many aspects to superconducting materials and many views that could be taken in discussing their properties. We shall not emphasize the electrical and magnetic properties of superconductors. For purposes of the present discussion, it is sufficient to note that in electrical conduction, the conduction electrons inside a metal are not bound to individual atoms, but rather are free to move through the metal over long distances as they contribute to the electric current.

In this essay, we shall emphasize that aspect of superconductors that contributes to their thermodynamic properties. We shall show that the change from the N state to the S state is a change of phase, just as the freezing of water is a change of phase from the liquid state to the solid state. However, as you will see, the N to S phase change in a superconductor is a very unusual kind of "freezing."

When the temperature of a material is lowered while the material remains in one phase, its properties will change gradually with temperature. For example, when the temperature of water is steadily lowered from 100°C to 0°C, its density will gradually change. On the other hand, there is a dramatic decrease in the density of water as it freezes at 0°C even though there is no change in its temperature during that process. This is the reason that water expands upon freezing. In analogy, the behavior of the resistance of a superconductor at the critical temperature, as shown in Figure 22E.1, is indicative of a phase transition.

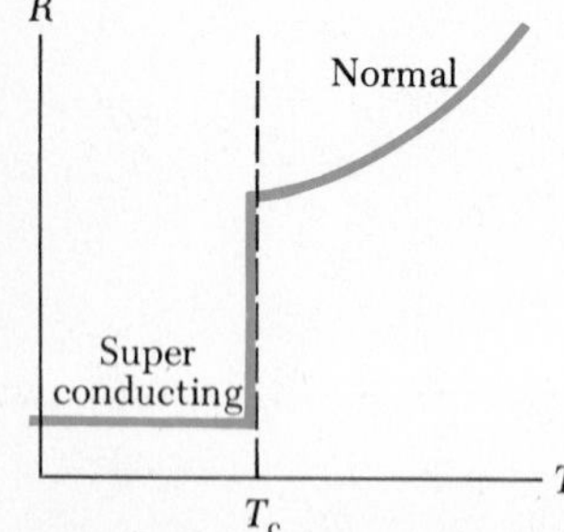

Figure 22E.1 Resistance versus temperature for a pure superconductor, such as mercury, lead, or aluminum. The metal behaves normally down to the critical temperature, T_c, below which the resistance drops to zero.

Many metals exhibit superconducting behavior, particularly those that have more than one or two valence electrons per atom. Moreover, alloys made from these metals also exhibit superconducting behavior. It is the valence electrons of the metal atoms that are the conduction electrons. Apparently the phenomenon of superconductivity does not depend on the detailed nature of any single metal, but rather requires some very general and wide-spread properties, such as the metal's having a sufficient number of conduction electrons. The explanation of what causes the N to S transition should likewise be a very general one.

Heike Kamerlingh-Onnes (1853–1926), a Dutch physicist, was the first to produce liquified helium in the early part of this century. Because helium is a liquid below 4.2 K, it may be used as a coolant for any material below this temperature. Kamerlingh-Onnes developed the helium liquifier because of his interest in the properties of other materials at those temperatures. As it turned out, he first observed the

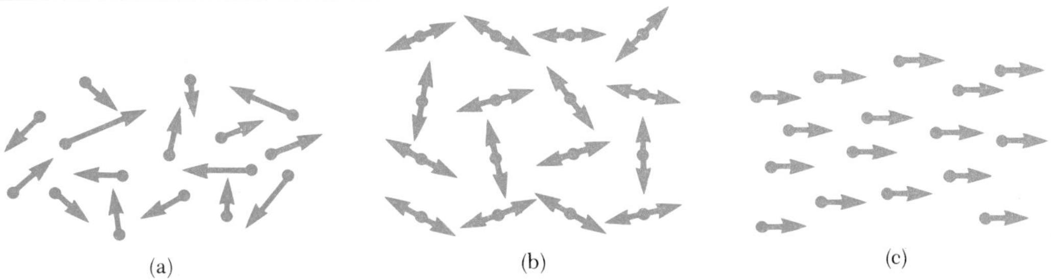

Figure 22E.2 (a) A fluid of particles. The particles are arranged neither in position nor in velocity. This represents a normal fluid. (b) An orderly arrangement of positions of particles. The particles vibrate around these positions, and these vibrations are not orderly. (c) An orderly arrangement of velocities of particles. The particles do not care where their positions are, as long as they maintain the exact same velocity.

phenomenon of superconductivity in mercury, which has (as fate would have it) a critical temperature of about 4.2 K. At that time, Kammerlingh-Onnes wished to test a proposed theory that predicted the behavior of metals near absolute zero. We shall first briefly discuss the shortcomings of this incorrect theory before describing some features of the correct one.

The conduction electrons in a metal can be viewed as a fluid of charged particles. Each electron moves inside its container (the block of metal) and bumps into neighboring electrons (via mutual electrical repulsion). The positions and motions of these particles are both disordered, as shown in Figure 22E.2a. An electric current is established when the fluid flows in a particular direction. (The motion of charges in a preferred direction constitutes a current.) Every fluid that was known early in this century formed a solid (a crystal lattice) at a sufficiently low temperature. Lattice ions in a solid do not flow but simply vibrate about some mean position, as if they are held in the lattice by attachment to springs.

So the incorrect theory is as follows. At very low temperatures, conduction electrons might freeze to form an electron lattice. Since they are not allowed to move through the metal in that state, the conduction current would be zero and the electrical resistance would be infinite. This is in complete contradiction with the experimental result. The experiment says that the resistance of a superconductor is *zero* at or below T_c. Thus, we must conclude that the electrons do not freeze into a lattice.

The correct explanation begins again with the electron fluid. Clearly, some sort of "freezing" must be taking place in a superconductor at or below T_c, but what is meant by *"freezing"* as used in this context? Ordinary freezing means that an orderly arrangement of the positions of particles is established—a place for each particle and each particle in its place. Note that each particle vibrates in place, but the vibrations are not orderly, as indicated in Figure 22E.2b. This is what happens when water becomes ice. On the other hand, this ordinary type of freezing *does not* occur for conduction electrons in a metal.

In order to gain some understanding of the type of "freezing" that occurs for conduction electrons in a superconductor, we must introduce a totally different type of order. The order we are referring to is an orderly arrangement of the velocities of the particles. Consider a system of particles whose positions are random, but all having the same velocity. Such a system is ordered in the sense that the relative positions of all the particles will remain the same as long as they all maintain the same velocity, as shown in Figure 22E.2c.

The following familiar example may be useful for understanding the concept and consequences of an orderly arrangement of velocities. On a moderately crowded highway, if a few cars travel much faster or much slower than the other cars, then traffic is greatly hindered, resulting in major traffic jams and perhaps an occasional collision. However, suppose all cars travel at nearly the same velocity. In this case, the traffic will flow very smoothly, regardless of the spacing between cars and regardless of their common velocity. Ideally, all cars should maintain exactly the same velocity in order to optimize the flow of traffic and eliminate collisions.

Although it is very difficult (if not impossible) for cars to maintain a common velocity on a highway, the situation is quite different for electrons in a superconductor. For a very subtle reason that requires an understanding of modern atomic physics, *individual* electrons in a superconductor are unable to all travel at the same velocity, but *pairs* of electrons are able to do so. In the superconducting state, the particles that have a common velocity are actually electron pairs. Above T_c, the center of mass of any pair of electrons moves randomly through the material. In fact, it is meaningless to refer to an electron pair in the normal state; they simply do not exist as pairs above T_c. Below T_c, however, electron pairs do exist, and *the center of mass of each pair travels at exactly the same velocity.*

The modern atomic physics you need to know comes later in the text. As you will learn, the Pauli Exclusion Principle forbids individual electrons from occupying the

During the first step, heat energy is transferred from the burning fuel to a water supply, which is converted into steam. In this process about 12% of the available energy is lost up the chimney. In the second step, thermal energy in the form of steam at high pressure and temperature passes through a turbine and is converted into mechanical energy. A well-designed turbine has an efficiency of about 47%. Steam, which leaves the turbine at a lower pressure, is then condensed into water and gives up heat in the process. Finally, in the third step, the turbine drives an electrical generator of very high efficiency, typically 99%. Hence, the overall efficiency is the product of the efficiencies of each step, which for the figures given becomes $(0.88)(0.47)(0.99) = 0.41$, or 41%. The thermal energy transferred to the cooling water amounts to about 47% of the initial fuel energy.

In the case of nuclear power plants, the steam generated by the nuclear reactor is at a lower temperature than that of a fossil-fuel plant. This is due primarily to material limitations in the reactor. Typical water moderated nuclear power plants have an overall efficiency of about 34%. High temperature gas-cooled reactors operate at temperatures and efficiencies comparable to fossil-fuel plants.

The waste heat from electrical power plants can be disposed of in various ways. The method shown in Figure 22.14 involves passing water from a river or lake through a condenser and returning it to that source at a higher temperature. This can raise the water temperature of the river or lake by several degrees, which can produce undesirable ecological effects, such as the increased growth of bacteria, undesirable blue-green algae, and pathogenic organisms. Fish and other marine life are also affected since they require oxygen, and the percentage of dissolved oxygen in the water decreases with increasing temperature. There is further demand for oxygen in the decomposition of organic matter, which also proceeds at a higher rate as the temperature increases.

A cooling tower at a reactor site in southern Washington. (From Jonathan Turk and Amos Turk, *Physical Science*, 2nd edition, Philadelphia, Saunders College Publishing, 1981.)

Cooling towers are also commonly used in disposing of waste heat. These towers usually use the heat to evaporate water, which is then released to the atmosphere. Cooling towers also present environmental problems since evaporated water can cause increased precipitation, fog, and ice. Another type of cooling tower is the dry cooling tower (nonevaporative), which transfers heat to the atmosphere by conduction. However, this type is more expensive and cannot cool to as low a temperature as an evaporative tower.

exact same state of motion (and spin) in an atom or even in a solid bar. Pairs of electrons are immune from this principle and acquire perfectly correlated motions in the superconducting metal.

We have attempted to explain, in very general terms, the meaning of the phase change upon going from the N state to the S state. It is a "freezing" of the substance (metal or fluid), not in position space, as with ice, but in velocity space, as with cars all traveling at exactly 55 mi/h down a highway.

Questions for thought:

1. Why do electrons wish to form pairs? (Why does any liquid wish to freeze?)
2. What mechanism produces electrical resistance in a normal metallic phase? Why is it absent from the superconducting metallic phase?

22.11 SUMMARY

The **first law of thermodynamics** is a generalization of the law of conservation of energy that includes heat transfer in any process.

Real processes proceed in an order governed by the second law of thermodynamics.

A **heat engine** is a device that converts thermal energy into other useful forms of energy. The net work done by a heat engine in carrying a substance through a cyclic process ($\Delta U = 0$) is given by

$$W = Q_h - Q_c \qquad (22.1)$$

Work done by a heat engine

where Q_h is the heat absorbed from a warmer reservoir and Q_c is the heat rejected to a cooler reservoir.

The **thermal efficiency,** e, of a heat engine is defined as the ratio of the net work done to the heat absorbed per cycle:

$$e = \frac{W}{Q_h} = 1 - \frac{Q_c}{Q_h} \qquad (22.2)$$

Thermal efficiency

The **second law of thermodynamics** can be stated in many ways:

Statements of the second law

1. No heat engine operating in a cycle can absorb thermal energy from a reservoir and perform an equal amount of work (Kelvin-Planck statement).
2. A perpetual-motion machine of the second kind is impossible to construct.
3. It is impossible to construct a cyclical machine whose sole effect is to transfer heat continuously from one body to another body at a higher temperature (Clausius statement).

Reversible process

A process is **reversible** if the system passes from the initial to the final state through a succession of equilibrium states. A process can be reversible only if it occurs quasi-statically.

Irreversible process

An **irreversible process** is one in which the system and its surroundings cannot be returned to their initial states. In such a process, the system passes from the initial to the final state through a series of nonequilibrium states.

The *efficiency of a heat engine* operating in the **Carnot cycle** is given by

Efficiency of a Carnot engine

$$e_c = 1 - \frac{T_c}{T_h} \qquad (22.4)$$

where T_c is the absolute temperature of the cold reservoir and T_h is the absolute temperature of the hot reservoir.

No real heat engine operating (irreversibly) between the temperatures T_c and T_h can be more efficient than an engine operating reversibly in a Carnot cycle between the same two temperatures.

The second law of thermodynamics states that when real (irreversible) processes occur, the degree of disorder in the system increases. When a process occurs in an isolated system, ordered energy is converted into disordered energy. The measure of disorder in a system is called **entropy,** S.

Clausius definition of change in entropy

The **change in entropy,** dS, of a system moving quasi-statically between two equilibrium states is given by

$$dS = \frac{đQ_r}{T} \qquad (22.8)$$

The change in entropy of a system moving reversibly between two equilibrium states is

$$\Delta S = \int_i^f \frac{đQ_r}{T} \qquad (22.9)$$

The value of ΔS is the same for all reversible paths connecting the initial and final states.

The change in entropy for any reversible, cyclic process is zero.

In any reversible process, the entropy of the universe remains constant.

The entropy of a system is a state function, that is, it depends on the state of the system. The change in entropy for a system undergoing a real (irreversible) process between two equilibrium states is the same as that of a reversible process between the same states.

In an irreversible process, the total entropy of an isolated system always increases. In general, the total entropy (and disorder) always increases in any irreversible process. Furthermore, the change in entropy of the universe is greater than zero for an irreversible process and is zero for a reversible process.

QUESTIONS

1. Distinguish clearly between temperature, heat, and internal energy.
2. When a sealed Thermos bottle full of hot coffee is shaken, what are the changes, if any, in (a) the temperature of the coffee and (b) its internal energy?
3. Use the first law of thermodynamics to explain why the total energy of an isolated system is always conserved.
4. Is it possible to convert internal energy to mechanical energy?
5. What are some factors that affect the efficiency of automobile engines?
6. The statement was made in this chapter that the first law says we cannot get more out of a process than we put in but the second law says that we cannot break even. Explain.
7. Is it possible to cool a room by leaving the door of a refrigerator open? What happens to the temperature of a room in which an air conditioner is left running on a table in the middle of the room?
8. In practical heat engines, which do we have more control of, the temperature of the hot reservoir or the temperature of the cold reservoir? Explain.

9. A steam-driven turbine is one major component of an electric power plant. Why is it advantageous to increase the temperature of the steam as much as possible?
10. Is it possible to construct a heat engine that creates no thermal pollution?
11. Electrical energy can be converted to heat energy with an efficiency of 100%. Why is this number misleading with regard to heating a home? That is, what other factors must be considered in comparing the cost of electric heating with the cost of hot air or hot water heating?
12. Discuss three common examples of natural processes that involve an increase in entropy. Be sure to account for all parts of each system under consideration.
13. Discuss the change in entropy of a gas that expands (a) at constant temperature and (b) adiabatically.

PROBLEMS

Section 22.1 Heat Engines and the Second Law of Thermodynamics

1. A heat engine absorbs 90 cal of heat and performs 25 J of work in each cycle. Find (a) the efficiency of the engine and (b) the heat expelled in each cycle.
2. A heat engine performs 200 J of work in each cycle and has an efficiency of 30%. For each cycle of operation, (a) how much heat is absorbed and (b) how much heat is expelled?
3. A refrigerator has a coefficient of performance equal to 5. If the refrigerator absorbs 30 cal of heat from a cold reservoir in each cycle, find (a) the work done in each cycle and (b) the heat expelled to the hot reservoir.
4. A particular engine has a power output of 5 kW and an efficiency of 25%. If the engine expels 2000 cal of heat in each cycle, find (a) the heat absorbed in each cycle and (b) the time for each cycle.
5. The heat absorbed by an engine is three times greater than the work it performs. (a) What is its thermal efficiency? (b) What fraction of the heat absorbed is expelled to the cold reservoir?
6. In each cycle of its operation, a certain refrigerator absorbs 25 cal from the cold reservoir and expels 32 cal. (a) What is the power required to operate the refrigerator if it works at 60 cycles/s? (b) What is the coefficient of performance of the refrigerator?
7. An engine absorbs 400 cal from a hot reservoir and expels 250 cal to a cold reservoir in each cycle. (a) What is the efficiency of the engine? (b) How much work is done in each cycle? (c) What is the power output of the engine if each cycle lasts for 0.3 s?

Section 22.3 The Carnot Engine

8. A heat engine operates between two reservoirs at temperatures of 20°C and 300°C. What is the maximum efficiency possible for this engine?
9. The efficiency of a Carnot engine is 30%. The engine absorbs 200 cal of heat per cycle from a hot reservoir at 500 K. Determine (a) the heat expelled per cycle and (b) the temperature of the cold reservoir.
10. A Carnot engine has a power output of 150 kW. The engine operates between two reservoirs at 20°C and 500°C. (a) How much heat energy is absorbed per hour? (b) How much heat energy is lost per hour?
11. A power plant has been proposed that would make use of the temperature gradient in the ocean. The system is to operate between 20°C (surface water temperature) and 5°C (water temperature at a depth of about 1 km). (a) What is the maximum efficiency of such a system? (b) If the power output of the plant is 75 MW, how much thermal energy is absorbed per hour? (c) In view of your results to (a), do you think such a system is worthwhile?
12. A heat engine operates in a Carnot cycle between 80°C and 350°C. It absorbs 5×10^3 cal of heat per cycle from the hot reservoir. The duration of each cycle is 1 s. (a) What is the maximum power output of this engine? (b) How much heat does it expel in each cycle?
13. One of the most efficient engines ever built operates between 430°C and 1870°C. Its actual efficiency is 42%. (a) What is its maximum theoretical efficiency? (b) How much power does the engine deliver if it absorbs 3.5×10^4 cal of heat each second?
14. An electrical generating plant has a power output of 500 MW. The plant uses steam at 200°C and exhausts water at 40°C. If the system operates with one half the maximum (Carnot) efficiency, (a) at what rate is heat expelled to the environment? (b) If the waste heat goes into a river whose flow rate is 1.2×10^6 kg/s, what is the rise in temperature of the river?
15. An air conditioner absorbs heat from its cooling coil at 13°C and expels heat to the outside at 30°C. (a) What is the *maximum* coefficient of performance of the air conditioner? (b) If the actual coefficient of performance is one third of the maximum value and if the air conditioner removes 2×10^3 cal of heat energy each second, what power must the motor deliver?
16. A heat pump powered by an electric motor absorbs heat from outside at 5°C and exhausts heat inside in the form of hot air at 40°C. (a) What is the maximum coefficient of performance of the heat pump? (b) If the actual coefficient of performance is 3.2, what *fraction* of the available work (electrical energy) is actually done?

17. An ideal gas is taken through a Carnot cycle. The isothermal expansion occurs at 250°C, and the isothermal compression takes place at 50°C. If the gas absorbs 300 cal of heat during the isothermal expansion, find (a) the heat expelled to the cold reservoir in each cycle and (b) the net work done by the gas in each cycle.

Section 22.5 The Gasoline Engine

18. A gasoline engine has a compression ratio of 6 and uses a gas with $\gamma = 1.4$. (a) What is the efficiency of the engine if it operates in an idealized Otto cycle? (b) If the actual efficiency is 15%, what fraction of the fuel is wasted as a result of friction and heat losses? (Assume complete combustion of the air-fuel mixture.)
19. A gasoline engine using an ideal diatomic gas ($\gamma = 1.4$) operates between temperature extremes of 300 K and 1500 K. Determine its compression ratio if it has an efficiency of 20%. Compare this efficiency to that of a Carnot engine operating between the same temperatures.

Section 22.8 Entropy

20. One mole of an ideal gas expands isothermally and quasi-statically to twice its initial volume. What is the change in entropy of the gas?
21. Calculate the change in entropy of 250 g of water when it is slowly heated from 20°C to 80°C. (*Hint:* Note that $đQ = mc\,dT$.)
22. An ice tray contains 500 g of water at 0°C. Calculate the change in entropy of the water as it freezes completely and slowly at 0°C.
23. One mole of an ideal monatomic gas is heated quasi-statically at constant volume from 300 K to 400 K. What is the change in entropy of the gas?
24. One kg of mercury is initially at −100°C. What is its change in entropy when heat is slowly added to raise its temperature to 100°C? (Mercury has a melting temperature of −39°C, a heat of fusion of 2.8 cal/g, and a specific heat of 0.033 cal/g · C°.)

Section 22.9 Entropy Changes in Irreversible Processes

25. If 800 cal of heat flows from a heat reservoir at 500 K to another reservoir at 300 K through a conducting metal rod, find the change in entropy of (a) the hot reservoir, (b) the cold reservoir, (c) the metal rod, and (d) the universe.
26. A 2-kg block moving with an initial speed of 5 m/s slides on a rough table and is stopped by the force of friction. Assuming the table and air remain at a temperature of 20°C, calculate the entropy change of the universe.
27. A 70-kg log falls from a height of 25 m into a lake. If the log, the lake, and the air are all at 300 K, find the change in entropy of the universe for this process.
28. A glass ampule of volume 150 cm³ contains 0.15 moles of an ideal gas. The ampule is broken in an evacuated vessel of volume 800 cm³. In this free expansion of the gas, find the change in entropy of the universe.
29. A 6-kg block of ice at 0°C is dropped into a lake at 27°C. Just after the ice has all melted, and before the ice water has had a chance to warm, what is the change in entropy of (a) the ice, (b) the lake, and (c) the universe?
30. A cyclic heat engine operates between two reservoirs at temperatures of 300 K and 500 K. In each cycle, the engine absorbs 700 J of heat from the hot reservoir and does 160 J of work. Find the entropy change in each cycle for (a) each reservoir, (b) the engine, and (c) the universe.
31. If 200 g of water at 20°C is mixed with 300 g of water at 75°C, find (a) the final equilibrium temperature of the mixture and (b) the change in entropy of the system.

GENERAL PROBLEMS

32. An ideal refrigerator (or heat pump) is equivalent to a Carnot engine running in reverse. That is, heat Q_c is absorbed from a cold reservoir and heat Q_h is rejected to a hot reservoir. (a) Show that the work that must be supplied to run the refrigerator is given by

$$W = \frac{T_h - T_c}{T_c} Q_2$$

(b) Show that the coefficient of performance of the ideal refrigerator is given by

$$\text{COP} = \frac{T_c}{T_h - T_c}$$

33. One mole of an ideal monatomic gas is taken through the cycle shown in Figure 22.15. The process *AB* is an isothermal expansion. Calculate (a) the net work done by the gas, (b) the heat added to the gas, (c) the heat expelled by the gas, and (d) the efficiency of the cycle.

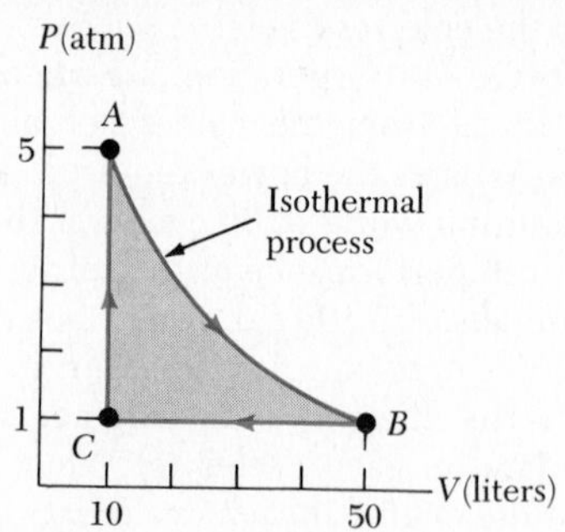

Figure 22.15 (Problem 33).

34. An athlete whose mass is 70 kg drinks 16 ounces (453.6 g) of refrigerated water. The water is at a temperature of 35°F. (a) Neglecting the temperature change of the body resulting from the water intake, (so that the body is regarded as a reservoir at 98.6°F), find the entropy increase of the entire system. (b) Assume that the entire body is cooled by the drink and that the average specific heat of a human is equal to the specific heat of liquid water. Neglecting any other heat transfers and any metabolic heat release, find the

athlete's temperature after drinking the cold water, given an initial body temperature of 98.6°F. Under *these* assumptions, what is the entropy increase of the entire system? Compare your result with that of (a).

35. Figure 22.16 represents n moles of an ideal monatomic gas being taken through a reversible cycle consisting of two isothermal processes at temperatures $3T_0$ and T_0 and two constant-volume processes. For each cycle, determine in terms of n, R, and T_0 (a) the net heat transferred to the gas and (b) the efficiency of an engine operating in this cycle.

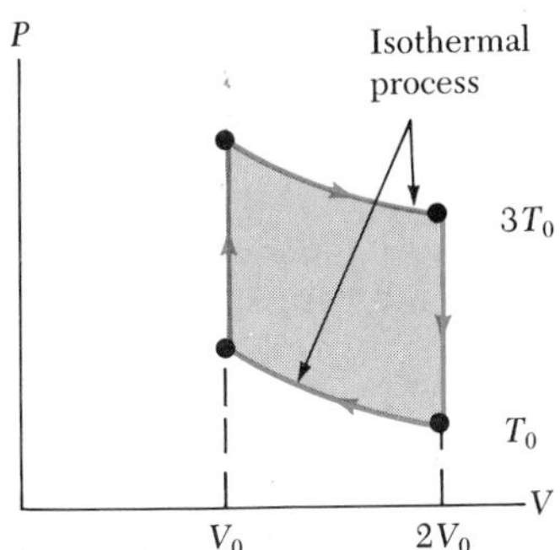

Figure 22.16 (Problem 35).

36. (a) Show that the work done by a system in any reversible cycle is given by $W = \oint T dS$, where the integral is over the closed path corresponding to the cyclic process. (*Hint:* Make use of the first law, $đQ = dU + P\,dV$, and the definition of the change in entropy as given by Equation 22.9.) (b) Construct an entropy-temperature plot for the Carnot cycle described in Figure 22.9. Identify the process associated with each line on the graph. What is the physical significance of the area enclosed by the cycle on this plot?

37. One mole of a monatomic ideal gas is taken through the reversible cycle shown in Figure 22.17. At point A, the pressure, volume, and temperature are P_0, V_0, and T_0, respectively. In terms of R and T_0, find (a) the total heat entering the system per cycle, (b) the total heat leaving the system per cycle, (c) the efficiency of an engine operating in this reversible cycle, and (d) the efficiency of an engine operating in a Carnot cycle between the same temperature extremes for this process.

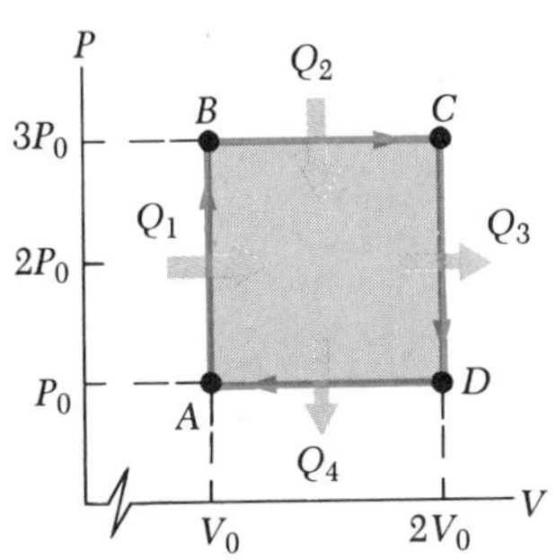

Figure 22.17 (Problem 37).

38. One mole of a diatomic ideal gas at an initial pressure of 4 atm and temperature of 300 K is carried through the following reversible cycle: (1) it expands isothermally until its volume is doubled; (2) it is compressed to its original volume at constant pressure; (3) it is compressed isothermally to a pressure of 4 atm; and (4) it expands at constant pressure to its original volume. (a) Make an accurate plot of the cyclic process on a PV diagram. (b) Calculate the work done by the gas per cycle. (c) Find the gross heat input and efficiency of the engine.

39. A system consisting of n moles of an ideal gas undergoes a reversible, *isobaric* process from a volume V_0 to a volume $3V_0$. Calculate the change in entropy of the gas. (*Hint:* Imagine that the system goes from the initial state to the final state first along an isotherm and then along an adiabatic curve, for which there is no change in entropy.)

40. An electrical power plant has an overall efficiency of 15%. The plant is to deliver 150 MW of power to a city, and its turbines use coal as the fuel. The burning coal produces steam at 190°C, which drives the turbines. This steam is then condensed into water at 25°C by passing it through cooling coils in contact with river water. (a) How many metric tons of coal does the plant consume each day (1 metric ton = 10^3 kg)? (b) What is the total cost of the fuel per year if the delivered price is $8/metric ton? (c) If the river water is delivered at 20°C, at what minimum rate must it flow over the cooling coils in order that its temperature not exceed 25°C? (*Note:* The heat of combustion of coal is 7.8×10^3 cal/g.)

•41. An idealized Diesel engine operates in a cycle known as the *air-standard Diesel cycle*, shown in Figure 22.18. Fuel is sprayed into the cylinder at the point of maximum compression, B. Combustion occurs during the expansion $B \rightarrow C$, which is approximated as an isobaric process. The rest of the cycle is the same as in the gasoline engine, described in Figure 22.11. Show that the efficiency of an engine operating in this idealized Diesel cycle is given by

$$e = 1 - \frac{1}{\gamma}\left(\frac{T_D - T_A}{T_C - T_B}\right)$$

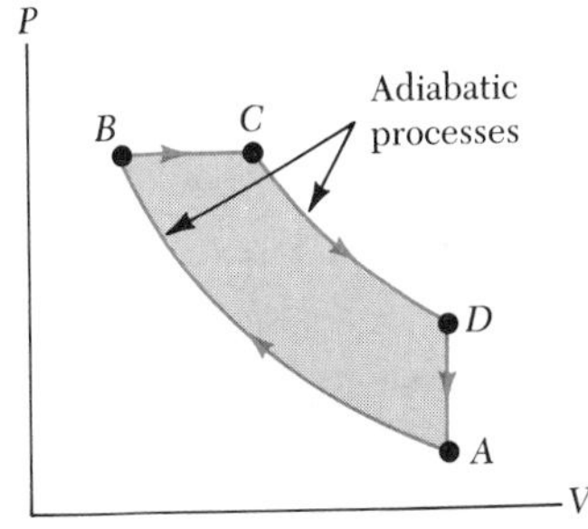

Figure 22.18 (Problem 41).

•**42.** One mole of an ideal gas ($\gamma = 1.4$) is carried through the Carnot cycle described in Figure 22.9. At point A, the pressure is 25 atm and the temperature is 600 K. At point C, the pressure is 1 atm and the temperature is 400 K. (a) Determine the pressures and volumes at points A, B, C, and D. (b) Calculate the net work done per cycle. (c) Determine the efficiency of an engine operating in this cycle.

•**43.** Consider one mole of an ideal gas which undergoes a quasi-static, reversible process in which the heat capacity C remains constant. Show that the pressure and volume of the gas obey the relation

$$PV^{\gamma'} = \text{constant, where} \quad \gamma' = \frac{C - C_p}{C - C_v}$$

(*Hint:* Start with the first law of thermodynamics, $đQ = dU + P\,dV$, and use the fact that $đQ = C\,dT$ and $dU = C_v\,dT$. Furthermore, note that $PV = RT$ applies to the gas, so it follows that $P\,dV + V\,dP = R\,dT$.)

•**44.** A typical human has a mass of 70 kg and produces about 2000 kcal (2×10^6 cal) of metabolic heat per day. (a) Find the rate of heat production in watts, and also express it in cal/h. (b) If none of the metabolic heat were lost, and assuming that the specific heat of the human body is 1 cal/g · °C, find the rate at which body temperature would rise. Give your answer in °C per hour and in °F per hour. (c) The result of (b) indicates that a human who is unable to transfer metabolic heat to the surroundings gets into serious trouble quickly. Assuming now that the heat transfer is rapid enough to maintain a constant body temperature, estimate the rate at which the entropy of the universe increases due to the release and transfer of metabolic heat. Express your result in cal/K/h. In making your estimate, assume that the given heat release is the only significant avenue of entropy production. Given this assumption, notice that there is no entropy increase of the body itself. (Why?) Use 20°C (68°F) for the temperature of the surroundings.

•**45.** A blacksmith plunges a hot iron horseshoe of mass 2 kg into a bucket containing 20 kg of water. The horseshoe is initially at a temperature of 600°C, and the water is initially at a temperature of 20°C. Assuming that no water is vaporized, find: (a) the final temperature of the water and the horseshoe, (b) the change of entropy of the horseshoe, (c) the change of entropy of the water, and (d) the overall change of entropy of the water and the horseshoe. (e) After a time interval, which is long compared with the time for the horseshoe to cool, the horseshoe and the water cool back to the temperature of the surroundings: 20°C. During this process, find the entropy changes of the water, the horseshoe, and the surroundings. (f) Using your results for (d) and (e), find the entropy change of the universe as a result of the entire sequence of events.

•**46.** Consider once more the situation described in Problem 45. A somewhat more realistic assumption is that water is heated to 100°C and vaporized in an amount sufficient to cool the horseshoe to 100°C, and thereafter the horseshoe equilibrates with the remaining water. Under this assumption: (a) How much water is vaporized? (b) What is the final temperature of the horseshoe and the remaining liquid water? (c) Identify where entropy changes occur and whether they are increases or decreases. (You need not give quantitative estimates for those changes.)

Appendix A

TABLE A.1 Conversion Factors

Length

	m	cm	km	in.	ft	mi
1 meter	1	10^2	10^{-3}	39.37	3.281	6.214×10^{-4}
1 centimeter	10^{-2}	1	10^{-5}	0.3937	3.281×10^{-2}	6.214×10^{-6}
1 kilometer	10^3	10^5	1	3.937×10^4	3.281×10^3	0.6214
1 inch	2.540×10^{-2}	2.540	2.540×10^{-5}	1	8.333×10^{-2}	1.578×10^{-5}
1 foot	0.3048	30.48	3.048×10^{-4}	12	1	1.894×10^{-4}
1 mile	1609	1.609×10^5	1.609	6.336×10^4	5280	1

Mass

	kg	g	slug	amu
1 kilogram	1	10^3	6.852×10^{-2}	6.024×10^{26}
1 gram	10^{-3}	1	6.852×10^{-5}	6.024×10^{23}
1 slug (lb/g)	14.59	1.459×10^4	1	8.789×10^{27}
1 atomic mass unit	1.660×10^{-27}	1.660×10^{-24}	1.137×10^{-28}	1

Time

	s	min	h	day	year
1 second	1	1.667×10^{-2}	2.778×10^{-4}	1.157×10^{-5}	3.169×10^{-8}
1 minute	60	1	1.667×10^{-2}	6.994×10^{-4}	1.901×10^{-6}
1 hour	3600	60	1	4.167×10^{-2}	1.141×10^{-4}
1 day	8.640×10^4	1440	24	1	2.738×10^{-3}
1 year (a)	3.156×10^7	5.259×10^5	8.766×10^3	365.2	1

Speed

	m/s	cm/s	ft/s	mi/h
1 meter/second	1	10^2	3.281	2.237
1 centimeter/second	10^{-2}	1	3.281×10^{-2}	2.237×10^{-2}
1 foot/second	0.3048	30.48	1	0.6818
1 mile/hour	0.4470	44.70	1.467	1

Note: 1 mi/min = 60 mi/h = 88 ft/s.

Force

	N	dyn	lb
1 newton	1	10^5	0.2248
1 dyne	10^{-5}	1	2.248×10^{-6}
1 pound	4.448	4.448×10^5	1

TABLE A.1 (Continued)

Work, Energy, Heat

	J	erg	ft · lb
1 joule	1	10^{7}	0.7376
1 erg	10^{-7}	1	7.376×10^{-8}
1 ft · lb	1.356	1.356×10^{7}	1
1 eV	1.602×10^{-19}	1.602×10^{-12}	1.182×10^{-19}
1 cal	4.186	4.186×10^{7}	3.087
1 Btu	1.055×10^{3}	1.055×10^{10}	7.779×10^{2}
1 kWh	3.600×10^{6}	3.600×10^{13}	2.655×10^{6}

	eV	cal	Btu	kWh
1 joule	6.242×10^{18}	0.2389	9.481×10^{-4}	2.778×10^{-7}
1 erg	6.242×10^{11}	2.389×10^{-8}	9.481×10^{-11}	2.778×10^{-14}
1 ft · lb	8.464×10^{18}	0.3239	1.285×10^{-3}	3.766×10^{-7}
1 eV	1	3.827×10^{-20}	1.519×10^{-22}	4.450×10^{-26}
1 cal	2.613×10^{19}	1	3.968×10^{-3}	1.163×10^{-6}
1 Btu	6.585×10^{21}	2.520×10^{2}	1	2.930×10^{-4}
1 kWh	2.247×10^{25}	8.601×10^{5}	3.413×10^{2}	1

Pressure

	N/m^2	dyn/cm^2	atm
1 newton/meter²	1	10	9.869×10^{-6}
1 dyne/centimeter²	10^{-1}	1	9.869×10^{-7}
1 atmosphere	1.013×10^{5}	1.013×10^{6}	1
1 centimeter mercury*	1.333×10^{3}	1.333×10^{4}	1.316×10^{-2}
1 pound/inch²	6.895×10^{3}	6.895×10^{4}	6.805×10^{-2}
1 pound/foot²	47.88	4.788×10^{2}	4.725×10^{-4}

	cm Hg	lb/in.²	lb/ft²
1 newton/meter²	7.501×10^{-4}	1.450×10^{-4}	2.089×10^{-2}
1 dyne/centimeter²	7.501×10^{-5}	1.450×10^{-5}	2.089×10^{-3}
1 atmosphere	76	14.70	2.116×10^{3}
1 centimeter mercury*	1	0.1943	27.85
1 pound/inch²	5.171	1	144
1 pound/foot²	3.591×10^{-2}	6.944×10^{-3}	1

* At 0°C and at a location where the acceleration due to gravity has its "standard" value, $9.80665\ m/s^2$.

TABLE A.2 Symbols, Dimensions, and Units of Physical Quantities

Quantity	Common Symbol	Unit*	Dimensions†	Unit in Terms of Basic SI Units
Acceleration	a	m/s^2	L/T^2	m/s^2
Angle	θ, ϕ	radian		
Angular acceleration	α	radian/s^2	T^{-2}	s^{-2}
Angular frequency	ω	radian/s	T^{-1}	s^{-1}
Angular momentum	L	kg · m^2/s	ML^2/T	kg · m^2/s
Angular velocity	ω	radian/s	T^{-1}	s^{-1}
Area	A	m^2	L^2	m^2
Atomic number	Z			
Capacitance	C	farad (F)(=C/V)	Q^{2}T^2/ML^2	A^2 · s^4/kg · m^2
Charge	q, Q, e	coulomb (C)	Q	A · s
Charge density				
Line	λ	C/m	Q/L	A · s/m
Surface	σ	C/m^2	Q/L^2	A · s/m^2
Volume	ρ	C/m^3	Q/L^3	A · s/m^3
Conductivity	σ	1/Ω · m	Q^{2}T/ML^3	A^2 · s^3/kg · m^3
Current	I	AMPERE	Q/T	A
Current density	J	A/m^2	Q/T^2	A/m^2
Density	ρ	kg/m^3	M/L^3	kg/m^3
Dielectric constant	κ			
Displacement	s	METER	L	m
Distance	d, h			
Length	ℓ, L			
Electric dipole moment	p	C · m	QL	A · s · m
Electric field	E	V/m	ML/QT^2	kg · m/A · s^3
Electric flux	Φ	V · m	ML^3/QT^2	kg · m^3/A · s^3
Electromotive force	$\mathcal{E}$	volt (V)	ML^2/QT^2	kg · m^2/A · s^3
Energy	E, U, K	joule (J)	ML^2/T^2	kg · m^2/s^2
Entropy	S	J/K	ML^2/T^2°K	kg · m^2/s^2 · K
Force	F	newton (N)	ML/T^2	kg · m/s^2
Frequency	f, v	hertz (Hz)	T^{-1}	s^{-1}
Heat	Q	joule (J)	ML^2/T^2	kg · m^2/s^2
Inductance	L	henry (H)	ML^2/Q^2	kg · m^2/A^2 · s^2
Magnetic dipole moment	$\mathfrak{M}$	N · m/T	QL^2/T	A · m^2
Magnetic field	B	tesla (T)(=Wb/m^2)	M/QT	kg/A · s^2
Magnetic flux	Φ_m	weber (Wb)	ML^2/QT	kg · m^2/A · s^2
Mass	m, M	KILOGRAM	M	kg
Molar specific heat	C	J/mole · K		kg · m^2/s^2 · kmole · K
Moment of inertia	I	kg · m^2	ML^2	kg · m^2
Momentum	p	kg · m/s	ML/T	kg · m/s
Period	T	s	T	s
Permeability of space	μ_o	N/A^2 (=H/m)	ML/Q^{2}T	kg · m/A^2 · s^2
Permittivity of space	ϵ_o	C^2/N · m^2 (= F/m)	Q^{2}T^2/ML^3	A^2 · s^4/kg · m^3
Potential (voltage)	V	volt (V)(=J/C)	ML^2/QT^2	kg · m^2/A · s^3
Power	P	watt (W)(=J/s)	ML^2/T^3	kg · m^2/s^3
Pressure	P, p	N/m^2	M/LT^2	kg/m · s^2
Resistance	R	ohm (Ω)(=V/A)	ML^2/Q^{2}T	kg · m^2/A^2 · s^3
Specific heat	c	J/kg · K	L^2/T^2°K	m^2/s^2 · K
Temperature	T	KELVIN	°K	K
Time	t	SECOND	T	s
Torque	τ	N · m	ML^2/T^2	kg · m^2/s^2
Velocity	v	m/s	L/T	m/s
Speed	v			

(Table continues)

TABLE A.2 (Continued)

Quantity	Common Symbol	Unit*	Dimensions†	Unit in Terms of Basic SI Units
Volume	V	m^3	L^3	m^3
Wavelength	λ	m	L	m
Work	W	joule (J)(=N · m)	ML^2/T^2	$kg \cdot m^2/s^2$

* The basic SI units are given in upper case letters.
† The symbols M, L, T, and Q denote mass, length, time, and charge, respectively.

TABLE A.3 Table of Selected Atomic Masses*

Atomic Number Z	Element	Symbol	Mass Number, A	Atomic Mass†	Percent Abundance, or Decay Mode (if radioactive)‡	Half-Life (if radioactive)
0	(Neutron)	n	1	1.008665	β^-	10.6 min
1	Hydrogen	H	1	1.007825	99.985	
	Deuterium	D	2	2.014102	0.015	
	Tritium	T	3	3.016049	β^-	12.33 yr
2	Helium	He	3	3.016029	0.00014	
			4	4.002603	≈100	
3	Lithium	Li	6	6.015123	7.5	
			7	7.016005	92.5	
4	Beryllium	Be	7	7.016930	EC, γ	53.3 days
			8	8.005305	2α	6.7×10^{-17} s
			9	9.012183	100	
5	Boron	B	10	10.012938	19.8	
			11	11.009305	80.2	
6	Carbon	C	11	11.011433	β^+, EC	20.4 min
			12	12.000000	98.89	
			13	13.003355	1.11	
			14	14.003242	β^-	5730 yr
7	Nitrogen	N	13	13.005739	β^+	9.96 min
			14	14.003074	99.63	
			15	15.000109	0.37	
8	Oxygen	O	15	15.003065	β^+, EC	122 s
			16	15.994915	99.76	
			18	17.999159	0.204	
9	Fluorine	F	19	18.998403	100	
10	Neon	Ne	20	19.992439	90.51	
			22	21.991384	9.22	
11	Sodium	Na	22	21.994435	β^+, EC, γ	2.602 yr
			23	22.989770	100	
			24	23.990964	β^-, γ	15.0 h
12	Magnesium	Mg	24	23.985045	78.99	
13	Aluminum	Al	27	26.981541	100	
14	Silicon	Si	28	27.976928	92.23	
			31	30.975364	β^-, γ	2.62 h
15	Phosphorus	P	31	30.973763	100	
			32	31.973908	β^-	14.28 days
16	Sulfur	S	32	31.972072	95.0	
			35	34.969033	β^-	87.4 days
17	Chlorine	Cl	35	34.968853	75.77	
			37	36.965903	24.23	

TABLE A.3 (Continued)

Atomic Number Z	Element	Symbol	Mass Number, A	Atomic Mass†	Percent Abundance, or Decay Mode (if radioactive)‡	Half-Life (if radioactive)
18	Argon	Ar	40	39.962383	99.60	
19	Potassium	K	39	38.963708	93.26	
			40	39.964000	β^-, EC, γ, β^+	1.28×10^9 yr
20	Calcium	Ca	40	39.962591	96.94	
21	Scandium	Sc	45	44.955914	100	
22	Titanium	Ti	48	47.947947	73.7	
23	Vanadium	V	51	50.943963	99.75	
24	Chromium	Cr	52	51.940510	83.79	
25	Manganese	Mn	55	54.938046	100	
26	Iron	Fe	56	55.934939	91.8	
27	Cobalt	Co	59	58.933198	100	
			60	59.933820	β^-, γ	5.271 yr
28	Nickel	Ni	58	57.935347	68.3	
			60	59.930789	26.1	
			64	63.927968	0.91	
29	Copper	Cu	63	62.929599	69.2	
			64	63.929766	β^-, β^+	12.7 h
			65	64.927792	30.8	
30	Zinc	Zn	64	63.929145	48.6	
			66	65.926035	27.9	
31	Gallium	Ga	69	68.925581	60.1	
32	Germanium	Ge	72	71.922080	27.4	
			74	73.921179	36.5	
33	Arsenic	As	75	74.921596	100	
34	Selenium	Se	80	79.916521	49.8	
35	Bromine	Br	79	78.918336	50.69	
36	Krypton	Kr	84	83.911506	57.0	
			89	88.917563	β^-	3.2 min
37	Rubidium	Rb	85	84.911800	72.17	
38	Strontium	Sr	86	85.909273	9.8	
			88	87.905625	82.6	
			90	89.907746	β^-	28.8 yr
39	Yttrium	Y	89	88.905856	100	
40	Zirconium	Zr	90	89.904708	51.5	
41	Niobium	Nb	93	92.906378	100	
42	Molybdenum	Mo	98	97.905405	24.1	
43	Technetium	Tc	98	97.907210	β^-, γ	4.2×10^6 yr
44	Ruthenium	Ru	102	101.904348	31.6	
45	Rhodium	Rh	103	102.90550	100	
46	Palladium	Pd	106	105.90348	27.3	
47	Silver	Ag	107	106.905095	51.83	
			109	108.904754	48.17	
48	Cadmium	Cd	114	113.903361	28.7	
49	Indium	In	115	114.90388	95.7; β^-	5.1×10^{14} yr
50	Tin	Sn	120	119.902199	32.4	
51	Antimony	Sb	121	120.903824	57.3	
52	Tellurium	Te	130	129.90623	34.5; β^-	2×10^{21} yr
53	Iodine	I	127	126.904477	100	
			131	130.906118	β^-, γ	8.04 days
54	Xenon	Xe	132	131.90415	26.9	
			136	135.90722	8.9	

TABLE A.3 (Continued)

Atomic Number Z	Element	Symbol	Mass Number, A	Atomic Mass†	Percent Abundance, or Decay Mode (if radioactive)‡	Half-Life (if radioactive)
55	Cesium	Cs	133	132.90543	100	
56	Barium	Ba	137	136.90582	11.2	
			138	137.90524	71.7	
			144	143.922673	β^-	11.9 s
57	Lanthanum	La	139	138.90636	99.911	
58	Cerium	Ce	140	139.90544	88.5	
59	Praseodymium	Pr	141	140.90766	100	
60	Neodymium	Nd	142	141.90773	27.2	
61	Promethium	Pm	145	144.91275	EC, α, γ	17.7 yr
62	Samarium	Sm	152	151.91974	26.6	
63	Europium	Eu	153	152.92124	52.1	
64	Gadolinium	Gd	158	157.92411	24.8	
65	Terbium	Tb	159	158.92535	100	
66	Dysprosium	Dy	164	163.92918	28.1	
67	Holmium	Ho	165	164.93033	100	
68	Erbium	Er	166	165.93031	33.4	
69	Thulium	Tm	169	168.93423	100	
70	Ytterbium	Yb	174	173.93887	31.6	
71	Lutecium	Lu	175	174.94079	97.39	
72	Hafnium	Hf	180	179.94656	35.2	
73	Tantalum	Ta	181	180.94801	99.988	
74	Tungsten (wolfram)	W	184	183.95095	30.7	
75	Rhenium	Re	187	186.95577	62.60, β^-	4×10^{10} yr
76	Osmium	Os	191	190.96094	β^-, γ	15.4 days
			192	191.96149	41.0	
77	Iridium	Ir	191	190.96060	37.3	
			193	192.96294	62.7	
78	Platinum	Pt	195	194.96479	33.8	
79	Gold	Au	197	196.96656	100	
80	Mercury	Hg	202	201.97063	29.8	
81	Thallium	Tl	205	204.97441	70.5	
82	Lead	Pb	210	209.990069	β^-	1.3 min
			204	203.973044	β^-, 1.48	1.4×10^{17} yr
			206	205.97446	24.1	
			207	206.97589	22.1	
			208	207.97664	52.3	
			210	209.98418	α, β^-, γ	22.3 yr
			211	210.98874	β^-, γ	36.1 min
			212	211.99188	β^-, γ	10.64 h
			214	213.99980	β^-, γ	26.8 min
83	Bismuth	Bi	209	208.98039	100	
			211	210.98726	α, β^-, γ	2.15 min
84	Polonium	Po	210	209.98286	α, γ	138.38 days
			214	213.99519	α, γ	164 μs
85	Astatine	At	218	218.00870	α, β^-	$\approx$2 s
86	Radon	Rn	222	222.017574	α, γ	3.8235 days
87	Francium	Fr	223	223.019734	α, β^-, γ	21.8 min
88	Radium	Ra	226	226.025406	α, γ	1.60×10^3 yr
			228	228.031069	β^-	5.76 yr

TABLE A.3 (Continued)

Atomic Number Z	Element	Symbol	Mass Number, A	Atomic Mass†	Percent Abundance, or Decay Mode (if radioactive)‡	Half-Life (if radioactive)
89	Actinium	Ac	227	227.027751	α, β^-, γ	21.773 yr
90	Thorium	Th	228	228.02873	α, γ	1.9131 yr
			232	232.038054	100, α, γ	1.41×10^{10} yr
91	Protactinium	Pa	231	231.035881	α, γ	3.28×10^4 yr
92	Uranium	U	232	232.03714	α, γ	72 yr
			233	233.039629	α, γ	1.592×10^5 yr
			235	235.043925	0.72; α, γ	7.038×10^8 yr
			236	236.045563	α, γ	2.342×10^7 yr
			238	238.050786	99.275; α, γ	4.468×10^9 yr
			239	239.054291	β^-, γ	23.5 min
93	Neptunium	Np	239	239.052932	β^-, γ	2.35 days
94	Plutonium	Pu	239	239.052158	α, γ	2.41×10^4 yr
95	Americium	Am	243	243.061374	α, γ	7.37×10^3 yr
96	Curium	Cm	245	245.065487	α, γ	8.5×10^3 yr
97	Berkelium	Bk	247	247.07003	α, γ	1.4×10^3 yr
98	Californium	Cf	249	249.074849	α, γ	351 yr
99	Einsteinium	Es	254	254.08802	α, γ, β^-	276 days
100	Fermium	Fm	253	253.08518	EC, α, γ	3.0 days
101	Mendelevium	Md	255	255.0911	EC, α	27 min
102	Nobelium	No	255	255.0933	EC, α	3.1 min
103	Lawrencium	Lr	257	257.0998	α	$\approx$35 s
104	Rutherfordium (?)	Rf	261	261.1087	α	1.1 min
105	Hahnium (?)	Ha	262	262.1138	α	0.7 min
106			263	263.1184	α	0.9 s
107			261	261	α	1–2 ms

° Data are taken from *Chart of the Nuclides,* 12th ed., General Electric, 1977, and from C. M. Lederer and V. S. Shirley, eds., *Table of Isotopes,* 7th ed., John Wiley & Sons, Inc., New York, 1978.

† The masses given in column (5) are those for the neutral atom, including the Z electrons.

‡ The process EC stands for "electron capture."

Appendix B
Mathematics Review

These appendices in mathematics are intended as a brief review of operations and methods. Early in this course, you should be totally familiar with basic algebraic techniques, analytic geometry, and trigonometry. The appendices on differential and integral calculus are more detailed and are intended for those students who have difficulties in applying calculus concepts to physical situations.

B.1 SCIENTIFIC NOTATION

Many quantities that scientists deal with often have very large or very small values. For example, the speed of light is about 300 000 000 m/s and the ink required to make the dot over an *i* in this textbook has a mass of about 0.000000001 kg. Obviously, it is very cumbersome to read, write, and keep track of numbers such as these. We avoid this problem by using a method dealing with powers of the number 10:

$$10^0 = 1$$

$$10^1 = 10$$

$$10^2 = 10 \times 10 = 100$$

$$10^3 = 10 \times 10 \times 10 = 1000$$

$$10^4 = 10 \times 10 \times 10 \times 10 = 10,000$$

$$10^5 = 10 \times 10 \times 10 \times 10 \times 10 = 100,000$$

and so on. The number of zeros corresponds to the power to which 10 is raised, called the **exponent** of 10. For example, the speed of light, 300 000 000 m/s, can be expressed as 3×10^8 m/s.

For numbers less than one, we note the following:

$$10^{-1} = \frac{1}{10} = 0.1$$

$$10^{-2} = \frac{1}{10 \times 10} = 0.01$$

$$10^{-3} = \frac{1}{10 \times 10 \times 10} = 0.001$$

$$10^{-4} = \frac{1}{10 \times 10 \times 10 \times 10} = 0.0001$$

$$10^{-5} = \frac{1}{10 \times 10 \times 10 \times 10 \times 10} = 0.00001$$

In these cases, the number of places the decimal point is to the left of the digit 1 equals the value of the (negative) exponent. Numbers that are expressed as some power of 10 multiplied by another number between 1 and 10 are said to be in **scientific notation.** For example, the scientific notation for 5 943 000 000 is 5.943×10^9 and that for 0.0000832 is 8.32×10^{-5}.

When numbers expressed in scientific notation are being multiplied, the following general rule is very useful:

$$10^n \times 10^m = 10^{n+m} \qquad \text{(B.1)}$$

where n and m can be *any* numbers (not necessarily integers). For example, $10^2 \times 10^5 = 10^7$. The rule also applies if one of the exponents is negative. For example, $10^3 \times 10^{-8} = 10^{-5}$.

When dividing numbers expressed in scientific notation, note that

$$\frac{10^n}{10^m} = 10^n \times 10^{-m} = 10^{n-m} \qquad \text{(B.2)}$$

EXERCISES

With help from the above rules, verify the answers to the following:

1. $86{,}400 = 8.64 \times 10^4$
2. $9{,}816{,}762.5 = 9.8167625 \times 10^6$
3. $0.0000000398 = 3.98 \times 10^{-8}$
4. $(4 \times 10^8)(9 \times 10^9) = 3.6 \times 10^{18}$
5. $(3 \times 10^7)(6 \times 10^{-12}) = 1.8 \times 10^{-4}$
6. $\dfrac{75 \times 10^{-11}}{5 \times 10^{-3}} = 1.5 \times 10^{-7}$
7. $\dfrac{(3 \times 10^6)(8 \times 10^{-2})}{(2 \times 10^{17})(6 \times 10^5)} = 2 \times 10^{-18}$

B.2 ALGEBRA

Some Basic Rules

When algebraic operations are performed, the laws of arithmetic apply. Symbols such as x, y, and z are usually used to represent quantities that are not specified, what are called the **unknowns.**

First, consider the equation

$$8x = 32$$

If we wish to solve for x, we can divide (or multiply) each side of the equation by the same factor without destroying the equality. In this case, if we divide both sides by 8, we have

$$\frac{8x}{8} = \frac{32}{8}$$

$$x = 4$$

Next consider the equation

$$x + 2 = 8$$

In this type of expression, we can add or subtract the same quantity from each side. If we subtract 2 from each side, we get

$$x + 2 - 2 = 8 - 2$$

$$x = 6$$

In general, if $x + a = b$, then $x = b - a$.

Now consider the equation

$$\frac{x}{5} = 9$$

If we multiply each side by 5, we are left with x on the left by itself and 45 on the right:

$$\left(\frac{x}{5}\right)(5) = 9 \times 5$$

$$x = 45$$

In all cases, *whatever operation is performed on the left side of the equality must also be performed on the right side.*

The following rules for multiplying, dividing, adding, and subtracting fractions should be recalled, where a, b, and c are three numbers:

	Rule	Example
Multiplying	$\left(\frac{a}{b}\right)\left(\frac{c}{d}\right) = \frac{ac}{bd}$	$\left(\frac{2}{3}\right)\left(\frac{4}{5}\right) = \frac{8}{15}$
Dividing	$\frac{(a/b)}{(c/d)} = \frac{ad}{bc}$	$\frac{2/3}{4/5} = \frac{(2)(5)}{(4)(3)} = \frac{10}{12}$
Adding	$\frac{a}{b} \pm \frac{c}{d} = \frac{ad \pm bc}{bd}$	$\frac{2}{3} - \frac{4}{5} = \frac{(2)(5) - (4)(3)}{(3)(5)} = -\frac{2}{15}$

EXERCISES

In the following exercises, solve for x:

	Answers
1. $a = \frac{1}{1 + x}$	$x = \frac{1 - a}{a}$
2. $3x - 5 = 13$	$x = 6$
3. $ax - 5 = bx + 2$	$x = \frac{7}{a - b}$
4. $\frac{5}{2x + 6} = \frac{3}{4x + 8}$	$x = -\frac{11}{7}$

Powers

When powers of a given quantity x are multiplied, the following rule applies:

$$x^n x^m = x^{n+m} \tag{B.3}$$

For example, $x^2 x^4 = x^{2+4} = x^6$.

When dividing the powers of a given quantity, note that

$$\frac{x^n}{x^m} = x^{n-m} \qquad \text{(B.4)}$$

For example, $x^8/x^2 = x^{8-2} = x^6$.

A power that is a fraction, such as $\frac{1}{3}$, corresponds to a root as follows:

$$x^{1/n} = \sqrt[n]{x} \qquad \text{(B.5)}$$

For example, $4^{1/3} = \sqrt[3]{4} = 1.5874$. (A scientific calculator is useful for such calculations.)

Finally, any quantity x^n that is raised to the mth power is

$$(x^n)^m = x^{nm} \qquad \text{(B.6)}$$

Table B.1 summarizes the rules of exponents.

TABLE B.1 Rules of Exponents

$x^0 = 1$
$x^1 = x$
$x^n x^m = x^{n+m}$
$x^n/x^m = x^{n-m}$
$x^{1/n} = \sqrt[n]{x}$
$(x^n)^m = x^{nm}$

EXERCISES

Verify the following:

1. $3^2 \times 3^3 = 243$
2. $x^5x^{-8} = x^{-3}$
3. $x^{10}/x^{-5} = x^{15}$
4. $5^{1/3} = 1.709975$ (Use your calculator.)
5. $60^{1/4} = 2.783158$ (Use your calculator.)
6. $(x^4)^3 = x^{12}$

Factoring

Some useful formulas for factoring an equation are

$ax + ay + az = a(x + y + x)$ common factor

$a^2 + 2ab + b^2 = (a + b)^2$ perfect square

$a^2 - b^2 = (a + b)(a - b)$ differences of squares

Quadratic Equations

The general form of a quadratic equation is

$$ax^2 + bx + c = 0 \qquad \text{(B.7)}$$

where x is the unknown quantity and a, b, and c are numerical factors referred to as **coefficients** of the equation. This equation has two roots, given by

$$x = \frac{-b \pm \sqrt{b^2 - 4ac}}{2a} \qquad \text{(B.8)}$$

If $b^2 \geq 4ac$, the roots will be real.

EXAMPLE 1

The equation $x^2 + 5x + 4 = 0$ has the following roots corresponding to the two signs of the square-root term:

$$x = \frac{-5 \pm \sqrt{5^2 - (4)(1)(4)}}{2(1)} = \frac{-5 \pm \sqrt{9}}{2} = \frac{-5 \pm 3}{2}$$

that is,

$$x_+ = \frac{-5 + 3}{2} = -1 \qquad x_- = \frac{-5 - 3}{2} = -4$$

where x_+ refers to the root corresponding to the positive sign and x_- refers to the root corresponding to the negative sign.

EXERCISES

Solve the following quadratic equations:

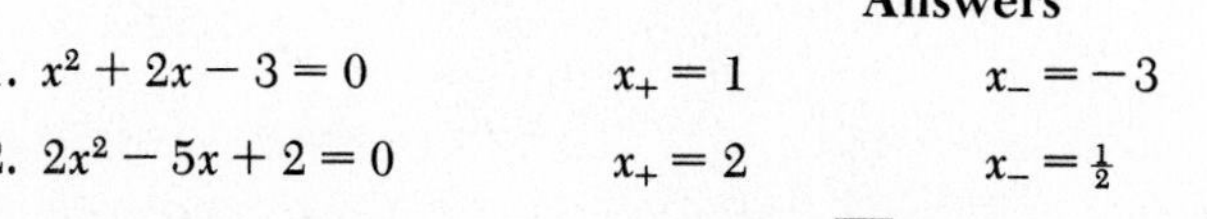

	Answers	
1. $x^2 + 2x - 3 = 0$	$x_+ = 1$	$x_- = -3$
2. $2x^2 - 5x + 2 = 0$	$x_+ = 2$	$x_- = \frac{1}{2}$
3. $2x^2 - 4x - 9 = 0$	$x_+ = 1 + \sqrt{22}/2$	$x_- = 1 - \sqrt{22}/2$

Linear Equations

A linear equation has the general form

$$y = ax + b \tag{B.9}$$

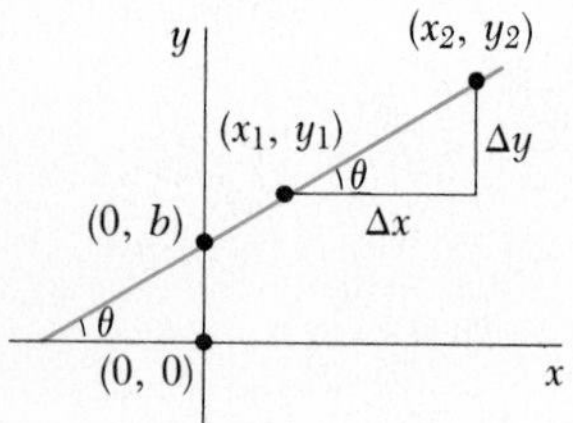

Figure B.1

where a and b are constants. This equation is referred to as being linear because the graph of y versus x is a straight line, as shown in Figure B.1. The constant b, called the **intercept**, represents the value of y at which the straight line intersects the y axis. The constant a is equal to the **slope** of the straight line and is also equal to the tangent of the angle that the line makes with the x axis. If any two points on the straight line are specified by the coordinates (x_1, y_1) and (x_2, y_2), as in Figure B.1, then the **slope** of the straight line can be expressed

$$\text{Slope} = \frac{y_2 - y_1}{x_2 - x_1} = \frac{\Delta y}{\Delta x} = \tan\theta \tag{B.10}$$

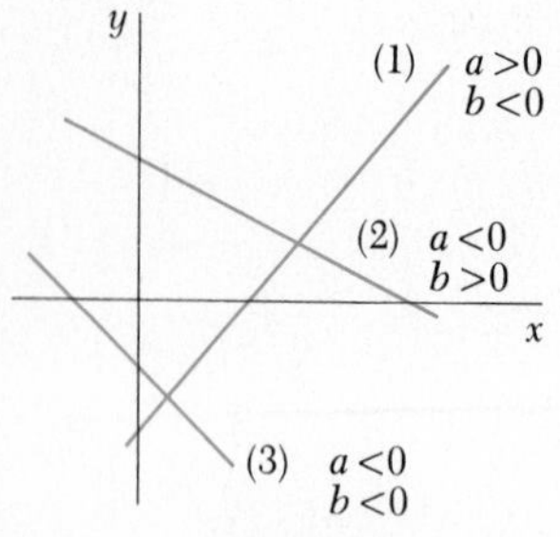

Figure B.2

Note that a and b can have either positive or negative values. If $a > 0$, the straight line has a *positive* slope, as in Figure B.1. If $a < 0$, the straight line has a *negative* slope. In Figure B.1, both a and b are positive. Three other possible situations are shown in Figure B.2: $a > 0, b < 0$; $a < 0, b > 0$; and $a < 0, b < 0$.

EXERCISES

1. Draw graphs of the following straight lines:
(a) $y = 5x + 3$ (b) $y = -2x + 4$ (c) $y = -3x - 6$

2. Find the slopes of the straight lines described in Exercise 1.
Answers: (a) 5, (b) -2, (c) -3

3. Find the slopes of the straight lines that pass through the following sets of points: (a) $(0, -4)$ and $(4, 2)$, (b) $(0, 0)$ and $(2, -5)$, and (c) $(-5, 2)$ and $(4, -2)$
Answers: (a) 3/2 (b) $-5/2$ (c) $-4/9$

Solving Simultaneous Linear Equations

Consider an equation such as $3x + 5y = 15$, which has two unknowns, x and y. Such an equation does not have a unique solution. That is, $(x = 0,\ y = 3)$, $(x = 5,\ y = 0)$, and $(x = 2,\ y = 9/5)$ are all solutions to this equation.

If a problem has two unknowns, a unique solution is possible only if we have *two* equations. In general, if a problem has n unknowns, its solution requires n equations. In order to solve two simultaneous equations involving two unknowns, x and y, we solve one of the equations for x in terms of y and substitute this expression into the other equation.

EXAMPLE 2

Solve the following two simultaneous equations:

$$(1)\ 5x + y = -8 \qquad (2)\ 2x - 2y = 4$$

Solution: From (2), $x = y + 2$. Substitution of this into (1) gives

$$5(y + 2) + y = -8$$
$$6y = -18$$
$$y = -3$$
$$x = y + 2 = -1$$

Alternate solution: Multiply each term in (1) by the factor 2 and add the result to (2):

$$10x + 2y = -16$$
$$\underline{2x - 2y = 4}$$
$$12x = -12$$
$$x = -1$$
$$y = x - 2 = -3$$

Two linear equations with two unknowns can also be solved by a graphical method. If the straight lines corresponding to the two equations are plotted in a conventional coordinate system, the intersection of the two lines represents the solution. For example, consider the two equations

$$x - y = 2$$
$$x - 2y = -1$$

These are plotted in Figure B.3. The intersection of the two lines has the coordinates $x = 5$, $y = 3$. This represents the solution to the equations. You should check this solution by the analytical technique discussed above.

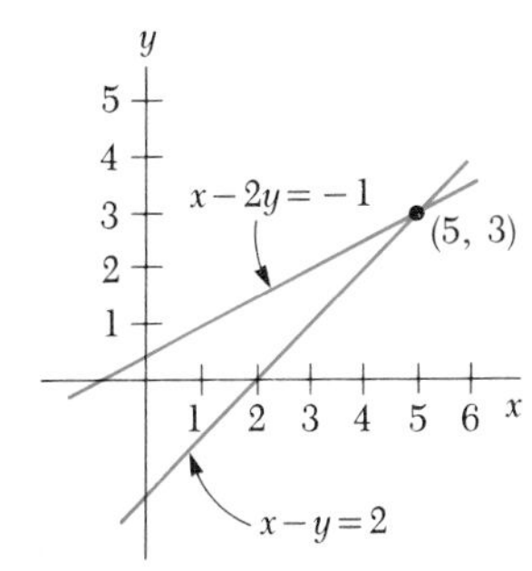

Figure B.3

EXERCISES

Solve the following pairs of simultaneous equations involving two unknowns:

		Answers
1.	$x + y = 8$ $x - y = 2$	$x = 5,\ y = 3$
2.	$98 - T = 10a$ $T - 49 = 5a$	$T = 65,\ a = 3.27$
3.	$6x + 2y = 6$ $8x - 4y = 28$	$x = 2,\ y = -3$

Logarithms

Suppose that a quantity x is expressed as a power of some quantity a:

$$x = a^y \tag{B.11}$$

The number y is called the **base** number. The **logarithm** of x with respect to the base a is equal to the exponent to which the base must be raised in order to satisfy the expression $x = a^y$:

$$y = \log_a x \tag{B.12}$$

Conversely, the **antilogarithm** of y is the number x:

$$x = \text{antilog}_a\, y \tag{B.13}$$

In practice, the two bases most often used are base 10, called the *common* logarithm base, and base $e = 2.718$ . . ., called the *natural* logarithm base. When common logarithms are used,

$$y = \log_{10} x \qquad (\text{or } x = 10^y) \tag{B.14}$$

When natural logarithms are used,

$$y = \ln_e x \qquad (\text{or } x = e^y) \tag{B.15}$$

For example, $\log_{10} 52 = 1.716$, so that $\text{antilog}_{10}\, 1.716 = 10^{1.716} = 52$. Likewise, $\ln_e 52 = 3.951$, so $\text{antiln}_e\, 3.951 = e^{3.951} = 52$.

In general, note that you can convert between base 10 and base e with the equality

$$\ln_e x = (2.302585) \log_{10} x \tag{B.16}$$

Finally, some useful properties of logarithms are as follows:

$$\log (ab) = \log a + \log b$$
$$\log (a/b) = \log a - \log b$$
$$\log (a^n) = n \log a$$
$$\ln e = 1$$
$$\ln e^a = a$$
$$\ln \left(\frac{1}{a}\right) = -\ln a$$

B.3 GEOMETRY

The **distance** d between two points whose coordinates are (x_1, y_1) and (x_2, y_2)

$$d = \sqrt{(x_2 - x_1)^2 + (y_2 - y_1)^2} \tag{B.17}$$

The **radian measure:** the arc length s of a circular arc (Fig. B.4) is proportional to the radius r for a fixed value of θ (in radians)

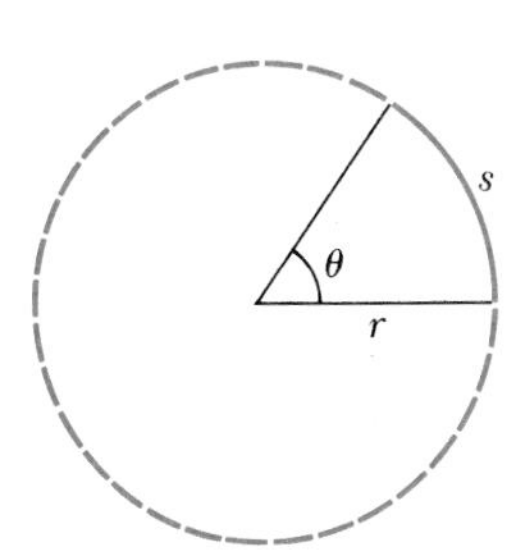

Figure B.4

$$s = r\theta \tag{B.18}$$

$$\theta = \frac{s}{r}$$

Table B.2 gives the areas and volumes for several geometric shapes used throughout this text:

The equation of a **straight line** (Fig. B.5) is given by

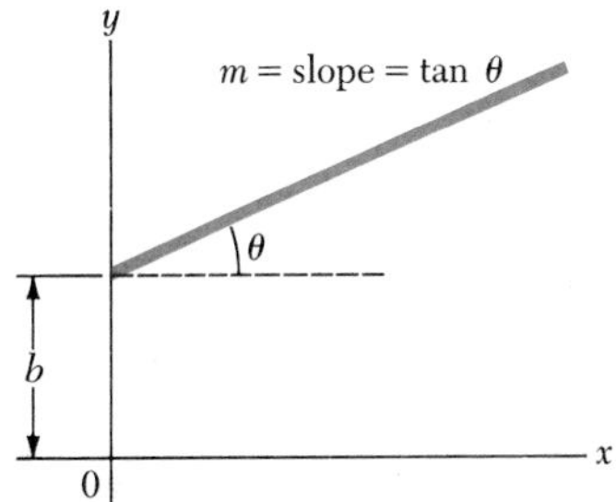

Figure B.5

$$y = mx + b \tag{B.19}$$

where b is the y intercept and m is the slope of the line.

The equation of a **circle** of radius R centered at the origin is

$$x^2 + y^2 = R^2 \tag{B.20}$$

The equation of an **ellipse** with the origin at its center (Fig. B.6) is

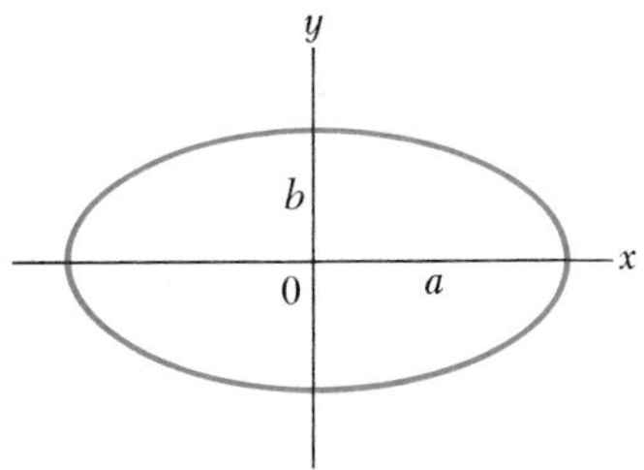

Figure B.6

$$\frac{x^2}{a^2} + \frac{y^2}{b^2} = 1 \tag{B.21}$$

TABLE B.2 Useful Information for Geometry

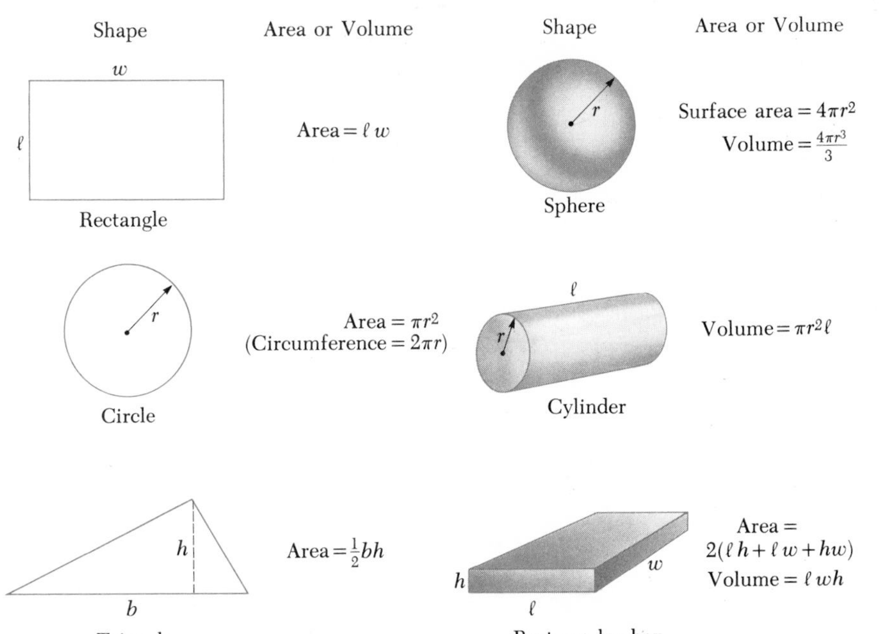

Shape	Area or Volume	Shape	Area or Volume
Rectangle (w, ℓ)	Area $= \ell w$	Sphere (r)	Surface area $= 4\pi r^2$ Volume $= \frac{4\pi r^3}{3}$
Circle (r)	Area $= \pi r^2$ (Circumference $= 2\pi r$)	Cylinder (r, ℓ)	Volume $= \pi r^2 \ell$
Triangle (h, b)	Area $= \frac{1}{2}bh$	Rectangular box (h, w, ℓ)	Area $=$ $2(\ell h + \ell w + hw)$ Volume $= \ell wh$

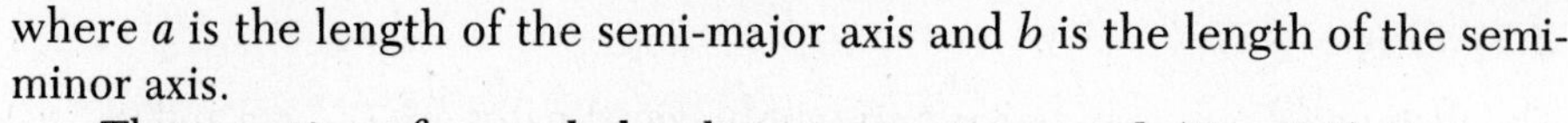
where a is the length of the semi-major axis and b is the length of the semi-minor axis.

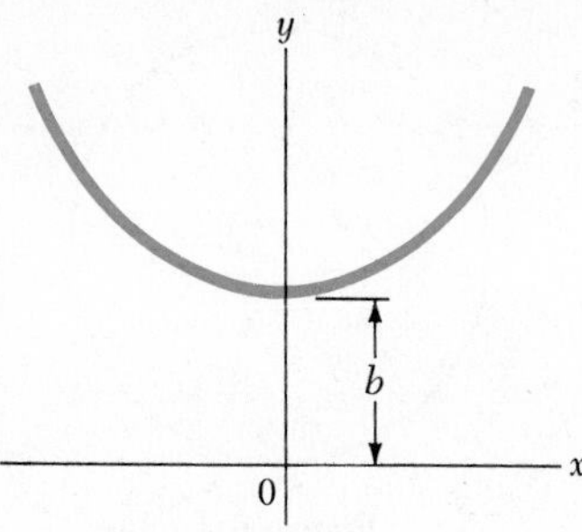

Figure B.7

The equation of a **parabola** whose vertex is at $y = b$ (Fig. B.7) is

$$y = ax^2 + b \tag{B.22}$$

The equation of a **rectangular hyperbola** (Fig. B.8) is

$$xy = \text{constant} \tag{B.23}$$

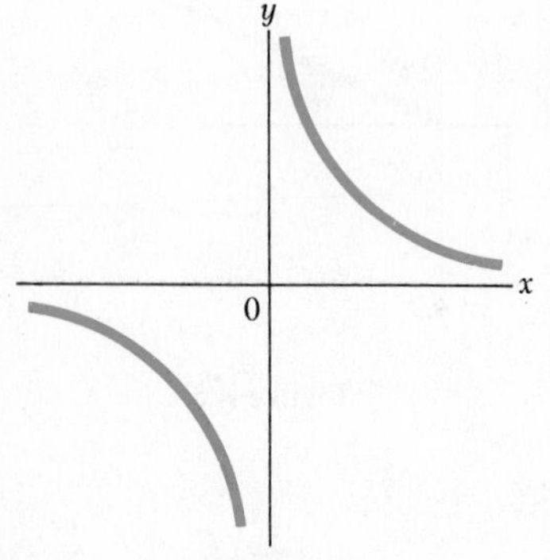

Figure B.8

B.4 TRIGONOMETRY

That portion of mathematics based on the special properties of the right triangle is called trigonometry. By definition, a right triangle is one containing a 90° angle. Consider the right triangle shown in Figure B.9, where side a is opposite the angle θ, side b is adjacent to the angle θ, and side c is the hypotenuse of the triangle. The three basic trigonometric functions defined by such a triangle are the sine (sin), cosine (cos), and tangent (tan) functions. In terms of the angle θ, these functions are defined by

$$\sin\theta \equiv \frac{\text{side opposite } \theta}{\text{hypotenuse}} = \frac{a}{c} \tag{B.24}$$

$$\cos\theta \equiv \frac{\text{side adjacent to } \theta}{\text{hypotenuse}} = \frac{b}{c} \tag{B.25}$$

$$\tan\theta \equiv \frac{\text{side opposite } \theta}{\text{side adjacent to } \theta} = \frac{a}{b} \tag{B.26}$$

The Pythagorean theorem provides the following relationship between the sides of a triangle:

$$c^2 = a^2 + b^2 \tag{B.27}$$

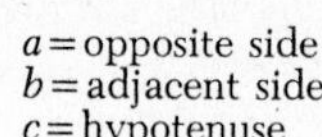

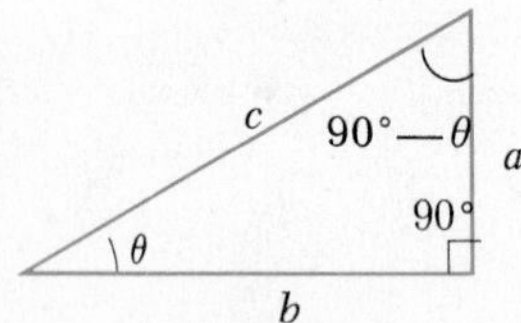

Figure B.9

From the above definitions and the Pythagorean theorem, it follows that

$$\sin^2\theta + \cos^2\theta = 1$$

$$\tan\theta = \frac{\sin\theta}{\cos\theta}$$

The cosecant, secant, and cotangent functions are defined by

$$\csc\theta \equiv \frac{1}{\sin\theta} \qquad \sec\theta \equiv \frac{1}{\cos\theta} \qquad \cot\theta \equiv \frac{1}{\tan\theta}$$

The relations below follow directly from the right triangle shown in Figure B.9:

$$\begin{cases} \sin\theta = \cos(90^\circ - \theta) \\ \cos\theta = \sin(90^\circ - \theta) \\ \cot\theta = \tan(90^\circ - \theta) \end{cases}$$

TABLE B.3 Some Trigonometric Identities

$\sin^2\theta + \cos^2\theta = 1$	$\csc^2\theta = 1 + \cot^2\theta$
$\sec^2\theta = 1 + \tan^2\theta$	$\sin^2\dfrac{\theta}{2} = \frac{1}{2}(1 - \cos\theta)$
$\sin 2\theta = 2\sin\theta\cos\theta$	$\cos^2\dfrac{\theta}{2} = \frac{1}{2}(1 + \cos\theta)$
$\cos 2\theta = \cos^2\theta - \sin^2\theta$	$1 - \cos\theta = 2\sin^2\dfrac{\theta}{2}$
$\tan 2\theta = \dfrac{2\tan\theta}{1 - \tan^2\theta}$	$\tan\dfrac{\theta}{2} = \sqrt{\dfrac{1-\cos\theta}{1+\cos\theta}}$
$\sin(A \pm B) = \sin A\cos B \pm \cos A\sin B$	
$\cos(A \pm B) = \cos A\cos B \mp \sin A\sin B$	
$\sin A \pm \sin B = 2\sin[\frac{1}{2}(A \pm B)]\cos[\frac{1}{2}(A \mp B)]$	
$\cos A + \cos B = 2\cos[\frac{1}{2}(A + B)]\cos[\frac{1}{2}(A - B)]$	
$\cos A - \cos B = 2\sin[\frac{1}{2}(A + B)]\sin[\frac{1}{2}(B - A)]$	

Some properties of trigonometric functions are as follows:

$$\begin{cases} \sin(-\theta) = -\sin\theta \\ \cos(-\theta) = \cos\theta \\ \tan(-\theta) = -\tan\theta \end{cases}$$

The following relations apply to *any* triangle as shown in Figure B.10:

$$\alpha + \beta + \gamma = 180°$$

Law of cosines $$\begin{cases} a^2 = b^2 + c^2 - 2bc\cos\alpha \\ b^2 = a^2 + c^2 - 2ac\cos\beta \\ c^2 = a^2 + b^2 - 2ab\cos\gamma \end{cases}$$

Law of sines $$\left\{ \frac{a}{\sin\alpha} = \frac{b}{\sin\beta} = \frac{c}{\sin\gamma} \right.$$

Table B.3 lists a number of useful trigonometric identities.

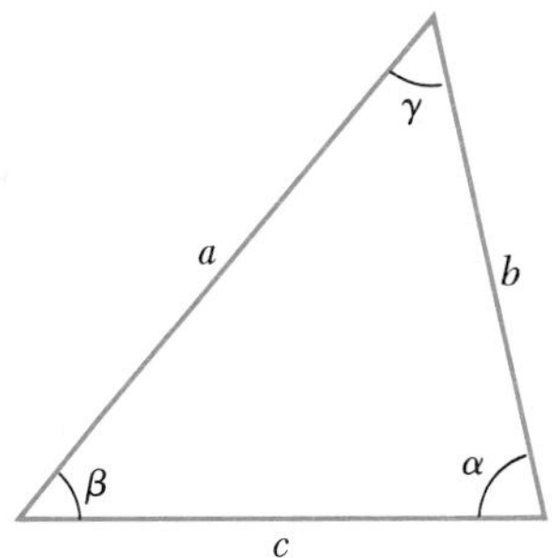

Figure B.10

EXAMPLE 3

Consider the right triangle in Figure B.11, in which $a = 2$, $b = 5$, and c is unknown. From the Pythagorean theorem, we have

$$c^2 = a^2 + b^2 = 2^2 + 5^2 = 4 + 25 = 29$$

$$c = \sqrt{29} = 5.39$$

To find the angle θ, note that

$$\tan\theta = \frac{a}{b} = \frac{2}{5} = 0.400$$

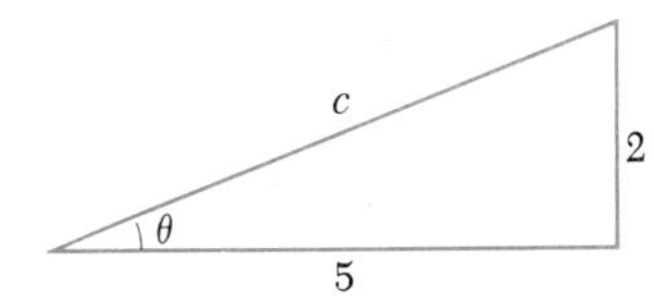

Figure B.11

From a table of functions or from a calculator, we have

$$\theta = \tan^{-1}(0.400) = 21.8°$$

where $\tan^{-1}(0.400)$ is the notation for "angle whose tangent is 0.400," sometimes written as arctan(0.400).

EXERCISES

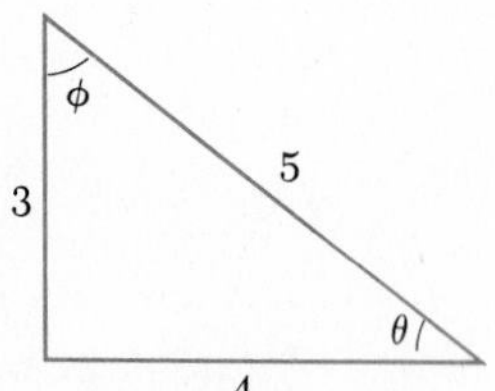

Figure B.12

1. In Figure B.12, find (a) the side opposite θ, (b) the side adjacent to ϕ, (c) $\cos\theta$, (d) $\sin\phi$, and (e) $\tan\phi$.
Answers: (a) 3, (b) 3, (c) $\frac{4}{5}$, (d) $\frac{4}{5}$, and (e) $\frac{4}{3}$

2. In a certain right triangle, the two sides that are perpendicular to each other are 5 m and 7 m long. What is the length of the third side of the triangle?
Answer: 8.60 m

3. A right triangle has a hypotenuse of length 3 m, and one of its angles is 30°. What is the length of (a) the side opposite the 30° angle and (b) the side adjacent to the 30° angle?
Answers: (a) 1.5 m and (b) 2.60 m

B.5 SERIES EXPANSIONS

$$(a+b)^n = a^n + \frac{n}{1!}a^{n-1}b + \frac{n(n-1)}{2!}a^{n-2}b^2 + \cdots$$

$$(1+x)^n = 1 + nx + \frac{n(n-1)}{2!}x^2 + \cdots$$

$$e^x = 1 + x + \frac{x^2}{2!} + \frac{x^3}{3!} + \cdots$$

$$\ln(1 \pm x) = \pm x - \tfrac{1}{2}x^2 \pm \tfrac{1}{3}x^3 - \cdots$$

$$\left.\begin{aligned} \sin x &= x - \frac{x^3}{3!} + \frac{x^5}{5!} - \cdots \\ \cos x &= 1 - \frac{x^2}{2!} + \frac{x^4}{4!} - \cdots \\ \tan x &= x + \frac{x^3}{3} + \frac{2x^5}{15} + \cdots \quad |x| < \pi/2 \end{aligned}\right\} x \text{ in radians}$$

For $x \ll 1$, the following approximations can be used:

$$(1+x)^n \approx 1 + nx \qquad \sin x \approx x$$

$$e^x \approx 1 + x \qquad \cos x \approx 1$$

$$\ln(1 \pm x) \approx \pm x \qquad \tan x \approx x$$

B.6 DIFFERENTIAL CALCULUS

In various branches of science, it is sometimes necessary to use the basic tools of calculus, first invented by Newton, to describe physical phenomena. The use of calculus is fundamental in the treatment of various problems in newtonian mechanics, electricity, and magnetism. In this section, we simply state some basic properties and "rules of thumb" that should be a useful review to the student.

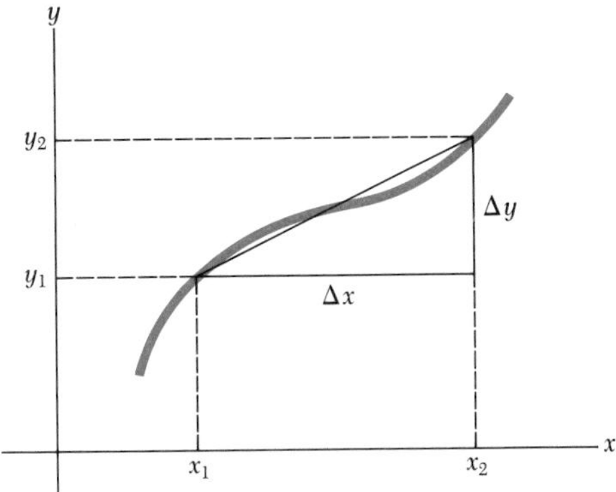

Figure B.13

First, a **function** must be specified that relates one variable to another (such as coordinate as a function of time). Suppose one of the variables is called y (the dependent variable), the other x (the independent variable). We might have a function relation such as

$$y(x) = ax^3 + bx^2 + cx + d$$

If a, b, c, and d are specified constants, then y can be calculated for any value of x. We usually deal with continuous functions, that is, those for which y varies "smoothly" with x.

The **derivative** of y with respect to x is defined as the limit of the slopes of chords drawn between two points on the y versus x curve as Δx approaches zero. Mathematically, we write this definition as

$$\frac{dy}{dx} = \lim_{\Delta x \to 0} \frac{\Delta y}{\Delta x} = \lim_{\Delta x \to 0} \frac{y(x + \Delta x) - y(x)}{\Delta x} \tag{B.28}$$

where Δy and Δx are defined as $\Delta x = x_2 - x_1$ and $\Delta y = y_2 - y_1$ (see Fig. B.13).

A useful expression to remember when $y(x) = ax^n$, where a is a *constant* and n is *any* positive or negative number (integer or fraction), is

$$\frac{dy}{dx} = nax^{n-1} \tag{B.29}$$

If $y(x)$ is a polynomial or algebraic function of x, we apply Equation B.29 to *each* term in the polynomial and take $da/dx = 0$. It is important to note that dy/dx *does not* mean dy divided by dx, but is simply a notation of the limiting process of the derivative as defined by Equation B.28. In Examples 4 through 7, we evaluate the derivatives of several well-behaved functions.

EXAMPLE 4

Suppose $y(x)$ (that is, y as a function of x) is given by

$$y(x) = ax^3 + bx + c$$

where a and b are constants. Then it follows that

$$y(x + \Delta x) = a(x + \Delta x)^3 + b(x + \Delta x) + c$$

$$y(x + \Delta x) = a(x^3 + 3x^2\,\Delta x + 3x\,\Delta x^2 + \Delta x^3) + b(x + \Delta x) + c$$

so

$$\Delta y = y(x + \Delta x) - y(x) = a(3x^2 \Delta x + 3x \Delta x^2 + \Delta x^3) + b \Delta x$$

Substituting this into Equation B.28 gives

$$\frac{dy}{dx} = \lim_{\Delta x \to 0} \frac{\Delta y}{\Delta x} = \lim_{\Delta x \to 0} [3ax^2 + 3x \Delta x + \Delta x^2] + b$$

$$\frac{dy}{dx} = 3ax^2 + b$$

EXAMPLE 5

$$y(x) = 8x^5 + 4x^3 + 2x + 7$$

Solution: Applying Equation B.29 to each term independently, and remembering that d/dx (constant) $= 0$, we have

$$\frac{dy}{dx} = 8(5)x^4 + 4(3)x^2 + 2(1)x^0 + 0$$

$$\frac{dy}{dx} = 40x^4 + 12x^2 + 2$$

Special Properties of the Derivative

A. **Derivative of the Product of Two Functions** If a function y is given by the product of two functions, say, $g(x)$ and $h(x)$, then the derivative of y is defined as

$$\frac{d}{dx} f(x) = \frac{d}{dx} [g(x)h(x)] = g \frac{dh}{dx} + h \frac{dg}{dx} \qquad \text{(B.30)}$$

B. **Derivative of the Sum of Two Functions** If a function y is equal to the sum of two functions, then the derivative of the sum is equal to the sum of the derivatives:

$$\frac{d}{dx} f(x) = \frac{d}{dx} [g(x) + h(x)] = \frac{dg}{dx} + \frac{dh}{dx} \qquad \text{(B.31)}$$

C. **Chain Rule of Differential Calculus** If $y = f(x)$ and x is a function of some other variable z, then dy/dx can be written as the product of two derivatives:

$$\frac{dy}{dx} = \frac{dy}{dz} \frac{dz}{dx} \qquad \text{(B.32)}$$

D. **The Second Derivative** The second derivative of y with respect to x is defined as the derivative of the function dy/dx (or, the derivative of the derivative). It is usually written

$$\frac{d^2y}{dx^2} = \frac{d}{dx}\left(\frac{dy}{dx}\right) \qquad \text{(B.33)}$$

EXAMPLE 6

Find the first derivative $y(x) = x^3/(x+1)^2$ with respect to x.

Solution: We can rewrite this function as $y(x) = x^3(x+1)^{-2}$ and apply Equation B.30 directly:

$$\frac{dy}{dx} = (x+1)^{-2}\frac{d}{dx}(x^3) + x^3\frac{d}{dx}(x+1)^{-2}$$

$$= (x+1)^{-2}\,3x^2 + x^3(-2)(x+1)^{-3}$$

$$\frac{dy}{dx} = \frac{3x^2}{(x+1)^2} - \frac{2x^3}{(x+1)^3}$$

EXAMPLE 7

A useful formula that follows from Equation B.30 is the derivative of the quotient of two functions. Show that the expression is given by

$$\frac{d}{dx}\left[\frac{g(x)}{h(x)}\right] = \frac{h\dfrac{dg}{dx} - g\dfrac{dh}{dx}}{h^2}$$

Solution: We can write the quotient as gh^{-1} and then apply Equations B.29 and B.30:

$$\frac{d}{dx}\left(\frac{g}{h}\right) = \frac{d}{dx}(gh^{-1}) = g\frac{d}{dx}(h^{-1}) + h^{-1}\frac{d}{dx}(g)$$

$$= -gh^{-2}\frac{dh}{dx} + h^{-1}\frac{dg}{dx}$$

$$= \frac{h\dfrac{dg}{dx} - g\dfrac{dh}{dx}}{h^2}$$

Some of the more commonly used derivatives of functions are listed in Table B.4.

TABLE B.4 Derivatives for Several Functions

$\frac{d}{dx}(a) = 0$	$\frac{d}{dx}(\tan ax) = a\sec^2 ax$
$\frac{d}{dx}(ax^n) = nax^{n-1}$	$\frac{d}{dx}(\cot ax) = -a\csc^2 ax$
$\frac{d}{dx}(e^{ax}) = ae^{ax}$	$\frac{d}{dx}(\sec x) = \tan x\sec x$
$\frac{d}{dx}(\sin ax) = a\cos ax$	$\frac{d}{dx}(\csc x) = -\cot x\csc x$
$\frac{d}{dx}(\cos ax) = -a\sin ax$	$\frac{d}{dx}(\ln ax) = \frac{a}{x}$

Note: The letters a and n are constants.

B.7 INTEGRAL CALCULUS

We think of integration as the inverse of differentiation. As an example, consider the expression

$$f(x) = \frac{dy}{dx} = 3ax^2 + b$$

which was the result of differentiating the function

$$y(x) = ax^3 + bx + c$$

in Example 4. We can write the first expression $dy = f(x)dx = (3ax^2 + b)dx$ and obtain $y(x)$ by "summing" over all values of x. Mathematically, we write this inverse operation

$$y(x) = \int f(x)dx$$

For the function $f(x)$ given above,

$$y(x) = \int (3ax^2 + b)dx = ax^3 + bx + c$$

where c is a constant of the integration. This type of integral is called an *indefinite integral* since its value depends on the choice of the constant c.

A general **indefinite integral** $I(x)$ is defined as

$$I(x) = \int f(x)dx \qquad \text{(B.34)}$$

where $f(x)$ is called the *integrand* and $f(x) = \dfrac{dI(x)}{dx}$.

For a *general continuous* function $f(x)$, the integral can be described as the area under the curve bounded by $f(x)$ and the x axis, between two specified values of x, say, x_1 and x_2, as in Figure B.14.

The area of the shaded element is approximately $f_i \Delta x_i$. If we sum all these area elements from x_1 to x_2 and take the limit of this sum as $\Delta x_i \to 0$, we obtain the *true* area under the curve bounded by $f(x)$ and x, between the limits x_1 and x_2:

$$\text{Area} = \lim_{\Delta x \to 0} \sum_i f_i(x)\Delta x_i = \int_{x_1}^{x_2} f(x)dx \qquad \text{(B.35)}$$

Integrals of the type defined by Equation B.35 are called **definite integrals.**

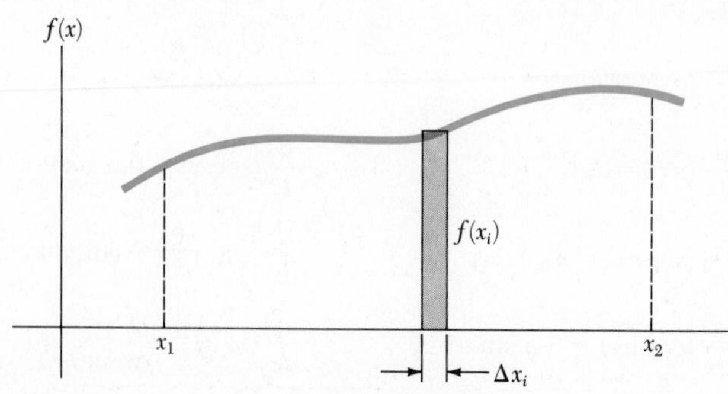

Figure B.14

One of the common types of integrals that arise in practical situations has the form

$$\int x^n\,dx = \frac{x^{n+1}}{n+1} + c \qquad (n \neq -1) \tag{B.36}$$

This result is obvious since differentiation of the right-hand side with respect to x gives $f(x) = x^n$ directly. If the limits of the integration are known, this integral becomes a *definite integral* and is written

$$\int_{x_1}^{x_2} x^n\,dx = \frac{x_2^{\,n+1} - x_1^{\,n+1}}{n+1} \qquad (n \neq -1) \tag{B.37}$$

EXAMPLES

1. $\displaystyle\int_0^a x^2\,dx = \frac{x^3}{3}\bigg]_0^a = \frac{a^3}{3}$

2. $\displaystyle\int_0^b x^{3/2}\,dx = \frac{x^{5/2}}{5/2}\bigg]_0^b = \frac{2}{5}b^{5/2}$

3. $\displaystyle\int_3^5 x\,dx = \frac{x^2}{2}\bigg]_3^5 = \frac{5^2-3^2}{2} = 8$

Partial Integration

Sometimes it is useful to apply the method of *partial integration* to evaluate certain integrals. The method uses the property that

$$\int u\,dv = uv - \int v\,du \tag{B.38}$$

where u and v are *carefully* chosen so as to reduce a complex integral to a simpler one. In many cases, several reductions have to be made. Consider the example

$$I(x) = \int x^2 e^x\,dx$$

This can be evaluated by integrating by parts twice. First, if we choose $u = x^2$, $v = e^x$, we get

$$\int x^2 e^x\,dx = \int x^2\,d(e^x) = x^2 e^x - 2\int e^x x\,dx + c_1$$

Now, in the second term, choose $u = x$, $v = e^x$, which gives

$$\int x^2 e^x\,dx = x^2 e^x - 2xe^x + 2\int e^x\,dx + c_1$$

or

$$\int x^2 e^x\,dx = x^2 e^x - 2xe^x + 2e^x + c_2$$

The Perfect Differential

Another useful method to remember is the use of the *perfect differential*. That is, we should sometimes look for a change of variable such that the differential of the function is the differential of the independent variable appearing in the integrand. For example, consider the integral

$$I(x) = \int \cos^2 x \sin x \, dx$$

This becomes easy to evaluate if we rewrite the differential as $d(\cos x) = -\sin x \, dx$. The integral then becomes

$$\int \cos^2 x \sin x \, dx = -\int \cos^2 x \, d(\cos x)$$

If we now change variables, letting $y = \cos x$, we get

$$\int \cos^2 x \sin x \, dx = -\int y^2 dy = -\frac{y^3}{3} + c = -\frac{\cos^3 x}{3} + c$$

TABLE B.5 Some Indefinite Integrals (an arbitrary constant should be added to each of these integrals)

$$\int x^n \, dx = \frac{x^{n+1}}{n+1} \quad \text{(provided } n \neq -1\text{)}$$

$$\int \frac{dx}{x} = \int x^{-1} \, dx = \ln x$$

$$\int \frac{dx}{a+bx} = \frac{1}{b} \ln(a+bx)$$

$$\int \frac{dx}{(a+bx)^2} = -\frac{1}{b(a+bx)}$$

$$\int \frac{dx}{a^2+x^2} = \frac{1}{a} \tan^{-1} \frac{x}{a}$$

$$\int \frac{dx}{a^2-x^2} = \frac{1}{2a} \ln \frac{a+x}{a-x} \quad (a^2 - x^2 > 0)$$

$$\int \frac{dx}{x^2-a^2} = \frac{1}{2a} \ln \frac{x-a}{x+a} \quad (x^2 - a^2 > 0)$$

$$\int \frac{x \, dx}{a^2 \pm x^2} = \pm \tfrac{1}{2} \ln(a^2 \pm x^2)$$

$$\int \frac{dx}{\sqrt{a^2-x^2}} = \sin^{-1} \frac{x}{a} = -\cos^{-1} \frac{x}{a} \quad (a^2 - x^2 > 0)$$

$$\int \frac{dx}{\sqrt{x^2 \pm a^2}} = \ln(x + \sqrt{x^2 \pm a^2})$$

$$\int \frac{x \, dx}{\sqrt{a^2-x^2}} = -\sqrt{a^2-x^2}$$

$$\int \frac{x \, dx}{\sqrt{x^2 \pm a^2}} = \sqrt{x^2 \pm a^2}$$

$$\int \sqrt{a^2-x^2} \, dx = \tfrac{1}{2}\left(x\sqrt{a^2-x^2} + a^2 \sin^{-1} \frac{x}{a}\right)$$

$$\int x\sqrt{a^2-x^2} \, dx = -\tfrac{1}{3}(a^2-x^2)^{3/2}$$

$$\int \sqrt{x^2 \pm a^2} \, dx = \tfrac{1}{2}[x\sqrt{x^2 \pm a^2} \pm a^2 \ln(x + \sqrt{x^2 \pm a^2})]$$

$$\int x\,(\sqrt{x^2 \pm a^2}) \, dx = \tfrac{1}{3}(x^2 \pm a^2)^{3/2}$$

$$\int e^{ax} \, dx = \frac{1}{a} e^{ax}$$

$$\int \ln ax \, dx = (x \ln ax) - x$$

$$\int xe^{ax} \, dx = \frac{e^{ax}}{a^2}(ax-1)$$

$$\int \frac{dx}{a+be^{cx}} = \frac{x}{a} - \frac{1}{ac} \ln(a+be^{cx})$$

$$\int \sin ax \, dx = -\frac{1}{a} \cos ax$$

$$\int \cos ax \, dx = \frac{1}{a} \sin ax$$

$$\int \tan ax \, dx = -\frac{1}{a} \ln(\cos ax) = \frac{1}{a} \ln(\sec ax)$$

$$\int \cot ax \, dx = \frac{1}{a} \ln(\sin ax)$$

$$\int \sec ax \, dx = \frac{1}{a} \ln(\sec ax + \tan ax) = \frac{1}{a} \ln\left[\tan\left(\frac{ax}{2} + \frac{\pi}{4}\right)\right]$$

$$\int \csc ax \, dx = \frac{1}{a} \ln(\csc ax - \cot ax) = \frac{1}{a} \ln\left(\tan \frac{ax}{2}\right)$$

$$\int \sin^2 ax \, dx = \frac{x}{2} - \frac{\sin 2ax}{4a}$$

$$\int \cos^2 ax \, dx = \frac{x}{2} + \frac{\sin 2ax}{4a}$$

$$\int \frac{dx}{\sin^2 ax} = -\frac{1}{a} \cot ax$$

$$\int \frac{dx}{\cos^2 ax} = \frac{1}{a} \tan ax$$

$$\int \tan^2 ax \, dx = \frac{1}{a}(\tan ax) - x$$

$$\int \cot^2 ax \, dx = -\frac{1}{a}(\cot ax) - x$$

$$\int \sin^{-1} ax \, dx = x(\sin^{-1} ax) + \frac{\sqrt{1-a^2x^2}}{a}$$

$$\int \cos^{-1} ax \, dx = x(\cos^{-1} ax) - \frac{\sqrt{1-a^2x^2}}{a}$$

$$\int \tan^{-1} ax \, dx = x(\tan^{-1} ax) - \frac{1}{2a} \ln(1+a^2x^2)$$

$$\int \cot^{-1} ax \, dx = x(\cot^{-1} ax) + \frac{1}{2a} \ln(1+a^2x^2)$$

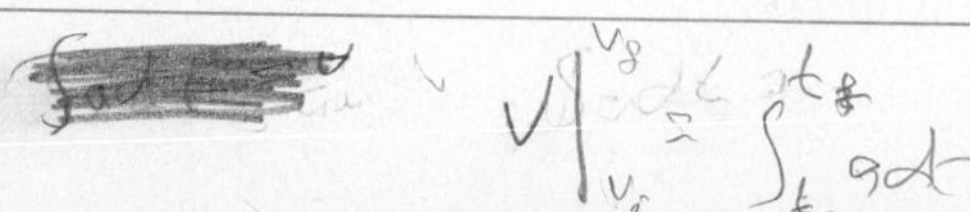

TABLE B.6 Gauss' Probability Integral and Related Integrals

$$I_0 = \int_0^\infty e^{-\alpha x^2}\,dx = \tfrac{1}{2}\sqrt{\frac{\pi}{\alpha}} \qquad \text{(Gauss' probability integral)}$$

$$I_1 = \int_0^\infty x e^{-\alpha x^2}\,dx = \frac{1}{2\alpha}$$

$$I_2 = \int_0^\infty x^2 e^{-\alpha x^2}\,dx = -\frac{dI_0}{d\alpha} = \tfrac{1}{4}\sqrt{\frac{\pi}{\alpha^3}}$$

$$I_3 = \int_0^\infty x^3 e^{-\alpha x^2}\,dx = -\frac{dI_1}{d\alpha} = \frac{1}{2\alpha^2}$$

$$I_4 = \int_0^\infty x^4 e^{-\alpha x^2}\,dx = \frac{d^2 I_0}{d\alpha^2} = \tfrac{3}{8}\sqrt{\frac{\pi}{\alpha^5}}$$

$$I_5 = \int_0^\infty x^5 e^{-\alpha x^2}\,dx = \frac{d^2 I_1}{d\alpha^2} = \frac{1}{\alpha^3}$$

$$\vdots$$

$$I_{2n} = (-1)^n \frac{d^n}{d\alpha^n} I_0$$

$$I_{2n+1} = (-1)^n \frac{d^n}{d\alpha^n} I_1$$

Table B.5 lists some useful indefinite integrals. Table B.6 gives Gauss' probability integral and other definite integrals. A more complete list can be found in various handbooks, such as *The Handbook of Chemistry and Physics*, CRC Press.

Appendix C
The Periodic Table

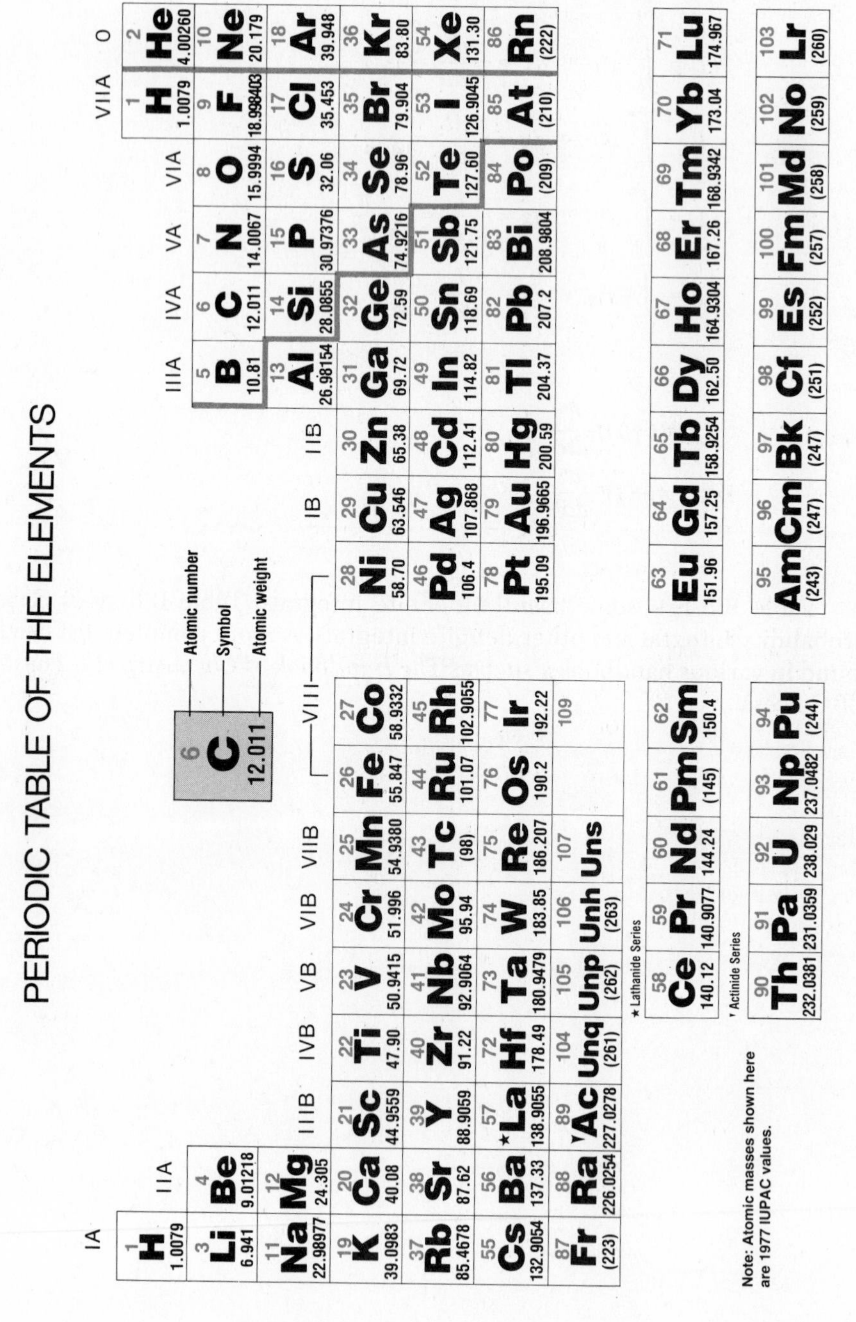

PERIODIC TABLE OF THE ELEMENTS

IA	IIA	IIIB	IVB	VB	VIB	VIIB	VIII	VIII	VIII	IB	IIB	IIIA	IVA	VA	VIA	VIIA	O
1 H 1.0079																1 H 1.0079	2 He 4.00260
3 Li 6.941	4 Be 9.01218											5 B 10.81	6 C 12.011	7 N 14.0067	8 O 15.9994	9 F 18.998403	10 Ne 20.179
11 Na 22.98977	12 Mg 24.305											13 Al 26.98154	14 Si 28.0855	15 P 30.97376	16 S 32.06	17 Cl 35.453	18 Ar 39.948
19 K 39.0983	20 Ca 40.08	21 Sc 44.9559	22 Ti 47.90	23 V 50.9415	24 Cr 51.996	25 Mn 54.9380	26 Fe 55.847	27 Co 58.9332	28 Ni 58.70	29 Cu 63.546	30 Zn 65.38	31 Ga 69.72	32 Ge 72.59	33 As 74.9216	34 Se 78.96	35 Br 79.904	36 Kr 83.80
37 Rb 85.4678	38 Sr 87.62	39 Y 88.9059	40 Zr 91.22	41 Nb 92.9064	42 Mo 95.94	43 Tc (98)	44 Ru 101.07	45 Rh 102.9055	46 Pd 106.4	47 Ag 107.868	48 Cd 112.41	49 In 114.82	50 Sn 118.69	51 Sb 121.75	52 Te 127.60	53 I 126.9045	54 Xe 131.30
55 Cs 132.9054	56 Ba 137.33	57 ★La 138.9055	72 Hf 178.49	73 Ta 180.9479	74 W 183.85	75 Re 186.207	76 Os 190.2	77 Ir 192.22	78 Pt 195.09	79 Au 196.9665	80 Hg 200.59	81 Tl 204.37	82 Pb 207.2	83 Bi 208.9804	84 Po (209)	85 At (210)	86 Rn (222)
87 Fr (223)	88 Ra 226.0254	89 ▾Ac 227.0278	104 Unq (261)	105 Unp (262)	106 Unh (263)	107 Uns		109									

Key: 6 C 12.011 — Atomic number, Symbol, Atomic weight

★ Lathanide Series

58 Ce 140.12	59 Pr 140.9077	60 Nd 144.24	61 Pm (145)	62 Sm 150.4	63 Eu 151.96	64 Gd 157.25	65 Tb 158.9254	66 Dy 162.50	67 Ho 164.9304	68 Er 167.26	69 Tm 168.9342	70 Yb 173.04	71 Lu 174.967

▾ Actinide Series

90 Th 232.0381	91 Pa 231.0359	92 U 238.029	93 Np 237.0482	94 Pu (244)	95 Am (243)	96 Cm (247)	97 Bk (247)	98 Cf (251)	99 Es (252)	100 Fm (257)	101 Md (258)	102 No (259)	103 Lr (260)

Note: Atomic masses shown here are 1977 IUPAC values.

Appendix D
SI Units

TABLE D.1 SI Base Units

Base Quantity	SI Base Unit Name	SI Base Unit Symbol
Length	Meter	m
Mass	Kilogram	kg
Time	Second	s
Electric current	Ampere	A
Temperature	Kelvin	K
Amount of substance	Mole	mol

Table D.2 Derived SI Units

Quantity	Name	Symbol	Expression in Terms of Base Units	Expression in Terms of Other SI Units
Plane angle	Radian	rad	m/m	
Frequency	Hertz	Hz	s^{-1}	
Force	Newton	N	$kg \cdot m/s^2$	J/m
Pressure	Pascal	Pa	$kg/m \cdot s^2$	N/m^2
Energy: work	Joule	J	$kg \cdot m^2/s^2$	$N \cdot m$
Power	Watt	W	$kg \cdot m^2/s^3$	J/s
Electric charge	Coulomb	C	$A \cdot s$	
Electric potential (emf)	Volt	V	$kg \cdot m^2/A \cdot s^3$	W/A
Capacitance	Farad	F	$A^2 \cdot s^4/kg \cdot m^2$	C/V
Electric resistance	Ohm	Ω	$kg \cdot m^2/A^2 \cdot s^3$	V/A
Magnetic flux	Weber	Wb	$kg \cdot m^2/A \cdot s^2$	$V \cdot s$
Magnetic field intensity	Tesla	T	$kg/A \cdot s^2$	Wb/m^2
Inductance	Henry	H	$kg \cdot m^2/A^2 \cdot s^3$	Wb/A

Answers to Odd-Numbered Problems

CHAPTER 1

1. 2.8 g/cm^3
3. 2.26×10^3 kg
5. (a) 9.83×10^{-16} g (b) 1.06×10^7 atoms
7. k cannot be found from this analysis.
9. L/T^3
11. 1.39×10^{-4} m^3
13. 7.46×10^{-4} m^3
15. 1.14×10^4 kg/m^3
17. 1.18×10^{17} kg/m^3
19. 2.87×10^8 s
21. 2.54×10^{22} atoms
23. Estimated at 3×10^9 beats.
25. Assuming two 6-packs per week per family and four people per family, we estimate 30 billion cans per year. Taking the mass of one can as 5 g, we estimate a total mass of 1.5×10^8 kg, corresponding to about 10^5 tons.
27. About 10^4 bricks. Estimating the area of one brick as 3 in. $\times$ 8 in. = 24 in.^{2} = 0.17 ft^2, and one wall as having an area of 12 ft $\times$ 30 ft = 360 ft^2 (for a total area of 1440 ft^2), we estimate a total number of $1440/0.17 \approx 10^4$ bricks.
29. 1.3×10^{10} lb, 4.2×10^7, assuming 300 net lb/head and 4 hamburgers/lb.
31. (a) 22 cm (b) 67.9 cm^2
33. (195.8 ± 1.4) cm^2

CHAPTER 2

1. (a) 8.6 m (b) (4.5 m, $-63°$) (c) (4.2 m, 135°)
3. $x = -2.75$ m, $y = -4.76$ m
5. (14.3 km, 65.2°)
7. (a) (b) $|\mathbf{A} + \mathbf{B}| = 8.39$ m

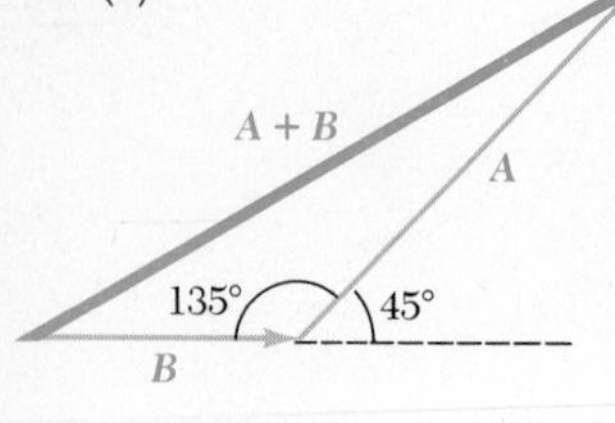

9. $|\mathbf{d}| = \sqrt{8^2 + 13^2} = 15.3$ m
$\theta = -58.4°$

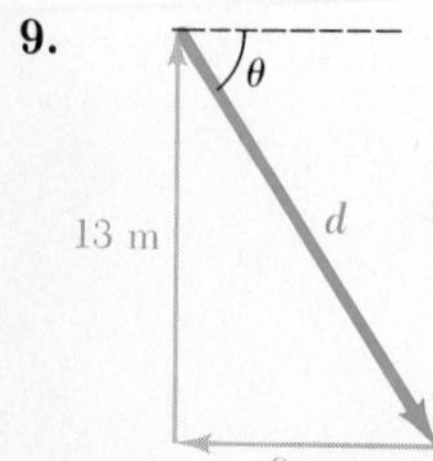

11. 47.2 units, $\theta = 122°$
13.

Quadrant	I	II	III	IV
x component	+	−	−	+
y component	+	+	−	−

15. (7.21 m, 56.3°)
17. (a) $\mathbf{A} + \mathbf{B} = 2\mathbf{i} - 6\mathbf{j}$ (b) $\mathbf{A} - \mathbf{B} = 4\mathbf{i} + 2\mathbf{j}$
(c) $|\mathbf{A} + \mathbf{B}| = 6.32$ (d) $|\mathbf{A} - \mathbf{B}| = 4.47$
(e) For $\mathbf{A} + \mathbf{B}$, $\theta = -71.6°$; for $\mathbf{A} - \mathbf{B}$, $\theta = 26.6°$.
19. (a) $\mathbf{r} = (-11.1\mathbf{i} + 6.40\mathbf{j})$ m
(b) $\mathbf{r} = (1.65\mathbf{i} + 2.86\mathbf{j})$ cm
(c) $\mathbf{r} = (-18.0\mathbf{i} - 12.6\mathbf{j})$ in.
21. 9.2 m west and 2.3 m north, *or* $\mathbf{R} = (-9.2\mathbf{i} + 2.3\mathbf{j})$ m
23. 1260 mi east and 386 mi north, *or* $\mathbf{R} = (1260\mathbf{i} + 386\mathbf{j})$ mi
25. $A_x = 2.6$ m, $B_x = 0$, $\mathbf{A} + \mathbf{B} = (2.6\mathbf{i} + 4.5\mathbf{j})$ m
$A_y = 1.5$ m, $B_y = 3$ m
27. (a) $\mathbf{A} = 8\mathbf{i} + 12\mathbf{j} - 4\mathbf{k}$
(b) $\mathbf{B} = \mathbf{A}/4 = 2\mathbf{i} + 3\mathbf{j} - \mathbf{k}$
(c) $\mathbf{C} = -3\mathbf{A} = -24\mathbf{i} - 36\mathbf{j} + 12\mathbf{k}$
29. (a) $|\mathbf{B}| = 7$ units, $0 = 217°$ (b) $C_x = -28$ units, $C_y = -91$ units
31. (a) $\mathbf{A} = -3\mathbf{i} + 2\mathbf{j}$ (b) 3.61, 146.3°
(c) $\mathbf{B} = 3\mathbf{i} - 6\mathbf{j}$
33. 5.83 N at $\theta = 149°$
35. 38.3 N in the positive y direction
37. (a) $F_x = 49.5$ N, $F_y = 27.1$ N
(b) 56.4 N at $\theta = 28.7°$
39. 240 m at $\theta = 237°$
41. (a) $\mathbf{r}_1 = (-3\mathbf{i} - 5\mathbf{j})$ m, $\mathbf{r}_2 = (-\mathbf{i} + 8\mathbf{j})$ m
(b) $\Delta\mathbf{r} = \mathbf{r}_2 - \mathbf{r}_1 = (2\mathbf{i} + 13\mathbf{j})$ m

CHAPTER 3

1. -3.89×10^{-2} m/s
3. (a) 1.92 km (b) 4.57 m/s
5. (a) 4 m/s (b) -4 m/s (c) zero (d) 2 m/s
7. (a) negative (b) positive (c) zero (d) zero
9. (b) 1.6 m/s
11. -2.5 m/s^2
13. (a) 4 m/s^2 (b) No. The acceleration is not necessarily constant, and so the average velocity cannot be evaluated. If the acceleration were constant, then $\bar{v} = 13$ m/s.
15. (b) 1 m/s^2 (c) 1.5 m/s^2
17. (a) -8 m/s^2 (b) -9 m/s (c) 7 m/s
19. (a) 0 (b) 6 m/s^2 (c) 825 m (d) 65 m/s
21. (a) 4 cm (b) 18 cm/s

23. (a) 12.7 m/s (b) −2.3 m/s
25. (a) 9.75 ft/s² (b) 3.08 s
27. (a) 24.5 s (b) 122 m
29. 24 s
31. (a) 3×10^{-10} s (b) 1.26×10^{-4} m
33. (a) -3.5×10^5 m/s² (b) 2.9×10^{-4} s
35. (a) 8.20 s (b) 134 m
37. (a) 96 ft/s (b) -3.07×10^3 ft/s², or 96g (c) 3.1×10^{-2} s
39. (a) 2.33 s (b) −32.8 m/s
41. (a) 17.2 m/s (b) 15.1 m
43. (a) 39.2 m (b) 17.9 m/s (c) −9.8 m/s²
45. (a) The velocity is constant; the acceleration is zero. (b) The velocity is directly proportional to the time; the acceleration is a constant.
47. (a) −6 m (b) 9 m (c) 3 m
49. (a) $15t^2$ (b) $5t^3$
51. (a) 0.75 s (b) −20 ft/s²
53. 1.0 s
55. (b) 4 m, 2 m/s (c) 1/3 s (d) −4 m, −10 m/s, −6 m/s²
57. (a) $(3t^2 - 18t + 6)$ cm/s (b) $(3 \pm \sqrt{7})$ s (c) $-6\sqrt{7}$ cm/s², $6\sqrt{7}$ cm/s² (d) −74 cm
59. (a) 6.46 s (b) 334 ft (c) $v_j = 103$ ft/s, $v_s = 89.5$ ft/s
61. (a) 79.3 ft/s (b) −113 ft/s (c) −16.7 ft/s
63. (a) 5410 ft (b) 361 ft/s
65. (a) 40.4 s (b) 1735 m (c) −184 m/s
67. (b) $(47/12)v$ (c) $(47/60)v$

CHAPTER 4

1. (a) $v_x = 2t$, $v_y = 4t$ (b) $x = t^2$, $y = 2t^2$ (c) $\sqrt{20}\, t$
3. (a) $\mathbf{v} = -12t\mathbf{j}$ m/s, $\mathbf{a} = -12\mathbf{j}$ m/s² (b) $\mathbf{r} = (3\mathbf{i} - 6\mathbf{j})$ m, $\mathbf{v} = -12\mathbf{j}$ m/s
5. (a) $\mathbf{v} = 4\mathbf{i}$ m/s (b) $x = 4$ m, $y = 6$ m
7. 2.70 m/s²
9. (a) 54.4 cm below the center of the target
11. 53.1°
13. (a) 14.7°, 75.4° (b) 10.4 s, 39.5 s
15. 80 m
17. (a) 12.6 m/s (b) 395 m/s² directed toward the center of rotation
19. $v = 10.5$ m/s, $a = 219$ m/s²
21. (a) 32 m/s² downward (b) 72 m/s² upward
23. (a) 13.0 m/s² toward the center (b) 6.24 m/s (c) 7.50 m/s² along $\mathbf{v}$
25. (a) 4 m/s² toward the center (b) $\sqrt{8}$ m/s
27. (a) 14.5° north of west (b) 194 km/h
29. 72 km/h, 56.3° north of east
31. 33.6 min (compared with 27.8 min)
33. (a) 41.7 m/s (b) 3.81 s (c) $v_x = 34.2$ m/s, $v_y = -13.4$ m/s, $v = 36.7$ m/s
35. (a) $v_x = 7.14$ m/s, $v_y = -12.1$ m/s (b) $t = 2.47$ s (c) $d = 4.90$ m
37. (b) $\mathbf{v} = -6 \sin 2t\mathbf{i} + 6 \cos 2t\mathbf{j}$ m/s, $\mathbf{a} = -12 \cos 2t\mathbf{i} - 12 \sin 2t\mathbf{j} = -4\mathbf{r}$ m/s² (c) $\dfrac{v^2}{r} = 12 \text{ m/s}^2 = |\mathbf{a}|$
39. (a) 0° (b) 9.8 m/s (c) linear motion along a vertical (d) 45.6° north of the vertical at 14 m/s (e) parabola
41. (a) 1.53×10^3 m (b) 36.2 s (c) 4.05 km
43. 0.139 m/s
45. (a) 36.9° east of south (b) 0.751 km
47. Less than 265 m or more than 3476 m
49. 7.52 m/s away from the quarterback
51. He will not reach safety by running off the edge horizontally; however, he will reach safety by using the long-jump technique.

CHAPTER 5

1. (a) 3 (b) 1.5 m/s²
3. (a) 534 N (b) 54.4 kg
5. 1.96×10^5 dynes, or 1.96 N
7. (a) 12 N (b) 3 m/s²
9. (a) $(4\mathbf{i} + 3\mathbf{j})$ m/s² (b) $(5.5\mathbf{i} + 2.6\mathbf{j})$ m/s²
11. 2 ft/s²
13. 8 N in the negative x direction
15. 6.4×10^3 N
17. (a) $F_x = 2.5$ N, $F_y = 5$ N (b) $F = 5.6$ N
19. (a) $T_1 = 31.5$ N, $T_2 = 37.5$ N, $T_3 = 49$ N (b) $T_1 = 113$ N, $T_2 = 56.6$ N, $T_3 = 98$ N
21. (a) 576 N (b) No; F would have to be infinitely large.
23. 3.73 m
25. (a) $T = 36.8$ N (b) $a = 2.45$ m/s² (c) 1.23 m
27. $a = \dfrac{F}{m_1 + m_2}$, $T = \dfrac{m_1}{m_1 + m_2} F$
29. $\mu_s = 0.38$, $\mu_k = 0.31$
31. (a) 16.3 N (b) 8.07 N
33. 0.458
35. (b) $T = 16.7$ N, $a = 0.69$ m/s²
37. (a) 1.78 m/s² (b) 0.368 (c) 9.37 N (d) 2.67 m/s
39. (a) 35.4 N (b) 0.601
41. (a) 0.55 (b) 0.25 m/s²
43. (a) 3.12 m/s² (b) 17.5 N
45. (a) $T_1 = T_2 = T_3 = $ mg/2, $T_4 = 3$ mg/2, $T_5 = $ mg (b) $F_A = T_1 = $ mg/2
47. (a) No (b) 80 lb
49. (a) friction between the two blocks (b) 34.7 N (c) 0.306
51. (b) 5.75 m/s² (c) $T_1 = 17.4$ N, $T_2 = 40.5$ N
53. (a) 20 lb (b) 12 lb (c) 18 ft/s² (d) the upper rope
55. (a) 1.02 m/s² (b) 2.04 N, 3.06 N, 4.08 N (c) 14 N between m_1 and m_2, 8 N between m_2 and m_3

57. (a) $mg\left(\frac{\sin\theta + \mu_s \cos\theta}{\cos\theta - \mu_s \sin\theta}\right)$
(b) $ma + mg\left(\frac{\sin\theta + \mu_k \cos\theta}{\cos\theta - \mu_k \sin\theta}\right)$
59. (a) $T_1 = 78.0$ N, $T_2 = 35.9$ N (b) 0.655
61. $T_A = 304$ N, $T_B = 290$ N, $T_C = 152$ N, $T_D = 138$ N
63. (b) 8 ft/s^2 (c) 20 lb
67. (a) $T_1 = 2mg/\sin\theta_1$, $T_2 = mg/\sin[\tan^{-1}(\frac{1}{2}\tan\theta_1)]$, $T_3 = 2mg/\tan\theta$, (b) $\theta_2 = \tan^{-1}(\frac{1}{2}\tan\theta_1)$
(c) $D = (L/5)(2\cos\theta_1 + 2\cos[\tan^{-1}(\frac{1}{2}\tan\theta_1)] + 1)$
69. (a) $T = m_2 g(m_1 M/[m_1 M + m_2(m_1 + M)])$
(b) $a = (m_2(M + m_1)/[m_1 M + m_2(M + m_1)])g$
(c) $A = (m_1 m_2/[m_1 M + m_2(m_1 + M)])g$
(d) $(a - A) = (M m_2/[m_1 M + m_2(m_1 + M)])g$

CHAPTER 6

1. 6.22×10^{-12} N
3. (a) friction (b) 0.128
5. (a) 5.56×10^3 m/s (b) 237 min (c) 1.47×10^3 N
7. (a) 8.32×10^{-8} N (b) 9.13×10^{22} m/s^2 (c) 6.61×10^{15} rev/s
9. (a) 2.49×10^4 N (b) 12.1 m/s
11. (a) 20.4 N (b) $a_t = 4.14$ m/s^2, $a_r = 32$ m/s^2 (c) 32.3 m/s^2
13. 2.42 m/s^2 in the forward direction
15. (a) 3.6 m/s^2 to the right (b) zero
17. (a) $N = m(g - a)$ (b) $N = m(g - a)$
19. (a) 1.47 N · s/m (b) 2.02×10^{-3} s (c) 2.94×10^{-2} N
21. (a) 0.61 rev/s (b) 0.77 m/s, 2.93 m/s^2
23. (a) 66.3 N (b) 36.6 N (c) 7.02 N
25. (a) $v_{max} = \sqrt{Rg\left(\frac{\tan\theta + \mu}{1 - \mu\tan\theta}\right)}$, $v_{min} = \sqrt{Rg\left(\frac{\tan\theta - \mu}{1 + \mu\tan\theta}\right)}$ (b) $\mu = \tan\theta$
(c) $v_{max} = 16.6$ m/s (37 mi/h), $v_{min} = 8.57$ m/s (19 mi/h)
27. (b) 2.54 s, 23.6 rev/min
29. $v = v_0 e^{-(b/m)t}$

CHAPTER 7

1. 5.88×10^3 J
3. (a) 317 J (b) −176 J (c) zero (d) zero (e) 141 J
5. (a) 2.94×10^5 J (b) -2.94×10^5 J
7. (a) 3 (b) 74.7°
9. 18.4
13. (a) 63.4° (b) 80.7° (c) 67.8°
15. (a) 7.5 J (b) 15 J (c) 7.5 J (d) 30 J
17. (b) −12 J
19. (a) 22.5 J (b) 90 J
21. (a) 51 J (b) 69 J
23. (a) 9×10^3 J (b) 300 N
25. (a) 1.94 m/s (b) 3.35 m/s (c) 3.87 m/s
27. (a) $v_0^2/2\mu_k g$ (b) 12.8 m
29. (a) 0.791 m/s (b) 0.531 m/s
31. (a) 63.9 J (b) −35.4 J (c) −9.51 J (d) 19.0 J
33. 829 N (186 lb)
35. (a) 0.41 m/s (b) 2.45×10^3 J
37. (a) 3920 W (5.25 hp) (b) 7.06×10^5 J
39. (a) 7.5×10^4 J (b) 2.50×10^4 W (33.5 hp) (c) 3.33×10^4 W (44.7 hp)
41. (a) 29.7 kW (b) 37.3 kW
43. 6.0 km/liter
45. (a) 980 J (b) −980 J (c) 24.5 W
47. (a) $\cos\alpha = A_x/A$, $\cos\beta = A_y/A$, $\cos\gamma = A_z/A$, where $A = (A_x^2 + A_y^2 + A_z^2)^{1/2}$
49. (a) 20 J (b) 6.71 m/s
51. (a) $kd/2mg$ (b) $kd/4mg$
57. (a) 2.7 m/s^2 (c) 4.04×10^3 N (d) 146 hp
59. (c) 7.29×10^7 J 1.97×10^4 W (d) 13.6%

CHAPTER 8

1. (a) $W_{OA} = 0$, $W_{AC} = -147$ J, and so $W_{OAC} = -147$ J (b) $W_{OB} = -147$ J, $W_{BC} = 0$, and so $W_{OBC} = -147$ J (c) $W_{OC} = -147$ J; the gravitational force is conservative.
3. (a) $W_{OAO} = -30$ J (b) $W_{OACO} = -51.2$ J (c) $W_{OCO} = -42.4$ J (d) Friction is a nonconservative force.
5. (a) 125 J (b) 50 J (c) 66.7 J (d) nonconservative, since W is path-dependent
7. (a) 70 J (b) −70 J (c) 6.83 m/s
9. (a) 15 J, 30 J (b) Yes. The total energy is not conserved since $E_i = 30$ J and $E_f = 20$ J
11. (a) −19.6 J (b) 39.2 J (c) zero
13. (a) 5.91 J (b) 3.47 m/s (c) 49.6 N (d) 0.816 m
15. (a) 31.3 m/s (b) 147 J (c) 4
17. (a) 0.225 J (b) 0.363 J (c) No. The normal force varies with position, and so the frictional force also varies.
19. (a) 8.33 m (b) −50 J (c) zero
21. (a) 8.85 m/s (b) 54.1%
23. (a) 9.90 m/s (b) −11.8 J (c) −11.8 J
25. (a) 0.180 J (b) 0.100 J
27. (a) $(2mgh/k)^{1/2}$ (b) 8.94 cm
29. (a) 588 N/m (b) 0.70 m/s
31. (a) $F_r = A/r^2$ (b) the gravitational force (A negative) and the electrostatic force (A positive or negative)
33. (a) zero at A, C, and E, positive at B, negative at D (b) unstable at A and E, stable at C

35.

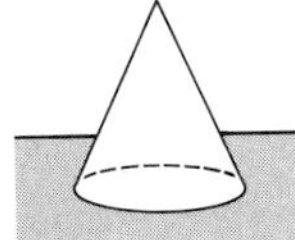

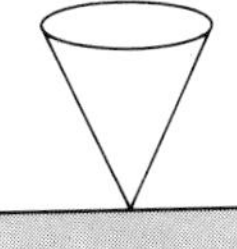

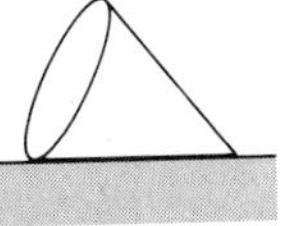

37. (a) 3.49 J, 676 J, 741 J
(b) 175 N, 338 N, 370 N (c) yes

39. (a) $\Delta U = -\dfrac{ax^2}{2} - \dfrac{bx^3}{3}$ (b) $\Delta U = -\dfrac{A}{\alpha}(1 - e^{\alpha x})$

41. 0.115
45. 1.07 m/s
47. 7.64 J
49. (a) 0.378 m (b) 2.30 m/s (c) 1.08 m

51. $y = \dfrac{mg}{k} + \sqrt{\left(\dfrac{mg}{k}\right)^2 + \dfrac{2mgh}{k}}$

CHAPTER 9

1. $p_x = 6$ kg · m/s, $p_y = -12$ kg · m/s, $p = 13.4$ kg · m/s
3. 1.70×10^4 kg · m/s in the northwesterly direction (b) 5.66×10^3 N
5. (a) 12 kg · m/s (b) 6 m/s (c) 4 m/s
7. (a) 1.35×10^4 kg · m/s (b) 9×10^3 N (c) 18×10^3 N
9. (a) quadrupled (b) $\sqrt{3}$ times its initial value
11. (a) 22.5 kg · m/s (b) 1.13×10^3 N
13. (a) 15.2 kg · m/s (b) 7.60×10^3 N
15. 6 m/s to the left
17. The boy moves westward with a speed of 2.46 m/s.
19. 2.68×10^{-20} m/s
21. 340 m/s
23. 6 kg
25. (a) 2.75 m/s (b) 6.75×10^4 J
27. (a) 0.284, or 28.4% (b) $K_n = 1.15 \times 10^{-13}$ J, $K_c = 0.45 \times 10^{-13}$ J
29. (a) −6.67 cm/s, 13.3 cm/s (b) 8/9
31. (b) and (c) are perfectly elastic
33. (a) 0.556 m/s (b) 11.1 J
35. (a) 24 cm/s (b) No. The earth recoils by a negligible amount.
37. $v = (2i - 1.8j)$ m/s
39. (a) $v_x = -9.3 \times 10^6$ m/s, $v_y = -8.3 \times 10^6$ m/s (b) 4.4×10^{-13} J
41. v(yellow) = 2.00 m/s, v(orange) = 3.46 m/s
43. 4.67×10^6 m (this point lies within the earth)
45. $(\frac{1}{3}, \frac{5}{3})$ m
47. (a) $v_c = (1.4i + 3.2j)$ m/s (b) $p = (7i + 16j)$ kg · m/s
49. $a_c = (i + 2j)$ m/s^2
51. 3×10^5 N
53. 1.42×10^4 m/s
55. (a) 2.04 m/s, south (b) 2.75 m/s to the south (c) 2.30 m/s at an angle 62° south of west
57. (a) $-2mv \sin\theta$ (b) zero (c) $(2mv \sin\theta/t)j$
59. 1.48×10^3 m/s
61. $x = \dfrac{2v_0^2}{9\mu g} - \dfrac{4}{9}d$
63. 108 N
65. (a) 6.93 m/s (b) 1.14 m
67. $\left(\dfrac{3Mg}{L}\right)x$

CHAPTER 10

1. 1.67 rad, or 95°
3. (a) 377 rad/s (b) 565 rad
5. (a) 5 rad/s^2 (b) 10 rad
7. (a) 1.99×10^{-7} rad/s (b) 2.6×10^{-6} rad/s
9. (a) 0.40 rad/s (b) 32 m/s^2 toward the center
11. (a) 8 rad/s (b) 16 m/s, $a_r = 128$ m/s^2, $a_t = 8$ m/s^2 (c) $\theta = 9$ rad
13. (a) 126 rad/s (b) 2.51 m/s (c) 953 m/s^2 (d) 15.1 m
15. (a) 143 kg · m^2 (b) 4.58×10^3 J
17. (a) 92 kg · m^2, 184 J (b) 6 m/s, 4 m/s, 8 m/s, 184 J
19. (a) $\frac{3}{2}MR^2$ (b) $\frac{7}{5}MR^2$
21. 3.2 N · m into the plane
23. (a) 12 kg · m^2 (b) 2.4 N · m (c) 43.8 rev
25. (a) $(2gh/[1 + I/(mR^2)])^{1/2}$ (b) $[2gh/(R^2 + I/m)]^{1/2}$
27. (a) 46.8 N (b) 0.234 kg · m^2 (c) 40 rad/s
29. (a) 2.0 N · m (b) 20 rad/s^2 (c) 4 m/s^2 (d) 40 rad/s (e) 8 m/s (f) 80 J (g) 80 J (h) 40 rad (i) 8 m
31. (a) $\dfrac{Mmg}{M + 4m}$ (b) $\dfrac{4mg}{M + 4m}$ (c) $\dfrac{1}{R}\sqrt{\dfrac{8mgh}{M + 4m}}$
33. (a) $m_1 g(m_1 + m_2 + I/R^2)^{-1}$
(b) $T_2 = m_1 m_2 g(m_1 + m_2 + I/R^2)^{-1}$, $T_1 = m_1 g(I + m_2 R^2)[I + (m_1 + m_2)R^2]^{-1}$
(c) $a = 3.12$ m/s^2, $T_1 = 26.7$ N, $T_2 = 9.37$ N
(d) $a = 5.6$ m/s^2, $T_1 = T_2 = 16.8$ N
37. (a) $\omega = \sqrt{3g/L}$ (b) $\alpha = 3g/2L$ (c) $a_x = \frac{3}{2}g$, $a_y = \frac{3}{4}g$ (d) $R_x = \frac{3}{2}Mg$, $R_y = \frac{1}{4}Mg$
39. (a) $0.707R$ (b) $0.289L$ (c) $0.632R$
41. (a) $h = (r^2\omega^2/2g)(m + \frac{1}{2}M)/m$
(b) $a = -[m/(m + \frac{1}{2}M)]g$
(c) $t = (r\omega_0/g)(m + \frac{1}{2}M)/m$

CHAPTER 11

1. (a) $5k$ (b) 135°
3. (a) $-6k$ (b) $-4i - 12j$ (c) $-2j + 6k$
5. (a) $-10k$ N · m (b) $8k$ N · m
9. 12.5 kg · m^2/s (out of the plane)

11. (a) $24k$ kg · m²/s (b) $-16k$ kg · m²/s
13. (a) mvd (out of the plane)
(b) $-2mvd$ (into the plane) (c) zero
15. (a) $L = md(v_0 + gt)k$ (b) $\tau = mgdk$
17. (a) 0.336 N · m (b) $L = 0.28v$ (c) 8.4 m/s²
19. (a) 0.367 kg · m²/s (b) 1.47 kg · m²/s
21. 7.35 rad/s
23. (a) 0.420 rad/s in the counterclockwise direction (b) 123 J
25. (a) 8.57 rad/s (b) increases by 234 J
(c) The student does work on the system.
27. (a) $a_c = \frac{2}{3}g \sin\theta$ (disk), $a_c = \frac{1}{2}g\sin\theta$ (hoop)
(b) $\frac{1}{3}\tan\theta$
29. (a) 500 J (b) 250 J (c) 750 J
31. (a) $(\frac{6}{5}gh)^{1/2}$ (b) $\frac{3}{5}g\sin\theta$

33. $\omega = \sqrt{\frac{10}{7}\frac{g}{r^2}(1-\cos\theta)(R-r)}$

35. (a) $\omega_f = 11.04$ rad/s (b) No
39. (a) $2.7(R-r)$

(b) $F_x = -\frac{10}{7}mg\left(\frac{2R+r}{R-r}\right)$ $F_y = -\frac{5}{7}mg$

41. $\frac{4}{3}\left(\frac{Fd}{M}\right)^{1/2}$

43. (a) $\frac{1}{3}\omega_0$ (b) $\frac{2}{3}$
45. (a) $\tau_x = yF_z - zF_y$, $\tau_y = zF_x - xF_z$, $\tau_z = xF_y - yF_x$
47. (a) v_0r_0/r (b) $T = (mv_0^2r_0^2)r^{-3}$

(c) $\frac{1}{2}mv_0^2\left(\frac{r_0^2}{r^2} - 1\right)$ (d) 4.5 m/s, 10.1 N, 0.45 J

49. (a) $F_y = \frac{W}{L}\left(d - \frac{ah}{g}\right)$

(b) $F_x = -306$ N, $F_y = 553$ N

CHAPTER 12

1. $F_1 + F_2 - W_1 - W_2 = 0$, $F_2\ell - W_1d_1 - W_2d_2 = 0$

3. $x = \frac{(W_1 + W)d + W_1\ell/2}{W_2}$

5. The y coordinate of the center of mass is 15.3 cm from the bottom. The x coordinate is 8 cm from the left side of the "tee."
7. $(\frac{1}{3}, \frac{5}{3})$ m
9. at the 75-cm mark
13. (a) 1.36 m from the front axle (b) 3560 N on each back tire and 4280 N on each front tire.
15. $N_a = 6.0 \times 10^5$ N, $N_b = 4.8 \times 10^5$ N
17. (b) $T = 17.3$ lb (c) $d = 0.76\ell$
19. (b) $T = 213$ N, $R_x = 184$ N, $R_y = 188$ N
21. (b) $T = 1.07 \times 10^3$ N, $R_x = 991$ N, $R_y = 497$ N
23. (a) $-268(x)$, 1300 N(y) (b) 0.324
25. (a) 180 N (b) 156 N
27. $T = 2710$ N, $R_x = 2650$ N
29. (a) $\mu_k = 0.57$, $\frac{6}{7}$ ft from the right corner
(b) $h = \frac{5}{3}$ ft

31. (a) $W = \frac{w}{2}\left(\frac{2\mu_s\sin\theta - \cos\theta}{\cos\theta - \mu_s\sin\theta}\right)$

(b) $R = (w + W)\sqrt{1+\mu_s^2}$, $F = \sqrt{W^2 + \mu_s^2(w+W)^2}$
33. (a) 133 N (b) $N_A = 429$ N, $N_B = 257$ N
(c) $R_x = 133$ N, $R_y = 257$ N

CHAPTER 13

1. (a) 1.5 Hz, 0.67 s (b) 4 m (c) π rad
(d) -4 m
3. (a) 4.3 cm (b) -5 cm/s (c) -17 cm/s²
(d) π s, 5 cm
5. (a) -14 cm/s, 16 cm/s² (b) 16 cm/s, 1.83s
(c) 32 cm/s², 1.05 s
7. (b) 6π cm/s, 0.33 s (c) $18\pi^2$, 0.5 s (d) 12 cm
9. 3.95 N/m
11. (a) 2.40 s (b) 0.417 Hz (c) 2.62 rad/s
(d) 0.23 s
13. (a) 0.4 m/s, 1.6 m/s² (b) 0.32 m/s, -0.96 m/s²
15. (a) 0.153 J (b) 0.783 m/s (c) 17.5 m/s²
17. (a) quadrupled (b) doubled (c) doubled
(d) no change
19. 2.6 cm
21. 0.158 Hz, 6.35 s
23. 106
25. increases by 1.78×10^{-3} s
27. 8.5×10^{-2} kg · m²
31. (a) 1 s (b) 5.09 cm
35. (a) $E = \frac{1}{2}mv^2 + mgL(1-\cos\theta)$

39. (a) 2 Mg, $T_p = Mg\left(1 + \frac{y}{L}\right)$

(b) $\frac{4\pi}{3}\sqrt{\frac{2L}{g}} = 2.68$ s

43. $\omega = \left(\frac{mgL + kh^2}{I}\right)^{1/2}$

45. (a) $\omega = \left(\frac{gd}{d^2 + L^2/12}\right)^{1/2}$ (b) 1.53 s

47. (a) $I = mL^2 + \frac{2}{5}mR^2$

(b) $T = 2\pi\sqrt{\frac{L}{g}}\left(1 + \frac{2}{5}\frac{R^2}{L^2}\right)^{1/2}$

49. (b) $-\frac{\pi}{\sqrt{g}}\frac{1}{2\rho a^2}\left(\frac{dM}{dt}\right) \times \left[L_0 - \left(\frac{1}{2\rho a^2}\right)\left(\frac{dM}{dt}\right)t\right]^{-1/2}$

(c) $\frac{2\pi}{\sqrt{g}}\left[L_0 - \left(\frac{1}{2\rho a^2}\right)\left(\frac{dM}{dt}\right)t\right]^{1/2}$

CHAPTER 14

1. 2.96×10^{-9} N
3. 4.62×10^{-8} N toward the center of the triangle

5. $F_x = Gm^2\left[\frac{2}{b^2} + \frac{3b}{(a^2+b^2)^{3/2}}\right]$,

$F_y = Gm^2\left[\frac{2}{a^2} + \frac{3a}{(a^2+b^2)^{3/2}}\right]$

7. $F_6 = (12.6i + 1.92j) \times 10^{-11}$ N, $F_6 = 12.7 \times 10^{-4}$ N
9. (a) $4\pi^2/GM_e = 9.89 \times 10^{-14}$ s^2/m^3 (b) 127 min
11. 9.37×10^6 m
13. 4.22×10^7 m
15. (a) -1.67×10^{-14} J (b) at the center of the triangle
17. $-20.95\,\frac{Gm^2}{a}$
19. 5.04×10^3 m/s
21. (a) 3.90×10^9 J (b) $|U|$ is halved, K is halved
23. (a) 1.87×10^{11} J (b) 103 kW
27. $Gm\lambda_0 L[d(L+d)]^{-1} + GmAL$ to the right
29. (a) 7.41×10^{-10} N (b) 1.04×10^{-8} N (c) 5.21×10^{-9} N
31. (a) $(GM/4R^3)^{1/2}$
    (b) $(g/R)^{1/2} = 1.57$ rad/s (0.249 rev/s)
33. (a) -2.34×10^{-10} N
    (b) 1.00×10^{-10} N along the positive x axis
35. (a) $k = \frac{GmM_e}{R_e^3}$, at $\frac{L}{2}$
    (b) $\frac{L}{2}\left(\frac{GM_e}{R_e^3}\right)^{1/2}$, at the middle of the tunnel
    (c) 311 m/s
37. (a) $v_1 = m_2\left[\frac{2G}{d(m_1+m_2)}\right]^{1/2}$,

$v_2 = m_1\left[\frac{2G}{d(m_1+m_2)}\right]^{1/2}$,

$v_{rel} = \left[\frac{2G(m_1+m_2)}{d}\right]^{1/2}$

    (b) $K_1 = 1.07 \times 10^{32}$ J, $K_2 = 2.67 \times 10^{31}$ J
39. (a) 7.34×10^{22} kg (b) 1.63×10^3 m/s (c) 1.32×10^{10} J
41. (a) $U = -\frac{3Gm}{R}\left(M + \frac{\sqrt{3}}{3}m\right)$
    (b) $v = \left(\frac{\sqrt{3}Gm}{3R} + \frac{GM}{R}\right)^{1/2}$
43. (a) $F = \frac{GMmd}{(R^2+d^2)^{3/2}}$ downward
    (b) $F = 0$ at the middle and $F \approx \frac{GMm}{d^2}$ for $d \gg R$

CHAPTER 15

1. 667 N
3. 9.52×10^{-6}
5. 1.40×10^7 N/m^2, 5.65×10^{-8} m^3
7. (a) 3.14×10^4 N (7060 lb)
   (b) 6.28×10^4 N (14 100 lb)
9. 0.11 kg
11. 4×10^{17} kg/m^2. Matter contains mostly free space.
13. 6.24×10^6 N/m^2
15. (b) $T = 1.07$
17. 1.62 m
19. 1.28×10^5 N/m^2, 2.68×10^4 N/m^2
25. (a) 7 cm (b) 2.8 kg
27. 0.439 kg
29. (a) 4.24 m/s (b) 17.0 m/s
31. (a) 0.83 m/s (lower), 3.3 m/s (upper)
    (b) 4.15×10^{-3} m^3/s
33. (a) 2.65 m/s (b) 2.31×10^4 N/m^2
35. 4.31×10^4 N
37. 4.9%
39. 1.40×10^8 N/m^2
51. (a) $s' = (4s)\tan^2\alpha$ (b) $s'/s = 4$
    (c) $\alpha = 26.6°$ or 0.464 rad
53. (a) $(\rho_1 h_1 + \rho_2 h_2)/(h_1 + h_2)$
    (b) $(\rho_1 h_1 + \rho_2 h_2)/\rho\omega$
    (c) $d' = d$ (d) $\Delta U = (\rho_2 - \rho_1)s^2 h_1 h_2 g$
57. (a) $T_\ell = \left[\frac{\rho_b - \rho_w}{\rho_b}\right] M_b g$
    (b) $T_u = T_\ell + \frac{1}{4}[\rho_c - \rho_w]\pi d^2 hg$
    (c) $T'_\ell = M_b g$; $T'_u = T'_\ell + \frac{1}{4}\rho_c \pi d^2 hg$
    (d) $T_\ell = 1.14 \times 10^4$ N; $T_u = 1.35 \times 10^4$ N
    $T'_\ell = 1.96 \times 10^4$ N; $T'_u = 2.20 \times 10^4$ N

CHAPTER 16

1. $y = \frac{6}{(x = 4.5t)^2 - 3}$
5. 25.8 m/s
7. 13.5 N
9. (b) 0.124 s
11. 0.319 m
13. (a) $y = A \sin k(x - vt)$
    (b) $y = A \sin \frac{2\pi}{\lambda}(x - vt)$
    (c) $y = A \sin 2\pi\left(\frac{x}{\lambda} - ft\right)$
    (d) $y = A \sin\left[2\pi f\left(\frac{x}{v} - t\right)\right]$
15. (a) 0.2 m (b) 4π rad/s (c) 5.03 m^{-1}
    (d) 1.25 m (e) 2.50 m/s (f) to the left
17. (a) $y = 0.08 \sin(7.85x + 6\pi t)$ m
    (b) $y = 0.08 \sin(7.85x + 6\pi t - 0.785)$ m
19. (a) 2370 cm/s^2 (b) 1675 cm/s^2
21. 1.07 kW
23. (a) remains constant (b) remains constant
    (c) remains constant (d) P is quadrupled
27. (a) 4.0 cm (b) π cm (c) 0.477 Hz
    (d) 2.09 s (e) to the right

29. (a) 179 m/s (b) 17.7 kW
31. 6.67 cm
33. (a) 25 m^{-1} (b) 12.0 rad/s (c) −0.34 rad (d) 0.467 s (e) 0.238 m
35. 20.2°C
37. (a) $0.7071(2\sqrt{\ell/g})$ (b) $\frac{1}{4}\ell$

CHAPTER 17

1. 4540 m/s
3. (a) N/m^2
5. 0.103 N/m^2, or 1.02×10^{-6} atm
7. 40 Hz
9. (a) zero (b) 3.86 N/m^2
11. -1.55×10^4 N/m^2
13. $s = (2.25 \times 10^{-8}) \cos(62.8x - 2.16 \times 10^4 t)$ m
15. 10 $\mu W/m^2$
17. (a) 3.75 W/m^2 (b) 0.6 W/m^2
19. 241 W
21. (a) 30 m (b) 9.49×10^5 m
23. 448 m/s
25. (a) 75 Hz (b) 0.948 m
27. (a) 56.5° (b) 2.11
29. 480 Hz
31. (a) 10 (b) 3 dB
33. 1204 Hz
35. (a) $[(f_r - f_e)/(f_r - f_e)]v$ (b) $\frac{1}{2}(v_p + v)T$
(c) $vt\left[\dfrac{f_e}{f_r + f_e}\right]$
(d) $v_p = 9.99$ m/s, $d_e = 52.07$ m/s, $d_r = 49.12$ m
37. (a) 0.948° (b)4.40°
39. 1.34×10^4 N
41. 34.3 W

CHAPTER 18

1. (a) 9.24 m (b) 600 Hz
3. $y_2 = 8 \sin[2\pi(0.1x - 80t - \frac{1}{6})]$
5. (a) 2π rad (b) 2 A
7. 0.522 m and 0.728 m measured from either speaker
9. (a) 4.24 cm (b) 6.00 cm (c) 6.00 cm (d) 0.5 cm, 1.5 cm, 2.5 cm
11. 25.1 m, 60 Hz
13. (a) 2.0 cm (b) 2.40 cm
15. (a) 0.60 m (b) 30 Hz
17. touch at the midpoint, pick at $L/4$
19. 0.786 Hz, 1.57 Hz, 2.36 Hz, 3.14 Hz
21. $n = 33$
23. nodes at 0, $\frac{8}{3}$ m, $\frac{16}{3}$ m, and 8 m; antinodes at $\frac{4}{3}$ m, 4 m, and $\frac{20}{3}$ m
25. 19 958 Hz
27. (a) $L_{open} = (n + 1)v/2f$, where n is the overtone number (0,1,2,3 . . .)
(b) $L_{closed(n)} = [2(1 + n) - 1]v/(4f)$
29. 0.504 m, 0.840 m
31. (a) 0.358 m (b) 0.717 m
35. 328 m/s
37. 1.16 m
39. (a) $f_n = [2(n + 1) - 1]v/(2L)$, where $n =$ (0,1,2,3 . . .) (b) $f_n = [2(n + 1) - 1]2v/L$
43. 3400 Hz
45. (a) $\dfrac{L_{closed}}{L_{open}} = \dfrac{3}{4}$
(b) $L_{open} = 0.664$ m $L_{closed} = 0.498$ m
47. (a) 133 Hz (b) 5.64 kg
49. (a) 6 m (b) 274 m/s
(c) $y = 0.04 \cos(1.05x) \sin(287t)$ m
51. 328 m/s
53. 5 Hz

CHAPTER 19

1. (a) 68.3 mm Hg (b) 131 mm Hg
3. (a) 42.9°C (b) 1.47 atm
5. (a) 832.3°F (b) 717.8 K
7. 37.0°C, 39.4°C
9. −40°C
11. 1.43 cm
13. 1.26 cm
15. 0.95 gal
17. (a) 1.35×10^{-2} cm (b) 6.75×10^{-4} cm
(c) 3.18×10^{-2} cm^3
19. 287°C, or 560 K
21. (a) 600 K, or 327°C (b) 1200 K, or 927°C
23. 53.3 lb/in.^{2}
25. 3.28×10^{13} molecules
27. (a) $R_o = 50\ \Omega$, $A = 1.55 \times 10^{-3}$ $(C°)^{-1}$
(b) 200°C
29. 3.28 cm
31. (a) 2.75×10^{-5} s (b) loses 16.6 s each week
35. 1.01×10^3 atm
37. (a) 24.5 m (b) 3.41×10^5 N/m^2
39. (b) 1.25 kg/m^3
41. (a) 0.58% (b) 0.018%
43. $T = 55.05$ N, $x = 42$ μm

CHAPTER 20

1. 2.69×10^4 cal
3. 80.6°C
5. 23.4°C
7. 63.9°C
9. 1.17×10^4 cal
11. 167 g
13. 1.21 liters
15. 9×10^6 cal/h
17. 3.91×10^{26} W
19. 63.8°C
21. 271 cal/s
23. 2.85×10^3 m
25. (a) 20.57°C (b) No. The change in potential energy and the heat absorbed are both proportional to the mass; hence the mass cancels.

27. (a) 9.89×10^{-3}C° (b) It goes into heating up the surface.
29. (a) 6.08×10^5 J (b) -4.05×10^5 J
31. (a) -87.9 J (b) 723 J
33. -420 J (heat leaves the system)
35. $A \to B(+++)$, $B \to C(0--)$, $C \to A(---)$
37. (a) 5.48×10^3 J (b) 5.48×10^3 J
39. (a) 3.08×10^{-2} m³ (b) -3.46×10^3 J (c) -3.46×10^3 J
41. (a) $-\frac{2}{3}P_oV_o$ (b) $-RT_o \ln 2$ (c) zero
43. (a) 60.3°C (b) 1.40×10^5 cal
45. (a) 1.52×10^{-2} J (b) 3.6 parts per million
47. (a) 725 cal/s (b) 12.4 min
49. 141 J
51. 5.75×10^3 J/s, or 5.75 kW
53. 0.654 cal/g · C°
55. (a) 26.9 liters (b) 8.43 liters/min
57. 12.2 h
59. (c) 1.47×10^{-4} J; 35 parts per billion (d) not a significant energy requirement

CHAPTER 21

1. 2.43×10^5 m²/s²
3. $\bar{F} = 8.0$ N, $P = 1.6$ N/m²
5. $\bar{F} = 0.943$ N, $P = 1.57$ N/m²
7. 2.54×10^3 m/s
9. (a) 731 m/s at 600 K 422 m/s at 200 K
11. (a) 2.28×10^3 J (b) 6.21×10^{-21} J
13. (a) 202 cal (b) 281 cal
15. (a) $C'_V = 8.96$ cal/K, $C'_P = 14.9$ cal/K (b) $C'_V = 14.9$ cal/K, $C'_P = 20.9$ cal/K
17. (a) 209 J (b) zero (c) 317 K
21. 10.1 atm, 756 K
23. 443 m/s
25. 4.33 s
27. 2.87%
33. (a) 558 m/s (b) 514 m/s (c) 456 m/s
35. $v_{mp} = 731$ m/s, $\bar{v} = 825$ m/s, $v_{rms} = 895$ m/s
37. (a) 3.4×10^4 molecules (b) 1.8×10^4 molecules
39. (a) 2.37×10^4 K (b) 1.06×10^3 K
41. (a) 3.21×10^{12} molecules (b) 7.78×10^5 m (c) 6.42×10^{-4} s⁻¹
43. (a) 1.96×10^{27} molecules/m³ (b) 1.84×10^{-9} m (c) 2.42×10^{11} s⁻¹
45. 346 m/s
49. (a) $3.65v$ (b) $3.99v$ (c) $3v$ (d) $106mv^2/V$ (e) $7.97\ mv^2$
51. (a) $C_V = aR$ (b) $C_P = (a+1)R$ (c) $a = 3.42$ (d) about 7
53. $N_v\,\Delta v \approx 1.4 \times 10^{21}$ molecules
59. (a) 5.18 (b) 0.518 (c) 5.18 (d) 0.139
61. (a) $n_1R/(\gamma_1 - 1) + n_2R/(\gamma_2 - 1)$ (b) $([n_1\gamma_1/(\gamma_1 - 1)] + [n_2\gamma_2/(\gamma_2 - 1)])R$

CHAPTER 22

1. (a) 6.64% (b) 84 cal
3. (a) 6 cal = 25.1 J (b) 36 cal
5. (a) 33% (b) 2/3
7. (a) 37.5% (b) 628 J (c) 2.09 kW
9. (a) 140 cal (b) 350 K
11. (a) 5.1% (b) 1.26×10^{12} cal/h
13. (a) 67.2% (b) 61.5 kW
15. (a) 16.8 (b) 1.49 kW
17. (a) 185 cal (b) 115 cal
19. 1.75, 80% Carnot efficiency
21. 46.6 cal/K
23. 3.59 J/K
25. (a) -6.70 J/K (b) 11.2 J/K (c) zero (d) 4.46 J/K
27. 57.2 J/K
29. (a) 7.33×10^3 J/K (b) -6.67×10^3 J/K (c) 0.66×10^3 J/K
31. (a) 53°C (b) 7.3 J/K
33. (a) 4.11×10^3 J (b) 1.42×10^4 J (c) 1.01×10^4 J (d) 28.9%
35. (a) $2nRT_o \ln 2$ (b) 0.273, or 27.3%
37. (a) $10.5RT_o$ (b) $-8.5RT_o$ (c) $4/21 \approx 19\%$ (d) $5/6 \approx 83\%$
39. $nR\left(\frac{\gamma}{\gamma - 1}\right) \ln 3$
45. (a) 299.3 K (b) -229.1 cal/K (c) 414.6 cal/K (d) 185.5 cal/K

Photo Credits

Chapter 1	p. 3, Photo Mill Valley, CA, University Science Books, 1982
Chapter 2	p. 17, Dr. Raymond A. Serway
Chapter 3	p. 32, Education Development Center, Newton, MA
Chapter 4	p. 55, Education Development, Center, Newton, MA
Chapter 5	p. 78, NASA
Chapter 6	p. 107, King's Island
Chapter 7	p. 122, Tennessee Valley Authority
Chapter 8	p. 145, Photo courtesy of Six Flags Great Adventure
Chapter 9	p. 107, Ken Ries, James Madison University, Harrisonburg, VA
Chapter 10	p. 194, Education Development Center, Newton, MA
Chapter 11	p. 217, David Leonardi
Chapter 13	p. 257, Griffith Observatory
Chapter 14	p. 282, NASA/JPL
Chapter 15	p. 310, H.E. Edgerton, M.I.T., Cambridge, MA
Chapter 16	p. 345, Jay Freedman
Chapter 17	p. 366, R. Steinway and Sons
Chapter 18	p. 384, Lyon & Healy Harps, Chicago
Chapter 19	p. 409, Corning Glass Works
Chapter 20	p. 425, Dr. Albert A. Bartlett, University of Colorado, Boulder
Chapter 21	p. 453, Lloyd Black
Chapter 22	p. 478, James Madison University

INDEX

Page numbers in *italics* indicate illustrations; page numbers followed by *n* indicate footnotes; page numbers followed by *t* indicate tables.